Linear Equations and Slope

The slope, m, of a line between two distinct points (x_1, y_1) and (x_2, y_2):

$$m = \frac{y_2 - y_1}{x_2 - x_1}, \quad x_2 - x_1 \neq 0$$

Standard form: $Ax + By = C$, $\quad A$ and B are not both zero

Horizontal line: $y = k$

Vertical line: $x = k$

Slope intercept form: $y = mx + b$

Point-slope formula: $y - y_1 = m(x - x_1)$

Midpoint Formula

Given two points (x_1, y_1) and (x_2, y_2), the midpoint is

$$\left(\frac{x_1 + x_2}{2}, \frac{y_1 + y_2}{2} \right)$$

Angles

Two angles are **complementary** if the sum of their measures is 90°.

Two angles are **supplementary** if the sum of their measures is 180°.

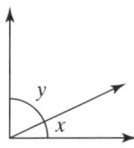 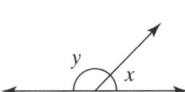

In the figure below, $\angle a$ and $\angle c$ are vertical angles and $\angle b$ and $\angle d$ are vertical angles. The measures of vertical angles are equal.

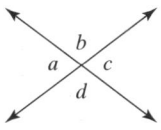

The sum of the measures of the angles of a triangle is 180°.

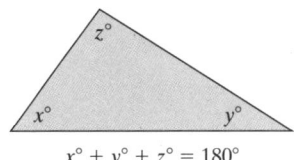

$$x° + y° + z° = 180°$$

Properties and Definitions of Exponents

Let a and b ($b \neq 0$) represent real numbers and m and n represent positive integers.

$$b^m b^n = b^{m+n}; \qquad \frac{b^m}{b^n} = b^{m-n}; \qquad (b^m)^n = b^{mn};$$

$$(ab)^m = a^m b^m; \qquad \left(\frac{a}{b} \right)^m = \frac{a^m}{b^m}; \qquad b^0 = 1; \qquad b^{-n} = \left(\frac{1}{b} \right)^n$$

Difference of Squares:

$$a^2 - b^2 = (a + b)(a - b)$$

Difference of Cubes:

$$a^3 - b^3 = (a - b)(a^2 + ab + b^2)$$

Sum of Cubes:

$$a^3 + b^3 = (a + b)(a^2 - ab + b^2)$$

Perfect Square Trinomials:

$$a^2 + 2ab + b^2 = (a + b)^2$$
$$a^2 - 2ab + b^2 = (a - b)^2$$

The Quadratic Formula

The solutions to $ax^2 + bx + c = 0$ $(a \neq 0)$ are given by

$$x = \frac{-b \pm \sqrt{b^2 - 4ac}}{2a}$$

The Vertex Formula

For $f(x) = ax^2 + bx + c$ $(a \neq 0)$, the vertex is

$$\left(\frac{-b}{2a}, \frac{4ac - b^2}{4a} \right) \quad \text{or} \quad \left(\frac{-b}{2a}, f\left(\frac{-b}{2a} \right) \right)$$

The Distance Formula

The distance between two points (x_1, y_1) and (x_2, y_2) is
$$d = \sqrt{(x_2 - x_1)^2 + (y_2 - y_1)^2}$$

The Standard Form of a Circle

$(x - h)^2 + (y - k)^2 = r^2$ with center (h, k) and radius r

Beginning and Intermediate
ALGEBRA

Second Edition

Julie Miller
Daytona Beach Community College

Molly O'Neill
Daytona Beach Community College

Nancy Hyde
Formerly of Broward Community College

With Contributions
by **Mitchel Levy**

 **Higher Education**

Boston   Burr Ridge, IL   Dubuque, IA   New York   San Francisco   St. Louis
Bangkok   Bogotá   Caracas   Kuala Lumpur   Lisbon   London   Madrid   Mexico City
Milan   Montreal   New Delhi   Santiago   Seoul   Singapore   Sydney   Taipei   Toronto

Higher Education

BEGINNING AND INTERMEDIATE ALGEBRA, SECOND EDITION

Published by McGraw-Hill, a business unit of The McGraw-Hill Companies, Inc., 1221 Avenue of the Americas, New York, NY 10020. Copyright © 2008 by The McGraw-Hill Companies, Inc. All rights reserved. No part of this publication may be reproduced or distributed in any form or by any means, or stored in a database or retrieval system, without the prior written consent of The McGraw-Hill Companies, Inc., including, but not limited to, in any network or other electronic storage or transmission, or broadcast for distance learning.

Some ancillaries, including electronic and print components, may not be available to customers outside the United States.

This book is printed on acid-free paper.

1 2 3 4 5 6 7 8 9 0 DOW/DOW 0 9 8 7 6

ISBN 978–0–07–305281–6
MHID 0–07–305281–7

ISBN 978–0–07–329793–4 (Annotated Instructor's Edition)
MHID 0–07–329793–3

Publisher: *Elizabeth J. Haefele*
Sponsoring Editor: *David Millage*
Director of Development: *David Dietz*
Developmental Editor: *Michelle Driscoll*
Marketing Manager: *Barbara Owca*
Senior Project Manager: *Vicki Krug*
Senior Production Supervisor: *Sherry L. Kane*
Lead Media Project Manager: *Stacy A. Patch*
Media Producer: *Amber M. Huebner*
Cover Design: *Laurie B. Janssen*
Interior Design: *Asylum Studios*
Lead Photo Research Coordinator: *Carrie K. Burger*
Supplement Producer: *Melissa M. Leick*
Compositor: *Techbooks*
Typeface: *10/12 Times Ten Roman*
Printer: *R. R. Donnelley Willard, OH*

Photo Credits: Page 3: © PhotoDisc/Getty R-F Website; p. 71: © Elena Rooraid/PhotoEdit; p. 79: © Vol. 44/Corbis CD; p. 91: © Corbis R-F Website; p. 154: © Vol. 44/Corbis CD; p. 163: © Judy Griesdieck/Corbis; p.173: © Corbis R-F Website; p. 187: © Robert Brenner/PhotoEdit; p. 210: © David Young-Wolff/PhotoEdit; p. 229: © Tony Freeman/PhotoEdit; p. 243: © Vol. 107/Corbis CD; p. 275: © Susan Van Etten/Photo Edit; p. 332: © Jeff Greenberg/PhotoEdit; p. 340: © Vol. 132/Corbis CD; p. 341: © Linda Waymire; p. 374: © Vol. 26/Corbis; p. 418: © Corbis Website; p. 501: © Paul Morris/Corbis; p. 536: © EyeWire/Getty Images Website; p. 568: © PhotoDisc Website; p. 611: Courtesy of NOAA; p. 621: © Vol. 59/Corbis CD; p. 639: © PhotoDisc Website; p. 680: © Quest/SPL/Photo Researchers, Inc.; p. 711: © The McGraw-Hill Companies, Inc./John Thoeming, photographer; p. 734: © Corbis Website; p. 787: © Tony Freeman/PhotoEdit; p. 794: © Vol. 145/Corbis CD; p. 798: © Nathan Benn/Corbis; p. 832: Corbis Website; p. 845: © Corbis Website; p. 877: © Mary Kate Denny/Photo Edit; p. 914: © Corbis; p. 933: © PhotoDisc/Getty R-F Website; p. 986: © Oliver Meckes/Photo Researchers, Inc.

Library of Congress Cataloging-in-Publication Data

Miller, Julie, 1962–
    Beginning and intermediate algebra. — 2nd ed. / Julie Miller, Molly O'Neill, Nancy Hyde.
      p. cm.
    Includes index.
    ISBN 978–0–07–305281–6 — ISBN 0–07–305281–7 (hard copy : alk. paper)
    1. Algebra—Textbooks. I. O'Neill, Molly, 1953– II. Hyde, Nancy. III. Title.
QA152.3.M57 2008
512.9—dc22
                      2006038338

www.mhhe.com

Contents

Chapter 10   More Equations and Inequalities   671

Chapter 11   Radicals and Complex Numbers   739

Chapter 12   Quadratic Equations and Functions   823

Dedication

To Suzanne and Nick

—Julie Miller

To my son and best friend, Stephen

—Molly O'Neill

In memory of my father, Harry Garvey

—Nancy Hyde

About the Authors

Julie Miller

Julie Miller has been on the faculty of the Mathematics Department at Daytona Beach Community College for 18 years, where she has taught developmental and upper-level courses. Prior to her work at DBCC, she worked as a software engineer for General Electric in the area of flight and radar simulation. Julie earned a bachelor of science in applied mathematics from Union College in Schenectady, New York, and a master of science in mathematics from the University of Florida. In addition to this textbook, she has authored several course supplements for college algebra, trigonometry, and precalculus, as well as several short works of fiction and nonfiction for young readers.

"My father is a medical researcher, and I got hooked on math and science when I was young and would visit his laboratory. I can remember using graph paper to plot data points for his experiments and doing simple calculations. He would then tell me what the peaks and features in the graph meant in the context of his experiment. I think that applications and hands-on experience made math come alive for me and I'd like to see math come alive for my students."

—Julie Miller

Molly O'Neill

Molly O'Neill is also from Daytona Beach Community College, where she has taught for 20 years in the Mathematics Department. She has taught a variety of courses from developmental mathematics to calculus. Before she came to Florida, Molly taught as an adjunct instructor at the University of Michigan–Dearborn, Eastern Michigan University, Wayne State University, and Oakland Community College. Molly earned a bachelor of science in mathematics and a master of arts and teaching from Western Michigan University in Kalamazoo, Michigan. Besides this textbook, she has authored several course supplements for college algebra, trigonometry, and precalculus and has reviewed texts for developmental mathematics.

"I differ from many of my colleagues in that math was not always easy for me. But in seventh grade I had a teacher who taught me that if I follow the rules of mathematics, even I could solve math problems. Once I understood this, I enjoyed math to the point of choosing it for my career. I now have the greatest job because I get to do math everyday and I have the opportunity to influence my students just as I was influenced. Authoring these texts has given me another avenue to reach even more students."

—Molly O'Neill

Nancy Hyde

Nancy Hyde served as a full-time faculty member of the Mathematics Department at Broward Community College for 24 years. During this time she taught the full spectrum of courses from developmental math through differential equations. She received a bachelor of science degree in math education from Florida State University and a master's degree in math education from Florida Atlantic University. She has conducted workshops and seminars for both students and teachers on the use of technology in the classroom. In addition to this textbook, she has authored a graphing calculator supplement for College Algebra.

"I grew up in Brevard County, Florida, with my father working at Cape Canaveral. I was always excited by mathematics and physics in relation to the space program. As I studied higher levels of mathematics I became more intrigued by its abstract nature and infinite possibilities. It is enjoyable and rewarding to convey this perspective to students while helping them to understand mathematics."

—Nancy Hyde

Mitchel Levy

Mitchel Levy of Broward Community College joined the team as the exercise consultant for the Miller/O'Neill/Hyde paperback series. Mitchel received his BA in mathematics in 1983 from the State University of New York at Albany and his MA in mathematical statistics from the University of Maryland, College Park in 1988. With over 17 years of teaching and extensive reviewing experience, Mitchel knows what makes exercise sets work for students. In 1987 he received the first annual "Excellence in Teaching" award for graduate teaching assistants at the University of Maryland. Mitchel was honored as the Broward Community College Professor of the year in 1994, and has co-coached the Broward math team to 3 state championships over 7 years.

"I love teaching all level of mathematics from Elementary Algebra through Calculus and Statistics."

—Mitchel Levy

Introducing . . .

The Miller/O'Neill/Hyde Series in Developmental Mathematics

Miller/O'Neill/Hyde casebound series

Beginning Algebra, 2/e

Intermediate Algebra, 2/e

Beginning and Intermediate Algebra, 2/e

Miller/O'Neill/Hyde worktext series

Basic College Mathematics

Introductory Algebra

Intermediate Algebra

Preface

From the Authors

First and foremost, we would like to thank the students and colleagues who have helped us prepare this text. The content and organization are based on a wealth of resources. Aside from an accumulation of our own notes and experiences as teachers, we recognize the influence of colleagues at Daytona Beach Community College as well as fellow presenters and attendees of national mathematics conferences and meetings. Perhaps our single greatest source of inspiration has been our students, who ask good, probing questions every day and challenge us to find new and better ways to convey mathematical concepts. We gratefully acknowledge the part that each has played in the writing of this book.

In designing the framework for this text, the time we have spent with our students has proved especially valuable. Over the years we have observed that students struggle consistently with certain topics. We have also come to know the influence of forces beyond the math, particularly motivational issues. An awareness of the various pitfalls has enabled us to tailor pedagogy and techniques that directly address students' needs and promote their success. These techniques and pedagogy are outlined here.

Active Classroom

First, we believe students retain more of what they learn when they are actively engaged in the classroom. Consequently, as we wrote each section of text, we also wrote accompanying worksheets called **Classroom Activities** to foster accountability and to encourage classroom participation. Classroom Activities resemble the examples that students encounter in the textbook. The activities can be assigned to individual students or to pairs or groups of students. Most of the activities have been tested in the classroom with our own students. In one class in particular, the introduction of Classroom Activities transformed a group of "clock watchers" into students who literally had to be ushered out of the classroom so that the next class could come in. The activities can be found in the *Instructor's Resource Manual*, which is available through MathZone.

Conceptual Support

While we believe students must practice basic skills to be successful in any mathematics class, we also believe concepts are important. To this end, we have included numerous writing questions and homework exercises that ask students to **"interpret the meaning in the context of the problem."** These questions make students stop and think, so they can process what they learn. In this way, students will learn underlying concepts. They will also form an understanding of what their answers mean in the contexts of the problems they solve.

Writing Style

Many students believe that reading a mathematics text is an exercise in futility. However, students who take the time to read the text and features may cast that notion aside. In particular, the **Tips** and **Avoiding Mistakes** boxes should prove especially enlightening. They offer the types of insights and hints that are usually only revealed during classroom lecture. On the whole, students should be very comfortable with the reading level, as the language and tone are consistent with those used daily within our own developmental mathematics classes.

Real-World Applications

Another critical component of the text is the inclusion of **contemporary real-world examples and applications.** We based examples and applications on information that students encounter daily when they turn on the news, read a magazine, or surf the Internet. We incorporated data for students to answer mathematical questions based on information in tables and graphs. When students encounter facts or information that is meaningful to them, they will relate better to the material and remember more of what they learn.

Study Skills

Many students in this course lack the basic study skills needed to be successful. Therefore, at the beginning of the homework exercises, we included a set of **Study Skills Exercises.** These exercises focus on one of nine areas: learning about the course, using the text, taking notes, completing homework assignments, test taking, time management, learning styles, preparing for a final exam, and defining **key terms.** Through completion of these exercises, students will be in a better position to pass the class and adopt techniques that will benefit them throughout their academic careers.

Language of Mathematics

Finally, for students to succeed in mathematics, they must be able to understand its language and notation. We place special emphasis on the skill of translating mathematical notation to English expressions and vice versa through **Translating Expressions Exercises.** These appear intermittently throughout the text. We also include key terms in the homework exercises and ask students to define these terms.

What Sets This Book Apart?

We believe that the thoughtfully designed pedagogy and contents of this textbook offer any willing student the opportunity to achieve success, opening the door to a wider world of possibilities.

While this textbook offers complete coverage of the beginning algebra and intermediate algebra curricula, there are several concepts that receive special emphasis.

Problem Recognition

Problem recognition is an important theme carried throughout this edition and is integrated into the textbook in a number of different ways. First, we developed **Problem Recognition Exercises** that appear in selected chapters. The purpose of the Problem Recognition Exercises is to present a collection of problems that may look very similar to students, upon first glance, but are actually quite different in the manner of their individual solutions. By completing these exercises, students gain a greater awareness in problem recognition—that is, identifying a particular problem type upon inspection and applying the appropriate method to solve it.

We have carefully selected the content areas that seem to present the greatest challenge for students in terms of their ability to differentiate among various problem types. For these topics (listed below) we developed Problem Recognition Exercises

> Addition and Subtraction of Signed Numbers (page 80)
> Properties of Exponents (page 376)
> Operations on Polynomials (page 410)
> Operations on Rational Expressions (page 528)
> Comparing Rational Equations and Rational Expressions (page 547)
> Equations and Inequalities (page 712)
> Operations on Radicals (page 788)
> Logarithmic and Exponential Forms (page 967)

The problem recognition theme is also threaded throughout the book within the exercise sets. In particular, Section 6.6, "General Factoring Summary," is worth special mention. While not formally labeled "Problem Recognition," each factoring exercise requires students to label the *type* of factoring problem presented before attempting to factor the polynomial (see page 469, directions for problems 7–74).

The concept of problem recognition is also applied in every chapter, on a more micro level. We looked for opportunities within the section-ending Practice Exercises to include exercises that involve comparison between problem types. See, for example,

> Section 5.6, exercises 2–13 on comparing addition versus multiplication of polynomials
> Section 8.3, exercises 18–29 on identifying function type as linear, quadratic or constant
> Section 10.2, exercises 13–18 and 39–42 on comparing the solutions to polynomial and rational equations versus inequalities
> Section 11.2, exercises 21–24 on the effect of the position of a negative sign within the base or exponent of an exponential expression

We also included Mixed Exercises that appear in many of the Practice Exercise sets to give students opportunities to practice exercises that are not lumped together by problem type.

We firmly believe that students who effectively learn how to distinguish between various problems and the methods to solve them will be better prepared for Intermediate Algebra and courses beyond.

Design

While the content of a textbook is obviously critical, we believe that the design of the page is equally vital. We have often heard instructors describe the importance of "white space" in the design of developmental math textbooks and the need for simplicity, to prevent distractions. All of these comments have factored heavily in the page layout of this textbook. For example, we left ample space between exercises in the section-ending Practice Exercise sets to make it easier for students to read and complete homework assignments.

Similarly, we developed design treatments within sections to present the content, including examples, definitions, and summary boxes, in an organized and reader-friendly way. We also limited the number of colors and photos in use, in an effort to avoid distraction on the pages. We believe these considerations directly reflect the needs of our students and will enable them to navigate through the content successfully.

Chapter R

Chapter R is a reference chapter. We designed it to help students reacquaint themselves with the fundamentals of fractions and geometry. This chapter also addresses study skills and helpful hints to use the resources provided in the text and its supplements.

Factoring

Many years ago, we experimented in the classroom with our approach to factoring. We began factoring trinomials with the general case first, that is, with leading coefficient not equal to 1. This gave the students one rule for all cases, and it provided us with an extra class day for practice and group work. Most importantly, this approach forces students always to consider the leading coefficient, whether it be 1 or some other number. Thus, when students take the product of the inner terms and the product of the outer terms, the factors of the leading coefficient always come into play.

While we recommend presenting trinomials with leading coefficient other than 1 first, we want to afford flexibility to the instructors using this textbook. Therefore, we have structured our chapter on Factoring, Chapter 6, to offer instructors the option of covering trinomials with leading coefficient of 1 first (Section 6.2), or to cover trinomials with leading coefficient other than 1 first (Sections 6.3 and 6.4).

For those instructors who have never tried the method of presenting trinomials with leading coefficient other than 1 first, we suggest giving it a try. We have heard other instructors, who were at first resistant to this approach, remark how well it has worked for their students. In fact, one instructor told us, "I was skeptical, but I had the best exam results I have ever seen in factoring as a result of this approach! It has a big thumbs up from me and students alike."

Identifying Equations and Inequalities

A student who completes intermediate algebra should be able to recognize and solve a variety of equations and inequalities; however, the skill of distinguishing different types of equations and inequalities is often overlooked. Chapter 10—More Equations and Inequalities—is designed as a synthesis chapter in which students are exposed to a variety of equations and inequalities appropriate at this level. Note that each section of Chapter 10 is a self-contained unit and can be introduced at the instructor's discretion at another location in the text.

Calculator Usage

The use of a scientific or a graphing calculator often inspires great debate among faculty who teach developmental mathematics. Our **Calculator Connections** boxes offer screen shots and some keystrokes to support applications where a calculator might enhance learning. Our approach is to use a calculator as a verification tool after analytical methods have been applied. The Calculator Connections boxes are self-contained units with accompanying exercises and can be employed or easily omitted at the recommendation of the instructor.

Calculator Exercises appear within the section-ending Practice Exercise sets where appropriate, but they are clearly noted as such and can be assigned or omitted at the instructor's discretion.

Listening to Students' and Instructors' Concerns

Our editorial staff has amassed the results of reviewer questionnaires, user diaries, focus groups, and symposia. We have consulted with a six-member panel of beginning and intermediate algebra instructors and their students on the development of this book. In addition, we have read hundreds of pages of reviews from instructors across the country. At McGraw-Hill symposia, faculty from across the United States gathered to discuss issues and trends in developmental mathematics. These efforts have involved hundreds of faculty and have explored issues such as content, readability, and even the aesthetics of page layout.

In our continuing efforts to improve our products, we invite you to contact us directly with your comments.

Julie Miller	Molly O'Neill	Nancy Hyde
millerj@dbcc.edu	oneillm@dbcc.edu	nhyde@montanasky.com

Acknowledgments and Reviewers

The development of this textbook would never have been possible without the creative ideas and constructive feedback offered by many reviewers. We are especially thankful to the following instructors for their valuable feedback and careful review of the manuscript.

Board of Advisors

Darla Aguilar, *Pima Community College*

Gloria Guerra, *St. Philip's College*

Sandee House, *Georgia Perimeter College–Clark Campus*

George Johnson, *St. Philip's College*

Wilene Leach, *Arkansas State University–Beebe Campus*

Richard Rupp, *Del Mar College*

Special thanks to Sharon Hamsa, Sharon Holmes, and Jack Haughn for their feedback during manuscript development.

In addition, we would like to thank the following manuscript reviewers of *Beginning Algebra,* Second Edition and *Intermediate Algebra,* Second Edition. Their suggestions were also helpful in developing the manuscript:

Beginning Algebra, Second Edition

Board of Advisors

Valarie Beaman-Hackle, *Macon State College*

Michael Everett, *Santa Ana College*

Teresa Hasenauer, *Indian River Community College*

Lori Holdren, *Lake City Community College*

Becky Hubiak, *Tidewater Community College*

Barbara Hughes, *San Jacinto College–Pasadena*

Cameron English, *Rio Hondo College*

Manuscript Reviewers

Cedric Atkins, *Charles Stewart Mott Community College*

David Bell, *Florida Community College*

Emilie Berglund, *Utah Valley State College*

Kirby Bunas, *Santa Rosa Junior College*

Annette Burden, *Youngstown State University*

Susan Caldiero, *Cosumnes River College*

Edie Carter, *Amarillo College*
Marcial Echenique, *Broward Community College*
Karen Estes, *St. Petersburg College*
Renu Gupta, *Louisiana State University–Alexandria*
Rodger Hergert, *Rock Valley College*
Linda Ho, *El Camino College*
Michelle Hollis, *Bowling Green Community College*
Laura Hoye, *Trident Technical College*
Randa Kress, *Idaho State University–Pocatello*
Donna Krichiver, *Johnson County Community College*
Karl Kruczek, *Northeastern State University*
Kathryn Lavelle, *Westchester Community College*
Jane Loftus, *Utah Valley State College*
Gary McCracken, *Shelton State Community College*
Karen Pagel, *Dona Ana Branch Community College*
Peter Remus, *Chicago State University*
Daniel Richbart, *Erie Community College–City*

Intermediate Algebra, Second Edition

Board of Advisors

David French, *Tidewater Community College*
Donna Gerken, *Miami-Dade College–Kendall*
Cynthia Harrison, *Baton Rouge Community College*
Mary Henderson, *Okaloosa–Walton College*
Jo Karen Hudson, *University of Central Arkansas*
Mike Kirby, *Tidewater Community College*

Manuscript Reviewers

Alina Coronel, *Miami-Dade College–Kendall*
Adrienne Goldstein, *Miami-Dade College*
Jane Gringauz, *Minneapolis Community and Technical College*
Kristin Humphrey, *Spokane Community College*
Patricia Jones, *Rogers State University*
Cheryl Kane, *University of Nebraska–Lincoln*
Kim Luna, *Eastern New Mexico University*
Cloyd Payne, *Owens Community College*
Nicole Pfeifer, *Spokane Community College*
Evalon St. John, *Rogers State University*
Jean Thornton, *Western Kentucky University*
Bettie Truitt, *Black Hawk College*
William Van Alstine, *Aiken Technical College*

Special thanks go to Doris McClellan-Lewis and Jon Weerts for preparing the Instructor's Solutions Manual and the Student's Solutions Manual and to Lauri Semarne and John Hunt for their work ensuring accuracy. Many, many thanks to Yolanda Davis, Andrea Hendricks, Patricia Jayne, and Cynthia Cruz for their work in the video series and to Kelly Jackson, an awesome teacher, for preparing the Instructor Notes.

Finally, we are forever grateful to the many people behind the scenes at McGraw-Hill, our publishing family. To Erin Brown, our lifeline on this project, without whom we'd be lost. To Liz Haefele for her passion for excellence and constant inspiration. To David Millage and Barb Owca for their countless hours of support and creative ideas promoting all of our efforts. To Jeff Huettman and Amber Huebner for the awesome technology so critical to student success, and finally to Vicki Krug for keeping the train on the track during production.

Most importantly, we give special thanks to all the students and instructors who use *Beginning and Intermediate Algebra*, Second Edition in their classes.

Julie Miller   Molly O'Neill   Nancy Hyde

A COMMITMENT TO ACCURACY

You have a right to expect an accurate textbook, and McGraw-Hill invests considerable time and effort to make sure that we deliver one. Listed below are the many steps we take to make sure this happens.

OUR ACCURACY VERIFICATION PROCESS

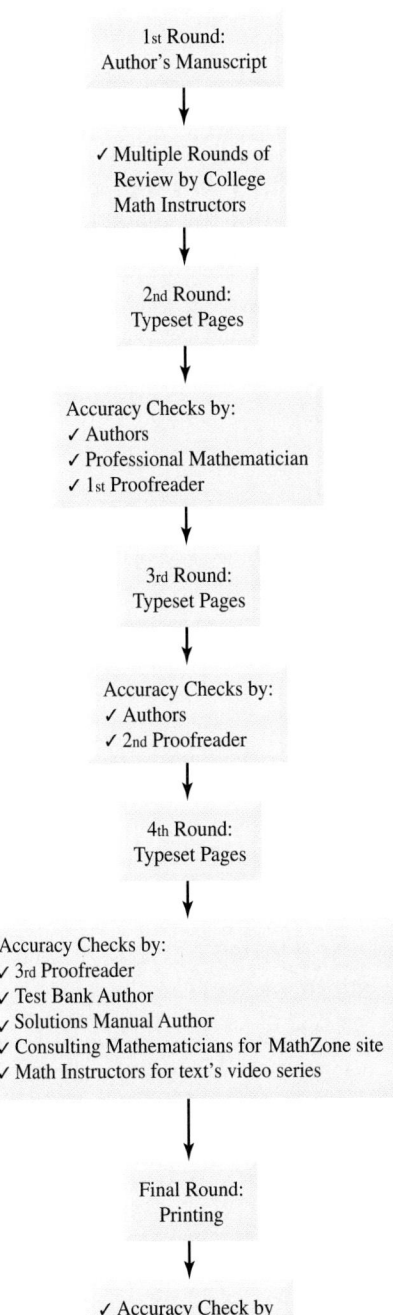

First Round

Step 1: Numerous **college math instructors** review the manuscript and report on any errors that they may find, and the authors make these corrections in their final manuscript.

Second Round

Step 2: Once the manuscript has been typeset, the **authors** check their manuscript against the first page proofs to ensure that all illustrations, graphs, examples, exercises, solutions, and answers have been correctly laid out on the pages, and that all notation is correctly used.

Step 3: An outside, **professional mathematician** works through every example and exercise in the page proofs to verify the accuracy of the answers.

Step 4: A **proofreader** adds a triple layer of accuracy assurance in the first pages by hunting for errors, then a second, corrected round of page proofs is produced.

Third Round

Step 5: The **author team** reviews the second round of page proofs for two reasons: (1) to make certain that any previous corrections were properly made, and (2) to look for any errors they might have missed on the first round.

Step 6: A **second proofreader** is added to the project to examine the new round of page proofs to double check the author team's work and to lend a fresh, critical eye to the book before the third round of paging.

Fourth Round

Step 7: A **third proofreader** inspects the third round of page proofs to verify that all previous corrections have been properly made and that there are no new or remaining errors.

Step 8: Meanwhile, in partnership with **independent mathematicians,** the text accuracy is verified from a variety of fresh perspectives:
- The **test bank author** checks for consistency and accuracy as they prepare the computerized test item file.
- The **solutions manual author** works every single exercise and verifies their answers, reporting any errors to the publisher.
- A **consulting group of mathematicians,** who write material for the text's MathZone site, notifies the publisher of any errors they encounter in the page proofs.
- A video production company employing **expert math instructors** for the text's videos will alert the publisher of any errors they might find in the page proofs.

Final Round

Step 9: The **project manager,** who has overseen the book from the beginning, performs a **fourth proofread** of the textbook during the printing process, providing a final accuracy review.

⇒ What results is a mathematics textbook that is as accurate and error-free as is humanly possible, and our authors and publishing staff are confident that our many layers of quality assurance have produced textbooks that are the leaders of the industry for their integrity and correctness.

Guided Tour

Chapter Opener

Each chapter opens with a puzzle relating to the content of the chapter. Section titles are clearly listed for easy reference.

Linear Equations and Inequalities

2

2.1 Addition, Subtraction, Multiplication, and Division Properties of Equality

2.2 Solving Linear Equations

2.3 Linear Equations: Clearing Fractions and Decimals

2.4 Applications of Linear Equations: Introduction to Problem Solving

2.5 Applications Involving Percents

2.6 Formulas and Applications of Geometry

2.7 Linear Inequalities

In this chapter we learn how to solve linear equations and inequalities in one variable. This important category of equations and inequalities is used in a variety of applications.

The list of words contains key terms used in this chapter. Search for them in the puzzle and in the text throughout the chapter. By the end of this chapter, you should be familiar with all of these terms.

Key Terms

compound	identity	notation
conditional	inequality	percent
consecutive	integer	property
contradiction	interest	set-builder
decimal	interval	solution
equation	linear	
fractions	literal	

113

Section 3.1	Rectangular Coordinate System

Concepts

1. Interpreting Graphs
2. Plotting Points in a Rectangular Coordinate System
3. Applications of Plotting and Identifying Points

1. Interpreting Graphs

Mathematics is a powerful tool used by scientists and has directly contributed to the highly technical world we live in. Applications of mathematics have led to advances in the sciences, business, computer technology, and medicine.

One fundamental application of mathematics is the graphical representation of numerical information (or **data**). For example, Table 3-1 represents the number of clients admitted to a drug and alcohol rehabilitation program over a 12-month period.

Table 3-1

Month		Number of Clients
Jan.	1	55
Feb.	2	62
March	3	64
April	4	60
May	5	70
June	6	73
July	7	77
Aug.	8	80
Sept.	9	80
Oct.	10	74
Nov.	11	85
Dec.	12	90

In table form, the information is difficult to picture and interpret. It appears that on a monthly basis, the number of clients fluctuates. However, when the data are represented in a graph, an upward trend is clear (Figure 3-1).

Number of Clients Admitted to Program

Month (x = 1 corresponds to Jan.)

Figure 3-1

Concepts

A list of important learning concepts is provided at the beginning of each section. Each concept corresponds to a heading within the section and within the exercises, making it easy for students to locate topics as they study or as they work through homework exercises.

Worked Examples

Examples are set off in boxes and organized so that students can easily follow the solutions. Explanations appear beside each step and color-coding is used, where appropriate. For additional step-by-step instruction, students can run the "e-Professors" in MathZone. The e-Professors are based on worked examples from the text and use the solution methodologies presented in the text.

Skill Practice Exercises

Every worked example is followed by one or more Skill Practice exercises. These exercises offer students an immediate opportunity to work problems that mirror the examples. Students can then check their work by referring to the answers at the bottom of the page.

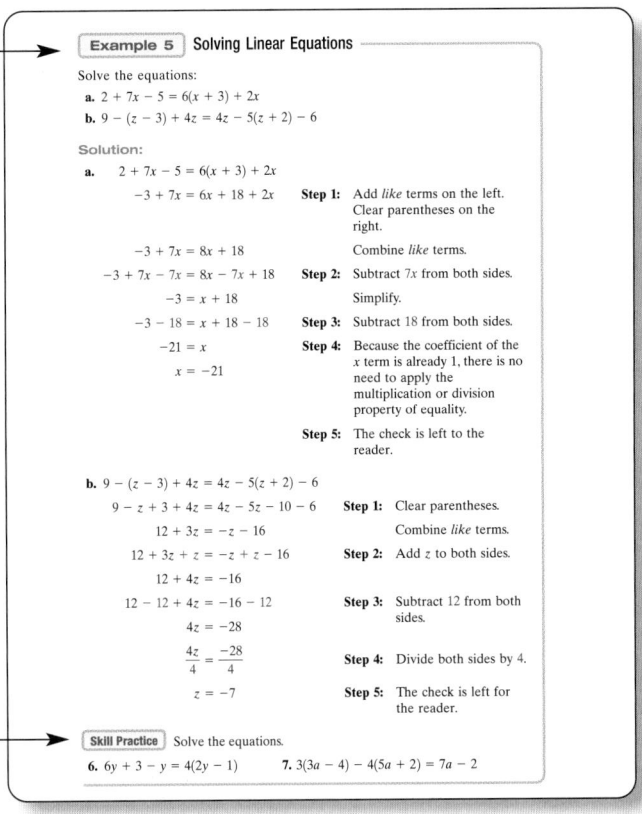

Example 5 Solving Linear Equations

Solve the equations:
a. $2 + 7x - 5 = 6(x + 3) + 2x$
b. $9 - (z - 3) + 4z = 4z - 5(z + 2) - 6$

Solution:

a.
$$2 + 7x - 5 = 6(x + 3) + 2x$$
$$-3 + 7x = 6x + 18 + 2x$$
Step 1: Add *like* terms on the left. Clear parentheses on the right.

$$-3 + 7x = 8x + 18$$
Combine *like* terms.

$$-3 + 7x - 7x = 8x - 7x + 18$$
Step 2: Subtract $7x$ from both sides.
$$-3 = x + 18$$
Simplify.

$$-3 - 18 = x + 18 - 18$$
Step 3: Subtract 18 from both sides.

$$-21 = x$$
Step 4: Because the coefficient of the x term is already 1, there is no need to apply the multiplication or division property of equality.
$$x = -21$$

Step 5: The check is left to the reader.

b.
$$9 - (z - 3) + 4z = 4z - 5(z + 2) - 6$$
$$9 - z + 3 + 4z = 4z - 5z - 10 - 6$$
Step 1: Clear parentheses.
$$12 + 3z = -z - 16$$
Combine *like* terms.

$$12 + 3z + z = -z + z - 16$$
Step 2: Add z to both sides.
$$12 + 4z = -16$$

$$12 - 12 + 4z = -16 - 12$$
Step 3: Subtract 12 from both sides.
$$4z = -28$$

$$\frac{4z}{4} = \frac{-28}{4}$$
Step 4: Divide both sides by 4.

$$z = -7$$
Step 5: The check is left for the reader.

Skill Practice Solve the equations.

6. $6y + 3 - y = 4(2y - 1)$ 7. $3(3a - 4) - 4(5a + 2) = 7a - 2$

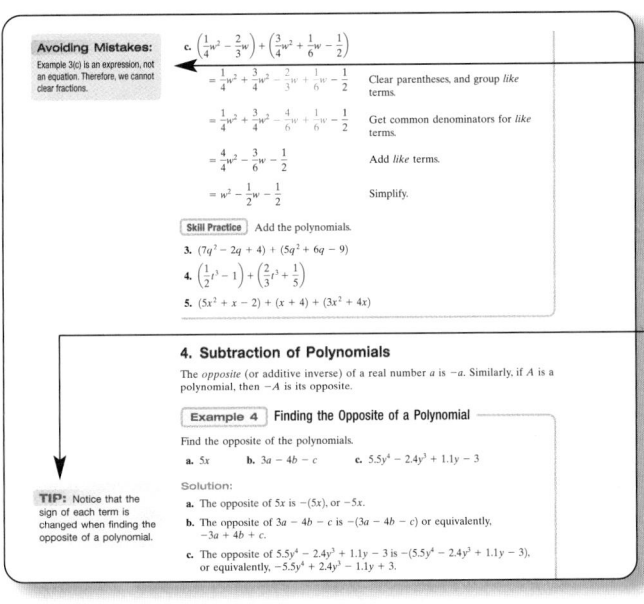

Avoiding Mistakes:
Example 3(c) is an expression, not an equation. Therefore, we cannot clear fractions.

c. $\left(\frac{1}{4}w^2 - \frac{2}{3}w\right) + \left(\frac{3}{4}w^2 + \frac{1}{6}w - \frac{1}{2}\right)$

$$= \frac{1}{4}w^2 + \frac{3}{4}w^2 - \frac{2}{3}w + \frac{1}{6}w - \frac{1}{2}$$
Clear parentheses, and group *like* terms.

$$= \frac{1}{4}w^2 + \frac{3}{4}w^2 - \frac{4}{6}w + \frac{1}{6}w - \frac{1}{2}$$
Get common denominators for *like* terms.

$$= \frac{4}{4}w^2 - \frac{3}{6}w - \frac{1}{2}$$
Add *like* terms.

$$= w^2 - \frac{1}{2}w - \frac{1}{2}$$
Simplify.

Skill Practice Add the polynomials.

3. $(7q^2 - 2q + 4) + (5q^2 + 6q - 9)$
4. $\left(\frac{1}{2}t^3 - 1\right) + \left(\frac{2}{3}t^3 + \frac{1}{5}\right)$
5. $(5x^2 + x - 2) + (x + 4) + (3x^2 + 4x)$

4. Subtraction of Polynomials

The *opposite* (or additive inverse) of a real number a is $-a$. Similarly, if A is a polynomial, then $-A$ is its opposite.

Example 4 Finding the Opposite of a Polynomial

Find the opposite of the polynomials.
a. $5x$ b. $3a - 4b - c$ c. $5.5y^4 - 2.4y^3 + 1.1y - 3$

Solution:
a. The opposite of $5x$ is $-(5x)$, or $-5x$.
b. The opposite of $3a - 4b - c$ is $-(3a - 4b - c)$ or equivalently, $-3a + 4b + c$.
c. The opposite of $5.5y^4 - 2.4y^3 + 1.1y - 3$ is $-(5.5y^4 - 2.4y^3 + 1.1y - 3)$, or equivalently, $-5.5y^4 + 2.4y^3 - 1.1y + 3$.

TIP: Notice that the sign of each term is changed when finding the opposite of a polynomial.

Avoiding Mistakes

Through notes labeled Avoiding Mistakes students are alerted to common errors and are shown methods to avoid them.

Tips

Tip boxes appear throughout the text and offer helpful hints and insight.

Problem Recognition Exercises

These exercises appear in selected chapters and cover topics that are inherently difficult for students. Problem recognition exercises give students the opportunity to practice the important, yet often overlooked, skill of distinguishing among different problem types.

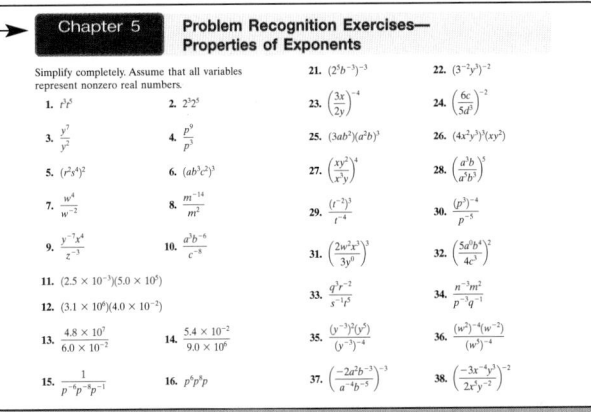

Chapter 5   Problem Recognition Exercises— Properties of Exponents

Simplify completely. Assume that all variables represent nonzero real numbers.

1. $r^3 t^5$
2. $2^3 2^5$
3. $\frac{y^7}{y^2}$
4. $\frac{p^9}{p^5}$
5. $(r^2 s^4)^2$
6. $(ab^3 c^2)^3$
7. $\frac{w^4}{w^{-2}}$
8. $\frac{m^{-14}}{m^2}$
9. $\frac{y^{-7} x^4}{z^{-3}}$
10. $\frac{a^3 b^{-6}}{c^{-8}}$
11. $(2.5 \times 10^{-3})(5.0 \times 10^5)$
12. $(3.1 \times 10^6)(4.0 \times 10^{-2})$
13. $\frac{4.8 \times 10^7}{6.0 \times 10^{-2}}$
14. $\frac{5.4 \times 10^{-2}}{9.0 \times 10^6}$
15. $\frac{1}{p^{-6} p^{-8} p^{-1}}$
16. $p^6 p^8 p$

21. $(2^3 b^{-3})^{-3}$
22. $(3^{-2} y^3)^{-2}$
23. $\left(\frac{3x}{2y}\right)^{-4}$
24. $\left(\frac{6c}{5d^3}\right)^{-2}$
25. $(3ab^2)(a^2 b)^3$
26. $(4x^2 y^3)^3 (xy^2)$
27. $\left(\frac{xy^2}{x^3 y}\right)^4$
28. $\left(\frac{a^3 b}{a^2 b^3}\right)^5$
29. $\frac{(t^{-2})^3}{t^{-4}}$
30. $\frac{(p^3)^{-4}}{p^{-5}}$
31. $\left(\frac{2w^2 x^3}{3y^0}\right)^3$
32. $\left(\frac{5a^0 b^4}{4c^3}\right)^2$
33. $\frac{q^3 r^{-2}}{s^{-1} t^5}$
34. $\frac{n^{-3} m^2}{p^{-3} q^{-1}}$
35. $\frac{(y^{-3})^2 (y^5)}{(y^{-3})^{-4}}$
36. $\frac{(w^2)^{-4} (w^{-2})}{(w^3)^{-4}}$
37. $\left(\frac{-2a^2 b^{-3}}{a^{-4} b^{-5}}\right)^{-3}$
38. $\left(\frac{-3x^{-4} y^3}{2x^3 y^{-2}}\right)^{-2}$

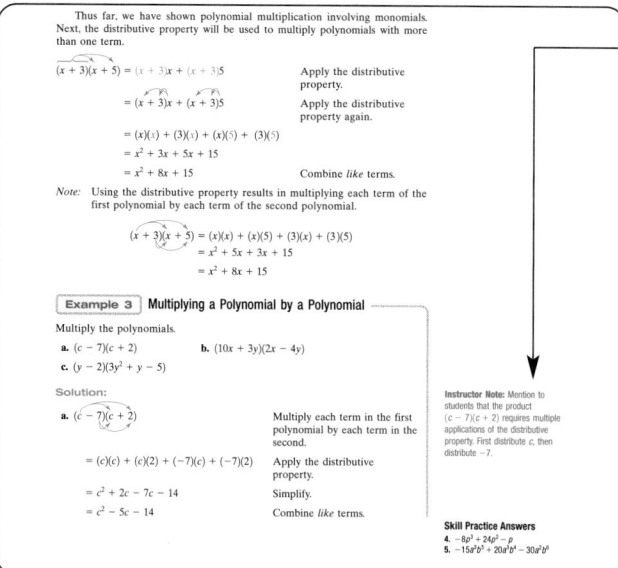

Thus far, we have shown polynomial multiplication involving monomials. Next, the distributive property will be used to multiply polynomials with more than one term.

$(x + 3)(x + 5) = (x + 3)x + (x + 3)5$ Apply the distributive property.

$= (x + 3)x + (x + 3)5$ Apply the distributive property again.

$= (x)(x) + (3)(x) + (x)(5) + (3)(5)$

$= x^2 + 3x + 5x + 15$

$= x^2 + 8x + 15$ Combine *like* terms.

Note: Using the distributive property results in multiplying each term of the first polynomial by each term of the second polynomial.

$(x + 3)(x + 5) = (x)(x) + (x)(5) + (3)(x) + (3)(5)$

$= x^2 + 5x + 3x + 15$

$= x^2 + 8x + 15$

Example 3   Multiplying a Polynomial by a Polynomial

Multiply the polynomials.

a. $(c - 7)(c + 2)$ **b.** $(10x + 3y)(2x - 4y)$

c. $(y - 2)(3y^2 + y - 5)$

Solution:

a. $(c - 7)(c + 2)$

$= (c)(c) + (c)(2) + (-7)(c) + (-7)(2)$ Apply the distributive property.

$= c^2 + 2c - 7c - 14$ Simplify.

$= c^2 - 5c - 14$ Combine *like* terms.

Instructor Note: Mention to students that the product $(c - 7)(c + 2)$ requires multiple applications of the distributive property. First distribute c, then distribute -7.

Skill Practice Answers
4. $-8p^3 + 24p^2 - p$
5. $-15a^3b^5 + 20a^3b^4 - 30a^2b^6$

Instructor Note (*AIE* only)

Instructor notes appear in the margins throughout each section of the *Annotated Instructor's Edition* (*AIE*). The notes may assist with lecture preparation in that they point out items that tend to confuse students, or lead students to err.

References to Classroom Activities (*AIE* only)

References are made to Classroom Activities at the beginning of each set of Practice Exercises in the *AIE*. The activities may be found in the *Instructor's Resource Manual*, which is available through MathZone and can be used during lecture or assigned for additional practice.

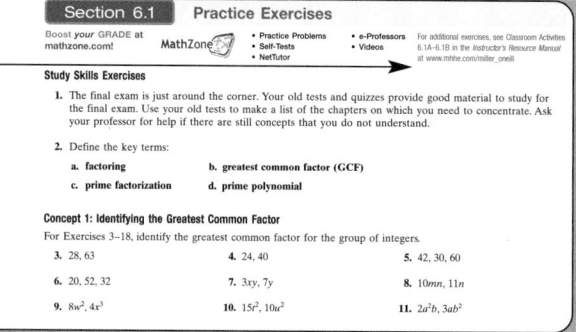

Section 6.1   Practice Exercises

Boost *your* GRADE at mathzone.com!   MathZone   • Practice Problems   • e-Professors   For additional exercises, see Classroom Activities
  • Self-Tests   • Videos   6.1A–6.1B in the *Instructor's Resource Manual*
  • NetTutor   at www.mhhe.com/miller_oneill

Study Skills Exercises

1. The final exam is just around the corner. Your old tests and quizzes provide good material to study for the final exam. Use your old tests to make a list of the chapters on which you need to concentrate. Ask your professor for help if there are still concepts that you do not understand.

2. Define the key terms:

   a. factoring   **b.** greatest common factor (GCF)

   c. prime factorization   **d.** prime polynomial

Concept 1: Identifying the Greatest Common Factor

For Exercises 3–18, identify the greatest common factor for the group of integers.

   3. 28, 63   **4.** 24, 40   **5.** 42, 30, 60

   6. 20, 52, 32   **7.** $3xy$, $7y$ **8.** $10mn$, $11n$

   9. $8w^2$, $4x^3$ **10.** $15t^2$, $10u^2$ **11.** $2a^2b$, $3ab^2$

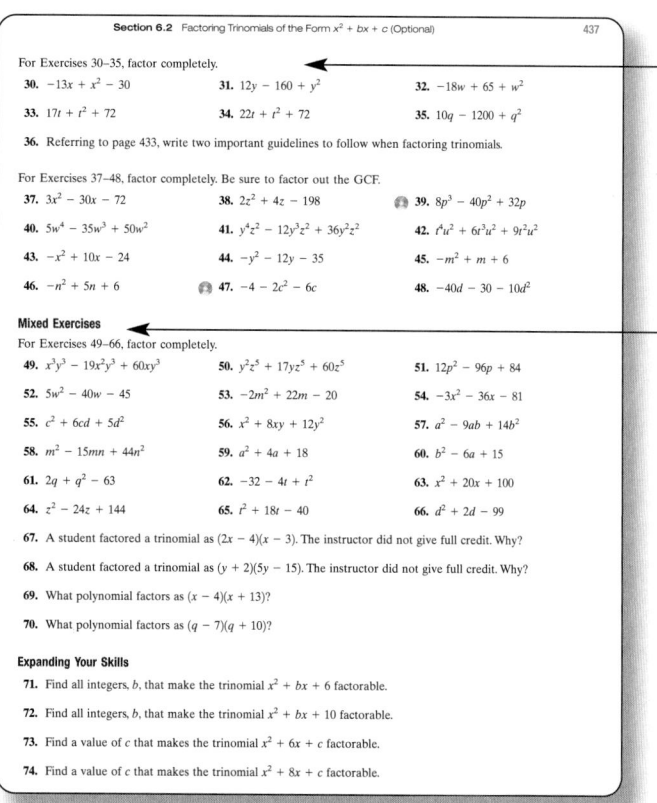

Section 6.2   Factoring Trinomials of the Form $x^2 + bx + c$ (Optional)   437

For Exercises 30–35, factor completely.

30. $-13x + x^2 - 30$ **31.** $12y - 160 + y^2$ **32.** $-18w + 65 + w^2$

33. $17t + t^2 + 72$ **34.** $22t + t^2 + 72$ **35.** $10q - 1200 + q^2$

36. Referring to page 433, write two important guidelines to follow when factoring trinomials.

For Exercises 37–48, factor completely. Be sure to factor out the GCF.

37. $3x^2 - 30x - 72$ **38.** $2z^2 + 4z - 198$ **39.** $8p^3 - 40p^2 + 32p$

40. $5w^4 - 35w^3 + 50w^2$ **41.** $y^4z^2 - 12y^3z^2 + 36y^2z^2$ **42.** $t^4u^2 + 6t^3u^2 + 9t^2u^2$

43. $-x^2 + 10x - 24$ **44.** $-y^2 - 12y - 35$ **45.** $-m^2 + m + 6$

46. $-n^2 + 5n + 6$ **47.** $-4 - 2c^2 - 6c$ **48.** $-40d - 30 - 10d^2$

Mixed Exercises

For Exercises 49–66, factor completely.

49. $x^3y^3 - 19x^2y^3 + 60xy^3$ **50.** $y^2z^5 + 17yz^5 + 60z^5$ **51.** $12p^2 - 96p + 84$

52. $5w^2 - 40w - 45$ **53.** $-2m^2 + 22m - 20$ **54.** $-3x^2 - 36x - 81$

55. $c^2 + 6cd + 5d^2$ **56.** $x^2 + 8xy + 12y^2$ **57.** $a^2 - 9ab + 14b^2$

58. $m^2 - 15mn + 44n^2$ **59.** $a^2 + 4a + 18$ **60.** $b^2 - 6a + 15$

61. $2q + q^2 - 63$ **62.** $-32 - 4t + t^2$ **63.** $x^2 + 20x + 100$

64. $z^2 - 24z + 144$ **65.** $t^2 + 18t - 40$ **66.** $d^2 + 2d - 99$

67. A student factored a trinomial as $(2x - 4)(x - 3)$. The instructor did not give full credit. Why?

68. A student factored a trinomial as $(y + 2)(5y - 15)$. The instructor did not give full credit. Why?

69. What polynomial factors as $(x - 4)(x + 13)$?

70. What polynomial factors as $(q - 7)(q + 10)$?

Expanding Your Skills

71. Find all integers, b, that make the trinomial $x^2 + bx + 6$ factorable.

72. Find all integers, b, that make the trinomial $x^2 + bx + 10$ factorable.

73. Find a value of c that makes the trinomial $x^2 + 6x + c$ factorable.

74. Find a value of c that makes the trinomial $x^2 + 8x + c$ factorable.

Practice Exercises

A variety of problem types appear in the section-ending Practice Exercises. Problem types are clearly labeled with either a heading or an icon for easy identification. References to MathZone are also found at the beginning of the Practice Exercises to remind students and instructors that additional help and practice problems are available. The core exercises for each section are organized by section concept. General references to examples are provided for blocks of core exercises. **Mixed Exercises** are also provided in some sections where no reference to concepts is offered.

Icon Key

The following key has been prepared for easy identification of "themed" exercises appearing within the Practice Exercises.

Student Edition

- Exercises Keyed to Video
- Calculator Exercises

AIE only

- Writing
- Translating Expressions
- Geometry

Study Skills Exercises appear at the beginning of selected exercise sets. They are designed to help students learn vocabulary and techniques to improve their study habits including exam preparation, note taking, and time management.

In the Practice Exercises, where appropriate, students are asked to define the **Key Terms** that are presented in the section. Assigning these exercises will help students to develop and expand their mathematical vocabularies.

Review Exercises also appear at the start of the Practice Exercises. The purpose of the Review Exercises is to help students retain their knowledge of concepts previously learned.

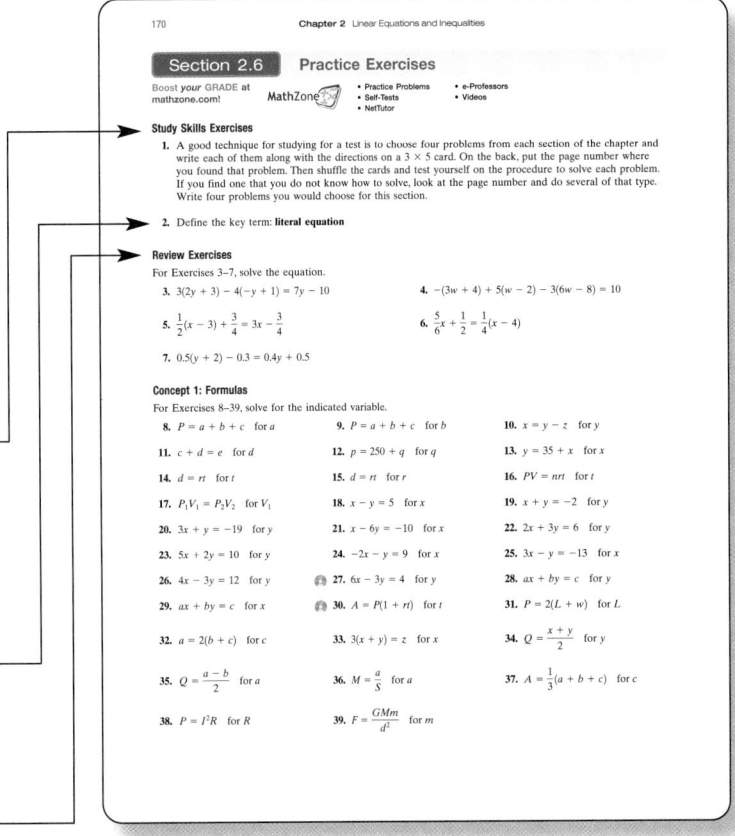

Writing Exercises offer students an opportunity to conceptualize and communicate their understanding of algebra. These, along with the **Translating Expressions Exercises**, enable students to strengthen their command of mathematical language and notation and to improve their reading and writing skills.

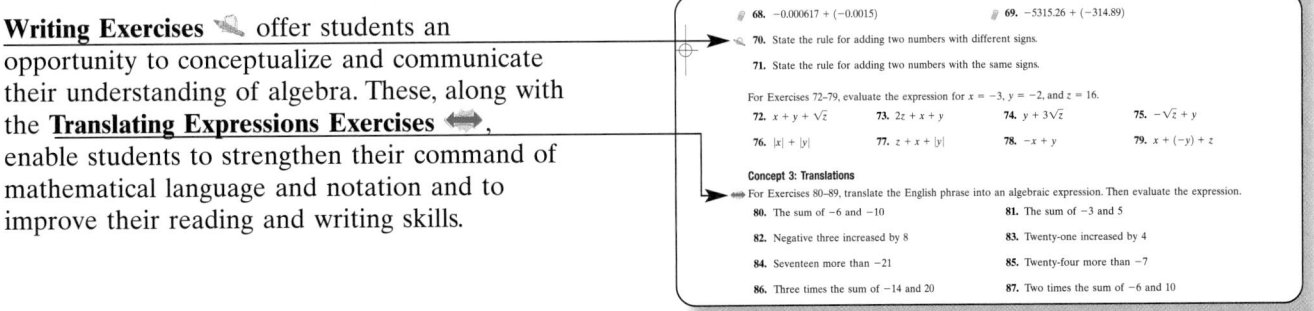

Geometry Exercises 🌐 appear throughout the Practice Exercises and encourage students to review and apply geometry concepts.

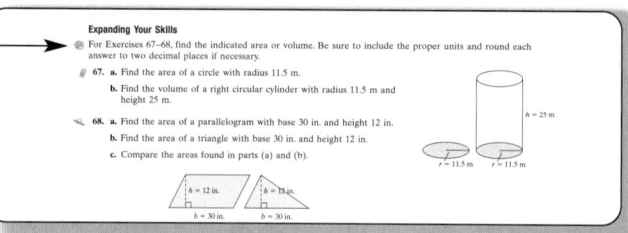

Calculator Exercises 🖩 signify situations where a calculator would provide assistance for time-consuming calculations. These exercises were carefully designed to demonstrate the types of situations in which a calculator is a handy tool rather than a "crutch."

Exercises Keyed to Video 🎥 are labeled with an icon to help students and instructors identify those exercises for which accompanying video instruction is available.

Applications based on real-world facts and figures motivate students and enable them to hone their problem-solving skills.

Expanding Your Skills exercises, found near the end of most Practice Exercises, challenge students' knowledge of the concepts presented.

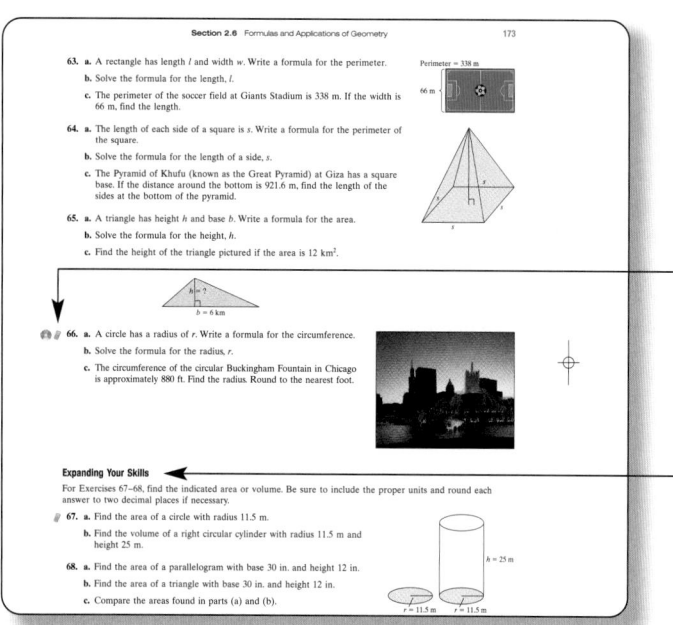

Optional Calculator Connections appear where appropriate and can be implemented at the instructor's discretion depending on the amount of emphasis placed on the calculator in the course. The Calculator Connections display keystrokes and provide an opportunity for students to apply the skill introduced.

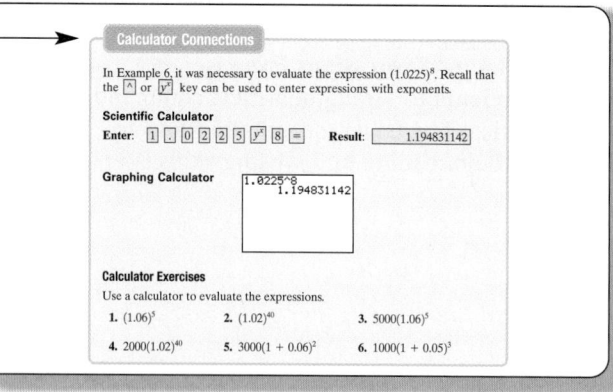

Chapter 6   SUMMARY

Section 6.1   Greatest Common Factor and Factoring by Grouping

Key Concepts

The **greatest common factor** (GCF) is the greatest factor common to all terms of a polynomial. To factor out the GCF from a polynomial, use the distributive property.

A four-term polynomial may be factorable by grouping.

Steps to Factoring by Grouping

1. Identify and factor out the GCF from all four terms.
2. Factor out the GCF from the first pair of terms. Factor out the GCF or its opposite from the second pair of terms.
3. If the two terms share a common binomial factor, factor out the binomial factor.

Examples

Example 1

$3x(a + b) - 5(a + b)$ Greatest common factor is $(a + b)$.

$= (a + b)(3x - 5)$

Example 2

$60xa - 30xb - 80ya + 40yb$

$= 10[6xa - 3xb - 8ya + 4yb]$ Factor out GCF.

$= 10[3x(2a - b) - 4y(2a - b)]$ Factor by grouping.

$= 10(2a - b)(3x - 4y)$

Chapter 6   Review Exercises

Section 6.1

For Exercises 1–4, identify the greatest common factor between each group of terms.

1. $15a^2b^4, 30a^3b, 9a^5b^3$
2. $3(x + 5), x(x + 5)$
3. $2c^3(3c - 5), 4c(3c - 5)$
4. $-2wyz, -4xyz$

For Exercises 5–10, factor out the greatest common factor.

5. $6x^2 + 2x^4 - 8x$
6. $11w^3y^3 - 44w^2y^5$
7. $-t^2 + 5t$
8. $-6u^2 - u$
9. $3b(b + 2) - 7(b + 2)$
10. $2(5x + 9) + 8x(5x + 9)$

26. When factoring a polynomial of the form $ax^2 - bx + c$, should the signs of the binomials be both positive, both negative, or different?

27. When factoring a polynomial of the form $ax^2 + bx + c$, should the signs of the binomials be both positive, both negative, or different?

28. When factoring a polynomial of the form $ax^2 + bx - c$, should the signs of the binomials be both positive, both negative, or different?

For Exercises 29–40, factor the trinomial using the trial-and-error method.

29. $2y^2 - 5y - 12$
30. $4w^2 - 5w - 6$
31. $10z^2 + 29z + 10$
32. $8z^2 + 6z - 9$

Chapter 6   Test

1. Factor out the GCF: $15x^4 - 3x + 6x^3$
2. Factor by grouping: $7a - 35 - a^2 + 5a$
3. Factor the trinomial: $6w^2 - 43w + 7$
4. Factor the difference of squares: $169 - p^2$
5. Factor the perfect square trinomial: $q^2 - 16q + 64$
6. Factor the sum of cubes: $8 + t^3$

For Exercises 7–25, factor completely.

7. $a^2 + 12a + 32$
8. $x^2 + x - 42$
9. $2y^2 - 17y + 8$
10. $6z^2 + 19z + 8$
11. $9t^2 - 100$
12. $v^2 - 81$
13. $3a^2 + 27ab + 54b^2$
14. $c^4 - 1$
15. $xy - 7x + 3y - 21$
16. $49 + p^2$
17. $-10u^2 + 30u - 20$
18. $12t^2 - 75$
19. $5y^2 - 50y + 125$
20. $21q^2 + 14q$
21. $2x^3 + x^2 - 8x - 4$
22. $y^3 - 125$

23. $m^2n^2 - 81$
24. $16a^2 - 64b^2$
25. $64x^3 - 27y^6$

For Exercises 26–30, solve the equation.

26. $(2x - 3)(x + 5) = 0$
27. $x^2 - 7x = 0$
28. $x^2 - 6x = 16$
29. $x(5x + 4) = 1$
30. $y^3 + 10y^2 - 9y - 90 = 0$

31. A tennis court has area of 312 yd². If the length is 2 yd more than twice the width, find the dimensions of the court.

32. The product of two consecutive odd integers is 35. Find the integers.

33. The height of a triangle is 5 in. less than the length of the base. The area is 42 in². Find the length of the base and the height of the triangle.

34. The hypotenuse of a right triangle is 2 ft less than three times the shorter leg. The longer leg is 3 ft less than three times the shorter leg. Find the length of the shorter leg.

Chapters 1–6   Cumulative Review Exercises

For Exercises 1–2, simplify completely.

1. $\dfrac{|4 - 25 \div (-5) \cdot 2|}{\sqrt{8^2 + 6^2}}$

2. Solve for t: $5 - 2(t + 4) = 3t + 12$

3. Solve for y: $3x - 2y = 8$

4. A child's piggy bank has $3.80 in quarters, dimes, and nickels. The number of nickels is two more than the number of quarters. The number of dimes is three less than the number of quarters. Find the number of each type of coin in the bank.

5. Solve the inequality. Graph the solution on a number line and write the solution set in interval notation.

$\dfrac{5}{12}x \le \dfrac{5}{3}$

8. Find an equation of the line passing through the point $(-3, 5)$ and having a slope of 3. Write the final answer in slope-intercept form.

9. Solve the system. $\begin{aligned} 2x - 3y &= 4 \\ 5x - 6y &= 13 \end{aligned}$

For Exercises 10–12, perform the indicated operations.

10. $2\left(\dfrac{1}{3}y^3 - \dfrac{3}{2}y^2 - 7\right) - \left(\dfrac{2}{3}y^3 + \dfrac{1}{2}y^2 + 5y\right)$

11. $(4p^2 - 5p - 1)(2p - 3)$

12. $(2w - 7)^2$

13. Divide using long division:
$(r^4 + 2r^3 - 5r + 1) \div (r - 3)$

14. $\dfrac{c^{12}c^{-5}}{c^3}$

End-of-Chapter Summary and Exercises

The **Summary**, located at the end of each chapter, outlines key concepts for each section and illustrates those concepts with examples. Following the Summary is a set of **Review Exercises** that are organized by section. A **Chapter Test** appears after each set of Review Exercises. Chapters 2–14 also include **Cumulative Reviews** that follow the Chapter Tests. These end-of-chapter materials provide students with ample opportunity to prepare for quizzes or exams.

Supplements

For the Instructor

Instructor's Resource Manual

The *Instructor's Resource Manual* (*IRM*), written by the authors, is a printable electronic supplement available through MathZone. The *IRM* includes discovery-based classroom activities, worksheets for drill and practice, materials for a student portfolio, and tips for implementing successful cooperative learning. Numerous classroom activities are available for each section of text and can be used as a complement to lecture or can be assigned for work outside of class. The activities are designed for group or individual work and take about 5–10 minutes each. With increasing demands on faculty schedules, these ready-made lessons offer a convenient means for both full-time and adjunct faculty to promote active learning in the classroom.

 **www.mathzone.com**

McGraw-Hill's **MathZone** is a complete online tutorial and course management system for mathematics and statistics, designed for greater ease of use than any other system available. Available with selected McGraw-Hill textbooks, the system enables instructors to **create and share courses and assignments** with colleagues and adjuncts with only a few clicks of the mouse. All assignments, questions, e-Professors, online tutoring, and video lectures are directly tied to **text-specific** materials.

MathZone courses are customized to your textbook, but you can edit questions and algorithms, import your own content, and **create** announcements and due dates for assignments.

MathZone has **automatic grading** and reporting of easy-to-assign, algorithmically generated homework, quizzing, and testing. All student activity within **MathZone** is automatically recorded and available to you through a **fully integrated grade book** that can be downloaded to Excel.

MathZone offers:

- **Practice exercises** based on the textbook and generated in an unlimited number for as much practice as needed to master any topic you study.
- **Videos** of classroom instructors giving lectures and showing you how to solve exercises from the textbook.
- **e-Professors** to take you through animated, step-by-step instructions (delivered via on-screen text and synchronized audio) for solving problems in the book, allowing you to digest each step at your own pace.
- **NetTutor,** which offers live, personalized tutoring via the Internet.

Instructor's Testing and Resource CD

This cross-platform CD-ROM provides a wealth of resources for the instructor. Among the supplements featured on the CD-ROM is a **computerized test bank** utilizing Brownstone Diploma® algorithm-based testing software to create customized exams quickly. This user-friendly program enables instructors to search for questions by topic, format, or difficulty level; to edit existing questions or to add new ones; and to scramble questions and answer keys for multiple versions of a single test. Hundreds of text-specific, open-ended, and multiple-choice questions are included in the question bank. Sample chapter tests are also provided.

ALEKS ALEKS (**A**ssessment and **LE**arning in **K**nowledge **S**paces) is an artificial-intelligence-based system for mathematics learning, available over the Web 24/7. Using unique adaptive questioning, ALEKS accurately assesses what topics each student knows and then determines exactly what each student is ready to learn next. ALEKS interacts with the students much as a skilled human tutor would, moving between explanation and practice as needed, correcting and analyzing errors, defining terms and changing topics on request, and helping them master the course content more quickly and easily. Moreover, the new ALEKS 3.0 now links to text-specific videos, multimedia tutorials, and textbook pages in PDF format. ALEKS also offers a robust classroom management system that enables instructors to monitor and direct student progress toward mastery of curricular goals. See www.highed.aleks.com.

New Connect2Developmental Mathematics Video Series!

Available on DVD and the MathZone website, these innovative videos bring essential Developmental Mathematics concepts to life! The videos take the concepts and place them in a real world setting so that students make the connection from what they learn in the classroom to real world experiences outside the classroom. Making use of 3D animations and lectures, Connect2Developmental Mathematics video series answers the age-old questions "Why is this important" and "When will I ever use it?" The videos cover topics from Arithmetic and Basic Mathematics through the Algebra sequence, mixing student-oriented themes and settings with basic theory.

Miller/O'Neill/Hyde Video Lectures on Digital Video Disk (DVD)

In the videos, qualified instructors work through selected problems from the textbook, following the solution methodology employed in the text. The video series is available on DVD or online as an assignable element of MathZone. The DVDs are closed-captioned for the hearing-impaired, are subtitled in Spanish, and meet the Americans with Disabilities Act Standards for Accessible Design. Instructors may use them as resources in a learning center, for online courses, and/or to provide extra help for students who require extra practice.

Annotated Instructor's Edition

In the *Annotated Instructor's Edition (AIE)*, **answers to all exercises and tests appear adjacent to each exercise,** in a color used *only* for annotations. The *AIE* also contains **Instructor Notes** that appear in the margin. The notes may assist with lecture preparation. Also found in the *AIE* are icons within the Practice Exercises that serve to guide instructors in their preparation of homework assignments and lessons.

Instructor's Solutions Manual

The *Instructor's Solutions Manual* provides comprehensive, worked-out solutions to all exercises in the Chapter Openers; the Practice Exercises; the Problem Recognition Exercises; the end-of-chapter Review Exercises; the Chapter Tests; and the Cumulative Review Exercises.

For the Student

 www.mathzone.com

McGraw-Hill's MathZone is a powerful Web-based tutorial for homework, quizzing, testing, and multimedia instruction. Also available in CD-ROM format, MathZone offers:

Practice exercises based on the text and generated in an unlimited quantity for as much practice as needed to master any objective

Video clips of classroom instructors showing how to solve exercises from the text, step by step

e-Professor animations that take the student through step-by-step instructions, delivered on-screen and narrated by a teacher on audio, for solving exercises from the textbook; the user controls the pace of the explanations and can review as needed

NetTutor, which offers personalized instruction by live tutors familiar with the textbook's objectives and problem-solving methods.

Every assignment, exercise, video lecture, and e-Professor is derived from the textbook.

Student's Solutions Manual

The *Student's Solutions Manual* provides comprehensive, worked-out solutions to the odd-numbered exercises in the Practice Exercise sets; the Problem Recognition Exercises, the end-of-chapter Review Exercises, the Chapter Tests, and the Cumulative Review Exercises. Answers to the odd- and even-numbered entries to the Chapter Opener Puzzles are also provided.

New Connect2Developmental Mathematics Video Series!

Available on DVD and the MathZone website, these innovative videos bring essential Developmental Mathematics concepts to life! The videos take the concepts and place them in a real world setting so that students make the connection from what they learn in the classroom to real world experiences outside the classroom. Making use of 3D animations and lectures, Connect2Developmental Mathematics video series answers the age-old questions "Why is this important" and "When will I ever use it?" The videos cover topics from Arithmetic and Basic Mathematics through the Algebra sequence, mixing student-oriented themes and settings with basic theory.

Video Lectures on Digital Video Disk

The video series is based on exercises from the textbook. Each presenter works through selected problems, following the solution methodology employed in the text. The video series is available on DVD or online as part of MathZone. The DVDs are closed-captioned for the hearing impaired, are subtitled in Spanish, and meet the Americans with Disabilities Act Standards for Accessible Design.

NetTutor

Available through MathZone, NetTutor is a revolutionary system that enables students to interact with a live tutor over the Web. NetTutor's Web-based, graphical chat capabilities enable students and tutors to use mathematical notation and even to draw graphs as they work through a problem together. Students can also submit questions and receive answers, browse previously answered questions, and view previous sessions. Tutors are familiar with the textbook's objectives and problem-solving styles.

Reference

Chapter R is a reference chapter that provides a review of the basic operations on fractions. It also offers background on some of the important facts from geometry that will be used later in the text.

As you work through Section R.3, see if you recognize any of these important shapes and figures. Match each figure number on the left with a letter on the right. Then fill in the blanks below with the appropriate letter. The letters will form a word to complete the puzzle.

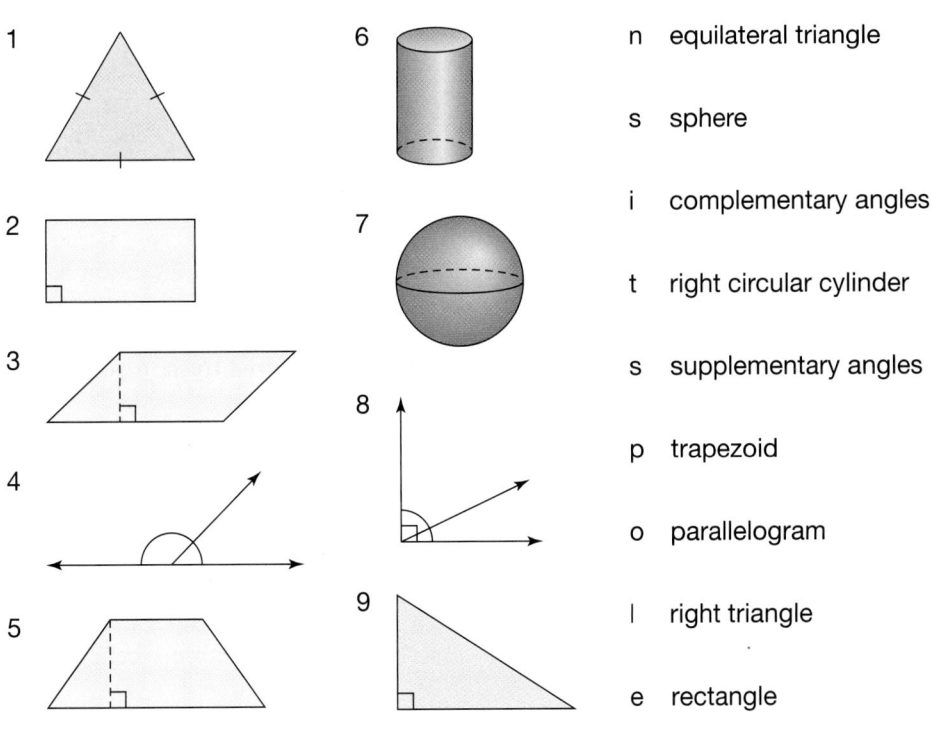

1
2
3
4
5
6
7
8
9

n   equilateral triangle

s   sphere

i   complementary angles

t   right circular cylinder

s   supplementary angles

p   trapezoid

o   parallelogram

l   right triangle

e   rectangle

Without geometry, life would be $\overline{}_{5}\ \overline{}_{3}\ \overline{}_{8}\ \overline{}_{1}\ \overline{}_{6}\ \overline{}_{9}\ \overline{}_{2}\ \overline{}_{7}\ \overline{}_{4}$

Study Tips

In taking a course in algebra, you are making a commitment to yourself, your instructor, and your classmates. Following some or all of the study tips listed here can help you be successful in this endeavor. The features of this text that will assist you are printed in blue.

1. Before the Course

- Purchase the necessary materials for the course before the course begins or on the first day.
- Obtain a three-ring binder to keep and organize your notes, homework, tests, and any other materials acquired in the class. We call this type of notebook a portfolio.
- Arrange your schedule so that you have enough time to attend class and to do homework. A common rule of thumb is to set aside at least 2 hours for homework for every hour spent in class. That is, if you are taking a 4 credit-hour course, plan on at least 8 hours a week for homework. If you experience difficulty in mathematics, plan for more time. A 4 credit-hour course will then take *at least* 12 hours each week—about the same as a part-time job.
- Communicate with your employer and family members the importance of your success in this course so that they can support you.
- Be sure to find out the type of calculator (if any) that your instructor requires.

2. During the Course

- Read the section in the text *before* the lecture to familiarize yourself with the material and terminology.
- Attend every class, and be on time.
- Take notes in class. Write down all of the examples that the instructor presents. Read the notes after class, and add any comments to make your notes clearer to you. Use a tape recorder to record the lecture if the instructor permits the recording of lectures.
- Ask questions in class.
- Read the section in the text *after* the lecture, and pay special attention to the Tip boxes and Avoiding Mistakes boxes.
- Do homework every night. Even if your class does not meet every day, you should still do some work every night to keep the material fresh in your mind.
- Check your homework with the answers that are supplied in the back of this text. Correct the exercises that do not match, and circle or star those that you cannot correct yourself. This way you can easily find them and ask your instructor the next day.
- Write the definition and give an example of each Key Term found at the beginning of the Practice Exercises.
- The Problem Recognition Exercises provide additional practice distinguishing among a variety of problem types. Sometimes the most difficult part of learning mathematics is retaining all that you learn. These exercises are excellent tools for retention of material.
- Form a study group with fellow students in your class, and exchange phone numbers. You will be surprised by how much you can learn by talking about mathematics with other students.

- If you use a calculator in your class, read the Calculator Connections boxes to learn how and when to use your calculator.
- Ask your instructor where you might obtain extra help if necessary.

3. Preparation for Exams

- Look over your homework. Pay special attention to the exercises you have circled or starred to be sure that you have learned that concept.
- Read through the Summary at the end of the chapter. Be sure that you understand each concept and example. If not, go to the section in the text and reread that section.
- Give yourself enough time to take the Chapter Test uninterrupted. Then check the answers. For each problem you answered incorrectly, go to the Review Exercises and do all of the problems that are similar.
- To prepare for the final exam, complete the Cumulative Review Exercises at the end of each chapter, starting with Chapter 2. If you complete the cumulative reviews after finishing each chapter, then you will be preparing for the final exam throughout the course. The Cumulative Review Exercises are another excellent tool for helping you retain material.

4. Where to Go for Help

- At the first sign of trouble, see your instructor. Most instructors have specific office hours set aside to help students. Don't wait until after you have failed an exam to seek assistance.
- Get a tutor. Most colleges and universities have free tutoring available.
- When your instructor and tutor are unavailable, use the Student Solutions Manual for step-by-step solutions to the odd-numbered problems in the exercise sets.
- Work with another student from your class.
- Work on the computer. Many mathematics tutorial programs and websites are available on the Internet, including the website that accompanies this text: www.mhhe.com/miller_oneill

Section R.1 Practice Exercises

Boost *your* GRADE at mathzone.com!

- Practice Problems
- Self-Tests
- NetTutor
- e-Professors
- Videos

Concept 1: Before the Course

1. To be motivated to complete a course, it is helpful to have a very clear reason for taking the course. List your goals for taking this course.

2. Budgeting enough time to do homework and to study for a class is one of the most important steps to success in a class. Use the weekly calendar to help you plan your time for your studies this week. Also write other obligations such as the time required for your job, for your family, for sleeping, and for eating. Be realistic when estimating the time for each activity.

Time	Mon.	Tues.	Wed.	Thurs.	Fri.	Sat.	Sun.
7–8							
8–9							
9–10							
10–11							
11–12							
12–1							
1–2							
2–3							
3–4							
4–5							
5–6							
6–7							
7–8							
8–9							
9–10							

3. Taking 12 credit hours is the equivalent of a full-time job. Often students try to work too many hours while taking classes at school.

   a. Write down how many hours you work per week and the number of credit hours you are taking this term.

   number of hours worked per week _____

   number of credit hours this term _____

   b. The next table gives a recommended limit to the number of hours you should work for the number of credit hours you are taking at school. (Keep in mind that other responsibilities in your life such as your family might also make it necessary to limit your hours at work even more.) How do your numbers from part (a) compare to those in the table? Are you working too many hours?

Number of credit hours	Maximum number of hours of work per week
3	40
6	30
9	20
12	10
15	0

4. It is important to establish a place where you can study—someplace that has few distractions and is readily available. Answer the questions about the space that you have chosen.

   a. Is there enough room to spread out books and paper to do homework?

   b. Is this space available anytime?

   c. Are there any distractions? Can you be interrupted?

   d. Is the furniture appropriate for studying? That is, is there a comfortable chair and good lighting?

5. Organization is an important ingredient to success. A calendar or pocket planner is a valuable resource for keeping track of assignments and test dates. Write the date of the first test in this class.

Concept 2: During the Course

6. Taking notes can help in many ways. Good notes provide examples for reference as you do your homework. Also, taking notes keeps your mind on track during the lecture and helps make you an active listener. Here are some tips to help you take better notes in class.

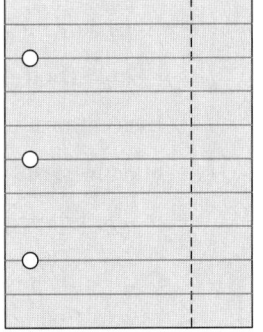

    a. In your next math class, take notes by drawing a vertical line about $\frac{3}{4}$ of the way across the paper as shown. On the left side, write down what your instructor puts on the board or overhead. On the right side, make your own comments about important words, procedures, or questions that you have.

    b. Revisit your notes as soon as possible after class to fill in the missing parts that you recall from lecture but did not have time to write down.

    c. Be sure that you label each page with the date, chapter, section, and topic. This will make it easier to study from when you study for a test.

7. Many instructors use a variety of styles to accommodate all types of learners. From the following list, try to identify the type of learner that best describes you.

Auditory Learner Do you learn best by listening to your instructor's lectures? Do you tape the lecture so that you can listen to it as many times as you need? Do you talk aloud when doing homework or study for a test?

Visual Learner Do you learn best by seeing problems worked out on the board? Do you understand better if there is a picture or illustration accompanying the problem? Do you take notes in class?

Tactile Learner Do you learn best with hands-on projects? Do you prefer having some sort of physical objects to manipulate? Do you prefer to move around the classroom in a lab situation?

Concept 3: Preparation for Exams

8. When taking a test, go through the test and work all the problems that you know first. Then go back and work on the problems that were more difficult. Give yourself a time limit for each problem (maybe 3 to 5 minutes the first time through the test). Circle the importance of each statement.

	not important	somewhat important	very important
a. Read through the entire test first.	1	2	3
b. If time allows, go back and check each problem.	1	2	3
c. Write out all steps instead of doing the work in your head.	1	2	3

9. One way to lessen test anxiety is to feel prepared for the exam. Check the ways that you think might be helpful in preparing for a test.

_____ Do the chapter test in the text.

_____ Get in a study group with your peers and go over the major topics.

_____ Write your own pretest.

_____ Use the online component that goes with your textbook.

_____ Write an outline of the major topics. Then find an example for each topic.

_____ Make flash cards with a definition, property, or rule on one side and an example on the other.

_____ Re-read your notes that you took in class.

_____ Write down the definitions of all key terms introduced in the chapter.

10. The following list gives symptoms of math anxiety.

- Experiencing loss of sleep and worrying about an upcoming exam.
- Mind becoming blank when answering a question or taking a test.
- Becoming nervous about asking questions in class.
- Experiencing anxiety that interferes with studying.
- Being afraid or embarrassed to let the instructor see your work.
- Becoming physically ill during a test.
- Having sweaty palms or shaking hands when asked a math question.

Have you experienced any of these symptoms? If so, how many and how often?

11. If you think that you have math anxiety, read the following list for some possible solutions. Check the activities that you can realistically try to help you overcome this problem.

_____ Read a book on math anxiety.

_____ Search the Web for help tips on handling math anxiety.

_____ See a counselor to discuss your anxiety.

_____ See your instructor to inform him or her about your situation.

_____ Evaluate your time management to see if you are trying to do too much. Then adjust your schedule accordingly.

Concept 4: Where to Go for Help

12. Does your college offer free tutoring? If so, write down the room number and the hours of the tutoring center.

13. Does your instructor have office hours? If so, write down your professor's office number and office hours.

14. Is there a supplement to your text? If so, find out its price and where you can get it.

15. Find out how to access the online tutoring available with this text.

Section R.2   Fractions

Concepts

1. Basic Definitions
2. Prime Factorization
3. Simplifying Fractions to Lowest Terms
4. Multiplying Fractions
5. Dividing Fractions
6. Adding and Subtracting Fractions
7. Operations on Mixed Numbers

1. Basic Definitions

The study of algebra involves many of the operations and procedures used in arithmetic. Therefore, we begin this text by reviewing the basic operations of addition, subtraction, multiplication, and division on fractions and mixed numbers.

In day-to-day life, the numbers we use for counting are

the **natural numbers:** $1, 2, 3, 4, \ldots$ and

the **whole numbers:** $0, 1, 2, 3, \ldots$.

Whole numbers are used to count the number of whole units in a quantity. A fraction is used to express part of a whole unit. If a child gains $2\frac{1}{2}$ lb, the child has gained two whole pounds plus a portion of a pound. To express the additional half pound mathematically, we may use the fraction, $\frac{1}{2}$.

A Fraction and Its Parts

Fractions are numbers of the form $\frac{a}{b}$, where $\frac{a}{b} = a \div b$ and b does not equal zero.

In the fraction $\frac{a}{b}$, the **numerator** is a, and the **denominator** is b.

The denominator of a fraction indicates how many equal parts divide the whole. The numerator indicates how many parts are being represented. For instance, suppose Jack wants to plant carrots in $\frac{2}{5}$ of a rectangular garden. He can divide the garden into five equal parts and use two of the parts for carrots (Figure R-1).

The shaded region represents $\frac{2}{5}$ of the garden.

Figure R-1

Definition of a Proper Fraction, an Improper Fraction, and a Mixed Number

1. If the numerator of a fraction is less than the denominator, the fraction is a **proper fraction**. A proper fraction represents a quantity that is less than a whole unit.

2. If the numerator of a fraction is greater than or equal to the denominator, then the fraction is an **improper fraction**. An improper fraction represents a quantity greater than or equal to a whole unit.

3. A **mixed number** is a whole number added to a proper fraction.

Proper Fractions: $\dfrac{3}{5}$ ⊘ $\dfrac{1}{8}$ ⊛

Improper Fractions: $\dfrac{7}{5}$ ⊘ ⊘ $\dfrac{8}{8}$ ⊛

Mixed Numbers: $1\frac{1}{5}$ ⊘ ⊘ $2\frac{3}{8}$ ⊛ ⊛ ⊛

2. Prime Factorization

To perform operations on fractions it is important to understand the concept of a factor. For example, when the numbers 2 and 6 are multiplied, the result (called the **product**) is 12.

$$2 \times 6 = 12$$

factors   product

The numbers 2 and 6 are said to be **factors** of 12. (In this context, we refer only to natural number factors.) The number 12 is said to be factored when it is written as the product of two or more natural numbers. For example, 12 can be factored in several ways:

$$12 = 1 \times 12 \qquad 12 = 2 \times 6 \qquad 12 = 3 \times 4 \qquad 12 = 2 \times 2 \times 3$$

A natural number greater than 1 that has only two factors, 1 and itself, is called a **prime number**. The first several prime numbers are 2, 3, 5, 7, 11, and 13. A natural number greater than 1 that is not prime is called a **composite number**. That is, a composite number has factors other than itself and 1. The first several composite numbers are 4, 6, 8, 9, 10, 12, 14, 15, and 16.

The number 1 is neither prime nor composite.

Example 1 | **Writing a Natural Number as a Product of Prime Factors**

Write each number as a product of prime factors.

a. 12   **b.** 30

Solution:

a. $12 = 2 \times 2 \times 3$ Divide 12 by prime numbers until only prime numbers are obtained.

$$\begin{array}{r} 2\overline{)12} \\ 2\overline{)6} \\ 3 \end{array}$$

Or use a factor tree

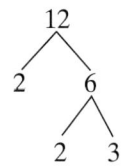

b. $30 = 2 \times 3 \times 5$

$$\begin{array}{r} 2\overline{)30} \\ 3\overline{)15} \\ 5 \end{array}$$

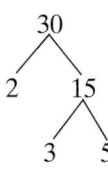

Skill Practice | Write the number as a product of prime factors.

1. 40   **2.** 60

3. Simplifying Fractions to Lowest Terms

The process of factoring numbers can be used to reduce or simplify fractions to lowest terms. A fractional portion of a whole can be represented by infinitely many fractions. For example, Figure R-2 shows that $\frac{1}{2}$ is equivalent to $\frac{2}{4}, \frac{3}{6}, \frac{4}{8}$, and so on.

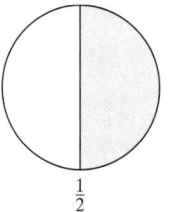
$\frac{1}{2}$

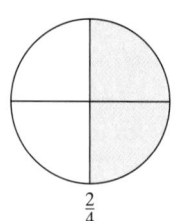
$\frac{2}{4}$

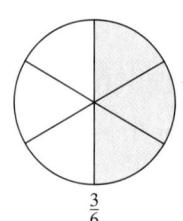
$\frac{3}{6}$

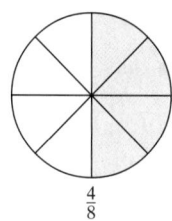
$\frac{4}{8}$

Skill Practice Answers
1. $2 \times 2 \times 2 \times 5$
2. $2 \times 2 \times 3 \times 5$

Figure R-2

The fraction $\frac{1}{2}$ is said to be in **lowest terms** because the numerator and denominator share no common factor other than 1.

To simplify a fraction to lowest terms, we use the following important principle.

The Fundamental Principle of Fractions

Consider the fraction $\frac{a}{b}$ and the nonzero number, c. Then,

$$\frac{a}{b} = \frac{a \div c}{b \div c}$$

The fundamental principle of fractions indicates that dividing both the numerator and denominator by the same nonzero number results in an equivalent fraction. For example, the numerator and denominator of the fraction $\frac{3}{6}$ both share a common factor of 3. To reduce $\frac{3}{6}$, we will divide both the numerator and denominator by the common factor, 3.

$$\frac{3}{6} = \frac{3 \div 3}{6 \div 3} = \frac{1}{2}$$

Before applying the fundamental principle of fractions, it is helpful to write the prime factorization of both the numerator and denominator. We can also use a slash, /, to denote division of common factors.

Example 2 Simplifying a Fraction to Lowest Terms

Simplify the fraction $\dfrac{30}{12}$ to lowest terms.

Solution:

From Example 1, we have $30 = 2 \times 3 \times 5$ and $12 = 2 \times 2 \times 3$. Hence,

$$\frac{30}{12} = \frac{2 \times 3 \times 5}{2 \times 2 \times 3} \qquad \text{Factor the numerator and denominator.}$$

$$= \frac{\overset{1}{2} \times \overset{1}{3} \times 5}{\underset{1}{2} \times 2 \times \underset{1}{3}} \qquad \begin{array}{l}\text{Divide the numerator and denominator by the} \\ \text{common factors, 2 and 3.}\end{array}$$

Multiply $1 \times 1 \times 5 = 5$.

$$= \frac{\overset{1}{2} \times \overset{1}{3} \times 5}{\underset{1}{2} \times 2 \times \underset{1}{3}} = \frac{5}{2}$$

Multiply $1 \times 2 \times 1 = 2$.

Skill Practice Simplify to lowest terms.

3. $\dfrac{20}{30}$

Skill Practice Answers

3. $\dfrac{2}{3}$

Simplifying a Fraction to Lowest Terms

Simplify $\dfrac{14}{42}$ to lowest terms.

Solution:

$$\frac{14}{42} = \frac{2 \times 7}{2 \times 3 \times 7}$$ Factor the numerator and denominator.

$$= \frac{\overset{1}{2} \times \overset{1}{7}}{\underset{1}{2} \times 3 \times \underset{1}{7}} = \frac{1}{3}$$

Multiply $1 \times 1 = 1$.

Multiply $1 \times 3 \times 1 = 3$.

Avoiding Mistakes:

In Example 3, the common factors 2 and 7 in the numerator and denominator simplify to 1. It is important to remember to write the factor of 1 in the numerator. The simplified form of the fraction is $\frac{1}{3}$.

Skill Practice   Simplify to lowest terms.

4. $\dfrac{32}{12}$

4. Multiplying Fractions

Multiplying Fractions

If b is not zero and d is not zero, then

$$\frac{a}{b} \times \frac{c}{d} = \frac{a \times c}{b \times d}$$

To multiply fractions, multiply the numerators and multiply the denominators.

Multiplying Fractions

Multiply the fractions: $\dfrac{1}{4} \times \dfrac{1}{2}$

Solution:

$$\frac{1}{4} \times \frac{1}{2} = \frac{1 \times 1}{4 \times 2} = \frac{1}{8}$$ Multiply the numerators. Multiply the denominators.

Notice that the product $\frac{1}{4} \times \frac{1}{2}$ represents a quantity that is $\frac{1}{4}$ of $\frac{1}{2}$. Taking $\frac{1}{4}$ of a quantity is equivalent to dividing the quantity by 4. One-half of a pie divided into four pieces leaves pieces that each represent $\frac{1}{8}$ of the pie (Figure R-3).

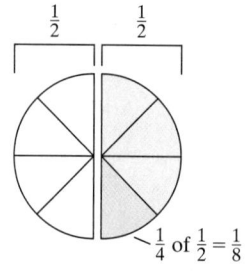

$$\frac{1}{4} \text{ of } \frac{1}{2} = \frac{1}{8}$$

Figure R-3

Skill Practice Answers

4. $\dfrac{8}{3}$

Skill Practice Multiply.

5. $\dfrac{2}{7} \times \dfrac{3}{4}$

Example 5   **Multiplying Fractions**

Multiply the fractions.

a. $\dfrac{7}{10} \times \dfrac{15}{14}$ **b.** $\dfrac{2}{13} \times \dfrac{13}{2}$ **c.** $5 \times \dfrac{1}{5}$

Solution:

a. $\dfrac{7}{10} \times \dfrac{15}{14} = \dfrac{7 \times 15}{10 \times 14}$ Multiply the numerators, and multiply the denominators.

Divide common factors in the numerator and denominator.

$= \dfrac{\overset{1}{7} \times 3 \times \overset{1}{5}}{2 \times \underset{1}{5} \times 2 \times \underset{1}{7}} = \dfrac{3}{4}$ Multiply $1 \times 3 \times 1 = 3$.

Multiply $2 \times 1 \times 2 \times 1 = 4$.

b. $\dfrac{2}{13} \times \dfrac{13}{2} = \dfrac{2 \times 13}{13 \times 2} = \dfrac{2 \times \overset{1}{13}}{\underset{1}{13} \times 2} = \dfrac{1}{1} = 1$ Multiply $1 \times 1 = 1$.

Multiply $1 \times 1 = 1$.

c. $5 \times \dfrac{1}{5} = \dfrac{5}{1} \times \dfrac{1}{5}$ The whole number 5 can be written as $\frac{5}{1}$.

$= \dfrac{\overset{1}{5} \times 1}{1 \times \underset{1}{5}} = \dfrac{1}{1} = 1$ Multiply and simplify to lowest terms.

Skill Practice Multiply.

6. $\dfrac{8}{9} \times \dfrac{3}{4}$ **7.** $\dfrac{4}{5} \times \dfrac{5}{4}$ **8.** $10 \times \dfrac{1}{10}$

---

5. Dividing Fractions

Before we divide fractions, we need to know how to find the reciprocal of a fraction. Notice from Example 5 that $\frac{2}{13} \times \frac{13}{2} = 1$ and $5 \times \frac{1}{5} = 1$. The numbers $\frac{2}{13}$ and $\frac{13}{2}$ are said to be reciprocals because their product is 1. Likewise the numbers 5 and $\frac{1}{5}$ are reciprocals.

Skill Practice Answers

5. $\dfrac{3}{14}$ **6.** $\dfrac{2}{3}$ **7.** 1   **8.** 1

The Reciprocal of a Number

Two nonzero numbers are **reciprocals** of each other if their product is 1. Therefore, the reciprocal of the fraction

$$\frac{a}{b} \text{ is } \frac{b}{a} \qquad \text{because} \qquad \frac{a}{b} \times \frac{b}{a} = 1$$

Number	Reciprocal	Product
$\dfrac{2}{13}$	$\dfrac{13}{2}$	$\dfrac{2}{13} \times \dfrac{13}{2} = 1$
$\dfrac{1}{8}$	$\dfrac{8}{1}$ (or equivalently 8)	$\dfrac{1}{8} \times 8 = 1$
$6\left(\text{or equivalently } \dfrac{6}{1}\right)$	$\dfrac{1}{6}$	$6 \times \dfrac{1}{6} = 1$

To understand the concept of dividing fractions, consider a pie that is half-eaten. Suppose the remaining half must be divided among three people, that is, $\frac{1}{2} \div 3$. However, dividing by 3 is equivalent to taking $\frac{1}{3}$ of the remaining $\frac{1}{2}$ of the pie (Figure R-4).

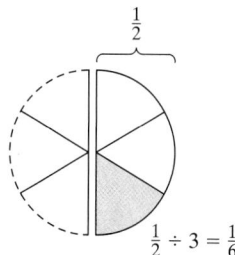

$$\frac{1}{2} \div 3 = \frac{1}{2} \cdot \frac{1}{3} = \frac{1}{6}$$

$$\frac{1}{2} \div 3 = \frac{1}{6}$$

Figure R-4

This example illustrates that dividing two numbers is equivalent to multiplying the first number by the reciprocal of the second number.

Division of Fractions

Let a, b, c, and d be numbers such that b, c, and d are not zero. Then,

$$\frac{a}{b} \div \frac{c}{d} = \frac{a}{b} \times \frac{d}{c}$$

multiply

reciprocal

To divide fractions, multiply the first fraction by the reciprocal of the second fraction.

| Example 6 | Dividing Fractions |

Divide the fractions.

a. $\dfrac{8}{5} \div \dfrac{3}{10}$ **b.** $\dfrac{12}{13} \div 6$

Solution:

a. $\dfrac{8}{5} \div \dfrac{3}{10} = \dfrac{8}{5} \times \dfrac{10}{3}$ Multiply by the reciprocal of $\frac{3}{10}$, which is $\frac{10}{3}$.

$= \dfrac{8 \times \overset{2}{\cancel{10}}}{\underset{1}{\cancel{5}} \times 3} = \dfrac{16}{3}$ Multiply and simplify to lowest terms.

b. $\dfrac{12}{13} \div 6 = \dfrac{12}{13} \div \dfrac{6}{1}$ Write the whole number 6 as $\frac{6}{1}$.

$= \dfrac{12}{13} \times \dfrac{1}{6}$ Multiply by the reciprocal of $\frac{6}{1}$, which is $\frac{1}{6}$.

$= \dfrac{\overset{2}{\cancel{12}} \times 1}{13 \times \underset{1}{\cancel{6}}} = \dfrac{2}{13}$ Multiply and simplify to lowest terms.

| Skill Practice | Divide.

9. $\dfrac{12}{25} \div \dfrac{8}{15}$ **10.** $\dfrac{9}{11} \div 3$

6. Adding and Subtracting Fractions

Adding and Subtracting Fractions

Two fractions can be added or subtracted if they have a common denominator. Let a, b, and c, be numbers such that b does not equal zero. Then,

$$\frac{a}{b} + \frac{c}{b} = \frac{a+c}{b} \quad \text{and} \quad \frac{a}{b} - \frac{c}{b} = \frac{a-c}{b}$$

To add or subtract fractions with the same denominator, add or subtract the numerators and write the result over the common denominator.

| Example 7 | Adding and Subtracting Fractions with the Same Denominator |

Add or subtract as indicated.

a. $\dfrac{1}{12} + \dfrac{7}{12}$ **b.** $\dfrac{13}{5} - \dfrac{3}{5}$

Skill Practice Answers

9. $\dfrac{9}{10}$ **10.** $\dfrac{3}{11}$

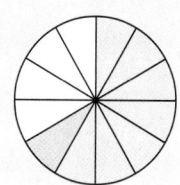

Solution:

a. $\dfrac{1}{12} + \dfrac{7}{12} = \dfrac{1+7}{12}$ Add the numerators.

$\qquad\qquad = \dfrac{8}{12}$

$\qquad\qquad = \dfrac{2}{3}$ Simplify to lowest terms.

b. $\dfrac{13}{5} - \dfrac{3}{5} = \dfrac{13-3}{5}$ Subtract the numerators.

$\qquad\qquad = \dfrac{10}{5}$ Simplify.

$\qquad\qquad = 2$ Simplify to lowest terms.

> **Skill Practice**   Add or subtract as indicated.

11. $\dfrac{2}{3} + \dfrac{5}{3}$ **12.** $\dfrac{5}{8} - \dfrac{1}{8}$

In Example 7, we added and subtracted fractions with the same denominators. To add or subtract fractions with different denominators, we must first become familiar with the idea of a least common multiple between two or more numbers. The **least common multiple (LCM)** of two numbers is the smallest whole number that is a multiple of each number. For example, the LCM of 6 and 9 is 18.

multiples of 6:   6, 12, 18, 24, 30, 36, . . .

multiples of 9:   9, 18, 27, 36, 45, 54, . . .

Listing the multiples of two or more given numbers can be a cumbersome way to find the LCM. Therefore, we offer the following method to find the LCM of two numbers.

Steps to Finding the LCM of Two Numbers

1. Write each number as a product of prime factors.
2. The LCM is the product of unique prime factors from *both* numbers. If a factor is repeated within the factorization of either number, use that factor the maximum number of times it appears in *either* factorization.

> **Example 8**   Finding the LCM of Two Numbers

Find the LCM of 9 and 15.

Solution:

$9 = 3 \times 3$ and $15 = 3 \times 5$ Factor the numbers.

$\text{LCM} = 3 \times 3 \times 5 = 45$ The LCM is the product of the factors of 3 and 5, where 3 is repeated twice.

> **Skill Practice**   Find the LCM.

13. 10 and 25

To add or subtract fractions with *different* denominators, we must first write each fraction as an equivalent fraction with a common denominator. A common denominator may be *any* common multiple of the denominators. However, we will use the least common denominator. The **least common denominator (LCD)** of two or more fractions is the LCM of the denominators of the fractions. The following steps outline the procedure to write a fraction as an equivalent fraction with a common denominator.

Writing Equivalent Fractions

To write a fraction as an equivalent fraction with a common denominator, multiply the numerator and denominator by the factors from the common denominator that are missing from the denominator of the original fraction.

Note: Multiplying the numerator and denominator by the *same* nonzero quantity will not change the value of the fraction.

Example 9 Writing Equivalent Fractions

a. Write the fractions $\frac{1}{9}$ and $\frac{1}{15}$ as equivalent fractions with the LCD as the denominator.

b. Subtract $\frac{1}{9} - \frac{1}{15}$.

Solution:

From Example 8, we know that the LCM for 9 and 15 is 45. Therefore, the LCD of $\frac{1}{9}$ and $\frac{1}{15}$ is 45.

a. $\dfrac{1}{9} = \dfrac{1 \times 5}{9 \times 5} = \dfrac{5}{45}$ Multiply numerator and denominator by 5. This creates a denominator of 45.

$\dfrac{1}{15} = \dfrac{1 \times 3}{15 \times 3} = \dfrac{3}{45}$ Multiply numerator and denominator by 3. This creates a denominator of 45.

b. $\dfrac{1}{9} - \dfrac{1}{15}$

$= \dfrac{5}{45} - \dfrac{3}{45}$ Write $\frac{1}{9}$ and $\frac{1}{15}$ as equivalent fractions with the same denominator.

$= \dfrac{2}{45}$ Subtract.

Skill Practice

14. Write the fractions $\frac{5}{8}$ and $\frac{5}{12}$ as equivalent fractions with the LCD as the denominator.

Example 10 Adding Fractions with Different Denominators

Suppose Nakeysha ate $\frac{1}{2}$ of an ice-cream pie, and her friend Carla ate $\frac{1}{3}$ of the pie. How much of the ice-cream pie was eaten?

Skill Practice Answers

14. $\dfrac{15}{24}, \dfrac{10}{24}$

Solution:

$$\frac{1}{2} + \frac{1}{3}$$ The LCD is $3 \times 2 = 6$.

$$= \frac{1 \times 3}{2 \times 3} + \frac{1 \times 2}{3 \times 2}$$ Multiply numerator and denominator by the missing factors.

$$= \frac{3}{6} + \frac{2}{6}$$

$$= \frac{5}{6}$$ Add the fractions.

Together, Nakeysha and Carla ate $\frac{5}{6}$ of the ice-cream pie.

By converting the fractions $\frac{1}{2}$ and $\frac{1}{3}$ to the same denominator, we are able to add *like*-size pieces of pie (Figure R-5).

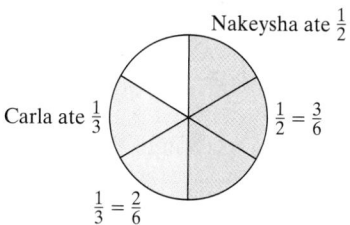

Nakeysha ate $\frac{1}{2}$

Carla ate $\frac{1}{3}$

$\frac{1}{2} = \frac{3}{6}$

$\frac{1}{3} = \frac{2}{6}$

Figure R-5

Skill Practice

15. Adam painted $\frac{1}{4}$ of a fence and Megan painted $\frac{1}{6}$ of the fence. How much of the fence was painted?

Example 11	**Adding and Subtracting Fractions**

Simplify: $\frac{1}{2} + \frac{5}{8} - \frac{11}{16}$.

Solution:

$$\frac{1}{2} + \frac{5}{8} - \frac{11}{16}$$ Factor the denominators: $2 = 2$, $8 = 2 \times 2 \times 2$, and $16 = 2 \times 2 \times 2 \times 2$. The LCD is $2 \times 2 \times 2 \times 2 = 16$. Next write the fractions as equivalent fractions with the LCD.

$$= \frac{1 \times 2 \times 2 \times 2}{2 \times 2 \times 2 \times 2} + \frac{5 \times 2}{8 \times 2} - \frac{11}{16}$$ Multiply numerators and denominators by the factors missing from each denominator.

$$= \frac{8}{16} + \frac{10}{16} - \frac{11}{16}$$

$$= \frac{8 + 10 - 11}{16}$$ Add and subtract the numerators.

$$= \frac{7}{16}$$ Simplify.

Skill Practice Answers

15. Together, Adam and Megan painted $\frac{5}{12}$ of the fence.

Skill Practice Add.

16. $\dfrac{2}{3} + \dfrac{1}{2} + \dfrac{5}{6}$

7. Operations on Mixed Numbers

Recall that a mixed number is a whole number added to a fraction. The number $3\frac{1}{2}$ represents the sum of three wholes plus a half, that is, $3\frac{1}{2} = 3 + \frac{1}{2}$. For this reason, any mixed number can be converted to an improper fraction by using addition.

$$3\frac{1}{2} = 3 + \frac{1}{2} = \frac{6}{2} + \frac{1}{2} = \frac{7}{2}$$

TIP: A shortcut to writing a mixed number as an improper fraction is to multiply the whole number by the denominator of the fraction. Then add this value to the numerator of the fraction, and write the result over the denominator.

$3\frac{1}{2} \longrightarrow$ Multiply the whole number by the denominator: $3 \times 2 = 6$.

Add the numerator: $6 + 1 = 7$.

Write the result over the denominator: $\frac{7}{2}$.

To add, subtract, multiply, or divide mixed numbers, we will first write the mixed number as an improper fraction.

Example 12   Operations on Mixed Numbers

Perform the indicated operations.

a. $5\frac{1}{3} - 2\frac{1}{4}$ **b.** $7\frac{1}{2} \div 3$

Solution:

a. $5\frac{1}{3} - 2\frac{1}{4}$

$= \dfrac{16}{3} - \dfrac{9}{4}$ Write the mixed numbers as improper fractions.

$= \dfrac{16 \times 4}{3 \times 4} - \dfrac{9 \times 3}{4 \times 3}$ The LCD is 12. Multiply numerators and denominators by the missing factors from the denominators.

$= \dfrac{64}{12} - \dfrac{27}{12}$

$= \dfrac{37}{12}$ or $3\dfrac{1}{12}$ Subtract the fractions.

Skill Practice Answers

16. 2

TIP: An improper fraction can also be written as a mixed number. Both answers are acceptable. Note that

$$\frac{37}{12} = \frac{36}{12} + \frac{1}{12} = 3 + \frac{1}{12}, \text{ or } 3\frac{1}{12}$$

This can easily be found by dividing.

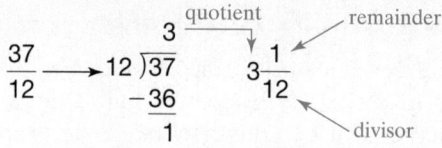

Avoiding Mistakes:

Remember that when dividing (or multiplying) fractions, a common denominator is not necessary.

b. $7\frac{1}{2} \div 3$

$= \frac{15}{2} \div \frac{3}{1}$ Write the mixed number and whole number as fractions.

$= \frac{\overset{5}{\cancel{15}}}{2} \times \frac{1}{\underset{1}{\cancel{3}}}$ Multiply by the reciprocal of $\frac{3}{1}$, which is $\frac{1}{3}$.

$= \frac{5}{2} \text{ or } 2\frac{1}{2}$ The answer may be written as an improper fraction or as a mixed number.

Skill Practice Answers

17. $\frac{17}{12}$ or $1\frac{5}{12}$ **18.** $\frac{22}{35}$

Skill Practice Perform the indicated operations.

17. $2\frac{3}{4} - 1\frac{1}{3}$ **18.** $3\frac{2}{3} \div 5\frac{5}{6}$

Section R.2   Practice Exercises

Boost *your* GRADE at mathzone.com!

MathZone

• Practice Problems
• Self-Tests
• NetTutor

• e-Professors
• Videos

Concept 1: Basic Definitions

For Exercises 1–8, identify the numerator and denominator of the fraction. Then determine if the fraction is a proper fraction or an improper fraction.

1. $\frac{7}{8}$ **2.** $\frac{2}{3}$ **3.** $\frac{9}{5}$ **4.** $\frac{5}{2}$

5. $\frac{6}{6}$ **6.** $\frac{4}{4}$ **7.** $\frac{12}{1}$ **8.** $\frac{5}{1}$

For Exercises 9–16, write a proper or improper fraction associated with the shaded region of each figure.

9. 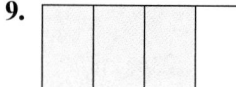   **10.**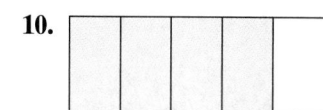

11.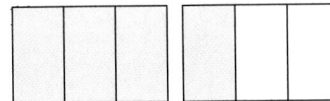

13.    **14.** 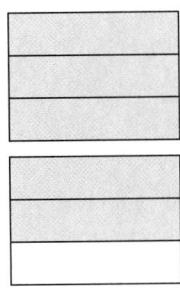   **15.** 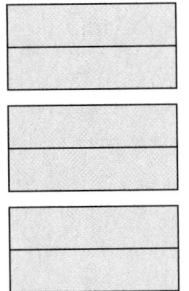   **16.**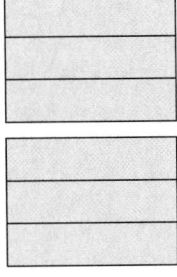

For Exercises 17–20, write both an improper fraction and a mixed number associated with the shaded region of each figure.

17. 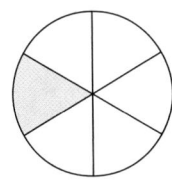   **18.** 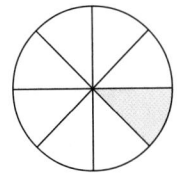   **19.** 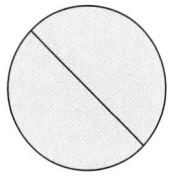   **20.**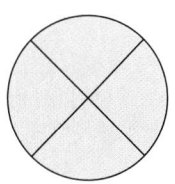

21. Explain the difference between the set of whole numbers and the set of natural numbers.

22. Explain the difference between a proper fraction and an improper fraction.

23. Write a fraction that simplifies to $\frac{1}{2}$. (Answers may vary.)

24. Write a fraction that simplifies to $\frac{1}{3}$. (Answers may vary.)

Concept 2: Prime Factorization

For Exercises 25–32, identify the number as either a prime number or a composite number.

25. 5   **26.** 9   **27.** 4   **28.** 2

29. 39   **30.** 23   **31.** 53   **32.** 51

For Exercises 33–40, write the number as a product of prime factors.

33. 36   **34.** 70   **35.** 42   **36.** 35

37. 110   **38.** 136   **39.** 135   **40.** 105

Concept 3: Simplifying Fractions to Lowest Terms

For Exercises 41–52, simplify each fraction to lowest terms.

41. $\frac{3}{15}$ **42.** $\frac{8}{12}$ **43.** $\frac{6}{16}$ **44.** $\frac{12}{20}$

45. $\dfrac{42}{48}$ **46.** $\dfrac{35}{80}$ **47.** $\dfrac{48}{64}$ **48.** $\dfrac{32}{48}$

49. $\dfrac{110}{176}$ **50.** $\dfrac{70}{120}$ **51.** $\dfrac{150}{200}$ **52.** $\dfrac{119}{210}$

Concepts 4–5: Multiplying and Dividing Fractions

For Exercises 53–54, determine if the statement is true or false. If it is false, rewrite as a true statement.

53. When multiplying or dividing fractions, it is necessary to have a common denominator.

54. When dividing two fractions, it is necessary to multiply the first fraction by the reciprocal of the second fraction.

For Exercises 55–66, multiply or divide as indicated.

55. $\dfrac{10}{13} \times \dfrac{26}{15}$ **56.** $\dfrac{15}{28} \times \dfrac{7}{9}$ **57.** $\dfrac{3}{7} \div \dfrac{9}{14}$ **58.** $\dfrac{7}{25} \div \dfrac{1}{5}$

59. $\dfrac{9}{10} \times 5$ **60.** $\dfrac{3}{7} \times 14$ **61.** $\dfrac{12}{5} \div 4$ **62.** $\dfrac{20}{6} \div 5$

63. $\dfrac{5}{2} \times \dfrac{10}{21} \times \dfrac{7}{5}$ **64.** $\dfrac{55}{9} \times \dfrac{18}{32} \times \dfrac{24}{11}$ **65.** $\dfrac{9}{100} \div \dfrac{13}{1000}$ **66.** $\dfrac{1000}{17} \div \dfrac{10}{3}$

67. Stephen's take-home pay is $1200 a month. If his rent is $\frac{1}{4}$ of his pay, how much is his rent?

68. Gus decides to save $\frac{1}{3}$ of his pay each month. If his monthly pay is $2112, how much will he save each month?

69. On a college basketball team, one-third of the team graduated with honors. If the team has 12 members, how many graduated with honors?

70. Shontell had only enough paper to print out $\frac{3}{5}$ of her book report before school. If the report is 10 pages long, how many pages did she print out?

71. Natalie has 4 yd of material with which she can make holiday aprons. If it takes $\frac{1}{2}$ yd of material per apron, how many aprons can she make?

72. There are 4 cups of oatmeal in a box. If each serving is $\frac{1}{3}$ of a cup, how many servings are contained in the box?

73. Gail buys 6 lb of mixed nuts to be divided into decorative jars that will each hold $\frac{3}{4}$ lb of nuts. How many jars will she be able to fill?

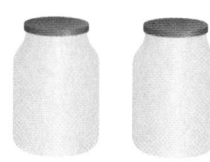

74. Troy has a $\frac{7}{8}$ in. nail that he must hammer into a board. Each strike of the hammer moves the nail $\frac{1}{16}$ in. into the board. How many strikes of the hammer must he make to drive the nail completely into the board?

Concept 6: Adding and Subtracting Fractions

For Exercises 75–78, add or subtract as indicated.

75. $\dfrac{5}{14} + \dfrac{1}{14}$ **76.** $\dfrac{9}{5} + \dfrac{1}{5}$ **77.** $\dfrac{17}{24} - \dfrac{5}{24}$ **78.** $\dfrac{11}{18} - \dfrac{5}{18}$

For Exercises 79–84, find the least common denominator for each pair of fractions.

79. $\frac{1}{6}, \frac{5}{24}$

80. $\frac{1}{12}, \frac{11}{30}$

81. $\frac{9}{20}, \frac{3}{8}$

82. $\frac{13}{24}, \frac{7}{40}$

83. $\frac{7}{10}, \frac{11}{45}$

84. $\frac{1}{20}, \frac{1}{30}$

For Exercises 85–100, add or subtract as indicated.

85. $\frac{1}{8} + \frac{3}{4}$

86. $\frac{3}{16} + \frac{1}{2}$

87. $\frac{3}{8} - \frac{3}{10}$

88. $\frac{12}{35} - \frac{1}{10}$

89. $\frac{7}{26} - \frac{2}{13}$

90. $\frac{11}{24} - \frac{5}{16}$

91. $\frac{7}{18} + \frac{5}{12}$

92. $\frac{3}{16} + \frac{9}{20}$

93. $\frac{3}{4} - \frac{1}{20}$

94. $\frac{1}{6} - \frac{1}{24}$

95. $\frac{5}{12} + \frac{5}{16}$

96. $\frac{3}{25} + \frac{8}{35}$

97. $\frac{1}{6} + \frac{3}{4} + \frac{5}{8}$

98. $\frac{1}{2} + \frac{2}{3} + \frac{5}{12}$

99. $\frac{4}{7} + \frac{1}{2} + \frac{3}{4}$

100. $\frac{9}{10} + \frac{4}{5} + \frac{3}{4}$

Concept 7: Operations on Mixed Numbers

For Exercises 101–123, perform the indicated operations.

101. $4\frac{3}{5} \div \frac{1}{10}$

102. $2\frac{4}{5} \div \frac{7}{11}$

103. $3\frac{1}{5} \times \frac{7}{8}$

104. $2\frac{1}{2} \times \frac{4}{5}$

105. $1\frac{2}{9} \div 6$

106. $2\frac{2}{5} \div \frac{2}{7}$

107. A board $26\frac{3}{8}$ in. long must be cut into three pieces of equal length. Find the length of each piece.

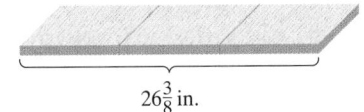

$26\frac{3}{8}$ in.

108. $2\frac{1}{8} + 1\frac{3}{8}$

109. $1\frac{3}{14} + 1\frac{1}{14}$

110. $1\frac{5}{6} - \frac{7}{8}$

111. $2\frac{1}{3} - \frac{5}{6}$

112. $1\frac{1}{6} + 3\frac{3}{4}$

113. $4\frac{1}{2} + 2\frac{2}{3}$

114. $1 - \frac{7}{8}$

115. $2 - \frac{3}{7}$

116. A futon, when set up as a sofa, measures $3\frac{5}{6}$ ft wide. When it is opened to be used as a bed, the width is increased by $1\frac{3}{4}$ ft. What is the total width of this bed?

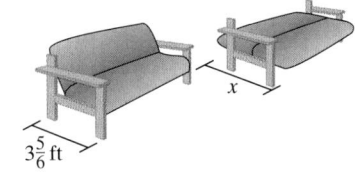

$3\frac{5}{6}$ ft

117. A plane trip from Orlando to Detroit takes $2\frac{3}{4}$ hr. If the plane traveled for $1\frac{1}{6}$ hr, how much time remains for the flight?

118. Antonio bought $\frac{1}{2}$ lb of turkey, $\frac{1}{3}$ lb of ham, and $\frac{3}{4}$ lb of roast beef. How much meat did Antonio buy?

119. José ordered two seafood platters for a party. One platter has $1\frac{1}{2}$ lb of shrimp, and the other has $\frac{3}{4}$ lb of shrimp. How many pounds does he have altogether?

120. Ayako took a trip to the store $5\frac{1}{2}$ miles away. If she rode the bus for $4\frac{5}{6}$ miles and walked the rest of the way, how far did she have to walk?

121. Average rainfall in Tampa, FL, for the month of November is $2\frac{3}{4}$ in. One stormy weekend $3\frac{1}{8}$ in. of rain fell. How many inches of rain over the monthly average is this?

122. Maria has 4 yd of material. If she sews a dress that requires $3\frac{1}{8}$ yd, how much material will she have left?

123. Pete started working out at the gym several months ago. His waist measured $38\frac{1}{2}$ in. when he began and is now $33\frac{3}{4}$ in. How many inches did he lose around his waist?

Section R.3   Introduction to Geometry

Concepts

1. Perimeter
2. Area
3. Volume
4. Angles
5. Triangles

1. Perimeter

In this section, we present several facts and formulas that may be used throughout the text in applications of geometry. One of the most important uses of geometry involves the measurement of objects of various shapes. We begin with an introduction to perimeter, area, and volume for several common shapes and objects.

Perimeter is defined as the distance around a figure. For example, if we were to put up a fence around a field, the perimeter would determine the amount of fencing. For a polygon (a closed figure constructed from line segments), the perimeter is the sum of the lengths of the sides. For a circle, the distance around the outside is called the **circumference**.

Rectangle	Square	Triangle	Circle
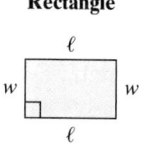	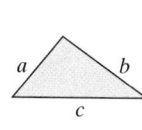	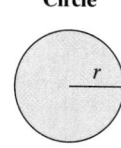	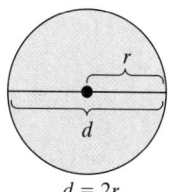
$P = 2\ell + 2w$	$P = 4s$	$P = a + b + c$	Circumference: $C = 2\pi r$

For a circle, r represents the length of a radius—the distance from the center to any point on the circle. The length of a diameter, d, of a circle is twice that of a radius. Thus, $d = 2r$. The number π is a constant equal to the ratio of the circumference of a circle divided by the length of a diameter, that is, $\pi = \frac{C}{d}$. The value of π is often approximated by 3.14 or $\frac{22}{7}$.

$d = 2r$

Example 1   Finding Perimeter and Circumference

Find the perimeter or circumference as indicated. Use 3.14 for π.

a. Perimeter of the polygon

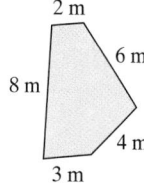

b. Perimeter of the rectangle

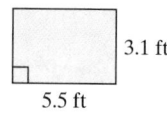

c. Circumference of the circle

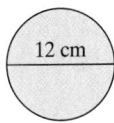

12 cm

Solution:

a. $P = (8\text{ m}) + (2\text{ m}) + (6\text{ m}) + (4\text{ m}) + (3\text{ m})$ Add the lengths of the sides.

  $= 23\text{ m}$ The perimeter is 23 m.

b. $P = 2\ell + 2w$

  $= 2(5.5\text{ ft}) + 2(3.1\text{ ft})$ Substitute $\ell = 5.5$ ft and
  $w = 3.1$ ft.

  $= 11\text{ ft} + 6.2\text{ ft}$

  $= 17.2\text{ ft}$ The perimeter is 17.2 ft.

c. $C = 2\pi r$

  $= 2(3.14)(6\text{ cm})$ Substitute 3.14 for π and
  $r = 6$ cm.

  $= 6.28(6\text{ cm})$

  $= 37.68\text{ cm}$ The circumference is 37.68 cm.

Skill Practice

1. Find the perimeter.

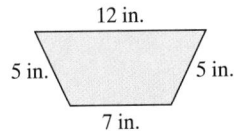

12 in.

5 in.   5 in.

7 in.

2. Find the circumference. Use 3.14 for π.

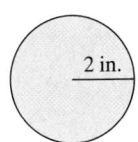

2 in.

3. Find the perimeter of the square.

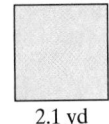

2.1 yd

> **TIP:** If a calculator is used to find the circumference of a circle, use the π key to get a more accurate answer than by using 3.14.

2. Area

The **area** of a geometric figure is the number of square units that can be enclosed within the figure. For example, the rectangle shown in Figure R-6 encloses 6 square inches (6 in.^{2}). In applications, we would find the area of a region if we were laying carpet or putting down sod for a lawn.

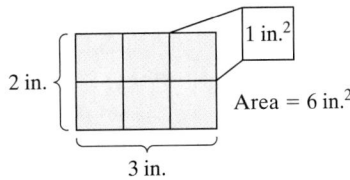

2 in.

3 in.

1 in.^{2}

Area = 6 in.^{2}

Figure R-6

Skill Practice Answers
1. 29 in.   **2.** 12.56 in.
3. 8.4 yd

The formulas used to compute the area for several common geometric shapes are given here:

Rectangle	Square	Parallelogram	Triangle	Trapezoid	Circle
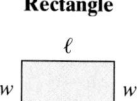	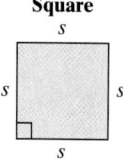	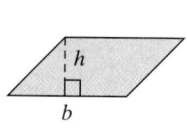	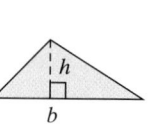	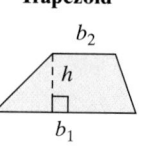	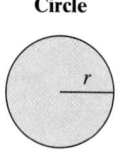
$A = \ell w$	$A = s^2$	$A = bh$	$A = \frac{1}{2}bh$	$A = \frac{1}{2}(b_1 + b_2)h$	$A = \pi r^2$

Example 2 Finding Area

Find the area enclosed by each figure.

a.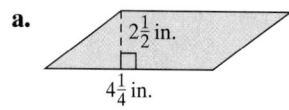

b.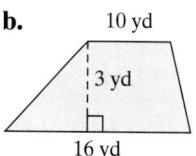

Solution:

a. $A = bh$ The figure is a parallelogram.

$\quad = (4\frac{1}{4} \text{ in.})(2\frac{1}{2} \text{ in.})$ Substitute $b = 4\frac{1}{4}$ in. and $h = 2\frac{1}{2}$ in.

$\quad = \left(\dfrac{17}{4} \text{ in.}\right)\left(\dfrac{5}{2} \text{ in.}\right)$

$\quad = \dfrac{85}{8} \text{ in.}^2$ or $10\frac{5}{8} \text{ in.}^2$

TIP: Notice that the units of area are given in square units such as square inches (in.^{2}), square feet (ft^2), square yards (yd^2), square centimeters (cm^2), and so on.

b. $A = \dfrac{1}{2}(b_1 + b_2)h$ The figure is a trapezoid.

$\quad = \dfrac{1}{2}(16 \text{ yd} + 10 \text{ yd})(3 \text{ yd})$ Substitute $b_1 = 16$ yd, $b_2 = 10$ yd, and $h = 3$ yd.

$\quad = \dfrac{1}{2}(26 \text{ yd})(3 \text{ yd})$

$\quad = (13 \text{ yd})(3 \text{ yd})$

$\quad = 39 \text{ yd}^2$ The area is 39 yd^2.

Skill Practice Find the area enclosed by the figure.

4. $\frac{3}{4}$ cm

$\frac{3}{4}$ cm

5.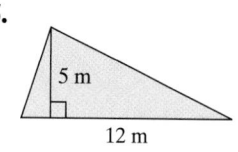

Skill Practice Answers

4. $\dfrac{9}{16}$ cm^2 **5.** 30 m^2

TIP: Notice that several of the formulas presented thus far involve multiple operations. The order in which we perform the arithmetic is called the **order of operations** and is covered in detail in Section 1.2. We will follow these guidelines in the order given below:

1. Perform operations within parentheses first.
2. Evaluate expressions with exponents.
3. Perform multiplication or division in order from left to right.
4. Perform addition or subtraction in order from left to right.

Example 3 | **Finding Area of a Circle**

Find the area of a circular fountain if the radius is 25 ft.
Use 3.14 for π.

Solution:

$A = \pi r^2$

$\quad = (3.14)(25 \text{ ft})^2$ Substitute 3.14 for π and $r = 25$ ft.

$\quad = (3.14)(625 \text{ ft}^2)$ Note: $(25 \text{ ft})^2 = (25 \text{ ft})(25 \text{ ft}) = 625 \text{ ft}^2$

$\quad = 1962.5 \text{ ft}^2$ The area of the fountain is 1962.5 ft^2.

Skill Practice | Find the area of the circular region. Use 3.14 for π.

6.

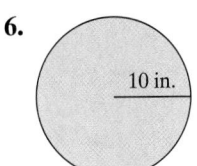

---

3. Volume

The **volume** of a solid is the number of cubic units that can be enclosed within a solid. The solid shown in Figure R-7 contains 18 cubic inches (18 in.^3). In applications, volume might refer to the amount of water in a swimming pool.

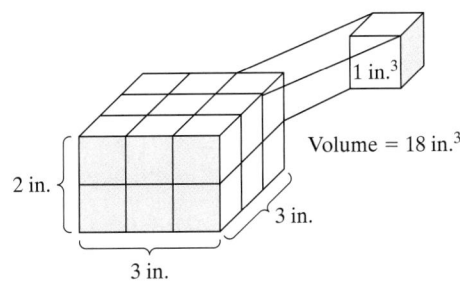

Figure R-7

The formulas used to compute the volume of several common solids are given here:

| **Rectangular Solid** | **Cube** | **Right Circular Cylinder** |

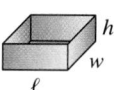 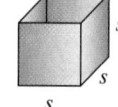

$$V = \ell wh \qquad V = s^3 \qquad V = \pi r^2 h$$

TIP: Notice that the volume formulas for the three figures above are given by the product of the area of the base and the height of the figure:

$$V = \ell wh \qquad V = s \cdot s \cdot s \qquad V = \pi r^2 h$$

Area of   Area of   Area of
Rectangular Base   Square Base   Circular Base

| **Right Circular Cone** | **Sphere** |

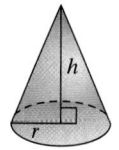 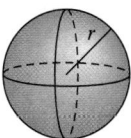

$$V = \tfrac{1}{3}\pi r^2 h \qquad\qquad V = \tfrac{4}{3}\pi r^3$$

Example 4 | **Finding Volume**

Find the volume of each object.

a.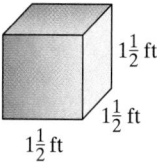

b.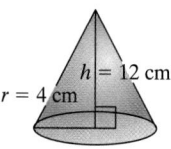

Solution:

a. $V = s^3$ The object is a cube.

$ = (1\tfrac{1}{2}\,\text{ft})^3$ Substitute $s = 1\tfrac{1}{2}$ ft.

$ = \left(\dfrac{3}{2}\,\text{ft}\right)^3$

$ = \left(\dfrac{3}{2}\,\text{ft}\right)\left(\dfrac{3}{2}\,\text{ft}\right)\left(\dfrac{3}{2}\,\text{ft}\right)$

$ = \dfrac{27}{8}\,\text{ft}^3, \text{ or } 3\tfrac{3}{8}\,\text{ft}^3$

TIP: Notice that the units of volume are cubic units such as cubic inches (in.^{3}), cubic feet (ft^3), cubic yards (yd^3), cubic centimeters (cm^3), and so on.

b. $V = \frac{1}{3}\pi r^2 h$ The object is a right circular cone.

$= \frac{1}{3}(3.14)(4 \text{ cm})^2(12 \text{ cm})$ Substitute 3.14 for π, $r = 4$ cm, and $h = 12$ cm.

$= \frac{1}{3}(3.14)(16 \text{ cm}^2)(12 \text{ cm})$

$= 200.96 \text{ cm}^3$

Skill Practice

7. Find the volume of the object.

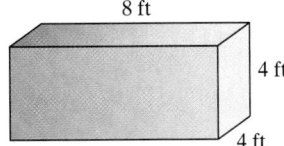

8 ft · 4 ft · 4 ft

8. Find the volume of the object. Use 3.14 for π. Round to the nearest whole unit.

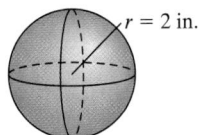

$r = 2$ in.

Example 5   **Finding Volume in an Application**

An underground gas tank is in the shape of a right circular cylinder.

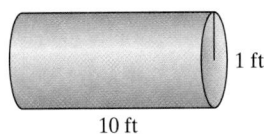

1 ft

10 ft

a. Find the volume of the tank. Use 3.14 for π.

b. Find the cost to fill the tank with gasoline if gasoline costs $9/ft^3.

Solution:

a. $V = \pi r^2 h$

$= (3.14)(1 \text{ ft})^2(10 \text{ ft})$ Substitute 3.14 for π, $r = 1$ ft, and $h = 10$ ft.

$= (3.14)(1 \text{ ft}^2)(10 \text{ ft})$

$= 31.4 \text{ ft}^3$ The tank holds 31.4 ft^3 of gasoline.

b. Cost $= (\$9/\text{ft}^3)(31.4 \text{ ft}^3)$

$= \$282.60$ It will cost \$282.60 to fill the tank.

Skill Practice Answers

7. 128 ft^3 **8.** 33 in.^{3}

Skill Practice

9a. Find the volume of soda in the can. Use 3.14 for π. Round
to the nearest whole unit.

b. Using your answer from part (a), how much soda is
contained in a six-pack?

12 cm
6 cm

4. Angles

Applications involving angles and their measure come up often in the study of
algebra, trigonometry, calculus, and applied sciences. The most common unit used
to measure an angle is the degree (°). Several angles and their corresponding
degree measure are shown in Figure R-8.

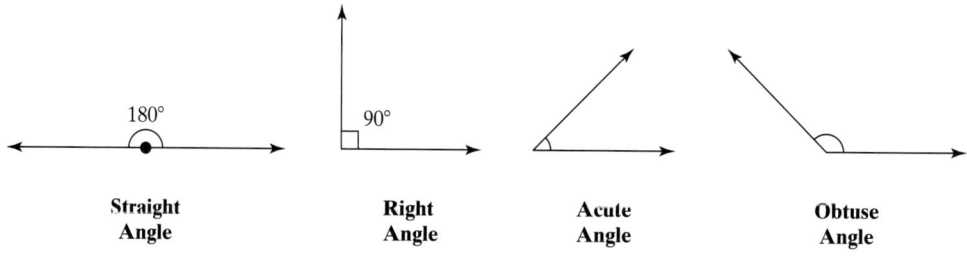

180°	90°		
Straight Angle	**Right Angle**	**Acute Angle**	**Obtuse Angle**

Figure R-8

- An angle that measures 90° is a **right angle** (right angles are often marked with
a square or corner symbol, ☐).
- An angle that measures 180° is called a **straight angle**.
- An angle that measures between 0° and 90° is called an **acute angle**.
- An angle that measures between 90° and 180° is called an **obtuse angle**.
- Two angles with the same measure are **equal angles** (or **congruent angles**).

The measure of an angle will be denoted by the symbol m written in front of
the angle. Therefore, the measure of $\angle A$ is denoted $m(\angle A)$.

- Two angles are said to be
complementary if the sum of
their measures is 90°.

- Two angles are said to be
supplementary if the sum of
their measures is 180°.

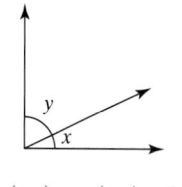
y
x

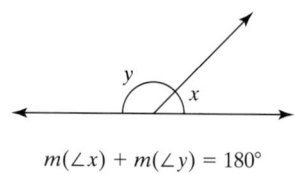
y
x

$m(\angle x) + m(\angle y) = 90°$ $m(\angle x) + m(\angle y) = 180°$

Skill Practice Answers
9a. 339 cm³   **b.** 2034 cm³

When two lines intersect, four angles are formed (Figure R-9). In Figure R-9, $\angle a$ and $\angle b$ are **vertical angles**. Another set of vertical angles is the pair $\angle c$ and $\angle d$. An important property of vertical angles is that the measures of two vertical angles are *equal*. In the figure, $m(\angle a) = m(\angle b)$ and $m(\angle c) = m(\angle d)$.

Parallel lines are lines that lie in the same plane and do not intersect. In Figure R-10, the lines L_1 and L_2 are parallel lines. If a line intersects two parallel lines, the line is called a **transversal**. In Figure R-10, the line m is a transversal and forms eight angles with the parallel lines L_1 and L_2.

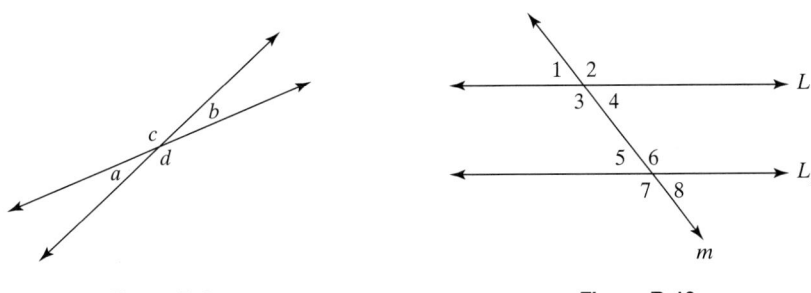

Figure R-9   Figure R-10

The angles 1–8 in Figure R-10 have special names and special properties.

Two angles that lie between the parallel lines (interior) and on alternate sides of the transversal are called **alternate interior angles**. Pairs of alternate interior angles are shown next.

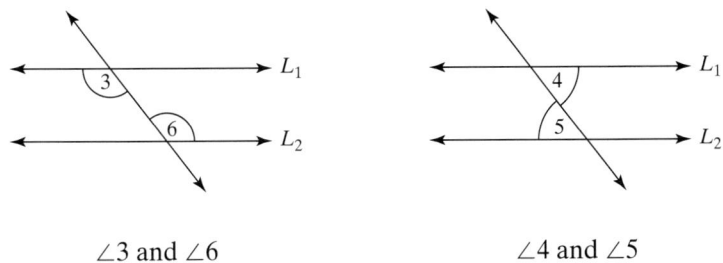

$\angle 3$ and $\angle 6$ $\angle 4$ and $\angle 5$

Alternate interior angles are equal in measure. Thus, $m(\angle 3) = m(\angle 6)$ and $m(\angle 4) = m(\angle 5)$.

Two angles that lie outside the parallel lines (exterior) and on alternate sides of the transversal are called **alternate exterior angles**. Pairs of alternate exterior angles are shown here.

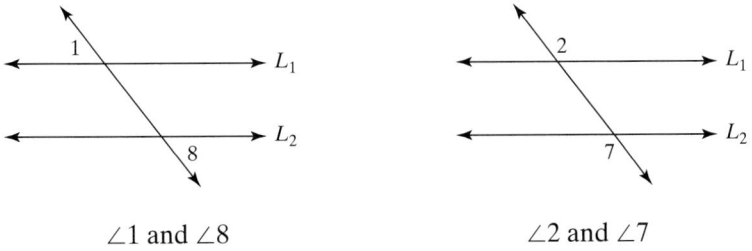

$\angle 1$ and $\angle 8$ $\angle 2$ and $\angle 7$

Alternate exterior angles are equal in measure. Thus, $m(\angle 1) = m(\angle 8)$ and $m(\angle 2) = m(\angle 7)$.

Two angles that lie on the same side of the transversal such that one is exterior and one is interior are called **corresponding angles**. Corresponding angles cannot be adjacent to each other. Pairs of corresponding angles are shown here in color.

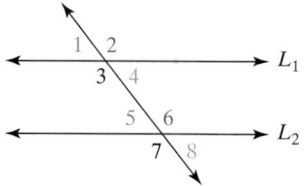

Corresponding angles are equal in measure. Thus, $m(\angle 1) = m(\angle 5)$, $m(\angle 2) = m(\angle 6)$, $m(\angle 3) = m(\angle 7)$, and $m(\angle 4) = m(\angle 8)$.

Example 6 **Finding Unknown Angles in a Diagram**

Find the measure of each angle and explain how the angle is related to the given angle of 70°.

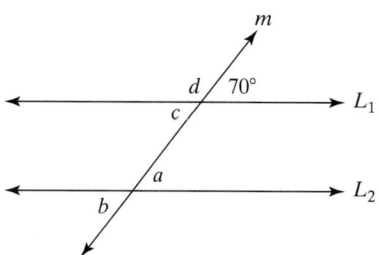

a. $\angle a$ **b.** $\angle b$ **c.** $\angle c$ **d.** $\angle d$

Solution:

a. $m(\angle a) = 70°$ $\angle a$ is a corresponding angle to the given angle of 70°.

b. $m(\angle b) = 70°$ $\angle b$ and the given angle of 70° are alternate exterior angles.

c. $m(\angle c) = 70°$ $\angle c$ and the given angle of 70° are vertical angles.

d. $m(\angle d) = 110°$ $\angle d$ is the supplement of the given angle of 70°.

Skill Practice

10. Refer to the figure. Assume that lines L_1 and L_2 are parallel. Given that $m(\angle 3) = 23°$, find:

    a. $m(\angle 2)$ **b.** $m(\angle 4)$

    c. $m(\angle 5)$ **d.** $m(\angle 8)$

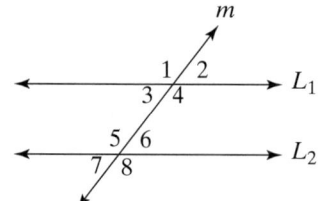

Skill Practice Answers

10a. 23°   **b.** 157°
  c. 157°   **d.** 157°

5. Triangles

Triangles are categorized by the measures of the angles (Figure R-11) and by the number of equal sides or angles (Figure R-12).

- An **acute triangle** is a triangle in which all three angles are acute.
- A **right triangle** is a triangle in which one angle is a right angle.
- An **obtuse triangle** is a triangle in which one angle is obtuse.

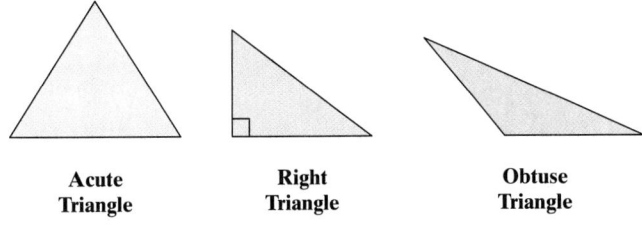

| Acute Triangle | Right Triangle | Obtuse Triangle |

Figure R-11

- An **equilateral triangle** is a triangle in which all three sides (and all three angles) are equal in measure.
- An **isosceles triangle** is a triangle in which two sides are equal in measure (the angles opposite the equal sides are also equal in measure).
- A **scalene triangle** is a triangle in which no sides (or angles) are equal in measure.

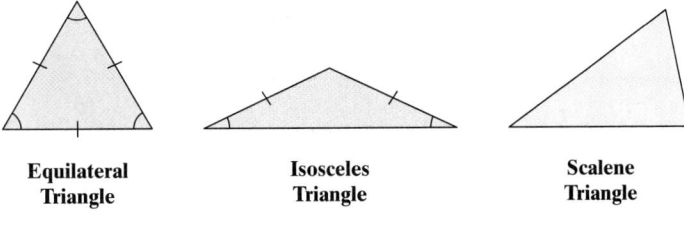

| Equilateral Triangle | Isosceles Triangle | Scalene Triangle |

Figure R-12

The following important property is true for all triangles.

Sum of the Angles in a Triangle

The sum of the measures of the angles of a triangle is 180°.

Example 7 | Finding Unknown Angles in a Diagram

Find the measure of each angle in the figure.

a. $\angle a$

b. $\angle b$

c. $\angle c$

d. $\angle d$

e. $\angle e$

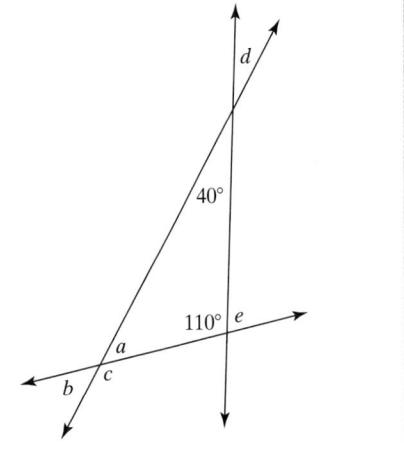

Solution:

a. $m(\angle a) = 30°$ The sum of the angles in a triangle is 180°.

b. $m(\angle b) = 30°$ $\angle a$ and $\angle b$ are vertical angles and are equal in measure.

c. $m(\angle c) = 150°$ $\angle c$ and $\angle a$ are supplementary angles ($\angle c$ and $\angle b$ are also supplementary).

d. $m(\angle d) = 40°$ $\angle d$ and the given angle of 40° are vertical angles.

e. $m(\angle e) = 70°$ $\angle e$ and the given angle of 110° are supplementary angles.

Skill Practice

11. Refer to the figure. Find the measure of the indicated angle.

a. $\angle a$ b. $\angle b$ c. $\angle c$

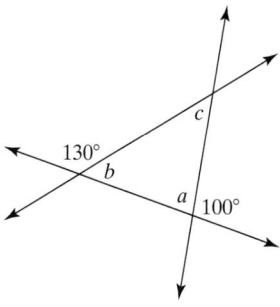

Skill Practice Answers

11a. 80°   **b.** 50°   **c.** 50°

Section R.3   Practice Exercises

Boost *your* **GRADE at mathzone.com!**

MathZone

• Practice Problems
• Self-Tests
• NetTutor
• e-Professors
• Videos

Concept 1: Perimeter

1. Identify which of the following units could be measures of perimeter.

a. Square inches (in.^{2})   b. Meters (m)   c. Cubic feet (ft^3)

d. Cubic meters (m^3)   e. Miles (mi)   f. Square centimeters (cm^2)

g. Square yards (yd^2)   h. Cubic inches (in.^{3})   i. Kilometers (km)

For Exercises 2–5, find the perimeter of each figure.

2.

6 m
10 m

3.
22 cm
32 cm

4.
4.3 mi
4.3 mi

5.
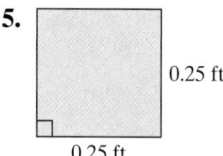
0.25 ft
0.25 ft

6. Identify which of the following units could be measures of circumference.

   a. Square inches (in.²)   **b.** Meters (m)   **c.** Cubic feet (ft³)

   d. Cubic meters (m³)   **e.** Miles (mi)   **f.** Square centimeters (cm²)

   g. Square yards (yd²)   **h.** Cubic inches (in.³)   **i.** Kilometers (km)

For Exercises 7–10, find the perimeter or circumference. Use 3.14 for π.

 7.
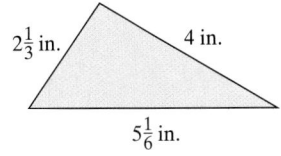

8.
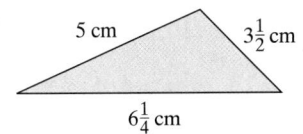

 9.
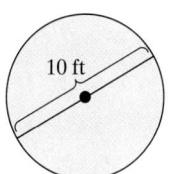

10.
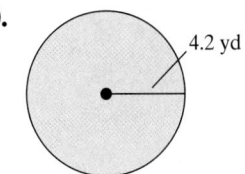

Concept 2: Area

11. Identify which of the following units could be measures of area.

   a. Square inches (in.²)   **b.** Meters (m)   **c.** Cubic feet (ft³)

   d. Cubic meters (m³)   **e.** Miles (mi)   **f.** Square centimeters (cm²)

   g. Square yards (yd²)   **h.** Cubic inches (in.³)   **i.** Kilometers (km)

For Exercises 12–25, find the area. Use 3.14 for π.

12.
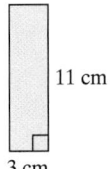

13.
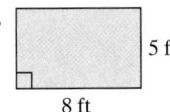

14.
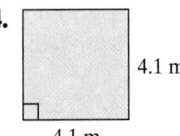

15.
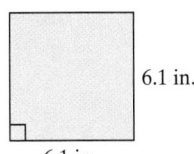

16.
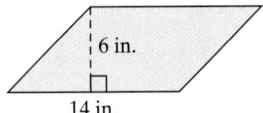

17.
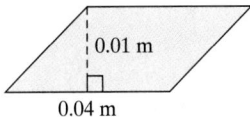

18.
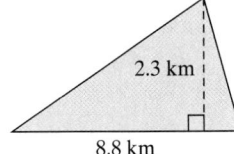

19.
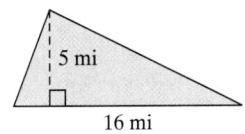

20.
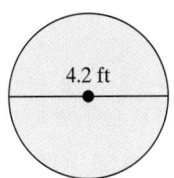

21.
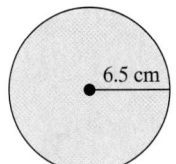

22.
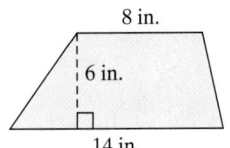

23.
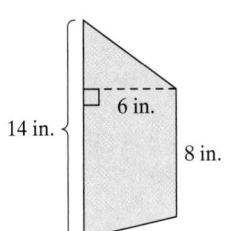

24.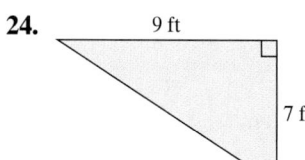

25.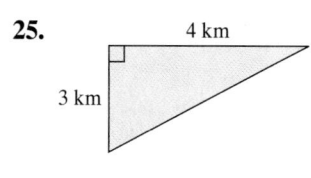

Concept 3: Volume

26. Identify which of the following units could be measures of volume.

   a. Square inches (in.²)   **b.** Meters (m)   **c.** Cubic feet (ft³)

   d. Cubic meters (m³)   **e.** Miles (mi)   **f.** Square centimeters (cm²)

   g. Square yards (yd²)   **h.** Cubic inches (in.³)   **i.** Kilometers (km)

For Exercises 27–34, find the volume of each figure. Use 3.14 for π.

27.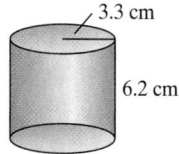

28.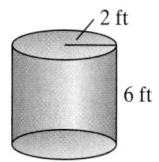

29.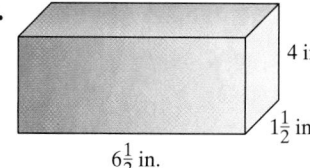

30.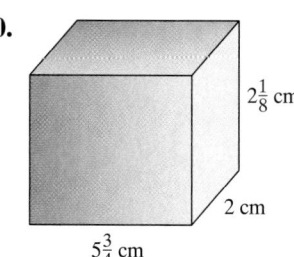

31.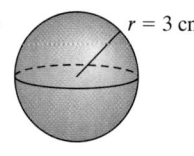

32.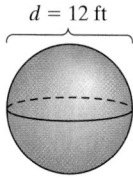

33.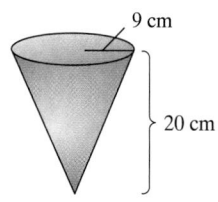

34.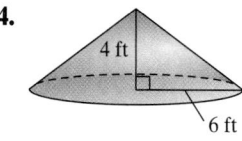

35. A florist sells balloons and needs to know how much helium to order. Each balloon is approximately spherical with a radius of 9 in. How much helium is needed to fill one balloon?

36. Find the volume of a spherical ball whose radius is 3 in. Use 3.14 for π.

37. Find the volume of a snow cone in the shape of a right circular cone whose radius is 3 cm and whose height is 12 cm. Use 3.14 for π.

38. A landscaping supply company has a pile of gravel in the shape of a right circular cone whose radius is 10 yd and whose height is 18 yd. Find the volume of the gravel. Use 3.14 for π.

39. Find the volume of a cube that is 3.2 ft on a side.

40. Find the volume of a cube that is 10.5 cm on a side.

Mixed Exercises: Perimeter, Area, and Volume

41. A wall measuring 20 ft by 8 ft can be painted for $50.

   a. What is the price per square foot? Round to the nearest cent.

   b. At this rate, how much would it cost to paint the remaining three walls that measure 20 ft by 8 ft, 16 ft by 8 ft, and 16 ft by 8 ft? Round to the nearest dollar.

42. Suppose it costs $320 to carpet a 16 ft by 12 ft room.

   a. What is the price per square foot? Round to the nearest cent.

   b. At this rate, how much would it cost to carpet a room that is 20 ft by 32 ft?

43. If you were to purchase fencing for a garden, would you measure the perimeter or area of the garden?

44. If you were to purchase sod (grass) for your front yard, would you measure the perimeter or area of the yard?

45. **a.** Find the area of a circular pizza that is 8 in. in diameter (the radius is 4 in.). Use 3.14 for π.

   b. Find the area of a circular pizza that is 12 in. in diameter (the radius is 6 in.).

   c. Assume that the 8-in. diameter and 12-in. diameter pizzas are both the same thickness. Which would provide more pizza, two 8-in. pizzas or one 12-in. pizza?

46. Find the area of a circular stained glass window that is 16 in. in diameter. Use 3.14 for π.

47. Find the volume of a soup can in the shape of a right circular cylinder if its radius is 3.2 cm and its height is 9 cm. Use 3.14 for π.

48. Find the volume of a coffee mug whose radius is 2.5 in. and whose height is 6 in. Use 3.14 for π.

Concept 4: Angles

For Exercises 49–56, answer True or False. If an answer is false, explain why.

49. The sum of the measures of two right angles equals the measure of a straight angle.

50. Two right angles are complementary.

51. Two right angles are supplementary.

52. Two acute angles cannot be supplementary.

53. Two obtuse angles cannot be supplementary.

54. An obtuse angle and an acute angle can be supplementary.

55. If a triangle is equilateral, then it is not scalene.

56. If a triangle is isosceles, then it is also scalene.

57. What angle is its own complement?

58. What angle is its own supplement?

59. If possible, find two acute angles that are supplementary.

60. If possible, find two acute angles that are complementary. Answers may vary.

61. If possible, find an obtuse angle and an acute angle that are supplementary. Answers may vary.

62. If possible, find two obtuse angles that are supplementary.

63. Refer to the figure.

    a. State all the pairs of vertical angles.

    b. State all the pairs of supplementary angles.

    c. If the measure of $\angle 4$ is 80°, find the measures of $\angle 1$, $\angle 2$, and $\angle 3$.

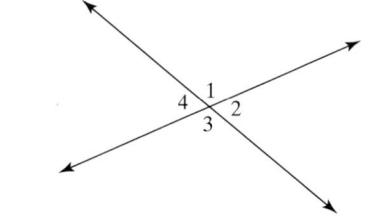

64. Refer to the figure.

    a. State all the pairs of vertical angles.

    b. State all the pairs of supplementary angles.

    c. If the measure of $\angle a$ is 25°, find the measures of $\angle b$, $\angle c$, and $\angle d$.

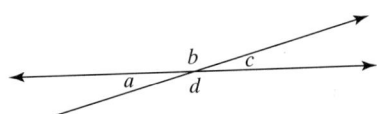

For Exercises 65–68, find the complement of each angle.

65. 33°   **66.** 87°   **67.** 12°   **68.** 45°

For Exercises 69–72, find the supplement of each angle.

69. 33°   **70.** 87°   **71.** 122°   **72.** 90°

For Exercises 73–80, refer to the figure. Assume that L_1 and L_2 are parallel lines cut by the transversal, n.

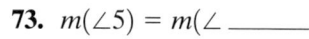

73. $m(\angle 5) = m(\angle \underline{\hspace{1cm}})$ Reason: Vertical angles have equal measures.

74. $m(\angle 5) = m(\angle \underline{\hspace{1cm}})$ Reason: Alternate interior angles have equal measures.

75. $m(\angle 5) = m(\angle \underline{\hspace{1cm}})$ Reason: Corresponding angles have equal measures.

76. $m(\angle 7) = m(\angle \underline{\hspace{1cm}})$ Reason: Corresponding angles have equal measures.

77. $m(\angle 7) = m(\angle \underline{\hspace{1cm}})$ Reason: Alternate exterior angles have equal measures.

78. $m(\angle 7) = m(\angle \underline{\hspace{1cm}})$ Reason: Vertical angles have equal measures.

79. $m(\angle 3) = m(\angle \underline{\hspace{1cm}})$ Reason: Alternate interior angles have equal measures.

80. $m(\angle 3) = m(\angle \underline{\hspace{1cm}})$ Reason: Vertical angles have equal measures.

81. Find the measures of angles a–g in the figure. Assume that L_1 and L_2 are parallel and that n is a transversal.

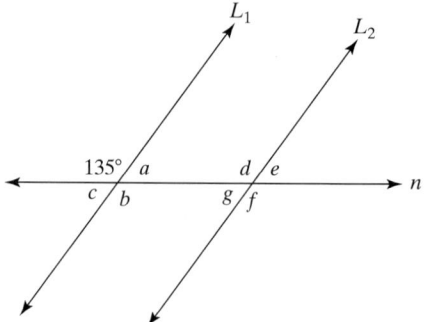

82. Find the measures of angles *a–g* in the figure. Assume that L_1 and L_2 are parallel and that *n* is a transversal.

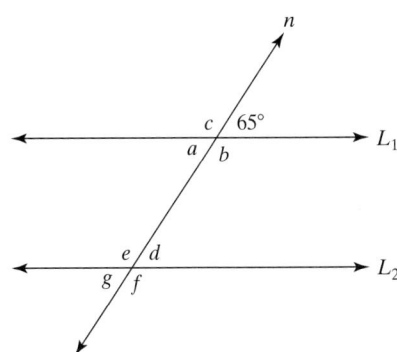

Concept 5: Triangles

For Exercises 83–86, identify the triangle as equilateral, isosceles, or scalene.

83.
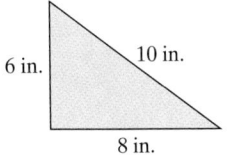

84.
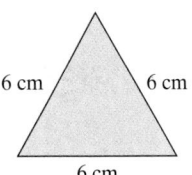

85.
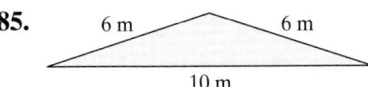

86.
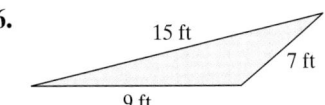

87. Can a triangle be both a right triangle and an obtuse triangle? Explain.

88. Can a triangle be both a right triangle and an isosceles triangle? Explain.

For Exercises 89–92, find the measures of the missing angles.

89.
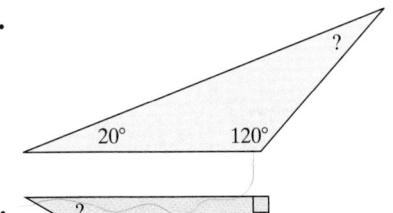

90.
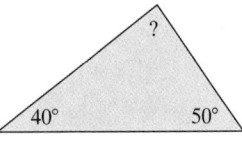

91.
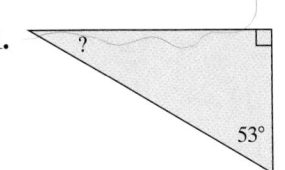

92.
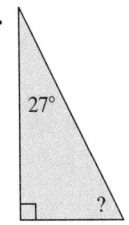

93. Refer to the figure. Find the measures of angles *a–j*.

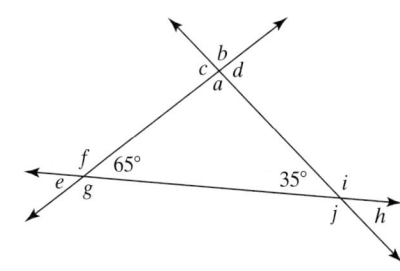

94. Refer to the figure. Find the measures of angles *a–j*.

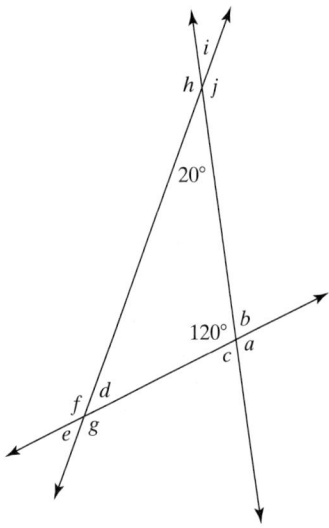

95. Refer to the figure. Find the measures of angles *a–k*. Assume that L_1 and L_2 are parallel.

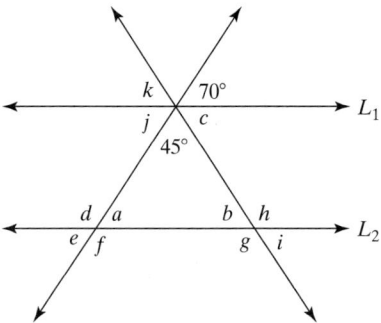

96. Refer to the figure. Find the measures of angles *a–k*. Assume that L_1 and L_2 are parallel.

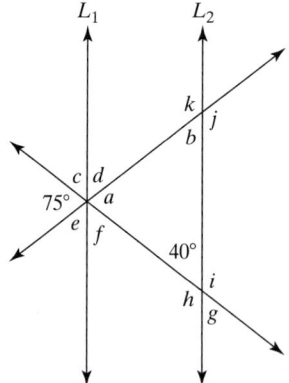

Expanding Your Skills

For Exercises 97–98, find the perimeter.

97.

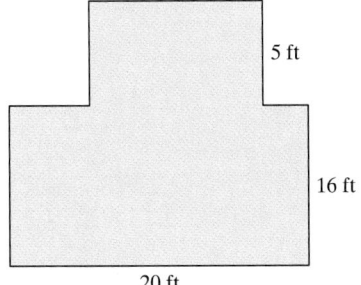

5 ft

16 ft

20 ft

98.

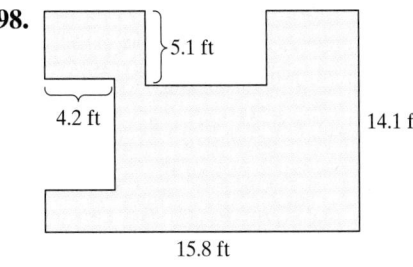

5.1 ft

4.2 ft

14.1 ft

15.8 ft

For Exercises 99–102, find the area of the shaded region. Use 3.14 for π.

99.

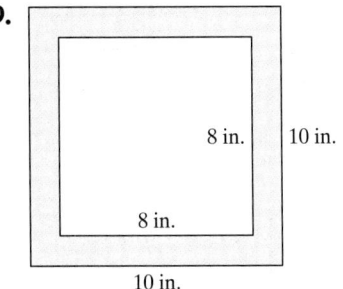

8 in.

10 in.

8 in.

10 in.

100.

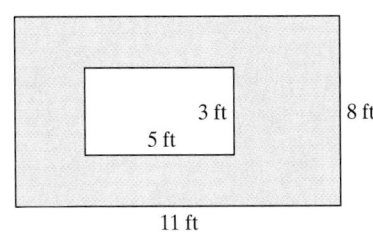

3 ft

5 ft

8 ft

11 ft

101.

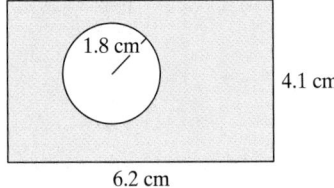

1.8 cm

4.1 cm

6.2 cm

102.

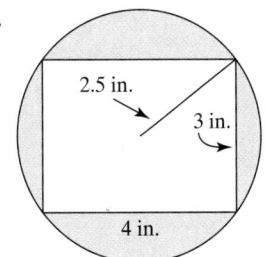

2.5 in.

3 in.

4 in.

The Set of Real Numbers

1.1 Sets of Numbers and the Real Number Line

1.2 Order of Operations

1.3 Addition of Real Numbers

1.4 Subtraction of Real Numbers

Problem Recognition Exercises—Addition and Subtraction of Signed Numbers

1.5 Multiplication and Division of Real Numbers

1.6 Properties of Real Numbers and Simplifying Expressions

In Chapter 1 we present operations on real numbers. To be successful in algebra, it is particularly important to understand the order of operations.

As you work through the chapter, pay attention to the order of operations. Then work through this puzzle. When you fill in the blanks, notice that negative signs may appear to the left of a horizontal number. They may also appear above a number written vertically.

Across

2. -12^2

5. $-3426 - 469 + 124 - 6421$

9. $-2.5 - 1.4 - 8.1 - 100$

11. $243 - |12 - 17| + |-600|$

12. $155 - 3(2 - 4^2)^2 - 4(1 - 3)$

Down

1. $(-12)^2$ 3. $120 - \dfrac{\sqrt{25 - 9}}{-2}$

4. $600 \div [2^3 + 7 - (4 + 1)]$

6. $-\dfrac{4}{3} - \left(-\dfrac{11}{6}\right) + 15 - \dfrac{1}{2}$

7. $84(-12) - (-10{,}625)$

8. $\dfrac{120 - \sqrt{25 - 9}}{-2}$

10. $(-2.5)(-1.4)(-8.1)(-100)$

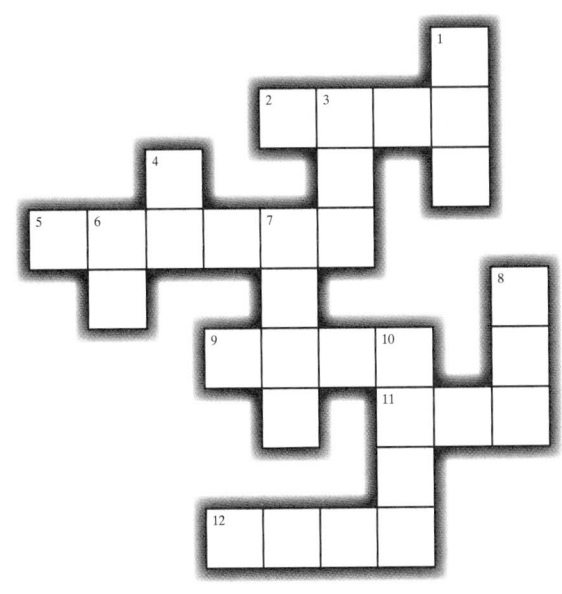

Sets of Numbers and the Real Number Line

1. Real Number Line

The numbers we work with on a day-to-day basis are all part of the set of **real numbers**. The real numbers encompass zero, all positive, and all negative numbers, including those represented by fractions and decimal numbers. The set of real numbers can be represented graphically on a horizontal number line with a point labeled as 0. Positive real numbers are graphed to the right of 0, and negative real numbers are graphed to the left. Zero is neither positive nor negative. Each point on the number line corresponds to exactly one real number. For this reason, this number line is called the **real number line** (Figure 1-1).

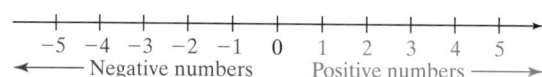

Figure 1-1

2. Plotting Points on the Number Line

> **Example 1**   Plotting Points on the Real Number Line

Plot the points on the real number line that represent the following real numbers.

a. -3 **b.** $\dfrac{3}{2}$ **c.** -4.8 **d.** $\dfrac{16}{5}$

Solution:

a. Because -3 is negative, it lies three units to the left of zero.

b. The fraction $\frac{3}{2}$ can be expressed as the mixed number $1\frac{1}{2}$, which lies halfway between 1 and 2 on the number line.

c. The negative number -4.8 lies $\frac{8}{10}$ units to the left of -4 on the number line.

d. The fraction $\frac{16}{5}$ can be expressed as the mixed number $3\frac{1}{5}$, which lies $\frac{1}{5}$ unit to the right of 3 on the number line.

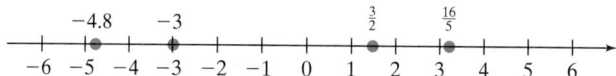

> **Skill Practice**

1. Plot the numbers on a real number line.
$\{-1, -2.5, \frac{3}{4}, 4\}$

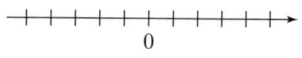

3. Set of Real Numbers

In mathematics, a well-defined collection of elements is called a set. "Well-defined" means the set is described in such a way that it is clear whether an

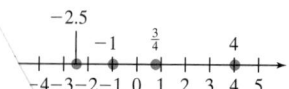

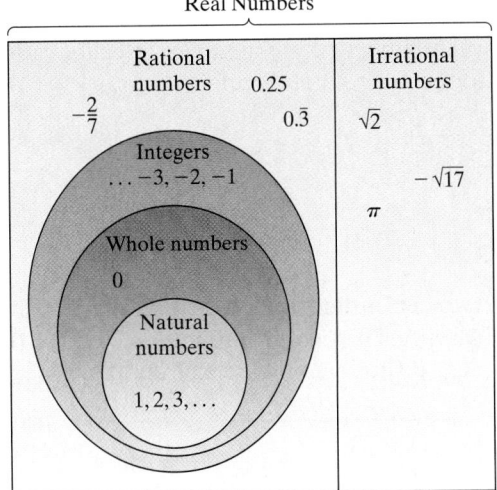

Figure 1-3

Example 3 | **Classifying Numbers by Set**

Check the set(s) to which each number belongs. The numbers may belong to more than one set.

	Natural Numbers	Whole Numbers	Integers	Rational Numbers	Irrational Numbers	Real Numbers
5						
$\frac{-47}{3}$						
1.48						
$\sqrt{7}$						
0						

Solution:

	Natural Numbers	Whole Numbers	Integers	Rational Numbers	Irrational Numbers	Real Numbers
5	✔	✔	✔	✔ (ratio of 5 and 1)		✔
$\frac{-47}{3}$				✔ (ratio of −47 and 3)		✔
1.48				✔ (ratio of 148 and 100)		✔
$\sqrt{7}$					✔	✔
0		✔	✔	✔ (ratio of 0 and 1)		✔

4. Inequalities

The relative size of two real numbers can be compared using the real number line. Suppose a and b represent two real numbers. We say that a is less than b, denoted $a < b$, if a lies to the left of b on the number line.

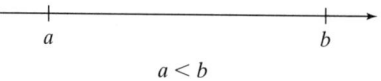

$a < b$

We say that a is greater than b, denoted $a > b$, if a lies to the right of b on the number line.

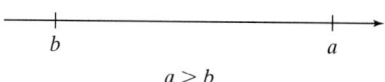

$a > b$

Table 1-1 summarizes the relational operators that compare two real numbers a and b.

Table 1-1

Mathematical Expression	Translation	Example
$a < b$	a is less than b.	$2 < 3$
$a > b$	a is greater than b.	$5 > 1$
$a \leq b$	a is less than or equal to b.	$4 \leq 4$
$a \geq b$	a is greater than or equal to b.	$10 \geq 9$
$a = b$	a is equal to b.	$6 = 6$
$a \neq b$	a is not equal to b.	$7 \neq 0$
$a \approx b$	a is approximately equal to b.	$2.3 \approx 2$

The symbols $<$, $>$, $\leq$, $\geq$, and $\neq$ are called inequality signs, and the expressions $a < b, a > b, a \leq b, a \geq b$, and $a \neq b$ are called **inequalities**.

Example 4   Ordering Real Numbers

The average temperatures (in degrees Celsius) for selected cities in the United States and Canada in January are shown in Table 1-2.

Table 1-2

City	Temp (°C)
Prince George, British Columbia	-12.1
Corpus Christi, Texas	13.4
Parkersburg, West Virginia	-0.9
San Jose, California	9.7
Juneau, Alaska	-5.7
New Bedford, Massachusetts	-0.2
Durham, North Carolina	4.2

Skill Practice Answers

6. integers, rational numbers, real numbers
7. rational numbers, real numbers
8. irrational numbers, real numbers
9. natural numbers, whole numbers, integers, rational numbers, real numbers
10. whole numbers, integers, rational numbers, real numbers

a. Plot a point on the real number line representing the temperature of each city.

b. Compare the temperatures between the following cities, and fill in the blank with the appropriate inequality sign: < or >.

Solution:

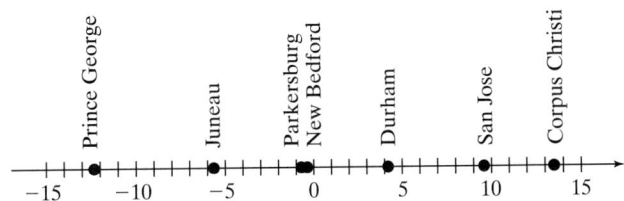

a. Temperature of San Jose $\boxed{<}$ temperature of Corpus Christi

b. Temperature of Juneau $\boxed{>}$ temperature of Prince George

c. Temperature of Parkersburg $\boxed{<}$ temperature of New Bedford

d. Temperature of Parkersburg $\boxed{>}$ temperature of Prince George

Skill Practice Based on the location of the numbers on the number line, fill in the blanks with the appropriate inequality sign: < or >.

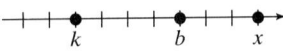

11. k _____ x **12.** x _____ b **13.** b _____ k

5. Opposite of a Real Number

To gain mastery of any algebraic skill, it is necessary to know the meaning of key definitions and key symbols. Two important definitions are the *opposite* of a real number and the *absolute value* of a real number.

> ### Definition of the Opposite of a Real Number
>
> Two numbers that are the same distance from 0 but on opposite sides of 0 on the number line are called **opposites** of each other. Symbolically, we denote the opposite of a real number a as $-a$.

Example 5 Finding the Opposite of a Real Number

a. Find the opposite of 5.

b. Find the opposite of $-\frac{4}{7}$.

c. Evaluate $-(0.46)$.

d. Evaluate $-\left(-\frac{11}{3}\right)$.

Skill Practice Answers

11. <   **12.** >   **13.** >

Solution:

 a. The opposite of 5 is -5.

 b. The opposite of $-\dfrac{4}{7}$ is $\dfrac{4}{7}$.

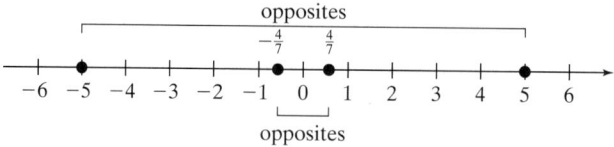

 c. $-(0.46) = -0.46$ The expression $-(0.46)$ represents the opposite of 0.46.

 d. $-\left(-\dfrac{11}{3}\right) = \dfrac{11}{3}$ The expression $-(-\tfrac{11}{3})$ represents the opposite of $-\tfrac{11}{3}$.

> **Skill Practice**

14. Find the opposite of 224.   **15.** Find the opposite of -3.4.

16. Evaluate $-(-22)$.   **17.** Evaluate $-\left(\dfrac{1}{5}\right)$.

6. Absolute Value of a Real Number

The concept of absolute value will be used to define the addition of real numbers in Section 1.3.

> **Informal Definition of the Absolute Value of a Real Number**
>
> The **absolute value** of a real number a, denoted $|a|$, is the distance between a and 0 on the number line.
>
> *Note:* The absolute value of any real number is nonnegative.

For example, $|3| = 3$ and $|-3| = 3$.

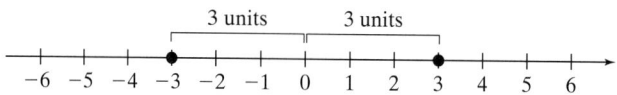

> **Example 6**   Finding the Absolute Value of a Real Number

Evaluate the absolute value expressions.

 a. $|-4|$ **b.** $\left|\tfrac{1}{2}\right|$ **c.** $|-6.2|$ **d.** $|0|$

Solution:

 a. $|-4| = 4$ -4 is 4 units from 0 on the number line.

b. $\left|\frac{1}{2}\right| = \frac{1}{2}$ $\frac{1}{2}$ is $\frac{1}{2}$ unit from 0 on the number line.

c. $|-6.2| = 6.2$ -6.2 is 6.2 units from 0 on the number line.

d. $|0| = 0$ 0 is 0 units from 0 on the number line.

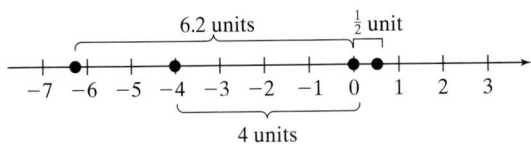

Skill Practice Evaluate.

18. $|14|$ **19.** $|-99|$

20. $|0|$ **21.** $\left|-\dfrac{7}{8}\right|$

The absolute value of a number a is its distance from zero on the number line. The definition of $|a|$ may also be given symbolically depending on whether a is negative or nonnegative.

Definition of the Absolute Value of a Real Number

Let a be a real number. Then

1. If a is nonnegative (that is, $a \geq 0$), then $|a| = a$.

2. If a is negative (that is, $a < 0$), then $|a| = -a$.

This definition states that if a is a nonnegative number, then $|a|$ equals a itself. If a is a negative number, then $|a|$ equals the opposite of a. For example,

$|9| = 9$ Because 9 is positive, then $|9|$ equals the number 9 itself.

$|-7| = 7$ Because -7 is negative, then $|-7|$ equals the opposite of -7, which is 7.

Example 7 **Comparing Absolute Value Expressions**

Determine if the statements are true or false.

a. $|3| \leq 3$ **b.** $-|5| = |-5|$

Solution:

a. $|3| \leq 3$ True. The symbol $\leq$ means "less than *or* equal to." Since $|3|$ is equal to 3, then $|3| \leq 3$ is a true statement.

b. $-|5| = |-5|$ False. On the left-hand side, $-|5|$ is the opposite of $|5|$. Hence $-|5| = -5$. On the right-hand side, $|-5| = 5$. Therefore, the original statement simplifies to $-5 = 5$, which is false.

Skill Practice True or False.

22. $-4 < -4$ **23.** $|-17| = 17$

Calculator Connections

Scientific and graphing calculators approximate irrational numbers by using rational numbers in the form of terminating decimals. For example, consider approximating π and $\sqrt{3}$:

Scientific Calculator

Enter: $\boxed{\pi}$ (or $\boxed{2^{nd}}$ $\boxed{\pi}$)   Result: 3.141592654

Enter: $\boxed{3}$ $\boxed{\sqrt{}}$ Result: 1.732050808

Graphing Calculator

Enter: $\boxed{2^{nd}}$ $\boxed{\pi}$ $\boxed{\text{ENTER}}$

Enter: $\boxed{2^{nd}}$ $\boxed{\sqrt{}}$ $\boxed{3}$ $\boxed{\text{ENTER}}$

```
π
          3.141592654
√(3)
          1.732050808
```

Note that when writing approximations, we use the symbol, $\approx$.

$$\pi \approx 3.141592654 \quad \text{and} \quad \sqrt{3} \approx 1.732050808$$

Section 1.1   Practice Exercises

Boost *your* GRADE at mathzone.com!

MathZone

- Practice Problems
- Self-Tests
- NetTutor
- e-Professors
- Videos

Study Skills Exercises

1. In this text, we will provide skills for you to enhance your learning experience. In the first four chapters, each set of Practice Exercises will begin with an activity that focuses on one of the following areas: learning about your course, using your text, taking notes, doing homework, and taking an exam. In subsequent chapters, we will insert skills pertaining to the specific material in the chapter. In Chapter 6 we will give tips on studying for the final exam.

Each activity requires only a few minutes and will help you to pass this class and become a better math student. Many of these skills can be carried over to other disciplines and help you to become a model college student. To begin, fill in the following information:

a. Instructor's name

b. Days of the week that the class meets

c. The room number in which the class meets

d. Is there a lab requirement for this course?

   If so, how often and what is the location of the lab?

e. Instructor's office number

f. Instructor's telephone number

g. Instructor's e-mail address

h. Instructor's office hours

2. Define the key terms:

a. real numbers

b. natural numbers

c. whole numbers

d. integers

e. rational numbers

f. irrational numbers

g. inequality

h. opposite

i. absolute value

j. real number line

Concept 2: Plotting Points on the Number Line

3. Plot the numbers on a real number line: $\{1, -2, -\pi, 0, -\frac{5}{2}, 5.1\}$

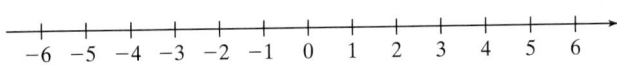

4. Plot the numbers on a real number line: $\{3, -4, \frac{1}{8}, -1.7, -\frac{4}{3}, 1.75\}$

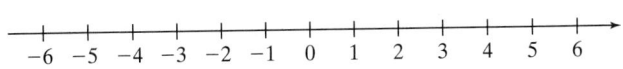

Concept 3: Set of Real Numbers

For Exercises 5–20, describe each number as (a) a terminating decimal, (b) a repeating decimal, or (c) a nonterminating, nonrepeating decimal.

5. 0.29

6. 3.8

7. $\frac{1}{9}$

8. $\frac{1}{3}$

9. $\frac{1}{8}$

10. $\frac{1}{5}$

11. 2π

12. 3π

13. -0.125

14. -3.24

15. -3

16. -6

17. $0.\overline{2}$

18. $0.\overline{6}$

19. $\sqrt{6}$

20. $\sqrt{10}$

21. List all of the numbers from Exercises 5–20 that are rational numbers.

22. List all of the numbers from Exercises 5–20 that are irrational numbers.

23. List three numbers that are real numbers but not rational numbers.

24. List three numbers that are real numbers but not irrational numbers.

25. List three numbers that are integers but not natural numbers.

26. List three numbers that are integers but not whole numbers.

27. List three numbers that are rational numbers but not integers.

For Exercises 28–33, let $A = \{-\frac{3}{2}, \sqrt{11}, -4, 0.\overline{6}, 0, \sqrt{7}, 1\}$

28. Are all of the numbers in set A real numbers?

29. List all of the rational numbers in set A.

30. List all of the whole numbers in set A.

31. List all of the natural numbers in set A.

32. List all of the irrational numbers in set A.

33. List all of the integers in set A.

34. Plot the real numbers of set A on a number line. (*Hint:* $\sqrt{11} \approx 3.3$ and $\sqrt{7} \approx 2.6$)

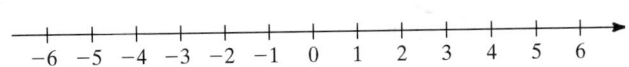

Concept 4: Inequalities

35. The LPGA Samsung World Championship of women's golf scores for selected players are given in the table. Compare the scores and fill in the blanks with the appropriate inequality sign: $<$ or $>$.

a. Kane's score _____ Pak's score.

b. Sorenstam's score _____ Davies' score.

c. Pak's score _____ McCurdy's score.

d. Kane's score _____ Davies' score.

LPGA Golfers	Final Score with Respect to Par
Annika Sorenstam	7
Laura Davies	−4
Lorie Kane	0
Cindy McCurdy	3
Se Ri Pak	−8

36. The elevations of selected cities in the United States are shown in the figure. Compare the elevations and fill in the blanks with the appropriate inequality sign: $<$ or $>$. (A negative number indicates that the city is below sea level.)

a. Elevation of Tucson _____ elevation of Cincinnati.

b. Elevation of New Orleans _____ elevation of Chicago.

c. Elevation of New Orleans _____ elevation of Houston.

d. Elevation of Chicago _____ elevation of Cincinnati.

Concept 5: Opposite of a Real Number

For Exercises 37–44, find the opposite of each number.

37. 18   **38.** 2   **39.** −6.1   **40.** −2.5

41. $-\dfrac{5}{8}$ **42.** $-\dfrac{1}{3}$ **43.** $\dfrac{7}{3}$ **44.** $\dfrac{1}{9}$

The opposite of a is denoted as $-a$. For Exercises 45–50, simplify.

45. $-(-3)$ **46.** $-(-5.1)$ **47.** $-\left(\dfrac{7}{3}\right)$ **48.** $-(-7)$

49. $-(-8)$ **50.** $-(36)$

Concept 6: Absolute Value of a Real Number

For Exercises 51–58, find the absolute value as indicated.

51. $|-2|$ **52.** $|-7|$ **53.** $|-1.5|$ **54.** $|-3.7|$

55. $-|-1.5|$ **56.** $-|-3.7|$ **57.** $\left|\dfrac{3}{2}\right|$ **58.** $\left|\dfrac{7}{4}\right|$

For Exercises 59–60, answer true or false. If a statement is false, explain why.

59. If n is positive, then $|n|$ is negative.   **60.** If m is negative, then $|m|$ is negative.

For Exercises 61–84, determine if the statements are true or false. Use the real number line to justify the answer.

61. $5 > 2$ **62.** $8 < 10$ **63.** $6 < 6$ **64.** $19 > 19$

65. $-7 \geq -7$ **66.** $-1 \leq -1$ **67.** $\dfrac{3}{2} \leq \dfrac{1}{6}$ **68.** $-\dfrac{1}{4} \geq -\dfrac{7}{8}$

69. $-5 > -2$ **70.** $6 < -10$ **71.** $8 \neq 8$ **72.** $10 \neq 10$

73. $|-2| \geq |-1|$ **74.** $|3| \leq |-1|$ **75.** $\left|-\dfrac{1}{9}\right| = \left|\dfrac{1}{9}\right|$ **76.** $\left|-\dfrac{1}{3}\right| = \left|\dfrac{1}{3}\right|$

77. $|7| \neq |-7|$ **78.** $|-13| \neq |13|$ **79.** $-1 < |-1|$ **80.** $-6 < |-6|$

81. $|-8| \geq |8|$ **82.** $|-11| \geq |11|$ **83.** $|-2| \leq |2|$ **84.** $|-21| \leq |21|$

Expanding Your Skills

85. For what numbers, a, is $-a$ positive?   **86.** For what numbers, a, is $|a| = a$?

Order of Operations

1. Variables and Expressions

A **variable** is a symbol or letter, such as x, y, and z, used to represent an unknown number. **Constants** are values that do not vary such as the numbers 3, -1.5, $\frac{2}{7}$, and π. An algebraic **expression** is a collection of variables and constants under algebraic operations. For example, $\frac{3}{x}$, $y + 7$, and $t - 1.4$ are algebraic expressions.

The symbols used to show the four basic operations of addition, subtraction, multiplication, and division are summarized in Table 1-3.

Concepts

1. Variables and Expressions
2. Evaluating Algebraic Expressions
3. Exponential Expressions
4. Square Roots
5. Order of Operations
6. Translations

Table 1-3

Operation	Symbols	Translation
Addition	$a + b$	**sum** of a and b a plus b b added to a b more than a a increased by b the total of a and b
Subtraction	$a - b$	**difference** of a and b a minus b b subtracted from a a decreased by b b less than a
Multiplication	$a \times b, a \cdot b, a(b), (a)b, (a)(b), ab$ (*Note:* We rarely use the notation $a \times b$ because the symbol, $\times$, might be confused with the variable, x.)	**product** of a and b a times b a multiplied by b
Division	$a \div b, \dfrac{a}{b}, a/b, b\overline{)a}$	**quotient** of a and b a divided by b b divided into a ratio of a and b a over b a per b

2. Evaluating Algebraic Expressions

The value of an algebraic expression depends on the values of the variables within the expression.

Example 1   Evaluating an Algebraic Expression

Evaluate the algebraic expression when $p = 4$ and $q = \frac{3}{4}$.

   a. $100 - p$ **b.** pq

Solution:

   a. $100 - p$

$$100 - (\quad)$$ When substituting a number for a variable, use parentheses.

$$= 100 - (4)$$ Substitute $p = 4$ in the parentheses.

$$= 96$$ Subtract.

   b. pq

$$= (\quad)(\quad)$$ When substituting a number for a variable, use parentheses.

$$= (4)\left(\frac{3}{4}\right)$$ Substitute $p = 4$ and $q = \frac{3}{4}$.

$$= \frac{4}{1} \cdot \frac{3}{4}$$ Write the whole number as a fraction.

$$= \frac{12}{4}$$ Multiply fractions.

$$= 3$$ Reduce to lowest terms.

Skill Practice   Evaluate the algebraic expressions when $x = 5$ and $y = 2$.

1a. xy **b.** $20 - y$

---

3. Exponential Expressions

In algebra, repeated multiplication can be expressed using exponents. The expression $4 \cdot 4 \cdot 4$ can be written as

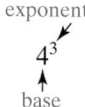

exponent

4^3

base

In the expression 4^3, 4 is the base, and 3 is the exponent, or power. The exponent indicates how many factors of the base to multiply.

Definition of b^n

Let b represent any real number and n represent a positive integer. Then,

$$b^n = \underbrace{b \cdot b \cdot b \cdot b \ldots \cdot b}_{n \text{ factors of } b}$$

b^n is read as "b to the nth power."

b is called the **base,** and n is called the **exponent,** or **power.**

b^2 is read as "b squared," and b^3 is read as "b cubed."

The exponent, n, is a count of the number of times the base, b, is used as a factor.

Example 2 | Evaluating Exponential Expressions

Translate the expression into words and then evaluate the expression.

a. 2^5 **b.** 5^2 **c.** $\left(\dfrac{3}{4}\right)^3$ **d.** 1^6

Solution:

a. The expression 2^5 is read as "two to the fifth power"
$2^5 = (2)(2)(2)(2)(2) = 32$.

b. The expression 5^2 is read as "five to the second power" or "five, squared"
$5^2 = (5)(5) = 25$.

c. The expression $\left(\frac{3}{4}\right)^3$ is read as "three-fourths to the third power" or "three-fourths, cubed"

$$\left(\frac{3}{4}\right)^3 = \left(\frac{3}{4}\right)\left(\frac{3}{4}\right)\left(\frac{3}{4}\right) = \frac{27}{64}$$

d. The expression 1^6 is read as "one to the sixth power"
$1^6 = (1)(1)(1)(1)(1)(1) = 1$.

Skill Practice Evaluate.

2. 4^3 **3.** 2^4 **4.** $\left(\dfrac{2}{3}\right)^2$ **5.** $(1)^7$

---

4. Square Roots

The inverse operation to squaring a number is to find its **square roots**. For example, finding a square root of 9 is equivalent to asking "what number(s) when squared equals 9?" The symbol, $\sqrt{}$ (called a radical sign), is used to find the *principal* square root of a number. By definition, the principal square root of a number is nonnegative. Therefore, $\sqrt{9}$ is the nonnegative number that when squared equals 9. Hence $\sqrt{9} = 3$ because 3 is nonnegative and $(3)^2 = 9$. Several more examples follow:

$$\sqrt{64} = 8 \qquad \text{Because } (8)^2 = 64$$

$$\sqrt{121} = 11 \qquad \text{Because } (11)^2 = 121$$

$$\sqrt{0} = 0 \qquad \text{Because } (0)^2 = 0$$

Skill Practice Answers

2. 64   **3.** 16

4. $\dfrac{4}{9}$ **5.** 1

$$\sqrt{\frac{1}{16}} = \frac{1}{4} \qquad \text{Because } \frac{1}{4} \cdot \frac{1}{4} = \frac{1}{16}$$

$$\sqrt{\frac{4}{9}} = \frac{2}{3} \qquad \text{Because } \frac{2}{3} \cdot \frac{2}{3} = \frac{4}{9}$$

TIP: To simplify square roots, it is advisable to become familiar with the following squares and square roots.

$0^2 = 0 \rightarrow \sqrt{0} = 0$	$7^2 = 49 \rightarrow \sqrt{49} = 7$
$1^2 = 1 \rightarrow \sqrt{1} = 1$	$8^2 = 64 \rightarrow \sqrt{64} = 8$
$2^2 = 4 \rightarrow \sqrt{4} = 2$	$9^2 = 81 \rightarrow \sqrt{81} = 9$
$3^2 = 9 \rightarrow \sqrt{9} = 3$	$10^2 = 100 \rightarrow \sqrt{100} = 10$
$4^2 = 16 \rightarrow \sqrt{16} = 4$	$11^2 = 121 \rightarrow \sqrt{121} = 11$
$5^2 = 25 \rightarrow \sqrt{25} = 5$	$12^2 = 144 \rightarrow \sqrt{144} = 12$
$6^2 = 36 \rightarrow \sqrt{36} = 6$	$13^2 = 169 \rightarrow \sqrt{169} = 13$

5. Order of Operations

When algebraic expressions contain numerous operations, it is important to evaluate the operations in the proper order. Parentheses ( ), brackets [ ], and braces { } are used for grouping numbers and algebraic expressions. It is important to recognize that operations within parentheses and other grouping symbols must be done first. Other grouping symbols include absolute value bars, radical signs, and fraction bars.

Order of Operations

1. Simplify expressions within parentheses and other grouping symbols first. These include absolute value bars, fraction bars, and radicals. If imbedded parentheses are present, start with the innermost parentheses.

2. Evaluate expressions involving exponents and radicals.

3. Perform multiplication or division in the order that they occur from left to right.

4. Perform addition or subtraction in the order that they occur from left to right.

Example 3 | **Applying the Order of Operations**

Simplify the expressions.

a. $17 - 3 \cdot 2 + 2^2$

b. $\frac{1}{2}\left(\frac{5}{6} - \frac{3}{4}\right)$

c. $25 - 12 \div 3 \cdot 4$

d. $6.2 - |-2.1| + \sqrt{15 - 6}$

e. $28 - 2[(6 - 3)^2 + 4]$

Solution:

a. $17 - 3 \cdot 2 + 2^2$

$\quad = 17 - 3 \cdot 2 + 4$ Simplify exponents.

$\quad = 17 - 6 + 4$ Multiply before adding or subtracting.

$\quad = 11 + 4$ Add or subtract from left to right.

$\quad = 15$

b. $\dfrac{1}{2}\left(\dfrac{5}{6} - \dfrac{3}{4}\right)$ Subtract fractions within the parentheses.

$\quad = \dfrac{1}{2}\left(\dfrac{10}{12} - \dfrac{9}{12}\right)$ The least common denominator is 12.

$\quad = \dfrac{1}{2}\left(\dfrac{1}{12}\right)$

$\quad = \dfrac{1}{24}$ Multiply fractions.

c. $25 - 12 \div 3 \cdot 4$ Multiply or divide in order from left to right.

$\quad = 25 - 4 \cdot 4$ Notice that the operation $12 \div 3$ is performed first (not $3 \cdot 4$).

$\quad = 25 - 16$ Multiply $4 \cdot 4$ before subtracting.

$\quad = 9$ Subtract.

d. $6.2 - |-2.1| + \sqrt{15 - 6}$

$\quad = 6.2 - |-2.1| + \sqrt{9}$ Simplify within the square root.

$\quad = 6.2 - (2.1) + 3$ Simplify the square root and absolute value.

$\quad = 4.1 + 3$ Add or subtract from left to right.

$\quad = 7.1$ Add.

e. $28 - 2[(6 - 3)^2 + 4]$

$\quad = 28 - 2[(3)^2 + 4]$ Simplify within the inner parentheses first.

$\quad = 28 - 2[(9) + 4]$ Simplify exponents.

$\quad = 28 - 2[13]$ Add within the square brackets.

$\quad = 28 - 26$ Multiply before subtracting.

$\quad = 2$ Subtract.

Skill Practice Simplify the expressions.

6. $14 - 3 \cdot 2$ **7.** $\dfrac{13}{4} - \dfrac{1}{4}(10 - 2)$ **8.** $1 + 2 \cdot 3^2 \div 6$

9. $|-20| - (7 - 2)$ **10.** $60 - 5[(6 - 3) + 2^2]$

Skill Practice Answers

6. 8   **7.** $\dfrac{5}{4}$ **8.** 4

9. 15   **10.** 25

6. Translations

| **Example 4** | Translating from English Form to Algebraic Form |

Translate each English phrase to an algebraic expression.

  a. The quotient of x and 5
  b. The difference of p and the square root of q
  c. Seven less than n
  d. Seven less n
  e. Eight more than the absolute value of w

Solution:

a. $\dfrac{x}{5}$ or $x \div 5$ — The quotient of x and 5

b. $p - \sqrt{q}$ — The difference of p and the square root of q

c. $n - 7$ — Seven less than n

d. $7 - n$ — Seven less n

e. $|w| + 8$ — Eight more than the absolute value of w

Avoiding Mistakes:

Recall that "*a* less than *b*" is translated as $b - a$. Therefore, the statement "seven less than n" must be translated as $n - 7$, not $7 - n$.

| **Skill Practice** | Translate each English phrase to an algebraic expression.

11. The product of 6 and y
12. The sum of b and the square root of c
13. Twelve less than x
14. Twelve less x
15. One more than two times x

| **Example 5** | Translating from English Form to Algebraic Form |

Translate each English phrase into an algebraic expression. Then evaluate the expression for $a = 6$, $b = 4$, and $c = 20$.

  a. The product of a and the square root of b
  b. Twice the sum of b and c
  c. The difference of twice a and b

Solution:

a. $a\sqrt{b}$ — The product of a and the square root of b.

$= (\ )\sqrt{(\ )}$ — Use parentheses to substitute a number for a variable.

$= (6)\sqrt{(4)}$ — Substitute $a = 6$ and $b = 4$.

$= 6 \cdot 2$ — Simplify the radical first.

$= 12$ — Multiply.

Skill Practice Answers

11. $6y$ **12.** $b + \sqrt{c}$
13. $x - 12$ **14.** $12 - x$
15. $2x + 1$

b. $2(b + c)$ Twice the sum of b and c. To compute "twice the sum of b and c," it is necessary to take the sum first and then multiply by 2. To ensure the proper order, the sum of b and c must be enclosed in parentheses. The proper translation is $2(b + c)$.

$= 2((\ ) + (\ ))$ Use parentheses to substitute a number for a variable.

$= 2((4) + (20))$ Substitute $b = 4$ and $c = 20$.

$= 2(24)$ Simplify within the parentheses first.

$= 48$ Multiply.

c. $2a - b$ The difference of twice a and b.

$= 2(\ ) - (\ )$ Use parentheses to substitute a number for a variable.

$= 2(6) - (4)$ Substitute $a = 6$ and $b = 4$.

$= 12 - 4$ Multiply first.

$= 8$ Subtract.

Skill Practice   Translate each English phrase to an algebraic expression. Then evaluate the expression for $x = 3, y = 9, z = 10$.

16. The quotient of the square root of y and x

17. One-half the sum of x and y

18. The difference of z and twice x

Calculator Connections

On a calculator, we enter exponents higher than the second power by using the key labeled $\boxed{y^x}$ or $\boxed{\wedge}$. For example, evaluate 2^4 and 10^6:

Scientific Calculator

Enter: $\boxed{2}$ $\boxed{y^x}$ $\boxed{4}$ $\boxed{=}$ **Result:** $\boxed{\qquad 16}$

Enter: $\boxed{10}$ $\boxed{y^x}$ $\boxed{6}$ $\boxed{=}$ **Result:** $\boxed{\qquad 1000000}$

Graphing Calculator

```
2^4
             16
10^6
       1000000
```

Most calculators also have the capability to enter several operations at once. However, it is important to note that fraction bars and radicals require user-defined parentheses to ensure that the proper order of operations is followed. For example, evaluate the following expressions on a calculator:

a. $130 - 2(5 - 1)^3$ **b.** $\dfrac{18 - 2}{11 - 9}$ **c.** $\sqrt{25 - 9}$

Skill Practice Answers

16. $\dfrac{\sqrt{y}}{x}; 1$ **17.** $\dfrac{1}{2}(x + y); 6$

18. $z - 2x; 4$

Scientific Calculator

Enter: $\boxed{130}\;\boxed{-}\;\boxed{2}\;\boxed{\times}\;\boxed{(\!(}\;\boxed{5}\;\boxed{-}\;\boxed{1}\;\boxed{)}\;\boxed{y^x}\;\boxed{3}\;\boxed{=}$ **Result:** $\boxed{2}$

Enter: $\boxed{(\!(}\;\boxed{18}\;\boxed{-}\;\boxed{2}\;\boxed{)}\;\boxed{\div}\;\boxed{(\!(}\;\boxed{11}\;\boxed{-}\;\boxed{9}\;\boxed{)}\;\boxed{=}$ **Result:** $\boxed{8}$

Enter: $\boxed{(\!(}\;\boxed{25}\;\boxed{-}\;\boxed{9}\;\boxed{)}\;\boxed{\sqrt{}}$ **Result:** $\boxed{4}$

Graphing Calculator

```
130-2*(5-1)^3
              2
(18-2)/(11-9)
              8
√(25-9)
              4
```

Calculator Exercises

Simplify the expressions without the use of a calculator. Then enter the expressions into the calculator to verify your answers.

1. $\dfrac{4+6}{8-3}$ **2.** $110 - 5(2+1) - 4$ **3.** $100 - 2(5-3)^3$

4. $3 + (4-1)^2$ **5.** $(12 - 6 + 1)^2$ **6.** $3 \cdot 8 - \sqrt{32 + 2^2}$

7. $\sqrt{18 - 2}$ **8.** $(4 \cdot 3 - 3 \cdot 3)^3$ **9.** $\dfrac{20 - 3^2}{26 - 2^2}$

Section 1.2   Practice Exercises

Study Skills Exercises

1. Sometimes you may run into a problem with homework or you find that you are having trouble keeping up with the pace of the class. A tutor can be a good resource. Answer the following questions.

   a. Does your college offer tutoring?

   b. Is it free?

   c. Where would you go to sign up for a tutor?

2. Define the key terms:

   a. variable   **b. constant**   **c. expression**   **d. base**

   e. exponent   **f. square root**   **g. order of operations**

Review Exercises

3. Which of the following are rational numbers. $\left\{ -4, 5.\overline{6}, \sqrt{29}, 0, \pi, 4.02, \dfrac{7}{9} \right\}$

4. Evaluate. $|-56|$ **5.** Evaluate. $|9.2|$

6. Find the opposite of 19.   **7.** Find the opposite of -34.2.

Concept 2: Evaluating Algebraic Expressions

For Exercises 8–19, evaluate the expressions for the given substitutions.

8. $y - 3$ when $y = 18$

9. $3q$ when $q = 5$

10. $\dfrac{15}{t}$ when $t = 5$

11. $8 + w$ when $w = 12$

12. $5 + 6d$ when $d = \dfrac{2}{3}$

13. $\dfrac{6}{5}h - 1$ when $h = 10$

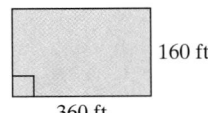

14. $2(c + 1) - 5$ when $c = 4$

15. $4(4x - 1)$ when $x = \dfrac{3}{4}$

16. $p^2 + \dfrac{2}{9}$ when $p = \dfrac{2}{3}$

17. $z^3 - \dfrac{2}{27}$ when $z = \dfrac{2}{3}$

18. $5(x + 2.3)$ when $x = 1.1$

19. $3(2.1 - y)$ when $y = 0.5$

20. The area of a rectangle may be computed as $A = \ell w$, where ℓ is the length of the rectangle and w is the width. Find the area for the rectangle shown.

21. The perimeter of the rectangle from Exercise 20 may be computed as $P = 2\ell + 2w$. Find the perimeter.

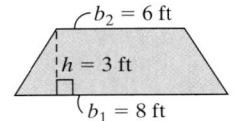

22. The area of a trapezoid is given by $A = \frac{1}{2}(b_1 + b_2)h$, where b_1 and b_2 are the lengths of the two parallel sides and h is the height. Find the area of the trapezoid with dimensions shown in the figure.

23. The volume of a rectangular solid is given by $V = \ell w h$, where ℓ is the length of the box, w is the width, and h is the height. Find the volume of the box shown in the figure.

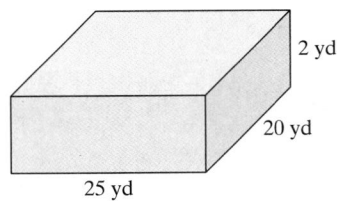

Concept 3: Exponential Expressions

For Exercises 24–31, write each of the products using exponents.

24. $\dfrac{1}{6} \cdot \dfrac{1}{6} \cdot \dfrac{1}{6} \cdot \dfrac{1}{6}$

25. $10 \cdot 10 \cdot 10 \cdot 10 \cdot 10 \cdot 10$

26. $a \cdot a \cdot a \cdot b \cdot b$

27. $7 \cdot x \cdot x \cdot y \cdot y$

28. $5c \cdot 5c \cdot 5c \cdot 5c \cdot 5c$

29. $3 \cdot w \cdot z \cdot z \cdot z \cdot z$

30. $8 \cdot y \cdot x \cdot x \cdot x \cdot x \cdot x$

31. $\dfrac{2}{3}t \cdot \dfrac{2}{3}t \cdot \dfrac{2}{3}t$

32. **a.** For the expression $5x^3$, what is the base for the exponent 3?

   b. Does 5 have an exponent? If so, what is it?

33. **a.** For the expression $2y^4$, what is the base for the exponent 4?

   b. Does 2 have an exponent? If so, what is it?

For Exercises 34–41, write each expression in expanded form using the definition of an exponent.

34. x^3

35. y^4

36. $(2b)^3$

37. $(8c)^2$

38. $10y^5$

39. x^2y^3

40. $2wz^2$

41. $3a^3b$

For Exercises 42–49, simplify the expressions.

42. 5^2

43. 4^3

44. $\left(\dfrac{1}{7}\right)^2$

45. $\left(\dfrac{1}{2}\right)^5$

46. $(0.25)^3$

47. $(0.8)^2$

48. 2^6

49. 13^2

Concept 4: Square Roots

For Exercises 50–61, simplify the square roots.

50. $\sqrt{81}$

51. $\sqrt{64}$

52. $\sqrt{4}$

53. $\sqrt{9}$

54. $\sqrt{100}$

55. $\sqrt{49}$

56. $\sqrt{16}$

57. $\sqrt{36}$

58. $\sqrt{\dfrac{1}{9}}$

59. $\sqrt{\dfrac{1}{64}}$

60. $\sqrt{\dfrac{25}{81}}$

61. $\sqrt{\dfrac{49}{100}}$

Concept 5: Order of Operations

For Exercises 62–89, use the order of operations to simplify the expressions.

62. $8 + 2 \cdot 6$

63. $7 + 3 \cdot 4$

64. $(8 + 2)6$

65. $(7 + 3)4$

66. $4 + 2 \div 2 \cdot 3 + 1$

67. $5 + 6 \cdot 2 \div 4 - 1$

68. $\dfrac{1}{4} \cdot \dfrac{2}{3} - \dfrac{1}{6}$

69. $\dfrac{3}{4} \cdot \dfrac{2}{3} + \dfrac{2}{3}$

70. $\dfrac{9}{8} - \dfrac{1}{3} \cdot \dfrac{3}{4}$

71. $\dfrac{11}{6} - \dfrac{3}{8} \cdot \dfrac{4}{3}$

72. $3[5 + 2(8 - 3)]$

73. $2[4 + 3(6 - 4)]$

74. $10 + |-6|$

75. $18 + |-3|$

76. $21 - |8 - 2|$

77. $12 - |6 - 1|$

78. $2^2 + \sqrt{9} \cdot 5$

79. $3^2 + \sqrt{16} \cdot 2$

80. $\sqrt{9 + 16} - 2$

81. $\sqrt{36 + 13} - 5$

82. $\dfrac{7 + 3(8 - 2)}{(7 + 3)(8 - 2)}$

83. $\dfrac{16 - 8 \div 4}{4 + 8 \div 4 - 2}$

84. $\dfrac{15 - 5(3 \cdot 2 - 4)}{10 - 2(4 \cdot 5 - 16)}$

85. $\dfrac{5(7 - 3) + 8(6 - 4)}{4[7 + 3(2 \cdot 9 - 8)]}$

86. $[4^2 \cdot (6 - 4) \div 8] + [7 \cdot (8 - 3)]$

87. $(18 \div \sqrt{4}) \cdot \{[(9^2 - 1) \div 2] - 15\}$

88. $48 - 13 \cdot 3 + [(50 - 7 \cdot 5) + 2]$

89. $80 \div 16 \cdot 2 + (6^2 - |-2|)$

Concept 6: Translations

For Exercises 90–101, translate each English phrase into an algebraic expression.

90. The product of 3 and x

91. The sum of b and 6

92. The quotient of x and 7

93. Four divided by k

94. The difference of 2 and a

95. Three subtracted from t

96. x more than twice y

97. Nine decreased by the product of 3 and p

98. Four times the sum of x and 12

99. Twice the difference of x and 3

100. Q less than 3

101. Fourteen less than t

For Exercises 102–111, use the order of operations to evaluate the expression when $x = 4$, $y = 2$, and $z = 10$.

102. $2y^3$

103. $3z^2$

104. $|z - 8|$

105. $|x - 3|$

106. $5\sqrt{x}$

107. $\sqrt{9 + x^2}$

108. $yz - x$

109. $z - xy$

110. xy^2

111. yx^2

For Exercises 112–123, translate each algebraic expression into an English phrase. (Answers may vary.)

112. $5 + r$

113. $18 - x$

114. $s - 14$

115. $y + 12$

116. xyz

117. $7x + 1$

118. 5^2

119. 6^3

120. $\sqrt{5}$

121. $\sqrt{10}$

122. 7^3

123. 10^2

124. Some students use the following common memorization device (mnemonic) to help them remember the order of operations: the acronym PEMDAS or **P**lease **E**xcuse **M**y **D**ear **A**unt **S**ally to remember **P**arentheses, **E**xponents, **M**ultiplication, **D**ivision, **A**ddition, and **S**ubtraction. The problem with this mnemonic is that it suggests that multiplication is done before division and similarly, it suggests that addition is performed before subtraction. Explain why following this acronym may give incorrect answers for the expressions:

a. $36 \div 4 \cdot 3$

b. $36 - 4 + 3$

125. If you use the acronym **P**lease **E**xcuse **M**y **D**ear **A**unt **S**ally to remember the order of operations, what must you keep in mind about the last four operations?

126. Explain why the acronym **P**lease **E**xcuse **D**r. **M**ichael **S**mith's **A**unt could also be used as a memory device for the order of operations.

Expanding Your Skills

For Exercises 127–130, use the order of operations to simplify the expressions.

127. $\dfrac{\sqrt{\frac{1}{9}} + \frac{2}{3}}{\sqrt{\frac{4}{25}} + \frac{3}{5}}$

128. $\dfrac{5 - \sqrt{9}}{\sqrt{\frac{4}{9}} + \frac{1}{3}}$

129. $\dfrac{|-2|}{|-10| - |2|}$

130. $\dfrac{|-4|^2}{2^2 + \sqrt{144}}$

Addition of Real Numbers

Section 1.3

1. Addition of Real Numbers and the Number Line

Concepts

Adding real numbers can be visualized on the number line. To add a positive number, move to the right on the number line. To add a negative number, move to the left on the number line. The following example may help to illustrate the process.

On a winter day in Detroit, suppose the temperature starts out at 5 degrees Fahrenheit (5°F) at noon, and then drops 12° two hours later when a cold front passes through. The resulting temperature can be represented by the expression

1. Addition of Real Numbers and the Number Line
2. Addition of Real Numbers
3. Translations
4. Applications Involving Addition of Real Numbers

$5° + (-12°)$. On the number line, start at 5 and count 12 units to the left (Figure 1-4). The resulting temperature at 2:00 P.M. is $-7°F$.

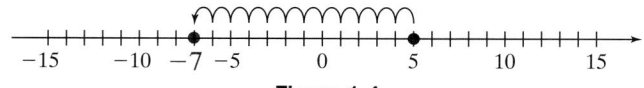

Figure 1-4

Example 1 **Using the Number Line to Add Real Numbers**

Use the number line to add the numbers.

a. $-5 + 2$ **b.** $-1 + (-4)$ **c.** $4 + (-7)$

Solution:

a. $-5 + 2 = -3$

Start at -5, and count 2 units to the right.

b. $-1 + (-4) = -5$

Start at -1, and count 4 units to the left.

c. $4 + (-7) = -3$

Start at 4, and count 7 units to the left.

Skill Practice Use the number line to add the numbers.

1. $-2 + (-3)$ **2.** $5 + (-6)$ **3.** $-2 + 4$

2. Addition of Real Numbers

When adding large numbers or numbers that involve fractions or decimals, counting units on the number line can be cumbersome. Study the following example to determine a pattern for adding two numbers with the *same* sign.

$1 + 4 = 5$

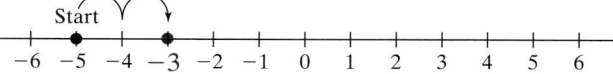

$-1 + (-4) = -5$

Adding Numbers with the *Same* Sign

To add two numbers with the *same* sign, add their absolute values and apply the common sign.

Skill Practice Answers

1. -5 **2.** -1 **3.** 2

Study the following example to determine a pattern for adding two numbers with *different* signs.

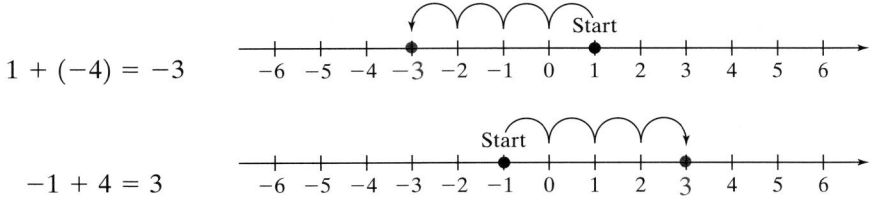

$1 + (-4) = -3$

$-1 + 4 = 3$

Adding Numbers with *Different* Signs

To add two numbers with *different* signs, subtract the smaller absolute value from the larger absolute value. Then apply the sign of the number having the larger absolute value.

Example 2   **Adding Real Numbers with the Same Sign**

Add.

a. $-12 + (-14)$ **b.** $7 + 63$ **c.** $-8.8 + (-3.7)$ **d.** $-\dfrac{4}{3} + \left(-\dfrac{6}{7}\right)$

Solution:

a. $-12 + (-14)$

$= -(12 + 14)$

common sign is negative

$= -26$

First find the absolute value of the addends.
$|-12| = 12$ and $|-14| = 14$.

Add their absolute values and apply the common sign (in this case, the common sign is negative).

The sum is -26.

b. $7 + 63$

$= + (7 + 63)$

common sign is positive

$= 70$

First find the absolute value of the addends.
$|7| = 7$ and $|63| = 63$.

Add their absolute values and apply the common sign (in this case, the common sign is positive).

The sum is 70.

c. $-8.8 + (-3.7)$

$= -(8.8 + 3.7)$

common sign is negative

$= -12.5$

First find the absolute value of the addends.
$|-8.8| = 8.8$ and $|-3.7| = 3.7$.

Add their absolute values and apply the common sign (in this case, the common sign is negative).

The sum is -12.5.

d. $-\dfrac{4}{3} + \left(-\dfrac{6}{7}\right)$ The least common denominator (LCD) is 21.

$= -\dfrac{4 \cdot 7}{3 \cdot 7} + \left(-\dfrac{6 \cdot 3}{7 \cdot 3}\right)$ Write each fraction with the LCD.

$= -\dfrac{28}{21} + \left(-\dfrac{18}{21}\right)$ Find the absolute value of the addends.

$$\left|-\dfrac{28}{21}\right| = \dfrac{28}{21} \text{ and } \left|-\dfrac{18}{21}\right| = \dfrac{18}{21}.$$

$= -\left(\dfrac{28}{21} + \dfrac{18}{21}\right)$ Add their absolute values and apply the common sign (in this case, the common sign is negative).

↑ common sign is negative

$= -\dfrac{46}{21}$ The sum is $-\dfrac{46}{21}$.

Skill Practice   Add the numbers.

4. $-5 + (-25)$ **5.** $7 + 12$

6. $-14.8 + (-9.7)$ **7.** $-\dfrac{1}{2} + \left(-\dfrac{5}{8}\right)$

Example 3   **Adding Real Numbers with Different Signs**

Add.

a. $12 + (-17)$ **b.** $-10.6 + 20.4$ **c.** $-8 + 8$ **d.** $\dfrac{2}{15} + \left(-\dfrac{4}{5}\right)$

Solution:

a. $12 + (-17)$ First find the absolute value of the addends. $|12| = 12$ and $|-17| = 17$.

     The absolute value of -17 is greater than the absolute value of 12. Therefore, the sum is negative.

$= -(17 - 12)$ Next, subtract the smaller absolute value from the larger absolute value.

↑ Apply the sign of the number with the larger absolute value.

$= -5$

b. $-10.6 + 20.4$ First find the absolute value of the addends. $|-10.6| = 10.6$ and $|20.4| = 20.4$.

     The absolute value of 20.4 is greater than the absolute value of -10.6. Therefore, the sum is positive.

$= +(20.4 - 10.6)$ Next, subtract the smaller absolute value from the larger absolute value.

↑ Apply the sign of the number with the larger absolute value.

$= 9.8$

Skill Practice Answers

4. -30 **5.** 19

6. -24.5 **7.** $-\dfrac{9}{8}$

c. $-8 + 8$

First find the absolute value of the addends. $|-8| = 8$ and $|8| = 8$.

$= (8 - 8)$

The absolute values are equal. Therefore, their difference is 0. The number zero is neither positive nor negative.

$= 0$

d. $\dfrac{2}{15} + \left(-\dfrac{4}{5}\right)$

The least common denominator is 15.

$= \dfrac{2}{15} + \left(-\dfrac{4 \cdot 3}{5 \cdot 3}\right)$

Write each fraction with the LCD.

$= \dfrac{2}{15} + \left(-\dfrac{12}{15}\right)$

Find the absolute value of the addends.

$$\left|\dfrac{2}{15}\right| = \dfrac{2}{15} \quad \text{and} \quad \left|-\dfrac{12}{15}\right| = \dfrac{12}{15}$$

The absolute value of $-\frac{12}{15}$ is greater than the absolute value of $\frac{2}{15}$. Therefore, the sum is negative.

$= -\left(\dfrac{12}{15} - \dfrac{2}{15}\right)$

Next, subtract the smaller absolute value from the larger absolute value.

↑ Apply the sign of the number with the larger absolute value.

$= -\dfrac{10}{15}$

Subtract.

$= -\dfrac{2}{3}$

Simplify by reducing to lowest terms. $-\dfrac{\overset{2}{\cancel{10}}}{\underset{3}{\cancel{15}}} = -\dfrac{2}{3}$

Skill Practice Add the numbers.

8. $-15 + 16$

9. $27.3 + (-18.1)$

10. $6.2 + (-6.2)$

11. $-\dfrac{9}{10} + \dfrac{2}{5}$

3. Translations

Example 4 **Translating Expressions Involving the Addition of Real Numbers**

Translate each English phrase into an algebraic expression. Then simplify the result.

a. The sum of -12, -8, 9, and -1

b. Negative three-tenths added to $-\frac{7}{8}$

c. The sum of -12 and its opposite

Solution:

a. $\underbrace{-12 + (-8)} + 9 + (-1)$

The sum of -12, -8, 9, and -1.

$= \underbrace{-20 + 9} + (-1)$

Add from left to right.

$= \underbrace{-11 + (-1)}$

$= -12$

b. $-\dfrac{7}{8} + \left(-\dfrac{3}{10}\right)$ Negative three-tenths added to $-\frac{7}{8}$.

$= -\dfrac{35}{40} + \left(-\dfrac{12}{40}\right)$ Get a common denominator.

$= -\dfrac{47}{40}$ The numbers have the same signs. Add their absolute values and keep the common sign. $-\left(\frac{35}{40} + \frac{12}{40}\right)$.

c. $-12 + (12)$ The sum of -12 and its opposite.

$= 0$ Add.

TIP: The sum of any number and its opposite is 0.

Skill Practice   Translate to an algebraic expression, and simplify the result.

12. The sum of -10, 4, and -6

13. The sum of $-\dfrac{9}{4}$ and $\dfrac{11}{3}$

14. -60 added to its opposite

---

4. Applications Involving Addition of Real Numbers

Example 5   Adding Real Numbers in Applications

a. A running back on a football team gains 4 yards (yd). On the next play, the quarterback is sacked and loses 13 yd. Write a mathematical expression to describe this situation and then simplify the result.

b. A student has $120 in her checking account. After depositing her paycheck of $215, she writes a check for $255 to cover her portion of the rent and another check for $294 to cover her car payment. Write a mathematical expression to describe this situation and then simplify the result.

Solution:

a. $4 + (-13)$ The loss of 13 yd can be interpreted as adding -13 yd.

$= -9$ The football team has a net loss of 9 yd.

b. $\underbrace{120 + 215} + (-255) + (-294)$ Writing a check is equivalent to adding a negative amount to the bank account.

$= \underbrace{335 + (-255)} + (-294)$ Use the order of operations. Add from left to right.

$= 80 + (-294)$

$= -214$ The student has overdrawn her account by $214.

Skill Practice

Skill Practice Answers

12. $-10 + 4 + (-6)$; -12

13. $-\dfrac{9}{4} + \dfrac{11}{3}; \dfrac{17}{12}$

14. $60 + (-60)$; 0

15. $32.00 + 2.15 + (-3.28)$; $30.87 per share

15. A share of GE stock was priced at $32.00 per share at the beginning of the month. After the first week, the price went up $2.15 per share. At the end of the second week it went down $3.28 per share. Write a mathematical expression to describe the price of the stock and find the price of the stock at the end of the 2-week period.

| Section 1.3 | **Practice Exercises** |

Study Skills Exercise

1. It is very important to attend class every day. Math is cumulative in nature, and you must master the material learned in the previous class to understand today's lesson. Because this is so important, many instructors tie attendance to the final grade. Write down the attendance policy for your class.

Review Exercises

Plot the points in set A on a number line. Then for Exercises 2–7 place the appropriate inequality ($<$, $>$) between the expressions.

$$A = \left\{-2, \frac{3}{4}, -\frac{5}{2}, 3, \frac{9}{2}, 1.6, 0\right\}$$

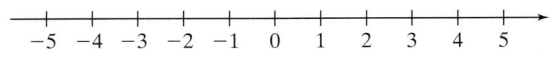

2. -2 ____ 0

3. $\dfrac{9}{2}$ ____ $\dfrac{3}{4}$

4. -2 ____ $-\dfrac{5}{2}$

5. 0 ____ $-\dfrac{5}{2}$

6. $\dfrac{3}{4}$ ____ 1.6

7. $\dfrac{3}{4}$ ____ $-\dfrac{5}{2}$

Concept 1: Addition of Real Numbers and the Number Line

For Exercises 8–15, add the numbers using the number line.

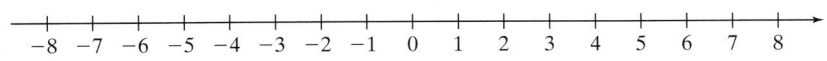

8. $-2 + (-4)$

9. $-3 + (-5)$

10. $-7 + 10$

11. $-2 + 9$

12. $6 + (-3)$

13. $8 + (-2)$

14. $2 + (-5)$

15. $7 + (-3)$

Concept 2: Addition of Real Numbers

For Exercises 16–43, add the integers.

16. $-19 + 2$

17. $-25 + 18$

18. $-4 + 11$

19. $-3 + 9$

20. $-16 + (-3)$

21. $-12 + (-23)$

22. $-2 + (-21)$

23. $-13 + (-1)$

24. $0 + (-5)$

25. $0 + (-4)$

26. $-3 + 0$

27. $-8 + 0$

28. $-16 + 16$

29. $11 + (-11)$

30. $41 + (-41)$

31. $-15 + 15$

32. $4 + (-9)$

33. $6 + (-9)$

34. $7 + (-2) + (-8)$

35. $2 + (-3) + (-6)$

36. $-17 + (-3) + 20$

37. $-9 + (-6) + 15$

38. $-3 + (-8) + (-12)$

39. $-8 + (-2) + (-13)$

40. $-42 + (-3) + 45 + (-6)$

41. $36 + (-3) + (-8) + (-25)$

42. $-5 + (-3) + (-7) + 4 + 8$

43. $-13 + (-1) + 5 + 2 + (-20)$

For Exercises 44–69, add the rational numbers.

44. $23.81 + (-2.51)$

45. $-9.23 + 10.53$

46. $-\dfrac{2}{7} + \dfrac{1}{14}$

47. $-\dfrac{1}{8} + \dfrac{5}{16}$

48. $\dfrac{2}{3} + \left(-\dfrac{5}{6}\right)$

49. $\dfrac{1}{2} + \left(-\dfrac{3}{4}\right)$

50. $-\dfrac{7}{8} + \left(-\dfrac{1}{16}\right)$

51. $-\dfrac{1}{9} + \left(-\dfrac{4}{3}\right)$

52. $-\dfrac{1}{4} + \dfrac{3}{10}$

53. $-\dfrac{7}{6} + \dfrac{7}{8}$

54. $-2.1 + \left(-\dfrac{3}{10}\right)$

55. $-8.3 + \left(-\dfrac{9}{10}\right)$

56. $\dfrac{3}{4} + (-0.5)$

57. $-\dfrac{3}{2} + 0.45$

58. $8.23 + (-8.23)$

59. $-7.5 + 7.5$

60. $-\dfrac{7}{8} + 0$

61. $0 + \left(-\dfrac{21}{22}\right)$

62. $-\dfrac{2}{3} + \left(-\dfrac{1}{9}\right) + 2$

63. $-\dfrac{1}{4} + \left(-\dfrac{3}{2}\right) + 2$

64. $-47.36 + 24.28$

65. $-0.015 + (0.0026)$

66. $516.816 + (-22.13)$

67. $87.02 + (-93.19)$

68. $-0.000617 + (-0.0015)$

69. $-5315.26 + (-314.89)$

70. State the rule for adding two numbers with different signs.

71. State the rule for adding two numbers with the same signs.

For Exercises 72–79, evaluate the expression for $x = -3$, $y = -2$, and $z = 16$.

72. $x + y + \sqrt{z}$

73. $2z + x + y$

74. $y + 3\sqrt{z}$

75. $-\sqrt{z} + y$

76. $|x| + |y|$

77. $z + x + |y|$

78. $-x + y$

79. $x + (-y) + z$

Concept 3: Translations

For Exercises 80–89, translate the English phrase into an algebraic expression. Then evaluate the expression.

80. The sum of -6 and -10

81. The sum of -3 and 5

82. Negative three increased by 8

83. Twenty-one increased by 4

84. Seventeen more than -21

85. Twenty-four more than -7

86. Three times the sum of -14 and 20

87. Two times the sum of -6 and 10

88. Five more than the sum of -7 and -2

89. Negative six more than the sum of 4 and -1

Concept 4: Applications Involving Addition of Real Numbers

90. The temperature in Minneapolis, Minnesota, began at $-5°F$ (5° below zero) at 6:00 A.M. By noon the temperature had risen 13°, and by the end of the day, the temperature had dropped 11° from its noon time high. Write an expression using addition that describes the changes in temperature during the day. Then evaluate the expression to give the temperature at the end of the day.

91. The temperature in Toronto, Ontario, Canada, began at 4°F. A cold front went through at noon, and the temperature dropped 9°. By 4:00 P.M. the temperature had risen 2° from its noon time low. Write an expression using addition that describes the changes in temperature during the day. Then evaluate the expression to give the temperature at 4:00 P.M.

92. During a football game, the Nebraska Cornhuskers lost 2 yd, gained 6 yd, and then lost 5 yd. Write an expression using addition that describes the team's total loss or gain and evaluate the expression.

93. During a football game, the University of Oklahoma's team gained 3 yd, lost 5 yd, and then gained 14 yd. Write an expression using addition that describes the team's total loss or gain and evaluate the expression.

94. Yoshima has $52.23 in her checking account. She writes a check for groceries for $52.95.

    a. Write an addition problem that expresses Yoshima's transaction.

    b. Is Yoshima's account overdrawn?

95. Mohammad has $40.02 in his checking account. He writes a check for a pair of shoes for $40.96.

    a. Write an addition problem that expresses Mohammad's transaction.

    b. Is Mohammad's account overdrawn?

96. In the game show *Jeopardy*, a contestant responds to six questions with the following outcomes:

$$+\$100, \ +\$200, \ -\$500, \ +\$300, \ +\$100, \ -\$200$$

    a. Write an expression using addition to describe the contestant's scoring activity.

    b. Evaluate the expression from part (a) to determine the contestant's final outcome.

97. A company that has been in business for 5 years has the following profit and loss record.

Year	Profit/Loss ($)
1	−50,000
2	−32,000
3	−5000
4	13,000
5	26,000

    a. Write an expression using addition to describe the company's profit/loss activity.

    b. Evaluate the expression from part (a) to determine the company's net profit or loss.

Section 1.4   Subtraction of Real Numbers

Concepts

1. Subtraction of Real Numbers
2. Translations
3. Applications Involving Subtraction
4. Applying the Order of Operations

1. Subtraction of Real Numbers

In Section 1.3, we learned the rules for adding real numbers. Subtraction of real numbers is defined in terms of the addition process. For example, consider the following subtraction problem and the corresponding addition problem:

$$6 - 4 = 2 \quad \Leftrightarrow \quad 6 + (-4) = 2$$

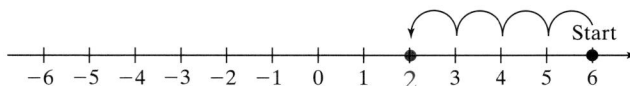

In each case, we start at 6 on the number line and move to the left 4 units. That is, adding the opposite of 4 produces the same result as subtracting 4. This is true in general. To subtract two real numbers, add the opposite of the second number to the first number.

Subtraction of Real Numbers

If a and b are real numbers, then $a - b = a + (-b)$

$$\left.\begin{array}{l} 10 - 4 = 10 + (-4) = 6 \\ -10 - 4 = -10 + (-4) = -14 \end{array}\right\} \quad \text{Subtracting 4 is the same as adding } -4.$$

$$\left.\begin{array}{l} 10 - (-4) = 10 + (4) = 14 \\ -10 - (-4) = -10 + (4) = -6 \end{array}\right\} \quad \text{Subtracting } -4 \text{ is the same as adding 4.}$$

Example 1	**Subtracting Integers**

Subtract the numbers.

  a. $4 - (-9)$ **b.** $-6 - 9$ **c.** $-11 - (-5)$ **d.** $7 - 10$

Solution:

  a. $4 - (-9)$

      $= 4 + (9) = 13$

      ↑   ↖ Take the opposite of -9.
  Change subtraction to addition.

  b. $-6 - 9$

      $= -6 + (-9) = -15$

      ↑   ↖ Take the opposite of 9.
  Change subtraction to addition.

  c. $-11 - (-5)$

      $= -11 + (5) = -6$

      ↑   ↖ Take the opposite of -5.
  Change subtraction to addition.

  d. $7 - 10$

      $= 7 + (-10) = -3$

      ↑   ↖ Take the opposite of 10.
  Change subtraction to addition.

Skill Practice	Subtract.

  1. $1 - (-3)$ **2.** $-2 - 2$ **3.** $-6 - (-11)$ **4.** $8 - 15$

Skill Practice Answers

1. 4   **2.** -4
3. 5   **4.** -7

Example 2 Subtracting Real Numbers

a. $\dfrac{3}{20} - \left(-\dfrac{4}{15}\right)$ **b.** $-2.3 - 6.04$

Solution:

a. $\dfrac{3}{20} - \left(-\dfrac{4}{15}\right)$ The least common denominator is 60.

$\dfrac{9}{60} - \left(-\dfrac{16}{60}\right)$ Write equivalent fractions with the LCD.

$\dfrac{9}{60} + \left(\dfrac{16}{60}\right)$ Write in terms of addition.

$\dfrac{25}{60}$ Add.

$\dfrac{5}{12}$ Reduce to lowest terms.

b. $-2.3 - 6.04$

$-2.3 + (-6.04)$ Write in terms of addition.

-8.34 Add.

Skill Practice Subtract.

5. $-\dfrac{1}{6} - \dfrac{7}{12}$ **6.** $7.5 - (-1.5)$

2. Translations

Example 3 Translating Expressions Involving Subtraction

Write an algebraic expression for each English phrase and then simplify the result.

a. The difference of -7 and -5

b. 12.4 subtracted from -4.7

c. -24 decreased by the sum of -10 and 13

d. Seven-fourths less than one-third

Solution:

a. $-7 - (-5)$ The difference of -7 and -5

$= -7 + (5)$ Rewrite subtraction in terms of addition.

$= -2$ Simplify.

b. $-4.7 - 12.4$ 12.4 subtracted from -4.7

$= -4.7 + (-12.4)$ Rewrite subtraction in terms of addition.

$= -17.1$ Simplify.

TIP: Recall that "b subtracted from a" is translated as $a - b$. In Example 3(b), -4.7 is written first and then 12.4 is subtracted.

Skill Practice Answers

5. $-\dfrac{3}{4}$ **6.** 9

c. $-24 - (-10 + 13)$ -24 decreased by the sum of -10 and 13

$= -24 - (3)$ Simplify inside parentheses.

$= -24 + (-3)$ Rewrite subtraction in terms of addition.

$= -27$ Simplify.

TIP: Parentheses must be used around the sum of -10 and 13 so that -24 is decreased by the entire quantity $(-10 + 13)$.

d. $\dfrac{1}{3} - \dfrac{7}{4}$ Seven-fourths less than one-third

$= \dfrac{1}{3} + \left(-\dfrac{7}{4}\right)$ Rewrite subtraction in terms of addition.

$= \dfrac{4}{12} + \left(-\dfrac{21}{12}\right)$ The common denominator is 12.

$= -\dfrac{17}{12} \quad \text{or} \quad -1\dfrac{5}{12}$

Skill Practice   Write an algebraic expression for each phrase and then simplify.

7. 8 less than -10

8. -72 subtracted from -82

9. The difference of 2.6 and -14.7

10. Two-fifths decreased by four-thirds

3. Applications Involving Subtraction

Example 4   **Using Subtraction of Real Numbers in an Application**

During one of his turns on *Jeopardy*, Harold selected the category "Show Tunes." He got the following results: the $200 question correct; the $400 question incorrect; the $600 question correct; the $800 question incorrect; and the $1000 question correct. Write an expression that determines Harold's score. Then simplify the expression to find his total winnings for that category.

Solution:

$200 - 400 + 600 - 800 + 1000$

$= 200 + (-400) + 600 + (-800) + 1000$ Rewrite subtraction in terms of addition.

$= -200 + 600 + (-800) + 1000$ Add from left to right.

$= 400 + (-800) + 1000$

$= -400 + 1000$

$= 600$ Harold won $600 in that category.

Skill Practice Answers

7. $-10 - 8; -18$
8. $-82 - (-72); -10$
9. $2.6 - (-14.7); 17.3$
10. $\dfrac{2}{5} - \dfrac{4}{3}; -\dfrac{14}{15}$

Skill Practice

11. During Harold's first round on *Jeopardy*, he got the $100, $200, and $400 questions correct but he got the $300 and $500 questions incorrect. Determine Harold's score for this round.

Example 5 Using Subtraction of Real Numbers in an Application

The highest recorded temperature in North America was 134°F, recorded on July 10, 1913, in Death Valley, California. The lowest temperature of −81°F was recorded on February 3, 1947, in Snag, Yukon, Canada.

Find the difference between the highest and lowest recorded temperatures in North America.

Solution:

$134 - (-81)$

$= 134 + (81)$ Rewrite subtraction in terms of addition.

$= 215$ Add.

The difference between the highest and lowest temperatures is 215°F.

Skill Practice

12. The record high temperature for the state of Montana occurred in 1937 and was 117°F. The record low occurred in 1954 and was −70°F. Find the difference between the highest and lowest temperatures.

4. Applying the Order of Operations

Example 6 Applying the Order of Operations

Simplify the expressions.

a. $-6 + \{10 - [7 - (-4)]\}$

b. $5 - \sqrt{35 - (-14)} - 2$

c. $\left(-\dfrac{5}{8} - \dfrac{2}{3}\right) - \left(\dfrac{1}{8} + 2\right)$

d. $-6 - |7 - 11| + (-3 + 7)^2$

Solution:

a. $-6 + \{10 - [7 - (-4)]\}$ Work inside the inner brackets first.

$= -6 + \{10 - [7 + (4)]\}$ Rewrite subtraction in terms of addition.

$= -6 + \{10 - (11)\}$ Simplify the expression inside braces.

$= -6 + \{10 + (-11)\}$ Rewrite subtraction in terms of addition.

$= -6 + (-1)$

$= -7$ Add.

Skill Practice Answers

11. −100, Harold lost $100.

12. 187°F

b. $5 - \sqrt{35 - (-14)} - 2$ Work inside the radical first.

$= 5 - \sqrt{35 + (14)} - 2$ Rewrite subtraction in terms of addition.

$= 5 - \sqrt{49} - 2$

$= 5 - 7 - 2$ Simplify the radical.

$= 5 + (-7) + (-2)$ Rewrite subtraction in terms of addition.

$= -2 + (-2)$ Add from left to right.

$= -4$

c. $\left(-\dfrac{5}{8} - \dfrac{2}{3}\right) - \left(\dfrac{1}{8} + 2\right)$ Work inside the parentheses first.

$= \left[-\dfrac{5}{8} + \left(-\dfrac{2}{3}\right)\right] - \left(\dfrac{1}{8} + 2\right)$ Rewrite subtraction in terms of addition.

$= \left[-\dfrac{15}{24} + \left(-\dfrac{16}{24}\right)\right] - \left(\dfrac{1}{8} + \dfrac{16}{8}\right)$ Get a common denominator in each parentheses.

$= \left(-\dfrac{31}{24}\right) - \left(\dfrac{17}{8}\right)$ Add fractions in each parentheses.

$= \left(-\dfrac{31}{24}\right) + \left(-\dfrac{17}{8}\right)$ Rewrite subtraction in terms of addition.

$= -\dfrac{31}{24} + \left(-\dfrac{51}{24}\right)$ Get a common denominator.

$= -\dfrac{82}{24}$ Add.

$= -\dfrac{41}{12}$ Reduce to lowest terms.

d. $-6 - |7 - 11| + (-3 + 7)^2$ Simplify within absolute value bars and parentheses first.

$= -6 - |7 + (-11)| + (-3 + 7)^2$ Rewrite subtraction in terms of addition.

$= -6 - |-4| + (4)^2$

$= -6 - (4) + 16$ Simplify absolute value and exponent.

$= -6 + (-4) + 16$ Rewrite subtraction in terms of addition.

$= -10 + 16$ Add from left to right.

$= 6$

Skill Practice   Simplify the expressions.

Skill Practice Answers
13. -12 **14.** 54
15. -1 **16.** 16

13. $\{-11 + [3 - (-4)]\} - 8$

14. $(12 - 5)^2 + \sqrt{4 - (-21)}$

15. $\left(-1 + \dfrac{1}{4}\right) - \left(\dfrac{3}{4} - \dfrac{1}{2}\right)$

16. $4 - 2|6 + (-8)| + (4)^2$

Calculator Connections

Most calculators can add, subtract, multiply, and divide signed numbers. It is important to note, however, that the key used for the negative sign is different from the key used for subtraction. On a scientific calculator, the $\boxed{+/-}$ key or $\boxed{+\bigcirc-}$ key is used to enter a negative number or to change the sign of an existing number. On a graphing calculator, the $\boxed{(-)}$ key is used. These keys should not be confused with the $\boxed{-}$ key which is used for subtraction. For example, try simplifying the following expressions.

a. $-7 + (-4) - 6$ **b.** $-3.1 - (-0.5) + 1.1$

Scientific Calculator

Enter: $\boxed{7}\ \boxed{+/-}\ \boxed{+}\ \boxed{(}\ \boxed{4}\ \boxed{+/-}\ \boxed{)}\ \boxed{-}\ \boxed{6}\ \boxed{=}$ **Result:** $\boxed{-17}$

Enter: $\boxed{3}\ \boxed{.}\ \boxed{1}\ \boxed{+/-}\ \boxed{-}\ \boxed{(}\ \boxed{0}\ \boxed{.}\ \boxed{5}\ \boxed{+/-}\ \boxed{)}$ **Result:** $\boxed{-1.5}$

$\boxed{+}\ \boxed{1}\ \boxed{.}\ \boxed{1}\ \boxed{=}$

Graphing Calculator

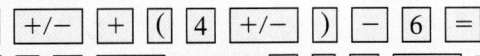

```
-7+(-4)-6
            -17
-3.1-(-0.5)+1.1
            -1.5
```

Calculator Exercises

Simplify the expression without the use of a calculator. Then use the calculator to verify your answer.

1. $-8 + (-5)$ **2.** $4 + (-5) + (-1)$ **3.** $627 - (-84)$

4. $-0.06 - 0.12$ **5.** $-3.2 + (-14.5)$ **6.** $-472 + (-518)$

7. $-12 - 9 + 4$ **8.** $209 - 108 + (-63)$

Section 1.4   Practice Exercises

Boost *your* GRADE at mathzone.com!    MathZone

- Practice Problems
- Self-Tests
- NetTutor

- e-Professors
- Videos

Study Skills Exercise

1. Some instructors allow the use of calculators. Does your instructor allow the use of a calculator? If so, what kind?

Will you be allowed to use a calculator on tests or just for occasional calculator problems in the text?

Helpful Hint: If you are not permitted to use a calculator on tests, it is good to do your homework in the same way, without the calculator.

Review Exercises

For Exercises 2–5, translate each English phrase into an algebraic expression.

2. The square root of 6

3. The square of x

4. Negative seven increased by 10

5. Two more than $-b$

For Exercises 6–9, simplify the expression.

6. $4^2 - 6 \div 2$

7. $1 + 36 \div 9 \cdot 2$

8. $14 - |10 - 6|$

9. $|-12 + 7|$

Concept 1: Subtraction of Real Numbers

For Exercises 10–17, fill in the blank to make each statement correct.

10. $5 - 3 = 5 + $ _____

11. $8 - 7 = 8 + $ _____

12. $-2 - 12 = -2 + $ _____

13. $-4 - 9 = -4 + $ _____

14. $7 - (-4) = 7 + $ _____

15. $13 - (-4) = 13 + $ _____

16. $-9 - (-3) = -9 + $ _____

17. $-15 - (-10) = -15 + $ _____

For Exercises 18–41, subtract the integers.

18. $3 - 5$

19. $9 - 12$

20. $3 - (-5)$

21. $9 - (-12)$

22. $-3 - 5$

23. $-9 - 12$

24. $-3 - (-5)$

25. $-9 - (-5)$

26. $23 - 17$

27. $14 - 2$

28. $23 - (-17)$

29. $14 - (-2)$

30. $-23 - 17$

31. $-14 - 2$

32. $-23 - (-17)$

33. $-14 - (-2)$

34. $-6 - 14$

35. $-9 - 12$

36. $-7 - 17$

37. $-8 - 21$

38. $13 - (-12)$

39. $20 - (-5)$

40. $-14 - (-9)$

41. $-21 - (-17)$

For Exercises 42–63, subtract the real numbers.

42. $-\dfrac{6}{5} - \dfrac{3}{10}$

43. $-\dfrac{2}{9} - \dfrac{5}{3}$

44. $\dfrac{3}{8} - \left(-\dfrac{4}{3}\right)$

45. $\dfrac{7}{10} - \left(-\dfrac{5}{6}\right)$

46. $\dfrac{1}{2} - \dfrac{1}{10}$

47. $\dfrac{2}{7} - \dfrac{3}{14}$

48. $-\dfrac{11}{12} - \left(-\dfrac{1}{4}\right)$

49. $-\dfrac{7}{8} - \left(-\dfrac{1}{6}\right)$

50. $6.8 - (-2.4)$

51. $7.2 - (-1.9)$

52. $3.1 - 8.82$

53. $1.8 - 9.59$

54. $-4 - 3 - 2 - 1$

55. $-10 - 9 - 8 - 7$

56. $6 - 8 - 2 - 10$

57. $20 - 50 - 10 - 5$

58. $-36.75 - 14.25$

59. $-84.21 - 112.16$

60. $-112.846 + (-13.03) - 47.312$

61. $-96.473 + (-36.02) - 16.617$

62. $0.085 - (-3.14) + (0.018)$

63. $0.00061 - (-0.00057) + (0.0014)$

Concept 2: Translations

For Exercises 64–73, translate each English phrase into an algebraic expression. Then evaluate the expression.

64. Six minus −7

65. Eighteen minus −1

66. Eighteen subtracted from 3

67. Twenty-one subtracted from 8

68. The difference of −5 and −11

69. The difference of −2 and −18

70. Negative thirteen subtracted from −1

71. Negative thirty-one subtracted from −19

72. Twenty less than −32

73. Seven less than −3

Concept 3: Applications Involving Subtraction

74. On the game, *Jeopardy*, Jasper selected the category "The Last." He got the first four questions correct (worth $200, $400, $600, and $800) but then missed the last question (worth $1000). Write an expression that determines Jasper's score. Then simplify the expression to find his total winnings for that category.

75. On Ethyl's turn in *Jeopardy*, she chose the category "Birds of a Feather." She already had $1200 when she selected a Double Jeopardy question. She wagered $500 but guessed incorrectly (therefore she lost $500). On her next turn, she got the $800 question correct. Write an expression that determines Ethyl's score. Then simplify the expression to find her total winnings.

76. In Ohio, the highest temperature ever recorded was 113°F and the lowest was −39°F. Find the difference between the highest and lowest temperatures. (*Source: Information Please Almanac*)

77. In Mississippi, the highest temperature ever recorded was 115°F and the lowest was −19°F. Find the difference between the highest and lowest temperatures. (*Source: Information Please Almanac*)

78. The highest mountain in the world is Mt. Everest, located in the Himalayas. Its height is 8848 meters (m) (29,028 ft). The lowest recorded depth in the ocean is located in the Marianas Trench in the Pacific Ocean. Its "height" relative to sea level is −11,033 m (−36,198 ft). Determine the difference in elevation, in meters, between the highest mountain in the world and the deepest ocean trench. (*Source: Information Please Almanac*)

79. The lowest point in North America is located in Death Valley, California, at an elevation of −282 ft (−86 m). The highest point in North America is Mt. McKinley, Alaska, at an elevation of 20,320 ft (6194 m). Find the difference in elevation, in feet, between the highest and lowest points in North America. (*Source: Information Please Almanac*)

Concept 4: Applying the Order of Operations

For Exercises 80–93, perform the indicated operations. Remember to perform addition or subtraction as they occur from left to right.

80. $6 + 8 − (−2) − 4 + 1$

81. $−3 − (−4) + 1 − 2 − 5$

82. $−1 − 7 + (−3) − 8 + 10$

83. $13 − 7 + 4 − 3 − (−1)$

84. $2 − (−8) + 7 + 3 − 15$

85. $8 − (−13) + 1 − 9$

86. $−6 + (−1) + (−8) + (−10)$

87. $−8 + (−3) + (−5) + (−2)$

88. $−6 − 1 − 8 − 10$

89. $-8 - 3 - 5 - 2$

90. $-\dfrac{13}{10} + \dfrac{8}{15} - \left(-\dfrac{2}{5}\right)$

91. $\dfrac{11}{14} - \left(-\dfrac{9}{7}\right) - \dfrac{3}{2}$

92. $\dfrac{2}{3} + \dfrac{5}{9} - \dfrac{4}{3} - \left(-\dfrac{1}{6}\right)$

93. $-\dfrac{9}{8} - \dfrac{1}{4} - \left(-\dfrac{5}{6}\right) + \dfrac{1}{8}$

For Exercises 94–101, evaluate the expressions for $a = -2$, $b = -6$, and $c = -1$.

94. $(a + b) - c$ **95.** $(a - b) + c$ **96.** $a - (b + c)$ **97.** $a + (b - c)$

98. $(a - b) - c$ **99.** $(a + b) + c$ **100.** $a - (b - c)$ **101.** $a + (b + c)$

For Exercises 102–107, evaluate the expression using the order of operations.

102. $\sqrt{29 + (-4)} - 7$ **103.** $8 - \sqrt{98 + (-3)} + 5$ **104.** $|10 + (-3)| - |-12 + (-6)|$

105. $|6 - 8| + |12 - 5|$ **106.** $\dfrac{3 - 4 + 5}{4 + (-2)}$ **107.** $\dfrac{12 - 14 + 6}{6 + (-2)}$

Chapter 1   Problem Recognition Exercises—Addition and Subtraction of Signed Numbers

This set of exercises allows you to practice recognizing the difference between a negative sign and a subtraction sign in the context of a problem.

1. State the rule for adding two negative numbers.

2. State the rule for adding a negative number to a positive number.

For Exercises 3–32, add or subtract as indicated.

3. $65 - 24$ **4.** $42 - 29$

5. $13 - (-18)$ **6.** $22 - (-24)$

7. $4.8 - 6.1$ **8.** $3.5 - 7.1$

9. $4 + (-20)$ **10.** $5 + (-12)$

11. $\dfrac{1}{3} - \dfrac{5}{12}$ **12.** $\dfrac{3}{8} - \dfrac{1}{12}$

13. $-32 - 4$ **14.** $-51 - 8$

15. $-6 + (-6)$ **16.** $-25 + (-25)$

17. $-4 - \left(-\dfrac{5}{6}\right)$ **18.** $-2 - \left(-\dfrac{2}{5}\right)$

19. $-60 + 55$ **20.** $-55 + 23$

21. $-18 - (-18)$ **22.** $-3 - (-3)$

23. $-3.5 - 4.2$ **24.** $-6.6 - 3.9$

25. $-90 + (-24)$ **26.** $-35 + (-21)$

27. $-14 + (-2) - 16$ **28.** $-25 + (-6) - 15$

29. $-42 + 12 + (-30)$ **30.** $-46 + 16 + (-40)$

31. $-10 - 8 - 6 - 4 - 2$

32. $-100 - 90 - 80 - 70 - 60$

For Exercises 33–34, evaluate the expression for $x = 3$ and $y = -5$.

33. $x - y$ **34.** $|x + y|$

For Exercises 35–36, write an algebraic expression for each English phrase and then simplify the result.

35. The sum of -8 and 20.

36. The difference of -11 and -2.

Multiplication and Division of Real Numbers

1. Multiplication of Real Numbers

Multiplication of real numbers can be interpreted as repeated addition.

Example 1	**Multiplying Real Numbers**

Multiply the real numbers by writing the expressions as repeated addition.

a. $3(4)$ **b.** $3(-4)$

Solution:

a. $3(4) = 4 + 4 + 4 = 12$ Add 3 groups of 4.

b. $3(-4) = -4 + (-4) + (-4) = -12$ Add 3 groups of -4.

The results from Example 1 suggest that the product of a positive number and a negative number is *negative*. Consider the following pattern of products.

$$
\begin{aligned}
4 \times 3 &= 12 \\
4 \times 2 &= 8 \\
4 \times 1 &= 4 \\
4 \times 0 &= 0 \\
4 \times -1 &= -4 \\
4 \times -2 &= -8 \\
4 \times -3 &= -12
\end{aligned}
$$

The pattern decreases by 4 with each row.

Thus, the product of a positive number and a negative number must be *negative* for the pattern to continue.

Now suppose we have a product of two negative numbers. To determine the sign, consider the following pattern of products.

$$
\begin{aligned}
-4 \times 3 &= -12 \\
-4 \times 2 &= -8 \\
-4 \times 1 &= -4 \\
-4 \times 0 &= 0 \\
-4 \times -1 &= 4 \\
-4 \times -2 &= 8 \\
-4 \times -3 &= 12
\end{aligned}
$$

The pattern increases by 4 with each row.

Thus, the product of two negative numbers must be *positive* for the pattern to continue.

We now summarize the rules for multiplying real numbers.

Multiplication of Real Numbers

1. The product of two real numbers with the *same* sign is positive.
2. The product of two real numbers with *different* signs is negative.
3. The product of any real number and zero is zero.

> **Example 2** **Multiplying Real Numbers**

Multiply the real numbers.

   a. $-8(-4)$ **b.** $-2.5(-1.7)$ **c.** $-7(10)$

   d. $\dfrac{1}{2}(-8)$ **e.** $0(-8.3)$ **f.** $-\dfrac{2}{7}\left(-\dfrac{7}{2}\right)$

Solution:

  a. $-8(-4) = 32$ *Same signs. Product is positive.*

  b. $-2.5(-1.7) = 4.25$

  c. $-7(10) = -70$ *Different signs. Product is negative.*

  d. $\dfrac{1}{2}(-8) = -4$

  e. $0(-8.3)$

     $= 0$ The product of any real number and zero is zero.

  f. $-\dfrac{2}{7}\left(-\dfrac{7}{2}\right)$

     $= \dfrac{14}{14}$ *Same signs.* Product is positive.

     $= 1$ Reduce to lowest terms.

> **Skill Practice** Multiply.
>
> **1.** $-9(3)$ **2.** $-6(-6)$ **3.** $0(-7)$
>
> **4.** $1.5(-8)$ **5.** $-\dfrac{4}{3}\left(-\dfrac{3}{4}\right)$ **6.** $(-1.2)(-4.8)$

Observe the pattern for repeated multiplications.

$$
\begin{array}{llll}
(-1)(-1) & (-1)(-1)(-1) & (-1)(-1)(-1)(-1) & (-1)(-1)(-1)(-1)(-1) \\
= 1 & = (1)(-1) & = (1)(-1)(-1) & = (1)(-1)(-1)(-1) \\
 & = -1 & = (-1)(-1) & = (-1)(-1)(-1) \\
 & & = 1 & = (1)(-1) \\
 & & & = -1
\end{array}
$$

The pattern demonstrated in these examples indicates that

- The product of an even number of negative factors is positive.
- The product of an odd number of negative factors is negative.

2. Exponential Expressions

Recall that for any real number b and any positive integer, n:

$$b^n = \underbrace{b \cdot b \cdot b \cdot b \ldots \cdot b}_{n \text{ factors of } b}$$

Be particularly careful when evaluating exponential expressions involving negative numbers. An exponential expression with a negative base is written with parentheses around the base, such as $(-2)^4$.

Skill Practice Answers

1. -27 **2.** 36 **3.** 0
4. -12 **5.** 1 **6.** 5.76

To evaluate $(-2)^4$, the base -2 is multiplied four times:

$$(-2)^4 = (-2)(-2)(-2)(-2) = 16$$

If parentheses are *not* used, the expression -2^4 has a different meaning:

- The expression -2^4 has a base of 2 (not -2) and can be interpreted as $-1 \cdot 2^4$. Hence,

$$-2^4 = -1(2)(2)(2)(2) = -16$$

- The expression -2^4 can also be interpreted as the opposite of 2^4. Hence,

$$-2^4 = -(2 \cdot 2 \cdot 2 \cdot 2) = -16$$

Example 3   **Evaluating Exponential Expressions**

Simplify.

a. $(-5)^2$ **b.** -5^2 **c.** $\left(-\dfrac{1}{2}\right)^3$ **d.** -0.4^3

Solution:

a. $(-5)^2 = (-5)(-5) = 25$ Multiply two factors of -5.

b. $-5^2 = -1(5)(5) = -25$ Multiply -1 by two factors of 5.

c. $\left(-\dfrac{1}{2}\right)^3 = \left(-\dfrac{1}{2}\right)\left(-\dfrac{1}{2}\right)\left(-\dfrac{1}{2}\right) = -\dfrac{1}{8}$ Multiply three factors of $-\dfrac{1}{2}$.

d. $-0.4^3 = -1(0.4)(0.4)(0.4) = -0.064$ Multiply -1 by three factors of 0.4.

Skill Practice   Simplify.

7. -7^2 **8.** $(-7)^2$ **9.** $(-3)^3$ **10.** -3^3

3. Division of Real Numbers

Two numbers are *reciprocals* if their product is 1. For example, $-\frac{2}{7}$ and $-\frac{7}{2}$ are reciprocals because $-\frac{2}{7}\left(-\frac{7}{2}\right) = 1$. Symbolically, if a is a nonzero real number, then the reciprocal of a is $\frac{1}{a}$ because $a \cdot \frac{1}{a} = 1$. This definition also implies that a number and its reciprocal have the same sign.

The Reciprocal of a Real Number

Let a be a nonzero real number. Then, the **reciprocal** of a is $\frac{1}{a}$.

Recall that to subtract two real numbers, we add the opposite of the second number to the first number. In a similar way, division of real numbers is defined in terms of multiplication. To divide two real numbers, we multiply the first number by the reciprocal of the second number.

Division of Real Numbers

Let a and b be real numbers such that $b \neq 0$. Then, $a \div b = a \cdot \frac{1}{b}$.

Skill Practice Answers

7. -49 **8.** 49

9. -27 **10.** -27

Consider the quotient $10 \div 5$. The reciprocal of 5 is $\frac{1}{5}$, so we have

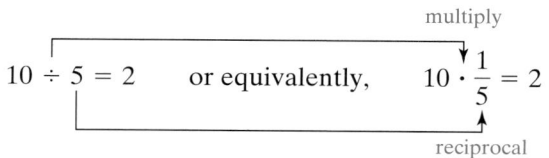

Because division of real numbers can be expressed in terms of multiplication, then the sign rules that apply to multiplication also apply to division.

Division of Real Numbers

1. The quotient of two real numbers with the *same* sign is positive.
2. The quotient of two real numbers with *different* signs is negative.

Example 4   **Dividing Real Numbers**

Divide the real numbers.

a. $200 \div (-10)$ b. $\dfrac{-48}{16}$ c. $\dfrac{-6.25}{-1.25}$ d. $\dfrac{-9}{-5}$

e. $15 \div -25$ f. $-\dfrac{3}{14} \div \dfrac{9}{7}$ g. $\dfrac{\frac{2}{5}}{-\frac{2}{5}}$

Solution:

a. $200 \div (-10) = -20$ *Different signs.* Quotient is negative.

b. $\dfrac{-48}{16} = -3$ *Different signs.* Quotient is negative.

c. $\dfrac{-6.25}{-1.25} = 5$ *Same signs.* Quotient is positive.

d. $\dfrac{-9}{-5} = \dfrac{9}{5}$ *Same signs.* Quotient is positive.

   Because 5 does not divide into 9 evenly, the answer can be left as a fraction.

e. $15 \div -25$ *Different signs.* Quotient is negative.

   $= \dfrac{15}{-25}$

   $= -\dfrac{3}{5}$

TIP: If the numerator and denominator of a fraction are both negative, then the quotient is positive. Therefore, $\frac{-9}{-5}$ can be simplified to $\frac{9}{5}$.

TIP: If the numerator and denominator of a fraction have opposite signs, then the quotient will be negative. Therefore, a fraction has the same value whether the negative sign is written in the numerator, in the denominator, or in front of the fraction.

$$\dfrac{-3}{5} = \dfrac{3}{-5} = -\dfrac{3}{5}$$

f. $-\dfrac{3}{14} \div \dfrac{9}{7}$ *Different signs.* Quotient is negative.

$= -\dfrac{3}{14} \cdot \dfrac{7}{9}$ Multiply by the reciprocal of $\frac{9}{7}$ which is $\frac{7}{9}$.

$= -\dfrac{21}{126}$ Multiply fractions.

$= -\dfrac{1}{6}$ Reduce to lowest terms.

g. $\dfrac{\frac{2}{5}}{-\frac{2}{5}}$ This is equivalent to $\frac{2}{5} \div \left(-\frac{2}{5}\right)$.

$= \dfrac{2}{5}\left(-\dfrac{5}{2}\right)$ Multiply by the reciprocal of $-\frac{2}{5}$, which is $-\frac{5}{2}$.

$= -1$ Multiply fractions.

Skill Practice Simplify.

11. $-14 \div 7$ **12.** $\dfrac{-20}{-5}$ **13.** $\dfrac{18}{-3}$ **14.** $\dfrac{-7}{-3}$

15. $\dfrac{-4}{-8}$ **16.** $\dfrac{3}{4} \div \left(\dfrac{-9}{16}\right)$ **17.** $\dfrac{-1}{1}$ **18.** $4.2 \div -0.2$

Multiplication can be used to check any division problem. If $\frac{a}{b} = c$, then $bc = a$ (provided that $b \ne 0$). For example,

$$\dfrac{8}{-4} = -2 \;\rightarrow\; \underline{\text{Check:}} \quad (-4)(-2) = 8 \;\checkmark$$

This relationship between multiplication and division can be used to investigate division problems involving the number zero.

1. The quotient of 0 and any nonzero number is 0. For example:

$$\dfrac{0}{6} = 0 \qquad \text{because } 6 \cdot 0 = 0 \;\checkmark$$

2. The quotient of any nonzero number and 0 is undefined. For example,

$$\dfrac{6}{0} = ?$$

Finding the quotient $\frac{6}{0}$ is equivalent to asking, "What number times zero will equal 6?" That is, $(0)(?) = 6$. No real number satisfies this condition. Therefore, we say that division by zero is undefined.

3. The quotient of 0 and 0 cannot be determined. Evaluating an expression of the form $\frac{0}{0} = ?$ is equivalent to asking, "What number times zero will equal 0?" That is, $(0)(?) = 0$. Any real number will satisfy this requirement; however, expressions involving $\frac{0}{0}$ are usually discussed in advanced mathematics courses.

Division Involving Zero

Let a represent a nonzero real number. Then,

1. $\dfrac{0}{a} = 0$ **2.** $\dfrac{a}{0}$ is undefined

Skill Practice Answers

11. -2 **12.** 4 **13.** -6
14. $\dfrac{7}{3}$ **15.** $\dfrac{1}{2}$ **16.** $-\dfrac{4}{3}$
17. -1 **18.** -21

4. Applying the Order of Operations

> **Example 5**   Applying the Order of Operations

Simplify the expressions.

a. $-36 \div (-27) \div (-9)$ **b.** $-8 + 8 \div (-2) \div (-6)$

c. $\dfrac{4 + \sqrt{30 - 5}}{-5 - 1}$ **d.** $\dfrac{24 - 2[-3 + (5 - 8)]^2}{2|-12 + 3|}$

Solution:

a. $\underbrace{-36 \div (-27)} \div (-9)$

$= \dfrac{-36}{-27} \div -9$ Divide from left to right.

$= \dfrac{4}{3} \div -9$ Reduce to lowest terms.

$= \dfrac{4}{3}\left(-\dfrac{1}{9}\right)$ Multiply by the reciprocal of -9, which is $-\frac{1}{9}$.

$= -\dfrac{4}{27}$ The product of two numbers with opposite signs is negative.

b. $-8 + 8 \div (-2) \div (-6)$

$= -8 + (-4) \div (-6)$ Perform division before addition.

$= -8 + \dfrac{4}{6}$ The quotient of -4 and -6 is positive $\frac{4}{6}$ or $\frac{2}{3}$.

$= -\dfrac{8}{1} + \dfrac{2}{3}$ Write -8 as a fraction.

$= -\dfrac{24}{3} + \dfrac{2}{3}$ Get a common denominator.

$= -\dfrac{22}{3}$ Add.

c. $\dfrac{4 + \sqrt{30 - 5}}{-5 - 1}$ Simplify numerator and denominator separately.

$= \dfrac{4 + \sqrt{25}}{-6}$ Simplify within the radical and simplify the denominator.

$= \dfrac{4 + 5}{-6}$ Simplify the radical.

$= \dfrac{9}{-6}$

$= \dfrac{3}{-2}$ or $-\dfrac{3}{2}$ Reduce to lowest terms. The quotient of two numbers with opposite signs is negative.

d. $\dfrac{24 - 2[-3 + (5 - 8)]^2}{2|-12 + 3|}$ Simplify numerator and denominator separately.

$= \dfrac{24 - 2[-3 + (-3)]^2}{2|-9|}$ Simplify within the inner parentheses and absolute value.

$= \dfrac{24 - 2[-6]^2}{2(9)}$ Simplify within brackets, [ ]. Simplify the absolute value.

$= \dfrac{24 - 2(36)}{2(9)}$ Simplify exponents.

$= \dfrac{24 - 72}{18}$ Perform multiplication before subtraction.

$= \dfrac{-48}{18}$ or $-\dfrac{8}{3}$ Reduce to lowest terms.

Skill Practice Simplify.

19. $-12 \div (-6) \div 2$

20. $-32 \div 4 \cdot 2 - 10$

21. $\dfrac{-4 - \sqrt{12 - 8}}{-6 - (-4)}$

22. $\dfrac{(3 + 15) - 1 - 2 + 11}{12 + 3(-4)}$

Example 6 **Evaluating an Algebraic Expression**

Given $y = -6$, evaluate the expressions.

a. y^2 **b.** $-y^2$

Solution:

a. y^2

$= (\ \ )^2$ When substituting a number for a variable, use parentheses.

$= (-6)^2$ Substitute $y = -6$.

$= 36$ Square -6, that is, $(-6)(-6) = 36$.

b. $-y^2$

$= -(\ \ )^2$ When substituting a number for a variable, use parentheses.

$= -(-6)^2$ Substitute $y = -6$.

$= -(36)$ Square -6.

$= -36$ Multiply by -1.

Skill Practice Given $a = -7$, evaluate the expressions.

23a. a^2 **b.** $-a^2$

Calculator Connections

Be particularly careful when raising a negative number to an even power on a calculator. For example, the expressions $(-4)^2$ and -4^2 have different values. That is, $(-4)^2 = 16$ and $-4^2 = -16$. Verify these expressions on a calculator.

Scientific Calculator

To evaluate $(-4)^2$

Enter: ⃞(⃞4 ⃞+/− ⃞) ⃞x²   Result: [16]

To evaluate -4^2 on a scientific calculator, it is important to square 4 first and then take its opposite.

Enter: ⃞4 ⃞x² ⃞+/−   Result: [−16]

Graphing Calculator

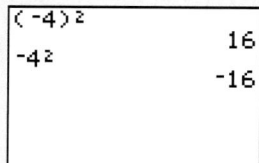

The graphing calculator allows for several methods of denoting the multiplication of two real numbers. For example, consider the product of -8 and 4.

Calculator Exercises

Simplify the expression without the use of a calculator. Then use the calculator to verify your answer.

1. $-6(5)$ 2. $\dfrac{-5.2}{2.6}$ 3. $(-5)(-5)(-5)(-5)$ 4. $(-5)^4$

5. -5^4 6. -2.4^2 7. $(-2.4)^2$

8. $(-1)(-1)(-1)$ 9. $\dfrac{-8.4}{-2.1}$ 10. $90 \div (-5)(2)$

Section 1.5   Practice Exercises

Study Skills Exercises

1. Look through the text, and write down the page number that contains:

    a. Avoiding Mistakes _____

    b. Tip box _____

    c. A key term (shown in bold) _____

2. Define the key term **reciprocal of a real number**.

Review Exercises

For Exercises 3–7, determine if the expression is true or false.

3. $6 + (-2) > -5 + 6$

4. $|-6| + |-14| \leq |-3| + |-17|$

5. $16 - |-2| \leq |-3| - (-11)$

6. $\sqrt{36} - |-6| > 0$

7. $\sqrt{9} + |-3| \leq 0$

Concept 1: Multiplication of Real Numbers

For each product in Exercises 8–11, write an equivalent addition problem.

8. $5(4)$

9. $2(6)$

10. $3(-2)$

11. $5(-6)$

For Exercises 12–19, multiply the real numbers.

12. $8(-7)$

13. $(-3) \cdot 4$

14. $(-6) \cdot 7$

15. $9(-5)$

16. $(-11)(-13)$

17. $(-5)(-26)$

18. $(-30)(-8)$

19. $(-16)(-8)$

Concept 2: Exponential Expressions

For Exercises 20–27, simplify the exponential expression.

20. $(-6)^2$

21. $(-10)^2$

22. -6^2

23. -10^2

24. $\left(-\dfrac{2}{3}\right)^3$

25. $\left(-\dfrac{5}{2}\right)^3$

26. $(-0.2)^4$

27. $(-0.1)^4$

Concept 3: Division of Real Numbers

For Exercises 28–33, show how multiplication can be used to check the division problems.

28. $\dfrac{14}{-2} = -7$

29. $\dfrac{-18}{-6} = 3$

30. $\dfrac{0}{-5} = 0$

31. $\dfrac{0}{-4} = 0$

32. $\dfrac{6}{0}$ is undefined

33. $\dfrac{-4}{0}$ is undefined

For Exercises 34–41, divide the real numbers.

34. $\dfrac{54}{-9}$

35. $\dfrac{-27}{3}$

36. $\dfrac{-100}{-10}$

37. $\dfrac{-120}{-40}$

38. $\dfrac{-14}{-7}$

39. $\dfrac{-21}{-3}$

40. $\dfrac{13}{-65}$

41. $\dfrac{7}{-77}$

Mixed Exercises

For Exercises 42–117, multiply or divide as indicated.

42. $2 \cdot 3$

43. $8 \cdot 6$

44. $2(-3)$

45. $8(-6)$

46. $(-2)3$

47. $(-8)6$

48. $(-2)(-3)$

49. $(-8)(-6)$

50. $24 \div 3$

51. $52 \div 2$

52. $24 \div (-3)$

53. $52 \div (-2)$

54. $(-24) \div 3$

55. $(-52) \div 2$

56. $(-24) \div (-3)$

57. $(-52) \div (-2)$

58. $-6 \cdot 0$

59. $-8 \cdot 0$

60. $-18 \div 0$

61. $-42 \div 0$

62. $0\left(-\dfrac{2}{5}\right)$

63. $0\left(-\dfrac{1}{8}\right)$

64. $0 \div \left(-\dfrac{1}{10}\right)$

65. $0 \div \left(\dfrac{4}{9}\right)$

66. $\dfrac{-9}{6}$

67. $\dfrac{-15}{10}$

68. $\dfrac{-30}{-100}$

69. $\dfrac{-250}{-1000}$

70. $\dfrac{26}{-13}$

71. $\dfrac{52}{-4}$

72. $1.72(-4.6)$

73. $361.3(-14.9)$

74. $-0.02(-4.6)$

75. $-0.06(-2.15)$

76. $\dfrac{14.4}{-2.4}$

77. $\dfrac{50.4}{-6.3}$

78. $\dfrac{-5.25}{-2.5}$

79. $\dfrac{-8.5}{-27.2}$

80. $(-3)^2$

81. $(-7)^2$

82. -3^2

83. -7^2

84. $\left(-\dfrac{4}{3}\right)^3$

85. $\left(-\dfrac{1}{5}\right)^3$

86. $(-0.2)^3$

87. $(-0.1)^6$

88. -0.2^4

89. -0.1^4

90. $87 \div (-3)$

91. $96 \div (-6)$

92. $-4(-12)$

93. $(-5)(-11)$

94. $2.8(-5.1)$

95. $(7.21)(-0.3)$

96. $(-6.8) \div (-0.02)$

97. $(-12.3) \div (-0.03)$

98. $\left(-\dfrac{2}{15}\right)\left(\dfrac{25}{3}\right)$

99. $\left(-\dfrac{5}{16}\right)\left(\dfrac{4}{9}\right)$

100. $\left(-\dfrac{7}{8}\right) \div \left(-\dfrac{9}{16}\right)$

101. $\left(-\dfrac{22}{23}\right) \div \left(-\dfrac{11}{3}\right)$

102. $(-2)(-5)(-3)$

103. $(-6)(-1)(-10)$

104. $(-8)(-4)(-1)(-3)$

105. $(-6)(-3)(-1)(-5)$

106. $100 \div (-10) \div (-5)$

107. $150 \div (-15) \div (-2)$

108. $-12 \div (-6) \div (-2)$

109. $-36 \div (-2) \div 6$

110. $\dfrac{2}{5} \cdot \dfrac{1}{3} \cdot \left(-\dfrac{10}{11}\right)$

111. $\left(-\dfrac{9}{8}\right) \cdot \left(-\dfrac{2}{3}\right) \cdot \left(1\dfrac{5}{12}\right)$

112. $\left(1\dfrac{1}{3}\right) \div 3 \div \left(-\dfrac{7}{9}\right)$

113. $-\dfrac{7}{8} \div \left(3\dfrac{1}{4}\right) \div (-2)$

114. $12 \div (-2)(4)$

115. $(-6) \cdot 7 \div (-2)$

116. $\left(-\dfrac{12}{5}\right) \div (-6) \cdot \left(-\dfrac{1}{8}\right)$

117. $10 \cdot \dfrac{1}{3} \div \dfrac{25}{6}$

Concept 4: Applying the Order of Operations

For Exercises 118–131, perform the indicated operations.

118. $8 - 2^3 \cdot 5 + 3 - (-6)$

119. $-14 \div (-7) - 8 \cdot 2 + 3^3$

120. $-(2 - 8)^2 \div (-6) \cdot 2$

121. $-(3 - 5)^2 \cdot 6 \div (-4)$

122. $\dfrac{6(-4) - 2(5 - 8)}{-6 - 3 - 5}$

123. $\dfrac{3(-4) - 5(9 - 11)}{-9 - 2 - 3}$

124. $\dfrac{-4 + 5}{(-2) \cdot 5 + 10}$

125. $\dfrac{-3 + 10}{2(-4) + 8}$

126. $|-5| - |-7|$

127. $|-8| - |-2|$

128. $-|-1| - |5|$

129. $-|-10| - |6|$

130. $\dfrac{|2 - 9| - |5 - 7|}{10 - 15}$

131. $\dfrac{|-2 + 6| - |3 - 5|}{13 - 11}$

For Exercises 132–139, evaluate the expression for $x = -2$, $y = -4$, and $z = 6$.

132. $x^2 - 2y$

133. $3y^2 - z$

134. $4(2x - z)$

135. $6(3x + y)$

136. $\dfrac{3x + 2y}{y}$

 **137.** $\dfrac{2z - y}{x}$

138. $\dfrac{x + 2y}{x - 2y}$

139. $\dfrac{x - z}{x^2 - z^2}$

140. Is the expression $\dfrac{10}{5x}$ equal to 10/5x? Explain.

141. Is the expression 10/(5x) equal to $\dfrac{10}{5x}$? Explain.

For Exercises 142–149, translate the English phrase into an algebraic expression. Then evaluate the expression.

142. The product of -3.75 and 0.3

143. The product of -0.4 and -1.258

144. The quotient of $\frac{16}{5}$ and $\left(-\frac{8}{9}\right)$

145. The quotient of $\left(-\frac{3}{14}\right)$ and $\frac{1}{7}$

146. The number -0.4 plus the quantity 6 times -0.42

147. The number 0.5 plus the quantity -2 times 0.125

148. The number $-\frac{1}{4}$ minus the quantity 6 times $-\frac{1}{3}$

149. Negative five minus the quantity $\left(-\frac{5}{6}\right)$ times $\frac{3}{8}$

150. For 3 weeks, Jim pays $2 a week for lottery tickets. Jim has one winning ticket for $3. Write an expression that describes his net gain or loss. How much money has Jim won or lost?

151. Stephanie pays $2 a week for 6 weeks for lottery tickets. Stephanie has one winning ticket for $5. Write an expression that describes her net gain or loss. How much money has Stephanie won or lost?

152. Evaluate the expressions in parts (a) and (b).

    a. $-4 - 3 - 2 - 1$

    b. $-4(-3)(-2)(-1)$

    c. Explain the difference between the operations in parts (a) and (b).

153. Evaluate the expressions in parts (a) and (b).

    a. $-10 - 9 - 8 - 7$

    b. $-10(-9)(-8)(-7)$

    c. Explain the difference between the operations in parts (a) and (b).

Section 1.6

Properties of Real Numbers and Simplifying Expressions

1. Commutative Properties of Real Numbers

When getting dressed in the morning, it makes no difference whether you put on your left shoe first and then your right shoe, or vice versa. This example illustrates a process in which the order does not affect the outcome. Such a process or operation is said to be *commutative*.

In algebra, the operations of addition and multiplication are commutative because the order in which we add or multiply two real numbers does not affect the result. For example,

$$10 + 5 = 5 + 10 \quad \text{and} \quad 10 \cdot 5 = 5 \cdot 10$$

Commutative Properties of Real Numbers

If a and b are real numbers, then

1. $a + b = b + a$ **commutative property of addition**
2. $ab = ba$ **commutative property of multiplication**

It is important to note that although the operations of addition and multiplication are commutative, subtraction and division are *not* commutative. For example,

$$10 - 5 \neq 5 - 10 \quad \text{and} \quad 10 \div 5 \neq 5 \div 10$$
$$5 \neq -5 \qquad\qquad 2 \neq \frac{1}{2}$$

Example 1 Applying the Commutative Property of Addition

Use the commutative property of addition to rewrite each expression.

a. $-3 + (-7)$ **b.** $3x^3 + 5x^4$

Solution:

a. $-3 + (-7) = -7 + (-3)$
b. $3x^3 + 5x^4 = 5x^4 + 3x^3$

Skill Practice Use the commutative property of addition to rewrite each expression.

1. $-5 + 9$ **2.** $7y + x$

Recall that subtraction is not a commutative operation. However, if we rewrite the difference of two numbers, $a - b$, as $a + (-b)$, we can apply the commutative property of addition. This is demonstrated in Example 2.

Skill Practice Answers
1. $9 + (-5)$ **2.** $x + 7y$

Example 2 Applying the Commutative Property of Addition

Rewrite the expression in terms of addition. Then apply the commutative property of addition.

a. $5a - 3b$ **b.** $z^2 - \dfrac{1}{4}$

Solution:

a. $5a - 3b$

$= 5a + (-3b)$ Rewrite subtraction as addition of $-3b$.

$= -3b + 5a$ Apply the commutative property of addition.

b. $z^2 - \dfrac{1}{4}$

$= z^2 + \left(-\dfrac{1}{4}\right)$ Rewrite subtraction as addition of $-\frac{1}{4}$.

$= -\dfrac{1}{4} + z^2$ Apply the commutative property of addition.

Skill Practice Rewrite each expression in terms of addition. Then apply the commutative property of addition.

3. $8m - 2n$ **4.** $\dfrac{1}{3}x - \dfrac{3}{4}$

In most cases, a detailed application of the commutative property will not be shown. Instead the process will be rewritten in one step. For example,

$$5a - 3b = -3b + 5a$$

Example 3 Applying the Commutative Property of Multiplication

Use the commutative property of multiplication to rewrite each expression.

a. $12(-6)$ **b.** $x \cdot 4$

Solution:

a. $12(-6) = -6(12)$

b. $x \cdot 4 = 4 \cdot x$ (or simply $4x$)

Skill Practice Use the commutative property of multiplication to rewrite each expression.

5. $-2(5)$ **6.** $y \cdot 6$

Skill Practice Answers

3. $8m + (-2n)$; $-2n + 8m$

4. $\dfrac{1}{3}x + \left(-\dfrac{3}{4}\right)$; $-\dfrac{3}{4} + \dfrac{1}{3}x$

5. $5(-2)$ **6.** $6y$

2. Associative Properties of Real Numbers

The associative property of real numbers states that the manner in which three or more real numbers are grouped under addition or multiplication will not affect the outcome. For example,

$$(5 + 10) + 2 = 5 + (10 + 2) \qquad \text{and} \qquad (5 \cdot 10)2 = 5(10 \cdot 2)$$
$$15 + 2 = 5 + 12 \qquad\qquad (50)2 = 5(20)$$
$$17 = 17 \qquad\qquad 100 = 100$$

Associative Properties of Real Numbers

If a, b, and c represent real numbers, then

1. $(a + b) + c = a + (b + c)$ associative property of addition

2. $(ab)c = a(bc)$ associative property of multiplication

Example 4   **Applying the Associative Property of Multiplication**

Use the associative property of multiplication to rewrite each expression. Then simplify the expression if possible.

a. $(5y)y$ **b.** $4(5z)$ **c.** $-\dfrac{3}{2}\left(-\dfrac{2}{3}w\right)$

Solution:

a. $(5y)y$

     $= 5(y \cdot y)$ Apply the associative property of multiplication.

     $= 5y^2$ Simplify.

b. $4(5z)$

     $= (4 \cdot 5)z$ Apply the associative property of multiplication.

     $= 20z$ Simplify.

c. $-\dfrac{3}{2}\left(-\dfrac{2}{3}w\right)$

     $= \left[-\dfrac{3}{2}\left(-\dfrac{2}{3}\right)\right]w$ Apply the associative property of multiplication.

     $= 1w$ Simplify.

     $= w$

Note: In most cases, a detailed application of the associative property will not be shown when multiplying two expressions. Instead, the process will be written in one step, such as

$$(5y)y = 5y^2, \quad 4(5z) = 20z, \quad \text{and} \quad -\dfrac{3}{2}\left(-\dfrac{2}{3}w\right) = w$$

Skill Practice Answers

7. $-3(z \cdot z)$; $-3z^2$

8. $(-2 \cdot 4)x$; $-8x$

9. $\left(\dfrac{5}{4} \cdot \dfrac{4}{5}\right)t$; t

Skill Practice   Use the associative property of multiplication to rewrite each expression. Simplify if possible.

7. $(-3z)z$ **8.** $-2(4x)$ **9.** $\dfrac{5}{4}\left(\dfrac{4}{5}t\right)$

3. Identity and Inverse Properties of Real Numbers

The number 0 has a special role under the operation of addition. Zero added to any real number does not change the number. Therefore, the number 0 is said to be the *additive identity* (also called the *identity element of addition*). For example,

$$-4 + 0 = -4 \qquad 0 + 5.7 = 5.7 \qquad 0 + \frac{3}{4} = \frac{3}{4}$$

The number 1 has a special role under the operation of multiplication. Any real number multiplied by 1 does not change the number. Therefore, the number 1 is said to be the *multiplicative identity* (also called the *identity element of multiplication*). For example,

$$(-8)1 = -8 \qquad 1(-2.85) = -2.85 \qquad 1\left(\frac{1}{5}\right) = \frac{1}{5}$$

Identity Properties of Real Numbers

If a is a real number, then

1. $a + 0 = 0 + a = a$ **identity property of addition**

2. $a \cdot 1 = 1 \cdot a = a$ **identity property of multiplication**

The sum of a number and its opposite equals 0. For example, $-12 + 12 = 0$. For any real number, a, the opposite of a (also called the *additive inverse* of a) is $-a$ and $a + (-a) = -a + a = 0$. The inverse property of addition states that the sum of any number and its additive inverse is the identity element of addition, 0. For example,

Number	Additive Inverse (Opposite)	Sum
9	-9	$9 + (-9) = 0$
-21.6	21.6	$-21.6 + 21.6 = 0$
$\dfrac{2}{7}$	$-\dfrac{2}{7}$	$\dfrac{2}{7} + \left(-\dfrac{2}{7}\right) = 0$

If b is a nonzero real number, then the reciprocal of b (also called the *multiplicative inverse* of b) is $\frac{1}{b}$. The inverse property of multiplication states that the product of b and its multiplicative inverse is the identity element of multiplication, 1. Symbolically, we have $b \cdot \frac{1}{b} = \frac{1}{b} \cdot b = 1$. For example,

Number	Multiplicative Inverse (Reciprocal)	Product
7	$\dfrac{1}{7}$	$7 \cdot \dfrac{1}{7} = 1$
3.14	$\dfrac{1}{3.14}$	$3.14\left(\dfrac{1}{3.14}\right) = 1$
$-\dfrac{3}{5}$	$-\dfrac{5}{3}$	$-\dfrac{3}{5}\left(-\dfrac{5}{3}\right) = 1$

Inverse Properties of Real Numbers

If a is a real number and b is a nonzero real number, then

   1. $a + (-a) = -a + a = 0$ **inverse property of addition**

   2. $b \cdot \dfrac{1}{b} = \dfrac{1}{b} \cdot b = 1$ **inverse property of multiplication**

4. Distributive Property of Multiplication over Addition

The operations of addition and multiplication are related by an important property called the **distributive property of multiplication over addition**. Consider the expression $6(2 + 3)$. The order of operations indicates that the sum $2 + 3$ is evaluated first, and then the result is multiplied by 6:

$$6(2 + 3)$$
$$= 6(5)$$
$$= 30$$

Notice that the same result is obtained if the factor of 6 is multiplied by each of the numbers 2 and 3, and then their products are added:

$6(2 + 3)$ The factor of 6 is *distributed* to the numbers 2 and 3.

$= 6(2) + 6(3)$
$= \quad 12 + 18$
$= \qquad 30$

The distributive property of multiplication over addition states that this is true in general.

TIP: The mathematical definition of the distributive property is consistent with the everyday meaning of the word *distribute*. To distribute means to "spread out from one to many." In the mathematical context, the factor a is distributed to both b and c in the parentheses.

Distributive Property of Multiplication over Addition

If a, b, and c are real numbers, then

$$a(b + c) = ab + ac \qquad \text{and} \qquad (b + c)a = ab + ac$$

Example 5   Applying the Distributive Property

Apply the distributive property: $2(a + 6b + 7)$

Solution:

$2(a + 6b + 7)$

$= 2(a + 6b + 7)$

$= 2(a) + 2(6b) + 2(7)$ Apply the distributive property.

$= 2a + 12b + 14$ Simplify.

TIP: Notice that the parentheses are removed after the distributive property is applied. Sometimes this is referred to as *clearing parentheses*.

Skill Practice

10. Apply the distributive property.

$7(x + 4y + z)$

Because the difference of two expressions $a - b$ can be written in terms of addition as $a + (-b)$, the distributive property can be applied when the operation of subtraction is present within the parentheses. For example,

$5(y - 7)$

$= 5[y + (-7)]$ Rewrite subtraction as addition of -7.

$= 5[y + (-7)]$ Apply the distributive property.

$= 5(y) + 5(-7)$
$= 5y + (-35),$ or $5y - 35$ Simplify.

Example 6   **Applying the Distributive Property**

Use the distributive property to rewrite each expression.

a. $-(-3a + 2b + 5c)$ **b.** $-6(2 - 4x)$

Solution:

a. $-(-3a + 2b + 5c)$

$= -1(-3a + 2b + 5c)$ The negative sign preceding the parentheses can be interpreted as taking the opposite of the quantity that follows or as
$-1(-3a + 2b + 5c)$

$= -1(-3a + 2b + 5c)$

$= -1(-3a) + (-1)(2b) + (-1)(5c)$ Apply the distributive property.

$= 3a + (-2b) + (-5c)$ Simplify.

$= 3a - 2b - 5c$

b. $-6(2 - 4x)$

$= -6[2 + (-4x)]$ Change subtraction to addition of $-4x$.

$= -6[2 + (-4x)]$

$= -6(2) + (-6)(-4x)$ Apply the distributive property. Notice that multiplying by -6 changes the signs of all terms to which it is applied.

$= -12 + 24x$ Simplify.

Skill Practice   Use the distributive property to rewrite each expression.

11. $-6(-3a + 7b)$ **12.** $-(12x + 8y - 3z)$

TIP: Notice that a negative factor preceding the parentheses changes the signs of all the terms to which it is multiplied.

$-1(-3a + 2b + 5c)$
$= +3a - 2b - 5c$

Skill Practice Answers

10. $7x + 28y + 7z$
11. $18a - 42b$
12. $-12x - 8y + 3z$

Note: In most cases, the distributive property will be applied without as much detail as shown in Examples 5 and 6. Instead, the distributive property will be applied in one step.

$$2(a + 6b + 7) \qquad -(3a + 2b + 5c) \qquad -6(2 - 4x)$$
$$1 \text{ step} = 2a + 12b + 14 \qquad 1 \text{ step} = -3a - 2b - 5c \qquad 1 \text{ step} = -12 + 24x$$

5. Simplifying Algebraic Expressions

An algebraic expression is the sum of one or more terms. A term is a constant or the product of a constant and one or more variables. For example, the expression

$$-7x^2 + xy - 100 \quad \text{or} \quad -7x^2 + xy + (-100)$$

consists of the terms $-7x^2$, xy, and -100.

The terms $-7x^2$ and xy are **variable terms** and the term -100 is called a **constant term**. It is important to distinguish between a term and the factors within a term. For example, the quantity xy is one term, and the values x and y are factors within the term. The constant factor in a term is called the *numerical coefficient* (or simply **coefficient**) of the term. In the terms $-7x^2$, xy, and -100, the coefficients are $-7, 1$, and -100, respectively.

Terms are said to be *like* terms if they each have the same variables and the corresponding variables are raised to the same powers. For example,

Like **Terms**		**Unlike Terms**		
$-3b$	and $5b$	$-5c$	and $7d$	(different variables)
$17xy$	and $-4xy$	$6xy$	and $3x$	(different variables)
$9p^2q^3$	and p^2q^3	$4p^2q^3$	and $8p^3q^2$	(different powers)
$5w$	and $2w$	$5w$	and 2	(different variables)
7	and 10	7	and $10a$	(different variables)

Example 7   Identifying Terms, Factors, Coefficients, and *Like* Terms

a. List the terms of the expression $5x^2 - 3x + 2$.

b. Identify the coefficient of the term $6yz^3$.

c. Which of the pairs are *like* terms: $8b, 3b^2$ $\quad$ or $\quad$ $4c^2d, -6c^2d$.

Solution:

a. The terms of the expression $5x^2 - 3x + 2$ are $5x^2$, $-3x$, and 2.

b. The coefficient of $6yz^3$ is 6.

c. $4c^2d$ and $-6c^2d$ are *like* terms.

Skill Practice

13. List the terms in the expression. $4xy - 9x^2 + 15$

14. Identify the coefficients of each term in the expression. $2a - 5b + c - 80$

15. Which of the pairs are *like* terms? $5x^3, 5x$ $\,$ or $\,$ $-7x^2, 11x^2$

Skill Practice Answers

13. $4xy, -9x^2, 15$
14. $2, -5, 1, -80$
15. $-7x^2$ and $11x^2$ are *like* terms.

6. Clearing Parentheses and Combining *Like* Terms

Two terms can be added or subtracted only if they are *like* terms. To add or subtract *like* terms, we use the distributive property as shown in the next example.

Example 8 **Using the Distributive Property to Add and Subtract *Like* Terms**

Add or subtract as indicated.

  a. $7x + 2x$ **b.** $-2p + 3p - p$

Solution:

  a. $7x + 2x$

      $= (7 + 2)x$ Apply the distributive property.

      $= 9x$ Simplify.

  b. $-2p + 3p - p$

      $= -2p + 3p - 1p$ Note that $-p$ equals $-1p$.

      $= (-2 + 3 - 1)p$ Apply the distributive property.

      $= (0)p$ Simplify.

      $= 0$

Skill Practice Simplify by adding *like* terms.

16. $8x + 3x$ **17.** $-6a + 4a + a$

Although the distributive property is used to add and subtract *like* terms, it is tedious to write each step. Observe that adding or subtracting *like* terms is a matter of combining the coefficients and leaving the variable factors unchanged. This can be shown in one step, a shortcut that we will use throughout the text. For example,

$$7x + 2x = 9x \qquad -2p + 3p - 1p = 0p = 0 \qquad -3a - 6a = -9a$$

Example 9 **Adding and Subtracting *Like* Terms**

  a. $3yz + 5 - 2yz + 9$ **b.** $1.2w^3 + 5.7w^3$

Solution:

  a. $3yz + 5 - 2yz + 9$

      $= 3yz - 2yz + 5 + 9$ Arrange *like* terms together. Notice that constants such as 5 and 9 are *like* terms.

      $= 1yz + 14$ Combine *like* terms.

      $= yz + 14$

Skill Practice Answers

16. $11x$ **17.** $-a$

b. $1.2w^3 + 5.7w^3$

$\quad = 6.9w^3$ Combine *like* terms.

Skill Practice   Simplify by adding *like* terms.

18. $4q + 3w - 2q - 3w + 2q$ **19.** $8x^2 + x + 5 + 4x - 3 + 2x^2$

Examples 10 and 11 illustrate how the distributive property is used to clear parentheses.

Example 10   Clearing Parentheses and Combining *Like* Terms

Simplify by clearing parentheses and combining *like* terms: $5 - 2(3x + 7)$

Solution:

$5 - 2(3x + 7)$ The order of operations indicates that we must perform multiplication before subtraction.

It is important to understand that a factor of -2 (not 2) will be multiplied to all terms within the parentheses. To see why this is so, we can rewrite the subtraction in terms of addition.

$= 5 + (-2)(3x + 7)$ Change subtraction to addition.

$= 5 + (-2)(3x + 7)$ A factor of -2 is to be distributed to terms in the parentheses.

$= 5 + (-2)(3x) + (-2)(7)$ Apply the distributive property.

$= 5 + (-6x) + (-14)$ Simplify.

$= 5 + (-14) + (-6x)$ Arrange *like* terms together.

$= -9 + (-6x)$ Combine *like* terms.

$= -9 - 6x$ Simplify by changing addition of the opposite to subtraction.

Skill Practice   Clear the parentheses and combine *like* terms.

20. $3(2x - 1) + 4(x + 3)$

Example 11   Clearing Parentheses and Combining *Like* Terms

Simplify by clearing parentheses and combining *like* terms.

a. $10(5y + 2) - 6(y - 1)$ **b.** $\frac{1}{4}(4k + 2) - \frac{1}{2}(6k + 1)$

c. $-(4s - 6t) - (3t + 5s) - 2s$

Skill Practice Answers

18. $4q$
19. $10x^2 + 5x + 2$
20. $10x + 9$

Solution:

a. $10(5y + 2) - 6(y - 1)$

$= 50y + 20 - 6y + 6$ Apply the distributive property. Notice that a factor of -6 is distributed through the second parentheses and changes the signs.

$= 50y - 6y + 20 + 6$ Arrange *like* terms together.

$= 44y + 26$ Combine *like* terms.

b. $\dfrac{1}{4}(4k + 2) - \dfrac{1}{2}(6k + 1)$

$= \dfrac{4}{4}k + \dfrac{2}{4} - \dfrac{6}{2}k - \dfrac{1}{2}$ Apply the distributive property. Notice that a factor of $-\frac{1}{2}$ is distributed through the second parentheses and changes the signs.

$= k + \dfrac{1}{2} - 3k - \dfrac{1}{2}$ Simplify fractions.

$= k - 3k + \dfrac{1}{2} - \dfrac{1}{2}$ Arrange *like* terms together.

$= -2k + 0$ Combine *like* terms.

$= -2k$

c. $-(4s - 6t) - (3t + 5s) - 2s$

$= -4s + 6t - 3t - 5s - 2s$ Apply the distributive property.

$= -4s - 5s - 2s + 6t - 3t$ Arrange *like* terms together.

$= -11s + 3t$ Combine *like* terms.

Skill Practice   Clear the parentheses and combine *like* terms.

21. $5(2y + 3) - 2(3y + 1)$ **22.** $\dfrac{1}{2}(8x + 1) + \dfrac{1}{3}(3x - 2)$

23. $-4(x + 2y) - (2x - y) - 5x$

Skill Practice Answers

21. $4y + 13$ **22.** $5x - \dfrac{1}{6}$

23. $-11x - 7y$

Section 1.6   Practice Exercises

Boost *your* GRADE at mathzone.com!

MathZone

- Practice Problems
- Self-Tests
- NetTutor
- e-Professors
- Videos

Study Skills Exercises

1. Write down the page number(s) for the Chapter Summary for this chapter. Describe one way in which you can use the Summary found at the end of each chapter.

2. Define the key terms:

     a. commutative properties   **b. associative properties**   **c. identity properties**

     d. inverse properties   **e. distributive property of multiplication over addition**

     f. variable term   **g. constant term**   **h. coefficient**   **i. *like* terms**

Review Exercises

For Exercises 3–18, perform the indicated operations.

3. $(-6) + 14$

4. $(-2) + 9$

5. $-13 - (-5)$

6. $-1 - (-19)$

7. $18 \div (-4)$

8. $-27 \div 5$

9. $-3 \cdot 0$

10. $0(-15)$

11. $\dfrac{1}{2} + \dfrac{3}{8}$

12. $\dfrac{7}{2} + \dfrac{5}{9}$

13. $\dfrac{25}{21} - \dfrac{6}{7}$

14. $\dfrac{8}{9} - \dfrac{1}{3}$

15. $\left(-\dfrac{3}{5}\right)\left(\dfrac{4}{27}\right)$

16. $\left(\dfrac{1}{6}\right)\left(-\dfrac{8}{3}\right)$

17. $\left(-\dfrac{11}{12}\right) \div \left(-\dfrac{5}{4}\right)$

18. $\left(-\dfrac{14}{15}\right) \div \left(-\dfrac{7}{5}\right)$

Concept 1: Commutative Properties of Real Numbers

For Exercises 19–26, rewrite each expression using the commutative property of addition or the commutative property of multiplication.

19. $5 + (-8)$

20. $7 + (-2)$

21. $8 + x$

22. $p + 11$

23. $5(4)$

24. $10(8)$

25. $x(-12)$

26. $y(-23)$

For Exercises 27–30, rewrite each expression using addition. Then apply the commutative property of addition.

27. $x - 3$

28. $y - 7$

29. $4p - 9$

30. $3m - 12$

Concept 2: Associative Properties of Real Numbers

For Exercises 31–38, use the associative property of multiplication to rewrite each expression. Then simplify the expression if possible.

31. $(4p)p$

32. $(-6y)y$

33. $-5(3x)$

34. $-12(4z)$

35. $\dfrac{6}{11}\left(\dfrac{11}{6}x\right)$

36. $\dfrac{3}{5}\left(\dfrac{5}{3}x\right)$

37. $-4\left(-\dfrac{1}{4}t\right)$

38. $-5\left(-\dfrac{1}{5}w\right)$

Concept 3: Identity and Inverse Properties of Real Numbers

39. What is another name for multiplicative inverse?

40. What is another name for additive inverse?

41. What is the additive identity?

42. What is the multiplicative identity?

Concept 4: The Distributive Property of Multiplication over Addition

For Exercises 43–64, use the distributive property to clear parentheses.

43. $6(5x + 1)$

44. $2(x + 7)$

45. $-2(a + 8)$

46. $-3(2z + 9)$

47. $3(5c - d)$

48. $4(w - 13z)$

49. $-7(y - 2)$

50. $-2(4x - 1)$

51. $-\dfrac{2}{3}(x - 6)$

52. $-\dfrac{1}{4}(2b - 8)$

53. $\dfrac{1}{3}(m - 3)$

54. $\dfrac{2}{5}(n - 5)$

55. $\dfrac{3}{8}(4 + 8s)$

56. $\dfrac{4}{9}(3 - 9t)$

57. $-(2p + 10)$

58. $-(7q + 1)$

59. $-(-3w - 5z)$ **60.** $-(-7a - b)$ **61.** $4(x + 2y - z)$ **62.** $-6(2a - b + c)$

63. $-(-6w + x - 3y)$ **64.** $-(-p - 5q - 10r)$

For Exercises 65–68, use the associative property or distributive property to clear parentheses.

65. $2(3 + x)$ **66.** $5(4 + y)$ **67.** $4(6z)$ **68.** $8(2p)$

For Exercises 69–77, match the statements with the properties of multiplication and addition.

69. $6 \cdot \dfrac{1}{6} = 1$

a. Commutative property of addition

b. Inverse property of multiplication

70. $7(4 \cdot 9) = (7 \cdot 4)9$

c. Commutative property of multiplication

71. $2(3 + k) = 6 + 2k$

d. Associative property of addition

e. Identity property of multiplication

72. $3 \cdot 7 = 7 \cdot 3$

f. Associative property of multiplication

73. $5 + (-5) = 0$

g. Inverse property of addition

74. $18 \cdot 1 = 18$

h. Identity property of addition

i. Distributive property of multiplication over addition

75. $(3 + 7) + 19 = 3 + (7 + 19)$

76. $23 + 6 = 6 + 23$

77. $3 + 0 = 3$

Concept 5: Simplifying Algebraic Expressions

For Exercises 78–81, list the terms and their coefficients for each expression.

78. $3xy - 6x^2 + y - 17$

Term	Coefficient

79. $2x - y + 18xy + 5$

Term	Coefficient

80. $x^4 - 10xy + 12 - y$

Term	Coefficient

81. $-x + 8y - 9x^2y - 3$

Term	Coefficient

82. Explain why $12x$ and $12x^2$ are not *like* terms.

83. Explain why $3x$ and $3xy$ are not *like* terms.

84. Explain why $7z$ and $\sqrt{13}z$ are *like* terms.

85. Explain why πx and $8x$ are *like* terms.

86. Write three different *like* terms.

87. Write three terms that are not *like*.

Concept 6: Clearing Parentheses and Combining *Like* Terms

For Exercises 88–97, simplify by combining *like* terms.

88. $5k - 10k - 12k + 16 + 7$

89. $-4p - 2p + 8p - 15 + 3$

90. $9x - 7x^2 + 12x + 14x^2$

91. $2y^2 - 8y + y - 5y^2 - 3y^2$

92. $4a + 2a^2 - 6a + 5 + 3a^2 - 2$

93. $8x^2 - 5x + 3 - 7 + 6x - x^2$

94. $\dfrac{1}{4}a + b - \dfrac{3}{4}a - 5b$

95. $\dfrac{2}{5} + 2t - \dfrac{3}{5} + t - \dfrac{6}{5}$

96. $2.8z - 8.1z + 6 - 15.2$

97. $2.4 - 8.4w - 2w + 0.9$

For Exercises 98–121, simplify by clearing parentheses and combining *like* terms.

98. $-3(2x - 4) + 10$

99. $-2(4a + 3) - 14$

100. $4(w + 3) - 12$

101. $5(2r + 6) - 30$

102. $5 - 3(x - 4)$

103. $4 - 2(3x + 8)$

104. $-3(2t + 4) + 8(2t - 4)$

105. $-5(5y + 9) + 3(3y + 6)$

106. $2(w - 5) - (2w + 8)$

107. $6(x + 3) - (6x - 5)$

108. $-\dfrac{1}{3}(6t + 9) + 10$

109. $-\dfrac{3}{4}(8 + 4q) + 7$

110. $10(5.1a - 3.1) + 4$

111. $100(-3.14p - 1.05) + 212$

112. $-4m + 2(m - 3) + 2m$

113. $-3b + 4(b + 2) - 8b$

114. $\dfrac{1}{2}(10q - 2) + \dfrac{1}{3}(2 - 3q)$

115. $\dfrac{1}{5}(15 - 4p) - \dfrac{1}{10}(10p + 5)$

116. $7n - 2(n - 3) - 6 + n$

117. $8k - 4(k - 1) + 7 - k$

118. $6(x + 3) - 12 - 4(x - 3)$

119. $5(y - 4) + 3 - 6(y - 7)$

120. $6.1(5.3z - 4.1) - 5.8$

121. $-3.6(1.7q - 4.2) + 14.6$

Expanding Your Skills

For Exercises 122–131, determine if the expressions are equivalent. If two expressions are not equivalent, state why.

122. $3a + b, b + 3a$

123. $4y + 1, 1 + 4y$

124. $2c + 7, 9c$

125. $5z + 4, 9z$

126. $5x - 3, 3 - 5x$

127. $6d - 7, 7 - 6d$

128. $5x - 3, -3 + 5x$

129. $8 - 2x, -2x + 8$

130. $5y + 6, 5 + 6y$

131. $7z - 2; 7 - 2z$

132. Which grouping of terms is easier computationally, $(14\frac{2}{7} + 2\frac{1}{3}) + \frac{2}{3}$ or $14\frac{2}{7} + (2\frac{1}{3} + \frac{2}{3})$?

133. Which grouping of terms is easier computationally, $(5\frac{1}{8} + 18\frac{2}{5}) + 1\frac{3}{5}$ or $5\frac{1}{8} + (18\frac{2}{5} + 1\frac{3}{5})$?

134. As a small child in school, the great mathematician Karl Friedrich Gauss (1777–1855) was said to have found the sum of the integers from 1 to 100 mentally:

$$1 + 2 + 3 + 4 + \cdots + 99 + 100$$

Rather than adding the numbers sequentially, he added the numbers in pairs:

$$(1 + 99) + (2 + 98) + (3 + 97) + \cdots$$

a. Use this technique to add the integers from 1 to 10.

$$1 + 2 + 3 + 4 + 5 + 6 + 7 + 8 + 9 + 10$$

b. Use this technique to add the integers from 1 to 20.

Chapter 1 SUMMARY

Section 1.1 Sets of Numbers and the Real Number Line

Key Concepts

Natural numbers: $\{1, 2, 3, \ldots\}$

Whole numbers: $\{0, 1, 2, 3, \ldots\}$

Integers: $\{\ldots -3, -2, -1, 0, 1, 2, 3, \ldots\}$

Rational numbers: The set of numbers that can be expressed in the form $\frac{p}{q}$, where p and q are integers and q does not equal 0.

Irrational numbers: The set of real numbers that are not rational.

Real numbers: The set of both the rational numbers and the irrational numbers.

$a < b$ "a is less than b."

$a > b$ "a is greater than b."

$a \leq b$ "a is less than or equal to b."

$a \geq b$ "a is greater than or equal to b."

Two numbers that are the same distance from zero but on opposite sides of zero on the number line are called **opposites**. The opposite of a is denoted $-a$.

The **absolute value** of a real number, a, denoted $|a|$, is the distance between a and 0 on the number line.

If $a \geq 0$, $|a| = a$

If $a < 0$, $|a| = -a$

Examples

Example 1

-5, 0, and 4 are integers.

$-\dfrac{5}{2}$, -0.5, and $0.\overline{3}$ are rational numbers.

$\sqrt{7}$, $-\sqrt{2}$, and π are irrational numbers.

Example 2

All real numbers can be located on a real number line.

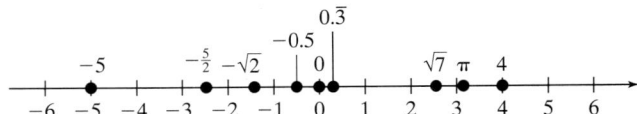

Example 3

$5 < 7$ "5 is less than 7."

$-2 > -10$ "-2 is greater than -10."

$y \leq 3.4$ "y is less than or equal to 3.4."

$x \geq \dfrac{1}{2}$ "x is greater than or equal to $\dfrac{1}{2}$."

Example 4

5 and -5 are opposites.

Example 5

$|7| = 7$

$|-7| = 7$

Section 1.2   Order of Operations

Key Concepts

A **variable** is a symbol or letter used to represent an unknown number.

A **constant** is a value that is not variable.

An algebraic **expression** is a collection of variables and constants under algebraic operations.

$$b^n = \underbrace{b \cdot b \cdot b \cdot b \ldots \cdot b}_{n \text{ factors of } b}$$ b is the **base**, n is the **exponent**

$\sqrt{x}$ is the positive **square root** of x.

The Order of Operations:

1. Simplify expressions within parentheses and other grouping symbols first.
2. Evaluate expressions involving exponents and radicals.
3. Do multiplication or division in the order that they occur from left to right.
4. Do addition or subtraction in the order that they occur from left to right.

Examples

Example 1

Variables: x, y, z, a, b

Constants: $2, -3, \pi$

Expressions: $2x + 5, 3a + b^2$

Example 2

$5^3 = 5 \cdot 5 \cdot 5 = 125$

Example 3

$\sqrt{49} = 7$

Example 4

$10 + 5(3 - 1)^2 - \sqrt{5 - 1}$

$= 10 + 5(2)^2 - \sqrt{4}$ Work within grouping symbols.

$= 10 + 5(4) - 2$ Simplify exponents.

$= 10 + 20 - 2$ Perform multiplication.

$= 30 - 2$ Add and subtract, left to right.

$= 28$

Section 1.3   Addition of Real Numbers

Key Concepts

Addition of Two Real Numbers:

Same Signs: Add the absolute values of the numbers and apply the common sign to the sum.

Different Signs: Subtract the smaller absolute value from the larger absolute value. Then apply the sign of the number having the larger absolute value.

Examples

Example 1

$-3 + (-4) = -7$

$-1.3 + (-9.1) = -10.4$

Example 2

$-5 + 7 = 2$

$\dfrac{2}{3} + \left(-\dfrac{7}{3}\right) = -\dfrac{5}{3}$

Section 1.4 Subtraction of Real Numbers

Key Concepts

Subtraction of Two Real Numbers:

Add the opposite of the second number to the first number. That is,

$a - b = a + (-b)$

Examples

Example 1

$7 - (-5) = 7 + (5) = 12$

$-3 - 5 = -3 + (-5) = -8$

$-11 - (-2) = -11 + (2) = -9$

Section 1.5 Multiplication and Division of Real Numbers

Key Concepts

Multiplication and Division of Two Real Numbers:

Same Signs:
Product is positive.
Quotient is positive.

Different Signs:
Product is negative.
Quotient is negative.

The **reciprocal** of a number a is $\dfrac{1}{a}$.

Multiplication and Division Involving Zero:

The product of any real number and 0 is 0.

The quotient of 0 and any nonzero real number is 0.

The quotient of any nonzero real number and 0 is undefined.

Examples

Example 1

$(-5)(-2) = 10$

$\dfrac{-20}{-4} = 5$

Example 2

$(-3)(7) = -21$

$\dfrac{-4}{8} = -\dfrac{1}{2}$

Example 3

The reciprocal of -6 is $-\dfrac{1}{6}$.

Example 4

$4 \cdot 0 = 0$

$0 \div 4 = 0$

$4 \div 0$ is undefined.

Section 1.6 Properties of Real Numbers and Simplifying Expressions

Key Concepts

The Properties of Real Numbers:

Commutative Property of Addition:

$a + b = b + a$

Associative Property of Addition:

$(a + b) + c = a + (b + c)$

Examples

Example 1

$-5 + (-7) = (-7) + (-5)$

Example 2

$(2 + 3) + 10 = 2 + (3 + 10)$

Identity Property of Addition: The number 0 is said to be the identity element for addition because:

$$0 + a = a \quad \text{and} \quad a + 0 = a$$

Inverse Property of Addition:

$$a + (-a) = 0 \quad \text{and} \quad -a + a = 0$$

Commutative Property of Multiplication:

$$ab = ba$$

Associative Property of Multiplication:

$$(ab)c = a(bc)$$

Identity Property of Multiplication: The number 1 is said to be the identity element for multiplication because:

$$1 \cdot a = a \quad \text{and} \quad a \cdot 1 = a$$

Inverse Property of Multiplication:

$$a \cdot \frac{1}{a} = 1 \quad \text{and} \quad \frac{1}{a} \cdot a = 1$$

The Distributive Property of Multiplication over Addition:

$$a(b + c) = ab + ac$$

A **term** is a constant or the product of a constant and one or more variables.

The **coefficient** of a term is the numerical factor of the term.

Like terms have the same variables, and the corresponding variables have the same powers.

Two terms can be added or subtracted if they are *like* terms. Sometimes it is necessary to clear parentheses before adding or subtracting *like* terms.

Example 3

$$0 + \frac{3}{4} = \frac{3}{4} \quad \text{and} \quad \frac{3}{4} + 0 = \frac{3}{4}$$

Example 4

$$1.5 + (-1.5) = 0 \quad \text{and} \quad -1.5 + 1.5 = 0$$

Example 5

$$(-3)(-4) = (-4)(-3)$$

Example 6

$$(2 \cdot 3)6 = 2(3 \cdot 6)$$

Example 7

$$1 \cdot 5 = 5 \quad \text{and} \quad 5 \cdot 1 = 5$$

Example 8

$$6 \cdot \frac{1}{6} = 1 \quad \text{and} \quad \frac{1}{6} \cdot 6 = 1$$

Example 9

$$2(x + 4y) = 2x + 8y$$

$$-(a + 6b - 5c) = -a - 6b + 5c$$

Example 10

$-2x$ is a term with coefficient -2.

yz^2 is a term with coefficient 1.

Example 11

$3x$ and $-5x$ are *like* terms.

$4a^2b$ and $4ab$ are not *like* terms.

Example 12

$$-4d + 12d + d = 9d$$

Example 13

$$-2w - 4(w - 2) + 3 \quad \text{Use the distributive property.}$$

$$= -2w - 4w + 8 + 3 \quad \text{Combine *like* terms.}$$

$$= -6w + 11$$

Chapter 1   Review Exercises

Section 1.1

1. Given the set $\{7, \frac{1}{3}, -4, 0, -\sqrt{3}, -0.\overline{2}, \pi, 1\}$,

   a. List the natural numbers.

   b. List the integers.

   c. List the whole numbers.

   d. List the rational numbers.

   e. List the irrational numbers.

   f. List the real numbers.

For Exercises 2–5, determine the absolute values.

2. $\left|\frac{1}{2}\right|$ 3. $|-6|$ 4. $|-\sqrt{7}|$ 5. $|0|$

For Exercises 6–13, identify whether the inequality is true or false.

6. $-6 > -1$ 7. $0 < -5$ 8. $-10 \leq 0$

9. $5 \neq -5$ 10. $7 \geq 7$ 11. $7 \geq -7$

12. $0 \leq -3$ 13. $-\frac{2}{3} \leq -\frac{2}{3}$

Section 1.2

For Exercises 14–23, translate the English phrases into algebraic expressions.

14. The product of x and $\frac{2}{3}$

15. The quotient of 7 and y

16. The sum of 2 and $3b$

17. The difference of a and 5

18. Two more than $5k$

19. Seven less than $13z$

20. The quotient of 6 and x, decreased by 18

21. The product of y and 3, increased by 12

22. Three-eighths subtracted from z

23. Five subtracted from two times p

For Exercises 24–29, simplify the expressions.

24. 6^3 25. 15^2 26. $\sqrt{36}$

27. $\left(\frac{1}{4}\right)^2$ 28. $\frac{1}{\sqrt{100}}$ 29. $\left(\frac{3}{2}\right)^3$

For Exercises 30–33, perform the indicated operations.

30. $15 - 7 \cdot 2 + 12$ 31. $|-11| + |5| - (7 - 2)$

32. $4^2 - (5 - 2)^2$ 33. $22 - 3(8 \div 4)^2$

Section 1.3

For Exercises 34–46, add the rational numbers.

34. $-6 + 8$ 35. $14 + (-10)$

36. $21 + (-6)$ 37. $-12 + (-5)$

38. $\frac{2}{7} + \left(-\frac{1}{9}\right)$ 39. $\left(-\frac{8}{11}\right) + \left(\frac{1}{2}\right)$

40. $\left(-\frac{1}{10}\right) + \left(-\frac{5}{6}\right)$ 41. $\left(-\frac{5}{2}\right) + \left(-\frac{1}{5}\right)$

42. $-8.17 + 6.02$ 43. $2.9 + (-7.18)$

44. $13 + (-2) + (-8)$ 45. $-5 + (-7) + 20$

46. $2 + 5 + (-8) + (-7) + 0 + 13 + (-1)$

47. Under what conditions will the expression $a + b$ be negative?

48. The high temperatures (in degrees Celsius) for the province of Alberta, Canada, during a week in January were $-8, -1, -4, -3, -4, 0$, and 7. What was the average high temperature for that week? Round to the nearest tenth of a degree.

Section 1.4

For Exercises 49–61, subtract the rational numbers.

49. $13 - 25$

50. $31 - (-2)$

51. $-8 - (-7)$

52. $-2 - 15$

53. $\left(-\dfrac{7}{9}\right) - \dfrac{5}{6}$

54. $\dfrac{1}{3} - \dfrac{9}{8}$

55. $7 - 8.2$

56. $-1.05 - 3.2$

57. $-16.1 - (-5.9)$

58. $7.09 - (-5)$

59. $\dfrac{11}{2} - \left(-\dfrac{1}{6}\right) - \dfrac{7}{3}$

60. $-\dfrac{4}{5} - \dfrac{7}{10} - \left(-\dfrac{13}{20}\right)$

61. $6 - 14 - (-1) - 10 - (-21) - 5$

62. Under what conditions will the expression $a - b$ be negative?

For Exercises 63–67, write an algebraic expression and simplify.

63. -18 subtracted from -7

64. The difference of -6 and 41

65. Seven decreased by 13

66. Five subtracted from the difference of 20 and -7

67. The sum of 6 and -12, decreased by 21

68. In Nevada, the highest temperature ever recorded was $125°F$ and the lowest was $-50°F$. Find the difference between the highest and lowest temperatures. (*Source: Information Please Almanac*)

Section 1.5

For Exercises 69–88, multiply or divide as indicated.

69. $10(-17)$

70. $(-7)13$

71. $(-52) \div 26$

72. $(-48) \div (-16)$

73. $\dfrac{7}{4} \div \left(-\dfrac{21}{2}\right)$

74. $\dfrac{2}{3}\left(-\dfrac{12}{11}\right)$

75. $-\dfrac{21}{5} \cdot 0$

76. $\dfrac{3}{4} \div 0$

77. $0 \div (-14)$

78. $\dfrac{0}{3} \cdot \dfrac{1}{8}$

79. $(-0.45)(-5)$

80. $(-2.1) \div (-0.07)$

81. $\dfrac{-21}{14}$

82. $\dfrac{-13}{-52}$

83. $(5)(-2)(3)$

84. $(-6)(-5)(15)$

85. $\left(-\dfrac{1}{2}\right)\left(\dfrac{7}{8}\right)\left(-\dfrac{4}{7}\right)$

86. $\left(\dfrac{12}{13}\right)\left(-\dfrac{1}{6}\right)\left(\dfrac{13}{14}\right)$

87. $40 \div 4 \div (-5)$

88. $\dfrac{10}{11} \div \dfrac{7}{11} \div \dfrac{5}{9}$

For Exercises 89–92, perform the indicated operations.

89. $9 - 4[-2(4 - 8) - 5(3 - 1)]$

90. $\dfrac{8(-3) - 6}{-7 - (-2)}$

91. $\dfrac{2}{3} - \left(\dfrac{3}{8} + \dfrac{5}{6}\right) \div \dfrac{5}{3}$

92. $5.4 - (0.3)^2 \div 0.09$

For Exercises 93–96, evaluate the expressions with the given substitutions.

93. $3(x + 2) \div y$ for $x = 4$ and $y = -9$

94. $a^2 - bc$ for $a = -6, b = 5,$ and $c = 2$

95. $w + xy - \sqrt{z}$

for $w = 12, x = 6, y = -5,$ and $z = 25$

96. $(u - v)^2 + (u^2 - v^2)$ for $u = 5$ and $v = -3$

97. In statistics, the formula $x = \mu + z\sigma$ is used to find cutoff values for data that follow a bell-shaped curve. Find x if $\mu = 100, z = -1.96,$ and $\sigma = 15$.

For Exercises 98–104, answer true or false. If a statement is false, explain why.

98. If n is positive, then $-n$ is negative.

99. If m is negative, then m^4 is negative.

100. If m is negative, then m^3 is negative.

101. If $m > 0$ and $n > 0$, then $mn > 0$.

102. If $p < 0$ and $q < 0$, then $pq < 0$.

103. A number and its reciprocal have the same signs.

104. A nonzero number and its opposite have different signs.

Section 1.6

For Exercises 105–112, answers may vary.

105. Give an example of the commutative property of addition.

106. Give an example of the associative property of addition.

107. Give an example of the inverse property of addition.

108. Give an example of the identity property of addition.

109. Give an example of the commutative property of multiplication.

110. Give an example of the associative property of multiplication.

111. Give an example of the inverse property of multiplication.

112. Give an example of the identity property of multiplication.

113. Explain why $5x - 2y$ is the same as $-2y + 5x$.

114. Explain why $3a - 9y$ is the same as $-9y + 3a$.

115. List the terms of the expression:
$3y + 10x - 12 + xy$

116. Identify the coefficients for the terms listed in Exercise 115.

117. Simplify each expression by combining *like* terms.

   a. $3a + 3b - 4b + 5a - 10$

   b. $-6p + 2q + 9 - 13q - p + 7$

118. Use the distributive property to clear the parentheses.

   a. $-2(4z + 9)$ **b.** $5(4w - 8y + 1)$

For Exercises 119–124, simplify the expressions.

119. $2p - (p + 5) + 3$

120. $6(h + 3) - 7h - 4$

121. $\frac{1}{2}(-6q) + q - 4\left(3q + \frac{1}{4}\right)$

122. $0.3b + 12(0.2 - 0.5b)$

123. $-4[2(x + 1) - (3x + 8)]$

124. $5[(7y - 3) + 3(y + 8)]$

Chapter 1   Test

1. Is $0.\overline{315}$ a rational number or irrational number? Explain your reasoning.

2. Plot the points on a number line: $|3|, 0, -2, 0.5,$ $\left|-\frac{3}{2}\right|, \sqrt{16}$.

$$\overset{}{\underset{0}{\vdash\!\!+\!\!+\!\!+\!\!+\!\!+\!\!+\!\!+\!\!+\!\!+\!\!\rightarrow}}$$

3. Use the number line in Exercise 2 to identify whether the statements are true or false.

   a. $|3| < -2$ **b.** $0 \leq \left|-\frac{3}{2}\right|$

   c. $-2 < 0.5$ **d.** $|3| \geq \left|-\frac{3}{2}\right|$

4. Use the definition of exponents to expand the expressions:

   a. $(4x)^3$ **b.** $4x^3$

5. a. Translate the expression into an English phrase: $2(a - b)$. (Answers may vary.)

   b. Translate the expression into an English phrase: $2a - b$. (Answers may vary.)

6. Translate the phrase into an algebraic expression: "The quotient of the principal square root of c and the square of d."

For Exercises 7–23, perform the indicated operations.

7. $18 + (-12)$

8. $-10 + (-9)$

9. $-15 - (-3)$

10. $21 - (-7)$

11. $-\dfrac{1}{8} + \left(-\dfrac{3}{4}\right)$

12. $-10.06 - (-14.72)$

13. $-14 + (-2) - 16$

14. $-84 \div 7$

15. $38 \div 0$

16. $7(-4)$

17. $-22 \cdot 0$

18. $(-16)(-2)(-1)(-3)$

19. $\dfrac{2}{5} \div \left(-\dfrac{7}{10}\right) \cdot \left(-\dfrac{7}{6}\right)$

20. $(8 - 10)\dfrac{3}{2} + (-5)$

21. $8 - [(2 - 4) - (8 - 9)]$

22. $\dfrac{\sqrt{5^2 - 4^2}}{|-12 + 3|}$

23. $\dfrac{|4 - 10|}{2 - 3(5 - 1)}$

24. Identify the property that justifies each statement.

    a. $6(-8) = (-8)6$ **b.** $5 + 0 = 5$

    c. $(2 + 3) + 4 = 2 + (3 + 4)$

    d. $\dfrac{1}{7} \cdot 7 = 1$ **e.** $8[7(-3)] = (8 \cdot 7)(-3)$

For Exercises 25–29, simplify the expression.

25. $-5x - 4y + 3 - 7x + 6y - 7$

26. $-3(4m + 8p - 7)$

27. $-4[2(5 - x) - 8x] - 9$

28. $4(p - 5) - (8p + 3)$

29. $\dfrac{1}{2}(12p - 4) + \dfrac{1}{3}(2 - 6p)$

For Exercises 30–33, evaluate the expressions given the values $x = 4$, $y = -3$, and $z = -7$.

30. $y^2 - x$

31. $3x - 2y$

32. $y(x - 2)$

33. $-y^2 - 4x + z$

For Exercises 34–35, translate the English statement to an algebraic expression. Then simplify the expression.

34. Subtract -4 from 12.

35. Find the difference of 6 and 8.

Linear Equations and Inequalities

2

In this chapter we learn how to solve linear equations and inequalities in one variable. This important category of equations and inequalities is used in a variety of applications.

The list of words contains key terms used in this chapter. Search for them in the puzzle and in the text throughout the chapter. By the end of this chapter, you should be familiar with all of these terms.

Key Terms

compound	identity	notation
conditional	inequality	percent
consecutive	integer	property
contradiction	interest	set-builder
decimal	interval	solution
equation	linear	
fractions	literal	

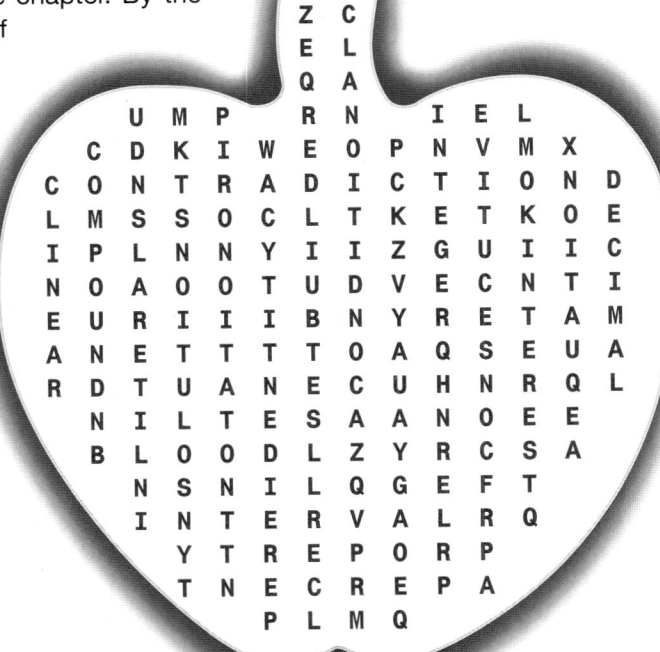

Section 2.1

Addition, Subtraction, Multiplication, and Division Properties of Equality

1. Definition of a Linear Equation in One Variable

An *equation* is a statement that indicates that two quantities are equal. The following are equations.

$$x = 5 \qquad y + 2 = 12 \qquad -4z = 28$$

All equations have an equal sign. Furthermore, notice that the equal sign separates the equation into two parts, the left-hand side and the right-hand side. A **solution to an equation** is a value of the variable that makes the equation a true statement. Substituting a solution to an equation for the variable makes the right-hand side equal to the left-hand side.

Equation	Solution	Check	
$x = 5$	5	$x = 5$ $5 = 5$ ✔	Substitute 5 for x. Right-hand side equals left-hand side.
$y + 2 = 12$	10	$y + 2 = 12$ $10 + 2 = 12$ ✔	Substitute 10 for y. Right-hand side equals left-hand side.
$-4z = 28$	-7	$-4z = 28$ $-4(-7) = 28$ ✔	Substitute -7 for z. Right-hand side equals left-hand side.

Example 1   **Determining Whether a Number is a Solution to an Equation**

Determine whether the given number is a solution to the equation.

a. $-6w + 14 = 4; \quad 3$ **b.** $4x + 7 = 5; \quad -\frac{1}{2}$

Solution:

a. $-6w + 14 = 4;$

$\quad -6(3) + 14 \overset{?}{=} 4$ Substitute 3 for w.

$\quad -18 + 14 \overset{?}{=} 4$ Simplify.

$\qquad\qquad -4 \neq 4$ Right-hand side does not equal left-hand side. Thus, 3 *is not a solution* to the equation $-6w + 14 = 4$.

b. $4x + 7 = 5$

$\quad 4\left(-\frac{1}{2}\right) + 7 \overset{?}{=} 5$ Substitute $-\frac{1}{2}$ for x.

$\quad -2 + 7 \overset{?}{=} 5$ Simplify.

$\qquad\qquad 5 = 5$ ✔   Right-hand side equals the left-hand side. Thus, $-\frac{1}{2}$ *is a solution* to the equation $4x + 7 = 5$.

Skill Practice Determine if the number given is a solution to the equation.

1. $-2y + 5 = 9$; -2 **2.** $4x - 1 = 7$; $\dfrac{1}{4}$

In the study of algebra, you will encounter a variety of equations. In this chapter, we will focus on a specific type of equation called a linear equation in one variable.

Definition of a Linear Equation in One Variable

Let a and b be real numbers such that $a \neq 0$. A **linear equation in one variable** is an equation that can be written in the form

$$ax + b = 0$$

Notice that a linear equation in one variable has only one variable. Furthermore, because the variable has an implied exponent of 1, a linear equation is sometimes called a *first-degree equation*.

linear equation in one variable	*not* a linear equation in one variable
$2x + 3 = 0$	$4x^2 + 8 = 0$ (exponent on x is not 1)
$\frac{1}{5}a + \frac{2}{7} = 0$	$\frac{3}{4}a + \frac{5}{8}b = 0$ (more than one variable)

2. Addition and Subtraction Properties of Equality

If two equations have the same solution then the equations are said to be equivalent. For example, the following equations are equivalent because the solution for each equation is 6.

Equivalent Equations:	Check the Solution 6:
$2x - 5 = 7$	$2(6) - 5 = 7 \Rightarrow 12 - 5 = 7$ ✔
$2x = 12$	$2(6) = 12 \Rightarrow \quad 12 = 12$ ✔
$x = 6$	$6 = 6 \Rightarrow \quad 6 = 6$ ✔

To solve a linear equation, $ax + b = 0$, the goal is to find *all* values of x that make the equation true. One general strategy for solving an equation is to rewrite it as an equivalent but simpler equation. This process is repeated until the equation can be written in the form $x =$ number. The addition and subtraction properties of equality help us do this.

Addition and Subtraction Properties of Equality

Let a, b, and c represent algebraic expressions.

1. Addition property of equality: If $a = b$,
then $a + c = b + c$

2. Subtraction property of equality: If $a = b$,
then $a - c = b - c$

The addition and subtraction properties of equality indicate that adding or subtracting the same quantity to each side of an equation results in an equivalent equation. This is true because if two quantities are increased or decreased by the same amount, then the resulting quantities will also be equal (Figure 2-1).

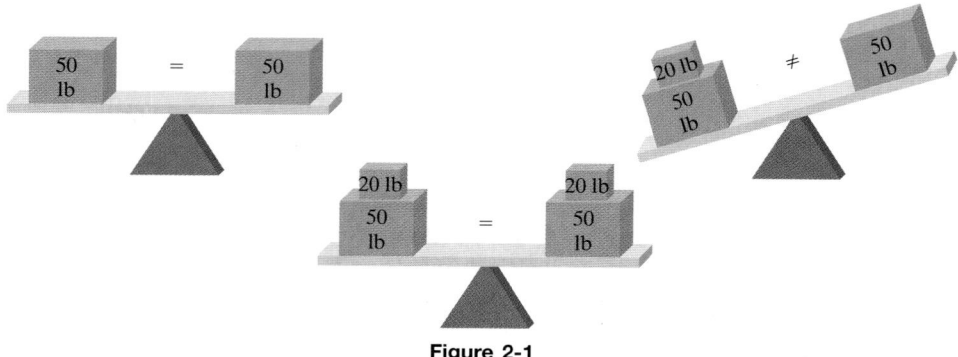

Figure 2-1

| **Example 2** | **Applying the Addition and Subtraction Properties of Equality** |

Solve the equations.

a. $p - 4 = 11$ **b.** $w + 5 = -2$ **c.** $\dfrac{9}{4} = q - \dfrac{3}{4}$ **d.** $-1.2 + z = 4.6$

Solution:

In each equation, the goal is to isolate the variable on one side of the equation. To accomplish this, we use the fact that the sum of a number and its opposite is zero and the difference of a number and itself is zero.

a. $p - 4 = 11$

$p - 4 + 4 = 11 + 4$ To isolate p, add 4 to both sides $(-4 + 4 = 0)$.

$p + 0 = 15$ Simplify.

$p = 15$

Check: $p - 4 = 11$ Check the solution by substituting $p = 15$ back in the original equation.

$15 - 4 \overset{?}{=} 11$

$11 = 11$ ✔   True

b. $w + 5 = -2$

$w + 5 - 5 = -2 - 5$ To isolate w, subtract 5 from both sides. $(5 - 5 = 0)$.

$w + 0 = -7$ Simplify.

$w = -7$

Check: $w + 5 = -2$ Check the solution by substituting $w = -7$ back in the original equation.

$-7 + 5 \overset{?}{=} -2$

$-2 = -2$ ✔   True

c.
$$\frac{9}{4} = q - \frac{3}{4}$$

$$\frac{9}{4} + \frac{3}{4} = q - \frac{3}{4} + \frac{3}{4}$$ To isolate q, add $\frac{3}{4}$ to both sides ($-\frac{3}{4} + \frac{3}{4} = 0$).

$$\frac{12}{4} = q + 0$$ Simplify.

$$3 = q \quad \text{or equivalently,} \quad q = 3$$

Check: $\dfrac{9}{4} = q - \dfrac{3}{4}$ Check the solution.

$$\frac{9}{4} \stackrel{?}{=} 3 - \frac{3}{4}$$ Substitute $q = 3$ in the original equation.

$$\frac{9}{4} \stackrel{?}{=} \frac{12}{4} - \frac{3}{4}$$ Get a common denominator.

$$\frac{9}{4} = \frac{9}{4} \checkmark$$ True

TIP: The variable may be isolated on either side of the equation.

d.
$$-1.2 + z = 4.6$$

$$-1.2 + 1.2 + z = 4.6 + 1.2$$ To isolate z, add 1.2 to both sides.

$$0 + z = 5.8$$

$$z = 5.8$$

Check: $-1.2 + z = 4.6$ Check the equation.

$$-1.2 + 5.8 \stackrel{?}{=} 4.6$$ Substitute $z = 5.8$ in the original equation.

$$4.6 = 4.6 \checkmark$$ True

Skill Practice   Solve the equations.

3. $v - 7 = 2$ **4.** $14 + c = -2$

5. $\dfrac{1}{4} = a - \dfrac{2}{3}$ **6.** $-8.1 + w = 11.5$

3. Multiplication and Division Properties of Equality

Adding or subtracting the same quantity to both sides of an equation results in an equivalent equation. In a similar way, multiplying or dividing both sides of an equation by the same nonzero quantity also results in an equivalent equation. This is stated formally as the multiplication and division properties of equality.

Multiplication and Division Properties of Equality

Let a, b, and c represent algebraic expressions.

   1. Multiplication property of equality: If $a = b$,

                                        then $ac = bc$

   2. Division property of equality: If $a = b$

                                        then $\dfrac{a}{c} = \dfrac{b}{c}$ (provided $c \neq 0$)

Skill Practice Answers

3. $v = 9$ **4.** $c = -16$

5. $a = \dfrac{11}{12}$ **6.** $w = 19.6$

To understand the multiplication property of equality, suppose we start with a true equation such as $10 = 10$. If both sides of the equation are multiplied by a constant such as 3, the result is also a true statement (Figure 2-2).

$$10 = 10$$

$$3 \cdot 10 = 3 \cdot 10$$

$$30 = 30$$

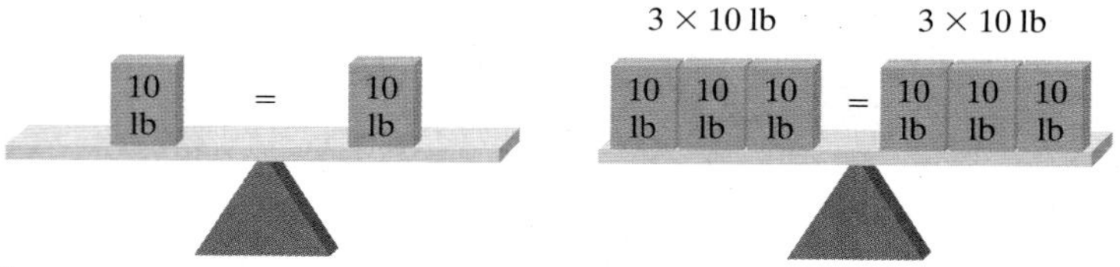

Figure 2-2

Similarly, if both sides of the equation are divided by a nonzero real number such as 2, the result is also a true statement (Figure 2-3).

$$10 = 10$$

$$\frac{10}{2} = \frac{10}{2}$$

$$5 = 5$$

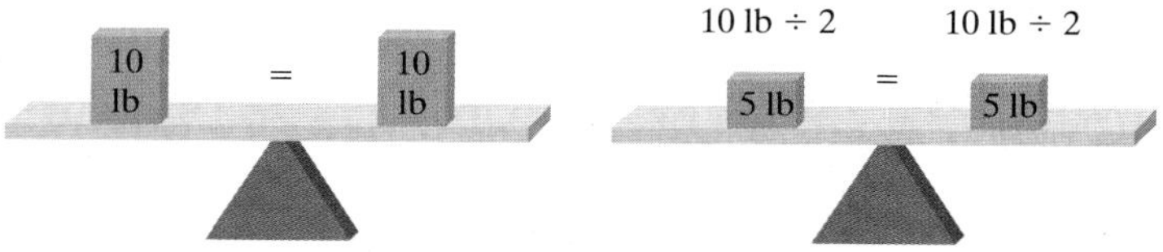

Figure 2-3

TIP: Recall that the product of a number and its reciprocal is 1. For example:

$$\frac{1}{5}(5) = 1$$

$$-\frac{7}{2}\left(-\frac{2}{7}\right) = 1$$

To solve an equation in the variable x, the goal is to write the equation in the form $x = $ number. In particular, notice that we desire the coefficient of x to be 1. That is, we want to write the equation as $1x = $ number. Therefore, to solve an equation such as $5x = 15$, we can multiply both sides of the equation by the reciprocal of the x-term coefficient. In this case, multiply both sides by the reciprocal of 5, which is $\frac{1}{5}$.

$$5x = 15$$

$$\frac{1}{5}(5x) = \frac{1}{5}(15) \qquad \text{Multiply by } \tfrac{1}{5}.$$

$$1x = 3 \qquad \text{The coefficient of the } x\text{-term is now 1.}$$

$$x = 3$$

TIP: Recall that the quotient of a nonzero real number and itself is 1. For example:

$$\frac{5}{5} = 1$$

$$\frac{-3.5}{-3.5} = 1$$

The division property of equality can also be used to solve the equation $5x = 15$ by dividing both sides by the coefficient of the x-term. In this case, divide both sides by 5 to make the coefficient of x equal to 1.

$$5x = 15$$

$$\frac{5x}{5} = \frac{15}{5} \qquad \text{Divide by 5.}$$

$$1x = 3 \qquad \text{The coefficient on the } x\text{-term is now 1.}$$

$$x = 3$$

| **Example 3** | **Applying the Multiplication and Division Properties of Equality** |

Solve the equations using the multiplication or division properties of equality.

a. $12x = 60$ **b.** $48 = -8w$ **c.** $-\dfrac{2}{9}q = \dfrac{1}{3}$

d. $-3.43 = -0.7z$ **e.** $\dfrac{d}{6} = -4$ **f.** $-x = 8$

Solution:

a. $12x = 60$

$$\frac{12x}{12} = \frac{60}{12}$$ To obtain a coefficient of 1 for the x-term, divide both sides by 12.

$1x = 5$ Simplify.

$x = 5$ Check: $12x = 60$

$12(5) \stackrel{?}{=} 60$

$60 = 60 \checkmark$ True

b. $48 = -8w$

$$\frac{48}{-8} = \frac{-8w}{-8}$$ To obtain a coefficient of 1 for the w-term, divide both sides by -8.

$-6 = 1w$ Simplify.

$-6 = w$ Check: $48w = -8w$

$48 \stackrel{?}{=} -8(-6)$

$48 = 48 \checkmark$ True

c. $$-\frac{2}{9}q = \frac{1}{3}$$

$$\left(-\frac{9}{2}\right)\left(-\frac{2}{9}q\right) = \frac{1}{3}\left(-\frac{9}{2}\right)$$ To obtain a coefficient of 1 for the q-term, multiply by the reciprocal of $-\frac{2}{9}$, which is $-\frac{9}{2}$.

$$1q = -\frac{3}{2}$$ Simplify. The product of a number and its reciprocal is 1.

$$q = -\frac{3}{2}$$ Check: $-\frac{2}{9}q = \frac{1}{3}$

$$-\frac{2}{9}\left(-\frac{3}{2}\right) \stackrel{?}{=} \frac{1}{3}$$

$$\frac{1}{3} = \frac{1}{3} \checkmark$$ True

TIP: When applying the multiplication or division properties of equality to obtain a coefficient of 1 for the variable term, we will generally use the following convention:

- If the coefficient of the variable term is expressed as a fraction, we usually multiply both sides by its reciprocal.
- If the coefficient of the variable term is an integer or decimal, we divide both sides by the coefficient itself.

d. $-3.43 = -0.7z$

$$\frac{-3.43}{-0.7} = \frac{-0.7z}{-0.7}$$ To obtain a coefficient of 1 for the z-term, divide by -0.7.

$$4.9 = 1z$$ Simplify.

$$4.9 = z$$

$$z = 4.9$$ Check: $-3.43 = -0.7z$

$$-3.43 \stackrel{?}{=} -0.7(4.9)$$

$$-3.43 = -3.43 \checkmark \quad \text{True}$$

e. $\dfrac{d}{6} = -4$

$$\frac{1}{6}d = -4$$ $\frac{d}{6}$ is equivalent to $\frac{1}{6}d$.

$$\frac{6}{1} \cdot \frac{1}{6}d = -4 \cdot \frac{6}{1}$$ To obtain a coefficient of 1 for the d-term, multiply by the reciprocal of $\frac{1}{6}$, which is $\frac{6}{1}$.

$$1d = -24$$ Simplify.

$$d = -24$$ Check: $\dfrac{d}{6} = -4$

$$\frac{-24}{6} \stackrel{?}{=} -4$$

$$-4 = -4 \checkmark \quad \text{True}$$

f. $-x = 8$ Note that $-x$ is equivalent to $-1 \cdot x$.

$$-1x = 8$$

$$\frac{-1x}{-1} = \frac{8}{-1}$$ To obtain a coefficient of 1 for the x-term, divide by -1.

$$x = -8$$ Check: $-x = 8$

$$-(-8) \stackrel{?}{=} 8$$

$$8 = 8 \checkmark \quad \text{True}$$

TIP: In Example 3(f), we could also have *multiplied* both sides by -1 to create a coefficient of 1 on the x-term.

$$-x = 8$$
$$(-1)(-x) = (-1)8$$
$$x = -8$$

Skill Practice Solve the equations.

7. $4x = -20$ **8.** $1005 = -5p$ **9.** $\dfrac{2}{3}a = \dfrac{1}{4}$

10. $6.82 = 2.2w$ **11.** $\dfrac{x}{5} = -8$ **12.** $-y = -11$

It is important to distinguish between cases where the addition or subtraction properties of equality should be used to isolate a variable versus those in which the multiplication or division properties of equality should be used. Remember the goal is to isolate the variable term and obtain a coefficient of 1. Compare the equations:

$$5 + x = 20 \quad \text{and} \quad 5x = 20$$

Skill Practice Answers

7. $x = -5$ **8.** $p = -201$

9. $a = \dfrac{3}{8}$ **10.** $w = 3.1$

11. $x = -40$ **12.** $y = 11$

In the first equation, the relationship between 5 and x is addition. Therefore, we want to reverse the process by subtracting 5 from both sides. In the second equation, the relationship between 5 and x is multiplication. To isolate x, we reverse the process by dividing by 5 or equivalently, multiplying by the reciprocal, $\frac{1}{5}$.

$$5 + x = 20 \qquad \text{and} \qquad 5x = 20$$

$$5 - 5 + x = 20 - 5 \qquad\qquad \frac{5x}{5} = \frac{20}{5}$$

$$x = 15 \qquad\qquad x = 4$$

4. Translations

| Example 4 | Translating to a Linear Equation

Write an algebraic equation to represent each English sentence. Then solve the equation.

a. The quotient of a number and 4 is 6.

b. The product of a number and 4 is 6.

c. Negative twelve is equal to the sum of -5 and a number.

d. The value 1.4 subtracted from a number is 5.7.

TIP: To help translate an English sentence to a mathematical equation, refer to page 53 for a table of equivalent expressions.

Solution:

For each case we will let x represent the unknown number.

a. The quotient of a number and 4 is 6.

$$\frac{x}{4} = 6$$

$$4 \cdot \frac{x}{4} = 4 \cdot 6 \qquad \text{Multiply both sides by 4.}$$

$$\frac{4}{1} \cdot \frac{x}{4} = 4 \cdot 6$$

$$x = 24$$

b. The product of a number and 4 is 6.

$$4x = 6$$

$$\frac{4x}{4} = \frac{6}{4} \qquad\qquad \text{Divide both sides by 4.}$$

$$x = \frac{3}{2}$$

c. Negative twelve is equal to the sum of -5 and a number.

$$-12 = -5 + x$$

$$-12 + 5 = -5 + 5 + x \qquad \text{Add 5 to both sides.}$$

$$-7 = x$$

d. The value 1.4 subtracted from a number is 5.7.

$$x - 1.4 = 5.7$$

$$x - 1.4 + 1.4 = 5.7 + 1.4 \qquad \text{Add } 1.4 \text{ to both sides.}$$

$$x = 7.1$$

Skill Practice Write an algebraic equation. Then solve the equation.

Skill Practice Answers

13. $x + 6 = -20$; $x = -26$
14. $-3x = -24$; $x = 8$
15. $13 = x - 5$; $x = 18$
16. $\dfrac{x}{6.3} = 1.2$; $x = 7.56$

13. The sum of a number and 6 is -20.

14. The product of a number and -3 is -24.

15. 13 is equal to 5 subtracted from a number.

16. The quotient of a number and 6.3 is 1.2.

Section 2.1   Practice Exercises

Boost *your* **GRADE at**
mathzone.com!

- Practice Problems
- Self-Tests
- NetTutor
- e-Professors
- Videos

Study Skills Exercises

1. After getting a test back, it is a good idea to correct the test so that you do not make the same errors again. One recommended approach is to use a clean sheet of paper, and divide the paper down the middle vertically as shown. For each problem that you missed on the test, rework the problem correctly on the left-hand side of the paper. Then give a written explanation on the right-hand side of the paper. To reinforce the correct procedure, do four more problems of that type.

   Take the time this week to make corrections from your last test.

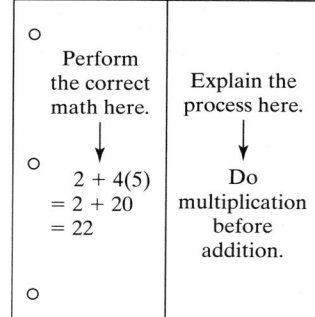

2. Define the key terms:

   a. linear equation in one variable   **b. solution to an equation**

   c. addition property of equality   **d. subtraction property of equality**

   e. multiplication property of equality   **f. division property of equality**

Concept 1: Definition of a Linear Equation in One Variable

For Exercises 3–6, identify the following as either an expression or an equation.

3. $x - 4 + 5x$ **4.** $8x + 2 = 7$ **5.** $9 = 2x - 4$ **6.** $3x^2 + x = -3$

7. Explain how to determine if a number is a solution to an equation.

For Exercises 8–13, determine whether the given number is a solution to the equation.

8. $x - 1 = 5$;   4   **9.** $x - 2 = 1$; -1 **10.** $5x = -10$; -2

11. $3x = 21$;   7   **12.** $3x + 9 = 3$; -2 **13.** $2x - 1 = -3$; -1

Concept 2: Addition and Subtraction Properties of Equality

For Exercises 14–33, solve the equations using the addition or subtraction property of equality. Be sure to check your answers.

14. $x + 6 = 5$

15. $x - 2 = 10$

16. $q - 14 = 6$

17. $w + 3 = -5$

18. $2 + m = -15$

19. $-6 + n = 10$

20. $-23 = y - 7$

21. $-9 = -21 + b$

22. $4 + c = 4$

23. $-13 + b = -13$

24. $4.1 = 2.8 + a$

25. $5.1 = -2.5 + y$

26. $5 = z - \dfrac{1}{2}$

27. $-7 = p + \dfrac{2}{3}$

28. $x + \dfrac{5}{2} = \dfrac{1}{2}$

29. $x - \dfrac{2}{3} = \dfrac{7}{3}$

30. $-6.02 + c = -8.15$

31. $p + 0.035 = -1.12$

32. $3.245 + t = -0.0225$

33. $-1.004 + k = 3.0589$

Concept 3: Multiplication and Division Properties of Equality

For Exercises 34–53, solve the equations using the multiplication or division property of equality. Be sure to check your answers.

34. $6x = 54$

35. $2w = 8$

36. $12 = -3p$

37. $6 = -2q$

38. $-5y = 0$

39. $-3k = 0$

40. $-\dfrac{y}{5} = 3$

41. $-\dfrac{z}{7} = 1$

42. $\dfrac{4}{5} = -t$

43. $-\dfrac{3}{7} = -h$

44. $\dfrac{2}{5}a = -4$

45. $\dfrac{3}{8}b = -9$

46. $-\dfrac{1}{5}b = -\dfrac{4}{5}$

47. $-\dfrac{3}{10}w = \dfrac{2}{5}$

48. $-41 = -x$

49. $32 = -y$

50. $3.81 = -0.03p$

51. $2.75 = -0.5q$

52. $5.82y = -15.132$

53. $-32.3x = -0.4522$

Concept 4: Translations

For Exercises 54–63, write an algebraic equation to represent each English sentence. (Let x represent the unknown number.) Then solve the equation.

54. The sum of negative eight and a number is forty-two.

55. The sum of thirty-one and a number is thirteen.

56. The difference of a number and negative six is eighteen.

57. The sum of negative twelve and a number is negative fifteen.

58. The product of a number and seven is the same as negative sixty-three.

59. The product of negative three and a number is the same as twenty-four.

60. The quotient of a number and twelve is one-third.

61. Eighteen is equal to the quotient of a number and two.

62. The sum of a number and $\frac{5}{8}$ is $\frac{13}{8}$.

63. The difference of a number and $\frac{2}{3}$ is $\frac{1}{3}$.

Mixed Exercises

For Exercises 64–91, solve the equation using the appropriate property of equality.

64. $a - 9 = 1$ **65.** $b - 2 = -4$ **66.** $-9x = 1$ **67.** $-2k = -4$

68. $-\dfrac{2}{3}h = 8$ **69.** $\dfrac{3}{4}p = 15$ **70.** $\dfrac{2}{3} + t = 8$ **71.** $\dfrac{3}{4} + y = 15$

72. $\dfrac{r}{3} = -12$ **73.** $\dfrac{d}{-4} = 5$ **74.** $k + 16 = 32$ **75.** $-18 = -9 + t$

76. $16k = 32$ **77.** $-18 = -9t$ **78.** $7 = -4q$ **79.** $-3s = 10$

80. $-4 + q = 7$ **81.** $s - 3 = 10$ **82.** $-\dfrac{1}{3}d = 12$ **83.** $-\dfrac{2}{5}m = 10$

84. $4 = \dfrac{1}{2} + z$ **85.** $3 = \dfrac{1}{4} + p$ **86.** $1.2y = 4.8$ **87.** $4.3w = 8.6$

88. $4.8 = 1.2 + y$ **89.** $8.6 = w - 4.3$ **90.** $0.0034 = y - 0.405$ **91.** $-0.98 = m + 1.0034$

For Exercises 92–99, determine if the equation is a linear equation in one variable. Answer yes or no.

92. $4p + 5 = 0$ **93.** $3x - 5y = 0$ **94.** $4 + 2a^2 = 5$ **95.** $-8t = 7$

96. $x - 4 = 9$ **97.** $2x^3 + y = 0$ **98.** $19b = -3$ **99.** $13 + x = 19$

Expanding Your Skills

For Exercises 100–105, construct an equation with the given solution. Answers will vary.

100. $y = 6$ **101.** $x = 2$ **102.** $p = -4$

103. $t = -10$ **104.** $a = 0$ **105.** $k = 1$

For Exercises 106–109, simplify by collecting the *like* terms. Then solve the equation.

106. $5x - 4x + 7 = 8 - 2$ **107.** $2 + 3 = 2y + 1 - y$

108. $6p - 3p = 15 + 6$ **109.** $12 - 20 = 2t + 2t$

Solving Linear Equations

1. Solving Linear Equations Involving Multiple Steps

In Section 2.1, we studied a one-step process to solve linear equations by using the addition, subtraction, multiplication, and division properties of equality. In the next example, we solve the equation $-2w - 7 = 11$. Solving this equation will require multiple steps. To understand the proper steps, always remember the ultimate goal—to isolate the variable. Therefore, we will first isolate the *term* containing the variable before dividing both sides by -2.

Concepts

1. Solving Linear Equations Involving Multiple Steps
2. Steps to Solve a Linear Equation in One Variable
3. Conditional Equations, Identities, and Contradictions

Example 1 Solving a Linear Equation

Solve the equation: $-2w - 7 = 11$

Solution:

$$-2w - 7 = 11$$

$$-2w - 7 + 7 = 11 + 7 \qquad \text{Add } 7 \text{ to both sides of the equation. This isolates the } w\text{-term.}$$

$$-2w = 18$$

$$\frac{-2w}{-2} = \frac{18}{-2} \qquad \text{Next, apply the division property of equality to obtain a coefficient of 1 for } w. \text{ Divide by } -2 \text{ on both sides.}$$

$$1w = -9$$

$$w = -9$$

Check:

$$-2w - 7 = 11$$

$$-2(-9) - 7 \stackrel{?}{=} 11 \qquad \text{Substitute } w = -9 \text{ in the original equation.}$$

$$18 - 7 \stackrel{?}{=} 11$$

$$11 = 11 \checkmark \qquad \text{True}$$

Skill Practice Solve.

1. $-5y + 25 = 10$

Example 2 Solving a Linear Equation

Solve the equation: $2 = \frac{1}{5}x + 3$

Skill Practice Answers

1. $y = 3$

Solution:

$$2 = \frac{1}{5}x + 3$$

$$2 - 3 = \frac{1}{5}x + 3 - 3 \qquad \text{Subtract 3 from both sides. This isolates the } x\text{-term.}$$

$$-1 = \frac{1}{5}x \qquad \text{Simplify.}$$

$$5(-1) = 5 \cdot \left(\frac{1}{5}x\right) \qquad \begin{array}{l}\text{Next, apply the multiplication property of} \\ \text{equality to obtain a coefficient of 1 for } x.\end{array}$$

$$-5 = 1x$$

$$-5 = x \qquad \text{Simplify. The answer checks in the original equation.}$$

Skill Practice Solve.

2. $2 = \frac{1}{2}a - 7$

In Example 3, the variable x appears on both sides of the equation. In this case, apply the addition or subtraction properties of equality to collect the variable terms on one side of the equation and the constant terms on the other side. Then use multiplication or division properties of equality to get a coefficient equal to 1.

Example 3 **Solving a Linear Equation**

Solve the equation: $6x - 4 = 2x - 8$

Solution:

$$6x - 4 = 2x - 8$$

$$6x - 2x - 4 = 2x - 2x - 8 \qquad \begin{array}{l}\text{Subtract } 2x \text{ from both sides leaving } 0x \text{ on} \\ \text{the right-hand side.}\end{array}$$

$$4x - 4 = 0x - 8 \qquad \text{Simplify.}$$

$$4x - 4 = -8 \qquad \begin{array}{l}\text{The } x\text{-terms have now been combined on} \\ \text{one side of the equation.}\end{array}$$

$$4x - 4 + 4 = -8 + 4 \qquad \begin{array}{l}\text{Add 4 to both sides of the equation. This} \\ \text{combines the constant terms on the } \textit{other} \\ \text{side of the equation.}\end{array}$$

$$4x = -4$$

$$\frac{4x}{4} = \frac{-4}{4} \qquad \begin{array}{l}\text{To obtain a coefficient of 1 for } x, \text{ divide} \\ \text{both sides of the equation by 4.}\end{array}$$

$$x = -1 \qquad \underline{\text{Check:}}$$

$$6x - 4 = 2x - 8$$

$$6(-1) - 4 \stackrel{?}{=} 2(-1) - 8$$

$$-6 - 4 \stackrel{?}{=} -2 - 8$$

Skill Practice Answers

2. $a = 18$

$$-10 = -10 \checkmark \qquad \text{True}$$

Skill Practice Solve.

3. $10x - 3 = 4x + 9$

TIP: It is important to note that the variable may be isolated on either side of the equation. We will solve the equation from Example 3 again, this time isolating the variable on the right-hand side.

$$6x - 4 = 2x - 8$$
$$6x - 6x - 4 = 2x - 6x - 8 \qquad \text{Subtract } 6x \text{ on both sides.}$$
$$0x - 4 = -4x - 8$$
$$-4 = -4x - 8$$
$$-4 + 8 = -4x - 8 + 8 \qquad \text{Add } 8 \text{ to both sides.}$$
$$4 = -4x$$
$$\frac{4}{-4} = \frac{-4x}{-4} \qquad \text{Divide both sides by } -4.$$
$$-1 = x \quad \text{or equivalently } x = -1$$

2. Steps to Solve a Linear Equation in One Variable

In some cases it is necessary to simplify both sides of a linear equation before applying the properties of equality. Therefore, we offer the following steps to solve a linear equation in one variable.

Steps to Solve a Linear Equation in One Variable

1. Simplify both sides of the equation.
   - Clear parentheses
   - Combine *like* terms
2. Use the addition or subtraction property of equality to collect the variable terms on one side of the equation.
3. Use the addition or subtraction property of equality to collect the constant terms on the *other* side of the equation.
4. Use the multiplication or division property of equality to make the coefficient of the variable term equal to 1.
5. Check your answer.

Example 4 Solving Linear Equations

Solve the equations:

a. $7 + 3 = 2(p - 3)$ **b.** $2.2y - 8.3 = 6.2y + 12.1$

Skill Practice Answers

3. $x = 2$

Solution:

a. $7 + 3 = 2(p - 3)$

$10 = 2p - 6$ **Step 1:** Simplify both sides of the equation by clearing parentheses and combining *like* terms.

                    Step 2: The variable terms are already on one side.

$10 + 6 = 2p - 6 + 6$ **Step 3:** Add 6 to both sides to collect the constant terms on the other side.

$16 = 2p$

$\dfrac{16}{2} = \dfrac{2p}{2}$ **Step 4:** Divide both sides by 2 to obtain a coefficient of 1 for p.

$8 = p$ **Step 5:** <u>Check:</u>

$7 + 3 = 2(p - 3)$

$10 \stackrel{?}{=} 2(8 - 3)$

$10 \stackrel{?}{=} 2(5)$

$10 = 10$ ✔ True

b. $2.2y - 8.3 = 6.2y + 12.1$ **Step 1:** The right- and left-hand sides are already simplified.

$2.2y - 2.2y - 8.3 = 6.2y - 2.2y + 12.1$ **Step 2:** Subtract $2.2y$ from both sides to collect the variable terms on one side of the equation.

$-8.3 = 4.0y + 12.1$

$-8.3 - 12.1 = 4.0y + 12.1 - 12.1$ **Step 3:** Subtract 12.1 from both sides to collect the constant terms on the other side.

$-20.4 = 4.0y$

$\dfrac{-20.4}{4.0} = \dfrac{4.0y}{4.0}$ **Step 4:** To obtain a coefficient of 1 for the y-term, divide both sides of the equation by 4.0.

$-5.1 = y$

$y = -5.1$ **Step 5:** <u>Check:</u>

$2.2y - 8.3 = 6.2y + 12.1$

$2.2(-5.1) - 8.3 \stackrel{?}{=} 6.2(-5.1) + 12.1$

$-11.22 - 8.3 \stackrel{?}{=} -31.62 + 12.1$

$-19.52 = -19.52$ ✔ True

Skill Practice Solve the equations.

4. $5 - 8 = -3(2x + 3)$ **5.** $1.5p + 2.3 = 3.5p - 1.9$

Skill Practice Answers

4. $x = -1$ **5.** $p = 2.1$

Example 5	Solving Linear Equations

Solve the equations:

a. $2 + 7x - 5 = 6(x + 3) + 2x$

b. $9 - (z - 3) + 4z = 4z - 5(z + 2) - 6$

Solution:

a.

$$2 + 7x - 5 = 6(x + 3) + 2x$$

$$-3 + 7x = 6x + 18 + 2x \qquad \textbf{Step 1:} \quad \text{Add } like \text{ terms on the left. Clear parentheses on the right.}$$

$$-3 + 7x = 8x + 18 \qquad\qquad \text{Combine } like \text{ terms.}$$

$$-3 + 7x - 7x = 8x - 7x + 18 \qquad \textbf{Step 2:} \quad \text{Subtract } 7x \text{ from both sides.}$$

$$-3 = x + 18 \qquad\qquad \text{Simplify.}$$

$$-3 - 18 = x + 18 - 18 \qquad \textbf{Step 3:} \quad \text{Subtract 18 from both sides.}$$

$$-21 = x \qquad\qquad \textbf{Step 4:} \quad \text{Because the coefficient of the } x \text{ term is already 1, there is no need to apply the multiplication or division property of equality.}$$

$$x = -21$$

Step 5: The check is left to the reader.

b. $9 - (z - 3) + 4z = 4z - 5(z + 2) - 6$

$$9 - z + 3 + 4z = 4z - 5z - 10 - 6 \qquad \textbf{Step 1:} \quad \text{Clear parentheses.}$$

$$12 + 3z = -z - 16 \qquad\qquad \text{Combine } like \text{ terms.}$$

$$12 + 3z + z = -z + z - 16 \qquad \textbf{Step 2:} \quad \text{Add } z \text{ to both sides.}$$

$$12 + 4z = -16$$

$$12 - 12 + 4z = -16 - 12 \qquad \textbf{Step 3:} \quad \text{Subtract 12 from both sides.}$$

$$4z = -28$$

$$\frac{4z}{4} = \frac{-28}{4} \qquad\qquad \textbf{Step 4:} \quad \text{Divide both sides by 4.}$$

$$z = -7 \qquad\qquad \textbf{Step 5:} \quad \text{The check is left to the reader.}$$

Skill Practice	Solve the equations.

6. $6y + 3 - y = 4(2y - 1)$ **7.** $3(3a - 4) - 4(5a + 2) = 7a - 2$

3. Conditional Equations, Identities, and Contradictions

The solutions to a linear equation are the values of x that make the equation a true statement. A linear equation has one unique solution. Some types of equations, however, have no solution while others have infinitely many solutions.

Skill Practice Answers

6. $y = \dfrac{7}{3}$ **7.** $a = -1$

I. Conditional Equations

An equation that is true for some values of the variable but false for other values is called a **conditional equation**. The equation $x + 4 = 6$, for example, is true on the condition that $x = 2$. For other values of x, the statement $x + 4 = 6$ is false.

II. Contradictions

Some equations have no solution, such as $x + 1 = x + 2$. There is no value of x, that when increased by 1 will equal the same value increased by 2. If we tried to solve the equation by subtracting x from both sides, we get the contradiction $1 = 2$. This indicates that the equation has no solution. An equation that has no solution is called a **contradiction**.

$$x + 1 = x + 2$$
$$x - x + 1 = x - x + 2$$
$$1 = 2 \quad \text{(contradiction)} \qquad \text{No solution}$$

III. Identities

An equation that has all real numbers as its solution set is called an **identity**. For example, consider the equation, $x + 4 = x + 4$. Because the left- and right-hand sides are identically equal, any real number substituted for x will result in equal quantities on both sides. If we subtract x from both sides of the equation, we get the identity $4 = 4$. In such a case, the solution is the set of all real numbers.

$$x + 4 = x + 4$$
$$x - x + 4 = x - x + 4$$
$$4 = 4 \quad \text{(identity)} \qquad \text{The solution is all real numbers.}$$

Example 6 | **Identifying Conditional Equations, Contradictions, and Identities**

Identify each equation as a conditional equation, a contradiction, or an identity. Then describe the solution.

a. $4k - 5 = 2(2k - 3) + 1$ **b.** $2(b - 4) = 2b - 7$ **c.** $3x + 7 = 2x - 5$

Solution:

a.
$$4k - 5 = 2(2k - 3) + 1$$
$$4k - 5 = 4k - 6 + 1 \qquad \text{Clear parentheses.}$$
$$4k - 5 = 4k - 5 \qquad \text{Combine } like \text{ terms.}$$
$$4k - 4k - 5 = 4k - 4k - 5 \qquad \text{Subtract } 4k \text{ from both sides.}$$
$$-5 = -5$$

This is an identity. The solution is all real numbers.

b.
$$2(b - 4) = 2b - 7$$
$$2b - 8 = 2b - 7 \qquad \text{Clear parentheses.}$$
$$2b - 2b - 8 = 2b - 2b - 7 \qquad \text{Subtract } 2b \text{ from both sides.}$$
$$-8 = -7 \quad \text{(Contradiction)}$$

This is a contradiction. There is no solution.

c.

$$3x + 7 = 2x - 5$$

$3x - 2x + 7 = 2x - 2x - 5$ Subtract $2x$ from both sides.

$x + 7 = -5$ Simplify.

$x + 7 - 7 = -5 - 7$ Subtract 7 from both sides.

$x = -12$

This is a conditional equation. The solution is $x = -12$. (The equation is true only on the condition that $x = -12$.)

Skill Practice   Solve. Then describe the solution.

8. $5x - 1 = 2x + 3(x - 3)$ **9.** $4(2t - 1) + t = 3(3t - 1) - 1$

10. $6(v - 2) = 4v - 2(v + 1)$

Skill Practice Answers

8. No solution—the equation is a contradiction.

9. All real numbers—the equation is an identity.

10. $v = \dfrac{5}{2}$; conditional equation

Section 2.2   Practice Exercises

Boost *your* GRADE at mathzone.com!

 MathZone

- Practice Problems
- Self-Tests
- NetTutor
- e-Professors
- Videos

Study Skills Exercises

1. Several topics are given here about taking notes. Which would you do first to help make the most of note-taking? Put them in order of importance to you by labeling them with the numbers 1–6.

_____ Read your notes after class to complete any abbreviations or incomplete sentences.

_____ Highlight important terms and definitions.

_____ Review your notes from the previous class.

_____ Bring pencils (more than one) and paper to class.

_____ Sit in class where you can clearly read the board and hear your instructor.

_____ Keep your focus on the instructor looking for phrases such as, "The most important point is . . ." and "Here is where the problem usually occurs."

2. Define the key terms:

     a. conditional equation   **b. contradiction**   **c. identity**

Review Exercises

For Exercises 3–6, simplify the expressions by clearing parentheses and combining *like* terms.

3. $5z + 2 - 7z - 3z$ **4.** $10 - 4w + 7w - 2 + w$

5. $-(-7p + 9) + (3p - 1)$ **6.** $8y - (2y + 3) - 19$

7. Explain the difference between simplifying an expression and solving an equation.

For Exercises 8–12, solve the equations using the addition, subtraction, multiplication, or division property of equality.

8. $5w = -30$

9. $-7y = 21$

10. $x + 8 = -15$

11. $z - 23 = -28$

12. $-\dfrac{9}{8} = -\dfrac{3}{4}k$

Concept 1: Solving Linear Equations Involving Multiple Steps

For Exercises 13–40, solve the equations using the steps outlined in the text.

13. $6z + 1 = 13$

14. $5x + 2 = -13$

15. $3y - 4 = 14$

16. $-7w - 5 = -19$

17. $-2p + 8 = 3$

18. $4q + 5 = 2$

19. $6 = 7m - 1$

20. $-9 = 4n - 1$

21. $-\dfrac{1}{2} - 4x = 8$

22. $2b - \dfrac{1}{4} = 5$

23. $0.2x + 3.1 = -5.3$

24. $-1.8 + 2.4a = -6.6$

25. $\dfrac{5}{8} = \dfrac{1}{4} - \dfrac{1}{2}p$

26. $\dfrac{6}{7} = \dfrac{1}{7} + \dfrac{5}{3}r$

27. $7w - 6w + 1 = 10 - 4$

28. $5v - 3 - 4v = 13$

29. $11h - 8 - 9h = -16$

30. $6u - 5 - 8u = -7$

31. $3a + 7 = 2a - 19$

32. $6b - 20 = 14 + 5b$

33. $-4r - 28 = -78 - r$

34. $-6x - 7 = -3 - 8x$

35. $-2z - 8 = -z$

36. $-7t + 4 = -6t$

37. $\dfrac{5}{6}x + \dfrac{2}{3} = -\dfrac{1}{6}x - \dfrac{5}{3}$

38. $\dfrac{3}{7}x - \dfrac{1}{4} = -\dfrac{4}{7}x - \dfrac{5}{4}$

39. $3y - 2 = 5y - 2$

40. $4 + 10t = -8t + 4$

Concept 2: Steps to Solve a Linear Equation in One Variable

For Exercises 41–58, solve the equations using the steps outlined in the text.

41. $3(2p - 4) = 15$

42. $4(t + 15) = 20$

43. $6(3x + 2) - 10 = -4$

44. $4(2k + 1) - 1 = 5$

45. $2(y - 3) - y = 6$

46. $4(w - 5) - 3w = 2$

47. $17(s + 3) = 4(s - 10) + 13$

48. $5(4 + p) = 3(3p - 1) - 9$

49. $6(3t - 4) + 10 = 5(t - 2) - (3t + 4)$

50. $-5y + 2(2y + 1) = 2(5y - 1) - 7$

51. $5 - 3(x + 2) = 5$

52. $1 - 6(2 - h) = 7$

53. $-2[(4p + 1) - (3p - 1)] = 5(3 - p) - 9$

54. $5 - (6k + 1) = 2[(5k - 3) - (k - 2)]$

55. $0.2w - 0.47 = 0.53 - 0.2(2w - 13)$

56. $0.4z - 0.15 = 0.65 - 0.3(6 - 2z)$

57. $3(-0.9n + 0.5) = -3.5n + 1.3$

58. $7(0.4m - 0.1) = 5.2m + 0.86$

Concept 3: Conditional Equations, Identities, and Contradictions

For Exercises 59–64, identify the equations as a conditional equation, a contradiction, or an identity. Then describe the solution.

59. $2(k - 7) = 2k - 13$

60. $5h + 4 = 5(h + 1) - 1$

61. $7x + 3 = 6(x - 2)$

62. $3y - 1 = 1 + 3y$

63. $3 - 5.2p = -5.2p + 3$

64. $2(q + 3) = 4q + q - 9$

65. A conditional linear equation has (choose one): One solution, no solution, or infinitely many solutions.

66. An equation that is a contradiction has (choose one): One solution, no solution, or infinitely many solutions.

67. An equation that is an identity has (choose one): One solution, no solution, or infinitely many solutions.

Mixed Exercises

For Exercises 68–91, find the solution, if possible.

68. $4p - 6 = 8 + 2p$

69. $\frac{1}{2}t - 2 = 3$

70. $2k + 9 = -8$

71. $3(y - 2) + 5 = 5$

72. $7(w - 2) = -14 - 3w$

73. $0.24 = 0.4m$

74. $2(x + 2) - 3 = 2x + 1$

75. $n + \frac{1}{4} = -\frac{1}{2}$

76. $0.5b = -23$

77. $3(2r + 1) = 6(r + 2) - 6$

78. $8 - 2q = 4$

79. $\frac{x}{7} - 3 = 1$

80. $2 - 4(y - 5) = -4$

81. $4 - 3(4p - 1) = -8$

82. $0.4(a + 20) = 6$

83. $2.2r - 12 = 3.4$

84. $10(2n + 1) - 6 = 20(n - 1) + 12$

85. $\frac{2}{5}y + 5 = -3$

86. $c + 0.123 = 2.328$

87. $4(2z + 3) = 8(z - 3) + 36$

88. $\frac{4}{5}t - 1 = \frac{1}{5}t + 5$

89. $6g - 8 = 4 - 3g$

90. $8 - (3q + 4) = 6 - q$

91. $6w - (8 + 2w) = 2(w - 4)$

Expanding Your Skills

92. Suppose $x = -5$ is a solution to the equation $x + a = 10$. Find the value of a.

93. Suppose $x = 6$ is a solution to the equation $x + a = -12$. Find the value of a.

94. Suppose $x = 3$ is a solution to the equation $ax = 12$. Find the value of a.

95. Suppose $x = 11$ is a solution to the equation $ax = 49.5$. Find the value of a.

96. Write an equation that is an identity. Answers may vary.

97. Write an equation that is a contradiction. Answers may vary.

Section 2.3   Linear Equations: Clearing Fractions and Decimals

1. Clearing Fractions and Decimals

Linear equations that contain fractions can be solved in different ways. The first procedure, illustrated here, uses the method outlined in Section 2.2.

$$\frac{5}{6}x - \frac{3}{4} = \frac{1}{3}$$

$$\frac{5}{6}x - \frac{3}{4} + \frac{3}{4} = \frac{1}{3} + \frac{3}{4} \qquad \text{To isolate the variable term, add } \tfrac{3}{4} \text{ to both sides.}$$

$$\frac{5}{6}x = \frac{4}{12} + \frac{9}{12} \qquad \text{Find the common denominator on the right-hand side.}$$

$$\frac{5}{6}x = \frac{13}{12} \qquad \text{Simplify.}$$

$$\frac{6}{5}\left(\frac{5}{6}x\right) = \frac{\overset{1}{6}}{5}\left(\frac{13}{\underset{2}{12}}\right) \qquad \text{Multiply by the reciprocal of } \tfrac{5}{6}, \text{ which is } \tfrac{6}{5}.$$

$$x = \frac{13}{10}$$

Sometimes it is simpler to solve an equation with fractions by eliminating the fractions first using a process called **clearing fractions**. To clear fractions in the equation $\frac{5}{6}x - \frac{3}{4} = \frac{1}{3}$, we can multiply both sides of the equation by the least common denominator (LCD) of all terms in the equation. In this case, the LCD of $\frac{5}{6}x$, $-\frac{3}{4}$, and $\frac{1}{3}$ is 12. Because each denominator in the equation is a factor of 12, we can simplify common factors to leave integer coefficients for each term.

Example 1   Solving a Linear Equation by Clearing Fractions

Solve the equation $\frac{5}{6}x - \frac{3}{4} = \frac{1}{3}$ by clearing fractions first.

Solution:

$$\frac{5}{6}x - \frac{3}{4} = \frac{1}{3}$$

$$12\left(\frac{5}{6}x - \frac{3}{4}\right) = 12\left(\frac{1}{3}\right) \qquad \text{Multiply both sides of the equation by the LCD, 12.}$$

$$\frac{\overset{2}{\cancel{12}}}{1}\left(\frac{5}{6}x\right) - \frac{\overset{3}{\cancel{12}}}{1}\left(\frac{3}{4}\right) = \frac{\overset{4}{\cancel{12}}}{1}\left(\frac{1}{3}\right) \qquad \text{Apply the distributive property (recall that } 12 = \tfrac{12}{1}\text{).}$$

$$2(5x) - 3(3) = 4(1) \qquad \text{Simplify common factors to clear the fractions.}$$

$$10x - 9 = 4$$

$$10x - 9 + 9 = 4 + 9 \qquad \text{Add 9 to both sides.}$$

$$10x = 13$$

$$\frac{10x}{10} = \frac{13}{10} \qquad \text{Divide both sides by 10.}$$

$$x = \frac{13}{10} \qquad \text{Simplify.}$$

TIP: Recall that the multiplication property of equality indicates that multiplying both sides of an equation by a nonzero constant results in an equivalent equation.

TIP: The fractions in this equation can be eliminated by multiplying both sides of the equation by *any* common multiple of the denominators. For example, try multiplying both sides of the equation by 24:

$$24\left(\frac{5}{6}x - \frac{3}{4}\right) = 24\left(\frac{1}{3}\right)$$

$$\overset{4}{\cancel{24}}\left(\frac{5}{6}x\right) - \overset{6}{\cancel{24}}\left(\frac{3}{4}\right) = \overset{8}{\cancel{24}}\left(\frac{1}{3}\right)$$

$$20x - 18 = 8$$

$$20x = 26$$

$$\frac{20x}{20} = \frac{26}{20}$$

$$x = \frac{13}{10}$$

Skill Practice Solve the equation by clearing fractions.

1. $\dfrac{2}{5}y + \dfrac{1}{2} = -\dfrac{7}{10}$

The same procedure used to clear fractions in an equation can be used to clear decimals. For example, consider the equation $0.05x + 0.25 = 0.2$. Because any terminating decimal can be written as a fraction, the equation can be interpreted as $\frac{5}{100}x + \frac{25}{100} = \frac{2}{10}$. A convenient common denominator for all terms in this equation is 100. Therefore, we can multiply the original equation by 100 to clear decimals.

Example 2 **Solving a Linear Equation by Clearing Decimals**

Solve the equation $0.05x + 0.25 = 0.2$ by clearing decimals first.

Solution:

$$0.05x + 0.25 = 0.2$$

$100(0.05x + 0.25) = 100(0.2)$ Multiply both sides of the equation by 100.

$100(0.05x) + 100(0.25) = 100(0.2)$ Apply the distributive property.

$5x + 25 = 20$ Simplify (decimals have been cleared).

$5x + 25 - 25 = 20 - 25$ Subtract 25 from both sides.

$5x = -5$

$\dfrac{5x}{5} = \dfrac{-5}{5}$ Divide both sides by 5.

$x = -1$ Simplify.

This equation can be checked by hand or by using a calculator.

$$0.05x + 0.25 = 0.2$$

$$0.05(-1) + 0.25 \overset{?}{=} 0.2$$

$$-0.05 + 0.25 \overset{?}{=} 0.2$$

$$0.2 = 0.2 ✔ \text{ True}$$

TIP: Notice that multiplying a decimal number by 100 has the effect of moving the decimal point two places to the right. Similarly, multiplying by 10 moves the decimal point one place to the right, multiplying by 1000 moves the decimal point three places to the right, and so on.

Skill Practice Answers

1. $y = -3$

> **Skill Practice**   Solve the equation by clearing fractions.
>
> **2.** $-0.12z - 0.5 = 1.3$

In this section, we combine the process for clearing fractions and decimals with the general strategies for solving linear equations. To solve a linear equation, it is important to follow the steps listed below.

Steps for Solving a Linear Equation in One Variable

1. Simplify both sides of the equation.
   - Clear parentheses
   - Consider clearing fractions or decimals (if any are present) by multiplying both sides of the equation by a common denominator of all terms.
   - Combine *like* terms
2. Use the addition or subtraction property of equality to collect the variable terms on one side of the equation.
3. Use the addition or subtraction property of equality to collect the constant terms on the other side of the equation.
4. Use the multiplication or division property of equality to make the coefficient of the variable term equal to 1.
5. Check your answer.

---

2. Solving Linear Equations with Fractions

> **Example 3**   Solving Linear Equations with Fractions
>
> **a.** $\dfrac{1}{6}x - \dfrac{2}{3} = \dfrac{1}{5}x - 1$ **b.** $\dfrac{1}{3}(x + 7) - \dfrac{1}{2}(x + 1) = 4$

Solution:

TIP: After clearing the parentheses, answer these two questions:

1. What is the LCD?
2. How many terms are in the equation? Then multiply each term by the LCD.

a.

$$\frac{1}{6}x - \frac{2}{3} = \frac{1}{5}x - 1$$

The LCD of $\frac{1}{6}x$, $-\frac{2}{3}$, and $\frac{1}{5}x$ is 30.

$$30\left(\frac{1}{6}x - \frac{2}{3}\right) = 30\left(\frac{1}{5}x - 1\right)$$

Multiply by the LCD, 30.

$$\frac{\overset{5}{\cancel{30}}}{1} \cdot \frac{1}{\cancel{6}}x - \frac{\overset{10}{\cancel{30}}}{1} \cdot \frac{2}{\cancel{3}} = \frac{\overset{6}{\cancel{30}}}{1} \cdot \frac{1}{\cancel{5}}x - 30(1)$$

Apply the distributive property (recall $30 = \frac{30}{1}$).

$$5x - 20 = 6x - 30$$

Clear fractions.

$$5x - 6x - 20 = 6x - 6x - 30$$

Subtract $6x$ from both sides.

$$-x - 20 = -30$$

$$-x - 20 + 20 = -30 + 20$$

Add 20 to both sides.

$$-x = -10$$

$$\frac{-x}{-1} = \frac{-10}{-1}$$

Divide both sides by -1.

$$x = 10$$

The check is left to the reader.

Skill Practice Answers

2. $z = -15$

b.
$$\frac{1}{3}(x + 7) - \frac{1}{2}(x + 1) = 4$$

$$\frac{1}{3}x + \frac{7}{3} - \frac{1}{2}x - \frac{1}{2} = 4 \qquad \text{Clear parentheses.}$$

$$6\left(\frac{1}{3}x + \frac{7}{3} - \frac{1}{2}x - \frac{1}{2}\right) = 6(4) \qquad \text{The LCD of } \frac{1}{3}x, \frac{7}{3}, -\frac{1}{2}x, \text{ and } -\frac{1}{2} \text{ is } 6.$$

$$\overset{2}{\cancel{6}} \cdot \frac{1}{3}x + \overset{2}{\cancel{6}} \cdot \frac{7}{3} + \overset{3}{\cancel{6}}\left(-\frac{1}{2}x\right) + \overset{3}{\cancel{6}}\left(-\frac{1}{2}\right) = 6(4) \qquad \begin{array}{l}\text{Apply the distributive} \\ \text{property.}\end{array}$$

$$2x + 14 - 3x - 3 = 24 \qquad \text{Multiply fractions.}$$

$$-x + 11 = 24 \qquad \text{Combine } like \text{ terms.}$$

$$-x + 11 - 11 = 24 - 11 \qquad \text{Subtract } 11.$$

$$-x = 13$$

$$\frac{-x}{-1} = \frac{13}{-1} \qquad \text{Divide by } -1.$$

$$x = -13 \qquad \begin{array}{l}\text{The check is left to} \\ \text{the reader.}\end{array}$$

Skill Practice Solve the equations.

3. $\frac{2}{5}x - \frac{1}{2} = \frac{7}{4} + \frac{3}{10}x$ **4.** $\frac{1}{5}(z + 1) + \frac{1}{4}(z + 3) = 2$

TIP: In Example 3(b) both parentheses and fractions are present within the equation. In such a case, we recommend that you clear parentheses first. Then clear the fractions.

Example 4 **Solving a Linear Equation with Fractions**

Solve. $\dfrac{x - 2}{5} - \dfrac{x - 4}{2} = 2$

Solution:

$$\frac{x - 2}{5} - \frac{x - 4}{2} = \frac{2}{1} \qquad \begin{array}{l}\text{The LCD of } \frac{x-2}{5}, \frac{x-4}{2}, \text{ and } \frac{2}{1} \\ \text{is } 10.\end{array}$$

$$10\left(\frac{x - 2}{5} - \frac{x - 4}{2}\right) = 10\left(\frac{2}{1}\right) \qquad \text{Multiply both sides by } 10.$$

$$\frac{\overset{2}{\cancel{10}}}{1} \cdot \left(\frac{x - 2}{\cancel{5}}\right) - \frac{\overset{5}{\cancel{10}}}{1} \cdot \left(\frac{x - 4}{\cancel{2}}\right) = \frac{10}{1} \cdot \left(\frac{2}{1}\right) \qquad \text{Apply the distributive property.}$$

$$2(x - 2) - 5(x - 4) = 20 \qquad \text{Clear fractions.}$$

$$2x - 4 - 5x + 20 = 20 \qquad \text{Apply the distributive property.}$$

$$-3x + 16 = 20 \qquad \text{Simplify both sides of the equation.}$$

$$-3x + 16 - 16 = 20 - 16 \qquad \text{Subtract } 16 \text{ from both sides.}$$

$$-3x = 4$$

$$\frac{-3x}{-3} = \frac{4}{-3} \qquad \text{Divide both sides by } -3.$$

$$x = -\frac{4}{3} \qquad \text{The check is left to the reader.}$$

Avoiding Mistakes:

In Example 4, several of the fractions in the equation have two terms in the numerator. It is important to enclose these fractions in parentheses when clearing fractions. In this way, we will remember to use the distributive property to multiply the factors shown in blue with both terms from the numerator of the fractions.

Skill Practice Answers

3. $x = \dfrac{45}{2}$ **4.** $z = \dfrac{7}{3}$

Skill Practice Solve the equation.

5. $\dfrac{x+1}{4} + \dfrac{x+2}{6} = 1$

---

3. Solving Linear Equations with Decimals

The process of **clearing decimals** is similar to clearing fractions in a linear equation. In this case, we multiply both sides of the equation by a convenient power of ten. We find the coefficient of the term within the equation that has the greatest number of digits following the decimal point. If the last nonzero digit in that coefficient is in the tenths place, we multiply by 10. If the coefficient is represented to the hundredths place, we multiply by 100, and so on.

Example 5 Solving Linear Equations Containing Decimals

Solve the equations by clearing decimals.

a. $2.5x + 3 = 1.7x - 6.6$ **b.** $0.2(x+4) - 0.45(x+9) = 12$

Solution:

a.

$$2.5x + 3 = 1.7x - 6.6$$

$$10(2.5x + 3) = 10(1.7x - 6.6)$$ Multiply both sides of the equation by 10.

$$25x + 30 = 17x - 66$$ Apply the distributive property.

$$25x - 17x + 30 = 17x - 17x - 66$$ Subtract $17x$ from both sides.

$$8x + 30 = -66$$

$$8x + 30 - 30 = -66 - 30$$ Subtract 30 from both sides.

$$8x = -96$$

$$\frac{8x}{8} = \frac{-96}{8}$$ Divide both sides by 8.

$$x = -12$$

TIP: The terms with the most digits following the decimal point are $-0.45x$ and -4.05. Each of these is written to the hundredths place. Therefore, we multiply both sides by 100.

b.

$$0.2(x+4) - 0.45(x+9) = 12$$

$$0.2x + 0.8 - 0.45x - 4.05 = 12$$ Clear parentheses first.

$$100(0.2x + 0.8 - 0.45x - 4.05) = 100(12)$$ Multiply both sides by 100.

$$20x + 80 - 45x - 405 = 1200$$ Apply the distributive property.

$$-25x - 325 = 1200$$ Simplify both sides.

$$-25x - 325 + 325 = 1200 + 325$$ Add 325 to both sides.

$$-25x = 1525$$

$$\frac{-25x}{-25} = \frac{1525}{-25}$$ Divide both sides by -25.

$$x = -61$$ The check is left to the reader.

Skill Practice Answers

5. $x = 1$

---

6. $1.2w + 3.5 = 2.1 + w$

7. $0.25(x + 2) - 0.15(x + 3) = 4$

Section 2.3   Practice Exercises

Study Skills Exercises

1. Instructors vary in what they emphasize on tests. For example, test material may come from the textbook, notes, handouts, or homework. What does your instructor emphasize?

2. Define the key terms:

   a. clearing fractions   **b. clearing decimals**

Review Exercises

For Exercises 3–6, solve the equation.

3. $5(x + 2) - 3 = 4x + 5$

4. $-2(2x - 4x) = 6 + 18$

5. $3(2y + 3) - 4(-y + 1) = 7y - 10$

6. $-(3w + 4) + 5(w - 2) - 3(6w - 8) = 10$

7. Solve the equation and describe the solution set: $7x + 2 = 7(x - 12)$

8. Solve the equation and describe the solution set: $2(3x - 6) = 3(2x - 4)$

Concept 1: Clearing Fractions and Decimals

For Exercises 9–14, determine which of the values could be used to clear fractions or decimals in the given equation.

9. $\dfrac{2}{3}x - \dfrac{1}{6} = \dfrac{x}{9}$

   Values: 6, 9, 12, 18, 24, 36

10. $\dfrac{1}{4}x - \dfrac{2}{7} = \dfrac{1}{2}x + 2$

   Values: 4, 7, 14, 21, 28, 42

11. $0.02x + 0.5 = 0.35x + 1.2$

   Values: 10; 100; 1000; 10,000

12. $0.003 - 0.002x = 0.1x$

   Values: 10; 100; 1000; 10,000

13. $\dfrac{1}{6}x + \dfrac{7}{10} = x$

   Values: 3, 6, 10, 30, 60

14. $2x - \dfrac{5}{2} = \dfrac{x}{3} - \dfrac{1}{4}$

   Values: 2, 3, 4, 6, 12, 24

Concept 2: Solving Linear Equations with Fractions

For Exercises 15–36, solve the equation.

15. $\dfrac{1}{2}x + 3 = 5$

16. $\dfrac{1}{3}y - 4 = 9$

17. $\dfrac{1}{6}y + 2 = \dfrac{5}{12}$

18. $\dfrac{2}{15}z + 3 = \dfrac{7}{5}$

19. $\frac{1}{3}q + \frac{3}{5} = \frac{1}{15}q - \frac{2}{5}$

20. $\frac{3}{7}x - 5 = \frac{24}{7}x + 7$

21. $\frac{12}{5}w + 7 = 31 - \frac{3}{5}w$

22. $-\frac{1}{9}p - \frac{5}{18} = -\frac{1}{6}p + \frac{1}{3}$

23. $\frac{1}{4}(3m - 4) - \frac{1}{5} = \frac{1}{4}m + \frac{3}{10}$

24. $\frac{1}{25}(20 - t) = \frac{4}{25}t - \frac{3}{5}$

25. $\frac{1}{6}(5s + 3) = \frac{1}{2}(s + 11)$

26. $\frac{1}{12}(4n - 3) = \frac{1}{12}n - \frac{3}{4}$

27. $\frac{2}{3}x + 4 = \frac{2}{3}x - 6$

28. $-\frac{1}{9}a + \frac{2}{9} = \frac{1}{3} - \frac{1}{9}a$

29. $\frac{1}{6}(2c - 1) = \frac{1}{3}c - \frac{1}{6}$

30. $\frac{3}{2}b - 1 = \frac{1}{8}(12b - 8)$

31. $\frac{2x + 1}{3} + \frac{x - 1}{3} = 5$

32. $\frac{4y - 2}{5} - \frac{y + 4}{5} = -3$

33. $\frac{3w - 2}{6} = 1 - \frac{w - 1}{3}$

34. $\frac{z - 7}{4} = \frac{6z - 1}{8} - 2$

35. $\frac{x + 3}{3} - \frac{x - 1}{2} = 4$

36. $\frac{5y - 1}{2} - \frac{y + 4}{5} = 1$

Concept 3: Solving Linear Equations with Decimals

For Exercises 37–54, solve the equation.

37. $9.2y - 4.3 = 50.9$

38. $-6.3x + 1.5 = -4.8$

39. $21.1w + 4.6 = 10.9w + 35.2$

40. $0.05z + 0.2 = 0.15z - 10.5$

41. $0.2p - 1.4 = 0.2(p - 7)$

42. $0.5(3q + 87) = 1.5q + 43.5$

43. $0.20x + 53.60 = x$

44. $z + 0.06z = 3816$

45. $0.15(90) + 0.05p = 0.10(90 + p)$

46. $0.25(60) + 0.10x = 0.15(60 + x)$

47. $0.40(y + 10) + 0.60y = 2$

48. $0.75(x - 2) + 0.25x = 0.5$

49. $0.4x + 0.2 = -3.6 - 0.6x$

50. $0.12x + 3 - 0.8x = 0.22x - 0.6$

51. $0.06(x - 0.5) = 0.06x + 0.01$

52. $0.125x = 0.025(5x + 1)$

53. $-3.5x + 1.3 = -0.3(9x - 5)$

54. $x + 4 = 2(0.4x + 1.3)$

Mixed Exercises

For Exercises 55–81, solve the equation.

55. $2b + 23 = 6b - 5$

56. $-x = 7$

57. $\frac{y}{4} = -2$

58. $10p - 9 + 2p - 3 = 8p - 18$

59. $0.5(2a - 3) - 0.1 = 0.4(6 + 2a)$

60. $-\frac{5}{9}w + \frac{11}{12} = \frac{23}{36}$

61. $-6x = 0$

62. $15.2q = -2.4q - 176$

63. $9.8h + 2 = 3.8h + 20$

64. $-k - 41 = 3 - k$

65. $\frac{1}{4}(x + 4) = \frac{1}{5}(2x + 3)$

66. $7y + 3(2y + 5) = 10y + 17$

67. $2z - 7 = 2(z - 13)$

68. $x - 17.8 = -21.3$

69. $\dfrac{4}{5}w = 10$

70. $5c + 25 = 20$

71. $4b - 8 - b = -3b + 2(3b - 4)$

72. $36 = 6z + 9$

73. $-3a + 1 = 19$

74. $-5(1 - x) + x = -(6 - 2x) + 6$

75. $3(4h - 2) - (5h - 8) = 8 - (2h + 3)$

76. $1.72w - 0.04w = 0.42$

77. $\dfrac{3}{8}t - \dfrac{5}{8} = \dfrac{1}{2}t + \dfrac{1}{8}$

78. $3(8x - 1) + 10 = 6(5 + 4x) - 23$

79. $\dfrac{2x - 1}{4} + \dfrac{3x + 2}{6} = 2$

80. $\dfrac{w - 4}{6} - \dfrac{3w - 1}{2} = -1$

81. $\dfrac{2k + 5}{4} = 2 - \dfrac{k + 2}{3}$

82. The sum of $\frac{2}{5}$ and twice a number is the same as the sum of $\frac{11}{5}$ and the number. Find the number.

83. The difference of three times a number and $\frac{5}{9}$ is the same as the sum of twice the number and $\frac{1}{9}$. Find the number.

84. The sum of twice a number and $\frac{3}{4}$ is the same as the difference of four times the number and $\frac{1}{8}$. Find the number.

85. The difference of a number and $-\frac{11}{12}$ is the same as the difference of three times the number and $\frac{1}{6}$. Find the number.

Expanding Your Skills

For Exercises 86–89, solve the equation.

86. $\dfrac{1}{2}a + 0.4 = -0.7 - \dfrac{3}{5}a$

87. $\dfrac{3}{4}c - 0.11 = 0.23(c - 5)$

88. $0.8 + \dfrac{7}{10}b = \dfrac{3}{2}b - 0.8$

89. $0.78 - \dfrac{1}{25}h = \dfrac{3}{5}h - 0.5$

Applications of Linear Equations: Introduction to Problem Solving

Section 2.4

1. Problem-Solving Strategies

Linear equations can be used to solve many real-world applications. However, with "word problems," students often do not know where to start. To help organize the problem-solving process, we offer the following guidelines:

Concepts

1. **Problem-Solving Strategies**
2. **Translations Involving Linear Equations**
3. **Consecutive Integer Problems**
4. **Applications of Linear Equations**
5. **Applications Involving Uniform Motion**

Problem-Solving Flowchart for Word Problems

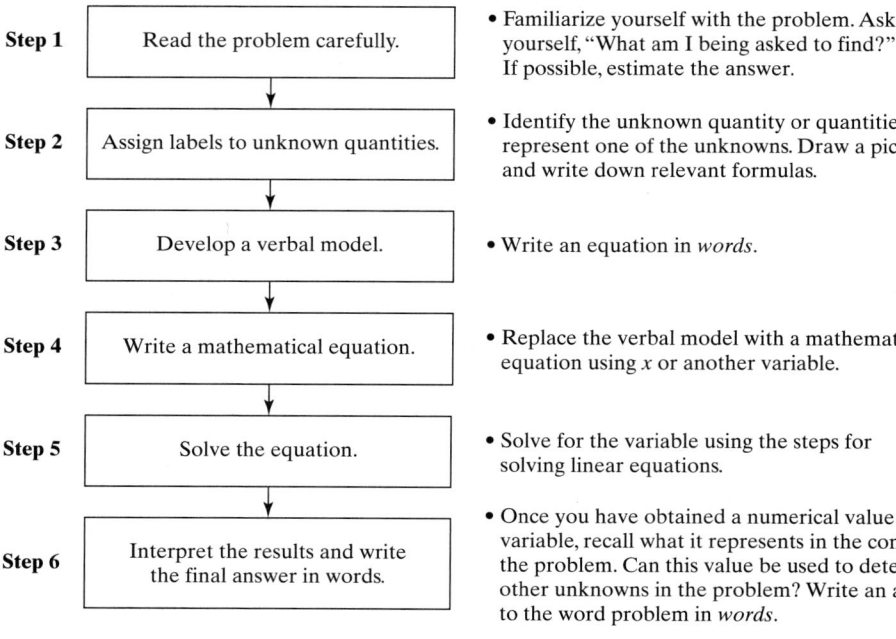

Step 1 — Read the problem carefully.	• Familiarize yourself with the problem. Ask yourself, "What am I being asked to find?" If possible, estimate the answer.
Step 2 — Assign labels to unknown quantities.	• Identify the unknown quantity or quantities. Let x represent one of the unknowns. Draw a picture and write down relevant formulas.
Step 3 — Develop a verbal model.	• Write an equation in *words*.
Step 4 — Write a mathematical equation.	• Replace the verbal model with a mathematical equation using x or another variable.
Step 5 — Solve the equation.	• Solve for the variable using the steps for solving linear equations.
Step 6 — Interpret the results and write the final answer in words.	• Once you have obtained a numerical value for the variable, recall what it represents in the context of the problem. Can this value be used to determine other unknowns in the problem? Write an answer to the word problem in *words*.

2. Translations Involving Linear Equations

We have already practiced translating an English sentence to a mathematical equation. Recall from Section 1.2 that several key words translate to the algebraic operations of addition, subtraction, multiplication, and division.

Addition: $a + b$	**Subtraction: $a - b$**
The sum of a and b	The difference of a and b
a plus b	a minus b
b added to a	b subtracted from a
b more than a	a decreased by b
a increased by b	b less than a
The total of a and b	

Multiplication: $a \cdot b$	**Division: $a \div b$**
The product of a and b	The quotient of a and b
a times b	a divided by b
a multiplied by b	b divided into a
	The ratio of a and b
	a over b
	a per b

| Example 1 | **Translating to a Linear Equation** |

The sum of a number and negative eleven is negative fifteen. Find the number.

Solution:

| | **Step 1:** | Read the problem. |
| Let x represent the unknown number. | **Step 2:** | Label the unknown. |

the sum of   is

(a number) $+ (-11) = (-15)$ **Step 3:** Develop a verbal model.

$x + (-11) = -15$ **Step 4:** Write an equation.

$x + (-11) + 11 = -15 + 11$ **Step 5:** Solve the equation.

$x = -4$

The number is -4.   **Step 6:** Write the final answer in words.

| Skill Practice |

1. The sum of a number and negative seven is twelve. Find the number.

| Example 2 | **Translating to a Linear Equation** |

Forty less than five times a number is fifty-two less than the number. Find the number.

Solution:

| | **Step 1:** | Read the problem. |
| Let x represent the unknown number. | **Step 2:** | Label the unknown. |

$\left(\begin{array}{c} 5 \text{ times} \\ \text{a number} \end{array} \right) - (40) = \left(\begin{array}{c} \text{the} \\ \text{number} \end{array} \right) - (52)$ **Step 3:** Develop a verbal model.

$5x \quad - 40 = \quad x \quad - 52$ **Step 4:** Write an equation.

$5x - 40 = x - 52$ **Step 5:** Solve the equation.

$5x - x - 40 = x - x - 52$

$4x - 40 = -52$

$4x - 40 + 40 = -52 + 40$

$4x = -12$

$\dfrac{4x}{4} = \dfrac{-12}{4}$

$x = -3$

The number is -3.   **Step 6:** Write the final answer in words.

Avoiding Mistakes:

It is important to remember that subtraction is not a commutative operation. Therefore, the order in which two real numbers are subtracted affects the outcome. The expression "forty less than five times a number" must be translated as: $5x - 40$ (not $40 - 5x$). Similarly, "fifty-two less than the number" must be translated as: $x - 52$ (not $52 - x$).

| Skill Practice |

2. Thirteen more than twice a number is five more than the number. Find the number.

Skill Practice Answers

1. The number is 19.
2. The number is -8.

| **Example 3** | **Translating to a Linear Equation** |

Twice the sum of a number and six is two more than three times the number. Find the number.

Solution:

Step 1:   Read the problem.

Let x represent the unknown number.   Step 2:   Label the unknown.

Step 3:   Develop a verbal model.

twice   the sum   is   2 more than
$$2 \quad (x + 6) \quad = \quad 3x + 2$$ Step 4:   Write an equation.
three times
a number

$$2(x + 6) = 3x + 2$$ Step 5:   Solve the equation.

$$2x + 12 = 3x + 2$$

$$2x - 2x + 12 = 3x - 2x + 2$$

$$12 = x + 2$$

$$12 - 2 = x + 2 - 2$$

$$10 = x$$

The number is 10.   Step 6:   Write the final answer in words.

Skill Practice

3. Three times the sum of a number and eight is four more than the number. Find the number.

3. Consecutive Integer Problems

The word *consecutive* means "following one after the other in order without gaps." The numbers 6, 7, 8 are examples of three **consecutive integers**. The numbers $-4, -2, 0, 2$ are examples of **consecutive even integers**. The numbers 23, 25, 27 are examples of **consecutive odd integers**.

Notice that any two consecutive integers differ by 1. Therefore, if x represents an integer, then $(x + 1)$ represents the next larger consecutive integer (Figure 2-4).

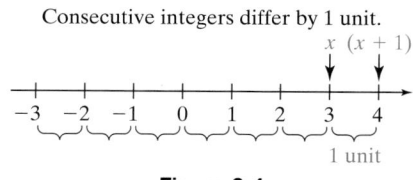

Figure 2-4

Any two consecutive even integers differ by 2. Therefore, if x represents an even integer, then $(x + 2)$ represents the next consecutive larger even integer (Figure 2-5).

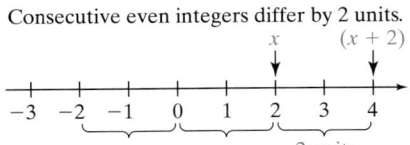

Consecutive even integers differ by 2 units.

Figure 2-5

Likewise, any two consecutive odd integers differ by 2. If x represents an odd integer, then $(x + 2)$ is the next larger odd integer (Figure 2-6).

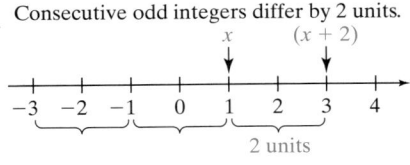

Consecutive odd integers differ by 2 units.

Figure 2-6

| **Example 4** | **Solving an Application Involving Consecutive Integers** |

The sum of two consecutive odd integers is -188. Find the integers.

Solution:

In this example we have two unknown integers. We can let x represent either of the unknowns.

                                      Step 1: Read the problem.

Suppose x represents the first odd integer.   **Step 2:** Label the variables.

Then $(x + 2)$ represents the second odd integer.

$$\left(\begin{array}{c}\text{first}\\\text{integer}\end{array}\right) + \left(\begin{array}{c}\text{second}\\\text{integer}\end{array}\right) = (\text{total})$$

Step 3: Write an equation in words.

$$x + (x + 2) = -188$$

Step 4: Write a mathematical equation.

$$x + (x + 2) = -188$$
$$2x + 2 = -188$$

Step 5: Solve for x.

$$2x + 2 - 2 = -188 - 2$$
$$2x = -190$$
$$\frac{2x}{2} = \frac{-190}{2}$$
$$x = -95$$

The first integer is $x = -95$.

The second integer is $x + 2 = -95 + 2 = -93$.

The two integers are -95 and -93.

Step 6: Interpret the results and write the answer in words.

Skill Practice

4. The sum of two consecutive even integers is 66. Find the integers.

Example 5   **Solving an Application Involving Consecutive Integers**

Ten times the smallest of three consecutive integers is twenty-two more than three times the sum of the integers. Find the integers.

Solution:

Step 1: Read the problem.

Let x represent the first integer.
$x + 1$ represents the second consecutive integer.
$x + 2$ represents the third consecutive integer.

Step 2: Label the variables.

$$\begin{pmatrix} 10 \text{ times} \\ \text{the first} \\ \text{integer} \end{pmatrix} = \begin{pmatrix} 3 \text{ times} \\ \text{the sum of} \\ \text{the integers} \end{pmatrix} + 22$$

Step 3: Write an equation in words.

$$10x = 3[(x) + (x + 1) + (x + 2)] + 22$$

the sum of the integers

Step 4: Write a mathematical equation.

$$10x = 3(x + x + 1 + x + 2) + 22$$

Step 5: Solve the equation.

$$10x = 3(3x + 3) + 22$$

Clear parentheses.

$$10x = 9x + 9 + 22$$

Combine *like* terms.

$$10x = 9x + 31$$

$$10x - 9x = 9x - 9x + 31$$

$$x = 31$$

Isolate the x-terms on one side.

The first integer is $x = 31$.

The second integer is $x + 1 = 31 + 1 = 32$.

The third integer is $x + 2 = 31 + 2 = 33$.

Step 6: Interpret the results and write the answer in words.

The three integers are 31, 32, and 33.

Skill Practice Answers

4. The integers are 32 and 34.

5. Five times the smallest of three consecutive integers is 117 more than the sum of the integers. Find the integers.

4. Applications of Linear Equations

Example 6 Using a Linear Equation in an Application

A carpenter cuts a 6-ft board in two pieces. One piece must be three times as long as the other. Find the length of each piece.

Solution:

In this problem, one piece must be three times as long as the other. Thus, if x represents the length of one piece, then $3x$ can represent the length of the other.

Step 1: Read the problem completely.

x represents the length of the smaller piece. $3x$ represents the length of the longer piece.

Step 2: Label the unknowns. Draw a figure.

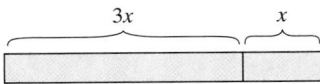

$$\begin{pmatrix} \text{length of} \\ \text{one piece} \end{pmatrix} + \begin{pmatrix} \text{length of} \\ \text{other piece} \end{pmatrix} = \begin{pmatrix} \text{total length} \\ \text{of the board} \end{pmatrix}$$

Step 3: Set up a verbal equation.

$$x \quad + \quad 3x \quad = \quad 6$$

Step 4: Write an equation.

$$4x = 6$$

Step 5: Solve the equation.

$$\frac{4x}{4} = \frac{6}{4}$$

$$x = 1.5$$

The smaller piece is $x = 1.5$ ft.

Step 6: Interpret the results.

The longer piece is $3x$ or $3(1.5 \text{ ft}) = 4.5$ ft.

6. A plumber cuts a 96-in. piece of pipe into two pieces. One piece is five times longer than the other piece. How long is each piece?

Example 7 Using a Linear Equation in an Application

The hit movies *Spider-Man* and *X-Men* together brought in $169.3 million during their opening weekends. *Spider-Man* earned $5.8 million more than twice what *X-Men* earned. How much revenue did each movie bring in during its opening weekend?

Skill Practice Answers

5. The integers are 60, 61, and 62.
6. One piece is 80 in. and the other is 16 in.

Solution:

In this example, we have two unknowns. The variable x can represent *either* quantity. However, the revenue from *Spider-Man* is given in terms of the revenue for *X-Men*.

Step 1:   Read the problem.

Let x represent the revenue for *X-Men*.

Step 2:   Label the variables.

Then $2x + 5.8$ represents the revenue for *Spider-Man*.

$$\left(\begin{array}{c}\text{Revenue from}\\ \text{X-Men}\end{array}\right) + \left(\begin{array}{c}\text{Revenue from}\\ \text{Spider-Man}\end{array}\right) = \left(\begin{array}{c}\text{Total}\\ \text{Revenue}\end{array}\right)$$

Step 3:   Set up a verbal equation.

$$x \qquad + \qquad 2x + 5.8 \qquad = \qquad 169.3$$

Step 4:   Write an equation.

$$3x + 5.8 = 169.3$$

Step 5:   Solve the equation.

$$3x + 5.8 - 5.8 = 169.3 - 5.8$$

$$3x = 163.5$$

$$\frac{3x}{3} = \frac{163.5}{3}$$

$$x = 54.5$$

Revenue from *X-Men*: $x = 54.5$

Step 6:   Interpret the results.

Revenue from *Spider-Man*: $2x + 5.8 = 2(54.5) + 5.8 = 114.8$

The revenue from *X-Men* was $54.5 million for its opening weekend. The revenue from *Spider-Man* was $114.8 million.

> **Skill Practice**
>
> **7.** There are 40 students in an algebra class. There are four more women than men. How many women and men are in the class?

5. Applications Involving Uniform Motion

The formula: (distance) = (rate)(time) or simply, $d = rt$, relates the distance traveled to the rate of travel and the time of travel.

For example, if a car travels at 60 mph for 3 hours, then

$$d = (60 \text{ mph})(3 \text{ hours})$$

$$= 180 \text{ miles}$$

If a car travels at 60 mph for x hours, then

$$d = (60 \text{ mph})(x \text{ hours})$$

$$= 60x \text{ miles}$$

Skill Practice Answers

7. There are 22 women and 18 men.

| Example 8 | Solving an Application Involving Distance, Rate, and Time |

One bicyclist rides 4 mph faster than another bicyclist. The faster rider takes 3 hr to complete a race, while the slower rider takes 4 hr. Find the speed for each rider.

Solution:

Step 1: Read the problem.

The problem is asking us to find the speed of each rider.

Let x represent the speed of the slower rider. Then $(x + 4)$ is the speed of the faster rider.

Step 2: Label the variables and organize the information given in the problem. A distance-rate-time chart may be helpful.

	Distance	Rate	Time
Faster rider	$3(x + 4)$	$x + 4$	3
Slower rider	$4(x)$	x	4

To complete the first column, we can use the relationship, $d = rt$.

faster rider's distance = (faster rate)(faster rider's time) = $(x + 4)(3)$

slower rider's distance = (slower rate)(slower rider's time) = $(x)(4)$

Because the riders are riding in the same race, their distances are equal.

$$\begin{pmatrix} \text{distance} \\ \text{by faster rider} \end{pmatrix} = \begin{pmatrix} \text{distance} \\ \text{by slower rider} \end{pmatrix}$$

Step 3: Set up a verbal model.

$$3(x + 4) = 4(x)$$

Step 4: Write a mathematical equation.

$$3x + 12 = 4x$$

Step 5: Solve the equation.

$$12 = x$$

Subtract $3x$ from both sides.

The variable x represents the slower rider's rate. The quantity $x + 4$ is the faster rider's rate. Thus, if $x = 12$, then $x + 4 = 16$.

The slower rider travels 12 mph and the faster rider travels 16 mph.

TIP: Check that the answer is reasonable. If the slower rider rides at 12 mph for 4 hr, he travels 48 mi. If the faster rider rides at 16 mph for 3 hr, he also travels 48 mi as expected.

Skill Practice

8. An express train travels 25 mph faster than a cargo train. It takes the express train 6 hr to travel a route, and it takes 9 hr for the cargo train to travel the same route. Find the speed of each train.

| Example 9 | Solving an Application Involving Distance, Rate, and Time |

Two families that live 270 miles apart plan to meet for an afternoon picnic. To share the driving, they want to meet somewhere between their two homes. Both families leave at 9.00 A.M., but one family averages 12 mph faster than the other family. If the families meet at the designated spot $2\frac{1}{2}$ hours later, determine

a. The average rate of speed for each family.

b. The distance each family traveled to the picnic.

Skill Practice Answers

8. The express train travels 75 mph, and the cargo train travels 50 mph.

Solution:

For simplicity, we will call the two families, Family A and Family B. Let Family A be the family that travels at the slower rate (Figure 2-7).

Step 1:   Read the problem and draw a sketch.

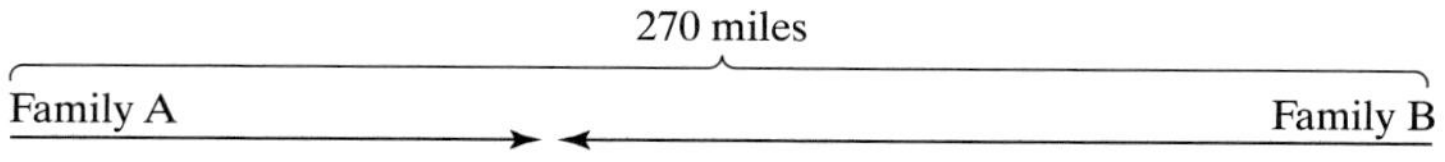

270 miles

Family A ————————————→ ←———————— Family B

Figure 2-7

Let x represent the rate of Family A.

Step 2:   Label the variables.

Then $(x + 12)$ is the rate of Family B.

	Distance	**Rate**	**Time**
Family A	$2.5x$	x	2.5
Family B	$2.5(x + 12)$	$x + 12$	2.5

To complete the first column, we can use the relationship $d = rt$.

The distance traveled by Family A is $(x)(2.5)$.

The distance traveled by Family B is $(x + 12)(2.5)$.

To set up an equation, recall that the total distance between the two families is given as 270 miles.

$$\left(\begin{array}{c}\text{distance} \\ \text{traveled by} \\ \text{Family A}\end{array}\right) + \left(\begin{array}{c}\text{distance} \\ \text{traveled by} \\ \text{Family B}\end{array}\right) = \left(\begin{array}{c}\text{total} \\ \text{distance}\end{array}\right)$$

Step 3:   Create a verbal equation.

$$2.5x \quad + \quad 2.5(x + 12) \quad = \quad 270$$

Step 4:   Write a mathematical equation.

$$2.5x + 2.5(x + 12) = 270$$
$$2.5x + 2.5x + 30 = 270$$

Step 5:   Solve for x.

$$5.0x + 30 = 270$$
$$5x = 240$$
$$\frac{5x}{5} = \frac{240}{5}$$
$$x = 48$$

a. Family A traveled 48 (mph).

Step 6:   Interpret the results and write the answer in words.

Family B traveled $x + 12 = 48 + 12 = 60$ (mph).

b. To compute the distance each family traveled, use $d = rt$:

Family A traveled: $(48 \text{ mph})(2.5 \text{ hr}) = 120$ miles

Family B traveled: $(60 \text{ mph})(2.5 \text{ hr}) = 150$ miles

Skill Practice

9. A Piper Cub airplane has an average air speed that is 10 mph faster than a Cessna 150 airplane. If the combined distance traveled by these two small planes is 690 miles after 3 hr, what is the average speed of each plane?

Skill Practice Answers

9. The Cessna's speed is 110 mph, and the Piper Cub's speed is 120 mph.

Section 2.4   Practice Exercises

Boost *your* GRADE at mathzone.com!

• Practice Problems   • e-Professors
• Self-Tests   • Videos
• NetTutor

Study Skills Exercises

1. After doing a section of homework, check the odd-numbered answers in the back of the text. Choose a method to identify the exercises that you got wrong or had trouble with (i.e., circle the number or put a star by the number). List some reasons why it is important to label these problems.

2. Define the key terms:

   a. consecutive integers   **b. consecutive even integers**   **c. consecutive odd integers**

Concept 2: Translations Involving Linear Equations

For Exercises 3–8, write an algebraic equation to represent the English sentence. Then solve the equation.

3. The sum of a number and sixteen is negative thirty-one. Find the number.

4. The sum of a number and negative twenty-one is fourteen. Find the number.

5. The difference of a number and six is negative three. Find the number.

6. The difference of a number and negative four is negative twelve. Find the number.

7. Sixteen less than a number is negative one. Find the number.

8. Ten less than a number is negative thirteen. Find the number.

For Exercises 9–18, use the problem-solving flowchart on page 142.

9. Six less than a number is −10. Find the number.

10. Fifteen less than a number is 41. Find the number.

11. Twice the sum of a number and seven is eight. Find the number.

12. Twice the sum of a number and negative two is sixteen. Find the number.

13. A number added to five is the same as twice the number. Find the number.

14. Three times a number is the same as the difference of twice the number and seven. Find the number.

15. The sum of six times a number and ten is equal to the difference of the number and fifteen. Find the number.

16. The difference of fourteen and three times a number is the same as the sum of the number and negative ten. Find the number.

17. If the difference of a number and four is tripled, the result is six more than the number. Find the number.

18. Twice the sum of a number and eleven is twenty-two less than three times the number. Find the number.

Concept 3: Consecutive Integer Problems

19. a. If x represents the smallest of three consecutive integers, write an expression to represent each of the next two consecutive integers.

   b. If x represents the largest of three consecutive integers, write an expression to represent each of the previous two consecutive integers.

20. a. If x represents the smallest of three consecutive odd integers, write an expression to represent each of the next two consecutive odd integers.

   b. If x represents the largest of three consecutive odd integers, write an expression to represent each of the previous two consecutive odd integers.

For Exercises 21–28, use the problem-solving flowchart from page 142.

21. The sum of two consecutive integers is -67. Find the integers.

22. The sum of two consecutive odd integers is 52. Find the integers.

23. The sum of two consecutive odd integers is 28. Find the integers.

24. The sum of three consecutive even integers is 66. Find the integers.

25. The sum of the page numbers on two facing pages in a book is 941. What are the page numbers?

26. Three raffle tickets are represented by three consecutive integers. If the sum of the three integers is 2,666,031, find the numbers.

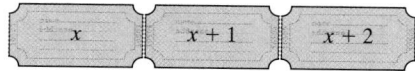

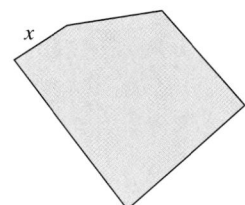

27. The perimeter of a pentagon (a five-sided polygon) is 80 in. The five sides are represented by consecutive integers. Find the measures of the sides.

28. The perimeter of a pentagon (a five-sided polygon) is 95 in. The five sides are represented by consecutive integers. Find the measures of the sides.

Concept 4: Applications of Linear Equations

For Exercises 29–40, use the problem-solving flowchart (page 142) to solve the problems.

29. Karen's age is 12 years more than Clarann's age. The sum of their ages is 58. Find their ages.

30. Maria's age is 15 years less than Orlando's age. The sum of their ages is 29. Find their ages.

31. For a recent year, 104 more Democrats than Republicans were in the U.S. House of Representatives. If the total number of representatives from these two parties in the House was 434, find the number of representatives from each party.

32. For a recent year, 12 more Republicans than Democrats were in the U.S. House of Representatives. If the House had a total of 434 representatives from these two parties, find the number of Democrats and the number of Republicans.

33. A board is 86 cm in length and must be cut so that one piece is 20 cm longer than the other piece. Find the length of each piece.

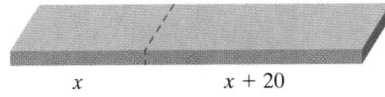

34. A rope is 54 in. in length and must be cut into two pieces. If one piece must be twice as long as the other, find the length of each piece.

35. Approximately 5.816 million people watch *The Oprah Winfrey Show*. This is 1.118 million more than watch *The Dr. Phil Show*. How many watch *The Dr. Phil Show*? (*Source: Neilson Media Research*)

36. Two of the largest Internet retailers are eBay and Amazon.com. Recently, the estimated U.S. sales of eBay were $0.1 billion less than twice the sales of Amazon.com. Given the total sales of $5.6 billion, determine the sales of eBay and Amazon.com.

37. The longest river in Africa is the Nile. It is 2455 km longer than the Congo River, also in Africa. The sum of the lengths of these rivers is 11,195 km. What is the length of each river?

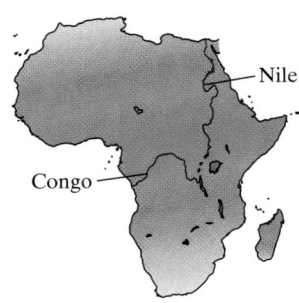

38. The average depth of the Gulf of Mexico is three times the depth of the Red Sea. The difference between the average depths is 1078 m. What is the average depth of the Gulf of Mexico and the average depth of the Red Sea?

39. Asia and Africa are the two largest continents in the world. The land area of Asia is approximately 14,514,000 km² larger than the land area of Africa. Together their total area is 74,644,000 km². Find the land area of Asia and the land area of Africa.

40. Mt. Everest, the highest mountain in the world, is 2654 m higher than Mt. McKinley, the highest mountain in the United States. If the sum of their heights is 15,042 m, find the height of each mountain.

Concept 5: Applications Involving Uniform Motion

41. A woman can hike 1 mph faster down a trail to Cochita Lake than she can on the return trip uphill. It takes her 2 hr to get to the lake and 3 hr to return. What is her speed hiking down to the lake?

	Distance	Rate	Time
Downhill to the lake			
Uphill from the lake			

42. A car travels 20 mph slower in a bad rain storm than in sunny weather. The car travels the same distance in 2 hr in sunny weather as it does in 3 hr in rainy weather. Find the speed of the car in sunny weather.

43. Hazel and Emilie fly from Atlanta to San Diego. The flight from Atlanta to San Diego is against the wind and takes 4 hr. The return flight with the wind takes 3.5 hr. If the wind speed is 40 mph, find the speed of the plane in still air.

44. A boat on the Potomac River travels the same distance downstream in $\frac{2}{3}$ hr as it does going upstream in 1 hr. If the speed of the current is 3 mph, find the speed of the boat in still water.

45. Two cars are 200 miles apart and traveling toward each other on the same road. They meet in 2 hr. One car is traveling 4 mph faster than the other. What is the speed of each car?

46. Two cars are 238 miles apart and traveling toward each other along the same road. They meet in 2 hr. One car is traveling 5 mph slower than the other. What is the speed of each car?

47. After Hurricane Katrina, a rescue vehicle leaves a station at noon and heads for New Orleans. An hour later a second vehicle traveling 10 mph faster leaves the same station. By 4:00 P.M., the first vehicle reaches its destination, and the second is still 10 miles away. How fast is each vehicle?

48. A truck leaves a truck stop at 9:00 A.M. and travels toward Sturgis, Wyoming. At 10:00 A.M., a motorcycle leaves the same truck stop and travels the same route. The motorcycle travels 15 mph faster than the truck. By noon, the truck has traveled 20 miles further than the motorcycle. How fast is each vehicle?

49. Two boats traveling the same direction leave a harbor at noon. After 2 hr, they are 40 miles apart. If one boat travels twice as fast as the other, find the rate of each boat.

50. Two canoes travel down a river, starting at 9:00 A.M. One canoe travels twice as fast as the other. After 3.5 hr, the canoes are 5.25 miles apart. Find the speed of each canoe.

Mixed Exercises

51. A number increased by 58 is -22. Find the number.

52. The sum of a number and -14 is -32. Find the number.

53. Three consecutive integers are such that three times the largest exceeds the sum of the two smaller integers by 47. Find the integers.

54. Four times the smallest of three consecutive odd integers is 236 more than the sum of the other two integers. Find the integers.

55. A boat on the Hudson River travels the same distance in $\frac{1}{2}$ hour downstream as it can going upstream for an hour. If the boat travels 12 mph in still water, what is the speed of the current?

56. A flight from Orlando to Phoenix takes 4 hr with the wind. The return flight against the wind takes $4\frac{1}{2}$ hr. If the plane flies at 500 mph in still air, find the speed of the wind. (Round to the nearest mile per hour.)

57. Five times the difference of a number and three is four less than four times the number. Find the number.

58. Three times the difference of a number and seven is one less than twice the number. Find the number.

59. In a recent year, the estimated earnings for Jennifer Lopez was $2.5 million more than half of the earnings for the band, U2. If the total earnings were $106 million, what were the earnings for Jennifer Lopez and U2? (*Source: Forbes*)

60. In a recent year, the best selling DVD was *The Lord of the Rings: The Fellowship of the Rings*. Another big seller was *Harry Potter and the Sorcerer's Stone*. There were 90 million fewer Harry Potter DVDs sold than Lord of the Rings. If the two movies totaled 424 million, how many DVDs of each were sold? (*Source: VSDA VidTrac*)

61. A boat in distress, 21 nautical miles from a marina, travels toward the marina at 3 knots (nautical miles per hour). A coast guard cruiser leaves the marina and travels toward the boat at 25 knots. How long will it take for the boats to reach each other?

62. An air traffic controller observes a plane heading from New York to San Francisco traveling at 450 mph. At the same time, another plane leaves San Francisco and travels 500 mph to New York. If the distance between the airports is 2850 miles, how long will it take for the planes to pass each other?

63. If three is added to five times a number, the result is forty-three more than the number. Find the number.

64. If seven is added to three times a number, the result is thirty-one more than the number.

65. The deepest point in the Pacific Ocean is 676 m more than twice the deepest point in the Arctic Ocean. If the deepest point in the Pacific is 10,920 m, how many meters is the deepest point in the Arctic Ocean?

66. The area of Greenland is 201,900 km^2 less than three times the area of New Guinea. What is the area of New Guinea if the area of Greenland is 2,175,600 km^2?

Applications Involving Percents

1. Solving Basic Percent Equations
2. Applications Involving Sales Tax
3. Applications Involving Simple Interest

1. Solving Basic Percent Equations

The word *percent* means "per hundred." For example:

Percent	Interpretation
63% of homes have a computer	63 out of 100 homes have a computer.
5% sales tax	5¢ in tax is charged for every 100¢ in merchandise.
15% commission	$15 is earned in commission for every $100 sold.

Percents come up in a variety of applications in day-to-day life. Many such applications follow the basic percent equation:

$$\text{amount} = (\text{percent})(\text{base}) \qquad \text{Basic percent equation}$$

In Example 1, we translate an English sentence into a percent equation.

Example 1   **Solving Basic Percent Equations**

a. What percent of 60 is 25.2?

b. 8.2 is 125% of what number?

c. 2% of 1500 is what number?

Solution:

a. Let x represent the unknown percent.

What percent of 60 is 25.2?

$$x \cdot 60 = 25.2$$

$$60x = 25.2$$

$$\frac{60x}{60} = \frac{25.2}{60}$$

$$x = 0.42, \text{ or } 42\%$$

25.2 is 42% of 60.

Step 1: Read the problem.
Step 2: Label the variables.
Step 3: Create a verbal model.

Step 4: Write a mathematical equation.

Step 5: Solve the equation.

Step 6: Interpret the results and write the answer in words.

Avoiding Mistakes:

Be sure to use the decimal form of a percentage within an equation.

$$125\% = 1.25$$

b. Let x represent the unknown number.

8.2 is 125% of what number?

$$8.2 = 1.25 \cdot x$$

$$8.2 = 1.25x$$

$$\frac{8.2}{1.25} = \frac{1.25x}{1.25}$$

$$6.56 = x$$

8.2 is 125% of 6.56.

Step 1: Read the problem.
Step 2: Label the variables.
Step 3: Create a verbal model.
Step 4: Write a mathematical equation.

Step 5: Solve the equation.

Step 6: Interpret the results and write the answer in words.

c. Let x represent the unknown number.

$$\underset{\underset{0.02}{\downarrow}}{2\%} \;\; \underset{}{\text{of}} \;\; \underset{\underset{1500}{\downarrow}}{1500} \;\; \underset{}{\text{is}} \;\; \underset{}{\text{what}} \;\; \underset{\underset{x}{\downarrow}}{\text{number?}}$$

$$0.02 \;\cdot\; 1500 \;=\; x$$

$$30 \;=\; x$$

30 is 2% of 1500.

Step 1: Read the problem.

Step 2: Label the variable.

Step 3: Create a verbal model.

Step 4: Write a mathematical equation.

Step 5: Solve the equation.

Step 6: Interpret the results.

Skill Practice

1. What percent of 85 is 11.9?

2. 279 is 90% of what number?

3. 115% of 82 is what number?

2. Applications Involving Sales Tax

One common use of percents is in computing **sales tax**.

$$\text{sales tax} = (\text{tax rate}) \cdot (\text{price of merchandise})$$

Example 2 Computing Sales Tax

A new digital camera costs $429.95.

a. Compute the sales tax if the tax rate is 4%.

b. Determine the total cost, including tax.

Solution:

a. Let x represent the amount of tax.

sales tax = (tax rate)(price of merchandise)

$x = (0.04)(\$429.95)$

$x = \$17.198$

$x = \$17.20$

The tax on the merchandise is $17.20.

Step 1: Read the problem.

Step 2: Label the variable.

Step 3: Write a verbal equation.

Step 4: Write a mathematical equation.

Step 5: Solve the equation.

Round to the nearest cent.

Step 6: Interpret the results.

b. The total cost is found by:

total cost = cost of merchandise + amount of tax

Therefore the total cost is: $429.95 + $17.20 = $447.15.

Skill Practice

4. Determine the total cost, including tax, of a portable CD player that sells for $89. Assume that the tax rate is 6%.

Skill Practice Answers

1. 14% **2.** 310
3. 94.3 **4.** $94.34

Example 3   Applying Percents

A video game is purchased for a total of $48.15, including sales tax. If the tax rate is 7%, find the original price of the video game before the sales tax.

Solution: **Step 1:**   Read the problem.

Let x represent the price of the video game.   **Step 2:**   Label variables.

$0.07x$ represents the amount of sales tax.

$$\left(\begin{array}{c}\text{original}\\\text{price}\end{array}\right) + \left(\begin{array}{c}\text{sales}\\\text{tax}\end{array}\right) = \left(\begin{array}{c}\text{total}\\\text{cost}\end{array}\right)$$

Step 3:   Write a verbal equation.

$$x \quad + \quad 0.07x \quad = \quad \$48.15$$

Step 4:   Write a mathematical equation.

$$1.07x = 48.15$$

Step 5:   Solve for x.

$$\frac{1.07x}{1.07} = \frac{48.15}{1.07}$$

Divide both sides by 1.07.

$$x = 45$$

Step 6:   Interpret the results and write the answer in words.

The original price was $45.

Skill Practice

5. The total price of a pair of jeans, including a 5% sales tax, is $27.30. Find the original price of the jeans.

3. Applications Involving Simple Interest

One important application of percents is in computing simple interest on a loan or on an investment.

Banks hold large quantities of money for their customers. However, because all bank customers are unlikely to withdraw all their money on a single day, a bank does not keep all the money in cash. Instead, it keeps some cash for day-to-day transactions but invests the remaining portion of the money. Because a bank uses its customer's money to make investments and because it wants to attract more customers, the bank pays interest on the money.

Simple interest is interest that is earned on principal (the original amount of money invested in an account). The following formula is used to compute simple interest:

$$\left(\begin{array}{c}\text{simple}\\\text{interest}\end{array}\right) = \left(\begin{array}{c}\text{principal}\\\text{invested}\end{array}\right)\left(\begin{array}{c}\text{annual}\\\text{interest rate}\end{array}\right)\left(\begin{array}{c}\text{time}\\\text{in years}\end{array}\right)$$

This formula is often written symbolically as $I = Prt$.

For example, to find the simple interest earned on $2000 invested at 7.5% interest for 3 years, we have

$$I = Prt$$

$$\text{Interest} = (\$2000)(0.075)(3)$$

$$= \$450$$

Skill Practice Answers

5. The original price was $26.00.

> **Example 4** **Computing Interest**

Wade borrows $12,000 for a new car.

  a. If he pays 7.5% simple interest over a 4-year period, how much total interest will he pay?

  b. Determine the total amount to be paid back.

Solution:

  a. Let *I* represent the amount of interest.

Step 1:	Read the problem.
Step 2:	Label the variables.

Interest = (Principal)(interest rate) (time in years)

Step 3:	Write an equation in words.

$I = (\$12{,}000)(0.075)(4)$

Step 4:	Write a mathematical equation.

$I = \$3600$

Step 5:	Solve the equation.

Wade will pay $3600 in interest.

Step 6:	Interpret the results.

  b. The total amount to be paid back includes the amount borrowed plus interest.

$$\$12{,}000 + \$3600 = \$15{,}600$$

> **Skill Practice**

  6. How much interest must be paid if $8000 is invested at a 5.5% simple interest rate for 2 years?

> **Example 5** **Applying Simple Interest**

Jorge wants to save money for his daughter's college education. If Jorge needs to have $4340 at the end of 4 years, how much money would he need to invest at a 6% simple interest rate?

Solution:

Let *P* represent the original amount invested.

$$\left(\begin{array}{c}\text{original} \\ \text{principal}\end{array}\right) + (\text{interest}) = (\text{total})$$

$$(P) \quad + \quad (Prt) \quad = (\text{total})$$

$$P \quad + \quad P(0.06)(4) = 4340$$

Step 1:	Read the problem.
Step 2:	Label the variables.
Step 3:	Write an equation in words.
	Recall that interest is computed by the formula $I = Prt$.

$$P + 0.24P = 4340$$

Step 4:	Write a mathematical equation.

$$1.24P = 4340$$

Step 5:	Solve the equation.

$$\frac{1.24P}{1.24} = \frac{4340}{1.24}$$

$$P = 3500$$

Skill Practice Answers

6. The interest is $880.

The original investment should be $3500.   **Step 6:** Interpret the results and write the answer in words.

Skill Practice

7. Cassandra invested some money in her bank account, and after 10 years at 4% simple interest, it has grown to $7700. What was the initial amount invested?

Skill Practice Answers

7. The initial investment was $5500.

Section 2.5   Practice Exercises

Study Skills Exercises

1. Go to the online service called MathZone that accompanies this text (www.mathzone.com). Name two features that this online service offers that can help you in this course.

2. Define the key terms:

   a. sales tax   **b. simple interest**

Review Exercises

3. List the six steps to solve an application.

For Exercises 4–5, use the steps for problem solving to solve these applications.

4. Find two consecutive integers such that 3 times the larger is the same as 45 more than the smaller. Find the numbers.

5. The height of the Great Pyramid of Giza is 17 m more than twice the height of the pyramid found in Saqqara. If the difference in their heights is 77 m, find the height of each pyramid.

Concept 1: Solving Basic Percent Equations

 For Exercises 6–17, find the missing values.

6. 45 is what percent of 360?

7. 338 is what percent of 520?

8. 544 is what percent of 640?

9. 576 is what percent of 800?

10. What is 0.5% of 150?

11. What is 9.5% of 616?

12. What is 142% of 740?

13. What is 156% of 280?

14. 177 is 20% of what number?

15. 126 is 15% of what number?

16. 275 is 12.5% of what number?

17. 594 is 45% of what number?

For Exercises 18–21, use the graph showing the distribution for leading forms of cancer in men. (*Source: Centers for Disease Control*)

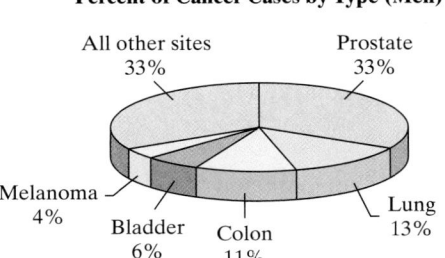

Percent of Cancer Cases by Type (Men)

18. If there are 700,000 cases of cancer in men in the United States, approximately how many are prostate cancer?

19. Approximately how many cases of lung cancer would be expected in 700,000 cancer cases among men in the United States.

20. There were 14,000 cases of cancer of the pancreas diagnosed out of 700,000 cancer cases. What percent is this?

21. There were 21,000 cases of leukemia diagnosed out of 700,000 cancer cases. What percent is this?

22. A Pioneer car CD/MP3 player costs $170. Circuit City has it on sale for 12% off with free installation.

    a. What is the discount on the CD/MP3 player?

    b. What is the sale price?

23. A laptop computer, originally selling for $899.00, is on sale for 10% off.

    a. What is the discount on the laptop?

    b. What is the sale price?

24. A Sony digital camera is on sale for $400.00. This price is 15% off the original price. What was the original price? Round to the nearest cent.

25. The *Star Wars III* DVD is on sale for $18. If this represents an 18% sale price, what was the original price of the DVD?

26. The original price of an Audio Jukebox was $250. It is on sale for $220. What percent discount does this represent?

27. During the holiday season, the Xbox 360 sold for $425.00 in stores. This product was in such demand that it sold for $800 online. What percent markup does this represent? (Round to the nearest whole percent.)

Concept 2: Applications Involving Sales Tax

For Exercises 28–37, solve for the unknown quantity.

28. A Craftsman drill is on sale for $99.99. If the sales tax rate is 7%, how much will Molly have to pay for the drill?

29. Patrick purchased four new tires that were regularly priced at $94.99 each, but are on sale for $20 off per tire. If the sales tax rate is 6%, how much will be charged to Patrick's VISA card?

30. The sales tax for a screwdriver set came to $1.04. If the sales tax rate is 6.5%, what was the price of the screwdriver?

31. The sales tax for a picture frame came to $1.32. If the sales tax rate is 5.5%, what was the price of the picture frame?

32. Sun Lei bought a laptop computer over the Internet for $1800. The total cost, including tax, came to $1890. What is the sales tax rate?

33. Jamie purchased a compact disc and paid $18.26. If the disc price is $16.99, what is the sales tax rate (round to the nearest tenth of a percent)?

34. When the Hendersons went to dinner, their total bill, including tax, was $43.74. If the sales tax rate is 8%, what was the original price of the dinner?

35. The admission to Walt Disney World costs $74.37 including taxes of 11%. What is the original price of a ticket?

36. A hotel room rented for five nights costs $706.25 including 13% in taxes. Find the original price of the room rental for the five nights. Then find the price per night.

37. The price of four CDs is $74.88, including a 4% sales tax. Find the original cost of a single CD assuming all CDs are the same price.

Concept 3: Applications Involving Simple Interest

For Exercises 38–47, solve these equations involving simple interest.

38. How much interest will Pam earn in 4 years if she invests $3000 in an account that pays 3.5% simple interest?

39. How much interest will Roxanne have to pay if she borrows $2000 for 2 years at a simple interest rate of 4%?

40. Bob borrowed some money for 1 year at 5% simple interest. If he had to pay back a total of $1260, how much did he originally borrow?

41. Mike borrowed some money for 2 years at 6% simple interest. If he had to pay back a total of $3640, how much did he originally borrow?

42. If $1500 grows to $1950 after 5 years, find the simple interest rate.

43. If $9000 grows to $10,440 in 2 years, find the simple interest rate.

44. A new bank offered simple interest loans at 11% for new customers. If a customer took out a loan for $2000 to be paid back in 18 months ($\frac{18}{12} = \frac{3}{2}$ years), find

    a. the interest on the loan

    b. the total amount that the customer owes (principal + interest).

45. Rafael has $3000 saved for a future trip to Europe. If he invests in an account that pays 4% simple interest, how much will he have after $2\frac{1}{2}$ years?

46. Perry is planning a vacation to Europe in 2 years. How much should he invest in a certificate of deposit that pays 3% simple interest to get the $3500 that he needs for the trip? Round to the nearest dollar.

47. Sherica invested in a mutual fund and at the end of 20 years she has $14,300 in her account. If the mutual fund returned an average yield of 8%, how much did she originally invest?

Expanding Your Skills

Percents are often used to represent rates. For example, salespeople may receive all or part of their salary as a percentage of their sales. This is called a **commission**. The **commission rate** is the percent of the sales that the salesperson receives in income.

$$\text{commission} = (\text{commission rate}) \cdot (\$ \text{ in merchandise sold})$$

48. The local car dealership pays its sales personnel a commission of 25% of the dealer profit on each car sold. The dealer made a profit of $18,250 on the cars Joëlle sold last month. What was her commission last month?

49. Dan sold a beachfront home for $650,000. If his commission rate is 4%, what did he earn on the sale of that home?

50. A salesperson at You Bought It discount store earns 3% commission on all appliances that he sells. If Geoff's commission for the month was $116.37, how much did he sell?

51. Anna makes a commission at an appliance store. In addition to her base salary, she earns a 2.5% commission on her sales. If Anna's commission for a month was $260, how much did she sell?

52. For selling software, Tom received a bonus commission based on sales over $500. If he received $180 in commission for selling a total of $2300 worth of software, what is his commission rate?

53. In addition to an hourly salary, Jessica earns a commission for selling ice cream bars at the beach. If she sells $708 worth of ice cream and receives a commission of $56.64, what is her commission rate?

54. Diane sells women's sportswear at a department store. She earns a regular salary and, as a bonus, she receives a commission of 4% on all sales over $200. If Diane earned an extra $25.80 last week in commission, how much merchandise did she sell over $200?

Formulas and Applications of Geometry

Section 2.6

1. Formulas

Concepts

1. Formulas
2. Geometry Applications

Literal equations are equations that contain several variables. A formula is a literal equation with a specific application. For example, the perimeter of a triangle (distance around the triangle) can be found by the formula $P = a + b + c$, where a, b, and c are the lengths of the sides (Figure 2-8).

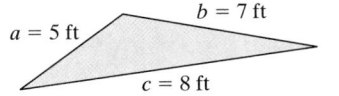

$$P = a + b + c$$
$$= 5 \text{ ft} + 7 \text{ ft} + 8 \text{ ft}$$
$$= 20 \text{ ft}$$

Figure 2-8

In this section, we will learn how to rewrite formulas to solve for a different variable within the formula. Suppose, for example, that the perimeter of a triangle is known and two of the sides are known (say, sides a and b). Then the third side, c, can be found by subtracting the lengths of the known sides from the perimeter (Figure 2-9).

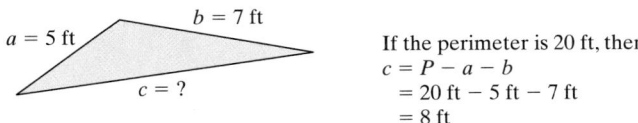

If the perimeter is 20 ft, then
$$c = P - a - b$$
$$= 20 \text{ ft} - 5 \text{ ft} - 7 \text{ ft}$$
$$= 8 \text{ ft}$$

Figure 2-9

To solve a formula for a different variable, we use the same properties of equality outlined in the earlier sections of this chapter. For example, consider the two equations $2x + 3 = 11$ and $wx + y = z$. Suppose we want to solve for x in each case:

$2x + 3 = 11$			$wx + y = z$	
$2x + 3 - 3 = 11 - 3$	Subtract 3.		$wx + y - y = z - y$	Subtract y.
$2x = 8$			$wx = z - y$	
$\dfrac{2x}{2} = \dfrac{8}{2}$	Divide by 2.		$\dfrac{wx}{w} = \dfrac{z - y}{w}$	Divide by w.
$x = 4$			$x = \dfrac{z - y}{w}$	

The equation on the left has only one variable and we are able to simplify the equation to find a numerical value for x. The equation on the right has multiple variables. Because we do not know the values of w, y, and z, we are not able to simplify further. The value of x is left as a formula in terms of w, y, and z.

Example 1   Solving Formulas for an Indicated Variable

Solve the formulas for the indicated variables.

a. $d = rt$ for t **b.** $5x + 2y = 12$ for y

Solution:

TIP: The original equation $d = rt$ represents the distance traveled, d, in terms of the rate of speed, r, and the time of travel, t.

The equation $t = \frac{d}{r}$ represents the same relationship among the variables, however the time of travel is expressed in terms of the distance and rate.

a. $d = rt$ for t The goal is to isolate the variable t.

$\dfrac{d}{r} = \dfrac{rt}{r}$ Because the relationship between r and t is multiplication, we reverse the process by dividing both sides by r.

$\dfrac{d}{r} = t$, or equivalently $t = \dfrac{d}{r}$

b. $5x + 2y = 12$ for y The goal is to solve for y.

$5x - 5x + 2y = 12 - 5x$ Subtract $5x$ from both sides to isolate the y-term.

$2y = -5x + 12$ $-5x + 12$ is the same as $12 - 5x$.

$\dfrac{2y}{2} = \dfrac{-5x + 12}{2}$ Divide both sides by 2 to isolate y.

$y = \dfrac{-5x + 12}{2}$

TIP: In the expression $\dfrac{-5x + 12}{2}$ do not try to divide the 2 into the 12. The divisor of 2 is dividing the entire quantity, $-5x + 12$ (not just the 12).

We may, however, apply the divisor to each term individually in the numerator. That is, $\dfrac{-5x + 12}{2}$ can be written in several different forms. Each is correct.

$$y = \dfrac{-5x + 12}{2} \quad \text{or} \quad y = \dfrac{-5x}{2} + \dfrac{12}{2} \;\Rightarrow\; y = -\dfrac{5x}{2} + 6$$

Skill Practice Solve for the indicated variable.

1. $A = lw$ for l **2.** $-2a + 4b = 7$ for a

Example 2 Solving Formulas for an Indicated Variable

The formula $C = \frac{5}{9}(F - 32)$ is used to find the temperature, C, in degrees Celsius for a given temperature expressed in degrees Fahrenheit, F. Solve the formula $C = \frac{5}{9}(F - 32)$ for F.

Solution:

$$C = \frac{5}{9}(F - 32)$$

$$C = \frac{5}{9}F - \frac{5}{9} \cdot 32 \qquad \text{Clear parentheses.}$$

$$C = \frac{5}{9}F - \frac{160}{9} \qquad \text{Multiply: } \frac{5}{9} \cdot \frac{32}{1} = \frac{160}{9}.$$

$$9(C) = 9\left(\frac{5}{9}F - \frac{160}{9}\right) \qquad \text{Multiply by the LCD to clear fractions.}$$

$$9C = \frac{9}{1} \cdot \frac{5}{9}F - \frac{9}{1} \cdot \frac{160}{9} \qquad \text{Apply the distributive property.}$$

$$9C = 5F - 160 \qquad \text{Simplify.}$$

$$9C + 160 = 5F - 160 + 160 \qquad \text{Add } 160 \text{ to both sides.}$$

$$9C + 160 = 5F$$

$$\frac{9C + 160}{5} = \frac{5F}{5} \qquad \text{Divide both sides by 5.}$$

$$\frac{9C + 160}{5} = F$$

The answer may be written in several forms:

$$F = \frac{9C + 160}{5} \quad \text{or} \quad F = \frac{9C}{5} + \frac{160}{5} \quad \Rightarrow \quad F = \frac{9}{5}C + 32$$

Skill Practice Solve the formula.

3. $M = \frac{1}{2}(a + b)$ for a

---

2. Geometry Applications

In Section R.3, we presented numerous facts and formulas related to geometry. Sometimes these are needed to solve applications in geometry.

Example 3 Solving a Geometry Application Involving Perimeter

The length of a rectangular lot is 1 m less than twice the width. If the perimeter is 190 m, find the length and width.

Skill Practice Answers

1. $l = \dfrac{A}{w}$

2. $a = \dfrac{4b - 7}{2}$ or $a = 2b - \dfrac{7}{2}$

3. $a = 2M - b$

Solution:

Step 1: Read the problem.

Let x represent the width of the rectangle.   **Step 2:** Label the variables.

Then $2x - 1$ represents the length.

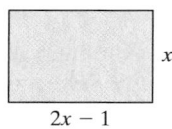

$$P = 2l + 2w$$ **Step 3:** Perimeter formula

$$190 = 2(2x - 1) + 2(x)$$ **Step 4:** Write an equation in terms of x.

$$190 = 4x - 2 + 2x$$ **Step 5:** Solve for x.

$$190 = 6x - 2$$

$$192 = 6x$$

$$\frac{192}{6} = \frac{6x}{6}$$

$$32 = x$$

The width is $x = 32$.

The length is $2x - 1 = 2(32) - 1 = 63$.   **Step 6:** Interpret the results and write the answer in words.

The width of the rectangular lot is 32 m and the length is 63 m.

Skill Practice

4. The length of a rectangle is 10 ft less than twice the width. If the perimeter is 178 ft, find the length and width.

Example 4   **Solving a Geometry Application Involving Complementary Angles**

Two complementary angles are drawn such that one angle is 4° more than seven times the other angle. Find the measure of each angle.

Solution: **Step 1:** Read the problem.

Let x represent the measure of one angle.   **Step 2:** Label the variables.

Then $7x + 4$ represents the measure of the other angle.
The angles are complementary, so their sum must be 90°.

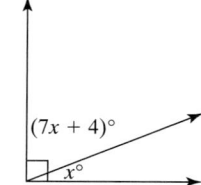

Skill Practice Answers

4. length is 56 ft, width is 33 ft

$$\left(\begin{array}{c}\text{Measure of}\\\text{first angle}\end{array}\right) + \left(\begin{array}{c}\text{measure of}\\\text{second angle}\end{array}\right) = 90°$$

Step 3: Create a verbal equation.

$$x \quad + \quad 7x + 4 \quad = 90$$

Step 4: Write a mathematical equation.

$$8x + 4 = 90$$

Step 5: Solve for x.

$$8x = 86$$

$$\frac{8x}{8} = \frac{86}{8}$$

$$x = 10.75$$

Step 6: Interpret the results and write the answer in words.

One angle is $x = 10.75$.

The other angle is $7x + 4 = 7(10.75) + 4 = 79.25$.

The angles are $10.75°$ and $79.25°$.

Skill Practice

5. Two supplementary angles are constructed so that the measure of one is $120°$ more than twice the other. Find the measures of the angles.

Example 5 Solving a Geometry Application

One angle in a triangle is twice as large as the smallest angle. The third angle is $10°$ more than seven times the smallest angle. Find the measure of each angle.

Solution:

Step 1: Read the problem.

Let x represent the measure of the smallest angle.

Step 2: Label the variables.

Then $2x$ and $7x + 10$ represent the measures of the other two angles.

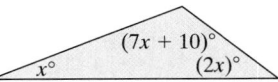

The sum of the angles must be $180°$.

Step 3: Create a verbal equation.

$$x + 2x + (7x + 10) = 180$$

Step 4: Write a mathematical equation.

$$x + 2x + 7x + 10 = 180$$

Step 5: Solve for x.

$$10x + 10 = 180$$

$$10x = 170$$

$$x = 17$$

Skill Practice Answers

5. $20°$ and $160°$

The smallest angle is $x = 17$.

The other angles are $2x = 2(17) = 34$

$$7x + 10 = 7(17) + 10 = 129$$

The angles are 17°, 34°, and 129°.

Step 6: Interpret the results and write the answer in words.

> **Skill Practice**
>
> 6. In a triangle, the measure of the largest angle is 80° greater than the measure of the smallest angle. The measure of the middle angle is twice that of the smallest. Find the measures of the angles.

Example 6 Solving a Geometry Application Involving Circumference

The circumference of a circle is 188.4 ft. Find the radius to the nearest tenth of a foot (Figure 2-10). Use 3.14 for π.

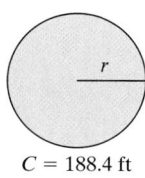

$C = 188.4$ ft

Figure 2-10

Solution:

$C = 2\pi r$ Use the formula for the circumference of a circle.

$188.4 = 2\pi r$ Substitute 188.4 for C.

$\dfrac{188.4}{2\pi} = \dfrac{2\pi r}{2\pi}$ Divide both sides by 2π.

$r = \dfrac{188.4}{2\pi}$

$r \approx \dfrac{188.4}{2(3.14)}$

$= 30.0$

The radius is approximately 30.0 ft.

> **Skill Practice**
>
> 7. The area of a triangle is 52 cm² and the height is 10 cm. Find the measure of the base of the triangle.

Skill Practice Answers

6. 25°, 50°, and 105°
7. The base measures 10.4 cm.

TIP: In Example 6, we could have solved the equation $C = 2\pi r$ for the variable r first before substituting the value of C.

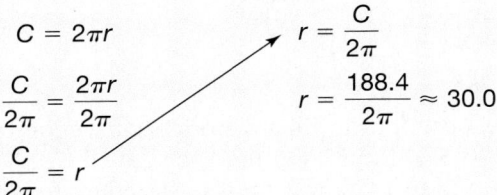

$$C = 2\pi r \qquad\qquad r = \frac{C}{2\pi}$$

$$\frac{C}{2\pi} = \frac{2\pi r}{2\pi} \qquad\qquad r = \frac{188.4}{2\pi} \approx 30.0$$

$$\frac{C}{2\pi} = r$$

Calculator Connections

In Example 6, we could obtain a more accurate result if we use the π key on the calculator.

Note that parentheses are required to divide 188.4 by the quantity 2π. This guarantees that the calculator follows the implied order of operations. Without parentheses, the calculator would divide 188.4 by 2 and then multiply the result by π.

Scientific Calculator

Enter: 1 8 8 . 4 ÷ (2 × π) =

Result: 29.98479128 correct

Enter: 1 8 8 . 4 ÷ 2 × π =

Result: 295.938028 incorrect

Graphing Calculator

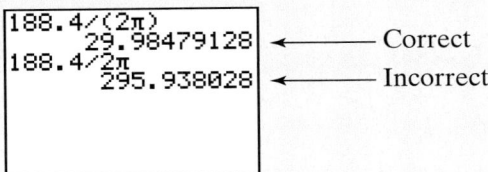

```
188.4/(2π)
          29.98479128
188.4/2π
          295.938028
```
← Correct
← Incorrect

Calculator Exercises

Approximate the expressions with a calculator. Round to three decimal places if necessary.

1. $\dfrac{880}{2\pi}$

2. $\dfrac{1600}{\pi(4)^2}$

3. $\dfrac{20}{(-0.05)(5)}$

4. $\dfrac{10}{0.5(6 + 4)}$

Section 2.6   Practice Exercises

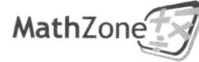
Study Skills Exercises

1. A good technique for studying for a test is to choose four problems from each section of the chapter and write each of them along with the directions on a 3×5 card. On the back, put the page number where you found that problem. Then shuffle the cards and test yourself on the procedure to solve each problem. If you find one that you do not know how to solve, look at the page number and do several of that type. Write four problems you would choose for this section.

2. Define the key term: **literal equation**

Review Exercises

For Exercises 3–7, solve the equation.

3. $3(2y + 3) - 4(-y + 1) = 7y - 10$

4. $-(3w + 4) + 5(w - 2) - 3(6w - 8) = 10$

5. $\dfrac{1}{2}(x - 3) + \dfrac{3}{4} = 3x - \dfrac{3}{4}$

6. $\dfrac{5}{6}x + \dfrac{1}{2} = \dfrac{1}{4}(x - 4)$

7. $0.5(y + 2) - 0.3 = 0.4y + 0.5$

Concept 1: Formulas

For Exercises 8–39, solve for the indicated variable.

8. $P = a + b + c$ for a

9. $P = a + b + c$ for b

10. $x = y - z$ for y

11. $c + d = e$ for d

12. $p = 250 + q$ for q

13. $y = 35 + x$ for x

14. $d = rt$ for t

15. $d = rt$ for r

16. $PV = nrt$ for t

17. $P_1V_1 = P_2V_2$ for V_1

18. $x - y = 5$ for x

19. $x + y = -2$ for y

20. $3x + y = -19$ for y

21. $x - 6y = -10$ for x

22. $2x + 3y = 6$ for y

23. $5x + 2y = 10$ for y

24. $-2x - y = 9$ for x

25. $3x - y = -13$ for x

26. $4x - 3y = 12$ for y

27. $6x - 3y = 4$ for y

28. $ax + by = c$ for y

29. $ax + by = c$ for x

30. $A = P(1 + rt)$ for t

31. $P = 2(L + w)$ for L

32. $a = 2(b + c)$ for c

33. $3(x + y) = z$ for x

34. $Q = \dfrac{x + y}{2}$ for y

35. $Q = \dfrac{a - b}{2}$ for a

36. $M = \dfrac{a}{S}$ for a

37. $A = \dfrac{1}{3}(a + b + c)$ for c

38. $P = I^2R$ for R

39. $F = \dfrac{GMm}{d^2}$ for m

Concept 2: Geometry Applications

For Exercises 40–60, use the problem-solving flowchart (page 142) from Section 2.4.

40. The perimeter of a rectangular garden is 24 ft. The length is 2 ft more than the width. Find the length and the width of the garden.

41. In a small rectangular wallet photo, the width is 7 cm less than the length. If the border (perimeter) of the photo is 34 cm, find the length and width.

42. A builder buys a rectangular lot of land such that the length is 5 m less than two times the width. If the perimeter is 590 m, find the length and the width.

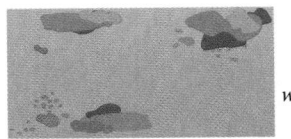

43. The perimeter of a rectangular pool is 140 yd. If the length is 10 yd more than the width, find the length and the width.

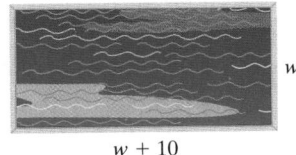

44. A triangular parking lot has two sides that are the same length and the third side is 5 m longer. If the perimeter is 71 m, find the lengths of the sides.

45. The perimeter of a triangle is 16 ft. One side is 3 ft longer than the shortest side. The third side is 1 ft longer than the shortest side. Find the lengths of all the sides.

46. The largest angle in a triangle is three times the smallest angle. The middle angle is two times the smallest angle. Given that the sum of the angles in a triangle is 180°, find the measure of each angle.

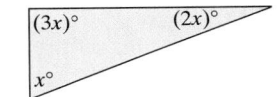

47. The smallest angle in a triangle measures 90° less than the largest angle. The middle angle measures 60° less than the largest angle. Find the measure of each angle.

48. The smallest angle in a triangle is half the largest angle. The middle angle measures 30° less than the largest angle. Find the measure of each angle.

49. The largest angle of a triangle is three times the middle angle. The smallest angle measures 10° less than the middle angle. Find the measure of each angle.

50. Find the value of x and the measure of each angle labeled in the figure.

51. Find the value of y and the measure of each angle labeled in the figure.

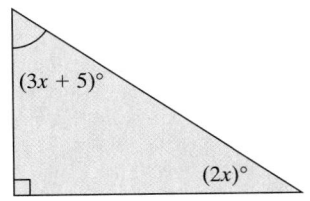

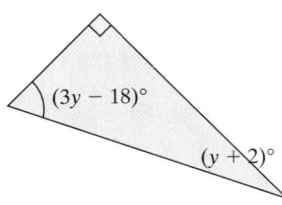

52. Sometimes memory devices are helpful for remembering mathematical facts. Recall that the sum of two complementary angles is 90°. That is, two complementary angles when added together form a right triangle or "corner." The words *Complementary* and *Corner* both start with the letter "*C*." Derive your own memory device for remembering that the sum of two supplementary angles is 180°.

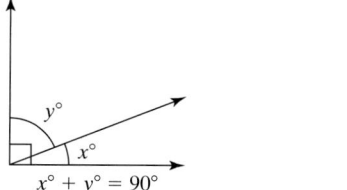

Complementary angles form a "Corner"   Supplementary angles . . .

53. Two angles are complementary. One angle is 20° less than the other angle. Find the measures of the angles.

54. Two angles are complementary. One angle is twice as large as the other angle. Find the measures of the angles.

55. Two angles are complementary. One angle is 4° less than three times the other angle. Find the measures of the angles.

56. Two angles are supplementary. One angle is three times as large as the other angle. Find the measures of the angles.

57. Two angles are supplementary. One angle is twice as large as the other angle. Find the measures of the angles.

58. Two angles are supplementary. One angle is 6° more than four times the other. Find the measures of the two angles.

59. Find the measures of the vertical angles labeled in the figure by first solving for x.

60. Find the measures of the vertical angles labeled in the figure by first solving for y.

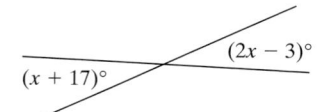

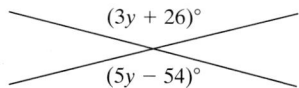

61. a. A rectangle has length l and width w. Write a formula for the area.

   b. Solve the formula for the width, w.

   c. The area of a rectangular volleyball court is 1740.5 ft² and the length is 59 ft. Find the width.

62. a. A parallelogram has height h and base b. Write a formula for the area.

   b. Solve the formula for the base, b.

   c. Find the base of the parallelogram pictured if the area is 40 m².

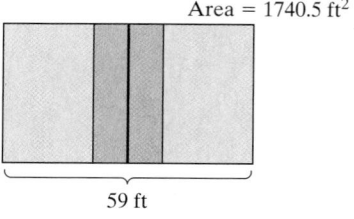

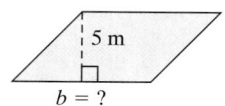

63. a. A rectangle has length *l* and width *w*. Write a formula for the perimeter.

b. Solve the formula for the length, *l*.

c. The perimeter of the soccer field at Giants Stadium is 338 m. If the width is 66 m, find the length.

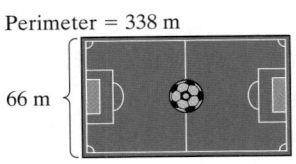

Perimeter = 338 m

66 m

64. a. The length of each side of a square is *s*. Write a formula for the perimeter of the square.

b. Solve the formula for the length of a side, *s*.

c. The Pyramid of Khufu (known as the Great Pyramid) at Giza has a square base. If the distance around the bottom is 921.6 m, find the length of the sides at the bottom of the pyramid.

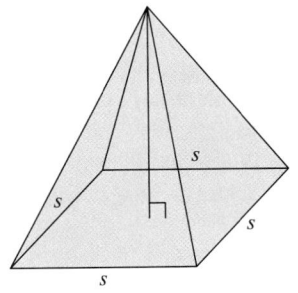

65. a. A triangle has height *h* and base *b*. Write a formula for the area.

b. Solve the formula for the height, *h*.

c. Find the height of the triangle pictured if the area is 12 km².

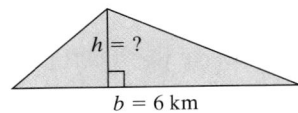

$h = ?$

$b = 6$ km

66. a. A circle has a radius of *r*. Write a formula for the circumference.

b. Solve the formula for the radius, *r*.

c. The circumference of the circular Buckingham Fountain in Chicago is approximately 880 ft. Find the radius. Round to the nearest foot.

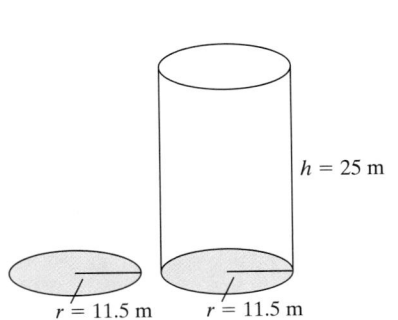

Expanding Your Skills

For Exercises 67–68, find the indicated area or volume. Be sure to include the proper units and round each answer to two decimal places if necessary.

67. a. Find the area of a circle with radius 11.5 m.

b. Find the volume of a right circular cylinder with radius 11.5 m and height 25 m.

68. a. Find the area of a parallelogram with base 30 in. and height 12 in.

b. Find the area of a triangle with base 30 in. and height 12 in.

c. Compare the areas found in parts (a) and (b).

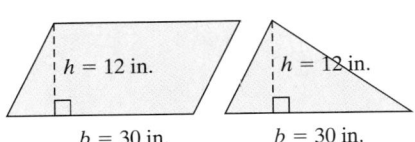

$h = 12$ in. $h = 12$ in.

$b = 30$ in. $b = 30$ in.

$h = 25$ m

$r = 11.5$ m $r = 11.5$ m

Linear Inequalities

1. Graphing Linear Inequalities

Recall that $a < b$ (equivalently $b > a$) means that a lies to the left of b on the number line. The statement $a > b$ (equivalently $b < a$) means that a lies to the right of b on the number line. If $a = b$, then a and b are represented by the same point on the number line.

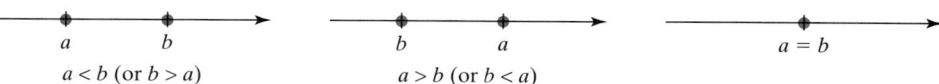

$a < b$ (or $b > a$) $a > b$ (or $b < a$)

A Linear Inequality in One Variable

A **linear inequality in one variable**, x, is defined as any relationship that can be written in one of the following forms:

$$ax + b < 0, \quad ax + b \le 0, \quad ax + b > 0, \quad \text{or} \quad ax + b \ge 0, \text{ where } a \ne 0.$$

The following inequalities are linear equalities in one variable.

$$2x - 3 < 0 \qquad -4z - 3 > 0 \qquad a \le 4 \qquad 5.2y \ge 10.4$$

The number line is a tool used to visualize the solution set of an equation or inequality. For example, the solution set to the equation $x = 2$ is {2} and may be graphed as a single point on the number line.

$x = 2$ $\begin{array}{c} \longleftarrow\!\!+\!\!+\!\!+\!\!+\!\!+\!\!+\!\!+\!\!+\!\bullet\!\!+\!\!+\!\!+\!\!+\!\!\longrightarrow \\ -6\ -5\ -4\ -3\ -2\ -1\ \ 0\ \ 1\ \ 2\ \ 3\ \ 4\ \ 5\ \ 6 \end{array}$

The solution set to an inequality is the set of real numbers that make the inequality a true statement. For example, the solution set to the inequality $x \ge 2$ is all real numbers 2 or greater. Because the solution set has an infinite number of values, we cannot list all of the individual solutions. However, we can graph the solution set on the number line.

$x \ge 2$ $\begin{array}{c} \longleftarrow\!\!+\!\!+\!\!+\!\!+\!\!+\!\!+\!\!+\!\!+\![\!\!=\!\!=\!\!=\!\!=\!\!\longrightarrow \\ -6\ -5\ -4\ -3\ -2\ -1\ \ 0\ \ 1\ \ 2\ \ 3\ \ 4\ \ 5\ \ 6 \end{array}$

The square bracket symbol, [, is used on the graph to indicate that the point $x = 2$ is included in the solution set. By convention, square brackets, either [or], are used to *include* a point on a graph. Parentheses, (or), are used to *exclude* a point on a graph.

The solution set of the inequality $x > 2$ includes the real numbers greater than 2 but not including 2. Therefore, a (symbol is used on the graph to indicate that $x = 2$ is not included.

$x > 2$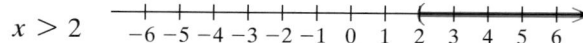

Example 1 | Graphing Linear Inequalities

Graph the solution sets.

a. $x > -1$ **b.** $c \le \dfrac{7}{3}$ **c.** $3 > y$

Solution:

a. $x > -1$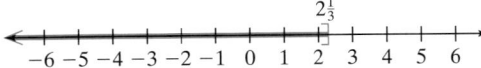

The solution set is the set of real numbers strictly greater than -1. Therefore, we graph the region on the number line to the right of -1. Because $x = -1$ is not included in the solution set, we use the (symbol at $x = -1$.

b. $c \le \frac{7}{3}$ is equivalent to $c \le 2\frac{1}{3}$.

The solution set is the set of real numbers less than or equal to $2\frac{1}{3}$. Therefore, graph the region on the number line to the left of and including $2\frac{1}{3}$. Use the symbol] to indicate that $c = 2\frac{1}{3}$ is included in the solution set.

c. $3 > y$ This inequality reads "3 is greater than y." This is equivalent to saying, "y is less than 3."

$y < 3$

The solution set is the set of real numbers less than 3. Therefore, graph the region on the number line to the left of 3. Use the symbol) to denote that the endpoint, 3, is not included in the solution.

Skill Practice | Graph the solution sets.

1. $y < 0$ **2.** $x \ge -\dfrac{5}{4}$ **3.** $5 \ge a$

TIP: Some textbooks use a closed circle or an open circle (● or ○) rather than a bracket or parenthesis to denote inclusion or exclusion of a value on the real number line. For example, the solution sets for the inequalities $x > -1$ and $c \le \frac{7}{3}$ are graphed here.

$x > -1$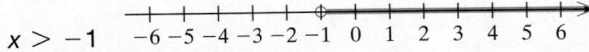

$c \le \frac{7}{3}$

A statement that involves more than one inequality is called a **compound inequality**. One type of compound inequality is used to indicate that one number is between two others. For example, the inequality $-2 < x < 5$ means that $-2 < x$ and $x < 5$. In words, this is easiest to understand if we read the variable first: x is greater than -2 and x is less than 5. The numbers satisfied by these two conditions are those between -2 and 5.

Skill Practice Answers

1.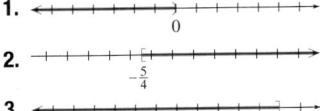

2.

3.

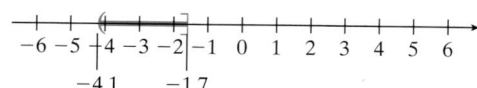

Example 2   Graphing a Compound Inequality

Graph the solution set of the inequality: $-4.1 < y \le -1.7$

Solution:

$-4.1 < y \le -1.7$ means that

$-4.1 < y$ and $y \le -1.7$

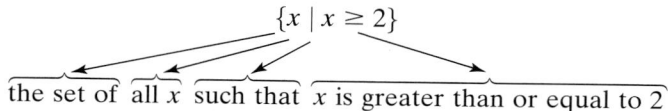

Shade the region of the number line greater than -4.1 and less than or equal to -1.7.

Skill Practice   Graph the solution set.

4. $0 \le y \le 8.5$

2. Set-Builder Notation and Interval Notation

Graphing the solution set to an inequality is one way to define the set. Two other methods are to use **set-builder notation** or **interval notation**.

Set-Builder Notation

The solution to the inequality $x \ge 2$ can be expressed in set-builder notation as follows:

$$\{x \mid x \ge 2\}$$

the set of   all x such that x is greater than or equal to 2

Interval Notation

To understand interval notation, first think of a number line extending infinitely far to the right and infinitely far to the left. Sometimes we use the infinity symbol, ∞, or negative infinity symbol, $-\infty$, to label the far right and far left ends of the number line (Figure 2-11).

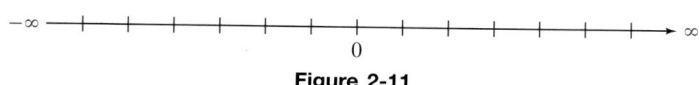

Figure 2-11

To express the solution set of an inequality in interval notation, sketch the graph first. Then use the endpoints to define the interval.

Inequality	**Graph**	**Interval Notation**
$x \ge 2$		$[2, \infty)$

$[2$, $\infty)$

The graph of the solution set $x \ge 2$ begins at 2 and extends infinitely far to the right. The corresponding interval notation begins at 2 and extends to ∞. Notice that a square bracket [is used at 2 for both the graph and the interval notation. A parenthesis is always used at ∞ and for $-\infty$, because there is no endpoint.

Skill Practice Answers

4.

Using Interval Notation

- The endpoints used in interval notation are always written from left to right. That is, the smaller number is written first, followed by a comma, followed by the larger number.
- A parenthesis, (or), indicates that an endpoint is excluded from the set.
- A square bracket, [or], indicates that an endpoint is included in the set.
- Parentheses, (and), are always used with $-\infty$ and ∞, respectively.

Example 3 | **Using Set-Builder Notation and Interval Notation**

Complete the chart.

Set-Builder Notation	Graph	Interval Notation
	$\leftarrow\!+\!+\!+\!)\!+\!+\!+\!+\!+\!+\!+\!+\!\rightarrow$ $-6\ -5\ -4\ -3\ -2\ -1\ \ 0\ \ 1\ \ 2\ \ 3\ \ 4\ \ 5\ \ 6$	
		$[-\frac{1}{2}, \infty)$
$\{y \mid -2 \le y < 4\}$		

Solution:

Set-Builder Notation	Graph	Interval Notation
$\{x \mid x < -3\}$	$\leftarrow\!+\!+\!+\!)\!+\!+\!+\!+\!+\!+\!+\!+\!\rightarrow$ $-6\ -5\ -4\ -3\ -2\ -1\ \ 0\ \ 1\ \ 2\ \ 3\ \ 4\ \ 5\ \ 6$	$(-\infty, -3)$
$\{x \mid x \ge -\frac{1}{2}\}$	$\leftarrow\!+\!+\!+\!+\!+\!+\![\!+\!+\!+\!+\!+\!+\!\rightarrow$ $-6\ -5\ -4\ -3\ -2\ -1\ 0\ \ 1\ \ 2\ \ 3\ \ 4\ \ 5\ \ 6$ $-\frac{1}{2}$	$[-\frac{1}{2}, \infty)$
$\{y \mid -2 \le y < 4\}$	$\leftarrow\!+\!+\!+\![\!+\!+\!+\!+\!)\!+\!+\!\rightarrow$ $-6\ -5\ -4\ -3\ -2\ -1\ 0\ \ 1\ \ 2\ \ 3\ \ 4\ \ 5\ \ 6$	$[-2, 4)$

Skill Practice | Graph the sets defined in Exercises 5–7, and give the set-builder notation and interval notation.

5. $\{x \mid x \ge -2\}$ **6.** $(-3, 1]$ **7.** $\leftarrow\!+\!+\!+\!+\!+\!+\!+\!+\!+\!)\!+\!+\!+\!+\!\rightarrow$
 2.1

---

3. Addition and Subtraction Properties of Inequality

Solving linear inequalities is similar to solving linear equations. Recall that adding or subtracting the same quantity to both sides of an equation results in an equivalent equation. The addition and subtraction properties of inequality state that the same is true for an inequality.

Skill Practice Answers

5. $[-2, \infty)$;
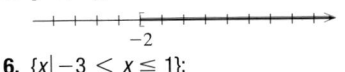

6. $\{x \mid -3 < x \le 1\}$;
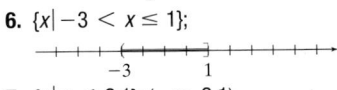

7. $\{x \mid x < 2.1\}$; $(-\infty, 2.1)$

Addition and Subtraction Properties of Inequality

Let a, b, and c represent real numbers.

1. ***Addition Property of Inequality:**   If $a < b$,

   then $a + c < b + c$

2. ***Subtraction Property of Inequality:**   If $a < b$,

   then $a - c < b - c$

*These properties may also be stated for $a \leq b, a > b$, and $a \geq b$.

To illustrate the addition property of inequality, consider the inequality $5 > 3$. If we add a real number such as 4 to both sides, the left-hand side will still be greater than the right-hand side (Figure 2-12).

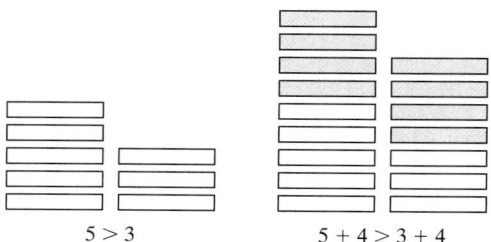

$5 > 3$ $5 + 4 > 3 + 4$

Figure 2-12

Example 4   Solving a Linear Inequality

Solve the inequality and graph the solution set. Express the solution set in set-builder notation and in interval notation.

$$-2p + 5 < -3p + 6$$

Solution:

$$-2p + 5 < -3p + 6$$

$$-2p + 3p + 5 < -3p + 3p + 6 \qquad \text{Addition property of inequality}$$
$$\text{(add } 3p \text{ to both sides)}$$

$$p + 5 < 6 \qquad\qquad\qquad \text{Simplify.}$$

$$p + 5 - 5 < 6 - 5 \qquad\qquad \text{Subtraction property of inequality}$$

$$p < 1$$

Graph:

$$\xleftarrow{\hspace{2cm}}\overset{\quad -6\;-5\;-4\;-3\;-2\;-1\;\;0\;\;1\;\;\;2\;\;\;3\;\;\;4\;\;\;5\;\;\;6}{\rule{6cm}{0.4pt}}$$

Set-builder notation: $\{p \,|\, p < 1\}$

Interval notation: $(-\infty, 1)$

Skill Practice   Solve the inequality. Graph the solution set and express in interval notation.

8. $2y - 5 < y - 11$

Skill Practice Answers

8.

$\xleftarrow{\hspace{1.5cm}}$
$\qquad -6$

$(-\infty, -6)$

TIP: The solution to an inequality gives a set of values that make the original inequality true. Therefore, you can test your final answer by using *test points*. That is, pick a value in the proposed solution set and verify that it makes the original inequality true. Furthermore, any test point picked outside the solution set should make the original inequality false. For example,

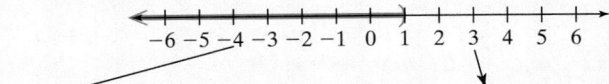

Pick $p = -4$ as an arbitrary test point within the proposed solution set.

Pick $p = 3$ as an arbitrary test point outside the proposed solution set.

$$-2p + 5 < -3p + 6$$
$$-2(-4) + 5 \overset{?}{<} -3(-4) + 6$$
$$8 + 5 \overset{?}{<} 12 + 6$$
$$13 < 18 \; \checkmark \quad \text{True}$$

$$-2p + 5 < -3p + 6$$
$$-2(3) + 5 \overset{?}{<} -3(3) + 6$$
$$-6 + 5 \overset{?}{<} -9 + 6$$
$$-1 < -3 \quad \text{False}$$

4. Multiplication and Division Properties of Inequality

Multiplying both sides of an equation by the same quantity results in an equivalent equation. However, the same is not always true for an inequality. If you multiply or divide an inequality by a negative quantity, the direction of the inequality symbol must be reversed.

For example, consider multiplying or dividing the inequality, $4 < 5$ by -1.

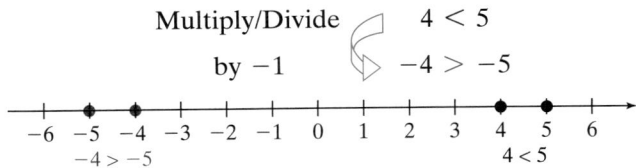

Multiply/Divide $4 < 5$
by -1 $-4 > -5$

$-4 > -5$ $4 < 5$

Figure 2-13

The number 4 lies to the left of 5 on the number line. However, -4 lies to the right of -5 (Figure 2-13). Changing the sign of two numbers changes their relative position on the number line. This is stated formally in the multiplication and division properties of inequality.

Multiplication and Division Properties of Inequality

Let a, b, and c represent real numbers. Then

*If c is positive and $a < b$, then $ac < bc$ and $\dfrac{a}{c} < \dfrac{b}{c}$.

*If c is negative and $a < b$, then $ac > bc$ and $\dfrac{a}{c} > \dfrac{b}{c}$.

The second statement indicates that if both sides of an inequality are multiplied or divided by a negative quantity, the inequality sign must be reversed.

*These properties may also be stated for $a \leq b, a > b$, and $a \geq b$.

Example 5	**Solving a Linear Inequality**

Solve the inequality $-5x - 3 \leq 12$. Graph the solution set and write the answer in interval notation.

Solution:

$$-5x - 3 \leq 12$$

$$-5x - 3 + 3 \leq 12 + 3 \qquad \text{Add 3 to both sides.}$$

$$-5x \leq 15$$

$$\frac{-5x}{-5} \geq \frac{15}{-5} \qquad \text{Divide by } -5. \text{ Reverse the direction of the inequality sign.}$$

$$x \geq -3$$

Interval notation: $[-3, \infty)$

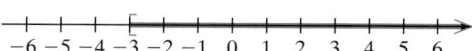

TIP: The inequality $-5x - 3 \leq 12$, could have been solved by isolating x on the right-hand side of the inequality. This would create a positive coefficient on the variable term and eliminate the need to divide by a negative number.

$$-5x - 3 \leq 12$$

$$-3 \leq 5x + 12$$

$$-15 \leq 5x \qquad \text{Notice that the coefficient of } x \text{ is positive.}$$

$$\frac{-15}{5} \leq \frac{5x}{5} \qquad \text{Do not reverse the inequality sign because we are dividing by a positive number.}$$

$$-3 \leq x, \text{ or equivalently, } x \geq -3$$

Skill Practice	Solve.

9. $-5p + 2 > 22$

Example 6	**Solving a Linear Inequality**

Solve the inequality. Graph the solution set and write the answer in interval notation.

$$1.4x + 4.5 - 0.2x < -0.3$$

Solution:

$$1.4x + 4.5 - 0.2x < -0.3$$

$$1.2x + 4.5 < -0.3 \qquad \text{Combine } like \text{ terms.}$$

$$1.2x + 4.5 - 4.5 < -0.3 - 4.5 \qquad \text{Subtract 4.5 from both sides.}$$

$$1.2x < -4.8 \qquad \text{Simplify.}$$

$$\frac{1.2x}{1.2} < \frac{-4.8}{1.2} \qquad \text{Divide by 1.2. The direction of the inequality sign is } not \text{ reversed because we divided by a positive number.}$$

$$x < -4$$

Interval notation: $(-\infty, -4)$

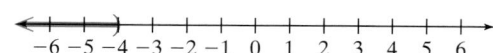

Skill Practice Answers

9.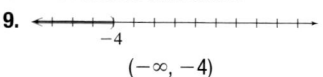

Skill Practice Solve.

10. $0.1x + 4.2 < 1.1x - 0.8$

Example 7 Solving Linear Inequalities

Solve the inequality $-\frac{1}{4}k + \frac{1}{6} \le 2 + \frac{2}{3}k$. Graph the solution set and write the answer in interval notation.

Solution:

$$-\frac{1}{4}k + \frac{1}{6} \le 2 + \frac{2}{3}k$$

$$12\left(-\frac{1}{4}k + \frac{1}{6}\right) \le 12\left(2 + \frac{2}{3}k\right)$$ Multiply both sides by 12 to clear fractions. (Because we multiplied by a positive number, the inequality sign is *not* reversed.)

$$\frac{12}{1}\left(-\frac{1}{4}k\right) + \frac{12}{1}\left(\frac{1}{6}\right) \le 12(2) + \frac{12}{1}\left(\frac{2}{3}k\right)$$ Apply the distributive property.

$$-3k + 2 \le 24 + 8k$$ Simplify.

$$-3k - 8k + 2 \le 24 + 8k - 8k$$ Subtract $8k$ from both sides.

$$-11k + 2 \le 24$$

$$-11k + 2 - 2 \le 24 - 2$$ Subtract 2 from both sides.

$$-11k \le 22$$

$$\frac{-11k}{-11} \ge \frac{22}{-11}$$ Divide both sides by -11. Reverse the inequality sign.

$$k \ge -2$$

Graph: —+—+—+—[—+—+—+—+—+—→
    -4 -3 -2 -1 0 1 2 3 4

Interval notation: $[-2, \infty)$

Skill Practice Solve.

11. $\frac{1}{5}t + 7 \le \frac{1}{2}t - 2$

---

5. Solving Inequalities of the Form $a < x < b$

To solve a compound inequality of the form $a < x < b$ we can work with the inequality as a three-part inequality and isolate the variable, x, as demonstrated in the next example.

 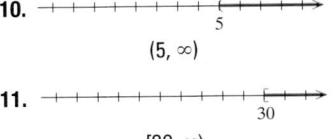

Example 8 Solving a Compound Inequality of the Form $a < x < b$

Solve the inequality: $-3 \leq 2x + 1 < 7$. Graph the solution and write the answer in interval notation.

Solution:

To solve the compound inequality $-3 \leq 2x + 1 < 7$ isolate the variable x in the middle. The operations performed on the middle portion of the inequality must also be performed on the left-hand side and right-hand side.

$$-3 \leq 2x + 1 < 7$$

$$-3 - 1 \leq 2x + 1 - 1 < 7 - 1 \qquad \text{Subtract 1 from all three parts of the inequality.}$$

$$-4 \leq 2x < 6 \qquad \text{Simplify.}$$

$$\frac{-4}{2} \leq \frac{2x}{2} < \frac{6}{2} \qquad \text{Divide by 2 in all three parts of the inequality.}$$

$$-2 \leq x < 3$$

Graph:

```
   ├──┼──┼──┼──┼──[──┼──┼──┼──┼──┼──┼──┼──>
  -6 -5 -4 -3 -2 -1  0  1  2  3  4  5  6
```

Interval notation: $[-2, 3)$

Skill Practice Solve.

12. $-3 \leq -5 + 2y < 11$

6. Applications of Linear Inequalities

Table 2-1 provides several commonly used translations to express inequalities.

Table 2-1

English Phrase	Mathematical Inequality
a is less than b	$a < b$
a is greater than b a exceeds b	$a > b$
a is less than or equal to b a is at most b a is no more than b	$a \leq b$
a is greater than or equal to b a is at least b a is no less than b	$a \geq b$

Example 9 Translating Expressions Involving Inequalities

Translate the English phrases into mathematical inequalities:

a. Claude's annual salary, s, is no more than $40,000.

b. A citizen must be at least 18 years old to vote. (Let a represent a citizen's age.)

c. An amusement park ride has a height requirement between 48 in. and 70 in. (Let h represent height in inches.)

Skill Practice Answers

12.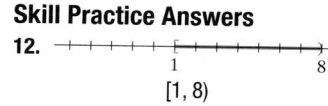
$[1, 8)$

Solution:

a. $s \leq 40{,}000$ Claude's annual salary, s, is no more than $40,000.

b. $a \geq 18$ A citizen must be at least 18 years old to vote.

c. $48 < h < 70$ An amusement park ride has a height requirement between 48 in. and 70 in.

Skill Practice   Translate the English phrase into a mathematical inequality.

13. Bill needs a score of at least 92 on the final exam. Let x represent Bill's score.

14. Fewer than 19 cars are in the parking lot. Let c represent the number of cars.

15. The heights, h, of women who wear petite size clothing are typically between 58 in. and 63 in., inclusive.

Linear inequalities are found in a variety of applications. See how the next example can help you determine the minimum grade you need on an exam to get an A in your math course.

Example 10   Solving an Application with Linear Inequalities

To earn an A in a math class, Alsha must average at least 90 on all of her tests. Suppose Alsha has scored 79, 86, 93, 90, and 95 on her first five math tests. Determine the minimum score she needs on her sixth test to get an A in the class.

Solution:

Let x represent the score on the sixth exam.   Label the variable.

$$\left(\begin{array}{c} \text{Average of} \\ \text{all tests} \end{array} \right) \geq 90$$ Create a verbal model.

$$\frac{79 + 86 + 93 + 90 + 95 + x}{6} \geq 90$$ The average score is found by taking the sum of the test scores and dividing by the number of scores.

$$\frac{443 + x}{6} \geq 90$$ Simplify.

$$6\left(\frac{443 + x}{6} \right) \geq (90)6$$ Multiply both sides by 6 to clear fractions.

$$443 + x \geq 540$$ Solve the inequality.

$$x \geq 97$$ Interpret the results.

Alsha must score at least 97 on her sixth exam to receive an A in the course.

Skill Practice

16. To get at least a B in math, Simon must average 80 on all tests. Suppose Simon has scored 60, 72, 98, and 85 on the first four tests. What score does he need on the fifth test to receive a B?

Skill Practice Answers

13. $x \geq 92$

14. $c < 19$

15. $58 \leq h \leq 63$

16. Simon needs at least 85.

Section 2.7   Practice Exercises

Study Skills Exercises

1. Find the page numbers for the Chapter Review Exercises, the Chapter Test, and the Cumulative Review Exercises for this chapter.

   Chapter Review Exercises _____   Chapter Test _____

   Cumulative Review Exercises _____

   Compare these features and state the advantages of each.

2. Define the key terms:

   a. linear inequality in one variable   **b. compound inequality**

   c. set-builder notation   **d. interval notation**

Review Problems

3. Solve the equation: $3(x + 2) - (2x - 7) = -(5x - 1) - 2(x + 6)$

4. Solve the equation: $6 - 8(x + 3) + 5x = 5x - (2x - 5) + 13$

Concepts 1–2: Graphing Linear Inequalities; Set-Builder Notation and Interval Notation

For Exercises 5–10, graph each inequality and write the solution set in interval notation.

Set-Builder Notation	Graph	Interval Notation
5. $\{x \mid x \geq 6\}$		
6. $\left\{x \mid \frac{1}{2} < x \leq 4\right\}$		
7. $\{x \mid x \leq 2.1\}$		
8. $\left\{x \mid x > \frac{7}{3}\right\}$		
9. $\{x \mid -2 < x \leq 7\}$		
10. $\{x \mid x < -5\}$		

For Exercises 11–16, write each set in set-builder notation and in interval notation.

Set-Builder Notation	Graph	Interval Notation
11.	$\frac{3}{4}$	
12.	-0.3	
13.	$-1 \quad 8$	
14.	0	
15.	-14	
16.	$0 \quad 9$	

For Exercises 17–22, graph each set and write the set in set-builder notation.

Set-Builder Notation	Graph	Interval Notation
17.		$[18, \infty)$
18.		$[-10, -2]$
19.		$(-\infty, -0.6)$
20.		$\left(-\infty, \dfrac{5}{3}\right)$
21.		$[-3.5, 7.1)$
22.		$[-10, \infty)$

Concepts 3–4: Properties of Inequality

For Exercises 23–30, solve the equation in part (a). For part (b), solve the inequality and graph the solution set.

23. a. $x + 3 = 6$ **24. a.** $y - 6 = 12$ **25. a.** $p - 4 = 9$ **26. a.** $k + 8 = 10$
 b. $x + 3 > 6$ **b.** $y - 6 \geq 12$ **b.** $p - 4 \leq 9$ **b.** $k + 8 < 10$

27. a. $4c = -12$ **28. a.** $5d = -35$ **29. a.** $-10z = 15$ **30. a.** $-2w = 14$
 b. $4c < -12$ **b.** $5d > -35$ **b.** $-10z \leq 15$ **b.** $-2w < 14$

Concept 5: Solving Inequalities of the Form $a < x < b$

For Exercises 31–36, graph the solution.

31. $-1 < y \leq 4$ **32.** $2.5 \leq t < 5.7$ **33.** $0 < x + 3 < 8$

34. $-2 \leq x - 4 \leq 3$ **35.** $8 \leq 4x \leq 24$ **36.** $-9 < 3x < 12$

Mixed Exercises

For Exercises 37–82, solve the inequality. Graph the solution set and write the set in interval notation.

37. $x + 5 \leq 6$ **38.** $y - 7 < 6$ **39.** $q - 7 > 3$

40. $r + 4 \geq -1$ **41.** $4 < 1 + z$ **42.** $3 > z - 6$

43. $2 \geq a - 6$ **44.** $7 \leq b + 12$ **45.** $3c > 6$

46. $4d \leq 12$

47. $-3c > 6$

48. $-4d \leq 12$

49. $-h \leq -14$

50. $-q > -7$

51. $12 \geq -\dfrac{x}{2}$

52. $6 < -\dfrac{m}{3}$

53. $-2 \leq p + 1 < 4$

54. $0 < k + 7 < 6$

55. $-3 < 6h - 3 < 12$

56. $-6 \leq 4a - 2 \leq 12$

57. $5 < \dfrac{1}{2}x < 6$

58. $-6 \leq 3x \leq 12$

59. $-5 \leq 4x - 1 < 15$

60. $-2 < \dfrac{1}{3}x - 2 \leq 2$

61. $0.6z \geq 54$

62. $-0.7w > 28$

63. $-\dfrac{2}{3}y < 6$

64. $\dfrac{3}{4}x \leq -12$

65. $-2x - 4 \leq 11$

66. $-3x + 1 > 0$

67. $-12 > 7x + 9$

68. $8 < 2x - 10$

69. $-7b - 3 \leq 2b$

70. $3t \geq 7t - 35$

71. $4n + 2 < 6n + 8$

72. $2w - 1 \leq 5w + 8$

73. $8 - 6(x - 3) > -4x + 12$

74. $3 - 4(h - 2) > -5h + 6$

75. $3(x + 1) - 2 \leq \dfrac{1}{2}(4x - 8)$

76. $8 - (2x - 5) \geq \dfrac{1}{3}(9x - 6)$

77. $\dfrac{7}{6}p + \dfrac{4}{3} \geq \dfrac{11}{6}p - \dfrac{7}{6}$

78. $\dfrac{1}{3}w - \dfrac{1}{2} \leq \dfrac{5}{6}w + \dfrac{1}{2}$

79. $\dfrac{1}{2}y - \dfrac{1}{4} > \dfrac{3}{4}y + 2$

80. $\dfrac{2}{5}t + \dfrac{1}{2} < \dfrac{1}{10}t - 1$

81. $-1.2a - 0.4 < -0.4a + 2$

82. $-0.4c + 1.2 > -2c - 0.4$

For Exercises 83–86, determine whether the given number is a solution to the inequality.

83. $-2x + 5 < 4$ $x = -2$

84. $-3y - 7 > 5$ $y = 6$

85. $4(p + 7) - 1 > 2 + p$ $p = 1$

86. $3 - k < 2(-1 + k)$ $k = 4$

Concept 6: Applications of Linear Inequalities

87. Let x represent a student's average in a math class. The grading scale is given here.

A	$93 \leq x \leq 100$
B+	$89 \leq x < 93$
B	$84 \leq x < 89$
C+	$80 \leq x < 84$
C	$75 \leq x < 80$
F	$0 \leq x < 75$

a. Write the range of scores corresponding to each letter grade in interval notation.

b. If Stephan's average is 84.01, what grade will he receive?

c. If Estella's average is 79.89, what grade will she receive?

88. Let x represent a student's average in a science class. The grading scale is given here.

A	$90 \leq x \leq 100$
B+	$86 \leq x < 90$
B	$80 \leq x < 86$
C+	$76 \leq x < 80$
C	$70 \leq x < 76$
D+	$66 \leq x < 70$
D	$60 \leq x < 66$
F	$0 \leq x < 60$

a. Write the range of scores corresponding to each letter grade in interval notation.

b. If Jacque's average is 89.99, what is her grade?

c. If Marc's average is 66.01, what is his grade?

For Exercises 89–99, translate the English phrase into a mathematical inequality.

89. The length of a fish, L, was at least 10 in.

90. Tasha's average test score, t, exceeded 90.

91. The wind speed, w, exceeded 75 mph.

92. The height of a cave, h, was no more than 2 ft.

93. The temperature of the water in Blue Spring, t, is no more than 72°F.

94. The temperature on the tennis court, t, was no less than 100°F.

95. The length of the hike, L, was no less than 8 km.

96. The depth, d, of a certain pool was at most 10 ft.

97. The amount of rain, a, in a recent storm was at most 2 in.

98. The snowfall, h, in Monroe County is between 2 inches and 5 inches.

99. The cost, c, of carpeting a room is between $300 and $400.

100. The average summer rainfall for Miami, Florida, for June, July, and August is 7.4 in. per month. If Miami receives 5.9 in. of rain in June and 6.1 in. in July, how much rain is required in August to exceed the 3-month summer average?

101. The average winter snowfall for Burlington, Vermont, for December, January, and February is 18.7 in. per month. If Burlington receives 22 in. of snow in December and 24 in. in January, how much snow is required in February to exceed the 3-month winter average?

102. An artist paints wooden birdhouses. She buys the birdhouses for $9 each. However, for large orders, the price per birdhouse is discounted by a percentage off the original price. Let x represent the number of birdhouses ordered. The corresponding discount is given in the table.

Size of Order	Discount
$x \leq 49$	0%
$50 \leq x \leq 99$	5%
$100 \leq x \leq 199$	10%
$x \geq 200$	20%

  a. If the artist places an order for 190 birdhouses, compute the total cost.

  b. Which costs more: 190 birdhouses or 200 birdhouses? Explain your answer.

103. A wholesaler sells T-shirts to a surf shop at $8 per shirt. However, for large orders, the price per shirt is discounted by a percentage off the original price. Let x represent the number of shirts ordered. The corresponding discount is given in the table.

Number of Shirts Ordered	Discount
$x \leq 24$	0%
$25 \leq x \leq 49$	2%
$50 \leq x \leq 99$	4%
$100 \leq x \leq 149$	6%
$x \geq 150$	8%

  a. If the surf shop orders 50 shirts, compute the total cost.

  b. Which costs more: 148 shirts or 150 shirts? Explain your answer.

104. Maggie sells lemonade at an art show. She has a fixed cost of $75 to cover the registration fee for the art show. In addition, her cost to produce each lemonade is $0.17. If x represents the number of lemonades, then the total cost to produce x lemonades is given by:

$$\text{Cost} = 75 + 0.17x$$

If Maggie sells each lemonade for $2, then her revenue (the amount she brings in) for selling x lemonades is given by:

$$\text{Revenue} = 2.00x$$

  a. Write an inequality that expresses the number of lemonades, x, that Maggie must sell to make a profit. Profit is realized when the revenue is greater than the cost (Revenue > Cost).

  b. Solve the inequality in part (a).

105. Two rental car companies rent subcompact cars at a discount. Company A rents for $14.95 per day plus 22 cents per mile. Company B rents for $18.95 a day plus 18 cents per mile. Let x represent the number of miles driven in one day.

  The cost to rent a subcompact car for one day from Company A is:

$$\text{Cost}_A = 14.95 + 0.22x$$

  The cost to rent a subcompact car for one day from Company B is:

$$\text{Cost}_B = 18.95 + 0.18x$$

  a. Write an inequality that expresses the number of miles, x, for which the daily cost to rent from Company A is less than the daily cost to rent from Company B.

  b. Solve the inequality in part (a).

Expanding Your Skills

For Exercises 106–111, solve the inequality. Graph the solution set and write the set in interval notation.

106. $3(x + 2) - (2x - 7) \leq (5x - 1) - 2(x + 6)$

107. $6 - 8(y + 3) + 5y > 5y - (2y - 5) + 13$

108. $-2 - \dfrac{w}{4} \leq \dfrac{1 + w}{3}$

109. $\dfrac{z - 3}{4} - 1 > \dfrac{z}{2}$

110. $-0.703 < 0.122p - 2.472$

111. $3.88 - 1.335t \geq 5.66$

Chapter 2   SUMMARY

Section 2.1   Addition, Subtraction, Multiplication, and Division Properties of Equality

Key Concepts

An equation is an algebraic statement that indicates two expressions are equal. A **solution to an equation** is a value of the variable that makes the equation a true statement. The set of all solutions to an equation is the solution set of the equation.

A **linear equation in one variable** can be written in the form $ax + b = 0$, where $a \neq 0$.

Addition Property of Equality:

If $a = b$, then $a + c = b + c$

Subtraction Property of Equality:

If $a = b$, then $a - c = b - c$

Multiplication Property of Equality:

If $a = b$, then $ac = bc$

Division Property of Equality:

If $a = b$, then $\dfrac{a}{c} = \dfrac{b}{c}$ $(c \neq 0)$

Examples

Example 1

$2x + 1 = 9$ is an equation with solution $x = 4$.

Check: $2(4) + 1 = 9$

$$8 + 1 = 9$$

$$9 = 9 \; \checkmark \quad \text{True}$$

Example 2

$$x - 5 = 12$$
$$x - 5 + 5 = 12 + 5$$
$$x = 17$$

Example 3

$$z + 1.44 = 2.33$$
$$z + 1.44 - 1.44 = 2.33 - 1.44$$
$$z = 0.89$$

Example 4

$$\frac{3}{4}x = 12$$

$$\frac{4}{3} \cdot \frac{3}{4}x = 12 \cdot \frac{4}{3}$$

$$x = 16$$

Example 5

$$16 = 8y$$
$$\frac{16}{8} = \frac{8y}{8}$$
$$2 = y$$

Section 2.2   Solving Linear Equations

Key Concepts

Steps for Solving a Linear Equation in One Variable:

1. Simplify both sides of the equation.
   - Clear parentheses
   - Combine *like* terms
2. Use the addition or subtraction property of equality to collect the variable terms on one side of the equation.
3. Use the addition or subtraction property of equality to collect the constant terms on the other side of the equation.
4. Use the multiplication or division property of equality to make the coefficient of the variable term equal to 1.
5. Check your answer.

A **conditional equation** is true for some values of the variable but is false for other values.

An equation that has all real numbers as its solution set is an **identity**.

An equation that has no solution is a **contradiction**.

Examples

Example 1

$$5y + 7 = 3(y - 1) + 2$$

$5y + 7 = 3y - 3 + 2$	Clear parentheses.
$5y + 7 = 3y - 1$	Combine *like* terms.
$2y + 7 = -1$	Isolate variable term.
$2y = -8$	Isolate constant term.
$y = -4$	Divide both sides by 2.

Check:

$$5(-4) + 7 \overset{?}{=} 3[(-4) - 1] + 2$$
$$-20 + 7 \overset{?}{=} 3(-5) + 2$$
$$-13 \overset{?}{=} -15 + 2$$
$$-13 = -13 \ ✔ \quad \text{True}$$

Example 2

$x + 5 = 7$ is a conditional equation because it is true only on the condition that $x = 2$.

Example 3

$$x + 4 = 2(x + 2) - x$$
$$x + 4 = 2x + 4 - x$$
$$x + 4 = x + 4$$
$$4 = 4 \quad \text{is an identity.}$$

The solution is all real numbers.

Example 4

$$y - 5 = 2(y + 3) - y$$
$$y - 5 = 2y + 6 - y$$
$$y - 5 = y + 6$$
$$-5 = 6 \quad \text{is a contradiction.}$$

There is no solution.

Section 2.3 Linear Equations: Clearing Fractions and Decimals

Key Concepts

To Clear Fractions or Decimals in an Equation:

Multiply both sides of an equation by an appropriate power of 10 to clear decimals.

Multiply both sides of the equation by a common denominator of all the fractions.

Examples

Example 1

$$-1.2x - 5.1 = 16.5$$

$$10(-1.2x - 5.1) = 10(16.5)$$ Multiply both sides by 10.

$$-12x - 51 = 165$$

$$-12x = 216$$

$$\frac{-12x}{-12} = \frac{216}{-12}$$

$$x = -18$$

Example 2

$$\frac{1}{2}x - 2 - \frac{3}{4}x = \frac{7}{4}$$

$$\frac{4}{1}\left(\frac{1}{2}x - 2 - \frac{3}{4}x\right) = \frac{4}{1}\left(\frac{7}{4}\right)$$ Multiply by the LCD.

$$2x - 8 - 3x = 7$$ Apply distributive property.

$$-x - 8 = 7$$ Combine *like* terms.

$$-x = 15$$ Add 8 to both sides.

$$x = -15$$ Divide by -1.

Section 2.4   Applications of Linear Equations: Introduction to Problem Solving

Key Concepts

Problem-Solving Steps for Word Problems:

1. Read the problem carefully.
2. Assign labels to unknown quantities.
3. Develop a verbal model.
4. Write a mathematical equation.
5. Solve the equation.
6. Interpret the results and write the answer in words.

Examples

Example 1

The perimeter of a triangle is 54 m. The lengths of the sides are represented by three consecutive even integers. Find the lengths of the three sides.

1. Read the problem.

2. Let x represent one side, $x + 2$ represent the second side, and $x + 4$ represent the third side.

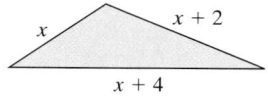

3. (First side) + (second side) + (third side) = perimeter

4. $x + (x + 2) + (x + 4) = 54$

5. $3x + 6 = 54$
   $3x = 48$
   $x = 16$

6. $x = 16$ represents the length of the shortest side. The lengths of the other sides are given by $x + 2 = 18$ and $x + 4 = 20$. The lengths of the three sides are 16 m, 18 m, and 20 m.

Section 2.5 Applications Involving Percents

Key Concepts

The following formula will help solve basic percent problems.

$$\text{amount} = (\text{percent})(\text{base})$$

One common use of percents is in computing **sales tax**.

Another use of percent is in computing **simple interest** using the formula:

$$\left(\begin{array}{c}\text{simple} \\ \text{interest}\end{array}\right) = \left(\begin{array}{c}\text{principal} \\ \text{invested}\end{array}\right)\left(\begin{array}{c}\text{annual} \\ \text{interest} \\ \text{rate}\end{array}\right)\left(\begin{array}{c}\text{time in} \\ \text{years}\end{array}\right)$$

or $I = Prt$.

Examples

Example 1

A dinette set costs $1260.00 after a 5% sales tax is included. What was the price before tax?

$$\left(\begin{array}{c}\text{price} \\ \text{before tax}\end{array}\right) + (\text{tax}) = \left(\begin{array}{c}\text{total} \\ \text{price}\end{array}\right)$$

$$x \quad\quad + 0.05x = 1260$$
$$1.05x = 1260$$
$$x = 1200$$

The dinette set cost $1200 before tax.

Example 2

John Li invests $5400 at 2.5% simple interest. How much interest does he make after 5 years?

$$I = Prt$$
$$I = (\$5400)(0.025)(5)$$
$$I = \$675$$

Section 2.6   Formulas and Applications of Geometry

Key Concepts

A **literal equation** is an equation that has more than one variable. Often such an equation can be manipulated to solve for different variables.

Formulas from Section R.4 can be used in applications involving geometry.

Examples

Example 1

$P = 2a + b$, solve for a.

$P - b = 2a + b - b$

$P - b = 2a$

$\dfrac{P - b}{2} = \dfrac{2a}{2}$

$\dfrac{P - b}{2} = a$ or $a = \dfrac{P - b}{2}$

Example 2

Find the length of a side of a square whose perimeter is 28 ft.

Use the formula $P = 4s$. Substitute 28 for P and solve:

$P = 4s$

$28 = 4s$

$7 = s$

The length of a side of the square is 7 ft.

Section 2.7   Linear Inequalities

Key Concepts

A **linear inequality in one variable**, x, is any relationship in the form: $ax + b < 0$, $ax + b > 0$, $ax + b \leq 0$, or $ax + b \geq 0$, where $a \neq 0$.

The solution set to an inequality can be expressed as a graph or in **set-builder notation** or in **interval notation**.

When graphing an inequality or when writing interval notation, a parenthesis, (or), is used to denote that an endpoint is *not included* in a solution set. A square bracket, [or], is used to show that an endpoint *is included* in a solution set. Parenthesis (or) are always used with $-\infty$ and ∞, respectively.

The inequality $a < x < b$ is used to show that x is greater than a and less than b. That is, x is *between* a and b.

Multiplying or dividing an inequality by a negative quantity requires the direction of the inequality sign to be reversed.

Examples

Example 1

$-2x + 6 \geq 14$

$-2x + 6 - 6 \geq 14 - 6$ Subtract 6.

$-2x \geq 8$ Simplify.

$\dfrac{-2x}{-2} \leq \dfrac{8}{-2}$ Divide by -2. Reverse the inequality sign.

$x \leq -4$

Set-builder notation: $\{x \mid x \leq -4\}$

Graph:

-4

Interval notation: $(-\infty, -4]$

Chapter 2   Review Exercises

Section 2.1

1. Label the following as either an expression or an equation:

   a. $3x + y = 10$ **b.** $9x + 10y - 2xy$

   c. $4(x + 3) = 12$ **d.** $-5x = 7$

2. Explain how to determine whether an equation is linear in one variable.

3. Identify which equations are linear.

   a. $4x^2 + 8 = -10$ **b.** $x + 18 = 72$

   c. $-3 + 2y^2 = 0$ **d.** $-4p - 5 = 6p$

4. For the equation, $4y + 9 = -3$, determine if the given numbers are solutions.

   a. $y = 3$ **b.** $y = 0$

   c. $y = -3$ **d.** $y = -2$

For Exercises 5–12, solve the equation using the addition property, subtraction property, multiplication property, or division property of equality.

5. $a + 6 = -2$ 6. $6 = z - 9$

7. $-\dfrac{3}{4} + k = \dfrac{9}{2}$ 8. $0.1r = 7$

9. $-5x = 21$ 10. $\dfrac{t}{3} = -20$

11. $-\dfrac{2}{5}k = \dfrac{4}{7}$ 12. $-m = -27$

13. The quotient of a number and negative six is equal to negative ten. Find the number.

14. The difference of a number and $-\frac{1}{8}$ is $\frac{5}{12}$. Find the number.

15. Four subtracted from a number is negative twelve. Find the number.

16. Six subtracted from a number is negative eight. Find the number.

Section 2.2

For Exercises 17–28, solve the equation.

17. $4d + 2 = 6$ 18. $5c - 6 = -9$

19. $-7c = -3c - 9$ 20. $-28 = 5w + 2$

21. $\dfrac{b}{3} + 1 = 0$ 22. $\dfrac{2}{3}h - 5 = 7$

23. $-3p + 7 = 5p + 1$ 24. $4t - 6 = -12t + 16$

25. $4a - 9 = 3(a - 3)$ 26. $3(2c + 5) = -2(c - 8)$

27. $7b + 3(b - 1) + 16 = 2(b + 8)$

28. $2 + (17 - x) + 2(x - 1) = 4(x + 2) - 8$

29. Explain the difference between an equation that is a contradiction and an equation that is an identity.

30. Label each equation as a conditional equation, a contradiction, or an identity.

    a. $x + 3 = 3 + x$ **b.** $3x - 19 = 2x + 1$

    c. $5x + 6 = 5x - 28$ **d.** $2x - 8 = 2(x - 4)$

    e. $-8x - 9 = -8(x - 9)$

Section 2.3

For Exercises 31–48, solve the equation.

31. $\dfrac{x}{8} - \dfrac{1}{4} = \dfrac{1}{2}$ 32. $\dfrac{y}{15} - \dfrac{2}{3} = \dfrac{4}{5}$

33. $\dfrac{4z + 7}{5} = z + 2$ 34. $\dfrac{5y - 3}{6} = 1 + y$

35. $\dfrac{1}{10}p - 3 = \dfrac{2}{5}p$ 36. $\dfrac{1}{4}y - \dfrac{3}{4} = \dfrac{1}{2}y + 1$

37. $-\dfrac{1}{4}(2 - 3t) = \dfrac{3}{4}$ 38. $\dfrac{2}{7}(w + 4) = \dfrac{1}{2}$

39. $17.3 - 2.7q = 10.55$

40. $4.9z + 4.6 = 3.2z - 2.2$

41. $5.74a + 9.28 = 2.24a - 5.42$

42. $62.84t - 123.66 = 4(2.36 + 2.4t)$

43. $0.05x + 0.10(24 - x) = 0.75(24)$

44. $0.20(x + 4) + 0.65x = 0.20(854)$

45. $100 - (t - 6) = -(t - 1)$

46. $3 - (x + 4) + 5 = 3x + 10 - 4x$

47. $5t - (2t + 14) = 3t - 14$

48. $9 - 6(2z + 1) = -3(4z - 1)$

Section 2.4

49. Twelve added to the sum of a number and two is forty-four. Find the number.

50. Twenty added to the sum of a number and six is thirty-seven. Find the number.

51. Three times a number is the same as the difference of twice the number and seven. Find the number.

52. Eight less than five times a number is forty-eight less than the number. Find the number.

53. Three times the largest of three consecutive even integers is 76 more than the sum of the other two integers. Find the integers.

54. Ten times the smallest of three consecutive integers is 213 more than the sum of the other two integers. Find the integers.

55. The perimeter of a triangle is 78 in. The lengths of the sides are represented by three consecutive integers. Find the lengths of the sides of the triangle.

56. The perimeter of a pentagon (a five-sided polygon) is 190 cm. The five sides are represented by consecutive integers. Find the measures of the sides.

57. The minimum salary for a major league baseball player in 1985 was $60,000. This was twice the minimum salary in 1980. What was the minimum salary in 1980?

58. The state of Indiana has approximately 2.1 million more people than Kentucky. Together their population totals 10.3 million. Approximately how many people are in each state?

59. A Cessna 182 airplane has an average air speed that is 50 mph slower than a Mooney airplane. If the combined distance traveled by these two small airplanes is 660 mi after 2 hr, what is the speed of each plane?

60. A bicyclist and a jogger leave a gym at the same time. The bicyclist rides 6 mph faster than the jogger can run. It takes the jogger 2 hr to complete the same distance completed by the bicyclist in 1 hr. Find the speed of each person.

Section 2.5

For Exercises 61–70, solve the problems involving percents.

61. What is 35% of 68?

62. What is 4% of 720?

63. 53.5 is what percent of 428?

64. 68.4 is what percent of 72?

65. 24 is 15% of what number?

66. 8.75 is 0.5% of what number?

67. A novel originally selling at $29.99 is on sale for 12% off. What is the sale price of the book? Round to the nearest cent.

68. What would be the total price (including tax) of a novel that sells for $26.39, if the sales tax rate is 7%?

69. Anna Tsao invested $3000 in an account paying 8% simple interest.

    a. How much interest will she earn in $3\frac{1}{2}$ years?

    b. What will her balance be at that time?

70. Eduardo invested money in an account earning 4% simple interest. At the end of 5 years, he had a total of $14,400. How much money did he originally invest?

Section 2.6

For Exercises 71–78, solve for the indicated variable.

71. $C = K - 273$ for K

72. $K = C + 273$ for C

73. $P = 4s$ for s **74.** $P = 3s$ for s

75. $y = mx + b$ for x **76.** $a + bx = c$ for x

77. $2x + 5y = -2$ for y

78. $4(a + b) = Q$ for b

For Exercises 79–85, use the appropriate geometry formula to solve the problem.

79. The volume of a cone is given by the formula $V = \dfrac{1}{3}\pi r^2 h$.

    a. Solve the formula for h.

    b. Find the height of a right circular cone whose volume is 47.8 in.^{3} and whose radius is 3 in. Round to the nearest tenth of an inch.

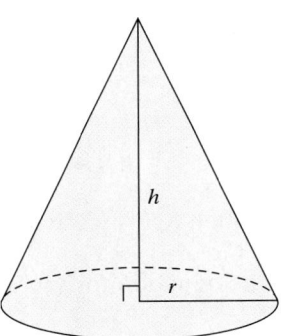

80. Find the height of a parallelogram whose area is 42 m^2 and whose base is 6 m.

81. The smallest angle of a triangle is 2° more than $\frac{1}{4}$ of the largest angle. The middle angle is 2° less than the largest angle. Find the measure of each angle.

82. One angle is 6° less than twice a second angle. If the two angles are complementary, what are their measures?

83. A rectangular window has a width 1 ft less than its length. The perimeter is 18 ft. Find the length and the width of the window.

84. Find the measure of the vertical angles by first solving for x.

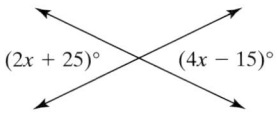

$(2x + 25)°$ $(4x - 15)°$

85. Find the measure of angle y.

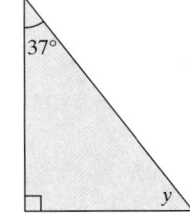

Section 2.7

86. Graph the inequalities and write the sets in interval notation.

    a. $\{x \mid x > -2\}$

    b. $\left\{ x \mid x \le \dfrac{1}{2} \right\}$

    c. $\{x \mid -1 < x \le 4\}$

87. A landscaper buys potted geraniums from a nursery at a price of $5 per plant. However, for large orders, the price per plant is discounted by a percentage off the original price. Let x represent the number of potted plants ordered. The corresponding discount is given in the table.

Number of Plants	Discount
$x \le 99$	0%
$100 \le x \le 199$	2%
$200 \le x \le 299$	4%
$x \ge 300$	6%

    a. Find the cost to purchase 130 plants.

    b. Which costs more, 300 plants or 295 plants? Explain your answer.

For Exercises 88–97, solve the inequality. Graph the solution set and express the answer in interval notation.

88. $c + 6 < 23$ ⟶

89. $3w - 4 > -5$ ⟶

90. $-2x - 7 \geq 5$ ⟶

91. $5(y + 2) \leq -4$ ⟶

92. $-\dfrac{3}{7}a \leq -21$ ⟶

93. $1.3 > 0.4t - 12.5$ ⟶

94. $4k + 23 < 7k - 31$ ⟶

95. $\dfrac{6}{5}h - \dfrac{1}{5} \leq \dfrac{3}{10} + h$ ⟶

96. $-6 < 2b \leq 14$ ⟶

97. $-2 \leq z + 4 \leq 9$ ⟶

98. The summer average rainfall for Bermuda for June, July, and August is 5.3 in. per month. If Bermuda receives 6.3 in. of rain in June and 7.1 in. in July, how much rain is required in August to exceed the 3-month summer average?

99. Reggie sells hot dogs at a ballpark. He has a fixed cost of $33 to use the concession stand at the park. In addition, the cost for each hot dog is $0.40. If x represents the number of hot dogs sold, then the total cost is given by

$$\text{Cost} = 33 + 0.40x$$

If Reggie sells each hot dog for $1.50, then his revenue (the amount he brings in) for selling x hot dogs is given by

$$\text{Revenue} = 1.50x$$

a. Write an inequality that expresses the number of hot dogs, x, that Reggie must sell to make a profit. Profit is realized when the revenue is greater than the cost (revenue > cost).

b. Solve the inequality in part (a).

Chapter 2   Test

1. Which of the equations have $x = -3$ as a solution?

   a. $4x + 1 = 10$

   b. $6(x - 1) = x - 21$

   c. $5x - 2 = 2x + 1$

   d. $\dfrac{1}{3}x + 1 = 0$

2. a. Simplify: $3x - 1 + 2x + 8$

   b. Solve: $3x - 1 = 2x + 8$

For Exercises 3 – 13, solve the equation.

3. $t + 3 = -13$

4. $8 = p - 4$

5. $\dfrac{t}{8} = -\dfrac{2}{9}$

6. $-3x + 5 = -2$

7. $2(p - 4) = p + 7$

8. $2 + d = 2 - 3(d - 5) - 2$

9. $\dfrac{3}{7} + \dfrac{2}{5}x = -\dfrac{1}{5}x + 1$

10. $3h + 1 = 3(h + 1)$

11. $\dfrac{3x + 1}{2} - \dfrac{4x - 3}{3} = 1$

12. $0.5c - 1.9 = 2.8 + 0.6c$

13. $-5(x + 2) + 8x = -2 + 3x - 8$

14. Solve the equation for y: $3x + y = -4$

15. Solve $C = 2\pi r$ for r.

16. 13% of what is 11.7?

17. One number is four plus one-half of another. The sum of the numbers is 31. Find the numbers.

18. The perimeter of a pentagon (a five-sided polygon) is 315 in. The five sides are represented by consecutive integers. Find the measures of the sides.

19. In the 1997–1998 season, a couple purchased two NHL hockey tickets and two NBA basketball tickets for $153.92. A hockey ticket cost $4.32 more than a basketball ticket. What were the prices of the individual tickets?

20. The total bill for a pair of basketball shoes (including sales tax) is $87.74. If the tax rate is 7%, find the cost of the shoes before tax.

21. Clarita borrowed money at a 6% simple interest rate. If she paid back a total of $8000 at the end of 10 years, how much did she originally borrow?

22. The length of a soccer field for international matches is 40 m less than twice its width. If the perimeter is 370 m, what are the dimensions of the field?

23. Given the triangle, find the measures of each angle by first solving for y.

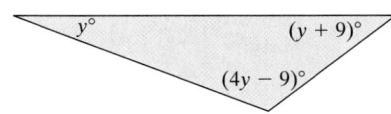

24. Two families leave their homes at the same time to meet for lunch. The families live 210 miles apart, and one family drives 5 mph slower than the other. If it takes them 2 hr to meet at a point between their homes, how fast does each family travel?

25. Two angles are complementary. One angle is 26° more than the other angle. What are the measures of the angles?

26. Graph the inequalities and write the sets in interval notation.

    a. $\{x \mid x < 0\}$

    b. $\{x \mid -2 \le x < 5\}$

For Exercises 27–29, solve the inequality. Graph the solution and write the solution set in interval notation.

27. $5x + 14 > -2x$

28. $2(3 - x) \ge 14$

29. $-13 \le 3p + 2 \le 5$

30. The average winter snowfall for Syracuse, New York, for December, January, and February is 27.5 in. per month. If Syracuse receives 24 in. of snow in December and 32 in. in January, how much snow is required in February to exceed the 3-month average?

Chapters 1–2   Cumulative Review Exercises

For Exercises 1–5, perform the indicated operations.

1. $\left| -\dfrac{1}{5} + \dfrac{7}{10} \right|$ **2.** $5 - 2[3 - (4 - 7)]$

3. $-\dfrac{2}{3} + \left(\dfrac{1}{2}\right)^2$ **4.** $-3^2 + (-5)^2$

5. $\sqrt{5 - (-20)} - 3^2$

For Exercises 6–9, translate the mathematical expressions and simplify the results.

6. The square root of the difference of five squared and nine

7. The sum of -14 and 12

8. List the terms of the expression:
$-7x^2y + 4xy - 6$

9. Simplify: $-4[2x - 3(x + 4)] + 5(x - 7)$

For Exercises 10–15, solve the equations.

10. $8t - 8 = 24$

11. $-2.5x - 5.2 = 12.8$

12. $-5(p - 3) + 2p = 3(5 - p)$

13. $\dfrac{x + 3}{5} + \dfrac{x - 2}{2} = 2$

14. $\dfrac{2}{9}x - \dfrac{1}{3} = x + \dfrac{1}{9}$

15. $-0.6w = 48$

16. The sum of two consecutive odd integers is 156. Find the integers.

17. The total bill for a man's three-piece suit (including sales tax) is $374.50. If the tax rate is 7%, find the cost of the suit before tax.

18. The area of a triangle is 41 cm². Find the height of the triangle if the base is 12 cm.

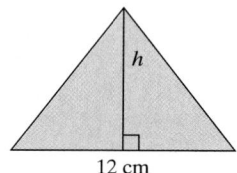

12 cm

For Exercises 19–20, solve the inequality. Graph the solution set on a number line and express the solution in interval notation.

19. $-3x - 3(x + 1) < 9$ ⟶

20. $-6 \leq 2x - 4 \leq 14$ ⟶

Graphing Linear Equations in Two Variables

3

In this chapter we study graphing and focus on the graphs of lines.

As you work through the chapter, pay attention to key terms. Then work through this puzzle. When you fill in the blanks, notice that hyphens also go in the boxes.

Across

1. (0, b)
3. one of four regions in the *xy*-plane
4. The point (0, 0)
5. horizontal axis
7. lines with the same slope and different *y*-intercepts
8. (a, 0)

Down

1. vertical axis
2. lines that intersect at a right angle
6. "slant" of a line

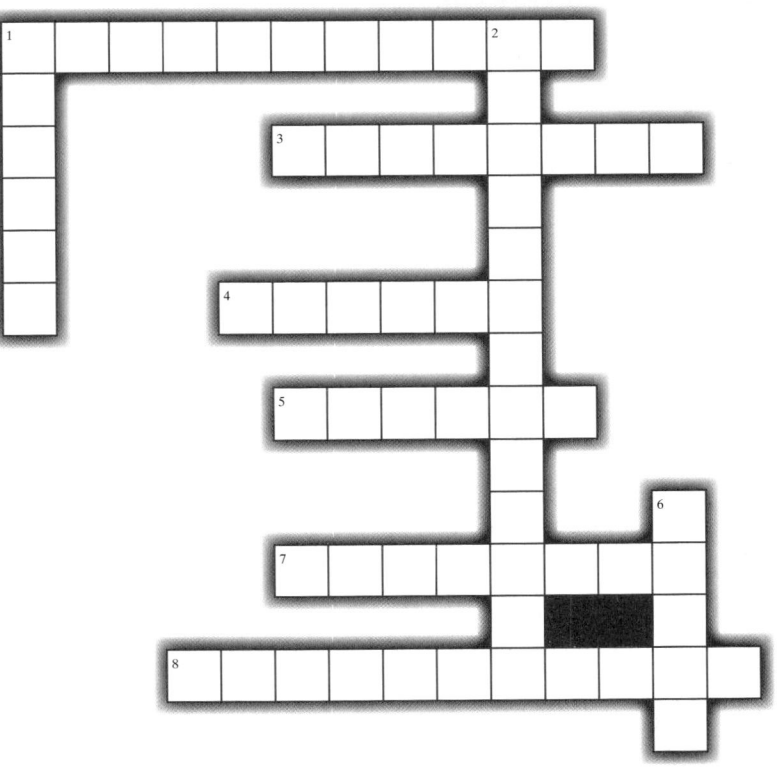

Rectangular Coordinate System

1. Interpreting Graphs

Mathematics is a powerful tool used by scientists and has directly contributed to the highly technical world we live in. Applications of mathematics have led to advances in the sciences, business, computer technology, and medicine.

One fundamental application of mathematics is the graphical representation of numerical information (or **data**). For example, Table 3-1 represents the number of clients admitted to a drug and alcohol rehabilitation program over a 12-month period.

Table 3-1

	Month	Number of Clients
Jan.	1	55
Feb.	2	62
March	3	64
April	4	60
May	5	70
June	6	73
July	7	77
Aug.	8	80
Sept.	9	80
Oct.	10	74
Nov.	11	85
Dec.	12	90

In table form, the information is difficult to picture and interpret. It appears that on a monthly basis, the number of clients fluctuates. However, when the data are represented in a graph, an upward trend is clear (Figure 3-1).

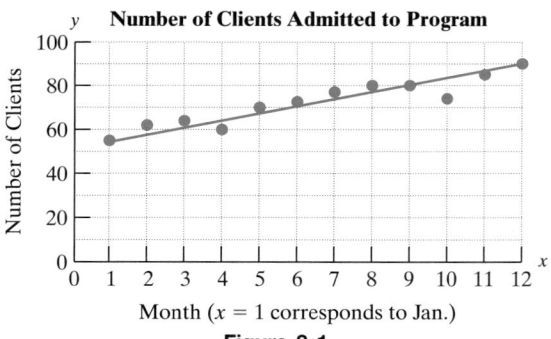

Month ($x = 1$ corresponds to Jan.)

Figure 3-1

From the increase in clients shown in this graph, management for the rehabilitation center might make plans for the future. If the trend continues, management might consider expanding its facilities and increasing its staff to accommodate the expected increase in clients.

Example 1	Interpreting a Graph

Refer to Figure 3-1 and Table 3-1.

  a. For which month was the number of clients the greatest?

  b. How many clients were served in the first month (January)?

  c. Which month corresponds to 60 clients served?

  d. Between which two months did the number of clients decrease?

  e. Between which two months did the number of clients remain the same?

Solution:

  a. Month 12 (December) corresponds to the highest point on the graph, which represents the most clients.

  b. In month 1 (January), there were 55 clients served.

  c. Month 4 (April).

  d. The number of clients decreased between months 3 and 4 and between months 9 and 10.

  e. The number of clients remained the same between months 8 and 9.

Skill Practice

1. Refer to Figure 3-1.

  a. How many clients were served in October?

  b. Which month corresponds to 70 clients served?

  c. What is the difference between the number of clients in month 12 and month 1?

  d. For which month was the number of clients served the least?

2. Plotting Points in a Rectangular Coordinate System

In Example 1, two variables are represented, time and the number of clients. To picture two variables, we use a graph with two number lines drawn at right angles to each other (Figure 3-2). This forms a **rectangular coordinate system**. The horizontal line is called the **x-axis**, and the vertical line is called the **y-axis**. The point where the lines intersect is called the **origin**. On the x-axis, the numbers to the right of the origin are positive and the numbers to the left are negative. On the y-axis, the numbers above the origin are positive and the numbers below are negative. The x- and y-axes divide the graphing area into four regions called **quadrants**.

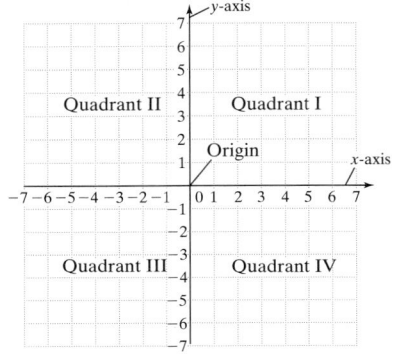

Figure 3-2

Skill Practice Answers

1a. 74
  b. Month 5, May
  c. 35
  d. Month 1, January

Points graphed in a rectangular coordinate system are defined by two numbers as an **ordered pair**, (x, y). The first number (called the first coordinate, or the abscissa) is the horizontal position from the origin. The second number (called the second coordinate, or the ordinate) is the vertical position from the origin. Example 2 shows how points are plotted in a rectangular coordinate system.

| **Example 2** | **Plotting Points in a Rectangular Coordinate System** |

Plot the points.

a. $(4, 5)$ **b.** $(-4, -5)$ **c.** $(-1, 3)$ **d.** $(3, -1)$

e. $\left(\dfrac{1}{2}, -\dfrac{7}{3}\right)$ **f.** $(-2, 0)$ **g.** $(0, 0)$ **h.** $(\pi, 1.1)$

Solution:

See Figure 3-3.

a. The ordered pair $(4, 5)$ indicates that $x = 4$ and $y = 5$. Beginning at the origin, move 4 units in the positive x-direction (4 units to the right), and from there move 5 units in the positive y-direction (5 units up). Then plot the point. The point is in Quadrant I.

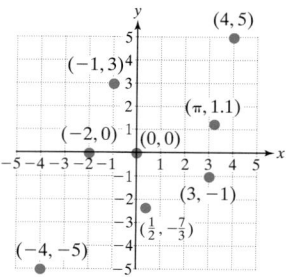

Figure 3-3

b. The ordered pair $(-4, -5)$ indicates that $x = -4$ and $y = -5$. Move 4 units in the negative x-direction (4 units to the left), and from there move 5 units in the negative y-direction (5 units down). Then plot the point. The point is in Quadrant III.

c. The ordered pair $(-1, 3)$ indicates that $x = -1$ and $y = 3$. Move 1 unit to the left and 3 units up. The point is in Quadrant II.

d. The ordered pair $(3, -1)$ indicates that $x = 3$ and $y = -1$. Move 3 units to the right and 1 unit down. The point is in Quadrant IV.

e. The improper fraction $-\dfrac{7}{3}$ can be written as the mixed number $-2\dfrac{1}{3}$. Therefore, to plot the point $\left(\dfrac{1}{2}, -\dfrac{7}{3}\right)$ move to the right $\dfrac{1}{2}$ unit, and down $2\dfrac{1}{3}$ units. The point is in Quadrant IV.

f. The point $(-2, 0)$ indicates $y = 0$. Therefore, the point is on the x-axis.

g. The point $(0, 0)$ is at the origin.

h. The irrational number, π, can be approximated as 3.14. Thus, the point $(\pi, 1.1)$ is located approximately 3.14 units to the right and 1.1 units up. The point is in Quadrant I.

TIP: Notice that changing the order of the x- and y-coordinates changes the location of the point. The point $(-1, 3)$ for example is in Quadrant II, whereas $(3, -1)$ is in Quadrant IV (Figure 3-3). This is why points are represented by *ordered* pairs. The order of the coordinates is important.

Skill Practice

2. Plot the points.

$$A(3, 4) \qquad B(-2, 2) \qquad C(4, 0) \qquad D\left(\frac{5}{2}, -\frac{1}{3}\right) \qquad E(-5, -2)$$

The effective use of graphs for mathematical models requires skill in identifying points and interpreting graphs.

Example 3 | **Identifying Points**

Refer to the figure, and give the coordinates of each point and the quadrant or axis where it is located.

a. P

b. Q

c. R

d. S

e. T

f. U

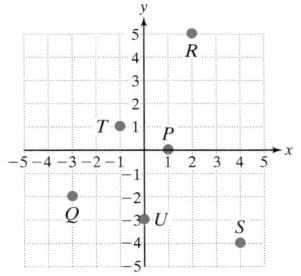

Solution:

a. P The coordinates are $(1, 0)$, and the point is on the x-axis.

b. Q The coordinates are $(-3, -2)$, and the point is in Quadrant III.

c. R The coordinates are $(2, 5)$, and the point is in Quadrant I.

d. S The coordinates are $(4, -4)$, and the point is in Quadrant IV.

e. T The coordinates are $(-1, 1)$, and the point is in Quadrant II.

f. U The coordinates are $(0, -3)$, and the point is on the y-axis.

Skill Practice

3. Give the coordinates of the labeled points, and state the quadrant or axis where the point is located.

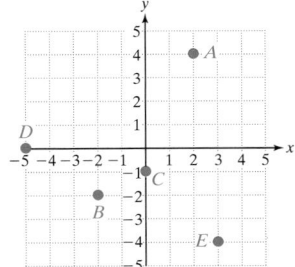

Skill Practice Answers

2.

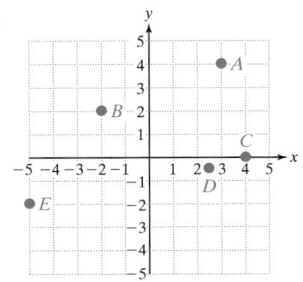

3. $A(2, 4)$; Quadrant I
   $B(-2, -2)$; Quadrant III
   $C(0, -1)$; y-axis
   $D(-5, 0)$; x-axis
   $E(3, -4)$; Quadrant IV

3. Applications of Plotting and Identifying Points

> **Example 4**   Plotting Points in an Application

The daily low temperatures (in degrees Fahrenheit) for one week in January for Sudbury, Ontario, Canada, are given in Table 3-2.

Table 3-2

Day Number, x	Temperature (°F), y
1	−3
2	−5
3	1
4	6
5	5
6	0
7	−4

a. Write an ordered pair for each row in the table using the day number as the x-coordinate and the temperature as the y-coordinate.

b. Plot the ordered pairs from part (a) on a rectangular coordinate system.

Solution:

a. $(1, -3)$
$(2, -5)$
$(3, 1)$
$(4, 6)$
$(5, 5)$
$(6, 0)$
$(7, -4)$

Each ordered pair represents the day number and the corresponding low temperature for that day.

b.

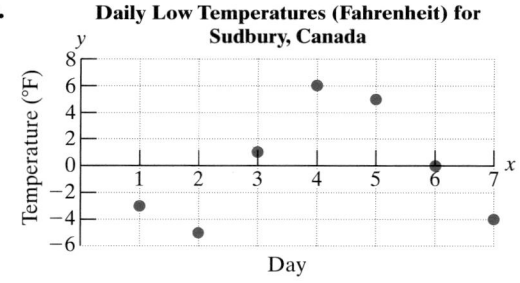

Daily Low Temperatures (Fahrenheit) for Sudbury, Canada

TIP: The graph in Example 4(b) shows only Quadrants I and IV because all of the x-coordinates are positive.

> **Skill Practice**

4. The table shows the number of homes sold in one town for a 6-month period. Plot the ordered pairs.

Month, x	Number Sold, y
1	20
2	25
3	28
4	40
5	45
6	30

Skill Practice Answers

4.

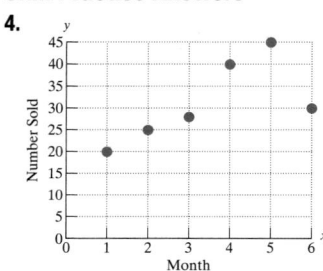

| Example 5 | **Determining Points from a Graph** |

A map of a national park is drawn so that the origin is placed at the ranger station (Figure 3-4). Four fire observation towers are located at points *A*, *B*, *C*, and *D*. Estimate the coordinates of the fire towers relative to the ranger station (all distances are in miles).

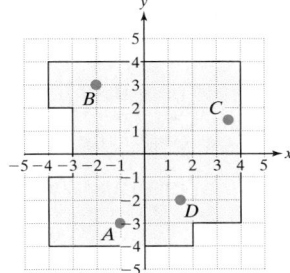

Figure 3-4

Solution:

Point *A*: $(-1, -3)$

Point *B*: $(-2, 3)$

Point *C*: $(3\frac{1}{2}, 1\frac{1}{2})$ or $(\frac{7}{2}, \frac{3}{2})$ or $(3.5, 1.5)$

Point *D*: $(1\frac{1}{2}, -2)$ or $(\frac{3}{2}, -2)$ or $(1.5, -2)$

Skill Practice

5. A map of a city is drawn so that the main office of a cell phone company is located at the origin. Cell towers are located at points *A*, *B*, *C*, and *D*. Estimate the coordinates of the towers.

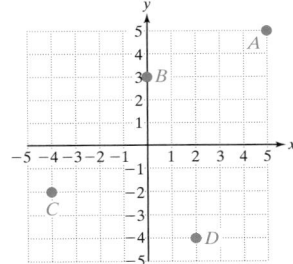

Skill Practice Answers

5. *A*(5, 5)
  B(0, 3)
  C(−4, −2)
  D(2, −4)

| Section 3.1 | **Practice Exercises** |

Study Skills Exercises

1. Before you proceed too much farther in Chapter 3, make your test corrections for the Chapter 2 test. See Exercise 1 of Section 2.1 for instructions.

2. Define the key terms:

    a. data   **b. ordered pair**   **c. origin**   **d. quadrant**

    e. rectangular coordinate system   **f. *x*-axis**   **g. *y*-axis**

Concept 1: Interpreting Graphs

For Exercises 3–6, refer to the graphs to answer the questions.

3. The number of patients served by a certain hospice care center is shown in the graph for the first 12 months after it opened.

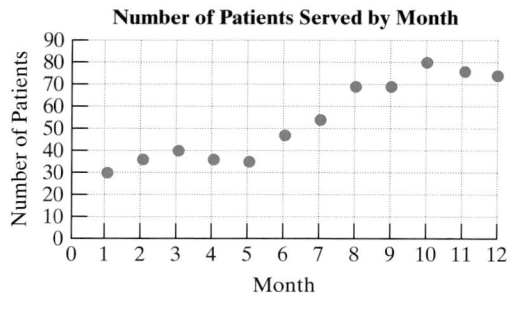

Number of Patients Served by Month

   a. For which month was the number of patients greatest?

   b. How many patients did the center serve in the first month?

   c. Between which months did the number of patients decrease?

   d. Between which two months did the number of patients remain the same?

   e. Which month corresponds to 40 patients served?

   f. Approximately how many patients were served during the 10th month?

4. The number of housing permits (in thousands) issued by a certain county in Texas between 1994 and 2003 is shown in the graph.

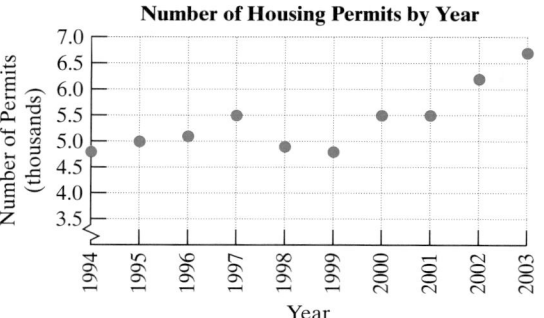

Number of Housing Permits by Year

   a. For which year was the number of permits greatest?

   b. How many permits did the county issue in 1997?

   c. Between which years did the number of permits decrease?

   d. Between which two consecutive years did the number of permits remain the same?

   e. Which year corresponds to 5000 permits issued?

5. The price per share of a stock (in dollars) over a period of 5 days is shown in the graph.

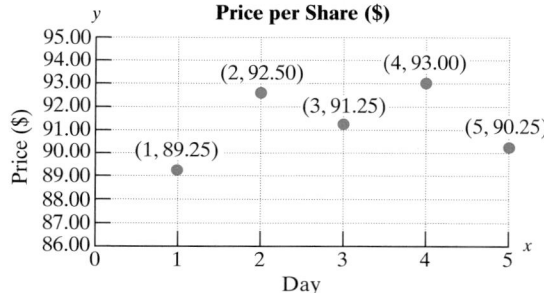

Price per Share ($)

   a. Interpret the meaning of the ordered pair (1, 89.25).

   b. What was the gain in price between day 3 and day 4?

   c. What was the loss in price between day 4 and day 5?

6. The price per share of a stock (in dollars) over a period of 5 days is shown in the graph.

   a. Interpret the meaning of the ordered pair (1, 10.125).

   b. What was the loss between day 4 and day 5?

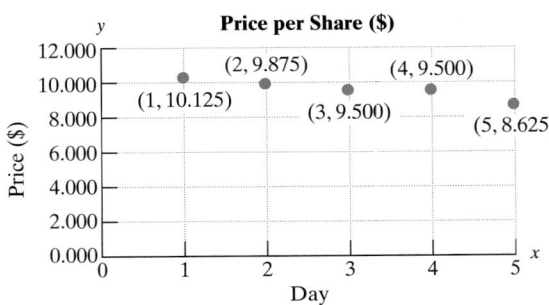

Concept 2: Plotting Points in a Rectangular Coordinate System

7. Plot the points on a rectangular coordinate system.

   a. (2, 6)   **b.** (6, 2)

   c. (0, −3)   **d.** (−3, 0)

8. Plot the points on a rectangular coordinate system.

   a. (−1, 2)   **b.** (2, −1)

   c. (0, 7)   **d.** (7, 0)

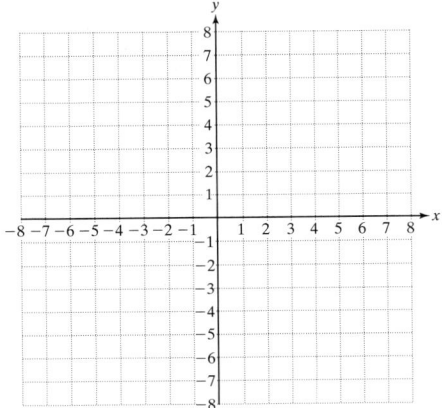

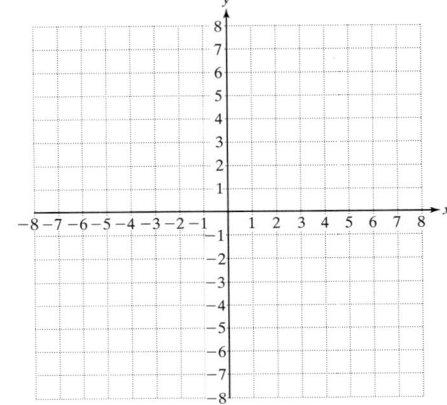

9. Plot the points on a rectangular coordinate system.

   a. (4, 5)   **b.** (−4, 5)

   c. (4, −5)   **d.** (−4, −5)

10. Plot the points on a rectangular coordinate system.

   a. (2, 3)   **b.** (−2, 3)

   c. (2, −3)   **d.** (−2, −3)

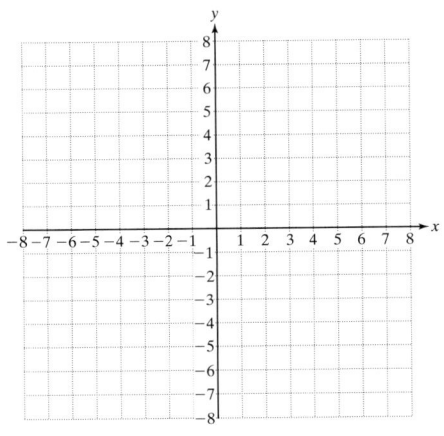

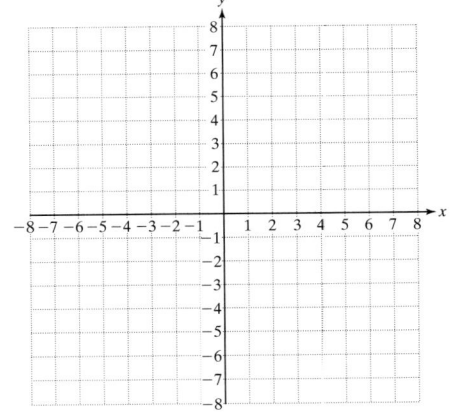

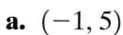

11. Plot the points on a rectangular coordinate system.

a. $(-1, 5)$

b. $(0, 4)$

c. $\left(-2, -\dfrac{3}{2}\right)$

d. $(2, -0.75)$

e. $(4, 2)$

f. $(-6.1, 0)$

g. $(0, 0)$

h. $(5, -5)$

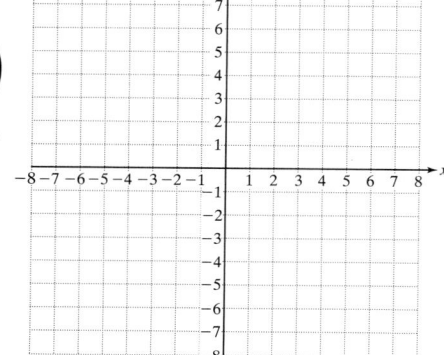

12. Plot the points on a rectangular coordinate system.

a. $(7, 0)$

b. $(-3, -1)$

c. $(0, 0)$

d. $(0, 1.5)$

e. $(6, 1)$

f. $\left(-\dfrac{1}{4}, 4\right)$

g. $\left(\dfrac{1}{4}, -4\right)$

h. $(-5, -6)$

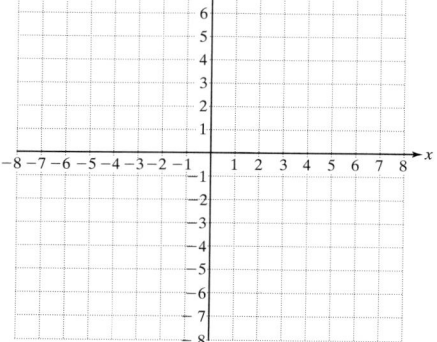

For Exercises 13–20, identify the quadrant in which the given point is found.

13. $(13, -2)$ **14.** $(25, 16)$ **15.** $(-8, 14)$ **16.** $(-82, -71)$

17. $(-5, -19)$ **18.** $(-31, 6)$ **19.** $\left(\dfrac{5}{2}, \dfrac{7}{4}\right)$ **20.** $(9, -40)$

21. What is the x-coordinate of a point on the y-axis?

22. What is the y-coordinate of a point on the x-axis?

23. Where is the point $\left(\dfrac{7}{8}, 0\right)$ located?   **24.** Where is the point $\left(0, \dfrac{6}{5}\right)$ located?

For Exercises 25–26, refer to the graph.

25. Estimate the coordinates of the points A, B, C, D, E, and F.

26. Estimate the coordinates of the points G, H, I, J, K, and L.

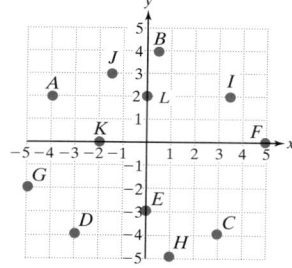

Concept 3: Applications of Plotting and Identifying Points

27. A movie theater has kept records of popcorn sales versus movie attendance.

a. Write the corresponding ordered pairs using the movie attendance as the x-variable and sales of popcorn as the y-variable. Interpret the meaning of the first ordered pair.

Movie Attendance (Number of People)	Sales of Popcorn ($)
250	225
175	193
315	330
220	209
450	570
400	480
190	185

b. Plot the data points on a rectangular coordinate system.

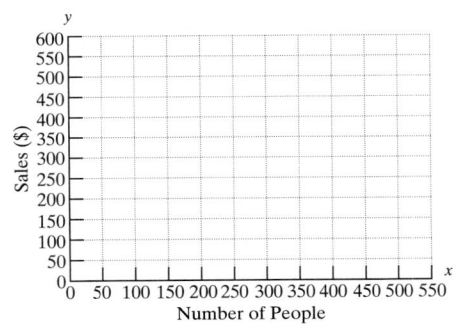

28. The age and systolic blood pressure (in millimeters of mercury, mm Hg) for eight different women are given in the table.

   a. Write the corresponding ordered pairs using each woman's age as the *x*-variable and the systolic blood pressure as the *y*-variable. Interpret the meaning of the first ordered pair.

   b. Plot the data points on a rectangular coordinate system.

Age (Years)	Systolic Blood Pressure (mm Hg)
57	149
41	120
71	158
36	115
64	151
25	110
40	118
77	165

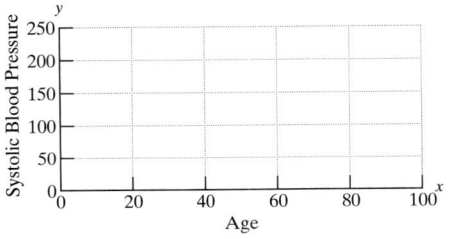

29. The income level defining the poverty line for an individual is given for selected years between 1980 and 2005. Let *x* represent the number of years since 1980. Let *y* represent the income defining the poverty level.

   (0, 4300)   (5, 5600)   (10, 6800)

   (15, 7900)   (20, 9000)   (25, 10,500)

   a. Interpret the meaning of the ordered pair (10, 6800).

   b. Plot the points on a rectangular coordinate system.

   (*Source: U.S. Department of the Census*)

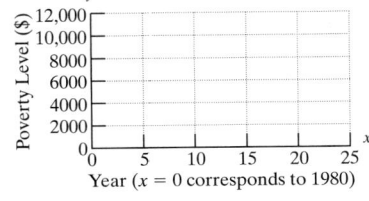

30. The following ordered pairs give the population of the U.S. colonies from 1700 to 1770. Let *x* represent the year, where *x* = 0 corresponds to 1700, *x* = 10 corresponds to 1710, and so on. Let *y* represent the population of the colonies.

   (0, 251000)   (10, 332000)   (20, 466000)

   (30, 629000)   (40, 906000)   (50, 1171000)

   (60, 1594000)   (70, 2148000)

   a. Interpret the meaning of the ordered pair (10, 332000).

   b. Plot the points on a rectangular coordinate system.

   (*Source: Information Please Almanac*)

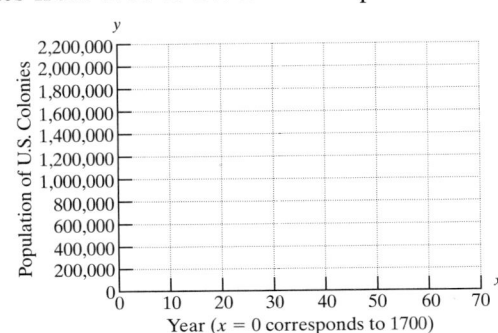

31. The following table shows the average temperature in degrees Celsius for Montreal, Quebec, Canada, by month.

   a. Write the corresponding ordered pairs, letting $x = 1$ correspond to the month of January.

   b. Plot the ordered pairs on a rectangular coordinate system.

Month, x		Temperature (°C), y
Jan.	1	−10.2
Feb.	2	−9.0
March	3	−2.5
April	4	5.7
May	5	13.0
June	6	18.3
July	7	20.9
Aug.	8	19.6
Sept.	9	14.8
Oct.	10	8.7
Nov.	11	2.0
Dec.	12	−6.9

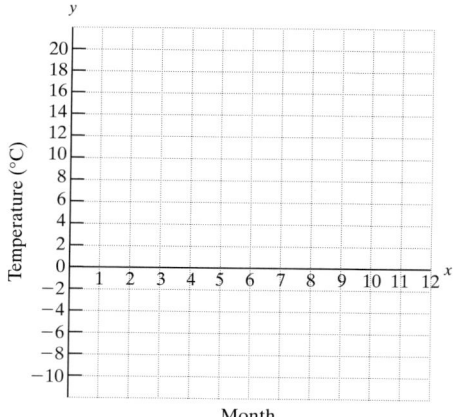

32. The table shows the average temperature in degrees Fahrenheit for Fairbanks, Alaska, by month.

   a. Write the corresponding ordered pairs, letting $x = 1$ correspond to the month of January.

   b. Plot the ordered pairs on a rectangular coordinate system.

Month, x		Temperature (°F), y
Jan.	1	−12.8
Feb.	2	−4.0
March	3	8.4
April	4	30.2
May	5	48.2
June	6	59.4
July	7	61.5
Aug.	8	56.7
Sept.	9	45.0
Oct.	10	25.0
Nov.	11	6.1
Dec.	12	−10.1

33. A map of a park is laid out with the Visitor Center located at the origin. Five visitors are in the park located at points A, B, C, D, and E. All distances are in meters.

   a. Estimate the coordinates of each hiker.

   b. How far apart are visitors C and D?

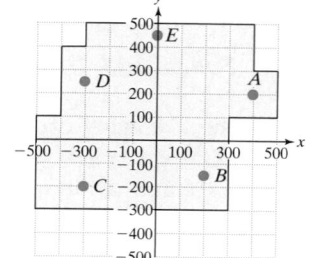

34. A townhouse has a sprinkler system in the backyard. With the water source at the origin, the sprinkler heads are located at points A, B, C, D, and E. All distances are in feet.

   a. Estimate the coordinates of each sprinkler head.

   b. How far is the distance from sprinkler head B to C?

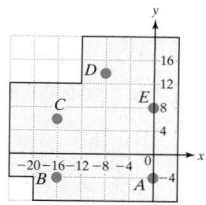

Linear Equations in Two Variables

1. Solutions to Linear Equations in Two Variables

Recall that an equation in the form $ax + b = 0$, where $a \neq 0$, is called a linear equation in one variable. A solution to such an equation is a value of x that makes the equation a true statement. For example, $3x + 6 = 0$ has a solution of $x = -2$.

In this section, we will look at linear equations in *two* variables.

Definition of a Linear Equation in Two Variables

Let A, B, and C be real numbers such that A and B are not both zero. Then, an equation that can be written in the form:

$$Ax + By = C$$

is called a **linear equation in two variables**.

The equation $x + y = 4$ is a linear equation in two variables. A solution to such an equation is an ordered pair (x, y) that makes the equation a true statement. Several solutions to the equation $x + y = 4$ are listed here:

Solution:	Check:
(x, y)	$x + y = 4$
$(2, 2)$	$(2) + (2) = 4$ ✔
$(1, 3)$	$(1) + (3) = 4$ ✔
$(4, 0)$	$(4) + (0) = 4$ ✔
$(-1, 5)$	$(-1) + (5) = 4$ ✔

By graphing these ordered pairs, we see that the solution points line up (Figure 3-5).

Notice that there are infinitely many solutions to the equation $x + y = 4$ so they cannot all be listed. Therefore, to visualize all solutions to the equation $x + y = 4$, we draw the line through the points in the graph. Every point on the line represents an ordered pair solution to the equation $x + y = 4$, and the line represents the set of *all* solutions to the equation.

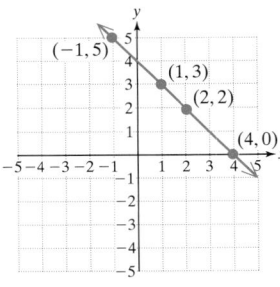

Figure 3-5

Example 1 Determining Solutions to a Linear Equation

For the linear equation, $4x - 5y = 8$, determine whether the given ordered pair is a solution.

a. $(2, 0)$ **b.** $(3, 1)$ **c.** $\left(1, -\dfrac{4}{5}\right)$

Solution:

a. $\quad 4x - 5y = 8$

$\quad\quad 4(2) - 5(0) \stackrel{?}{=} 8$ $\quad\quad$ Substitute $x = 2$ and $y = 0$.

$\quad\quad\quad\quad 8 - 0 = 8$ ✔ (true) $\quad\quad$ The ordered pair $(2, 0)$ is a solution.

b. $4x - 5y = 8$

$4(3) - 5(1) \overset{?}{=} 8$ Substitute $x = 3$ and $y = 1$.

$12 - 5 \neq 8$ The ordered pair $(3, 1)$ is *not* a solution.

c. $4x - 5y = 8$

$4(1) - 5\left(-\dfrac{4}{5}\right) \overset{?}{=} 8$ Substitute $x = 1$ and $y = -\dfrac{4}{5}$.

$4 + 4 = 8$ ✔ (true)   The ordered pair $\left(1, -\dfrac{4}{5}\right)$ is a solution.

> **Skill Practice**
>
> **1.** Given the equation $3x - 2y = -12$, determine whether the given ordered pair is a solution.
>
>   **a.** $(-2, 3)$ **b.** $(4, 0)$ **c.** $\left(1, \dfrac{15}{2}\right)$

2. Graphing Linear Equations in Two Variables by Plotting Points

The word *linear* means "relating to or resembling a line." It is not surprising then that the solution set for any linear equation in two variables forms a line in a rectangular coordinate system. Because two points determine a line, to graph a linear equation it is sufficient to find two solution points and draw the line between them. We will find three solution points and use the third point as a check point. This process is demonstrated in Example 2.

> **Example 2**   **Graphing a Linear Equation**
>
> Graph the equation $x - 2y = 8$.
>
> **Solution:**
>
> We will find three ordered pairs that are solutions to $x - 2y = 8$. To find the ordered pairs, choose arbitrary values of x or y, such as those shown in the table. Then complete the table to find the corresponding ordered pairs.

x	y
2		$\longrightarrow (2, \quad)$
	−1	$\longrightarrow (\ , -1)$
0		$\longrightarrow (0, \quad)$

From the first row, substitute $x = 2$:

$x - 2y = 8$

$(2) - 2y = 8$

$-2y = 6$

$y = -3$

From the second row, substitute $y = -1$:

$x - 2y = 8$

$x - 2(-1) = 8$

$x + 2 = 8$

$x = 6$

From the third row, substitute $x = 0$:

$x - 2y = 8$

$(0) - 2y = 8$

$-2y = 8$

$y = -4$

Skill Practice Answers

1a. Yes   **b.** No   **c.** Yes

The completed table is shown below with the corresponding ordered pairs.

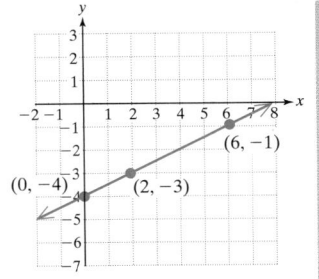

x	y
2	−3	→ (2, −3)
6	−1	→ (6, −1)
0	−4	→ (0, −4)

Figure 3-6

To graph the equation, plot the three solutions and draw the line through the points (Figure 3-6).

Skill Practice

2. Graph the equation $2x + y = 6$.

TIP: Only two points are needed to graph a line. However, in Example 2, we found a third ordered pair, (0, −4). Notice that this point "lines up" with the other two points. If the three points do not line up, then we know that a mistake was made in solving for at least one of the ordered pairs.

In Example 2, the original values for x and y given in the table were picked arbitrarily by the authors. It is important to note, however, that once you pick an arbitrary value for x, the corresponding y-value is determined by the equation. Similarly, once you pick an arbitrary value for y, the x-value is determined by the equation.

Example 3   Graphing a Linear Equation

Graph the equation $4x + 3y = 15$.

Solution:

We will find three ordered pairs that are solutions to the equation $4x + 3y = 15$. In the table, we have selected arbitrary values for x and y and must complete the ordered pairs. Notice that in this case, we are choosing zero for x and zero for y to illustrate that the resulting equation is often easy to solve.

x	y
0		→ (0,)
	0	→ (, 0)
3		→ (3,)

From the first row, substitute $x = 0$:

$$4x + 3y = 15$$
$$4(0) + 3y = 15$$
$$3y = 15$$
$$y = 5$$

From the second row, substitute $y = 0$:

$$4x + 3y = 15$$
$$4x + 3(0) = 15$$
$$4x = 15$$
$$x = \frac{15}{4} \text{ or } 3\frac{3}{4}$$

From the third row, substitute $x = 3$:

$$4x + 3y = 15$$
$$4(3) + 3y = 15$$
$$12 + 3y = 15$$
$$3y = 3$$
$$y = 1$$

Skill Practice Answers

2. $2x + y = 6$

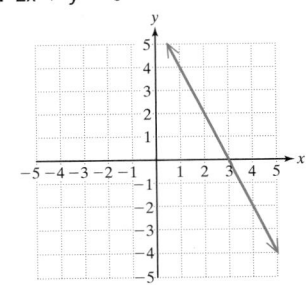

The completed table is shown with the corresponding ordered pairs.

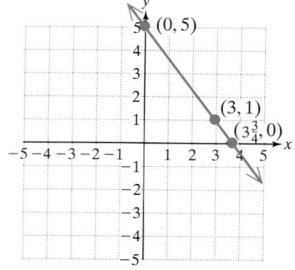

Figure 3-7

x	y	
0	5	$\longrightarrow (0, 5)$
$3\frac{3}{4}$	0	$\longrightarrow (3\frac{3}{4}, 0)$
3	1	$\longrightarrow (3, 1)$

To graph the equation, plot the three solutions and draw the line through the points (Figure 3-7).

Skill Practice

3. Graph the equation $2x + 3y = 12$.

Example 4   **Graphing a Linear Equation**

Graph the line $y = -\frac{1}{3}x + 1$.

Solution:

Because the y-variable is isolated in the equation, it is easy to substitute a value for x and simplify the right-hand side to find y. Since any number for x can be picked, choose numbers that are multiples of 3 that will simplify easily when multiplied by $-\frac{1}{3}$.

x	y
3	
0	
−3	

$$y = -\frac{1}{3}x + 1$$

Let $x = 3$:

$$y = -\frac{1}{3}(3) + 1$$
$$y = -1 + 1$$
$$y = 0$$

Let $x = 0$:

$$y = -\frac{1}{3}(0) + 1$$
$$y = 0 + 1$$
$$y = 1$$

Let $x = -3$:

$$y = -\frac{1}{3}(-3) + 1$$
$$y = 1 + 1$$
$$y = 2$$

x	y
3	0
0	1
−3	2

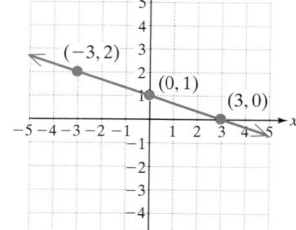

Figure 3-8

The line through the three ordered pairs $(3, 0)$, $(0, 1)$, and $(-3, 2)$ is shown in Figure 3-8. The line represents the set of all solutions to the equation $y = -\frac{1}{3}x + 1$.

Skill Practice

4. Graph the line $y = \frac{1}{2}x + 3$.

Skill Practice Answers

3. $2x + 3y = 12$

4. $y = \frac{1}{2}x + 3$

3. *x*- and *y*-Intercepts

The x- and y-intercepts are the points where the graph intersects the x- and y-axes, respectively. From Example 4, we see that the x-intercept is at the point $(3, 0)$ and the y-intercept is at the point $(0, 1)$. Notice that an x-intercept is a point on the x-axis and must have a y-coordinate of 0. Likewise, a y-intercept is a point on the y-axis and has an x-coordinate of 0.

> ***Definition of *x*- and *y*-Intercepts**
>
> An **x-intercept** is a point $(a, 0)$ where a graph intersects the x-axis.
>
> A **y-intercept** is a point $(0, b)$ where a graph intersects the y-axis.

Although any two points may be used to graph a line, in some cases it is convenient to use the x- and y-intercepts of the line. To find the x- and y-intercepts of any two-variable equation in x and y, follow these steps:

> **Finding *x*- and *y*-Intercepts**
>
> **Step 1.** Find the x-intercept(s) by substituting $y = 0$ into the equation and solving for x.
>
> **Step 2.** Find the y-intercept(s) by substituting $x = 0$ into the equation and solving for y.

| **Example 5** | **Finding the *x*- and *y*-Intercepts of a Line** |

Given the equation $-3x + 2y = 8$,

  a. Find the x-intercept.   **b.** Find the y-intercept.

  c. Graph the equation.

Solution:

a. To find the x-intercept, substitute $y = 0$.

$$-3x + 2y = 8$$
$$-3x + 2(0) = 8$$
$$-3x \quad\quad = 8$$
$$\frac{-3x}{-3} = \frac{8}{-3}$$
$$x = -\frac{8}{3}$$

The x-intercept is $\left(-\frac{8}{3}, 0\right)$.

b. To find the y-intercept, substitute $x = 0$.

$$-3x + 2y = 8$$
$$-3(0) + 2y = 8$$
$$2y = 8$$
$$y = 4$$

The y-intercept is $(0, 4)$.

---

*In some applications, an x-intercept is defined as the x-coordinate of a point of intersection that a graph makes with the x-axis. For example, if an x-intercept is at the point $(3, 0)$, it is sometimes stated simply as 3 (the y-coordinate is assumed to be 0). Similarly, a y-intercept is sometimes defined as the y-coordinate of a point of intersection that a graph makes with the y-axis. For example, if a y-intercept is at the point $(0, 7)$, it may be stated simply as 7 (the x-coordinate is assumed to be 0).

TIP: A third point such as $(-2, 1)$, can be plotted to check the line.

c. The line through the ordered pairs $\left(-\frac{8}{3}, 0\right)$ and $(0, 4)$ is shown in Figure 3-9. Note that the point $\left(-\frac{8}{3}, 0\right)$ can be written as $\left(-2\frac{2}{3}, 0\right)$.

The line represents the set of all solutions to the equation $-3x + 2y = 8$.

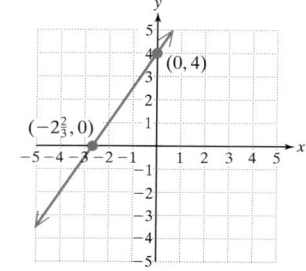

Figure 3-9

Skill Practice

5. Given the equation $4x - 3y = -9$,
   a. Find the x-intercept.
   b. Find the y-intercept.
   c. Graph the equation.

Example 6   **Finding the *x*- and *y*-Intercepts of a Line**

Given the equation $4x + 5y = 0$,

a. Find the x-intercept.   b. Find the y-intercept.

c. Graph the line.

Solution:

a. To find the x-intercept, substitute $y = 0$.

$$4x + 5y = 0$$
$$4x + 5(0) = 0$$
$$4x = 0$$
$$x = 0$$

The x-intercept is $(0, 0)$.

b. To find the y-intercept, substitute $x = 0$.

$$4x + 5y = 0$$
$$4(0) + 5y = 0$$
$$5y = 0$$
$$y = 0$$

The y-intercept is $(0, 0)$.

c. Because the x-intercept and the y-intercept are the same point (the origin), one or more additional points are needed to graph the line. In the table, we have arbitrarily selected additional values for x and y to find two more points on the line.

x	y
-5	
	2

Skill Practice Answers

5a. $\left(-\dfrac{9}{4}, 0\right)$ **b.** $(0, 3)$

c. $4x - 3y = -9$

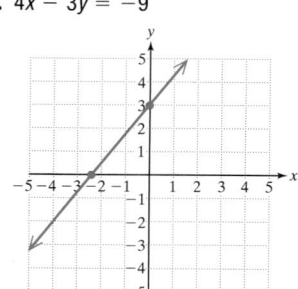

Let $x = -5$: $4x + 5y = 0$
$$4(-5) + 5y = 0$$
$$-20 + 5y = 0$$
$$5y = 20$$
$$y = 4$$

Let $y = 2$: $4x + 5y = 0$
$$4x + 5(2) = 0$$
$$4x + 10 = 0$$
$$4x = -10$$
$$x = -\frac{10}{4}$$
$$x = -\frac{5}{2}$$

x	y
-5	4
$-\frac{5}{2}$	2

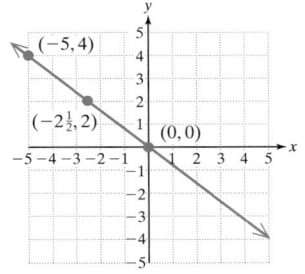

(−5,4)

(−2½,2)   (0,0)

The line through the ordered pairs $(0, 0)$, $(-5, 4)$, and $(-\frac{5}{2}, 2)$ is shown in Figure 3-10. Note that the point $(-\frac{5}{2}, 2)$ can be written as $(-2\frac{1}{2}, 2)$.

The line represents the set of all solutions to the equation $4x + 5y = 0$.

Figure 3-10

Skill Practice

6. Given the equation $2x - 3y = 0$,
   a. Find the x-intercept.
   b. Find the y-intercept.
   c. Graph the line. (*Hint:* You may need to find an additional point.)

4. Horizontal and Vertical Lines

Recall that a linear equation can be written in the form $Ax + By = C$, where A and B are not both zero. However, if A or B is 0, then the line is either parallel to the x-axis (horizontal) or parallel to the y-axis (vertical), respectively.

Definitions of Vertical and Horizontal Lines

1. A **vertical line** is a line that can be written in the form, $x = k$, where k is a constant.

2. A **horizontal line** is a line that can be written in the form, $y = k$, where k is a constant.

Example 7 **Graphing a Horizontal Line**

Graph the line $y = 3$.

Solution:

Because this equation is in the form $y = k$, the line is horizontal and must cross the y-axis at $y = 3$ (Figure 3-11).

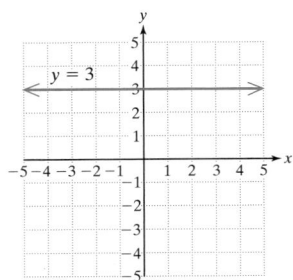

$y = 3$

Figure 3-11

Skill Practice Answers

6a. $(0, 0)$ **b.** $(0, 0)$

c. $2x - 3y = 0$

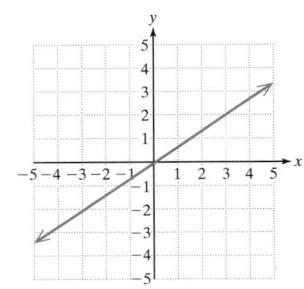

Alternative Solution:

Create a table of values for the equation $y = 3$. The choice for the y-coordinate must be 3, but x can be any real number.

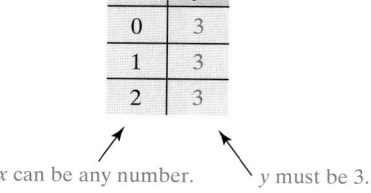

x	y
0	3
1	3
2	3

x can be any number.   *y must be 3.*

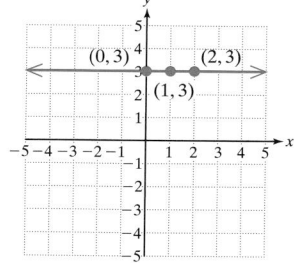

TIP: Notice that a horizontal line has a y-intercept, but does not have an x-intercept (unless the horizontal line is the x-axis itself).

Skill Practice

7. Graph the line $y = -2$.

Example 8 **Graphing a Vertical Line**

Graph the line $4x = -8$.

Solution:

Because the equation does not have a y-variable, we can solve the equation for x.

$$4x = -8 \quad \text{is equivalent to} \quad x = -2$$

This equation is in the form $x = k$, indicating that the line is vertical and must cross the x-axis at $x = -2$ (Figure 3-12).

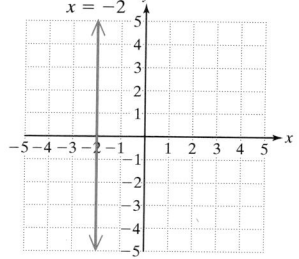

Figure 3-12

Alternative Solution:

Create a table of values for the equation $x = -2$. The choice for the x-coordinate must be -2, but y can be any real number.

x	y
-2	0
-2	3
-2	-4

x must be −2.   *y can be any number.*

TIP: Notice that a vertical line has an x-intercept but does not have a y-intercept (unless the vertical line is the y-axis itself).

Skill Practice

8. Graph the line $x = 4$.

Skill Practice Answers

7. $y = -2$

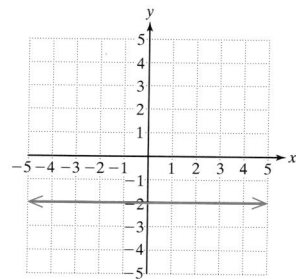

8. $x = 4$

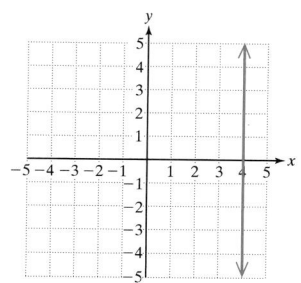

Calculator Connections

A viewing window of a graphing calculator shows a portion of a rectangular coordinate system. The standard viewing window for many calculators shows the x-axis between -10 and 10 and the y-axis between -10 and 10 (Figure 3-13). Furthermore, the scale defined by the "tick" marks on both the x- and y-axes is usually set to 1.

The "Standard Viewing Window"

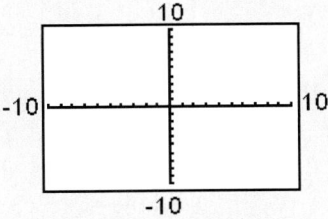

Figure 3-13

To graph an equation in x and y on a graphing calculator, the equation must be written with the y-variable isolated. Therefore, the equation $x + 3y = 3$ must first be written as $y = -\frac{1}{3}x + 1$ before it can be entered into a graphing calculator. To enter the equation $y = -\frac{1}{3}x + 1$, use parentheses around the fraction $\frac{1}{3}$. The *Graph* option displays the graph of the line.

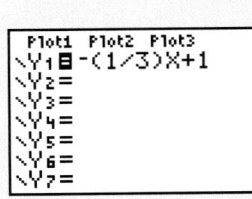

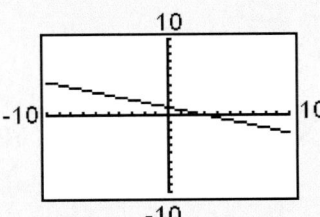

Sometimes the standard viewing window does not provide an adequate display for the graph of an equation. For example, the graph of $y = -x + 15$ is visible only in a small portion of the upper right corner of the standard viewing window.

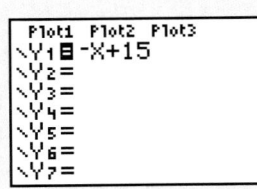

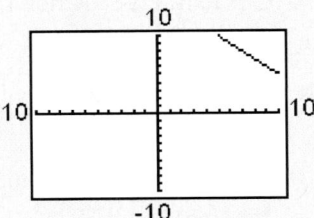

To see where this line crosses the x- and y-axes, we can change the viewing window to accommodate larger values of x and y. Most calculators have a *Range* feature or *Window* feature that allows the user to change the minimum and maximum x- and y-values.

To get a better picture of the equation $y = -x + 15$, change the minimum x-value to -10 and the maximum x-value to 20. Similarly, use a minimum y-value of -10 and a maximum y-value of 20.

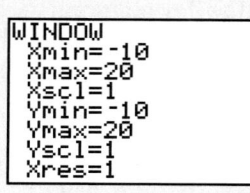

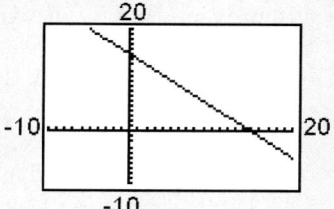

Calculator Exercises

Graph the equations on the standard viewing window.

1. $y = -2x + 5$

2. $y = 3x - 1$

3. $y = \dfrac{1}{2}x - \dfrac{7}{2}$

4. $y = -\dfrac{3}{4}x + \dfrac{5}{3}$

5. $4x - 7y = 21$

6. $2x + 3y = 12$

Graph the equations on the given viewing window.

7. $y = 3x + 15$

   Window: $-10 \le x \le 10$
             $-5 \le y \le 20$

8. $y = -2x - 25$

   Window: $-30 \le x \le 30$
             $-30 \le y \le 30$

   Xscl = 3 (sets the x-axis tick marks to increments of 3)

   Yscl = 3 (sets the y-axis tick marks to increments of 3)

9. $y = -0.2x + 0.04$

   Window: $-0.1 \le x \le 0.3$
             $-0.1 \le y \le 0.1$

   Xscl = 0.01 (sets the x-axis tick marks to increments of 0.01)

   Yscl = 0.01 (sets the y-axis tick marks to increments of 0.01)

10. $y = 0.3x - 0.5$

   Window: $-1 \le x \le 3$
             $-1 \le y \le 1$

   Xscl = 0.1 (sets the x-axis tick marks to increments of 0.1)

   Yscl = 0.1 (sets the y-axis tick marks to increments of 0.1)

Section 3.2   Practice Exercises

Study Skills Exercises

1. Check your progress by answering these questions.

Yes _____ No _____   Did you have sufficient time to study for the test on Chapter 2? If not, what could you have done to create more time for studying?

Yes _____ No _____   Did you work all of the assigned homework problems in Chapter 2?

Yes _____ No _____   If you encountered difficulty, did you see your instructor or tutor for help?

Yes _____ No _____   Have you taken advantage of the textbook supplements such as the *Student Solutions Manual* and MathZone?

2. Define the key terms:

    a. horizontal line   **b. linear equation in two variables**   **c. vertical line**

    d. *x*-intercept   **e. *y*-intercept**

Review Exercises

For Exercises 3–8, refer to the figure to give the coordinates of the labeled points, and state the quadrant or axis where the point is located.

3. A   **4.** B   **5.** C

6. D   **7.** E   **8.** F

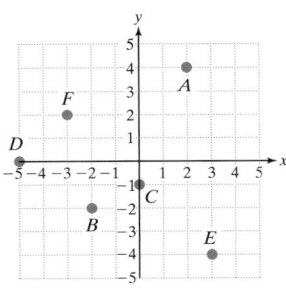

Concept 1: Solutions to Linear Equations in Two Variables

For Exercises 9–17, determine if the given ordered pair is a solution to the equation.

9. $x - y = 6$; $(8, 2)$

10. $y = 3x - 2$; $(1, 1)$

11. $y = -\dfrac{1}{3}x + 3$; $(-3, 4)$

12. $y = -\dfrac{5}{2}x + 5$; $(-2, 0)$

13. $4x + 5y = 20$; $(-5, -4)$

14. $y = 7$; $(0, 7)$

15. $y = -2$; $(-2, 6)$

16. $x = 1$; $(0, 1)$

17. $x = -5$; $(-5, 6)$

Concept 2: Graphing Linear Equations in Two Variables by Plotting Points

For Exercises 18–33, complete each table, and graph the corresponding ordered pairs. Draw the line defined by the points to represent all solutions to the equation.

18. $x + y = 3$

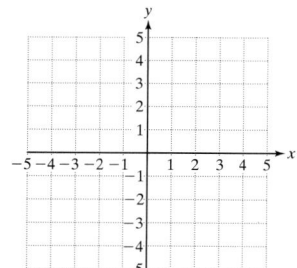

x	y
2	
	3
−1	
	0

19. $x + y = -2$

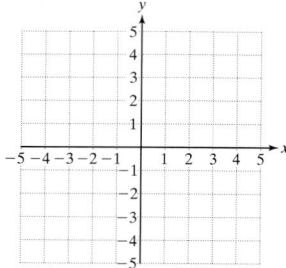

x	y
1	
	0
−3	
	2

20. $y = 5x + 1$

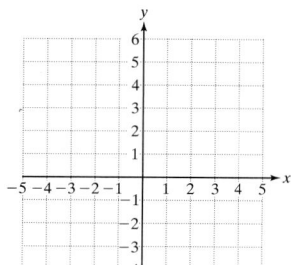

x	y
1	
	1
−1	

21. $y = -3x - 3$

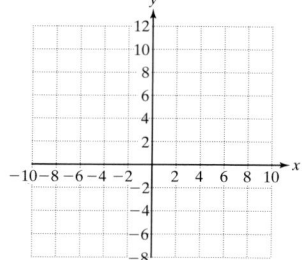

x	y
−2	
	0
−4	

22. $2x - 3y = 6$

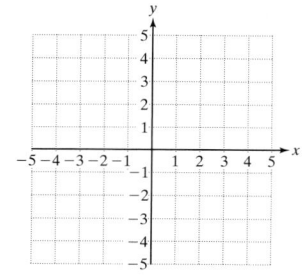

x	y
0	
	0
2	

23. $4x + 2y = 8$

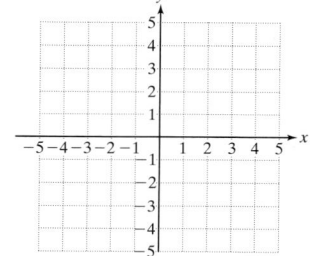

x	y
0	
	0
3	

24. $y = \dfrac{2}{7}x - 5$

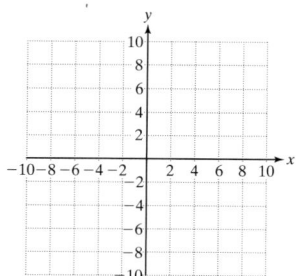

x	y
7	
−7	
0	

25. $y = -\dfrac{3}{5}x - 2$

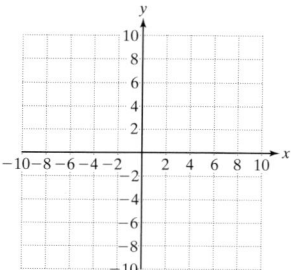

x	y
0	
5	
10	

26. $5x + 3y = 12$

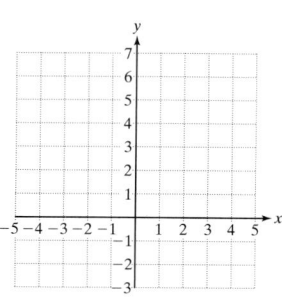

x	y
1	
	4
−1	

27. $4x - 3y = 6$

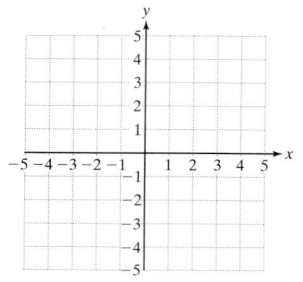

x	y
−2	
	4
0	

28. $y = 3$

x	y
2	
0	
-1	

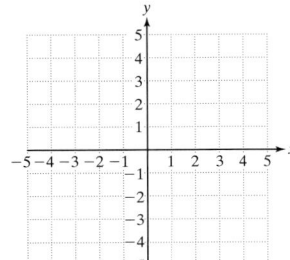

29. $y = -2$

x	y
0	
-3	
5	

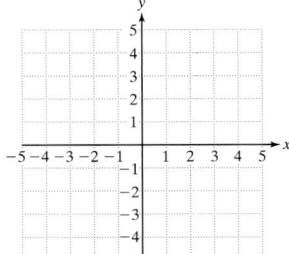

30. $x = -4$

x	y
	1
	-2
	4

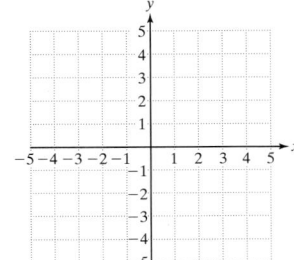

31. $x = \frac{3}{2}$

x	y
	-1
	2
	-3

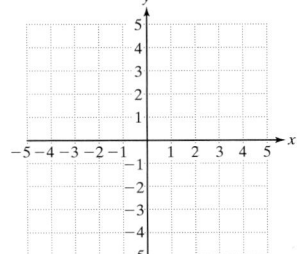

32. $y = -3.4x + 5.8$

x	y
0	
1	
2	

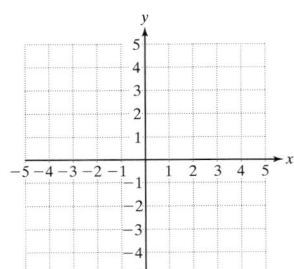

33. $y = -1.2x + 4.6$

x	y
0	
1	
2	

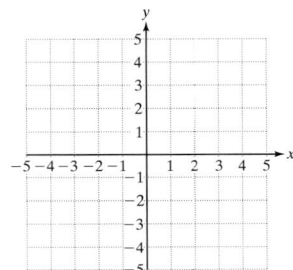

For Exercises 34–47, graph the lines by making a table of at least three ordered pairs and plotting the points.

34. $x = y + 2$

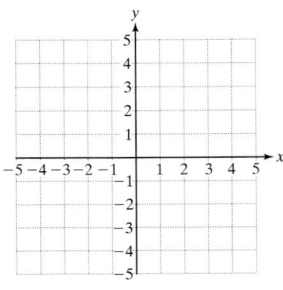

35. $x - y = 4$

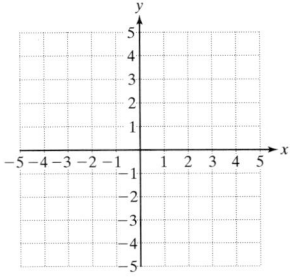

36. $-3x + y = -6$

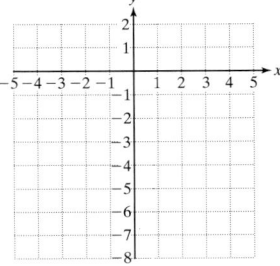

37. $2x - 5y = 10$

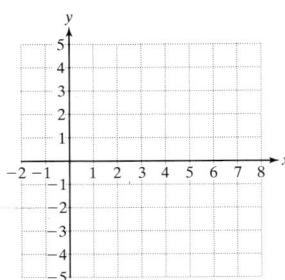

38. $y = 4x$

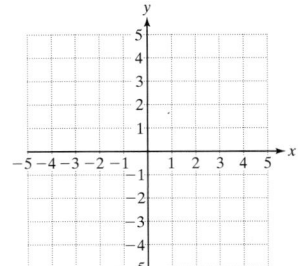

39. $y = -2x$

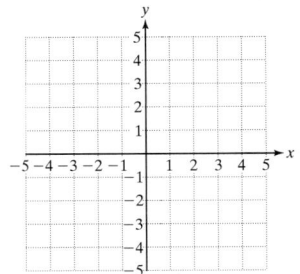

40. $y = -\dfrac{1}{2}x + 3$

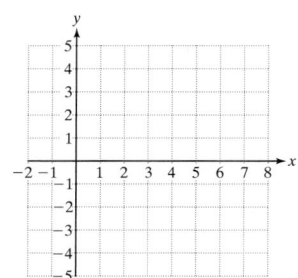

41. $y = \dfrac{1}{4}x - 2$

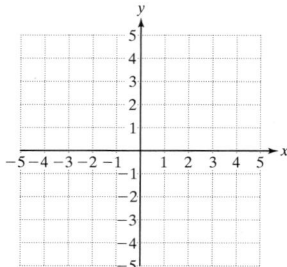

42. $x + y = 0$

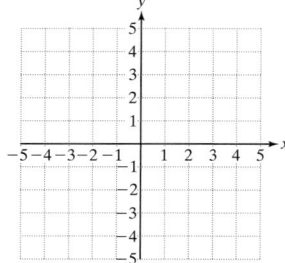

43. $-x + y = 0$

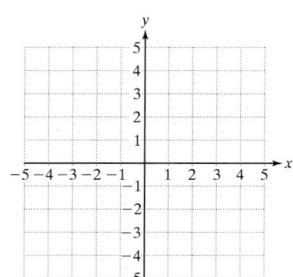

44. $2x + 3y = 8$

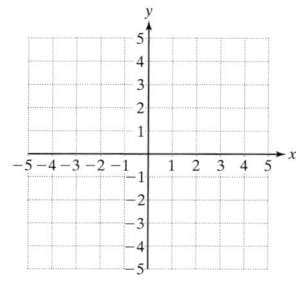

45. $4x - 5y = 15$

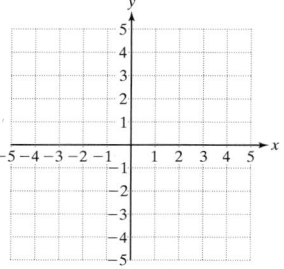

46. $50x - 40y = 200$

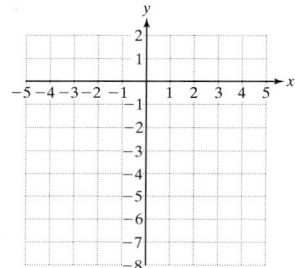

47. $-30x - 20y = 60$

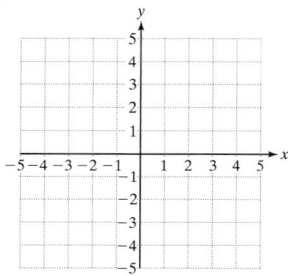

Concept 3: *x*- and *y*-Intercepts

48. The *x*-intercept is on which axis?

49. The *y*-intercept is on which axis?

For Exercises 50–53, estimate the coordinates of the *x*- and *y*-intercepts.

50.

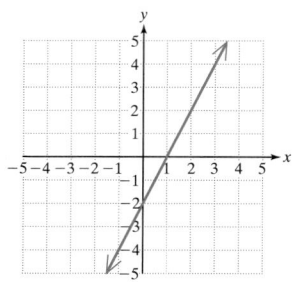

51.

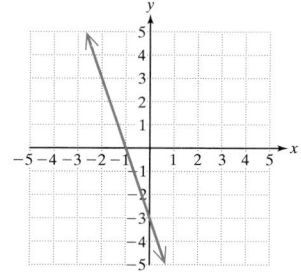

52.

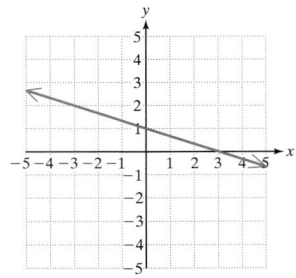

53.

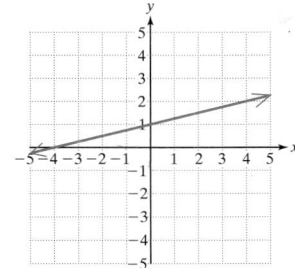

For Exercises 54–67, find the *x*- and *y*-intercepts (if they exist), and graph the line.

54. $5x + y = 5$

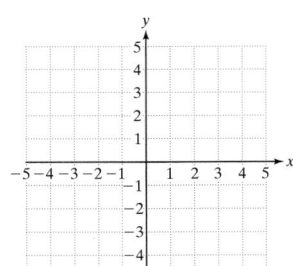

55. $x - 3y = -9$

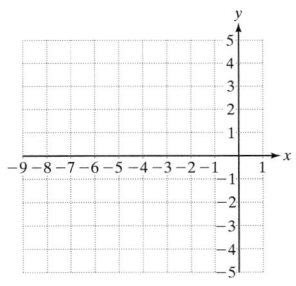

56. $y = \frac{2}{3}x - 1$

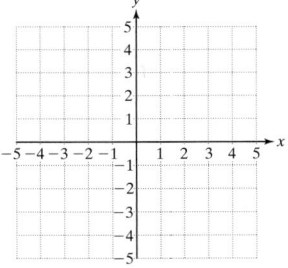

57. $y = -\frac{3}{4}x + 2$

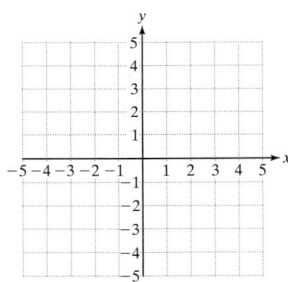

58. $x - 3 = y$

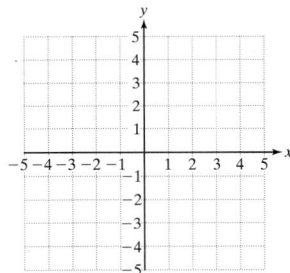

59. $2x + 8 = y$

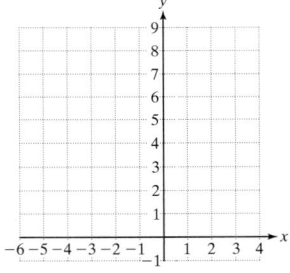

60. $-3x + y = 0$

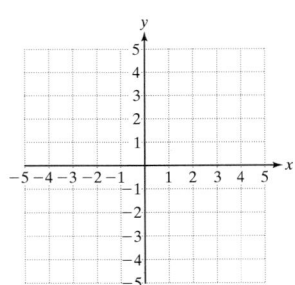

61. $2x - 2y = 0$

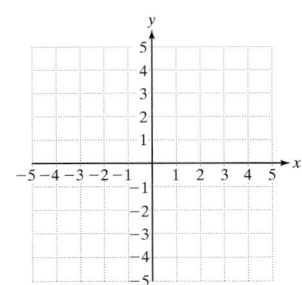

62. $25y = 10x + 100$

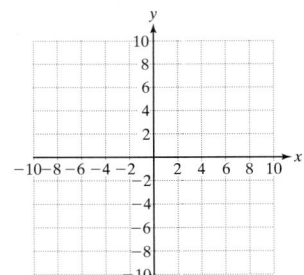

63. $20x = -40y + 200$

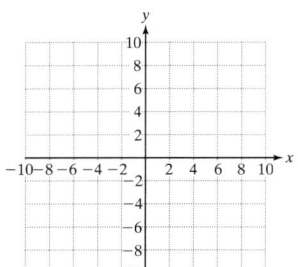

64. $1.2x - 2.4y = 3.6$

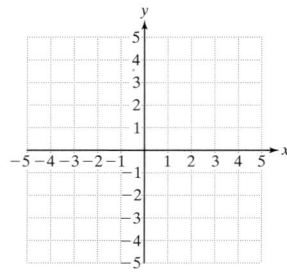

65. $-8.1x - 10.8y = 16.2$

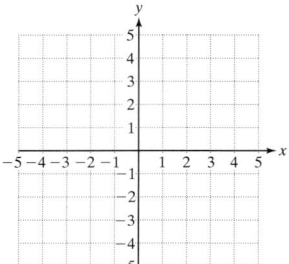

66. $x = 2y$

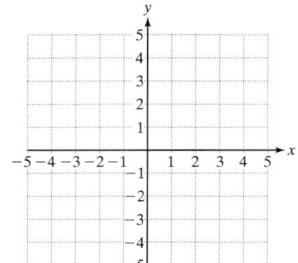

67. $x = -5y$

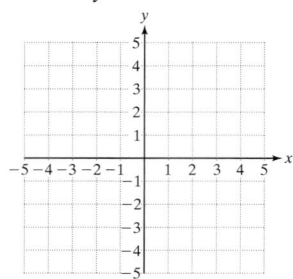

Concept 4: Horizontal and Vertical Lines

68. Explain why not every line has both an x- and a y-intercept.

69. Which of the lines will have only one intercept?

   a. $2x - 3y = 6$ **b.** $x = 5$ **c.** $2y = 8$ **d.** $-x + y = 0$

70. Which of the lines will have only one intercept?

   a. $y = 2$ **b.** $x + y = 0$ **c.** $2x - 10 = 2$ **d.** $x + 4y = 8$

For Exercises 71–74, answer true or false. If the statement is false, rewrite it to be true.

71. The line $x = 3$ is horizontal.

72. The line $y = -4$ is horizontal.

73. A line parallel to the y-axis is vertical.

74. A line perpendicular to the x-axis is vertical.

For Exercises 75–86,

a. Identify as representing a horizontal or vertical line.   **b.** Graph the line.

c. Identify the x- and y-intercepts if they exist.

75. $x = 3$

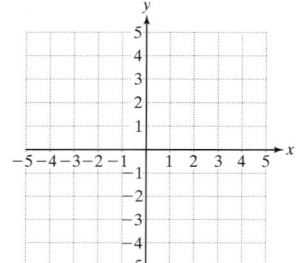

76. $y = -1$

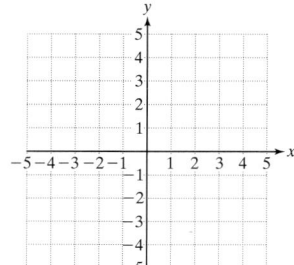

77. $-2y = 8$

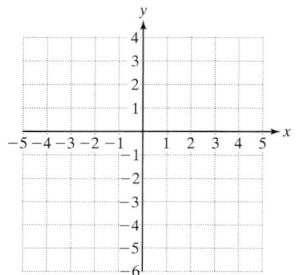

78. $5x = 20$

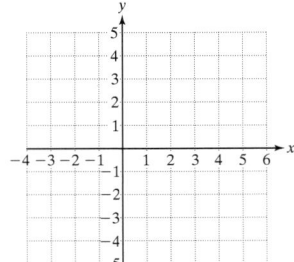

79. $x + 3 = 7$

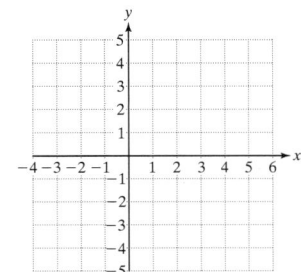

80. $y - 8 = -13$

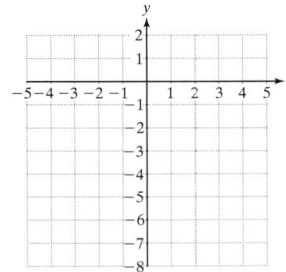

81. $3y = 0$

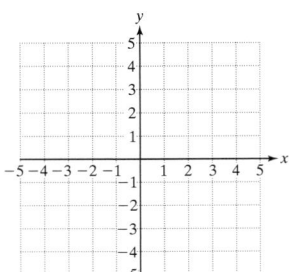

82. $5x = 0$

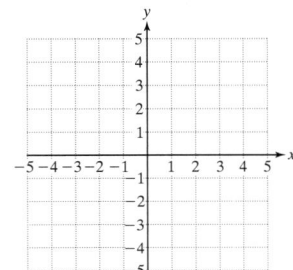

83. $2x + 7 = 10$

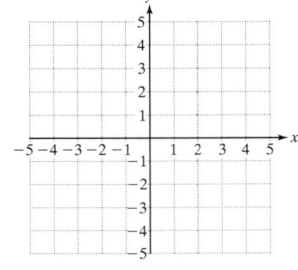

84. $-3y + 2 = 9$

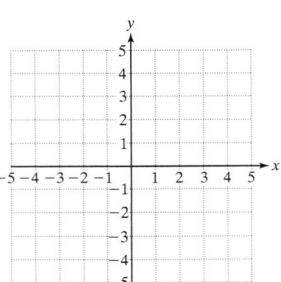

85. $9 = 3 + 4y$

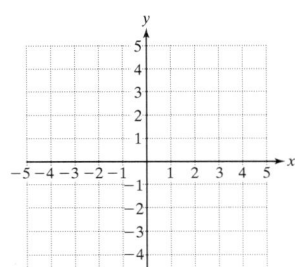

86. $7 = -2x - 5$

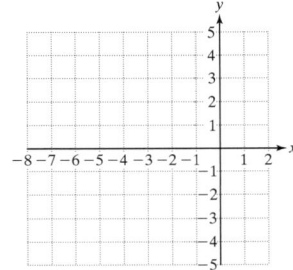

Expanding Your Skills

87. The students in the ninth grade at Atlantic High School pick up aluminum cans to be recycled. The current value of aluminum is $0.69 per pound. If the students pay $20 to rent a truck to haul the cans, then the following equation expresses the amount of money that they earn, y, given the number of pounds of aluminum, x.

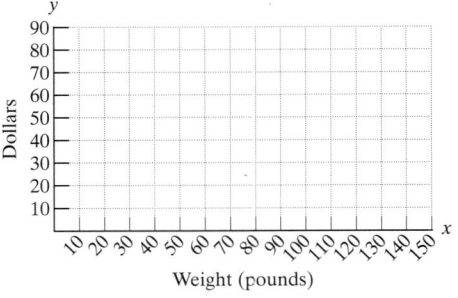

$$y = 0.69x - 20 \quad (x \geq 0)$$

a. Let $x = 55$ and solve for y.

b. Let $y = 80.05$ and solve for x.

c. Write the ordered pairs from parts (a) and (b), and interpret their meaning in the context of the problem.

d. Graph the ordered pairs and the line defined by the points.

88. The store "CDs R US" sells all compact disks for $13.99. The following equation represents the revenue, y, (in dollars) generated by selling x CDs.

$$y = 13.99x \quad (x \geq 0)$$

a. Find y when $x = 13$.

b. Find x when $y = 279.80$.

c. Write the ordered pairs from parts (a) and (b), and interpret their meaning in the context of the problem.

d. Graph the ordered pairs and the line defined by the points.

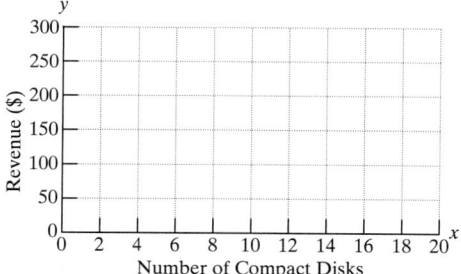

89. The value of a car depreciates once it is driven off of the dealer's lot. For a Hyundai Accent, the value of the car is given by the equation $y = -1531x + 11{,}599 \ (x \geq 0)$ where y is the value of the car in dollars x years after its purchase. (*Source: Kelly Blue Book*)

a. Find y when $x = 1$.

b. Find x when $y = 7006$.

c. Write the ordered pairs from parts (a) and (b), and interpret their meaning in the context of the problem.

Section 3.3   Slope of a Line

Concepts

1. Introduction to Slope
2. Slope Formula
3. Parallel and Perpendicular Lines
4. Applications of Slope

1. Introduction to Slope

The x- and y-intercepts represent the points where a line crosses the x- and y-axes. Another important feature of a line is its slope. Geometrically, the slope of a line measures the "steepness" of the line. For example, two ski runs are depicted by the lines in Figure 3-14.

Beginner's Hill   Daredevil Hill

Figure 3-14

By visual inspection, Daredevil Hill is "steeper" than Beginner's Hill. To measure the slope of a line quantitatively, consider two points on the line. The slope of the line is the ratio of the vertical change (change in y) between the two points and the horizontal change (change in x). As a memory device, we might think of the slope of a line as "rise over run." See Figure 3-15.

$$\text{Slope} = \frac{\text{change in } y}{\text{change in } x} = \frac{\text{rise}}{\text{run}}$$

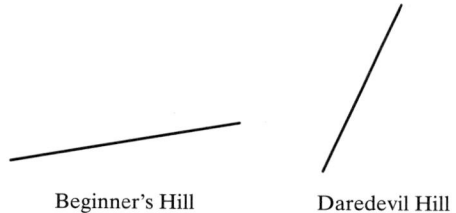

Figure 3-15

To move from point A to point B on Beginner's Hill, rise 2 ft and move to the right 6 ft (Figure 3-16).

To move from point A to point B on Daredevil Hill, rise 12 ft and move to the right 6 ft (Figure 3-17).

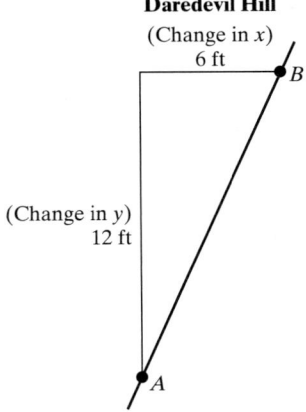

Daredevil Hill
(Change in x)
6 ft

(Change in y)
12 ft

Figure 3-17

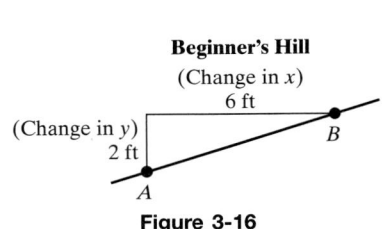

Beginner's Hill
(Change in x)
6 ft

(Change in y)
2 ft

Figure 3-16

$$\text{Slope} = \frac{\text{change in } y}{\text{change in } x} = \frac{2 \text{ ft}}{6 \text{ ft}} = \frac{1}{3}$$

$$\text{Slope} = \frac{\text{change in } y}{\text{change in } x} = \frac{12 \text{ ft}}{6 \text{ ft}} = \frac{2}{1} = 2$$

The slope of Daredevil Hill is greater than the slope of Beginner's Hill, confirming the observation that Daredevil Hill is steeper. On Daredevil Hill, there is a 12-ft change in elevation for every 6 ft of horizontal distance (a 2:1 ratio). On Beginner's Hill there is only a 2-ft change in elevation for every 6 ft of horizontal distance (a 1:3 ratio).

Example 1   **Finding Slope in an Application**

Find the slope of the ramp up the stairs.

Solution:

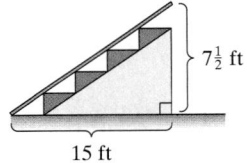

$7\frac{1}{2}$ ft

15 ft

$$\text{Slope} = \frac{\text{change in } y}{\text{change in } x} = \frac{7\frac{1}{2} \text{ ft}}{15 \text{ ft}}$$

$$\frac{7\frac{1}{2}}{15} = \frac{\frac{15}{2}}{\frac{15}{1}} \qquad \text{Write the mixed number as an improper fraction.}$$

$$\frac{15}{2} \cdot \frac{1}{15} = \frac{1}{2} \qquad \text{Multiply by the reciprocal and simplify.}$$

The slope is $\frac{1}{2}$.

Skill Practice

1. Calculate the slope of the aircraft's takeoff path.

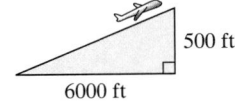

500 ft

6000 ft

Skill Practice Answers

1. $\dfrac{500}{6000} = \dfrac{1}{12}$

2. Slope Formula

The slope of a line may be found using any two points on the line—call these points (x_1, y_1) and (x_2, y_2). The change in y between the points can be found by taking the difference of the y-values: $y_2 - y_1$. The change in x can be found by taking the difference of the x-values in the same order: $x_2 - x_1$ (Figure 3-18).

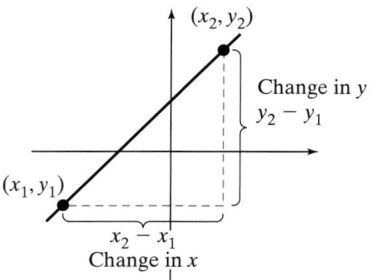

Figure 3-18

The slope of a line is often symbolized by the letter m and is given by the following formula.

Definition of the Slope of a Line

The **slope** of a line passing through the distinct points (x_1, y_1) and (x_2, y_2) is

$$m = \frac{y_2 - y_1}{x_2 - x_1} \quad \text{provided } x_2 - x_1 \neq 0$$

Example 2   Finding the Slope of a Line Given Two Points

Find the slope of the line through the points $(-1, 3)$ and $(-4, -2)$.

Solution:

To use the slope formula, first label the coordinates of each point and then substitute the coordinates into the slope formula.

$$\underset{(x_1, y_1)}{(-1, 3)} \quad \text{and} \quad \underset{(x_2, y_2)}{(-4, -2)} \qquad \text{Label the points.}$$

$$m = \frac{y_2 - y_1}{x_2 - x_1} = \frac{(-2) - (3)}{(-4) - (-1)} \qquad \text{Apply the slope formula.}$$

$$= \frac{-5}{-3}; \quad \text{hence, } m = \frac{5}{3} \qquad \text{Simplify to lowest terms.}$$

The slope of the line can be verified from the graph (Figure 3-19).

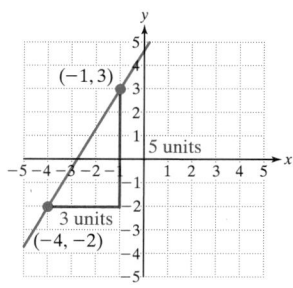

Figure 3-19

Skill Practice   Find the slope of the line through the given points.

2. $(-5, 2)$ and $(1, 3)$

TIP: The slope formula is not dependent on which point is labeled (x_1, y_1) and which point is labeled (x_2, y_2). In Example 2, reversing the order in which the points are labeled results in the same slope.

$$\underset{(x_2,\, y_2)}{(-1, 3)} \quad \text{and} \quad \underset{(x_1,\, y_1)}{(-4, -2)} \qquad \text{Label the points.}$$

$$m = \frac{(3) - (-2)}{(-1) - (-4)} = \frac{5}{3} \qquad \text{Apply the slope formula.}$$

When you apply the slope formula, you will see that the slope of a line may be positive, negative, zero, or undefined.

- Lines that increase, or rise, from left to right have a positive slope.
- Lines that decrease, or fall, from left to right have a negative slope.
- Horizontal lines have a slope of zero.
- Vertical lines have an undefined slope.

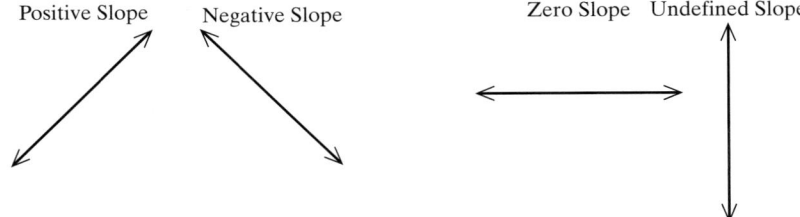

Positive Slope   Negative Slope   Zero Slope   Undefined Slope

---

Example 3 **Finding the Slope of a Line Given Two Points**

Find the slope of the line passing through the points $(-5, 0)$ and $(2, -3)$.

Solution:

$$\underset{(x_1,\, y_1)}{(-5, 0)} \quad \text{and} \quad \underset{(x_2,\, y_2)}{(2, -3)} \qquad \text{Label the points.}$$

$$m = \frac{y_2 - y_1}{x_2 - x_1} = \frac{(-3) - (0)}{(2) - (-5)} \qquad \text{Apply the slope formula.}$$

$$= \frac{-3}{7} \quad \text{or} \quad -\frac{3}{7} \qquad \text{Simplify.}$$

By graphing the points $(-5, 0)$ and $(2, -3)$, we can verify that the slope is $-\frac{3}{7}$ (Figure 3-20). Notice that the line slopes downward from left to right.

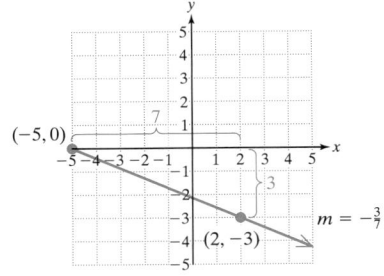

Figure 3-20

Skill Practice Find the slope of the line through the given points.

3. $(0, -8)$ and $(-2, -2)$

Skill Practice Answers

3. -3

| **Example 4** | **Determining the Slope of a Horizontal and Vertical Line** |

a. Find the slope of the line passing through the points $(2, -1)$ and $(2, 4)$.

b. Find the slope of the line passing through the points $(3, -2)$ and $(-4, -2)$.

Solution:

a. $\underset{(x_1, y_1)}{(2, -1)}$ and $\underset{(x_2, y_2)}{(2, 4)}$ Label the points.

$$m = \frac{y_2 - y_1}{x_2 - x_1} = \frac{(4) - (-1)}{(2) - (2)}$$ Apply the slope formula.

$$m = \frac{5}{0} \quad \text{Undefined}$$

Because the slope, m, is undefined, we expect the points to form a vertical line as shown in Figure 3-21.

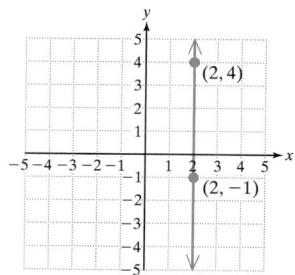

Figure 3-21

b. $\underset{(x_1, y_1)}{(3, -2)}$ and $\underset{(x_2, y_2)}{(-4, -2)}$ Label the points.

$$m = \frac{y_2 - y_1}{x_2 - x_1} = \frac{(-2) - (-2)}{(-4) - (3)}$$ Apply the slope formula.

$$m = \frac{-2 + 2}{-4 - 3} = \frac{0}{-7} = 0$$

Because the slope is 0, we expect the points to form a horizontal line, as shown in Figure 3-22.

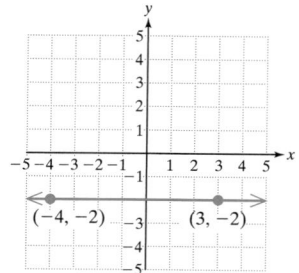

Figure 3-22

| **Skill Practice** |

4. Find the slope of the line through the given points.

    a. $(5, 6)$ and $(5, -2)$ **b.** $(3, 8)$ and $(-5, 8)$

Skill Practice Answers

4a. Undefined   **b.** 0

3. Parallel and Perpendicular Lines

Lines in the same plane that do not intersect are called **parallel lines**. Parallel lines have the same slope and different y-intercepts (Figure 3-23).

Lines that intersect at a right angle are **perpendicular lines**. If two lines are perpendicular, then the slope of one line is the opposite of the reciprocal of the slope of the other line (provided neither line is vertical) (Figure 3-24).

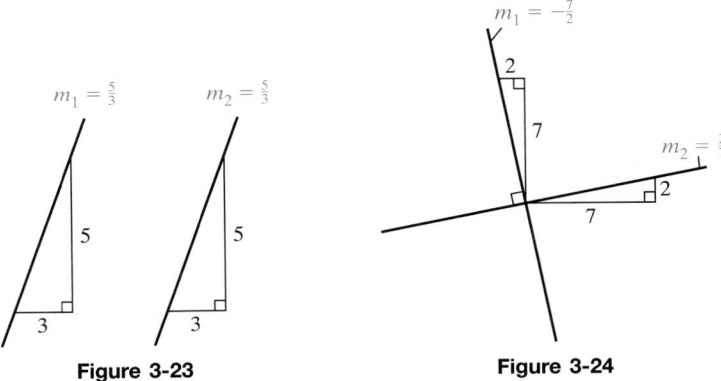

Figure 3-23   **Figure 3-24**

Slopes of Parallel Lines

If m_1 and m_2 represent the slopes of two parallel (nonvertical) lines, then

$$m_1 = m_2.$$

See Figure 3-23.

Slopes of Perpendicular Lines

If $m_1 \neq 0$ and $m_2 \neq 0$ represent the slopes of two perpendicular lines, then

$$m_1 = -\frac{1}{m_2} \text{ or equivalently, } m_1 m_2 = -1. \text{ See Figure 3-24.}$$

Example 5 **Determining the Slope of Parallel and Perpendicular Lines**

Suppose a given line has a slope of $-\frac{1}{4}$.

   a. Find the slope of a line parallel to the given line.

   b. Find the slope of a line perpendicular to the given line.

Solution:

   a. Parallel lines must have the same slope. The slope of a line parallel to the given line is $m = -\dfrac{1}{4}$.

   b. Perpendicular lines must have opposite and reciprocal slopes. The slope of a line perpendicular to the given line is $m = +\dfrac{4}{1}$ or simply, $m = 4$.

5. A given line has a slope of $\frac{5}{3}$.

   a. Find the slope of a line parallel to the given line.

   b. Find the slope of a line perpendicular to the given line.

4. Applications of Slope

In many applications, the interpretation of slope refers to the *rate of change* of the *y*-variable to the *x*-variable.

| Example 6 | Interpreting Slope in an Application

Mario earns $10.00/hr working for a landscaping company. Shannelle earns $15.00/hr working for an in-home nursing agency. Figure 3-25 shows their total earnings versus the number of hours they work.

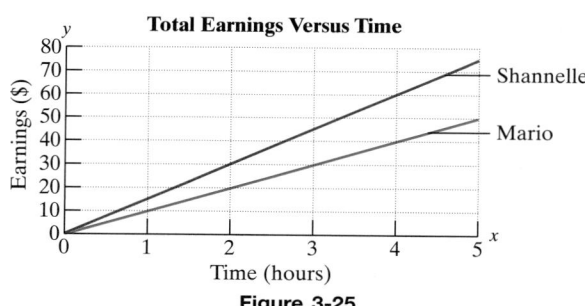

Figure 3-25

   a. Find the slope of the line representing Mario's earnings.

   b. Find the slope of the line representing Shannelle's earnings.

Solution:

   a. After 1 hr, Mario earns $10. After 2 hr, he earns $20, and so on. For each 1-hr change in time, there is a $10 increase in wages. The slope of the line representing Mario's earnings is $10/hr.

   b. After 1 hr, Shannelle earns $15. After 2 hr, she earns $30, and so on. For each 1-hr change in time, there is a $15 increase in wages. The slope of the line representing Shannelle's earnings is $15/hr.

| Skill Practice |

6. Anita invested in a stock that went up $50 in 4 months. David invested in a stock that went up $15 in 2 months. Let the variable *y* represent the amount that a stock is worth. Let *x* represent the time in months that a stock is invested.

   a. Find the slope of the line represented by Anita's earnings.

   b. Find the slope of the line representing David's earnings.

TIP: To find the slope for Example 6(a) we can pick points from the graph, such as (1, 10) and (2, 20). Then use the slope formula to calculate the slope.

$$m = \frac{y_2 - y_1}{x_2 - x_1} \text{ becomes}$$

$$m = \frac{20 - 10}{2 - 1} = \frac{10}{1} = 10$$

Skill Practice Answers

5a. $\frac{5}{3}$ **b.** $-\frac{3}{5}$

6a. $\dfrac{\$50}{4 \text{ months}}$ or $12.50/month

b. $\dfrac{\$15}{2 \text{ months}}$ or $7.50/month

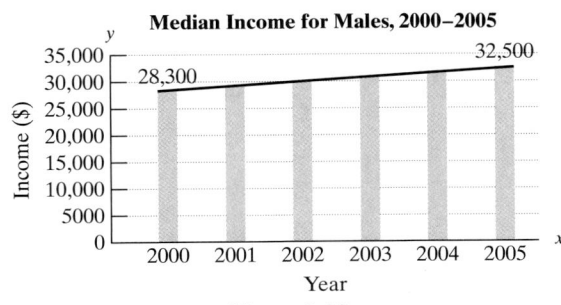

| Example 7 | Interpreting Slope in an Application |

Figure 3-26 shows the annual median income for males in the United States between 2000 and 2005. The trend is approximately linear. Find the slope of the line and interpret the meaning of the slope in the context of this problem.

Median Income for Males, 2000–2005

Figure 3-26
(Source: U.S. Department of the Census)

Solution:

To determine the slope, we need to know two points on the line. From the graph, the median income for males in 2000 was $28,300. This corresponds to the ordered pair (2000, 28300). In 2005, the median income was $32,500. This corresponds to the ordered pair (2005, 32500).

$$(2000, 28300) \quad \text{and} \quad (2005, 32500)$$
$$(x_1, y_1) \qquad\qquad\quad (x_2, y_2) \qquad\qquad \text{Label the points.}$$

$$m = \frac{y_2 - y_1}{x_2 - x_1} = \frac{32{,}500 - 28{,}300}{2005 - 2000} \qquad \text{Apply the slope formula.}$$

$$= \frac{4200}{5} \text{ or } 840 \qquad\qquad \text{Simplify.}$$

The slope indicates that the median income for males in the United States increased at a rate of approximately $840 per year between 2000 and 2005.

Skill Practice

7. In the year 2000, the population of Alaska was approximately 630,000. By 2005, it had grown to 670,000. Use the ordered pairs (2000, 630000) and (2005, 670000) to determine the slope of the line through the points. Then interpret the meaning in the context of this problem.

Skill Practice Answers

7. $m = 8000$; The population of Alaska is increasing by 8000 people per year.

Section 3.3   Practice Exercises

Study Skills Exercises

1. Make up a practice test for yourself. Use examples or exercises from the text. Be sure to cover each concept that was presented.

2. Define the key terms:

    a. parallel lines   **b. perpendicular lines**   **c. slope**

Review Exercises

For Exercises 3–8, find the *x*- and *y*-intercepts (if they exist). Then graph the lines.

3. $x - 3y = 6$

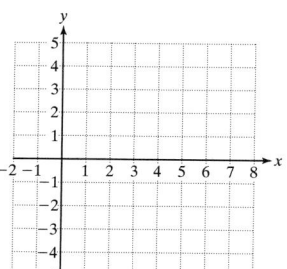

4. $x - 5 = 2$

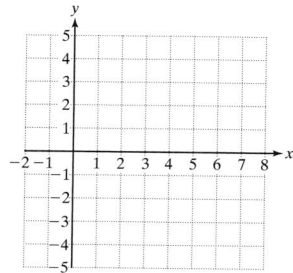

5. $y = \dfrac{2}{3}x$

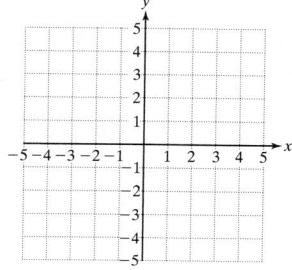

6. $2y - 3 = 0$

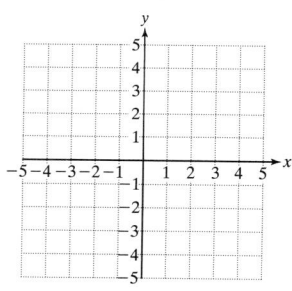

7. $4x + y = 8$

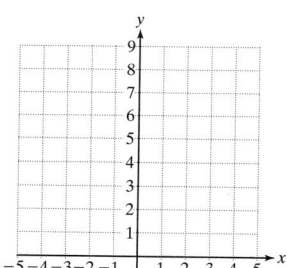

8. $2x = 4y$

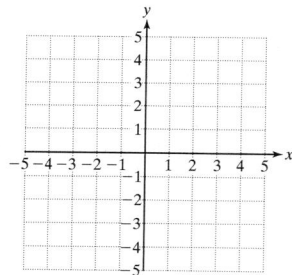

Concept 1: Introduction to Slope

9. Determine the pitch (slope) of the roof.

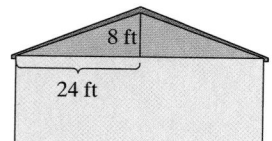

10. Determine the slope of the stairs.

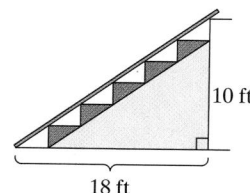

11. Determine the slope of the ramp.

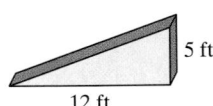

5 ft

12 ft

12. Determine the slope of the treadmill.

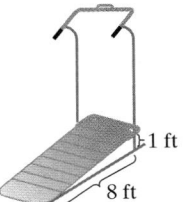

1 ft

8 ft

Concept 2: Slope Formula

For Exercises 13–16, fill in the blank with the appropriate term: zero, negative, positive, or undefined.

13. The slope of a line parallel to the *y*-axis is _____.

14. The slope of a horizontal line is _____.

15. The slope of a line that rises from left to right is _____.

16. The slope of a line that falls from left to right is _____.

For Exercises 17–24, label the lines as having a positive, negative, zero, or undefined slope.

17.

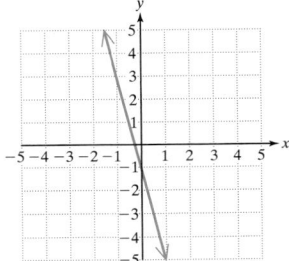

18.

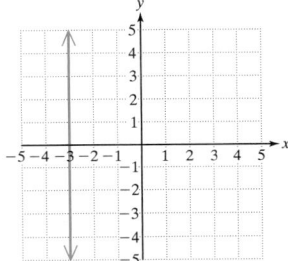

19.

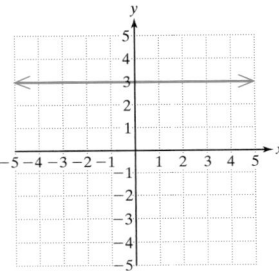

20.

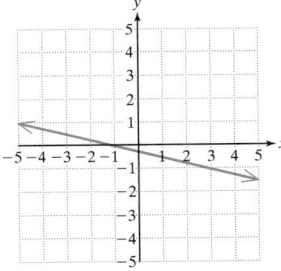

21.

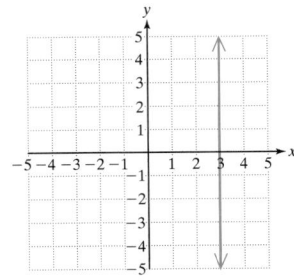

22.

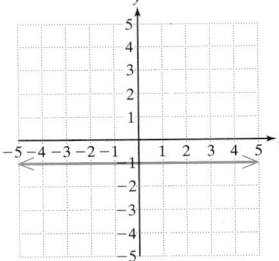

23.

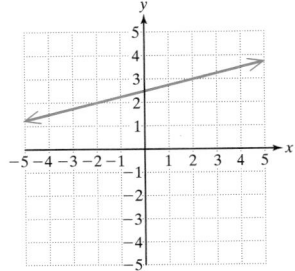

24.

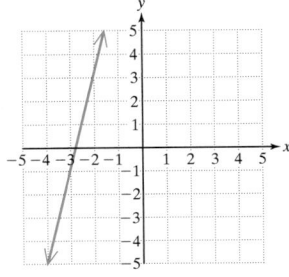

For Exercises 25–32, determine the slope by using the slope formula and any two points on the line. Check your answer by drawing a right triangle and labeling the "rise" and "run."

25.
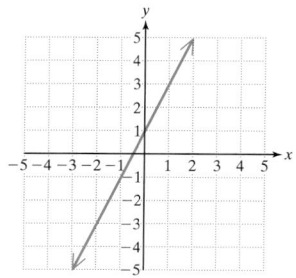

26.
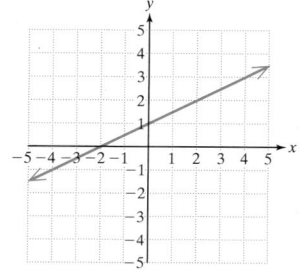

27.
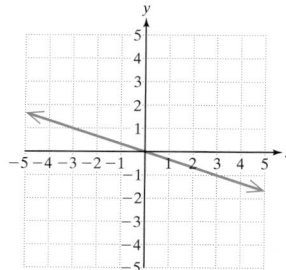

28.
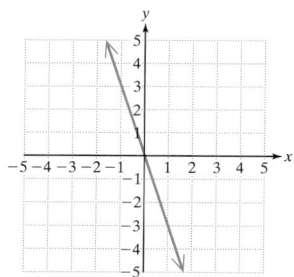

29.
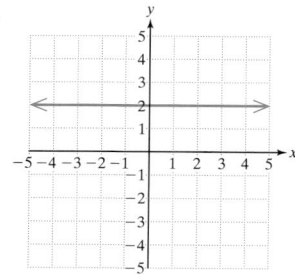

30.
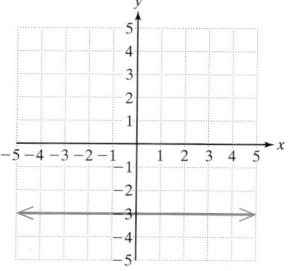

31.
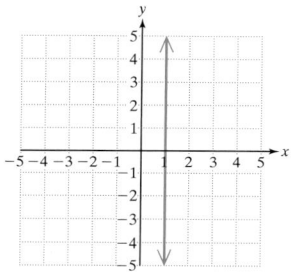

32.
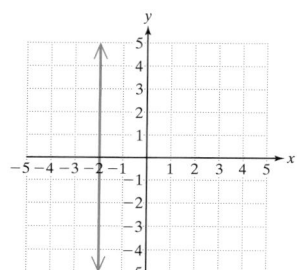

For Exercises 33–50, find the slope of the line that passes through the two points.

33. $(2, 4)$ and $(-1, 3)$

34. $(0, 4)$ and $(3, 0)$

35. $(-2, 3)$ and $(-1, 0)$

36. $(-3, -4)$ and $(1, -5)$

37. $(1, 5)$ and $(-4, 2)$

38. $(-6, -1)$ and $(-2, -3)$

39. $(5, 3)$ and $(-2, 3)$

40. $(0, -1)$ and $(-4, -1)$

41. $(2, -7)$ and $(2, 5)$

42. $(-4, 3)$ and $(-4, -4)$

43. $\left(\dfrac{1}{2}, \dfrac{3}{5}\right)$ and $\left(\dfrac{1}{4}, -\dfrac{4}{5}\right)$

44. $\left(-\dfrac{2}{7}, \dfrac{1}{3}\right)$ and $\left(\dfrac{8}{7}, -\dfrac{5}{6}\right)$

45. $(3, -1)$ and $(-5, 6)$

46. $(-6, 5)$ and $(-10, 4)$

47. $(6.8, -3.4)$ and $(-3.2, 1.1)$

48. $(-3.15, 8.25)$ and $(6.85, -4.25)$

49. $(1994, 3.5)$ and $(2000, 2.6)$

50. $(1988, 4.65)$ and $(1998, 9.25)$

Concept 3: Parallel and Perpendicular Lines

For Exercises 51–58, information regarding the slope of a line is given.

a. Determine the slope of a line parallel to the given line.

b. Determine the slope of a line perpendicular to the given line.

51. $m = -2$ **52.** $m = \dfrac{2}{3}$

53. $m = 0$ **54.** The slope is undefined.

55. $m = \dfrac{4}{5}$ **56.** $m = -4$

57. The slope is undefined.   **58.** $m = 0$

For Exercises 59–66, find the slopes of the lines l_1 and l_2 determined by the two given points. Then identify whether l_1 and l_2 are parallel, perpendicular, or neither.

59. l_1: $(2, 4)$ and $(-1, -2)$
l_2: $(1, 7)$ and $(0, 5)$

60. l_1: $(0, 0)$ and $(-2, 4)$
l_2: $(1, -5)$ and $(-1, -1)$

61. l_1: $(1, 9)$ and $(0, 4)$
l_2: $(5, 2)$ and $(10, 1)$

62. l_1: $(3, -4)$ and $(-1, -8)$
l_2: $(5, -5)$ and $(-2, 2)$

63. l_1: $(4, 4)$ and $(0, 3)$
l_2: $(1, 7)$ and $(-1, -1)$

64. l_1: $(3, 5)$ and $(-2, -5)$
l_2: $(2, 0)$ and $(-4, -3)$

65. l_1: $(3.1, 6.3)$ and $(3.1, -5.7)$
l_2: $(1.2, 4.7)$ and $(1.2, -5.3)$

66. l_1: $(4.5, -6.7)$ and $(-2.3, -6.7)$
l_2: $(-2.2, 6.7)$ and $(-1.4, 6.7)$

Concept 4: Applications of Slope

67. In 1980, there were 304 thousand male inmates in federal and state prisons. By 2005, the number increased to 1479 thousand.

Let x represent the year, and let y represent the number of prisons (in thousands).

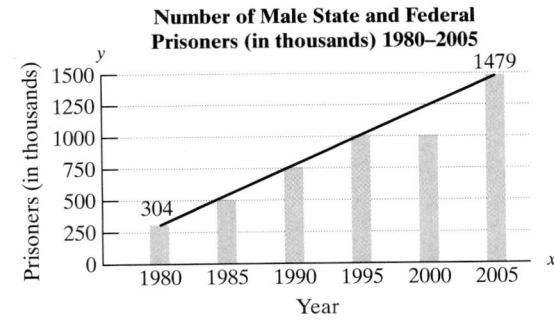

Number of Male State and Federal Prisoners (in thousands) 1980–2005

(Source: U.S. Bureau of Justice Statistics)

a. Using the ordered pairs $(1980, 304)$ and $(2005, 1479)$, find the slope of the line.

b. Interpret the slope in the context of this problem.

68. In the year 1980, there were 12 thousand female inmates in federal and state prisons. By 2005, the number increased to 102 thousand.

Let x represent the year, and let y represent the number of prisoners (in thousands).

a. Using the ordered pairs (1980, 12) and (2005, 102), find the slope of the line.

b. Interpret the slope in the context of this problem.

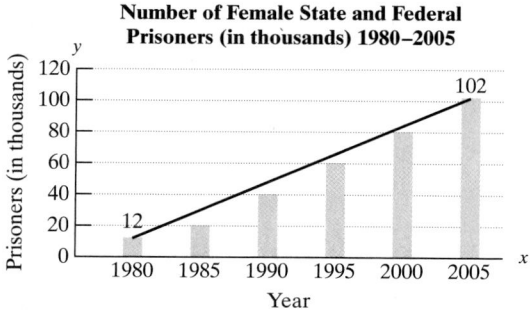

Number of Female State and Federal Prisoners (in thousands) 1980–2005

(Source: U.S. Bureau of Justice Statistics)

69. The following graph shows the median income for females in the United States between 2000 and 2005.

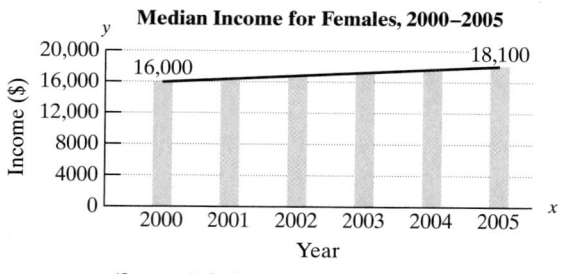

Median Income for Females, 2000–2005

(Source: U.S. Department of the Census)

a. Find the slope of the line and interpret the meaning in the context of this problem.

b. Compare the slopes for the rise in median income per year for women and for men (see Example 7). Based on the slopes of the two lines, will the median income for women ever catch up to the median income for men? Explain.

70. Jorge is paid by the hour according to the equation

$P = 11.50x$ P is his total pay (in dollars) and x is the number of hours worked.

a. How much money will Jorge earn if he works 20 hr?

b. How much money will Jorge earn if he works 21 hr?

c. How much money will Jorge earn if he works 22 hr?

d. What is the slope of the line? Interpret the meaning of the slope in the context of this problem.

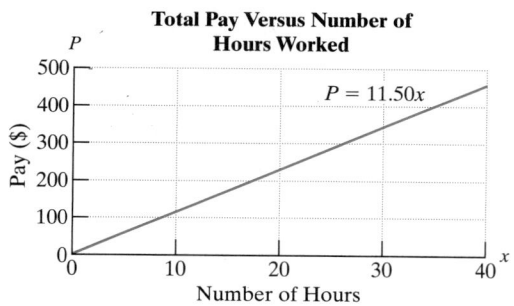

Total Pay Versus Number of Hours Worked

71. The distance, d (in miles), between a lightning strike and an observer is given by the equation $d = 0.2t$, where t is the time (in seconds) between seeing lightning and hearing thunder.

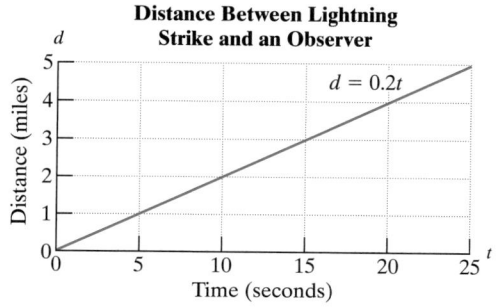

Distance Between Lightning Strike and an Observer

a. If an observer counts 5 sec between seeing lightning and hearing thunder, how far away was the lightning strike?

b. If an observer counts 10 sec between seeing lightning and hearing thunder, how far away was the lightning strike?

c. If an observer counts 15 sec between seeing lightning and hearing thunder, how far away was the lightning strike?

d. What is the slope of the line? Interpret the meaning of the slope in the context of this problem.

Mixed Exercises

For Exercises 72–77, determine the slope of the line passing through points *A* and *B*.

72. Point *A* is located 3 units up and 4 units to the right of point *B*.

73. Point *A* is located 2 units up and 5 units to the left of point *B*.

74. Point *A* is located 3 units up and 3 units to the left of point *B*.

75. Point *A* is located 2 units down and 2 units to the left of point *B*.

76. Point *A* is located 5 units to the right of point *B*.

77. Point *A* is located 3 units down from point *B*.

78. Graph the line through the point $(1, -2)$ having slope $\frac{2}{3}$. Then give two other points on the line.

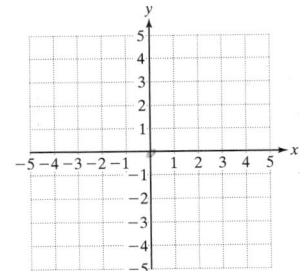

79. Graph the line through the point $(-2, -3)$ having slope $\frac{3}{4}$. Then give two other points on the line.

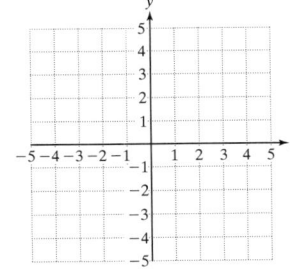

80. Graph the line through the point $(2, 2)$ having slope -3. Then give two other points on the line.

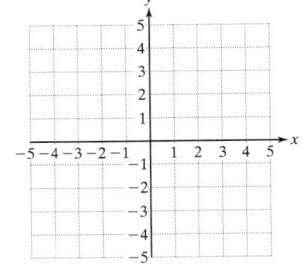

81. Graph the line through the point $(-1, 3)$ having slope -2. Then give two other points on the line.

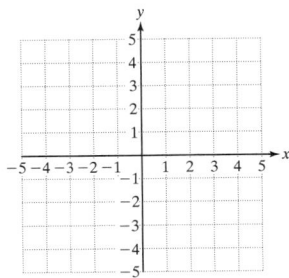

For Exercises 82–89, draw a line as indicated. Answers may vary.

82. Draw a line with a positive slope and a positive *y*-intercept.

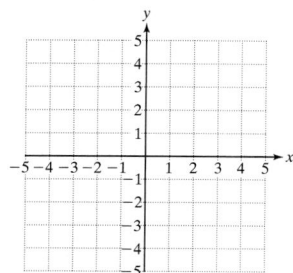

83. Draw a line with a positive slope and a negative *y*-intercept.

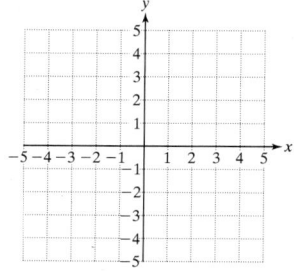

84. Draw a line with a negative slope and a negative *y*-intercept.

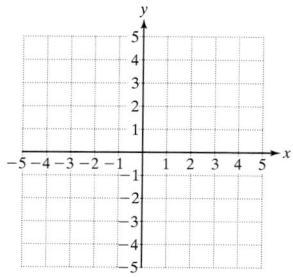

85. Draw a line with a negative slope and positive *y*-intercept.

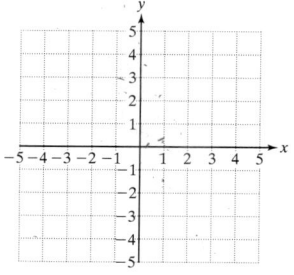

86. Draw a line with a zero slope and a positive *y*-intercept.

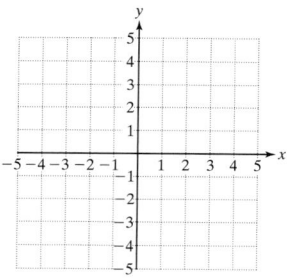

87. Draw a line with a zero slope and a negative *y*-intercept.

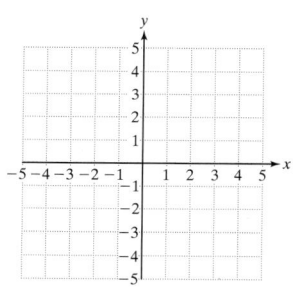

88. Draw a line with undefined slope and a negative *x*-intercept.

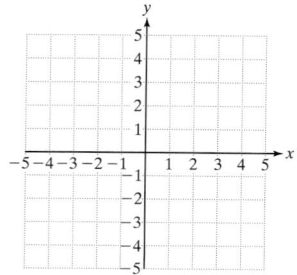

89. Draw a line with undefined slope and a positive *x*-intercept.

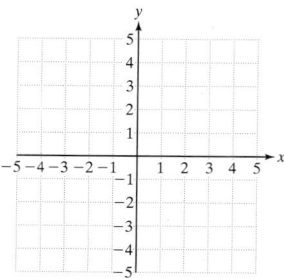

Expanding Your Skills

90. Find the slope between the points $(a + b, 4m - n)$ and $(a - b, m + 2n)$,

91. Find the slope between the points $(3c - d, s + t)$ and $(c - 2d, s - t)$.

92. Find the x-intercept of the line $ax + by = c$.

93. Find the y-intercept of the line $ax + by = c$.

94. Find another point on the line that contains the point $(2, -1)$ and has a slope of $\frac{2}{5}$.

95. Find another point on the line that contains the point $(-3, 4)$ and has a slope of $\frac{1}{4}$.

Slope-Intercept Form of a Line

1. Slope-Intercept Form of a Line

In Section 3.2, we learned that an equation of the form $Ax + By = C$ (where A and B are not both zero) represents a line in a rectangular coordinate system. An equation of a line written in this way is in **standard form**. In this section, we will learn a new form, called **slope-intercept form**, which is useful in determining the slope and y-intercept of a line.

Let $(0, b)$ represent the y-intercept of a line. Let (x, y) represent any other point on the line. Then the slope of the line can be found as follows:

Let $(0, b)$ represent (x_1, y_1), and let (x, y) represent (x_2, y_2). Apply the slope formula.

Concepts

1. Slope-Intercept Form of a Line
2. Graphing a Line from Its Slope and y-Intercept
3. Determining Whether Two Lines Are Parallel, Perpendicular, or Neither
4. Writing an Equation of a Line Given Its Slope and y-Intercept

$$m = \frac{(y_2 - y_1)}{(x_2 - x_1)} \rightarrow m = \frac{y - b}{x - 0} \qquad \text{Apply the slope formula.}$$

$$m = \frac{y - b}{x} \qquad \text{Simplify.}$$

$$mx = \left(\frac{y - b}{x}\right)x \qquad \text{Multiply by } x \text{ to clear fractions.}$$

$$mx = y - b$$

$$mx + b = y - b + b \qquad \text{To isolate } y, \text{ add } b \text{ to both sides.}$$

$$mx + b = y \quad \text{or} \quad y = mx + b \qquad \text{The equation is in slope-intercept form.}$$

Slope-Intercept Form of a Line

$y = mx + b$ is the slope-intercept form of a line.

m is the slope and the point $(0, b)$ is the y-intercept.

> **Example 1** Identifying the Slope and *y*-Intercept of a Line

For each equation, identify the slope and *y*-intercept.

a. $y = 3x - 1$ **b.** $y = 4x$ **c.** $y = -2.7x + 5$ **d.** $y = 5$

Solution:

Each equation is written in slope-intercept form, $y = mx + b$. The slope is the coefficient of *x*, and the *y*-intercept is determined by the constant term.

a. $y = 3x - 1$ The slope is 3.   The *y*-intercept is $(0, -1)$.

b. $y = 4x$ can be written as
$y = 4x + 0$.   The slope is 4.   The *y*-intercept is $(0, 0)$.

c. $y = -2.7x + 5$ The slope is -2.7.   The *y*-intercept is $(0, 5)$.

d. $y = 5$ can be written as
$y = 0x + 5$.   The slope is 0.   The *y*-intercept is $(0, 5)$.

> **Skill Practice** Identify the slope and the *y*-intercept.

1. $y = 4x + 6$ **2.** $y = -\dfrac{3}{4}x$ **3.** $y = 3.56x - 4.27$ **4.** $y = -7$

Given the equation of a line, we can write the equation in slope-intercept form by solving the equation for the *y*-variable. This is demonstrated in Example 2.

> **Example 2** Identifying the Slope and *y*-Intercept of a Line

Given the line $-5x - 2y = 6$,

a. Write the slope-intercept form of the line.
b. Identify the slope and *y*-intercept.

Solution:

a. Write the equation in slope-intercept form, $y = mx + b$, by solving for *y*.

$$-5x - 2y = 6$$
$$-2y = 5x + 6 \quad \text{Add } 5x \text{ to both sides.}$$
$$\frac{-2y}{-2} = \frac{5x}{-2} + \frac{6}{-2} \quad \text{Divide both sides by } -2.$$
$$y = -\frac{5}{2}x - 3 \quad \text{Slope-intercept form}$$

b. The slope is $-\frac{5}{2}$, and the *y*-intercept is $(0, -3)$.

> **Skill Practice**

5. Given the equation of the line $2x - 6y = -3$,
a. Write the slope-intercept form of the line.
b. Identify the slope and the *y*-intercept.

Skill Practice Answers

1. slope: 4; *y*-intercept: $(0, 6)$
2. slope: $-\dfrac{3}{4}$; *y*-intercept: $(0, 0)$
3. slope: 3.56; *y*-intercept: $(0, -4.27)$
4. slope: 0; *y*-intercept: $(0, -7)$
5a. $y = \dfrac{1}{3}x + \dfrac{1}{2}$
b. slope is $\dfrac{1}{3}$; *y*-intercept is: $\left(0, \dfrac{1}{2}\right)$

2. Graphing a Line from Its Slope and *y*-Intercept

Slope-intercept form is a useful tool to graph a line. The *y*-intercept is a known point on the line. The slope indicates the direction of the line and can be used to find a second point. Using slope-intercept form to graph a line is demonstrated in Example 3.

Example 3 Graphing a Line Using the Slope and *y*-Intercept

Graph the line $y = -\frac{5}{2}x - 3$ by using the slope and *y*-intercept.

Solution:

First plot the *y*-intercept, $(0, -3)$.

The slope, $m = -\frac{5}{2}$ can be written as

$$m = \frac{-5}{2} \quad \longleftarrow \text{The change in } y \text{ is } -5.$$
$$\quad \longleftarrow \text{The change in } x \text{ is } 2.$$

To find a second point on the line, start at the *y*-intercept and move down 5 units and to the right 2 units. Then draw the line through the two points (Figure 3-27).

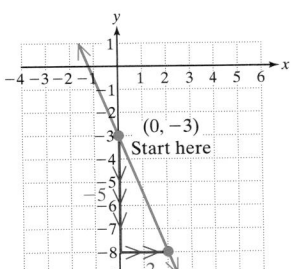

Figure 3-27

Similarly, the slope can be written as

$$m = \frac{5}{-2} \quad \longleftarrow \text{The change in } y \text{ is } 5.$$
$$\quad \longleftarrow \text{The change in } x \text{ is } -2.$$

To find a second point, start at the *y*-intercept and move up 5 units and to the left 2 units. Then draw the line through the two points (Figure 3-28).

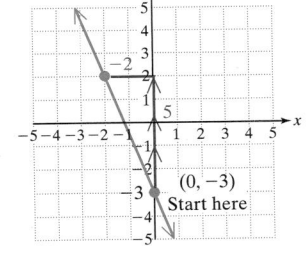

Figure 3-28

Skill Practice

6. Graph the line $y = 2x - 3$ by using the slope and the *y*-intercept.

Example 4 Graphing a Line from Its Slope and *y*-Intercept

Graph the line $y = 4x$ by using the slope and *y*-intercept.

Solution:

The line can be written as $y = 4x + 0$. Therefore, we can plot the *y*-intercept at $(0, 0)$. The slope $m = 4$ can be written as

$$m = \frac{4}{1} \quad \begin{array}{l} \longleftarrow \text{The change in } y \text{ is } 4. \\ \longleftarrow \text{The change in } x \text{ is } 1. \end{array}$$

Skill Practice Answers

6. $y = 2x - 3$

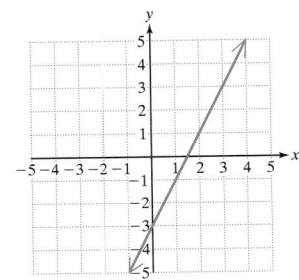

To find a second point on the line, start at the *y*-intercept and move up 4 units and to the right 1 unit. Then draw the line through the two points (Figure 3-29).

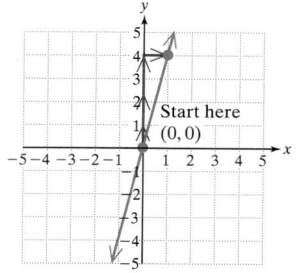

Figure 3-29

Skill Practice

7. Graph the line by using the slope and the *y*-intercept.

$$y = -\frac{1}{4}x$$

3. Determining Whether Two Lines Are Parallel, Perpendicular, or Neither

The slope-intercept form provides a means to find the slope of a line by inspection. Furthermore, if the slopes of two lines are known, then we can compare the slopes to determine if the lines are parallel, perpendicular, or neither parallel nor perpendicular. (Recall that two distinct nonvertical lines are parallel if their slopes are equal. Two lines are perpendicular if the slope of one line is the opposite of the reciprocal of the slope of the other line.)

Example 5   **Determining If Two Lines Are Parallel, Perpendicular, or Neither**

For each pair of lines, determine if they are parallel, perpendicular, or neither.

a. $l_1:$ $y = 3x - 5$ **b.** $l_1:$ $x - 3y = -9$

  $l_2:$ $y = 3x + 1$ $l_2:$ $3x = -y + 4$

c. $l_1:$ $y = \frac{3}{2}x + 2$ **d.** $l_1:$ $x = 2$

  $l_2:$ $y = \frac{2}{3}x + 1$ $l_2:$ $2y = 8$

Solution:

a. $l_1:$ $y = 3x - 5$ The slope of l_1 is 3.

  $l_2:$ $y = 3x + 1$ The slope of l_2 is 3.

Because the slopes are the same, the lines are parallel.

b. First write the equation of each line in slope-intercept form.

$l_1:$ $x - 3y = -9$ $l_2:$ $3x = -y + 4$

  $-3y = -x - 9$ $3x + y = 4$

  $\dfrac{-3y}{-3} = \dfrac{-x}{-3} - \dfrac{9}{-3}$ $y = -3x + 4$

  $y = \dfrac{1}{3}x + 3$

$l_1:$ $y = \frac{1}{3}x + 3$ The slope of l_1 is $\frac{1}{3}$.

$l_2:$ $y = -3x + 4$ The slope of l_2 is -3.

The slope of $\frac{1}{3}$ is the opposite of the reciprocal of -3. Therefore, the lines are perpendicular.

Skill Practice Answers

7. $y = -\dfrac{1}{4}x$

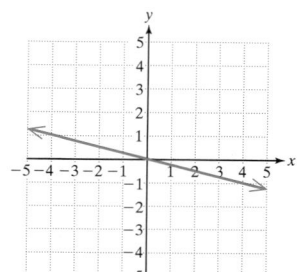

c. l_1: $y = \frac{3}{2}x + 2$ The slope of l_1 is $\frac{3}{2}$.

    l_2: $y = \frac{2}{3}x + 1$ The slope of l_2 is $\frac{2}{3}$.

The slopes are not the same. Therefore, the lines are not parallel. The values of the slopes are reciprocals, but they are not opposite in sign. Therefore, the lines are not perpendicular. The lines are neither parallel nor perpendicular.

d. The equation $x = 2$ represents a vertical line because the equation is in the form $x = k$.

The equation $2y = 8$ can be simplified to $y = 4$, which represents a horizontal line.

In this example, we do not need to analyze the slopes because vertical lines and horizontal lines are perpendicular.

Skill Practice   For each pair of lines determine if they are parallel, perpendicular, or neither.

8. $y = 3x - 5$

    $y = -3x - 15$

9. $x - 5y = 10$

    $5x - 1 = -y$

10. $y = \dfrac{5}{6}x - \dfrac{1}{2}$

     $y = \dfrac{5}{6}x + \dfrac{1}{2}$

11. $y = -5$

     $x = 6$

4. Writing an Equation of a Line Given Its Slope and y-Intercept

The slope-intercept form of a line can be used to write an equation of a line when the slope is known and the y-intercept is known.

Example 6   **Writing an Equation of a Line Using Slope-Intercept Form**

Write an equation of the line with a slope of $\frac{2}{3}$ and y-intercept $(0, 8)$.

Solution:

The slope is given as $m = \frac{2}{3}$, and the y-intercept $(0, b)$ is given as $(0, 8)$. Substitute the values $m = \frac{2}{3}$ and $b = 8$, into the slope-intercept form of a line.

$$y = mx + b$$
$$y = \frac{2}{3}x + 8$$

Skill Practice

12. Write an equation of the line with slope of -4 and y-intercept $(0, -10)$.

Skill Practice Answers

8. Neither   **9.** Perpendicular
10. Parallel   **11.** Perpendicular
12. $y = -4x - 10$

Example 7	**Writing an Equation of a Line Given Its Slope and *y*-Intercept**

Write an equation of the line with a slope of -2 that passes through the origin.

Solution:

The slope is given as $m = -2$. Furthermore, the line passes through the origin. Therefore, the *y*-intercept is $(0, 0)$ and the corresponding value of *b* is 0. The slope-intercept form of the line becomes:

$$y = mx + b$$

$$y = -2x + 0 \text{ or equivalently } y = -2x.$$

Skill Practice Answers

13. $y = \dfrac{6}{5}x$

Skill Practice

13. Write an equation of the line with a slope of $\frac{6}{5}$ that passes through the origin.

Calculator Connections

In Example 5(b) we found that the lines $y = \frac{1}{3}x + 3$ and $y = -3x + 4$ are perpendicular. We can verify our results by graphing the lines on a graphing calculator.

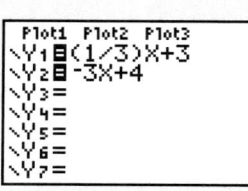 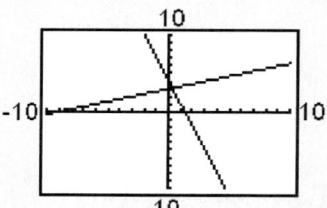

Notice that the lines do not appear perpendicular in the calculator display. That is, they do not appear to form a right angle at the point of intersection. Because many calculators have a rectangular screen, the standard viewing window is elongated in the horizontal direction. To eliminate this distortion, try using a *ZSquare* option. This feature will set the viewing window so that equal distances on the display denote an equal number of units on the graph.

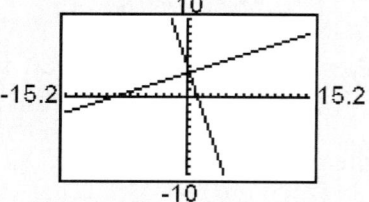

Calculator Exercises

For each pair of lines, determine if the lines are parallel, perpendicular, or neither. Then use a square viewing window to graph the lines on a graphing calculator to verify your results.

1. $x + y = 1$

   $x - y = -3$

2. $3x + y = -2$

   $6x + 2y = 6$

3. $2x - y = 4$

   $3x + 2y = 4$

4. Graph the lines: $y = x + 1$ and $y = 0.99x + 3$. Are these lines parallel? Explain.

5. Graph the lines $y = -2x - 1$ and $y = -2x - 0.99$. Are these lines the same? Explain.

Study Skills Exercises

1. When taking a test, go through the test and do all the problems that you know first. Then go back and work on the problems that were more difficult. Give yourself a time limit for how much time you spend on each problem (maybe 3 to 5 minutes the first time through). Circle the importance of each statement.

	not important	somewhat important	very important
a. Read through the entire test first.	1	2	3
b. If time allows, go back and check each problem.	1	2	3
c. Write out all steps instead of doing the work in your head.	1	2	3

2. Define the key terms:

    a. slope-intercept form of a line   **b. standard form of a line**

Review Exercises

For Exercises 3–9, determine the x- and y-intercepts, if they exist.

3. $x - 5y = 10$ **4.** $3x + y = -12$ **5.** $3y = -9$ **6.** $2 + y = 5$

7. $-4x = 6y$ **8.** $-x + 3 = 8$ **9.** $5x = 20$

Concept 1: Slope-Intercept Form of a Line

For Exercises 10–29, identify the slope and y-intercept, if they exist.

10. $y = -2x + 3$ **11.** $y = \dfrac{2}{3}x + 5$ **12.** $y = x - 2$

13. $y = -x + 6$ **14.** $y = -x$ **15.** $y = -4x$

16. $y = \dfrac{3}{4}x - 1$ **17.** $y = x - \dfrac{5}{3}$ **18.** $2x - 5y = 4$

19. $3x + 2y = 9$ **20.** $3x - y = 5$ **21.** $7x - 3y = -6$

22. $x + y = 6$ **23.** $x - y = 1$ **24.** $x + 6 = 8$

25. $-4 + x = 1$ **26.** $-8y = 2$ **27.** $1 - y = 9$

28. $3y - 2x = 0$ **29.** $5x = 6y$

Concept 2: Graphing a Line from Its Slope and *y*-Intercept

For Exercises 30–33, graph the line using the slope and *y*-intercept.

30. Graph the line through the point $(0, 2)$, having a slope of -4.

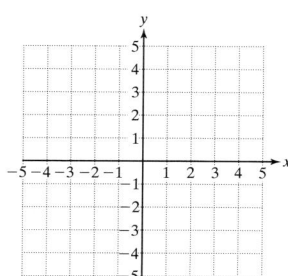

31. Graph the line through the point $(0, -1)$, having a slope of -3.

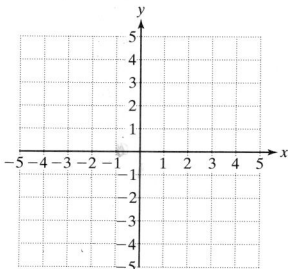

32. Graph the line through the point $(0, -5)$, having a slope of $\frac{3}{2}$.

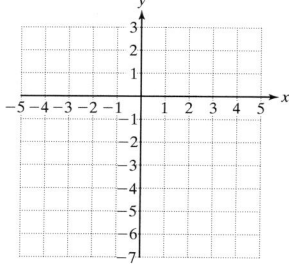

33. Graph the line through the point $(0, 3)$, having a slope of $-\frac{1}{4}$.

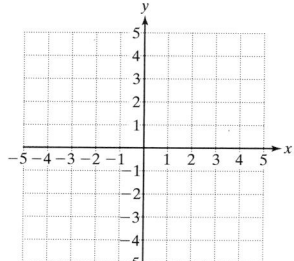

For Exercises 34–39, match the equation with the graph (a–f) by identifying if the slope is positive or negative and if the *y*-intercept is positive, negative, or zero.

34. $y = 2x + 3$

35. $y = -3x - 2$

36. $y = -\frac{1}{3}x + 3$

37. $y = \frac{1}{2}x - 2$

38. $y = x$

39. $y = -2x$

a.

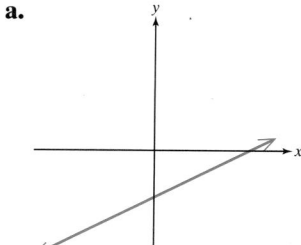

b.

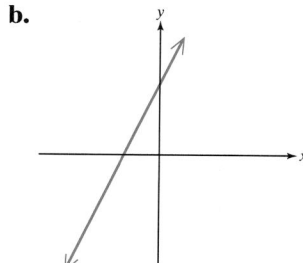

c.

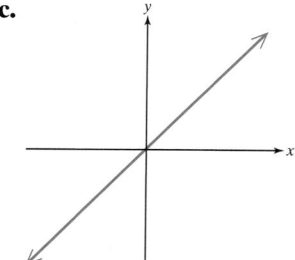

d.

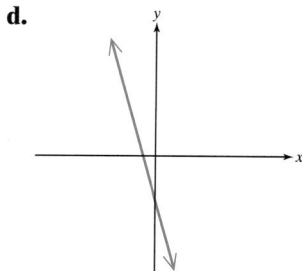

e.

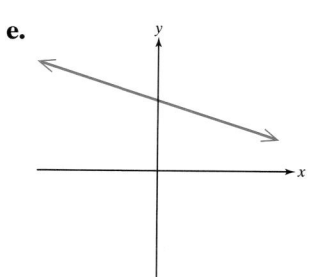

f.

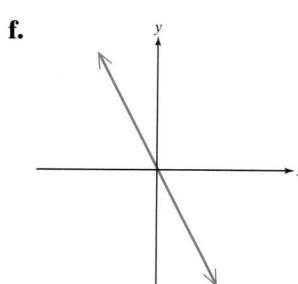

For Exercises 40–55, write each equation in slope-intercept form (if possible) and graph the line.

40. $x - 2y = 6$

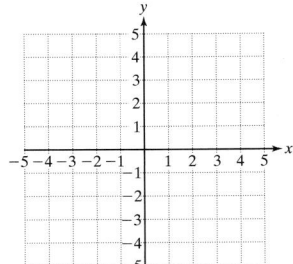

41. $5x - 2y = 2$

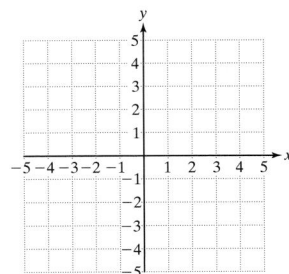

42. $2x + y = 9$

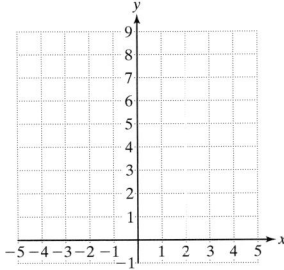

43. $-6x + y = 8$

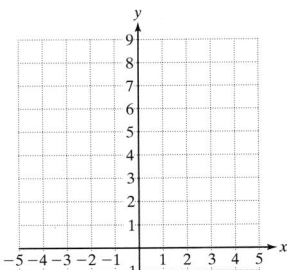

44. $2x = -4y + 6$

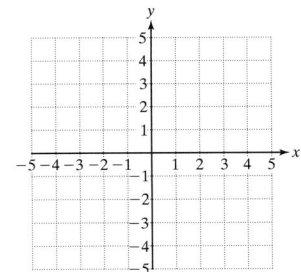

45. $3x = y - 7$

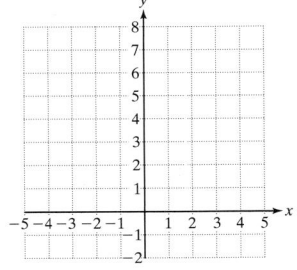

46. $x + y = 0$

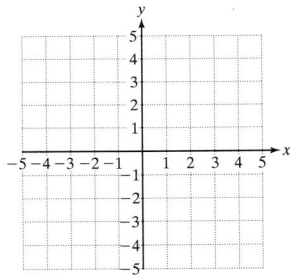

47. $x - y = 0$

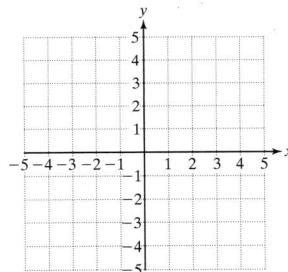

48. $5y = 4x$

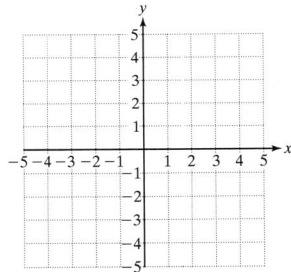

49. $-2x = 5y$

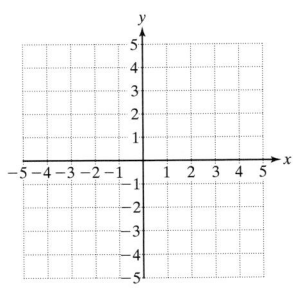

50. $3y + 2 = 0$

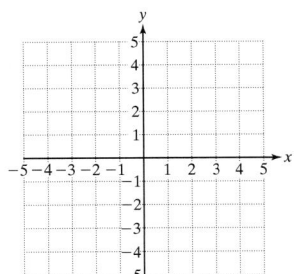

51. $1 + 5y = 6$

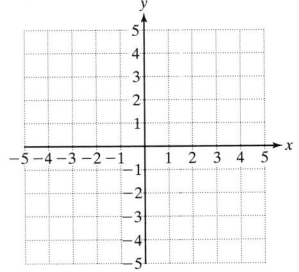

52. $3x + 1 = 7$

53. $-2x - 5 = 1$

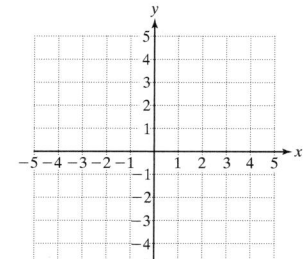
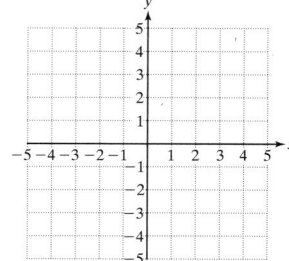

54. $\dfrac{1}{2}x + \dfrac{1}{4}y = \dfrac{1}{2}$

55. $\dfrac{1}{3}x - \dfrac{1}{6}y = \dfrac{1}{2}$

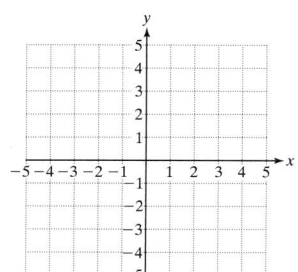
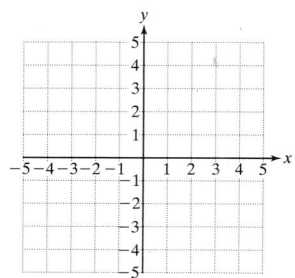

Concept 3: Determining Whether Two Lines Are Parallel, Perpendicular, or Neither

For Exercises 56–61, let m_1 and m_2 represent the slopes of two lines. Determine if the lines are parallel, perpendicular, or neither.

56. $m_1 = -2$, $m_2 = \dfrac{1}{2}$

57. $m_1 = \dfrac{2}{3}$, $m_2 = \dfrac{3}{2}$

58. $m_1 = 1$, $m_2 = \dfrac{4}{4}$

59. $m_1 = \dfrac{3}{4}$, $m_2 = -\dfrac{8}{6}$

60. $m_1 = \dfrac{2}{7}$, $m_2 = -\dfrac{2}{7}$

61. $m_1 = 5$, $m_2 = 5$

For Exercises 62–77, determine if the lines l_1 and l_2 are parallel, perpendicular, or neither.

62. l_1: $y = -2x - 3$

  l_2: $y = \dfrac{1}{2}x + 4$

63. l_1: $y = \dfrac{4}{3}x - 2$

  l_2: $y = -\dfrac{3}{4}x + 6$

64. l_1: $y = \dfrac{4}{5}x - \dfrac{1}{2}$

  l_2: $y = \dfrac{5}{4}x - \dfrac{2}{3}$

65. l_1: $y = \dfrac{1}{5}x + 1$

  l_2: $y = 5x - 3$

66. l_1: $y = -9x + 6$

  l_2: $y = -9x - 1$

67. l_1: $y = 4x - 1$

  l_2: $y = 4x + \dfrac{1}{2}$

68. l_1: $x = 3$

  l_2: $y = \dfrac{7}{4}$

69. l_1: $y = \dfrac{2}{3}$

  l_2: $x = 6$

70. l_1: $2x = 4$

  l_2: $6 = x$

71. l_1: $2y = 7$

  l_2: $y = 4$

72. l_1: $2x + 3y = 6$

  l_2: $3x - 2y = 12$

73. l_1: $4x + 5y = 20$

  l_2: $5x - 4y = 60$

74. l_1: $4x + 2y = 6$

  l_2: $4x + 8y = 16$

75. l_1: $3x + y = 5$

  l_2: $x + 3y = 18$

76. l_1: $y = \dfrac{1}{5}x - 3$

  l_2: $2x - 10y = 20$

77. l_1: $y = \dfrac{1}{3}x + 2$

  l_2: $-x + 3y = 12$

Concept 4: Writing an Equation of a Line Given Its Slope and *y*-Intercept

For Exercises 78–87, write an equation of the line given the following information. Write the answer in slope-intercept form if possible.

78. The slope is $-\frac{1}{3}$, and the *y*-intercept is $(0, 2)$.

79. The slope is $\frac{2}{3}$, and the *y*-intercept is $(0, -1)$.

80. The slope is 10, and the *y*-intercept is $(0, -19)$.

81. The slope is -14, and the *y*-intercept is $(0, 2)$.

82. The slope is 0, and the *y*-intercept is -11.

83. The slope is 0, and the *y*-intercept is $\frac{6}{7}$.

84. The slope is 5, and the line passes through the origin.

85. The slope is -3, and the line passes through the origin.

86. The slope is 6, and the line passes through the point $(0, -2)$.

87. The slope is -4, and the line passes through the point $(0, -3)$.

Expanding Your Skills

88. The cost for a rental car is \$49.95 per day plus a flat fee of \$31.95 for insurance. The equation, $C = 49.95x + 31.95$ represents the total cost, C (in dollars), to rent the car for x days.

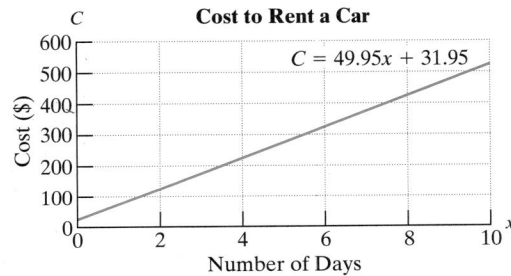

Cost to Rent a Car

$C = 49.95x + 31.95$

a. Identify the slope. Interpret the meaning of the slope in the context of this problem.

b. Identify the C-intercept. Interpret the meaning of the C-intercept in the context of this problem.

c. Use the equation to determine how much it would cost to rent the car for 1 week.

89. A phone bill is determined each month by a \$16.95 flat fee plus \$0.10/min of long distance. The equation $C = 0.10x + 16.95$ represents the total monthly cost, C, for x minutes of long distance.

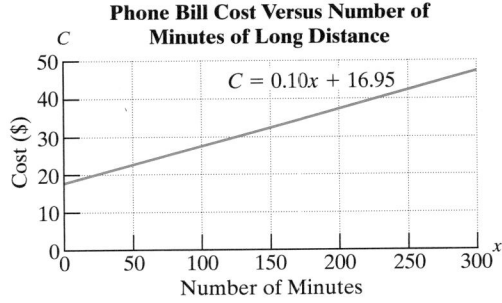

Phone Bill Cost Versus Number of Minutes of Long Distance

$C = 0.10x + 16.95$

a. Identify the slope. Interpret the meaning of the slope in the context of this problem.

b. Identify the C-intercept. Interpret the meaning of the C-intercept in the context of this problem.

c. Use the equation to determine the total cost of 234 min of long distance.

90. A linear equation is said to be written in standard form if it can be written as $Ax + By = C$, where A and B are not both zero. Write the equation $Ax + By = C$ in slope-intercept form to show that the slope is given by the ratio, $-\frac{A}{B}$. $(B \neq 0.)$

For Exercises 91–94, use the result of Exercise 90 to find the slope of the line.

91. $2x + 5y = 8$

92. $6x + 7y = -9$

93. $4x - 3y = -5$

94. $11x - 8y = 4$

Section 3.5 Point-Slope Formula

1. Writing an Equation of a Line Using the Point-Slope Formula

In Section 3.4, the slope-intercept form of a line was used as a tool to construct an equation of a line. Another useful tool to determine an equation of a line is the point-slope formula. The point-slope formula can be derived from the slope formula as follows:

Suppose a line passes through a given point (x_1, y_1) and has slope m. If (x, y) is any other point on the line, then:

$$m = \frac{y - y_1}{x - x_1} \qquad \text{Slope formula}$$

$$m(x - x_1) = \frac{y - y_1}{x - x_1}(x - x_1) \qquad \text{Clear fractions}$$

$$m(x - x_1) = y - y_1$$

or

$$y - y_1 = m(x - x_1) \qquad \text{Point-slope formula}$$

> **Point-Slope Formula**
>
> The **point-slope formula** is given by
>
> $$y - y_1 = m(x - x_1)$$
>
> where m is the slope of the line and (x_1, y_1) is a known point on the line.

Example 1 demonstrates how to use the point-slope formula to find an equation of a line when a point on the line and slope are given.

Example 1   Writing an Equation of a Line Using the Point-Slope Formula

Use the point-slope formula to find an equation of the line having a slope of 3 and passing through the point $(-2, -4)$. Write the answer in slope-intercept form.

Solution:

The slope of the line is given: $m = 3$.

A point on the line is given: $(x_1, y_1) = (-2, -4)$.

The point-slope formula:

$$y - y_1 = m(x - x_1)$$

$$y - (-4) = 3[x - (-2)] \qquad \text{Substitute } m = 3, x_1 = -2, \text{ and } y_1 = -4.$$

$$y + 4 = 3(x + 2) \qquad \text{Simplify. Because the final answer is required in slope-intercept form, simplify the equation and solve for } y.$$

$$y + 4 = 3x + 6 \qquad \text{Apply the distributive property.}$$

$$y = 3x + 6 - 4 \qquad \text{Subtract 4 from both sides.}$$

$$y = 3x + 2 \qquad \text{Slope-intercept form}$$

1. Write an equation of the line passing through the point $(-1, 5)$ and having slope -4.

The equation $y = 3x + 2$ from Example 1 is graphed in Figure 3-30. Notice that the line does indeed pass through the point $(-2, -4)$.

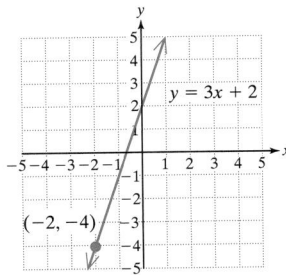

Figure 3-30

2. Writing an Equation of a Line Through Two Points

Example 2 is similar to Example 1; however, the slope must first be found from two given points.

Example 2 | **Writing an Equation of a Line Through Two Points**

Use the point-slope formula to find an equation of the line passing through the points $(-2, 5)$ and $(4, -1)$. Write the final answer in slope-intercept form.

Solution:

Given two points on a line, the slope can be found with the slope formula.

$$\underset{(x_1, y_1)}{(-2, 5)} \quad \text{and} \quad \underset{(x_2, y_2)}{(4, -1)} \quad \text{Label the points.}$$

$$m = \frac{y_2 - y_1}{x_2 - x_1} = \frac{(-1) - (5)}{(4) - (-2)} = \frac{-6}{6} = -1$$

To apply the point-slope formula, use the slope, $m = -1$ and either given point. We will choose the point $(-2, 5)$ as (x_1, y_1).

$$y - y_1 = m(x - x_1)$$

$$y - 5 = -1[x - (-2)] \quad \text{Substitute } m = -1, x_1 = -2, \text{ and } y_1 = 5.$$

$$y - 5 = -1(x + 2) \quad \text{Simplify.}$$

$$y - 5 = -x - 2$$

$$y = -x + 3 \quad \text{Slope-intercept form}$$

TIP: The point-slope formula can be applied using either given point for (x_1, y_1). In Example 2, using the point $(4, -1)$ for (x_1, y_1) produces the same result.

$$y - y_1 = m(x - x_1)$$
$$y - (-1) = -1(x - 4)$$
$$y + 1 = -x + 4$$
$$y = -x + 3$$

2. Use the point-slope formula to write an equation of the line passing through the two points $(1, -1)$ and $(-1, -5)$. Write the final answer in slope-intercept form.

Skill Practice Answers

1. $y = -4x + 1$
2. $y = 2x - 3$

The solution to Example 2 can be checked by graphing the line $y = -x + 3$ using the slope and y-intercept. Notice that the line passes through the points $(-2, 5)$ and $(4, -1)$ as expected. See Figure 3-31.

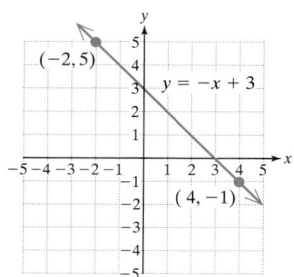

Figure 3-31

3. Writing an Equation of a Line Parallel or Perpendicular to Another Line

| **Example 3** | **Writing an Equation of a Line Parallel to Another Line** |

Use the point-slope formula to find an equation of the line passing through the point $(-1, 0)$ and parallel to the line $y = -4x + 3$. Write the final answer in slope-intercept form.

Solution:

Figure 3-32 shows the line $y = -4x + 3$ (pictured in black) and a line parallel to it (pictured in blue) that passes through the point $(-1, 0)$. The equation of the given line, $y = -4x + 3$, is written in slope-intercept form, and its slope is easily identified as -4. The line parallel to the given line must also have a slope of -4.

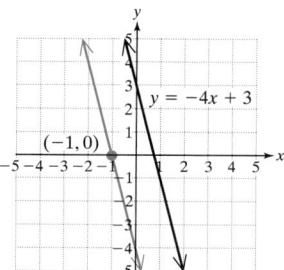

Figure 3-32

Apply the point-slope formula using $m = -4$ and the point $(x_1, y_1) = (-1, 0)$.

$$y - y_1 = m(x - x_1)$$
$$y - 0 = -4[x - (-1)]$$
$$y = -4(x + 1)$$
$$y = -4x - 4$$

3. Use the point-slope formula to write an equation of the line passing through the point $(8, 2)$ and parallel to the line $y = \frac{3}{4}x - \frac{1}{2}$.

Example 4 **Writing an Equation of a Line Perpendicular to Another Line**

Use the point-slope formula to find an equation of the line passing through the point $(-3, 1)$ and perpendicular to the line $3x + y = -2$. Write the final answer in slope-intercept form.

Solution:

The given line can be written in slope-intercept form as $y = -3x - 2$. The slope of this line is -3. Therefore, the slope of a line perpendicular to the given line is $\frac{1}{3}$.

Apply the point-slope formula with $m = \frac{1}{3}$, and $(x_1, y_1) = (-3, 1)$.

$y - y_1 = m(x - x_1)$ Point-slope formula

$y - (1) = \frac{1}{3}[x - (-3)]$ Substitute $m = \frac{1}{3}$, $x_1 = -3$, and $y_1 = 1$.

$y - 1 = \frac{1}{3}(x + 3)$ To write the final answer in slope-intercept form, simplify the equation and solve for y.

$y - 1 = \frac{1}{3}x + 1$ Apply the distributive property.

$y = \frac{1}{3}x + 2$ Add 1 to both sides.

A sketch of the perpendicular lines $y = \frac{1}{3}x + 2$ and $y = -3x - 2$ is shown in Figure 3-33. Notice that the line $y = \frac{1}{3}x + 2$ passes through the point $(-3, 1)$.

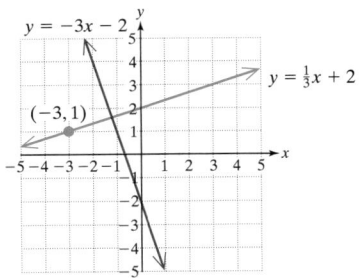

Figure 3-33

4. Write an equation of the line passing through the point $(10, 4)$ and perpendicular to the line $x + 2y = 1$.

4. Different Forms of Linear Equations: A Summary

A linear equation can be written in several different forms as summarized in Table 3-3.

Skill Practice Answers

3. $y = \dfrac{3}{4}x - 4$

4. $y = 2x - 16$

Table 3-3

Form	Example	Comments
Standard Form $Ax + By = C$	$4x + 2y = 8$	A and B must not both be zero.
Horizontal Line $y = k$ (k is constant)	$y = 4$	The slope is zero, and the y-intercept is $(0, k)$.
Vertical Line $x = k$ (k is constant)	$x = -1$	The slope is undefined, and the x-intercept is $(k, 0)$.
Slope-Intercept Form $y = mx + b$ the slope is m y-intercept is $(0, b)$	$y = -3x + 7$ Slope $= -3$ y-intercept is $(0, 7)$	Solving a linear equation for y results in slope-intercept form. The coefficient of the x-term is the slope, and the constant defines the location of the y-intercept.
Point-Slope Formula $y - y_1 = m(x - x_1)$	$m = -3$ $(x_1, y_1) = (4, 2)$ $y - 2 = -3(x - 4)$	This formula is typically used to build an equation of a line when a point on the line is known and the slope of the line is known.

Although standard form and slope-intercept form can be used to express an equation of a line, often the slope-intercept form is used to give a *unique* representation of the line. For example, the following linear equations are all written in standard form, yet they each define the same line.

$$2x + 5y = 10$$
$$-4x - 10y = -20$$
$$6x + 15y = 30$$
$$\frac{2}{5}x + y = 2$$

The line can be written uniquely in slope-intercept form as: $y = -\frac{2}{5}x + 2$.

Although it is important to understand and apply slope-intercept form and the point-slope formula, they are not necessarily applicable to all problems, particularly when dealing with a horizontal or vertical line.

Example 5   Writing an Equation of a Line

Find an equation of the line passing through the point $(2, -4)$ and perpendicular to the x-axis.

Solution:

Because the line is perpendicular to the x-axis, the line must be vertical. Recall that all vertical lines can be written in the form $x = k$, where k is a constant. A quick sketch can help find the value of the constant. See Figure 3-34.

Because the line must pass through a point whose x-coordinate is 2, then the equation of the line must be $x = 2$.

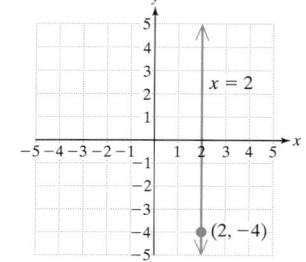

Figure 3-34

Skill Practice

5. Write an equation for the vertical line that passes through the point $(-7, 2)$.

Section 3.5   Practice Exercises

Study Skills Exercises

1. Prepare a one-page summary sheet with the most important information that you need for the test. On the day of the test, look at this sheet several times to refresh your memory instead of trying to memorize new information.

2. Define the key term: **point-slope formula**

Review Exercises

For Exercises 3–6, graph the equations.

3. $2x - 3y = -3$

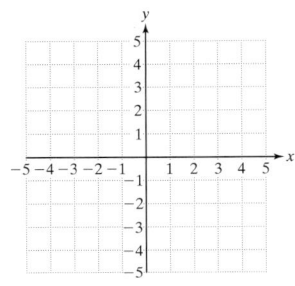

4. $y = -2x$

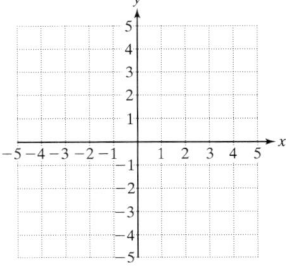

5. $3 - y = 9$

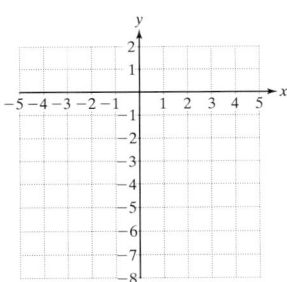

6. $y = \frac{4}{5}x$

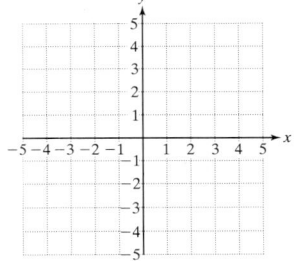

For Exercises 7–10, find the slope of the line that passes through the given points.

7. $(1, -3)$ and $(2, 6)$

8. $(2, -4)$ and $(-2, 4)$

9. $(-2, 5)$ and $(5, 5)$

10. $(6.1, 2.5)$ and $(6.1, -1.5)$

Concept 1: Writing an Equation of a Line Using the Point-Slope Formula

For Exercises 11–22, use the point-slope formula (if possible) to write an equation of the line given the following information.

11. The slope is 3, and the line passes through the point $(-2, 1)$.

12. The slope is -2, and the line passes through the point $(1, -5)$.

13. The slope is -4, and the line passes through the point $(-3, -2)$.

14. The slope is 5, and the line passes through the point $(-1, -3)$.

15. The slope is $-\frac{1}{2}$, and the line passes through $(-1, 0)$.

16. The slope is $-\frac{3}{4}$, and the line passes through $(2, 0)$.

17. The slope is $\frac{1}{4}$, and the line passes through the point $(-8, 6)$.

18. The slope is $\frac{2}{5}$, and the line passes through the point $(-5, 4)$.

19. The slope is 4.5, and the line passes through the point $(5.2, -2.2)$.

20. The slope is -3.6, and the line passes through the point $(10.0, 8.2)$.

21. The slope is 0, and the line passes through the point $(3, -2)$.

22. The slope is 0, and the line passes through the point $(0, 5)$.

Concept 2: Writing an Equation of a Line through Two Points

For Exercises 23–26, find an equation of the line through the given points. Write the final answer in slope-intercept form.

23.

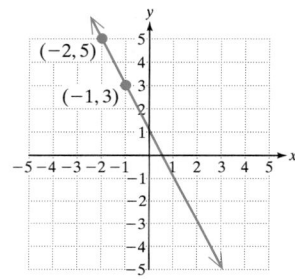

24.

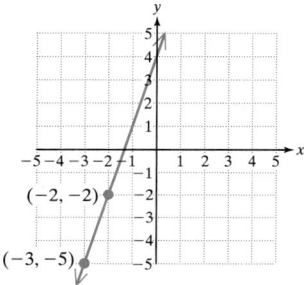

25.

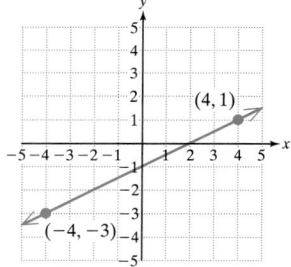

26.

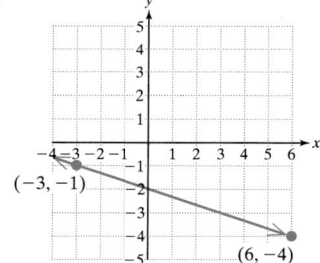

For Exercises 27–32, use the point-slope formula to write an equation of the line given the following information.

27. The line passes through the points $(-2, -6)$ and $(1, 0)$.

28. The line passes through the points $(-2, 5)$ and $(0, 1)$.

29. The line passes through the points $(1, -3)$ and $(-7, 2)$.

30. The line passes through the points $(0, -4)$ and $(-1, -3)$.

31. The line passes through the points $(2.2, -3.3)$ and $(12.2, -5.3)$.

32. The line passes through the points $(4.7, -2.2)$ and $(-0.3, 6.8)$.

Concept 3: Writing an Equation of a Line Parallel or Perpendicular to Another Line

For Exercises 33–44, use the point-slope formula to write an equation of the line given the following information.

33. The line passes through the point $(-3, 1)$ and is parallel to the line $y = 4x + 3$.

34. The line passes through the point $(4, -1)$ and is parallel to the line $y = 3x + 1$.

35. The line passes through the point $(4, 0)$ and is parallel to the line $3x + 2y = 8$.

36. The line passes through the point $(2, 0)$ and is parallel to the line $5x + 3y = 6$.

37. The line passes through the point $(-5, 2)$ and is perpendicular to the line $y = \frac{1}{2}x + 3$.

38. The line passes through the point $(-2, -2)$ and is perpendicular to the line $y = \frac{1}{3}x - 5$.

39. The line passes through the point $(0, -6)$ and is perpendicular to the line $-5x + y = 4$.

40. The line passes through the point $(0, -8)$ and is perpendicular to the line $2x - y = 5$.

41. The line passes through the point $(4, 4)$ and is parallel to the line $3x - y = 6$.

42. The line passes through the point $(-1, -7)$ and is parallel to the line $5x + y = -5$.

43. The line passes through the point $(-2, -1)$ and is perpendicular to $4x - y = 3$.

44. The line passes through the point $(-4, 3)$ and is perpendicular to $3x - 2y = 9$.

Concept 4: Different Forms of Linear Equations: A Summary

For Exercises 45–50, match the form or formula on the left with its name on the right.

45. $x = k$ i. Standard form

46. $y = mx + b$ ii. Point-slope formula

47. $m = \dfrac{y_2 - y_1}{x_2 - x_1}$ iii. Horizontal line

48. $y - y_1 = m(x - x_1)$ iv. Vertical line

49. $y = k$ v. Slope-intercept form

50. $Ax + By = C$ vi. Slope formula

For Exercises 51–62, find an equation for the line given the following information.

51. The line passes through the point $(3, 1)$ and is parallel to the line $y = -4$. See the figure.

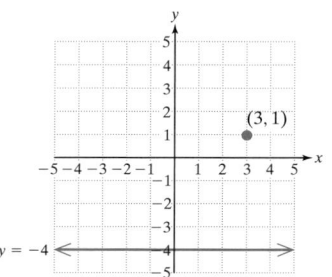

52. The line passes through the point $(-1, 1)$ and is parallel to the line $y = 2$. See the figure.

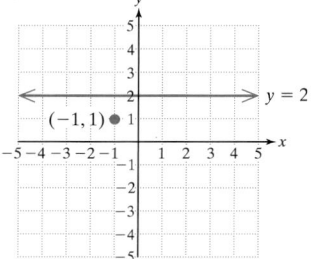

53. The line passes through the point $(2, 6)$ and is perpendicular to the line $y = 1$. (*Hint:* Sketch the line first.)

54. The line passes through the point $(0, 3)$ and is perpendicular to the line $y = -5$. (*Hint:* Sketch the line first.)

55. The line passes through the point $\left(\frac{5}{2}, \frac{1}{2}\right)$ and is parallel to the line $x = 4$.

56. The line passes through the point $\left(-6, \frac{2}{3}\right)$ and is parallel to the line $x = -2$.

57. The line passes through the point $(2, 2)$ and is perpendicular to the line $x = 0$.

58. The line passes through the point $(5, -2)$ and is perpendicular to the line $x = 0$.

59. The slope is undefined, and the line passes through the point $(-6, -3)$.

60. The slope is undefined, and the line passes through the point $(2, -1)$.

61. The line passes through the points $(-4, 0)$ and $(-4, 3)$.

62. The line passes through the points $(1, 3)$ and $(1, -4)$.

Expanding Your Skills

63. The following table represents the percentage of females, y, who smoked for selected years. Let x represent the number of years after 1965. Let y represent percentage of women who smoked.

Year		Percentage
1965	$x = 0$	33.9
1975	$x = 10$	32.1
1985	$x = 20$	27.9
1995	$x = 30$	23.4

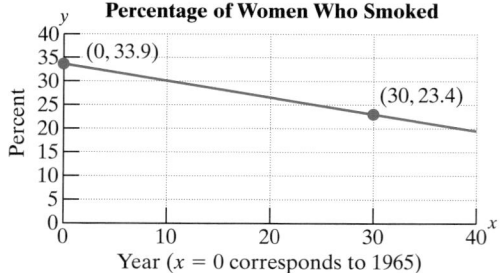

(Source: U.S. National Center for Health Statistics.)

a. Find the slope of the line between the points $(0, 33.9)$ and $(30, 23.4)$.

b. Find an equation of the line between the points $(0, 33.9)$ and $(30, 23.4)$. Write the answer in slope-intercept form.

c. Use the equation from part (b) to estimate the percentage of women who smoked in the year 2000.

64. The following table represents the median selling price, y, of new privately owned one-family houses sold in the Midwest from 1980 to 2005. Let x represent the number of years after 1980. Let y represent price in thousands of dollars. (Source: U.S. Bureau of Census.)

Year		Price (in $ thousands)
1980	$x = 0$	67
1985	$x = 5$	80
1990	$x = 10$	108
1995	$x = 15$	134
2000	$x = 20$	167
2005	$x = 25$	185

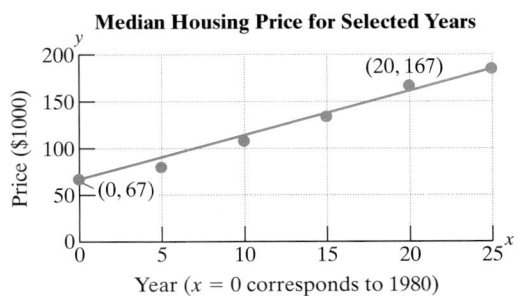

Median Housing Price for Selected Years

a. Find the slope of the line between the points $(0, 67)$ and $(20, 167)$.

b. Find an equation of the line between the points $(0, 67)$ and $(20, 167)$. Write the answer in slope-intercept form.

c. Use the equation from part (b) to estimate the median price of a one-family house sold in the Midwest in the year 2008.

Applications of Linear Equations

1. Interpreting a Linear Equation in Two Variables

Concepts

Linear equations can often be used to describe (or model) the relationship between two variables in a real-world event. In an xy-coordinate system, the variable being predicted by the mathematical equation is called the **dependent variable** (or response variable) and is represented by y. The variable used to make the prediction is called the **independent variable** (or predictor variable) and is represented by x.

1. Interpreting a Linear Equation in Two Variables

2. Writing a Linear Equation Using Observed Data Points

3. Writing a Linear Equation Given a Fixed Value and a Rate of Change

Example 1 Interpreting a Linear Equation

The cost, y, of a speeding ticket (in dollars) is given by $y = 10x + 100$, where $x > 0$ is the number of miles per hour over the speed limit.

a. Which is the independent variable?

b. Which variable is the dependent variable?

c. What is the slope of the line?

d. Interpret the meaning of the slope in terms of cost and the number of miles per hour over the speed limit.

e. Graph the line.

Solution:

a. The independent variable is the number of miles over the speed limit and is represented by x.

b. The dependent variable is the cost of the speeding ticket and is represented by y. The cost of the ticket *depends* on the number of miles per hour over the speed limit.

c. The equation is written in slope-intercept form where $m = 10$.

d. The slope $m = 10$ or $\frac{10}{1}$ indicates that there is a $10 increase in the cost of the speeding ticket for every 1 mph over the speed limit.

e.

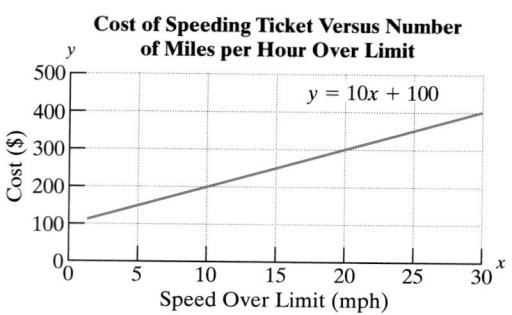

Cost of Speeding Ticket Versus Number of Miles per Hour Over Limit

$y = 10x + 100$

Skill Practice

1. The cost, y, for a local move by a small moving company is given by $y = 60x + 100$, where $x > 0$ is the number of hours required for the move.
   a. Which variable is the independent variable?
   b. Which variable is the dependent variable?
   c. What is the slope of the line?
   d. Interpret the meaning of the slope in terms of cost and the number of hours it takes to move.
   e. Graph the line.

Example 2   **Interpreting a Linear Equation**

The total number of crimes in the United States decreased from the year 1990 to 2006 (Figure 3-35). The decrease followed a trend that is approximately linear and can be represented by the linear equation:

$$N = -0.28x + 14.8$$ where N is the number of crimes (in millions) and x is the number of years since 1990.

Skill Practice Answers

1a. Number of hours, x
   b. Cost, y **c.** 60
   d. There is an increase in cost of $60 for each hour of the move.
   e.

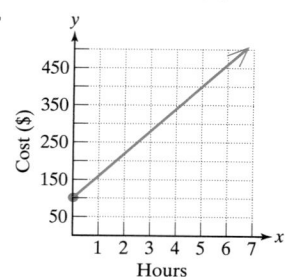

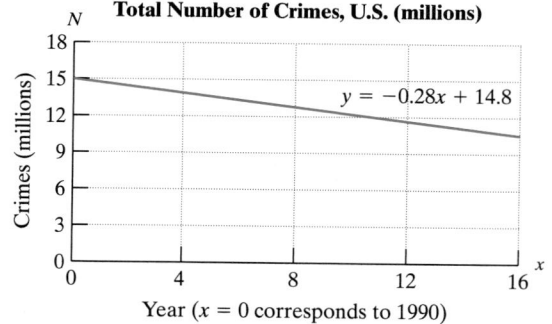

Total Number of Crimes, U.S. (millions)

$y = -0.28x + 14.8$

Year ($x = 0$ corresponds to 1990)

Figure 3-35

(Source: FBI, Uniform Crime Reports)

a. Which is the independent variable?

b. Which is the dependent variable?

c. Use the equation to predict the number of crimes in 2000.

d. What is the N-intercept of the line? Interpret the meaning of the N-intercept in terms of the number of crimes and the year.

e. What is the slope of the line? Interpret the meaning of the slope in terms of the number of crimes and the year.

f. Is it possible for this linear trend to continue indefinitely?

g. From the equation, determine the value of the x-intercept. Round to the nearest whole unit. Interpret the meaning of the x-intercept in terms of the number of crimes and the year. Realistically, is it possible for this linear trend to continue to its x-intercept?

Solution:

a. The number of years since 1990 is the independent variable. It is represented by x.

b. The number of reported crimes is the dependent variable. It is represented by N.

c. The year 2000 is 10 years after 1990. Therefore, substitute $x = 10$ into the linear equation.

$$N = -0.28x + 14.8$$

$$N = -0.28(10) + 14.8 \qquad \text{Substitute } x = 10.$$

$$= 12$$

The number of reported crimes in the U.S. in 2000 was approximately 12 million.

d. Notice that the variable N is playing the role of y (the dependent variable). The equation is written in slope-intercept form, $N = -0.28x + 14.8$. The N-intercept is $(0, 14.8)$ and indicates that in the year $x = 0$ (1990), the number of reported crimes in the United States was approximately 14.8 million.

e. From the slope intercept form of the line, $N = -0.28x + 14.8$, the slope is -0.28 or equivalently, $\frac{-0.28}{1}$. The slope indicates that the number of crimes decreased by 0.28 million per year during this time period.

f. It is not possible for the linear trend to continue indefinitely because eventually the number of crimes would reduce to a negative number.

g. To find the x-intercept, substitute $N = 0$.

$$N = -0.28x + 14.8$$

$$0 = -0.28x + 14.8 \qquad \text{Substitute } N = 0.$$

$$-14.8 = -0.28x \qquad \text{Subtract 14.8 from both sides.}$$

$$\frac{-14.8}{-0.28} = \frac{-0.28x}{-0.28} \qquad \text{Divide both sides by } -0.28.$$

$$53 \approx x$$

The x-intercept is $(53, 0)$ and indicates that approximately 53 years after 1990 (the year 2043), the number of crimes will be 0. Although the concept of having zero reported crimes is appealing, it is not realistic. This shows that the linear trend will not continue indefinitely.

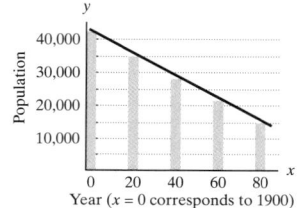

Skill Practice

2. The number of tigers, y, in India decreased between the years 1900 to 1980 according to the linear equation $y = -350x + 42,000$, where x is the number of years since 1900. (*Source:* Environmental Investigation Agency)

   a. Which is the independent variable?

   b. Which is the dependent variable?

   c. Use the equation to predict the number of tigers in 1960.

   d. What is the slope of the line? Interpret the meaning of the slope in terms of the number of tigers and the year.

   e. Find the x-intercept. Interpret the meaning of the x-intercept in terms of the number of tigers.

2. Writing a Linear Equation Using Observed Data Points

Example 3 Writing a Linear Equation from Observed Data Points

In the 1990s sales for SUVs increased in the United States until higher gas prices eventually slowed the trend. Let x represent the number of years since 1994, and let y represent the yearly number of SUVs sold (in millions) in the United States. From Figure 3-36, we see that SUV sales followed a trend that was approximately linear.

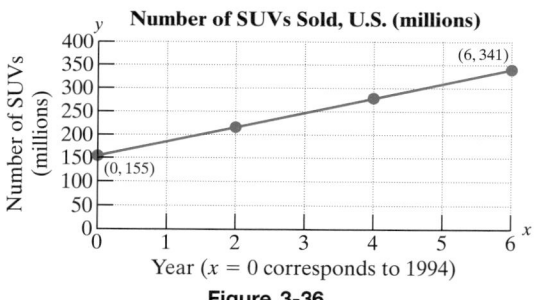

Figure 3-36

   a. Use the data points from Figure 3-36 to find a linear equation that represents the number of SUVs sold in the U.S. x years after 1994.

   b. Use the linear equation found in part (a) to estimate the number of SUVs sold for the year 1998.

Solution:

   a. From the graph, the two given data values are $(0, 155)$ and $(6, 341)$. From these points we can find the slope of the line.

$$(0, 155) \qquad (6, 341)$$
$$(x_1, y_1) \qquad (x_2, y_2) \qquad\qquad \text{Label the points.}$$

Skill Practice Answers

2a. The number of years since 1900, x
  b. The number of tigers, y
  c. 21,000 tigers
  d. -350; The number of tigers decreases by 350 per year.
  e. (0, 120) If the trend continues, there will be 0 tigers in the year 2020.

$$m = \frac{y_2 - y_1}{x_2 - x_1} = \frac{341 - 155}{6 - 0}$$

$$= \frac{186}{6} = \frac{31}{1} = 31$$

The slope is 31 and indicates that there has been an increase of 31 million SUVs sold per year in the United States.

With $m = 31$ and the y-intercept given as $(0, 155)$, we have the following linear equation in slope-intercept form:

$$y = 31x + 155$$

b. The year 1998 is 4 years after the base year of 1994. Therefore, substitute $x = 4$ into the linear equation to approximate the number of SUVs sold in 1998.

$$y = 31x + 155$$

$$y = 31(4) + 155 \qquad \text{Substitute } x = 4.$$

$$= 279$$

The number of SUVs sold in the U.S. in 1998 was approximately 279 million.

Skill Practice

3. Soft drink sales at a concession stand at a softball stadium have increased linearly over the course of the summer softball season.

   a. Use the given data points to find a linear equation that relates the sales, y, to week number, x.

   b. Use the equation to predict the number of soft drinks sold in week 10.

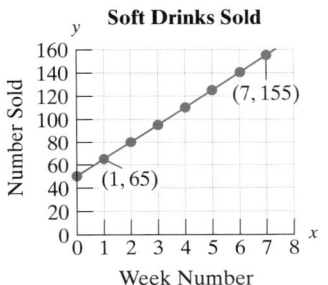

---

3. Writing a Linear Equation Given a Fixed Value and a Rate of Change

Another way to look at the equation $y = mx + b$ is to identify the term mx as the variable term and the term b as the constant term. The value of the term mx will change with the value of x (this is why the slope, m, is called a *rate of change*). However, the term b will remain constant regardless of the value of x. With these ideas in mind, we can write a linear equation if the rate of change and the constant are known.

Example 4 Finding a Linear Equation

A small word-processing business has a fixed monthly cost of \$6000 (this includes rent, utilities, and salaries of its employees). The business has a variable cost of \$5 per project (this includes mostly paper and computer supplies).

   a. Write a linear equation to compute the total cost, y, for 1 month if x projects are completed.

   b. Use the equation to compute the cost to run the business for 1 month if 800 projects are completed.

Skill Practice Answers

3a. $y = 15x + 50$
   b. 200 soft drinks

Solution:

a. $6000 is the constant (fixed) monthly cost. The variable cost is $5 per project. If the slope, m, is replaced with 5, and b is replaced with 6000, then the equation $y = mx + b$ becomes

$y = 5x + 6000$ where y represents the total cost of completing x projects in 1 month

b. Because x represents the number of projects, substitute $x = 800$.

$$y = 5(800) + 6000$$
$$= 4000 + 6000$$
$$= 10,000$$

The total cost of completing 800 projects is $10,000.

Skill Practice

4. The commission for buying stock at a discount brokerage firm is $12.95 plus $.02 per share of stock.

    a. Write a linear equation to compute the commission, y, on a purchase of x shares of stock.

    b. Find the commission that would be charged to purchase 250 shares of stock.

Calculator Connections

In Example 2, the equation $N = -0.28x + 14.8$ was used to represent the number of crimes, N (in millions), in the United States versus the number of years, x, since 1990. The equation is based on data between 1990 and 2006. This corresponds to x-values between 0 and 16. To graph the equation on a graphing calculator, the viewing window can be set for x between 0 and 16. The calculator interprets the values of N as the y-variable. We set the viewing window to accommodate N-values up to 20.

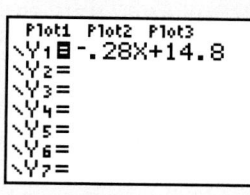
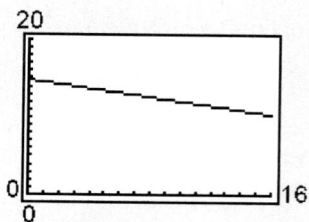

An *Eval* feature can be used to find solutions to an equation by evaluating the value of y for user-defined values of x. For example, entering a value of 10 for x results in 12 for y. This indicates that for $x = 10$ (the year 2000) there were approximately 12 million reported crimes in the United States.

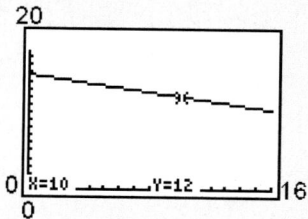

Calculator Exercises

Use a graphing calculator to graph the lines on an appropriate viewing window. Evaluate the equations at the given values of x.

1. $y = -4.6x + 27.1$ at $x = 3$

2. $y = -3.6x - 42.3$ at $x = 0$

3. $y = 40x + 105$ at $x = 6$

4. $y = 20x - 65$ at $x = 8$

Section 3.6   Practice Exercises

Study Skills Exercises

1. On test day, take a look at any formulas or important points that you had to memorize before you enter the classroom. Then when you sit down to take your test, write these formulas on the test or scrap paper. This is called a memory dump. Write down the formulas from Chapter 3.

2. Define the key terms:

   a. dependent variable   **b. independent variable**

Review Exercises

For Exercises 3–7, find the x- and y-intercepts of the lines, if possible.

3. $5x + 6y = 30$

4. $3x + 4y = 1$

5. $y = -2x - 4$

6. $y = 5x$

7. $y = -9$

Concept 1: Interpreting a Linear Equation in Two Variables

8. The electric bill charge for a certain utility company is $0.095 per kilowatt-hour. The total cost, y, depends on the number of kilowatt-hours, x, according to the equation

$$y = 0.095x$$

   a. Which variable is the independent variable?

   b. Which variable is the dependent variable?

   c. Determine the cost of using 1000 kilowatt-hours.

   d. Determine the cost of using 2000 kilowatt-hours.

   e. Determine the y-intercept. Interpret the meaning of the y-intercept in the context of this problem.

   f. Determine the slope. Interpret the meaning of the slope in the context of this problem.

9. The minimum hourly wage, y (in dollars/hour), in the United States since 1960 can be approximated by the equation $y = 0.10x + 0.82$, where x represents the number of years since 1960 ($x = 0$ corresponds to 1960, $x = 1$ corresponds to 1961, and so on).

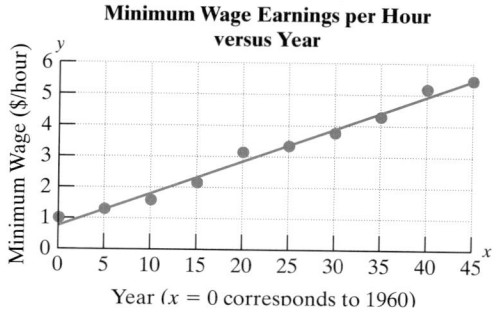

Minimum Wage Earnings per Hour versus Year

Year ($x = 0$ corresponds to 1960)

a. Which variable is the independent variable?

b. Which variable is the dependent variable?

c. Approximate the minimum wage in 1985.

d. Use the equation to predict the minimum wage in 2010.

e. Find the y-intercept. Interpret the meaning of the y-intercept in the context of this problem.

f. Find the slope. Interpret the meaning of the slope in the context of this problem.

10. The graph depicts the rise in the number of jail inmates in the United States since 1995. Two linear equations are given: one to describe the number of female inmates and one to describe the number of male inmates by year.

Let y represent the number of inmates (in thousands). Let x represent the number of years since 1995.

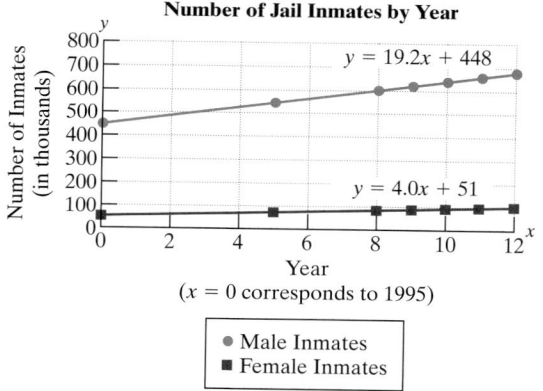

Number of Jail Inmates by Year

$y = 19.2x + 448$

$y = 4.0x + 51$

Year
($x = 0$ corresponds to 1995)

● Male Inmates
■ Female Inmates

(Source: U.S. Bureau of Justice Statistics)

a. What is the slope of the line representing the number of female inmates? Interpret the meaning of the slope in the context of this problem.

b. What is the slope of the line representing the number of male inmates? Interpret the meaning of the slope in the context of this problem.

c. Which group, males or females, has the larger slope? What does this imply about the rise in the number of male and female prisoners?

11. The following graph shows the number of points scored by Shaquille O'Neal and by Allen Iverson according to the number of minutes played for several games. Two linear equations are given: one to describe the number of points scored by O'Neal and one to describe the number of points scored by Iverson. In both equations, y represents the number of points scored and x represents the number of minutes played.

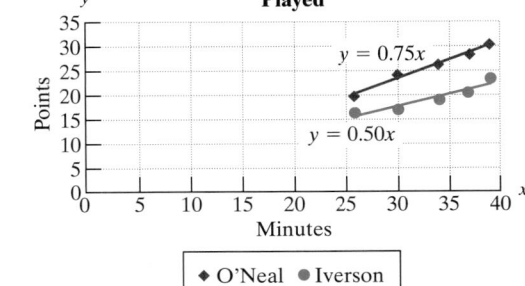

a. What is the slope of the line representing the number of points scored by O'Neal? Interpret the meaning of the slope in the context of this problem.

b. What is the slope of the line representing the number of points scored by Iverson? Interpret the meaning of the slope in the context of this problem.

c. According to these linear equations, approximately how many points would each player expect to score if he played for 36 minutes? Round to the nearest point.

12. The water bill charge for a certain utility company is $4.20 per 1000 gallons used. The total cost, y, depends on the number of thousands of gallons of water used, x, according to the equation

$$y = 4.20x \quad x \geq 0$$

a. Determine the cost of using 3000 gallons. (*Hint:* $x = 3$.)

b. Determine the cost of using 5000 gallons.

c. Determine the y-intercept. Interpret the meaning of the y-intercept in the context of this problem.

d. Determine the slope. Interpret the meaning of the slope in the context of this problem.

13. The average daily temperature in January for cities along the eastern seaboard of the United States and Canada generally decreases for cities farther north. A city's latitude in the northern hemisphere is a measure of how far north it is on the globe.

The average temperature, y, (measured in degrees Fahrenheit) can be described by the equation

$$y = -2.333x + 124.0 \quad \text{where } x \text{ is the latitude of the city.}$$

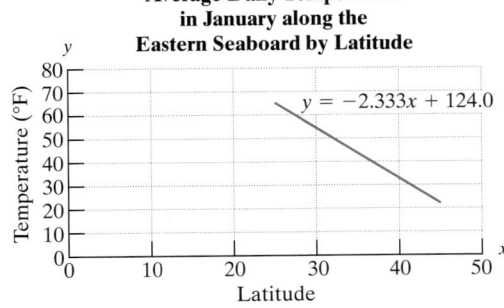

(Source: U.S. National Oceanic and Atmospheric Administration.)

City	x Latitude (°N)	y Average Daily Temperature (°F)
Jacksonville, FL	30.3	52.4
Miami, FL	25.8	67.2
Atlanta, GA	33.8	41.0
Baltimore, MD	39.3	31.8
Boston, MA	42.3	28.6
Atlantic City, NJ	39.4	30.9
New York, NY	40.7	31.5
Portland, ME	43.7	20.8
Charlotte, NC	35.2	39.3
Norfolk, VA	36.9	39.1

a. Which variable is the dependent variable?

b. Which variable is the independent variable?

c. Use the equation to predict the average daily temperature in January for Philadelphia, Pennsylvania, whose latitude is 40.0°N. Round to one decimal place.

d. Use the equation to predict the average daily temperature in January for Edmundston, New Brunswick, Canada, whose latitude is 47.4°N. Round to one decimal place.

e. What is the slope of the line? Interpret the meaning of the slope in terms of the latitude and temperature.

f. From the equation, determine the value of the x-intercept. Round to one decimal place. Interpret the meaning of the x-intercept in terms of latitude and temperature.

Concept 2: Writing a Linear Equation Using Observed Data Points

14. The average length of stay for community hospitals has been decreasing in the United States from 1980 to 2005. Let x represent the number of years since 1980. Let y represent the average length of a hospital stay in days.

a. Find a linear equation that relates the average length of hospital stays versus the year.

b. Use the linear equation found in part (a) to predict the average length of stay in community hospitals in the year 2010. Round to the nearest day.

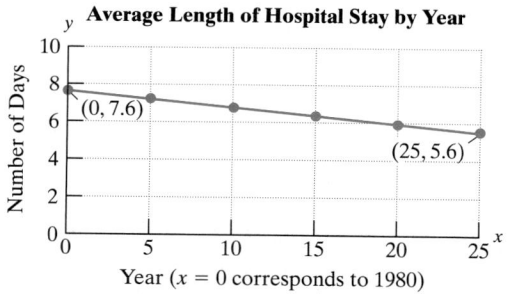

15. The graph shows the average height for boys based on age. Let x represent a boy's age, and let y represent his height (in inches).

a. Find a linear equation that represents the height of a boy versus his age.

b. Use the linear equation found in part (a) to predict the average height of a 5-year-old boy.

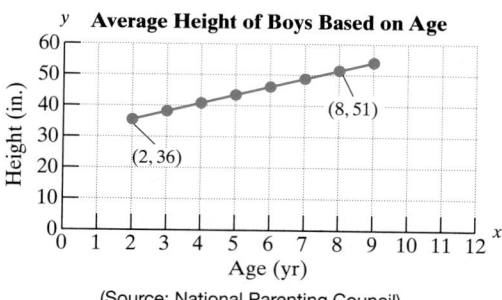

16. The figure depicts a relationship between a person's height, y (in inches), and the length of the person's arm, x (measured in inches from shoulder to wrist).

a. Use the points $(17, 57.75)$ and $(24, 82.25)$ to find a linear equation relating height to arm length.

b. What is the slope of the line? Interpret the slope in the context of this problem.

c. Use the equation from part (a) to estimate the height of a person whose arm length is 21.5 in.

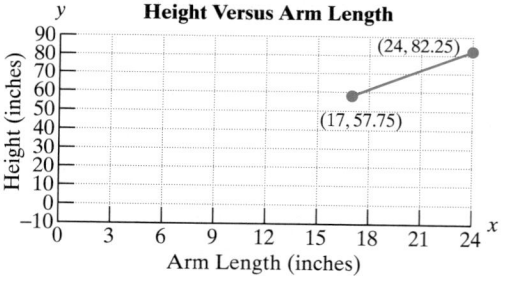

17. In a certain city, the time required to commute to work, y (in minutes), by car is related linearly to the distance traveled, x (in miles).

a. Use the points $(5, 12)$ and $(16, 34)$ to find a linear equation relating the commute time to work to the distance traveled.

b. What is the slope of the line? Interpret the slope in the context of this problem.

c. Use the equation from part (a) to find the time required to commute to work for a motorist who lives 18 miles away.

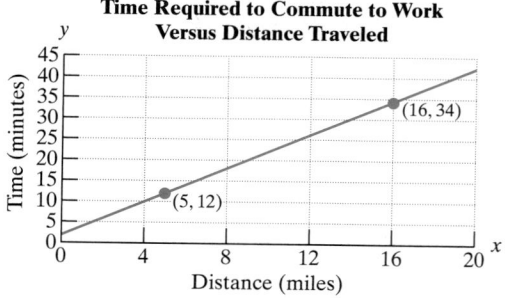

Concept 3: Writing a Linear Equation Given a Fixed Value and a Rate of Change

18. The cost to rent a car, y, for 1 day is $20 plus $0.25 per mile.

   a. Write a linear equation to compute the cost, y, of driving a car x miles for 1 day.

   b. Use the equation to compute the cost of driving 258 miles in the rental car for 1 day.

19. A phone bill is determined each month by a $18.95 flat fee plus $0.08 per minute of long distance.

   a. Write a linear equation to compute the monthly cost of a phone bill, y, if x minutes of long distance are used.

   b. Use the equation to compute the phone bill for a month in which 1 hr and 27 min of long distance was used.

20. A tennis instructor charges a student $25 per lesson plus a one-time court fee of $20.

   a. Write a linear equation to compute the total cost, y, for x tennis lessons.

   b. What is the total cost to a student who takes 20 tennis lessons?

21. The cost to rent a 10 ft by 10 ft storage space is $90 per month plus a nonrefundable deposit of $105.

   a. Write a linear equation to compute the cost, y, of renting a 10 ft by 10 ft space for x months.

   b. What is the cost of renting such a storage space for 1 year (12 months)?

22. A business has a fixed monthly cost of $1200. In addition, the business has a variable cost of $35 for each item produced.

   a. Write a linear equation to compute the total cost, y, for 1 month if x items are produced.

   b. Use the equation to compute the cost for 1 month if 100 items are produced.

23. An air-conditioning and heating company has a fixed monthly cost of $5000. Furthermore, each service call costs the company $25.

   a. Write a linear equation to compute the total cost, y, for 1 month if x service calls are made.

   b. Use the equation to compute the cost for 1 month if 150 service calls are made.

24. A bakery that specializes in bread rents a booth at a flea market. The daily cost to rent the booth is $100. Each loaf of bread costs the bakery $0.80 to produce.

   a. Write a linear equation to compute the total cost, y, for 1 day if x loaves of bread are produced.

   b. Use the equation to compute the cost for 1 day if 200 loaves of bread are produced.

25. A beverage company rents a booth at an art show to sell lemonade. The daily cost to rent a booth is $35. Each lemonade costs $0.50 to produce.

   a. Write a linear equation to compute the total cost, y, for 1 day if x lemonades are produced.

   b. Use the equation to compute the cost for 1 day if 350 lemonades are produced.

Chapter 3   SUMMARY

Section 3.1   Rectangular Coordinate System

Key Concepts

Graphical representation of numerical **data** is often helpful to study problems in real-world applications.

A **rectangular coordinate system** is made up of a horizontal line called the **x-axis** and a vertical line called the **y-axis**. The point where the lines meet is the **origin**. The four regions of the plane are called **quadrants**.

The point (x, y) is an **ordered pair**. The first element in the ordered pair is the point's horizontal position from the origin. The second element in the ordered pair is the point's vertical position from the origin.

Example

Example 1

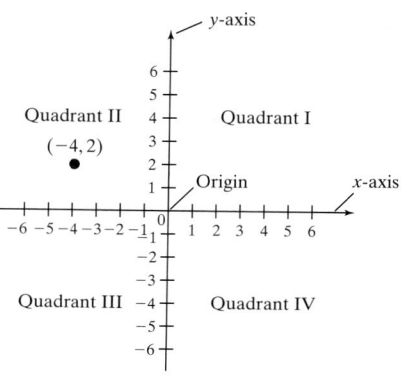

Section 3.2   Linear Equations in Two Variables

Key Concepts

An equation written in the form $Ax + By = C$ (where A and B are not both zero) is a **linear equation in two variables**.

A solution to a linear equation in x and y is an ordered pair (x, y) that makes the equation a true statement. The graph of the set of all solutions of a linear equation in two variables is a line in a rectangular coordinate system.

A linear equation can be graphed by finding at least two solutions and graphing the line through the points.

Examples

Example 1

Graph the equation $2x + y = 2$.

Select arbitrary values of x or y such as those shown in the table. Then complete the table to find the corresponding ordered pairs.

x	y	
0	2	→ $(0, 2)$
−1	4	→ $(−1, 4)$
1	0	→ $(1, 0)$

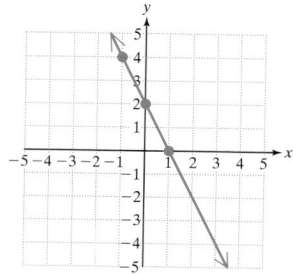

Example 2

For the line $2x + y = 2$, the x-intercept is $(1, 0)$ and the y-intercept is $(0, 2)$.

An **x-intercept** of a graph is a point $(a, 0)$ where the graph intersects the x-axis.

A **y-intercept** of a graph is a point $(0, b)$ where the graph intersects the y-axis.

A **vertical line** can be written in the form $x = k$.
A **horizontal line** can be written in the form $y = k$.

Example 3

$x = 3$ is a vertical line

$y = 3$ is a horizontal line

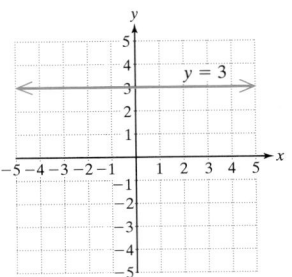

Section 3.3 — Slope of a Line

Key Concepts

The **slope**, m, of a line between two points (x_1, y_1) and (x_2, y_2) is given by

$$m = \frac{y_2 - y_1}{x_2 - x_1} \quad \text{or} \quad \frac{\text{change in } y}{\text{change in } x}$$

The slope of a line may be positive, negative, zero, or undefined.

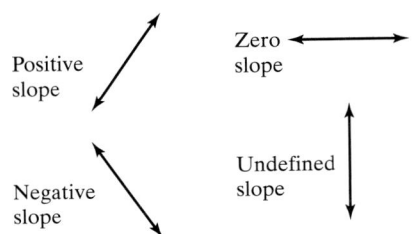

If m_1 and m_2 represent the slopes of two **parallel lines** (nonvertical), then $m_1 = m_2$.

If $m_1 \neq 0$ and $m_2 \neq 0$ represent the slopes of two nonvertical **perpendicular lines**, then

$$m_1 = -\frac{1}{m_2} \quad \text{or equivalently,} \quad m_1 m_2 = -1.$$

Examples

Example 1

Find the slope of the line between $(1, -5)$ and $(-3, 7)$.

$$m = \frac{7 - (-5)}{-3 - 1} = \frac{12}{-4} = -3$$

Example 2

The slope of the line $y = -2$ is 0 because the line is horizontal.

Example 3

The slope of the line $x = 4$ is undefined because the line is vertical.

Example 4

The slopes of two distinct lines are given. Determine whether the lines are parallel, perpendicular, or neither.

a. $m_1 = -7$ and $m_2 = -7$ Parallel

b. $m_1 = -\dfrac{1}{5}$ and $m_2 = 5$ Perpendicular

c. $m_1 = -\dfrac{3}{2}$ and $m_2 = -\dfrac{2}{3}$ Neither

Section 3.4   Slope-Intercept Form of a Line

Key Concepts

The **slope-intercept form** of a line is

$$y = mx + b$$

where m is the slope of the line and $(0, b)$ is the y-intercept.

Slope-intercept form is used to identify the slope and y-intercept of a line when the equation is given.

Slope-intercept form can also be used to graph a line.

Examples

Example 1

Find the slope and y-intercept.

$$7x - 2y = 4$$
$$-2y = -7x + 4 \qquad \text{Solve for } y.$$
$$\frac{-2y}{-2} = \frac{-7x}{-2} + \frac{4}{-2}$$
$$y = \frac{7}{2}x - 2$$

The slope is $\frac{7}{2}$. The y-intercept is $(0, -2)$.

Example 2

Graph the line.

$$y = \frac{7}{2}x - 2$$

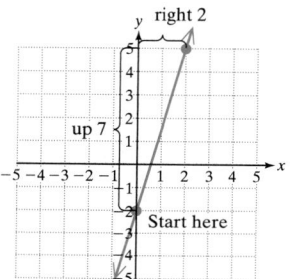

Section 3.5   Point-Slope Formula

Key Concepts

The **point-slope formula** is used primarily to construct an equation of a line given a point and the slope.

Equations of Lines—A Summary:

Standard form: $Ax + By = C$
Horizontal line: $y = k$
Vertical line: $x = k$
Slope-intercept form: $y = mx + b$
Point-slope formula: $y - y_1 = m(x - x_1)$

Examples

Example 1

Find an equation of the line passing through the point $(6, -4)$ and having a slope of $-\frac{1}{2}$.

Label the given information:
$m = -\frac{1}{2}$ and $(x_1, y_1) = (6, -4)$

$$y - y_1 = m(x - x_1)$$
$$y - (-4) = -\frac{1}{2}(x - 6)$$
$$y + 4 = -\frac{1}{2}x + 3$$
$$y = -\frac{1}{2}x - 1$$

Section 3.6   Applications of Linear Equations

Key Concepts

Linear equations can often be used to describe or model the relationship between variables in a real-world event. In such applications, the slope may be interpreted as a rate of change.

The two variables used in an application are called the independent and the dependent variables. The **independent variable** is used to make a prediction and is represented on the horizontal axis. The **dependent variable** is the response to the value of the independent variable and is represented on the vertical axis.

Examples

Example 1

The number of drug-related arrests for a small city has been growing approximately linearly since 1980.

Let y represent the number of drug arrests, and let x represent the number of years after 1980.

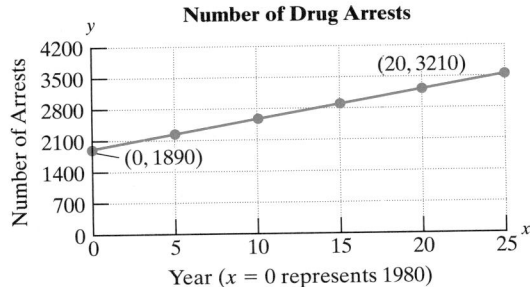

Number of Drug Arrests

a. Use the ordered pairs $(0, 1890)$ and $(20, 3210)$ to find an equation of the line shown in the graph.

$$m = \frac{y_2 - y_1}{x_2 - x_1} = \frac{3210 - 1890}{20 - 0}$$

$$= \frac{1320}{20} = 66$$

The slope is 66, indicating that the number of drug arrests is increasing at a rate of 66 per year. $m = 66$, and the y-intercept is $(0, 1890)$. Hence:

$$y = mx + b \implies y = 66x + 1890$$

b. Use the equation in part (a) to predict the number of drug-related arrests in the year 2010. (The year 2010 is 30 years after 1980. Hence, $x = 30$.)

$$y = 66(30) + 1890$$

$$y = 3870$$

The number of drug arrests is predicted to be 3870 by the year 2010.

Chapter 3 — Review Exercises

Section 3.1

1. Graph the points on a rectangular coordinate system.

   a. $\left(\dfrac{1}{2}, 5\right)$ b. $(-1, 4)$ c. $(2, -1)$

   d. $(0, 3)$ e. $(0, 0)$ f. $\left(-\dfrac{8}{5}, 0\right)$

   g. $(-2, -5)$ h. $(3, 1)$

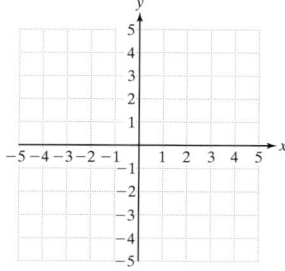

2. Estimate the coordinates of the points A, B, C, D, E, and F.

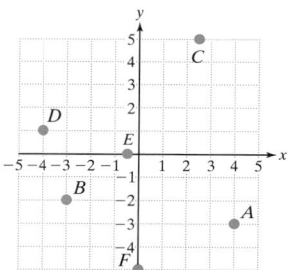

For Exercises 3–8, determine the quadrant in which the given point is found.

3. $(-2, -10)$ 4. $(-4, 6)$

5. $(3, -5)$ 6. $\left(\dfrac{1}{2}, \dfrac{7}{5}\right)$

7. $(\pi, -2.7)$ 8. $(-1.2, -6.8)$

9. On which axis is the point $(2, 0)$ found?

10. On which axis is the point $(0, -3)$ found?

11. The price per share of a stock (in dollars) over a period of 5 days is shown in the graph.

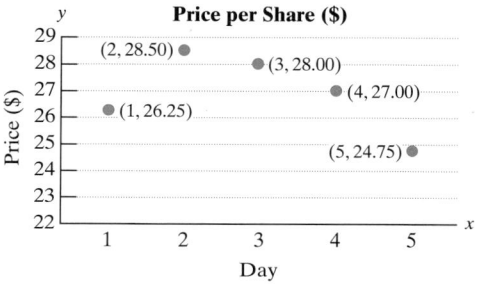

   a. Interpret the meaning of the ordered pair $(1, 26.25)$.

   b. On which day was the price the highest?

   c. What was the increase in price between day 1 and day 2?

12. The number of space shuttle launches for selected years is given by the ordered pairs. Let x represent the number of years since 1995. Let y represent the number of launches.

   $(1, 7)$ $(2, 8)$ $(3, 5)$ $(4, 3)$

   $(5, 5)$ $(6, 6)$ $(7, 5)$ $(8, 1)$

   a. Interpret the meaning of the ordered pair $(8, 1)$.

   b. Plot the points on a rectangular coordinate system.

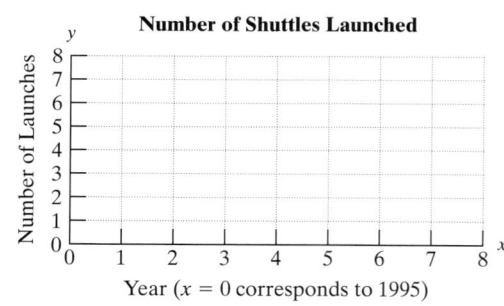

Section 3.2

For Exercises 13–16, determine if the given ordered pair is a solution to the equation.

13. $5x - 3y = 12$; $(0, 4)$

14. $2x - 4y = -6$; $(3, 0)$

15. $y = \dfrac{1}{3}x - 2$; $(9, 1)$

16. $y = -\dfrac{2}{5}x + 1$; $(-10, 5)$

For Exercises 17–20, complete the table and graph the corresponding ordered pairs. Graph the line through the points to represent all solutions to the equation.

17. $3x - y = 5$

x	y
2	
	4
1	

18. $\dfrac{1}{2}x + 3y = 6$

x	y
	2
-2	
	3

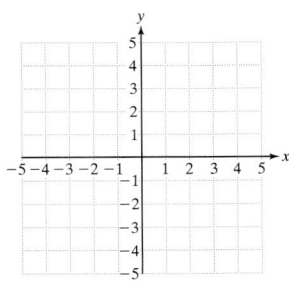

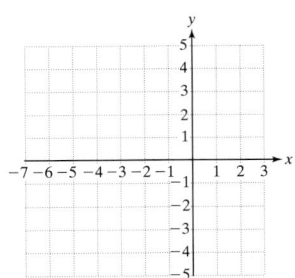

19. $y = \dfrac{2}{3}x - 1$

x	y
0	
3	
-6	

20. $y = -2x - 3$

x	y
0	
-3	
1	

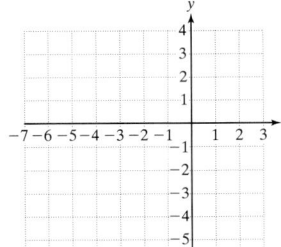

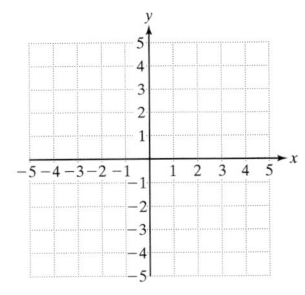

For Exercises 21–24, graph the line.

21. $x + 2y = 4$

22. $x - y = 5$

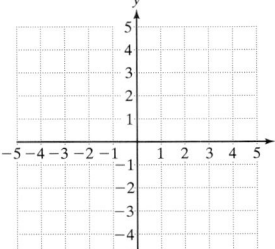

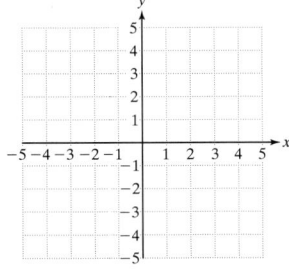

23. $y = 3x - 2$

24. $y = \dfrac{1}{4}x$

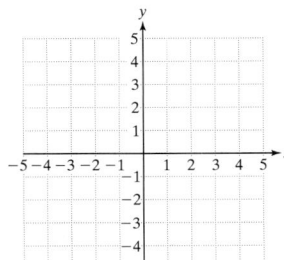

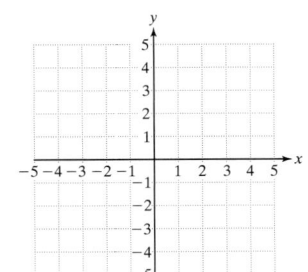

For Exercises 25–28, identify the line as horizontal or vertical. Then graph the line.

25. $3x - 2 = 10$

26. $2x + 1 = -2$

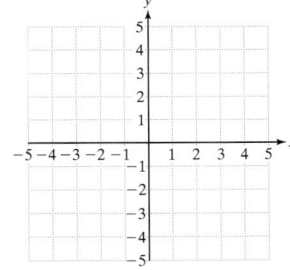

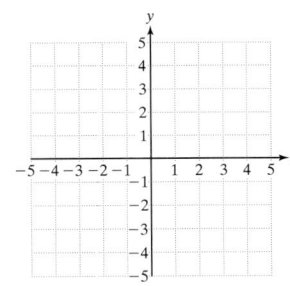

27. $6y + 1 = 13$

28. $5y - 1 = 14$

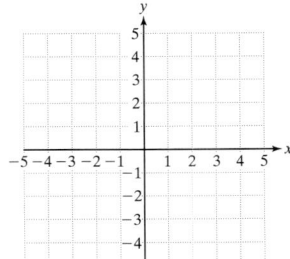

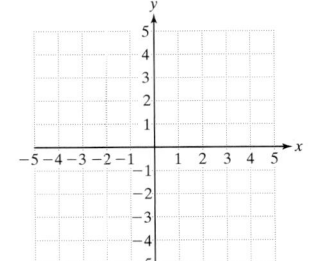

For Exercises 29–36, find the x- and y-intercepts if they exist.

29. $-4x + 8y = 12$　　　**30.** $2x + y = 6$

31. $y = 8x$　　　**32.** $5x - y = 0$

33. $6y = -24$　　　**34.** $2y - 3 = 1$

35. $2x + 5 = 0$　　　**36.** $-3x + 1 = 0$

Section 3.3

37. What is the slope of the ladder leaning up against the wall?

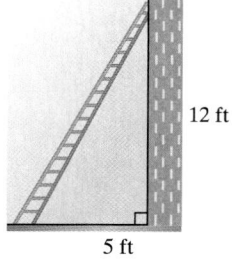
12 ft

5 ft

38. Point A is located 4 units down and 2 units to the right of point B. What is the slope of the line through points A and B?

39. Determine the slope of the line that passes through the points $(7, -9)$ and $(-5, -1)$.

40. Determine the slope of the line that has x- and y-intercepts of $(-1, 0)$ and $(0, 8)$.

41. Determine the slope of the line that passes through the points $(3, 0)$ and $(3, -7)$.

42. Determine the slope of the horizontal line $y = -1$.

43. A given line has a slope of -5.

   a. What is the slope of a line parallel to the given line?

   b. What is the slope of a line perpendicular to the given line?

44. A given line has a slope of 0.

   a. What is the slope of a line parallel to the given line?

   b. What is the slope of a line perpendicular to the given line?

For Exercises 45–48, find the slopes of the lines l_1 and l_2 from the two given points. Then determine whether l_1 and l_2 are parallel, perpendicular, or neither.

45. l_1:　$(3, 7)$ and $(0, 5)$

   l_2:　$(6, 3)$ and $(-3, -3)$

46. l_1:　$(-2, 1)$ and $(-1, 9)$

   l_2:　$(0, -6)$ and $(2, 10)$

47. l_1:　$(0, \frac{5}{6})$ and $(2, 0)$

   l_2:　$(0, \frac{6}{5})$ and $(-\frac{1}{2}, 0)$

48. l_1:　$(1, 1)$ and $(1, -8)$

   l_2:　$(4, -5)$ and $(7, -5)$

Section 3.4

For Exercises 49–54, write each equation in slope-intercept form. Identify the slope and the y-intercept.

49. $5x - 2y = 10$　　　**50.** $3x + 4y = 12$

51. $x - 3y = 0$　　　**52.** $5y - 8 = 4$

53. $2y = -5$　　　**54.** $y - x = 0$

For Exercises 55–59, determine whether the lines l_1 and l_2 are parallel, perpendicular, or neither.

55. l_1:　$y = \frac{3}{5}x + 3$　　　**56.** l_1:　$2x - 5y = 10$

   l_2:　$y = \frac{5}{3}x + 1$　　　 l_2:　$5x + 2y = 20$

57. l_1:　$3x + 2y = 6$　　　**58.** l_1:　$y = \frac{1}{4}x - 3$

   l_2:　$-6x - 4y = 4$　　　 l_2:　$-x + 4y = 8$

59. l_1:　$2x = 4$

   l_2:　$y = 6$

60. Write an equation of the line whose slope is $-\frac{4}{3}$ and whose y-intercept is $(0, -1)$.

61. Write an equation of the line that passes through the origin and has a slope of 5.

Section 3.5

62. Write a linear equation in two variables in slope-intercept form. (Answers may vary.)

63. Write a linear equation in two variables in standard form. (Answers may vary.)

64. Write the slope formula to find the slope of the line between the points (x_1, y_1) and (x_2, y_2).

65. Write the point-slope formula.

66. Write an equation of a vertical line (answers may vary).

67. Write an equation of a horizontal line (answers may vary).

For Exercises 68–73, use the point-slope formula to write an equation of a line given the following information.

68. The slope is -6, and the line passes through the point $(-1, 8)$.

69. The slope is $\frac{2}{3}$, and the line passes through the point $(5, 5)$.

70. The line passes through the points $(0, -4)$ and $(8, -2)$.

71. The line passes through the points $(2, -5)$ and $(8, -5)$.

72. The line passes through the point $(5, 12)$ and is perpendicular to the line $y = -\frac{5}{6}x - 3$.

73. The line passes through the point $(-6, 7)$ and is parallel to the line $4x - y = 0$.

Section 3.6

74. The graph shows the average height for girls based on age (*Source:* National Parenting Council). Let x represent a girl's age, and let y represent her height (in inches).

$$y = 2.4x + 31$$

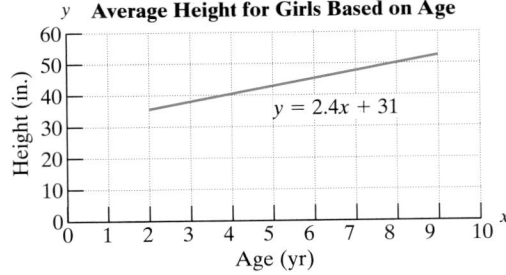

a. Which variable is the independent variable?

b. Which variable is the dependent variable?

c. Use the equation to estimate the average height of a 7-year-old girl.

d. What is the slope of the line? Interpret the meaning of the slope in the context of the problem.

75. The number of drug prescriptions has increased between 1995 and 2007 (see graph). Let x represent the number of years since 1995. Let y represent the number of prescriptions (in millions).

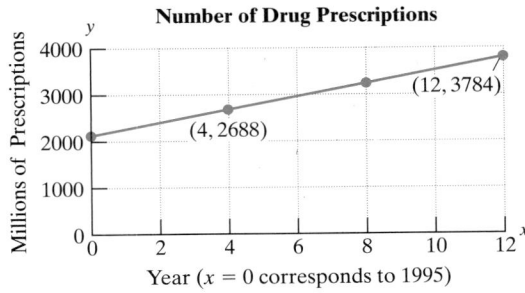

a. Using the ordered pairs $(4, 2688)$ and $(12, 3784)$ find the slope of the line.

b. Interpret the meaning of the slope in the context of this problem.

c. Find a linear equation that represents the number of prescriptions, y, versus the year, x.

d. Predict the number of prescriptions for the year 2008.

76. The amount of money that U.S. consumers had in outstanding automobile loans beginning at year 1992 is shown in the graph. Let x represent the number of years since 1992. Let y represent the total debt in auto loans (in billions of dollars).

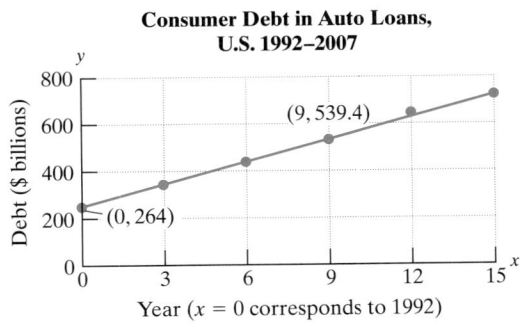

a. Find a linear equation that represents the total debt in auto loans, y, versus the year, x.

b. Use the linear equation found in part (a) to predict the debt in the year 2010.

77. A water purification company charges $20 per month and a $55 installation fee.

a. Write a linear equation to compute the total cost, y, of renting this system for x months.

b. Use the equation from part (a) to determine the total cost to rent the system for 9 months.

78. A small cleaning company has a fixed monthly cost of $700 and a variable cost of $8 per service call.

a. Write a linear equation to compute the total cost, y, of making x service calls in one month.

b. Use the equation from part (a) to determine the total cost of making 80 service calls.

Chapter 3   Test

1. In which quadrant is the given point found?

a. $\left(-\dfrac{7}{2}, 4\right)$ **b.** $(4.6, -2)$ **c.** $(-37, -45)$

2. What is the y-coordinate for a point on the x-axis?

3. What is the x-coordinate for a point on the y-axis?

4. The following table depicts a boy's height versus his age. Let x represent the boy's age and y represent his height.

Age (years), x	Height (inches), y
5	46
7	50
9	55
11	60

a. Write the data as ordered pairs and interpret the meaning of the first ordered pair.

b. Graph the ordered pairs on a rectangular coordinate system.

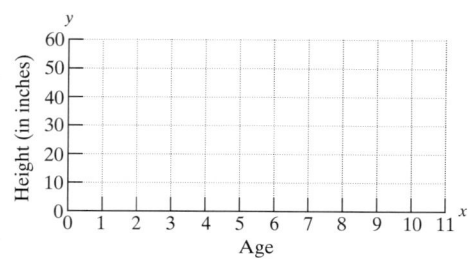

c. From the graph and table estimate the boy's height at age 10.

d. The data appear to follow an upward trend up to the boy's teenage years. Do you think this trend will continue? Would it be reasonable to use these data to predict the boy's height at age 25?

5. Determine whether the ordered pair is a solution to the equation $2x - y = 6$

a. $(0, 6)$ **b.** $(4, 2)$

c. $(3, 0)$ **d.** $\left(\dfrac{9}{2}, 3\right)$

6. Given the equation $y = \frac{1}{4}x - 2$, complete the table. Plot the ordered pairs and graph the line through the points to represent the set of all solutions to the equation.

x	y
0	
4	
6	

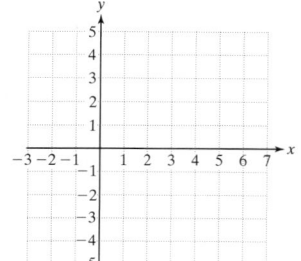

7. If x represents an adult's age, then the person's maximum recommended heart rate, y, during exercise is approximated by the equation

$$y = 220 - x \quad (x \geq 18)$$

   a. Use the equation to find the maximum recommended heart rate for a person who is 18 years old.

   b. Use the equation to complete the following ordered pairs: $(20, \quad), (30, \quad), (40, \quad),$ $(50, \quad), (60, \quad).$

For Exercises 8–9, determine whether the equation represents a horizontal or vertical line. Then graph the line.

8. $-6y = 18$ **9.** $5x + 1 = 8$

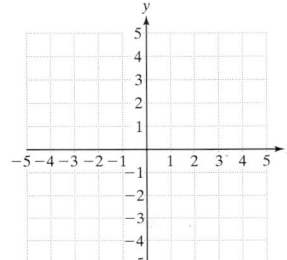 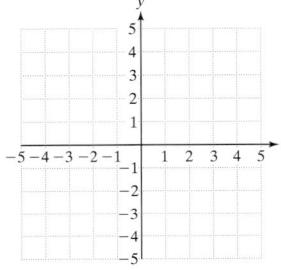

10. Find the x-intercept and the y-intercept of the line $-4x + 3y = 6$.

11. What is the slope of the hill?

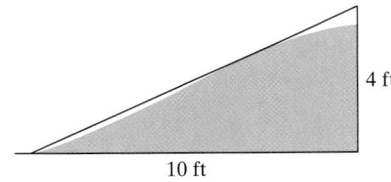

4 ft

10 ft

12. a. Find the slope of the line that passes through the points $(-2, 0)$ and $(-5, -1)$.

   b. Find the slope of the line $4x - 3y = 9$.

13. a. What is the slope of a line parallel to the line $x + 4y = -16$?

   b. What is the slope of a line perpendicular to the line $x + 4y = -16$?

14. a. What is the slope of the line $x = 5$?

   b. What is the slope of the line $y = -3$?

For Exercises 15–18, find the x- and y-intercepts if they exist, and graph the lines.

15. $y = 8x + 2$

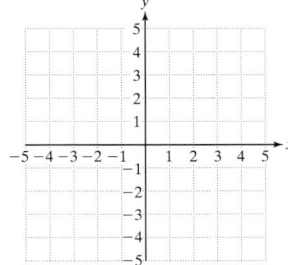

16. $2x + 9y = 0$

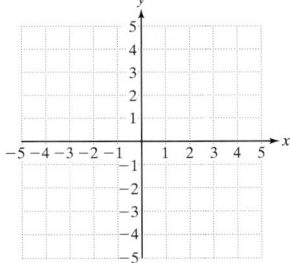

17. $x - 3 = 0$

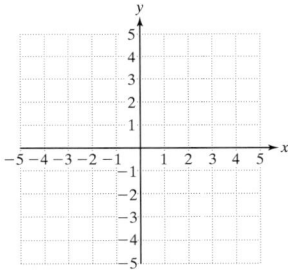

18. $-4y = 12$

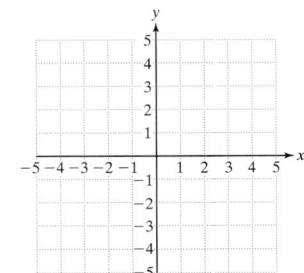

19. Determine whether the lines l_1 and l_2 are parallel, perpendicular, or neither.

$l_1: \ 2y = 3x - 3 \qquad l_2: \ 4x = -6y + 1$

20. Write an equation of the line that has y-intercept $(0, \frac{1}{2})$ and slope $\frac{1}{4}$.

21. Write an equation of the line that passes through the points $(2, 8)$, and $(4, 1)$.

22. Write an equation of the line that passes through the point $(2, -6)$ and is parallel to the x-axis.

23. Write an equation of the line that passes through the point $(3, 0)$ and is parallel to the line $2x + 6y = -5$.

24. Write an equation of the line that passes through the point $(-3, -1)$ and is perpendicular to the line $x + 3y = 9$.

25. Hurricane Floyd dumped rain at an average rate of $\frac{3}{4}$ in./hr on Southport, NC. Further inland, in Lumberton, NC, the storm dropped $\frac{1}{2}$ in. of rain per hour. The following graph depicts the total amount of rainfall (in inches) versus the time (in hours) for both locations in North Carolina.

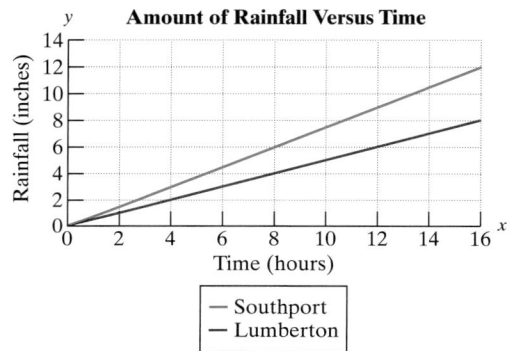

a. What is the slope of the line representing the rainfall for Southport?

b. What is the slope of the line representing the rainfall for Lumberton?

26. To attend a state fair, the cost is $10 per person to cover exhibits and musical entertainment. There is an additional cost of $1.50 per ride.

    a. Write an equation that gives the total cost, y, of visiting the state fair and going on x rides.

    b. Use the equation from part (a) to determine the cost of going to the state fair and going on 10 rides.

27. The number of medical doctors for selected years is shown in the graph. Let x represent the number of years since 1980, and let y represent the number of medical doctors (in thousands) in the United States.

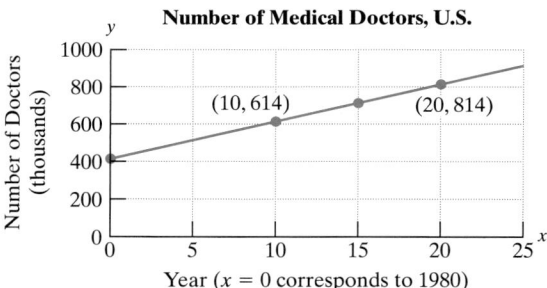

    a. Find the slope of the line shown in the graph. Interpret the meaning of the slope in the context of this problem.

    b. Find an equation of the line.

    c. Use the equation from part (b) to predict the number of medical doctors in the United States for the year 2010.

Chapters 1–3   Cumulative Review Exercises

1. Identify the numbers as rational or irrational.

    a. -3 **b.** $\dfrac{5}{4}$ **c.** $\sqrt{10}$ **d.** 0

2. Write the opposite and the absolute value for each number.

    a. $-\dfrac{2}{3}$ **b.** 5.3

3. Simplify the expression using the order of operations: $32 \div 2 \cdot 4 + 5$

4. Add: $3 + (-8) + 2 + (-10)$

5. Subtract: $16 - 5 - (-7)$

For Exercises 6–7, translate the English phrase into an algebraic expression. Then evaluate the expression.

6. The quotient of $\dfrac{3}{4}$ and $-\dfrac{7}{8}$.

7. The product of -2.1 and -6.

8. Name the property that is illustrated by the following statement. $6 + (8 + 2) = (6 + 8) + 2$

For Exercises 9–12, solve the equation.

9. $6x - 10 = 14$ **10.** $3(m + 2) - 3 = 2m + 8$

11. $\dfrac{2}{3}y - \dfrac{1}{6} = y + \dfrac{4}{3}$ **12.** $1.7z + 2 = -2(0.3z + 1.3)$

13. The area of Texas is $267{,}277$ mi^2. If this is 712 mi^2 less than 29 times the area of Maine, find the area of Maine.

14. For the formula $3a + b = c$, solve for a.

15. Graph the line $-6x + 2y = 0$.

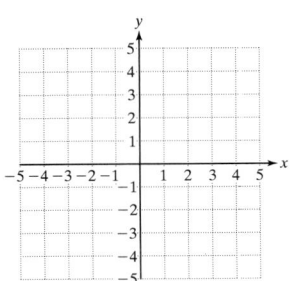

16. Find the x- and y-intercepts of $-2x + 4y = 4$.

17. Write the equation in slope-intercept form. Then identify the slope and the y-intercept.
$3x + 2y = -12$

18. Explain why the line $2x + 3 = 5$ has only one intercept.

19. Find an equation of a line passing through $(2, -5)$ with slope -3.

20. Find an equation of the line passing through $(0, 6)$ and $(-3, 4)$.

Systems of Linear Equations in Two Variables

4

This chapter is devoted to solving systems of linear equations. Applications of systems of equations involve two or more variables subject to two or more constraints. For example:

At a movie theater, one group of students bought three drinks and two small popcorns for a total of $13.00 (excluding tax). Another group bought five drinks and three popcorns for $20.50. There are two unknown quantities in this scenario: the cost per drink, and the cost per popcorn.

Fill in the blanks below using trial and error to determine the cost per drink and the cost per popcorn. You have the correct answer if both equations are true.

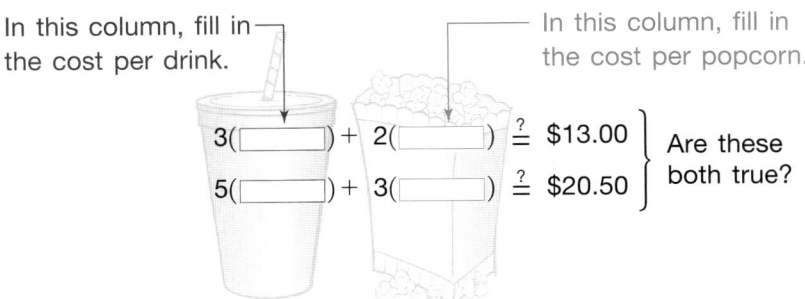

In this column, fill in the cost per drink.

In this column, fill in the cost per popcorn.

$$3(\boxed{}) + 2(\boxed{}) \overset{?}{=} \$13.00$$
$$5(\boxed{}) + 3(\boxed{}) \overset{?}{=} \$20.50$$

Are these both true?

If you have trouble with this puzzle, don't fret. Later in the chapter, we'll use the power of algebra to set up and solve a system of equations that will take the guess work away!

Section 4.1 — Solving Systems of Equations by the Graphing Method

Concepts

1. Determining Solutions to a System of Linear Equations
2. Dependent and Inconsistent Systems of Linear Equations
3. Solving Systems of Linear Equations by Graphing

1. Determining Solutions to a System of Linear Equations

Recall from Section 3.2 that a linear equation in two variables has an infinite number of solutions. The set of all solutions to a linear equation forms a line in a rectangular coordinate system. Two or more linear equations form a **system of linear equations**. For example, here are three systems of equations:

$$x - 3y = -5 \qquad\qquad y = \tfrac{1}{4}x - \tfrac{3}{4} \qquad\qquad 5a + b = 4$$
$$2x + 4y = 10 \qquad -2x + 8y = -6 \qquad -10a - 2b = 8$$

A **solution to a system of linear equations** is an ordered pair that is a solution to both individual equations in the system.

Example 1　Determining Solutions to a System of Linear Equations

Determine whether the ordered pairs are solutions to the system.

$$x + y = 4$$
$$-2x + y = -5$$

a. $(3, 1)$　　　**b.** $(0, 4)$

Solution:

a. Substitute the ordered pair $(3, 1)$ into both equations:

$$x + y = 4 \longrightarrow (3) + (1) \overset{?}{=} 4 \ ✔ \quad \text{True}$$
$$-2x + y = -5 \longrightarrow -2(3) + (1) \overset{?}{=} -5 \ ✔ \quad \text{True}$$

Because the ordered pair $(3, 1)$ is a solution to both equations, it is a solution to the *system* of equations.

b. Substitute the ordered pair $(0, 4)$ into both equations.

$$x + y = 4 \longrightarrow (0) + (4) \overset{?}{=} 4 \ ✔ \quad \text{True}$$
$$-2x + y = -5 \longrightarrow -2(0) + (4) \overset{?}{=} -5 \quad \text{False}$$

Because the ordered pair $(0, 4)$ is not a solution to the second equation, it is *not* a solution to the system of equations.

Skill Practice

1. Determine whether the ordered pairs are solutions to the system.

$$5x - 2y = 24$$
$$2x + y = 6$$

a. $(6, 3)$　　　**b.** $(4, -2)$

Avoiding Mistakes:

It is important to test an ordered pair in *both* equations to determine if the ordered pair is a solution.

Skill Practice Answers

1a. No
　b. Yes

A solution to a system of two linear equations may be interpreted graphically as a point of intersection between the two lines. Using slope-intercept form to graph the lines from Example 1, we have

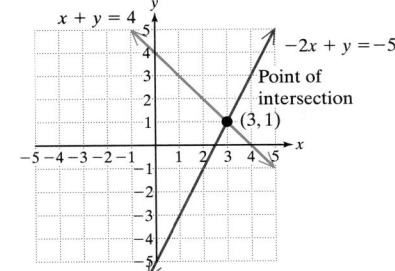

$$x + y = 4 \longrightarrow y = -x + 4$$

$$-2x + y = -5 \longrightarrow y = 2x - 5$$

Notice that the lines intersect at $(3, 1)$ (Figure 4-1).

Figure 4-1

2. Dependent and Inconsistent Systems of Linear Equations

When two lines are drawn in a rectangular coordinate system, three geometric relationships are possible:

1. Two lines may intersect at *exactly one point.*

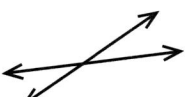

2. Two lines may intersect at *no point.* This occurs if the lines are parallel.

3. Two lines may intersect at *infinitely many points* along the line. This occurs if the equations represent the same line (the lines coincide).

If a system of linear equations has one or more solutions, the system is **consistent**. If a system of linear equations has no solution, it is **inconsistent**.

If two equations represent the same line, then all points along the line are solutions to the system of equations. In such a case, the system is characterized as a **dependent system**. An **independent system** is one in which the two equations represent different lines.

Solutions to Systems of Linear Equations in Two Variables

One Unique Solution	No Solution	Infinitely Many Solutions

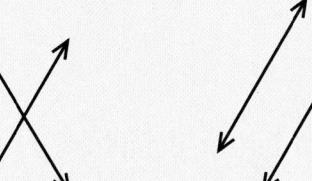
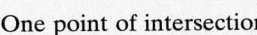
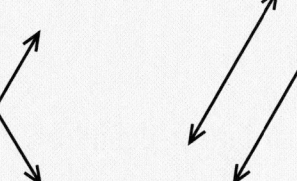

One point of intersection	Parallel lines	Coinciding lines

- System is consistent.
- System is independent.

- System is inconsistent.
- System is independent.

- System is consistent.
- System is dependent.

3. Solving Systems of Linear Equations by Graphing

One way to find a solution to a system of equations is to graph the equations and find the point (or points) of intersection. This is called the *graphing method* to solve a system of equations.

Example 2 Solving a System of Linear Equations by Graphing

Solve the system by the graphing method. $y = 2x$

$$y = 2$$

Solution:

The equation $y = 2x$ is written in slope-intercept form as $y = 2x + 0$. The line passes through the origin, with a slope of 2.

The line $y = 2$ is a horizontal line and has a slope of 0.

Because the lines have different slopes, the lines must be different and non-parallel. From this, we know that the lines must intersect at exactly one point. Graph the lines to find the point of intersection (Figure 4-2).

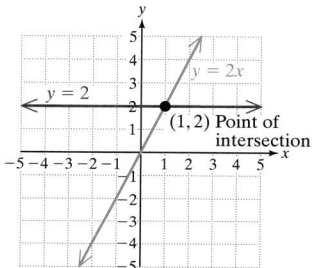

Figure 4-2

The point $(1, 2)$ appears to be the point of intersection. This can be confirmed by substituting $x = 1$ and $y = 2$ into both original equations.

$$y = 2x \longrightarrow (2) \stackrel{?}{=} 2(1) \ \checkmark \quad \text{True}$$

$$y = 2 \longrightarrow (2) \stackrel{?}{=} 2 \ \checkmark \quad \text{True}$$

The solution is $(1, 2)$.

Skill Practice Solve the system by graphing.

2. $y = -3x$
  $x = -1$

Skill Practice Answers

2.

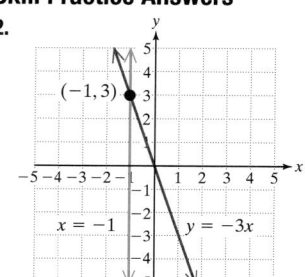

Example 3 Solving a System of Linear Equations by Graphing

Solve the system by the graphing method.

$$x - 2y = -2$$

$$-3x + 2y = 6$$

Solution:

To graph each equation, write the equation in slope-intercept form: $y = mx + b$.

Equation 1	**Equation 2**
$x - 2y = -2$	$-3x + 2y = 6$
$-2y = -x - 2$	$2y = 3x + 6$
$\dfrac{-2y}{-2} = \dfrac{-x}{-2} - \dfrac{2}{-2}$	$\dfrac{2y}{2} = \dfrac{3x}{2} + \dfrac{6}{2}$
$y = \dfrac{1}{2}x + 1$	$y = \dfrac{3}{2}x + 3$

From their slope-intercept forms, we see that the lines have different slopes, indicating that the lines are different and nonparallel. Therefore, the lines must intersect at exactly one point. Graph the lines to find that point (Figure 4-3).

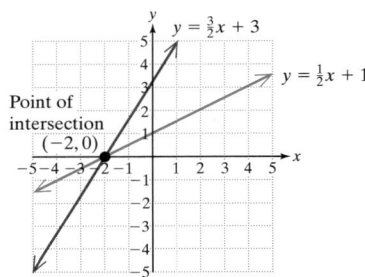

Figure 4-3

The point $(-2, 0)$ appears to be the point of intersection. This can be confirmed by substituting $x = -2$ and $y = 0$ into both equations.

$$x - 2y = -2 \longrightarrow (-2) - 2(0) \stackrel{?}{=} -2 \ ✔ \quad \text{True}$$

$$-3x + 2y = 6 \longrightarrow -3(-2) + 2(0) \stackrel{?}{=} 6 \ ✔ \quad \text{True}$$

The solution is $(-2, 0)$.

Skill Practice   Solve the system by graphing.

3. $y = 2x - 3$

  $6x + 2y = 4$

TIP: In Examples 2 and 3, the lines could also have been graphed by using the x- and y-intercepts or by using a table of points. However, the advantage of writing the equations in slope-intercept form is that we can compare the slopes and y-intercepts of each line.

1. If the slopes differ, the lines are different and nonparallel and must cross in exactly one point.
2. If the slopes are the same and the y-intercepts are different, the lines are parallel and will not intersect.
3. If the slopes are the same and the y-intercepts are the same, the two equations represent the same line.

Skill Practice Answers

3.

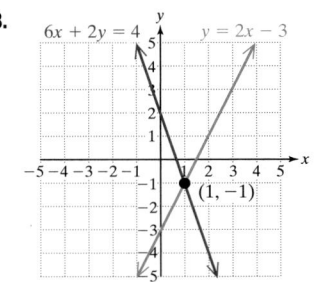

Example 4   Graphing an Inconsistent System

Solve the system by graphing.

$$-x + 3y = -6$$
$$6y = 2x + 6$$

Solution:

To graph the lines, write each equation in slope-intercept form.

Equation 1	**Equation 2**
$-x + 3y = -6$	$6y = 2x + 6$
$3y = x - 6$	
$\dfrac{3y}{3} = \dfrac{x}{3} - \dfrac{6}{3}$	$\dfrac{6y}{6} = \dfrac{2x}{6} + \dfrac{6}{6}$
$y = \dfrac{1}{3}x - 2$	$y = \dfrac{1}{3}x + 1$

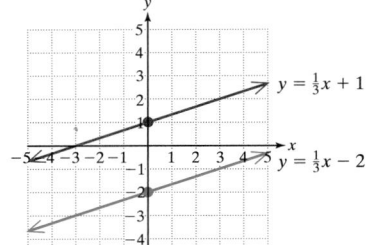

Figure 4-4

Because the lines have the same slope but different y-intercepts, they are parallel (Figure 4-4). Two parallel lines do not intersect, which implies that the system has no solution. The system is inconsistent.

Skill Practice   Solve the system by graphing.

4. $4x + y = 8$
      $y = -4x + 3$

Example 5   Graphing a Dependent System

Solve the system by graphing.

$$x + 4y = 8$$
$$y = -\frac{1}{4}x + 2$$

Solution:

Write the first equation in slope-intercept form. The second equation is already in slope-intercept form.

Skill Practice Answers

4.

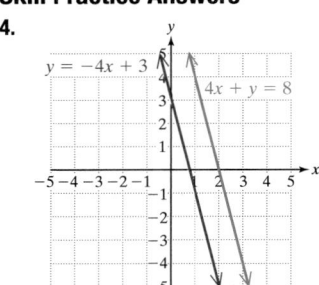

No solution, the lines are parallel.
The system is inconsistent.

Equation 1	**Equation 2**
$x + 4y = 8$	$y = -\dfrac{1}{4}x + 2$
$4y = -x + 8$	
$\dfrac{4y}{4} = \dfrac{-x}{4} + \dfrac{8}{4}$	
$y = -\dfrac{1}{4}x + 2$	

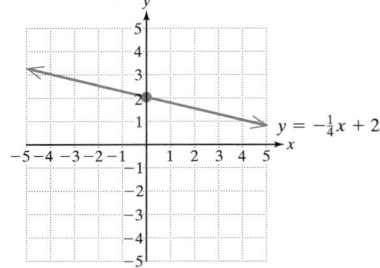

Figure 4-5

Notice that the slope-intercept forms of the two lines are identical. Therefore, the equations represent the same line (Figure 4-5). The system is dependent, and the solution to the system of equations is the set of all points on the line.

Because the ordered pairs in the solution set cannot all be listed, we can write the solution in set-builder notation. Furthermore, the equations $x + 4y = 8$ and $y = -\frac{1}{4}x + 2$ represent the same line. Therefore, the solution set may be written as $\{(x, y) \mid x + 4y = 8\}$ or as $\{(x, y) \mid y = -\frac{1}{4}x + 2\}$.

Skill Practice   Solve the system by graphing.

5. $x - 3y = 4$

$y = \dfrac{1}{3}x - \dfrac{4}{3}$

Calculator Connections

The solution to a system of equations can be found by using either a *Trace* feature or an *Intersect* feature on a graphing calculator to find the point of intersection.

For example, consider the system:

$$-2x + y = 6$$

$$5x + y = -1$$

First graph the equations together on the same viewing window. Recall that to enter the equations into the calculator, the equations must be written with the y-variable isolated.

$$-2x + y = 6 \xrightarrow{\text{isolate } y} y = 2x + 6$$

$$5x + y = -1 \longrightarrow y = -5x - 1$$

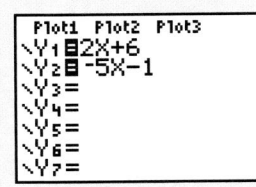 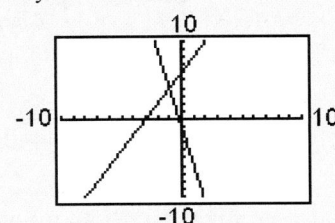

By inspection of the graph, it appears that the solution is $(-1, 4)$. The *Trace* option on the calculator may come close to $(-1, 4)$ but may not show the exact solution (Figure 4-6). However, an *Intersect* feature on a graphing calculator may provide the exact solution (Figure 4-7). See your user's manual for further details.

Using *Trace*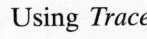

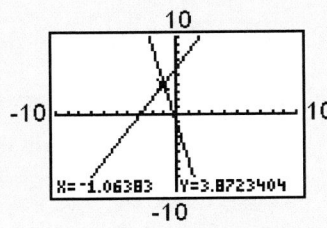

Figure 4-6

Using *Intersect*

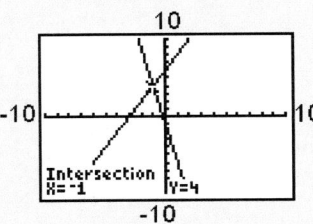

Figure 4-7

Skill Practice Answers

5.

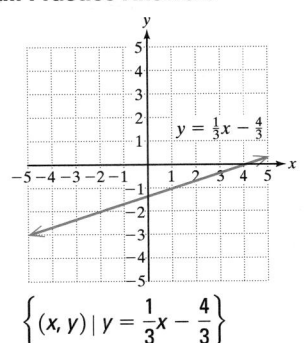

$$\left\{ (x, y) \mid y = \frac{1}{3}x - \frac{4}{3} \right\}$$

The system is dependent.

Calculator Exercises

Use a graphing calculator to graph each linear equation on the same viewing window. Use a *Trace* or *Intersect* feature to find the point(s) of intersection.

1. $y = 2x - 3$
$y = -4x + 9$

2. $y = -\dfrac{1}{2}x + 2$
$y = \dfrac{1}{3}x - 3$

3. $x + y = 4$ (Example 1)
$-2x + y = -5$

4. $x - 2y = -2$ (Example 3)
$-3x + 2y = 6$

5. $-x + 3y = -6$ (Example 4)
$6y = 2x + 6$

6. $x + 4y = 8$ (Example 5)
$y = -\dfrac{1}{4}x + 2$

Section 4.1   Practice Exercises

Boost *your* GRADE at mathzone.com!

- Practice Problems
- Self-Tests
- NetTutor
- e-Professors
- Videos

Study Skills Exercises

1. Figure out your grade at this point. Are you earning the grade that you want? If not, maybe organizing a study group would help.

In a study group, check the activities that you might try to help you learn and understand the material.

_____ Quiz each other by asking each other questions.

_____ Practice teaching each other.

_____ Share and compare class notes.

_____ Support and encourage each other.

_____ Work together on exercises and sample problems.

2. Define the key terms:

a. system of linear equations

b. solution to a system of linear equations

c. consistent system

d. inconsistent system

e. dependent system

f. independent system

Concept 1: Determining Solutions to a System of Linear Equations

For Exercises 3–10, determine if the given point is a solution to the system.

3. $3x - y = 7$ $\quad(2, -1)$
$x - 2y = 4$

4. $x - y = 3$ $\quad(4, 1)$
$x + y = 5$

5. $4y = -3x + 12$ $\quad(0, 4)$
$y = \dfrac{2}{3}x - 4$

6. $y = -\dfrac{1}{3}x + 2$ $\quad(9, -1)$
$x = 2y + 6$

7. $3x - 6y = 9$ $\quad\left(4, \dfrac{1}{2}\right)$
$x - 2y = 3$

8. $x - y = 4$ $\quad(6, 2)$
$3x - 3y = 12$

9. $\dfrac{1}{3}x = \dfrac{2}{5}y - \dfrac{4}{5}$ $\quad(0, 2)$
$\dfrac{3}{4}x + \dfrac{1}{2}y = 2$

10. $\dfrac{1}{4}x + \dfrac{1}{2}y = \dfrac{3}{2}$ $\quad(4, 1)$
$y = \dfrac{3}{2}x - 6$

For Exercises 11–14, match the graph of the system of equations with the appropriate description of the solution.

11.

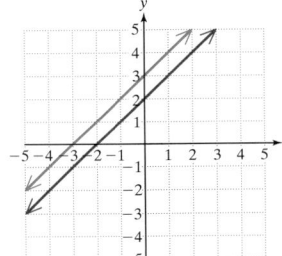

a. The solution is $(1, 3)$.

b. No solution.

c. There are infinitely many solutions.

d. The solution is $(0, 0)$.

12.

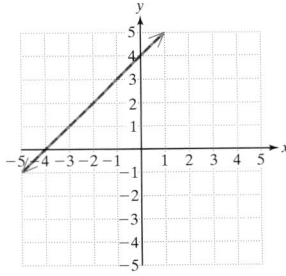

13.

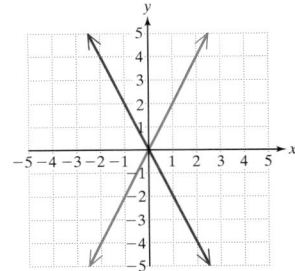

14.

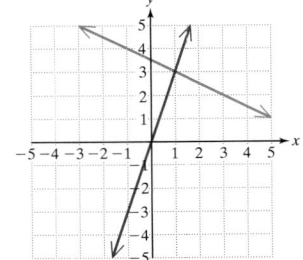

Concept 2: Dependent and Inconsistent Systems of Linear Equations

15. Graph each system of equations.

a. $y = 2x - 3$
$y = 2x + 5$

b. $y = 2x + 1$
$y = 4x - 5$

c. $y = 3x - 5$
$y = 3x - 5$

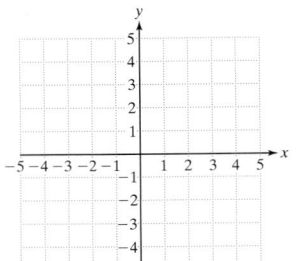

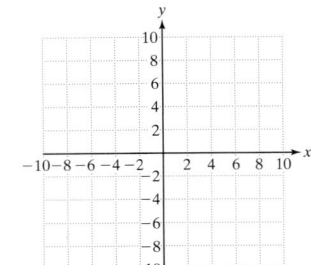

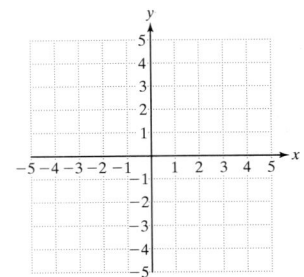

For Exercises 16–26, determine which system of equations (a, b, or c) makes the statement true. (*Hint:* Refer to the graphs from Exercise 15.)

16. The lines are parallel.   A

17. The lines coincide.

18. The lines intersect at exactly one point.   b

19. The system is inconsistent.

20. The system is dependent.   C

21. The lines have the same slope but different y-intercepts.

22. The lines have the same slope and same y-intercept.   C

23. The lines have different slopes.

24. The system has exactly one solution.   b

25. The system has infinitely many solutions.

26. The system has no solution.   A

a. $y = 2x - 3$
  $y = 2x + 5$

b. $y = 2x + 1$
  $y = 4x - 5$

c. $y = 3x - 5$
  $y = 3x - 5$

Concept 3: Solving Systems of Linear Equations by Graphing

For Exercises 27–52, solve the systems by graphing. If a system does not have a unique solution, identify the system as inconsistent or dependent.

27. $y = -x + 4$
  $y = x - 2$

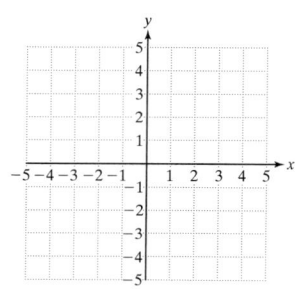

28. $y = 3x + 2$
  $y = 2x$

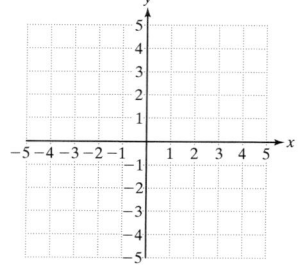

29. $2x + y = 0$
  $3x + y = 1$

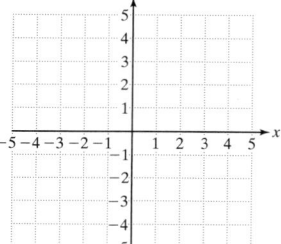

30. $x + y = -1$
  $2x - y = -5$

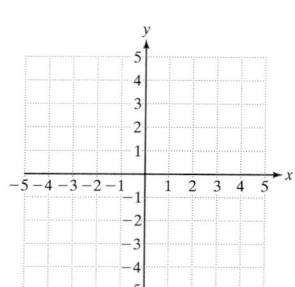

31. $2x + y = 6$
  $x = 1$

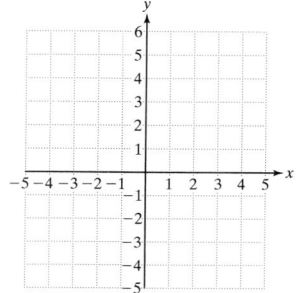

32. $4x + 3y = 9$
  $x = 3$

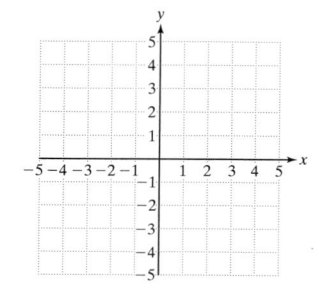

33. $-6x - 3y = 0$
$4x + 2y = 4$

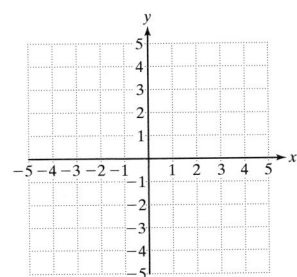

34. $2x - 6y = 12$
$-3x + 9y = 12$

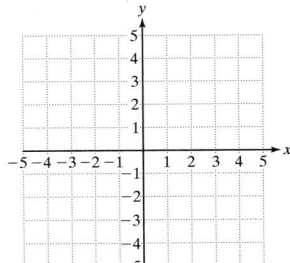

35. $-2x + y = 3$
$6x - 3y = -9$

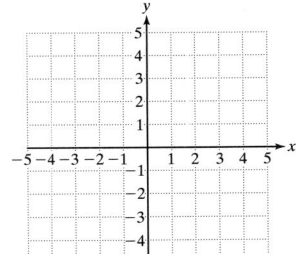

36. $x + 3y = 0$
$-2x - 6y = 0$

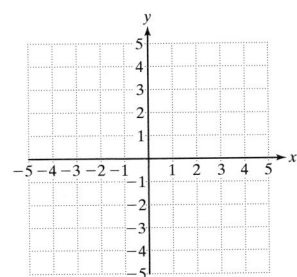

37. $y = 6$
$2x + 3y = 12$

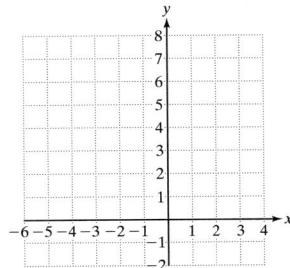

38. $y = -2$
$x - 2y = 10$

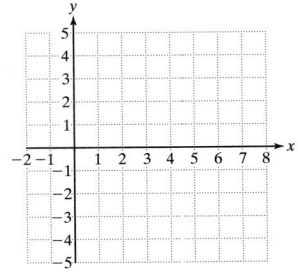

39. $-5x + 3y = -9$
$y = \dfrac{5}{3}x - 3$

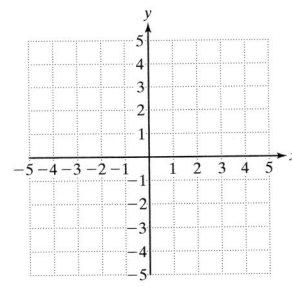

40. $4x + 2y = 6$
$y = -2x + 3$

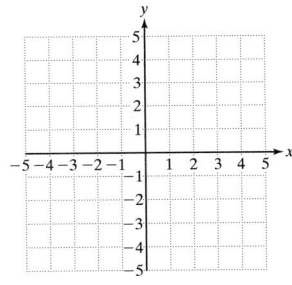

41. $x = 4 + y$
$3y = -3x$

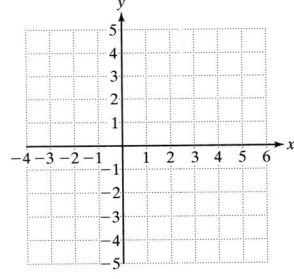

42. $3y = 4x$
$x - y = -1$

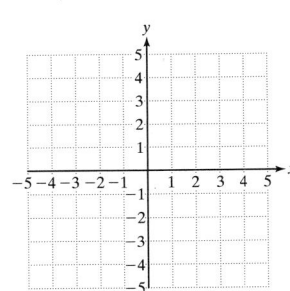

43. $-x + y = 3$
$4y = 4x + 6$

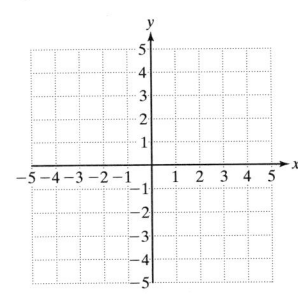

44. $x - y = 4$
$3y = 3x + 6$

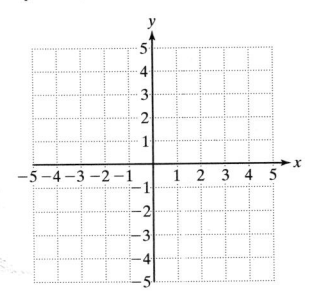

45. $x = 4$

$2y = 4$

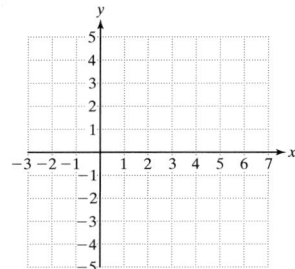

46. $-3x = 6$

$y = 2$

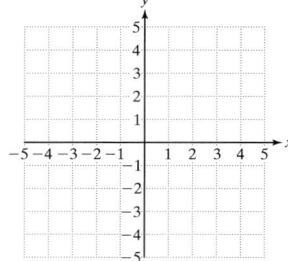

47. $2x + 3y = 8$

$-4x - 6y = 6$

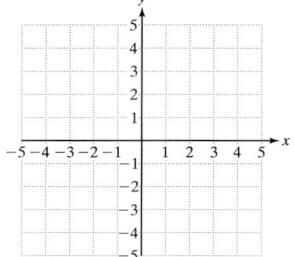

48. $4x + 4y = 8$

$5x + 5y = 5$

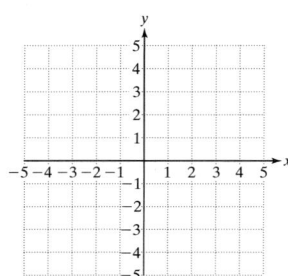

49. $2x + y = 4$

$4x - 2y = -4$

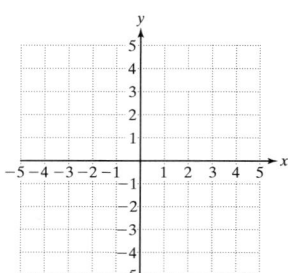

50. $6x + 6y = 3$

$2x - y = 4$

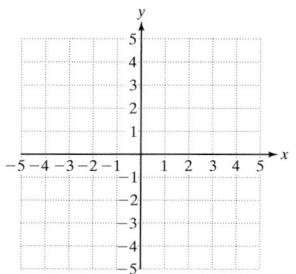

51. $y = 0.5x + 2$

$-x + 2y = 4$

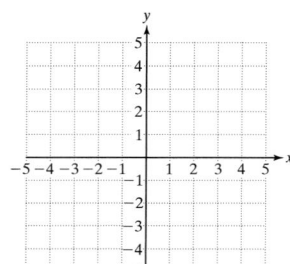

52. $3x - 4y = 6$

$-6x + 8y = -12$

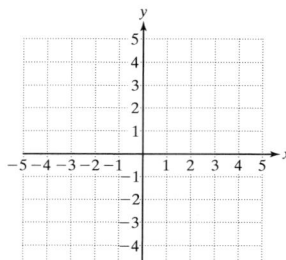

53. Two tennis instructors have two different fee schedules. Owen charges $25 per lesson plus a one-time court fee of $20 at the tennis club. Joan charges $30 per lesson but does not require a court fee. The total cost, y, depends on the number of lessons, x, according to the equations

Owen: $y = 25x + 20$

Joan: $y = 30x$

From the graph, determine the number of lessons for which the total cost is the same for both instructors.

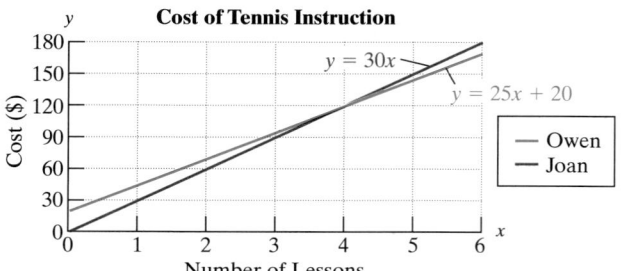

54. The cost to rent a 10 ft by 10 ft storage space is different for two different storage companies. The Storage Bin charges $90 per month plus a nonrefundable deposit of $120. AAA Storage charges $110 per month with no deposit. The total cost, y, to rent a 10 ft by 10 ft space depends on the number of months, x, according to the equations

The Storage Bin: $y = 90x + 120$

AAA Storage: $y = 110x$

From the graph, determine the number of months required for which the cost to rent space is equal for both companies.

Total Cost for Storage

$y = 110x$

$y = 90x + 120$

— (The Storage Bin)
— (AAA Storage)

For the systems graphed in Exercises 55–56, explain why the ordered pair cannot be a solution to the system of equations.

55. $(-3, 1)$

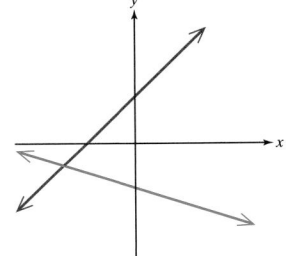

56. $(-1, -4)$

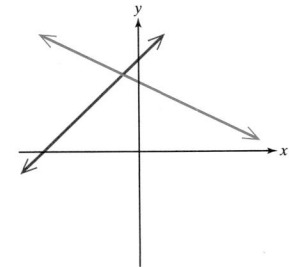

Expanding Your Skills

57. Write a system of linear equations whose solution is $(2, 1)$.

58. Write a system of linear equations whose solution is $(1, 4)$.

59. One equation in a system of linear equations is $x + y = 4$. Write a second equation such that the system will have no solution. (Answers may vary.)

60. One equation in a system of linear equations is $x - y = 3$. Write a second equation such that the system will have infinitely many solutions. (Answers may vary.)

Section 4.2

Solving Systems of Equations by the Substitution Method

1. Solving Systems of Linear Equations by the Substitution Method

In Section 4.1, we used the graphing method to find the solution set to a system of equations. However, sometimes it is difficult to determine the solution using this method because of limitations in the accuracy of the graph. This is particularly true when the coordinates of a solution are not integer values or when the solution is a point not sufficiently close to the origin. Identifying the coordinates of the point $\left(\frac{3}{17}, -\frac{23}{9}\right)$ or $(-251, 8349)$ for example, might be difficult from a graph.

In this section and the next, we will cover two algebraic methods to solve a system of equations that do not require graphing. The first method, called the *substitution method*, is demonstrated in Examples 1–5.

Example 1   Solving a System of Linear Equations Using the Substitution Method

Solve by using the substitution method.

$$x = 2y - 3$$
$$-4x + 3y = 2$$

Solution:

The variable x has been isolated in the first equation. The quantity $2y - 3$ is equal to x and therefore can be substituted for x in the second equation. This leaves the second equation in terms of y only.

First equation: $x = \underline{2y - 3}$

Second equation: $-4x + 3y = 2$

$-4(2y - 3) + 3y = 2$ This equation now contains only one variable.

$-8y + 12 + 3y = 2$ Solve the resulting equation.

$-5y + 12 = 2$

$-5y = -10$

$y = 2$

To find x, substitute $y = 2$ back into the first equation.

$$x = 2y - 3$$
$$x = 2(2) - 3$$
$$x = 1$$

Check the ordered pair $(1, 2)$ in both original equations.

$$x = 2y - 3 \longrightarrow 1 \overset{?}{=} 2(2) - 3 \ ✔ \quad \text{True}$$
$$-4x + 3y = 2 \longrightarrow -4(1) + 3(2) \overset{?}{=} 2 \ ✔ \quad \text{True}$$

The solution is $(1, 2)$ because it checks in both original equations.

1. Solve by using the substitution method.

$$y = -2x + 4$$
$$3x - y = -4$$

In Example 1, we eliminated the x-variable from the second equation by substituting an equivalent expression for x. The resulting equation was relatively simple to solve because it had only one variable. This is the premise of the substitution method.

The substitution method can be summarized as follows.

> **Solving a System of Equations by the Substitution Method**
>
> **1.** Isolate one of the variables from one equation.
>
> **2.** Substitute the quantity found in step 1 into the other equation.
>
> **3.** Solve the resulting equation.
>
> **4.** Substitute the value found in step 3 back into the equation in step 1 to find the value of the remaining variable.
>
> **5.** Check the solution in both original equations and write the answer as an ordered pair.

Example 2 **Solving a System of Linear Equations Using the Substitution Method**

Solve the system using the substitution method.

$$x + y = 4$$
$$-5x + 3y = -12$$

Solution:

The x- or y-variable in the first equation is easy to isolate because the coefficients are both 1. While either variable can be isolated, we arbitrarily choose the x-variable.

$x + y = 4 \longrightarrow x = 4 - y$ **Step 1:** Solve the first equation for x.

$-5(4 - y) + 3y = -12$ **Step 2:** Substitute $4 - y$ for x in the other equation.

$-20 + 5y + 3y = -12$ **Step 3:** Solve for y.

$-20 + 8y = -12$

$8y = 8$

$y = 1$

$x = 4 - y$ **Step 4:** Substitute $y = 1$ into the equation $x = 4 - y$.

$x = 4 - 1$

$x = 3$

Step 5: Check the ordered pair $(3, 1)$ in both original equations.

$$x + y = 4 \qquad\qquad (3) + (1) \stackrel{?}{=} 4 \checkmark \text{ True}$$

$$-5x + 3y = -12 \qquad -5(3) + 3(1) \stackrel{?}{=} -12 \checkmark \text{ True}$$

The solution is $(3, 1)$ because it checks in both original equations.

Skill Practice Solve the system by the substitution method.

2. $2x + 3y = -2$

$\quad -x + y = 1$

TIP: The solution to a system of linear equations can be confirmed by graphing. The system from Example 2 is graphed here.

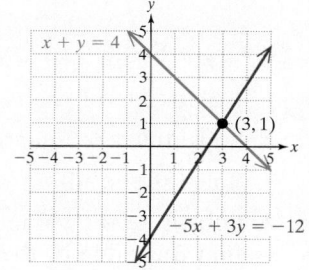

Example 3 **Solving a System of Linear Equations Using the Substitution Method**

Solve the system by using the substitution method.

$$3x + 5y = 17$$
$$2x - y = -6$$

Solution:

The y-variable in the second equation is the easiest variable to isolate because its coefficient is -1.

$$3x + 5y = 17$$

$$2x - y = -6 \longrightarrow -y = -2x - 6$$

$$y = \underline{2x + 6} \qquad \textbf{Step 1:} \quad \text{Solve the second equation for } y.$$

$$3x + 5(2x + 6) = 17 \qquad \textbf{Step 2:} \quad \text{Substitute the quantity } 2x + 6 \text{ for } y \text{ in the other equation.}$$

$$3x + 10x + 30 = 17 \qquad \textbf{Step 3:} \quad \text{Solve for } x.$$

$$13x + 30 = 17$$

$$13x = 17 - 30$$

$$13x = -13$$

$$x = -1$$

Avoiding Mistakes:

Do not substitute $y = 2x + 6$ into the same equation from which it came. This mistake will result in an identity:

$$2x - y = -6$$

$$2x - (2x + 6) = -6$$

$$2x - 2x - 6 = -6$$

$$-6 = -6$$

Skill Practice Answers

2. $(-1, 0)$

$y = 2x + 6$

$y = 2(-1) + 6$

$y = -2 + 6$

$y = 4$

Step 4: Substitute $x = -1$ into the equation $y = 2x + 6$.

Step 5: The ordered pair $(-1, 4)$ can be checked in the original equations to verify the answer.

$3x + 5y = 17 \longrightarrow 3(-1) + 5(4) \overset{?}{=} 17 \longrightarrow -3 + 20 = 17$ ✔ True

$2x - y = -6 \longrightarrow 2(-1) - (4) \overset{?}{=} -6 \longrightarrow -2 - 4 = -6$ ✔ True

The solution is $(-1, 4)$.

Skill Practice Solve the system by the substitution method.

3. $x + 4y = 11$

$2x - 5y = -4$

Recall from Section 4.1, that a system of linear equations may represent two parallel lines. In such a case, there is no solution to the system.

Example 4 Solving an Inconsistent System Using Substitution

Solve the system by using the substitution method.

$$2x + 3y = 6$$
$$y = -\tfrac{2}{3}x + 4$$

Solution:

$2x + 3y = 6$

$y = -\tfrac{2}{3}x + 4$ **Step 1:** The variable y is already isolated in the second equation.

$2x + 3(-\tfrac{2}{3}x + 4) = 6$ **Step 2:** Substitute $y = -\tfrac{2}{3}x + 4$ from the second equation into the first equation.

$2x - 2x + 12 = 6$ **Step 3:** Solve the resulting equation.

$12 = 6$ (contradiction)

The equation results in a contradiction. There are no values of x and y that will make 12 equal to 6. Therefore, there is no solution, and the system is inconsistent.

Skill Practice Solve the system by the substitution method.

4. $y = -\dfrac{1}{2}x + 3$

$2x + 4y = 5$

TIP: The answer to Example 3 can be verified by writing each equation in slope-intercept form and graphing the lines.

Equation 1

$$2x + 3y = 6$$

$$3y = -2x + 6$$

$$\frac{3y}{3} = \frac{-2x}{3} + \frac{6}{3}$$

$$y = -\frac{2}{3}x + 2$$

Equation 2

$$y = -\frac{2}{3}x + 4$$

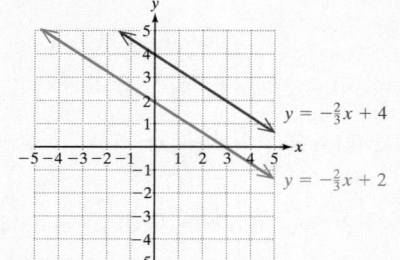

The equations indicate that the lines have the same slope but different y-intercepts. Therefore, the lines must be parallel. There is no point of intersection, indicating that the system has no solution.

Recall that a system of two linear equations may represent the same line. In such a case, the solution is the set of all points on the line.

Example 5   **Solving a Dependent System Using Substitution**

Solve the system by using the substitution method.

$$\frac{1}{2}x - \frac{1}{4}y = 1$$

$$6x - 3y = 12$$

Solution:

$$\frac{1}{2}x - \frac{1}{4}y = 1$$

$$6x - 3y = 12$$

To make the first equation easier to work with, we have the option of clearing fractions.

$$\frac{1}{2}x - \frac{1}{4}y = 1 \xrightarrow{\text{multiply by 4}} 4\left(\frac{1}{2}x\right) - 4\left(\frac{1}{4}y\right) = 4(1) \longrightarrow 2x - y = 4$$

Now the system becomes:

$$2x - y = 4$$
$$6x - 3y = 12$$

The y-variable in the first equation is the easiest to isolate because its coefficient is -1.

$$2x - y = 4 \xrightarrow{\text{solve for } y} -y = -2x + 4 \rightarrow y = 2x - 4$$
$$6x - 3y = 12$$

Step 1: Isolate one of the variables.

$$6x - 3(2x - 4) = 12$$

Step 2: Substitute $y = 2x - 4$ from the first equation into the second equation.

$$6x - 6x + 12 = 12 \qquad \textbf{Step 3:} \quad \text{Solve the resulting equation.}$$

$$12 = 12 \quad \text{(identity)}$$

Because the equation produces an identity, all values of x make this equation true. Thus, x can be any real number. Substituting any real number, x, into the equation $y = 2x - 4$ produces an ordered pair on the line $y = 2x - 4$. Hence, the solution set to the system of equations is the set of all ordered pairs on the line $y = 2x - 4$. This can be written as $\{(x, y) | y = 2x - 4\}$. The system is dependent.

Skill Practice Solve the system by the substitution method.

5. $2x + \dfrac{1}{3}y = -\dfrac{1}{3}$

$12x + 2y = -2$

TIP: The solution to Example 5 can be verified by writing each equation in slope-intercept form and graphing the lines.

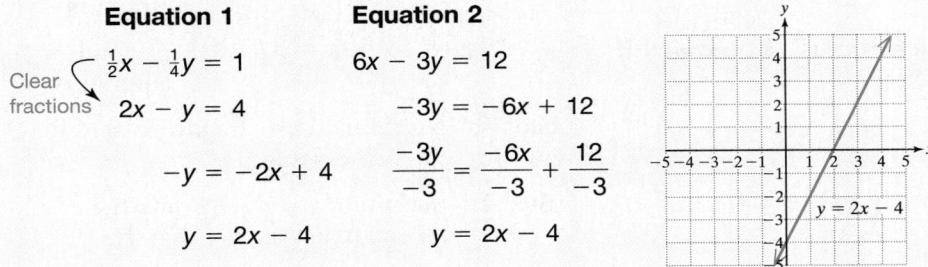

Equation 1	Equation 2
$\frac{1}{2}x - \frac{1}{4}y = 1$	$6x - 3y = 12$
Clear fractions $2x - y = 4$	$-3y = -6x + 12$
$-y = -2x + 4$	$\dfrac{-3y}{-3} = \dfrac{-6x}{-3} + \dfrac{12}{-3}$
$y = 2x - 4$	$y = 2x - 4$

Notice that the slope-intercept forms for both equations are identical. The equations represent the same line, indicating that the system is dependent. Each point on the line is a solution to the system of equations.

2. Solutions to Systems of Linear Equations: A Summary

The following summary reviews the three different geometric relationships between two lines and the solutions to the corresponding systems of equations.

Solutions to a System of Two Linear Equations

1. The lines may intersect at one point (yielding one unique solution).

2. The lines may be parallel and have no point of intersection (yielding no solution). This is detected algebraically when a contradiction (false statement) is obtained (for example, $0 = -3$ and $12 = 6$).

3. The lines may be the same and intersect at all points on the line (yielding an infinite number of solutions). This is detected algebraically when an identity is obtained (for example, $0 = 0$ and $12 = 12$).

Skill Practice Answers

5. Infinitely many solutions:
$\{(x, y) | 12x + 2y = -2\}$

3. Applications of the Substitution Method

In Chapter 2, we solved word problems using one linear equation and one variable. In this chapter, we investigate application problems with two unknowns. In such a case, we can use two variables to represent the unknown quantities. However, if two variables are used, we must write a system of *two* distinct equations.

Example 6 Applying the Substitution Method

One number is 3 more than 4 times another. Their sum is 133. Find the numbers.

Solution:

We can use two variables to represent the two unknown numbers.

Let x represent one number.
Let y represent the other number.   Label the variables.

We must now write two equations. Each of the first two sentences gives a relationship between x and y:

One number is 3 more than 4 times another. $\longrightarrow x = 4y + 3$ (first equation)

Their sum is 133. $\longrightarrow x + y = 133$ (second equation)

Step 1: Notice that x is already isolated in the first equation.

$(4y + 3) + y = 133$ **Step 2:** Substitute $x = 4y + 3$ into the second equation, $x + y = 133$.

$5y + 3 = 133$ **Step 3:** Solve the resulting equation.
$5y = 130$
$y = 26$

$x = 4y + 3$
$x = 4(26) + 3$ **Step 4:** To solve for x, substitute $y = 26$ into the equation $x = 4y + 3$.
$x = 104 + 3$
$x = 107$

One number is 26, and the other is 107.

Skill Practice

6. One number is 16 more than another. Their sum is 92. Use a system of equations to find the numbers.

Skill Practice Answers

6. One number is 38, and the other number is 54.

Example 7 Using the Substitution Method
in a Geometry Application

Two angles are supplementary. The measure of one angle is 15° more than twice the measure of the other angle. Find the measures of the two angles.

Solution:

Let x represent the measure of one angle.
Let y represent the measure of the other angle.

The sum of the measures of supplementary angles is 180° $\longrightarrow$ $x + y = 180$

The measure of one angle is 15° more than
twice the other angle $\longrightarrow$ $x = 2y + 15$

$$x + y = 180$$
$$x = 2y + 15$$

Step 1: The x-variable in the second equation is already isolated.

$$(2y + 15) + y = 180$$

Step 2: Substitute $x = 2y + 15$ from the second equation into the first equation.

$$2y + 15 + y = 180$$

Step 3: Solve the resulting equation.

$$3y + 15 = 180$$
$$3y = 165$$
$$y = 55$$

$$x = 2y + 15$$

Step 4: Substitute $y = 55$ into the equation $x = 2y + 15$.

$$x = 2(55) + 15$$
$$x = 110 + 15$$
$$x = 125$$

One angle is 55°, and the other is 125°.

TIP: Check that the angles 55° and 125° meet the conditions of Example 7.
- Because $55° + 125° = 180°$, the angles are supplementary. ✔
- The angle 125° is 15° more than twice 55°: $125° = 2(55°) + 15°$. ✔

Skill Practice

7. The measure of one angle is 2° less than 3 times the measure of another angle. The angles are complementary. Use a system of equations to find the measures of the two angles.

Skill Practice Answers
7. The measures of the angles are 23° and 67°.

Section 4.2   **Practice Exercises**

Boost *your* GRADE at mathzone.com!

 MathZone

- Practice Problems
- Self-Tests
- NetTutor
- e-Professors
- Videos

Review Exercises

For Exercises 1–6, write each pair of lines in slope-intercept form. Then identify whether the lines intersect in exactly one point or if the lines are parallel or coinciding.

1. $2x - y = 4$
$-2y = -4x + 8$

2. $x - 2y = 5$
$3x = 6y + 15$

3. $2x + 3y = 6$
$x - y = 5$

4. $x - y = -1$

$x + 2y = 4$

5. $2x = \dfrac{1}{2}y + 2$

$4x - y = 13$

6. $4y = 3x$

$3x - 4y = 15$

Concept 1: Solving Systems of Linear Equations by the Substitution Method

For Exercises 7–10, solve each system using the substitution method.

7. $3x + 2y = -3$

$y = 2x - 12$

8. $4x - 3y = -19$

$y = -2x + 13$

9. $x = -4y + 16$

$3x + 5y = 20$

10. $x = -y + 3$

$-2x + y = 6$

11. Given the system: $4x - 2y = -6$

$3x + y = 8$

    a. Which variable from which equation is easiest to isolate and why?

    b. Solve the system using the substitution method.

12. Given the system: $x - 5y = 2$

$11x + 13y = 22$

    a. Which variable from which equation is easiest to isolate and why?

    b. Solve the system using the substitution method.

For Exercises 13–48, solve each system using the substitution method.

13. $x - 3y = -1$

$2x = 4y + 2$

14. $3x + y = 1$

$2y = x + 9$

15. $-2x + 5y = 5$

$x - 4y = -10$

16. $3x - 7y = -2$

$2x + y = 27$

17. $4x - y = -1$

$2x + 4y = 13$

18. $5x - 3y = -2$

$10x - y = 1$

19. $4x - 3y = 11$

$x = 5$

20. $y = -3x - 9$

$y = 12$

21. $4x = 8y + 4$

$5x - 3y = 5$

22. $3y = 6x - 6$

$-3x + y = -4$

23. $x - 3y = -11$

$6x - y = 2$

24. $-2x - y = 9$

$x + 7y = 15$

25. $3x + 2y = -1$

$\dfrac{3}{2}x + y = 4$

26. $5x - 2y = 6$

$-\dfrac{5}{2}x + y = 5$

27. $10x - 30y = -10$

$2x - 6y = -2$

28. $3x + 6y = 6$

$-6x - 12y = -12$

29. $2x + y = 3$

$y = -7$

30. $-3x = 2y + 23$

$x = -1$

31. $x + 2y = -2$

$4x = -2y - 17$

32. $x + y = 1$

$2x - y = -2$

33. $y = -\dfrac{1}{2}x - 4$

$y = 4x - 13$

34. $y = \dfrac{2}{3}x - 3$

$y = 6x - 19$

35. $y = \dfrac{1}{2}x + 8$

$y - 4 = -2(x + 3)$

36. $x = -2y + 7$

$x - 3 = 6(y + 2)$

37. $3x + 2y = 4$

$2x - 3y = -6$

38. $4x + 3y = 4$

$-2x + 5y = -2$

39. $y = 0.25x + 1$

$-x + 4y = 4$

40. $y = 0.75x - 3$

$-3x + 4y = -12$

41. $11x + 6y = 17$

$5x - 4y = 1$

42. $3x - 8y = 7$

$10x - 5y = 45$

43. $x + 2y = 4$

$4y = -2x - 8$

44. $-y = x - 6$

$2x + 2y = 4$

45. $\dfrac{1}{3}(2x + y) = 1$

$x + y = 4$

46. $2(x - y) = 4$

$3x + y = 10$

47. $\dfrac{x}{3} + \dfrac{y}{2} = -4$

$x - 3y = 6$

48. $x - 2y = -5$

$\dfrac{2x}{3} + \dfrac{y}{3} = 0$

Concept 3: Applications of the Substitution Method

For Exercises 49–58, set up a system of linear equations and solve for the indicated quantities.

49. Two numbers have a sum of 106. One number is 10 less than the other. Find the numbers.

50. Two positive numbers have a difference of 8. The larger number is 2 less than 3 times the smaller number. Find the numbers.

51. The difference between two positive numbers is 26. The larger number is three times the smaller. Find the numbers.

52. The sum of two numbers is 956. One number is 94 less than 6 times the other. Find the numbers.

53. Two angles are supplementary. One angle is 15° more than 10 times the other angle. Find the measure of each angle.

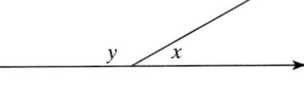

54. Two angles are complementary. One angle is 1° less than 6 times the other angle. Find the measure of each angle.

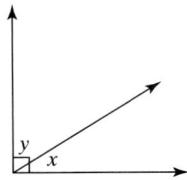

55. Two angles are complementary. One angle is 10° more than 3 times the other angle. Find the measure of each angle.

56. Two angles are supplementary. One angle is 5° less than twice the other angle. Find the measure of each angle.

57. In a right triangle, one of the acute angles is 6° less than the other acute angle. Find the measure of each acute angle.

58. In a right triangle, one of the acute angles is 9° less than twice the other acute angle. Find the measure of each acute angle.

Expanding Your Skills

59. The following system of equations is dependent and has infinitely many solutions. Find three ordered pairs that are solutions to the system of equations.

$$y = 2x + 3$$
$$-4x + 2y = 6$$

60. The following system of equations is dependent and has infinitely many solutions. Find three ordered pairs that are solutions to the system of equations.

$$y = -x + 1$$
$$2x + 2y = 2$$

Section 4.3 — Solving Systems of Equations by the Addition Method

1. Solving Systems of Linear Equations by the Addition Method

Thus far in Chapter 4 we have used the graphing method and the substitution method to solve a system of linear equations in two variables. In this section, we present another algebraic method to solve a system of linear equations, called the *addition method* (sometimes called the *elimination method*). The purpose of the addition method is to eliminate one variable.

Example 1   Solving a System of Linear Equations Using the Addition Method

Solve the system using the addition method.

$$x + y = -2$$
$$x - y = -6$$

Solution:

Notice that the coefficients of the y-variables are opposites:

Coefficient is 1.

$$x + 1y = -2$$
$$x - 1y = -6$$

Coefficient is -1.

Because the coefficients of the y-variables are opposites, we can add the two equations to eliminate the y-variable.

$$x + y = -2$$
$$\underline{x - y = -6}$$
$$2x \quad\ \ = -8 \leftarrow \text{After adding the equations, we have one equation and one variable.}$$

$$2x = -8 \qquad \text{Solve the resulting equation.}$$
$$x = -4$$

To find the value of y, substitute $x = -4$ into *either* of the original equations.

$$x + y = -2 \qquad \text{First equation}$$
$$(-4) + y = -2$$
$$y = -2 + 4$$
$$y = 2$$

The solution is $(-4, 2)$.

TIP: Notice that the value $x = -4$ could have been substituted into the second equation to obtain the same value for y.

$$x - y = -6$$
$$(-4) - y = -6$$
$$-y = -6 + 4$$
$$-y = -2$$
$$y = 2$$

<u>Check:</u>

$$x + y = -2 \longrightarrow (-4) + (2) \overset{?}{=} -2 \longrightarrow -2 = -2 \ \checkmark \quad \text{True}$$
$$x - y = -6 \longrightarrow (-4) - (2) \overset{?}{=} -6 \longrightarrow -6 = -6 \ \checkmark \quad \text{True}$$

Skill Practice Solve the system using the addition method.

1. $x + y = 13$

$2x - y = 2$

It is important to note that the addition method works on the premise that the two equations have *opposite* values for the coefficients of one of the variables. Sometimes it is necessary to manipulate the original equations to create two coefficients that are opposites. This is accomplished by multiplying one or both equations by an appropriate constant. The process is outlined as follows.

Solving a System of Equations by the Addition Method

1. Write both equations in standard form: $Ax + By = C$.

2. Clear fractions or decimals (optional).

3. Multiply one or both equations by nonzero constants to create opposite coefficients for one of the variables.

4. Add the equations from step 3 to eliminate one variable.

5. Solve for the remaining variable.

6. Substitute the known value from step 5 into one of the original equations to solve for the other variable.

7. Check the solution in both equations.

Example 2 **Solving a System of Linear Equations Using the Addition Method**

Solve the system using the addition method.

$$3x + 5y = 17$$

$$2x - y = -6$$

Solution:

$3x + 5y = 17$ **Step 1:** Both equations are already written in standard form.

$2x - y = -6$ **Step 2:** There are no fractions or decimals.

Notice that neither the coefficients of x nor the coefficients of y are opposites. However, multiplying the second equation by 5 creates the term $-5y$ in the second equation. This is the opposite of the term $+5y$ in the first equation.

$3x + 5y = 17$ $3x + 5y = 17$ **Step 3:** Multiply the second
$2x - y = -6$ $\xrightarrow{\text{Multiply by 5}}$ $10x - 5y = -30$ equation by 5.
                        $\overline{13x \quad\quad = -13}$ **Step 4:** Add the equations.

$13x = -13$ **Step 5:** Solve the equation.

$x = -1$

Step 6: Substitute $x = -1$ into one of the original equations.

$3x + 5y = 17$ First equation

$3(-1) + 5y = 17$

$5y = 20$

$y = 4$

Skill Practice Answers

1. $(5, 8)$

The solution is $(-1, 4)$.   **Step 7:** Check the solution in both original equations.

Check:

$$3x + 5y = 17 \longrightarrow 3(-1) + 5(4) \overset{?}{=} 17 \longrightarrow -3 + 20 = 17 ✔ \quad \text{True}$$

$$2x - y = -6 \longrightarrow 2(-1) - (4) \overset{?}{=} -6 \longrightarrow -2 - 4 = -6 ✔ \quad \text{True}$$

> **Skill Practice** Solve the system using the addition method.
>
> **2.** $4x + 3y = 3$
>
> $x - 2y = 9$

In Example 3, the system of equations uses the variables a and b instead of x and y. In such a case, we will write the solution as an ordered pair with the variables written in alphabetical order, such as (a, b).

Example 3 Solving a System of Linear Equations Using the Addition Method

Solve the system using the addition method.

$$5b = 7a + 8$$
$$-4a - 2b = -10$$

Solution:

Step 1: Write the equations in standard form.

The first equation becomes: $5b = 7a + 8 \longrightarrow -7a + 5b = 8$

The system becomes: $-7a + 5b = 8$

$-4a - 2b = -10$

Step 2: There are no fractions or decimals.

Step 3: We need to obtain opposite coefficients on either the a or b term.

Notice that neither the coefficients of a nor the coefficients of b are opposites. However, it is possible to change the coefficients of b to 10 and -10 (this is because the LCM of 5 and 2 is 10). This is accomplished by multiplying the first equation by 2 and the second equation by 5.

$$-7a + 5b = 8 \xrightarrow{\text{Multiply by 2}} -14a + 10b = 16$$
$$-4a - 2b = -10 \xrightarrow{\text{Multiply by 5}} \underline{-20a - 10b = -50}$$
$$-34a \qquad = -34$$

Step 4: Add the equations.

$$-34a = -34$$

$$\frac{-34a}{-34} = \frac{-34}{-34}$$

Step 5: Solve the resulting equation.

$$a = 1$$

$5b = 7a + 8$ First equation

$5b = 7(1) + 8$

$5b = 15$

$b = 3$

The solution is $(1, 3)$.

Check:

$$5b = 7a + 8 \longrightarrow 5(3) \overset{?}{=} 7(1) + 8 \longrightarrow 15 = 7 + 8 \checkmark \text{ True}$$

$$-4a - 2b = -10 \longrightarrow -4(1) - 2(3) \overset{?}{=} -10 \longrightarrow -4 - 6 = -10 \checkmark \text{ True}$$

Step 6: Substitute $a = 1$ into one of the original equations.

Step 7: Check the solution in the original equations.

Skill Practice   Solve the system using the addition method.

3. $8n = 4 - 5m$

  $7m + 6n = -10$

Example 4   **Solving a System of Linear Equations Using the Addition Method**

Solve the system using the addition method.

$$3(x - 10) = 7y + 11$$
$$-2(x - y) = 2x - 18$$

Solution:

Step 1: Write the equations in standard form.

$$3(x - 10) = 7y + 11 \xrightarrow{\text{Clear parentheses}} 3x - 30 = 7y + 11 \xrightarrow{\text{Write as } Ax + By = C} 3x - 7y = 41$$

$$-2(x - y) = 2x - 18 \longrightarrow -2x + 2y = 2x - 18 \longrightarrow -4x + 2y = -18$$

Step 2: There are no fractions or decimals.

Notice that neither the coefficients of x nor the coefficients of y are opposites. However, it is possible to change the coefficients of x to 12 and -12 (notice that 12 is the LCM of 3 and 4). This is accomplished by multiplying the first equation by 4 and the second equation by 3.

$$3x - 7y = 41 \xrightarrow{\text{Multiply by 4}} 12x - 28y = 164$$
$$-4x + 2y = -18 \xrightarrow[\text{Multiply by 3}]{} \underline{-12x + 6y = -54}$$
$$-22y = 110$$

Step 3: Create opposite coefficients of x.

Step 4: Add the equations.

$$-22y = 110$$

$$\frac{-22y}{-22} = \frac{110}{-22}$$

$$y = -5$$

Step 5: Solve the resulting equation.

$3x - 7y = 41$ First equation

Step 6: Substitute $y = -5$ into one of the equations.

$3x - 7(-5) = 41$

$3x + 35 = 41$

$3x = 6$

$x = 2$

The solution is $(2, -5)$.

Step 7: Check the solution in the original equations.

Check:

$3(x - 10) = 7y + 11 \longrightarrow 3(2 - 10) \overset{?}{=} 7(-5) + 11 \longrightarrow -24 = -24 \checkmark$ True

$-2(x - y) = 2x - 18 \longrightarrow -2[2 - (-5)] \overset{?}{=} 2(2) - 18 \longrightarrow -14 = -14 \checkmark$ True

Skill Practice Solve the system using the addition method.

4. $x = 5(y + 2) - x$

$ 7y = 3(x - 5)$

TIP: When using the addition method, it makes no difference which variable is eliminated. In Example 4 we eliminated x. However, we could easily have eliminated y by changing the coefficients of y to -14 and 14. This would be accomplished by multiplying the first equation by 2 and the second equation by 7.

$$3x - 7y = 41 \xrightarrow{\text{Multiply by 2}} 6x - 14y = 82$$

$$-4x + 2y = -18 \xrightarrow[\text{Multiply by 7}]{} \underline{-28x + 14y = -126}$$

$$-22x = -44$$

Because $-22x = -44$, then $x = 2$. Substituting $x = 2$ into either original equation yields $y = -5$.

Example 5 **Solving a System of Linear Equations Using the Addition Method**

Solve the system using the addition method.

$$34x - 22y = 4$$

$$17x - 88y = -19$$

Solution:

The equations are already in standard form. There are no fractions or decimals to clear.

$$34x - 22y = 4 \xrightarrow{} 34x - 22y = 4$$

$$17x - 88y = -19 \xrightarrow[\text{Multiply by } -2]{} \underline{-34x + 176y = 38}$$

$$154y = 42$$

$$154y = 42$$

$$\frac{154y}{154} = \frac{42}{154}$$

$$y = \frac{3}{11}$$

To find the value of x, we normally substitute y into one of the original equations and solve for x. In this example, we will show an alternative approach for finding x. By repeating the addition method, this time eliminating y, we can solve for x. This approach enables us to avoid substitution of the fractional value for y.

$$34x - 22y = 4 \xrightarrow{\text{Multiply by } -4} -136x + 88y = -16$$
$$17x - 88y = -19 \xrightarrow{ } \underline{\quad 17x - 88y = -19 \quad}$$
$$-119x = -35$$

$$-119x = -35 \qquad \text{Solve for } x.$$

$$\frac{-119x}{-119} = \frac{-35}{-119}$$

$$x = \frac{5}{17} \qquad \text{Simplify.}$$

The solution is $\left(\frac{5}{17}, \frac{3}{11}\right)$. These values can be checked in the original equations:

$$34x - 22y = 4 \qquad\qquad 17x - 88y = -19$$

$$34\left(\frac{5}{17}\right) - 22\left(\frac{3}{11}\right) \stackrel{?}{=} 4 \qquad\qquad 17\left(\frac{5}{17}\right) - 88\left(\frac{3}{11}\right) \stackrel{?}{=} -19$$

$$10 - 6 = 4 \checkmark \text{ True} \qquad\qquad 5 - 24 = -19 \checkmark \text{ True}$$

Skill Practice Solve the system using the addition method.

5. $15x - 16y = 1$
$\ 45x + 4y = 16$

Example 6 Solving an Inconsistent System of Linear Equations

Solve the system using the addition method.

$$2x - 5y = 10$$
$$\frac{1}{2}x = 1 + \frac{5}{4}y$$

Solution:

$$2x - 5y = 10 \xrightarrow{} 2x - 5y = 10$$
$$\frac{1}{2}x = 1 + \frac{5}{4}y \xrightarrow{} \frac{1}{2}x - \frac{5}{4}y = 1 \qquad \textbf{Step 1:} \text{ Write the equations in standard form.}$$

Step 2: Multiply both sides of the second equation by 4 to clear fractions.

$$\frac{1}{2}x - \frac{5}{4}y = 1 \xrightarrow{} 4\left(\frac{1}{2}x - \frac{5}{4}y\right) = 4(1) \xrightarrow{} 2x - 5y = 4$$

Skill Practice Answers

5. $\left(\frac{1}{3}, \frac{1}{4}\right)$

Now the system becomes
$$2x - 5y = 10$$
$$2x - 5y = 4$$

To make either the *x*-coefficients or *y*-coefficients opposites, multiply either equation by -1.

$$2x - 5y = 10 \xrightarrow{\text{Multiply by } -1} -2x + 5y = -10 \qquad \textbf{Step 3:} \quad \text{Create opposite coefficients.}$$

$$2x - 5y = 4 \xrightarrow{\hspace{2cm}} \underline{2x - 5y = 4}$$
$$0 = -6 \qquad \textbf{Step 4:} \quad \text{Add the equations.}$$

Because the equation results in a contradiction, there is no solution, and the system of equations is inconsistent. Writing each line in slope-intercept form verifies that the lines are parallel (Figure 4-8).

$$2x - 5y = 10 \xrightarrow{\text{Slope-intercept form}} y = \frac{2}{5}x - 2$$

$$\frac{1}{2}x = 1 + \frac{5}{4}y \xrightarrow{\text{Slope-intercept form}} y = \frac{2}{5}x - \frac{4}{5}$$

There is no solution.

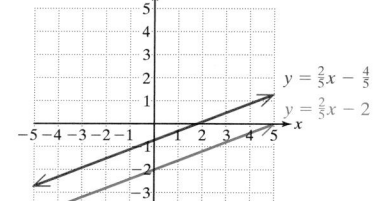

Figure 4-8

Skill Practice   Solve the system using the addition method.

6. $\dfrac{2}{3}x = 2 + \dfrac{3}{4}y$

   $8x - 9y = 6$

Example 7   Solving a Dependent System of Linear Equations

Solve the system by the addition method.
$$3x - y = 4$$
$$2y = 6x - 8$$

Solution:

$$3x - y = 4 \xrightarrow{\hspace{2cm}} 3x - y = 4 \qquad \textbf{Step 1:} \quad \text{Write the equations in standard form.}$$

$$2y = 6x - 8 \longrightarrow -6x + 2y = -8 \qquad \textbf{Step 2:} \quad \text{There are no fractions or decimals.}$$

Notice that the equations differ exactly by a factor of -2, which indicates that these two equations represent the same line. Multiply the first equation by 2 to create opposite coefficients for the variables.

$$3x - y = 4 \xrightarrow{\text{Multiply by 2}} 6x - 2y = 8 \qquad \textbf{Step 3:} \quad \text{Create opposite coefficients.}$$

$$-6x + 2y = -8 \xrightarrow{\hspace{2cm}} \underline{-6x + 2y = -8}$$
$$0 = 0 \qquad \textbf{Step 4:} \quad \text{Add the equations.}$$

Skill Practice Answers

6. No solution.

Because the resulting equation is an identity, the original equations represent the same line. This can be confirmed by writing each equation in slope-intercept form.

$$3x - y = 4 \longrightarrow -y = -3x + 4 \longrightarrow y = 3x - 4$$

$$-6x + 2y = -8 \longrightarrow 2y = 6x - 8 \longrightarrow y = 3x - 4$$

The solution is the set of all points on the line, or equivalently, $\{(x, y) \mid 3x - y = 4\}$.

Skill Practice Solve the system by using the addition method.

7. $3x = 3y + 15$

$2x = 10 + 2y$

2. Summary of Methods for Solving Linear Equations in Two Variables

If no method of solving a system of linear equations is specified, you may use the method of your choice. However, we recommend the following guidelines:

1. If one of the equations is written with a variable isolated, the substitution method is a good choice. For example:

$$2x + 5y = 2 \qquad \text{or} \qquad y = \frac{1}{3}x - 2$$
$$x = y - 6 \qquad\qquad x - 6y = 9$$

2. If both equations are written in standard form, $Ax + By = C$, where none of the variables has coefficients of 1 or -1, then the addition method is a good choice.

$$4x + 5y = 12$$
$$5x + 3y = 15$$

3. If both equations are written in standard form, $Ax + By = C$, and at least one variable has a coefficient of 1 or -1, then either the substitution method or the addition method is a good choice.

Skill Practice Answers

7. $\{(x, y) \mid 3x = 3y + 15\}$

Section 4.3	**Practice Exercises**

Boost *your* GRADE at mathzone.com!

MathZone

- Practice Problems
- Self-Tests
- NetTutor
- e-Professors
- Videos

Study Skills Exercise

1. Now that you have learned three methods of solving a system of linear equations with two variables, choose a system and solve it all three ways. There are two advantages to this. One is to check your answer (you should get the same answer using all three methods). The second advantage is to show you which method is the easiest for you to use.

Solve the system by using the graphing method, the substitution method, and the addition method.

$$2x + y = -7$$
$$x - 10 = 4y$$

Review Exercises

For Exercises 2–5, check to see if the given ordered pair is a solution to the system.

2. $x + y = 8$ $(5, 3)$

$y = x - 2$

3. $3x = y + 1$ $(3, 2)$ yes

$-x + 2y = 0$ no

4. $3x + 2y = 14$ $(5, -2)$

$5x - 2y = 29$

5. $x = 2y - 11$ $(-3, 4)$

$-x + 5y = 23$ yes

Concept 1: Solving Systems of Linear Equations by the Addition Method

For Exercises 6–7, answer as true or false.

6. Given the system

$$5x - 4y = 1$$
$$7x - 2y = 5$$

a. To eliminate the y-variable using the addition method, multiply the second equation by 2.

b. To eliminate the x-variable, multiply the first equation by 7 and the second equation by -5.

7. Given the system

$$3x + 5y = -1$$
$$9x - 8y = -26$$

a. To eliminate the x-variable using the addition method, multiply the first equation by -3.

b. To eliminate the y-variable, multiply the first equation by 8 and the second equation by -5.

8. Given the system

$$3x - 4y = 2$$
$$17x + y = 35$$

a. Which variable, x or y, is easier to eliminate using the addition method?

b. Solve the system using the addition method.

9. Given the system

$$-2x + 5y = -15$$
$$6x - 7y = 21$$

a. Which variable, x or y, is easier to eliminate using the addition method?

b. Solve the system using the addition method.

For Exercises 10–21, solve the systems using the addition method.

10. $x + 2y = 8$

$5x - 2y = 4$

11. $2x - 3y = 11$

$-4x + 3y = -19$

12. $a + b = 3$

$3a + b = 13$

13. $-2u + 6v = 10$

$-2u + v = -5$

14. $-3x + y = 1$

$-6x - 2y = -2$

15. $5m - 2n = 4$

$3m + n = 9$

16. $3x - 5y = 13$
$x - 2y = 5$

17. $7a + 2b = -1$
$3a - 4b = 19$

18. $6c - 2d = -2$
$5c + 3d = 17$

19. $2s + 3t = -1$
$5s - 2t = 7$

20. $6y - 4z = -2$
$4y + 6z = 42$

21. $4k - 2r = -4$
$2k + 4r = -32$

22. In solving a system of equations, suppose you get the statement $0 = 5$. How many solutions will the system have? What can you say about the graphs of these equations?

23. In solving a system of equations, suppose you get the statement $0 = 0$. How many solutions will the system have? What can you say about the graphs of these equations?

24. In solving a system of equations, suppose you get the statement $3 = 3$. How many solutions will the system have? What can you say about the graphs of these equations?

25. In solving a system of equations, suppose you get the statement $2 = -5$. How many solutions will the system have? What can you say about the graphs of these equations?

For Exercises 26–37, solve the system of equations using the addition method.

26. $-2x + y = -5$
$8x - 4y = 12$

27. $x - 3y = 2$
$-5x + 15y = 10$

28. $x + 2y = 2$
$-3x - 6y = -6$

29. $4x - 3y = 6$
$-12x + 9y = -18$

30. $3a + 2b = 11$
$7a - 3b = -5$

31. $4y + 5z = -2$
$5y - 3z = 16$

32. $3x - 5y = 7$
$5x - 2y = -1$

33. $4s + 3t = 9$
$3s + 4t = 12$

34. $2(x + 1) = -3y + 9$
$3x - 10 = -4y$

35. $-3(x - 2) + 7y = 5$
$5y = 2x$

36. $4x - 5y = 0$
$8(x - 1) = 10y$

37. $y = 2x + 1$
$-3(2x - y) = 0$

Concept 2: Summary of Methods for Solving Linear Equations in Two Variables

For Exercises 38–57, solve the system of equations by either the addition method or the substitution method.

38. $5x - 2y = 4$
$y = -3x + 9$

39. $-x = 8y + 5$
$4x - 3y = -20$

40. $0.1x + 0.1y = 0.6$
$0.1x - 0.1y = 0.1$

41. $0.1x + 0.1y = 0.2$
$0.1x - 0.1y = 0.3$

42. $3x = 5y - 9$
$2y = 3x + 3$

43. $10x - 5 = 3y$
$2x + 3y = 1$

44. $y = -5x - 5$
$6x - 3 = -3y$

45. $4x + 5y = -2$
$6x = -10y + 2$

46. $x = -\dfrac{1}{2}$
$6x - 5y = -8$

47. $4x - 2y = 1$
$y = 3$

48. $0.02x + 0.04y = 0.12$
$0.03x - 0.05y = -0.15$

49. $-0.03x + 0.01y = 0.01$
$-0.06x - 0.02y = -0.02$

50. $8x - 16y = 24$

   $2x - 4y = 0$

51. $y = -\dfrac{1}{2}x - 5$

   $2x + 4y = -8$

52. $\dfrac{m}{2} + \dfrac{n}{5} = \dfrac{13}{10}$

   $3(m - n) = m - 10$

53. $\dfrac{a}{4} - \dfrac{3b}{2} = \dfrac{15}{2}$

   $\dfrac{1}{5}(a + 2b) = -2$

54. $2(m - 3n) = m + 4$

   $3m + 8 = 5m - n$

55. $m - 3n = 10$

   $3(m + 4n) = -12$

56. $9a - 2b = 8$

   $6(3a + 1) = 4b + 22$

57. $a = 5 + 2b$

   $3(a - 2b) = 15$

For Exercises 58–61, set up a system of linear equations, and solve for the indicated quantities.

58. The sum of two positive numbers is 26. Their difference is 14. Find the numbers.

59. The difference of two positive numbers is 2. The sum of the numbers is 36. Find the numbers.

60. Eight times the smaller of two numbers plus 2 times the larger number is 44. Three times the smaller number minus 2 times the larger number is zero. Find the numbers.

61. Six times the smaller of two numbers minus the larger number is -9. Ten times the smaller number plus five times the larger number is 5. Find the numbers.

For Exercises 62–65, use the addition method to eliminate the x-variable to solve for y. Then use the addition method to eliminate the y-variable to solve for x. Write the solution as an ordered pair.

62. $2x + 3y = 6$

   $x - y = 5$

63. $6x + 6y = 8$

   $9x - 18y = -3$

64. $2x - 5y = 4$

   $3x - 3y = 4$

65. $6x - 5y = 7$

   $4x - 6y = 7$

For Exercises 66–68, solve the system by using each of the three methods: (a) the graphing method, (b) the substitution method, and (c) the addition method.

66. $2x + y = 1$

   $-4x - 2y = -2$

67. $3x + y = 6$

   $-2x + 2y = 4$

68. $2x - 2y = 6$

   $5y = 5x + 5$

Expanding Your Skills

69. Explain why a system of linear equations cannot have exactly two solutions.

70. The solution to the following system of linear equations is $(1, 2)$. Find A and B.

$$Ax + 3y = 8$$
$$x + By = -7$$

71. The solution to the following system of linear equations is $(-3, 4)$. Find A and B.

$$4x + Ay = -32$$
$$Bx + 6y = 18$$

Applications of Linear Equations in Two Variables

1. Applications Involving Cost

In Sections 2.4–2.6, we solved several applied problems by setting up a linear equation in one variable. When solving an application that involves two unknowns, sometimes it is convenient to use a system of linear equations in two variables.

| Example 1 | Using a System of Linear Equations Involving Cost |

At a movie theater a couple buys one large popcorn and two drinks for $5.75. A group of teenagers buys two large popcorns and five drinks for $13.00. Find the cost of one large popcorn and the cost of one drink.

Solution:

In this application we have two unknowns, which we can represent by x and y.

Let x represent the cost of one large popcorn.
Let y represent the cost of one drink.

We must now write two equations. Each of the first two sentences in the problem gives a relationship between x and y:

$$\begin{pmatrix} \text{Cost of 1} \\ \text{large popcorn} \end{pmatrix} + \begin{pmatrix} \text{cost of 2} \\ \text{drinks} \end{pmatrix} = \begin{pmatrix} \text{total} \\ \text{cost} \end{pmatrix} \longrightarrow x + 2y = 5.75$$

$$\begin{pmatrix} \text{Cost of 2} \\ \text{large popcorns} \end{pmatrix} + \begin{pmatrix} \text{cost of 5} \\ \text{drinks} \end{pmatrix} = \begin{pmatrix} \text{total} \\ \text{cost} \end{pmatrix} \longrightarrow 2x + 5y = 13.00$$

To solve this system, we may either use the substitution method or the addition method. We will use the substitution method by solving for x in the first equation.

$$x + 2y = 5.75 \longrightarrow x = -2y + 5.75 \qquad \text{Isolate } x \text{ in the first equation.}$$

$$2x + 5y = 13.00$$

$$2(-2y + 5.75) + 5y = 13.00 \qquad \text{Substitute } x = -2y + 5.75 \text{ into the other equation.}$$

$$-4y + 11.50 + 5y = 13.00 \qquad \text{Solve for } y.$$

$$y + 11.50 = 13.00$$

$$y = 1.50$$

$$x = -2y + 5.75$$

$$x = -2(1.50) + 5.75 \qquad \text{Substitute } y = 1.50 \text{ into the equation}$$

$$x = -3.00 + 5.75 \qquad x = -2y + 5.75.$$

$$x = 2.75$$

The cost of one large popcorn is $2.75 and the cost of one drink is $1.50.

Check by verifying that the solutions meet the specified conditions.

$$1 \text{ popcorn} + 2 \text{ drinks} = 1(\$2.75) + 2(\$1.50) = \$5.75 \ ✔ \ \text{True}$$

$$2 \text{ popcorns} + 5 \text{ drinks} = 2(\$2.75) + 5(\$1.50) = \$13.00 \ ✔ \ \text{True}$$

Skill Practice

1. Lynn went to a fast-food restaurant and spent $9.00. She purchased 4 hamburgers and 5 orders of fries. The next day, Ricardo went to the same restaurant and purchased 10 hamburgers and 7 orders of fries. He spent $18.10. Use a system of equations to determine the cost of a burger and the cost of an order of fries.

2. Applications Involving Principal and Interest

In Section 2.5, we learned that simple interest is interest computed on the principal amount of money invested (or borrowed). Simple interest, I, is found by using the formula

$$I = Prt \qquad \text{where } P \text{ is the principal,}$$
$$r \text{ is the annual interest rate, and}$$
$$t \text{ is the time in years.}$$

If the amount of time is taken to be 1 year, we have: $I = Pr(1)$ or simply $I = Pr$.

In Example 2, we apply the concept of simple interest to two accounts to produce a desired amount of interest after 1 year.

Example 2 **Using a System of Linear Equations Involving Investments**

Joanne has a total of $6000 to deposit in two accounts. One account earns 3.5% simple interest and the other earns 2.5% simple interest. If the total amount of interest at the end of 1 year is $195, find the amount she deposited in each account.

Solution:

Let x represent the principal deposited in the 2.5% account.
Let y represent the principal deposited in the 3.5% account.

	2.5% Account	3.5% Account	Total
Principal	x	y	6000
Interest $(I = Pr)$	$0.025x$	$0.035y$	195

Each row of the table yields an equation in x and y:

$$\begin{pmatrix} \text{Principal} \\ \text{invested} \\ \text{at } 2.5\% \end{pmatrix} + \begin{pmatrix} \text{principal} \\ \text{invested} \\ \text{at } 3.5\% \end{pmatrix} = \begin{pmatrix} \text{total} \\ \text{principal} \end{pmatrix} \longrightarrow x + y = 6000$$

$$\begin{pmatrix} \text{Interest} \\ \text{earned} \\ \text{at } 2.5\% \end{pmatrix} + \begin{pmatrix} \text{interest} \\ \text{earned} \\ \text{at } 3.5\% \end{pmatrix} = \begin{pmatrix} \text{total} \\ \text{interest} \end{pmatrix} \longrightarrow 0.025x + 0.035y = 195$$

Skill Practice Answers

1. The cost of a burger is $1.25 and the cost of an order of fries is $0.80.

We will choose the addition method to solve the system of equations. First multiply the second equation by 1000 to clear decimals.

$$x + y = 6000 \xrightarrow{\hspace{3cm}} x + y = 6000$$
$$0.025x + 0.035y = 195 \xrightarrow[\text{Multiply by 1000}]{} 25x + 35y = 195{,}000$$

$$x + y = 6000 \xrightarrow[\hspace{1cm}]{\text{Multiply by } -25} -25x - 25y = -150{,}000$$
$$25x + 35y = 195{,}000 \xrightarrow{\hspace{2cm}} \underline{25x + 35y = 195{,}000}$$
$$10y = 45{,}000$$

$$\frac{10y}{10} = \frac{45{,}000}{10}$$

$y = 4500$ The amount invested in the 3.5% account is $4500.

$x + y = 6000$ Substitute $y = 4500$ into the equation $x + y = 6000$.

$x + 4500 = 6000$

$x = 1500$ The amount invested in the 2.5% account is $1500.

Joanne deposited $1500 in the 2.5% account and $4500 in the 3.5% account.

To check the solution, verify that the conditions of the problem have been met.

1. The sum of $1500 and $4500 is $6000 as desired. ✔

2. The interest earned on $1500 at 2.5% is: 0.025($1500) = $37.5
   The interest earned on $4500 at 3.5% is: 0.035($4500) = $157.5
   Total interest: $195.00 ✔

Skill Practice

2. Addie has a total of $8000 in two accounts. One pays 5% interest, and the other pays 6.5% interest. At the end of one year, she earned $475 interest. Use a system of equations to determine the amount invested in each account.

---

3. Applications Involving Mixtures

Example 3 **Using a System of Linear Equations in a Mixture Application**

A 10% alcohol solution is mixed with a 40% alcohol solution to produce 30 L of a 20% alcohol solution. Find the number of liters of 10% solution and the number of liters of 40% solution required for this mixture.

Solution:

Each solution contains a percentage of alcohol plus some other mixing agent such as water. Before we set up a system of equations to model this situation,

Skill Practice Answers

2. $3000 is invested at 5%, and $5000 is invested at 6.5%.

it is helpful to have background understanding of the problem. In Figure 4-9, the liquid depicted in blue is pure alcohol and the liquid shown in gray is the mixing agent (such as water). Together these liquids form a solution. (Realistically the mixture may not separate as shown, but this image may be helpful for your understanding.)

Let x represent the number of liters of 10% solution.
Let y represent the number of liters of 40% solution.

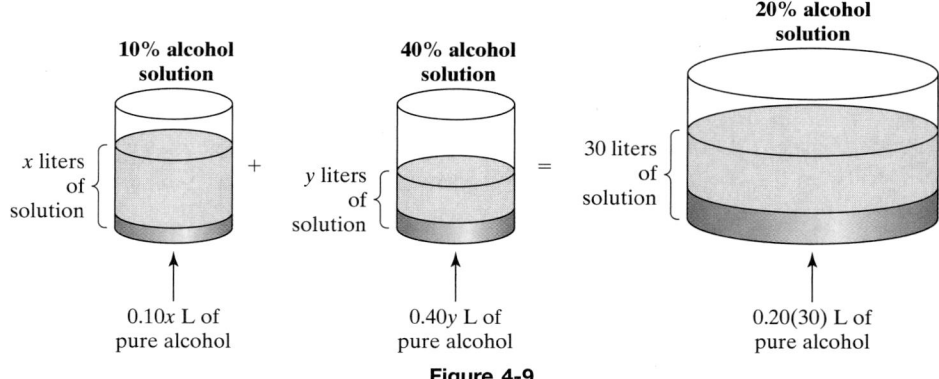

Figure 4-9

The information given in the statement of the problem can be organized in a chart.

	10% Alcohol	40% Alcohol	20% Alcohol
Number of liters of solution	x	y	30
Number of liters of pure alcohol	$0.10x$	$0.40y$	$0.20(30) = 6$

From the first row, we have

$$\left(\begin{array}{c}\text{Amount of}\\ 10\%\text{ solution}\end{array}\right) + \left(\begin{array}{c}\text{amount of}\\ 40\%\text{ solution}\end{array}\right) = \left(\begin{array}{c}\text{total amount}\\ \text{of }20\%\text{ solution}\end{array}\right) \rightarrow x + y = 30$$

From the second row, we have

$$\left(\begin{array}{c}\text{Amount of}\\ \text{alcohol in}\\ 10\%\text{ solution}\end{array}\right) + \left(\begin{array}{c}\text{amount of}\\ \text{alcohol in}\\ 40\%\text{ solution}\end{array}\right) = \left(\begin{array}{c}\text{total amount of}\\ \text{alcohol in}\\ 20\%\text{ solution}\end{array}\right) \rightarrow 0.10x + 0.40y = 6$$

We will solve the system with the addition method by first clearing decimals.

$$
\begin{array}{llll}
x + y = 30 & \xrightarrow{} & x + y = 30 & \xrightarrow{\text{Multiply by }-1} & -x - y = -30 \\
0.10x + 0.40y = 6 & \xrightarrow[\text{Multiply by 10}]{} & x + 4y = 60 & \xrightarrow{} & \underline{x + 4y = 60} \\
& & & & 3y = 30
\end{array}
$$

$$3y = 30 \qquad \text{After eliminating the } x\text{-variable, solve for } y.$$
$$y = 10 \qquad \text{10 L of 40\% solution is needed.}$$

$$x + y = 30 \qquad \text{Substitute } y = 10 \text{ into either of the original equations.}$$
$$x + (10) = 30$$
$$x = 20 \qquad \text{20 L of 10\% solution is needed.}$$

10 L of 40% solution must be mixed with 20 L of 10% solution.

Skill Practice

3. How many ounces of 20% and 35% acid solution should be mixed together to obtain 15 ounces of 30% acid solution?

4. Applications Involving Distance, Rate, and Time

The following formula relates the distance traveled to the rate and time of travel.

$$d = rt \qquad \text{distance} = \text{rate} \cdot \text{time}$$

For example, if a car travels at 60 mph for 3 hours, then

$$d = (60 \text{ mph})(3 \text{ hours})$$
$$= 180 \text{ miles}$$

If a car travels at 60 mph for x hours, then

$$d = (60 \text{ mph})(x \text{ hours})$$
$$= 60x \text{ miles}$$

The relationship $d = rt$ is used in Example 4.

Example 4 Using a System of Linear Equations in a Distance, Rate, Time Application

A plane travels with a tail wind from Kansas City, Missouri, to Denver, Colorado, a distance of 600 miles in 2 hours. The return trip against a head wind takes 3 hours. Find the speed of the plane in still air, and find the speed of the wind.

Solution:

Let p represent the speed of the plane in still air.
Let w represent the speed of the wind.

Notice that when the plane travels with the wind, the net speed is $p + w$.
When the plane travels against the wind, the net speed is $p - w$.

The information given in the problem can be organized in a chart.

	Distance	Rate	Time
With a tail wind	600	$p + w$	2
Against a head wind	600	$p - w$	3

Skill Practice Answers

3. 10 ounces of the 35% solution, and 5 ounces of the 20% solution.

To set up two equations in p and w, recall that $d = rt$.

From the first row, we have

$$\begin{pmatrix} \text{Distance} \\ \text{with the wind} \end{pmatrix} = \begin{pmatrix} \text{rate with} \\ \text{the wind} \end{pmatrix}\begin{pmatrix} \text{time traveled} \\ \text{with the wind} \end{pmatrix} \longrightarrow 600 = (p + w) \cdot 2$$

From the second row, we have

$$\begin{pmatrix} \text{Distance} \\ \text{against the wind} \end{pmatrix} = \begin{pmatrix} \text{rate against} \\ \text{the wind} \end{pmatrix}\begin{pmatrix} \text{time traveled} \\ \text{against the wind} \end{pmatrix} \longrightarrow 600 = (p - w) \cdot 3$$

Using the distributive property to clear parentheses produces the following system:

$$2p + 2w = 600$$
$$3p - 3w = 600$$

The coefficients on the w-variable can be changed to 6 and -6 by multiplying the first equation by 3 and the second equation by 2.

$$
\begin{array}{l}
2p + 2w = 600 \quad \xrightarrow{\text{Multiply by 3}} \quad 6p + 6w = 1800 \\
3p - 3w = 600 \quad \xrightarrow{\text{Multiply by 2}} \quad 6p - 6w = 1200 \\
\hline
\qquad\qquad\qquad\qquad\qquad\qquad\qquad\quad 12p \quad\;\;\; = 3000
\end{array}
$$

$$12p = 3000$$

$$\frac{12p}{12} = \frac{3000}{12}$$

$$p = 250 \qquad \text{The speed of the plane in still air is 250 mph.}$$

> **TIP:** To create opposite coefficients on the w-variables, we could have divided the first equation by 2 and divided the second equation by 3:
>
> $$
> \begin{array}{l}
> 2p + 2w = 600 \quad \xrightarrow{\text{Divide by 2}} \quad p + w = 300 \\
> 3p - 3w = 600 \quad \xrightarrow{\text{Divide by 3}} \quad p - w = 200 \\
> \hline
> \qquad\qquad\qquad\qquad\qquad\qquad\quad 2p \qquad\;\; = 500 \\
> \qquad\qquad\qquad\qquad\qquad\qquad\quad\;\; p = 250
> \end{array}
> $$

$$2p + 2w = 600 \qquad \text{Substitute } p = 250 \text{ into the first equation.}$$
$$2(250) + 2w = 600$$
$$500 + 2w = 600$$
$$2w = 100$$
$$w = 50 \qquad \text{The speed of the wind is 50 mph.}$$

The speed of the plane in still air is 250 mph. The speed of the wind is 50 mph.

Skill Practice

4. Dan and Cheryl paddled their canoe 40 miles in 5 hours with the current and 16 miles in 8 hours against the current. Find the speed of the current and the speed of the canoe in still water.

5. Miscellaneous Mixture Applications

| Example 5 | Solving a Miscellaneous Mixture Application

At the start of business each day, Petersen's Bakery in St. Augustine has twice the number of $1 bills as $5 bills in the cash register. If the register holds a total of $175 in $1 and $5 bills at the start of the day, how many of each type of bill does it have?

Solution:

First label the unknown quantities.

Let x represent the number of $1 bills.
Let y represent the number of $5 bills.

We must now write two equations that relate x and y.

Since the number of $1 bills is twice the number of $5 bills,
we have: $\longrightarrow x = 2y$

Since the total value in the register is $175, we have:

$$\begin{pmatrix} \text{Value of} \\ \text{the \$1 bills} \end{pmatrix} + \begin{pmatrix} \text{value of} \\ \text{the \$5 bills} \end{pmatrix} = \begin{pmatrix} \text{total} \\ \text{value} \end{pmatrix} \longrightarrow 1x + 5y = 175$$

The system can be written as: $x = 2y$
$$x + 5y = 175$$

In the first equation, the value of x is isolated. Therefore, the substitution method is a good choice to solve the system of equations.

$x + 5y = 175$	Second equation
$(2y) + 5y = 175$	Substitute $x = 2y$ into the second equation.
$7y = 175$	Solve the resulting equation.
$\dfrac{7y}{7} = \dfrac{175}{7}$	
$y = 25$	There are twenty-five $5 bills.

$x = 2y$	Now find x by substituting the known value of y.
$= 2(25)$	Substitute $y = 25$.
$= 50$	There are fifty $1 bills.

The cash register holds twenty-five $5 bills and fifty $1 bills.

| Skill Practice |

5. A postal worker has four times as many 39¢ stamps as 3¢ stamps. If the total value of the stamps is $15.90, how many of each type of stamp are there?

Section 4.4   Practice Exercises

Boost *your* GRADE at
mathzone.com!

• Practice Problems   • e-Professors
• Self-Tests   • Videos
• NetTutor

Review Exercises

For Exercises 1–4, solve each system of equations by three different methods:

   a. Graphing method   **b.** Substitution method   **c.** Addition method

1. $-2x + y = 6$
    $2x + y = 2$

2. $x - y = 2$
    $x + y = 6$

3. $y = -2x + 6$
    $4x - 2y = 8$

4. $2x = y + 4$
    $4x = 2y + 8$

For Exercises 5–8, set up a system of linear equations in two variables to solve for the unknown quantities.

5. One number is eight more than twice another. Their sum is 20. Find the numbers.

6. The difference of two positive numbers is 264. The larger number is three times the smaller number. Find the numbers.

7. Two angles are complementary. The measure of one angle is 10° less than nine times the measure of the other. Find the measure of each angle.

8. Two angles are supplementary. The measure of one angle is 9° more than twice the measure of the other angle. Find the measure of each angle.

Concept 1: Applications Involving Cost

9. Kent bought three tapes and two CDs for $62.50. Demond bought one tape and four CDs for $72.50. Find the cost of one tape and the cost of one CD.

10. Tanya bought three adult tickets and one child's ticket to a movie for $23.00. Li bought two adult tickets and five children's tickets for $30.50. Find the cost of one adult ticket and the cost of one children's ticket.

11. Linda bought 100 shares of a technology stock and 200 shares of a mutual fund for $3800. Her sister, Sandy, bought 300 shares of technology stock and 50 shares of a mutual fund for $5350. Find the cost per share of the technology stock, and the cost per share of the mutual fund.

12. Two videos and three DVDs can be rented for $19.15. Four videos and one DVD can be rented for $17.35. Find the cost to rent one video and the cost to rent one DVD.

Concept 2: Applications Involving Principal and Interest

13. Shanelle invested $10,000, and at the end of 1 year, she received $805 in interest. She invested part of the money in an account earning 10% simple interest and the remaining money in an account earning 7% simple interest. How much did she invest in each account?

	10% Account	7% Account	Total
Principal invested			
Interest earned			

14. $12,000 was invested in two accounts, one earning 12% simple interest and the other earning 8% simple interest. If the total interest at the end of 1 year was $1240, how much was invested in each account?

	12% Account	8% Account	Total
Principal invested			
Interest earned			

15. Troy borrowed a total of $12,000 in two different loans to help pay for his new Chevy Silverado. One loan charges 9% simple interest, and the other charges 6% simple interest. If he is charged $810 in interest after 1 year, find the amount borrowed at each rate.

16. Blake has a total of $4000 to invest in two accounts. One account earns 2% simple interest, and the other earns 5% simple interest. How much should be invested in both accounts to earn exactly $155 at the end of 1 year?

Concept 3: Applications Involving Mixtures

17. How much 50% disinfectant solution must be mixed with a 40% disinfectant solution to produce 25 gal of a 46% disinfectant solution?

	50% Mixture	40% Mixture	46% Mixture
Amount of solution			
Amount of disinfectant			

18. How many gallons of 20% antifreeze solution and a 10% antifreeze solution must be mixed to obtain 40 gal of a 16% antifreeze solution?

19 How much 45% disinfectant solution must be mixed with a 30% disinfectant solution to produce 20 gal of a 39% disinfectant solution?

20. How many gallons of a 25% antifreeze solution and a 15% antifreeze solution must be mixed to obtain 15 gal of a 23% antifreeze solution?

Concept 4: Applications Involving Distance, Rate, and Time

21. It takes a boat 2 hr to go 16 miles downstream with the current and 4 hr to return against the current. Find the speed of the boat in still water and the speed of the current.

	Distance	Rate	Time
Downstream			
Return			

22. A boat takes 1.5 hr to go 12 miles upstream against the current. It can go 24 miles downstream with the current in the same amount of time. Find the speed of the current and the speed of the boat in still water.

23. A plane can fly 960 miles with the wind in 3 hr. It takes the same amount of time to fly 840 miles against the wind. What is the speed of the plane in still air and the speed of the wind?

24. A plane flies 720 miles with the wind in 3 hr. The return trip takes 4 hr. What is the speed of the wind and the speed of the plane in still air?

Concept 5: Miscellaneous Mixture Applications

25. Debi has $2.80 in a collection of dimes and nickels. The number of nickels is five more than the number of dimes. Find the number of each type of coin.

26. A child is collecting state quarters and new $1 coins. If she has a total of 25 coins, and the number of quarters is nine more than the number of dollar coins, how many of each type of coin does she have?

27. In the 1961–1962 NBA basketball season, Wilt Chamberlain of the Philadelphia Warriors made 2432 baskets. Some of the baskets were free throws (worth 1 point each) and some were field goals (worth 2 points each). The number of field goals was 762 more than the number of free throws.

   a. How many field goals did he make and how many free throws did he make?

   b. What was the total number of points scored?

   c. If Wilt Chamberlain played 80 games during this season, what was the average number of points per game?

28. In the 1971–1972 NBA basketball season, Kareem Abdul-Jabbar of the Milwaukee Bucks made 1663 baskets. Some of the baskets were free throws (worth 1 point each) and some were field goals (worth 2 points each). The number of field goals he scored was 151 more than twice the number of free throws.

   a. How many field goals did he make and how many free throws did he make?

   b. What was the total number of points scored?

   c. If Kareem Abdul-Jabbar played 81 games during this season, what was the average number of points per game?

29. A small plane can fly 350 miles with a tailwind in $1\frac{3}{4}$ hours. In the same amount of time, the same plane can travel only 210 miles with a headwind. What is the speed of the plane in still air and the speed of the wind?

30. A plane takes 2 hr to travel 1000 miles with the wind. It can travel only 880 miles against the wind in the same time. Find the speed of the wind and the speed of the plane in still air.

31. At the holidays, Erica likes to sell a candy/nut mixture to her neighbors. She wants to combine candy that costs $1.80 per pound with nuts that cost $1.20 per pound. If Erica needs 20 lb of mixture that will sell for $1.56 per pound, how many pounds of candy and how many pounds of nuts should she use?

32. Mary Lee's natural food store sells a combination of teas. The most popular is a mixture of a tea that sells for $3.00 per pound with one that sells for $4.00 per pound. If she needs 40 lb of tea that will sell for $3.65 per pound, how many pounds of each tea should she use?

33. A total of $60,000 is invested in two accounts, one that earns 5.5% simple interest, and one that earns 6.5% simple interest. If the total interest at the end of 1 year is $3750, find the amount invested in each account.

34. Jacques borrows a total of $15,000. Part of the money is borrowed from a bank that charges 12% simple interest per year. Jacques borrows the remaining part of the money from his sister and promises to pay her 7% simple interest per year. If Jacques' total interest for the year is $1475, find the amount he borrowed from each source.

35. Miracle-Gro All-Purpose Plant Food contains 15% nitrogen. Green Light Super Bloom contains 12% nitrogen. How much Miracle-Gro and how much Green Light fertilizer must be mixed to obtain 60 oz of a mixture that is 13% nitrogen?

36. A textile manufacturer wants to combine a mixture of 20% dye with a mixture that is 50% dye to form 200 gal of a mixture that is 42.5% dye. How much of the 20% and 50% dye mixtures should he use?

37. In the 1994 Super Bowl, the Dallas Cowboys scored four more points than twice the number of points scored by the Buffalo Bills. If the total number of points scored by both teams was 43, find the number of points scored by each team.

38. In the 1973 Super Bowl, the Miami Dolphins scored twice as many points as the Washington Redskins. If the total number of points scored by both teams was 21, find the number of points scored by each team.

Expanding Your Skills

39. In a survey conducted among 500 college students, 340 said that the campus lacked adequate lighting. If $\frac{4}{5}$ of the women and $\frac{1}{2}$ of the men said that they thought the campus lacked adequate lighting, how many men and how many women were in the survey?

40. A thousand people were surveyed in southern California, and 445 said that they worked out at least three times a week. If $\frac{1}{2}$ of the women and $\frac{3}{8}$ of the men said that they worked out at least three times a week, how many men and how many women were in the survey?

41. During a 1-hour television program, there were 22 commercials. Some commercials were 15 sec and some were 30 sec long. Find the number of 15-sec commercials and the number of 30-sec commercials if the total playing time for commercials was 9.5 min.

Chapter 4   SUMMARY

Section 4.1   Solving Systems of Equations by the Graphing Method

Key Concepts

A **system of two linear equations** can be solved by graphing.

A **solution to a system of linear equations** is an ordered pair that satisfies each equation in the system. Graphically, this represents a point of intersection of the lines.

There may be one solution, infinitely many solutions, or no solution.

One solution   Infinitely many solutions   No solution
Consistent   Consistent   Inconsistent
Independent   Dependent   Independent

A system of equations is **consistent** if there is at least one solution. A system is **inconsistent** if there is no solution.

A linear system in x and y is **dependent** if two equations represent the same line. The solution set is the set of all points on the line.

If two linear equations represent different lines, then the system of equations is **independent**.

Examples

Example 1

Solve by graphing. $x + y = 3$
$$2x - y = 0$$

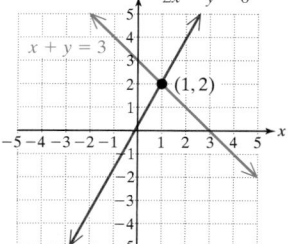

The solution is $(1, 2)$.

Example 2

Solve by graphing. $3x - 2y = 2$
$$-6x + 4y = 4$$

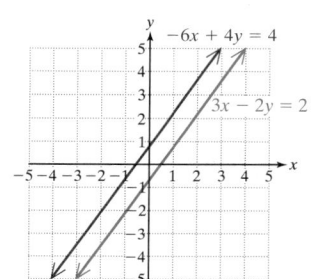

There is no solution. The system is inconsistent.

Example 3

Solve by graphing. $x + 2y = 2$
$$-3x - 6y = -6$$

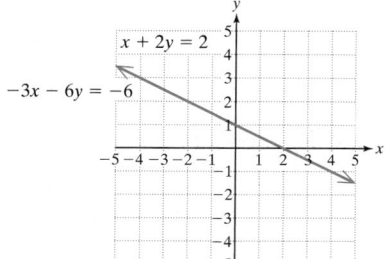

The system is dependent, and the solution set consists of all points on the line, given by
$$\{(x, y) \mid x + 2y = 2\}$$

Section 4.2	Solving Systems of Equations by the Substitution Method

Key Concepts

Steps to Solve a System of Equations by the Substitution Method:

1. Isolate one of the variables from one equation.
2. Substitute the quantity found in step 1 into the other equation.
3. Solve the resulting equation.
4. Substitute the value found in step 3 back into the equation in step 1 to find the remaining variable.
5. Check the solution in both original equations and write the answer as an ordered pair.

An inconsistent system has no solution and is detected algebraically by a contradiction (such as $0 = 3$). When graphed, the lines are parallel.

If two linear equations represent the same line, the system is dependent. This is detected algebraically by an identity (such as $0 = 0$).

Examples

Example 1

Solve by the substitution method.

$$x + 4y = -11$$
$$3x - 2y = -5$$

Isolate x in the first equation: $x = -4y - 11$
Substitute into the second equation.

$$3(-4y - 11) - 2y = -5 \qquad \text{Solve the equation.}$$

$$-12y - 33 - 2y = -5$$
$$-14y = 28$$
$$y = -2$$

$$\begin{aligned} & && \text{Substitute} \\ x &= -4y - 11 && y = -2. \\ x &= -4(-2) - 11 && \text{Solve for } x. \\ x &= -3 \end{aligned}$$

The solution is $(-3, -2)$ and checks in both original equations.

Example 2

Solve by the substitution method.

$$3x + y = 4$$
$$-6x - 2y = 2$$

Isolate y in the first equation: $y = -3x + 4$.
Substitute into the second equation.

$$-6x - 2(-3x + 4) = 2$$
$$-6x + 6x - 8 = 2$$
$$-8 = 2 \qquad \text{Contradiction}$$

The system is inconsistent and has no solution.

Example 3

Solve by the substitution method.

$$y = x + 2 \qquad y \text{ is already isolated.}$$
$$x - y = -2$$

$$x - (x + 2) = -2 \qquad \text{Substitute } y = x + 2 \text{ into the}$$
$$x - x - 2 = -2 \qquad \text{second equation.}$$
$$-2 = -2 \qquad \text{Identity}$$

The system is dependent. The solution set is all points on the line $y = x + 2$ or $\{(x, y) | y = x + 2\}$.

Section 4.3

Solving Systems of Equations by the Addition Method

Key Concepts

Solving a System of Linear Equations by the Addition Method:

1. Write both equations in standard form: $Ax + By = C$.
2. Clear fractions or decimals (optional).
3. Multiply one or both equations by a nonzero constant to create opposite coefficients for one of the variables.
4. Add the equations to eliminate one variable.
5. Solve for the remaining variable.
6. Substitute the known value into one of the original equations to solve for the other variable.
7. Check the solution in both equations.

Examples

Example 1

Solve by using the addition method.

$$5x = -4y - 7 \qquad \text{Write the first equation in}$$
$$6x - 3y = 15 \qquad \text{standard form.}$$

$$5x + 4y = -7 \xrightarrow{\text{Multiply by 3}} 15x + 12y = -21$$
$$6x - 3y = 15 \xrightarrow{\text{Multiply by 4}} \underline{24x - 12y = 60}$$
$$\qquad\qquad\qquad\qquad 39x \qquad\quad = 39$$
$$\qquad\qquad\qquad\qquad\qquad x = 1$$

$$5x = -4y - 7$$
$$5(1) = -4y - 7$$
$$5 = -4y - 7$$
$$12 = -4y$$
$$-3 = y \qquad \text{The solution is } (1, -3) \text{ and checks in both original equations.}$$

Section 4.4

Applications of Linear Equations in Two Variables

Examples

Example 1

A riverboat travels 36 miles with the current to a marina in 2 hr. The return trip takes 3 hr against the current. Find the speed of the current and the speed of the boat in still water.

Let x represent the speed of the boat in still water.
Let y represent the speed of the current.

	Distance	Rate	Time
Against current	36	$x - y$	3
With current	36	$x + y$	2

Distance = (rate)(time)

$$36 = (x - y) \cdot 3 \longrightarrow 36 = 3x - 3y$$
$$36 = (x + y) \cdot 2 \longrightarrow 36 = 2x - 2y$$

$$36 = 3x - 3y \xrightarrow{\text{Multiply by 2}} 72 = 6x - 6y$$
$$36 = 2x + 2y \xrightarrow{\text{Multiply by 3}} \underline{108 = 6x + 6y}$$
$$\qquad\qquad\qquad\qquad 180 = 12x$$
$$\qquad\qquad\qquad\qquad 15 = x$$

$$36 = 2(15) + 2y$$
$$36 = 30 + 2y$$
$$6 = 2y$$
$$3 = y$$

The speed of the boat in still water is 15 mph, and the speed of the current is 3 mph.

Example 2

Diane invests $15,000 more in an account earning 8% simple interest than in an account earning 5% simple interest. If the total interest after 1 year is $1850, how much was invested in each account?

	8%	5%	Total
Principal	x	y	
Interest	$0.08x$	$0.05y$	1850

$$x = y + 15{,}000$$

$$0.08x + 0.05y = 1850$$

Substitute $x = y + 15{,}000$ into the second equation:

$$0.08(y + 15{,}000) + 0.05y = 1850$$

$$0.08y + 1200 + 0.05y = 1850$$

$$0.13y + 1200 = 1850$$

$$0.13y = 650$$

$$\frac{0.13y}{0.13} = \frac{650}{0.13}$$

$$y = 5000$$

$$x = y + 15{,}000$$

$$x = 5000 + 15{,}000$$

$$x = 20{,}000$$

The amount invested in the 8% account is $20,000, and the amount invested at 5% is $5000.

Chapter 4   Review Exercises

Section 4.1

For Exercises 1–4, determine if the ordered pair is a solution to the system.

1. $x - 4y = -4$ $(4, 2)$
$x + 2y = 8$

2. $x - 6y = 6$ $(12, 1)$
$-x + y = 4$

3. $3x + y = 9$ $(1, 3)$
$y = 3$

4. $2x - y = 8$ $(2, -4)$
$x = 2$

For Exercises 5–10, identify whether the system represents intersecting lines, parallel lines, or coinciding lines by comparing their slopes and y-intercepts.

5. $y = -\dfrac{1}{2}x + 4$
$y = x - 1$

6. $y = -3x + 4$
$y = 3x + 4$

7. $y = -\dfrac{4}{7}x + 3$
$y = -\dfrac{4}{7}x - 5$

8. $y = 5x - 3$
$y = \dfrac{1}{5}x - 3$

9. $y = 9x - 2$
$9x - y = 2$

10. $x = -5$
$y = 2$

For Exercises 11–18, solve the systems by graphing. If a system does not have a unique solution, identify the system as inconsistent or dependent.

11. $y = -\dfrac{2}{3}x - 2$
$-x + 3y = -6$

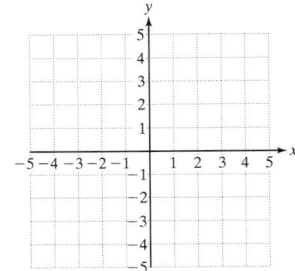

12. $y = -2x - 1$
$2x + 3y = 5$

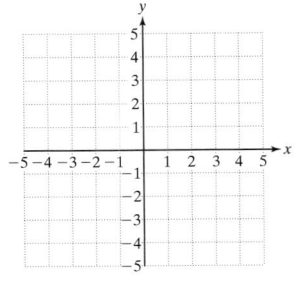

13. $4x = -2y + 10$
$2x + y = 5$

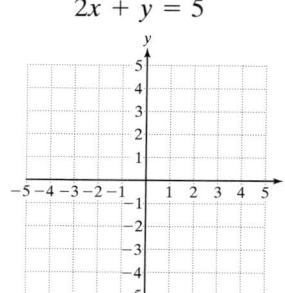

14. $10y = 2x - 10$
$-x + 5y = -5$

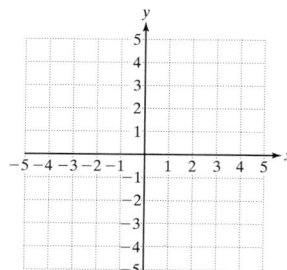

15. $6x - 3y = 9$
$y = -1$

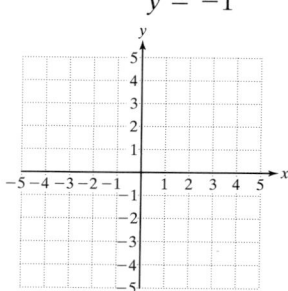

16. $5x + y = -11$
$x = -1$

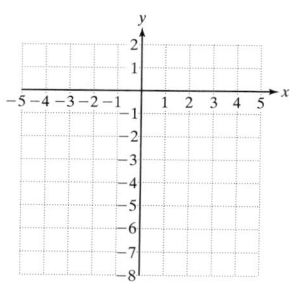

17. $x - 7y = 14$
$-2x + 14y = 14$

18. $y = -5x + 6$
$10x + 2y = 6$

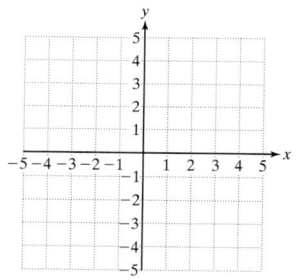

19. A rental car company rents a compact car for $20 a day plus $0.25 per mile. A midsize car rents for $30 a day plus $0.20 per mile.

The cost, y_c, to rent a compact car for one day is given by the equation:

$y_c = 20 + 0.25x$ where x is the number of miles driven

The cost, y_m, to rent a midsize car for one day is given by the equation:

$y_m = 30 + 0.20x$ where x is the number of miles driven

Find the number of miles at which the cost to rent either car would be the same, and confirm your answer with the graph.

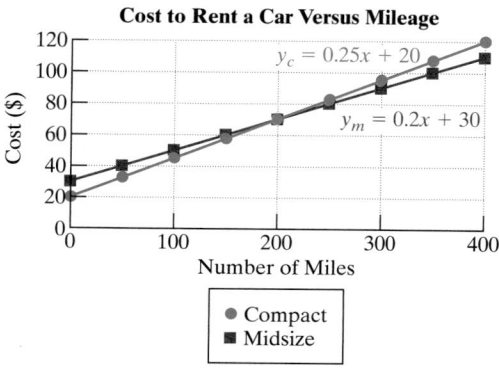

Section 4.2

For Exercises 20–23, solve the systems using the substitution method.

20. $6x + y = 2$
$y = 3x - 4$

21. $2x + 3y = -5$
$x = y - 5$

22. $2x + 6y = 10$
$x = -3y + 6$

23. $4x + 2y = 4$
$y = -2x + 2$

24. Given the system:

$$x + 2y = 11$$
$$5x + 4y = 40$$

a. Which variable from which equation is easiest to isolate and why?

b. Solve the system using the substitution method.

25. Given the system:

$$4x - 3y = 9$$
$$2x + y = 12$$

a. Which variable from which equation is easiest to isolate and why?

b. Solve the system using the substitution method.

For Exercises 26–29, solve the systems using the substitution method.

26. $3x - 2y = 23$
$\ x + 5y = -15$

27. $\ x + 5y = 20$
$\ 3x + 2y = 8$

28. $\ x - 3y = 9$
$\ 5x - 15y = 45$

29. $-3x + y = 15$
$\ 6x - 2y = 12$

30. The difference of two positive numbers is 42. The larger number is 2 more than 6 times the smaller number. Find the numbers.

31. In a right triangle, one of the acute angles is 6° less than the other acute angle. Find the measure of each acute angle.

32. Two angles are supplementary. One angle measures 14° less than two times the other angle. Find the measure of each angle.

Section 4.3

33. Explain the process for solving a system of two equations using the addition method.

34. Given the system:
$$3x - 5y = 1$$
$$2x - y = -4$$

a. Which variable, x or y, is easier to eliminate using the addition method? (Answers may vary.)

b. Solve the system using the addition method.

35. Given the system:
$$9x - 2y = 14$$
$$4x + 3y = 14$$

a. Which variable, x or y, is easier to eliminate using the addition method? (Answers may vary.)

b. Solve the system using the addition method.

For Exercises 36–43, solve the systems using the addition method.

36. $2x + 3y = 1$
$\ x - 2y = 4$

37. $\ x + 3y = 0$
$\ -3x - 10y = -2$

38. $8(x + 1) = -6y + 6$
$\ 10x = 9y - 8$

39. $12x = 5(y + 1)$
$\ 5y = -1 - 4x$

40. $-4x - 6y = -2$
$\ 6x + 9y = 3$

41. $-8x - 4y = 16$
$\ 10x + 5y = 5$

42. $\dfrac{1}{2}x - \dfrac{3}{4}y = -\dfrac{1}{2}$
$\ \dfrac{1}{3}x + y = -\dfrac{10}{3}$

43. $0.5x - 0.2y = 0.5$
$\ 0.4x + 0.7y = 0.4$

44. Given the system:
$$4x + 9y = -7$$
$$y = 2x - 13$$

a. Which method would you choose to solve the system, the substitution method or the addition method? Explain your choice. (Answers may vary.)

b. Solve the system.

45. Given the system:
$$5x - 8y = -2$$
$$3x - y = -5$$

a. Which method would you choose to solve the system, the substitution method or the addition method? Explain your choice. (Answers may vary.)

b. Solve the system.

Section 4.4

46. Miami Metrozoo charges $11.50 for adult admission and $6.75 for children under 12. The total bill before tax for a school group of 60 people is $443. How many adults and how many children were admitted?

47. Emillo invested $20,000, and at the end of 1 year he received $1525 in interest. If he invested part of the money at 5% simple interest and the remaining money at 8% simple interest, how much did he invest in each account?

48. To produce a 16% alcohol solution, a chemist mixes a 20% alcohol solution and a 14% alcohol solution. How much 20% solution and how much 14% solution must be used to produce 15 L of a 16% alcohol solution?

49. A boat travels 80 miles downstream with the current in 4 hr and 80 miles upstream against the current in 5 hr. Find the speed of the current and the speed of the boat in still water.

50. Suzanne has a collection of new quarters and new $1 coins. She has four more quarters than dollar coins and the total value of the coins is $4.75. How many of each coin does Suzanne have?

51. At Conseco Fieldhouse, home of the Indiana Pacers, the total cost of a soft drink and a hot dog is $8.00. The price of the hot dog is $1.00 more than the cost of the soft drink. Find the cost of a soft drink and the cost of a hot dog.

52. In a recent election 5700 votes were cast and 3675 voters voted for the winning candidate. If

$\frac{5}{8}$ of the women and $\frac{2}{3}$ of the men voted for the winning candidate, how many men and how many women voted?

53. Ray played two rounds of golf at Pebble Beach for a total score of 154. If his score in the second round is 10 more than his score in the first round, find the scores for each round.

Chapter 4   Test

1. Write each line in slope-intercept form. Then determine if the lines represent intersecting lines, parallel lines, or coinciding lines.

$$5x + 2y = -6$$

$$-\frac{5}{2}x - y = -3$$

For Exercises 2–3 solve the system by graphing.

2. $y = 2x - 4$

$-2x + 3y = 0$

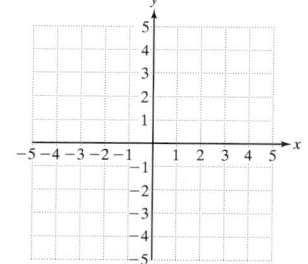

3. $2x + 4y = 6$

$2y - 3 = -x$

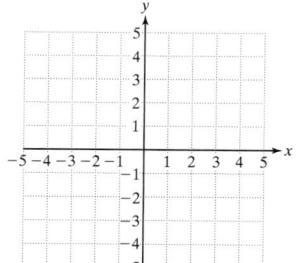

4. Solve the system using the substitution method.

$$x = 5y - 2$$

$$2x + y = -4$$

5. In the 2005 WNBA (basketball) season, the league's leading scorer was Sheryl Swoopes from the Houston Comets. Swoopes scored 17 points more than the second leading scorer, Lauren Jackson from the Seattle Storm. Together they scored a total of 1211 points. How many points did each player score?

6. Solve the system using the addition method.

$$3x - 6y = 8$$

$$2x + 3y = 3$$

7. How many milliliters of a 50% acid solution and how many milliliters of a 20% acid solution must be mixed to produce 36 mL of a 30% acid solution?

8. a. How many solutions does a system of two linear equations have if the equations represent parallel lines?

b. How many solutions does a system of two linear equations have if the equations represent coinciding lines?

c. How many solutions does a system of two linear equations have if the equations represent intersecting lines?

For Exercises 9–14, solve the systems using any method.

9. $\dfrac{1}{3}x + y = \dfrac{7}{3}$

$x = \dfrac{3}{2}y - 11$

10. $2(x - 6) = y$

$2x - \dfrac{1}{2}y = x + 5$

11. $3x - 4y = 29$

$2x + 5y = -19$

12. $2x = 6y - 14$

$2y = 3 - x$

13. $-0.25x - 0.05y = 0.2$

$10x + 2y = -8$

14. $3(x + y) = -2y - 7$

$-3y = 10 - 4x$

15. At *Best Buy*, Latrell buys four CDs and two DVDs for $54 from the sale rack. Kendra buys two CDs and three DVDs from the same rack for $49. What is the price per CD and the price per DVD?

16. The cost to ride the trolley one-way in San Diego is $2.25. Kelly and Hazel had to buy eight tickets for their group.

   a. What was the total amount of money required?

   b. Kelly and Hazel had only quarters and $1 bills. They also determined that they used twice as many quarters as $1 bills. How many quarters and how many $1 bills did they use?

17. Suppose a total of $5000 is borrowed from two different loans. One loan charges 10% simple interest, and the other charges 8% simple interest. How much was borrowed at each rate if $424 in interest is charged at the end of 1 year?

18. During the first 13 years of his football career, Jerry Rice scored a total of 166 touchdowns. One touchdown was scored on a kickoff return, and the remaining 165 were scored rushing or receiving. The number of receiving touchdowns he scored was 5 more than 15 times the number of rushing touchdowns he scored. How many receiving touchdowns and how many rushing touchdowns did he score?

19. A plane travels 880 miles in 2 hr against the wind and 1000 miles in 2 hr with the same wind. Find the speed of the plane in still air and the speed of the wind.

20. The number of calories in a piece of cake is 20 less than 3 times the number of calories in a scoop of ice cream. Together, the cake and ice cream have 460 calories. How many calories are in each?

21. A police force has 240 officers. If there are 116 more men than women, find the number of men and the number of women on the force.

Chapters 1–4   Cumulative Review Exercises

1. Simplify.

$$\dfrac{|2 - 5| + 10 \div 2 + 3}{\sqrt{10^2 - 8^2}}$$

2. Solve for x: $\frac{1}{3}x - \frac{3}{4} = \frac{1}{2}(x + 2)$

3. Solve for a: $-4(a + 3) + 2 = -5(a + 1) + a$

4. Solve for y: $3x - 2y = 6$

5. Solve for z. Graph the solution set on a number line and write the solution in interval notation:

$$-2(3z + 1) \le 5(z - 3) + 10$$

6. The largest angle in a triangle is 110°. Of the remaining two angles, one is 4° less than the other angle. Find the measure of the three angles.

7. Two hikers start at opposite ends of an 18-mile trail and walk toward each other. One hiker walks predominately down hill and averages 2 mph faster than the other hiker. Find the average rate of each hiker if they meet in 3 hr.

8. Jesse Ventura became the 38th governor of Minnesota by receiving 37% of the votes. If approximately 2,060,000 votes were cast, how many did Mr. Ventura get?

9. The YMCA wants to raise $2500 for its summer program for disadvantaged children. If the YMCA has already raised $900, what percent of its goal has been achieved?

10. Two angles are complementary. One angle measures 17° more than the other angle. Find the measure of each angle.

11. Solve for x: $z = \dfrac{x - m}{5}$

12. Solve for y: $2x - 3y = 6$

13. The slope of a given line is $-\frac{2}{3}$.

   a. What is the slope of a line parallel to the given line?

   b. What is the slope of a line perpendicular to the given line?

14. Find an equation of the line passing through the point $(2, -3)$ and having a slope of -3. Write the final answer in slope-intercept form.

15. Sketch the following equations on the same graph.

   a. $2x + 5y = 10$

   b. $2y = 4$

   c. Find the point of intersection and check the solution in each equation.

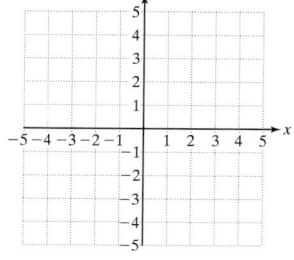

16. Solve the system of equations by using the substitution method.

$$2x + 5y = 10$$
$$2y = 4$$

17. Graph the line $2x + y = 3$.

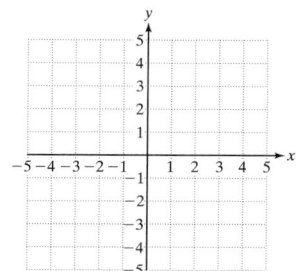

18. How many gallons of a 15% antifreeze solution should be mixed with a 60% antifreeze solution to produce 60 gal of a 45% antifreeze solution?

19. Use a system of linear equations to solve for x and y.

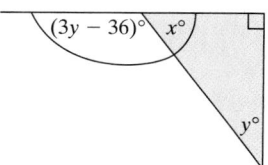

20. In 1920, the average speed for the winner of the Indianapolis 500 car race was 88.6 mph. By 1990, the average speed of the winner was 186.0 mph.

   a. Find the slope of the line shown in the figure. Round to one decimal place.

   b. Interpret the meaning of the slope in the context of this problem.

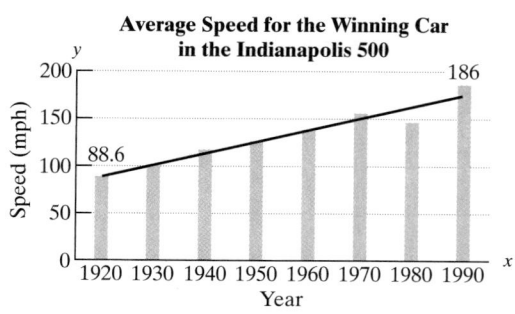

Polynomials and Properties of Exponents

5

In this chapter we introduce the concept of a polynomial and learn how to add, subtract, multiply, and divide polynomials. In addition, we cover the fundamental properties and applications of expressions involving exponents, including scientific notation.

Complete the fill-it-in puzzle using terms from this chapter. Write the words left to right and down.

Key Terms

base
binomial
coefficient
conjugates
degree
exponent
leading

magnitude
monomial
polynomial
power
reciprocal
trinomial

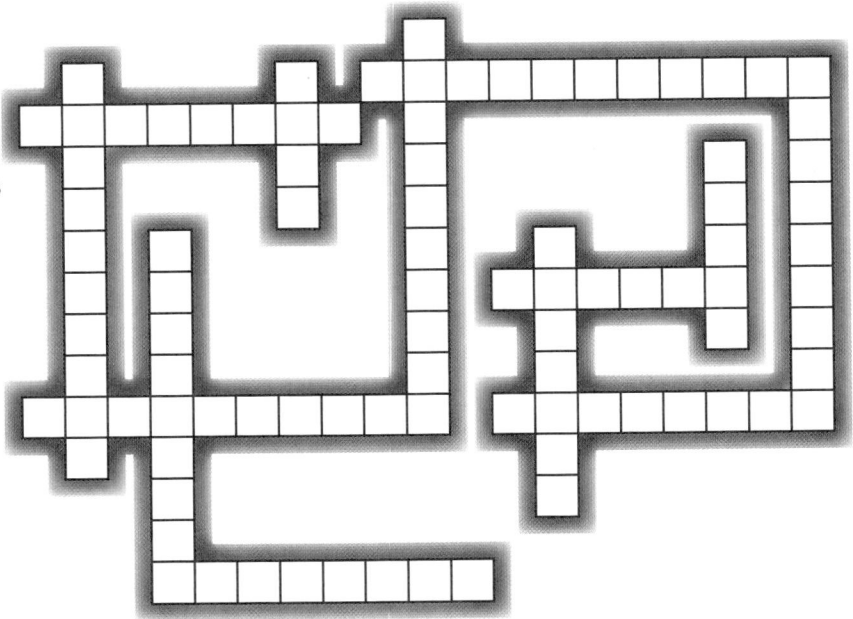

Exponents: Multiplying and Dividing Common Bases

1. Review of Exponential Notation

Recall that an **exponent** is used to show repeated multiplication of the **base**.

Definition of b^n

Let b represent any real number and n represent a positive integer. Then,

$$b^n = \underbrace{b \cdot b \cdot b \cdot b \ldots \cdot b}_{n \text{ factors of } b}$$

Example 1   Evaluating Expressions with Exponents

For each expression, identify the exponent and base. Then evaluate the expression.

   a. 6^2 **b.** $\left(-\dfrac{1}{2}\right)^3$ **c.** 0.8^4

Solution:

Expression	Base	Exponent	Result
a. 6^2	6	2	$(6)(6) = 36$
b. $\left(-\dfrac{1}{2}\right)^3$	$-\dfrac{1}{2}$	3	$\left(-\dfrac{1}{2}\right)\left(-\dfrac{1}{2}\right)\left(-\dfrac{1}{2}\right) = -\dfrac{1}{8}$
c. 0.8^4	0.8	4	$(0.8)(0.8)(0.8)(0.8) = 0.4096$

Skill Practice   For each expression, identify the base and exponent.

   1. 8^3 **2.** $(-4)^2$ **3.** 0.02^4

Note that if no exponent is explicitly written for an expression, then the expression has an implied exponent of 1. For example,

$$x = x^1$$
$$y = y^1$$
$$5 = 5^1$$

2. Evaluating Expressions with Exponents

Recall from Section 1.2 that particular care must be taken when evaluating exponential expressions involving negative numbers. An exponential expression with a negative base is written with parentheses around the base, such as $(-3)^2$.

      To evaluate $(-3)^2$, we have: $(-3)^2 = (-3)(-3) = 9$

If no parentheses are present, the expression -3^2, is the *opposite* of 3^2, or equivalently, $-1 \cdot 3^2$.

     Hence: $-3^2 = -1(3^2) = -1(3)(3) = -9$

Skill Practice Answers

1. Base 8; exponent 3
2. Base -4; exponent 2
3. Base 0.02; exponent 4

| **Example 2** | **Evaluating Expressions with Exponents** |

Evaluate each expression.

a. -5^4 **b.** $(-5)^4$ **c.** $(-0.2)^3$ **d.** -0.2^3

Solution:

a. -5^4

$$= -1 \cdot 5^4 \qquad \text{5 is the base to the power of 4.}$$

$$= -1 \cdot 5 \cdot 5 \cdot 5 \cdot 5 \qquad \text{Multiply } -1 \text{ times four factors of 5.}$$

$$= -625$$

b. $(-5)^4$

$$= (-5)(-5)(-5)(-5) \qquad \text{Parentheses indicate that } -5 \text{ is the base to the power of 4.}$$

$$= 625 \qquad \text{Multiply four factors of } -5.$$

c. $(-0.2)^3$ Parentheses indicate that -0.2 is the base to the power of 3.

$$= (-0.2)(-0.2)(-0.2) \qquad \text{Multiply three factors of } -0.2.$$

$$= -0.008$$

d. -0.2^3

$$= -1 \cdot 0.2^3 \qquad \text{0.2 is the base to the power of 3.}$$

$$= -1 \cdot 0.2 \cdot 0.2 \cdot 0.2 \qquad \text{Multiply } -1 \text{ times three factors of 0.2.}$$

$$= -0.008$$

| **Skill Practice** | Evaluate.

4. -2^4 **5.** $(-2)^4$ **6.** $(-0.1)^3$ **7.** -0.1^3

| **Example 3** | **Evaluating Expressions with Exponents** |

Evaluate each expression for $a = 2$ and $b = -3$.

a. $5a^2$ **b.** $(5a)^2$ **c.** $5ab^2$ **d.** $(b + a)^2$

Solution:

a. $5a^2$

$$= 5(\ )^2 \qquad \text{Use parentheses to substitute a number for a variable.}$$

$$= 5(2)^2 \qquad \text{Substitute } a = 2.$$

$$= 5(4) \qquad \text{Simplify.}$$

$$= 20$$

b. $(5a)^2$

$$= [5(\ )]^2 \qquad \text{Use parentheses to substitute a number for a variable.}$$

$$= [5(2)]^2 \qquad \text{Substitute } a = 2.$$

$$= (10)^2 \qquad \text{Simplify inside the parentheses first.}$$

$$= 100$$

Skill Practice Answers

4. -16 **5.** 16
6. -0.001 **7.** -0.001

c. $5ab^2$

$$= 5(2)(-3)^2 \qquad \text{Substitute } a = 2, b = -3.$$

$$= 5(2)(9) \qquad\quad \text{Simplify exponents first.}$$

$$= 90 \qquad\qquad\;\; \text{Multiply.}$$

TIP: In the expression $5ab^2$, the exponent, 2, applies only to the variable b. The constant 5 and the variable a both have an implied exponent of 1.

d. $(b + a)^2$

$$= [(-3) + (2)]^2 \qquad \text{Substitute } b = -3 \text{ and } a = 2.$$

$$= (-1)^2 \qquad\qquad\;\; \text{Simplify within the parentheses first.}$$

$$= 1$$

Avoiding Mistakes:

Be sure to follow the order of operations. In Example 3(d), it would be incorrect to square the terms within the parentheses before adding.

Skill Practice Evaluate each expression for $x = -2$ and $y = 5$.

8. $6x^2$ **9.** $(6x)^2$ **10.** $-2xy^2$ **11.** $(x - y)^2$

3. Multiplying and Dividing Common Bases

In this section, we investigate the effect of multiplying or dividing two quantities with the same base. For example, consider the expressions: x^5x^2 and $\frac{x^5}{x^2}$. Simplifying each expression, we have:

$$x^5x^2 = (x \cdot x \cdot x \cdot x \cdot x)(x \cdot x) = \overbrace{x \cdot x \cdot x \cdot x \cdot x \cdot x \cdot x}^{7 \text{ factors of } x} = x^7$$

$$\frac{x^5}{x^2} = \frac{x \cdot x \cdot x \cdot \cancel{x} \cdot \cancel{x}}{\cancel{x} \cdot \cancel{x}} = \frac{x \cdot x \cdot x}{1} = x^3$$

These examples suggest that to multiply two quantities with the same base, we add the exponents. To divide two quantities with the same base, we subtract the exponent in the denominator from the exponent in the numerator. These rules are stated formally as Properties 1 and 2 of exponents, respectively.

Multiplication of Like Bases

Assume that $a \neq 0$ is a real number and that m and n represent positive integers. Then,

$$\text{Property 1: } \quad a^m a^n = a^{m+n}$$

Division of Like Bases

Assume that $a \neq 0$ is a real number and that m and n represent positive integers such that $m > n$. Then,

$$\text{Property 2: } \quad \frac{a^m}{a^n} = a^{m-n}$$

Skill Practice Answers

8. 24 **9.** 144
10. 100 **11.** 49

Example 4   **Simplifying Expressions with Exponents**

Simplify the expressions.

a. w^3w^4 **b.** 2^32^4 **c.** $\dfrac{t^6}{t^4}$ **d.** $\dfrac{5^6}{5^4}$ **e.** $\dfrac{z^4z^5}{z^3}$ **f.** $\dfrac{10^7}{10^2 \cdot 10}$

Solution:

a. w^3w^4

$\qquad (w \cdot w \cdot w)(w \cdot w \cdot w \cdot w)$

$\quad = w^{3+4}$ Add the exponents.

$\quad = w^7$

b. 2^32^4

$\qquad (2 \cdot 2 \cdot 2)(2 \cdot 2 \cdot 2 \cdot 2)$

$\quad = 2^{3+4}$ Add the exponents (the base is unchanged).

$\quad = 2^7$ or 128

c. $\dfrac{t^6}{t^4}$

$\qquad \dfrac{\cancel{t} \cdot \cancel{t} \cdot \cancel{t} \cdot \cancel{t} \cdot t \cdot t}{\cancel{t} \cdot \cancel{t} \cdot \cancel{t} \cdot \cancel{t}}$

$\quad = t^{6-4}$ Subtract the exponents.

$\quad = t^2$

d. $\dfrac{5^6}{5^4}$

$\qquad \dfrac{\cancel{5} \cdot \cancel{5} \cdot \cancel{5} \cdot \cancel{5} \cdot 5 \cdot 5}{\cancel{5} \cdot \cancel{5} \cdot \cancel{5} \cdot \cancel{5}}$

$\quad = 5^{6-4}$ Subtract the exponents (the base is unchanged).

$\quad = 5^2$ or 25

e. $\dfrac{z^4z^5}{z^3}$

$\quad = \dfrac{z^{4+5}}{z^3}$ Add the exponents in the numerator (the base is unchanged).

$\quad = \dfrac{z^9}{z^3}$

$\quad = z^{9-3}$ Subtract the exponents.

$\quad = z^6$

f. $\dfrac{10^7}{10^2 \cdot 10}$

$\quad = \dfrac{10^7}{10^2 \cdot 10^1}$ Note that 10 is equivalent to 10^1.

$\quad = \dfrac{10^7}{10^{2+1}}$ Add the exponents in the denominator (the base is unchanged).

$\quad = \dfrac{10^7}{10^3}$

$\quad = 10^{7-3}$ Subtract the exponents.

$\quad = 10^4$ or 10,000 Simplify.

Avoiding Mistakes:

When we use Property 1 to add exponents, the base does not change. In Example 4(b), we have $2^32^4 = 2^7$.

Skill Practice Simplify the expressions.

12. $q^4 \cdot q^8$ **13.** $8^5 \cdot 8^{10}$ **14.** $\dfrac{y^{15}}{y^8}$ **15.** $\dfrac{2^{15}}{2^8}$

16. $\dfrac{a^3 a^8}{a^7}$ **17.** $\dfrac{3^3 \cdot 3^8}{3^7}$

4. Simplifying Expressions with Exponents

Example 5 Simplifying Expressions with Exponents

Use the commutative and associative properties of real numbers and the properties of exponents to simplify the expressions.

a. $(3p^2q^4)(2pq^5)$ **b.** $\dfrac{16w^9z^3}{3w^8z}$

Solution:

a. $(3p^2q^4)(2pq^5)$

$= (3 \cdot 2)(p^2p)(q^4q^5)$ Apply the associative and commutative properties of multiplication to group coefficients and common bases.

$= (3 \cdot 2)p^{2+1}q^{4+5}$ Add the exponents when multiplying common bases.

$= 6p^3q^9$ Simplify.

b. $\dfrac{16w^9z^3}{3w^8z}$

$= \left(\dfrac{16}{3}\right)\left(\dfrac{w^9}{w^8}\right)\left(\dfrac{z^3}{z}\right)$ Group like coefficients and factors.

$= \left(\dfrac{16}{3}\right)w^{9-8}z^{3-1}$ Subtract the exponents when dividing common bases.

$= \left(\dfrac{16}{3}\right)wz^2$ or $\dfrac{16wz^2}{3}$ Simplify.

Skill Practice Simplify the expressions.

18. $(4x^2y^3)(3x^5y^7)$ **19.** $\dfrac{3^5x^4y^7}{3^2xy^3}$

5. Applications of Exponents

Recall that **simple interest** on an investment or loan is computed by the formula $I = Prt$, where P is the amount of principal, r is the interest rate, and t is the time in years. Simple interest is based only on the original principal. However, in most day-to-day applications, the interest computed on money invested or

Skill Practice Answers

12. q^{12} **13.** 8^{15}
14. y^7 **15.** 2^7 or 128
16. a^4 **17.** 3^4 or 81
18. $12x^7y^{10}$
19. $3^3x^3y^4$ or $27x^3y^4$

borrowed is compound interest. **Compound interest** is computed on the original principal and on the interest already accrued.

Suppose $1000 is invested at 8% interest for 3 years. Compare the total amount in the account if the money earns simple interest versus if the interest is compounded annually.

Simple Interest

The simple interest earned is given by $I = Prt$

$$= (1000)(0.08)(3)$$

$$= \$240$$

Thus, the total amount in the account after 3 years is $1240.

Compound Interest (Annual)

To compute interest compounded annually over a period of 3 years, compute the interest earned in the first year. Then add the principal plus the interest earned in the first year. This value then becomes the principal on which to base the interest earned in the second year. We repeat this process, finding the interest for the second and third years based on the principal and interest earned in the preceding years. This process is outlined using a table.

Year	Interest Earned $I = Prt$	Total Amount in the Account
First year	$I = (\$1000)(0.08)(1) = \$80$	$1000 + $80 = $1080
Second year	$I = (\$1080)(0.08)(1) = \$86.40$	$1080 + $86.40 = $1166.40
Third year	$I = (\$1166.40)(0.08)(1) \approx \$93.31$	$1166.40 + 93.31 = **$1259.71**

The total amount in the account found by compounding interest annually is $1259.71.

The difference in the account balance for interest compounded annually versus for simple interest is $1259.71 − $1240 = $19.71.

The total amount, A, in an account earning compound annual interest can be computed quickly using the following formula:

$A = P(1 + r)^t$ where P is the amount of principal, r is the annual interest rate (expressed in decimal form), and t is the number of years.

For example, for $1000 invested at 8% interest compounded annually for 3 years, we have $P = 1000$, $r = 0.08$, and $t = 3$.

$$A = P(1 + r)^t$$

$$A = 1000(1 + 0.08)^3$$

$$= 1000(1.08)^3$$

$$= 1000(1.259712)$$

$$= 1259.712$$

Rounding to the nearest cent, we have $A = \$1259.71$, as expected.

Example 6 Using Exponents in an Application

Find the amount in an account after 8 years if the initial investment is $7000, invested at 2.25% interest compounded annually.

Solution:

Identify the values for each variable.

$P = 7000$

$r = 0.0225$ — Note that the decimal form of a percent is used for calculations.

$t = 8$

$$A = P(1 + r)^t$$
$$= 7000(1 + 0.0225)^8 \quad \text{Substitute.}$$
$$= 7000(1.0225)^8 \quad \text{Simplify inside the parentheses.}$$
$$\approx 7000(1.194831142) \quad \text{Approximate } (1.0225)^8.$$
$$= 8363.82 \quad \text{Multiply (round to the nearest cent).}$$

The amount in the account after 8 years is $8363.82.

Skill Practice

20. Find the amount in an account after 3 years if the initial investment is $4000 invested at 5% interest compounded annually.

Calculator Connections

In Example 6, it was necessary to evaluate the expression $(1.0225)^8$. Recall that the $\boxed{\wedge}$ or $\boxed{y^x}$ key can be used to enter expressions with exponents.

Scientific Calculator

Enter: $\boxed{1}\ \boxed{.}\ \boxed{0}\ \boxed{2}\ \boxed{2}\ \boxed{5}\ \boxed{y^x}\ \boxed{8}\ \boxed{=}$ **Result:** $\boxed{1.194831142}$

Graphing Calculator

```
1.0225^8
         1.194831142
```

Calculator Exercises

Use a calculator to evaluate the expressions.

1. $(1.06)^5$

2. $(1.02)^{40}$

3. $5000(1.06)^5$

4. $2000(1.02)^{40}$

5. $3000(1 + 0.06)^2$

6. $1000(1 + 0.05)^3$

Skill Practice Answers

20. $4630.50

Section 5.1 Practice Exercises

For this exercise set, assume all variables represent nonzero real numbers.

Study Skills Exercise

1. Define the key terms:

   a. exponent   b. base   c. simple interest   d. compound interest

Concept 1: Review of Exponential Notation

For Exercises 2–9, identify the base and the exponent.

2. c^3

3. x^4

4. 5^2

5. 3^5

6. $(-4)^8$

7. $(-1)^4$

8. x

9. q

10. What base corresponds to the exponent 5 in the expression $x^3y^5z^2$?

11. What base corresponds to the exponent 2 in the expression w^3v^2?

12. What base corresponds to the exponent 6 in the expression $4x^6$?

13. What base corresponds to the exponent 3 in the expression $2y^3$?

For Exercises 14–21, write the expression using exponents.

14. $(4n)(4n)(4n)$

15. $(-6b)(-6b)$

16. $4 \cdot n \cdot n \cdot n$

17. $-6 \cdot b \cdot b$

18. $(x - 5)(x - 5)(x - 5)$

19. $(y + 2)(y + 2)(y + 2)(y + 2)$

20. $\dfrac{4}{x \cdot x \cdot x \cdot x \cdot x}$

21. $\dfrac{-2}{t \cdot t \cdot t}$

Concept 2: Evaluating Expressions with Exponents

For Exercises 22–29, evaluate the two expressions and compare the answers. Do the expressions have the same value?

22. -5^2 and $(-5)^2$

23. -3^4 and $(-3)^4$

24. -2^5 and $(-2)^5$

25. -5^3 and $(-5)^3$

26. $\left(\dfrac{1}{2}\right)^3$ and $\dfrac{1}{2^3}$

27. $\left(\dfrac{1}{5}\right)^2$ and $\dfrac{1}{5^2}$

28. $\left(\dfrac{3}{10}\right)^2$ and $(0.3)^2$

29. $\left(\dfrac{7}{10}\right)^3$ and $(0.7)^3$

For Exercises 30–39, evaluate the expressions.

30. 16^1

31. 20^1

32. $(-1)^{21}$

33. $(-1)^{30}$

34. 0^6

35. 0^4

36. $\left(-\dfrac{1}{3}\right)^2$

37. $\left(-\dfrac{1}{4}\right)^3$

38. $-\left(\dfrac{2}{5}\right)^2$

39. $-\left(\dfrac{3}{5}\right)^2$

For Exercises 40–47, simplify using the order of operations.

40. $3 \cdot 2^4$

41. $2 \cdot 0^5$

42. $-4(-1)^7$

43. $-3(-1)^4$

44. $6^2 - 3^3$

45. $4^3 + 2^3$

46. $2 \cdot 3^2 + 4 \cdot 2^3$

47. $6^2 - 3 \cdot 1^3$

For Exercises 48–59, evaluate each expression for $a = -4$ and $b = 5$.

48. $-4b^2$

49. $5a^2$

50. $(-4b)^2$

51. $(5a)^2$

52. $(a + b)^2$

53. $(a - b)^2$

54. $a^2 + 2ab + b^2$

55. $a^2 - 2ab + b^2$

56. $-10ab^2$

57. $-6a^3b$

58. $-10a^2b$

59. $-a^2b$

Concept 3: Multiplying and Dividing Common Bases

60. Expand the following expressions first. Then simplify using exponents.

    a. $x^4 \cdot x^3$

    b. $5^4 \cdot 5^3$

61. Expand the following expressions first. Then simplify using exponents.

    a. $y^2 \cdot y^4$

    b. $3^2 \cdot 3^4$

For Exercises 62–73, simplify the expressions. Write the answers in exponent form.

62. z^5z^3

63. w^4w^7

64. $a \cdot a^8$

65. p^4p

66. $4^5 \cdot 4^9$

67. $6^7 \cdot 6^5$

68. $\left(\dfrac{2}{3}\right)^3\left(\dfrac{2}{3}\right)$

69. $\left(\dfrac{1}{x}\right)\left(\dfrac{1}{x}\right)^2$

70. $c^5c^2c^7$

71. $b^7b^2b^8$

72. $x \cdot x^4 \cdot x^{10} \cdot x^3$

73. $z^7 \cdot z^{11} \cdot z^{60} \cdot z$

74. Expand the following expressions. Then simplify.

    a. $\dfrac{p^8}{p^3}$

    b. $\dfrac{8^8}{8^3}$

75. Expand the following expressions. Then simplify.

    a. $\dfrac{w^5}{w^2}$

    b. $\dfrac{4^5}{4^2}$

For Exercises 76–93, simplify the expressions. Write the answers in exponent form.

76. $\dfrac{x^8}{x^6}$

77. $\dfrac{z^5}{z^4}$

78. $\dfrac{a^{10}}{a}$

79. $\dfrac{b^{12}}{b}$

80. $\dfrac{7^{13}}{7^6}$

81. $\dfrac{2^6}{2^4}$

82. $\dfrac{5^8}{5}$

83. $\dfrac{3^5}{3}$

84. $\dfrac{y^{13}}{y^{12}}$

85. $\dfrac{w^7}{w^6}$

86. $\dfrac{h^3 h^8}{h^7}$

87. $\dfrac{n^5 n^4}{n^2}$

88. $\dfrac{7^2 \cdot 7^6}{7}$

89. $\dfrac{5^3 \cdot 5^8}{5}$

90. $\dfrac{10^{20}}{10^3 \cdot 10^8}$

91. $\dfrac{3^{15}}{3^2 \cdot 3^{10}}$

92. $\dfrac{z^3 z^{11}}{z^4 z^6}$

93. $\dfrac{w^{12} w^2}{w^4 w^5}$

Concept 4: Simplifying Expressions with Exponents

For Exercises 94–109, use the commutative and associative properties of real numbers and the properties of exponents to simplify the expressions.

94. $(5a^2 b)(8a^3 b^4)$

95. $(10xy^3)(3x^4 y)$

96. $(r^6 s^4)(13r^2 s)$

97. $(6p^2 q^8)(7p^5 q^3)$

98. $\left(\dfrac{2}{3}m^{13}n^8\right)(24m^7 n^2)$

99. $\left(\dfrac{1}{4}c^6 d^6\right)(28c^2 d^7)$

100. $\dfrac{14c^4 d^5}{7c^3 d}$

101. $\dfrac{36h^5 k^2}{9h^3 k}$

102. $\dfrac{2x^3 y^5}{8xy^3}$

103. $\dfrac{13w^8 z^3}{26w^2 z}$

104. $\dfrac{25h^3 jk^5}{12h^2 k}$

105. $\dfrac{15m^5 np^{12}}{4mp^9}$

106. $(-4p^6 q^8 r^4)(2pqr^2)$

107. $(-5a^4 bc)(-10a^2 b)$

108. $\dfrac{-12s^2 tu^3}{4su^2}$

109. $\dfrac{15w^5 x^{10} y^3}{-15w^4 x}$

Concept 5: Applications of Exponents

Use the formula $A = P(1 + r)^t$ for Exercises 110–113.

110. Find the amount in an account after 2 years if the initial investment is $5000, invested at 7% interest compounded annually.

111. Find the amount in an account after 5 years if the initial investment is $2000, invested at 4% interest compounded annually.

112. Find the amount in an account after 3 years if the initial investment is $4000, invested at 6% interest compounded annually.

113. Find the amount in an account after 4 years if the initial investment is $10,000, invested at 5% interest compounded annually.

For Exercises 114–117, use the geometry formulas found in Section R.3.

114. Find the area of the pizza shown in the figure. Round to the nearest square inch.

115. Find the area of a circular pool 50 ft in diameter. Round to the nearest square foot.

116. Find the volume of the sphere shown in the figure. Round to the nearest cubic centimeter.

117. Find the volume of a spherical balloon that is 8 in. in diameter. Round to the nearest cubic inch.

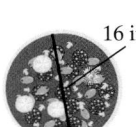

16 in.

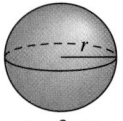

$r = 3$ cm

Expanding Your Skills

For Exercises 118–125, simplify the expressions using the addition or subtraction rules of exponents. Assume that a, b, m, and n represent positive integers.

118. $x^n x^{n+1}$ **119.** $y^a y^{2a}$ **120.** $p^{3m+5} p^{-m-2}$ **121.** $q^{4b-3} q^{-4b+4}$

122. $\dfrac{z^{b+1}}{z^b}$ **123.** $\dfrac{w^{5n+3}}{w^{2n}}$ **124.** $\dfrac{r^{3a+3}}{r^{3a}}$ **125.** $\dfrac{t^{3+2m}}{t^{2m}}$

Section 5.2 More Properties of Exponents

Concepts

1. **Power Rule for Exponents**
2. **The Properties**
   $(ab)^m = a^m b^m$ and
   $\left(\dfrac{a}{b}\right)^m = \dfrac{a^m}{b^m}$
3. **Simplifying Expressions with Exponents**

1. Power Rule for Exponents

The expression $(x^2)^3$ indicates that the quantity x^2 is cubed.

$$(x^2)^3 = (x^2)(x^2)(x^2) = (x \cdot x)(x \cdot x)(x \cdot x) = x^6$$

From this example, it appears that to raise a base to successive powers, we multiply the exponents and leave the base unchanged. This is stated formally as the power rule for exponents.

> **Power Rule for Exponents**
>
> Assume that $a \neq 0$ is a real number and that m and n represent positive integers. Then,
>
> $$\text{Property 3:}\quad (a^m)^n = a^{m \cdot n}$$

Example 1 Simplifying Expressions with Exponents

Simplify the expressions.

a. $(s^4)^2$ **b.** $(3^4)^2$ **c.** $(x^2 x^5)^4$

Solution:

a. $(s^4)^2$

$\quad = s^{4 \cdot 2}$ Multiply exponents (the base is unchanged).

$\quad = s^8$

b. $(3^4)^2$

$\quad = 3^{4 \cdot 2}$ Multiply exponents (the base is unchanged).

$\quad = 3^8$ or 6561

c. $(x^2 x^5)^4$

$\quad = (x^7)^4$ Simplify inside the parentheses by adding exponents.

$\quad = x^{7 \cdot 4}$ Multiply exponents (the base is unchanged).

$\quad = x^{28}$

Skill Practice Simplify the expressions.

1. $(y^3)^5$ **2.** $(2^8)^{10}$ **3.** $(q^5 q^4)^3$

Skill Practice Answers

1. y^{15} **2.** 2^{80} **3.** q^{27}

2. The Properties $(ab)^m = a^m b^m$ and $\left(\dfrac{a}{b}\right)^m = \dfrac{a^m}{b^m}$

Consider the following expressions and their simplified forms:

$$(xy)^3 = (xy)(xy)(xy) = (x \cdot x \cdot x)(y \cdot y \cdot y) = x^3 y^3$$

$$\left(\frac{x}{y}\right)^3 = \left(\frac{x}{y}\right)\left(\frac{x}{y}\right)\left(\frac{x}{y}\right) = \left(\frac{x \cdot x \cdot x}{y \cdot y \cdot y}\right) = \frac{x^3}{y^3}$$

The expressions were simplified using the commutative and associative properties of multiplication. The simplified forms for each expression could have been reached in one step by applying the exponent to each factor inside the parentheses.

> **Power of a Product and Power of a Quotient**
>
> Assume that a and b are real numbers such that $b \neq 0$. Let m represent a positive integer. Then,
>
> Property 4: $(ab)^m = a^m b^m$
>
> Property 5: $\left(\dfrac{a}{b}\right)^m = \dfrac{a^m}{b^m}$

> **Avoiding Mistakes:**
>
> The power rule of exponents can be applied to a product of bases but in general cannot be applied to a sum or difference of bases.
>
> $$(ab)^n = a^n b^n$$
>
> but $(a + b)^n \neq a^n + b^n$

Applying these properties of exponents, we have

$$(xy)^3 = x^3 y^3 \qquad \text{and} \qquad \left(\frac{x}{y}\right)^3 = \frac{x^3}{y^3}$$

3. Simplifying Expressions with Exponents

| **Example 2** | Simplifying Expressions with Exponents |

Simplify the expressions.

a. $(-2xyz)^4$ **b.** $(5x^2 y^7)^3$ **c.** $\left(\dfrac{2}{5}\right)^3$ **d.** $\left(\dfrac{1}{3xy^4}\right)^2$

Solution:

a. $(-2xyz)^4$

$= (-2)^4 x^4 y^4 z^4$ or $16x^4 y^4 z^4$ Raise each factor within parentheses to the fourth power.

b. $(5x^2 y^7)^3$

$= 5^3 (x^2)^3 (y^7)^3$ Raise each factor within parentheses to the third power.

$= 125x^6 y^{21}$ Multiply exponents and simplify.

c. $\left(\dfrac{2}{5}\right)^3$

$= \dfrac{(2)^3}{(5)^3}$ Apply the exponent to each factor in parentheses.

$= \dfrac{8}{125}$ Simplify.

d. $\left(\dfrac{1}{3xy^4}\right)^2$

$= \dfrac{1^2}{3^2x^2(y^4)^2}$ Square each factor within parentheses.

$= \dfrac{1}{9x^2y^8}$ Multiply exponents and simplify.

Skill Practice   Simplify the expressions.

4. $(3x^4y^{10})^3$ **5.** $(-5ab)^3$ **6.** $\left(\dfrac{4}{3}\right)^4$ **7.** $\left(\dfrac{2x^3}{y^5}\right)^2$

The properties of exponents can be used along with the properties of real numbers to simplify complicated expressions.

Example 3   **Simplifying Expressions with Exponents**

Simplify the expressions.

a. $\dfrac{(x^2)^6(x^3)}{(x^7)^2}$ **b.** $(3cd^2)(2cd^3)^3$ **c.** $\left(\dfrac{x^7yz^4}{8xz^3}\right)^2$

Solution:

a. $\dfrac{(x^2)^6(x^3)}{(x^7)^2}$ Clear parentheses by applying the power rule.

$= \dfrac{x^{2\cdot6}x^3}{x^{7\cdot2}}$ Multiply exponents.

$= \dfrac{x^{12}x^3}{x^{14}}$

$= \dfrac{x^{12+3}}{x^{14}}$ Add exponents in the numerator.

$= \dfrac{x^{15}}{x^{14}}$

$= x^{15-14}$ Subtract exponents.

$= x$ Simplify.

b. $(3cd^2)(2cd^3)^3$ Clear parentheses by applying the power rule.

$= 3cd^2 \cdot 2^3c^3d^9$ Raise each factor in the second parentheses to the third power.

$= 3 \cdot 2^3cc^3d^2d^9$ Group like factors.

$= 3 \cdot 8c^{1+3}d^{2+9}$ Add exponents from like factors.

$= 24c^4d^{11}$ Simplify.

Skill Practice Answers

4. $27x^{12}y^{30}$ **5.** $-125a^3b^3$

6. $\dfrac{256}{81}$ **7.** $\dfrac{4x^6}{y^{10}}$

c. $\left(\dfrac{x^7yz^4}{8xz^3}\right)^2$

$=\left(\dfrac{x^{7-1}yz^{4-3}}{8}\right)^2$ Simplify inside the parentheses by subtracting exponents from like factors.

$=\left(\dfrac{x^6yz}{8}\right)^2$

$=\dfrac{(x^6)^2y^2z^2}{8^2}$ Apply the power rule of exponents.

$=\dfrac{x^{12}y^2z^2}{64}$

Skill Practice Simplify the expressions.

8. $\dfrac{k^5 \cdot k^8}{(k^2)^4}$

9. $(x^4y^8z^{10})^4(xyz^3)^3$

10. $\left(\dfrac{2w^2xy^4}{6xy^3}\right)^2$

Skill Practice Answers

8. k^5 **9.** $x^{19}y^{35}z^{49}$

10. $\dfrac{w^4y^2}{9}$

Section 5.2   Practice Exercises

For this exercise set assume all variables represent nonzero real numbers.

Review Exercises

For Exercises 1–8, simplify.

1. $4^2 \cdot 4^7$

2. $5^8 \cdot 5^3 \cdot 5$

3. $a^{13} \cdot a \cdot a^6$

4. $y^{14}y^3$

5. $\dfrac{d^{13}d}{d^5}$

6. $\dfrac{3^8 \cdot 3}{3^2}$

7. $\dfrac{7^{11}}{7^5}$

8. $\dfrac{z^4}{z^3}$

9. Explain when to add exponents versus when to subtract exponents.

10. Explain when to add exponents versus when to multiply exponents.

Concept 1: Power Rule for Exponents

For Exercises 11–22, simplify and write answers in exponent form.

11. $(5^3)^4$

12. $(2^8)^7$

13. $(12^3)^2$

14. $(6^4)^4$

15. $(y^7)^2$

16. $(z^6)^4$

17. $(w^5)^5$

18. $(t^3)^6$

19. $(a^2a^4)^6$

20. $(z \cdot z^3)^2$

21. $(y^3y^4)^2$

22. $(w^5w)^4$

23. Evaluate the two expressions and compare the answers: $(2^2)^3$ and $(2^3)^2$.

24. Evaluate the two expressions and compare the answers: $(4^4)^2$ and $(4^2)^4$.

25. Evaluate the two expressions and compare the answers. Which expression is greater? Why?

$$2^{(2^4)} \quad \text{and} \quad (2^2)^4$$

26. Evaluate the two expressions and compare the answers. Which expression is greater? Why?

$$3^{(2^4)} \quad \text{and} \quad (3^2)^4$$

Concept 3: Simplifying Expressions with Exponents

For Exercises 27–42, use the appropriate property to clear the parentheses.

27. $(5w)^2$

28. $(4y)^3$

29. $(srt)^4$

30. $(wxy)^6$

31. $\left(\dfrac{2}{r}\right)^4$

32. $\left(\dfrac{1}{t}\right)^8$

33. $\left(\dfrac{x}{y}\right)^5$

34. $\left(\dfrac{w}{z}\right)^7$

35. $(-3a)^4$

36. $(2x)^5$

37. $(-3abc)^3$

38. $(-5xyz)^2$

39. $\left(-\dfrac{4}{x}\right)^3$

40. $\left(-\dfrac{1}{w}\right)^4$

41. $\left(-\dfrac{a}{b}\right)^2$

42. $\left(-\dfrac{r}{s}\right)^3$

For Exercises 43–76, simplify the expressions.

43. $(6u^2v^4)^3$

44. $(3a^5b^2)^6$

45. $5(x^2y)^4$

46. $18(u^3v^4)^2$

47. $(-h^4)^7$

48. $(-k^6)^3$

49. $(-m^2)^6$

50. $(-n^3)^8$

51. $\left(\dfrac{4}{rs^4}\right)^5$

52. $\left(\dfrac{2}{h^7k}\right)^3$

53. $\left(\dfrac{3p}{q^3}\right)^5$

54. $\left(\dfrac{5x^2}{y^3}\right)^4$

55. $\dfrac{y^8(y^3)^4}{(y^2)^3}$

56. $\dfrac{(w^3)^2(w^4)^5}{(w^4)^2}$

57. $(x^2)^5(x^3)^7$

58. $(y^3)^4(y^2)^5$

59. $(a^2b)^3(a^4b^3)^5$

60. $(c^3d^5)^2(cd^3)^3$

61. $(-2p^2q^4)^4$

62. $(-7x^4y^5)^2$

63. $(-m^7n^3)^5$

64. $(-a^3b^6)^7$

65. $\dfrac{(5a^3b)^4(a^2b)^4}{(5ab)^2}$

66. $\dfrac{(6s^3)^2(s^4t^5)^2}{(3s^4t^2)^2}$

67. $\dfrac{(21x^5y)(2x^8y^4)}{14xy}$

68. $\dfrac{(4u^3v^3)(9u^4v)}{12u^5v^2}$

69. $\left(\dfrac{2c^3d^4}{3c^2d}\right)^2$

70. $\left(\dfrac{x^3y^5z}{5xy^2}\right)^2$

71. $(2c^3d^2)^5\left(\dfrac{c^6d^8}{4c^2d}\right)^3$

72. $\left(\dfrac{s^5t^6}{2s^2t}\right)^2(10s^3t^3)^2$

73. $\left(\dfrac{-3a^3b}{c^2}\right)^3$

74. $\left(\dfrac{-4x^2}{y^4z}\right)^3$

75. $\dfrac{(-8b^6)^2(b^3)^5}{4b}$

76. $\dfrac{(-6a^2)^2(a^3)^4}{9a}$

Expanding Your Skills

For Exercises 77–84, simplify the expressions using the addition or subtraction properties of exponents. Assume that a, b, m, and n represent positive integers.

77. $(x^m)^2$

78. $(y^3)^n$

79. $(5a^{2n})^3$

80. $(3b^4)^m$

81. $\left(\dfrac{m^2}{n^3}\right)^b$

82. $\left(\dfrac{x^5}{y^3}\right)^m$

83. $\left(\dfrac{3a^3}{5b^4}\right)^n$

84. $\left(\dfrac{4m^6}{3n^2}\right)^b$

Definitions of b^0 and b^{-n}

Section 5.3

In Sections 5.1 and 5.2, we learned several rules that allow us to manipulate expressions containing *positive* integer exponents. In this section, we present definitions that can be used to simplify expressions with negative exponents or with an exponent of zero.

Concepts

1. Definition of b^0
2. Definition of b^{-n}
3. Properties of Integer Exponents: A Summary
4. Simplifying Expressions with Exponents

1. Definition of b^0

To begin, consider the following pattern.

$3^3 = 27$ divide by 3
$3^2 = 9$ divide by 3
$3^1 = 3$ divide by 3
$3^0 = 1$

As the exponents decrease by 1, the resulting expressions are divided by 3.

For the pattern to continue, we define $3^0 = 1$.

This pattern suggests that we should define an expression with a zero exponent as follows.

Definition of b^0

Let b be a nonzero real number. Then, $b^0 = 1$.

Avoiding Mistakes:

$b^0 = 1$ provided that b is not zero. Therefore, the expression 0^0 cannot be simplified by this rule.

Example 1   Simplifying Expressions with a Zero Exponent

Simplify.

   a. 4^0 **b.** $(-4)^0$ **c.** -4^0

   d. z^0 **e.** $-4z^0$ **f.** $(4z)^0$

Solution:

   a. $4^0 = 1$ By definition

   b. $(-4)^0 = 1$ By definition

   c. $-4^0 = -1 \cdot 4^0 = -1 \cdot 1 = -1$ The exponent 0 applies only to 4.

   d. $z^0 = 1$ By definition

   e. $-4z^0 = -4 \cdot z^0 = -4 \cdot 1 = -4$ The exponent 0 applies only to z.

   f. $(4z)^0 = 1$ The parentheses indicate that the exponent, 0, applies to both factors 4 and z.

Skill Practice   Evaluate the expressions. Assume all variables represent nonzero real numbers.

   1. 7^0 **2.** $(-7)^0$ **3.** -5^0

   4. $(4 + -8)^0$ **5.** $2x^0$ **6.** $(2x)^0$

Skill Practice Answers

1. 1   **2.** 1   **3.** -1
4. 1   **5.** 2   **6.** 1

The definition of b^0 is consistent with the other properties of exponents learned thus far. For example, we know that $1 = \frac{5^3}{5^3}$. If we subtract exponents, the result is 5^0.

subtract exponents

$$1 = \frac{5^3}{5^3} = 5^{3-3} = 5^0.$$ Therefore, 5^0 must be defined as 1.

2. Definition of b^{-n}

To understand the concept of a *negative* exponent, consider the following pattern.

$3^3 = 27$
$\quad\quad$ divide by 3
$3^2 = 9$
$\quad\quad$ divide by 3   As the exponents decrease by
$3^1 = 3$
$\quad\quad$ divide by 3   1, the resulting expressions are
$3^0 = 1$
$\quad\quad$ divide by 3   divided by 3.
$\quad\quad$ divide by 3

$3^{-1} = \dfrac{1}{3}$ ←——————— For the pattern to continue, we define $3^{-1} = \dfrac{1}{3^1} = \dfrac{1}{3}$.

$3^{-2} = \dfrac{1}{9}$ ←——————— For the pattern to continue, we define $3^{-2} = \dfrac{1}{3^2} = \dfrac{1}{9}$.

$3^{-3} = \dfrac{1}{27}$ ←——————— For the pattern to continue, we define $3^{-3} = \dfrac{1}{3^3} = \dfrac{1}{27}$.

This pattern suggests that $3^{-n} = \frac{1}{3^n}$ for all integers, n. In general, we have the following definition involving negative exponents.

Definition of b^{-n}

Let n be an integer and b be a nonzero real number. Then,

$$b^{-n} = \left(\frac{1}{b}\right)^n \quad \text{or} \quad \frac{1}{b^n}$$

The definition of b^{-n} implies that to evaluate b^{-n}, take the reciprocal of the base and change the sign of the exponent.

change the sign of   change the sign of
the exponent   the exponent

$$4^{-2} = \left(\frac{1}{4}\right)^2 \quad \text{or} \quad \frac{1}{4^2} \qquad\qquad \left(\frac{a}{b}\right)^{-n} = \left(\frac{b}{a}\right)^n \quad \text{or} \quad \frac{b^n}{a^n}$$

reciprocal   reciprocal
of the base   of the base

Example 2   **Simplifying Expressions with Negative Exponents**

Simplify.

a. c^{-3} $\qquad\qquad$ **b.** 5^{-1} $\qquad$ **c.** $(-3)^{-4}$ $\qquad$ **d.** $\left(\dfrac{1}{4}\right)^{-2}$

e. $\left(-\dfrac{3}{5}\right)^{-3}$ $\qquad$ **f.** $\dfrac{1}{y^{-5}}$ $\qquad$ **g.** $(5x)^{-3}$ $\qquad$ **h.** $5x^{-3}$

Solution:

a. $c^{-3} = \dfrac{1}{c^3}$ By definition

b. $5^{-1} = \dfrac{1}{5^1}$ By definition

   $= \dfrac{1}{5}$ Simplify.

c. $(-3)^{-4} = \dfrac{1}{(-3)^4}$ The base is -3 and must be enclosed in parentheses.

   $= \dfrac{1}{81}$ Simplify. Note that $(-3)^4 = (-3)(-3)(-3)(-3) = 81$.

d. $\left(\dfrac{1}{4}\right)^{-2} = 4^2$ Take the reciprocal of the base, and change the sign of the exponent.

   $= 16$ Simplify.

e. $\left(-\dfrac{3}{5}\right)^{-3} = \left(-\dfrac{5}{3}\right)^3$ Take the reciprocal of the base, and change the sign of the exponent.

   $= -\dfrac{125}{27}$ Simplify.

f. $\dfrac{1}{y^{-5}} = \left(\dfrac{1}{y}\right)^{-5}$ Apply the power of a quotient rule from Section 5.2.

   $= (y)^5$ Take the reciprocal of the base, and change the sign of the exponent.

   $= y^5$

g. $(5x)^{-3} = \left(\dfrac{1}{5x}\right)^3$ Take the reciprocal of the base, and change the sign of the exponent.

   $= \dfrac{(1)^3}{(5x)^3}$ Apply the exponent of 3 to each factor within parentheses.

   $= \dfrac{1}{125x^3}$ Simplify.

h. $5x^{-3} = 5 \cdot x^{-3}$ Note that the exponent, -3, applies only to x.

   $= 5 \cdot \dfrac{1}{x^3}$ Rewrite x^{-3} as $\dfrac{1}{x^3}$.

   $= \dfrac{5}{x^3}$ Multiply.

Skill Practice   Evaluate.

7. 3^{-3} **8.** x^{-1} **9.** $(-5)^{-2}$ **10.** $\left(\dfrac{1}{5}\right)^{-3}$

11. $\left(-\dfrac{5}{3}\right)^{-3}$ **12.** $\dfrac{2}{z^{-4}}$ **13.** $(2w)^{-4}$ **14.** $2w^{-4}$

Skill Practice Answers

7. $\dfrac{1}{27}$ **8.** $\dfrac{1}{x}$

9. $\dfrac{1}{25}$ **10.** 125

11. $-\dfrac{27}{125}$ **12.** $2z^4$

13. $\dfrac{1}{16w^4}$ **14.** $\dfrac{2}{w^4}$

It is important to note that the definition of b^{-n} is consistent with the other properties of exponents learned thus far. For example, consider the expression

$$\frac{x^4}{x^7} = \frac{\cancel{x} \cdot \cancel{x} \cdot \cancel{x} \cdot \cancel{x}}{\cancel{x} \cdot \cancel{x} \cdot \cancel{x} \cdot \cancel{x} \cdot x \cdot x \cdot x} = \frac{1}{x^3}$$

subtract exponents

By subtracting exponents, we have $\dfrac{x^4}{x^7} = x^{4-7} = x^{-3}$

Hence, $x^{-3} = \dfrac{1}{x^3}$

3. Properties of Integer Exponents: A Summary

The definitions of b^0 and b^{-n} allow us to extend the properties of exponents learned in Sections 5.1 and 5.2 to include integer exponents. These are summarized in Table 5-1.

Table 5-1

Properties of Integer Exponents Assume that a and b are real numbers ($b \neq 0$) and that m and n represent integers.		
Property	**Example**	**Details/Notes**
Multiplication of Like Bases 1. $b^m b^n = b^{m+n}$	$b^2 b^4 = b^{2+4} = b^6$	$b^2 b^4 = (b \cdot b)(b \cdot b \cdot b \cdot b) = b^6$
Division of Like Bases 2. $\dfrac{b^m}{b^n} = b^{m-n}$	$\dfrac{b^5}{b^2} = b^{5-2} = b^3$	$\dfrac{b^5}{b^2} = \dfrac{\cancel{b} \cdot \cancel{b} \cdot b \cdot b \cdot b}{\cancel{b} \cdot \cancel{b}} = b^3$
The Power Rule 3. $(b^m)^n = b^{m \cdot n}$	$(b^4)^2 = b^{4 \cdot 2} = b^8$	$(b^4)^2 = (b \cdot b \cdot b \cdot b)(b \cdot b \cdot b \cdot b) = b^8$
Power of a Product 4. $(ab)^m = a^m b^m$	$(ab)^3 = a^3 b^3$	$(ab)^3 = (ab)(ab)(ab)$ $= (a \cdot a \cdot a)(b \cdot b \cdot b) = a^3 b^3$
Power of a Quotient 5. $\left(\dfrac{a}{b}\right)^m = \dfrac{a^m}{b^m}$	$\left(\dfrac{a}{b}\right)^3 = \dfrac{a^3}{b^3}$	$\left(\dfrac{a}{b}\right)^3 = \left(\dfrac{a}{b}\right)\left(\dfrac{a}{b}\right)\left(\dfrac{a}{b}\right) = \dfrac{a \cdot a \cdot a}{b \cdot b \cdot b} = \dfrac{a^3}{b^3}$
Definitions Assume that b is a real number ($b \neq 0$) and that n represents an integer.		
Definition	**Example**	**Details/Notes**
$b^0 = 1$	$(4)^0 = 1$	Any nonzero quantity raised to the zero power equals 1.
$b^{-n} = \left(\dfrac{1}{b}\right)^n = \dfrac{1}{b^n}$	$b^{-5} = \left(\dfrac{1}{b}\right)^5 = \dfrac{1}{b^5}$	To simplify a negative exponent, take the reciprocal of the base and make the exponent positive.

4. Simplifying Expressions with Exponents

Example 3   Simplifying Expressions with Exponents

Simplify the following expressions. Write the answers with positive exponents only. Assume all variables are nonzero.

a. $\dfrac{a^3 b^{-2}}{c^{-5}}$ **b.** $\dfrac{x^2 x^{-7}}{x^3}$ **c.** $\dfrac{z^2}{w^{-4} w^4 z^{-8}}$

d. $(-4ab^{-2})^{-3}$ **e.** $\left(\dfrac{2p^{-4} q^3}{5p^2 q}\right)^{-1}$

Solution:

a. $\dfrac{a^3 b^{-2}}{c^{-5}}$

$\quad = \dfrac{a^3}{1} \cdot \dfrac{b^{-2}}{1} \cdot \dfrac{1}{c^{-5}}$

$\quad = \dfrac{a^3}{1} \cdot \dfrac{1}{b^2} \cdot \dfrac{c^5}{1}$ Simplify negative exponents.

$\quad = \dfrac{a^3 c^5}{b^2}$ Multiply.

b. $\dfrac{x^2 x^{-7}}{x^3}$

$\quad = \dfrac{x^{2+(-7)}}{x^3}$ Add the exponents in the numerator.

$\quad = \dfrac{x^{-5}}{x^3}$ Simplify.

$\quad = x^{-5-3}$ Subtract the exponents.

$\quad = x^{-8}$

$\quad = \dfrac{1}{x^8}$ Simplify the negative exponent.

c. $\dfrac{z^2}{w^{-4} w^4 z^{-8}}$

$\quad = \dfrac{z^2}{w^{-4+4} z^{-8}}$ Add the exponents in the denominator.

$\quad = \dfrac{z^2}{w^0 z^{-8}}$

$\quad = \dfrac{z^2}{(1)z^{-8}}$ Recall that $w^0 = 1$.

$\quad = z^{2-(-8)}$ Subtract the exponents.

$\quad = z^{10}$ Simplify.

d. $(-4ab^{-2})^{-3}$

$$= (-4)^{-3}a^{-3}(b^{-2})^{-3} \qquad \text{Apply the power rule of exponents.}$$

$$= (-4)^{-3}a^{-3}b^{6}$$

$$= \frac{1}{(-4)^3} \cdot \frac{1}{a^3} \cdot b^6 \qquad \text{Simplify the negative exponents.}$$

$$= \frac{1}{-64} \cdot \frac{1}{a^3} \cdot b^6 \qquad \text{Simplify.}$$

$$= -\frac{b^6}{64a^3} \qquad \text{Multiply fractions.}$$

TIP: Example 3(e) can also be simplified by clearing parentheses first:

$$\frac{2^{-1}p^4q^{-3}}{5^{-1}p^{-2}q^{-1}}$$

e. $\left(\dfrac{2p^{-4}q^3}{5p^2q}\right)^{-1}$ The negative exponent outside the parentheses can be eliminated by taking the reciprocal of the quantity within the parentheses.

$$= \left(\frac{5p^2q}{2p^{-4}q^3}\right)^{1} \qquad \text{Take the reciprocal of the base and make the outer exponent positive.}$$

$$= \frac{5p^2q}{2p^{-4}q^3}$$

$$= \frac{5p^{2-(-4)}q^{1-3}}{2} \qquad \text{Subtract the exponents.}$$

$$= \frac{5p^6q^{-2}}{2} \qquad \text{Simplify.}$$

$$= \frac{5p^6}{2} \cdot \frac{1}{q^2} \qquad \text{Simplify the negative exponent.}$$

$$= \frac{5p^6}{2q^2} \qquad \text{Simplify.}$$

Skill Practice Simplify the expressions. Use positive exponents only. Assume all variables are nonzero.

15. $\dfrac{x^{-6}}{y^4z^{-8}}$ **16.** $\dfrac{x^5 \cdot x^{-8}}{x^2}$ **17.** $\dfrac{a^7 \cdot a^{-4}}{a^{-3}}$

18. $(x^{-2} \cdot y^3)^{-2}$ **19.** $\left(\dfrac{3x^{-3}y^{-2}}{4xy^{-3}}\right)^{-2}$

Example 4 **Simplifying an Expression with Exponents**

Simplify the expression $2^{-1} + 3^{-1} + 5^0$. Write the answer with positive exponents only.

Skill Practice Answers

15. $\dfrac{z^8}{y^4x^6}$ **16.** $\dfrac{1}{x^5}$ **17.** a^6

18. $\dfrac{x^4}{y^6}$ **19.** $\dfrac{16x^8}{9y^2}$

Solution:

$2^{-1} + 3^{-1} + 5^0$

$= \dfrac{1}{2} + \dfrac{1}{3} + 1$ Simplify negative exponents. Simplify $5^0 = 1$.

$= \dfrac{3}{6} + \dfrac{2}{6} + \dfrac{6}{6}$ Get a common denominator.

$= \dfrac{11}{6}$ Simplify.

Skill Practice   Simplify the expression.

20. $2^{-1} + 4^{-2} + 3^0$

Skill Practice Answers

20. $\dfrac{25}{16}$

Section 5.3   Practice Exercises

Boost *your* GRADE at
mathzone.com!

 MathZone

- Practice Problems
- Self-Tests
- NetTutor
- e-Professors
- Videos

For this set of exercises, assume all variables represent nonzero real numbers.

Study Skills Exercise

1. To help you remember the properties of exponents, write them on 3 × 5 cards. On each card, write a property on one side and an example using that property on the other side. Keep these cards with you, and when you have a spare moment (such as waiting at the doctor's office), pull out these cards and go over the properties.

Review Exercises

For Exercises 2–11, simplify the expressions.

2. $b^3 b^8$

3. $c^7 c^2$

4. $\dfrac{x^6}{x^2}$

5. $\dfrac{y^9}{y^8}$

6. $\dfrac{9^4 \cdot 9^8}{9}$

7. $\dfrac{3^{14}}{3^3 \cdot 3^5}$

8. $(6ab^3c^2)^5$

9. $(7w^7z^2)^4$

10. $\left(\dfrac{s^2 t^5}{4}\right)^3$

11. $\left(\dfrac{5k^3}{h^7}\right)^2$

Concept 1: Definition of b^0

12. Simplify.

   a. 8^0 **b.** $\dfrac{8^4}{8^4}$

13. Simplify.

   a. d^0 **b.** $\dfrac{d^3}{d^3}$

For Exercises 14–27, simplify the expression.

14. p^0 **15.** k^0 **16.** 5^0 **17.** 2^0

18. -4^0 **19.** -1^0 **20.** $(-6)^0$ **21.** $(-2)^0$

22. $(8x)^0$ **23.** $(-3y^3)^0$ **24.** $-7x^0$ **25.** $6y^0$

26. ab^0 **27.** $-pq^0$

Concept 2: Definition of b^{-n}

28. Simplify and write the answers with positive exponents.

    a. t^{-5} **b.** $\dfrac{t^3}{t^8}$

29. Simplify and write the answers with positive exponents.

    a. 4^{-3} **b.** $\dfrac{4^2}{4^5}$

For Exercises 30–49, simplify.

30. $\left(\dfrac{2}{7}\right)^{-3}$ **31.** $\left(\dfrac{5}{4}\right)^{-1}$ **32.** $\left(-\dfrac{1}{5}\right)^{-2}$ **33.** $\left(-\dfrac{1}{3}\right)^{-3}$

34. a^{-3} **35.** c^{-5} **36.** 12^{-1} **37.** 4^{-2}

38. $(4b)^{-2}$ **39.** $(3z)^{-1}$ **40.** $6x^{-2}$ **41.** $7y^{-1}$

42. $(-8)^{-2}$ **43.** -8^{-2} **44.** -5^{-3} **45.** $(-w)^{-2}$

46. $(-t)^{-3}$ **47.** $(-r)^{-5}$ **48.** $\dfrac{1}{a^{-5}}$ **49.** $\dfrac{1}{b^{-6}}$

Concept 3: Properties of Integer Exponents: A Summary

50. Explain what is wrong with the following logic. $\dfrac{x^4}{x^{-6}} = x^{4-6} = x^{-2}$

51. Explain what is wrong with the following logic. $\dfrac{y^5}{y^{-3}} = y^{5-3} = y^2$

52. Explain what is wrong with the following logic. $2a^{-3} = \dfrac{1}{2a^3}$

53. Explain what is wrong with the following logic. $5b^{-2} = \dfrac{1}{5b^2}$

Concept 4: Simplifying Expressions with Exponents

For Exercises 54–95, simplify the expression. Write the answer with positive exponents only.

54. $x^{-8}x^4$ **55.** s^5s^{-6} **56.** $a^{-8}a^8$ **57.** q^3q^{-3}

58. $y^{17}y^{-13}$ **59.** $b^{20}b^{-14}$ **60.** $(m^{-6}n^9)^3$ **61.** $(c^4d^{-5})^{-2}$

62. $(-3j^{-5}k^6)^4$ **63.** $(6xy^{-11})^{-3}$ **64.** $\dfrac{p^3}{p^9}$ **65.** $\dfrac{q^2}{q^{10}}$

66. $\dfrac{r^{-5}}{r^{-2}}$

67. $\dfrac{u^{-2}}{u^{-6}}$

68. $\dfrac{a^2}{a^{-6}}$

69. $\dfrac{p^3}{p^{-5}}$

70. $\dfrac{y^{-2}}{y^6}$

71. $\dfrac{s^{-4}}{s^3}$

72. $\dfrac{7^3}{7^2 \cdot 7^8}$

73. $\dfrac{3^4 \cdot 3}{3^7}$

74. $\dfrac{a^2 a}{a^3}$

75. $\dfrac{t^5}{t^2 t^3}$

76. $\dfrac{a^{-1}b^2}{a^3 b^8}$

77. $\dfrac{k^{-4}h^{-1}}{k^6 h}$

78. $\dfrac{w^{-8}(w^2)^{-5}}{w^3}$

79. $\dfrac{p^2 p^{-7}}{(p^2)^3}$

80. $\dfrac{3^{-2}}{3}$

81. $\dfrac{5^{-1}}{5}$

82. $\left(\dfrac{p^{-1}q^5}{p^{-6}}\right)^0$

83. $\left(\dfrac{ab^{-4}}{a^{-5}}\right)^0$

84. $(8x^3 y^0)^{-2}$

85. $(3u^2 v^0)^{-3}$

86. $(-8y^{-12})(2y^{16}z^{-2})$

87. $(5p^{-2}q^5)(-2p^{-4}q^{-1})$

88. $\dfrac{-18a^{10}b^6}{108a^{-2}b^6}$

89. $\dfrac{-35x^{-4}y^{-3}}{-21x^2 y^{-3}}$

90. $\dfrac{(-4c^{12}d^7)^2}{(5c^{-3}d^{10})^{-1}}$

91. $\dfrac{(s^3 t^{-2})^4}{(3s^{-4}t^6)^{-2}}$

92. $\left(\dfrac{2}{p^6 p^3}\right)^{-3}$

93. $\left(\dfrac{5x}{x^7}\right)^{-2}$

94. $\left(\dfrac{5cd^{-3}}{10d^5}\right)^{-1}$

95. $\left(\dfrac{4m^{10}n^4}{2m^{12}n^{-2}}\right)^{-1}$

For Exercises 96–105, simplify the expression.

96. $5^{-1} + 2^{-2}$

97. $4^{-2} + 8^{-1}$

98. $10^0 - 10^{-1}$

99. $3^0 - 3^{-2}$

100. $2^{-2} + 1^{-2}$

101. $4^{-1} + 8^{-1}$

102. $4 \cdot 5^0 - 2 \cdot 3^{-1}$

103. $2 \cdot 4^0 - 3 \cdot 4^{-1}$

104. $5 \cdot 2^{-3} + 2 \cdot 4^{-1}$

105. $2 \cdot 3^{-2} + 3 \cdot 9^{-1}$

Scientific Notation

1. Introduction to Scientific Notation

In many applications in mathematics, it is necessary to work with very large or very small numbers. For example, the number of movie tickets sold in the United States in 2004 is estimated to be 1,500,000,000. The weight of a flea is approximately 0.00066 lb. To avoid writing numerous zeros in very large or small numbers, scientific notation was devised as a shortcut. Scientific notation is useful when performing calculations and when comparing the relative sizes of very large or very small numbers.

Concepts

1. Introduction to Scientific Notation
2. Writing Numbers in Scientific Notation
3. Writing Numbers without Scientific Notation
4. Multiplying and Dividing Numbers in Scientific Notation
5. Applications of Scientific Notation

The principle behind scientific notation is to use a power of 10 to express the magnitude of the number. Consider the following powers of 10:

$$10^0 = 1$$

$$10^1 = 10 \qquad 10^{-1} = \frac{1}{10^1} = \frac{1}{10} = 0.1$$

$$10^2 = 100 \qquad 10^{-2} = \frac{1}{10^2} = \frac{1}{100} = 0.01$$

$$10^3 = 1000 \qquad 10^{-3} = \frac{1}{10^3} = \frac{1}{1000} = 0.001$$

$$10^4 = 10{,}000 \qquad 10^{-4} = \frac{1}{10^4} = \frac{1}{10{,}000} = 0.0001$$

In the base-10 numbering system, each place value to the left and right of the decimal point represents a different power of 10 (Figure 5-1).

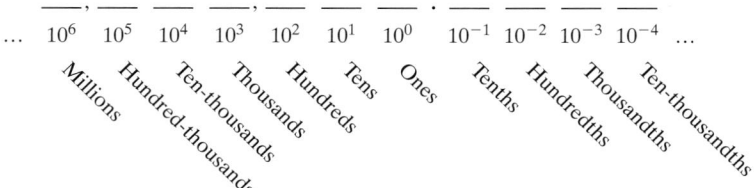

Figure 5-1

Therefore, a number such as 4000 can be written as 4.0×1000, or equivalently, 4.0×10^3. Similarly, the number 0.07 can be written as 7.0×0.01, or equivalently, 7.0×10^{-2}.

Definition of a Number Written in Scientific Notation

A number expressed in the form: $a \times 10^n$, where $1 \le |a| < 10$ and n is an integer is said to be written in **scientific notation**.

The numbers 4.0×10^3 and 7.0×10^{-2} are both expressed in scientific notation. To write a positive number in scientific notation, we apply the following guidelines:

1. Move the decimal point so that its new location is to the right of the first nonzero digit. The number should now be greater than or equal to 1, but less than 10. Count the number of places that the decimal point is moved.
2. If the original number is *large* (greater than or equal to 10), use the number of places the decimal point was moved as a *positive* power of 10.

$$450{,}000 = 4.5 \times 100{,}000 = 4.5 \times 10^5$$

5 places

3. If the original number is *small* (between 0 and 1), use the number of places the decimal point was moved as a *negative* power of 10.

$$0.0002 = 2.0 \times 0.0001 = 2.0 \times 10^{-4}$$

4 places

4. If the original number is greater than or equal to 1 but less than 10, use 0 as the power of 10.

$7.592 = 7.592 \times 10^0$ *Note:* A number between 1 and 10 is seldom written in scientific notation.

2. Writing Numbers in Scientific Notation

Example 1   **Writing Numbers in Scientific Notation**

Write the numbers in scientific notation.

  a. 53,000   **b.** 0.00053

Solution:

a. $53,000. = 5.3 \times 10^4$ To write 53,000 in scientific notation, the decimal point must be moved four places to the left. Because 53,000 is larger than 10, a *positive* power of 10 is used.

b. $0.00053 = 5.3 \times 10^{-4}$ To write 0.00053 in scientific notation, the decimal point must be moved four places to the right. Because 0.00053 is less than 1, a *negative* power of 10 is used.

Skill Practice   Write the numbers in scientific notation.

  1. 175,000,000   **2.** 0.000005

Example 2   **Writing Numbers in Scientific Notation**

Write the numerical values in scientific notation.

  a. The number of movie tickets sold in the United States in 2004 is estimated to be 1,500,000,000.

  b. The weight of a flea is approximately 0.00066 lb.

  c. The temperature on a January day in Fargo dropped to $-43°F$.

  d. A bench is 8.2 ft long.

Solution:

  a. $1,500,000,000 = 1.5 \times 10^9$ **b.** $0.00066 \text{ lb} = 6.6 \times 10^{-4} \text{ lb}$

  c. $-43°F = -4.3 \times 10^1 \text{ °F}$ **d.** $8.2 \text{ ft} = 8.2 \times 10^0 \text{ ft}$

Skill Practice   Write the numbers in scientific notation.

  3. The population of the Earth is approximately 6,400,000,000.

  4. The weight of a grain of salt is approximately 0.000002 ounces.

  5. The lowest recorded temperature is $-128.6°F$.

  6. A gallon of water weighs 8.33 lb.

Skill Practice Answers

1. 1.75×10^8
2. 5.0×10^{-6}
3. 6.4×10^9
4. 2.0×10^{-6} oz
5. -1.286×10^2 °F
6. 8.33×10^0 lb

TIP: For a number written in scientific notation, the power of 10 is sometimes called the **order of magnitude** (or simply the magnitude) of the number.

- The order of magnitude of the number of movie tickets sold in 2004 in the United States is 10^9 (billions of dollars).
- The mass of a flea is on the order of 10^{-4} lb (ten-thousandths of a pound).

3. Writing Numbers without Scientific Notation

Example 3 Writing Numbers Without Scientific Notation

Write the numerical values without scientific notation.

a. The mass of a proton is approximately 1.67×10^{-24} g.

b. The "nearby" star Vega is approximately 1.552×10^{14} miles from Earth.

Solution:

a. 1.67×10^{-24} g = 0.000 000 000 000 000 000 000 001 67 g

Because the power of 10 is negative, the value of 1.67×10^{-24} is a decimal number between 0 and 1. Move the decimal point 24 places to the *left*.

b. 1.552×10^{14} miles = 155,200,000,000,000 miles

Because the power of 10 is a positive integer, the value of 1.552×10^{14} is a large number greater than 10. Move the decimal point 14 places to the *right*.

Skill Practice Write the numerical values without scientific notation.

7. The probability of winning the California Super Lotto Jackpot is 5.5×10^{-8}.

8. The Sun's mass is 2×10^{30} kilograms.

4. Multiplying and Dividing Numbers in Scientific Notation

To multiply or divide two numbers in scientific notation, use the commutative and associative properties of multiplication to group the powers of 10. For example,

$$400 \times 2000 = (4 \times 10^2)(2 \times 10^3) = (4 \cdot 2) \times (10^2 \cdot 10^3) = 8 \times 10^5$$

$$\frac{0.00054}{150} = \frac{5.4 \times 10^{-4}}{1.5 \times 10^2} = \left(\frac{5.4}{1.5}\right) \times \left(\frac{10^{-4}}{10^2}\right) = 3.6 \times 10^{-6}$$

Example 4 Multiplying and Dividing Numbers in Scientific Notation

a. $(8.7 \times 10^4)(2.5 \times 10^{-12})$

b. $\dfrac{4.25 \times 10^{13}}{8.5 \times 10^{-2}}$

Solution:

a. $(8.7 \times 10^4)(2.5 \times 10^{-12})$

$= (8.7 \cdot 2.5) \times (10^4 \cdot 10^{-12})$ Commutative and associative properties of multiplication

Skill Practice Answers

7. 0.000 000 055

8. 2,000,000,000,000,000,000,000,000,000,000

$$= 21.75 \times 10^{-8}$$ The number 21.75 is not in proper scientific notation because 21.75 is not between 1 and 10.

$$= (2.175 \times 10^1) \times 10^{-8}$$ Rewrite 21.75 as 2.175×10^1.

$$= 2.175 \times (10^1 \times 10^{-8})$$ Associative property of multiplication

$$= 2.175 \times 10^{-7}$$ Simplify.

b. $\dfrac{4.25 \times 10^{13}}{8.5 \times 10^{-2}}$

$$= \left(\frac{4.25}{8.5}\right) \times \left(\frac{10^{13}}{10^{-2}}\right)$$ Commutative and associative properties

$$= 0.5 \times 10^{15}$$ The number 0.5×10^{15} is not in proper scientific notation because 0.5 is not between 1 and 10.

$$= (5.0 \times 10^{-1}) \times 10^{15}$$ Rewrite 0.5 as 5.0×10^{-1}.

$$= 5.0 \times (10^{-1} \times 10^{15})$$ Associative property of multiplication

$$= 5.0 \times 10^{14}$$ Simplify.

| Skill Practice | Multiply.

9. $(7 \times 10^5)(5 \times 10^3)$

10. $\dfrac{1 \times 10^{-2}}{4 \times 10^{-7}}$

5. Applications of Scientific Notation

| Example 5 | Applying Scientific Notation

If a spacecraft travels at 1.6×10^4 mph, how long will it take the craft to travel to Mars if the distance is approximately 8.0×10^7 miles?

Solution:

Since $d = rt$, then $t = \dfrac{d}{r}$

$$t = \frac{8.0 \times 10^7 \text{ miles}}{1.6 \times 10^4 \text{ miles/hour}}$$

$$= \left(\frac{8.0}{1.6}\right) \times \left(\frac{10^7}{10^4}\right) \text{hr}$$

$$= 5.0 \times 10^3 \text{ hr}$$

The time required to travel to Mars is approximately $5.0 \times 10^3 = 5000$ hr, or 208 days.

| Skill Practice |

11. The Earth's orbit is approximately 9.49×10^{11} meters. The Earth travels through its orbit at a speed of approximately 2.6×10^9 meters per day. How many days does it take for the Earth to complete the orbit?

Skill Practice Answers

9. 3.5×10^9
10. 2.5×10^4
11. Approximately 3.65×10^2 or 365 days.

Calculator Connections

Both scientific and graphing calculators can perform calculations involving numbers written in scientific notation. Most calculators use an $\boxed{EE}$ key or an $\boxed{EXP}$ key to enter the power of 10.

Scientific Calculator

Enter: 2.7 $\boxed{EE}$ (or $\boxed{EXP}$) 5 $\boxed{=}$ **Result:** | 270000 |

Enter: 7.1 $\boxed{EE}$ (or $\boxed{EXP}$) 3 $\boxed{+/-}$ $\boxed{=}$ **Result:** | 0.0071 |

Graphing Calculator

```
2.7E5
              270000
7.1E-3
                .0071
```

We recommend that you use parentheses to enclose each number written in scientific notation when performing calculations. Try using your calculator to perform the calculations from Example 4.

TIP: Note that the symbol E on the calculator screen means $\times\ 10$. Therefore, 5 E 14 means 5×10^{14}.

a. $(8.7 \times 10^4)(2.5 \times 10^{-12})$ **b.** $\dfrac{4.25 \times 10^{13}}{8.5 \times 10^{-2}}$

Scientific Calculator

Enter: $\boxed{(}$ 8.7 $\boxed{EE}$ 4 $\boxed{)}$ $\boxed{\times}$ $\boxed{(}$ 2.5 $\boxed{EE}$ 12 $\boxed{+/-}$ $\boxed{)}$ $\boxed{=}$

Result: | 0.000000218 |

Enter: $\boxed{(}$ 4.25 $\boxed{EE}$ 13 $\boxed{)}$ $\boxed{\div}$ $\boxed{(}$ 8.5 $\boxed{EE}$ 2 $\boxed{+/-}$ $\boxed{)}$ $\boxed{=}$

Result: | 5E14 |

Notice that the answer to part (b) is shown on the calculator in scientific notation. The calculator does not have enough room to display 14 zeros.

Graphing Calculator

```
(8.7E4)*(2.5E-12
)
            2.175E-7
(4.25E13)/(8.5E-
2)
               5E14
```

Calculator Exercises

Use a calculator to perform the indicated operations.

1. $(5.2 \times 10^6)(4.6 \times 10^{-3})$ **2.** $(2.19 \times 10^{-8})(7.84 \times 10^{-4})$

3. $\dfrac{4.76 \times 10^{-5}}{2.38 \times 10^9}$ **4.** $\dfrac{8.5 \times 10^4}{4.0 \times 10^{-1}}$

5. $\dfrac{(9.6 \times 10^7)(4.0 \times 10^{-3})}{2.0 \times 10^{-2}}$ **6.** $\dfrac{(5.0 \times 10^{-12})(6.4 \times 10^{-5})}{(1.6 \times 10^{-8})(4.0 \times 10^2)}$

Section 5.4   Practice Exercises

- Practice Problems
- Self-Tests
- NetTutor
- e-Professors
- Videos

Study Skills Exercise

1. Define the key terms:

   a. scientific notation   **b. order of magnitude**

Review Exercises

For Exercises 2–13, simplify the expression. Assume all variables represent nonzero real numbers.

2. $a^3 a^{-4}$

3. $b^5 b^8$

4. $10^3 \cdot 10^{-4}$

5. $10^5 \cdot 10^8$

6. $\dfrac{x^3}{x^6}$

7. $\dfrac{y^2}{y^7}$

8. $\dfrac{10^3}{10^6}$

9. $\dfrac{10^2}{10^7}$

10. $\dfrac{z^9 z^4}{z^3}$

11. $\dfrac{w^{-2} w^5}{w^{-1}}$

12. $\dfrac{10^9 \cdot 10^4}{10^3}$

13. $\dfrac{10^{-2} \cdot 10^5}{10^{-1}}$

Concept 1: Introduction to Scientific Notation

For Exercises 14–21, simplify the expressions.

14. 10^3

15. 10^5

16. 10^{-2}

17. 10^{-4}

18. 10^0

19. 10^1

20. 10^{-1}

21. 10^2

Concept 2: Writing Numbers in Scientific Notation

22. Explain how scientific notation might be valuable in studying astronomy. Answers may vary.

23. Explain how you would write the number 0.000 000 000 23 in scientific notation.

24. Explain how you would write the number 23,000,000,000,000 in scientific notation.

For Exercises 25–36, write the number in scientific notation.

25. 50,000

26. 900,000

27. 208,000

28. 420,000,000

29. 6,010,000

30. 75,000

31. 0.000008

32. 0.003

33. 0.000125

34. 0.00000025

35. 0.006708

36. 0.02004

For Exercises 37–42, write the numbers in scientific notation.

37. The mass of a proton is approximately 0.000 000 000 000 000 000 000 0017 g.

38. The total combined salaries of the president, vice president, senators, and representatives of the United States federal government is approximately $85,000,000.

39. The Bill Gates Foundation has over $27,000,000,000 from which it makes contributions to global charities.

40. One gram is equivalent to 0.0035 oz.

41. In the world's largest tanker disaster, *Amoco Cadiz* spilled 68,000,000 gal of oil off Portsall, France, causing widespread environmental damage over 100 miles of Brittany coast.

42. The human heart pumps about 1400 L of blood per day. That would mean that it pumps approximately 10,000,000 L per year.

Concept 3: Writing Numbers without Scientific Notation

43. Explain how you would write the number 3.1×10^{-9} without scientific notation.

44. Explain how you would write the number 3.1×10^{9} without scientific notation.

For Exercises 45–60, write the numbers without scientific notation.

45. 5×10^{-5}

46. 2×10^{-7}

47. 2.8×10^{3}

48. 9.1×10^{6}

49. 6.03×10^{-4}

50. 7.01×10^{-3}

51. 2.4×10^{6}

52. 3.1×10^{4}

53. 1.9×10^{-2}

54. 2.8×10^{-6}

55. 7.032×10^{3}

56. 8.205×10^{2}

57. One picogram (pg) is equal to 1×10^{-12} g.

58. A nanometer (nm) is approximately 3.94×10^{-8} in.

59. A normal diet contains between 1.6×10^{3} Cal and 2.8×10^{3} Cal per day.

60. The total land area of Texas is approximately 2.62×10^{5} square miles.

Concept 4: Multiplying and Dividing Numbers in Scientific Notation

For Exercises 61–80, multiply or divide as indicated. Write the answers in scientific notation.

61. $(2.5 \times 10^{6})(2.0 \times 10^{-2})$

62. $(2.0 \times 10^{-7})(3.0 \times 10^{13})$

63. $(1.2 \times 10^{4})(3 \times 10^{7})$

64. $(3.2 \times 10^{-3})(2.5 \times 10^{8})$

65. $\dfrac{7.7 \times 10^{6}}{3.5 \times 10^{2}}$

66. $\dfrac{9.5 \times 10^{11}}{1.9 \times 10^{3}}$

67. $\dfrac{9.0 \times 10^{-6}}{4.0 \times 10^{7}}$

68. $\dfrac{7.0 \times 10^{-2}}{5.0 \times 10^{9}}$

69. $(8.0 \times 10^{10})(4.0 \times 10^{3})$

70. $(6.0 \times 10^{-4})(3.0 \times 10^{-2})$

71. $(3.2 \times 10^{-4})(7.6 \times 10^{-7})$

72. $(5.9 \times 10^{12})(3.6 \times 10^{9})$

73. $\dfrac{2.1 \times 10^{11}}{7.0 \times 10^{-3}}$

74. $\dfrac{1.6 \times 10^{14}}{8.0 \times 10^{-5}}$

75. $\dfrac{5.7 \times 10^{-2}}{9.5 \times 10^{-8}}$

76. $\dfrac{2.72 \times 10^{-6}}{6.8 \times 10^{-4}}$

77. $6{,}000{,}000{,}000 \times 0.0000000023$

78. $0.000055 \times 40{,}000$

79. $\dfrac{0.0000000003}{6000}$

80. $\dfrac{420{,}000}{0.0000021}$

Concept 5: Applications of Scientific Notation

81. If a piece of paper is 3.0×10^{-3} in. thick, how thick is a stack of 1.25×10^{3} pieces of paper?

82. A box of staples contains 5.0×10^{3} staples and weighs 15 oz. How much does one staple weigh? Write your answer in scientific notation.

83. Bill Gates owned approximately 1,100,000,000 shares of Microsoft stock. If the stock price was $27 per share, how much was Bill Gates' stock worth?

84. A state lottery had a jackpot of $\$5.2 \times 10^{7}$. This week the winner was a group of office employees that included 13 people. How much would each person receive?

85. Dinosaurs became extinct about 65 million years ago.

    a. Write the number 65 million in scientific notation.   **b.** How many days is 65 million years?

    c. How many hours is 65 million years?   **d.** How many seconds is 65 million years?

86. The Earth is 111,600,000 km from the Sun.

    a. Write the number 111,600,000 in scientific notation.

    b. If there are 1000 m in a kilometer, how many meters is the Earth from the Sun?

    c. If there are 100 cm in a meter, how many centimeters is the Earth from the Sun?

87. For a recent year, the U.S. national debt was about $7,500,000,000,000. If the U.S. population was approximately 300,000,000, how much would each person pay to relieve the country of its debt?

88. The area of Japan is approximately 380,000 km^{2}. If the population is about 129,200,000, what is the density (people per square kilometer) of Japan?

89. The longest paper clip chain is 2.961×10^{3} ft. If the chain took 2.3688×10^{5} clips, what is the length of each paper clip in feet?

90. A sports figure had a 5-year (60-month) contract. If he makes $\$1.5 \times 10^{6}$ per month, how much will he make in the 5-year period?

Chapter 5   Problem Recognition Exercises— Properties of Exponents

Simplify completely. Assume that all variables represent nonzero real numbers.

1. $t^3 t^5$

2. $2^3 2^5$

3. $\dfrac{y^7}{y^2}$

4. $\dfrac{p^9}{p^3}$

5. $(r^2 s^4)^2$

6. $(ab^3 c^2)^3$

7. $\dfrac{w^4}{w^{-2}}$

8. $\dfrac{m^{-14}}{m^2}$

9. $\dfrac{y^{-7} x^4}{z^{-3}}$

10. $\dfrac{a^3 b^{-6}}{c^{-8}}$

11. $(2.5 \times 10^{-3})(5.0 \times 10^5)$

12. $(3.1 \times 10^6)(4.0 \times 10^{-2})$

13. $\dfrac{4.8 \times 10^7}{6.0 \times 10^{-2}}$

14. $\dfrac{5.4 \times 10^{-2}}{9.0 \times 10^6}$

15. $\dfrac{1}{p^{-6} p^{-8} p^{-1}}$

16. $p^6 p^8 p$

17. $\dfrac{v^9}{v^{11}}$

18. $(c^5 d^4)^{10}$

19. $\left(\dfrac{1}{2}\right)^{-1} + \left(\dfrac{1}{3}\right)^0$

20. $\left(\dfrac{1}{4}\right)^0 - \left(\dfrac{1}{5}\right)^{-1}$

21. $(2^5 b^{-3})^{-3}$

22. $(3^{-2} y^3)^{-2}$

23. $\left(\dfrac{3x}{2y}\right)^{-4}$

24. $\left(\dfrac{6c}{5d^3}\right)^{-2}$

25. $(3ab^2)(a^2 b)^3$

26. $(4x^2 y^3)^3 (xy^2)$

27. $\left(\dfrac{xy^2}{x^3 y}\right)^4$

28. $\left(\dfrac{a^3 b}{a^5 b^3}\right)^5$

29. $\dfrac{(t^{-2})^3}{t^{-4}}$

30. $\dfrac{(p^3)^{-4}}{p^{-5}}$

31. $\left(\dfrac{2w^2 x^3}{3y^0}\right)^3$

32. $\left(\dfrac{5a^0 b^4}{4c^3}\right)^2$

33. $\dfrac{q^3 r^{-2}}{s^{-1} t^5}$

34. $\dfrac{n^{-3} m^2}{p^{-3} q^{-1}}$

35. $\dfrac{(y^{-3})^2 (y^5)}{(y^{-3})^{-4}}$

36. $\dfrac{(w^2)^{-4} (w^{-2})}{(w^5)^{-4}}$

37. $\left(\dfrac{-2a^2 b^{-3}}{a^{-4} b^{-5}}\right)^{-3}$

38. $\left(\dfrac{-3x^{-4} y^3}{2x^5 y^{-2}}\right)^{-2}$

39. $(5h^{-2} k^0)^3 (5k^{-2})^{-4}$

40. $(6m^3 n^{-5})^{-4} (6m^0 n^{-2})^5$

Section 5.5   Addition and Subtraction of Polynomials

Concepts

1. Introduction to Polynomials
2. Applications of Polynomials
3. Addition of Polynomials
4. Subtraction of Polynomials
5. Polynomials and Applications to Geometry

1. Introduction to Polynomials

One commonly used algebraic expression is called a polynomial. A **polynomial** in one variable, x, is defined as a sum of terms of the form ax^n, where a is a real number and the exponent, n, is a nonnegative integer. For each term, a is called the **coefficient**, and n is called the **degree of the term**. For example,

Term (Expressed in the Form ax^n)	Coefficient	Degree
$-12z^7$	-12	7
$x^3 \rightarrow$ rewrite as $1x^3$	1	3
$10w \rightarrow$ rewrite as $10w^1$	10	1
$7 \rightarrow$ rewrite as $7x^0$	7	0

If a polynomial has exactly one term, it is categorized as a **monomial**. A two-term polynomial is called a **binomial**, and a three-term polynomial is called a **trinomial**. Usually the terms of a polynomial are written in descending order according to degree. The term with highest degree is called the **leading term**, and its coefficient is called the **leading coefficient**. The **degree of a polynomial** is the largest degree of all of its terms. Thus, the leading term determines the degree of the polynomial.

	Expression	Descending Order	Leading Coefficient	Degree of Polynomial
Monomials	$-3x^4$	$-3x^4$	-3	4
	17	17	17	0
Binomials	$4y^3 - 6y^5$	$-6y^5 + 4y^3$	-6	5
	$\dfrac{1}{2} - \dfrac{1}{4}c$	$-\dfrac{1}{4}c + \dfrac{1}{2}$	$-\dfrac{1}{4}$	1
Trinomials	$4p - 3p^3 + 8p^6$	$8p^6 - 3p^3 + 4p$	8	6
	$7a^4 - 1.2a^8 + 3a^3$	$-1.2a^8 + 7a^4 + 3a^3$	-1.2	8

Example 1 Identifying the Parts of a Polynomial

Given: $4.5a - 2.7a^{10} + 1.6 - 3.7a^5$

a. List the terms of the polynomial, and state the coefficient and degree of each term.

b. Write the polynomial in descending order.

c. State the degree of the polynomial and the leading coefficient.

Solution:

a. term: $4.5a$ coefficient: 4.5   degree: 1

term: $-2.7a^{10}$ coefficient: -2.7 degree: 10

term: 1.6   coefficient: 1.6   degree: 0

term: $-3.7a^5$ coefficient: -3.7 degree: 5

b. $-2.7a^{10} - 3.7a^5 + 4.5a + 1.6$

c. The degree of the polynomial is 10 and the leading coefficient is -2.7.

Skill Practice

1. Given $5x^3 - 7 + 8x^4$,

a. Write the polynomial in descending order.

b. State the degree of each term.

c. State the leading coefficient.

Polynomials may have more than one variable. In such a case, the degree of a term is the sum of the exponents of the variables contained in the term. For example, the term, $32x^2y^5z$, has degree 8 because the exponents applied to x, y, and z are 2, 5, and 1, respectively. The following polynomial has a degree of 11 because the highest degree of its terms is 11.

$$32x^2y^5z \quad - \quad 2x^3y \quad + \quad 2x^2yz^8 \quad + \quad 7$$

degree   degree   degree   degree
  8   4   11   0

Skill Practice Answers

1a. $8x^4 + 5x^3 - 7$
b. $4, 3, 0$
c. 8

2. Applications of Polynomials

> **Example 2**   Using Polynomials in an Application

A child throws a ball upward and the height of the ball, h (in feet), can be computed by the following equation:

$$h = -16t^2 + 64t + 2$$ where t is the time (in seconds) after the ball is released.

a. Find the height of the ball after 0.5 sec, 1 sec, and 1.5 sec.

b. Find the height of the ball at the time of release.

Solution:

a. $h = -16t^2 + 64t + 2$

$= -16(0.5)^2 + 64(0.5) + 2$ Substitute $t = 0.5$.

$= -16(0.25) + 32 + 2$

$= -4 + 32 + 2$

$= 30$ The height of the ball after 0.5 sec is 30 ft.

$h = -16t^2 + 64t + 2$

$= -16(1)^2 + 64(1) + 2$ Substitute $t = 1$.

$= -16(1) + 64 + 2$

$= -16 + 64 + 2$

$= 50$ The height of the ball after 1 sec is 50 ft.

$h = -16t^2 + 64t + 2$

$= -16(1.5)^2 + 64(1.5) + 2$ Substitute $t = 1.5$.

$= -16(2.25) + 96 + 2$

$= -36 + 96 + 2$

$= 62$ The height of the ball after 1.5 sec is 62 ft.

b. $h = -16t^2 + 64t + 2$

$= -16(0)^2 + 64(0) + 2$ At the time of release, $t = 0$.

$= 0 + 0 + 2$

$= 2$ The height of the ball at the time of release is 2 ft.

> **Skill Practice**

2. The number of cases, C (in millions), of soft drinks sold by a company the first 10 years it is in business is given by

$$C = -3t^2 + 25t + 150 \ (0 \le t \le 10)$$

where t is the number of years the company has been in business.

a. Find the number of cases sold after 2 years.

b. Find the number of cases sold after 5 years.

Skill Practice Answers

2a. 188 million
  b. 200 million

3. Addition of Polynomials

Recall that two terms are said to be *like* terms if they each have the same variables, and the corresponding variables are raised to the same powers.

Like Terms: $3x^2, -7x^2$ $\quad -5yz^3, yz^3$

Unlike Terms: $9z^2, 12z^6$ $\quad \dfrac{1}{3}w^6, \dfrac{2}{5}p^6 \quad\quad 4y, 7$

Recall that the distributive property is used to add or subtract *like* terms. For example,

$3x^2 + 9x^2 - 2x^2$

$= (3 + 9 - 2)x^2 \quad$ Apply the distributive property.

$= (10)x^2 \quad\quad$ Simplify.

$= 10x^2$

Example 3 Adding Polynomials

Add the polynomials.

a. $3x^2y + 5x^2y$

b. $(-3c^3 + 5c^2 - 7c) + (11c^3 + 6c^2 + 3)$

c. $\left(\dfrac{1}{4}w^2 - \dfrac{2}{3}w\right) + \left(\dfrac{3}{4}w^2 + \dfrac{1}{6}w - \dfrac{1}{2}\right)$

Solution:

a. $3x^2y + 5x^2y$

$= (3 + 5)x^2y \quad$ Apply the distributive property.

$= (8)x^2y$

$= 8x^2y \quad\quad$ Simplify.

b. $(-3c^3 + 5c^2 - 7c) + (11c^3 + 6c^2 + 3)$

$= -3c^3 + 11c^3 + 5c^2 + 6c^2 - 7c + 3 \quad$ Clear parentheses, and group *like* terms.

$= 8c^3 + 11c^2 - 7c + 3 \quad$ Combine *like* terms.

TIP: Although the distributive property is used to combine *like* terms, the process is simplified by combining the coefficients of *like* terms.

TIP: Polynomials can also be added by combining *like* terms in columns. The sum of the polynomials from Example 3(b) is shown here.

$\begin{array}{r} -3c^3 + 5c^2 - 7c + 0 \\ + 11c^3 + 6c^2 + 0c + 3 \\ \hline 8c^3 + 11c^2 - 7c + 3 \end{array}$

Place holders such as 0 and 0c may be used to help line up *like* terms.

c. $\left(\dfrac{1}{4}w^2 - \dfrac{2}{3}w\right) + \left(\dfrac{3}{4}w^2 + \dfrac{1}{6}w - \dfrac{1}{2}\right)$

$= \dfrac{1}{4}w^2 + \dfrac{3}{4}w^2 - \dfrac{2}{3}w + \dfrac{1}{6}w - \dfrac{1}{2}$ Clear parentheses, and group *like* terms.

$= \dfrac{1}{4}w^2 + \dfrac{3}{4}w^2 - \dfrac{4}{6}w + \dfrac{1}{6}w - \dfrac{1}{2}$ Get common denominators for *like* terms.

$= \dfrac{4}{4}w^2 - \dfrac{3}{6}w - \dfrac{1}{2}$ Add *like* terms.

$= w^2 - \dfrac{1}{2}w - \dfrac{1}{2}$ Simplify.

Skill Practice   Add the polynomials.

3. $(7q^2 - 2q + 4) + (5q^2 + 6q - 9)$

4. $\left(\dfrac{1}{2}t^3 - 1\right) + \left(\dfrac{2}{3}t^3 + \dfrac{1}{5}\right)$

5. $(5x^2 + x - 2) + (x + 4) + (3x^2 + 4x)$

4. Subtraction of Polynomials

The *opposite* (or additive inverse) of a real number a is $-a$. Similarly, if A is a polynomial, then $-A$ is its opposite.

Example 4   **Finding the Opposite of a Polynomial**

Find the opposite of the polynomials.

a. $5x$ **b.** $3a - 4b - c$ **c.** $5.5y^4 - 2.4y^3 + 1.1y - 3$

Solution:

TIP: Notice that the sign of each term is changed when finding the opposite of a polynomial.

a. The opposite of $5x$ is $-(5x)$, or $-5x$.

b. The opposite of $3a - 4b - c$ is $-(3a - 4b - c)$ or equivalently, $-3a + 4b + c$.

c. The opposite of $5.5y^4 - 2.4y^3 + 1.1y - 3$ is $-(5.5y^4 - 2.4y^3 + 1.1y - 3)$, or equivalently, $-5.5y^4 + 2.4y^3 - 1.1y + 3$.

Skill Practice   Find the opposite of the polynomials.

6. $x - 3$ **7.** $3y^2 - 2xy + 6x + 2$ **8.** $\dfrac{1}{3}x^3 - \dfrac{1}{2}x + \dfrac{5}{3}$

Skill Practice Answers

3. $12q^2 + 4q - 5$

4. $\dfrac{7}{6}t^3 - \dfrac{4}{5}$

5. $8x^2 + 6x + 2$

6. $-x + 3$

7. $-3y^2 + 2xy - 6x - 2$

8. $-\dfrac{1}{3}x^3 + \dfrac{1}{2}x - \dfrac{5}{3}$

Subtraction of two polynomials is similar to subtracting real numbers. Add the opposite of the second polynomial to the first polynomial.

Definition of Subtraction of Polynomials

If A and B are polynomials, then $A - B = A + (-B)$.

Example 5	**Subtracting Polynomials**

Subtract the polynomials.

a. $(-4p^4 + 5p^2 - 3) - (11p^2 + 4p - 6)$

b. $(a^2 - 2ab + 7b^2) - (-8a^2 - 6ab + 2b^2)$

Solution:

a. $(-4p^4 + 5p^2 - 3) - (11p^2 + 4p - 6)$

$= (-4p^4 + 5p^2 - 3) + (-11p^2 - 4p + 6)$ Add the opposite of the second polynomial.

$= -4p^4 + 5p^2 - 11p^2 - 4p - 3 + 6$ Group *like* terms.

$= -4p^4 - 6p^2 - 4p + 3$ Combine *like* terms.

TIP: Two polynomials can also be subtracted in columns by adding the opposite of the second polynomial to the first polynomial. Place holders (shown in red) may be used to help line up *like* terms.

$$\begin{array}{l} -4p^4 + 0p^3 + 5p^2 + 0p - 3 \\ \underline{-(0p^4 + 0p^3 + 11p^2 + 4p - 6)} \end{array} \xrightarrow{\text{add the opposite}} + \begin{array}{l} -4p^4 + 0p^3 + 5p^2 + 0p - 3 \\ \underline{-0p^4 - 0p^3 - 11p^2 - 4p + 6} \\ -4p^4 - 6p^2 - 4p + 3 \end{array}$$

---

b. $(a^2 - 2ab + 7b^2) - (-8a^2 - 6ab + 2b^2)$

$= (a^2 - 2ab + 7b^2) + (8a^2 + 6ab - 2b^2)$ Add the opposite of the second polynomial.

$= a^2 + 8a^2 - 2ab + 6ab + 7b^2 - 2b^2$ Group *like* terms.

$= 9a^2 + 4ab + 5b^2$ Combine *like* terms.

TIP: Recall that $a - b = a + (-b)$, or equivalently, $a + -1b$. Therefore, subtraction of polynomials can be simplified by applying the distributive property to clear parentheses.

$(a^2 - 2ab + 7b^2) - (-8a^2 - 6ab + 2b^2)$

$= a^2 - 2ab + 7b^2 - 1(-8a^2 - 6ab + 2b^2)$

$= a^2 - 2ab + 7b^2 + 8a^2 + 6ab - 2b^2$ Apply the distributive property.

$= a^2 + 8a^2 - 2ab + 6ab + 7b^2 - 2b^2$ Group *like* terms.

$= 9a^2 + 4ab + 5b^2$ Combine *like* terms.

Skill Practice Subtract the polynomials.

9. $(x^2 + 3x - 2) - (4x^2 + 6x + 1)$

10. $(-3y^2 + xy + 2x^2) - (-2y^2 - 3xy - 8x^2)$

Example 6 Subtracting Polynomials

Subtract $\frac{1}{3}t^4 + \frac{1}{2}t^2$ from $t^2 - 4$, and simplify the result.

Solution:

To subtract a from b, we write $b - a$. Thus, to subtract $\overbrace{\frac{1}{3}t^4 + \frac{1}{2}t^2}^{a}$ from $\overbrace{t^2 - 4}^{b}$, we have

$$\overset{b}{(t^2 - 4)} \overset{-}{-} \overset{a}{\left(\frac{1}{3}t^4 + \frac{1}{2}t^2\right)}$$

$$= t^2 - 4 - \frac{1}{3}t^4 - \frac{1}{2}t^2 \qquad \text{Apply the distributive property.}$$

$$= -\frac{1}{3}t^4 + t^2 - \frac{1}{2}t^2 - 4 \qquad \text{Group } \textit{like} \text{ terms in descending order.}$$

$$= -\frac{1}{3}t^4 + \frac{2}{2}t^2 - \frac{1}{2}t^2 - 4 \qquad \text{The } t^2\text{-terms are the only } \textit{like} \text{ terms.}$$
$$\qquad\qquad\qquad\qquad\qquad \text{Get a common denominator for the } t^2\text{-terms.}$$

$$= -\frac{1}{3}t^4 + \frac{1}{2}t^2 - 4 \qquad \text{Add } \textit{like} \text{ terms.}$$

Skill Practice

11. Subtract $\frac{1}{2}y^2 - \frac{1}{3}y$ from $y^3 + 3y^2 - y$.

---

5. Polynomials and Applications to Geometry

Example 7 Application of Polynomials in Geometry

If the perimeter of the triangle in Figure 5-2 can be represented by the polynomial $2x^2 + 5x + 6$, find a polynomial that represents the length of the missing side.

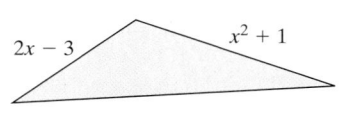

Figure 5-2

Skill Practice Answers

9. $-3x^2 - 3x - 3$

10. $-y^2 + 4xy + 10x^2$

11. $y^3 + \frac{5}{2}y^2 - \frac{2}{3}y$

Solution:

The missing side of the triangle can be found by subtracting the sum of the two known sides from the perimeter.

$$\begin{pmatrix} \text{Length} \\ \text{of missing} \\ \text{side} \end{pmatrix} = (\text{perimeter}) - \begin{pmatrix} \text{sum of the} \\ \text{two known sides} \end{pmatrix}$$

$$\begin{pmatrix} \text{Length} \\ \text{of missing} \\ \text{side} \end{pmatrix} = (2x^2 + 5x + 6) - [(2x - 3) + (x^2 + 1)]$$

$= 2x^2 + 5x + 6 - [2x - 3 + x^2 + 1]$	Clear inner parentheses.
$= 2x^2 + 5x + 6 - (x^2 + 2x - 2)$	Combine *like* terms within [ ].
$= 2x^2 + 5x + 6 - x^2 - 2x + 2$	Apply the distributive property.
$= 2x^2 - x^2 + 5x - 2x + 6 + 2$	Group *like* terms.
$= x^2 + 3x + 8$	Combine *like* terms.

The polynomial $x^2 + 3x + 8$ represents the length of the missing side.

Skill Practice

12. If the perimeter of the triangle is represented by the polynomial $6x - 9$, find the polynomial that represents the missing side.

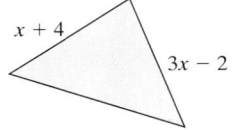

Skill Practice Answers

12. $2x - 11$

Section 5.5 Practice Exercises

Boost *your* GRADE at mathzone.com!

MathZone

- Practice Problems
- Self-Tests
- NetTutor
- e-Professors
- Videos

Study Skills Exercise

1. Define the key terms:

   a. polynomial   b. coefficient   c. degree of term

   d. monomial   e. binomial   f. trinomial

   g. leading term   h. leading coefficient   i. degree of polynomial

Review Exercises

For Exercises 2–7, simplify the expression. Assume all variables represent nonzero real numbers.

2. $\dfrac{p^3 \cdot 4p}{p^2}$

3. $(3x)(5x^{-4})$

4. $(6y^{-3})(2y^9)$

5. $\dfrac{8t^{-6}}{4t^{-2}}$

6. $\dfrac{8^3 \cdot 8^{-4}}{8^{-2} \cdot 8^6}$

7. $\dfrac{3^4 \cdot 3^{-8}}{3^{12} \cdot 3^{-4}}$

8. Explain the difference between 3.0×10^7 and 3^7.

9. Explain the difference between 4.0×10^{-2} and 4^{-2}.

10. Write the polynomial in descending order. $10 - 8a - a^3 + 2a^2 + a^5$

Concept 1: Introduction to Polynomials

11. Write the polynomial in descending order:

$$6 + 7x^2 - 7x^4 + 9x$$

12. Write the polynomial in descending order:

$$\frac{1}{2}y + y^2 - 12y^4 + y^3 - 6$$

For Exercises 13–24, categorize the expression as a monomial, a binomial, or a trinomial. Then identify the coefficient and degree of the leading term.

13. $10a^2 + 5a$

14. $7z + 13z^2 - 15$

15. $6x^2$

16. 9

17. $2t - t^4 - 5t$

18. $7x + 2$

19. $12y^4 - 3y + 1$

20. $5bc^2$

21. 23

22. $4 - 2c$

23. $-32xyz$

24. $w^4 - w^2$

Concept 2: Applications of Polynomials

25. A ball is dropped off a building and the height of the ball, h (in feet), can be computed by the equation:

$$h = -16t^2 + 150 \qquad \text{where } t \text{ is the time (in seconds) after the ball is released.}$$

   a. Find the height of the ball after 1 sec, 1.5 sec, and 2 sec.

   b. Find the height of the building by determining the height at the time of release.

26. An object is dropped off a building and the height of the object, h (in meters), can be computed by the equation:

$$h = -4.9t^2 + 45 \qquad \text{where } t \text{ is the time (in seconds) after the object is released.}$$

   a. Find the height of the object after 1 sec, 1.5 sec, and 2 sec.

   b. Find the height of the building by determining the height at the time of release.

27. A small business produces candles. The equation $P = -0.02x^2 + 10x - 60$ gives the profit, P in dollars, for x boxes of candles produced.

   a. Find the profit when $x = 100$, 250, and 300.

   b. What is P when $x = 0$. What does this number mean in the context of the problem?

28. An engineering student has a radio controlled model airplane. The student flies the plane over an open field and has a programmed flight path for landing. Four hundred horizontal feet from the landing point, the plane begins its descent (see figure). The flight path for landing can be modeled by:

$$h = -\frac{1}{320{,}000}x^3 + \frac{3}{1600}x^2$$ where h is the altitude of the plane in feet and x is the horizontal distance from the landing point.

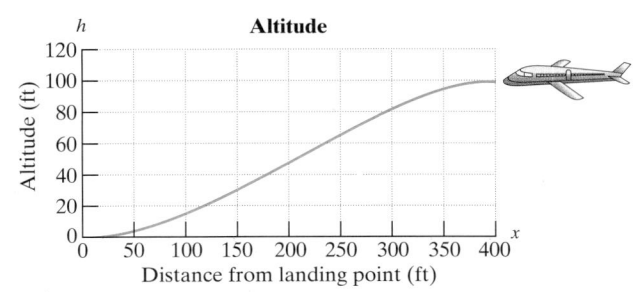

a. Find the altitude when the plane is 400 horizontal feet from the landing point.

b. Find the altitude when the plane is 100 horizontal feet from the landing point.

Concept 3: Addition of Polynomials

29. Explain why the terms $3x$ and $3x^2$ are not *like* terms.

30. Explain why the terms $4w^3$ and $4z^3$ are not *like* terms.

For Exercises 31–46, add the polynomials.

31. $23x^2y + 12x^2y$

32. $-5ab^3 + 17ab^3$

33. $(6y + 3x) + (4y - 3x)$

34. $(2z - 5h) + (-3z + h)$

35. $3b^2 + (5b^2 - 9)$

36. $4c + (3 - 10c)$

37. $(7y^2 + 2y - 9) + (-3y^2 - y)$

38. $(-3w^2 + 4w - 6) + (5w^2 + 2)$

39. $(6a + 2b - 5c) + (-2a - 2b - 3c)$

40. $(-13x + 5y + 10z) + (-3x - 3y + 2z)$

41. $\left(\frac{2}{5}a + \frac{1}{4}b - \frac{5}{6}\right) + \left(\frac{3}{5}a - \frac{3}{4}b - \frac{7}{6}\right)$

42. $\left(\frac{5}{9}x + \frac{1}{10}y\right) + \left(-\frac{4}{9}x + \frac{3}{10}y\right)$

43. $\left(z - \frac{8}{3}\right) + \left(\frac{4}{3}z^2 - z + 1\right)$

44. $\left(-\frac{7}{5}r + 1\right) + \left(-\frac{3}{5}r^2 + \frac{7}{5}r + 1\right)$

45. $(7.9t^3 + 2.6t - 1.1) + (-3.4t^2 + 3.4t - 3.1)$

46. $(0.34y^2 + 1.23) + (3.42y - 7.56)$

Concept 4: Subtraction of Polynomials

For Exercises 47–54, find the opposite of each polynomial.

47. $4h - 5$

48. $5k - 12$

49. $-2m^2 + 3m - 15$

50. $-n^2 - 6n + 9$

51. $3v^3 + 5v^2 + 10v + 22$

52. $7u^4 + 3v^2 + 17$

53. $-9t^4 - 8t - 39$

54. $-5r^5 - 3r^3 - r - 23$

For Exercises 55–74, subtract the polynomials.

55. $4a^3b^2 - 12a^3b^2$

56. $5yz^4 - 14yz^4$

57. $-32x^3 - 21x^3$

58. $-23c^5 - 12c^5$

59. $(7a - 7) - (12a - 4)$

60. $(4x + 3v) - (-3x + v)$

61. $(4k + 3) - (-12k - 6)$

62. $(3h - 15) - (8h - 13)$

63. $25s - (23s - 14)$

64. $3x^2 - (-x^2 - 12)$

65. $(5t^2 - 3t - 2) - (2t^2 + t + 1)$

66. $(k^2 + 2k + 1) - (3k^2 - 6k + 2)$

67. $(10r - 6s + 2t) - (12r - 3s - t)$

68. $(a - 14b + 7c) - (-3a - 8b + 2c)$

69. $\left(\dfrac{7}{8}x + \dfrac{2}{3}y - \dfrac{3}{10}\right) - \left(\dfrac{1}{8}x + \dfrac{1}{3}y\right)$

70. $\left(r - \dfrac{1}{12}s\right) - \left(\dfrac{1}{2}r - \dfrac{5}{12}s - \dfrac{4}{11}\right)$

71. $\left(\dfrac{2}{3}h^2 - \dfrac{1}{5}h - \dfrac{3}{4}\right) - \left(\dfrac{4}{3}h^2 - \dfrac{4}{5}h + \dfrac{7}{4}\right)$

72. $\left(\dfrac{3}{8}p^3 - \dfrac{5}{7}p^2 - \dfrac{2}{5}\right) - \left(\dfrac{5}{8}p^3 - \dfrac{2}{7}p^2 + \dfrac{7}{5}\right)$

73. $(4.5x^4 - 3.1x^2 - 6.7) - (2.1x^4 + 4.4x)$

74. $(1.3c^3 + 4.8) - (4.3c^2 - 2c - 2.2)$

75. Find the difference of $(4b^3 + 6b - 7)$ and $(-12b^2 + 11b + 5)$.

76. Find the difference of $(-5y^2 + 3y - 21)$ and $(-4y^2 - 5y + 23)$.

77. Subtract $(3x^3 - 5x + 10)$ from $(-2x^2 + 6x - 21)$.

78. Subtract $(7a^5 - 2a^3 - 5a)$ from $(3a^5 - 9a^2 + 3a - 8)$.

Concept 5: Polynomials and Applications to Geometry

79. Find a polynomial that represents the perimeter of the figure.

80. Find a polynomial that represents the perimeter of the figure.

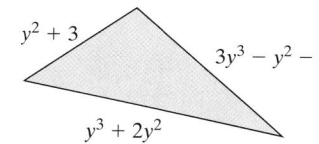

81. If the perimeter of the figure can be represented by the polynomial $5a^2 - 2a + 1$, find a polynomial that represents the length of the missing side.

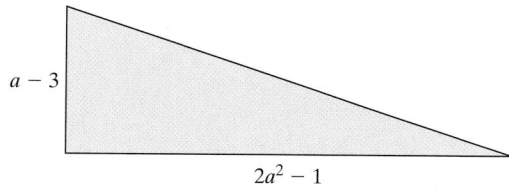

82. If the perimeter of the figure can be represented by the polynomial $6w^3 - 2w - 3$, find a polynomial that represents the length of the missing side.

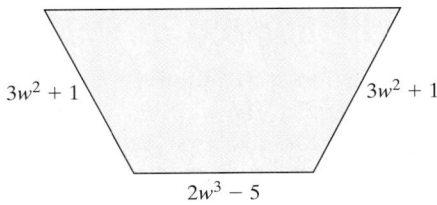

$3w^2 + 1$ $3w^2 + 1$

$2w^3 - 5$

Mixed Exercises

For Exercises 83–98, perform the indicated operation.

83. $(2ab^2 + 9a^2b) + (7ab^2 - 3ab + 7a^2b)$

84. $(8x^2y - 3xy - 6xy^2) + (3x^2y - 12xy)$

85. $(4z^5 + z^3 - 3z + 13) - (-z^4 - 8z^3 + 15)$

86. $(-15t^4 - 23t^2 + 16t) - (21t^3 + 18t^2 + t)$

87. $(9x^4 + 2x^3 - x + 5) + (9x^3 - 3x^2 + 8x + 3) - (7x^4 - x + 12)$

88. $(-6y^3 - 9y^2 + 23) - (7y^2 + 2y - 11) + (3y^3 - 25)$

89. $(5w^2 - 3w + 2) + (-4w + 6) - (7w^2 - 10)$

90. $(10u^3 - 5u^2 + 4) - (2u^3 + 5u^2 + u) - (u^3 - 3u + 9)$

91. $(7p^2q - 3pq^2) - (8p^2q + pq) + (4pq - pq^2)$

92. $(12c^2d - 2cd + 8cd^2) - (-c^2d + 4cd) - (5cd - 2cd^2)$

93. $(5x - 2x^3) + (2x^3 - 5x)$

94. $(p^2 - 4p + 2) - (2 + p^2 - 4p)$

95. $(2a^2b - 4ab + ab^2) - (-5ab^2 + 2a^2b + ab)$

96. $-3xy + 7xy^2 + 5x^2y + (3x^2y - 8xy - 11xy^2)$

97. $[(3y^2 - 5y) - (2y^2 + y - 1)] + (10y^2 - 4y - 5)$

98. $(12c^3 - 5c^2 - 2c) + [(7c^3 - 2c^2 + c) - (4c^3 + 4c)]$

Expanding Your Skills

99. Write a binomial of degree 3. (Answers may vary.)

100. Write a trinomial of degree 6. (Answers may vary.)

101. Write a monomial of degree 5. (Answers may vary.)

102. Write a monomial of degree 1. (Answers may vary.)

103. Write a trinomial with the leading coefficient -6. (Answers may vary.)

104. Write a binomial with the leading coefficient 13. (Answers may vary.)

Section 5.6　Multiplication of Polynomials

Concepts

1. Multiplication of Polynomials
2. Special Case Products: Difference of Squares and Perfect Square Trinomials
3. Applications to Geometry

1. Multiplication of Polynomials

The properties of exponents covered in Sections 5.1–5.3 can be used to simplify many algebraic expressions including the multiplication of monomials. To multiply monomials, first use the associative and commutative properties of multiplication to group coefficients and like bases. Then simplify the result by using the properties of exponents.

Example 1　Multiplying Monomials

Multiply.

a. $(3x^4)(4x^2)$　　　　**b.** $(-4c^5d)(2c^2d^3e)$　　　　**c.** $\left(\frac{1}{3}a^4b^3\right)\left(\frac{3}{4}b^7\right)$

Solution:

a. $(3x^4)(4x^2)$

$\qquad = (3 \cdot 4)(x^4x^2)$　　　　Group coefficients and like bases.

$\qquad = 12x^6$　　　　Add the exponents and simplify.

b. $(-4c^5d)(2c^2d^3e)$

$\qquad = (-4 \cdot 2)(c^5c^2)(dd^3)(e)$　　　　Group coefficients and like bases.

$\qquad = -8c^7d^4e$　　　　Simplify.

c. $\left(\frac{1}{3}a^4b^3\right)\left(\frac{3}{4}b^7\right)$

$\qquad = \left(\frac{1}{3} \cdot \frac{3}{4}\right)(a^4)(b^3b^7)$　　　　Group coefficients and like bases.

$\qquad = \frac{1}{4}a^4b^{10}$　　　　Simplify.

Skill Practice　Multiply.

1. $-5y(6y^3)$　　　　**2.** $7x^2y(-2x^3y^4)$　　　　**3.** $\left(-\frac{1}{2}w^2\right)\left(\frac{2}{3}wy^3\right)$

The distributive property is used to multiply polynomials: $a(b + c) = ab + ac$.

Example 2　Multiplying a Polynomial by a Monomial

Multiply.

a. $2t(4t - 3)$　　　　**b.** $-3a^2\left(-4a^2 + 2a - \frac{1}{3}\right)$

Skill Practice Answers

1. $-30y^4$
2. $-14x^5y^5$
3. $-\frac{1}{3}w^3y^3$

Solution:

a. $2t(4t - 3)$　　　　Multiply each term of the polynomial by $2t$.

$\qquad = (2t)(4t) + 2t(-3)$　　　　Apply the distributive property.

$\qquad = 8t^2 - 6t$　　　　Simplify each term.

b. $-3a^2\left(-4a^2 + 2a - \dfrac{1}{3}\right)$ Multiply each term of the polynomial by $-3a^2$.

$= (-3a^2)(-4a^2) + (-3a^2)(2a) + (-3a^2)\left(-\dfrac{1}{3}\right)$ Apply the distributive property.

$= 12a^4 - 6a^3 + a^2$ Simplify each term.

Skill Practice   Multiply.

4. $-4p(2p^2 - 6p + \frac{1}{4})$ **5.** $5a^2b^3(-3ab^2 + 4ab - 6b^3)$

Thus far, we have shown polynomial multiplication involving monomials. Next, the distributive property will be used to multiply polynomials with more than one term.

$(x + 3)(x + 5) = (x + 3)x + (x + 3)5$ Apply the distributive property.

$= (x + 3)x + (x + 3)5$ Apply the distributive property again.

$= (x)(x) + (3)(x) + (x)(5) + (3)(5)$

$= x^2 + 3x + 5x + 15$

$= x^2 + 8x + 15$ Combine *like* terms.

Note: Using the distributive property results in multiplying each term of the first polynomial by each term of the second polynomial.

$(x + 3)(x + 5) = (x)(x) + (x)(5) + (3)(x) + (3)(5)$

$= x^2 + 5x + 3x + 15$

$= x^2 + 8x + 15$

Example 3   **Multiplying a Polynomial by a Polynomial**

Multiply the polynomials.

a. $(c - 7)(c + 2)$ **b.** $(10x + 3y)(2x - 4y)$

c. $(y - 2)(3y^2 + y - 5)$

Solution:

a. $(c - 7)(c + 2)$ Multiply each term in the first polynomial by each term in the second.

$= (c)(c) + (c)(2) + (-7)(c) + (-7)(2)$ Apply the distributive property.

$= c^2 + 2c - 7c - 14$ Simplify.

$= c^2 - 5c - 14$ Combine *like* terms.

Skill Practice Answers

4. $-8p^3 + 24p^2 - p$
5. $-15a^3b^5 + 20a^3b^4 - 30a^2b^6$

TIP: Notice that the product of two *binomials* equals the sum of the products of the **F**irst terms, the **O**uter terms, the **I**nner terms, and the **L**ast terms. The acronym, **FOIL** (First Outer Inner Last) can be used as a memory device to multiply two binomials.

Outer terms

First terms

First   Outer   Inner   Last

$$(c - 7)(c + 2) = (c)(c) + (c)(2) + (-7)(c) + (-7)(2)$$

Inner terms

$$= c^2 + 2c - 7c - 14$$

Last terms

$$= c^2 - 5c - 14$$

b. $(10x + 3y)(2x - 4y)$

Multiply each term in the first polynomial by each term in the second.

$$= (10x)(2x) + (10x)(-4y) + (3y)(2x) + (3y)(-4y)$$

Apply the distributive property.

$$= 20x^2 - 40xy + 6xy - 12y^2$$

Simplify each term.

$$= 20x^2 - 34xy - 12y^2$$

Combine *like* terms.

c. $(y - 2)(3y^2 + y - 5)$

Multiply each term in the first polynomial by each term in the second.

$$= (y)(3y^2) + (y)(y) + (y)(-5) + (-2)(3y^2) + (-2)(y) + (-2)(-5)$$

$$= 3y^3 + y^2 - 5y - 6y^2 - 2y + 10$$

Simplify each term.

$$= 3y^3 - 5y^2 - 7y + 10$$

Combine *like* terms.

Avoiding Mistakes:

It is important to note that the acronym **FOIL** does not apply to Example 3(c) because the product does not involve two binomials.

TIP: Multiplication of polynomials can be performed vertically by a process similar to column multiplication of real numbers. For example,

$$
\begin{array}{r}
235 \\
\times \ 21 \\
\hline
235 \\
4700 \\
\hline
4935
\end{array}
\qquad
\begin{array}{r}
3y^2 + y - 5 \\
\times \quad y - 2 \\
\hline
-6y^2 - 2y + 10 \\
3y^3 + y^2 - 5y + 0 \\
\hline
3y^3 - 5y^2 - 7y + 10
\end{array}
$$

Note: When multiplying by the column method, it is important to *align like* terms vertically before adding terms.

Skill Practice Multiply and simplify.

6. $(x + 2)(x + 8)$ **7.** $(4a - 3c)(5a - 2c)$ **8.** $(2y + 4)(3y^2 - 5y + 2)$

---

2. Special Case Products: Difference of Squares and Perfect Square Trinomials

In some cases the product of two binomials takes on a special pattern.

I. The first special case occurs when multiplying the sum and difference of the same two terms. For example,

$(2x + 3)(2x - 3)$

$= 4x^2 - 6x + 6x - 9$

$= 4x^2 - 9$

Notice that the middle terms are opposites. This leaves only the difference between the square of the first term and the square of the second term. For this reason, the product is called a *difference of squares*.

Note: The sum and difference of the same two terms are called **conjugates**. Thus, the expressions $2x + 3$ and $2x - 3$ are conjugates of each other.

II. The second special case involves the square of a binomial. For example,

$(3x + 7)^2$

$= (3x + 7)(3x + 7)$

$= 9x^2 + 21x + 21x + 49$

$= 9x^2 + \quad 42x \quad + 49$

$\qquad \uparrow \qquad \uparrow \qquad \uparrow$

$= (3x)^2 + 2(3x)(7) + (7)^2$

When squaring a binomial, the product will be a trinomial called a *perfect square trinomial*. The first and third terms are formed by squaring each term of the binomial. The middle term equals twice the product of the terms in the binomial.

Note: The expression $(3x - 7)^2$ also expands to a perfect square trinomial, but the middle term will be negative:

$(3x - 7)(3x - 7) = 9x^2 - 21x - 21x + 49 = 9x^2 - 42x + 49$

Special Case Product Formulas

1. $(a + b)(a - b) = a^2 - b^2$ The product is called a **difference of squares**.

2. $\left. \begin{array}{l} (a + b)^2 = a^2 + 2ab + b^2 \\ (a - b)^2 = a^2 - 2ab + b^2 \end{array} \right\}$ The product is called a **perfect square trinomial**.

You should become familiar with these special case products because they will be used again in the next chapter to factor polynomials.

Example 4 Finding Special Products

Multiply the conjugates.

a. $(x - 9)(x + 9)$ **b.** $\left(\dfrac{1}{2}p - 6\right)\left(\dfrac{1}{2}p + 6\right)$

Skill Practice Answers
6. $x^2 + 10x + 16$
7. $20a^2 - 23ac + 6c^2$
8. $6y^3 + 2y^2 - 16y + 8$

TIP: The product of two conjugates can be checked by applying the distributive property:

$(x - 9)(x + 9)$

$= x^2 + 9x - 9x - 81$

$= x^2 - 81$

Solution:

a. $(x - 9)(x + 9)$ Apply the formula: $(a + b)(a - b) = a^2 - b^2$.

$a^2 - b^2$

$= (x)^2 - (9)^2$ Substitute $a = x$ and $b = 9$.

$= x^2 - 81$

b. $\left(\dfrac{1}{2}p - 6\right)\left(\dfrac{1}{2}p + 6\right)$ Apply the formula: $(a + b)(a - b) = a^2 - b^2$.

$a^2 - b^2$

$= \left(\dfrac{1}{2}p\right)^2 - (6)^2$ Substitute $a = \dfrac{1}{2}p$ and $b = 6$.

$= \dfrac{1}{4}p^2 - 36$ Simplify each term.

Skill Practice Multiply the conjugates.

9. $(a + 7)(a - 7)$ **10.** $(\tfrac{4}{5}x - 10)(\tfrac{4}{5}x + 10)$

Example 5   Finding Special Products

Square the binomials.

a. $(3w - 4)^2$ **b.** $(5x^2 + 2)^2$

Solution:

a. $(3w - 4)^2$ Apply the formula:
$(a - b)^2 = a^2 - 2ab + b^2$.

$a^2 - 2ab + b^2$

$= (3w)^2 - 2(3w)(4) + (4)^2$ Substitute $a = 3w$ and $b = 4$.

$= 9w^2 - 24w + 16$ Simplify each term.

TIP: The square of a binomial can be checked by explicitly writing the product of the two binomials and applying the distributive property:

$(3w - 4)^2 = (3w - 4)(3w - 4) = 9w^2 - 12w - 12w + 16$
$= 9w^2 - 24w + 16$

Avoiding Mistakes:

The property for squaring two factors is different than the property of squaring two terms: $(ab)^2 = a^2b^2$ but $(a + b)^2 = a^2 + 2ab + b^2$

b. $(5x^2 + 2)^2$ Apply the formula:
$(a + b)^2 = a^2 + 2ab + b^2$.

$a^2 + 2ab + b^2$

$= (5x^2)^2 + 2(5x^2)(2) + (2)^2$ Substitute $a = 5x^2$ and $b = 2$.

$= 25x^4 + 20x^2 + 4$ Simplify each term.

Skill Practice Square the binomials.

11. $(x + 3)^2$ **12.** $(3c^2 - 4)^2$

Skill Practice Answers

9. $a^2 - 49$ **10.** $\tfrac{16}{25}x^2 - 100$
11. $x^2 + 6x + 9$
12. $9c^4 - 24c^2 + 16$

3. Applications to Geometry

Example 6   Using Special Case Products in an Application of Geometry

Find a polynomial that represents the volume of the cube (Figure 5-3).

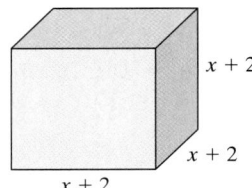

$x + 2$

$x + 2$

$x + 2$

Figure 5-3

Solution:

$$\text{Volume} = (\text{length})(\text{width})(\text{height})$$

$$V = (x + 2)(x + 2)(x + 2) \quad \text{or} \quad V = (x + 2)^3$$

To expand $(x + 2)(x + 2)(x + 2)$, multiply the first two factors. Then multiply the result by the last factor.

$$V = \underbrace{(x + 2)(x + 2)}(x + 2)$$

$$= (x^2 + 4x + 4)(x + 2)$$

> **TIP:** $(x + 2)(x + 2) = (x + 2)^2$ and results in a perfect square trinomial.
>
> $(x + 2)^2 = (x)^2 + 2(x)(2) + (2)^2$
>
> $\qquad\quad = x^2 + 4x + 4$

$$= (x^2)(x) + (x^2)(2) + (4x)(x) + (4x)(2) + (4)(x) + (4)(2)$$
Apply the distributive property.

$$= x^3 + 2x^2 + 4x^2 + 8x + 4x + 8$$
Group *like* terms.

$$= x^3 + 6x^2 + 12x + 8$$
Combine *like* terms.

The volume of the cube can be represented by

$$V = (x + 2)^3 = x^3 + 6x^2 + 12x + 8.$$

Skill Practice

13. Find the polynomial that represents the area of the square.

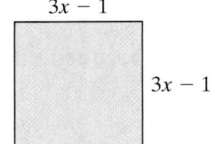

$3x - 1$

$3x - 1$

> **Example 7** **Using the Product of Polynomials in Geometry**

Find a polynomial that represents the area of the triangle (Figure 5-4).

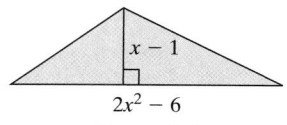

Figure 5-4

Solution:

The area of a triangle is found by $A = \frac{1}{2}bh$. In this case, $b = 2x^2 - 6$ and $h = x - 1$; therefore, $A = \frac{1}{2}(2x^2 - 6)(x - 1)$.

$A = \dfrac{1}{2}(2x^2 - 6)(x - 1)$ Apply the distributive property.

$= (x^2 - 3)(x - 1)$

$= (x^2 - 3)(x - 1)$ Multiply each term in the first polynomial by each term in the second.

$= (x^2)(x) + (x^2)(-1) + (-3)(x) + (-3)(-1)$ Apply the distributive property.

$= x^3 - x^2 - 3x + 3$

The area of the triangle can be represented by the polynomial $x^3 - x^2 - 3x + 3$.

> **Skill Practice**

14. Find the polynomial that represents the area of the triangle.

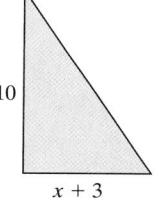

$4x - 10$

$x + 3$

Skill Practice Answers

14. The area of the triangle can be represented by $2x^2 + x - 15$

Section 5.6   Practice Exercises

Boost *your* GRADE at mathzone.com!

- Practice Problems
- Self-Tests
- NetTutor
- e-Professors
- Videos

Study Skills Exercise

1. Define the key terms:

     a. conjugates   **b. difference of squares**   **c. perfect square trinomial**

Review Exercises

For Exercises 2–13, simplify the expressions (if possible).

2. $4x + 5x$ **3.** $2y^2 - 4y^2$ **4.** $(4x)(5x)$ **5.** $(2y^2)(-4y^2)$

6. $-5a^3b - 2a^3b$ **7.** $7uvw^2 + uvw^2$ **8.** $(-5a^3b)(-2a^3b)$ **9.** $(7uvw^2)(uvw^2)$

10. $-c + 4c^2$ **11.** $3t + 3t^3$ **12.** $(-c)(4c^2)$ **13.** $(3t)(3t^3)$

Concept 1: Multiplication of Polynomials

For Exercises 14–22, multiply the expressions.

14. $8(4x)$

15. $-2(6y)$

16. $-10(5z)$

17. $7(3p)$

18. $(x^{10})(4x^3)$

19. $(a^{13}b^4)(12ab^4)$

20. $(4m^3n^7)(-3m^6n)$

21. $(2c^7d)(-c^3d^{11})$

22. $(-5u^2v)(-8u^3v^2)$

For Exercises 23–52, multiply the polynomials.

23. $8pq(2pq - 3p + 5q)$

24. $5ab(2ab + 6a - 3b)$

25. $(k^2 - 13k - 6)(-4k)$

26. $(h^2 + 5h - 12)(-2h)$

27. $-15pq(3p^2 + p^3q^2 - 2q)$

28. $-4u^2v(2u - 5uv^3 + v)$

29. $(y - 10)(y + 9)$

30. $(x + 5)(x - 6)$

31. $(m - 12)(m - 2)$

32. $(n - 7)(n - 2)$

33. $(2p - 3)(p + 1)$

34. $(4q + 11)(q - 5)$

35. $(3w + 5)(4w + 3)$

36. $(3z - 4)(2z - 1)$

37. $(-p + 1)(p - 11)$

38. $(-y - 7)(y - 10)$

39. $(6x - 1)(2x + 5)$

40. $(3x + 7)(x - 8)$

41. $(4a - 9)(2a - 1)$

42. $(3b + 5)(b - 5)$

43. $(3t - 7)(3t + 1)$

44. $(5w - 2)(2w - 5)$

45. $(3x + 4)(x + 8)$

46. $(7y + 1)(3y + 5)$

47. $(5s + 3)(s^2 + s - 2)$

48. $(t - 4)(2t^2 - t + 6)$

49. $(3w - 2)(9w^2 + 6w + 4)$

50. $(z + 5)(z^2 - 5z + 25)$

51. $(p^2 + p - 5)(p^2 + 4p - 1)$

52. $(-x^2 - 2x + 4)(x^2 + 2x - 6)$

Concept 2: Special Case Products: Difference of Squares and Perfect Square Trinomials

For Exercises 53–64, multiply the conjugates.

53. $(3a - 4b)(3a + 4b)$

54. $(5y + 7x)(5y - 7x)$

55. $(9k + 6)(9k - 6)$

56. $(2h - 5)(2h + 5)$

57. $\left(\dfrac{1}{2} - t\right)\left(\dfrac{1}{2} + t\right)$

58. $\left(r + \dfrac{1}{4}\right)\left(r - \dfrac{1}{4}\right)$

59. $(u^3 + 5v)(u^3 - 5v)$

60. $(8w^2 - x)(8w^2 + x)$

61. $(2 - 3a)(2 + 3a)$

62. $(1 - 4x^2)(1 + 4x^2)$

63. $\left(\dfrac{2}{3} - p\right)\left(\dfrac{2}{3} + p\right)$

64. $\left(\dfrac{1}{8} - q\right)\left(\dfrac{1}{8} + q\right)$

For Exercises 65–76, square the binomials.

65. $(a + b)^2$

66. $(a - b)^2$

67. $(x - y)^2$

68. $(x + y)^2$

69. $(2c + 5)^2$

70. $(5d - 9)^2$

71. $(3t^2 - 4s)^2$

72. $(u^2 + 4v)^2$

73. $(7 - t)^2$

74. $(4 + w)^2$

75. $(3 + 4q)^2$

76. $(2 - 3b)^2$

77. **a.** Evaluate $(2 + 4)^2$ by working within the parentheses first.

   b. Evaluate $2^2 + 4^2$.

   c. Compare the answers to parts (a) and (b) and make a conjecture about $(a + b)^2$ and $a^2 + b^2$.

78. **a.** Evaluate $(6 - 5)^2$ by working within the parentheses first.

   b. Evaluate $6^2 - 5^2$.

   c. Compare the answers to parts (a) and (b) and make a conjecture about $(a - b)^2$ and $a^2 - b^2$.

79. **a.** Simplify $(3x + y)^2$.

   b. Simplify $(3xy)^2$.

   c. Compare the answers to parts (a) and (b) to make a conjecture about $(a + b)^2$ and $(ab)^2$.

Concept 3: Applications to Geometry

80. Find a polynomial expression that represents the area of the rectangle shown in the figure.

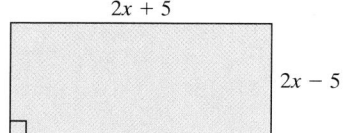

81. Find a polynomial expression that represents the area of the rectangle shown in the figure.

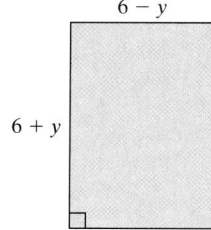

82. Find a polynomial expression that represents the area of the square shown in the figure.

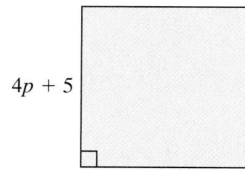

83. Find a polynomial expression that represents the area of the square shown in the figure.

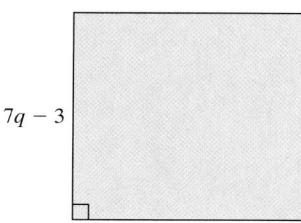

84. Find a polynomial that represents the area of the triangle shown in the figure.

   (Recall: $A = \frac{1}{2}bh$)

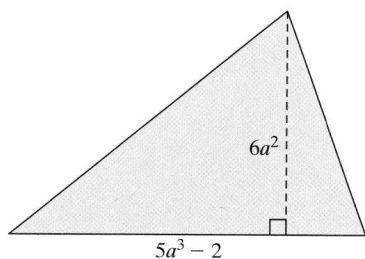

85. Find a polynomial that represents the area of the triangle shown in the figure.

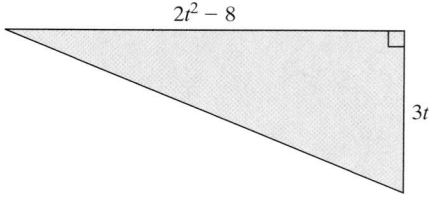

86. Find a polynomial that represents the volume of the cube shown in the figure.

(Recall: $V = s^3$)

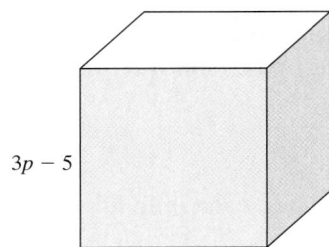

$3p - 5$

87. Find a polynomial that represents the volume of the rectangular solid shown in the figure.

(Recall: $V = lwh$)

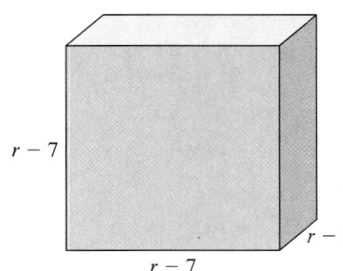

$r - 7$

$r - 1$

$r - 7$

Mixed Exercises

For Exercises 88–117, multiply the expressions.

88. $(7x + y)(7x - y)$

89. $(9w - 4z)(9w + 4z)$

90. $(5s + 3t)^2$

91. $(5s - 3t)^2$

92. $(7x - 3y)(3x - 8y)$

93. $(5a - 4b)(2a - b)$

94. $\left(\dfrac{2}{3}t + 2\right)(3t + 4)$

95. $\left(\dfrac{1}{5}s + 6\right)(5s - 3)$

96. $(5z + 3)(z^2 + 4z - 1)$

97. $(2k - 5)(2k^2 + 3k + 5)$

98. $(3a - 2)(5a + 1 + 2a^2)$

99. $(u + 4)(2 - 3u + u^2)$

100. $(y^2 + 2y + 4)(y - 5)$

101. $(w^2 - w + 6)(w + 2)$

102. $\left(\dfrac{1}{3}m - n\right)^2$

103. $\left(\dfrac{2}{5}p - q\right)^2$

104. $(2x + 3)(x^3 - x^2 + 4x + 1)$

105. $(4y - 1)(2y^3 + 3y^2 - y - 3)$

106. $(4y - 8.1)(4y + 8.1)$

107. $(2h + 2.7)(2h - 2.7)$

108. $(3c^2 + 4)(7c^2 - 8)$

109. $(5k^3 - 9)(k^3 - 2)$

110. $(3.1x + 4.5)^2$

111. $(2.5y + 1.1)^2$

112. $(k - 4)^3$

113. $(h + 3)^3$

114. $(5x + 3)^3$

115. $(2a - 4)^3$

116. $(y^2 + 2y + 1)(2y^2 - y + 3)$

117. $(2w^2 - w - 5)(3w^2 + 2w + 1)$

Expanding Your Skills

For Exercises 118–121, multiply the expressions containing more than two factors.

118. $2a(3a - 4)(a + 5)$

119. $5x(x + 2)(6x - 1)$

120. $(x - 3)(2x + 1)(x - 4)$

121. $(y - 2)(2y - 3)(y + 3)$

122. What binomial when multiplied by $(3x + 5)$ will produce a product of $6x^2 - 11x - 35$? [*Hint:* Let the quantity $(a + b)$ represent the unknown binomial.] Then find a and b such that $(3x + 5)(a + b) = 6x^2 - 11x - 35$.

123. What binomial when multiplied by $(2x - 4)$ will produce a product of $2x^2 + 8x - 24$?

For Exercises 124–126, determine what values of k would create a perfect square trinomial.

124. $x^2 + kx + 25$

125. $w^2 + kw + 9$

126. $a^2 + ka + 16$

Section 5.7   Division of Polynomials

Concepts

1. Division by a Monomial
2. Long Division
3. Synthetic Division

Division of polynomials will be presented in this section as two separate cases. The first case illustrates division by a monomial divisor. The second case illustrates long division by a polynomial with two or more terms.

1. Division by a Monomial

To divide a polynomial by a monomial, divide each individual term in the polynomial by the divisor and simplify the result.

Dividing a Polynomial by a Monomial

If a, b, and c are polynomials such that $c \neq 0$, then

$$\frac{a + b}{c} = \frac{a}{c} + \frac{b}{c} \qquad \text{Similarly,} \qquad \frac{a - b}{c} = \frac{a}{c} - \frac{b}{c}$$

Example 1   **Dividing a Polynomial by a Monomial**

Divide the polynomials.

a. $\dfrac{5a^3 - 10a^2 + 20a}{5a}$ **b.** $(12y^2z^3 - 15yz^2 + 6y^2z) \div (-6y^2z)$

Solution:

a. $\dfrac{5a^3 - 10a^2 + 20a}{5a}$

$= \dfrac{5a^3}{5a} - \dfrac{10a^2}{5a} + \dfrac{20a}{5a}$ Divide each term in the numerator by $5a$.

$= a^2 - 2a + 4$ Simplify each term using the properties of exponents.

b. $(12y^2z^3 - 15yz^2 + 6y^2z) \div (-6y^2z)$

$= \dfrac{12y^2z^3 - 15yz^2 + 6y^2z}{-6y^2z}$

$= \dfrac{12y^2z^3}{-6y^2z} - \dfrac{15yz^2}{-6y^2z} + \dfrac{6y^2z}{-6y^2z}$ Divide each term by $-6y^2z$.

$= -2z^2 + \dfrac{5z}{2y} - 1$ Simplify each term.

Skill Practice   Divide the polynomials.

1. $(36a^4 - 48a^3 + 12a^2) \div (6a^3)$ **2.** $\dfrac{-15x^3y^4 + 25x^2y^3 - 5xy^2}{-5xy^2}$

Skill Practice Answers

1. $6a - 8 + \dfrac{2}{a}$

2. $3x^2y^2 - 5xy + 1$

2. Long Division

If the divisor has two or more terms, a *long division* process similar to the division of real numbers is used. Take a minute to review the long division process for real numbers by dividing 5074 by 31.

$$
\begin{array}{r}
31\overline{)5074}
\end{array}
\qquad
\begin{array}{r}
1 \\
31\overline{)5074} \\
-31 \\
\hline
19
\end{array}
\quad \text{Subtract.}
$$

$$
\begin{array}{r}
16 \\
31\overline{)5074} \\
-31 \\
\hline
197 \\
-186 \\
\hline
11
\end{array}
$$

Bring down next column and repeat the process.

Subtract.

$$
\begin{array}{r}
163 \\
31\overline{)5074} \\
-31 \\
\hline
197 \\
-186 \\
\hline
114 \\
-93 \\
\hline
21
\end{array}
$$

Quotient

Remainder

Therefore, $5074 \div 31 = 163\frac{21}{31}$.

A similar procedure is used for long division of polynomials, as shown in Example 2.

Example 2 **Using Long Division to Divide Polynomials**

Divide the polynomials using long division: $(2x^2 - x + 3) \div (x - 3)$

Solution:

$$x - 3\overline{)2x^2 - x + 3}$$

Divide the leading term in the dividend by the leading term in the divisor.

$$\frac{2x^2}{x} = 2x$$

This is the first term in the quotient.

$$
\begin{array}{r}
2x \\
x - 3\overline{)2x^2 - x + 3} \\
-(2x^2 - 6x)
\end{array}
$$

Multiply $2x$ by the divisor $2x(x - 3) = 2x^2 - 6x$ and subtract the result.

$$
\begin{array}{r}
2x \\
x - 3\overline{)2x^2 - x + 3} \\
-2x^2 + 6x \\
\hline
5x
\end{array}
$$

Subtract the quantity $2x^2 - 6x$. To do this, add the opposite.

$$
\begin{array}{r}
2x + 5 \\
x - 3\overline{)2x^2 - x + 3} \\
-2x^2 + 6x \\
\hline
5x + 3
\end{array}
$$

Bring down the next column, and repeat the process.

Divide the leading term by x: $(5x)/x = 5$. Place 5 in the quotient.

$$\begin{array}{r} 2x + 5 \\ x - 3\overline{)2x^2 - x + 3} \\ -2x^2 + 6x \\ \hline 5x + 3 \\ -(5x - 15) \end{array}$$

Multiply the divisor by 5: $5(x - 3) = 5x - 15$ and subtract the result.

$$\begin{array}{r} 2x + 5 \\ x - 3\overline{)2x^2 - x + 3} \\ -2x^2 + 6x \\ \hline 5x + 3 \\ -5x + 15 \\ \hline 18 \end{array}$$

Subtract the quantity $5x - 15$ by adding the opposite.

The remainder is 18.

Summary:

The quotient is $2x + 5$
The remainder is   18
The divisor is $x - 3$
The dividend is $2x^2 - x + 3$

The solution to a long division problem is usually written in the form:

$$\text{quotient} + \frac{\text{remainder}}{\text{divisor}}$$

Hence,

$$(2x^2 - x + 3) \div (x - 3) = 2x + 5 + \frac{18}{x - 3}$$

Skill Practice Divide the polynomials using long division.

3. $(3x^2 + 2x - 5) \div (x + 2)$

The division of polynomials can be checked in the same fashion as the division of real numbers. To check Example 2, we have

$$\text{Dividend} = (\text{divisor})(\text{quotient}) + \text{remainder}$$

$$2x^2 - x + 3 \overset{?}{=} (x - 3)(2x + 5) + (18)$$

$$\overset{?}{=} 2x^2 + 5x - 6x - 15 + (18)$$

$$= 2x^2 - x + 3 \ ✔$$

Example 3 Using Long Division to Divide Polynomials

Divide the polynomials using long division: $(2w^3 + 8w^2 - 16) \div (2w + 4)$

Solution:

First note that the dividend has a missing power of w and can be written as $2w^3 + 8w^2 + 0w - 16$. The term $0w$ is a place holder for the missing term. It is helpful to use the place holder to keep the powers of w lined up.

$$\begin{array}{r} w^2 \\ 2w + 4\overline{)2w^3 + 8w^2 + 0w - 16} \\ -(2w^3 + 4w^2) \end{array}$$

Divide $2w^3 \div 2w = w^2$. This is the first term of the quotient.

Then multiply $w^2(2w + 4) = 2w^3 + 4w^2$.

Skill Practice Answers

3. $3x - 4 + \dfrac{3}{x + 2}$

$$\begin{array}{r} w^2 \\ 2w + 4\overline{)2w^3 + 8w^2 + 0w - 16} \\ \underline{-2w^3 - 4w^2} \\ 4w^2 + 0w \end{array}$$

— Subtract by adding the opposite.

— Bring down the next column, and repeat the process.

$$\begin{array}{r} w^2 + 2w \\ 2w + 4\overline{)2w^3 + 8w^2 + 0w - 16} \\ \underline{-2w^3 - 4w^2} \\ 4w^2 + 0w \\ -(4w^2 + 8w) \end{array}$$

— Divide $4w^2$ by the leading term in the divisor. $4w^2 \div 2w = 2w$. Place $2w$ in the quotient.

— Multiply $2w(2w + 4) = 4w^2 + 8w$.

$$\begin{array}{r} w^2 + 2w \\ 2w + 4\overline{)2w^3 + 8w^2 + 0w - 16} \\ \underline{-2w^3 - 4w^2} \\ 4w^2 + 0w \\ \underline{-4w^2 - 8w} \\ -8w - 16 \end{array}$$

— Subtract by adding the opposite.

— Bring down the next column and repeat.

$$\begin{array}{r} w^2 + 2w - 4 \\ 2w + 4\overline{)2w^3 + 8w^2 + 0w - 16} \\ \underline{-2w^3 - 4w^2} \\ 4w^2 + 0w \\ \underline{-4w^2 - 8w} \\ -8w - 16 \\ -(-8w - 16) \end{array}$$

— Divide $-8w$ by the leading term in the divisor. $-8w \div 2w = -4$. Place -4 in the quotient.

— Multiply $-4(2w + 4) = -8w - 16$.

$$\begin{array}{r} w^2 + 2w - 4 \\ 2w + 4\overline{)2w^3 + 8w^2 + 0w - 16} \\ \underline{-2w^3 - 4w^2} \\ 4w^2 + 0w \\ \underline{-4w^2 - 8w} \\ -8w - 16 \\ \underline{8w + 16} \\ 0 \end{array}$$

— Subtract by adding the opposite.

— The remainder is 0.

The quotient is $w^2 + 2w - 4$, and the remainder is 0.

Skill Practice Divide the polynomials using long division.

4. $\dfrac{4x^3 - 11x - 5}{2x + 1}$

In Example 3, the remainder is zero. Therefore, we say that $2w + 4$ divides *evenly* into $2w^3 + 8w^2 - 16$. For this reason, the divisor and quotient are factors of $2w^3 + 8w^2 - 16$. To check, we have

$$\text{Dividend} = (\text{divisor})(\text{quotient}) + \text{remainder}$$

$$2w^3 + 8w^2 - 16 \stackrel{?}{=} (2w + 4)(w^2 + 2w - 4) + 0$$

$$\stackrel{?}{=} 2w^3 + 4w^2 - 8w + 4w^2 + 8w - 16$$

$$= 2w^3 + 8w^2 - 16 ✔$$

Skill Practice Answers

4. $2x^2 - x - 5$

Example 4 Using Long Division to Divide Polynomials

Divide the polynomials using long division.

$$\frac{2y + y^4 - 5}{1 + y^2}$$

Solution:

First note that both the dividend and divisor should be written in descending order:

$$\frac{y^4 + 2y - 5}{y^2 + 1}$$

Also note that the dividend and the divisor have missing powers of y. Leave place holders.

$$y^2 + 0y + 1 \overline{)y^4 + 0y^3 + 0y^2 + 2y - 5}$$

$$\begin{array}{r} y^2 \\ y^2 + 0y + 1 \overline{)y^4 + 0y^3 + 0y^2 + 2y - 5} \\ -(y^4 + 0y^3 + y^2) \end{array}$$
  Divide $y^4 \div y^2 = y^2$. This is the first term of the quotient.

  Multiply $y^2(y^2 + 0y + 1) = y^4 + 0y^3 + y^2$

$$\begin{array}{r} y^2 \\ y^2 + 0y + 1 \overline{)y^4 + 0y^3 + 0y^2 + 2y - 5} \\ \underline{-y^4 - 0y^3 - y^2} \\ -y^2 + 2y - 5 \end{array}$$
  Subtract by adding the opposite.

  Bring down the next columns.

$$\begin{array}{r} y^2 \quad -1 \\ y^2 + 0y + 1 \overline{)y^4 + 0y^3 + 0y^2 + 2y - 5} \\ \underline{-y^4 - 0y^3 - y^2} \\ -y^2 + 2y - 5 \\ -(-y^2 - 0y - 1) \end{array}$$
  Divide $-y^2 \div y^2 = -1$.

  Multiply $-1(y^2 + 0y + 1) = -y^2 - 0y - 1$.

$$\begin{array}{r} y^2 \quad -1 \\ y^2 + 0y + 1 \overline{)y^4 + 0y^3 + 0y^2 + 2y - 5} \\ \underline{-y^4 - 0y^3 - y^2} \\ -y^2 + 2y - 5 \\ \underline{y^2 + 0y + 1} \\ 2y - 4 \end{array}$$
  Subtract by adding the opposite.

  Remainder

Therefore, $\dfrac{y^4 + 2y - 5}{y^2 + 1} = y^2 - 1 + \dfrac{2y - 4}{y^2 + 1}$.

Skill Practice Divide the polynomials using long division.

5. $(4 - x^2 + x^3) \div (2 + x^2)$

Skill Practice Answers

5. $x - 1 + \dfrac{-2x + 6}{x^2 + 2}$

Example 5	Determining Whether Long Division Is Necessary

Determine whether long division is necessary for each division of polynomials.

a. $\dfrac{2p^5 - 8p^4 + 4p - 16}{p^2 - 2p + 1}$

b. $\dfrac{2p^5 - 8p^4 + 4p - 16}{2p^2}$

c. $(3z^3 - 5z^2 + 10) \div (15z^3)$

d. $(3z^3 - 5z^2 + 10) \div (3z + 1)$

TIP: Recall that

- Long division is used when the divisor has *two or more terms.*
- If the divisor has *one term,* then divide each term in the dividend by the monomial divisor.

Solution:

a. $\dfrac{2p^5 - 8p^4 + 4p - 16}{p^2 - 2p + 1}$ The divisor has three terms. Use long division.

b. $\dfrac{2p^5 - 8p^4 + 4p - 16}{2p^2}$ The divisor has one term. No long division.

c. $(3z^3 - 5z^2 + 10) \div (15z^3)$ The divisor has one term. No long division.

d. $(3z^3 - 5z^2 + 10) \div (3z + 1)$ The divisor has two terms. Use long division.

Skill Practice	Determine whether long division is necessary for each division of polynomials.

6. $\dfrac{6x^3 - x^2 + 3x - 5}{2x + 3}$

7. $\dfrac{4x^4 - 3x^2}{2x^2}$

8. $(4y^3 - 3y^2 + 1) \div (7y)$

9. $(p^2 - p - 12) \div (p - 4)$

Skill Practice Answers

6. Long division
7. No long division
8. No long division
9. Long division

3. Synthetic Division

In this section we introduced the process of long division to divide two polynomials. Next, we will learn another technique, called **synthetic division**, to divide two polynomials. Synthetic division may be used when dividing a polynomial by a first-degree divisor of the form $x - r$, where r is a constant. Synthetic division is considered a "shortcut" because it uses the coefficients of the divisor and dividend without writing the variables.

Consider dividing the polynomials $(3x^2 - 14x - 10) \div (x - 2)$.

$$
\begin{array}{r}
3x - 8 \\
x - 2 \overline{\smash{)}\, 3x^2 - 14x - 10} \\
\underline{-(3x^2 - 6x)} \\
-8x - 10 \\
\underline{-(-8x + 16)} \\
-26
\end{array}
$$

First note that the divisor $x - 2$ is in the form $x - r$, where $r = 2$. Hence synthetic division can also be used to find the quotient and remainder.

Step 1: Write the value of r in a box. $\longrightarrow$

$$\underline{2|\;3\;\;-14\;\;-10}$$
$$3$$

$\longleftarrow$ **Step 2:** Write the coefficients of the dividend to the right of the box.

Step 3: Skip a line and draw a horizontal line below the list of coefficients.

Step 4: Bring down the leading coefficient from the dividend and write it below the line.

Step 5: Multiply the value of r by the number below the line $(2 \times 3 = 6)$. Write the result in the next column above the line.

$$\underline{2|\;3\;\;-14\;\;-10}$$
$$6$$
$$3\;\;\;\;-8$$

Step 6: Add the numbers in the column above the line $(-14 + 6)$, and write the result below the line.

Repeat steps 5 and 6 until all columns have been completed.

Step 7: To get the final result, we use the numbers below the line. The number in the last column is the remainder. The other numbers are the coefficients of the quotient.

$$\underline{2|\;3\;\;-14\;\;-10}$$
$$6\;\;-16$$
$$3\;\;\;\;-8\;\;\boxed{-26}$$

Quotient: $3x - 8$,   remainder $= -26$

A box is usually drawn around the remainder.

The degree of the quotient will always be 1 less than that of the dividend. Because the dividend is a second-degree polynomial, the quotient will be a first-degree polynomial. In this case, the quotient is $3x - 8$ and the remainder is -26.

Example 6   **Using Synthetic Division to Divide Polynomials**

Divide the polynomials $(5x + 4x^3 - 6 + x^4) \div (x + 3)$ by using synthetic division.

Solution:

As with long division, the terms of the dividend and divisor should be written in descending order. Furthermore, missing powers must be accounted for by using placeholders (shown here in bold).

$$5x + 4x^3 - 6 + x^4$$
$$= x^4 + 4x^3 + \mathbf{0}x^2 + 5x - 6$$

To use synthetic division, the divisor must be in the form $(x - r)$. The divisor $x + 3$ can be written as $x - (-3)$. Hence, $r = -3$.

Step 1: Write the value of r in a box. ⟶ **Step 2:** Write the coefficients of the dividend to the right of the box.

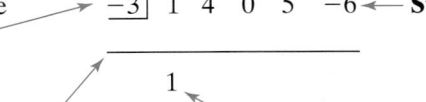

$$-3 | \quad 1 \quad 4 \quad 0 \quad 5 \quad -6$$
$$\overline{ \quad 1}$$

Step 3: Skip a line and draw a horizontal line below the list of coefficients.

Step 4: Bring down the leading coefficient from the dividend and write it below the line.

Step 5: Multiply the value of r by the number below the line $(-3 \times 1 = -3)$. Write the result in the next column above the line.

$$-3 | \quad 1 \quad 4 \quad 0 \quad 5 \quad -6$$
$$ \quad -3$$
$$\overline{ \quad 1 \quad 1}$$

Step 6: Add the numbers in the column above the line: $4 + (-3) = 1$.

Repeat steps 5 and 6:

$$-3 | \quad 1 \quad 4 \quad 0 \quad 5 \quad -6$$
$$ \quad -3 \; -3 \quad 9 \; -42$$
$$\overline{ \quad 1 \quad 1 \; -3 \; 14 \; \boxed{-48}}$$

⟵ remainder
constant
x-term coefficient
x^2-term coefficient
x^3-term coefficient

The quotient is

$x^3 + x^2 - 3x + 14$.

The remainder is -48.

The solution is $x^3 + x^2 - 3x + 14 + \dfrac{-48}{x + 3}$

Skill Practice Divide the polynomials by using synthetic division. Identify the quotient and the remainder.

10. $(5y^2 - 4y + 2y^3 - 5) \div (y + 3)$

TIP: It is interesting to compare the long division process to the synthetic division process. For Example 6, long division is shown on the left, and synthetic division is shown on the right. Notice that the same pattern of coefficients used in long division appears in the synthetic division process.

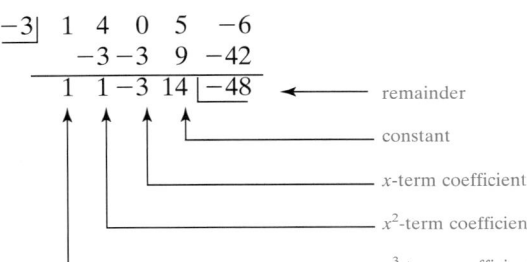

$$\begin{array}{r} x^3 + x^2 - 3x + 14 \\ x + 3 \overline{)\, x^4 + 4x^3 + 0x^2 + 5x - 6} \\ \underline{-(x^4 + 3x^3)} \\ x^3 + 0x^2 \\ \underline{-(x^3 + 3x^2)} \\ -3x^2 + 5x \\ \underline{-(-3x^2 - 9x)} \\ 14x - 6 \\ \underline{-(14x + 42)} \\ -48 \end{array}$$

$$-3 | \quad 1 \quad 4 \quad 0 \quad 5 \quad -6$$
$$ \quad -3 \; -3 \quad 9 \; -42$$
$$\overline{ \quad 1 \quad 1 \; -3 \; 14 \; \boxed{-48}}$$

$x^3 \quad x^2 \quad x$ constant remainder

Quotient: $x^3 + x^2 - 3x + 14$
Remainder: -48

Skill Practice Answers

10. Quotient: $2y^2 - y - 1$;
remainder: -2

Example 7	Using Synthetic Division to Divide Polynomials

Divide the polynomials by using synthetic division. Identify the quotient and remainder.

a. $(2m^7 - 3m^5 + 4m^4 - m + 8) \div (m + 2)$

b. $(p^4 - 81) \div (p - 3)$

Solution:

a. Insert placeholders (bold) for missing powers of m.

$(2m^7 - 3m^5 + 4m^4 - m + 8) \div (m + 2)$

$(2m^7 + \mathbf{0m^6} - 3m^5 + 4m^4 + \mathbf{0m^3} + \mathbf{0m^2} - m + 8) \div (m + 2)$

Because $m + 2$ can be written as $m - (-2), r = -2$.

$$
\begin{array}{r|rrrrrrrr}
-2 & 2 & 0 & -3 & 4 & 0 & 0 & -1 & 8 \\
   &   & -4 & 8 & -10 & 12 & -24 & 48 & -94 \\
\hline
   & 2 & -4 & 5 & -6 & 12 & -24 & 47 & \boxed{-86}
\end{array}
$$

Quotient: $2m^6 - 4m^5 + 5m^4 - 6m^3 + 12m^2 - 24m + 47$

Remainder: -86

The quotient is 1 degree less than dividend.

The solution is $2m^6 - 4m^5 + 5m^4 - 6m^3 + 12m^2 - 24m + 47 + \dfrac{-86}{m + 2}$.

b. $(p^4 - 81) \div (p - 3)$

$(p^4 + \mathbf{0p^3} + \mathbf{0p^2} + \mathbf{0p} - 81) \div (p - 3)$ Insert placeholders (bold) for missing powers of p.

$$
\begin{array}{r|rrrrr}
3 & 1 & 0 & 0 & 0 & -81 \\
  &   & 3 & 9 & 27 & 81 \\
\hline
  & 1 & 3 & 9 & 27 & \boxed{0}
\end{array}
$$

Quotient: $p^3 + 3p^2 + 9p + 27$

Remainder: 0

The solution is $p^3 + 3p^2 + 9p + 27$.

Skill Practice	Divide the polynomials by using synthetic division. Identify the quotient and the remainder.

11. $(4c^4 - 3c^2 - 6c - 3) \div (c - 2)$ **12.** $(x^3 + 1) \div (x + 1)$

Skill Practice Answers

11. Quotient: $4c^3 + 8c^2 + 13c + 20$;
remainder: 37

12. Quotient: $x^2 - x + 1$;
remainder: 0

Review Exercises

For Exercises 1–10, perform the indicated operations.

1. $(6z^5 - 2z^3 + z - 6) - (10z^4 + 2z^3 + z^2 + z)$

2. $(7a^2 + a - 6) + (2a^2 + 5a + 11)$

3. $(10x + y)(x - 3y)$

4. $8b^2(2b^2 - 5b + 12)$

5. $(10x + y) + (x - 3y)$

6. $(2w^3 + 5)^2$

7. $\left(\frac{4}{3}y^2 - \frac{1}{2}y + \frac{3}{8}\right) - \left(\frac{1}{3}y^2 + \frac{1}{4}y - \frac{1}{8}\right)$

8. $\left(\frac{7}{8}w - 1\right)\left(\frac{7}{8}w + 1\right)$

9. $(a + 3)(a^2 - 3a + 9)$

10. $(2x + 1)(5x - 3)$

Concept 1: Division by a Monomial

11. There are two methods for dividing polynomials. Explain when long division is used.

12. Explain how to check a polynomial division problem.

13. a. Divide $\dfrac{15t^3 + 18t^2}{3t}$

14. a. Divide $(-9y^4 + 6y^2 - y) \div (3y)$

    b. Check by multiplying the quotient by the divisor.

    b. Check by multiplying the quotient by the divisor.

For Exercises 15–30, divide the polynomials.

15. $(6a^2 + 4a - 14) \div (2)$

16. $\dfrac{4b^2 + 16b - 12}{4}$

17. $\dfrac{-5x^2 - 20x + 5}{-5}$

18. $\dfrac{-3y^3 + 12y - 6}{-3}$

19. $\dfrac{3p^3 - p^2}{p}$

20. $(7q^4 + 5q^2) \div q$

21. $(4m^2 + 8m) \div 4m^2$

22. $\dfrac{n^2 - 8}{n}$

23. $\dfrac{14y^4 - 7y^3 + 21y^2}{-7y^2}$

24. $(25a^5 - 5a^4 + 15a^3 - 5a) \div (-5a)$

25. $(4x^3 - 24x^2 - x + 8) \div (4x)$

26. $\dfrac{20w^3 + 15w^2 - w + 5}{10w}$

27. $\dfrac{-a^3b^2 + a^2b^2 - ab^3}{-a^2b^2}$

28. $(3x^4y^3 - x^2y^2 - xy^3) \div (-x^2y^2)$

29. $(6t^4 - 2t^3 + 3t^2 - t + 4) \div (2t^3)$

30. $\dfrac{2y^3 - 2y^2 + 3y - 9}{2y^2}$

Concept 2: Long Division

31. a. Divide $(z^2 + 7z + 11) \div (z + 5)$

   b. Check by multiplying the quotient by the divisor and adding the remainder.

32. a. Divide $\dfrac{2w^2 - 7w + 3}{w - 4}$

   b. Check by multiplying the quotient by the divisor and adding the remainder.

For Exercises 33–54, divide the polynomials.

33. $\dfrac{t^2 + 4t + 3}{t + 1}$

34. $(3x^2 + 8x + 4) \div (x + 2)$

35. $(7b^2 - 3b - 4) \div (b - 1)$

36. $\dfrac{w^2 - w - 2}{w - 2}$

37. $\dfrac{5k^2 - 29k - 6}{5k + 1}$

38. $(4y^2 + 25y - 21) \div (4y - 3)$

39. $(4p^3 + 12p^2 + p - 12) \div (2p + 3)$

40. $\dfrac{12a^3 - 2a^2 - 17a - 5}{3a + 1}$

41. $\dfrac{-k - 6 + k^2}{1 + k}$

42. $(1 + h^2 + 3h) \div (2 + h)$

43. $(4x^3 - 8x^2 + 15x - 16) \div (2x - 3)$

44. $\dfrac{3b^3 + b^2 + 17b - 49}{3b - 5}$

45. $\dfrac{9 + a^2}{a + 3}$

46. $(3 + m^2) \div (m + 3)$

47. $(4x^3 - 3x - 26) \div (x - 2)$

48. $(4y^3 + y + 1) \div (2y + 1)$

49. $(w^4 + 5w^3 - 5w^2 - 15w + 7) \div (w^2 - 3)$

50. $\dfrac{p^4 - p^3 - 4p^2 - 2p - 15}{p^2 + 2}$

51. $\dfrac{2n^4 + 5n^3 - 11n^2 - 20n + 12}{2n^2 + 3n - 2}$

52. $(6y^4 - 5y^3 - 8y^2 + 16y - 8) \div (2y^2 - 3y + 2)$

53. $(5x^3 - 4x - 9) \div (5x^2 + 5x + 1)$

54. $\dfrac{3a^3 - 5a + 16}{3a^2 - 6a + 7}$

55. Show that $(x^3 - 8) \div (x - 2)$ is *not* $(x^2 + 4)$.

56. Explain why $(y^3 + 27) \div (y + 3)$ is *not* $(y^2 + 9)$.

Concept 3: Synthetic Division

57. Explain the conditions under which you may use synthetic division to divide polynomials.

58. Can synthetic division be used to divide $(4x^4 + 3x^3 - 7x + 9)$ by $(2x + 5)$? Explain why or why not.

59. Can synthetic division be used to divide $(6x^5 - 3x^2 + 2x - 14)$ by $(x^2 - 3)$? Explain why or why not.

60. Can synthetic division be used to divide $(3x^4 - x + 1)$ by $(x - 5)$? Explain why or why not.

61. The following table represents the result of a synthetic division.

$$5\underline{|}\ \ 1\ \ -2\ \ -4\ \ \ 3$$
$$\underline{\ \ \ \ \ \ \ 5\ \ \ 15\ \ 55}$$
$$1\ \ \ \ 3\ \ \ 11\ \ \ \underline{|58}$$

Use x as the variable.

a. Identify the divisor.

b. Identify the quotient.

c. Identify the remainder.

62. The following table represents the result of a synthetic division.

$$-2\underline{|}\ \ 2\ \ \ \ 3\ \ \ 0\ \ -1\ \ \ \ 6$$
$$\underline{\ \ \ \ \ \ \ -4\ \ \ 2\ \ -4\ \ \ 10}$$
$$2\ \ -1\ \ \ 2\ \ -5\ \ \ \underline{|16}$$

Use x as the variable.

a. Identify the divisor.

b. Identify the quotient.

c. Identify the remainder.

For Exercises 63–74, divide by using synthetic division. Check your answer by multiplication.

63. $(x^2 - 2x - 48) \div (x - 8)$

64. $(x^2 - 4x - 12) \div (x - 6)$

65. $(t^2 - 3t - 4) \div (t + 1)$

66. $(h^2 + 7h + 12) \div (h + 3)$

67. $(5y^2 + 5y + 1) \div (y - 1)$

68. $(3w^2 + w - 5) \div (w + 2)$

69. $(3 + 7y^2 - 4y + 3y^3) \div (y + 3)$

70. $(2z - 2z^2 + z^3 - 5) \div (z + 3)$

71. $(x^3 - 3x^2 + 4) \div (x - 2)$

72. $(3y^4 - 25y^2 - 18) \div (y - 3)$

73. $(4w^4 - w^2 + 6w - 3) \div \left(w - \dfrac{1}{2}\right)$

74. $(-12y^4 - 5y^3 - y^2 + y + 3) \div \left(y + \dfrac{3}{4}\right)$

Mixed Exercises

For Exercises 75–86, determine which method to use to divide the polynomials: monomial division or long division. Then use that method to divide the polynomials.

75. $\dfrac{9a^3 + 12a^2}{3a}$

76. $\dfrac{3y^2 + 17y - 12}{y + 6}$

77. $(p^3 + p^2 - 4p - 4) \div (p^2 - p - 2)$

78. $(q^3 + 1) \div (q + 1)$

79. $\dfrac{t^4 + t^2 - 16}{t + 2}$

80. $\dfrac{-8m^5 - 4m^3 + 4m^2}{-2m^2}$

81. $(w^4 + w^2 - 5) \div (w^2 - 2)$

82. $(2k^2 + 9k + 7) \div (k + 1)$

83. $\dfrac{n^3 - 64}{n - 4}$

84. $\dfrac{15s^2 + 34s + 28}{5s + 3}$

85. $(9r^3 - 12r^2 + 9) \div (-3r^2)$

86. $(6x^4 - 16x^3 + 15x^2 - 5x + 10) \div (3x + 1)$

Expanding Your Skills

87. Given $P(x) = 4x^3 + 10x^2 - 8x - 20$,

a. Evaluate $P(-4)$.

b. Divide. $(4x^3 + 10x^2 - 8x - 20) \div (x + 4)$

c. Compare the value found in part (a) to the remainder found in part (b).

88. Given $P(x) = -3x^3 - 12x^2 + 5x - 8$,

    a. Evaluate $P(-6)$.

    b. Divide. $(-3x^3 - 12x^2 + 5x - 8) \div (x + 6)$

    c. Compare the value found in part (a) to the remainder found in part (b).

For Exercises 89–96, divide the polynomials and note any patterns.

89. $(x^2 - 1) \div (x - 1)$ **90.** $(x^3 - 1) \div (x - 1)$ **91.** $(x^4 - 1) \div (x - 1)$

92. $(x^5 - 1) \div (x - 1)$ **93.** $x^2 \div (x - 1)$ **94.** $x^3 \div (x - 1)$

95. $x^4 \div (x - 1)$ **96.** $x^5 \div (x - 1)$

Chapter 5   Problem Recognition Exercises—Operations on Polynomials

Perform the indicated operations and simplify.

1. $(2x - 4)(x^2 - 2x + 3)$ **2.** $(3y^2 + 8)(-y^2 - 4)$

3. $(2x - 4) + (x^2 - 2x + 3)$

4. $(3y^2 + 8) - (-y^2 - 4)$ **5.** $(6y - 7)^2$

6. $(3z + 2)^2$ **7.** $(6y - 7)(6y + 7)$

8. $(3z + 2)(3z - 2)$

9. $(-2x^4 - 6x^3 + 8x^2) \div (2x^2)$

10. $(-15m^3 + 12m^2 - 3m) \div (-3m)$

11. $(4x + y)^2$ **12.** $(2a + b)^2$

13. $(4xy)^2$ **14.** $(2ab)^2$

15. $(m^3 - 4m^2 - 6) - (3m^2 + 7m) + (-m^3 - 9m + 6)$

16. $(n^4 + 2n^2 - 3n) + (4n^2 + 2n - 1) - (4n^5 + 6n - 3)$

17. $(8x^3 + 2x + 6) \div (x - 2)$

18. $(-4x^3 + 2x^2 - 5) \div (x - 3)$

19. $(2x - y)(3x^2 + 4xy - y^2)$

20. $(3a + b)(2a^2 - ab + 2b^2)$

21. $(x + y^2)(x^2 - xy^2 + y^4)$

22. $(m^2 + 1)(m^4 - m^2 + 1)$

23. $(a^2 + 2b) - (a^2 - 2b)$ **24.** $(y^3 - 6z) - (y^3 + 6z)$

25. $(a^2 + 2b)(a^2 - 2b)$ **26.** $(y^3 - 6z)(y^3 + 6z)$

27. $(8u + 3v)^2$ **28.** $(2p - t)^2$

29. $\dfrac{8p^2 + 4p - 6}{2p - 1}$ **30.** $\dfrac{4v^2 - 8v + 8}{2v + 3}$

31. $\dfrac{12x^3y^7}{3xy^5}$ **32.** $\dfrac{-18p^2q^4}{2pq^3}$

33. $(2a - 9)(5a - 6)$ **34.** $(7a + 1)(4a - 3)$

35. $\left(\dfrac{3}{7}x - \dfrac{1}{2}\right)\left(\dfrac{3}{7}x + \dfrac{1}{2}\right)$ **36.** $\left(\dfrac{2}{5}y + \dfrac{4}{3}\right)\left(\dfrac{2}{5}y - \dfrac{4}{3}\right)$

37. $\left(\dfrac{1}{9}x^3 + \dfrac{2}{3}x^2 + \dfrac{1}{6}x - 3\right) - \left(\dfrac{4}{3}x^3 + \dfrac{1}{9}x^2 + \dfrac{2}{3}x + 1\right)$

38. $\left(\dfrac{1}{10}y^2 - \dfrac{3}{5}y - \dfrac{1}{15}\right) - \left(\dfrac{7}{5}y^2 + \dfrac{3}{10}y - \dfrac{1}{3}\right)$

39. $(0.05x^2 - 0.16x - 0.75) + (1.25x^2 - 0.14x + 0.25)$

40. $(1.6w^3 + 2.8w + 6.1) + (3.4w^3 - 4.1w^2 - 7.3)$

Chapter 5 SUMMARY

Section 5.1 Exponents: Multiplying and Dividing Common Bases

Key Concepts

Definition

$$b^n = \underbrace{b \cdot b \cdot b \cdot b \cdots b}_{n \text{ factors of } b}$$

b is the base,
n is the exponent

Multiplying Common Bases

$$a^m a^n = a^{m+n}$$

Dividing Common Bases

$$\frac{a^m}{a^n} = a^{m-n} \quad (a \neq 0)$$

Examples

Example 1

$3^4 = 3 \cdot 3 \cdot 3 \cdot 3 = 81$ 3 is the base,
                                    4 is the exponent

Example 2

Compare: $(-5)^2$ versus -5^2

versus $\begin{cases} (-5)^2 = (-5)(-5) = 25 \\ \\ -5^2 = -1(5^2) = -1(5)(5) = -25 \end{cases}$

Example 3

Simplify: $x^3 \cdot x^4 \cdot x^2 \cdot x = x^{10}$

Example 4

Simplify: $\dfrac{c^4 d^{10}}{cd^5} = c^{4-1} d^{10-5} = c^3 d^5$

Section 5.2 More Properties of Exponents

Key Concepts

Power Rule for Exponents

$$(a^m)^n = a^{mn} \quad (a \neq 0, m, n \text{ positive integers})$$

Power of a Product and Power of a Quotient

Assume m and n are positive integers and a and b are real numbers where $b \neq 0$.

$$(ab)^m = a^m b^m \quad \text{and} \quad \left(\frac{a}{b}\right)^m = \frac{a^m}{b^m}$$

Examples

Example 1

Simplify: $(x^4)^5 = x^{20}$

Example 2

Simplify: $(4uv^2)^3 = 4^3 u^3 (v^2)^3 = 64u^3 v^6$

Example 3

Simplify: $\left(\dfrac{p^5 q^3}{5pq^2}\right)^2 = \left(\dfrac{p^{5-1} q^{3-2}}{5}\right)^2 = \left(\dfrac{p^4 q}{5}\right)^2$

$$= \frac{p^8 q^2}{25}$$

Section 5.3 Definitions of b^0 and b^{-n}

Key Concepts

Definitions

If b is a real number such that $b \neq 0$ and n is an integer, then:

1. $b^0 = 1$

2. $b^{-n} = \left(\dfrac{1}{b}\right)^n = \dfrac{1}{b^n}$

Examples

Example 1

Simplify: $4^0 = 1$

Example 2

Simplify: $y^{-7} = \dfrac{1}{y^7}$

Example 3

Simplify: $\left(\dfrac{2a^3 b}{a^{-2} c^{-4}}\right)^{-2}$

$$= \left(\frac{2a^{3-(-2)}b}{c^{-4}}\right)^{-2}$$

$$= \left(\frac{2a^5 b}{c^{-4}}\right)^{-2} = \frac{2^{-2} a^{-10} b^{-2}}{c^8}$$

$$= \frac{1}{2^2 a^{10} b^2 c^8}$$

$$= \frac{1}{4 a^{10} b^2 c^8}$$

Section 5.4 Scientific Notation

Key Concepts

A number written in **scientific notation** is expressed in the form:

$a \times 10^n$ where $1 \leq |a| < 10$ and n is an integer. The value 10^n is sometimes called the **order of magnitude** or simply the magnitude of the number.

Examples

Example 1

Write the numbers in scientific notation:

$35{,}000 = 3.5 \times 10^4$

$0.000\,000\,548 = 5.48 \times 10^{-7}$

Example 2

Multiply: $(3.5 \times 10^4)(2.0 \times 10^{-6})$

$$= 7.0 \times 10^{-2}$$

Example 3

Divide: $\dfrac{8.4 \times 10^{-9}}{2.1 \times 10^3} = 4.0 \times 10^{-9-3} = 4.0 \times 10^{-12}$

Section 5.5 — Addition and Subtraction of Polynomials

Key Concepts

A **polynomial** in one variable is a finite sum of terms of the form ax^n, where a is a real number and the exponent, n, is a nonnegative integer. For each term, a is called the **coefficient** of the term and n is the **degree of the term**. The term with highest degree is the **leading term**, and its coefficient is called the **leading coefficient**. The **degree of the polynomial** is the largest degree of all its terms.

To add or subtract polynomials, add or subtract *like* terms.

Examples

Example 1

Given: $4x^5 - 8x^3 + 9x - 5$

Coefficients of each term:	$4, -8, 9, -5$
Degree of each term:	$5, 3, 1, 0$
Leading term:	$4x^5$
Leading coefficient:	4
Degree of polynomial:	5

Example 2

Perform the indicated operations:

$$(2x^4 - 5x^3 + 1) - (x^4 + 3) + (x^3 - 4x - 7)$$
$$= 2x^4 - 5x^3 + 1 - x^4 - 3 + x^3 - 4x - 7$$
$$= 2x^4 - x^4 - 5x^3 + x^3 - 4x + 1 - 3 - 7$$
$$= x^4 - 4x^3 - 4x - 9$$

Section 5.6 — Multiplication of Polynomials

Key Concepts

Multiplying Monomials

Use the commutative and associative properties of multiplication to group coefficients and like bases.

Multiplying Polynomials

Multiply each term in the first polynomial by each term in the second polynomial.

Product of Conjugates

Results in a **difference of squares**

$$(a + b)(a - b) = a^2 - b^2$$

Square of a Binomial

Results in a **perfect square trinomial**

$$(a + b)^2 = a^2 + 2ab + b^2$$
$$(a - b)^2 = a^2 - 2ab + b^2$$

Examples

Example 1

Multiply: $(5a^2b)(-2ab^3)$
$$= (5 \cdot -2)(a^2a)(bb^3)$$
$$= -10a^3b^4$$

Example 2

Multiply: $(x - 2)(3x^2 - 4x + 11)$
$$= 3x^3 - 4x^2 + 11x - 6x^2 + 8x - 22$$
$$= 3x^3 - 10x^2 + 19x - 22$$

Example 3

Multiply: $(3w - 4v)(3w + 4v)$
$$= (3w)^2 - (4v)^2$$
$$= 9w^2 - 16v^2$$

Example 4

Multiply: $(5c - 8d)^2$
$$= (5c)^2 - 2(5c)(8d) + (8d)^2$$
$$= 25c^2 - 80cd + 64d^2$$

| Section 5.7 | **Division of Polynomials** |

Key Concepts

Division of Polynomials

1. Division by a monomial, use the properties:

$$\frac{a+b}{c} = \frac{a}{c} + \frac{b}{c} \quad \text{and} \quad \frac{a-b}{c} = \frac{a}{c} - \frac{b}{c}$$

2. If the divisor has more than one term, use long division.

3. **Synthetic division** may be used to divide a polynomial by a binomial in the form $x - r$, where r is a constant.

Examples

Example 1

Divide: $\dfrac{-3x^2 - 6x + 9}{-3x}$

$$= \frac{-3x^2}{-3x} - \frac{6x}{-3x} + \frac{9}{-3x}$$

$$= x + 2 - \frac{3}{x}$$

Example 2

Divide: $(3x^2 - 5x + 1) \div (x + 2)$

$$\begin{array}{r} 3x - 11 \\ x+2\overline{)3x^2 - 5x + 1} \\ -\underline{(3x^2 + 6x)} \\ -11x + 1 \\ -\underline{(-11x - 22)} \\ 23 \end{array}$$

$$3x - 11 + \frac{23}{x+2}$$

Example 3

$(3x^2 - 5x + 1) \div (x + 2)$

$$\begin{array}{r} -2\,\underline{|\,3 \quad -5 \quad 1} \\ -6 \quad 22 \\ \hline 3 \quad -11 \quad \underline{|\,23} \end{array}$$

Answer: $3x - 11 + \dfrac{23}{x+2}$

| Chapter 5 | **Review Exercises** |

Section 5.1

For Exercises 1–4, identify the base and the exponent.

1. 5^3 **2.** x^4 **3.** $(-2)^0$ **4.** y

5. Evaluate the expressions.

   a. 6^2 **b.** $(-6)^2$ **c.** -6^2

6. Evaluate the expressions.

   a. 4^3 **b.** $(-4)^3$ **c.** -4^3

For Exercises 7–18, simplify and write the answers in exponent form. Assume that all variables represent nonzero real numbers.

7. $5^3 \cdot 5^{10}$ **8.** $a^7 a^4$

9. $x \cdot x^6 \cdot x^2$ **10.** $6^3 \cdot 6 \cdot 6^5$

11. $\dfrac{10^7}{10^4}$ **12.** $\dfrac{y^{14}}{y^8}$

13. $\dfrac{b^9}{b}$

14. $\dfrac{7^8}{7}$

15. $\dfrac{k^2k^3}{k^4}$

16. $\dfrac{8^4 \cdot 8^7}{8^{11}}$

17. $\dfrac{2^8 \cdot 2^{10}}{2^3 \cdot 2^7}$

18. $\dfrac{q^3q^{12}}{qq^8}$

19. Explain why $2^2 \cdot 4^4$ does *not* equal 8^6.

20. Explain why $\frac{10^5}{5^2}$ does *not* equal 2^3.

For Exercises 21–22, use the formula

$$A = P(1 + r)^t$$

21. Find the amount in an account after 3 years if the initial investment is $6000, invested at 6% interest compounded annually.

22. Find the amount in an account after 2 years if the initial investment is $20,000, invested at 5% interest compounded annually.

Section 5.2

For Exercises 23–40, simplify the expressions. Write the answers in exponent form. Assume all variables represent nonzero real numbers.

23. $(7^3)^4$

24. $(c^2)^6$

25. $(p^4p^2)^3$

26. $(9^5 \cdot 9^2)^4$

27. $\left(\dfrac{a}{b}\right)^2$

28. $\left(\dfrac{1}{3}\right)^4$

29. $\left(\dfrac{5}{c^2d^5}\right)^2$

30. $\left(-\dfrac{m^2}{4n^6}\right)^5$

31. $(2ab^2)^4$

32. $(-x^7y)^2$

33. $\left(\dfrac{-3x^3}{5y^2z}\right)^3$

34. $\left(\dfrac{r^3}{s^2t^6}\right)^5$

35. $\dfrac{a^4(a^2)^8}{(a^3)^3}$

36. $\dfrac{(8^3)^4 \cdot 8^{10}}{(8^4)^5}$

37. $\dfrac{(4h^2k)^2(h^3k)^4}{(2hk^3)^2}$

38. $\dfrac{(p^3q)^3(2p^2q^4)^4}{(8p)(pq^3)^2}$

39. $\left(\dfrac{2x^4y^3}{4xy^2}\right)^2$

40. $\left(\dfrac{a^4b^6}{ab^4}\right)^3$

Section 5.3

For Exercises 41–62, simplify the expressions. Assume all variables represent nonzero real numbers.

41. 8^0

42. $(-b)^0$

43. 1^0

44. $-x^0$

45. $2y^0$

46. $(2y)^0$

47. z^{-5}

48. 10^{-4}

49. $(6a)^{-2}$

50. $6a^{-2}$

51. $4^0 + 4^{-2}$

52. $9^{-1} + 9^0$

53. $t^{-6}t^{-2}$

54. r^8r^{-9}

55. $\dfrac{12x^{-2}y^3}{6x^4y^{-4}}$

56. $\dfrac{8ab^{-3}c^0}{10a^{-5}b^{-4}c^{-1}}$

57. $(-2m^2n^{-4})^{-4}$

58. $(3u^{-5}v^2)^{-3}$

59. $\dfrac{(k^{-6})^{-2}(k^3)}{5k^{-6}k^0}$

60. $\dfrac{(3h)^{-2}(h^{-5})^{-3}}{h^{-4}h^8}$

61. $2 \cdot 3^{-1} - 6^{-1}$

62. $2^{-1} - 2^{-2} + 2^0$

Section 5.4

63. Write the numbers in scientific notation.

   a. In a recent year there were 97,000,000 packages of M&Ms sold in the United States.

   b. The thickness of a piece of paper is 0.0042 in.

   c. The area of the Pacific Ocean is 166,241,000 km^2.

64. Write the numbers without scientific notation.

   a. A pH of 10 means the hydrogen ion concentration is 1×10^{-10} units.

   b. When it was released, *The Lord of the Rings: The Fellowship of the Ring* sold 2.573×10^8 DVDs.

   c. A fund-raising event for neurospinal research raised 2.56×10^5.

For Exercises 65–68, perform the indicated operations. Write the answers in scientific notation.

65. $(4.1 \times 10^{-6})(2.3 \times 10^{11})$

66. $\dfrac{9.3 \times 10^3}{6.0 \times 10^{-7}}$

67. $\dfrac{2000}{0.000008}$

68. $(0.000078)(21,000,000)$

69. Use your calculator to evaluate 5^{20}. Why is scientific notation necessary on your calculator to express the answer?

70. Use your calculator to evaluate $(0.4)^{30}$. Why is scientific notation necessary on your calculator to express the answer?

71. The average distance between the Earth and Sun is 9.3×10^7 miles.

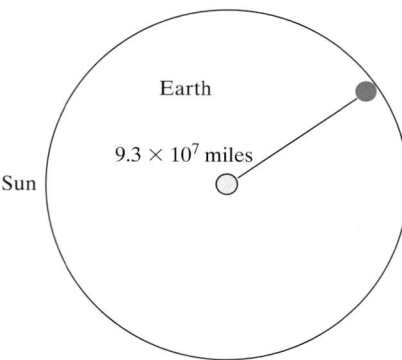

a. If the Earth's orbit is approximated by a circle, find the total distance the Earth travels around the Sun in one orbit. (*Hint:* The circumference of a circle is given by $C = 2\pi r$.) Express the answer in scientific notation.

b. If the Earth makes one complete trip around the Sun in 1 year (365 days $= 8.76 \times 10^3$ hr), find the average speed that the Earth travels around the Sun in miles per hour. Express the answer in scientific notation.

72. The average distance between the planet Mercury and the Sun is 3.6×10^7 miles.

a. If Mercury's orbit is approximated by a circle, find the total distance Mercury travels around the Sun in one orbit. (*Hint:* The circumference of a circle is given by $C = 2\pi r$.) Express the answer in scientific notation.

b. If Mercury makes one complete trip around the Sun in 88 days (2.112×10^3 hr), find the average speed that Mercury travels around the Sun in miles per hour. Express the answer in scientific notation.

Section 5.5

73. For the polynomial $7x^4 - x + 6$

a. Classify as a monomial, a binomial, or a trinomial.

b. Identify the degree of the polynomial.

c. Identify the leading coefficient.

74. For the polynomial $2y^3 - 5y^7$

a. Classify as a monomial, a binomial, or a trinomial.

b. Identify the degree of the polynomial.

c. Identify the leading coefficient.

For Exercises 75–80, add or subtract as indicated.

75. $(4x + 2) + (3x - 5)$

76. $(7y^2 - 11y - 6) - (8y^2 + 3y - 4)$

77. $(9a^2 - 6) - (-5a^2 + 2a)$

78. $(8w^4 - 6w + 3) + (2w^4 + 2w^3 - w + 1)$

79. $\left(5x^3 - \dfrac{1}{4}x^2 + \dfrac{5}{8}x + 2\right) + \left(\dfrac{5}{2}x^3 + \dfrac{1}{2}x^2 - \dfrac{1}{8}x\right)$

80. $(-0.02b^5 + b^4 - 0.7b + 0.3) +$ $(0.03b^5 - 0.1b^3 + b + 0.03)$

81. Subtract $(9x^2 + 4x + 6)$ from $(7x^2 - 5x)$.

82. Find the difference of $(x^2 - 5x - 3)$ and $(6x^2 + 4x + 9)$.

83. Write a trinomial of degree 2 with a leading coefficient of -5. (Answers may vary.)

84. Write a binomial of degree 6 with leading coefficient 6. (Answers may vary.)

85. Find a polynomial that represents the perimeter of the given rectangle.

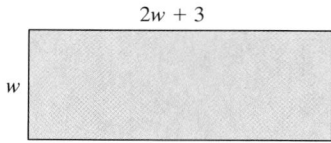

Section 5.6

For Exercises 86–101, multiply the expressions.

86. $(25x^4y^3)(-3x^2y)$ **87.** $(9a^6)(2a^2b^4)$

88. $5c(3c^3 - 7c + 5)$ **89.** $(x^2 + 5x - 3)(-2x)$

90. $(5k - 4)(k + 1)$ **91.** $(4t - 1)(5t + 2)$

92. $(q + 8)(6q - 1)$ **93.** $(2a - 6)(a + 5)$

94. $\left(7a + \dfrac{1}{2}\right)^2$ **95.** $(b - 4)^2$

96. $(4p^2 + 6p + 9)(2p - 3)$

97. $(2w - 1)(-w^2 - 3w - 4)$

98. $(b - 4)(b + 4)$

99. $\left(\dfrac{1}{3}r^4 - s^2\right)\left(\dfrac{1}{3}r^4 + s^2\right)$

100. $(-7z^2 + 6)^2$

101. $(2h + 3)(h^4 - h^3 + h^2 - h + 1)$

102. Find a polynomial that represents the area of the given rectangle.

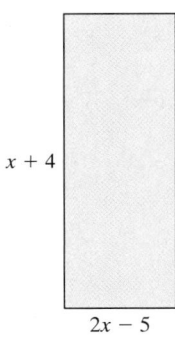

$x + 4$

$2x - 5$

Section 5.7

For Exercises 103–117, divide the polynomials.

103. $\dfrac{20y^3 - 10y^2}{5y}$

104. $(18a^3b^2 - 9a^2b - 27ab^2) \div 9ab$

105. $(12x^4 - 8x^3 + 4x^2) \div (-4x^2)$

106. $\dfrac{10z^7w^4 - 15z^3w^2 - 20zw}{-20z^2w}$

107. $\dfrac{6m^3 + 5m^2 - 6m}{6m}$ **108.** $\dfrac{18n^4 - 6n^3 + 12n^2}{3n^2}$

109. $\dfrac{x^2 + 7x + 10}{x + 5}$ **110.** $(2t^2 + t - 10) \div (t - 2)$

111. $(2p^2 + p - 16) \div (2p + 7)$

112. $\dfrac{5a^2 + 27a - 22}{5a - 3}$ **113.** $\dfrac{b^3 - 125}{b - 5}$

114. $(z^3 + 4z^2 + 5z + 20) \div (5 + z^2)$

115. $(-3y - 4y^3 + 5y^2 + y^4 + 2) \div (y^2 + 3)$

116. $(3t^4 - 8t^3 + t^2 - 4t - 5) \div (3t^2 + t + 1)$

117. $\dfrac{2w^4 + w^3 + 4w - 3}{2w^2 - w + 3}$

Chapter 5   Test

Assume all variables represent nonzero real numbers.

1. Expand the expression using the definition of exponents, then simplify: $\dfrac{3^4 \cdot 3^3}{3^6}$

For Exercises 2–11, simplify the expression. Write the answer with positive exponents only.

2. $9^5 \cdot 9$ **3.** $\dfrac{q^{10}}{q^2}$

4. $(3a^2b)^3$ **5.** $\left(\dfrac{2x}{y^3}\right)^4$

6. $(-7)^0$

7. c^{-3}

8. $\dfrac{14^3 \cdot 14^9}{14^{10} \cdot 14}$

9. $\dfrac{(s^2t)^3(7s^4t)^4}{(7s^2t^3)^2}$

10. $(2a^0b^{-6})^2$

11. $\left(\dfrac{6a^{-5}b}{8ab^{-2}}\right)^{-2}$

12. a. Write the number in scientific notation: 43,000,000,000

b. Write the number without scientific notation: 5.6×10^{-6}

13. The average amount of water flowing over Niagara Falls is 1.68×10^5 m³/min.

a. How many cubic meters of water flow over the falls in one day?

b. How many cubic meters of water flow over the falls in one year?

14. Write the polynomial in descending order: $4x + 5x^3 - 7x^2 + 11$.

a. Identify the degree of the polynomial.

b. Identify the leading coefficient of the polynomial.

15. Perform the indicated operations.

$(7w^2 - 11w - 6) + (8w^2 + 3w + 4) - (-9w^2 - 5w + 2)$

16. Subtract $(3x^2 - 5x^3 + 2x)$ from $(10x^3 - 4x^2 + 1)$.

For Exercises 17–23, multiply the polynomials.

17. $-2x^3(5x^2 + x - 15)$

18. $(4a - 3)(2a - 1)$

19. $(4y - 5)(y^2 - 5y + 3)$

20. $(2 + 3b)(2 - 3b)$

21. $(5z - 6)^2$

22. $(10 - 3w)(10 + 3w)$

23. $(y^2 - 5y + 2)(y - 6)$

24. Find the perimeter and the area of the rectangle shown in the figure.

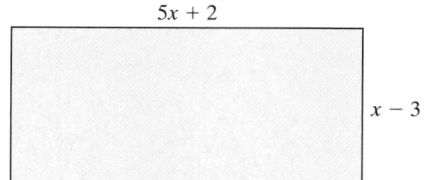

For Exercises 25–27, divide:

25. $(-12x^8 + x^6 - 8x^3) \div (4x^2)$

26. $\dfrac{2y^2 - 13y + 21}{y - 3}$

27. $(2w^3 - 2w - 5w^2 + 5) \div (2w + 3)$

28. $(2x^4 + x^3 - 4x + 1) \div (x^2 - 2)$

Chapters 1–5   Cumulative Review Exercises

For Exercises 1–2, simplify completely.

1. $-5 - \dfrac{1}{2}[4 - 3(-7)]$

2. $|-3^2 + 5|$

3. Translate the phrase into a mathematical expression and simplify:

The difference of the square of five and the square root of four.

4. Solve for x: $\dfrac{1}{2}(x - 6) + \dfrac{2}{3} = \dfrac{1}{4}x$

5. Solve for y: $-2y - 3 = -5(y - 1) + 3y$

6. For a point in a rectangular coordinate system, in which quadrant are both the x- and y-coordinates negative?

7. For a point in a rectangular coordinate system, on which axis is the x-coordinate zero and the y-coordinate nonzero?

8. In a triangle, one angle measures $23°$ more than the smallest angle. The third angle measures $10°$ more than the sum of the other two angles. Find the measure of each angle.

9. A snow storm lasts for 9 hr and dumps snow at a rate of $1\frac{1}{2}$ in./hr. If there was already 6 in. of snow on the ground before the storm, the snow depth is given by the equation:

$y = \dfrac{3}{2}x + 6$ where y is the snow depth in inches and $x \geq 0$ is the time in hours.

a. Find the snow depth after 4 hr.

b. Find the snow depth at the end of the storm.

c. How long had it snowed when the total depth of snow was $14\frac{1}{4}$ in.?

d. Graph the line.

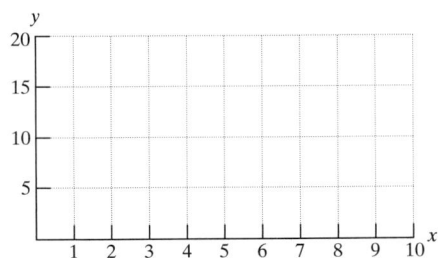

10. Solve the system of equations.

$$5x + 3y = -3$$
$$3x + 2y = -1$$

11. Solve the inequality. Graph the solution set on the real number line and express the solution in interval notation. $2 - 3(2x + 4) \leq -2x - (x - 5)$

For Exercises 12–15, perform the indicated operations.

12. $(2x^2 + 3x - 7) - (-3x^2 + 12x + 8)$

13. $(2y + 3z)(-y - 5z)$

14. $(4t - 3)^2$

15. $\left(\dfrac{2}{5}a + \dfrac{1}{3}\right)\left(\dfrac{2}{5}a - \dfrac{1}{3}\right)$

For Exercises 16–17, divide the polynomials.

16. $(12a^4b^3 - 6a^2b^2 + 3ab) \div (-3ab)$

17. $\dfrac{4m^3 - 5m + 2}{m - 2}$

For Exercises 18–19, use the properties of exponents to simplify the expressions. Write the answers with positive exponents only. Assume all variables represent nonzero real numbers.

18. $\left(\dfrac{2c^2d^4}{8cd^6}\right)^2$

19. $\dfrac{10a^{-2}b^{-3}}{5a^0b^{-6}}$

20. Perform the indicated operations, and write the final answer in scientific notation.

$$\dfrac{(8.2 \times 10^{-2})(6.8 \times 10^{-6})}{2.0 \times 10^{-5}}$$

Factoring Polynomials

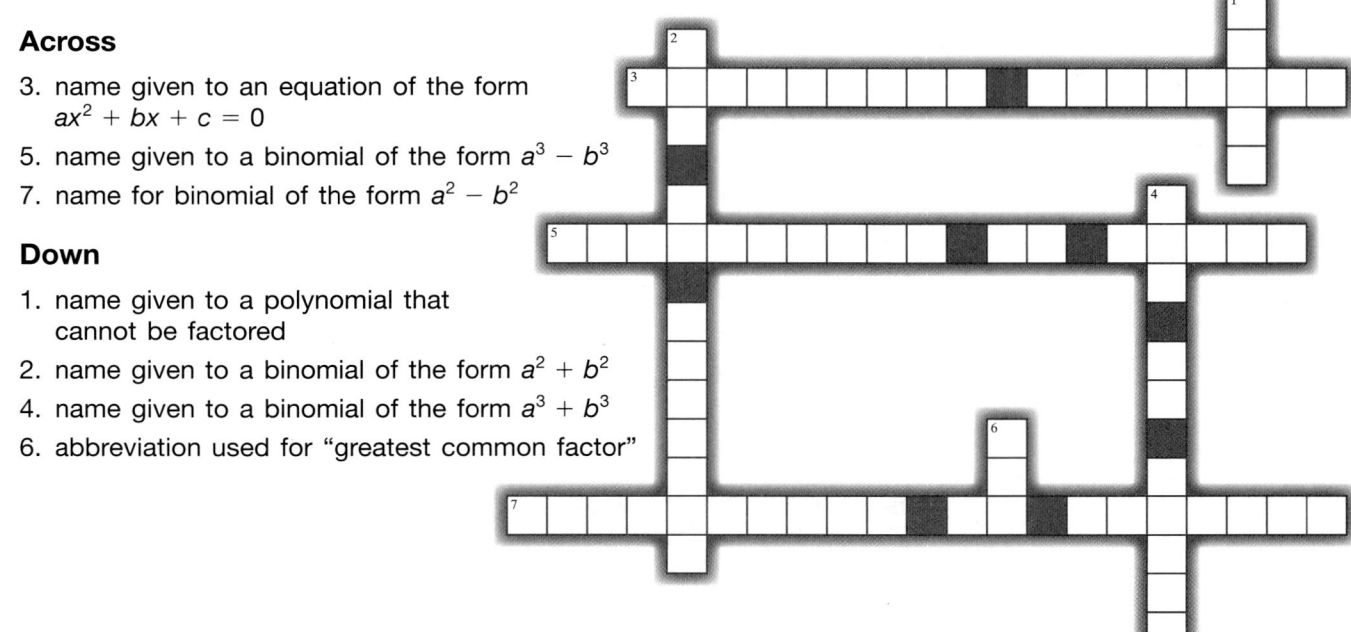

6

Chapter 6 is devoted to a mathematical operation called factoring. The applications of factoring are far-reaching, and in this chapter, we use factoring as a tool to solve a type of equation called a quadratic equation.

As you work through the chapter, pay attention to key terms. Then work through this puzzle.

Across

3. name given to an equation of the form $ax^2 + bx + c = 0$

5. name given to a binomial of the form $a^3 - b^3$

7. name for binomial of the form $a^2 - b^2$

Down

1. name given to a polynomial that cannot be factored

2. name given to a binomial of the form $a^2 + b^2$

4. name given to a binomial of the form $a^3 + b^3$

6. abbreviation used for "greatest common factor"

Greatest Common Factor and Factoring by Grouping

1. Identifying the Greatest Common Factor

Chapter 6 is devoted to a mathematical operation called **factoring**. To factor an integer means to write the integer as a product of two or more integers. To factor a polynomial means to express the polynomial as a product of two or more polynomials.

In the product $2 \cdot 5 = 10$, for example, 2 and 5 are factors of 10.

In the product $(3x + 4)(2x - 1) = 6x^2 + 5x - 4$, the quantities $(3x + 4)$ and $(2x - 1)$ are factors of $6x^2 + 5x - 4$.

We begin our study of factoring by factoring integers. The number 20, for example, can be factored as $1 \cdot 20, 2 \cdot 10, 4 \cdot 5,$ or $2 \cdot 2 \cdot 5$. The product $2 \cdot 2 \cdot 5$ (or equivalently $2^2 \cdot 5$) consists only of prime numbers and is called the **prime factorization**.

The **greatest common factor** (denoted **GCF**) of two or more integers is the greatest factor common to each integer. To find the greatest common factor of two integers, it is often helpful to express the numbers as a product of prime factors as shown in the next example.

Example 1   Identifying the GCF

Find the greatest common factor.

    a. 24 and 36   **b.** 105, 40, and 60

Solution:

First find the prime factorization of each number. Then find the product of common factors.

a.
$$2\underline{|24} \qquad 2\underline{|36}$$
$$2\underline{|12} \qquad 2\underline{|18}$$
$$2\underline{|6} \qquad 3\underline{|9}$$
$$3 \qquad\quad 3$$

Factors of 24 $= 2 \cdot 2 \cdot 2 \cdot 3$

Factors of 36 $= 2 \cdot 2 \cdot 3 \cdot 3$

The numbers 24 and 36 share two factors of 2 and one factor of 3. Therefore, the greatest common factor is $2 \cdot 2 \cdot 3 = 12$.

b.
$$5\underline{|105} \qquad 5\underline{|40} \qquad 5\underline{|60}$$
$$3\underline{|21} \qquad 2\underline{|8} \qquad 3\underline{|12}$$
$$7 \qquad\quad 2\underline{|4} \qquad 2\underline{|4}$$
$$2 \qquad\quad 2$$

Factors of 105 $= 3 \cdot 7 \cdot 5$

Factors of 40 $= 2 \cdot 2 \cdot 2 \cdot 5$

Factors of 60 $= 2 \cdot 2 \cdot 3 \cdot 5$

The greatest common factor is 5.

Skill Practice

   1. Find the GCF of 12 and 20.   **2.** Find the GCF of 45, 75, and 30.

Skill Practice Answers

1. 4   **2.** 15

In Example 2, we find the greatest common factor of two or more variable terms.

Example 2 | **Identifying the Greatest Common Factor**

Find the GCF among each group of terms.

a. $7x^3, 14x^2, 21x^4$ **b.** $15a^4b, 25a^3b^2$ **c.** $8c^2d^7e, 6c^3d^4$

Solution:

List the factors of each term.

a. $7x^3 = 7 \cdot x \cdot x \cdot x$

$14x^2 = 2 \cdot 7 \cdot x \cdot x$ The GCF is $7x^2$.

$21x^4 = 3 \cdot 7 \cdot x \cdot x \cdot x \cdot x$

b. $15a^4b = 3 \cdot 5 \cdot a \cdot a \cdot a \cdot a \cdot b$

$25a^3b^2 = 5 \cdot 5 \cdot a \cdot a \cdot a \cdot b \cdot b$ The GCF is $5a^3b$.

TIP: Notice that the expressions $15a^4b$ and $25a^3b^2$ share factors of 5, a, and b. The GCF is the product of the common factors, where each factor is raised to the lowest power to which it occurs in all the original expressions.

$15a^4b = 3 \cdot 5a^4b$ Lowest power of 5 is 1: 5^1
$25a^3b^2 = 5^2a^3b^2$ Lowest power of a is 3: a^3 The GCF is $5a^3b$.
 Lowest power of b is 1: b^1

---

c. $8c^2d^7e = 2^3c^2d^7e$
$6c^3d^4 = 2 \cdot 3c^3d^4$ The common factors are 2, c, and d.

The lowest power of 2 is 1: 2^1
The lowest power of c is 2: c^2 The GCF is $2c^2d^4$.
The lowest power of d is 4: d^4

Skill Practice Find the GCF.

3. $10z^3, 15z^5, 40z$ **4.** $6w^3y^5, 21w^4y^2$ **5.** $9m^2np^8, 15n^4p^5$

Sometimes polynomials share a common binomial factor as shown in Example 3.

Example 3 | **Finding the Greatest Common Binomial Factor**

Find the greatest common factor between the terms: $3x(a + b)$ and $2y(a + b)$

Solution:

$3x(a + b)$
$2y(a + b)$ The only common factor is the binomial $(a + b)$. The GCF is $(a + b)$.

Skill Practice Answers

3. $5z$ **4.** $3w^3y^2$ **5.** $3np^5$

Skill Practice

6. Find the greatest common binomial factor of the terms:
$$a(x + 2) \text{ and } b(x + 2)$$

2. Factoring out the Greatest Common Factor

The process of factoring a polynomial is the reverse process of multiplying polynomials. Both operations use the distributive property: $ab + ac = a(b + c)$.

Multiply

$$5y(y^2 + 3y + 1) = 5y(y^2) + 5y(3y) + 5y(1)$$
$$= 5y^3 + 15y^2 + 5y$$

Factor

$$5y^3 + 15y^2 + 15y = 5y(y^2) + 5y(3y) + 5y(1)$$
$$= 5y(y^2 + 3y + 1)$$

Steps to Factor out the Greatest Common Factor

1. Identify the GCF of all terms of the polynomial.
2. Write each term as the product of the GCF and another factor.
3. Use the distributive property to remove the GCF.

Note: To check the factorization, multiply the polynomials to remove parentheses.

Example 4 Factoring out the Greatest Common Factor

Factor out the GCF.

  a. $4x - 20$ **b.** $6w^2 + 3w$

  c. $15y^3 + 12y^4$ **d.** $9a^4b - 18a^5b + 27a^6b$

Solution:

  a. $4x - 20$ The GCF is 4.

      $= 4(x) - 4(5)$ Write each term as the product of the GCF and another factor.

      $= 4(x - 5)$ Use the distributive property to factor out the GCF.

TIP: Any factoring problem can be checked by multiplying the factors:

$$\underline{\text{Check}}: 4(x - 5) = 4x - 20 \checkmark$$

Skill Practice Answers

6. $(x + 2)$

b. $6w^2 + 3w$ The GCF is $3w$.

$\quad = 3w(2w) + 3w(1)$ Write each term as the product of $3w$ and another factor.

$\quad = 3w(2w + 1)$ Use the distributive property to factor out the GCF.

Check: $3w(2w + 1) = 6w^2 + 3w$ ✔

c. $15y^3 + 12y^4$ The GCF is $3y^3$.

$\quad = 3y^3(5) + 3y^3(4y)$ Write each term as the product of $3y^3$ and another factor.

$\quad = 3y^3(5 + 4y)$ Use the distributive property to factor out the GCF.

Check: $3y^3(5 + 4y) = 15y^3 + 12y^4$ ✔

d. $9a^4b - 18a^5b + 27a^6b$ The GCF is $9a^4b$.

$\quad = 9a^4b(1) - 9a^4b(2a) + 9a^4b(3a^2)$ Write each term as the product of $9a^4b$ and another factor.

$\quad = 9a^4b(1 - 2a + 3a^2)$ Use the distributive property to factor out the GCF.

Check: $9a^4b(1 - 2a + 3a^2) = 9a^4b - 18a^5b + 27a^6b$ ✔

Skill Practice	Factor out the GCF.

7. $6w + 18$ **8.** $21m^3 - 7m^2$ **9.** $9y^2 - 6y^5$ **10.** $50s^3t - 40st^2 + 10st$

The greatest common factor of the polynomial $2x + 5y$ is 1. If we factor out the GCF, we have $1(2x + 5y)$. A polynomial whose only factors are itself and 1 is called a **prime polynomial**. A prime polynomial cannot be factored further.

3. Factoring out a Negative Factor

Sometimes it is advantageous to factor out the *opposite* of the GCF when the leading coefficient of the polynomial is negative. This is demonstrated in the next example. Notice that this *changes the signs* of the remaining terms inside the parentheses.

Example 5	Factoring out a Negative Factor

a. Factor out -3 from the polynomial $-3x^2 + 6x - 33$.

b. Factor out the quantity $-4pq$ from the polynomial $-12p^3q - 8p^2q^2 + 4pq^3$.

Solution:

a. $-3x^2 + 6x - 33$

The GCF is 3. However, in this case, we will factor out the *opposite* of the GCF, -3.

$= -3(x^2) + (-3)(-2x) + (-3)(11)$

Write each term as the product of -3 and another factor.

$= -3[x^2 + (-2x) + 11]$

Factor out -3.

$= -3(x^2 - 2x + 11)$

Simplify. Notice that each sign within the trinomial has changed.

Check: $-3(x^2 - 2x + 11) = -3x^2 + 6x - 33$ ✔

Check by multiplying.

b. $-12p^3q - 8p^2q^2 + 4pq^3$

The GCF is $4pq$. However, in this case, we will factor out the *opposite* of the GCF, $-4pq$.

$= -4pq(3p^2) + (-4pq)(2pq) + (-4pq)(-q^2)$

Write each term as the product of $-4pq$ and another factor.

$= -4pq[3p^2 + 2pq + (-q^2)]$

Factor out $-4pq$. Notice that each sign within the trinomial has changed.

$= -4pq(3p^2 + 2pq - q^2)$

To verify that this is the correct factorization and that the signs are correct, multiply the factors.

Check: $-4pq(3p^2 + 2pq - q^2) = -12p^3q - 8p^2q^2 + 4pq^3$ ✔

Skill Practice

11. Factor out the opposite of the GCF: $-2x^2 - 10x + 16$

12. Factor out $-5xy$ from the polynomial: $-10x^2y + 5xy - 15xy^2$

4. Factoring out a Binomial Factor

The distributive property can also be used to factor out a common factor that consists of more than one term as shown in Example 6.

Example 6   Factoring out a Binomial Factor

Factor out the GCF: $2w(x + 3) - 5(x + 3)$

Skill Practice Answers

11. $-2(x^2 + 5x - 8)$
12. $-5xy(2x - 1 + 3y)$

Solution:

$2w(x + 3) - 5(x + 3)$ The greatest common factor is the quantity $(x + 3)$.

$= (x + 3)(2w) - (x + 3)(5)$ Write each term as the product of $(x + 3)$ and another factor.

$= (x + 3)(2w - 5)$ Use the distributive property to factor out the GCF.

| Skill Practice | Factor out the GCF.

13. $8y(a + b) + 9(a + b)$

5. Factoring by Grouping

When two binomials are multiplied, the product before simplifying contains four terms. For example:

$$(x + 4)(3a + 2b) = (x + 4)(3a) + (x + 4)(2b)$$
$$= (x + 4)(3a) + (x + 4)(2b)$$
$$= 3ax + 12a + 2bx + 8b$$

In Example 7, we learn how to reverse this process. That is, given a four-term polynomial, we will factor it as a product of two binomials. The process is called *factoring by grouping*.

| Example 7 | **Factoring by Grouping**

Factor by grouping: $3ax + 12a + 2bx + 8b$

Solution:

$3ax + 12a + 2bx + 8b$ **Step 1:** Identify and factor out the GCF from all four terms. In this case, the GCF is 1.

$= 3ax + 12a \mathrel{\vdots} + 2bx + 8b$ Group the first pair of terms and the second pair of terms.

$= 3a(x + 4) + 2b(x + 4)$ **Step 2:** Factor out the GCF from each pair of terms. *Note:* The two terms now share a common binomial factor of $(x + 4)$.

$= (x + 4)(3a + 2b)$ **Step 3:** Factor out the common binomial factor.

Check: $(x + 4)(3a + 2b) = 3ax + 2bx + 12a + 8b$ ✔

Note: Step 2 results in two terms with a common binomial factor. If the two binomials are different, step 3 cannot be performed. In such a case, the original polynomial may not be factorable by grouping, or different pairs of terms may need to be grouped and inspected.

Skill Practice Answers

13. $(a + b)(8y + 9)$

> **Skill Practice**   Factor by grouping.
>
> **14.** $5x + 10y + ax + 2ay$

> **TIP:** One frequently asked question when factoring is whether the order can be switched between the factors. The answer is yes. Because multiplication is commutative, the order in which the factors are written does not matter.
>
> $$(x + 4)(3a + 2b) = (3a + 2b)(x + 4)$$

Steps to Factoring by Grouping

To factor a four-term polynomial by grouping:

1. Identify and factor out the GCF from all four terms.

2. Factor out the GCF from the first pair of terms. Factor out the GCF from the second pair of terms. (Sometimes it is necessary to factor out the opposite of the GCF.)

3. If the two terms share a common binomial factor, factor out the binomial factor.

Example 8   **Factoring by Grouping**

Factor the polynomials by grouping.

   a. $ax + ay - bx - by$ **b.** $16w^4 - 40w^3 - 12w^2 + 30w$

Solution:

a. $ax + ay - bx - by$ **Step 1:** Identify and factor out the GCF from all four terms. In this case, the GCF is 1.

    $= ax + ay \,\vdots\, - bx - by$ Group the first pair of terms and the second pair of terms.

    $= a(x + y) - b(x + y)$ **Step 2:** Factor out a from the first pair of terms.

                          Factor out $-b$ from the second pair of terms. (This causes sign changes within the second parentheses. The terms in parentheses now match.)

    $= (x + y)(a - b)$ **Step 3:** Factor out the common binomial factor.

   Check: $(x + y)(a - b) = x(a) + x(-b) + y(a) + y(-b)$

                        $= ax - bx + ay - by$ ✔

Avoiding Mistakes:

In step 2, the expression $a(x + y) - b(x + y)$ is not yet factored because it is a *difference*, not a product. To factor the expression, you must carry it one step further.

   $a(x + y) - b(x + y)$

   $= (x + y)(a - b)$

The factored form must be represented as a product.

b. $16w^4 - 40w^3 - 12w^2 + 30w$

$= 2w(8w^3 - 20w^2 - 6w + 15)$

$= 2w[8w^3 - 20w^2 \vdots - 6w + 15]$

$= 2w[4w^2(2w - 5) - 3(2w - 5)]$

$= 2w[(2w - 5)(4w^2 - 3)]$

$= 2w(2w - 5)(4w^2 - 3)$

Step 1: Identify and factor out the GCF from all four terms. In this case, the GCF is $2w$.

Group the first pair of terms and the second pair of terms.

Step 2: Factor out $4w^2$ from the first pair of terms.

Factor out -3 from the second pair of terms. (This causes sign changes within the second parentheses. The terms in parentheses now match.)

Step 3: Factor out the common binomial factor.

Skill Practice Factor by grouping.

15. $tu - tv - 2u + 2v$ **16.** $3ab^2 + 6b^2 - 12ab - 24b$

Skill Practice Answers

15. $(u - v)(t - 2)$
16. $3b(a + 2)(b - 4)$

Section 6.1   Practice Exercises

Study Skills Exercises

1. The final exam is just around the corner. Your old tests and quizzes provide good material to study for the final exam. Use your old tests to make a list of the chapters on which you need to concentrate. Ask your professor for help if there are still concepts that you do not understand.

2. Define the key terms:

    a. factoring

    c. prime factorization

    b. greatest common factor (GCF)

    d. prime polynomial

Concept 1: Identifying the Greatest Common Factor

For Exercises 3–18, identify the greatest common factor for the group of integers.

  3. 28, 63

  6. 20, 52, 32

  9. $8w^2, 4x^3$

  4. 24, 40

  7. $3xy, 7y$

  10. $15t^2, 10u^2$

  5. 42, 30, 60

  8. $10mn, 11n$

  11. $2a^2b, 3ab^2$

12. $3x^3y^2, 5xy^4$ **13.** $12w^3z, 16w^2z$ **14.** $20cd, 15c^3d$

15. $8x^3y^4z^2, 12xy^5z^4, 6x^2y^8z^3$ **16.** $15r^2s^2t^5, 5r^3s^4t^3, 30r^4s^3t^2$ **17.** $7(x-y), 9(x-y)$

18. $(2a-b), 3(2a-b)$

Concept 2: Factoring out the Greatest Common Factor

19. a. Use the distributive property to multiply $3(x-2y)$.

  b. Use the distributive property to factor $3x-6y$.

20. a. Use the distributive property to multiply $a^2(5a+b)$.

  b. Use the distributive property to factor $5a^3 + a^2b$.

For Exercises 21–40, factor out the GCF.

21. $4p + 12$ **22.** $3q - 15$ **23.** $5c^2 - 10c + 15$ **24.** $16d^3 + 24d^2 + 32d$

25. $x^5 + x^3$ **26.** $y^2 - y^3$ **27.** $t^4 - 4t + 8t^2$ **28.** $7r^3 - r^5 + r^4$

29. $2ab + 4a^3b$ **30.** $5u^3v^2 - 5uv$ **31.** $38x^2y - 19x^2y^4$ **32.** $100a^5b^3 + 16a^2b$

33. $6x^3y^5 - 18xy^9z$ **34.** $15mp^7q^4 + 12m^4q^3$ **35.** $5 + 7y^3$ **36.** $w^3 - 5u^3v^2$

37. $42p^3q^2 + 14pq^2 - 7p^4q^4$ **38.** $8m^2n^3 - 24m^2n^2 + 4m^3n$

39. $t^5 + 2rt^3 - 3t^4 + 4r^2t^2$ **40.** $u^2v + 5u^3v^2 - 2u^2 + 8uv$

Concept 3: Factoring out a Negative Factor

41. For the polynomial $-2x^3 - 4x^2 + 8x$ **42.** For the polynomial $-9y^5 + 3y^3 - 12y$

  a. Factor out $-2x$.   **b.** Factor out $2x$.   **a.** Factor out $-3y$.   **b.** Factor out $3y$.

43. Factor out -1 from the polynomial $-8t^2 - 9t - 2$.

44. Factor out -1 from the polynomial $-6x^3 - 2x - 5$.

For Exercises 45–50, factor out the opposite of the greatest common factor.

45. $-15p^3 - 30p^2$ **46.** $-24m^3 - 12m^4$ **47.** $-q^4 + 2q^2 - 9q$

48. $-r^3 + 9r^2 - 5r$ **49.** $-7x - 6y - 2z$ **50.** $-4a + 5b - c$

Concept 4: Factoring out a Binomial Factor

For Exercises 51–56, factor out the GCF.

51. $13(a+6) - 4b(a+6)$ **52.** $7(x^2+1) - y(x^2+1)$ **53.** $8v(w^2-2) + (w^2-2)$

54. $t(r+2) + (r+2)$ **55.** $21x(x+3) + 7x^2(x+3)$ **56.** $5y^3(y-2) - 15y(y-2)$

Concept 5: Factoring by Grouping

For Exercises 57–76, factor by grouping.

57. $8a^2 - 4ab + 6ac - 3bc$

58. $4x^3 + 3x^2y + 4xy^2 + 3y^3$

59. $3q + 3p + qr + pr$

60. $xy - xz + 7y - 7z$

61. $6x^2 + 3x + 4x + 2$

62. $4y^2 + 8y + 7y + 14$

63. $2t^2 + 6t - 5t - 15$

64. $2p^2 - p - 6p + 3$

65. $6y^2 - 2y - 9y + 3$

66. $5a^2 + 30a - 2a - 12$

67. $b^4 + b^3 - 4b - 4$

68. $8w^5 + 12w^2 - 10w^3 - 15$

69. $3j^2k + 15k + j^2 + 5$

70. $2ab^2 - 6ac + b^2 - 3c$

71. $14w^6x^6 + 7w^6 - 2x^6 - 1$

72. $18p^4q - 9p^5 - 2q + p$

73. $ay + bx + by + ax$
(*Hint:* Rearrange the terms.)

74. $2c + 3ay + ac + 6y$

75. $vw^2 - 3 + w - 3wv$

76. $2x^2 + 6m + 12 + x^2m$

For Exercises 77–82, factor out the GCF first. Then factor by grouping.

77. $15x^4 + 15x^2y^2 + 10x^3y + 10xy^3$

78. $2a^3b - 4a^2b + 32ab - 64b$

79. $4abx - 4b^2x - 4ab + 4b^2$

80. $p^2q - pq^2 - rp^2q + rpq^2$

81. $6st^2 - 18st - 6t^4 + 18t^3$

82. $15j^3 - 10j^2k - 15j^2k^2 + 10jk^3$

83. The formula $P = 2l + 2w$ represents the perimeter, P, of a rectangle given the length, l, and the width, w. Factor out the GCF, and write an equivalent formula in factored form.

84. The formula $P = 2a + 2b$ represents the perimeter, P, of a parallelogram given the base, b, and an adjacent side, a. Factor out the GCF, and write an equivalent formula in factored form.

85. The formula $S = 2\pi r^2 + 2\pi rh$ represents the surface area, S, of a cylinder with radius, r, and height, h. Factor out the GCF, and write an equivalent formula in factored form.

86. The formula $A = P + Prt$ represents the total amount of money, A, in an account that earns simple interest at a rate, r, for t years. Factor out the GCF, and write an equivalent formula in factored form.

Expanding Your Skills

87. Factor out $\dfrac{1}{7}$ from $\dfrac{1}{7}x^2 + \dfrac{3}{7}x - \dfrac{5}{7}$.

88. Factor out $\dfrac{1}{5}$ from $\dfrac{6}{5}y^2 - \dfrac{4}{5}y + \dfrac{1}{5}$.

89. Factor out $\dfrac{1}{4}$ from $\dfrac{5}{4}w^2 + \dfrac{3}{4}w + \dfrac{9}{4}$.

90. Factor out $\dfrac{1}{6}$ from $\dfrac{1}{6}p^2 - \dfrac{3}{6}p + \dfrac{5}{6}$.

91. Write a polynomial that has a GCF of $3x$. (Answers may vary.)

92. Write a polynomial that has a GCF of $7y$. (Answers may vary.)

93. Write a polynomial that has a GCF of $4p^2q$. (Answers may vary.)

94. Write a polynomial that has a GCF of $2ab^2$. (Answers may vary.)

| Section 6.2 | **Factoring Trinomials of the Form $x^2 + bx + c$ (Optional)** |

Concept

1. Factoring Trinomials with a Leading Coefficient of 1

1. Factoring Trinomials with a Leading Coefficient of 1

In Section 5.6, we learned how to multiply two binomials. We also saw that such a product often results in a trinomial. For example,

Product of   Product of
first terms   last terms

$$(x + 3)(x + 7) = x^2 + \underline{7x + 3x} + 21 = x^2 + 10x + 21$$

Sum of products of inner
terms and outer terms

In this section, we want to reverse the process. That is, given a trinomial, we want to *factor* it as a product of two binomials. In particular, we begin our study with the case in which a trinomial has a leading coefficient of 1.

Consider the trinomial $x^2 + bx + c$. To produce a leading term of x^2, we can construct binomials of the form $(x + \quad)(x + \quad)$. The remaining terms can be satisfied by two integers, p and q, whose product is c and whose sum is b.

Factors of c

$$x^2 + bx + c = (x + p)(x + q) = x^2 + qx + px + pq$$
$$= x^2 + \underline{(q + p)}x + \underline{pq}$$

Sum = b Product = c

This process is demonstrated in Example 1.

| Example 1 | **Factoring a Trinomial of the Form $x^2 + bx + c$** |

Factor: $x^2 + 4x - 45$

Solution:

$x^2 + 4x - 45 = (x + \square)(x + \square)$ The product $x \cdot x = x^2$.

We must fill in the blanks with two integers whose product is -45 and whose sum is 4. The factors must have opposite signs to produce a negative product. The possible factorizations of -45 are:

Product = -45	Sum
$-1 \cdot 45$	44
$-3 \cdot 15$	12
$-5 \cdot 9$	4
$-9 \cdot 5$	-4
$-15 \cdot 3$	-12
$-45 \cdot 1$	-44

$$x^2 + 4x - 45 = (x + \Box)(x + \Box)$$
$$= (x + (-5))(x + 9) \qquad \text{Fill in the blanks with } -5 \text{ and } 9,$$
$$= (x - 5)(x + 9) \qquad \text{Factored form}$$

<u>Check:</u>
$$(x - 5)(x + 9) = x^2 + 9x - 5x - 45$$
$$= x^2 + 4x - 45 \ \checkmark$$

Skill Practice   Factor.

1. $x^2 - 5x - 14$

One frequently asked question is whether the order of factors can be reversed. The answer is yes because multiplication of polynomials is a commutative operation. Therefore, in Example 1, we can express the factorization as $(x - 5)(x + 9)$ or as $(x + 9)(x - 5)$.

Example 2   **Factoring a Trinomial of the Form $x^2 + bx + c$**

Factor: $w^2 - 15w + 50$

Solution:
$$w^2 - 15w + 50 = (w + \Box)(w + \Box) \qquad \text{The product } w \cdot w = w^2.$$

Find two integers whose product is 50 and whose sum is -15. To form a positive product, the factors must be either both positive or both negative. The sum must be negative, so we will choose negative factors of 50.

<u>Product = 50</u>	<u>Sum</u>
$(-1)(-50)$	-51
$(-2)(-25)$	-27
$(-5)(-10)$	-15

$$w^2 - 15w + 50 = (w + \Box)(w + \Box)$$
$$= (w + (-5))(w + (-10))$$
$$= (w - 5)(w - 10) \qquad \text{Factored form}$$

Skill Practice   Factor.

2. $z^2 - 16z + 48$

Practice will help you become proficient in factoring polynomials. As you do your homework, keep these important guidelines in mind:

- To factor a trinomial, write the trinomial in descending order such as $x^2 + bx + c$.
- For all factoring problems, always factor out the GCF from all terms first.

Furthermore, we offer the following rules for determining the signs within the binomial factors.

Skill Practice Answers
1. $(x - 7)(x + 2)$
2. $(z - 4)(z - 12)$

Sign Rules for Factoring Trinomials

Given the trinomial $x^2 + bx + c$, the signs within the binomial factors are determined as follows:

1. If c is *positive*, then the signs in the binomials must be the same (either both positive or both negative). The correct choice is determined by the middle term. If the middle term is positive, then both signs must be positive. If the middle term is negative, then both signs must be negative.

c is positive

$x^2 + 6x + 8$

$(x + 2)(x + 4)$

Same signs

c is positive

$x^2 - 6x + 8$

$(x - 2)(x - 4)$

Same signs

2. If c is *negative*, then the signs in the binomials must be different.

c is negative

$x^2 + 2x - 35$

$(x + 7)(x - 5)$

Different signs

c is negative

$x^2 - 2x - 35$

$(x - 7)(x + 5)$

Different signs

Example 3 **Factoring Trinomials**

Factor.

a. $-8p - 48 + p^2$ **b.** $-40t - 30t^2 + 10t^3$

c. $-a^2 + 6a - 8$ **d.** $2c^2 + 22cd + 60d^2$

Solution:

a. $-8p - 48 + p^2$

$= p^2 - 8p - 48$ Write in descending order.

$= (p \quad \square)(p \quad \square)$ Find two integers whose product is -48 and whose sum is -8. The numbers are -12 and 4.

$= (p - 12)(p + 4)$ Factored form

b. $-40t - 30t^2 + 10t^3$

$= 10t^3 - 30t^2 - 40t$ Write in descending order.

$= 10t(t^2 - 3t - 4)$ Factor out the GCF.

$= 10t(t \quad \square)(t \quad \square)$ Find two integers whose product is -4 and whose sum is -3. The numbers are -4 and 1.

$= 10t(t - 4)(t + 1)$ Factored form

c. $-a^2 + 6a - 8$ It is generally easier to factor a trinomial with a *positive* leading coefficient. Therefore, we will factor out -1 from all terms.

$= -1(a^2 - 6a + 8)$

$= -1(a \quad \square)(a \quad \square)$ Find two integers whose product is 8 and whose sum is -6. The numbers are -4 and -2.

$= -1(a - 4)(a - 2)$

TIP: Recall that factoring out -1 from a polynomial changes the signs of the terms within parentheses.

d. $2c^2 + 22cd + 60d^2$

$= 2(c^2 + 11cd + 30d^2)$ Factor out the GCF.

$= 2(c \quad \square d)(c \quad \square d)$ Notice that the second pair of terms has a factor of d. This will produce a product of d^2.

$= 2(c + 5d)(c + 6d)$ Find two integers whose product is 30 and whose sum is 11. The numbers are 5 and 6.

Skill Practice   Factor.

3. $-5w + w^2 - 6$ **4.** $30y^3 + 2y^4 + 112y^2$

5. $-x^2 + x + 12$ **6.** $3a^2 - 15ab + 12b^2$

To factor a trinomial of the form $x^2 + bx + c$, we must find two integers whose product is c and whose sum is b. If no such integers exist, then the trinomial is not factorable and is called a **prime polynomial**.

Example 4   **Factoring Trinomials**

Factor: $x^2 - 13x + 8$

Solution:

$x^2 - 13x + 8$ The trinomial is in descending order. The GCF is 1.

$= (x \quad \square)(x \quad \square)$ Find two integers whose product is 8 and whose sum is -13. No such integers exist.

The trinomial $x^2 - 13x + 8$ is prime.

Skill Practice   Factor.

7. $x^2 - 7x + 28$

Skill Practice Answers

3. $(w - 6)(w + 1)$
4. $2y^2(y + 8)(y + 7)$
5. $-(x - 4)(x + 3)$
6. $3(a - b)(a - 4b)$
7. Prime

Section 6.2   Practice Exercises

Boost *your* GRADE at mathzone.com!

MathZone

- Practice Problems   • e-Professors
- Self-Tests   • Videos
- NetTutor

Study Skills Exercises

1. Sometimes the problems on a test do not appear in the same order as the concepts appear in the text. In order to better prepare for a test, try to practice on problems taken from the book and placed in random order. Choose 30 problems from various chapters, randomize the order, and use them to review for the test. Repeat the process several times for additional practice.

2. Define the key term **prime polynomial**.

Review Exercises

For Exercises 3–6, factor completely.

3. $4x^3y^7 - 12x^4y^5 + 8xy^8$

4. $9a^6b^3 - 27a^3b^6 - 3a^2b^2$

5. $ax + 2bx - 5a - 10b$

6. $m^2 - mx - 3pm + 3px$

Concept 1: Factoring Trinomials with a Leading Coefficient of 1

For Exercises 7–20, factor completely.

7. $x^2 + 10x + 16$ **8.** $y^2 + 18y + 80$ **9.** $z^2 - 11z + 18$

10. $w^2 - 7w + 12$ **11.** $z^2 - 9z + 18$ **12.** $w^2 + 4w - 12$

13. $p^2 + 3p - 40$ **14.** $a^2 - 10a + 9$ **15.** $t^2 + 6t - 40$

16. $m^2 - 12m + 11$ **17.** $x^2 - 3x + 20$ **18.** $y^2 + 6y + 18$

19. $n^2 + 8n + 16$ **20.** $v^2 + 10v + 25$

For Exercises 21–24, assume that b and c represent positive integers.

21. When factoring a polynomial of the form $x^2 + bx + c$, pick an appropriate combination of signs.

    a. (+)(+)   **b.** (−)(−)   **c.** (+)(−)

22. When factoring a polynomial of the form $x^2 + bx - c$, pick an appropriate combination of signs.

    a. (+)(+)   **b.** (−)(−)   **c.** (+)(−)

23. When factoring a polynomial of the form $x^2 - bx - c$, pick an appropriate combination of signs.

    a. (+)(+)   **b.** (−)(−)   **c.** (+)(−)

24. When factoring a polynomial of the form $x^2 - bx + c$, pick an appropriate combination of signs.

    a. (+)(+)   **b.** (−)(−)   **c.** (+)(−)

25. Which is the correct factorization of $y^2 - y - 12$?

    $(y - 4)(y + 3)$ or $(y + 3)(y - 4)$

26. Which is the correct factorization of $x^2 + 14x + 13$?

    $(x + 13)(x + 1)$ or $(x + 1)(x + 13)$

27. Which is the correct factorization of $w^2 + 2w + 1$?

    $(w + 1)(w + 1)$ or $(w + 1)^2$

28. Which is the correct factorization of $z^2 - 4z + 4$?

    $(z - 2)(z - 2)$ or $(z - 2)^2$

29. In what order should a trinomial be written before attempting to factor it?

For Exercises 30–35, factor completely.

30. $-13x + x^2 - 30$

31. $12y - 160 + y^2$

32. $-18w + 65 + w^2$

33. $17t + t^2 + 72$

34. $22t + t^2 + 72$

35. $10q - 1200 + q^2$

36. Referring to page 433, write two important guidelines to follow when factoring trinomials.

For Exercises 37–48, factor completely. Be sure to factor out the GCF.

37. $3x^2 - 30x - 72$

38. $2z^2 + 4z - 198$

39. $8p^3 - 40p^2 + 32p$

40. $5w^4 - 35w^3 + 50w^2$

41. $y^4z^2 - 12y^3z^2 + 36y^2z^2$

42. $t^4u^2 + 6t^3u^2 + 9t^2u^2$

43. $-x^2 + 10x - 24$

44. $-y^2 - 12y - 35$

45. $-m^2 + m + 6$

46. $-n^2 + 5n + 6$

47. $-4 - 2c^2 - 6c$

48. $-40d - 30 - 10d^2$

Mixed Exercises

For Exercises 49–66, factor completely.

49. $x^3y^3 - 19x^2y^3 + 60xy^3$

50. $y^2z^5 + 17yz^5 + 60z^5$

51. $12p^2 - 96p + 84$

52. $5w^2 - 40w - 45$

53. $-2m^2 + 22m - 20$

54. $-3x^2 - 36x - 81$

55. $c^2 + 6cd + 5d^2$

56. $x^2 + 8xy + 12y^2$

57. $a^2 - 9ab + 14b^2$

58. $m^2 - 15mn + 44n^2$

59. $a^2 + 4a + 18$

60. $b^2 - 6a + 15$

61. $2q + q^2 - 63$

62. $-32 - 4t + t^2$

63. $x^2 + 20x + 100$

64. $z^2 - 24z + 144$

65. $t^2 + 18t - 40$

66. $d^2 + 2d - 99$

67. A student factored a trinomial as $(2x - 4)(x - 3)$. The instructor did not give full credit. Why?

68. A student factored a trinomial as $(y + 2)(5y - 15)$. The instructor did not give full credit. Why?

69. What polynomial factors as $(x - 4)(x + 13)$?

70. What polynomial factors as $(q - 7)(q + 10)$?

Expanding Your Skills

71. Find all integers, b, that make the trinomial $x^2 + bx + 6$ factorable.

72. Find all integers, b, that make the trinomial $x^2 + bx + 10$ factorable.

73. Find a value of c that makes the trinomial $x^2 + 6x + c$ factorable.

74. Find a value of c that makes the trinomial $x^2 + 8x + c$ factorable.

Factoring Trinomials: Trial-and-Error Method

In Section 6.2, we learned how to factor trinomials of the form $x^2 + bx + c$. These trinomials have a leading coefficient of 1. In this section and the next, we will consider the more general case in which the leading coefficient may be *any* integer. That is, we will factor trinomials of the form $ax^2 + bx + c$ (where $a \neq 0$). The method presented in this section is called the trial-and-error method.

1. Factoring Trinomials by the Trial-and-Error Method

To understand the basis of factoring trinomials of the form $ax^2 + bx + c$, first consider the multiplication of two binomials:

$$\overset{\text{Product of } 2 \cdot 1}{} \quad \overset{\text{Product of } 3 \cdot 2}{}$$

$$(2x + 3)(1x + 2) = 2x^2 + \underbrace{4x + 3x}_{\substack{\text{Sum of products of inner} \\ \text{terms and outer terms}}} + 6 = 2x^2 + 7x + 6$$

To factor the trinomial, $2x^2 + 7x + 6$, this operation is reversed. Hence,

$$2x^2 + 7x + 6 = \overset{\text{Factors of 2}}{\underset{\text{Factors of 6}}{(\square x \quad \square)(\square x \quad \square)}}$$

We need to fill in the blanks so that the product of the first terms in the binomials is $2x^2$ and the product of the last terms in the binomials is 6. Furthermore, the factors of $2x^2$ and 6 must be chosen so that the sum of the products of the inner terms and outer terms equals $7x$.

To produce the product $2x^2$, we might try the factors $2x$ and x within the binomials:

$$(2x \quad \square)(x \quad \square)$$

To produce a product of 6, the remaining terms in the binomials must either both be positive or both be negative. To produce a positive middle term, we will try positive factors of 6 in the remaining blanks until the correct product is found. The possibilities are $1 \cdot 6, 2 \cdot 3, 3 \cdot 2,$ and $6 \cdot 1$.

$(2x + 1)(x + 6) = 2x^2 + 12x + 1x + 6 = 2x^2 + 13x + 6$ Wrong middle term

$(2x + 2)(x + 3) = 2x^2 + 6x + 2x + 6 = 2x^2 + 8x + 6$ Wrong middle term

$(2x + 3)(x + 2) = 2x^2 + 4x + 3x + 6 = 2x^2 + 7x + 6$ Correct!

$(2x + 6)(x + 1) = 2x^2 + 2x + 6x + 6 = 2x^2 + 8x + 6$ Wrong middle term

The correct factorization of $2x^2 + 7x + 6$ is $(2x + 3)(x + 2)$. ✔

As this example shows, we factor a trinomial of the form $ax^2 + bx + c$ by shuffling the factors of a and c within the binomials until the correct product is obtained. However, sometimes it is not necessary to test all the possible combinations of factors.

In the previous example, the GCF of the original trinomial is 1. Therefore, any binomial factor that shares a common factor *greater than 1* does not need to be considered. In this case, the possibilities $(2x + 2)(x + 3)$ and $(2x + 6)(x + 1)$ cannot work.

$$\underbrace{(2x + 2)(x + 3)}_{\substack{\text{Common} \\ \text{factor of 2}}} \qquad \underbrace{(2x + 6)(x + 1)}_{\substack{\text{Common} \\ \text{factor of 2}}}$$

The steps to factor a trinomial by the trial-and-error method are outlined in the following box.

Trial-and-Error Method to Factor $ax^2 + bx + c$

1. Factor out the GCF.

2. List all pairs of positive factors of a and pairs of positive factors of c. Consider the reverse order for one of the lists of factors.

3. Construct two binomials of the form:

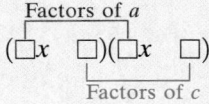

Test each combination of factors and signs until the correct product is found. If no combination of factors produces the correct product, the trinomial cannot be factored further and is a **prime polynomial**.

Before we begin our next example, keep these two important guidelines in mind:

- For any factoring problem you encounter, always factor out the GCF from all terms first.
- To factor a trinomial, write the trinomial in the form $ax^2 + bx + c$.

Example 1   Factoring a Trinomial by the Trial-and-Error Method

Factor the trinomial by the trial-and-error method: $10x^2 - 9x - 1$

Solution:

$10x^2 - 9x - 1$ **Step 1:** Factor out the GCF from all terms. In this case, the GCF is 1.

The trinomial is written in the form $ax^2 + bx + c$.

To factor $10x^2 - 9x - 1$, two binomials must be constructed in the form:

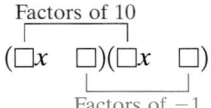

Step 2: To produce the product $10x^2$, we might try $5x$ and $2x$, or $10x$ and $1x$. To produce a product of -1, we will try the factors $(1)(-1)$ and $(-1)(1)$.

Step 3: Construct all possible binomial factors using different combinations of the factors of $10x^2$ and -1.

$(5x + 1)(2x - 1) = 10x^2 - 5x + 2x - 1 = 10x^2 - 3x - 1$ Wrong middle term

$(5x - 1)(2x + 1) = 10x^2 + 5x - 2x - 1 = 10x^2 + 3x - 1$ Wrong middle term

Because the numbers 1 and -1 did not produce the correct trinomial when coupled with $5x$ and $2x$, try using $10x$ and $1x$.

$$(10x - 1)(1x + 1) = 10x^2 + 10x - 1x - 1 = 10x^2 + 9x - 1 \qquad \text{Wrong middle term}$$

$$(10x + 1)(1x - 1) = 10x^2 - 10x + 1x - 1 = 10x^2 - 9x - 1 \qquad \text{Correct!}$$

Hence, $10x^2 - 9x - 1 = (10x + 1)(x - 1)$.

Skill Practice   Factor using the trial-and-error method.

1. $3b^2 + 8b + 4$

In Example 1, the factors of -1 must have opposite signs to produce a negative product. Therefore, one binomial factor is a sum and one is a difference. Determining the correct signs is an important aspect of factoring trinomials. We suggest the following guidelines:

Sign Rules for the Trial-and-Error Method

Given the trinomial $ax^2 + bx + c, (a > 0)$, the signs can be determined as follows:

1. If c is *positive*, then the signs in the binomials must be the same (either both positive or both negative). The correct choice is determined by the middle term. If the middle term is positive, then both signs must be positive. If the middle term is negative, then both signs must be negative.

$$\overset{c \text{ is positive}}{20x^2 + 43x + 21} \qquad \overset{c \text{ is positive}}{20x^2 - 43x + 21}$$

$$\underset{\text{Same signs}}{(4x + 3)(5x + 7)} \qquad \underset{\text{Same signs}}{(4x - 3)(5x - 7)}$$

2. If c is *negative*, then the signs in the binomial must be different. The middle term in the trinomial determines which factor gets the positive sign and which gets the negative sign.

$$\overset{c \text{ is negative}}{x^2 + 3x - 28} \qquad \overset{c \text{ is negative}}{x^2 - 3x - 28}$$

$$\underset{\text{Different signs}}{(x + 7)(x - 4)} \qquad \underset{\text{Different signs}}{(x - 7)(x + 4)}$$

Example 2   Factoring a Trinomial

Factor the trinomial: $13y - 6 + 8y^2$

Solution:

$13y - 6 + 8y^2$

$= 8y^2 + 13y - 6$ \qquad Write the polynomial in descending order.

$(\square y \quad \square)(\square y \quad \square)$ \qquad **Step 1:** The GCF is 1.

Skill Practice Answers

1. $(3b + 2)(b + 2)$

Factors of 8	Factors of 6	
$1 \cdot 8$	$1 \cdot 6$	**Step 2:** List the positive factors of 8 and positive factors of 6. Consider the reverse order in only one list of factors.
$2 \cdot 4$	$2 \cdot 3$	
	$3 \cdot 2$ $\}$ (reverse order)	
	$6 \cdot 1$	

$(2y \quad 1)(4y \quad 6)$

$(2y \quad 2)(4y \quad 3)$

$(2y \quad 3)(4y \quad 2)$

$(2y \quad 6)(4y \quad 1)$

$(1y \quad 1)(8y \quad 6)$

$(1y \quad 3)(8y \quad 2)$

Step 3: Construct all possible binomial factors using different combinations of the factors of 8 and 6.

Without regard to signs, these factorizations cannot work because the terms in the binomials share a common factor greater than 1.

Test the remaining factorizations. Keep in mind that to produce a product of -6, the signs within the parentheses must be opposite (one positive and one negative). Also, the sum of the products of the inner terms and outer terms must be combined to form $13y$.

$(1y \quad 6)(8y \quad 1)$ *Incorrect.*   Wrong middle term. Regardless of the signs, the product of inner terms, $48y$, and the product of outer terms, $1y$, cannot be combined to form the middle term $13y$.

$(1y \quad 2)(8y \quad 3)$ *Correct.*   The terms $16y$ and $3y$ can be combined to form the middle term $13y$, provided the signs are applied correctly. We require $+16y$ and $-3y$.

Hence, the correct factorization of $8y^2 + 13y - 6$ is $(y + 2)(8y - 3)$.

Skill Practice Factor.

2. $-25w + 6w^2 + 4$

2. Identifying GCF and Factoring Trinomials

Remember that the first step in any factoring problem is to remove the GCF. By removing the GCF, the remaining terms of the trinomial will be simpler and may have smaller coefficients.

Example 3   Factoring a Trinomial by the Trial-and-Error Method

Factor the trinomial by the trial-and-error method: $40x^3 - 104x^2 + 10x$

Solution:

$40x^3 - 104x^2 + 10x$

$= 2x(20x^2 - 52x + 5)$ **Step 1:** The GCF is $2x$.

TIP: Notice that when the GCF, $2x$, is removed from the original trinomial, the new trinomial has smaller coefficients. This makes the factoring process simpler. It is easier to list the factors of 20 and 5 rather than the factors of 40 and 10.

$= 2x(20x^2 - 52x + 5)$

$= 2x(\square x \quad \square)(\square x \quad \square)$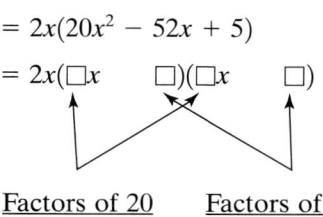

Step 2: List the factors of 20 and factors of 5. Consider the reverse order in one list of factors.

Factors of 20	Factors of 5
$1 \cdot 20$	$1 \cdot 5$
$2 \cdot 10$	$5 \cdot 1$
$4 \cdot 5$	

Step 3: Construct all possible binomial factors using different combinations of the factors of 20 and factors of 5. The signs in the parentheses must both be negative.

$= 2x(1x - 1)(20x - 5)$

$= 2x(2x - 1)(10x - 5)$ *Incorrect.*   The binomials contain a GCF greater than 1.

$= 2x(4x - 1)(5x - 5)$

$= 2x(1x - 5)(20x - 1)$ *Incorrect.*   Wrong middle term.

$\qquad 2x(x - 5)(20x - 1)$
$\qquad = 2x(20x^2 - 1x - 100x + 5)$
$\qquad = 2x(20x^2 - 101x + 5)$

$= 2x(4x - 5)(5x - 1)$ *Incorrect.*   Wrong middle term.

$\qquad 2x(4x - 5)(5x - 1)$
$\qquad = 2x(20x^2 - 4x - 25x + 5)$
$\qquad = 2x(20x^2 - 29x + 5)$

$= 2x(2x - 5)(10x - 1)$ *Correct.* $2x(2x - 5)(10x - 1)$
$\qquad = 2x(20x^2 - 2x - 50x + 5)$
$\qquad = 2x(20x^2 - 52x + 5)$
$\qquad = 40x^3 - 104x^2 + 10x$

The correct factorization is $2x(2x - 5)(10x - 1)$.

Skill Practice   Factor.

3. $8t^3 + 38t^2 + 24t$

Often it is easier to factor a trinomial when the leading coefficient is positive. If the leading coefficient is negative, consider factoring out the opposite of the GCF.

Example 4   **Factoring a Trinomial by the Trial-and-Error Method**

Factor: $-45x^2 - 3xy + 18y^2$

Solution:

$-45x^2 - 3xy + 18y^2$

$= -3(15x^2 + xy - 6y^2)$ **Step 1:** Factor out -3 to make the leading term positive.

$= -3(\square x \quad \square y)(\square x \quad \square y)$ **Step 2:** List the factors of 15 and 6.

Factors of 15	Factors of 6	$-3(15x^2 + xy - 6y^2)$
$1 \cdot 15$	$1 \cdot 6$	
$3 \cdot 5$	$2 \cdot 3$	
	$3 \cdot 2$	
	$6 \cdot 1$	**Step 3:** We will construct all binomial combinations, without regard to signs first.

$-3(x \quad y)(15x \quad 6y)$
$-3(x \quad 2y)(15x \quad 3y)$ *Incorrect.* The binomials contain a common factor.
$-3(3x \quad 3y)(5x \quad 2y)$
$-3(3x \quad 6y)(5x \quad y)$

Test the remaining factorizations. The signs within parentheses must be opposite to produce a product of $-6y^2$. Also, the sum of the products of the inner terms and outer terms must be combined to form $1xy$.

$-3(x \quad 3y)(15x \quad 2y)$ *Incorrect.* Regardless of signs, $45xy$ and $2xy$ cannot be combined to equal xy.

$-3(x \quad 6y)(15x \quad y)$ *Incorrect.* Regardless of signs, $90xy$ and xy cannot be combined to equal xy.

$-3(3x \quad y)(5x \quad 6y)$ *Incorrect.* Regardless of signs, $5xy$ and $18xy$ cannot be combined to equal xy.

$-3(3x \quad 2y)(5x \quad 3y)$ *Correct.* The terms $10xy$ and $9xy$ can be combined to form xy provided that the signs are applied correctly. We require $10xy$ and $-9xy$.

$-3(3x + 2y)(5x - 3y)$ Factored form

TIP: Do not forget to write the GCF in the final answer.

Skill Practice Factor.

4. $-4x^2 + 26xy - 40y^2$

3. Factoring Perfect Square Trinomials

Recall from Section 5.6 that the square of a binomial always results in a **perfect square trinomial**.

$$(a + b)^2 = (a + b)(a + b) \xrightarrow{\text{Multiply}} = a^2 + 2ab + b^2$$
$$(a - b)^2 = (a - b)(a - b) \xrightarrow{\text{Multiply}} = a^2 - 2ab + b^2$$

For example, $(3x + 5)^2 = (3x)^2 + 2(3x)(5) + (5)^2$
$$= 9x^2 + 30x + 25 \text{ (perfect square trinomial)}$$

We now want to reverse this process by factoring a perfect square trinomial. The trial-and-error method can always be used; however, if we recognize the pattern for a perfect square trinomial, we can use one of the following formulas to reach a quick solution.

Skill Practice Answers

4. $-2(2x - 5y)(x - 4y)$

> ### Factored Form of a Perfect Square Trinomial
>
> $$a^2 + 2ab + b^2 = (a + b)^2$$
> $$a^2 - 2ab + b^2 = (a - b)^2$$

For example, $9x^2 + 30x + 25$ is a perfect square trinomial with $a = 3x$ and $b = 5$. Therefore, it factors as

$$9x^2 + 30x + 25 = (3x)^2 + 2(3x)(5) + (5)^2 = (3x + 5)^2$$
$$a^2 \ + 2 \ (a) \ (b) + (b)^2 = (a \ + \ b)^2$$

To apply the factored form of a perfect square trinomial, we must first be sure that the trinomial is indeed a perfect square trinomial.

> ### Checking for a Perfect Square Trinomial
>
> **1.** Determine whether the first and third terms are both perfect squares and have positive coefficients.
>
> **2.** If this is the case, identify a and b, and determine if the middle term equals $2ab$.

| **Example 5** | **Factoring Perfect Square Trinomials** |

Factor the trinomials completely.

a. $x^2 + 14x + 49$ **b.** $25y^2 - 20y + 4$

c. $18c^3 - 48c^2d + 32cd^2$ **d.** $5w^2 + 50w + 45$

Solution:

a. $x^2 + 14x + 49$ The GCF is 1.

- The first and third terms are positive.

- The first term is a perfect square: $x^2 = (x)^2$.

Perfect squares

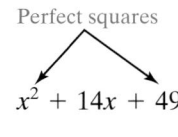

$x^2 + 14x + 49$

- The third term is a perfect square: $49 = (7)^2$.

- The middle term is twice the product of x and 7: $14x = 2(x)(7)$

$= (x)^2 + 2(x)(7) + (7)^2$ Hence, the trinomial is in the form $a^2 + 2ab + b^2$, where $a = x$ and $b = 7$.

$= (x + 7)^2$ Factor as $(a + b)^2$.

b. $25y^2 - 20y + 4$

Perfect squares

$$25y^2 - 20y + 4$$

$$= (5y)^2 - 2(5y)(2) + (2)^2$$

$$= (5y - 2)^2$$

The GCF is 1.

- The first and third terms are positive.

- The first term is a perfect square: $25y^2 = (5y)^2$.

- The third term is a perfect square: $4 = (2)^2$.

- In the middle: $20y = 2(5y)(2)$

Factor as $(a - b)^2$.

c. $18c^3 - 48c^2d + 32cd^2$

$$= 2c(9c^2 - 24cd + 16d^2)$$

Perfect squares

$$= 2c(9c^2 - 24cd + 16d^2)$$

$$= 2c[(3c)^2 - 2(3c)(4d) + (4d)^2]$$

$$= 2c(3c - 4d)^2$$

The GCF is $2c$.

- The first and third terms are positive.

- The first term is a perfect square: $9c^2 = (3c)^2$.

- The third term is a perfect square: $16d^2 = (4d)^2$.

- In the middle: $24cd = 2(3c)(4d)$

Factor as $(a - b)^2$.

d. $5w^2 + 50w + 45$

$$= 5(w^2 + 10w + 9)$$

Perfect squares

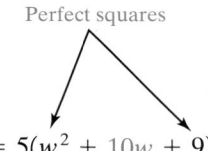

$$= 5(w^2 + 10w + 9)$$

$$= 5(w + 9)(w + 1)$$

The GCF is 5.

The first and third terms are perfect squares.

$$w^2 = (w)^2 \quad \text{and} \quad 9 = (3)^2$$

However, the middle term is not 2 times the product of w and 3. Therefore, this is not a perfect square trinomial.

$$10w \neq 2(w)(3)$$

To factor, use the trial-and-error method.

| Skill Practice | Factor completely.

5. $x^2 - 6x + 9$

6. $81w^2 + 72w + 16$

7. $5z^2 + 20z + 20$

8. $4x^2 + 13x + 9$

Skill Practice Answers

5. $(x - 3)^2$ **6.** $(9w + 4)^2$

7. $5(z + 2)^2$ **8.** $(4x + 9)(x + 1)$

> **TIP:** To help you identify a perfect square trinomial, we recommend that you familiarize yourself with the first several perfect squares.
>
> $(1)^2 = 1$ $(6)^2 = 36$ $(11)^2 = 121$
> $(2)^2 = 4$ $(7)^2 = 49$ $(12)^2 = 144$
> $(3)^2 = 9$ $(8)^2 = 64$ $(13)^2 = 169$
> $(4)^2 = 16$ $(9)^2 = 81$ $(14)^2 = 196$
> $(5)^2 = 25$ $(10)^2 = 100$ $(15)^2 = 225$
>
> If you do not recognize that a trinomial is a perfect square trinomial, you may still use the trial-and-error method to factor it.

Recall that a prime polynomial is a polynomial whose only factors are itself and 1. Not every trinomial is factorable by the methods presented in this text.

Example 6 | **Factoring a Trinomial by the Trial-and-Error Method**

Factor the trinomial by the trial-and-error method: $14p^2 - 56p + 21$

Solution:

$14p^2 - 56p + 21$

$= 7(2p^2 - 8p + 3)$ **Step 1:** The GCF is 7. Now factor the trinomial $2p^2 - 8p + 3$

$= 7(2p^2 - 8p + 3)$

$= 7(1p\ \square)(2p\ \square)$ **Step 2:** List the factors of 2 and the factors of 3.

Factors of 2	Factors of 3
$1 \cdot 2$	$1 \cdot 3$
	$3 \cdot 1$

Step 3: Construct all possible binomial factors using different combinations of the factors of 2 and 3. Because the third term in the trinomial is positive, both signs in the binomial must be the same. Because the middle term coefficient is negative, both signs will be negative.

$7(p - 1)(2p - 3) = 7(2p^2 - 3p - 2p + 3)$
$= 7(2p^2 - 5p + 3)$ *Incorrect.*   Wrong middle term.

$7(p - 3)(2p - 1) = 7(2p^2 - p - 6p + 3)$
$= 7(2p^2 - 7p + 3)$ *Incorrect.*   Wrong middle term.

None of the combinations of factors results in the correct product. Therefore, the polynomial $2p^2 - 8p + 3$ is prime and cannot be factored further.

The factored form of $14p^2 - 56p + 21$ is $7(2p^2 - 8p + 3)$.

Skill Practice Factor.

9. $3a^2 + a + 4$

Skill Practice Answers
9. Prime

Section 6.3   **Practice Exercises**

Study Skills Exercises

1. In addition to studying the material for a test, here are some other activities that people use when preparing for a test. Circle the importance of each statement.

	not important	somewhat important	very important
a. Get a good night's sleep the night before the test.	1	2	3
b. Eat a good meal before the test.	1	2	3
c. Wear comfortable clothes on the day of the test.	1	2	3
d. Arrive early to class on the day of the test.	1	2	3

2. Define the key terms:

   a. prime polynomial   **b. perfect square trinomial**

Review Exercises

For Exercises 3–8, factor completely.

3. $21a^2b^2 + 12ab^2 - 15a^2b$

4. $5uv^2 - 10u^2v + 25u^2v^2$

5. $mn - m - 2n + 2$

6. $5x - 10 - xy + 2y$

7. $6a^2 - 30a - 84$

8. $10b^2 + 20b - 240$

Concept 1: Factoring Trinomials by the Trial-and-Error Method

For Exercises 9–11, assume a, b, and c represent positive integers.

9. When factoring a polynomial of the form $ax^2 + bx + c$, pick an appropriate combination of signs.

   a. $(\ +\ )(\ +\ )$ **b.** $(\ -\ )(\ -\ )$ **c.** $(\ +\ )(\ -\ )$

10. When factoring a polynomial of the form $ax^2 - bx - c$, pick an appropriate combination of signs.

   a. $(\ +\ )(\ +\ )$ **b.** $(\ -\ )(\ -\ )$ **c.** $(\ +\ )(\ -\ )$

11. When factoring a polynomial of the form $ax^2 - bx + c$, pick an appropriate combination of signs.

   a. $(\ +\ )(\ +\ )$ **b.** $(\ -\ )(\ -\ )$ **c.** $(\ +\ )(\ -\ )$

For Exercises 12–29, factor completely by using the trial-and-error method.

12. $2y^2 - 3y - 2$

13. $2w^2 + 5w - 3$

14. $3n^2 + 13n + 4$

15. $2a^2 + 7a + 6$

16. $5x^2 - 14x - 3$

17. $7y^2 + 9y - 10$

18. $12c^2 - 5c - 2$

19. $6z^2 + z - 12$

20. $-12 + 10w^2 + 37w$

21. $-10 + 10p^2 + 21p$

22. $-5q - 6 + 6q^2$

23. $17a - 2 + 3a^2$

24. $6b - 23 + 4b^2$

25. $8 + 7x^2 - 18x$

26. $-8 + 25m^2 - 10m$

27. $8q^2 + 31q - 4$

28. $6y^2 - 19xy - 20x^2$

29. $12y^2 - 73yz + 6z^2$

Concept 2: Identifying GCF and Factoring Trinomials

For Exercises 30–39, factor completely. Be sure to factor out the GCF first.

30. $2m^2 - 12m - 80$

31. $3c^2 - 33c + 72$

32. $2y^5 + 13y^4 + 6y^3$

33. $3u^8 - 13u^7 + 4u^6$

34. $-a^2 - 15a + 34$

35. $-x^2 - 7x - 10$

36. $-12u^3 - 22u^2 + 20u$

37. $-18z^4 + 15z^3 + 12z^2$

38. $80m^2 - 100mp - 30p^2$

39. $60w^2 + 550wz - 500z^2$

Concept 3: Factoring Perfect Square Trinomials

40. Multiply. $(3x + 5)^2$

41. Multiply. $(2y - 7)^2$

42. a. Which trinomial is a perfect square trinomial? $x^2 + 4x + 4$ or $x^2 + 5x + 4$

   b. Factor the trinomials from part (a).

43. a. Which trinomial is a perfect square trinomial? $x^2 + 13x + 36$ or $x^2 + 12x + 36$

   b. Factor the trinomials from part (a).

For Exercises 44–55, factor completely. (*Hint:* Look for the pattern of a perfect square trinomial.)

44. $x^2 + 18x + 81$

45. $y^2 - 8y + 16$

46. $25z^2 - 20z + 4$

47. $36p^2 + 60p + 25$

48. $49a^2 + 42ab + 9b^2$

49. $25m^2 - 30mn + 9n^2$

50. $-2y + y^2 + 1$

51. $4 + w^2 - 4w$

52. $80z^2 + 120z + 45$

53. $36p^2 - 24p + 4$

54. $9y^2 + 78x + 25$

55. $4y^2 + 20y + 9$

Mixed Exercises

For Exercises 56–97, factor the trinomial completely.

56. $20z - 18 - 2z^2$

57. $25t - 5t^2 - 30$

58. $42 - 13q + q^2$

59. $-5w - 24 + w^2$

60. $6t^2 + 7t - 3$

61. $4p^2 - 9p + 2$

62. $4m^2 - 20m + 25$

63. $16r^2 + 24r + 9$

64. $5c^2 - c + 2$

65. $7s^2 + 2s + 9$

66. $6x^2 - 19xy + 10y^2$

67. $15p^2 + pq - 2q^2$

68. $12m^2 + 11mn - 5n^2$

69. $4a^2 + 5ab - 6b^2$

70. $6r^2 + rs - 2s^2$

71. $18x^2 - 9xy - 2y^2$

72. $4s^2 - 8st + t^2$

73. $6u^2 - 10uv + 5v^2$

74. $10t^2 - 23t - 5$

75. $16n^2 + 14n + 3$

76. $14w^2 + 13w - 12$

77. $12x^2 - 16x + 5$

78. $x^2 + 7x - 18$

79. $y^2 - 6y - 40$

80. $a^2 - 10a - 24$

81. $b^2 + 6b - 7$

82. $r^2 + 5r - 24$

83. $t^2 + 20t + 100$

84. $x^2 + 9xy + 20y^2$

85. $p^2 - 13pq + 36q^2$

86. $v^2 + 2v + 15$

87. $x^2 - x - 1$

88. $a^2 + 21ab + 20b^2$ **89.** $x^2 - 17xy - 18y^2$ **90.** $t^2 - 10t + 21$ **91.** $z^2 - 15z + 36$

92. $5d^3 + 3d^2 - 10d$ **93.** $3y^3 - y^2 + 12y$ ● **94.** $4b^3 - 4b^2 - 80b$ **95.** $2w^2 + 20w + 42$

96. $x^2y^2 - 13xy^2 + 30y^2$ **97.** $p^2q^2 - 14pq^2 + 33q^2$

Expanding Your Skills

Each pair of trinomials looks similar but differs by one sign. Factor each trinomial, and see how their factored forms differ.

98. **a.** $x^2 - 10x - 24$

b. $x^2 - 10x + 24$

99. **a.** $x^2 - 13x - 30$

b. $x^2 - 13x + 30$

100. **a.** $x^2 - 5x - 6$

b. $x^2 - 5x + 6$

101. **a.** $x^2 - 10x + 9$

b. $x^2 + 10x + 9$

For Exercises 102–105, factor completely.

102. $x^4 + 10x^2 + 9$

103. $y^4 + 4y^2 - 21$

104. $w^4 + 2w^2 - 15$

105. $p^4 - 13p^2 + 40$

Factoring Trinomials: AC-Method

Section 6.4

In Section 6.2, we factored trinomials with a leading coefficient of 1. In Section 6.3, we learned the trial-and-error method to factor the more general case in which the leading coefficient is any integer. In this section, we provide an alternative method to factor trinomials, called the ac-method.

Concepts

1. Factoring Trinomials by the AC-Method
2. Factoring Perfect Square Trinomials

1. Factoring Trinomials by the AC-Method

The product of two binomials results in a four-term expression that can sometimes be simplified to a trinomial. To factor the trinomial, we want to reverse the process.

<u>Multiply</u>:

Multiply the binomials.   Add the middle terms.

$(2x + 3)(x + 2) = \longrightarrow 2x^2 + 4x + 3x + 6 = \longrightarrow 2x^2 + 7x + 6$

<u>Factor</u>:

$2x^2 + 7x + 6 = \longrightarrow 2x^2 + 4x + 3x + 6 = \longrightarrow (2x + 3)(x + 2)$

Rewrite the middle term as   Factor by grouping.
a sum or difference of terms.

To factor a trinomial, $ax^2 + bx + c$, by the ac-method, we rewrite the middle term, bx, as a sum or difference of terms. The goal is to produce a four-term polynomial that can be factored by grouping. The process is outlined as follows.

> **AC-Method Factor $ax^2 + bx + c$ ($a \neq 0$)**
>
> 1. Multiply the coefficients of the first and last terms (ac).
> 2. Find two integers whose product is ac and whose sum is b. (If no pair of integers can be found, then the trinomial cannot be factored further and is a **prime polynomial**.)
> 3. Rewrite the middle term, bx, as the sum of two terms whose coefficients are the integers found in step 2.
> 4. Factor by grouping.

The ac-method for factoring trinomials is illustrated in Example 1. However, before we begin, keep these two important guidelines in mind:

- For any factoring problem you encounter, always factor out the GCF from all terms first.
- To factor a trinomial, write the trinomial in the form $ax^2 + bx + c$.

| Example 1 | **Factoring a Trinomial by the AC-Method** |

Factor the trinomial by the ac-method: $2x^2 + 7x + 6$

Solution:

$2x^2 + 7x + 6$

Factor out the GCF from all terms. In this case, the GCF is 1.

$2x^2 + 7x + 6$

Step 1: The trinomial is written in the form $ax^2 + bx + c$.

$a = 2, b = 7, c = 6$

Find the product $ac = (2)(6) = 12$.

12	12
$1 \cdot 12$	$(-1)(-12)$
$2 \cdot 6$	$(-2)(-6)$
$3 \cdot 4$	$(-3)(-4)$

Step 2: List all factors of ac and search for the pair whose sum equals the value of b. That is, list the factors of 12 and find the pair whose sum equals 7.

The numbers 3 and 4 satisfy both conditions: $3 \cdot 4 = 12$ and $3 + 4 = 7$.

$2x^2 + 7x + 6$

$= 2x^2 + 3x + 4x + 6$

Step 3: Write the middle term of the trinomial as the sum of two terms whose coefficients are the selected pair of numbers: 3 and 4.

$= 2x^2 + 3x \mid + 4x + 6$ **Step 4:** Factor by grouping.

$= x(2x + 3) + 2(2x + 3)$

$= (2x + 3)(x + 2)$

$\underline{\text{Check:}} (2x + 3)(x + 2) = 2x^2 + 4x + 3x + 6$
$= 2x^2 + 7x + 6 \checkmark$

| **Skill Practice** | Factor by the ac-method.

1. $2x^2 + 5x + 3$

TIP: One frequently asked question is whether the order matters when we rewrite the middle term of the trinomial as two terms (step 3). The answer is no. From the previous example, the two middle terms in step 3 could have been reversed to obtain the same result:

$$2x^2 + 7x + 6$$
$$= 2x^2 + 4x + 3x + 6$$
$$= 2x(x + 2) + 3(x + 2)$$
$$= (x + 2)(2x + 3)$$

This example also points out that the order in which two factors are written does not matter. The expression $(x + 2)(2x + 3)$ is equivalent to $(2x + 3)(x + 2)$ because multiplication is a commutative operation.

Example 2 | **Factoring Trinomials by the AC-Method**

Factor the trinomial by the ac-method: $-2x + 8x^2 - 3$

Solution:

$-2x + 8x^2 - 3$ First rewrite the polynomial in the form $ax^2 + bx + c$.

$= 8x^2 - 2x - 3$ The GCF is 1.

$a = 8, b = -2, c = -3$ **Step 1:** Find the product $ac = (8)(-3) = -24$.

-24	-24
$-1 \cdot 24$	$-24 \cdot 1$
$-2 \cdot 12$	$-12 \cdot 2$
$-3 \cdot 8$	$-8 \cdot 3$
$-4 \cdot 6$	$-6 \cdot 4$

Step 2: List all the factors of -24 and find the pair of factors whose sum equals -2.

The numbers -6 and 4 satisfy both conditions: $(-6)(4) = -24$ and $-6 + 4 = -2$.

$= 8x^2 - 2x - 3$ **Step 3:** Write the middle term of the trinomial as two terms whose coefficients are the selected pair of numbers, -6 and 4.

$= 8x^2 - 6x + 4x - 3$

$= 8x^2 - 6x \mid + 4x - 3$ **Step 4:** Factor by grouping.

$= 2x(4x - 3) + 1(4x - 3)$

$= (4x - 3)(2x + 1)$

Check: $(4x - 3)(2x + 1) = 8x^2 + 4x - 6x - 3$
$= 8x^2 - 2x - 3$ ✔

Skill Practice Factor by the ac-method.

2. $13w + 6w^2 + 6$

Avoiding Mistakes:

Before factoring a trinomial, be sure to write the trinomial in descending order. That is, write it in the form $ax^2 + bx + c$.

Skill Practice Answers
2. $(2w + 3)(3w + 2)$

Example 3	**Factoring a Trinomial by the AC-Method**

Factor the trinomial by the ac-method: $10x^3 - 85x^2 + 105x$

Solution:

$10x^3 - 85x^2 + 105x$ — The GCF is $5x$.

$= 5x(2x^2 - 17x + 21)$ — The trinomial is in the form $ax^2 + bx + c$.

$a = 2, b = -17, c = 21$ — **Step 1:** Find the product $ac = (2)(21) = 42$.

42	42
$1 \cdot 42$	$(-1)(-42)$
$2 \cdot 21$	$(-2)(-21)$
$3 \cdot 14$	$(-3)(-14)$
$6 \cdot 7$	$(-6)(-7)$

Step 2: List all the factors of 42 and find the pair whose sum equals -17.

The numbers -3 and -14 satisfy both conditions: $(-3)(-14) = 42$ and $-3 + (-14) = -17$.

$= 5x(2x^2 - 17x + 21)$ — **Step 3:** Write the middle term of the trinomial as two terms whose coefficients are the selected pair of numbers, -3 and -14.

$= 5x(2x^2 - 3x - 14x + 21)$

$= 5x(2x^2 - 3x \mid - 14x + 21)$ — **Step 4:** Factor by grouping.

$= 5x[x(2x - 3) - 7(2x - 3)]$

$= 5x[(2x - 3)(x - 7)]$

$= 5x(2x - 3)(x - 7)$

Avoiding Mistakes:

Be sure to bring down the GCF in each successive step as you factor.

TIP: Notice when the GCF is removed from the original trinomial, the new trinomial has smaller coefficients. This makes the factoring process simpler because the product ac is smaller. It is much easier to list the factors of 42 than the factors of 1050.

Original trinomial	**With the GCF factored out**
$10x^3 - 85x^2 + 105x$	$5x(2x^2 - 17x + 21)$
$ac = (10)(105) = 1050$	$ac = (2)(21) = 42$

Skill Practice	Factor by the ac-method.

3. $9y^3 - 30y^2 + 24y$

In most cases, it is easier to factor a trinomial with a positive leading coefficient.

Example 4	**Factoring a Trinomial by the AC-Method**

Factor: $-18x^2 + 21xy + 15y^2$

Skill Practice Answers

3. $3y(3y - 4)(y - 2)$

Solution:

$-18x^2 + 21xy + 15y^2$

$= -3(6x^2 - 7xy - 5y^2)$ Factor out -3 to make the leading term positive.

Step 1: The product $ac = (6)(-5) = -30$.

Step 2: The numbers -10 and 3 have a product of -30 and a sum of -7.

$= -3[6x^2 - 10xy + 3xy - 5y^2]$ **Step 3:** Rewrite the middle term, $-7xy$ as $-10xy + 3xy$.

$= -3[6x^2 - 10xy \mid + 3xy - 5y^2]$ **Step 4:** Factor by grouping.

$= -3[2x(3x - 5y) + y(3x - 5y)]$

$= -3(3x - 5y)(2x + y)$ Factored form

TIP: Do not forget to write the GCF in the final answer.

Skill Practice Factor.

4. $-8x^2 - 8xy + 30y^2$

2. Factoring Perfect Square Trinomials

Recall from Section 5.6 that the square of a binomial always results in a **perfect square trinomial**.

$(a + b)^2 = (a + b)(a + b) \xrightarrow{\text{Multiply}} = a^2 + 2ab + b^2$

$(a - b)^2 = (a - b)(a - b) \xrightarrow{\text{Multiply}} = a^2 - 2ab + b^2$

For example, $(3x + 5)^2 = (3x)^2 + 2(3x)(5) + (5)^2$

$= 9x^2 + 30x + 25$ (perfect square trinomial)

We now want to reverse this process by factoring a perfect square trinomial. The ac-method or the trial-and-error method can always be used; however, if we recognize the pattern for a perfect square trinomial, we can use one of the following formulas to reach a quick solution.

> **Factored Form of a Perfect Square Trinomial**
>
> $a^2 + 2ab + b^2 = (a + b)^2$
>
> $a^2 - 2ab + b^2 = (a - b)^2$

For example, $9x^2 + 30x + 25$ is a perfect square trinomial with $a = 3x$ and $b = 5$. Therefore, it factors as

$9x^2 + 30x + 25 = (3x)^2 + 2(3x)(5) + (5)^2 = (3x + 5)^2$
$ a^2 + 2\ (a)\ (b) + (b)^2 = (a\ +\ b)^2$

To apply the factored form of a perfect square trinomial, we must first be sure that the trinomial is indeed a perfect square trinomial.

Skill Practice Answers
4. $-2(2x - 3y)(2x + 5y)$

> ### Checking for a Perfect Square Trinomial
> 1. Determine whether the first and third terms are both perfect squares and have positive coefficients.
> 2. If this is the case, identify a and b, and determine if the middle term equals $2ab$.

Example 5	**Factoring Perfect Square Trinomials**

Factor the trinomials completely.

a. $x^2 + 10x + 25$ **b.** $49y^2 - 28y + 4$

c. $8w^3 - 24w^2z + 18wz^2$ **d.** $2x^2 + 52x + 50$

Solution:

a. $x^2 + 10x + 25$ The GCF is 1.

- The first and third terms are positive.

Perfect squares

$= x^2 + 10x + 25$

- The first term is a perfect square: $x^2 = (x)^2$

- The third term is a perfect square: $25 = (5)^2$

$= (x)^2 + 2(x)(5) + (5)^2$

- The middle term is twice the product of x and 5: $10x = 2(x)(5)$.

$= (x + 5)^2$

This is a perfect square trinomial with $a = x$ and $b = 5$. Factor as $a^2 + 2ab + b^2 = (a + b)^2$.

b. $49y^2 - 28y + 4$ The GCF is 1.

- The first and third terms are positive.

Perfect squares

$= 49y^2 - 28y + 4$

- The first term is a perfect square: $49y^2 = (7y)^2$

- The third term is a perfect square: $4 = (2)^2$

$= (7y)^2 - 2(7y)(2) + (2)^2$

- In the middle: $28y = 2(7y)(2)$.

$= (7y - 2)^2$

This is a perfect square trinomial with $a = 7y$ and $b = 2$. Factor as $a^2 - 2ab + b^2 = (a - b)^2$.

c. $8w^3 - 24w^2z + 18wz^2$

$= 2w(4w^2 - 12wz + 9z^2)$ The GCF is $2w$.

Perfect squares

- $4w^2$ is a perfect square: $4w^2 = (2w)^2$

- $9z^2$ is a perfect square: $9z^2 = (3z)^2$

$= 2w[(2w)^2 - 2(2w)(3z) + (3z)^2]$ • In the middle: $12wz = 2(2w)(3z)$

$= 2w(2w - 3z)^2$ This is a perfect square trinomial with $a = 2w$ and $b = 3z$. Factor as $a^2 - 2ab + b^2 = (a - b)^2$.

d. $2x^2 + 52x + 50$

$= 2(x^2 + 26x + 25)$ The GCF is 2.

        ↑

Middle term does not fit
the pattern for a perfect
square trinomial.

• The first and third terms are perfect squares. $x^2 = (x)^2$ and $25 = (5)^2$.

• However, the middle term is *not* 2 times the product of x and 5. Therefore, this is not a perfect square trinomial. $26x \neq 2(x)(5)$

$= 2(x + 25)(x + 1)$ Factor by using either the ac-method or the trial-and-error method.

Skill Practice   Factor.

5. $y^2 + 20y + 100$ **6.** $16x^2 - 24x + 9$

7. $27w^3 + 18w^2 + 3w$ **8.** $9x^2 + 15x + 4$

TIP: To help you identify a perfect square trinomial, we recommend that you familiarize yourself with the first several perfect squares.

$(1)^2 = 1$	$(6)^2 = 36$	$(11)^2 = 121$
$(2)^2 = 4$	$(7)^2 = 49$	$(12)^2 = 144$
$(3)^2 = 9$	$(8)^2 = 64$	$(13)^2 = 169$
$(4)^2 = 16$	$(9)^2 = 81$	$(14)^2 = 196$
$(5)^2 = 25$	$(10)^2 = 100$	$(15)^2 = 225$

If you do not recognize that a trinomial is a perfect square trinomial, you may still use the trial-and-error method or ac-method to factor it.

Recall that a prime polynomial is a polynomial whose only factors are itself and 1. It also should be noted that not every trinomial is factorable by the methods presented in this text.

Example 6   **Factoring a Trinomial by the AC-Method**

Factor the trinomial by the ac-method: $2p^2 - 8p + 3$

Solution:

$2p^2 - 8p + 3$ **Step 1:** The GCF is 1.

                   Step 2: The product $ac = 6$.

Skill Practice Answers

5. $(y + 10)^2$ **6.** $(4x - 3)^2$
7. $3w(3w + 1)^2$
8. $(3x + 1)(3x + 4)$

Step 3: List the factors of 6. Notice that no pair of factors has a sum of -8. Therefore, the trinomial cannot be factored.

$$\begin{array}{cc} 6 & 6 \\ 1 \cdot 6 & (-1)(-6) \\ 2 \cdot 3 & (-2)(-3) \end{array}$$

The trinomial $2p^2 - 8p + 3$ is a prime polynomial.

Skill Practice Factor.

Skill Practice Answer

9. The trinomial is prime.

9. $4x^2 + 5x + 2$

Section 6.4   Practice Exercises

Boost *your* GRADE at mathzone.com!

MathZone

• Practice Problems   • e-Professors
• Self-Tests   • Videos
• NetTutor

Study Skills Exercise

1. Define the key terms:

   a. prime polynomial   **b. perfect square trinomial**

Review Exercises

For Exercises 2–4, factor completely.

2. $5x(x - 2) - 2(x - 2)$ **3.** $8(y + 5) + 9y(y + 5)$ **4.** $6ab + 24b - 12a - 48$

Concept 1: Factoring Trinomials by the AC-Method

For Exercises 5–12, find the pair of integers whose product and sum are given.

5. Product: 12   Sum: 13   **6.** Product: 12   Sum: 7

7. Product: 8   Sum: -9 **8.** Product: -4 Sum: -3

9. Product: -20 Sum: 1   **10.** Product: -6 Sum: -1

11. Product: -18 Sum: 7    **12.** Product: -72 Sum: -6

For Exercises 13–30, factor the trinomials using the ac-method.

13. $3x^2 + 13x + 4$ **14.** $2y^2 + 7y + 6$ **15.** $4w^2 - 9w + 2$

16. $2p^2 - 3p - 2$ **17.** $2m^2 + 5m - 3$ **18.** $6n^2 + 7n - 3$

19. $8k^2 - 6k - 9$ **20.** $9h^2 - 12h + 4$ **21.** $4k^2 - 20k + 25$

22. $16h^2 + 24h + 9$ **23.** $5x^2 + x + 7$ **24.** $4y^2 - y + 2$

25. $10 + 9z^2 - 21z$

26. $13x + 4x^2 - 12$

27. $50y + 24 + 14y^2$

28. $-24 + 10w + 4w^2$

29. $12y^2 + 8yz - 15z^2$

30. $20a^2 + 3ab - 9b^2$

Concept 2: Factoring Perfect Square Trinomials

31. Multiply. $(5y - 7)^2$

32. Multiply. $(3x + 4)^2$

33. a. Which trinomial is a perfect square trinomial? $4x^2 - 25x + 25$ or $4x^2 - 20x + 25$

    b. Factor the trinomials from part (a).

34. a. Which trinomial is a perfect square trinomial? $9x^2 + 12x + 4$ or $9x^2 + 15x + 4$

    b. Factor the trinomials from part (a).

For Exercises 35–46, factor completely. (*Hint:* Look for the pattern of a perfect square trinomial.)

35. $t^2 - 16t + 64$

36. $n^2 + 18n + 81$

37. $49q^2 - 28q + 4$

38. $64y^2 - 80y + 25$

39. $36x + 4x^2 + 81$

40. $42p + 9p^2 + 49$

41. $32x^2 + 80xy + 50y^2$

42. $12w^2 - 36wz + 27z^2$

43. $4c^2 + 8c + 4$

44. $36x^2 - 24x + 4$

45. $9y^2 + 15y + 4$

46. $4w^2 + 20w + 9$

Mixed Exercises

For Exercises 47–79, factor completely.

47. $20p^2 - 19p + 3$

48. $4p^2 + 5pq - 6q^2$

49. $6u^2 - 19uv + 10v^2$

50. $15m^2 + mn - 2n^2$

51. $12a^2 + 11ab - 5b^2$

52. $3r^2 - rs - 14s^2$

53. $3h^2 + 19hk - 14k^2$

54. $2x^2 - 13xy + y^2$

55. $3p^2 + 20pq - q^2$

56. $3 - 14z + 16z^2$

57. $10w + 1 + 16w^2$

58. $b^2 + 16 - 8b$

59. $1 + q^2 - 2q$

60. $25x - 5x^2 - 30$

61. $20a - 18 - 2a^2$

62. $-6 - t + t^2$

63. $-6 + m + m^2$

64. $72x^2 + 18x - 2$

65. $20y^2 - 78y - 8$

66. $p^3 - 6p^2 - 27p$

67. $w^5 - 11w^4 + 28w^3$

68. $3x^3 + 10x^2 + 7x$

69. $4r^3 + 3r^2 - 10r$

70. $2p^3 - 38p^2 + 120p$

71. $4q^3 - 4q^2 - 80q$

72. $x^2y^2 + 14x^2y + 33x^2$

73. $a^2b^2 + 13ab^2 + 30b^2$

74. $-k^2 - 7k - 10$

75. $-m^2 - 15m + 34$

76. $-3n^2 - 3n + 90$

77. $-2h^2 + 28h - 90$

78. $x^4 - 7x^2 + 10$

79. $m^4 + 10m^2 + 21$

80. Is the expression $(2x + 4)(x - 7)$ factored completely? Explain why or why not.

81. Is the expression $(3x + 1)(5x - 10)$ factored completely? Explain why or why not.

Section 6.5   Factoring Binomials

1. Factoring a Difference of Squares

Up to this point, we have learned several methods of factoring, including:

- Factoring out the greatest common factor from a polynomial
- Factoring a four-term polynomial by grouping
- Recognizing and factoring perfect square trinomials
- Factoring trinomials by the ac-method or by the trial-and-error method

In this section, we will learn how to factor binomials. We begin by factoring a difference of squares. Recall from Section 5.6 that the product of two conjugates results in a **difference of squares**:

$$(a + b)(a - b) = a^2 - b^2$$

Therefore, to factor a difference of squares, the process is reversed. Identify a and b and construct the conjugate factors.

Factored Form of a Difference of Squares
$$a^2 - b^2 = (a + b)(a - b)$$

In addition to recognizing numbers that are perfect squares, it is helpful to recognize that a variable expression is a perfect square if its exponent is a multiple of 2. For example:

Perfect Squares

$$x^2 = (x)^2$$
$$x^4 = (x^2)^2$$
$$x^6 = (x^3)^2$$
$$x^8 = (x^4)^2$$
$$x^{10} = (x^5)^2$$

Example 1   Factoring Differences of Squares

Factor the binomials.

a. $y^2 - 25$ **b.** $49s^2 - 4t^4$ **c.** $18w^2z - 2z$

Solution:

a. $y^2 - 25$ The binomial is a difference of squares.

$= (y)^2 - (5)^2$ Write in the form: $a^2 - b^2$, where $a = y$, $b = 5$.

$= (y + 5)(y - 5)$ Factor as $(a + b)(a - b)$.

b. $49s^2 - 4t^4$ The binomial is a difference of squares.

$= (7s)^2 - (2t^2)^2$ Write in the form $a^2 - b^2$, where $a = 7s$ and $b = 2t^2$.

$= (7s + 2t^2)(7s - 2t^2)$ Factor as $(a + b)(a - b)$.

c. $18w^2z - 2z$ The GCF is $2z$.

$= 2z(9w^2 - 1)$ $(9w^2 - 1)$ is a difference of squares.

$= 2z[(3w)^2 - (1)^2]$ Write in the form: $a^2 - b^2$, where $a = 3w$, $b = 1$.

$= 2z(3w + 1)(3w - 1)$ Factor as $(a + b)(a - b)$.

| **Skill Practice** | Factor completely.

1. $a^2 - 64$ **2.** $25q^2 - 49w^2$ **3.** $98m^3n - 50mn$

The difference of squares $a^2 - b^2$ factors as $(a - b)(a + b)$. However, the *sum* of squares is not factorable.

Sum of Squares

Suppose a and b have no common factors. Then the **sum of squares** $a^2 + b^2$ is *not* factorable over the real numbers.

That is, $a^2 + b^2$ is prime over the real numbers.

To see why $a^2 + b^2$ is not factorable, consider the product of binomials:

$(a + b)(a - b) = a^2 - b^2$ Wrong sign

$(a + b)(a + b) = a^2 + 2ab + b^2$ Wrong middle term

$(a - b)(a - b) = a^2 - 2ab + b^2$ Wrong middle term

After exhausting all possibilities, we see that if a and b share no common factors, then the sum of squares $a^2 + b^2$ is a prime polynomial.

| **Example 2** | **Factoring Binomials**

Factor the binomials, if possible.

a. $p^2 - 9$ **b.** $p^2 + 9$

Solution:

a. $p^2 - 9$ Difference of squares

$= (p - 3)(p + 3)$ Factor as $a^2 - b^2 = (a - b)(a + b)$.

b. $p^2 + 9$ Sum of squares

Prime (cannot be factored)

| **Skill Practice** | Factor the binomials, if possible.

4. $t^2 - 144$ **5.** $t^2 + 144$

Skill Practice Answers

1. $(a + 8)(a - 8)$
2. $(5q + 7w)(5q - 7w)$
3. $2mn(7m + 5)(7m - 5)$
4. $(t - 12)(t + 12)$
5. Prime

2. Factoring a Sum or Difference of Cubes

A binomial $a^2 - b^2$ is a difference of squares and can be factored as $(a - b)(a + b)$. Furthermore, if a and b share no common factors, then a sum of squares $a^2 + b^2$ is not factorable over the real numbers. In this section, we will learn that both a difference of cubes, $a^3 - b^3$, and a sum of cubes, $a^3 + b^3$, are factorable.

Factoring a Sum or Difference of Cubes

Sum of Cubes: $a^3 + b^3 = (a + b)(a^2 - ab + b^2)$

Difference of Cubes: $a^3 - b^3 = (a - b)(a^2 + ab + b^2)$

Multiplication can be used to confirm the formulas for factoring a sum or difference of cubes:

$$(a + b)(a^2 - ab + b^2) = a^3 - a^2b + ab^2 + a^2b - ab^2 + b^3 = a^3 + b^3 \checkmark$$
$$(a - b)(a^2 + ab + b^2) = a^3 + a^2b + ab^2 - a^2b - ab^2 - b^3 = a^3 - b^3 \checkmark$$

To help you remember the formulas for factoring a sum or difference of cubes, keep the following guidelines in mind:

- The factored form is the product of a binomial and a trinomial.
- The first and third terms in the trinomial are the squares of the terms within the binomial factor.
- Without regard to signs, the middle term in the trinomial is the product of terms in the binomial factor.

Square the first term of the binomial.   Product of terms in the binomial

$$x^3 + 8 = (x)^3 + (2)^3 = (x + 2)[(x)^2 - (x)(2) + (2)^2]$$

Square the last term of the binomial.

- The sign within the binomial factor is the same as the sign of the original binomial.
- The first and third terms in the trinomial are always positive.
- The sign of the middle term in the trinomial is opposite the sign within the binomial.

Same sign   Positive

$$x^3 + 8 = (x)^3 + (2)^3 = (x + 2)[(x)^2 - (x)(2) + (2)^2]$$

Opposite signs

To help you recognize a sum or difference of cubes, we recommend that you familiarize yourself with the first several perfect cubes:

Perfect Cube	Perfect Cube
$1 = (1)^3$	$216 = (6)^3$
$8 = (2)^3$	$343 = (7)^3$
$27 = (3)^3$	$512 = (8)^3$
$64 = (4)^3$	$729 = (9)^3$
$125 = (5)^3$	$1000 = (10)^3$

It is also helpful to recognize that a variable expression is a perfect cube if its exponent is a multiple of 3. For example,

Perfect Cube

$$x^3 = (x)^3$$

$$x^6 = (x^2)^3$$

$$x^9 = (x^3)^3$$

$$x^{12} = (x^4)^3$$

Example 3 **Factoring a Sum of Cubes**

Factor: $w^3 + 64$

Solution:

$w^3 + 64$	w^3 and 64 are perfect cubes.
$= (w)^3 + (4)^3$	Write as $a^3 + b^3$, where $a = w$, $b = 4$.
$a^3 + b^3 = (a + b)(a^2 - ab + b^2)$	Apply the formula for a sum of cubes.
$(w)^3 + (4)^3 = (w + 4)[(w)^2 - (w)(4) + (4)^2]$	
$\qquad\qquad = (w + 4)(w^2 - 4w + 16)$	Simplify.

Skill Practice Factor.

6. $p^3 + 125$

Example 4 **Factoring a Difference of Cubes**

Factor: $27p^3 - q^6$

Solution:

$27p^3 - q^6$	$27p^3$ and q^6 are perfect cubes.
$(3p)^3 - (q^2)^3$	Write as $a^3 - b^3$, where $a = 3p$, $b = q^2$.
$a^3 - b^3 = (a - b)(a^2 + ab + b^2)$	Apply the formula for a difference of cubes.
$(3p)^3 - (q^2)^3 = (3p - q^2)[(3p)^2 + (3p)(q^2) + (q^2)^2]$	
$\qquad\qquad = (3p - q^2)(9p^2 + 3pq^2 + q^4)$	Simplify.

Skill Practice Factor.

7. $8y^3 - 27z^6$

Skill Practice Answers

6. $(p + 5)(p^2 - 5p + 25)$
7. $(2y - 3z^2)(4y^2 + 6yz^2 + 9z^4)$

3. Factoring Binomials: A Summary

After removing the GCF, the next step in any factoring problem is to recognize what type of pattern it follows. Exponents that are divisible by 2 are perfect squares and those divisible by 3 are perfect cubes. The formulas for factoring binomials are summarized in the following box:

Factoring Binomials

1. Difference of Squares: $a^2 - b^2 = (a + b)(a - b)$
2. Difference of Cubes: $a^3 - b^3 = (a - b)(a^2 + ab + b^2)$
3. Sum of Cubes: $a^3 + b^3 = (a + b)(a^2 - ab + b^2)$

Example 5   Factoring Binomials

Factor completely.

a. $27y^3 + 1$ **b.** $m^2 - \dfrac{1}{4}$ **c.** $3y^4 - 48$ **d.** $z^6 - 8w^3$

Solution:

a. $27y^3 + 1$

Sum of cubes: $27y^3 = (3y)^3$ and $1 = (1)^3$.

$= (3y)^3 + (1)^3$

Write as $a^3 + b^3$, where $a = 3y$ and $b = 1$.

$= (3y + 1)((3y)^2 - (3y)(1) + (1)^2)$

Apply the formula $a^3 + b^3 = (a + b)(a^2 - ab + b^2)$.

$= (3y + 1)(9y^2 - 3y + 1)$

Simplify.

b. $m^2 - \dfrac{1}{4}$

Difference of squares

$= (m)^2 - \left(\dfrac{1}{2}\right)^2$

Write as $a^2 - b^2$, where $a = m$ and $b = \frac{1}{2}$.

$= \left(m + \dfrac{1}{2}\right)\left(m - \dfrac{1}{2}\right)$

Apply the formula $a^2 - b^2 = (a + b)(a - b)$.

c. $3y^4 - 48$

$= 3(y^4 - 16)$

Factor out the GCF. The binomial is a difference of squares.

$= 3[(y^2)^2 - (4)^2]$

Write as $a^2 - b^2$, where $a = y^2$ and $b = 4$.

$= 3(y^2 + 4)(y^2 - 4)$

Apply the formula $a^2 - b^2 = (a + b)(a - b)$.

$y^2 + 4$ is a sum of squares and cannot be factored.

$= 3(y^2 + 4)(y + 2)(y - 2)$

$y^2 - 4$ is a difference of squares and can be factored further.

d. $z^6 - 8w^3$

Difference of cubes: $z^6 = (z^2)^3$ and $8w^3 = (2w)^3$

$= (z^2)^3 - (2w)^3$

Write as $a^3 - b^3$, where $a = z^2$ and $b = 2w$.

$= (z^2 - 2w)[(z^2)^2 + (z^2)(2w) + (2w)^2]$

Apply the formula $a^3 - b^3 = (a - b)(a^2 + ab + b^2)$.

$= (z^2 - 2w)(z^4 + 2z^2w + 4w^2)$

Simplify.

Each of the factorizations in Example 3 can be checked by multiplying.

Skill Practice Factor completely.

8. $1000x^3 + 1$ **9.** $25p^2 - \dfrac{1}{9}$ **10.** $2x^4 - 2$ **11.** $27a^6 - b^3$

Some factoring problems require more than one method of factoring. In general, when factoring a polynomial, be sure to factor completely.

Example 6 **Factoring Polynomials**

Factor completely.

a. $w^4 - 81$ **b.** $4x^3 + 4x^2 - 25x - 25$

Solution:

a. $w^4 - 81$

The GCF is 1. $w^4 - 81$ is a difference of squares.

$= (w^2)^2 - (9)^2$

Write in the form: $a^2 - b^2$, where $a = w^2$, $b = 9$.

$= (w^2 + 9)(w^2 - 9)$

Factor as $(a + b)(a - b)$.

$= (w^2 + 9)(w + 3)(w - 3)$

Note that $w^2 - 9$ can be factored further as a difference of squares. (The binomial $w^2 + 9$ is a sum of squares and cannot be factored further.)

b. $4x^3 + 4x^2 - 25x - 25$

The GCF is 1.

$= 4x^3 + 4x^2 - 25x - 25$

The polynomial has four terms. Factor by grouping.

$= 4x^2(x + 1) - 25(x + 1)$

$= (x + 1)(4x^2 - 25)$

$4x^2 - 25$ is a difference of squares.

$= (x + 1)(2x + 5)(2x - 5)$

Skill Practice Factor completely, if possible.

12. $y^4 - 625$ **13.** $x^3 + 6x^2 - 4x - 24$

Skill Practice Answers

8. $(10x + 1)(100x^2 - 10x + 1)$

9. $\left(5p - \dfrac{1}{3}\right)\left(5p + \dfrac{1}{3}\right)$

10. $2(x^2 + 1)(x - 1)(x + 1)$

11. $(3a^2 - b)(9a^4 + 3a^2b + b^2)$

12. $(y^2 + 25)(y + 5)(y - 5)$

13. $(x + 6)(x + 2)(x - 2)$

Example 7 Factoring Binomials

Factor the binomial $x^6 - y^6$ as

  a. A difference of cubes   **b.** A difference of squares

Solution:

a. $x^6 - y^6$

    Difference of cubes

$= (x^2)^3 - (y^2)^3$ Write as $a^3 - b^3$, where $a = x^2$ and $b = y^2$.

$= (x^2 - y^2)[(x^2)^2 + (x^2)(y^2) + (y^2)^2]$ Apply the formula $a^3 - b^3 = (a - b)(a^2 + ab + b^2)$.

$= (x^2 - y^2)(x^4 + x^2y^2 + y^4)$ Factor $x^2 - y^2$ as a difference of squares.

$= (x + y)(x - y)(x^4 + x^2y^2 + y^4)$

b. $x^6 - y^6$

    Difference of squares

$= (x^3)^2 - (y^3)^2$ Write as $a^2 - b^2$, where $a = x^3$ and $b = y^3$.

$= (x^3 + y^3)(x^3 - y^3)$ Apply the formula $a^2 - b^2 = (a + b)(a - b)$.

  Sum of cubes   Difference of cubes   Factor $x^3 + y^3$ as a sum of cubes.

                                   Factor $x^3 - y^3$ as a difference of cubes.

$= (x + y)(x^2 - xy + y^2)(x - y)(x^2 + xy + y^2)$

Notice that the expressions x^6 and y^6 are both perfect squares and perfect cubes because both exponents are multiples of 2 and of 3. Consequently, $x^6 - y^6$ can be factored initially as either the difference of squares or as the difference of cubes. In such a case, it is recommended that you factor the expression as a difference of squares first because it factors more completely into polynomials of lower degree. Hence,

$$x^6 - y^6 = (x + y)(x^2 - xy + y^2)(x - y)(x^2 + xy + y^2)$$

Skill Practice Factor completely.

14. $z^6 - 64$

Section 6.5	**Practice Exercises**

Study Skills Exercise

1. Define the key terms:

    a. difference of squares   **b. sum of squares**   **c. difference of cubes**   **d. sum of cubes**

Review Exercises

For Exercises 2–10, factor completely.

2. $3x^2 + x - 10$

3. $6x^2 - 17x + 5$

4. $6a^2b + 3a^3b$

5. $15x^2y^5 - 10xy^6$

6. $5p^2q + 20p^2 - 3pq - 12p$

7. $ax + ab - 6x - 6b$

8. $2xy - 3x - 4y + 6$

9. $25x^2 + 30x + 9$

10. $81y^2 - 36y + 4$

Concept 1: Factoring a Difference of Squares

11. What binomial factors as $(x - 5)(x + 5)$?

12. What binomial factors as $(n - 3)(n + 3)$?

13. What binomial factors as $(2p - 3q)(2p + 3q)$?

14. What binomial factors as $(7x - 4y)(7x + 4y)$?

For Exercises 15–34, factor the binomials completely.

15. $x^2 - 36$

16. $r^2 - 81$

17. $w^2 - 100$

18. $t^2 - 49$

19. $4a^2 - 121b^2$

20. $9x^2 - y^2$

21. $49m^2 - 16n^2$

22. $100a^2 - 49b^2$

23. $9q^2 + 16$

24. $36 + s^2$

25. $y^2 - 4z^2$

26. $b^2 - 144c^2$

27. $a^2 - b^4$

28. $y^4 - x^2$

29. $25p^2q^2 - 1$

30. $81s^2t^2 - 1$

31. $c^6 - 25$

32. $z^6 - 4$

33. $25 - 16t^2$

34. $64 - h^2$

Concept 2: Factoring a Sum or Difference of Cubes

35. Identify the expressions that are perfect cubes:

$$\{x^3, 8, 9, y^6, a^4, b^2, 3p^3, 27q^3, w^{12}, r^3s^6\}$$

36. Identify the expressions that are perfect cubes:

$$\{z^9, -81, 30, 8, 6x^3, y^{15}, 27a^3, b^2, p^3q^2, -1\}$$

37. From memory, write the formula to factor a sum of cubes:

$$a^3 + b^3 = \underline{\hspace{2in}}$$

38. From memory, write the formula to factor a difference of cubes:

$$a^3 - b^3 = \underline{\hspace{2in}}$$

For Exercises 39–58, factor the sums or differences of cubes.

39. $y^3 - 8$

40. $x^3 + 27$

41. $1 - p^3$

42. $q^3 + 1$

43. $w^3 + 64$

44. $8 - t^3$

45. $x^3 - 1000$

46. $8y^3 - 27$

47. $64t^3 + 1$

48. $125r^3 + 1$

49. $1000a^3 + 27$

50. $216b^3 - 125$

51. $n^3 - \dfrac{1}{8}$

52. $\dfrac{8}{27} + m^6$

53. $a^3 + b^6$

54. $u^6 - v^3$

55. $x^9 + 64y^3$

56. $125w^3 - z^9$

57. $25m^{12} + 16$

58. $36p^6 + 49q^4$

Concept 3: Factoring Binomials: A Summary

For Exercises 59–90, factor completely.

59. $x^4 - 4$

60. $b^4 - 25$

61. $a^2 + 9$

62. $w^2 + 36$

63. $t^3 + 64$

64. $u^3 + 27$

65. $g^3 - 4$

66. $h^3 - 25$

67. $4b^3 + 108$

68. $3c^3 - 24$

69. $5p^2 - 125$

70. $2q^4 - 8$

71. $\dfrac{1}{64} - 8h^3$

72. $\dfrac{1}{125} + k^6$

73. $x^4 - 16$

74. $p^4 - 81$

75. $q^6 - 64$

76. $a^6 - 1$

77. $\dfrac{4x^2}{9} - w^2$

78. $\dfrac{16y^2}{25} - x^2$

79. $2x^3 + 3x^2 - 2x - 3$

80. $3x^3 + x^2 - 12x - 4$

81. $16x^4 - y^4$

82. $1 - t^4$

83. $81y^4 - 16$

84. $u^4 - 256$

85. $81k^2 + 30k + 1$

86. $9h^2 - 15h + 4$

87. $k^3 + 4k^2 - 9k - 36$

88. $w^3 - 2w^2 - 4w + 8$

89. $2t^3 - 10t^2 - 2t + 10$

90. $9a^3 + 27a^2 - 4a - 12$

Expanding Your Skills

For Exercises 91–98, factor the difference of squares.

91. $(y - 3)^2 - 9$

92. $(x - 2)^2 - 4$

93. $(2p + 1)^2 - 36$

94. $(4q + 3)^2 - 25$

95. $16 - (t + 2)^2$

96. $81 - (a + 5)^2$

97. $100 - (2b - 5)^2$

98. $49 - (3k - 7)^2$

Section 6.6   General Factoring Summary

Concepts

1. Factoring Strategy
2. Mixed Practice

1. Factoring Strategy

To factor a polynomial, remember always to look for the greatest common factor first. Then identify the number of terms and the type of factoring problem the polynomial represents.

Factoring Strategy

1. Factor out the GCF (Section 6.1).

2. Identify whether the polynomial has two terms, three terms, or more than three terms.

3. If the polynomial has more than three terms, try factoring by grouping (Section 6.1).

4. If the polynomial has three terms, check first for a perfect square trinomial. Otherwise, factor the trinomial with the ac-method or the trial-and-error method (Sections 6.3 or 6.4, respectively).

5. If the polynomial has two terms, determine if it fits the pattern for
   - A difference of squares: $a^2 - b^2 = (a - b)(a + b)$
   - A sum of squares: $a^2 + b^2$ prime
   - A difference of cubes: $a^3 - b^3 = (a - b)(a^2 + ab + b^2)$
   - A sum of cubes: $a^3 + b^3 = (a + b)(a^2 - ab + b^2)$ (Section 6.5)

6. Be sure to factor the polynomial completely.

7. Check by multiplying.

2. Mixed Practice

Example 1 **Factoring Polynomials**

Factor out the GCF and identify the number of terms and type of factoring pattern represented by the polynomial. Then factor the polynomial completely.

a. $abx^2 - 3ax + 5bx - 15$ **b.** $20y^2 - 110y - 210$

c. $4p^3 + 20p^2 + 25p$ **d.** $w^3 + 1000$

e. $t^4 - \dfrac{1}{16}$ **f.** $y^3 - 5y^2 - 4y + 20$

Solution:

a. $abx^2 - 3ax + 5bx - 15$ The GCF is 1. The polynomial has four terms. Therefore, factor by grouping.

$abx^2 - 3ax \mid + 5bx - 15$

$= ax(bx - 3) + 5(bx - 3)$

$= (bx - 3)(ax + 5)$

Check: $(bx - 3)(ax + 5) = abx^2 + 5bx - 3ax - 15$ ✔

b. $20y^2 - 110y - 210$ The GCF is 10. The polynomial has three terms. The trinomial is not a perfect square trinomial. Use either the ac-method or the trial-and-error method.

$= 10(2y^2 - 11y - 21)$

$= 10(2y + 3)(y - 7)$

Check: $10(2y + 3)(y - 7) = 10(2y^2 - 14y + 3y - 21)$

$= 10(2y^2 - 11y - 21)$

$= 20y^2 - 110y - 210$ ✔

c. $4p^3 + 20p^2 + 25p$

$= p(4p^2 + 20p + 25)$

The GCF is p. The polynomial has three terms and is a perfect square trinomial, $a^2 + 2ab + b^2$, where $a = 2p$ and $b = 5$.

$= p(2p + 5)^2$

Apply the formula $a^2 + 2ab + b^2 = (a + b)^2$.

Check: $p(2p + 5)^2 = p[(2p)^2 + 2(2p)(5) + (5)^2]$

$= p(4p^2 + 20p + 25)$

$= 4p^3 + 20p^2 + 25p$ ✔

d. $w^3 + 1000$

$= (w)^3 + (10)^3$

The GCF is 1. The polynomial has two terms. The binomial is a sum of cubes, $a^3 + b^3$, where $a = w$ and $b = 10$.

$= (w + 10)(w^2 - 10w + 100)$

Apply the formula
$a^3 + b^3 = (a + b)(a^2 - ab + b^2)$.

Check: $(w + 10)(w^2 - 10w + 100) = w^3 - \cancel{10w^2} + \cancel{100w} + \cancel{10w^2} - \cancel{100w} + 1000$

$= w^3 + 100$ ✔

e. $t^4 - \dfrac{1}{16}$

The GCF is 1. The binomial is a difference of squares, $a^2 - b^2$.

$= \left(t^2 - \dfrac{1}{4}\right)\left(t^2 + \dfrac{1}{4}\right)$

Apply the formula
$a^2 - b^2 = (a - b)(a + b)$.

$= \left(t - \dfrac{1}{2}\right)\left(t + \dfrac{1}{2}\right)\left(t^2 + \dfrac{1}{4}\right)$

The binomial $t^2 - \frac{1}{4}$ is also a difference of squares.

Check: $(t - \frac{1}{2})(t + \frac{1}{2})(t^2 + \frac{1}{4}) = (t^2 + \cancel{\frac{1}{2}t} - \cancel{\frac{1}{2}t} - \frac{1}{4})(t^2 + \frac{1}{4})$

$= (t^2 - \frac{1}{4})(t^2 + \frac{1}{4})$

$= t^4 + \cancel{\frac{1}{4}t^2} - \cancel{\frac{1}{4}t^2} - \frac{1}{16}$

$= t^4 - \frac{1}{16}$ ✔

f. $y^3 - 5y^2 - 4y + 20$

The GCF is 1. The polynomial has four terms. Factor by grouping.

$= y^3 - 5y^2 \,\vert\, -4y + 20$

$= y^2(y - 5) - 4(y - 5)$

$= (y - 5)(y^2 - 4)$

$= (y - 5)(y - 2)(y + 2)$

The expression $y^2 - 4$ is a difference of squares and can be factored further as $(y - 2)(y + 2)$.

Check: $(y - 5)(y - 2)(y + 2) = (y - 5)(y^2 - \cancel{2y} + \cancel{2y} - 4)$

$= (y - 5)(y^2 - 4)$

$= (y^3 - 4y - 5y^2 + 20)$

$= y^3 - 5y^2 - 4y + 20$ ✔

Skill Practice   Factor completely.

1. $3ac + 6bc + ad + 2bd$

2. $18y^3 - 36y^2 + 10y$

3. $25t^2 - 20t + 4$

4. $4x^3 + 32y^3$

5. $w^4 - \dfrac{1}{81}$

6. $p^3 + 7p^2 - 9p - 63$

Skill Practice Answers

1. $(a + 2b)(3c + d)$
2. $2y(3y - 1)(3y - 5)$
3. $(5t - 2)^2$
4. $4(x + 2y)(x^2 - 2xy + 4y^2)$
5. $(w - \frac{1}{3})(w + \frac{1}{3})(w^2 + \frac{1}{9})$
6. $(p - 3)(p + 3)(p + 7)$

Section 6.6	**Practice Exercises**

Study Skills Exercise

1. This section summarizes the different methods of factoring that we have learned. In your own words, write a strategy for factoring any expression completely. Write it as if you were explaining the factoring process to another person.

Concept 1: Factoring Strategy

2. What is meant by a prime factor?

3. When factoring a trinomial, what pattern do you look for first before using the ac-method or trial-and-error method?

4. What is the first step in factoring any polynomial?

5. When factoring a binomial, what patterns can you look for?

6. What technique should be considered when factoring a four-term polynomial?

Concept 2: Mixed Practice

For Exercises 7–74,

a. Factor out the GCF from each polynomial. Then identify the category in which the polynomial best fits. Choose from
- difference of squares
- sum of squares
- difference of cubes
- sum of cubes
- trinomial (perfect square trinomial)
- trinomial (nonperfect square trinomial)
- four terms-grouping
- none of these

b. Factor the polynomial completely.

7. $2a^2 - 162$

8. $y^2 + 4y + 3$

9. $6w^2 - 6w$

10. $16z^4 - 81$

11. $3t^2 + 13t + 4$

12. $5r^3 + 5$

13. $3ac + ad - 3bc - bd$

14. $x^3 - 125$

15. $y^3 + 8$

16. $7p^2 - 29p + 4$

17. $3q^2 - 9q - 12$

18. $-2x^2 + 8x - 8$

19. $18a^2 + 12a$

20. $54 - 2y^3$

21. $4t^2 - 100$

22. $4t^2 - 31t - 8$

23. $10c^2 + 10c + 10$

24. $2xw - 10x + 3yw - 15y$

25. $x^3 + 0.001$

26. $4q^2 - 9$

27. $64 + 16k + k^2$

28. $s^2t + 5t + 6s^2 + 30$

29. $2x^2 + 2x - xy - y$

30. $w^3 + y^3$

31. $a^3 - c^3$

32. $3y^2 + y + 1$

33. $c^2 + 8c + 9$

34. $a^2 + 2a + 1$

35. $b^2 + 10b + 25$ **36.** $-t^2 - 4t + 32$ **37.** $-p^3 - 5p^2 - 4p$ **38.** $x^2 y^2 - 49$

39. $6x^2 - 21x - 45$ **40.** $20y^2 - 14y + 2$ **41.** $5a^2 bc^3 - 7abc^2$

42. $8a^2 - 50$ **43.** $t^2 + 2t - 63$ **44.** $b^2 + 2b - 80$

45. $ab + ay - b^2 - by$ **46.** $6x^3 y^4 + 3x^2 y^5$ **47.** $14u^2 - 11uv + 2v^2$

48. $9p^2 - 36pq + 4q^2$ **49.** $4q^2 - 8q - 6$ **50.** $9w^2 + 3w - 15$

51. $9m^2 + 16n^2$ **52.** $5b^2 - 30b + 45$ **53.** $6r^2 + 11r + 3$

54. $4s^2 + 4s - 15$ **55.** $16a^4 - 1$ **56.** $p^3 + p^2 c - 9p - 9c$

57. $81u^2 - 90uv + 25v^2$ **58.** $4x^2 + 16$ **59.** $x^2 - 5x - 6$

60. $q^2 + q - 7$ **61.** $2ax - 6ay + 4bx - 12by$ **62.** $8m^3 - 10m^2 - 3m$

63. $21x^4 y + 41x^3 y + 10x^2 y$ **64.** $2m^4 - 128$ **65.** $8uv - 6u + 12v - 9$

66. $4t^2 - 20t + st - 5s$ **67.** $12x^2 - 12x + 3$ **68.** $p^2 + 2pq + q^2$

69. $6n^3 + 5n^2 - 4n$ **70.** $4k^3 + 4k^2 - 3k$ **71.** $64 - y^2$

72. $36b - b^3$ **73.** $b^2 - 4b + 10$ **74.** $y^2 + 6y + 8$

Expanding Your Skills

For Exercises 75–78, factor completely.

75. $\dfrac{64}{125} p^3 - \dfrac{1}{8} q^3$ **76.** $\dfrac{1}{1000} r^3 + \dfrac{8}{27} s^3$ **77.** $a^{12} + b^{12}$ **78.** $a^9 - b^9$

Use Exercises 79–80 to investigate the relationship between division and factoring.

79. **a.** Use long division to divide $x^3 - 8$ by $(x - 2)$.

  b. Factor $x^3 - 8$.

80. **a.** Use long division to divide $y^3 + 27$ by $(y + 3)$.

  b. Factor $y^3 + 27$.

81. What trinomial multiplied by $(x - 2)$ gives a difference of cubes?

82. What trinomial multiplied by $(p + 3)$ gives a sum of cubes?

83. Write a binomial that when multiplied by $(4x^2 - 2x + 1)$ produces a sum of cubes.

84. Write a binomial that when multiplied by $(9y^2 + 15y + 25)$ produces a difference of cubes.

85. a. Write a polynomial that represents the area of the shaded region in the figure.

 b. Factor the expression from part (a).

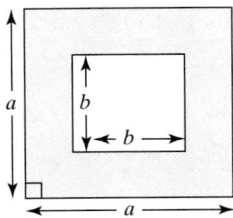

86. a. Write a polynomial that represents the area of the shaded region in the figure.

 b. Factor the expression from part (a).

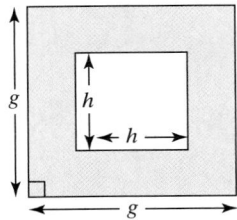

87. Use a difference of squares to find the product 67×73. [*Hint:* Write the product as $(70 - 3)(70 + 3)$.]

88. Use a difference of squares to find the product 85×75.

Solving Equations Using the Zero Product Rule   Section 6.7

1. Definition of a Quadratic Equation

In Section 2.1, we solved linear equations in one variable. These are equations of the form $ax + b = 0$ $(a \neq 0)$. A linear equation in one variable is sometimes called a first-degree polynomial equation because the highest degree of all its terms is 1. A second-degree polynomial equation in one variable is called a quadratic equation.

Concepts

1. Definition of a Quadratic Equation
2. Zero Product Rule
3. Solving Equations Using the Zero Product Rule
4. Applications of Quadratic Equations
5. Pythagorean Theorem

Definition of a Quadratic Equation in One Variable

If a, b, and c are real numbers such that $a \neq 0$, then a **quadratic equation** is an equation that can be written in the form

$$ax^2 + bx + c = 0$$

The following equations are quadratic because they can each be written in the form $ax^2 + bx + c = 0$, $(a \neq 0)$.

$$-4x^2 + 4x = 1 \qquad x(x - 2) = 3 \qquad (x - 4)(x + 4) = 9$$
$$-4x^2 + 4x - 1 = 0 \qquad x^2 - 2x = 3 \qquad x^2 - 16 = 9$$
$$x^2 - 2x - 3 = 0 \qquad x^2 - 25 = 0$$
$$x^2 + 0x - 25 = 0$$

2. Zero Product Rule

One method for solving a quadratic equation is to factor the equation and apply the zero product rule. The **zero product rule** states that if the product of two factors is zero, then one or both of its factors is zero.

> **Zero Product Rule**
>
> $$\text{If } ab = 0, \text{ then } a = 0 \text{ or } b = 0.$$

For example, the quadratic equation $x^2 - x - 12 = 0$ can be written in factored form as $(x - 4)(x + 3) = 0$. By the zero product rule, one or both factors must be zero. Hence, either $x - 4 = 0$ or $x + 3 = 0$. Therefore, to solve the quadratic equation, set each factor equal to zero and solve for x.

$$(x - 4)(x + 3) = 0 \qquad \text{Apply the zero product rule.}$$
$$x - 4 = 0 \quad \text{or} \quad x + 3 = 0 \qquad \text{Set each factor equal to zero.}$$
$$x = 4 \quad \text{or} \quad x = -3 \qquad \text{Solve each equation for } x.$$

3. Solving Equations Using the Zero Product Rule

Quadratic equations, like linear equations, arise in many applications in mathematics, science, and business. The following steps summarize the factoring method for solving a quadratic equation.

> **Steps for Solving a Quadratic Equation by Factoring**
>
> 1. Write the equation in the form: $ax^2 + bx + c = 0$.
> 2. Factor the equation completely.
> 3. Apply the zero product rule. That is, set each factor equal to zero, and solve the resulting equations.
>
> *Note:* The solution(s) found in step 3 may be checked by substitution in the original equation.

> **Example 1** **Solving Quadratic Equations**

Solve the quadratic equations.

a. $2x^2 - 9x = 5$ **b.** $4x^2 + 24x = 0$ **c.** $5x(5x + 2) = 10x + 9$

Solution:

a. $2x^2 - 9x = 5$

$\quad 2x^2 - 9x - 5 = 0 \qquad$ Write the equation in the form $ax^2 + bx + c = 0$.

$\quad (2x + 1)(x - 5) = 0 \qquad$ Factor the polynomial completely.

$\quad 2x + 1 = 0 \quad \text{or} \quad x - 5 = 0 \qquad$ Set each factor equal to zero.

$\quad\quad 2x = -1 \quad \text{or} \quad\quad x = 5 \qquad$ Solve each equation.

$\quad\quad\quad x = -\dfrac{1}{2} \quad \text{or} \quad\quad x = 5$

$\underline{\text{Check: } x = -\dfrac{1}{2}}$ $\underline{\text{Check: } x = 5}$

$2x^2 - 9x = 5$ $2x^2 - 9x = 5$

$2\left(-\dfrac{1}{2}\right)^2 - 9\left(-\dfrac{1}{2}\right) \overset{?}{=} 5$ $2(5)^2 - 9(5) \overset{?}{=} 5$

$2\left(\dfrac{1}{4}\right) + \dfrac{9}{2} \overset{?}{=} 5$ $2(25) - 45 \overset{?}{=} 5$

$\dfrac{1}{2} + \dfrac{9}{2} \overset{?}{=} 5$ $50 - 45 = 5 ✔$

$\dfrac{10}{2} = 5 ✔$

b. $4x^2 + 24x = 0$ The equation is already in the form $ax^2 + bx + c = 0$. (Note that $c = 0$.)

$4x(x + 6) = 0$ Factor completely.

$4x = 0$ or $x + 6 = 0$ Set each factor equal to zero.

$x = 0$ or $x = -6$ Each solution checks in the original equation.

c. $5x(5x + 2) = 10x + 9$

$25x^2 + 10x = 10x + 9$ Clear parentheses.

$25x^2 + 10x - 10x - 9 = 0$ Set the equation equal to zero.

$25x^2 - 9 = 0$ The equation is in the form $ax^2 + bx + c = 0$. (Note that $b = 0$.)

$(5x - 3)(5x + 3) = 0$ Factor completely.

$5x - 3 = 0$ or $5x + 3 = 0$ Set each factor equal to zero.

$5x = 3$ or $5x = -3$ Solve each equation.

$\dfrac{5x}{5} = \dfrac{3}{5}$ or $\dfrac{5x}{5} = \dfrac{-3}{5}$

$x = \dfrac{3}{5}$ or $x = -\dfrac{3}{5}$ Each solution checks in the original equation.

Skill Practice   Solve the quadratic equations.

1. $2y^2 + 19y = -24$

2. $5s^2 = 45$

3. $4z(z + 3) = 4z + 5$

The zero product rule can be used to solve higher degree polynomial equations provided the equations can be set to zero and written in factored form.

Skill Practice Answers

1. $y = -8,\ y = -\dfrac{3}{2}$

2. $s = 3,\ s = -3$

3. $z = -\dfrac{5}{2},\ z = \dfrac{1}{2}$

Example 2 **Solving Higher Degree Polynomial Equations**

Solve the equations.

a. $-6(y + 3)(y - 5)(2y + 7) = 0$ **b.** $w^3 + 5w^2 - 9w - 45 = 0$

Solution:

a. $-6(y + 3)(y - 5)(2y + 7) = 0$ The equation is already in factored form and equal to zero.

Set each factor equal to zero.

Solve each equation for y.

$-6 \neq 0$ or $y + 3 = 0$ or $y - 5 = 0$ or $2y + 7 = 0$

No solution, $y = -3$ or $y = 5$ or $y = -\dfrac{7}{2}$

Notice that when the constant factor is set equal to zero, the result is a contradiction $-6 = 0$. The constant factor does not produce a solution to the equation. Therefore, the only solutions are $y = -3$, $y = 5$, and $y = -\frac{7}{2}$. Each solution can be checked in the original equation.

b. $w^3 + 5w^2 - 9w - 45 = 0$ This is a higher degree polynomial equation.

$w^3 + 5w^2 \mid - 9w - 45 = 0$ The equation is already set equal to zero. Now factor.

$w^2(w + 5) - 9(w + 5) = 0$ Because there are four terms,
$(w + 5)(w^2 - 9) = 0$ try factoring by grouping.

$(w + 5)(w - 3)(w + 3) = 0$ $w^2 - 9$ is a difference of squares and can be factored further.

$w + 5 = 0$ or $w - 3 = 0$ or $w + 3 = 0$ Set each factor equal to zero.

$w = -5$ or $w = 3$ or $w = -3$ Solve each equation.

Each solution checks in the original equation.

Skill Practice   Solve the equations.

4. $5(p - 4)(p + 7)(2p - 9) = 0$
5. $x^3 + x^2 - 6x = 0$

4. Applications of Quadratic Equations

Example 3   **Using a Quadratic Equation in a Geometry Application**

A rectangular sign has an area of 40 ft^2. If the width is 3 feet shorter than the length, what are the dimensions of the sign?

Skill Practice Answers

4. $p = 4$, $p = -7$, $p = \dfrac{9}{2}$

5. $x = 0$, $x = -3$, $x = 2$

Solution:

Let x represent the length of the sign. Then $x - 3$ represents the width (Figure 6-1).

Label the variables.

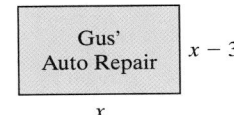

Figure 6-1

The problem gives information about the length of the sides and about the area. Therefore, we can form a relationship by using the formula for the area of a rectangle.

$A = l \cdot w$	Area equals length times width.
$40 = x(x - 3)$	Set up an algebraic equation.
$40 = x^2 - 3x$	Clear parentheses.
$0 = x^2 - 3x - 40$	Write the equation in the form, $ax^2 + bx + c = 0$.
$0 = (x - 8)(x + 5)$	Factor the equation.
$0 = x - 8$ or $0 = x + 5$	Set each factor equal to zero.
$8 = x$ or $-5 \cancel{=} x$	Because x represents the length of a rectangle, reject the negative solution.

The variable x represents the length of the sign. Thus, the length is 8 ft.

The expression $x - 3$ represents the width. The width is 8 ft − 3 ft, or 5 ft.

Skill Practice

6. The length of a rectangle is 5 ft more than the width. The area is 36 ft^2.

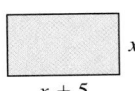

x

$x + 5$

Find the length and width of the rectangle.

Example 4 **Translating to a Quadratic Equation**

The product of two consecutive integers is 48 more than the larger integer. Find the integers.

Solution:

Let x represent the first (smaller) integer.

Then $x + 1$ represents the second (larger) integer.   Label the variables.

(First integer)(second integer) = (second integer) + 48   Verbal model

$$x(x + 1) = (x + 1) + 48$$ Algebraic equation

$$x^2 + x = x + 49$$ Simplify.

$$x^2 + x - x - 49 = 0$$ Set the equation equal to zero.

Skill Practice Answers

6. The width is 4 ft, and the length is 9 ft.

$$x^2 - 49 = 0$$

$$(x - 7)(x + 7) = 0 \qquad\qquad \text{Factor.}$$

$$x - 7 = 0 \quad \text{or} \quad x + 7 = 0 \qquad \text{Set each factor equal}$$
to zero.

$$x = 7 \quad \text{or} \quad x = -7 \qquad \text{Solve for } x.$$

Recall that x represents the smaller integer. Therefore, there are two possibilities for the pairs of consecutive integers.

If $x = 7$, then the larger integer is $x + 1$ or $7 + 1 = 8$.

If $x = -7$, then the larger integer is $x + 1$ or $-7 + 1 = -6$.

The integers are 7 and 8, or -7 and -6.

Skill Practice

7. The product of two consecutive odd integers is 9 more than ten times the smaller integer. Find the pairs of integers.

Example 5 **Using a Quadratic Equation in an Application**

A stone is dropped off a 64-ft cliff and falls into the ocean below. The height of the stone above sea level is given by the equation

$$h = -16t^2 + 64 \qquad \text{where } h \text{ is the stone's height in feet, and } t \text{ is the time in seconds.}$$

Find the time required for the stone to hit the water.

Solution:

When the stone hits the water, its height is zero. Therefore, substitute $h = 0$ into the equation.

$$h = -16t^2 + 64 \qquad\qquad \text{The equation is quadratic.}$$

$$0 = -16t^2 + 64 \qquad\qquad \text{Substitute } h = 0.$$

$$0 = -16(t^2 - 4) \qquad\qquad \text{Factor out the GCF.}$$

$$0 = -16(t - 2)(t + 2) \qquad\quad \text{Factor as a difference of squares.}$$

$$-16 \neq 0 \quad \text{or} \quad t - 2 = 0 \quad \text{or} \quad t + 2 = 0 \qquad \text{Set each factor to zero.}$$

$$\text{No solution,} \qquad t = 2 \quad \text{or} \qquad t = -2 \quad \text{Solve for } t.$$

The negative value of t is rejected because the stone cannot fall for a negative time. Therefore, the stone hits the water after 2 seconds.

Skill Practice

8. An object is launched into the air from the ground and its height is given by $h = -16t^2 + 144t$, where h is the height in feet after t seconds. Find the time required for the object to hit the ground.

Skill Practice Answers

7. The pairs of consecutive odd integers are 9 and 11, or -1 and 1.
8. 9 seconds

In Example 5, we can analyze the path of the stone as it falls from the cliff. Compute the height values at various times between 0 and 2 seconds (Table 6-1 and Table 6-2). The ordered pairs can be graphed where t is used in place of x and h is used in place of y.

Table 6-1

Time, t (sec)	Height, h (ft)
0.0	
0.5	
1.0	
1.5	
2.0	

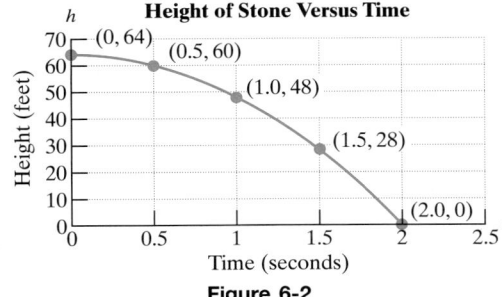

$h = -16(0.0)^2 + 64 = 64$

$h = -16(0.5)^2 + 64 = 60$

$h = -16(1.0)^2 + 64 = 48$

$h = -16(1.5)^2 + 64 = 28$

$h = -16(2.0)^2 + 64 = 0$

Table 6-2

Time, t (sec)	Height, h (ft)
0.0	64
0.5	60
1.0	48
1.5	28
2.0	0

The graph of the height of the stone versus time is shown in Figure 6-2.

Height of Stone Versus Time

(0, 64), (0.5, 60), (1.0, 48), (1.5, 28), (2.0, 0)

Height (feet) vs. Time (seconds)

Figure 6-2

5. Pythagorean Theorem

Recall that a right triangle is a triangle that contains a 90° angle. Furthermore, the sum of the squares of the two legs (the shorter sides) of a right triangle equals the square of the hypotenuse (the longest side). This important fact is known as the Pythagorean theorem. The Pythagorean theorem is an enduring landmark of mathematical history from which many mathematical ideas have been built. Although the theorem is named after Pythagoras (sixth century B.C.E.), a Greek mathematician and philosopher, it is thought that the ancient Babylonians were familiar with the principle more than a thousand years earlier.

For the right triangle shown in Figure 6-3, the **Pythagorean theorem** is stated as:

$$a^2 + b^2 = c^2$$

In this formula, a and b are the legs of the right triangle and c is the hypotenuse. Notice that the hypotenuse is the longest side of the right triangle and is opposite the 90° angle.

The triangle shown below is a right triangle. Notice that the lengths of the sides satisfy the Pythagorean theorem.

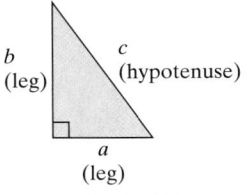

b (leg) c (hypotenuse)

a (leg)

Figure 6-3

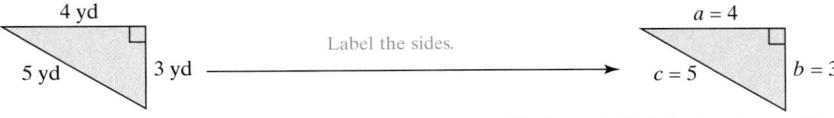

The shorter sides are labeled as a and b.

4 yd

5 yd   3 yd

Label the sides.

$a = 4$

$c = 5$ $b = 3$

The longest side is the hypotenuse. It is always labeled as c.

$$a^2 + b^2 = c^2 \qquad \text{Apply the Pythagorean theorem.}$$

$$(4)^2 + (3)^2 = (5)^2 \qquad \text{Substitute } a = 4, b = 3, \text{ and } c = 5.$$

$$16 + 9 = 25$$

$$25 = 25 \text{ ✔}$$

Example 6 **Applying the Pythagorean Theorem**

Find the length of the missing side of the right triangle.

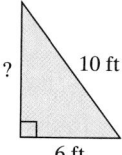

Solution:

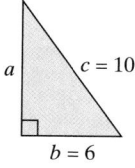

Label the triangle.

$$a^2 + b^2 = c^2 \qquad \text{Apply the Pythagorean theorem.}$$

$$a^2 + 6^2 = 10^2 \qquad \text{Substitute } b = 6 \text{ and } c = 10.$$

$$a^2 + 36 = 100 \qquad \text{Simplify.}$$

The equation is quadratic. Set the equation equal to zero.

$$a^2 + 36 - 100 = 100 - 100 \qquad \text{Subtract 100 from both sides.}$$

$$a^2 - 64 = 0$$

$$(a + 8)(a - 8) = 0 \qquad \text{Factor.}$$

$$a + 8 = 0 \quad \text{or} \quad a - 8 = 0 \qquad \text{Set each factor equal to zero.}$$

$$a \cancel{=} -8 \quad \text{or} \quad a = 8 \qquad \text{Because } x \text{ represents the length of a side of a triangle, reject the negative solution.}$$

The third side is 8 ft.

Skill Practice

9. Find the length of the missing side.

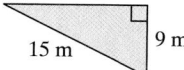

Example 7 **Using a Quadratic Equation in an Application**

A 13-ft board is used as a ramp to unload furniture off a loading platform. If the distance between the top of the board and the ground is 7 ft less than the distance between the bottom of the board and the base of the platform, find both distances.

Skill Practice Answers

9. The length of the third side is 12 m.

Solution:

Let x represent the distance between the bottom of the board and the base of the platform. Then $x - 7$ represents the distance between the top of the board and the ground (Figure 6-4).

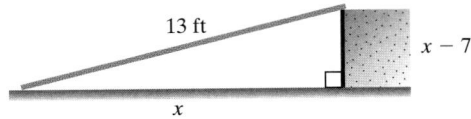

Figure 6-4

$$a^2 + b^2 = c^2 \qquad \text{Pythagorean theorem}$$

$$x^2 + (x - 7)^2 = (13)^2$$

$$x^2 + [(x)^2 - 2(x)(7) + (7)^2] = 169$$

$$x^2 + x^2 - 14x + 49 = 169$$

$$2x^2 - 14x + 49 = 169 \qquad \text{Combine } like \text{ terms.}$$

$$2x^2 - 14x + 49 - 169 = 169 - 169 \qquad \text{Set the equation equal to zero.}$$

$$2x^2 - 14x - 120 = 0 \qquad \text{Write the equation in the form } ax^2 + bx + c = 0.$$

$$2(x^2 - 7x - 60) = 0 \qquad \text{Factor.}$$

$$2(x - 12)(x + 5) = 0$$

$$2 \neq 0 \quad \text{or} \quad x - 12 = 0 \quad \text{or} \quad x + 5 = 0 \qquad \text{Set each factor equal to zero.}$$

$$x = 12 \quad \text{or} \quad x \neq -5 \qquad \text{Solve both equations for } x.$$

> **Avoiding Mistakes:**
>
> Recall that the square of a binomial results in a perfect square trinomial.
>
> $(a - b)^2 = a^2 - 2ab + b^2$
>
> Don't forget the middle term.

Recall that x represents the distance between the bottom of the board and the base of the platform. We reject the negative value of x because a distance cannot be negative. Therefore, the distance between the bottom of the board and the base of the platform is 12 ft. The distance between the top of the board and the ground is $x - 7 = 5$ ft.

> **Skill Practice**

10. A 5-yd ladder leans against a wall. The distance from the bottom of the wall to the top of the ladder is 1 yd more than the distance from the bottom of the wall to the bottom of the ladder. Find both distances.

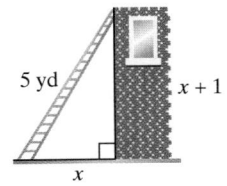

Section 6.7   Practice Exercises

Study Skills Exercise

1. Define the key terms:

    a. quadratic equation   **b.** Pythagorean theorem   **c.** zero product rule

Review Exercises

For Exercises 2–7, factor completely.

2. $6a - 8 - 3ab + 4b$

3. $4b^2 - 44b + 120$

4. $8u^2v^2 - 4uv$

5. $3x^2 + 10x - 8$

6. $3h^2 - 75$

7. $4x^2 + 16y^2$

Concept 1: Definition of a Quadratic Equation

For Exercises 8–13, identify the equations as linear, quadratic, or neither.

8. $4 - 5x = 0$

9. $5x^3 + 2 = 0$

10. $3x - 6x^2 = 0$

11. $1 - x + 2x^2 = 0$

12. $7x^4 + 8 = 0$

13. $3x + 2 = 0$

Concept 2: Zero Product Rule

For Exercises 14–21, solve the equations using the zero product rule.

14. $(x - 5)(x + 1) = 0$

15. $(x + 3)(x - 1) = 0$

16. $(3x - 2)(3x + 2) = 0$

17. $(2x - 7)(2x + 7) = 0$

18. $2(x - 7)(x - 7) = 0$

19. $3(x + 5)(x + 5) = 0$

20. $x(x - 4)(2x + 3) = 0$

21. $x(3x + 1)(x + 1) = 0$

22. For a quadratic equation of the form $ax^2 + bx + c = 0$, what must be done before applying the zero product rule?

23. What are the requirements needed to use the zero product rule to solve a quadratic equation or higher degree polynomial equation?

Concept 3: Solving Equations Using the Zero Product Rule

For Exercises 24–67, solve the equations.

24. $p^2 - 2p - 15 = 0$

25. $y^2 - 7y - 8 = 0$

26. $z^2 + 10z - 24 = 0$

27. $w^2 - 10w + 16 = 0$

28. $2q^2 - 7q - 4 = 0$

29. $4x^2 - 11x - 3 = 0$

30. $0 = 9x^2 - 4$

31. $4a^2 - 49 = 0$

32. $2k^2 - 28k + 96 = 0$

33. $0 = 2t^2 + 20t + 50$

34. $0 = 2m^3 - 5m^2 - 12m$

35. $3n^3 + 4n^2 + n = 0$

36. $(3p + 1)(p - 3)(p + 6) = 0$

37. $(2x - 1)(x - 10)(x + 7) = 0$

38. $x^3 - 16x = 0$

39. $t^3 - 36t = 0$

40. $3x^2 - 14x - 5 = 0$

41. $2y^2 + 3y - 9 = 0$

42. $16m^2 = 9$

43. $9n^2 = 1$

44. $2y^3 + 14y^2 = -20y$

45. $3d^3 - 6d^2 = 24d$

46. $5t - 2(t - 7) = 0$

47. $8h = 5(h - 9) + 6$

48. $2c(c - 8) = -30$

49. $3q(q - 3) = 12$

50. $b^3 = -4b^2 - 4b$

51. $x^3 + 36x = 12x^2$

52. $3(a^2 + 2a) = 2a^2 - 9$

53. $9(k - 1) = -4k^2$

54. $2n(n + 2) = 6$

55. $3p(p - 1) = 18$

56. $x(2x + 5) - 1 = 2x^2 + 3x + 2$

57. $3z(z - 2) - z = 3z^2 + 4$

58. $27q^2 = 9q$

59. $21w^2 = 14w$

60. $3(c^2 - 2c) = 0$

61. $2(4d^2 + d) = 0$

62. $y^3 - 3y^2 - 4y + 12 = 0$

63. $t^3 + 2t^2 - 16t - 32 = 0$

64. $(x - 1)(x + 2) = 18$

65. $(w + 5)(w - 3) = 20$

66. $(p + 2)(p + 3) = 1 - p$

67. $(k - 6)(k - 1) = -k - 2$

Concept 4: Applications of Quadratic Equations

68. If eleven is added to the square of a number, the result is sixty. Find all such numbers.

69. If a number is added to two times its square, the result is thirty-six. Find all such numbers.

70. If twelve is added to six times a number, the result is twenty-eight less than the square of the number. Find all such numbers.

71. The square of a number is equal to twenty more than the number. Find all such numbers.

72. The product of two consecutive odd integers is sixty-three. Find all such integers.

73. The product of two consecutive even integers is forty-eight. Find all such integers.

74. The sum of the squares of two consecutive integers is one more than ten times the larger number. Find all such integers.

75. The sum of the squares of two consecutive integers is nine less than ten times the sum of the integers. Find all such integers.

76. The length of a rectangular room is 5 yd more than the width. If 300 yd² of carpeting cover the room, what are the dimensions of the room?

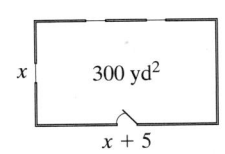

77. The width of a rectangular painting is 2 in. less than the length. The area is 120 in.² Find the length and width.

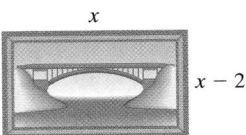

78. The width of a rectangular slab of concrete is 3 m less than the length. The area is 28 m².

    a. What are the dimensions of the rectangle?

    b. What is the perimeter of the rectangle?

79. The width of a rectangular picture is 7 in. less than the length. The area of the picture is 78 in.²

    a. What are the dimensions of the rectangle?

    b. What is the perimeter of the rectangle?

80. The base of a triangle is 1 ft less than twice the height. The area is 14 ft². Find the base and height of the triangle.

81. The height of a triangle is 5 cm less than 3 times the base. The area is 125 cm². Find the base and height of the triangle.

82. In a physics experiment, a ball is dropped off a 144-ft platform. The height of the ball above the ground is given by the equation

$$h = -16t^2 + 144$$

    where h is the ball's height in feet, and t is the time in seconds after the ball is dropped ($t \geq 0$).

Find the time required for the ball to hit the ground. (*Hint:* Let $h = 0$.)

83. A stone is dropped off a 256-ft cliff. The height of the stone above the ground is given by the equation

$$h = -16t^2 + 256$$

    where h is the stone's height in feet, and t is the time in seconds after the stone is dropped ($t \geq 0$).

Find the time required for the stone to hit the ground.

84. An object is shot straight up into the air from ground level with an initial speed of 24 ft/sec. The height of the object (in feet) is given by the equation

$$h = -16t^2 + 24t$$

    where t is the time in seconds after launch ($t \geq 0$).

Find the time(s) when the object is at ground level.

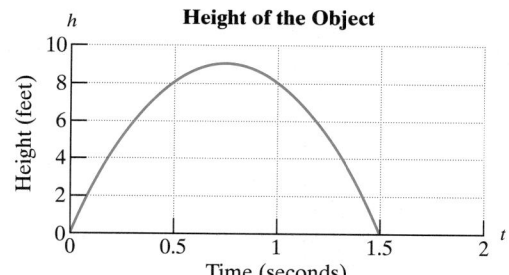

85. A rocket is launched straight up into the air from the ground with initial speed of 64 ft/sec. The height of the rocket (in feet) is given by the equation

$$h = -16t^2 + 64t$$

    where t is the time in seconds after launch ($t \geq 0$).

Find the time(s) when the ball is at ground level.

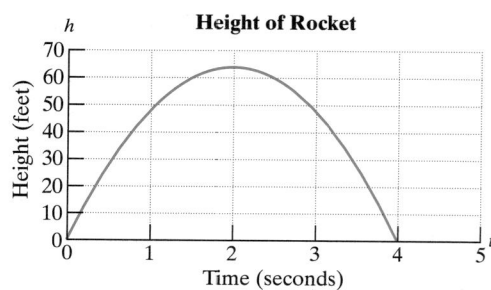

Concept 5: Pythagorean Theorem

86. Sketch a right triangle and label the sides with the words *leg* and *hypotenuse*.

87. State the Pythagorean theorem.

For Exercises 88–91, find the length of the missing side of the right triangle.

88.

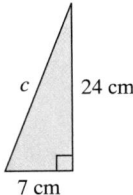

c 24 cm

7 cm

89.

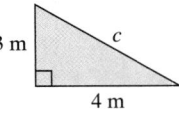

3 m *c* 4 m

90.

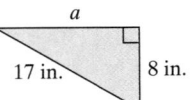

a 17 in. 8 in.

91.

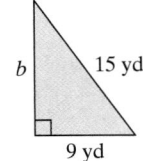

b 15 yd 9 yd

92. Find the length of the supporting brace.

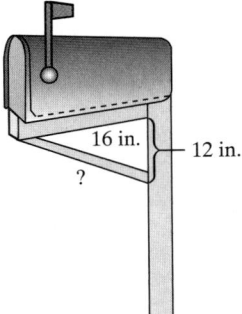

16 in. 12 in. ?

93. Find the height of the airplane above the ground.

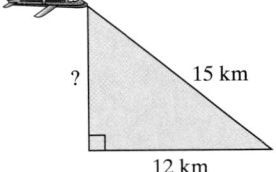

? 15 km 12 km

94. A 17-ft ladder rests against the side of a house. The distance between the top of the ladder and the ground is 7 ft more than the distance between the base of the ladder and the bottom of the house. Find both distances.

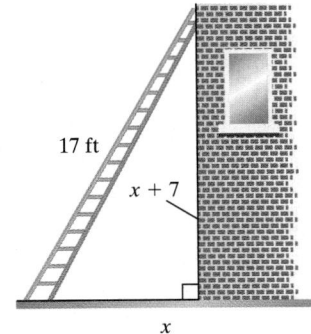

17 ft

x + 7

x

95. Darcy holds the end of a kite string 3 ft (1 yd) off the ground and wants to estimate the height of the kite. Her friend Jenna is 24 yd away from her, standing directly under the kite as shown in the figure. If Darcy has 30 yd of string out, find the height of the kite (ignore the sag in the string).

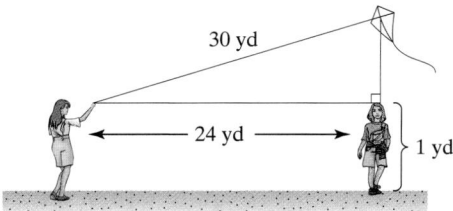

96. Two boats leave a marina. One travels east, and the other travels south. After 30 min, the second boat has traveled 1 mile farther than the first boat and the distance between the boats is 5 miles. Find the distance each boat traveled.

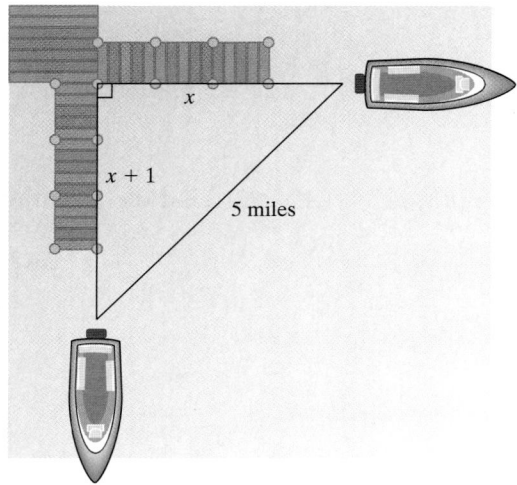

97. One leg of a right triangle is 4 m less than the hypotenuse. The other leg is 2 m less than the hypotenuse. Find the length of the hypotenuse.

98. The longer leg of a right triangle is 1 cm less than twice the shorter leg. The hypotenuse is 1 cm greater than twice the shorter leg. Find the length of the shorter leg.

Chapter 6 SUMMARY

Section 6.1 Greatest Common Factor and Factoring by Grouping

Key Concepts

The **greatest common factor** (GCF) is the greatest factor common to all terms of a polynomial. To factor out the GCF from a polynomial, use the distributive property.

A four-term polynomial may be factorable by grouping.

Steps to Factoring by Grouping

1. Identify and factor out the GCF from all four terms.
2. Factor out the GCF from the first pair of terms. Factor out the GCF or its opposite from the second pair of terms.
3. If the two terms share a common binomial factor, factor out the binomial factor.

Examples

Example 1

$3x(a + b) - 5(a + b)$ Greatest common factor is $(a + b)$.

$= (a + b)(3x - 5)$

Example 2

$60xa - 30xb - 80ya + 40yb$

$= 10[6xa - 3xb - 8ya + 4yb]$ Factor out GCF.

$= 10[3x(2a - b) - 4y(2a - b)]$ Factor by grouping.

$= 10(2a - b)(3x - 4y)$

Section 6.2 Factoring Trinomials of the Form $x^2 + bx + c$ (Optional)

Key Concepts

Factoring a Trinomial with a Leading Coefficient of 1:

A trinomial of the form $x^2 + bx + c$ factors as

$$x^2 + bx + c = (x \ \Box)(x \ \Box)$$

where the remaining terms are given by two integers whose product is c and whose sum is b.

Examples

Example 1

Factor: $x^2 - 14x + 45$

$x^2 - 14x + 45$ The integers -5 and -9 have a product of 45 and a sum of -14.

$= (x \ \Box)(x \ \Box)$

$= (x - 5)(x - 9)$

Section 6.3 Factoring Trinomials: Trial-and-Error Method

Key Concepts

Trial-and-Error Method for Factoring Trinomials in the Form $ax^2 + bx + c$ (where $a \neq 0$)

1. Factor out the GCF from all terms.
2. List the pairs of factors of a and the pairs of factors of c. Consider the reverse order in one of the lists.
3. Construct two binomials of the form

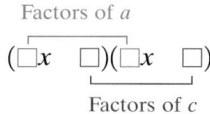

Factors of a

$(\square x \quad \square)(\square x \quad \square)$

Factors of c

4. Test each combination of factors and signs until the product forms the correct trinomial.
5. If no combination of factors produces the correct product, then the trinomial is prime.

The factored form of a **perfect square trinomial** is the square of a binomial:

$a^2 + 2ab + b^2 = (a + b)^2$

$a^2 - 2ab + b^2 = (a - b)^2$

Examples

Example 1

$10y^2 + 35y - 20$

$= 5(2y^2 + 7y - 4)$

The pairs of factors of 2 are: $2 \cdot 1$
The pairs of factors of -4 are:

$$-1(4) \qquad 1(-4)$$
$$-2(2) \qquad 2(-2)$$
$$-4(1) \qquad 4(-1)$$

$(2y - 2)(y + 2) = 2y^2 + 2y - 4$	No
$(2y - 4)(y + 1) = 2y^2 - 2y - 4$	No
$(2y + 1)(y - 4) = 2y^2 - 7y - 4$	No
$(2y + 2)(y - 2) = 2y^2 - 2y - 4$	No
$(2y + 4)(y - 1) = 2y^2 + 2y - 4$	No
$(2y - 1)(y + 4) = 2y^2 + 7y - 4$	Yes

Example 2

$9w^2 - 30wz + 25z^2$

$= (3w)^2 - 2(3w)(5z) + (5z)^2$

$= (3w - 5z)^2$

Section 6.4 Factoring Trinomials: AC-Method

Key Concepts

AC-Method for Factoring Trinomials of the Form
$ax^2 + bx + c$ (where $a \neq 0$)

1. Factor out the GCF from all terms.
2. Find the product ac.
3. Find two integers whose product is ac and whose sum is b. (If no pair of integers can be found, then the trinomial is prime.)
4. Rewrite the middle term (bx) as the sum of two terms whose coefficients are the numbers found in step 3.
5. Factor the polynomial by grouping.

The factored form of a **perfect square trinomial** is the square of a binomial:

$a^2 + 2ab + b^2 = (a + b)^2$
$a^2 - 2ab + b^2 = (a - b)^2$

Examples

Example 1

$10y^2 + 35y - 20$

$= 5(2y^2 + 7y - 4)$ First factor out GCF.

Identify the product $ac = (2)(-4) = -8$.

Find two integers whose product is -8 and whose sum is 7. The numbers are 8 and -1.

$5[2y^2 + 8y - 1y - 4]$

$= 5[2y(y + 4) - 1(y + 4)]$

$= 5(y + 4)(2y - 1)$

Example 2

Factor: $25y^2 + 10y + 1$

$= (5y)^2 + 2(5y)(1) + (1)^2$

$= (5y + 1)^2$

Section 6.5 Factoring Binomials

Key Concepts

Factoring a Difference of Squares

$a^2 - b^2 = (a - b)(a + b)$

Factoring a Sum or Difference of Cubes

$a^3 - b^3 = (a - b)(a^2 + ab + b^2)$

$a^3 + b^3 = (a + b)(a^2 - ab + b^2)$

Examples

Example 1

$25z^2 - 4y^2$

$= (5z - 2y)(5z + 2y)$

Example 2

$m^3 - 64$

$= (m)^3 - (4)^3$

$= (m - 4)(m^2 + 4m + 16)$

Example 3

$x^6 + 8y^3$

$= (x^2)^3 + (2y)^3$

$= (x^2 + 2y)(x^4 - 2x^2y + 4y^2)$

Section 6.6   General Factoring Summary

Key Concepts

Factoring Strategy

1. Factor out the greatest common factor, GCF.
2. Identify whether the polynomial has two terms, three terms, or more than three terms.
3. If the polynomial has more than three terms, try factoring by grouping (Section 6.1).
4. If the polynomial has three terms, check first for a perfect square trinomial. Otherwise, factor the trinomial with the trial-and-error method or ac-method (Sections 6.3 and 6.4, respectively).
5. If the polynomial has two terms, determine if it fits the pattern for a difference of squares, difference of cubes, or sum of cubes (Section 6.5).
6. Be sure to factor completely. Check that no factor can be factored further.
7. Check by multiplying.

Examples

Example 1

$$y^3 + 2y^2 - 9y - 18$$
$$= y^3 + 2y^2 \,\vert\, -9y - 18$$
$$= y^2(y + 2) - 9(y + 2)$$
$$= (y + 2)(y^2 - 9)$$
$$= (y + 2)(y - 3)(y + 3)$$

Example 2

$$9x^3 + 9x^2 - 4x$$
$$= x(9x^2 + 9x - 4) \qquad \text{Factor out the GCF.}$$
$$= x(3x + 4)(3x - 1) \qquad \text{Factor the trinomial.}$$

Example 3

$$5x^3 + 5$$
$$= 5(x^3 + 1) \qquad \text{Factor out the GCF.}$$
$$= 5(x + 1)(x^2 - x + 1) \qquad \text{Factor the sum of cubes.}$$

Section 6.7   Solving Equations Using the Zero Product Rule

Key Concepts

An equation of the form $ax^2 + bx + c = 0$, where $a \neq 0$, is a **quadratic equation**.

The zero product rule states that if $ab = 0$, then either $a = 0$ or $b = 0$. The zero product rule can be used to solve a quadratic equation or a higher degree polynomial equation that is factored and set to zero.

Examples

Example 1

The equation $2x^2 - 17x + 30 = 0$ is a quadratic equation.

Example 2

$$3w(w - 4)(2w + 1) = 0$$
$$3w = 0 \quad \text{or} \quad w - 4 = 0 \quad \text{or} \quad 2w + 1 = 0$$
$$w = 0 \quad \text{or} \quad w = 4 \quad \text{or} \quad w = -\frac{1}{2}$$

Example 3

$$4x^2 = 34x - 60$$
$$4x^2 - 34x + 60 = 0$$
$$2(2x^2 - 17x + 30) = 0$$
$$2(2x - 5)(x - 6) = 0$$
$$2 \neq 0 \quad \text{or} \quad 2x - 5 = 0 \quad \text{or} \quad x - 6 = 0$$
$$x = \frac{5}{2} \quad \text{or} \quad x = 6$$

Chapter 6   Review Exercises

Section 6.1

For Exercises 1–4, identify the greatest common factor between each group of terms.

1. $15a^2b^4, 30a^3b, 9a^5b^3$ **2.** $3(x + 5), x(x + 5)$

3. $2c^3(3c - 5), 4c(3c - 5)$ **4.** $-2wyz, -4xyz$

For Exercises 5–10, factor out the greatest common factor.

5. $6x^2 + 2x^4 - 8x$ **6.** $11w^3y^3 - 44w^2y^5$

7. $-t^2 + 5t$ **8.** $-6u^2 - u$

9. $3b(b + 2) - 7(b + 2)$

10. $2(5x + 9) + 8x(5x + 9)$

For Exercises 11–14, factor by grouping.

11. $7w^2 + 14w + wb + 2b$

12. $b^2 - 2b + yb - 2y$

13. $60y^2 - 45y - 12y + 9$

14. $6a - 3a^2 - 2ab + a^2b$

Section 6.2

For Exercises 15–24, factor completely.

15. $x^2 - 10x + 21$ **16.** $y^2 - 19y + 88$

17. $-6z + z^2 - 72$ **18.** $-39 + q^2 - 10q$

19. $3p^2w + 36pw + 60w$ **20.** $2m^4 + 26m^3 + 80m^2$

21. $-t^2 + 10t - 16$ **22.** $-w^2 - w + 20$

23. $a^2 + 12ab + 11b^2$ **24.** $c^2 - 3cd - 18d^2$

Section 6.3

For Exercises 25–28, let a, b, and c represent positive integers.

25. When factoring a polynomial of the form $ax^2 - bx - c$, should the signs of the binomials be both positive, both negative, or different?

26. When factoring a polynomial of the form $ax^2 - bx + c$, should the signs of the binomials be both positive, both negative, or different?

27. When factoring a polynomial of the form $ax^2 + bx + c$, should the signs of the binomials be both positive, both negative, or different?

28. When factoring a polynomial of the form $ax^2 + bx - c$, should the signs of the binomials be both positive, both negative, or different?

For Exercises 29–40, factor the trinomial using the trial-and-error method.

29. $2y^2 - 5y - 12$ **30.** $4w^2 - 5w - 6$

31. $10z^2 + 29z + 10$ **32.** $8z^2 + 6z - 9$

33. $2p^2 - 5p + 1$ **34.** $5r^2 - 3r + 7$

35. $10w^2 - 60w - 270$ **36.** $3y^2 - 18y - 48$

37. $9c^2 - 30cd + 25d^2$ **38.** $x^2 + 12x + 36$

39. $v^4 - 2v^2 - 3$ **40.** $x^4 + 7x^2 + 10$

41. In Exercises 29–40, which trinomials are perfect square trinomials?

Section 6.4

For Exercises 42–43, find a pair of integers whose product and sum are given.

42. Product: -5 sum: 4

43. Product: 15   sum: -8

For Exercises 44–57, factor the trinomial using the ac-method.

44. $3c^2 - 5c - 2$ **45.** $4y^2 + 13y + 3$

46. $t^2 + 13t + 12$ **47.** $4x^3 + 17x^2 - 15x$

48. $w^3 + 4w^2 - 5w$ **49.** $p^2 - 8pq + 15q^2$

50. $40v^2 + 22v - 6$ **51.** $40s^2 + 30s - 100$

52. $a^3b - 10a^2b^2 + 24ab^3$ **53.** $2z^6 + 8z^5 - 42z^4$

54. $3m + 9m^2 - 2$ **55.** $10 + 6p^2 + 19p$

56. $49x^2 + 140x + 100$ **57.** $9w^2 - 6wz + z^2$

58. In Exercises 42–57, which trinomials are perfect square trinomials?

Section 6.5

For Exercises 59–62, write the formula to factor each binomial, if possible.

59. $a^2 - b^2$ **60.** $a^2 + b^2$

61. $a^3 + b^3$ **62.** $a^3 - b^3$

For Exercises 63–78, factor completely.

63. $a^2 - 49$ **64.** $d^2 - 64$

65. $100 - 81t^2$ **66.** $4 - 25k^2$

67. $x^2 + 16$ **68.** $y^2 + 121$

69. $64 + a^3$ **70.** $125 - b^3$

71. $p^6 + 8$ **72.** $q^6 - \dfrac{1}{27}$

73. $6x^3 - 48$ **74.** $7y^3 + 7$

75. $2c^4 - 18$ **76.** $72x^2 - 2y^2$

77. $p^3 + 3p^2 - 16p - 48$ **78.** $4k - 8 - k^3 + 2k^2$

Section 6.6

For Exercises 79–94, factor completely using the factoring strategy found on page 467.

79. $6y^2 - 11y - 2$ **80.** $3p^2 - 6p + 3$

81. $x^3 - 36x$ **82.** $k^2 - 13k + 42$

83. $7ac - 14ad - bc + 2bd$

84. $q^4 - 64q$ **85.** $8h^2 + 20$

86. $2t^2 + t + 3$ **87.** $m^2 - 8m$

88. $x^3 + 4x^2 - x - 4$

89. $12s^3t - 45s^2t^2 - 12st^3$ **90.** $5p^4q - 20q^3$

91. $18a^2 + 39a - 15$ **92.** $w^4 + w^3 - 56w^2$

93. $8n + n^4$ **94.** $14m^3 - 14$

Section 6.7

95. For which of the following equations can the zero product rule be applied directly? Explain.

$$(x - 3)(2x + 1) = 0 \quad \text{or} \quad (x - 3)(2x + 1) = 6$$

For Exercises 96–111, solve the equation using the zero product rule.

96. $(4x - 1)(3x + 2) = 0$

97. $(a - 9)(2a - 1) = 0$

98. $3w(w + 3)(5w + 2) = 0$

99. $6u(u - 7)(4u - 9) = 0$

100. $7k^2 - 9k - 10 = 0$

101. $4h^2 - 23h - 6 = 0$

102. $q^2 - 144 = 0$ **103.** $r^2 = 25$

104. $5v^2 - v = 0$ **105.** $x(x - 6) = -8$

106. $36t^2 + 60t = -25$ **107.** $9s^2 + 12s = -4$

108. $3(y^2 + 4) = 20y$ **109.** $2(p^2 - 66) = -13p$

110. $2y^3 - 18y^2 = -28y$ **111.** $x^3 - 4x = 0$

112. The base of a parallelogram is 1 ft longer than twice the height. If the area is 78 ft², what are the base and height of the parallelogram?

113. A ball is tossed into the air from ground level with initial speed of 16 ft/sec. The height of the ball is given by the equation.

$$h = -16x^2 + 16x \quad (x \geq 0) \qquad \text{where } h \text{ is the ball's height in feet, and } x \text{ is the time in seconds}$$

Find the time(s) when the ball is at ground level.

114. Find the length of the ramp.

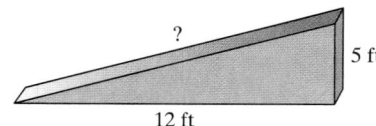

115. A right triangle has one leg that is 2 ft longer than the other leg. The hypotenuse is 2 ft less than twice the shorter leg. Find the length of all sides of the triangle.

116. If the square of a number is subtracted from 60, the result is -4. Find all such numbers.

117. The product of two consecutive integers is 44 more than 14 times their sum.

118. The base of a triangle is 1 m longer than twice the height. If the area of the triangle is 18 m^2, find the base and height.

Chapter 6   Test

1. Factor out the GCF: $15x^4 - 3x + 6x^3$

2. Factor by grouping: $7a - 35 - a^2 + 5a$

3. Factor the trinomial: $6w^2 - 43w + 7$

4. Factor the difference of squares: $169 - p^2$

5. Factor the perfect square trinomial: $q^2 - 16q + 64$

6. Factor the sum of cubes: $8 + t^3$

For Exercises 7–25, factor completely.

7. $a^2 + 12a + 32$

8. $x^2 + x - 42$

9. $2y^2 - 17y + 8$

10. $6z^2 + 19z + 8$

11. $9t^2 - 100$

12. $v^2 - 81$

13. $3a^2 + 27ab + 54b^2$

14. $c^4 - 1$

15. $xy - 7x + 3y - 21$

16. $49 + p^2$

17. $-10u^2 + 30u - 20$

18. $12t^2 - 75$

19. $5y^2 - 50y + 125$

20. $21q^2 + 14q$

21. $2x^3 + x^2 - 8x - 4$

22. $y^3 - 125$

23. $m^2n^2 - 81$

24. $16a^2 - 64b^2$

25. $64x^3 - 27y^6$

For Exercises 26–30, solve the equation.

26. $(2x - 3)(x + 5) = 0$

27. $x^2 - 7x = 0$

28. $x^2 - 6x = 16$

29. $x(5x + 4) = 1$

30. $y^3 + 10y^2 - 9y - 90 = 0$

31. A tennis court has an area of 312 yd^2. If the length is 2 yd more than twice the width, find the dimensions of the court.

32. The product of two consecutive odd integers is 35. Find the integers.

33. The height of a triangle is 5 in. less than the length of the base. The area is 42 in^2. Find the length of the base and the height of the triangle.

34. The hypotenuse of a right triangle is 2 ft less than three times the shorter leg. The longer leg is 3 ft less than three times the shorter leg. Find the length of the shorter leg.

Chapters 1–6 Cumulative Review Exercises

For Exercises 1–2, simplify completely.

1. $\dfrac{|4 - 25 \div (-5) \cdot 2|}{\sqrt{8^2 + 6^2}}$

2. Solve for t: $5 - 2(t + 4) = 3t + 12$

3. Solve for y: $3x - 2y = 8$

4. A child's piggy bank has \$3.80 in quarters, dimes, and nickels. The number of nickels is two more than the number of quarters. The number of dimes is three less than the number of quarters. Find the number of each type of coin in the bank.

5. Solve the inequality. Graph the solution on a number line and write the solution set in interval notation.

$$-\frac{5}{12}x \le \frac{5}{3} \qquad \longrightarrow$$

6. Given the equation $y = x + 4$

   a. Is the equation linear?

   b. Identify the slope.

   c. Identify the y-intercept.

   d. Identify the x-intercept.

   e. Graph the line.

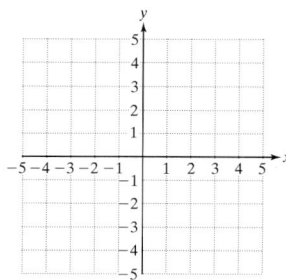

7. Consider the equation, $x = 5$.

   a. Does the equation present a horizontal or vertical line?

   b. Determine the slope of the line, if it exists.

   c. Identify the x-intercept, if it exists.

   d. Identify the y-intercept, if it exists.

8. Find an equation of the line passing through the point $(-3, 5)$ and having a slope of 3. Write the final answer in slope-intercept form.

9. Solve the system. $\begin{array}{l} 2x - 3y = 4 \\ 5x - 6y = 13 \end{array}$

For Exercises 10–12, perform the indicated operations.

10. $2\left(\dfrac{1}{3}y^3 - \dfrac{3}{2}y^2 - 7\right) - \left(\dfrac{2}{3}y^3 + \dfrac{1}{2}y^2 + 5y\right)$

11. $(4p^2 - 5p - 1)(2p - 3)$

12. $(2w - 7)^2$

13. Divide using long division: $(r^4 + 2r^3 - 5r + 1) \div (r - 3)$

14. $\dfrac{c^{12}c^{-5}}{c^3}$

15. Divide. Write the final answer in scientific notation: $\dfrac{8.0 \times 10^{-3}}{5.0 \times 10^{-6}}$

For Exercises 16–19, factor completely.

16. $w^4 - 16$

17. $2ax + 10bx - 3ya - 15yb$

18. $4x^2 - 8x - 5$

19. $y^2 - 27$

For Exercise 20, solve the equation.

20. $4x(2x - 1)(x + 5) = 0$

Rational Expressions

7

In Chapter 7 we define a rational expression as the ratio of two polynomials. Then we will learn how to add, subtract, multiply, and divide rational expressions. Then we study equations and applications involving rational expressions.

Working with rational expressions is similar to working with fractions so it is worthwhile to review operations of fractions first.

Fill in the appropriate operations $(+, -, \cdot, \div)$ into the equations to make them true. Use each of the symbols $+, -, \cdot, \div$ in each column only once.

Column 1	Column 2
$\frac{3}{2} \ \Box \ \frac{1}{2} = 2$	$\frac{3}{10} \ \Box \ \frac{5}{6} \ \Box \ \frac{1}{8} = \frac{1}{8}$
$\frac{3}{4} \ \Box \ \frac{1}{8} = 6$	$\frac{1}{2} \ \Box \ \frac{1}{2} \ \Box \ \frac{3}{4} = \frac{7}{6}$
$\frac{5}{7} \ \Box \ \frac{14}{3} = \frac{10}{3}$	
$\frac{2}{3} \ \Box \ \frac{1}{2} = \frac{1}{6}$	

Introduction to Rational Expressions

1. Definition of a Rational Expression

In Section 1.1, we defined a rational number as the ratio of two integers, $\frac{p}{q}$, where $q \neq 0$.

$$\text{Examples of rational numbers:} \quad \frac{2}{3}, \ -\frac{1}{5}, \ 9$$

In a similar way, we define a **rational expression** as the ratio of two polynomials, $\frac{p}{q}$, where $q \neq 0$.

$$\text{Examples of rational expressions:} \quad \frac{3x-6}{x^2-4}, \ \frac{3}{4}, \ \frac{6r^5+2r}{7}$$

2. Evaluating Rational Expressions

Example 1   Evaluating Rational Expressions

Evaluate the rational expression (if possible) for the given values of x: $\dfrac{12}{x-3}$

a. $x = 0$ **b.** $x = 1$ **c.** $x = -3$ **d.** $x = 3$

Solution:

Substitute the given value for the variable. Then use the order of operations to simplify.

a. $\dfrac{12}{x-3}$

$\dfrac{12}{(0)-3}$ Substitute $x = 0$.

$= \dfrac{12}{-3}$

$= -4$

b. $\dfrac{12}{x-3}$

$\dfrac{12}{(1)-3}$ Substitute $x = 1$.

$= \dfrac{12}{-2}$

$= -6$

c. $\dfrac{12}{x-3}$

$\dfrac{12}{(-3)-3}$ Substitute $x = -3$.

$= \dfrac{12}{-6}$

$= -2$

d. $\dfrac{12}{x-3}$

$\dfrac{12}{(3)-3}$ Substitute $x = 3$.

$= \dfrac{12}{0}$ Undefined. Recall that division by zero is undefined.

Skill Practice

1. Evaluate the expression for the given values of x.

$$\frac{x-3}{x+5}$$

a. $x = 2$ **b.** $x = 0$ **c.** $x = 3$ **d.** $x = -5$

Skill Practice Answers

1a. $-\dfrac{1}{7}$ **b.** $-\dfrac{3}{5}$
c. 0   **d.** Undefined

3. Domain of a Rational Expression

In Example 1(d), the expression $12/(x - 3)$ is undefined for $x = 3$. The fact that a rational expression may be defined for some values of the variable but not for others leads us to an important concept called the domain of an expression.

> **Informal Definition of the Domain of an Algebraic Expression**
>
> Given an algebraic expression, the **domain** of the expression is the set of real numbers that when substituted for the variable makes the expression result in a real number.

For a rational expression, the domain will consist of the real numbers that when substituted into the expression does not make the denominator equal to zero. Therefore, in Example 1(d), because $12/(x - 3)$ is undefined for $x = 3$, the domain is all real numbers except 3. We write this in set-builder notation as

$$\{x \mid x \text{ is a real number and } x \neq 3\}$$

> **Steps to Find the Domain of a Rational Expression**
>
> 1. Set the denominator equal to zero and solve the resulting equation.
> 2. The domain is the set of real numbers *excluding* the values found in step 1.

| **Example 2** | **Finding the Domain of Rational Expressions** |

Find the domain of the expressions.

a. $\dfrac{y - 3}{2y + 7}$ **b.** $\dfrac{-5}{x}$ **c.** $\dfrac{a + 10}{a^2 - 25}$ **d.** $\dfrac{2x^3 + 5}{x^2 + 9}$

Solution:

a. $\dfrac{y - 3}{2y + 7}$

$2y + 7 = 0$ Set the denominator equal to zero.

$2y = -7$ Solve the equation.

$\dfrac{2y}{2} = \dfrac{-7}{2}$

$y = -\dfrac{7}{2}$ The domain is the set of real numbers except $-\frac{7}{2}$.

Domain: $\{y \mid y \text{ is a real number and } y \neq -\frac{7}{2}\}$

b. $\dfrac{-5}{x}$

$x = 0$ Set the denominator equal to zero.

Domain: $\{x \mid x \text{ is a real number and } x \neq 0\}$ The domain is the set of real numbers except 0.

c. $\dfrac{a + 10}{a^2 - 25}$

$a^2 - 25 = 0$ — Set the denominator equal to zero. The equation is quadratic.

$(a - 5)(a + 5) = 0$ — Factor the equation.

$a - 5 = 0$ or $a + 5 = 0$ — Set each factor equal to zero.

$a = 5$ or $a = -5$ — The domain is the set of real numbers except 5 and -5.

Domain: $\{a \mid a \text{ is a real number and } a \neq 5, a \neq -5\}$

d. $\dfrac{2x^3 + 5}{x^2 + 9}$

The quantity x^2 cannot be negative for any real number, x, so the denominator $x^2 + 9$ cannot equal zero. Therefore, no numbers are excluded from the domain. The domain is the set of all real numbers.

Skill Practice Find the domain.

2. $\dfrac{a + 2}{2a - 8}$ **3.** $\dfrac{t}{t^2 + 10t}$ **4.** $\dfrac{w - 4}{w^2 - 9}$ **5.** $\dfrac{8}{z^4 + 1}$

4. Simplifying Rational Expressions to Lowest Terms

In many cases, it is advantageous to simplify or reduce a fraction to lowest terms. The same is true for rational expressions.

The method for simplifying rational expressions mirrors the process for simplifying fractions. In each case, factor the numerator and denominator. Common factors in the numerator and denominator form a ratio of 1 and can be reduced.

Simplifying a fraction: $\dfrac{21}{35} \xrightarrow{\text{factor}} \dfrac{3 \cdot \overset{1}{\cancel{7}}}{5 \cdot \cancel{7}} = \dfrac{3}{5} \cdot (1) = \dfrac{3}{5}$

Simplifying a rational expression: $\dfrac{2x - 6}{x^2 - 9} \xrightarrow{\text{factor}} \dfrac{2\overset{1}{\cancel{(x - 3)}}}{(x + 3)\cancel{(x - 3)}} = \dfrac{2}{(x + 3)} (1) = \dfrac{2}{x + 3}$

Informally, to simplify a rational expression to lowest terms, we reduce common factors whose ratio is 1. Formally, this is accomplished by applying the fundamental principle of rational expressions.

Fundamental Principle of Rational Expressions

Let p, q, and r represent polynomials. Then

$$\frac{pr}{qr} = \frac{p}{q} \text{ for } q \neq 0 \text{ and } r \neq 0$$

Skill Practice Answers

2. $\{a \mid a \text{ is a real number and } a \neq 4\}$
3. $\{t \mid t \text{ is a real number and } t \neq 0, t \neq -10\}$
4. $\{w \mid w \text{ is a real number and } w \neq 3, w \neq -3\}$
5. The set of all real numbers

| **Example 3** | **Simplifying a Rational Expression to Lowest Terms** |

Given the expression: $\dfrac{2p - 14}{p^2 - 49}$

a. Factor the numerator and denominator.

b. Determine the domain of the expression.

c. Simplify the expression to lowest terms.

Solution:

a. $\dfrac{2p - 14}{p^2 - 49}$

$= \dfrac{2(p - 7)}{(p + 7)(p - 7)}$ Factor out the GCF in the numerator.
Factor the denominator as a difference of squares.

b. $(p + 7)(p - 7) = 0$ To find the domain restrictions, set the denominator equal to zero. The equation is quadratic.

$p + 7 = 0$ or $p - 7 = 0$ Set each factor equal to 0.

$p = -7$ or $p = 7$ The domain is all real numbers except -7 and 7.

Domain: $\{p \mid p \text{ is a real number and } p \neq -7, p \neq 7\}$

c. $\dfrac{2(\overset{1}{\cancel{p - 7}})}{(p + 7)(\cancel{p - 7})}$ Reduce common factors whose ratio is 1.

$= \dfrac{2}{p + 7}$ (provided $p \neq 7$ and $p \neq -7$)

> **Avoiding Mistakes:**
>
> The domain of a rational expression is always determined *before* simplifying the expression to lowest terms.

| **Skill Practice** |

6. Given: $\dfrac{5z + 25}{z^2 + 3z - 10}$

a. Factor the numerator and the denominator.

b. Determine the domain of the expression.

c. Simplify the rational expression to lowest terms.

In Example 3, it is important to note that the expressions

$$\frac{2p - 14}{p^2 - 49} \quad \text{and} \quad \frac{2}{p + 7}$$

are equal for all values of p that make each expression a real number. Therefore,

$$\frac{2p - 14}{p^2 - 49} = \frac{2}{p + 7}$$

for all values of p except $p = 7$ and $p = -7$. (At $p = 7$ and $p = -7$, the original expression is undefined.) This is why the domain of an expression is always determined before the expression is simplified.

Skill Practice Answers

6a. $\dfrac{5(z + 5)}{(z + 5)(z - 2)}$

b. $\{z \mid z \text{ is a real number and } z \neq -5, z \neq 2\}$

c. $\dfrac{5}{z - 2}$

From this point forward, we will write statements of equality between two rational expressions with the assumption that they are equal for all values of the variable for which each expression is defined.

Example 4   Simplifying Rational Expressions to Lowest Terms

Simplify the rational expressions to lowest terms.

a. $\dfrac{18a^4}{9a^5}$ **b.** $\dfrac{2c - 8}{10c^2 - 80c + 160}$

Solution:

a. $\dfrac{18a^4}{9a^5}$

$= \dfrac{2 \cdot 3 \cdot 3 \cdot a \cdot a \cdot a \cdot a}{3 \cdot 3 \cdot a \cdot a \cdot a \cdot a \cdot a}$ Factor the numerator and denominator.

$= \dfrac{2 \cdot \overset{1}{(3 \cdot 3 \cdot a \cdot a \cdot a \cdot a)}}{(3 \cdot 3 \cdot a \cdot a \cdot a \cdot a) \cdot a}$ Reduce common factors to lowest terms.

$= \dfrac{2}{a}$

b. $\dfrac{2c - 8}{10c^2 - 80c + 160}$

$= \dfrac{2(c - 4)}{10(c^2 - 8c + 16)}$ Factor out the GCF.

$= \dfrac{2(c - 4)}{10(c - 4)^2}$ The denominator is a perfect square trinomial.

$= \dfrac{\overset{1}{2(c - 4)}}{2 \cdot 5(c - 4)(c - 4)}$ Reduce common factors whose ratio is 1.

$= \dfrac{1}{5(c - 4)}$

> **Avoiding Mistakes:**
>
> Given the expression
>
> $$\dfrac{2c - 8}{10c^2 - 80c + 160}$$
>
> do not be tempted to reduce before factoring. The terms $2c$ and $10c^2$ cannot be "canceled" because they are *terms*, not factors.
>
> Similarly, the -8 in the numerator cannot be "canceled" with the $-80c$ or 160 in the denominator because they are terms, not factors.

Skill Practice   Simplify to lowest terms.

7. $\dfrac{15q^3}{9q^2}$ **8.** $\dfrac{x^2 - 1}{2x^2 - x - 3}$

The process to simplify a rational expression to lowest terms is based on the identity property of multiplication. Therefore, this process applies only to factors (remember that factors are multiplied). For example,

$$\dfrac{3x}{3y} = \dfrac{\overset{1}{3} \cdot x}{3 \cdot y} = 1 \cdot \dfrac{x}{y} = \dfrac{x}{y}$$

$\underset{\text{Simplify}}{\uparrow}$

Skill Practice Answers

7. $\dfrac{5q}{3}$ **8.** $\dfrac{x - 1}{2x - 3}$

Terms that are added or subtracted cannot be reduced to lowest terms. For example,

$$\frac{x + 3}{y + 3}$$

↑

Cannot be simplified

The objective of simplifying a rational expression to lowest terms is to create an equivalent expression that is simpler to use. Consider the rational expression from Example 4(b) in its original form and in its simplified form. If we choose an arbitrary value of c from the domain of the original expression and substitute that value into each expression, we see that the simplified form is easier to evaluate. For example, substitute $c = 3$:

	Original Expression	**Simplified Expression**
	$\dfrac{2c - 8}{10c^2 - 80c + 160}$	$\dfrac{1}{5(c - 4)}$
Substitute $c = 3$.	$= \dfrac{2(3) - 8}{10(3)^2 - 80(3) + 160}$	$= \dfrac{1}{5(3 - 4)}$
	$= \dfrac{6 - 8}{10(9) - 240 + 160}$	$= \dfrac{1}{5(-1)}$
	$= \dfrac{-2}{90 - 240 + 160}$	$= -\dfrac{1}{5}$
	$= \dfrac{-2}{10} \quad$ or $\quad -\dfrac{1}{5}$	

5. Simplifying a Ratio of −1

When two factors are identical in the numerator and denominator, they form a ratio of 1 and can be reduced. Sometimes we encounter two factors that are opposites and form a ratio of −1. For example,

Simplified Form **Details/Notes**

$\dfrac{-5}{5} = -1$ The ratio of a number and its opposite is −1.

$\dfrac{100}{-100} = -1$ The ratio of a number and its opposite is −1.

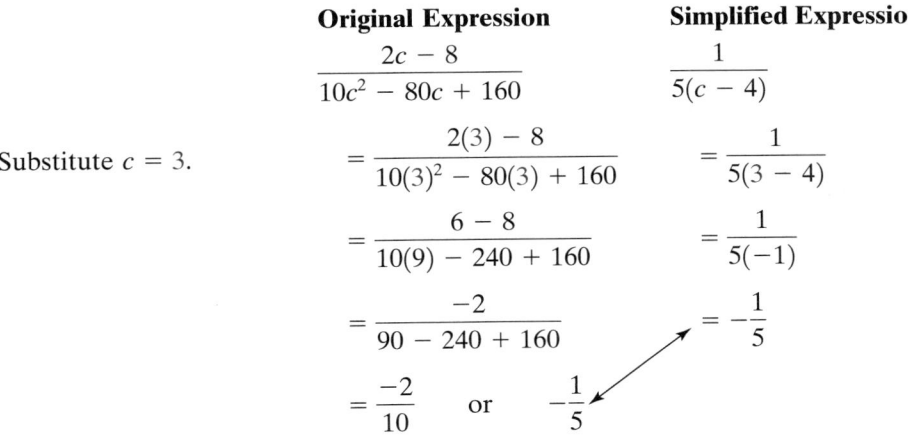

$\dfrac{x + 7}{-x - 7} = -1$ $\qquad \dfrac{x + 7}{-x - 7} = \dfrac{x + 7}{-1(x + 7)} = \dfrac{\overset{1}{\cancel{x + 7}}}{-1(\cancel{x + 7})} = \dfrac{1}{-1} = -1$

Factor out −1

$\dfrac{2 - x}{x - 2} = -1$ $\qquad \dfrac{2 - x}{x - 2} = \dfrac{-1(-2 + x)}{x - 2} = \dfrac{-1(\cancel{x - 2})}{\cancel{x - 2}} = \dfrac{-1}{1} = -1$

Recognizing factors that are opposites is useful when simplifying rational expressions.

Example 5   Simplifying Rational Expressions to Lowest Terms

Simplify the rational expressions to lowest terms.

a. $\dfrac{3c - 3d}{d - c}$ b. $\dfrac{5 - y}{y^2 - 25}$

Solution:

a. $\dfrac{3c - 3d}{d - c}$

$= \dfrac{3(c - d)}{d - c}$ Factor the numerator and denominator.

Notice that $(c - d)$ and $(d - c)$ are opposites and form a ratio of -1.

$= \dfrac{3(\overset{-1}{\cancel{c - d}})}{\cancel{d - c}}$ <u>Details:</u> $\dfrac{3(c - d)}{d - c} = \dfrac{3(c - d)}{-1(-d + c)} = \dfrac{3(c - d)}{-1(c - d)}$

$= 3(-1)$

$= -3$

b. $\dfrac{5 - y}{y^2 - 25}$

$= \dfrac{5 - y}{(y - 5)(y + 5)}$ Factor the numerator and denominator.

Notice that $5 - y$ and $y - 5$ are opposites and form a ratio of -1.

$= \dfrac{\overset{-1}{\cancel{5 - y}}}{\cancel{(y - 5)}(y + 5)}$ <u>Details:</u> $\dfrac{5 - y}{(y - 5)(y + 5)} = \dfrac{-1(-5 + y)}{(y - 5)(y + 5)}$

$= \dfrac{-1(y - 5)}{(y - 5)(y + 5)}$

$= \dfrac{-1}{y + 5}$ or $\dfrac{1}{-(y + 5)}$ or $-\dfrac{1}{y + 5}$

TIP: It is important to recognize that a rational expression may be written in several equivalent forms, particularly when a negative factor is present. For example, since two numbers with opposite signs form a negative quotient, the number $-\frac{3}{4}$ can be written as $\frac{-3}{4}$ or as $\frac{3}{-4}$. The $-$ sign can be written in the numerator, in the denominator, or out in front of the fraction.

For this reason, the expression in Example 5(b) can be written in a variety of forms.

Skill Practice Answers

9. $\dfrac{-1}{a + b}$ or $\dfrac{1}{-(a + b)}$ or $-\dfrac{1}{a + b}$

10. $\dfrac{-5}{y + 4}$ or $\dfrac{5}{-(y + 4)}$ or $-\dfrac{5}{y + 4}$

Skill Practice   Simplify to lowest terms.

9. $\dfrac{b - a}{a^2 - b^2}$ 10. $\dfrac{20 - 5y}{y^2 - 16}$

Section 7.1   Practice Exercises

Study Skills Exercises

1. Review Section R.2 in this text. Write an example of how to simplify (reduce) a fraction, multiply two fractions, divide two fractions, add two fractions, and subtract two fractions. Then as you learn about rational expressions, compare the operations on rational expressions with those on fractions. This is a great place to use 3×5 cards again. Write an example of an operation with fractions on one side and the same operation with rational expressions on the other side.

2. Define the key terms:

a. rational expression   **b. domain**

Concept 1: Definition of a Rational Expression

3. a. What is a rational number?

b. What is a rational expression?

4. a. Write an example of a rational number. (Answers will vary.)

b. Write an example of a rational expression. (Answers will vary.)

Concept 2: Evaluating Rational Expressions

For Exercises 5–11, substitute the given number into the expression and simplify (if possible).

5. $\dfrac{1}{x - 6}$ let $x = -2$

6. $\dfrac{w - 10}{w + 6}$ let $w = 0$

7. $\dfrac{w - 4}{2w + 8}$ let $w = 0$

8. $\dfrac{y - 8}{2y^2 + y - 1}$ let $y = 8$

9. $\dfrac{y + 3}{3y^2 - 25y - 18}$ let $y = -3$

10. $\dfrac{(a - 7)(a + 1)}{(a - 2)(a + 5)}$ let $a = 2$

11. $\dfrac{(a + 4)(a + 1)}{(a - 4)(a - 1)}$ let $a = 1$

12. A bicyclist rides 24 miles against a wind and returns 24 miles with the same wind. His average speed for the return trip traveling with the wind is 8 mph faster than his speed going out against the wind. If x represents the bicyclist's speed going out against the wind, then the total time, t, required for the round trip is given by

$$t = \frac{24}{x} + \frac{24}{x + 8} \qquad \text{where } x > 0 \text{ and } t \text{ is measured in hours.}$$

a. Find the time required for the round trip if the cyclist rides 12 mph against the wind.

b. Find the time required for the round trip if the cyclist rides 24 mph against the wind.

13. The manufacturer of mountain bikes has a fixed cost of $56,000, plus a variable cost of $140 per bike. The average cost per bike, y (in dollars), is given by the equation:

$$y = \frac{56,000 + 140x}{x}$$ where x represents the number of bikes produced and $x > 0$.

a. Find the average cost per bike if the manufacturer produces 1000 bikes.

b. Find the average cost per bike if the manufacturer produces 2000 bikes.

c. Find the average cost per bike if the manufacturer produces 10,000 bikes.

Concept 3: Domain of a Rational Expression

For Exercises 14–23, write the domain.

14. $\dfrac{5}{k + 2}$

15. $\dfrac{-3}{h - 4}$

16. $\dfrac{x + 5}{(2x - 5)(x + 8)}$

17. $\dfrac{4y + 1}{(3y + 7)(y + 3)}$

18. $\dfrac{b + 12}{b^2 + 5b + 6}$

19. $\dfrac{c - 11}{c^2 - 5c - 6}$

20. $\dfrac{x - 4}{x^2 + 9}$

21. $\dfrac{x + 1}{x^2 + 4}$

22. $\dfrac{y^2 - y - 12}{12}$

23. $\dfrac{z^2 + 10z + 9}{9}$

24. Construct a rational expression that is undefined for $x = 2$. (Answers will vary.)

25. Construct a rational expression that is undefined for $x = 5$. (Answers will vary.)

26. Construct a rational expression that is undefined for $x = -3$ and $x = 7$. (Answers will vary.)

27. Construct a rational expression that is undefined for $x = -1$ and $x = 4$. (Answers will vary.)

28. Evaluate the expressions for $x = 4$.

a. $\dfrac{5x + 5}{x^2 - 1}$ **b.** $\dfrac{5}{x - 1}$

29. Evaluate the expressions for $x = 3$.

a. $\dfrac{2x^2 - 4x - 6}{2x^2 - 18}$ **b.** $\dfrac{x + 1}{x + 3}$

30. Evaluate the expressions for $x = -1$.

a. $\dfrac{3x^2 - 2x - 1}{6x^2 - 7x - 3}$ **b.** $\dfrac{x - 1}{2x - 3}$

31. Evaluate the expressions for $x = 4$.

a. $\dfrac{(x + 5)^2}{x^2 + 6x + 5}$ **b.** $\dfrac{x + 5}{x + 1}$

Concept 4: Simplifying Rational Expressions to Lowest Terms

For Exercises 32–41,

a. Write the domain in set-builder notation.

b. Simplify the expression to lowest terms.

32. $\dfrac{3y + 6}{6y + 12}$

33. $\dfrac{8x - 8}{4x - 4}$

34. $\dfrac{t^2 - 1}{t + 1}$

35. $\dfrac{r^2 - 4}{r - 2}$

36. $\dfrac{7w}{21w^2 - 35w}$ **37.** $\dfrac{12a^2}{24a^2 - 18a}$ **38.** $\dfrac{9x^2 - 4}{6x + 4}$ **39.** $\dfrac{8b - 20}{4b^2 - 25}$

40. $\dfrac{a^2 + 3a - 10}{a^2 + a - 6}$ **41.** $\dfrac{t^2 + 3t - 10}{t^2 + t - 20}$

For Exercises 42–83, simplify the expression to lowest terms.

42. $\dfrac{7b^2}{21b}$ **43.** $\dfrac{15c^3}{3c^5}$ **44.** $\dfrac{18st^5}{12st^3}$ **45.** $\dfrac{20a^4b^2}{25ab^2}$

46. $\dfrac{-24x^2y^5z}{8xy^4z^3}$ **47.** $\dfrac{60rs^4t^2}{-12r^4s^2t^3}$ **48.** $\dfrac{3(y + 2)}{6(y + 2)}$ **49.** $\dfrac{8(x - 1)}{4(x - 1)}$

50. $\dfrac{(p - 3)(p + 5)}{(p + 5)(p + 4)}$ **51.** $\dfrac{(c + 4)(c - 1)}{(c + 4)(c + 2)}$ **52.** $\dfrac{(m + 11)}{4(m + 11)(m - 11)}$ **53.** $\dfrac{(n - 7)}{9(n + 2)(n - 7)}$

54. $\dfrac{x(2x + 1)^2}{4x^3(2x + 1)}$ **55.** $\dfrac{(p + 2)(p - 3)^4}{(p + 2)^2(p - 3)^2}$ **56.** $\dfrac{5}{20a - 25}$ **57.** $\dfrac{7}{14c - 21}$

58. $\dfrac{4w - 8}{w^2 - 4}$ **59.** $\dfrac{3x + 15}{x^2 - 25}$ **60.** $\dfrac{3x^2 - 6x}{9xy + 18x}$ **61.** $\dfrac{6p^2 + 12p}{2pq - 4p}$

62. $\dfrac{2x + 4}{x^2 - 3x - 10}$ **63.** $\dfrac{5z + 15}{z^2 - 4z - 21}$ **64.** $\dfrac{a^2 - 49}{a - 7}$ **65.** $\dfrac{b^2 - 64}{b - 8}$

66. $\dfrac{q^2 + 25}{q + 5}$ **67.** $\dfrac{r^2 + 36}{r + 6}$ **68.** $\dfrac{y^2 + 6y + 9}{2y^2 + y - 15}$ **69.** $\dfrac{h^2 + h - 6}{h^2 + 2h - 8}$

70. $\dfrac{3x^2 + 7x - 6}{x^2 + 7x + 12}$ **71.** $\dfrac{x^2 - 5x - 14}{2x^2 - x - 10}$ **72.** $\dfrac{5q^2 + 5}{q^4 - 1}$ **73.** $\dfrac{4t^2 + 16}{t^4 - 16}$

74. $\dfrac{ac - ad + 2bc - 2bd}{2ac + ad + 4bc + 2bd}$ (*Hint:* Factor by grouping.) **75.** $\dfrac{3pr - ps - 3qr + qs}{3pr - ps + 3qr - qs}$ (*Hint:* Factor by grouping.)

76. $\dfrac{2t^2 - 3t}{2t^4 - 13t^3 + 15t^2}$ **77.** $\dfrac{4m^3 + 3m^2}{4m^3 + 7m^2 + 3m}$

78. $\dfrac{49p^2 - 28pq + 4q^2}{14p - 4q}$ **79.** $\dfrac{3x - 3y}{2x^2 - 4xy + 2y^2}$

80. $\dfrac{5x^3 + 4x^2 - 45x - 36}{x^2 - 9}$ **81.** $\dfrac{x^2 - 1}{ax^3 - bx^2 - ax + b}$

82. $\dfrac{2x^2 - xy - 3y^2}{2x^2 - 11xy + 12y^2}$ **83.** $\dfrac{2c^2 + cd - d^2}{5c^2 + 3cd - 2d^2}$

Concept 5: Simplifying a Ratio of −1

84. What is the relationship between $x - 2$ and $2 - x$?

85. What is the relationship between $w + p$ and $-w - p$?

For Exercises 86–99, simplify to lowest terms.

86. $\dfrac{x-5}{5-x}$

87. $\dfrac{8-p}{p-8}$

88. $\dfrac{-4-y}{4+y}$

89. $\dfrac{z+10}{-z-10}$

90. $\dfrac{3y-6}{12-6y}$

91. $\dfrac{4q-4}{12-12q}$

92. $\dfrac{2m-7n}{7n-2m}$

93. $\dfrac{3a^2-5}{5-3a^2}$

94. $\dfrac{k+5}{5-k}$

95. $\dfrac{2+n}{2-n}$

96. $\dfrac{10x-12}{10x+12}$

97. $\dfrac{4t-16}{16+4t}$

98. $\dfrac{x^2-x-12}{16-x^2}$

99. $\dfrac{49-b^2}{b^2-10b+21}$

Expanding Your Skills

For Exercises 100–103, factor and simplify to lowest terms.

100. $\dfrac{w^3-8}{w^2+2w+4}$

101. $\dfrac{y^3+27}{y^2-3y+9}$

102. $\dfrac{z^2-16}{z^3-64}$

103. $\dfrac{x^2-25}{x^3+125}$

Section 7.2 Multiplication and Division of Rational Expressions

Concepts

1. Multiplication of Rational Expressions
2. Division of Rational Expressions

1. Multiplication of Rational Expressions

Recall from Section R.2 that to multiply fractions, we multiply the numerators and multiply the denominators. The same is true for multiplying rational expressions.

> **Multiplication of Rational Expressions**
>
> Let p, q, r, and s represent polynomials, such that $q \neq 0, s \neq 0$. Then,
>
> $$\frac{p}{q} \cdot \frac{r}{s} = \frac{pr}{qs}$$

For example:

Multiply the Fractions

$$\frac{2}{3} \cdot \frac{5}{7} = \frac{10}{21}$$

Multiply the Rational Expressions

$$\frac{2x}{3y} \cdot \frac{5z}{7} = \frac{10xz}{21y}$$

Sometimes it is possible to simplify a ratio of common factors to 1 *before* multiplying. To do so, we must first factor the numerators and denominators of each fraction.

$$\frac{15}{14} \cdot \frac{21}{10} = \frac{3 \cdot \overset{1}{\cancel{5}}}{2 \cdot \cancel{7}} \cdot \frac{3 \cdot \overset{1}{\cancel{7}}}{2 \cdot \cancel{5}} = \frac{9}{4}$$

The same process is also used to multiply rational expressions.

Steps to Multiply Rational Expressions

1. Factor the numerators and denominators of all rational expressions.
2. Simplify the ratios of common factors to 1 or -1.
3. Multiply the remaining factors in the numerator, and multiply the remaining factors in the denominator.

Example 1 **Multiplying Rational Expressions**

Multiply.

a. $\dfrac{5a^2b}{2} \cdot \dfrac{6a}{10b}$ **b.** $\dfrac{3c - 3d}{6c} \cdot \dfrac{2}{c^2 - d^2}$ **c.** $\dfrac{35 - 5x}{5x + 5} \cdot \dfrac{x^2 + 5x + 4}{x^2 - 49}$

Solution:

a. $\dfrac{5a^2b}{2} \cdot \dfrac{6a}{10b}$

$= \dfrac{5 \cdot a \cdot a \cdot b}{2} \cdot \dfrac{2 \cdot 3 \cdot a}{2 \cdot 5 \cdot b}$ Factor into prime factors.

$= \dfrac{\overset{1}{\cancel{5}} \cdot a \cdot a \cdot \overset{1}{\cancel{b}}}{2} \cdot \dfrac{\overset{1}{\cancel{2}} \cdot 3 \cdot a}{\cancel{2} \cdot \cancel{5} \cdot \cancel{b}}$ Simplify.

$= \dfrac{3a^3}{2}$ Multiply remaining factors.

b. $\dfrac{3c - 3d}{6c} \cdot \dfrac{2}{c^2 - d^2}$

$= \dfrac{3(c - d)}{2 \cdot 3 \cdot c} \cdot \dfrac{2}{(c - d)(c + d)}$ Factor into prime factors.

$= \dfrac{\overset{1}{\cancel{3}}(\overset{1}{\cancel{c - d}})}{\cancel{2} \cdot \cancel{3} \cdot c} \cdot \dfrac{\overset{1}{\cancel{2}}}{(\cancel{c - d})(c + d)}$ Simplify.

$= \dfrac{1}{c(c + d)}$

c. $\dfrac{35 - 5x}{5x + 5} \cdot \dfrac{x^2 + 5x + 4}{x^2 - 49}$

$= \dfrac{5(7 - x)}{5(x + 1)} \cdot \dfrac{(x + 4)(x + 1)}{(x - 7)(x + 7)}$ Factor the numerators and denominators completely.

$= \dfrac{\overset{1}{\cancel{5}}(\overset{-1}{\cancel{7 - x}})}{\cancel{5}(\cancel{x + 1})} \cdot \dfrac{(x + 4)(\overset{1}{\cancel{x + 1}})}{(\cancel{x - 7})(x + 7)}$ Simplify the ratios of common factors to 1 or -1.

$= \dfrac{-1(x + 4)}{x + 7}$

$= \dfrac{-(x + 4)}{x + 7}$ or $\dfrac{x + 4}{-(x + 7)}$ or $-\dfrac{x + 4}{x + 7}$

Avoiding Mistakes:

If all the factors in the numerator reduce to a ratio of 1, do not forget to write the factor of 1 in the numerator.

TIP: The ratio $\frac{7 - x}{x - 7} = -1$ because $7 - x$ and $x - 7$ are opposites.

Skill Practice Multiply.

1. $\dfrac{7a}{3b} \cdot \dfrac{15b}{14a^2}$ **2.** $\dfrac{4x - 8}{x + 6} \cdot \dfrac{x^2 + 6x}{2x}$

3. $\dfrac{p^2 + 4p + 3}{5p + 10} \cdot \dfrac{p^2 - p - 6}{9 - p^2}$

2. Division of Rational Expressions

Recall that to divide fractions, multiply the first fraction by the reciprocal of the second.

$$\dfrac{21}{10} \div \dfrac{49}{15} \xrightarrow[\text{of the second fraction}]{\text{Multiply by the reciprocal}} \dfrac{21}{10} \cdot \dfrac{15}{49} \xrightarrow{\text{Factor}} \dfrac{3 \cdot \overset{1}{\cancel{7}}}{2 \cdot \cancel{5}} \cdot \dfrac{3 \cdot \overset{1}{\cancel{5}}}{\cancel{7} \cdot 7} = \dfrac{9}{14}$$

The same process is used to divide rational expressions.

> **Division of Rational Expressions**
>
> Let p, q, r, and s represent polynomials, such that $q \neq 0, r \neq 0, s \neq 0$. Then,
>
> $$\dfrac{p}{q} \div \dfrac{r}{s} = \dfrac{p}{q} \cdot \dfrac{s}{r} = \dfrac{ps}{qr}$$

Example 2 Dividing Rational Expressions

Divide.

a. $\dfrac{5t - 15}{2} \div \dfrac{t^2 - 9}{10}$ **b.** $\dfrac{p^2 - 11p + 30}{10p^2 - 250} \div \dfrac{30p - 5p^2}{2p + 4}$ **c.** $\dfrac{\dfrac{3x}{4y}}{\dfrac{5x}{6y}}$

Solution:

a. $\dfrac{5t - 15}{2} \div \dfrac{t^2 - 9}{10}$

$= \dfrac{5t - 15}{2} \cdot \dfrac{10}{t^2 - 9}$ Multiply the first fraction by the reciprocal of the second.

$= \dfrac{5(t - 3)}{2} \cdot \dfrac{2 \cdot 5}{(t - 3)(t + 3)}$ Factor each polynomial.

$= \dfrac{5\overset{1}{\cancel{(t - 3)}}}{\cancel{2}} \cdot \dfrac{\cancel{2} \cdot 5}{\overset{1}{\cancel{(t - 3)}}(t + 3)}$ Reduce common factors.

$= \dfrac{25}{t + 3}$

b. $\dfrac{p^2 - 11p + 30}{10p^2 - 250} \div \dfrac{30p - 5p^2}{2p + 4}$

$= \dfrac{p^2 - 11p + 30}{10p^2 - 250} \cdot \dfrac{2p + 4}{30p - 5p^2}$ Multiply the first fraction by the reciprocal of the second.

Factor the trinomial.
$p^2 - 11p + 30 = (p - 5)(p - 6)$

Factor out the GCF.
$2p + 4 = 2(p + 2)$

$= \dfrac{(p - 5)(p - 6)}{2 \cdot 5(p - 5)(p + 5)} \cdot \dfrac{2(p + 2)}{5p(6 - p)}$ Factor out the GCF. Then factor the difference of squares.
$10p^2 - 250 = 10(p^2 - 25)$
$\qquad\qquad\quad = 2 \cdot 5(p - 5)(p + 5)$

Factor out the GCF.
$30p - 5p^2 = 5p(6 - p)$

$= \dfrac{\overset{1}{\cancel{(p-5)}}\overset{-1}{\cancel{(p-6)}}}{\cancel{2} \cdot 5\cancel{(p-5)}(p + 5)} \cdot \dfrac{\overset{1}{\cancel{2}}(p + 2)}{5p\cancel{(6-p)}}$ Reduce common factors.

$= -\dfrac{(p + 2)}{25p(p + 5)}$

c. $\dfrac{\dfrac{3x}{4y}}{\dfrac{5x}{6y}}$ ←————— This fraction bar denotes division ($\div$).

$= \dfrac{3x}{4y} \div \dfrac{5x}{6y}$

$= \dfrac{3x}{4y} \cdot \dfrac{6y}{5x}$ Multiply by the reciprocal of the second fraction.

$= \dfrac{3 \cdot \overset{1}{\cancel{x}}}{\cancel{2} \cdot 2 \cdot \cancel{y}} \cdot \dfrac{\overset{1}{\cancel{2}} \cdot 3 \cdot \overset{1}{\cancel{y}}}{5 \cdot \cancel{x}}$ Reduce common factors.

$= \dfrac{9}{10}$

Skill Practice Divide the rational expressions.

4. $\dfrac{7y - 14}{y + 1} \div \dfrac{y^2 + 2y - 8}{2y + 2}$ **5.** $\dfrac{4x^2 - 9}{2x^2 - x - 3} \div \dfrac{20x + 30}{x^2 + 7x + 6}$

6. $\dfrac{\dfrac{a^3 b}{9c}}{\dfrac{4ab}{3c^3}}$

TIP: $(p - 6)$ and $(6 - p)$ are opposites and form a ratio of -1.

$\dfrac{p - 6}{6 - p} = \dfrac{p - 6}{-1(-6 + p)}$

$\qquad = \dfrac{p - 6}{-1(p - 6)} = -1$

TIP: A fraction with one or more rational expressions in its numerator or denominator is called a *complex fraction*, for example,

$\dfrac{\dfrac{3x}{4y}}{\dfrac{5x}{6y}}$

Skill Practice Answers

4. $\dfrac{14}{y + 4}$ **5.** $\dfrac{x + 6}{10}$ **6.** $\dfrac{a^2 c^2}{12}$

Section 7.2   Practice Exercises

Boost your GRADE at mathzone.com!

- Practice Problems
- Self-Tests
- NetTutor
- e-Professors
- Videos

Review Exercises

1. Explain the difference between multiplying the fractions $\dfrac{2}{3} \cdot \dfrac{5}{9}$ and dividing the fractions $\dfrac{2}{3} \div \dfrac{5}{9}$.

For Exercises 2–9, multiply or divide the fractions.

2. $\dfrac{3}{5} \cdot \dfrac{1}{2}$

3. $\dfrac{6}{7} \cdot \dfrac{5}{12}$

4. $\dfrac{3}{4} \div \dfrac{3}{8}$

5. $\dfrac{18}{5} \div \dfrac{2}{5}$

6. $6 \cdot \dfrac{5}{12}$

7. $\dfrac{7}{25} \cdot 5$

8. $\dfrac{\frac{21}{4}}{\frac{7}{5}}$

9. $\dfrac{\frac{9}{2}}{\frac{3}{4}}$

Concept 1: Multiplication of Rational Expressions

For Exercises 10–21, multiply.

10. $\dfrac{2xy}{5x^2} \cdot \dfrac{15}{4y}$

11. $\dfrac{7s}{t^2} \cdot \dfrac{t^2}{14s^2}$

12. $\dfrac{6x^3}{9x^6y^2} \cdot \dfrac{18x^4y^7}{4y}$

13. $\dfrac{10a^2b}{15b^2} \cdot \dfrac{30b}{2a^3}$

14. $\dfrac{4x - 24}{20x} \cdot \dfrac{5x}{8}$

15. $\dfrac{5a + 20}{a} \cdot \dfrac{3a}{10}$

16. $\dfrac{3y + 18}{y^2} \cdot \dfrac{4y}{6y + 36}$

17. $\dfrac{2p - 4}{6p} \cdot \dfrac{4p^2}{8p - 16}$

18. $\dfrac{10}{2 - a} \cdot \dfrac{a - 2}{16}$

19. $\dfrac{b - 3}{6} \cdot \dfrac{20}{3 - b}$

20. $\dfrac{b^2 - a^2}{a - b} \cdot \dfrac{a}{a^2 - ab}$

21. $\dfrac{(x - y)^2}{x^2 + xy} \cdot \dfrac{x}{y - x}$

Concept 2: Division of Rational Expressions

For Exercises 22–35, divide.

22. $\dfrac{4x}{7y} \div \dfrac{2x^2}{21xy}$

23. $\dfrac{6cd}{5d^2} \div \dfrac{8c^3}{10d}$

24. $\dfrac{8m^4n^5}{5n^6} \div \dfrac{24mn}{15m^3}$

25. $\dfrac{10a^3b}{3a} \div \dfrac{5b}{9ab}$

26. $\dfrac{4a + 12}{6a - 18} \div \dfrac{3a + 9}{5a - 15}$

27. $\dfrac{8b - 16}{3b + 3} \div \dfrac{5b - 10}{2b + 2}$

28. $\dfrac{3x - 21}{6x^2 - 42x} \div \dfrac{7}{12x}$

29. $\dfrac{4a^2 - 4a}{9a - 9} \div \dfrac{5}{12a}$

30. $\dfrac{m^2 - n^2}{9} \div \dfrac{3n - 3m}{27m}$

31. $\dfrac{9 - b^2}{15b + 15} \div \dfrac{b - 3}{5b}$

32. $\dfrac{3p + 4q}{p^2 + 4pq + 4q^2} \div \dfrac{4}{p + 2q}$

33. $\dfrac{x^2 + 2xy + y^2}{2x - y} \div \dfrac{x + y}{5}$

34. $\dfrac{p^2 - 2p - 3}{p^2 - p - 6} \div \dfrac{p^2 - 1}{p^2 + 2p}$

35. $\dfrac{4t^2 - 1}{t^2 - 5t} \div \dfrac{2t^2 + 5t + 2}{t^2 - 3t - 10}$

Mixed Exercises

For Exercises 36–61, multiply or divide as indicated.

36. $(w + 3) \cdot \dfrac{w}{2w^2 + 5w - 3}$

37. $\dfrac{5t + 1}{5t^2 - 31t + 6} \cdot (t - 6)$

38. $\dfrac{\dfrac{5t - 10}{12}}{\dfrac{4t - 8}{8}}$

39. $\dfrac{\dfrac{6m + 6}{5}}{\dfrac{3m + 3}{10}}$

40. $\dfrac{q + 1}{5q^2 - 28q - 12} \cdot (5q + 2)$

41. $(r - 5) \cdot \dfrac{4r}{2r^2 - 7r - 15}$

42. $\dfrac{2a^2 + 13a - 24}{8a - 12} \div (a + 8)$

43. $\dfrac{3y^2 + 20y - 7}{5y + 35} \div (3y - 1)$

44. $\dfrac{y^2 + 5y - 36}{y^2 - 2y - 8} \cdot \dfrac{y + 2}{y - 6}$

45. $\dfrac{z^2 - 11z + 28}{z - 1} \cdot \dfrac{z + 1}{z^2 - 6z - 7}$

46. $\dfrac{t^2 + 4t - 5}{t^2 + 7t + 10} \cdot \dfrac{t + 4}{t - 1}$

47. $\dfrac{p^2 - 3p + 2}{p^2 - 4p + 3} \cdot \dfrac{p + 1}{p - 2}$

48. $(5t - 1) \div \dfrac{5t^2 + 9t - 2}{3t + 8}$

49. $(2q - 3) \div \dfrac{2q^2 + 5q - 12}{q - 7}$

50. $\dfrac{x^2 + 2x - 3}{x^2 - 3x + 2} \cdot \dfrac{x^2 + 2x - 8}{x^2 + 4x + 3}$

51. $\dfrac{y^2 + y - 12}{y^2 - y - 20} \cdot \dfrac{y^2 + y - 30}{y^2 - 2y - 3}$

52. $\dfrac{\dfrac{w^2 - 6w + 9}{8}}{\dfrac{9 - w^2}{4w + 12}}$

53. $\dfrac{\dfrac{p^2 - 6p + 8}{24}}{\dfrac{16 - p^2}{6p + 6}}$

54. $\dfrac{k^2 + 3k + 2}{k^2 + 5k + 4} \div \dfrac{k^2 + 5k + 6}{k^2 + 10k + 24}$

55. $\dfrac{4h^2 - 5h + 1}{h^2 + h - 2} \div \dfrac{6h^2 - 7h + 2}{2h^2 + 3h - 2}$

56. $\dfrac{ax + a + bx + b}{2x^2 + 4x + 2} \cdot \dfrac{4x + 4}{a^2 + ab}$

57. $\dfrac{3my + 9m + ny + 3n}{9m^2 + 6mn + n^2} \cdot \dfrac{30m + 10n}{5y^2 + 15y}$

58. $\dfrac{y^4 - 1}{2y^2 - 3y + 1} \div \dfrac{2y^2 + 2}{8y^2 - 4y}$

59. $\dfrac{x^4 - 16}{6x^2 + 24} \div \dfrac{x^2 - 2x}{3x}$

60. $\dfrac{x^2 - xy - 2y^2}{x + 2y} \div \dfrac{x^2 - 4xy + 4y^2}{x^2 - 4y^2}$

61. $\dfrac{4m^2 - 4mn - 3n^2}{8m^2 - 18n^2} \div \dfrac{3m + 3n}{6m^2 + 15mn + 9n^2}$

For Exercises 62–67, multiply or divide as indicated.

62. $\dfrac{b^3 - 3b^2 + 4b - 12}{b^4 - 16} \cdot \dfrac{3b^2 + 5b - 2}{3b^2 - 10b + 3} \div \dfrac{3}{6b - 12}$

63. $\dfrac{x^2 - 25}{3x^2 + 3xy} \cdot \dfrac{x^2 + 4x + xy + 4y}{x^2 + 9x + 20} \div \dfrac{x - 5}{x}$

64. $\dfrac{a^2 - 5a}{a^2 + 7a + 12} \div \dfrac{a^3 - 7a^2 + 10a}{a^2 + 9a + 18} \div \dfrac{a + 6}{a + 4}$

65. $\dfrac{t^2 + t - 2}{t^2 + 5t + 6} \div \dfrac{t - 1}{t} \div \dfrac{5t - 5}{t + 3}$

66. $\dfrac{p^3 - q^3}{p - q} \cdot \dfrac{p + q}{2p^2 + 2pq + 2q^2}$

67. $\dfrac{r^3 + s^3}{r - s} \div \dfrac{r^2 + 2rs + s^2}{r^2 - s^2}$

Section 7.3   Least Common Denominator

1. Writing Equivalent Rational Expressions

In Sections 7.1 and 7.2, we learned how to simplify, multiply, and divide rational expressions. Our next goal is to add and subtract rational expressions. As with fractions, rational expressions can be added or subtracted only if they have the same denominator. Therefore, we must first learn how to identify a common denominator between two or more rational expressions. Then we must learn how to convert a rational expression into an equivalent rational expression with the indicated denominator.

Using the identity property of multiplication, we know that for $q \neq 0$ and $r \neq 0$,

$$\frac{p}{q} = \frac{p}{q} \cdot 1 = \frac{p}{q} \cdot \frac{r}{r} = \frac{pr}{qr}$$

This principle is used to convert a rational expression into an equivalent expression with a different denominator. For example, $\frac{1}{2}$ can be converted into an equivalent expression with a denominator of 12 as follows:

$$\frac{1}{2} = \frac{1}{2} \cdot \frac{6}{6} = \frac{1 \cdot 6}{2 \cdot 6} = \frac{6}{12}$$

In this example, we multiplied $\frac{1}{2}$ by a convenient form of 1. The ratio $\frac{6}{6}$ was chosen so that the product produced a new denominator of 12. Notice that multiplying $\frac{1}{2}$ by $\frac{6}{6}$ is equivalent to multiplying the numerator and denominator of the original expression by 6. In general, if the numerator and denominator of a rational expression are both multiplied by the same nonzero quantity, the value of the expression remains unchanged.

Example 1   Creating Equivalent Fractions

Convert each expression into an equivalent expression with the indicated denominator.

a. $\dfrac{5}{x} = \dfrac{}{xyz}$ **b.** $\dfrac{7}{5p^2} = \dfrac{}{20p^6}$

c. $\dfrac{w}{w + 5} = \dfrac{}{(w + 5)(w - 2)}$ **d.** $\dfrac{1}{5} = \dfrac{}{5x + 20}$

Solution:

a. $\dfrac{5}{x} = \dfrac{}{xyz}$

$\dfrac{5 \cdot yz}{x \cdot yz} = \dfrac{5yz}{xyz}$

To convert $\frac{5}{x}$ to an equivalent fraction with a denominator of xyz, multiply both numerator and denominator by the missing factor of yz.

TIP: When multiplying both the numerator and denominator of the fraction $\frac{5}{x}$ by yz, we are actually multiplying the fraction by 1. This is because $\frac{yz}{yz} = 1$. Hence,

$$\frac{5}{x} = \frac{5}{x} \cdot 1 = \frac{5}{x} \cdot \left(\frac{yz}{yz}\right) = \frac{5 \cdot yz}{x \cdot yz} = \frac{5yz}{xyz}$$

b. $\dfrac{7}{5p^2} = \dfrac{}{20p^6}$

Multiply the numerator and denominator of the fraction by the missing factor of $4p^4$.

$$\frac{7 \cdot 4p^4}{5p^2 \cdot 4p^4} = \frac{28p^4}{20p^6}$$

c. $\dfrac{w}{w+5} = \dfrac{}{(w+5)(w-2)}$

Multiply numerator and denominator by the missing factor of $(w-2)$.

$$\frac{w}{w+5} = \frac{w \cdot (w-2)}{(w+5) \cdot (w-2)} \qquad \text{or} \qquad \frac{w^2 - 2w}{(w+5)(w-2)}$$

d. $\dfrac{1}{5} = \dfrac{}{5x+20}$

$$\frac{1}{5} = \frac{}{5(x+4)}$$

Factor the denominator of the second expression. The factor missing from the denominator of the first expression is $(x+4)$.

Multiply numerator and denominator by $(x+4)$.

$$\frac{1}{5} = \frac{1 \cdot (x+4)}{5 \cdot (x+4)} = \frac{x+4}{5(x+4)}$$

> **TIP:** Notice that in Example 1(c) we multiplied the polynomials in the numerator but left the denominator in factored form. This convention is followed because when we add and subtract rational expressions in the next section, the terms in the numerators must be combined.

Skill Practice Convert each expression to an equivalent expression with the indicated denominator.

1. $\dfrac{12}{a} = \dfrac{}{abc}$

2. $\dfrac{6}{7y} = \dfrac{}{14y^3}$

3. $\dfrac{x}{x+4} = \dfrac{}{(x+6)(x+4)}$

4. $\dfrac{1}{8} = \dfrac{}{8y-16}$

2. Least Common Denominator

Recall from Section R.2 that to add or subtract fractions, the fractions must have a common denominator. The same is true for rational expressions. In this section, we present a method to find the least common denominator of two rational expressions.

The **least common denominator (LCD)** of two or more rational expressions is defined as the least common multiple of the denominators. For example, consider the fractions $\frac{1}{20}$ and $\frac{1}{8}$. By inspection, you can probably see that the least common denominator is 40. To understand why, find the prime factorization of both denominators:

$$20 = 2^2 \cdot 5 \qquad \text{and} \qquad 8 = 2^3$$

A common multiple of 20 and 8 must be a multiple of 5, a multiple of 2^2, and a multiple of 2^3. However, any number that is a multiple of $2^3 = 8$ is automatically a multiple of $2^2 = 4$. Therefore, it is sufficient to construct the least common denominator as the product of unique prime factors, in which each factor is raised to its highest power.

$$\text{The LCD of } \frac{1}{20} \text{ and } \frac{1}{8} \text{ is } 2^3 \cdot 5 = 40.$$

Skill Practice Answers

1. $\dfrac{12}{a} = \dfrac{12bc}{abc}$ **2.** $\dfrac{6}{7y} = \dfrac{12y^2}{14y^3}$

3. $\dfrac{x}{x+4} = \dfrac{x^2 + 6x}{(x+4)(x+6)}$

4. $\dfrac{1}{8} = \dfrac{y-2}{8(y-2)}$

> **Steps to Find the Least Common Denominator of Two or More Rational Expressions**
>
> **1.** Factor all denominators completely.
>
> **2.** The LCD is the product of unique factors from the denominators, in which each factor is raised to the highest power to which it appears in any denominator.

Example 2   Finding the Least Common Denominator of Rational Expressions

Find the LCD of the following sets of rational expressions.

a. $\dfrac{5}{14}; \dfrac{3}{49}; \dfrac{1}{8}$ **b.** $\dfrac{5}{3x^2z}; \dfrac{7}{x^5y^3}$ **c.** $\dfrac{a+b}{a^2-25}; \dfrac{1}{2a-10}$

d. $\dfrac{x-5}{x^2-2x}; \dfrac{1}{x^2-4x+4}$

Solution:

a. $\dfrac{5}{14}; \dfrac{3}{49}; \dfrac{1}{8}$

$= \dfrac{5}{2\cdot7}; \dfrac{3}{7^2}; \dfrac{1}{2^3}$ **Step 1:** Factor the denominators.

The LCD is $2^3 \cdot 7^2 = 392$.   **Step 2:** The LCD is the product of unique factors, each raised to its highest power.

b. $\dfrac{5}{3x^2z}; \dfrac{7}{x^5y^3}$

$= \dfrac{5}{3x^2z}; \dfrac{7}{x^5y^3}$ **Step 1:** The denominators are already factored.

The LCD is $3x^5y^3z$.   **Step 2:** The LCD is the product of unique factors, each raised to its highest power.

c. $\dfrac{a+b}{a^2-25}; \dfrac{1}{2a-10}$

$= \dfrac{a+b}{(a-5)(a+5)}; \dfrac{1}{2(a-5)}$ **Step 1:** Factor the denominators.

The LCD is $2(a-5)(a+5)$.   **Step 2:** The LCD is the product of unique factors, each raised to its highest power.

d. $\dfrac{x-5}{x^2-2x}; \dfrac{1}{x^2-4x+4}$

$= \dfrac{x-5}{x(x-2)}; \dfrac{1}{(x-2)^2}$ **Step 1:** Factor the denominators.

The LCD is $x(x-2)^2$.   **Step 2:** The LCD is the product of unique factors, each raised to its highest power.

Skill Practice Find the LCD for each set of expressions.

5. $\dfrac{3}{8}; \dfrac{7}{10}; \dfrac{1}{15}$

6. $\dfrac{1}{5a^3b^2}; \dfrac{1}{10a^4b}$

7. $\dfrac{x}{x^2 - 16}; \dfrac{2}{3x + 12}$

8. $\dfrac{6}{t^2 + 5t - 14}; \dfrac{8}{t^2 - 3t + 2}$

---

3. Writing Rational Expressions with the Least Common Denominator

To add or subtract two rational expressions, the expressions must have the same denominator. Therefore, we must first practice the skill of converting each rational expression into an equivalent expression with the LCD as its denominator. The process is as follows: Identify the LCD for the two expressions. Then, multiply the numerator and denominator of each fraction by the factors from the LCD that are missing from the original denominators.

Example 3 **Converting to the Least Common Denominator**

Find the LCD of each pair of rational expressions. Then convert each expression to an equivalent fraction with the denominator equal to the LCD.

a. $\dfrac{3}{2ab}; \dfrac{6}{5a^2}$

b. $\dfrac{4}{x + 1}; \dfrac{7}{x - 4}$

c. $\dfrac{w + 2}{w^2 - w - 12}; \dfrac{1}{w^2 - 9}$

Solution:

a. $\dfrac{3}{2ab}; \dfrac{6}{5a^2}$ The LCD is $10a^2b$.

$$\dfrac{3}{2ab} = \dfrac{3 \cdot 5a}{2ab \cdot 5a} = \dfrac{15a}{10a^2b}$$ The first expression is missing the factor $5a$ from the denominator.

$$\dfrac{6}{5a^2} = \dfrac{6 \cdot 2b}{5a^2 \cdot 2b} = \dfrac{12b}{10a^2b}$$ The second expression is missing the factor $2b$ from the denominator.

b. $\dfrac{4}{x + 1}; \dfrac{7}{x - 4}$ The LCD is $(x + 1)(x - 4)$.

$$\dfrac{4}{x + 1} = \dfrac{4(x - 4)}{(x + 1)(x - 4)} = \dfrac{4x - 16}{(x + 1)(x - 4)}$$ The first expression is missing the factor $(x - 4)$ from the denominator.

$$\dfrac{7}{x - 4} = \dfrac{7(x + 1)}{(x - 4)(x + 1)} = \dfrac{7x + 7}{(x - 4)(x + 1)}$$ The second expression is missing the factor $(x + 1)$ from the denominator.

Skill Practice Answers

5. 120   **6.** $10a^4b^2$
7. $3(x - 4)(x + 4)$
8. $(t + 7)(t - 2)(t - 1)$

c. $\dfrac{w+2}{w^2-w-12}; \dfrac{1}{w^2-9}$ To find the LCD, factor each denominator.

$\dfrac{w+2}{(w-4)(w+3)}; \dfrac{1}{(w-3)(w+3)}$ The LCD is $(w-4)(w+3)(w-3)$.

$\dfrac{w+2}{(w-4)(w+3)} = \dfrac{(w+2)(w-3)}{(w-4)(w+3)(w-3)}$ The first expression is missing the factor $(w-3)$ from the denominator.

$= \dfrac{w^2-w-6}{(w-4)(w+3)(w-3)}$

$\dfrac{1}{(w-3)(w+3)} = \dfrac{1(w-4)}{(w-3)(w+3)(w-4)}$ The second expression is missing the factor $(w-4)$ from the denominator.

$= \dfrac{w-4}{(w-3)(w+3)(w-4)}$

Skill Practice   For each pair of expressions, find the LCD, and then convert each expression to an equivalent fraction with the denominator equal to the LCD.

9. $\dfrac{2}{rs^2}; \dfrac{-1}{r^3s}$ **10.** $\dfrac{5}{x-3}; \dfrac{x}{x+1}$ **11.** $\dfrac{z}{z^2-4}; \dfrac{-3}{z^2-z-2}$

Example 4   **Converting to the Least Common Denominator**

Find the LCD of the expressions $\dfrac{3}{x-7}$ and $\dfrac{1}{7-x}$.

Solution:

Notice that the expressions $x-7$ and $7-x$ are opposites and differ by a factor of -1. Therefore, we may use either $x-7$ or $7-x$ as a common denominator. Each case is detailed in the following conversions.

Converting to the Denominator $x-7$

$\dfrac{3}{x-7}; \dfrac{1}{7-x}$ Leave the first fraction unchanged because it has the desired LCD.

$\dfrac{1}{7-x} = \dfrac{(-1)1}{(-1)(7-x)}$ Multiply the *second* rational expression by the ratio $\frac{-1}{-1}$ to change its denominator to $x-7$.

$= \dfrac{-1}{-7+x}$ Apply the distributive property.

$= \dfrac{-1}{x-7}$

TIP: In Example 4, the expressions

$\dfrac{3}{x-7}$ and $\dfrac{1}{7-x}$

have opposite factors in the denominators. In such a case, you do not need to include *both* factors in the LCD.

Skill Practice Answers

9. $\dfrac{2r^2}{r^3s^2}; \dfrac{-s}{r^3s^2}$

10. $\dfrac{5x+5}{(x-3)(x+1)};$ $\dfrac{x^2-3x}{(x+1)(x-3)}$

11. $\dfrac{z^2+z}{(z-2)(z+2)(z+1)};$ $\dfrac{-3z-6}{(z-2)(z+2)(z+1)}$

Converting to the Denominator $7 - x$

$\dfrac{3}{x - 7}; \dfrac{1}{7 - x}$ Leave the second fraction unchanged because it has the desired LCD.

$\dfrac{3}{x - 7} = \dfrac{(-1)3}{(-1)(x - 7)};$ Multiply the first rational expression by the ratio $\frac{-1}{-1}$ to change its denominator to $7 - x$.

$= \dfrac{-3}{-x + 7}$ Apply the distributive property.

$= \dfrac{-3}{7 - x}$

Skill Practice Answers

12a. The LCD is $(w - 2)$ or $(2 - w)$.

b. $\dfrac{9}{w - 2} = \dfrac{9}{w - 2};$

$\dfrac{11}{2 - w} = \dfrac{-11}{w - 2}$

or

$\dfrac{9}{w - 2} = \dfrac{-9}{2 - w};$

$\dfrac{11}{2 - w} = \dfrac{11}{2 - w}$

> **Skill Practice**
>
> **12a.** Find the LCD of the expressions.
>
> **b.** Then convert each expression to an equivalent fraction with denominator equal to the LCD.
>
> $$\dfrac{9}{w - 2}; \dfrac{11}{2 - w}$$

Section 7.3   Practice Exercises

Study Skills Exercise

1. Define the key term **least common denominator**.

Review Exercises

2. Evaluate the expression for the given values of x. $\dfrac{2x}{x + 5}$

     a. $x = 1$ **b.** $x = 5$ **c.** $x = -5$

For Exercises 3–4, write the domain in set-builder notation. Then reduce the expression to lowest terms.

3. $\dfrac{3x + 3}{5x^2 - 5}$ 4. $\dfrac{x + 2}{x^2 - 3x - 10}$

For Exercises 5–8, multiply or divide as indicated.

5. $\dfrac{a + 3}{a + 7} \cdot \dfrac{a^2 + 3a - 10}{a^2 + a - 6}$ 6. $\dfrac{6(a + 2b)}{2(a - 3b)} \cdot \dfrac{4(a + 3b)(a - 3b)}{9(a + 2b)(a - 2b)}$

7. $\dfrac{16y^2}{9y + 36} \div \dfrac{8y^3}{3y + 12}$ 8. $\dfrac{5b^2 + 6b + 1}{b^2 + 5b + 6} \div (5b + 1)$

Concept 1: Writing Equivalent Rational Expressions

For Exercises 9–24, convert the expressions into equivalent expressions with the indicated denominator.

9. $\dfrac{6}{7} = \dfrac{}{42}$

10. $\dfrac{4}{9} = \dfrac{}{72}$

11. $\dfrac{2}{13} = \dfrac{}{39}$

12. $\dfrac{1}{8} = \dfrac{}{64}$

13. $\dfrac{3}{p^2 q} = \dfrac{}{5p^3 q}$

14. $\dfrac{2}{3rs} = \dfrac{}{18rs^3}$

15. $\dfrac{2x}{yz} = \dfrac{}{6y^2 z^4}$

16. $\dfrac{8a}{b^2 c} = \dfrac{}{2b^4 c^5}$

17. $\dfrac{w + 6}{w - 7} = \dfrac{}{(w - 7)(w + 2)}$

18. $\dfrac{z - 1}{z + 1} = \dfrac{}{(z + 1)(z - 3)}$

19. $\dfrac{-4}{z - 3} = \dfrac{}{5z - 15}$

20. $\dfrac{-8}{3a + 2} = \dfrac{}{12a + 8}$

21. $\dfrac{5}{x + 2} = \dfrac{}{x^2 - 4}$

22. $\dfrac{3y}{4y - 5} = \dfrac{}{16y^2 - 25}$

23. $\dfrac{6}{x - 3} = \dfrac{}{3 - x}$

24. $\dfrac{2}{a - 9} = \dfrac{}{9 - a}$

25. Which of the expressions are equivalent to $-\dfrac{5}{x - 3}$? Circle all that apply.

   a. $\dfrac{-5}{x - 3}$ **b.** $\dfrac{5}{-x + 3}$ **c.** $\dfrac{5}{3 - x}$ **d.** $\dfrac{5}{-(x - 3)}$

26. Which of the expressions are equivalent to $\dfrac{4 - a}{6}$? Circle all that apply.

   a. $\dfrac{a - 4}{-6}$ **b.** $\dfrac{a - 4}{6}$ **c.** $\dfrac{-(4 - a)}{-6}$ **d.** $-\dfrac{a - 4}{6}$

Concept 2: Least Common Denominator

27. Explain why the least common denominator of $\frac{1}{x^3}$, $\frac{1}{x^5}$, and $\frac{1}{x^4}$ is x^5.

28. Explain why the least common denominator of $\frac{2}{y^3}$, $\frac{9}{y^6}$, and $\frac{4}{y^5}$ is y^6.

29. Explain why the least common denominator of

$$\dfrac{1}{x + 3} \quad \text{and} \quad \dfrac{3}{x - 2}$$

is $(x + 3)(x - 2)$.

30. Explain why the least common denominator of

$$\dfrac{7}{y - 8} \quad \text{and} \quad \dfrac{3}{y + 1}$$

is $(y - 8)(y + 1)$.

31. Explain why a common denominator of

$$\dfrac{b + 1}{b - 1} \quad \text{and} \quad \dfrac{b}{1 - b}$$

could be either $(b - 1)$ or $(1 - b)$.

32. Explain why a common denominator of

$$\dfrac{1}{6 - t} \quad \text{and} \quad \dfrac{t}{t - 6}$$

could be either $(6 - t)$ or $(t - 6)$.

For Exercises 33–50, identify the LCD.

33. $\dfrac{4}{15}$; $\dfrac{5}{9}$

34. $\dfrac{7}{12}$; $\dfrac{1}{18}$

35. $\dfrac{3}{16}$; $\dfrac{1}{4}$

36. $\dfrac{1}{2}$; $\dfrac{11}{12}$

37. $\dfrac{1}{7}$; $\dfrac{2}{9}$

38. $\dfrac{2}{3}$; $\dfrac{5}{8}$

39. $\dfrac{1}{3x^2y}$; $\dfrac{8}{9xy^3}$

40. $\dfrac{5}{2a^4b^2}$; $\dfrac{1}{8ab^3}$

41. $\dfrac{6}{w^2}$; $\dfrac{7}{y}$

42. $\dfrac{2}{r}$; $\dfrac{3}{s^2}$

43. $\dfrac{p}{(p+3)(p-1)}$; $\dfrac{2}{(p+3)(p+2)}$

44. $\dfrac{6}{(q+4)(q-4)}$; $\dfrac{q^2}{(q+1)(q+4)}$

45. $\dfrac{7}{3t(t+1)}$; $\dfrac{10t}{9(t+1)^2}$

46. $\dfrac{13x}{15(x-1)^2}$; $\dfrac{5}{3x(x-1)}$

47. $\dfrac{y}{y^2-4}$; $\dfrac{3y}{y^2+5y+6}$

48. $\dfrac{4}{w^2-3w+2}$; $\dfrac{w}{w^2-4}$

49. $\dfrac{5}{3-x}$; $\dfrac{7}{x-3}$

50. $\dfrac{4}{x-6}$; $\dfrac{9}{6-x}$

Concept 3: Writing Rational Expressions with the Least Common Denominator

For Exercises 51–74, find the LCD. Then convert each expression to an equivalent expression with the denominator equal to the LCD.

51. $\dfrac{6}{5x^2}$; $\dfrac{1}{x}$

52. $\dfrac{3}{y}$; $\dfrac{7}{9y^2}$

53. $\dfrac{4}{5x^2}$; $\dfrac{y}{6x^3}$

54. $\dfrac{3}{15b^2}$; $\dfrac{c}{3b^2}$

55. $\dfrac{5}{6a^2b}$; $\dfrac{a}{12b}$

56. $\dfrac{x}{15y^2}$; $\dfrac{y}{5xy}$

57. $\dfrac{6}{m+4}$; $\dfrac{3}{m-1}$

58. $\dfrac{3}{n-5}$; $\dfrac{7}{n+2}$

59. $\dfrac{6}{2x-5}$; $\dfrac{1}{x+3}$

60. $\dfrac{4}{m+3}$; $\dfrac{-3}{5m+1}$

61. $\dfrac{6}{(w+3)(w-8)}$; $\dfrac{w}{(w-8)(w+1)}$

62. $\dfrac{t}{(t+2)(t+12)}$; $\dfrac{18}{(t-2)(t+2)}$

63. $\dfrac{6p}{p^2-4}$; $\dfrac{3}{p^2+4p+4}$

64. $\dfrac{5}{q^2-6q+9}$; $\dfrac{q}{q^2-9}$

65. $\dfrac{1}{a-4}$; $\dfrac{a}{4-a}$

66. $\dfrac{3b}{2b-5}$; $\dfrac{2b}{5-2b}$

67. $\dfrac{4}{x-7}$; $\dfrac{y}{14-2x}$

68. $\dfrac{4}{3x-15}$; $\dfrac{z}{5-x}$

69. $\dfrac{1}{a+b}$; $\dfrac{6}{-a-b}$

70. $\dfrac{p}{-q-8}$; $\dfrac{1}{q+8}$

71. $\dfrac{-3}{24y+8}$; $\dfrac{5}{18y+6}$

72. $\dfrac{r}{10r+5}$; $\dfrac{2}{16r+8}$

73. $\dfrac{3}{5z}$; $\dfrac{1}{z+4}$

74. $\dfrac{-1}{4a-8}$; $\dfrac{5}{4a}$

Expanding Your Skills

For Exercises 75–78, find the LCD. Then convert each expression to an equivalent expression with the denominator equal to the LCD.

75. $\dfrac{z}{z^2 + 9z + 14}$; $\dfrac{-3z}{z^2 + 10z + 21}$; $\dfrac{5}{z^2 + 5z + 6}$

76. $\dfrac{6}{w^2 - 3w - 4}$; $\dfrac{1}{w^2 + 6w + 5}$; $\dfrac{-9w}{w^2 + w - 20}$

77. $\dfrac{3}{p^3 - 8}$; $\dfrac{p}{p^2 - 4}$; $\dfrac{5p}{p^2 + 2p + 4}$

78. $\dfrac{7}{q^3 + 125}$; $\dfrac{q}{q^2 - 25}$; $\dfrac{12}{q^2 - 5q + 25}$

| Section 7.4 | Addition and Subtraction of Rational Expressions |

Concepts

1. Addition and Subtraction of Rational Expressions with the Same Denominator
2. Addition and Subtraction of Rational Expressions with Different Denominators
3. Using Rational Expressions in Translations

1. Addition and Subtraction of Rational Expressions with the Same Denominator

To add or subtract rational expressions, the expressions must have the same denominator. As with fractions, we add or subtract rational expressions with the same denominator by combining the terms in the numerator and then writing the result over the common denominator. Then, if possible, we simplify the expression to lowest terms.

Addition and Subtraction of Rational Expressions

Let p, q, and r represent polynomials where $q \neq 0$. Then,

1. $\dfrac{p}{q} + \dfrac{r}{q} = \dfrac{p + r}{q}$

2. $\dfrac{p}{q} - \dfrac{r}{q} = \dfrac{p - r}{q}$

Example 1 **Adding and Subtracting Rational Expressions with a Common Denominator**

Add or subtract as indicated.

a. $\dfrac{1}{12} + \dfrac{7}{12}$

b. $\dfrac{2}{5p} - \dfrac{7}{5p}$

c. $\dfrac{2}{3d + 5} + \dfrac{7d}{3d + 5}$

d. $\dfrac{x^2}{x - 3} - \dfrac{-5x + 24}{x - 3}$

Solution:

a. $\dfrac{1}{12} + \dfrac{7}{12}$

The fractions have the same denominator.

$= \dfrac{1 + 7}{12}$

Add the terms in the numerators, and write the result over the common denominator.

$= \dfrac{8}{12}$

$= \dfrac{2}{3}$

Simplify to lowest terms.

b. $\dfrac{2}{5p} - \dfrac{7}{5p}$

The rational expressions have the same denominator.

$= \dfrac{2 - 7}{5p}$

Subtract the terms in the numerators, and write the result over the common denominator.

$= \dfrac{-5}{5p}$

$= \dfrac{(-\overset{-1}{\cancel{5}})}{\cancel{5}p}$

Simplify to lowest terms.

$= -\dfrac{1}{p}$

c. $\dfrac{2}{3d + 5} + \dfrac{7d}{3d + 5}$

The rational expressions have the same denominator.

$= \dfrac{2 + 7d}{3d + 5}$

Add the terms in the numerators, and write the result over the common denominator.

$= \dfrac{7d + 2}{3d + 5}$

Because the numerator and denominator share no common factors, the expression is in lowest terms.

d. $\dfrac{x^2}{x - 3} - \dfrac{-5x + 24}{x - 3}$

The rational expressions have the same denominator.

$= \dfrac{x^2 - (-5x + 24)}{x - 3}$

Subtract the terms in the numerators, and write the result over the common denominator.

$= \dfrac{x^2 + 5x - 24}{x - 3}$

Simplify the numerator.

$= \dfrac{(x + 8)(x - 3)}{(x - 3)}$

Factor the numerator and denominator to determine if the rational expression can be simplified.

$= \dfrac{(x + 8)(\cancel{x - 3})^{1}}{(\cancel{x - 3})}$

Simplify to lowest terms.

$= x + 8$

Avoiding Mistakes:

When subtracting rational expressions, use parentheses to group the terms in the numerator that follow the subtraction sign. This will help you remember to apply the distributive property.

Skill Practice   Add or subtract as indicated.

1. $\dfrac{3}{14} + \dfrac{4}{14}$ **2.** $\dfrac{3}{7d} - \dfrac{6}{7d}$

3. $\dfrac{x^2 + 2}{x + 3} + \dfrac{4x + 1}{x + 3}$ **4.** $\dfrac{4t - 9}{2t + 1} - \dfrac{t - 5}{2t + 1}$

2. Addition and Subtraction of Rational Expressions with Different Denominators

To add or subtract two rational expressions with unlike denominators, we must convert the expressions to equivalent expressions with the same denominator. For example, consider adding

$$\frac{1}{10} + \frac{12}{5y}$$

The LCD is $10y$. For each expression, identify the factors from the LCD that are missing from the denominator. Then multiply the numerator and denominator of the expression by the missing factor(s).

$$\underset{\substack{\text{missing} \\ y}}{\frac{1}{10}} + \underset{\substack{\text{missing} \\ 2}}{\frac{12}{5y}}$$

$$= \frac{1 \cdot y}{10 \cdot y} + \frac{12 \cdot 2}{5y \cdot 2}$$

$$= \frac{y}{10y} + \frac{24}{10y} \qquad \text{The rational expressions now have the same denominators.}$$

$$= \frac{y + 24}{10y} \qquad \text{Add the numerators.}$$

Avoiding Mistakes:

In the expression $\frac{y + 24}{10y}$, notice that you cannot reduce the 24 and 10 because 24 is not a factor in the numerator. It is a term. Only factors can be reduced.

After successfully adding or subtracting two rational expressions, always check to see if the final answer is simplified. If necessary, factor the numerator and denominator, and reduce common factors. The expression

$$\frac{y + 24}{10y}$$

is in lowest terms because the numerator and denominator do not share any common factors.

Steps to Add or Subtract Rational Expressions

1. Factor the denominators of each rational expression.

2. Identify the LCD.

3. Rewrite each rational expression as an equivalent expression with the LCD as its denominator.

4. Add or subtract the numerators, and write the result over the common denominator.

5. Simplify to lowest terms.

Skill Practice Answers

1. $\dfrac{1}{2}$ **2.** $-\dfrac{3}{7d}$

3. $x + 1$ **4.** $\dfrac{3t - 4}{2t + 1}$

| Example 2 | **Adding and Subtracting Rational Expressions with Unlike Denominators** |

Add or subtract as indicated.

a. $\dfrac{4}{7k} - \dfrac{3}{k^2}$ **b.** $\dfrac{2q - 4}{3} - \dfrac{q + 1}{2}$ **c.** $\dfrac{1}{x - 5} + \dfrac{-10}{x^2 - 25}$

Solution:

a. $\dfrac{4}{7k} - \dfrac{3}{k^2}$

Step 1: The denominators are already factored.

Step 2: The LCD is $7k^2$.

$= \dfrac{4 \cdot k}{7k \cdot k} - \dfrac{3 \cdot 7}{k^2 \cdot 7}$

Step 3: Write each expression with the LCD.

$= \dfrac{4k}{7k^2} - \dfrac{21}{7k^2}$

$= \dfrac{4k - 21}{7k^2}$

Step 4: Subtract the numerators, and write the result over the LCD.

Step 5: The expression is in lowest terms because the numerator and denominator share no common factors.

Avoiding Mistakes:

Do not reduce after rewriting the fractions with the LCD. You will revert back to the original expression.

b. $\dfrac{2q - 4}{3} - \dfrac{q + 1}{2}$

Step 1: The denominators are already factored.

Step 2: The LCD is 6.

$= \dfrac{2(2q - 4)}{2 \cdot 3} - \dfrac{3(q + 1)}{3 \cdot 2}$

Step 3: Write each expression with the LCD.

$= \dfrac{2(2q - 4) - 3(q + 1)}{6}$

Step 4: Subtract the numerators, and write the result over the LCD.

$= \dfrac{4q - 8 - 3q - 3}{6}$

$= \dfrac{q - 11}{6}$

Step 5: The expression is in lowest terms because the numerator and denominator share no common factors.

c. $\dfrac{1}{x-5} + \dfrac{-10}{x^2-25}$

$= \dfrac{1}{x-5} + \dfrac{-10}{(x-5)(x+5)}$ **Step 1:** Factor the denominators.

Step 2: The LCD is $(x-5)(x+5)$.

$= \dfrac{1(x+5)}{(x-5)(x+5)} + \dfrac{-10}{(x-5)(x+5)}$ **Step 3:** Write each expression with the LCD.

$= \dfrac{1(x+5) + (-10)}{(x-5)(x+5)}$ **Step 4:** Add the numerators, and write the result over the LCD.

$= \dfrac{x+5-10}{(x-5)(x+5)}$

$= \dfrac{x - 5}{(x - 5)(x+5)}$ **Step 5:** Simplify.

$= \dfrac{1}{x+5}$

Skill Practice Add or subtract as indicated.

5. $\dfrac{4}{3x} + \dfrac{1}{2x^2}$ **6.** $\dfrac{q}{12} - \dfrac{q-2}{4}$ **7.** $\dfrac{1}{x-4} + \dfrac{-8}{x^2-16}$

Example 3 **Subtracting Rational Expressions with Different Denominators**

Subtract the rational expressions. $\dfrac{p+2}{p-1} - \dfrac{2}{p+6} - \dfrac{14}{p^2+5p-6}$

Solution:

$\dfrac{p+2}{p-1} - \dfrac{2}{p+6} - \dfrac{14}{p^2+5p-6}$

$= \dfrac{p+2}{p-1} - \dfrac{2}{p+6} - \dfrac{14}{(p-1)(p+6)}$ **Step 1:** Factor the denominators.

Step 2: The LCD is $(p-1)(p+6)$.

Step 3: Write each expression with the LCD.

$= \dfrac{(p+2)(p+6)}{(p-1)(p+6)} - \dfrac{2(p-1)}{(p+6)(p-1)} - \dfrac{14}{(p-1)(p+6)}$

$= \dfrac{(p+2)(p+6) - 2(p-1) - 14}{(p-1)(p+6)}$ **Step 4:** Combine the numerators, and write the result over the LCD.

Skill Practice Answers

5. $\dfrac{8x+3}{6x^2}$ **6.** $\dfrac{-q+3}{6}$

7. $\dfrac{1}{x+4}$

$$= \frac{p^2 + 6p + 2p + 12 - 2p + 2 - 14}{(p - 1)(p + 6)}$$

Step 5: Clear parentheses in the numerator.

$$= \frac{p^2 + 6p}{(p - 1)(p + 6)}$$

Combine *like* terms.

$$= \frac{p(p + 6)}{(p - 1)(p + 6)}$$

Factor the numerator to determine if the expression is in lowest terms.

$$= \frac{p(\overset{1}{\cancel{p + 6}})}{(p - 1)(\cancel{p + 6})}$$

Simplify to lowest terms.

$$= \frac{p}{p - 1}$$

Skill Practice Subtract.

8. $\dfrac{2y}{y - 1} - \dfrac{1}{y} - \dfrac{2y + 1}{y^2 - y}$

When the denominators of two rational expressions are opposites, we can produce identical denominators by multiplying one of the expressions by the ratio $\frac{-1}{-1}$. This is demonstrated in Example 4.

Example 4 Adding Rational Expressions with Different Denominators

Add the rational expressions. $\dfrac{1}{d - 7} + \dfrac{5}{7 - d}$

Solution:

$$\frac{1}{d - 7} + \frac{5}{7 - d}$$

The expressions $d - 7$ and $7 - d$ are opposites and differ by a factor of -1. Therefore, multiply the numerator and denominator of *either* expression by -1 to obtain a common denominator.

$$= \frac{1}{d - 7} + \frac{(-1)5}{(-1)(7 - d)}$$

Note that $-1(7 - d) = -7 + d$ or $d - 7$.

$$= \frac{1}{d - 7} + \frac{-5}{d - 7}$$

Simplify.

$$= \frac{1 + (-5)}{d - 7}$$

Add the terms in the numerators, and write the result over the common denominator.

$$= \frac{-4}{d - 7}$$

Skill Practice Add.

9. $\dfrac{3}{p - 8} + \dfrac{1}{8 - p}$

Skill Practice Answers

8. $\dfrac{2y - 3}{y - 1}$ **9.** $\dfrac{2}{p - 8}$ or $\dfrac{-2}{8 - p}$

3. Using Rational Expressions in Translations

Example 5 Using Rational Expressions in Translations

Translate the English phrase into a mathematical expression. Then simplify by combining the rational expressions.

The difference of the reciprocal of x and the quotient of x and 3

Solution:

The difference of the reciprocal of x and the quotient of x and 3

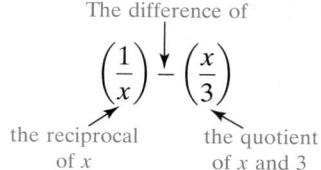

The difference of

$$\left(\frac{1}{x}\right) - \left(\frac{x}{3}\right)$$

the reciprocal   the quotient
of x of x and 3

$$\frac{1}{x} - \frac{x}{3} \qquad \text{The LCD is } 3x.$$

$$= \frac{3 \cdot 1}{3 \cdot x} - \frac{x \cdot x}{3 \cdot x} \qquad \text{Write each expression over the LCD.}$$

$$= \frac{3 - x^2}{3x} \qquad \text{Subtract the numerators.}$$

Skill Practice Translate the English phrase into a mathematical expression. Then simplify by combining the rational expressions.

10. The sum of 1 and the quotient of 2 and a

Skill Practice Answers

10. $1 + \dfrac{2}{a}; \dfrac{a + 2}{a}$

Section 7.4	**Practice Exercises**

Review Exercises

1. Write the domain of the expression. $\dfrac{x + 4}{x^2 - 36}$

2. For the rational expression $\dfrac{x^2 - 4x - 5}{x^2 - 7x + 10}$

    a. Find the value of the expression (if possible) when $x = 0, 1, -1, 2,$ and 5.

    b. Factor the denominator and identify the domain. Write the domain in set-builder notation.

    c. Reduce the expression to lowest terms.

3. For the rational expression $\dfrac{a^2 + a - 2}{a^2 - 4a - 12}$

   a. Find the value of the expression (if possible) when $a = 0, 1, -2, 2$, and 6.

   b. Factor the denominator, and identify the domain. Write the domain in set-builder notation.

   c. Reduce the expression to lowest terms.

For Exercises 4–5, multiply or divide as indicated.

4. $\dfrac{2b^2 - b - 3}{2b^2 - 3b - 9} \div \dfrac{b^2 - 1}{4b + 6}$

5. $\dfrac{6t - 1}{5t - 30} \cdot \dfrac{10t - 25}{2t^2 - 3t - 5}$

Concept 1: Addition and Subtraction of Rational Expressions with the Same Denominator

For Exercises 6–27, add or subtract the expressions with like denominators as indicated.

6. $\dfrac{7}{8} + \dfrac{3}{8}$

7. $\dfrac{1}{3} + \dfrac{7}{3}$

8. $\dfrac{9}{16} - \dfrac{3}{16}$

9. $\dfrac{14}{15} - \dfrac{4}{15}$

10. $\dfrac{5a}{a + 2} - \dfrac{3a - 4}{a + 2}$

11. $\dfrac{2b}{b - 3} - \dfrac{b - 9}{b - 3}$

12. $\dfrac{5c}{c + 6} + \dfrac{30}{c + 6}$

13. $\dfrac{12}{2 + d} + \dfrac{6d}{2 + d}$

14. $\dfrac{5}{t - 8} - \dfrac{2t + 1}{t - 8}$

15. $\dfrac{7p + 1}{2p + 1} - \dfrac{p - 4}{2p + 1}$

16. $\dfrac{9x^2}{3x - 7} - \dfrac{49}{3x - 7}$

17. $\dfrac{4w^2}{2w - 1} - \dfrac{1}{2w - 1}$

18. $\dfrac{m^2}{m + 5} + \dfrac{10m + 25}{m + 5}$

19. $\dfrac{k^2}{k - 3} - \dfrac{6k - 9}{k - 3}$

20. $\dfrac{2a}{a + 2} + \dfrac{4}{a + 2}$

21. $\dfrac{5b}{b + 4} + \dfrac{20}{b + 4}$

22. $\dfrac{x^2}{x + 5} - \dfrac{25}{x + 5}$

23. $\dfrac{y^2}{y - 7} - \dfrac{49}{y - 7}$

24. $\dfrac{r}{r^2 + 3r + 2} + \dfrac{2}{r^2 + 3r + 2}$

25. $\dfrac{x}{x^2 - x - 12} - \dfrac{4}{x^2 - x - 12}$

26. $\dfrac{1}{3y^2 + 22y + 7} - \dfrac{-3y}{3y^2 + 22y + 7}$

27. $\dfrac{5}{2x^2 + 13x + 20} + \dfrac{2x}{2x^2 + 13x + 20}$

For Exercises 28–29, find an expression that represents the perimeter of the figure (assume that $x > 0, y > 0$, and $t > 0$).

28.

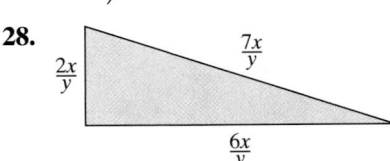

29.

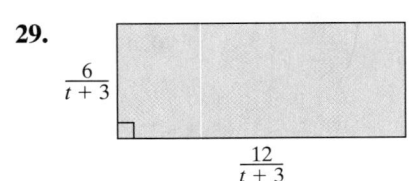

Concept 2: Addition and Subtraction of Rational Expressions with Different Denominators

For Exercises 30–71, add or subtract the expressions with unlike denominators as indicated.

30. $\dfrac{5}{4} + \dfrac{3}{2a}$

31. $\dfrac{11}{6p} + \dfrac{-7}{4p}$

32. $\dfrac{4}{5xy^3} + \dfrac{2x}{15y^2}$

33. $\dfrac{5}{3a^2b} + \dfrac{-7}{6b^2}$

34. $\dfrac{2}{s^3t^3} - \dfrac{3}{s^4t}$

35. $\dfrac{1}{p^2q} - \dfrac{2}{pq^3}$

36. $\dfrac{z}{3z - 9} - \dfrac{z - 2}{z - 3}$

37. $\dfrac{3w - 8}{2w - 4} - \dfrac{w - 3}{w - 2}$

38. $\dfrac{5}{a + 1} + \dfrac{4}{3a + 3}$

39. $\dfrac{2}{c - 4} + \dfrac{1}{5c - 20}$

40. $\dfrac{k}{k^2 - 9} - \dfrac{4}{k - 3}$

41. $\dfrac{7}{h + 5} - \dfrac{2h - 3}{h^2 - 25}$

42. $\dfrac{3a - 7}{6a + 10} - \dfrac{10}{3a^2 + 5a}$

43. $\dfrac{k + 2}{8k} - \dfrac{3 - k}{12k}$

44. $\dfrac{10}{3x - 7} + \dfrac{5}{7 - 3x}$

45. $\dfrac{8}{2w - 1} + \dfrac{4}{1 - 2w}$

46. $\dfrac{6a}{a^2 - b^2} + \dfrac{2a}{a^2 + ab}$

47. $\dfrac{7x}{x^2 + 2xy + y^2} + \dfrac{3x}{x^2 + xy}$

48. $\dfrac{p}{3} - \dfrac{4p - 1}{-3}$

49. $\dfrac{r}{7} - \dfrac{r - 5}{-7}$

50. $\dfrac{4n}{n - 8} - \dfrac{2n - 1}{8 - n}$

51. $\dfrac{m}{m - 2} - \dfrac{3m + 1}{2 - m}$

52. $\dfrac{5}{x} + \dfrac{3}{x + 2}$

53. $\dfrac{6}{y - 1} + \dfrac{9}{y}$

54. $\dfrac{5}{p - 3} - \dfrac{2}{p - 1}$

55. $\dfrac{1}{7x} + \dfrac{5}{2y^2}$

56. $\dfrac{y}{4y + 2} + \dfrac{3y}{6y + 3}$

57. $\dfrac{4}{q^2 - 2q} - \dfrac{5}{3q - 6}$

58. $\dfrac{4w}{w^2 + 2w - 3} + \dfrac{2}{1 - w}$

59. $\dfrac{z - 23}{z^2 - z - 20} - \dfrac{2}{5 - z}$

60. $\dfrac{3a - 8}{a^2 - 5a + 6} + \dfrac{a + 2}{a^2 - 6a + 8}$

61. $\dfrac{3b + 5}{b^2 + 4b + 3} + \dfrac{-b + 5}{b^2 + 2b - 3}$

62. $\dfrac{3x}{x^2 + x - 6} + \dfrac{x}{x^2 + 5x + 6}$

63. $\dfrac{x}{x^2 + 5x + 4} - \dfrac{2x}{x^2 - 2x - 3}$

64. $\dfrac{3y}{2y^2 - y - 1} - \dfrac{4y}{2y^2 - 7y - 4}$

65. $\dfrac{5}{6y^2 - 7y - 3} + \dfrac{4y}{3y^2 + 4y + 1}$

66. $\dfrac{3}{2p - 1} - \dfrac{4p + 4}{4p^2 - 1}$

67. $\dfrac{1}{3q - 2} - \dfrac{6q + 4}{9q^2 - 4}$

68. $\dfrac{m}{m + n} - \dfrac{m}{m - n} + \dfrac{1}{m^2 - n^2}$

69. $\dfrac{x}{x + y} - \dfrac{2xy}{x^2 - y^2} + \dfrac{y}{x - y}$

70. $\dfrac{2}{a + b} + \dfrac{2}{a - b} - \dfrac{4a}{a^2 - b^2}$

71. $\dfrac{-2x}{x^2 - y^2} + \dfrac{1}{x + y} - \dfrac{1}{x - y}$

For Exercises 72–73, find an expression that represents the perimeter of the figure (assume that $x > 0$ and $t > 0$).

72.

$\dfrac{2}{x + 3}$

$\dfrac{1}{x + 2}$

73.

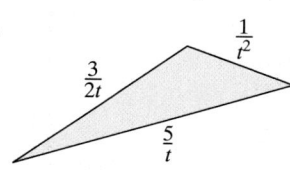

$\dfrac{1}{t^2}$

$\dfrac{3}{2t}$

$\dfrac{5}{t}$

Concept 3: Using Rational Expressions in Translations

74. Let a number be represented by n. Write the reciprocal of n.

75. Write the reciprocal of the sum of a number and 6.

76. Write the quotient of 5 and the sum of a number and 2.

77. Let a number be represented by p. Write the quotient of 12 and p.

For Exercises 78–81, translate the English phrases into algebraic expressions. Then simplify by combining the rational expressions.

78. The sum of a number and the quantity seven times the reciprocal of the number.

79. The sum of a number and the quantity five times the reciprocal of the number.

80. The difference of the reciprocal of n and the quotient of 2 and n.

81. The difference of the reciprocal of m and the quotient of $3m$ and 7.

Expanding Your Skills

For Exercises 82–87, perform the indicated operations.

82. $\dfrac{-3}{w^3 + 27} - \dfrac{1}{w^2 - 9}$

83. $\dfrac{m}{m^3 - 1} + \dfrac{1}{(m - 1)^2}$

84. $\dfrac{2p}{p^2 + 5p + 6} - \dfrac{p + 1}{p^2 + 2p - 3} + \dfrac{3}{p^2 + p - 2}$

85. $\dfrac{3t}{8t^2 + 2t - 1} - \dfrac{5t}{2t^2 - 9t - 5} + \dfrac{2}{4t^2 - 21t + 5}$

86. $\dfrac{3m}{m^2 + 3m - 10} + \dfrac{5}{4 - 2m} - \dfrac{1}{m + 5}$

87. $\dfrac{2n}{3n^2 - 8n - 3} + \dfrac{1}{6 - 2n} - \dfrac{3}{3n + 1}$

For Exercises 88–91, simplify by applying the order of operations.

88. $\left(\dfrac{2}{k + 1} + 3\right)\left(\dfrac{k + 1}{4k + 7}\right)$

89. $\left(\dfrac{p + 1}{3p + 4}\right)\left(\dfrac{1}{p + 1} + 2\right)$

90. $\left(\dfrac{1}{10a} - \dfrac{b}{10a^2}\right) \div \left(\dfrac{1}{10} - \dfrac{b}{10a}\right)$

91. $\left(\dfrac{1}{2m} + \dfrac{n}{2m^2}\right) \div \left(\dfrac{1}{4} + \dfrac{n}{4m}\right)$

Chapter 7

Problem Recognition Exercises— Operations on Rational Expressions

In Sections 7.1–7.4, we learned how to simplify, add, subtract, multiply, and divide rational expressions. The procedure for each operation is different, and it takes considerable practice to determine the correct method to apply for a given problem. The following review exercises give you the opportunity to practice the specific techniques for simplifying rational expressions.

1. Subtract. $\dfrac{5}{3x+1} - \dfrac{2x-4}{3x+1}$

2. Divide. $\dfrac{\dfrac{w+1}{w^2-16}}{\dfrac{w+1}{w+4}}$

3. Multiply. $\dfrac{3}{y} \cdot \dfrac{y^2-5y}{6y-9}$

4. Add. $\dfrac{-1}{x+3} + \dfrac{2}{2x-1}$

5. Simplify. $\dfrac{x-9}{9x-x^2}$

6. Add and subtract. $\dfrac{1}{p} - \dfrac{3}{p^2+3p} + \dfrac{p}{3p+9}$

7. Divide. $\dfrac{c^2+5c+6}{c^2+c-2} \div \dfrac{c}{c-1}$

8. Multiply. $\dfrac{2x^2-5x-3}{x^2-9} \cdot \dfrac{x^2+6x+9}{10x+5}$

9. Simplify. $\dfrac{6a^2b^3}{72ab^7c}$

10. Subtract. $\dfrac{2a}{a+b} - \dfrac{b}{a-b} - \dfrac{-4ab}{a^2-b^2}$

11. Divide. $\dfrac{p^2+10pq+25q^2}{p^2+6pq+5q^2} \div \dfrac{10p+50q}{2p^2-2q^2}$

12. Add. $\dfrac{3k-8}{k-5} + \dfrac{k-12}{k-5}$

13. Simplify. $\dfrac{20x^2+10x}{4x^3+4x^2+x}$

14. Multiply.

$\dfrac{w^2-81}{w^2+10w+9} \cdot \dfrac{w^2+w+2zw+2z}{w^2-9w+zw-9z}$

15. Divide. $\dfrac{8x^2-18x-5}{4x^2-25} \div \dfrac{4x^2-11x-3}{3x-9}$

16. Simplify. $\dfrac{xy+7x+5y+35}{x^2+ax+5x+5a}$

17. Subtract. $\dfrac{a}{a^2-9} - \dfrac{3}{6a-18}$

18. Add. $\dfrac{4}{y^2-36} + \dfrac{2}{y^2-4y-12}$

19. Multiply. $(t^2+5t-24)\left(\dfrac{t+8}{t-3}\right)$

20. Simplify. $\dfrac{6b^2-7b-10}{b-2}$

For Exercises 21–22, determine the domain.

21. $\dfrac{x-3}{x+1}$

22. $\dfrac{x-2}{x^2-9}$

Complex Fractions

1. Simplifying Complex Fractions (Method I)

A **complex fraction** is a fraction whose numerator or denominator contains one or more rational expressions. For example,

$$\frac{\dfrac{1}{ab}}{\dfrac{2}{b}} \quad \text{and} \quad \frac{1 + \dfrac{3}{4} - \dfrac{1}{6}}{\dfrac{1}{2} + \dfrac{1}{3}}$$

are complex fractions.

Two methods will be presented to simplify complex fractions. The first method (Method I) follows the order of operations to simplify the numerator and denominator separately before dividing. The process is summarized as follows.

> ### Steps to Simplify a Complex Fraction (Method I)
>
> 1. Add or subtract expressions in the numerator to form a single fraction. Add or subtract expressions in the denominator to form a single fraction.
>
> 2. Divide the rational expressions from step 1 by multiplying the numerator of the complex fraction by the reciprocal of the denominator of the complex fraction.
>
> 3. Simplify to lowest terms if possible.

Example 1   Simplifying Complex Fractions (Method I)

Simplify the expression. $\dfrac{\dfrac{1}{ab}}{\dfrac{2}{b}}$

Solution:

> **Step 1:** The numerator and denominator of the complex fraction are already single fractions.

$\dfrac{\dfrac{1}{ab}}{\dfrac{2}{b}}$ ⟵   This fraction bar denotes division ($\div$).

$= \dfrac{1}{ab} \div \dfrac{2}{b}$

$= \dfrac{1}{ab} \cdot \dfrac{b}{2}$ **Step 2:** Multiply the numerator of the complex fraction by the reciprocal of $\frac{2}{b}$, which is $\frac{b}{2}$.

$= \dfrac{1}{a\cancel{b}} \cdot \dfrac{\overset{1}{\cancel{b}}}{2}$ **Step 3:** Reduce common factors and simplify.

$= \dfrac{1}{2a}$

Simplify the expression.

1. $\dfrac{\dfrac{6x}{y}}{\dfrac{9}{2y}}$

Sometimes it is necessary to simplify the numerator and denominator of a complex fraction before the division can be performed. This is illustrated in the next example.

Example 2   Simplifying Complex Fractions (Method I)

Simplify the expression. $\dfrac{1 + \dfrac{3}{4} - \dfrac{1}{6}}{\dfrac{1}{2} + \dfrac{1}{3}}$

Solution:

$\dfrac{1 + \dfrac{3}{4} - \dfrac{1}{6}}{\dfrac{1}{2} + \dfrac{1}{3}}$

Step 1: Combine fractions in the numerator and denominator separately.

$= \dfrac{1 \cdot \dfrac{12}{12} + \dfrac{3}{4} \cdot \dfrac{3}{3} - \dfrac{1}{6} \cdot \dfrac{2}{2}}{\dfrac{1}{2} \cdot \dfrac{3}{3} + \dfrac{1}{3} \cdot \dfrac{2}{2}}$

The LCD in the numerator is 12. The LCD in the denominator is 6.

$= \dfrac{\dfrac{12}{12} + \dfrac{9}{12} - \dfrac{2}{12}}{\dfrac{3}{6} + \dfrac{2}{6}}$

$= \dfrac{\dfrac{19}{12}}{\dfrac{5}{6}}$

Form a single fraction in the numerator and in the denominator.

$= \dfrac{19}{\underset{2}{12}} \cdot \dfrac{\overset{1}{6}}{5}$

Step 2: Multiply by the reciprocal of $\frac{5}{6}$, which is $\frac{6}{5}$.

$= \dfrac{19}{10}$

Step 3: Simplify.

Simplify the expression.

2. $\dfrac{\dfrac{3}{4} - \dfrac{1}{6} + 2}{\dfrac{1}{3} + \dfrac{1}{2}}$

Skill Practice Answers

1. $\dfrac{4x}{3}$

2. $\dfrac{31}{10}$

| Example 3 | Simplifying Complex Fractions (Method I) |

Simplify the expression.

$$\dfrac{\dfrac{1}{x} + \dfrac{1}{y}}{x - \dfrac{y^2}{x}}$$

Solution:

$$\dfrac{\dfrac{1}{x} + \dfrac{1}{y}}{x - \dfrac{y^2}{x}}$$

The LCD in the numerator is xy. The LCD in the denominator is x.

$$= \dfrac{\dfrac{1 \cdot y}{x \cdot y} + \dfrac{1 \cdot x}{y \cdot x}}{\dfrac{x \cdot x}{1 \cdot x} - \dfrac{y^2}{x}}$$

Rewrite the expressions using common denominators.

$$= \dfrac{\dfrac{y}{xy} + \dfrac{x}{xy}}{\dfrac{x^2}{x} - \dfrac{y^2}{x}}$$

$$= \dfrac{\dfrac{y + x}{xy}}{\dfrac{x^2 - y^2}{x}}$$

Form single fractions in the numerator and denominator.

$$= \dfrac{y + x}{xy} \cdot \dfrac{x}{x^2 - y^2}$$

Multiply by the reciprocal of the denominator.

$$= \dfrac{\overset{1}{\cancel{y + x}}}{xy} \cdot \dfrac{\overset{1}{x}}{\cancel{(x + y)}(x - y)}$$

Factor and reduce. Note that $(y + x) = (x + y)$.

$$= \dfrac{1}{y(x - y)}$$

Simplify.

| Skill Practice | Simplify the expression.

3. $\dfrac{1 - \dfrac{q}{p}}{\dfrac{p}{q} - \dfrac{q}{p}}$

2. Simplifying Complex Fractions (Method II)

We will now simplify the expressions from Examples 2 and 3 again using a second method to simplify complex fractions (Method II). Recall that multiplying the numerator and denominator of a rational expression by the same quantity does not change the value of the expression because we are multiplying by a number equivalent to 1. This is the basis for Method II.

> **Steps to Simplifying a Complex Fraction (Method II)**
>
> **1.** Multiply the numerator and denominator of the complex fraction by the LCD of *all* individual fractions within the expression.
>
> **2.** Apply the distributive property, and simplify the numerator and denominator.
>
> **3.** Simplify to lowest terms if possible.

Example 4   **Simplifying Complex Fractions (Method II)**

Simplify the expression. $\dfrac{1 + \dfrac{3}{4} - \dfrac{1}{6}}{\dfrac{1}{2} + \dfrac{1}{3}}$

Solution:

TIP: In step 1, we are multiplying the original expression by $\frac{12}{12}$, which equals 1.

$$\dfrac{1 + \dfrac{3}{4} - \dfrac{1}{6}}{\dfrac{1}{2} + \dfrac{1}{3}}$$

The LCD of the expressions $1, \frac{3}{4}, \frac{1}{6}, \frac{1}{2}$, and $\frac{1}{3}$ is 12.

$$= \dfrac{12\left(1 + \dfrac{3}{4} - \dfrac{1}{6}\right)}{12\left(\dfrac{1}{2} + \dfrac{1}{3}\right)}$$

Step 1: Multiply the numerator and denominator of the complex fraction by 12.

$$= \dfrac{12 \cdot 1 + 12 \cdot \dfrac{3}{4} - 12 \cdot \dfrac{1}{6}}{12 \cdot \dfrac{1}{2} + 12 \cdot \dfrac{1}{3}}$$

Step 2: Apply the distributive property.

$$= \dfrac{12 \cdot 1 + \overset{3}{12} \cdot \dfrac{3}{4} - \overset{2}{12} \cdot \dfrac{1}{6}}{\overset{6}{12} \cdot \dfrac{1}{2} + \overset{4}{12} \cdot \dfrac{1}{3}}$$

Simplify each term.

$$= \dfrac{12 + 9 - 2}{6 + 4}$$

$$= \dfrac{19}{10}$$

Step 3: Simplify.

Skill Practice   Simplify the expression.

4. $\dfrac{1 - \dfrac{3}{5}}{\dfrac{1}{4} - \dfrac{7}{10} + 1}$

Skill Practice Answers

4. $\dfrac{8}{11}$

Example 5 Simplifying a Complex Fraction (Method II)

Simplify the expression. $\dfrac{\dfrac{1}{x}+\dfrac{1}{y}}{x-\dfrac{y^2}{x}}$

Solution:

$\dfrac{\dfrac{1}{x}+\dfrac{1}{y}}{x-\dfrac{y^2}{x}}$ The LCD of the expressions $\frac{1}{x}, \frac{1}{y}, x$, and $\frac{y^2}{x}$ is xy.

$=\dfrac{xy\left(\dfrac{1}{x}+\dfrac{1}{y}\right)}{xy\left(x-\dfrac{y^2}{x}\right)}$ **Step 1:** Multiply numerator and denominator of the complex fraction by xy.

$=\dfrac{xy\cdot\dfrac{1}{x}+xy\cdot\dfrac{1}{y}}{xy\cdot x-xy\cdot\dfrac{y^2}{x}}$ **Step 2:** Apply the distributive property, and simplify each term.

$=\dfrac{y+x}{x^2y-y^3}$

$=\dfrac{y+x}{y(x^2-y^2)}$ **Step 3:** Factor completely, and reduce common factors.

$=\dfrac{\overset{1}{\cancel{y+x}}}{y\cancel{(x+y)}(x-y)}$ Note that $(y+x)=(x+y)$.

$=\dfrac{1}{y(x-y)}$

Skill Practice Simplify the expression.

5. $\dfrac{\dfrac{z}{3}-\dfrac{3}{z}}{1+\dfrac{3}{z}}$

Example 6 Simplifying a Complex Fraction (Method II)

Simplify the expression. $\dfrac{\dfrac{1}{k+1}-1}{\dfrac{1}{k+1}+1}$

Skill Practice Answers

5. $\dfrac{z-3}{3}$

Solution:

$$\frac{\dfrac{1}{k+1}-1}{\dfrac{1}{k+1}+1}$$

The LCD of $\dfrac{1}{k+1}$ and 1 is $(k+1)$.

$$=\frac{(k+1)\left(\dfrac{1}{k+1}-1\right)}{(k+1)\left(\dfrac{1}{k+1}+1\right)}$$

Step 1: Multiply numerator and denominator of the complex fraction by $(k+1)$.

$$=\frac{(\overset{1}{\cancel{k+1}})\cdot\dfrac{1}{(\cancel{k+1})}-(k+1)\cdot 1}{(\overset{1}{\cancel{k+1}})\cdot\dfrac{1}{(\cancel{k+1})}+(k+1)\cdot 1}$$

Step 2: Apply the distributive property.

$$=\frac{1-(k+1)}{1+(k+1)}$$

Simplify.

$$=\frac{1-k-1}{1+k+1}$$

$$=\frac{-k}{k+2}$$

Step 3: The expression is already in lowest terms.

> **Skill Practice**   Simplify the expression.

6. $\dfrac{\dfrac{4}{p-3}+1}{1+\dfrac{2}{p-3}}$

Skill Practice Answers

6. $\dfrac{p+1}{p-1}$

Section 7.5   Practice Exercises

Study Skills Exercise

1. Define the key term **complex fraction**.

Review Exercises

For Exercises 2–3, write the domain in set-builder notation, and simplify the expression.

2. $\dfrac{y(2y+9)}{y^2(2y+9)}$

3. $\dfrac{a+5}{2a^2+7a-15}$

For Exercises 4–9, perform the indicated operations.

4. $\dfrac{2}{w-2}+\dfrac{3}{w}$

5. $\dfrac{6}{5}-\dfrac{3}{5k-10}$

6. $\dfrac{p^2+2p}{2p-1}\cdot\dfrac{10p^2-5p}{12p^3+24p^2}$

7. $\dfrac{x^2 - 2xy + y^2}{x^4 - y^4} \div \dfrac{3x^2y - 3xy^2}{x^2 + y^2}$

8. $\left(\dfrac{1}{z} - \dfrac{1}{2z}\right) \div \left(\dfrac{1}{2} + \dfrac{1}{2z}\right)$

9. $\left(\dfrac{2}{3a^2} - \dfrac{3}{b}\right) \div \left(\dfrac{5}{ab} - 4\right)$

Concepts 1–2: Simplifying Complex Fractions (Methods I and II)

For Exercises 10–35, simplify the complex fractions.

10. $\dfrac{\dfrac{7}{18y}}{\dfrac{2}{9}}$

11. $\dfrac{\dfrac{a^2}{2a - 3}}{\dfrac{5a}{8a - 12}}$

12. $\dfrac{\dfrac{3x + 2y}{2y}}{\dfrac{6x + 4y}{2}}$

13. $\dfrac{\dfrac{2x - 10}{4}}{\dfrac{x^2 - 5x}{3x}}$

14. $\dfrac{\dfrac{8a^4b^3}{3c}}{\dfrac{a^7b^2}{9c}}$

15. $\dfrac{\dfrac{12x^2}{5y}}{\dfrac{8x^6}{9y^2}}$

16. $\dfrac{\dfrac{4r^3s}{t^5}}{\dfrac{2s^7}{r^2t^9}}$

17. $\dfrac{\dfrac{5p^4q}{w^4}}{\dfrac{10p^2}{qw^2}}$

18. $\dfrac{\dfrac{1}{8} + \dfrac{4}{3}}{\dfrac{1}{2} - \dfrac{5}{12}}$

19. $\dfrac{\dfrac{8}{9} - \dfrac{1}{3}}{\dfrac{7}{6} + \dfrac{1}{9}}$

20. $\dfrac{\dfrac{1}{h} + \dfrac{1}{k}}{\dfrac{1}{hk}}$

21. $\dfrac{\dfrac{1}{b} + 1}{\dfrac{1}{b}}$

22. $\dfrac{\dfrac{n + 1}{n^2 - 9}}{\dfrac{2}{n + 3}}$

23. $\dfrac{\dfrac{5}{k - 5}}{\dfrac{k + 1}{k^2 - 25}}$

24. $\dfrac{2 + \dfrac{1}{x}}{4 + \dfrac{1}{x}}$

25. $\dfrac{6 + \dfrac{6}{k}}{1 + \dfrac{1}{k}}$

26. $\dfrac{\dfrac{m}{7} - \dfrac{7}{m}}{\dfrac{1}{7} + \dfrac{1}{m}}$

27. $\dfrac{\dfrac{2}{p} + \dfrac{p}{2}}{\dfrac{p}{3} - \dfrac{3}{p}}$

28. $\dfrac{\dfrac{1}{5} - \dfrac{1}{y}}{\dfrac{7}{10} + \dfrac{1}{y^2}}$

29. $\dfrac{\dfrac{1}{m^2} + \dfrac{2}{3}}{\dfrac{1}{m} - \dfrac{5}{6}}$

30. $\dfrac{\dfrac{8}{a + 4} + 2}{\dfrac{12}{a + 4} - 2}$

31. $\dfrac{\dfrac{2}{w + 1} + 3}{\dfrac{3}{w + 1} + 4}$

32. $\dfrac{1 - \dfrac{4}{t^2}}{1 - \dfrac{2}{t} - \dfrac{8}{t^2}}$

33. $\dfrac{1 - \dfrac{9}{p^2}}{1 - \dfrac{1}{p} - \dfrac{6}{p^2}}$

34. $\dfrac{t + 4 + \dfrac{3}{t}}{t - 4 - \dfrac{5}{t}}$

35. $\dfrac{\dfrac{9}{4m} + \dfrac{9}{2m^2}}{\dfrac{3}{2} + \dfrac{3}{m}}$

For Exercises 36–39, translate the English phrases into algebraic expressions. Then simplify the expressions.

36. The sum of one-half and two-thirds, divided by five.

37. The quotient of ten and the difference of two-fifths and one-fourth.

38. The quotient of three and the sum of two-thirds and three-fourths.

39. The difference of three-fifths and one-half, divided by four.

40. In electronics, resistors oppose the flow of current. For two resistors in parallel, the total resistance is given by

$$R = \dfrac{1}{\dfrac{1}{R_1} + \dfrac{1}{R_2}}$$

a. Find the total resistance if $R_1 = 2\ \Omega$ (ohms) and $R_2 = 3\ \Omega$.

b. Find the total resistance if $R_1 = 10\ \Omega$ and $R_2 = 15\ \Omega$.

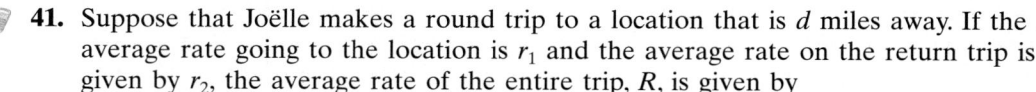 **41.** Suppose that Joëlle makes a round trip to a location that is d miles away. If the average rate going to the location is r_1 and the average rate on the return trip is given by r_2, the average rate of the entire trip, R, is given by

$$R = \dfrac{2d}{\dfrac{d}{r_1} + \dfrac{d}{r_2}}$$

a. Find the average rate of a trip to a destination 30 miles away when the average rate going there was 60 mph and the average rate returning home was 45 mph. (Round to the nearest tenth of a mile per hour.)

b. Find the average rate of a trip to a destination that is 50 miles away if the driver travels at the same rates as in part (a). (Round to the nearest tenth of a mile per hour.)

c. Compare your answers from parts (a) and (b) and explain the results in the context of the problem.

Expanding Your Skills

For Exercises 42–45, simplify the complex fractions using either method.

42. $\dfrac{\dfrac{1}{z^2 - 9} + \dfrac{2}{z + 3}}{\dfrac{3}{z - 3}}$

43. $\dfrac{\dfrac{5}{w^2 - 25} - \dfrac{3}{w + 5}}{\dfrac{4}{w - 5}}$

44. $\dfrac{\dfrac{2}{x - 1} + 2}{\dfrac{2}{x + 1} - 2}$

45. $\dfrac{\dfrac{1}{y - 3} + 1}{\dfrac{2}{y + 3} - 1}$

For Exercises 46–48, simplify the complex fractions. (*Hint:* Use the order of operations and begin with the fraction on the lower right.)

46. $1 + \dfrac{1}{1 + 1}$

47. $1 + \dfrac{1}{1 + \dfrac{1}{1 + 1}}$

48. $1 + \dfrac{1}{1 + \dfrac{1}{1 + \dfrac{1}{1 + 1}}}$

Rational Equations

1. Introduction to Rational Equations

Thus far we have studied two specific types of equations in one variable: linear equations and quadratic equations. Recall,

$$ax + b = 0, \text{ where } a \neq 0, \text{ is a } \textbf{linear equation.}$$

$$ax^2 + bx + c = 0, \text{ where } a \neq 0, \text{ is a } \textbf{quadratic equation.}$$

We will now study another type of equation called a rational equation.

> **Definition of a Rational Equation**
>
> An equation with one or more rational expressions is called a **rational equation.**

The following equations are rational equations:

$$\frac{y}{2} + \frac{y}{4} = 6 \qquad \frac{1}{x} + \frac{1}{3} = \frac{5}{6} \qquad \frac{6}{t^2 - 7t + 12} + \frac{2t}{t - 3} = \frac{3t}{t - 4}$$

To understand the process of solving a rational equation, first review the process of clearing fractions from Section 2.3.

Example 1 Solving a Rational Equation

Solve. $\dfrac{y}{2} + \dfrac{y}{4} = 6$

Solution:

$$\frac{y}{2} + \frac{y}{4} = 6 \qquad \text{The LCD of all terms in the equation is } 4.$$

$$4\left(\frac{y}{2} + \frac{y}{4}\right) = 4(6) \qquad \text{Multiply both sides of the equation by 4 to clear fractions.}$$

$$4 \cdot \frac{y}{2} + 4 \cdot \frac{y}{4} = 4(6) \qquad \text{Apply the distributive property.}$$

$$2y + y = 24 \qquad \text{Clear fractions.}$$

$$3y = 24 \qquad \text{Solve the resulting equation (linear).}$$

$$y = 8$$

$$\underline{\text{Check:}} \quad \frac{y}{2} + \frac{y}{4} = 6$$

$$\frac{(8)}{2} + \frac{(8)}{4} \stackrel{?}{=} 6$$

$$4 + 2 \stackrel{?}{=} 6$$

$$6 = 6 \checkmark$$

| Skill Practice | Solve the equation.

1. $\dfrac{t}{5} - \dfrac{t}{4} = 2$

2. Solving Rational Equations

The same process of clearing fractions is used to solve rational equations when variables are present in the denominator.

| Example 2 | **Solving a Rational Equation**

Solve the equation. $\dfrac{x+1}{x} + \dfrac{1}{3} = \dfrac{5}{6}$

Solution:

$$\dfrac{x+1}{x} + \dfrac{1}{3} = \dfrac{5}{6} \qquad \text{The LCD of all the expressions is } 6x.$$

$$6x \cdot \left(\dfrac{x+1}{x} + \dfrac{1}{3}\right) = 6x \cdot \left(\dfrac{5}{6}\right) \qquad \text{Multiply by the LCD.}$$

$$6x \cdot \left(\dfrac{x+1}{x}\right) + 6x \cdot \left(\dfrac{1}{3}\right) = 6x \cdot \left(\dfrac{5}{6}\right) \qquad \text{Apply the distributive property.}$$

$$6(x+1) + 2x = 5x \qquad \text{Clear fractions.}$$

$$6x + 6 + 2x = 5x \qquad \text{Solve the resulting equation.}$$

$$8x + 6 = 5x$$

$$3x = -6$$

$$x = -2$$

Check: $\dfrac{x+1}{x} + \dfrac{1}{3} = \dfrac{5}{6}$

$$\dfrac{(-2)+1}{(-2)} + \dfrac{1}{3} \overset{?}{=} \dfrac{5}{6}$$

$$\dfrac{-1}{-2} + \dfrac{1}{3} \overset{?}{=} \dfrac{5}{6}$$

$$\dfrac{1}{2} + \dfrac{1}{3} \overset{?}{=} \dfrac{5}{6}$$

$$\dfrac{3}{6} + \dfrac{2}{6} = \dfrac{5}{6} \checkmark$$

| Skill Practice | Solve the equation.

2. $\dfrac{3}{4} + \dfrac{5+a}{a} = \dfrac{1}{2}$

Skill Practice Answers

1. $t = -40$ **2.** $a = -4$

Example 3 Solving a Rational Equation

Solve the equation. $1 + \dfrac{3a}{a - 2} = \dfrac{6}{a - 2}$

Solution:

$$1 + \frac{3a}{a - 2} = \frac{6}{a - 2}$$ The LCD of all the expressions is $a - 2$.

$$(a - 2)\left(1 + \frac{3a}{a - 2}\right) = (a - 2)\left(\frac{6}{a - 2}\right)$$ Multiply by the LCD.

$$(a - 2)1 + (a - 2)\left(\frac{3a}{a - 2}\right) = (a - 2)\left(\frac{6}{a - 2}\right)$$ Apply the distributive property.

$$a - 2 + 3a = 6$$ Solve the resulting equation (linear).

$$4a - 2 = 6$$

$$4a = 8$$

$$a = 2$$

$$\underline{\text{Check:}} \quad 1 + \frac{3a}{a - 2} = \frac{6}{a - 2}$$

$$1 + \frac{3(2)}{(2) - 2} \stackrel{?}{=} \frac{6}{(2) - 2}$$

$$1 + \frac{6}{0} \stackrel{?}{=} \frac{6}{0}$$

The denominator is 0 when $a = 2$.

Because the value $a = 2$ makes the denominator zero in one (or more) of the rational expressions within the equation, the equation is undefined for $a = 2$. That is, $a = 2$ is not in the domain of the equation, therefore, it is an extraneous solution. No other potential solutions exist for the equation.

The equation $1 + \dfrac{3a}{a - 2} = \dfrac{6}{a - 2}$ has no solution.

Skill Practice Solve the equation.

3. $\dfrac{x}{x + 1} - 2 = \dfrac{-1}{x + 1}$

Examples 1–3 show that the steps to solve a rational equation mirror the process of clearing fractions from Section 2.3. However, there is one significant difference. The solutions of a rational equation must not make the denominator equal to zero for any expression within the equation. When $a = 2$ is substituted into the expression

$$\frac{3a}{a - 2} \qquad \text{or} \qquad \frac{6}{a - 2}$$

Skill Practice Answers

3. No solution; ($x = -1$ does not check.)

the denominator is zero and the expression is undefined. Hence, $a = 2$ cannot be a solution to the equation

$$1 + \frac{3a}{a - 2} = \frac{6}{a - 2}$$

The steps to solve a rational equation are summarized as follows.

> **Steps to Solve a Rational Equation**
>
> **1.** Factor the denominators of all rational expressions.
> **2.** Identify the LCD of all expressions in the equation.
> **3.** Multiply both sides of the equation by the LCD.
> **4.** Solve the resulting equation.
> **5.** Check potential solutions in the original equation.

Example 4   **Solving Rational Equations**

Solve the equations.

a. $1 - \dfrac{4}{p} = -\dfrac{3}{p^2}$ **b.** $\dfrac{6}{t^2 - 7t + 12} + \dfrac{2t}{t - 3} = \dfrac{3t}{t - 4}$

Solution:

a. $1 - \dfrac{4}{p} = -\dfrac{3}{p^2}$ **Step 1:** The denominators are already factored.

Step 2: The LCD of all expressions is p^2.

$$p^2\left(1 - \frac{4}{p}\right) = p^2\left(-\frac{3}{p^2}\right)$$ **Step 3:** Multiply by the LCD.

$$p^2(1) - p^2\left(\frac{4}{p}\right) = p^2\left(-\frac{3}{p^2}\right)$$ Apply the distributive property.

$$p^2 - 4p = -3$$ **Step 4:** Solve the resulting equation (quadratic).

$$p^2 - 4p + 3 = 0$$ Set the equation equal to zero and factor.

$$(p - 3)(p - 1) = 0$$

$$p - 3 = 0 \quad \text{or} \quad p - 1 = 0$$ Set each factor equal to zero.

$$p = 3 \quad \text{or} \quad p = 1$$ **Step 5:** Check: $p = 3$ Check: $p = 1$

$$1 - \frac{4}{p} = -\frac{3}{p^2} \qquad 1 - \frac{4}{p} = -\frac{3}{p^2}$$

$$1 - \frac{4}{(3)} \stackrel{?}{=} -\frac{3}{(3)^2} \qquad 1 - \frac{4}{(1)} \stackrel{?}{=} -\frac{3}{(1)^2}$$

$$\frac{3}{3} - \frac{4}{3} \stackrel{?}{=} -\frac{3}{9} \qquad 1 - 4 \stackrel{?}{=} -3$$

Both solutions $p = 3$ and $p = 1$ check. $-\dfrac{1}{3} = -\dfrac{1}{3} ✔ \qquad -3 = -3 ✔$

b. $\dfrac{6}{t^2 - 7t + 12} + \dfrac{2t}{t - 3} = \dfrac{3t}{t - 4}$

$\dfrac{6}{(t - 3)(t - 4)} + \dfrac{2t}{t - 3} = \dfrac{3t}{t - 4}$

Step 1: Factor the denominators.

Step 2: The LCD is $(t - 3)(t - 4)$.

Step 3: Multiply by the LCD on both sides.

$$(t - 3)(t - 4)\left(\dfrac{6}{(t - 3)(t - 4)} + \dfrac{2t}{t - 3}\right) = (t - 3)(t - 4)\left(\dfrac{3t}{t - 4}\right)$$

$$(t - 3)(t - 4)\left(\dfrac{6}{(t - 3)(t - 4)}\right) + (t - 3)(t - 4)\left(\dfrac{2t}{t - 3}\right) = (t - 3)(t - 4)\left(\dfrac{3t}{t - 4}\right)$$

$$6 + 2t(t - 4) = 3t(t - 3)$$

$6 + 2t^2 - 8t = 3t^2 - 9t$

$0 = 3t^2 - 2t^2 - 9t + 8t - 6$

$0 = t^2 - t - 6$

$0 = (t - 3)(t + 2)$

$t - 3 = 0$ or $t + 2 = 0$

$t = 3$ or $t = -2$

Step 4: Solve the resulting equation.

Because the resulting equation is quadratic, set the equation equal to zero and factor.

Set each factor equal to zero.

Step 5: Check the potential solutions in the original equation.

Check: $t = 3$

$t = 3$ cannot be a solution to the equation because it will make the denominator zero in the original equation.

$\dfrac{6}{t^2 - 7t + 12} + \dfrac{2t}{t - 3} = \dfrac{3t}{t - 4}$

$\dfrac{6}{(3)^2 - 7(3) + 12} + \dfrac{2(3)}{(3) - 3} \overset{?}{=} \dfrac{3(3)}{(3) - 4}$

$\dfrac{6}{0} + \dfrac{6}{0} \overset{?}{=} \dfrac{9}{-1}$

zero in the denominator

Check: $t = -2$

$\dfrac{6}{t^2 - 7t + 12} + \dfrac{2t}{t - 3} = \dfrac{3t}{t - 4}$

$\dfrac{6}{(-2)^2 - 7(-2) + 12} + \dfrac{2(-2)}{(-2) - 3} \overset{?}{=} \dfrac{3(-2)}{(-2) - 4}$

$\dfrac{6}{4 + 14 + 12} + \dfrac{-4}{-5} \overset{?}{=} \dfrac{-6}{-6}$

$\dfrac{6}{30} + \dfrac{4}{5} \overset{?}{=} 1$

$\dfrac{1}{5} + \dfrac{4}{5} = 1$ ✔

$t = -2$ is a solution.

The only solution is $t = -2$.

TIP: Note that $t = 3$ and $t = 4$ are not defined in the original expressions. Therefore, they cannot be solutions to the original equation.

TIP: $t = 3$ is not a solution because it is not in the domain of the equation.

Skill Practice Solve the equations.

4. $\dfrac{z}{2} - \dfrac{1}{2z} = \dfrac{12}{z}$

5. $\dfrac{-8}{x^2 + 6x + 8} + \dfrac{x}{x + 4} = \dfrac{2}{x + 2}$

Skill Practice Answers

4. $z = 5$ or $z = -5$
5. $x = 4$; ($x = -4$ does not check.)

| **Example 5** | Translating to a Rational Equation |

Ten times the reciprocal of a number is added to four. The result is equal to the quotient of twenty-two and the number. Find the number.

Solution:

Let x represent the number.

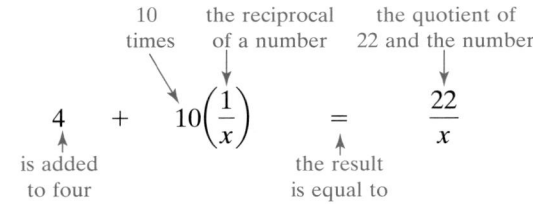

$$4 + \frac{10}{x} = \frac{22}{x}$$ **Step 1:** The denominators are already factored.

Step 2: The LCD is x.

$$x\left(4 + \frac{10}{x}\right) = x\left(\frac{22}{x}\right)$$ **Step 3:** Multiply both sides by the LCD.

$$4x + 10 = 22$$ Apply the distributive property.

$$4x = 12$$ **Step 4:** Solve the resulting equation (linear).

$x = 3$ is a potential solution.   **Step 5:** Substituting $x = 3$ into the original equation verifies that it is a solution.

The number is 3.

Skill Practice

6. The quotient of ten and a number is two less than four times the reciprocal of the number. Find the number.

3. Solving Formulas Involving Rational Equations

A rational equation may have more than one variable. To solve for a specific variable within a rational equation, we can still apply principles of clearing fractions.

| **Example 6** | Solving a Formula Involving a Rational Equation |

Solve for k. $F = \dfrac{ma}{k}$

Skill Practice Answers

6. The number is -3.

Solution:

To solve for k, we must clear fractions so that k appears in the numerator.

$$F = \frac{ma}{k} \qquad \text{The LCD is } k.$$

$$k \cdot (F) = k \cdot \left(\frac{ma}{k}\right) \qquad \text{Multiply both sides of the equation by the LCD.}$$

$$kF = ma \qquad \text{Clear fractions.}$$

$$\frac{kF}{F} = \frac{ma}{F} \qquad \text{Divide both sides by } F.$$

$$k = \frac{ma}{F}$$

> **Skill Practice**
>
> **7.** Solve for t. $C = \dfrac{rt}{d}$

Example 7 | **Solving a Formula Involving a Rational Equation**

Solve for b. $h = \dfrac{2A}{B + b}$

Solution:

To solve for b, we must clear fractions so that b appears in the numerator.

$$h = \frac{2A}{B + b} \qquad \text{The LCD is } (B + b).$$

$$h(B + b) = \left(\frac{2A}{B + b}\right) \cdot (B + b) \qquad \text{Multiply both sides of the equation by the LCD.}$$

$$hB + hb = 2A \qquad \text{Apply the distributive property.}$$

$$hb = 2A - hB \qquad \text{Subtract } hB \text{ from both sides to isolate the } b \text{ term.}$$

$$\frac{hb}{h} = \frac{2A - hB}{h} \qquad \text{Divide by } h.$$

$$b = \frac{2A - hB}{h}$$

Avoiding Mistakes:

Algebra is case-sensitive. The variables B and b represent different values.

> **Skill Practice**
>
> **8.** Solve the formula for x. $y = \dfrac{3}{x - 2}$

Skill Practice Answers

7. $t = \dfrac{Cd}{r}$

8. $x = \dfrac{3 + 2y}{y}$ or $x = \dfrac{3}{y} + 2$

TIP: The solution to Example 7 can be written in several forms. The quantity

$$\frac{2A - hB}{h}$$

can be left as a single rational expression or can be split into two fractions and simplified.

$$b = \frac{2A - hB}{h} = \frac{2A}{h} - \frac{hB}{h} = \frac{2A}{h} - B$$

Example 8   Solving a Formula Involving a Rational Expression

Solve for z. $y = \dfrac{x - z}{x + z}$

Solution:

To solve for z, we must clear fractions so that z appears in the numerator only.

$$y = \frac{x - z}{x + z} \qquad \text{LCD is } (x + z).$$

$$y(x + z) = \left(\frac{x - z}{x + z}\right)(x + z) \qquad \begin{array}{l}\text{Multiply both sides of the equation by}\\ \text{the LCD.}\end{array}$$

$$yx + yz = x - z \qquad \text{Apply the distributive property.}$$

$$yz + z = x - yx \qquad \text{Collect } z \text{ terms on one side of the equation.}$$

$$z(y + 1) = x - yx \qquad \text{Factor out a } z.$$

$$z = \frac{x - yx}{y + 1} \qquad \text{Divide by } y + 1 \text{ to solve for } z.$$

Skill Practice

Skill Practice Answers

9. $h = \dfrac{xa}{b - x}$ or $\dfrac{-ax}{x - b}$

9. Solve for h. $\dfrac{b}{x} = \dfrac{a}{h} + 1$

Section 7.6   Practice Exercises

- Practice Problems
- Self-Tests
- NetTutor
- e-Professors
- Videos

Study Skills Exercise

1. Define the key terms:

    a. linear equation   **b. quadratic equation**   **c. rational equation**

Review Exercises

For Exercises 2–7, perform the indicated operations.

2. $\dfrac{2}{x-3} - \dfrac{3}{x^2-x-6}$

3. $\dfrac{2x-6}{4x^2+7x-2} \div \dfrac{x^2-5x+6}{x^2-4}$

4. $\dfrac{2y}{y-3} + \dfrac{4}{y^2-9}$

5. $\dfrac{h-\dfrac{1}{h}}{\dfrac{1}{5}-\dfrac{1}{5h}}$

6. $\dfrac{w-4}{w^2-9} \cdot \dfrac{w-3}{w^2-8w+16}$

7. $1 + \dfrac{1}{x} - \dfrac{12}{x^2}$

Concept 1: Introduction to Rational Equations

For Exercises 8–13, solve the equations by first clearing the fractions.

8. $\dfrac{1}{3}z + \dfrac{2}{3} = -2z + 10$

9. $\dfrac{5}{2} + \dfrac{1}{2}b = 5 - \dfrac{1}{3}b$

10. $\dfrac{3}{2}p + \dfrac{1}{3} = \dfrac{2p-3}{4}$

11. $\dfrac{5}{3} - \dfrac{1}{6}k = \dfrac{3k+5}{4}$

12. $\dfrac{2x-3}{4} + \dfrac{9}{10} = \dfrac{x}{5}$

13. $\dfrac{4y+2}{3} - \dfrac{7}{6} = -\dfrac{y}{6}$

14. For the equation

$$\frac{1}{w} - \frac{1}{2} = -\frac{1}{4}$$

  a. Identify the domain of the equation.

  b. Identify the LCD of all the denominators of the equation.

  c. Solve the equation.

15. For the equation

$$\frac{3}{z} - \frac{4}{5} = -\frac{1}{5}$$

  a. Identify the domain of the equation.

  b. Identify the LCD of all the denominators of the equation.

  c. Solve the equation.

16. For the equation

$$\frac{x+1}{x^2+2x-3} = \frac{1}{x+3} - \frac{1}{x-1}$$

  a. Identify the domain of the equation.

  b. Identify the LCD of all the denominators of the equation.

  c. Solve the equation.

17. For the equation

$$\frac{10}{x-2} - \frac{40}{x^2+x-6} = \frac{12}{x+3}$$

  a. Identify the domain of the equation.

  b. Identify the LCD of all the denominators of the equation.

  c. Solve the equation.

Concept 2: Solving Rational Equations

For Exercises 18–47, solve the equations.

18. $\dfrac{1}{8} = \dfrac{3}{5} + \dfrac{5}{y}$

19. $\dfrac{2}{7} - \dfrac{1}{x} = \dfrac{2}{3}$

20. $\dfrac{4}{t} = \dfrac{3}{t} + \dfrac{1}{8}$

21. $\dfrac{9}{b} - \dfrac{8}{b} = \dfrac{1}{4}$

22. $\dfrac{5}{6x} + \dfrac{7}{x} = 1$

23. $\dfrac{14}{3x} - \dfrac{5}{x} = 2$

24. $1 - \dfrac{2}{y} = \dfrac{3}{y^2}$

25. $1 - \dfrac{2}{m} = \dfrac{8}{m^2}$

26. $\dfrac{a+1}{a} = 1 + \dfrac{a-2}{2a}$

27. $\dfrac{7b - 4}{5b} = \dfrac{9}{5} - \dfrac{4}{b}$

28. $\dfrac{w}{5} - \dfrac{w + 3}{w} = -\dfrac{3}{w}$

29. $\dfrac{t}{12} + \dfrac{t + 3}{3t} = \dfrac{1}{t}$

30. $\dfrac{2}{m + 3} = \dfrac{5}{4m + 12} - \dfrac{3}{8}$

31. $\dfrac{2}{4n - 4} - \dfrac{7}{4} = \dfrac{-3}{n - 1}$

32. $\dfrac{p}{p - 4} - 5 = \dfrac{4}{p - 4}$

33. $\dfrac{-5}{q + 5} = \dfrac{q}{q + 5} + 2$

34. $\dfrac{2t}{t + 2} - 2 = \dfrac{t - 8}{t + 2}$

35. $\dfrac{4w}{w - 3} - 3 = \dfrac{3w - 1}{w - 3}$

36. $\dfrac{x^2 - x}{x - 2} = \dfrac{12}{x - 2}$

37. $\dfrac{x^2 + 9}{x + 4} = \dfrac{-10x}{x + 4}$

38. $\dfrac{x^2 + 3x}{x - 1} = \dfrac{4}{x - 1}$

39. $\dfrac{2x^2 - 21}{2x - 3} = \dfrac{-11x}{2x - 3}$

40. $\dfrac{2x}{x + 4} - \dfrac{8}{x - 4} = \dfrac{2x^2 + 32}{x^2 - 16}$

41. $\dfrac{4x}{x + 3} - \dfrac{12}{x - 3} = \dfrac{4x^2 + 36}{x^2 - 9}$

42. $\dfrac{x}{x + 6} = \dfrac{72}{x^2 - 36} + 4$

43. $\dfrac{y}{y + 4} = \dfrac{32}{y^2 - 16} + 3$

44. $\dfrac{x}{3x - 3} - \dfrac{3}{x - 2} = \dfrac{7}{x^2 - 3x + 2}$

45. $\dfrac{a}{5a + 10} - \dfrac{1}{a - 5} = \dfrac{7}{a^2 - 3a - 10}$

46. $\dfrac{y - 2}{y - 3} = \dfrac{-1}{y^2 - 7y + 12} + \dfrac{3y}{y - 4}$

47. $\dfrac{6}{w + 1} - \dfrac{w}{w + 5} = \dfrac{16}{w^2 + 6w + 5}$

For Exercises 48–51, translate to a rational equation and solve.

48. The reciprocal of a number is added to three. The result is the quotient of 25 and the number. Find the number.

49. The difference of three and the reciprocal of a number is equal to the quotient of 20 and the number. Find the number.

50. If a number added to five is divided by the difference of the number and two, the result is three-fourths. Find the number.

51. If twice a number added to three is divided by the number plus one, the result is three-halves. Find the number.

Concept 3: Solving Formulas Involving Rational Equations

For Exercises 52–69, solve for the indicated variable.

52. $K = \dfrac{ma}{F}$ for m

53. $K = \dfrac{ma}{F}$ for a

54. $K = \dfrac{IR}{E}$ for E

55. $K = \dfrac{IR}{E}$ for R

56. $I = \dfrac{E}{R + r}$ for R

57. $I = \dfrac{E}{R + r}$ for r

58. $h = \dfrac{2A}{B + b}$ for B

59. $\dfrac{C}{\pi r} = 2$ for r

60. $\dfrac{V}{\pi h} = r^2$ for h

61. $\dfrac{V}{lw} = h$ for w

62. $x = \dfrac{at + b}{t}$ for t

63. $\dfrac{T + mf}{m} = g$ for m

64. $\dfrac{x - y}{xy} = z$ for x

65. $\dfrac{w - n}{wn} = P$ for w

66. $a + b = \dfrac{2A}{h}$ for h

67. $1 + rt = \dfrac{A}{P}$ for P

68. $\dfrac{1}{R} = \dfrac{1}{R_1} + \dfrac{1}{R_2}$ for R

69. $\dfrac{b + a}{ab} = \dfrac{1}{f}$ for b

Chapter 7	Problem Recognition Exercises—Comparing Rational Equations and Rational Expressions

Often adding or subtracting rational expressions is confused with solving rational equations. When adding rational expressions, we combine the terms to simplify the expression. When solving an equation, we clear the fractions and find numerical solutions, if possible. Both processes begin with finding the LCD, but the LCD is used differently in each process. Compare these two examples.

Example 1:

Add. $\dfrac{4}{x} + \dfrac{x}{3}$ (The LCD is $3x$.)

$= \dfrac{3}{3} \cdot \left(\dfrac{4}{x}\right) + \left(\dfrac{x}{3}\right) \cdot \dfrac{x}{x}$

$= \dfrac{12}{3x} + \dfrac{x^2}{3x}$

$= \dfrac{12 + x^2}{3x}$ The final answer is a rational expression.

Example 2:

Solve. $\dfrac{4}{x} + \dfrac{x}{3} = -\dfrac{8}{3}$ (The LCD is $3x$.)

$\dfrac{3x}{1}\left(\dfrac{4}{x} + \dfrac{x}{3}\right) = \dfrac{3x}{1}\left(-\dfrac{8}{3}\right)$

$12 + x^2 = -8x$

$x^2 + 8x + 12 = 0$

$(x + 2)(x + 6) = 0$

$x + 2 = 0 \text{ or } x - 2 = 0$

$x = -2 \text{ or } x = 2$ The final answers are numbers.

For Exercises 1–12, solve the equation or simplify the expression by combining the terms.

1. $\dfrac{y}{2y + 4} - \dfrac{2}{y^2 + 2y}$

2. $\dfrac{1}{x + 2} + 2 = \dfrac{x + 11}{x + 2}$

3. $\dfrac{5t}{2} - \dfrac{t - 2}{3} = 5$

4. $3 - \dfrac{2}{a - 5}$

5. $\dfrac{7}{6p^2} + \dfrac{2}{9p} + \dfrac{1}{3p^2}$

6. $\dfrac{3b}{b + 1} - \dfrac{2b}{b - 1}$

7. $4 + \dfrac{2}{h - 3} = 5$

8. $\dfrac{2}{w + 1} + \dfrac{3}{(w + 1)^2}$

9. $\dfrac{1}{x - 6} - \dfrac{3}{x^2 - 6x} = \dfrac{4}{x}$

10. $\dfrac{3}{m} - \dfrac{6}{5} = -\dfrac{3}{m}$

11. $\dfrac{7}{2x + 2} + \dfrac{3x}{4x + 4}$

12. $\dfrac{10}{2t - 1} - 1 = \dfrac{t}{2t - 1}$

Section 7.7   Applications of Rational Equations and Proportions

1. Solving Proportions

In this section, we look at how rational equations can be used to solve a variety of applications. The first type of rational equation that will be applied is called a proportion.

> **Definition of a Proportion**
>
> An equation that equates two ratios or rates is called a **proportion**. Thus, for $b \neq 0$ and $d \neq 0$, $\frac{a}{b} = \frac{c}{d}$ is a proportion.

A proportion can be solved by multiplying both sides of the equation by the LCD and clearing fractions.

Example 1   Solving a Proportion

Solve the proportion. $\dfrac{3}{11} = \dfrac{123}{w}$

Solution:

$$\frac{3}{11} = \frac{123}{w} \qquad \text{The LCD is } 11w.$$

$$11w\left(\frac{3}{11}\right) = 11w\left(\frac{123}{w}\right) \qquad \text{Multiply by the LCD and clear fractions.}$$

$$3w = 11 \cdot 123 \qquad \text{Solve the resulting equation (linear).}$$

$$3w = 1353$$

$$\frac{3w}{3} = \frac{1353}{3}$$

$$w = 451$$

$$\underline{\text{Check}}: w = 451$$

$$\frac{3}{11} = \frac{123}{w}$$

$$\frac{3}{11} \overset{?}{=} \frac{123}{(451)}$$

$$\frac{3}{11} = \frac{3}{11} \checkmark \qquad \text{Simplify to lowest terms.}$$

Skill Practice   Solve the proportion.

1. $\dfrac{10}{b} = \dfrac{2}{33}$

Skill Practice Answers

1. $b = 165$

TIP: The cross products of any proportion are equal. That is, for $b \neq 0$ and $d \neq 0$, the proportion $\frac{a}{b} = \frac{c}{d}$ is equivalent to $ad = bc$. Some rational equations are proportions and can be solved by equating the cross products. Consider the proportion from Example 1:

$$\frac{3}{11} \diagup\!\!\!\!\diagup \frac{123}{w}$$

$$3 \cdot w = 11 \cdot 123 \qquad \text{Equate the cross products.}$$

$$3w = 1353 \qquad \text{Solve the resulting equation.}$$

$$\frac{3w}{3} = \frac{1353}{3}$$

$$w = 451$$

2. Applications of Proportions and Similar Triangles

| Example 2 | Using a Proportion in an Application |

For a recent year, the population of Alabama was approximately 4.2 million. At that time, Alabama had seven representatives in the U.S. House of Representatives. In the same year, North Carolina had a population of approximately 7.2 million. If representation in the House is based on population in equal proportions for each state, how many representatives did North Carolina have?

Solution:

Let x represent the number of representatives for North Carolina.

Set up a proportion by writing two equivalent ratios.

$$\boxed{\begin{array}{c}\text{Population of Alabama}\\\hline\text{number of representatives}\end{array}} \rightarrow \frac{4.2}{7} = \frac{7.2}{x} \leftarrow \boxed{\begin{array}{c}\text{Population of North Carolina}\\\hline\text{number of representatives}\end{array}}$$

$$\frac{4.2}{7} = \frac{7.2}{x}$$

$$7x \cdot \frac{4.2}{7} = 7x \cdot \frac{7.2}{x} \qquad \text{Multiply by the LCD, } 7x.$$

$$4.2x = (7.2)(7) \qquad \text{Solve the resulting equation (linear).}$$

$$4.2x = 50.4$$

$$\frac{4.2x}{4.2} = \frac{50.4}{4.2}$$

$$x = 12 \qquad \text{North Carolina had 12 representatives.}$$

TIP: The equation from Example 2 could have been solved by first equating the cross products:

$$\frac{4.2}{7} \diagup\!\!\!\!\diagup \frac{7.2}{x}$$

$$4.2x = (7.2)(7)$$

$$4.2x = 50.4$$

$$x = 12$$

Skill Practice

2. A college keeps the ratio of students to faculty at 105 to 2. If the student population at the college is 1575, how many faculty members are needed?

Skill Practice Answers

2. 30 faculty members are needed.

Proportions are used in geometry with **similar triangles**. Two triangles are said to be similar if their corresponding angles have equal measures. In such a case, the lengths of the corresponding sides are proportional. The triangles in Figure 7-1 are similar. Therefore, the following ratios are equivalent.

$$\frac{a}{x} = \frac{b}{y} = \frac{c}{z}$$

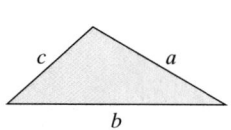 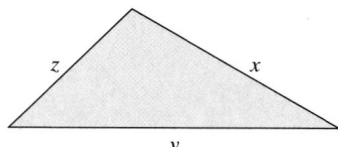

Figure 7-1

Example 3 Using Similar Triangles to Find an Unknown Side in a Triangle

The triangles in Figure 7-2 are similar.

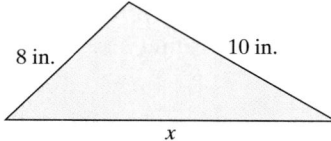 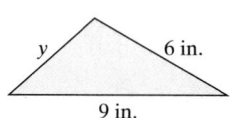

Figure 7-2

a. Solve for x. **b.** Solve for y.

Solution:

a. The lengths of the upper right sides of the triangles are given. These form a known ratio of $\frac{10}{6}$. Because the triangles are similar, the ratio of the other corresponding sides must be equal to $\frac{10}{6}$. To solve for x, we have:

Bottom side from large triangle	→	x	$=$	10 in.	←	Right side from large triangle
Bottom side from small triangle	→	9 in.		6 in.	←	Right side from small triangle

$$\frac{x}{9} = \frac{10}{6} \qquad \text{The LCD is 18.}$$

$$18 \cdot \left(\frac{x}{9}\right) = 18 \cdot \left(\frac{10}{6}\right) \qquad \text{Multiply by the LCD.}$$

$$2x = 30 \qquad \text{Clear fractions.}$$

$$x = 15 \qquad \text{Divide by 2.}$$

The length of side x is 15 in.

b. To solve for y, the ratio of the upper left sides of the triangles must equal $\frac{10}{6}$.

Left side from large triangle	$\rightarrow$	$\dfrac{8 \text{ in.}}{y}$	$=$	$\dfrac{10 \text{ in.}}{6 \text{ in.}}$	$\leftarrow$	Right side from large triangle
Left side from small triangle	$\rightarrow$				$\leftarrow$	Right side from small triangle

$$\frac{8}{y} = \frac{10}{6} \qquad \text{The LCD is } 6y.$$

$$6y \cdot \left(\frac{8}{y}\right) = 6y \cdot \left(\frac{10}{6}\right) \qquad \text{Multiply by the LCD.}$$

$$48 = 10y \qquad \text{Clear fractions.}$$

$$\frac{48}{10} = \frac{10y}{10}$$

$$4.8 = y$$

The length of side y is 4.8 in.

Skill Practice

3. The two triangles shown are similar triangles. Solve for the lengths of the missing sides.

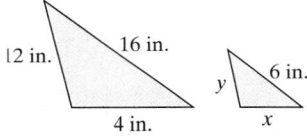

Example 4 **Using Similar Triangles in an Application**

The shadow cast by a yardstick is 2 ft long. The shadow cast by a tree is 11 ft long. Find the height of the tree.

Solution:

Let x represent the height of the tree.   Label the variable.

We will assume that the measurements were taken at the same time of day. Therefore, the angle of the sun is the same on both objects, and we can set up similar triangles (Figure 7-3).

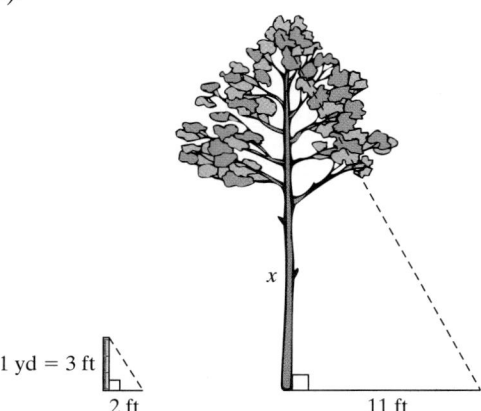

Figure 7-3

Skill Practice Answers

3. $x = 1.5$ in., and $y = 4.5$ in.

Create a verbal model.

Height of yardstick	$\longrightarrow$	$\dfrac{3 \text{ ft}}{2 \text{ ft}}$ = $\dfrac{x}{11 \text{ ft}}$	$\longleftarrow$	Height of tree
Length of yardstick's shadow	$\longrightarrow$		$\longleftarrow$	Length of tree's shadow

$$\frac{3}{2} = \frac{x}{11}$$ Write a mathematical equation.

$$\overset{11}{\cancel{22}} \cdot \left(\frac{3}{2}\right) = \left(\frac{x}{\cancel{11}}\right) \cdot \overset{2}{\cancel{22}}$$ Multiply by the LCD.

$$33 = 2x$$ Solve the equation.

$$\frac{33}{2} = \frac{2x}{2}$$

$$16.5 = x$$ Interpret the results and write the answer in words.

The tree is 16.5 ft high.

Skill Practice

4. The sun casts a 3.2-ft shadow of a 6-ft man. At the same time, the sun casts an 80-ft shadow of a building. How tall is the building?

3. Distance, Rate, and Time Applications

In Sections 2.4 and 4.4, we presented applications involving the relationship among the variables distance, rate, and time. Recall that $d = rt$.

Example 5   **Using a Rational Equation in a Distance, Rate, and Time Application**

A small plane flies 440 miles with the wind from Memphis, Tennessee, to Oklahoma City, Oklahoma. In the same amount of time, the plane flies 340 miles against the wind from Oklahoma City to Little Rock, Arkansas (see Figure 7-4). If the wind speed is 30 mph, find the speed of the plane in still air.

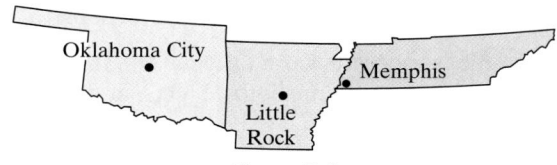

Figure 7-4

Solution:

Let x represent the speed of the plane in still air.

Organize the given information in a chart.

	Distance	Rate	Time
With the wind	440	$x + 30$	$\dfrac{440}{x + 30}$
Against the wind	340	$x - 30$	$\dfrac{340}{x - 30}$

Because $d = rt$, then $t = \dfrac{d}{r}$.

Skill Practice Answers

4. The building is 150 ft tall.

The plane travels with the wind for the same amount of time as it travels against the wind, so we can equate the two expressions for time.

$$\begin{pmatrix} \text{Time with} \\ \text{the wind} \end{pmatrix} = \begin{pmatrix} \text{time against} \\ \text{the wind} \end{pmatrix}$$

$$\frac{440}{x + 30} = \frac{340}{x - 30}$$

The LCD is $(x + 30)(x - 30)$.

$$\cancel{(x + 30)}(x - 30) \cdot \frac{440}{\cancel{x + 30}} = (x + 30)\cancel{(x - 30)} \cdot \frac{340}{\cancel{x - 30}}$$

$$440(x - 30) = 340(x + 30)$$

$$440x - 13{,}200 = 340x + 10{,}200$$

Solve the resulting linear equation.

$$100x = 23{,}400$$

$$x = 234$$

The plane's speed in still air is 234 mph.

> **TIP:** The equation
> $$\frac{440}{x + 30} = \frac{340}{x - 30}$$
> is a proportion. The fractions can also be cleared by equating the cross products.
> $$\frac{440}{x + 30} \diagdown\diagup \frac{340}{x - 30}$$
> $$440(x - 30) = 340(x + 30)$$

Skill Practice

5. Alison paddles her kayak in a river where the current of the water is 2 mph. She can paddle 20 miles with the current in the same time that she can paddle 10 miles against the current. Find the speed of the kayak in still water.

Example 6 **Using a Rational Equation in a Distance, Rate, and Time Application**

A motorist drives 100 miles between two cities in a bad rainstorm. For the return trip in sunny weather, she averages 10 mph faster and takes $\frac{1}{2}$ hour less time. Find the average speed of the motorist in the rainstorm and in sunny weather.

Solution:

Let x represent the motorist's speed during the rain.

Then $x + 10$ represents the speed in sunny weather.

	Distance	Rate	Time
Trip during rainstorm	100	x	$\dfrac{100}{x}$
Trip during sunny weather	100	$x + 10$	$\dfrac{100}{x + 10}$

Because $d = rt$, then $t = \dfrac{d}{r}$.

Because the same distance is traveled in $\frac{1}{2}$ hr less time, the difference between the time of the trip during the rainstorm and the time during sunny weather is $\frac{1}{2}$ hr.

$$\left(\begin{matrix}\text{Time during}\\\text{the rainstorm}\end{matrix}\right) - \left(\begin{matrix}\text{time during}\\\text{sunny weather}\end{matrix}\right) = \left(\frac{1}{2}\,\text{hr}\right) \qquad \text{Verbal model}$$

$$\frac{100}{x} - \frac{100}{x+10} = \frac{1}{2} \qquad \begin{matrix}\text{Mathematical}\\\text{equation}\end{matrix}$$

$$2x(x+10)\left(\frac{100}{x} - \frac{100}{x+10}\right) = 2x(x+10)\left(\frac{1}{2}\right) \qquad \begin{matrix}\text{Multiply by}\\\text{the LCD.}\end{matrix}$$

$$2x(x+10)\left(\frac{100}{x}\right) - 2x(x+10)\left(\frac{100}{x+10}\right) = 2x(x+10)\left(\frac{1}{2}\right) \qquad \begin{matrix}\text{Apply the}\\\text{distributive}\\\text{property.}\end{matrix}$$

$$200(x+10) - 200x = x(x+10) \qquad \text{Clear fractions.}$$

$$200x + 2000 - 200x = x^2 + 10x \qquad \begin{matrix}\text{Solve the}\\\text{resulting}\\\text{equation}\\\text{(quadratic).}\end{matrix}$$

$$2000 = x^2 + 10x$$

$$0 = x^2 + 10x - 2000 \qquad \begin{matrix}\text{Set the}\\\text{equation equal}\\\text{to zero.}\end{matrix}$$

$$0 = (x-40)(x+50) \qquad \text{Factor.}$$

$$x = 40 \qquad \text{or} \qquad x = -50$$

Because a rate of speed cannot be negative, reject $x = -50$. Therefore, the speed of the motorist in the rainstorm is 40 mph. Because $x + 10 = 40 + 10 = 50$, the average speed for the return trip in sunny weather is 50 mph.

Skill Practice

6. Harley rode his mountain bike 12 miles to the top of the mountain and the same distance back down. His speed going up was 8 mph slower than coming down. The ride up took 2 hours longer than coming down. Find his speeds.

4. Work Applications

Suppose Winston can paint a room in 2 hours. Then he paints $\frac{1}{2}$ room per hour. Suppose Clyde can paint the room in 4 hours. Then he paints $\frac{1}{4}$ room per hour. In general, we can define a work rate as follows:

Work rate: $\frac{1}{t}$ jobs per hour, where t is the total time required to complete the job.

Furthermore, if we multiply a work rate by the time an individual works, we compute the portion of the job completed. That is,

Portion of job completed = (Work rate)(Time)

Therefore, in 3 hours, Clyde would paint $(\frac{1}{4}$ room/hr$)(3$ hr$) = \frac{3}{4}$ room. We use this basic principle to solve equations involving "work."

Skill Practice Answers

6. Uphill speed was 4 mph; downhill speed was 12 mph.

| Example 7 | Using a Rational Equation in a "Work" Application |

A new printing press can print the morning edition in 2 hours, whereas the old printer required 4 hours. How long would it take to print the morning edition if both printers were working together?

Solution:

Let x represent the time required for both printers working together to complete the job.

One method to approach this problem is to determine the portion of the job that each printer can complete in 1 hour and extend that rate to the portion of the job completed in x hours.

- The old printer can perform the job in 4 hours. Therefore, it completes $\frac{1}{4}$ of the job in 1 hour and $\frac{1}{4}x$ jobs in x hours.
- The new printer can perform the job in 2 hours. Therefore, it completes $\frac{1}{2}$ of the job in 1 hour and $\frac{1}{2}x$ jobs in x hours.

	Work Rate	**Time**	**Portion of Job Completed**
Old printer	$\dfrac{1 \text{ job}}{4 \text{ hr}}$	x hours	$\dfrac{1}{4}x$
New printer	$\dfrac{1 \text{ job}}{2 \text{ hr}}$	x hours	$\dfrac{1}{2}x$

The sum of the portions of the job completed by each printer must equal one whole job.

$$\begin{pmatrix} \text{Portion of job} \\ \text{completed by} \\ \text{old printer} \end{pmatrix} + \begin{pmatrix} \text{portion of job} \\ \text{completed by} \\ \text{new printer} \end{pmatrix} = \begin{pmatrix} 1 \\ \text{whole} \\ \text{job} \end{pmatrix}$$

$\dfrac{1}{4}x + \dfrac{1}{2}x = 1$ The LCD is 4.

$4\left(\dfrac{1}{4}x + \dfrac{1}{2}x\right) = 4(1)$ Multiply by the LCD.

$4 \cdot \dfrac{1}{4}x + 4 \cdot \dfrac{1}{2}x = 4 \cdot 1$ Apply the distributive property.

$x + 2x = 4$ Solve the resulting linear equation.

$3x = 4$

$x = \dfrac{4}{3}$ or $x = 1\dfrac{1}{3}$ The time required to print the morning edition using both printers is $1\frac{1}{3}$ hr.

| Skill Practice |

7. The computer at a bank can process and prepare the bank statements in 30 hours. A new faster computer can do the job in 20 hours. If the bank uses both computers together, how long will it take to process the statements?

Skill Practice Answers

7. 12 hours

Section 7.7   Practice Exercises

Study Skills Exercise

1. Define the key terms:

  a. proportion   **b. similar triangles**

Review Exercises

For Exercises 2–8, determine whether each of the following is an equation or an expression. If it is an equation, solve it. If it is an expression, perform the indicated operation.

2. $\dfrac{b}{5} + 3 = 9$

3. $\dfrac{m}{m-1} - \dfrac{2}{m+3}$

4. $\dfrac{2}{a+5} + \dfrac{5}{a^2-25}$

5. $\dfrac{3y+6}{20} \div \dfrac{4y+8}{8}$

6. $\dfrac{z^2+z}{24} \cdot \dfrac{8}{z+1}$

7. $\dfrac{3}{p+3} = \dfrac{12p+19}{p^2+7p+12} - \dfrac{5}{p+4}$

8. $\dfrac{\dfrac{1}{t^2} + \dfrac{2}{3}}{\dfrac{1}{t} - \dfrac{5}{6}}$

Concept 1: Solving Proportions

For Exercises 9–22, solve the proportions.

9. $\dfrac{8}{5} = \dfrac{152}{p}$

10. $\dfrac{6}{7} = \dfrac{96}{y}$

11. $\dfrac{19}{76} = \dfrac{z}{4}$

12. $\dfrac{15}{135} = \dfrac{w}{9}$

13. $\dfrac{5}{3} = \dfrac{a}{8}$

14. $\dfrac{b}{14} = \dfrac{3}{8}$

15. $\dfrac{2}{1.9} = \dfrac{x}{38}$

16. $\dfrac{16}{1.3} = \dfrac{30}{p}$

17. $\dfrac{y+1}{2y} = \dfrac{2}{3}$

18. $\dfrac{w-2}{4w} = \dfrac{1}{6}$

19. $\dfrac{9}{2z-1} = \dfrac{3}{z}$

20. $\dfrac{1}{t} = \dfrac{1}{4-t}$

21. $\dfrac{8}{9a-1} = \dfrac{5}{3a+2}$

22. $\dfrac{4p+1}{3} = \dfrac{2p-5}{6}$

23. Charles' law describes the relationship between the initial and final temperature and volume of a gas held at a constant pressure.

$$\frac{V_i}{V_f} = \frac{T_i}{T_f}$$

  a. Solve the equation for V_f.   **b.** Solve the equation for T_f.

24. The relationship between the area, height, and base of a triangle is given by the proportion

$$\frac{A}{b} = \frac{h}{2},$$ where A is area, b is the base, and h is the height.

a. Solve the equation for A. **b.** Solve the equation for b.

Concept 2: Applications of Proportions and Similar Triangles

For Exercises 25–32, solve using proportions.

25. Toni drives her Honda Civic 132 miles on the highway on 4 gallons of gas. At this rate how many miles can she drive on 9 gallons of gas?

26. Tim takes his pulse for 10 seconds and counts 12 beats. How many beats per minute is this?

27. Suppose a household of 4 people produces 128 lb of garbage in one week. At this rate, how many pounds will 48 people produce in one week?

28. Property tax on a $180,000 house is $4000. At this rate, how much tax would be paid on a $216,000 home?

29. Martin won an election by a ratio 5 to 4. If he received 5420 votes, how many votes did his opponent receive?

30. Cooking oatmeal requires 1 cup of water for every $\frac{1}{2}$ cup of oats. How many cups of water will be required for $\frac{3}{4}$ cup of oats?

31. A map has a scale of 75 miles/in. If two cities measure 3.5 in. apart, how many miles does this represent?

32. A map has a scale of 50 miles/in. If two cities measure 6.5 in. apart, how many miles does this represent?

33. $\triangle ABC$ is similar to $\triangle DEF$.

    a. Find the length of $\overline{EF}$.

    b. Find the length of $\overline{DF}$.

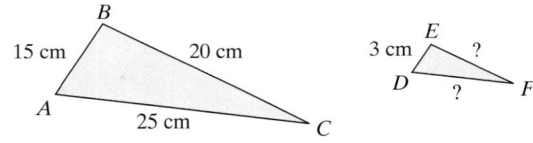

34. Figure $ABCD$ is similar to Figure $EFGH$.

    a. Find the length of $\overline{EH}$.

    b. Find the length of $\overline{AB}$.

    c. Find the length of $\overline{BC}$.

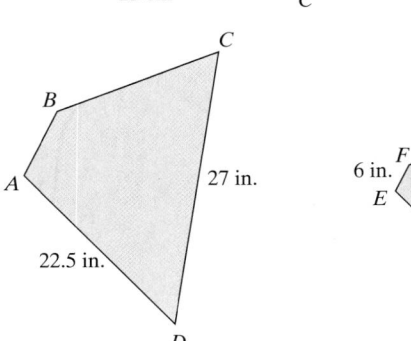

35. Solve for x and y.

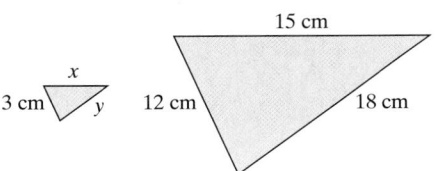

36. Solve for x and y.

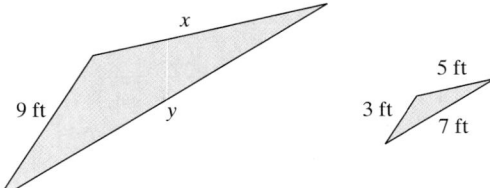

37. To estimate the height of a light pole, a mathematics student measures the length of a shadow cast by a meterstick and the length of the shadow cast by the light pole. Find the height of the light pole (see figure).

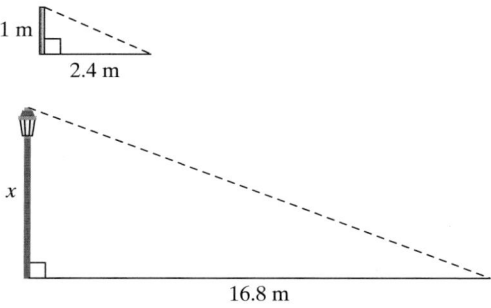

38. To estimate the height of a building, a student measures the length of a shadow cast by a yardstick and the length of the shadow cast by the building (see figure). Find the height of the building.

39. A 6-ft-tall man standing 54 ft from a light post casts an 18-ft shadow. What is the height of the light post?

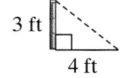

40. For a science project at school, a student must measure the height of a tree. The student measures the length of the shadow of the tree and then measures the length of the shadow cast by a yardstick. Use similar triangles to find the height of the tree.

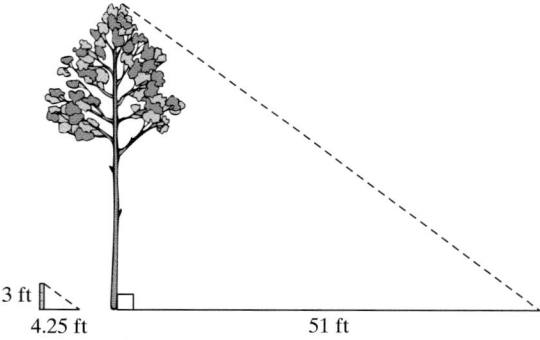

Concept 3: Distance, Rate, and Time Applications

41. A boat travels 54 miles upstream against the current in the same amount of time it takes to travel 66 miles downstream with the current. If the boat has the speed of 20 mph in still water, what is the speed of the current? (Use $t = \frac{d}{r}$ to complete the table.)

	Distance	Rate	Time
With the current (downstream)			
Against the current (upstream)			

42. A fisherman travels 9 miles downstream with the current in the same time that he travels 3 miles upstream against the current. If the speed of the current is 6 mph, what is the speed at which the fisherman travels in still water?

43. A plane flies 630 miles with the wind in the same time that it takes to fly 455 miles against the wind. If this plane flies at the rate of 217 mph in still air, what is the speed of the wind? (Use $t = \frac{d}{r}$ to complete the table.)

	Distance	Rate	Time
With the wind			
Against the wind			

44. A plane flies 370 miles with the wind in the same time that it takes to fly 290 miles against the wind. If the speed of the wind is 20 mph, what is the speed of the plane in still air?

45. Devon can cross-country ski 5 km/hr faster than his sister Shanelle. Devon skis 45 km in the same time Shanelle skis 30 km. Find their speeds.

46. Brooke walks 2 km/hr slower than her older sister Adrianna. Brooke can walk 12 km in the same amount of time that Adriana can walk 18 km. Find their speeds.

47. One motorist travels 15 mph faster than another. The slower driver takes 2 hours longer to travel 360 mi than the faster driver. What are the speeds of the two motorists?

48. A train travels 180 miles in 1 hour less time than a bus traveling the same distance. If the speed of the bus is 15 mph slower than the speed of the train, find the speed of the train and the bus.

49. Kendra flew 900 miles to Cincinnati, Ohio. When she returned she traveled 30 mph slower. If it took her 1 hour longer on her return flight, what were her speeds going to and returning from Cincinnati?

50. A plane flew 500 km from Atlanta, Georgia, to Louisville, Kentucky. When returning to Atlanta, the flight took $\frac{1}{2}$ hr less time. If the rate to Louisville was 50 km/hr slower than the rate returning, find the two rates in km/hr.

51. Sergio rode his bike 4 miles. Then he got a flat tire and had to walk back 4 miles. It took him 1 hr longer to walk than it did to ride. If his rate walking was 9 mph less than his rate riding, find the two rates.

52. Amber jogs 10 km in $\frac{3}{4}$ hr less than she can walk the same distance. If her walking rate is 3 km/hr less than her jogging rate, find her rates jogging and walking (in km/hr).

Concept 4: Work Applications

53. If it takes a person 2 hr to paint a room, what fraction of the room would be painted in 1 hr?

54. If it takes a copier 3 hr to complete a job, what fraction of the job would be completed in 1 hr?

55. If the cold-water faucet is left on, the sink will fill in 10 min. If the hot-water faucet is left on, the sink will fill in 12 min. How long would it take the sink to fill if both faucets are left on?

56. The CUT-IT-OUT lawn mowing company consists of two people: Tina and Bill. If Tina cuts a lawn by herself, she can do it in 4 hr. If Bill cuts the same lawn himself, it takes him an hour longer than Tina. How long would it take them if they worked together?

57. A manuscript needs to be printed. One printer can do the job in 50 min, and another printer can do the job in 40 min. How long would it take if both printers were used?

58. A pump can empty a small pond in 4 hr. Another more efficient pump can do the job in 3 hr. How long would it take to empty the pond if both pumps were used?

59. Tim and Al are bricklayers. Tim can construct an outdoor grill in 5 days. If Al helps Tim, they can build it in only 2 days. How long would it take Al to build the grill alone?

60. Norma is a new and inexperienced secretary. It takes her 3 hr to prepare a mailing. If her boss helps her, the mailing can be completed in 1 hr. How long would it take the boss to do the job by herself?

61. A pipe can fill a reservoir in 16 hr. A drainage pipe can drain the reservoir in 24 hr. How long would it take to fill the reservoir if the drainage pipe were left open by mistake? (*Hint:* The rate at which water drains should be negative.)

62. A hole in the bottom of a child's plastic swimming pool can drain the pool in 60 min. If the pool had no hole, a hose could fill the pool in 40 min. How long would it take the hose to fill the pool with the hole?

63. A new copy machine works three times faster than the older model. It takes 12 minutes to complete a job when both machines are working together. Find the time required for the new copy machine to complete the job by itself.

64. A cold-water faucet can fill a tub twice as fast as a hot-water faucet. Together it takes 12 min to fill the tub. How long will it take to fill the tub using only the cold-water faucet?

Expanding Your Skills

For Exercises 65–68, solve using proportions.

65. The ratio of smokers to nonsmokers in a restaurant is 2 to 7. There are 100 more nonsmokers than smokers. How many smokers and nonsmokers are in the restaurant?

66. The ratio of fiction to nonfiction books sold in a bookstore is 5 to 3. One week there are 180 more fiction books sold than nonfiction. Find the number of fiction and nonfiction books sold during that week.

67. There are 440 students attending a biology lecture. The ratio of male to female students at the lecture is 6 to 5. How many men and women are attending the lecture?

68. The ratio of dogs to cats at the humane society is 5 to 8. There are a total of 650 dogs and cats. How many dogs and how many cats are at the humane society?

Chapter 7   SUMMARY

Section 7.1   Introduction to Rational Expressions

Key Concepts

A **rational expression** is a ratio of the form $\frac{p}{q}$ where p and q are polynomials and $q \neq 0$.

The **domain** of an algebraic expression is the set of real numbers that when substituted for the variable makes the expression result in a real number. For a rational expression, the domain is all real numbers except those that make the denominator zero.

Simplifying a Rational Expression to Lowest Terms

Factor the numerator and denominator completely, and reduce factors whose ratio is equal to 1 or to -1. A rational expression written in lowest terms will still have the same restrictions on the domain as the original expression.

Examples

Example 1

$$\frac{x + 2}{x^2 - 5x - 14} \quad \text{is a rational expression.}$$

Example 2

To find the domain of $\dfrac{x + 2}{x^2 - 5x - 14}$ factor the denominator: $\dfrac{x + 2}{(x + 2)(x - 7)}$

The domain is $\{x \mid x$ is a real number and $x \neq -2, x \neq 7\}$.

Example 3

Simplify to lowest terms. $\dfrac{x + 2}{x^2 - 5x - 14}$

$$\frac{\overset{1}{\cancel{x + 2}}}{\cancel{(x + 2)}(x - 7)} \quad \text{Simplify.}$$

$$= \frac{1}{x - 7} \quad \text{(provided } x \neq 7, x \neq -2\text{).}$$

Section 7.2   Multiplication and Division of Rational Expressions

Key Concepts

Multiplying Rational Expressions

Factor the numerator and denominator completely. Then reduce factors whose ratio is 1 or -1.

Examples

Example 1

Multiply. $\dfrac{b^2 - a^2}{a^2 - 2ab + b^2} \cdot \dfrac{a^2 - 3ab + 2b^2}{2a + 2b}$

$$= \frac{\overset{-1}{\cancel{(b - a)}}\overset{1}{\cancel{(b + a)}}}{\cancel{(a - b)}\cancel{(a - b)}} \cdot \frac{(a - 2b)\overset{1}{\cancel{(a - b)}}}{2\cancel{(a + b)}}$$

$$= -\frac{a - 2b}{2} \quad \text{or} \quad \frac{2b - a}{2}$$

Dividing Rational Expressions

Multiply the first expression by the reciprocal of the second expression. That is, for $q \neq 0$, $r \neq 0$, and $s \neq 0$,

$$\frac{p}{q} \div \frac{r}{s} = \frac{p}{q} \cdot \frac{s}{r}$$

Example 2

Divide. $\dfrac{2c^2d^5}{15e^4} \div \dfrac{6c^4d^3}{20e}$

$$= \frac{2c^2d^5}{15e^4} \cdot \frac{20e}{6c^4d^3}$$

$$= \frac{40c^2d^5e}{90c^4d^3e^4}$$

$$= \frac{4d^2}{9c^2e^3}$$

Section 7.3 Least Common Denominator

Key Concepts

Converting a Rational Expression to an Equivalent Expression with a Different Denominator

Multiply numerator and denominator of the rational expression by the missing factors necessary to create the desired denominator.

Examples

Example 1

Convert $\dfrac{-3}{x-2}$ to an equivalent expression with the indicated denominator:

$$\frac{-3}{x-2} = \frac{}{5x^2 - 20}$$

$$\frac{-3}{x-2} = \frac{}{5(x^2 - 4)} \qquad \text{Factor.}$$

$$\frac{-3}{x-2} = \frac{}{5(x-2)(x+2)}$$

Multiply numerator and denominator by the missing factors from the denominator.

$$\frac{-3 \cdot 5(x+2)}{(x-2) \cdot 5(x+2)} = \frac{-15x - 30}{5(x-2)(x+2)}$$

Finding the Least Common Denominator (LCD) of Two or More Rational Expressions

1. Factor all denominators completely.
2. The LCD is the product of unique factors from the denominators, where each factor is raised to its highest power.

Example 2

Identify the LCD. $\dfrac{1}{8x^3y^2z}; \dfrac{5}{6xy^4}$

1. Write the denominators as a product of prime factors:

$$\frac{1}{2^3x^3y^2z}; \frac{5}{2 \cdot 3xy^4}$$

2. The LCD is $2^3 3x^3y^4z$ or $24x^3y^4z$

Section 7.4	Addition and Subtraction of Rational Expressions

Key Concepts

To add or subtract rational expressions, the expressions must have the same denominator.

Steps to Add or Subtract Rational Expressions

1. Factor the denominators of each rational expression.
2. Identify the LCD.
3. Rewrite each rational expression as an equivalent expression with the LCD as its denominator.
4. Add or subtract the numerators, and write the result over the common denominator.
5. Simplify.

Examples

Example 1

Subtract. $\dfrac{c}{c^2 - c - 12} - \dfrac{1}{2c - 8}$

$$= \dfrac{c}{(c - 4)(c + 3)} - \dfrac{1}{2(c - 4)}$$

The LCD is $2(c - 4)(c + 3)$.

$$= \dfrac{2c}{2(c - 4)(c + 3)} - \dfrac{1(c + 3)}{2(c - 4)(c + 3)}$$

$$= \dfrac{2c - (c + 3)}{2(c - 4)(c + 3)}$$

$$= \dfrac{2c - c - 3}{2(c - 4)(c + 3)} = \dfrac{c - 3}{2(c - 4)(c + 3)}$$

Section 7.5	Complex Fractions

Key Concepts

Complex fractions can be simplified by using Method I or Method II.

Method I

1. Simplify the numerator and denominator of the complex fraction separately to form a single fraction in the numerator and a single fraction in the denominator.
2. Perform the division represented by the complex fraction. (Multiply the numerator of the complex fraction by the reciprocal of the denominator of the complex fraction.)
3. Simplify to lowest terms, if possible.

Examples

Example 1

Simplify. $\dfrac{1 - \dfrac{4}{w^2}}{1 - \dfrac{1}{w} - \dfrac{6}{w^2}} = \dfrac{\dfrac{w^2}{w^2} - \dfrac{4}{w^2}}{\dfrac{w^2}{w^2} - \dfrac{w}{w^2} - \dfrac{6}{w^2}}$

$$= \dfrac{\dfrac{w^2 - 4}{w^2}}{\dfrac{w^2 - w - 6}{w^2}} = \dfrac{w^2 - 4}{w^2} \cdot \dfrac{w^2}{w^2 - w - 6}$$

$$= \dfrac{(w - 2)(w + 2)}{w^2} \cdot \dfrac{w^2}{(w - 3)(w + 2)}$$

$$= \dfrac{w - 2}{w - 3}$$

Method II

1. Multiply the numerator and denominator of the complex fraction by the LCD of all individual fractions within the expression.
2. Apply the distributive property, and simplify the result.
3. Simplify to lowest terms, if possible.

Example 2

Simplify.

$$\frac{1 - \dfrac{4}{w^2}}{1 - \dfrac{1}{w} - \dfrac{6}{w^2}} = \frac{w^2\left(1 - \dfrac{4}{w^2}\right)}{w^2\left(1 - \dfrac{1}{w} - \dfrac{6}{w^2}\right)}$$

$$= \frac{w^2 - 4}{w^2 - w - 6} = \frac{(w-2)(w+2)}{(w-3)(w+2)}$$

$$= \frac{w-2}{w-3}$$

Section 7.6   Rational Equations

Key Concepts

An equation with one or more rational expressions is called a **rational equation**.

Steps to Solve a Rational Equation

1. Factor the denominators of all rational expressions.
2. Identify the LCD of all expressions in the equation.
3. Multiply both sides of the equation by the LCD.
4. Solve the resulting equation.
5. Check each potential solution in the original equation.

Examples

Example 1

Solve. $\dfrac{1}{w} - \dfrac{1}{2w-1} = \dfrac{-2w}{2w-1}$

The LCD is $w(2w - 1)$.

$$w(2w-1)\frac{1}{w} - w(2w-1)\frac{1}{2w-1}$$

$$= w(2w-1)\frac{-2w}{2w-1}$$

$$(2w-1)(1) - w(1) = w(-2w)$$

$$2w - 1 - w = -2w^2 \qquad \text{Quadratic equation}$$

$$2w^2 + w - 1 = 0$$

$$(2w-1)(w+1) = 0$$

$$w \neq \tfrac{1}{2} \qquad \text{or} \qquad w = -1$$

Does not check.   Checks.

Example 2

Solve for I. $q = \dfrac{VQ}{I}$

$$I \cdot q = \frac{VQ}{I} \cdot I$$

$$Iq = VQ$$

$$I = \frac{VQ}{q}$$

| Section 7.7 | **Applications of Rational Equations and Proportions** |

Key Concepts

Solving Proportions

An equation that equates two rates or ratios is called a **proportion**:

$$\frac{a}{b} = \frac{c}{d} \quad (b \neq 0, d \neq 0)$$

To solve a proportion, multiply both sides of the equation by the LCD.

Examples 2 and 3 give applications of rational equations.

1. Applications involving $d = rt$
   (distance = rate · time) and work.

Example 2

Two cars travel from Los Angeles to Las Vegas. One car travels an average of 8 mph faster than the other car. If the faster car travels 189 miles in the same time as the slower car travels 165 miles, what is the average speed of each car?

Let r represent the speed of the slower car.
Let $r + 8$ represent the speed of the faster car.

	Distance	Rate	Time
Slower car	165	r	$\dfrac{165}{r}$
Faster car	189	$r + 8$	$\dfrac{189}{r + 8}$

$$\frac{165}{r} = \frac{189}{r + 8}$$

$$165(r + 8) = 189r$$

$$165r + 1320 = 189r$$

$$1320 = 24r$$

$$55 = r$$

The slower car travels 55 mph, and the faster car travels $55 + 8 = 63$ mph.

Examples

Example 1

A 90-g serving of a particular ice cream contains 10 g of fat. How much fat does 400 g of the same ice cream contain?

$$\frac{10 \text{ g fat}}{90 \text{ g ice cream}} = \frac{x \text{ grams fat}}{400 \text{ g ice cream}}$$

$$\frac{10}{90} = \frac{x}{400}$$

$$\overset{40}{3600} \cdot \left(\frac{10}{90}\right) = \left(\frac{x}{400}\right) \cdot \overset{9}{3600}$$

$$400 = 9x$$

$$x = \frac{400}{9} \approx 44.4 \text{ g}$$

Example 3

Beth and Cecelia have a house cleaning business. Beth can clean a particular house in 5 hr by herself. Cecelia can clean the same house in 4 hr. How long would it take if they cleaned the house together?

Let x be the number of hours it takes for both Beth and Cecelia to clean the house.

Beth can clean $\frac{1}{5}$ of the house in an hour and $\frac{1}{5}x$ of the house in x hours.
Cecelia can clean $\frac{1}{4}$ of the house in an hour and $\frac{1}{4}x$ of the house in x hours.

$$\frac{1}{5}x + \frac{1}{4}x = 1 \qquad \text{Together they clean one whole house.}$$

$$20\left(\frac{1}{5}x + \frac{1}{4}x\right) = (1)20$$

$$4x + 5x = 20$$

$$9x = 20$$

$$x = \frac{20}{9}, \text{ or } 2\tfrac{2}{9} \text{ hr working together.}$$

Chapter 7   Review Exercises

Section 7.1

1. For the rational expression $\dfrac{t-2}{t+9}$

   a. Evaluate the expression (if possible) for $t = 0, 1, 2, -3, -9$

   b. Write the domain of the expression in set-builder notation.

2. For the rational expression $\dfrac{k+1}{k-5}$

   a. Evaluate the expression for $k = 0, 1, 5, -1, -2$

   b. Write the domain of the expression in set-builder notation.

3. Which of the rational expressions are equal to -1 for all values of x for which the expressions are defined?

   a. $\dfrac{2-x}{x-2}$ **b.** $\dfrac{x-5}{x+5}$

   c. $\dfrac{-x-7}{x+7}$ **d.** $\dfrac{x^2-4}{4-x^2}$

For Exercises 4–13, write the domain in set-builder notation. Then simplify the expressions to lowest terms.

4. $\dfrac{x-3}{(2x-5)(x-3)}$ 5. $\dfrac{h+7}{(3h+1)(h+7)}$

6. $\dfrac{4a^2+7a-2}{a^2-4}$ 7. $\dfrac{2w^2+11w+12}{w^2-16}$

8. $\dfrac{z^2-4z}{8-2z}$ 9. $\dfrac{15-3k}{2k^2-10k}$

10. $\dfrac{2b^2+4b-6}{4b+12}$ 11. $\dfrac{3m^2-12m-15}{9m+9}$

12. $\dfrac{n+3}{n^2+6n+9}$ 13. $\dfrac{p+7}{p^2+14p+49}$

Section 7.2

For Exercises 14–27, multiply or divide as indicated.

14. $\dfrac{3y^3}{3y-6} \cdot \dfrac{y-2}{y}$ 15. $\dfrac{2u+10}{u} \cdot \dfrac{u^3}{4u+20}$

16. $\dfrac{11}{v-2} \cdot \dfrac{2v^2-8}{22}$ 17. $\dfrac{8}{x^2-25} \cdot \dfrac{3x+15}{16}$

18. $\dfrac{4c^2+4c}{c^2-25} \div \dfrac{8c}{c^2-5c}$ 19. $\dfrac{q^2-5q+6}{2q+4} \div \dfrac{2q-6}{q+2}$

20. $\left(\dfrac{-2t}{t+1}\right)(t^2-4t-5)$ 21. $(s^2-6s+8)\left(\dfrac{4s}{s-2}\right)$

22. $\dfrac{\dfrac{a^2+5a+1}{7a-7}}{\dfrac{a^2+5a+1}{a-1}}$ 23. $\dfrac{\dfrac{n^2+n+1}{n^2-4}}{\dfrac{n^2+n+1}{n+2}}$

24. $\dfrac{5h^2-6h+1}{h^2-1} \div \dfrac{16h^2-9}{4h^2+7h+3} \cdot \dfrac{3-4h}{30h-6}$

25. $\dfrac{3m-3}{6m^2+18m+12} \cdot \dfrac{2m^2-8}{m^2-3m+2} \div \dfrac{m+3}{m+1}$

26. $\dfrac{x-2}{x^2-3x-18} \cdot \dfrac{6-x}{x^2-4}$

27. $\dfrac{4y^2-1}{1+2y} \div \dfrac{y^2-4y-5}{5-y}$

Section 7.3

For Exercises 28–33, convert each expression into an equivalent expression with the indicated denominator.

28. $\dfrac{x+1}{x-2} = \dfrac{}{5x-10}$ 29. $\dfrac{y+2}{y-3} = \dfrac{}{2y-6}$

30. $\dfrac{6}{w} = \dfrac{}{w^2-4w}$ 31. $\dfrac{2}{r} = \dfrac{}{r^2+3r}$

32. $\dfrac{s-2}{s+4} = \dfrac{}{s^2-16}$ 33. $\dfrac{u+1}{u+6} = \dfrac{}{u^2-36}$

For Exercises 34–41, identify the LCD.

34. $\dfrac{2}{a^2bc^2}; \dfrac{5}{ab^3}$ **35.** $\dfrac{6x}{y^2z}; \dfrac{3}{xy^2z^4}$

36. $\dfrac{5}{p+2}; \dfrac{p}{p-4}$ **37.** $\dfrac{6}{q}; \dfrac{1}{q+8}$

38. $\dfrac{8}{m^2-16}; \dfrac{7}{m^2-m-12}$

39. $\dfrac{6}{n^2-9}; \dfrac{5}{n^2-n-6}$

40. $\dfrac{4}{2t-5}; \dfrac{5}{5-2t}$ **41.** $\dfrac{-2}{3k-1}; \dfrac{6}{1-3k}$

42. State two possible LCDs that could be used to add the fractions.

$$\dfrac{7}{c-2} + \dfrac{4}{2-c}$$

43. State two possible LCDs that could be used to subtract the fractions.

$$\dfrac{10}{3-x} - \dfrac{5}{x-3}$$

Section 7.4

For Exercises 44–55, add or subtract as indicated.

44. $\dfrac{h+3}{h+1} + \dfrac{h-1}{h+1}$ **45.** $\dfrac{b-6}{b-2} + \dfrac{b+2}{b-2}$

46. $\dfrac{a^2}{a-5} - \dfrac{25}{a-5}$ **47.** $\dfrac{x^2}{x+7} - \dfrac{49}{x+7}$

48. $\dfrac{y}{y^2-81} + \dfrac{2}{9-y}$ **49.** $\dfrac{3}{4-t^2} + \dfrac{t}{2-t}$

50. $\dfrac{4}{3m} - \dfrac{1}{m+2}$ **51.** $\dfrac{5}{2r+12} - \dfrac{1}{r}$

52. $\dfrac{4p}{p^2+6p+5} - \dfrac{3p}{p^2+5p+4}$

53. $\dfrac{3q}{q^2+7q+10} - \dfrac{2q}{q^2+6q+8}$

54. $\dfrac{1}{h} + \dfrac{h}{2h+4} - \dfrac{2}{h^2+2h}$

55. $\dfrac{x}{3x+9} - \dfrac{3}{x^2+3x} + \dfrac{1}{x}$

Section 7.5

For Exercises 56–63, simplify the complex fractions.

56. $\dfrac{\dfrac{a-4}{3}}{\dfrac{a-2}{3}}$ **57.** $\dfrac{\dfrac{z+5}{6z}}{\dfrac{2z+10}{3}}$

58. $\dfrac{\dfrac{2-3w}{2}}{\dfrac{2}{w}-3}$ **59.** $\dfrac{\dfrac{2}{y}+6}{\dfrac{3y+1}{4}}$

60. $\dfrac{\dfrac{y}{x}-\dfrac{x}{y}}{\dfrac{1}{x}+\dfrac{1}{y}}$ **61.** $\dfrac{\dfrac{b}{a}-\dfrac{a}{b}}{\dfrac{1}{b}-\dfrac{1}{a}}$

62. $\dfrac{\dfrac{6}{p+2}+4}{\dfrac{8}{p+2}-4}$ **63.** $\dfrac{\dfrac{25}{k+5}+5}{\dfrac{5}{k+5}-5}$

Section 7.6

For Exercises 64–71, solve the equations.

64. $\dfrac{2}{x} + \dfrac{1}{2} = \dfrac{1}{4}$ **65.** $\dfrac{1}{y} + \dfrac{3}{4} = \dfrac{1}{4}$

66. $\dfrac{2}{h-2} + 1 = \dfrac{h}{h+2}$ **67.** $\dfrac{w}{w-1} = \dfrac{3}{w+1} + 1$

68. $\dfrac{t+1}{3} - \dfrac{t-1}{6} = \dfrac{1}{6}$

69. $\dfrac{4p-4}{p^2+5p-14} + \dfrac{2}{p+7} = \dfrac{1}{p-2}$

70. $\dfrac{y+1}{y+3} = \dfrac{y^2-11y}{y^2+y-6} - \dfrac{y-3}{y-2}$

71. $\dfrac{1}{z+2} = \dfrac{4}{z^2-4} - \dfrac{1}{z-2}$

72. Four times a number is added to 5. The sum is then divided by 6. The result is $\frac{7}{2}$. Find the number.

73. Solve the formula $\dfrac{V}{h} = \dfrac{\pi r^2}{3}$ for h.

74. Solve the formula $\dfrac{A}{b} = \dfrac{h}{2}$ for b.

Section 7.7

For Exercises 75–76, solve the proportions.

75. $\dfrac{m+2}{8} = \dfrac{m}{3}$ **76.** $\dfrac{12}{a} = \dfrac{5}{8}$

77. A bag of popcorn states that it contains 4 g of fat per serving. If a serving is 2 oz, how many grams of fat are in a 5-oz bag?

78. Bud goes 10 mph faster on his Harley Davidson motorcycle than Ed goes on his Honda motorcycle. If Bud travels 105 miles in the same time that

Ed travels 90 miles, what are the rates of the two bikers?

79. There are two pumps set up to fill a small swimming pool. One pump takes 24 min by itself to fill the pool, but the other takes 56 min by itself. How long would it take if both pumps work together?

80. Consider the similar triangles shown here. Find the values of x and y.

Chapter 7   Test

For Exercises 1–2,

a. Write the domain in set-builder notation.

b. Reduce the rational expression to lowest terms.

1. $\dfrac{5(x-2)(x+1)}{30(2-x)}$ **2.** $\dfrac{7a^2 - 42a}{a^3 - 4a^2 - 12a}$

3. Identify the rational expressions that are equal to -1 for all values of x for which the expression is defined.

a. $\dfrac{x+4}{x-4}$ **b.** $\dfrac{7-2x}{2x-7}$

c. $\dfrac{9x^2+16}{-9x^2-16}$ **d.** $-\dfrac{x+5}{x+5}$

For Exercises 4–9, perform the indicated operation.

4. $\dfrac{2}{y^2+4y+3} + \dfrac{1}{3y+9}$

5. $\dfrac{9-b^2}{5b+15} \div \dfrac{b-3}{b+3} \div (9b^2+28b+3)$

6. $\dfrac{w^2-4w}{w^2-8w+16} \cdot \dfrac{w-4}{w^2+w}$

7. $\dfrac{t}{t-2} - \dfrac{8}{t^2-4}$

8. $\dfrac{1}{3x+4} + \dfrac{2}{3x^2-2x-8} + \dfrac{x}{x-2}$

9. $\dfrac{1 - \dfrac{4}{m}}{m - \dfrac{16}{m}}$

For Exercises 10–13, solve the equation.

10. $\dfrac{3}{a} + \dfrac{5}{2} = \dfrac{7}{a}$

11. $\dfrac{p}{p-1} - \dfrac{1}{p} = \dfrac{2p^2 - 1}{p^2 - p}$

12. $\dfrac{3}{c-2} - \dfrac{1}{c+1} = \dfrac{7}{c^2 - c - 2}$

13. $\dfrac{4x}{x-4} = 3 + \dfrac{16}{x-4}$

14. Solve the formula $\dfrac{C}{2} = \dfrac{A}{r}$ for r.

15. If $\frac{3}{2}$ is added to the reciprocal of a number the result is $\frac{2}{5}$ times the reciprocal of that number. Find the number.

16. Solve the proportion.
$$\frac{y+7}{-4} = \frac{1}{4}$$

17. A recipe for vegetable soup calls for $\frac{1}{2}$ cup of carrots for six servings. How many cups of carrots are needed to prepare 15 servings?

18. A motorboat can travel 28 miles downstream in the same amount of time as it can travel 18 miles upstream. Find the speed of the current if the boat can travel 23 mph in still water.

19. One printer requires 3 hr to do a job and a second printer requires 6 hr to do the same job. If they worked together, how long would it take to complete the task?

20. Consider the similar triangles shown here. Find the values of a and b.

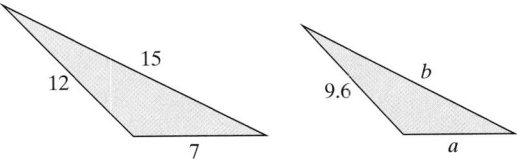

21. Find the LCD of the following pairs of rational expressions.

a. $\dfrac{x}{3(x+3)}; \dfrac{7}{5(x+3)}$ **b.** $\dfrac{-2}{3x^2 y}; \dfrac{4}{xy^2}$

Chapters 1–7   Cumulative Review Exercises

For Exercises 1–2, simplify completely.

1. $\left(\dfrac{1}{2}\right)^{-4} + 2^4$ **2.** $|3 - 5| + |-2 + 7|$

3. Solve for y: $\dfrac{1}{2} - \dfrac{3}{4}(y - 1) = \dfrac{5}{12}$

4. Complete the table.

Set-Builder Notation	Graph	Interval Notation
$\{x \mid x \geq -1\}$		
		$(-\infty, 5)$

5. The perimeter of a rectangular swimming pool is 104 m. The length is 1 m more than twice the width. Find the length and width.

6. The height of a triangle is 2 in. less than the base. The area is 40 in.^{2} Find the base and height of the triangle.

7. Simplify. $\left(\dfrac{4x^{-1}y^{-2}}{z^4}\right)^{-2}(2y^{-1}z^3)^3$

8. The length and width of a rectangle are given in terms of x.

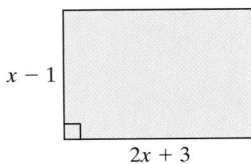

a. Write a polynomial that represents the perimeter of the rectangle.

b. Write a polynomial that represents the area of the rectangle.

9. Factor completely: $25x^2 - 30x + 9$

10. Factor. $10cd + 5d - 6c - 3$

11. Determine the domain of the expression.

$$\frac{x + 3}{(x - 5)(2x + 1)}$$

12. Simplify to lowest terms.

$$\frac{x^2 - 9}{x^2 + 8x + 15}$$

13. Divide. $\dfrac{2x - 6}{x^2 - 16} \div \dfrac{10x^2 - 90}{x^2 - x - 12}$

14. Simplify.

$$\frac{\dfrac{3}{4} - \dfrac{1}{x}}{\dfrac{1}{3x} - \dfrac{1}{4}}$$

15. Solve.

$$\frac{7}{y^2 - 4} = \frac{3}{y - 2} + \frac{2}{y + 2}$$

16. Solve the proportion.

$$\frac{2b - 5}{6} = \frac{4b}{7}$$

17. Determine the x- and y-intercepts.

   a. $-2x + 4y = 8$ **b.** $y = 5x$

18. Determine the slope

   a. of the line containing the points $(0, -6)$ and $(-5, 1)$

   b. of the line $y = -\dfrac{2}{3}x - 6$

   c. of a line parallel to a line having a slope of 4.

   d. of a line perpendicular to a line having a slope of 4.

19. Find an equation of a line passing through the point $(1, 2)$ and having a slope of 5. Write the answer in slope-intercept form.

20. A group of teenagers buys 2 large popcorns and 6 drinks at the movie theater for $16. A couple buys 1 large popcorn and 2 drinks for $6.50. Find the price for 1 large popcorn and the price for 1 drink.

Introduction to Relations and Functions

8

8.1 Introduction to Relations

8.2 Introduction to Functions

8.3 Graphs of Functions

8.4 Variation

In this chapter we introduce the concept of a function. In general terms, a function defines how one variable depends on one or more other variables. The words in the puzzle are key terms found in this chapter.

Across

1. A type of variation such that as one variable increases, the other increases.

4. A type of variation such that as one variable increases, the other variable decreases.

5. A set of ordered pairs such that for every element in the domain, there corresponds exactly one element in the range.

7. A set of ordered pairs.

Down

1. The set of first coordinates of a set of ordered pairs.

2. The shape of the graph of a quadratic function.

3. A function whose graph is a horizontal line.

6. A function whose graph is a line that is not vertical or horizontal.

7. The set of second coordinates of a set of ordered pairs.

Section 8.1   Introduction to Relations

1. Domain and Range of a Relation

In many naturally occurring phenomena, two variables may be linked by some type of relationship. For instance, an archeologist finds the bones of a woman at an excavation site. One of the bones is a femur. The femur is the large bone in the thigh attached to the knee and hip. Table 8-1 shows a correspondence between the length of a woman's femur and her height.

Table 8-1

Length of Femur (cm) x	Height (in.) y		Ordered Pair
45.5	65.5	→	(45.5, 65.5)
48.2	68.0	→	(48.2, 68.0)
41.8	62.2	→	(41.8, 62.2)
46.0	66.0	→	(46.0, 66.0)
50.4	70.0	→	(50.4, 70.0)

Each data point from Table 8-1 may be represented as an ordered pair. In this case, the first value represents the length of a woman's femur and the second, the woman's height. The set of ordered pairs {(45.5, 65.5), (48.2, 68.0), (41.8, 62.2), (46.0, 66.0), (50.4, 70.0)} defines a relation between femur length and height.

Definition of a Relation in *x* and *y*

Any set of ordered pairs (x, y) is called a **relation in *x* and *y***. Furthermore,

- The set of first components in the ordered pairs is called the **domain of the relation**.
- The set of second components in the ordered pairs is called the **range of the relation**.

Example 1   Finding the Domain and Range of a Relation

Find the domain and range of the relation linking the length of a woman's femur to her height {(45.5, 65.5), (48.2, 68.0), (41.8, 62.2), (46.0, 66.0), (50.4, 70.0)}.

Solution:

Domain: {45.5, 48.2, 41.8, 46.0, 50.4}   Set of first coordinates

Range: {65.5, 68.0, 62.2, 66.0, 70.0}   Set of second coordinates

Skill Practice

1. Find the domain and range of the relation.

$$\left\{ (0, 0), (-8, 4), \left(\frac{1}{2}, 1\right), (-3, 4), (-8, 0) \right\}$$

Skill Practice Answers

1. Domain $\left\{ 0, -8, \frac{1}{2}, -3 \right\}$, range {0, 4, 1}

The *x*- and *y*-components that constitute the ordered pairs in a relation do not need to be numerical. For example, Table 8-2 depicts five states in the United States and the corresponding number of representatives in the House of Representatives as of July 2005.

Table 8-2

State *x*	Number of Representatives *y*
Alabama	7
California	53
Colorado	7
Florida	25
Kansas	4

These data define a relation:

{(Alabama, 7), (California, 53), (Colorado, 7), (Florida, 25), (Kansas, 4)}

Example 2 **Finding the Domain and Range of a Relation**

Find the domain and range of the relation

{(Alabama, 7), (California, 53), (Colorado, 7), (Florida, 25), (Kansas, 4)}

Solution:

Domain: {Alabama, California, Colorado, Florida, Kansas}

Range: {7, 53, 25, 4}   (*Note:* The element 7 is not listed twice.)

Skill Practice

2. The table gives the longevity for four types of animals. Write the ordered pairs (*x*, *y*) indicated by this relation, and state the domain and range.

Animal, *x*	Longevity (years), *y*
Bear	22.5
Cat	11
Deer	12.5
Dog	11

A relation may consist of a finite number of ordered pairs or an infinite number of ordered pairs. Furthermore, a relation may be defined by several different methods: by a list of ordered pairs, by a correspondence between the domain and range, by a graph, or by an equation.

Skill Practice Answers

2. {(Bear, 22.5), (Cat, 11), (Deer, 12.5), (Dog, 11)}; domain: {Bear, Cat, Deer, Dog}, range: {22.5, 11, 12.5}

- A relation may be defined as a set of ordered pairs.

$$\{(1, 2), (-3, 4), (1, -4), (3, 4)\}$$

- A relation may be defined by a correspondence (Figure 8-1). The corresponding ordered pairs are $\{(1, 2), (1, -4), (-3, 4), (3, 4)\}$.

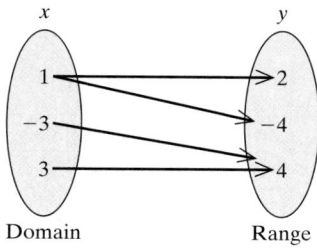

Domain   Range

Figure 8-1

- A relation may be defined by a graph (Figure 8-2). The corresponding ordered pairs are $\{(1, 2), (-3, 4), (1, -4), (3, 4)\}$.

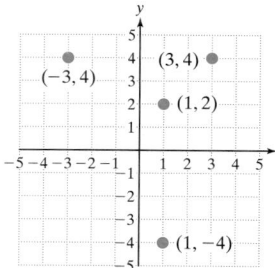

Figure 8-2

- A relation may be expressed by an equation such as $x = y^2$. The solutions to this equation define an infinite set of ordered pairs of the form $\{(x, y) | x = y^2\}$. The solutions can also be represented by a graph in a rectangular coordinate system (Figure 8-3).

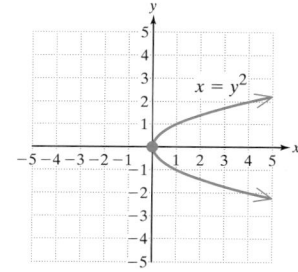

Figure 8-3

Example 3   **Finding the Domain and Range of a Relation**

Find the domain and range of the relations:

Solution:

a.

Domain: $\{3, 2, -7\}$

Range: $\{-9\}$

b.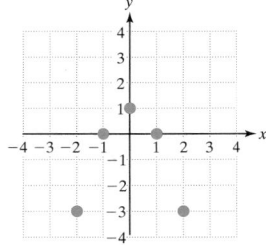

The domain elements are the x-coordinates of the points, and the range elements are the y-coordinates.

Domain: $\{-2, -1, 0, 1, 2\}$

Range: $\{-3, 0, 1\}$

c.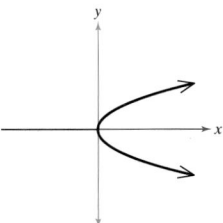

The domain consists of an infinite number of x-values extending from -8 to 8 (shown in red). The range consists of all y-values from -5 to 5 (shown in blue). Thus, the domain and range must be expressed in set-builder notation or in interval notation.

Domain: $\{x \mid x$ is a real number and $-8 \leq x \leq 8\}$ or $[-8, 8]$

Range: $\{y \mid y$ is a real number and $-5 \leq y \leq 5\}$ or $[-5, 5]$

d. $x = y^2$

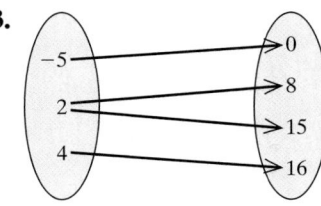

The arrows on the curve indicate that the graph extends infinitely far up and to the right and infinitely far down and to the right.

Domain: $\{x \mid x$ is a real number and $x \geq 0\}$ or $[0, \infty)$

Range: $\{y \mid y$ is any real number$\}$ or $(-\infty, \infty)$

Skill Practice   Find the domain and range of the relations.

3.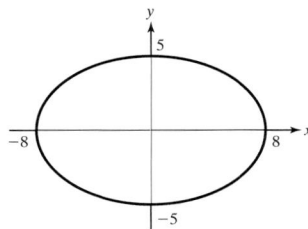

4.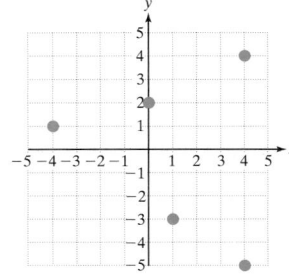

5.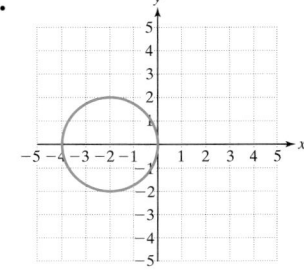

6. Find the domain and range of the relation $x = -|y|$ whose graph is shown here. Express the answer in interval notation.

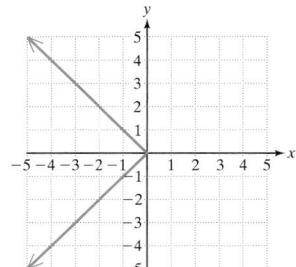

Skill Practice Answers

3. Domain $\{-5, 2, 4\}$, range $\{0, 8, 15, 16\}$
4. Domain $\{-4, 0, 1, 4\}$, range $\{-5, -3, 1, 2, 4\}$
5. Domain: $\{x \mid x$ is a real number and $-4 \leq x \leq 0\}$ or $[-4, 0]$, range: $\{y \mid y$ is a real number and $-2 \leq y \leq 2\}$ or $[-2, 2]$
6. Domain: $(-\infty, 0]$, range: $(-\infty, \infty)$

2. Applications Involving Relations

> **Example 4** **Analyzing a Relation**

The data in Table 8-3 depict the length of a woman's femur and her corresponding height. Based on these data, a forensics specialist or archeologist can find a linear relationship between height y and femur length x:

$$y = 0.906x + 24.3 \qquad 40 \le x \le 55$$

From this type of relationship, the height of a woman can be inferred based on skeletal remains.

Table 8-3

Length of Femur (cm) x	Height (in.) y
45.5	65.5
48.2	68.0
41.8	62.2
46.0	66.0
50.4	70.0

a. Find the height of a woman whose femur is 46.0 cm.

b. Find the height of a woman whose femur is 51.0 cm.

c. Why is the domain restricted to $40 \le x \le 55$?

Solution:

a. $y = 0.906x + 24.3$
$\quad = 0.906(46.0) + 24.3$ Substitute $x = 46.0$ cm.
$\quad = 65.976$ The woman is approximately 66.0 in. tall.

b. $y = 0.906x + 24.3$
$\quad = 0.906(51.0) + 24.3$ Substitute $x = 51.0$ cm.
$\quad = 70.506$ The woman is approximately 70.5 in. tall.

c. The domain restricts femur length to values between 40 cm and 55 cm inclusive. These values are within the normal lengths for an adult female and are in the proximity of the observed data (Figure 8-4).

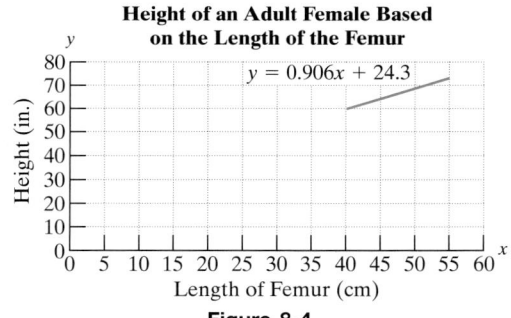

Height of an Adult Female Based on the Length of the Femur

$y = 0.906x + 24.3$

Length of Femur (cm)

Figure 8-4

> **Skill Practice**

7. The linear equation, $y = -0.014x + 64.5$, relates the weight of a car, x, (in pounds) to its gas mileage, y, (in mpg).

   a. Find the gas mileage in miles per gallon for a car weighing 2550 lb.

   b. Find the gas mileage for a car weighing 2850 lb.

Skill Practice Answers

7a. 28.8 mpg   **b.** 24.6 mpg

Section 8.1   Practice Exercises

Study Skills Exercise

1. Define the key terms.

    a. relation in *x* and *y*   **b. domain of a relation**   **c. range of a relation**

Concept 1: Domain and Range of a Relation

2. Given the relation $\{(2, 4), (-3, 1), (4, 4)\}$, give the domain and range.

For Exercises 3–6, write each relation as a set of ordered pairs.

3.

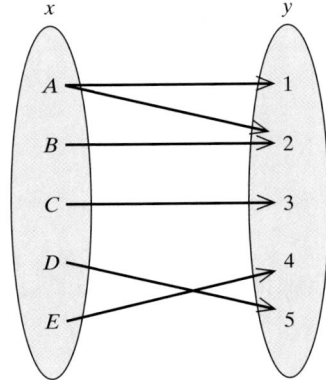

4.

State, *x*	Year of Statehood, *y*
Maine	1820
Nebraska	1823
Utah	1847
Hawaii	1959
Alaska	1959

5.

Memory Stick	Price
64 MB	$37.99
128 MB	$42.99
256 MB	$49.99
512 MB	$74.99

6.

x	*y*
0	3
−2	$\frac{1}{2}$
5	10
−7	1
−2	8
5	1

7. List the domain and range of Exercise 3.

8. List the domain and range of Exercise 4.

9. List the domain and range of Exercise 5.

10. List the domain and range of Exercise 6.

For Exercises 11–24, find the domain and range of the relations. Use interval notation where appropriate.

11.

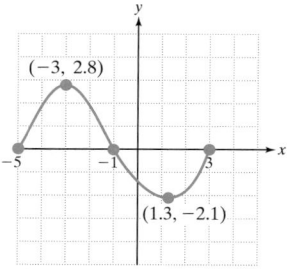

12.

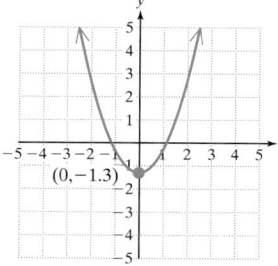

13.

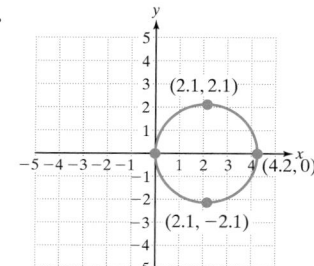

14.

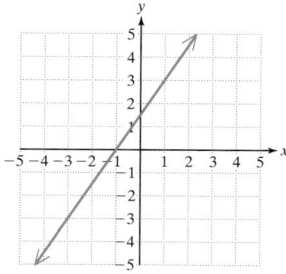

15.

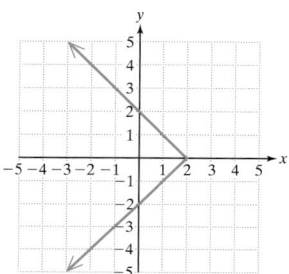

16.

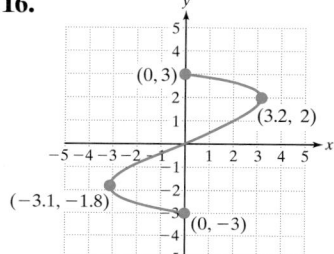

17. *Hint:* The open circle indicates that the point is not included in the relation.

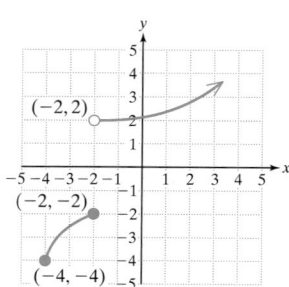

18.

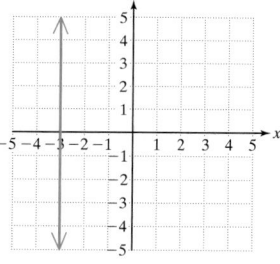

19.

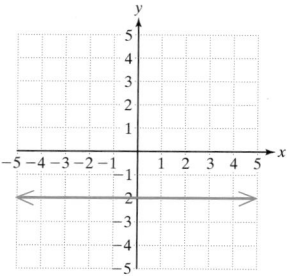

20.

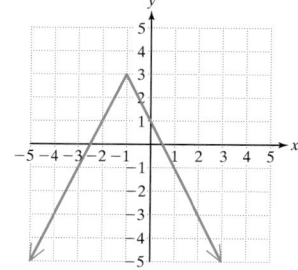

21.

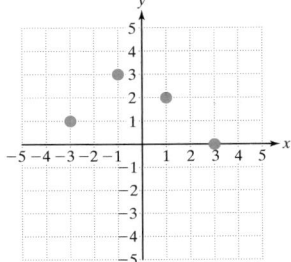

22.

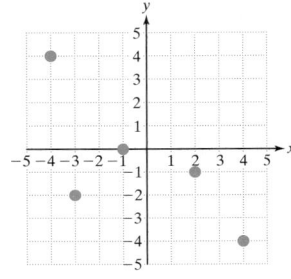

23.

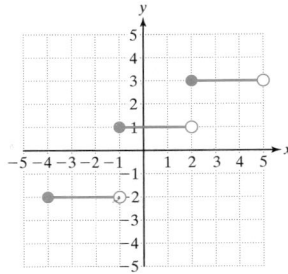

24.

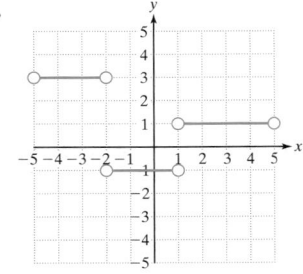

Concept 2: Applications Involving Relations

25. The table gives a relation between the month of the year and the average precipitation for that month for Miami, Florida.

   a. What is the range element corresponding to April?

   b. What is the range element corresponding to June?

   c. Which element in the domain corresponds to the least value in the range?

   d. Complete the ordered pair: (, 2.66)

   e. Complete the ordered pair: (Sept.,)

   f. What is the domain of this relation?

Month x	Precipitation (in.) y	Month x	Precipitation (in.) y
Jan.	2.01	July	5.70
Feb.	2.08	Aug.	7.58
Mar.	2.39	Sept.	7.63
Apr.	2.85	Oct.	5.64
May	6.21	Nov.	2.66
June	9.33	Dec.	1.83

Source: U.S. National Oceanic and Atmospheric Administration

26. The table gives a relation between a person's age and the person's maximum recommended heart rate.

   a. What is the domain?

   b. What is the range?

   c. The range element 200 corresponds to what element in the domain?

   d. Complete the ordered pair: (50,   )

   e. Complete the ordered pair: (, 190)

Age (years) x	Maximum Recommended Heart Rate (Beats per Minute) y
20	200
30	190
40	180
50	170
60	160

27. The population of Canada, y, (in millions) can be approximated by the relation $y = 0.146x + 31$, where x represents the number of years since 2000.

   a. Approximate the population of Canada in the year 2006.

   b. In what year will the population of Canada reach approximately 32,752,000?

28. As of April 2006, the world record times for selected women's track and field events are shown in the table.

The women's world record time y (in seconds) required to run x meters can be approximated by the relation $y = -10.78 + 0.159x$.

a. Predict the time required for a 500-m race.

b. Use this model to predict the time for a 1000-m race. Is this value exactly the same as the data value given in the table? Explain.

Distance (m)	Time (sec)	Winner's Name and Country
100	10.49	Florence Griffith Joyner (United States)
200	21.34	Florence Griffith Joyner (United States)
400	47.60	Marita Koch (East Germany)
800	113.28	Jarmila Kratochvilova (Czechoslovakia)
1000	148.98	Svetlana Masterkova (Russia)
1500	230.46	Qu Yunxia (China)

Expanding Your Skills

29. a. Define a relation with four ordered pairs such that the first element of the ordered pair is the name of a friend and the second element is your friend's place of birth.

b. State the domain and range of this relation.

30. a. Define a relation with four ordered pairs such that the first element is a state and the second element is its capital.

b. State the domain and range of this relation.

31. Use a mathematical equation to define a relation whose second component y is 1 less than 2 times the first component x.

32. Use a mathematical equation to define a relation whose second component y is 3 more than the first component x.

33. Use a mathematical equation to define a relation whose second component is the square of the first component.

34. Use a mathematical equation to define a relation whose second component is one-fourth the first component.

Section 8.2   Introduction to Functions

Concepts

1. Definition of a Function
2. Vertical Line Test
3. Function Notation
4. Finding Function Values from a Graph
5. Domain of a Function

1. Definition of a Function

In this section we introduce a special type of relation called a function.

Definition of a Function

Given a relation in x and y, we say "y is a **function** of x" if for every element x in the domain, there corresponds exactly one element y in the range.

To understand the difference between a relation that is a function and a relation that is not a function, consider Example 1.

Example 1 **Determining Whether a Relation Is a Function**

Determine which of the relations define y as a function of x.

a.

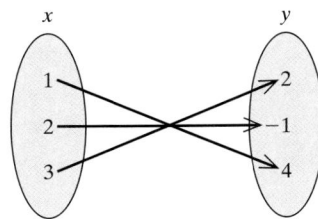

b.

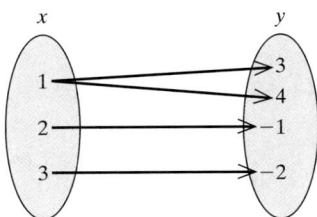

c.

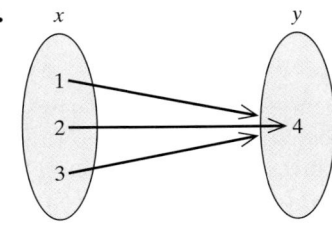

Solution:

a. This relation is defined by the set of ordered pairs $\{(1, 4), (2, -1), (3, 2)\}$.

Notice that for each x in the domain there is only one corresponding y in the range. Therefore, this relation is a function.

When $x = 1$, there is only one possibility for y: $y = 4$

When $x = 2$, there is only one possibility for y: $y = -1$

When $x = 3$, there is only one possibility for y: $y = 2$

b. This relation is defined by the set of ordered pairs

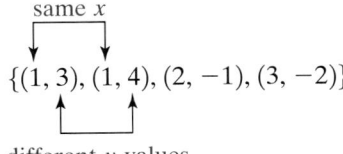

When $x = 1$, there are *two* possible range elements: $y = 3$ and $y = 4$. Therefore, this relation is *not* a function.

c. This relation is defined by the set of ordered pairs $\{(1, 4), (2, 4), (3, 4)\}$.

When $x = 1$, there is only one possibility for y: $y = 4$

When $x = 2$, there is only one possibility for y: $y = 4$

When $x = 3$, there is only one possibility for y: $y = 4$

Because each value of x in the domain has only one corresponding y value, this relation is a function.

Skill Practice   Determine if the relations define *y* as a function of *x*.

1.

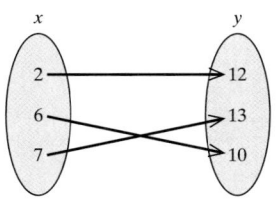

2. {(4, 2), (−5, 4), (0, 0), (8, 4)}

3. {(−1, 6), (8, 9), (−1, 4), (−3, 10)}

2. Vertical Line Test

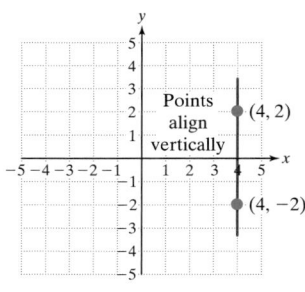

A relation that is not a function has at least one domain element *x* paired with more than one range value *y*. For example, the ordered pairs (4, 2) and (4, −2) do not constitute a function because two different *y*-values correspond to the same *x*. These two points are aligned vertically in the *xy*-plane, and a vertical line drawn through one point also intersects the other point. Thus, if a vertical line drawn through a graph of a relation intersects the graph in more than one point, the relation cannot be a function. This idea is stated formally as the **vertical line test**.

> **The Vertical Line Test**
>
> Consider a relation defined by a set of points (*x*, *y*) in a rectangular coordinate system. The graph defines *y* as a function of *x* if no vertical line intersects the graph in more than one point.

The vertical line test also implies that if any vertical line drawn through the graph of a relation intersects the relation in more than one point, then the relation does *not define y as a function of x*.

The vertical line test can be demonstrated by graphing the ordered pairs from the relations in Example 1.

a. {(1, 4), (2, −1), (3, 2)}   **b.** {(1, 3), (1, 4), (2, −1), (3, −2)}

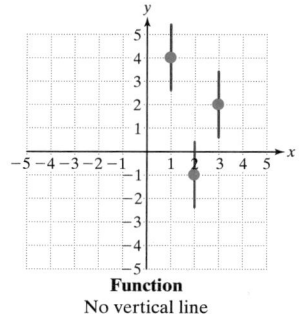

Function
No vertical line
intersects more than once.

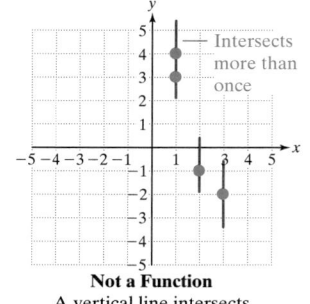

Not a Function
A vertical line intersects
in more than one point.

Skill Practice Answers
1. Yes   **2.** Yes   **3.** No

| Example 2 | **Using the Vertical Line Test**

Use the vertical line test to determine whether the following relations define y as a function of x.

a.

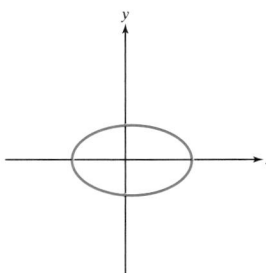

b.

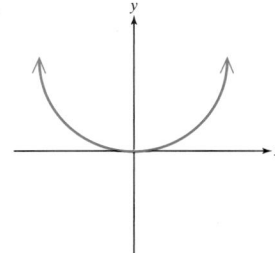

Solution:

a.

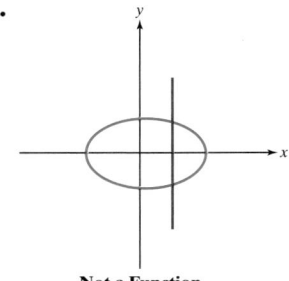

Not a Function
A vertical line intersects
in more than one point.

b.

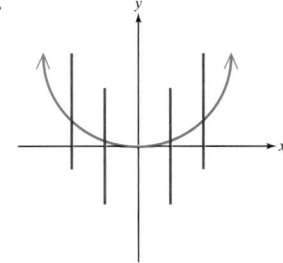

Function
No vertical line intersects
in more than one point.

| Skill Practice | Use the vertical line test to determine whether the relations define y as a function of x.

4.

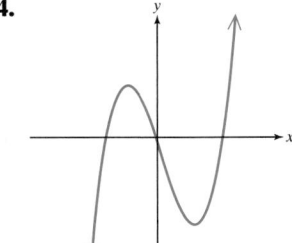

5.

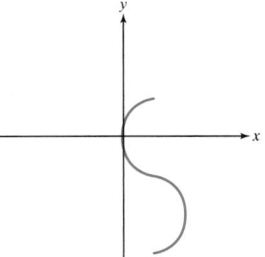

3. Function Notation

A function is defined as a relation with the added restriction that each value in the domain must have only one corresponding y-value in the range. In mathematics, functions are often given by rules or equations to define the relationship between two or more variables. For example, the equation $y = 2x$ defines the set of ordered pairs such that the y-value is twice the x-value.

When a function is defined by an equation, we often use **function notation**. For example, the equation $y = 2x$ can be written in function notation as

Skill Practice Answers
4. Yes   **5.** No

$$f(x) = 2x$$ where f is the name of the function, x is an input value from the domain of the function, and $f(x)$ is the function value (or y-value) corresponding to x

The notation $f(x)$ is read as "f of x" or "the value of the function f at x."

A function may be evaluated at different values of x by substituting x-values from the domain into the function. For example, to evaluate the function defined by $f(x) = 2x$ at $x = 5$, substitute $x = 5$ into the function.

$$f(x) = 2x$$
$$f(5) = 2(5)$$
$$f(5) = 10$$

Thus, when $x = 5$, the corresponding function value is 10. We say "f of 5 is 10" or "f at 5 is 10."

The names of functions are often given by either lowercase or uppercase letters, such as f, g, h, p, K, and M.

Example 3 Evaluating a Function

Given the function defined by $g(x) = \frac{1}{2}x - 1$, find the function values.

a. $g(0)$ **b.** $g(2)$ **c.** $g(4)$ **d.** $g(-2)$

Solution:

a. $g(x) = \dfrac{1}{2}x - 1$

$g(0) = \dfrac{1}{2}(0) - 1$ Substitute 0 for x.

$= 0 - 1$

$= -1$ We say, "g of 0 is -1." This is equivalent to the ordered pair $(0, -1)$.

b. $g(x) = \dfrac{1}{2}x - 1$

$g(2) = \dfrac{1}{2}(2) - 1$

$= 1 - 1$

$= 0$ We say "g of 2 is 0." This is equivalent to the ordered pair $(2, 0)$.

c. $g(x) = \dfrac{1}{2}x - 1$

$g(4) = \dfrac{1}{2}(4) - 1$

$= 2 - 1$

$= 1$ We say "g of 4 is 1." This is equivalent to the ordered pair $(4, 1)$.

d. $g(x) = \dfrac{1}{2}x - 1$

$g(-2) = \dfrac{1}{2}(-2) - 1$

$= -1 - 1$

$= -2$ We say "g of -2 is -2." This is equivalent to the ordered pair $(-2, -2)$.

Notice that $g(0)$, $g(2)$, $g(4)$, and $g(-2)$ correspond to the ordered pairs $(0, -1)$, $(2, 0)$, $(4, 1)$, and $(-2, -2)$. In the graph, these points "line up." The graph of *all* ordered pairs defined by this function is a line with a slope of $\frac{1}{2}$ and y-intercept of $(0, -1)$ (Figure 8-5). This should not be surprising because the function defined by $g(x) = \frac{1}{2}x - 1$ is equivalent to $y = \frac{1}{2}x - 1$.

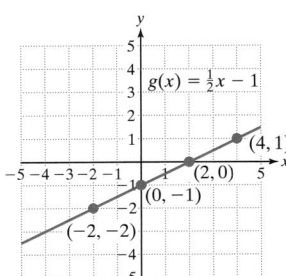

Figure 8-5

Skill Practice

6. Given the function defined by $f(x) = -2x - 3$, find the function values.

    a. $f(1)$ **b.** $f(0)$ **c.** $f(-3)$ **d.** $f\left(\dfrac{1}{2}\right)$

Calculator Connections

The values of $g(x)$ in Example 3 can be found using a *Table* feature.

$$Y_1 = \tfrac{1}{2}x - 1$$

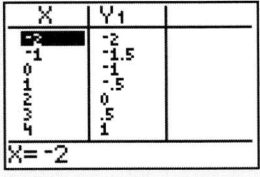

Function values can also be evaluated by using a *Value* (or *Eval*) feature. The value of $g(4)$ is shown here.

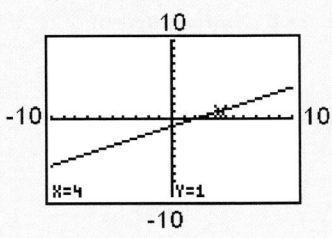

A function may be evaluated at numerical values or at algebraic expressions, as shown in Example 4.

Skill Practice Answers

6a. -5 **b.** -3 **c.** 3
  d. -4

Example 4   Evaluating Functions

Given the functions defined by $f(x) = x^2 - 2x$ and $g(x) = 3x + 5$, find the function values.

    a. $f(t)$ **b.** $g(w + 4)$ **c.** $f(-t)$

Solution:

   a. $f(x) = x^2 - 2x$

       $f(t) = (t)^2 - 2(t)$ Substitute $x = t$ for all values of x in the function.

          $= t^2 - 2t$ Simplify.

   b. $g(x) = 3x + 5$

       $g(w + 4) = 3(w + 4) + 5$ Substitute $x = w + 4$ for all values of x in the function.

             $= 3w + 12 + 5$

             $= 3w + 17$ Simplify.

   c. $f(x) = x^2 - 2x$ Substitute $-t$ for x.

       $f(-t) = (-t)^2 - 2(-t)$

          $= t^2 + 2t$ Simplify.

Skill Practice

7. Given the function defined by $g(x) = 4x - 3$, find the function values.

    a. $g(a)$ **b.** $g(x + h)$ **c.** $g(-x)$

4. Finding Function Values from a Graph

We can find function values by looking at a graph of the function. The value of $f(a)$ refers to the y-coordinate of a point with x-coordinate a.

Example 5   Finding Function Values from a Graph

Consider the function pictured in Figure 8-6.

   a. Find $h(-1)$.

   b. Find $h(2)$.

   c. For what value of x is $h(x) = 3$?

   d. For what values of x is $h(x) = 0$?

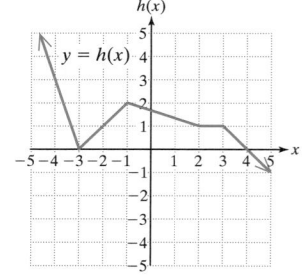

Figure 8-6

Solution:

   a. $h(-1) = 2$ This corresponds to the ordered pair $(-1, 2)$.

   b. $h(2) = 1$ This corresponds to the ordered pair $(2, 1)$.

   c. $h(x) = 3$ for $x = -4$ This corresponds to the ordered pair $(-4, 3)$.

   d. $h(x) = 0$ for $x = -3$ and $x = 4$ These are the ordered pairs $(-3, 0)$ and $(4, 0)$.

Skill Practice Answers

7a. $4a - 3$ **b.** $4x + 4h - 3$
   c. $-4x - 3$

Skill Practice Refer to the function graphed here.

8. Find $f(0)$.

9. Find $f(-2)$.

10. For what value(s) of x is $f(x) = 0$?

11. For what value(s) of x is $f(x) = -4$?

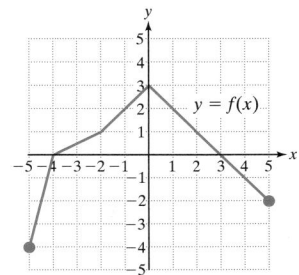

5. Domain of a Function

A function is a relation, and it is often necessary to determine its domain and range. Consider a function defined by the equation $y = f(x)$. The **domain** of f is the set of all x-values that when substituted into the function, produce a real number. The **range** of f is the set of all y-values corresponding to the values of x in the domain.

To find the domain of a function defined by $y = f(x)$, keep these guidelines in mind.

- Exclude values of x that make the denominator of a fraction zero.
- Exclude values of x that make a negative value within a square root.

Example 6 Finding the Domain of a Function

Find the domain of the functions. Write the answers in interval notation.

a. $f(x) = \dfrac{x + 7}{2x - 1}$ **b.** $h(x) = \dfrac{x - 4}{x^2 + 9}$

c. $k(t) = \sqrt{t + 4}$ **d.** $g(t) = t^2 - 3t$

Solution:

a. The function will be undefined when the denominator is zero, that is, when

$$2x - 1 = 0$$

$$2x = 1$$

$$x = \frac{1}{2} \qquad \text{The value } x = \tfrac{1}{2} \text{ must be } excluded \text{ from the domain.}$$

$$\text{Interval notation: } \left(-\infty, \frac{1}{2}\right) \cup \left(\frac{1}{2}, \infty\right)$$

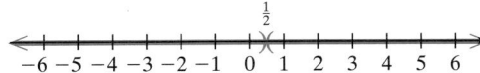

b. The quantity x^2 is greater than or equal to 0 for all real numbers x, and the number 9 is positive. Therefore, the sum $x^2 + 9$ must be *positive* for all real numbers x. The denominator of $h(x) = (x - 4)/(x^2 + 9)$ will never be zero; the domain is the set of all real numbers.

Interval notation: $(-\infty, \infty)$

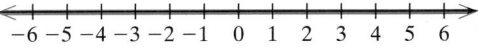

Skill Practice Answers

8. 3   **9.** 1
10. $x = -4$ and $x = 3$
11. $x = -5$

  c. The function defined by $k(t) = \sqrt{t + 4}$ will not be a real number when $t + 4$ is negative; hence the domain is the set of all t-values that make $t + 4$ *greater than or equal to zero*:

$$t + 4 \geq 0$$
$$t \geq -4$$

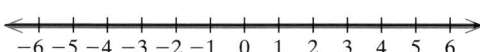

  Interval notation: $[-4, \infty)$

  d. The function defined by $g(t) = t^2 - 3t$ has no restrictions on its domain because any real number substituted for t will produce a real number. The domain is the set of all real numbers.

  Interval notation: $(-\infty, \infty)$

Skill Practice Write the domain of the functions in interval notation.

Skill Practice Answers
12. $(-\infty, 9) \cup (9, \infty)$
13. $(-\infty, \infty)$
14. $[2, \infty)$
15. $(-\infty, \infty)$

12. $f(x) = \dfrac{2x + 1}{x - 9}$ **13.** $p(x) = \dfrac{-5}{4x^2 + 1}$

14. $g(x) = \sqrt{x - 2}$ **15.** $h(x) = x + 6$

Section 8.2 Practice Exercises

Boost *your* GRADE at mathzone.com!

• Practice Problems • e-Professors
• Self-Tests   • Videos
• NetTutor

Study Skills Exercise

1. Define the key terms.

  a. function   **b. function notation**   **c. domain**   **d. range**   **e. vertical line test**

Review Exercises

For Exercises 2–4, **a.** write the relation as a set of ordered pairs, **b.** identify the domain, and **c.** identify the range.

2.

Parent, x	Child, y
Kevin	Katie
Kevin	Kira
Kathleen	Katie
Kathleen	Kira

3.

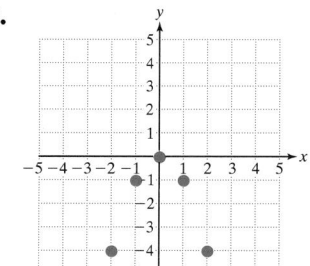

4.

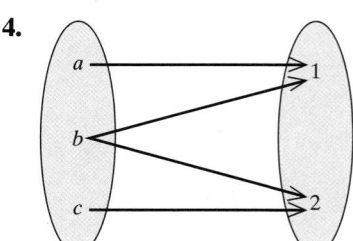

Concept 1: Definition of a Function

For Exercises 5–10, determine if the relation defines y as a function of x.

5.
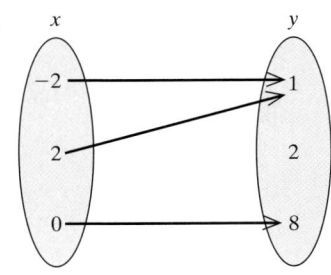

6.
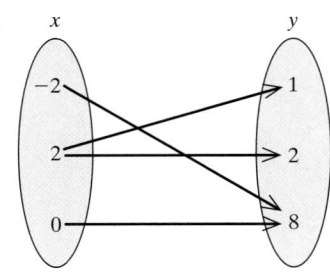

7.
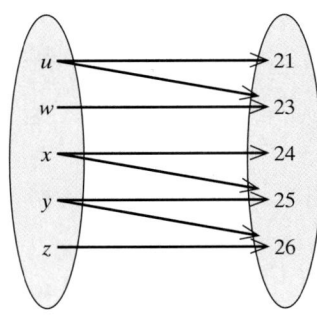

8.
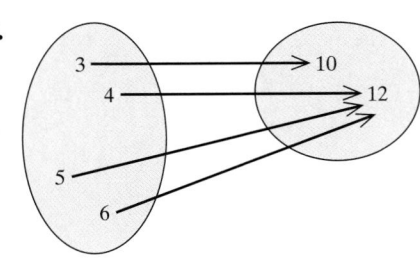

9. $\{(1, 2), (3, 4), (5, 4), (-9, 3)\}$

10. $\left\{(0, -1.1), \left(\frac{1}{2}, 8\right), (1.1, 8), \left(4, \frac{1}{2}\right)\right\}$

Concept 2: Vertical Line Test

For Exercises 11–16, use the vertical line test to determine whether the relation defines y as a function of x.

11.
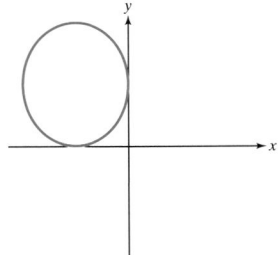

12.
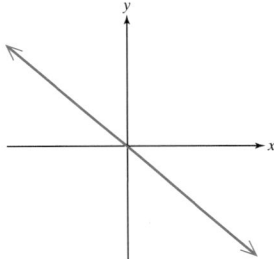

13.
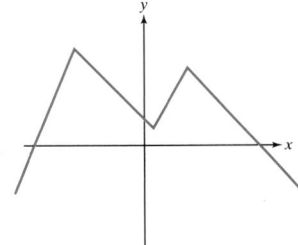

14.
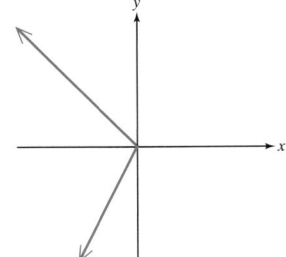

15.
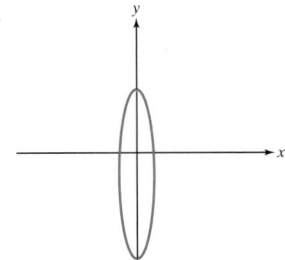

16.
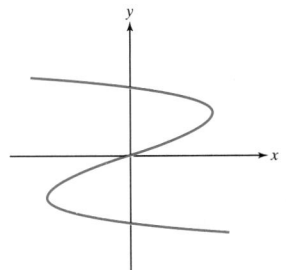

Concept 3: Function Notation

Consider the functions defined by $f(x) = 6x - 2$, $g(x) = -x^2 - 4x + 1$, $h(x) = 7$, and $k(x) = |x - 2|$. For Exercises 17–48, find the following.

17. $g(2)$ 　　　　 **18.** $k(2)$ 　　　　 **19.** $g(0)$ 　　　　 **20.** $h(0)$

21. $k(0)$ 　　　　 **22.** $f(0)$ 　　　　 **23.** $f(t)$ 　　　　 **24.** $g(a)$

25. $h(u)$

26. $k(v)$

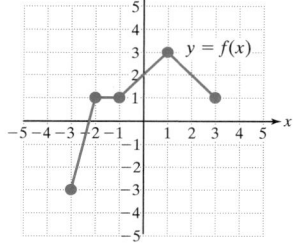 **27.** $g(-3)$

28. $h(-5)$

29. $k(-2)$

30. $f(-6)$

31. $f(x + 1)$

32. $h(x + 1)$

33. $g(2x)$

34. $k(x - 3)$

35. $g(-\pi)$

36. $g(x + h)$

37. $h(a + b)$

38. $f(x + h)$

39. $f(-a)$

40. $g(-b)$

41. $k(-c)$

42. $h(-x)$

43. $f\left(\dfrac{1}{2}\right)$

44. $g\left(\dfrac{1}{4}\right)$

45. $h\left(\dfrac{1}{7}\right)$

46. $k\left(\dfrac{3}{2}\right)$

47. $f(-2.8)$

48. $k(-5.4)$

Consider the functions $p = \{(\frac{1}{2}, 6), (2, -7), (1, 0), (3, 2\pi)\}$ and $q = \{(6, 4), (2, -5), (\frac{3}{4}, \frac{1}{5}), (0, 9)\}$. For Exercises 49–56, find the function values.

49. $p(2)$

50. $p(1)$

51. $p(3)$

52. $p\left(\dfrac{1}{2}\right)$

53. $q(2)$

54. $q\left(\dfrac{3}{4}\right)$

55. $q(6)$

56. $q(0)$

Concept 4: Finding Function Values from a Graph

57. The graph of $y = f(x)$ is given.

    a. Find $f(0)$.

    b. Find $f(3)$.

    c. Find $f(-2)$.

    d. For what value(s) of x is $f(x) = -3$?

    e. For what value(s) of x is $f(x) = 3$?

    f. Write the domain of f.

    g. Write the range of f.

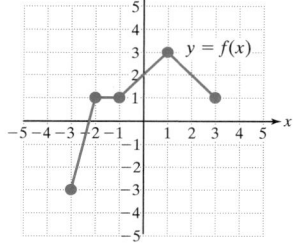

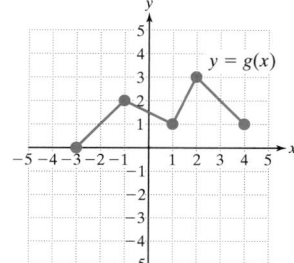 **58.** The graph of $y = g(x)$ is given.

    a. Find $g(-1)$.

    b. Find $g(1)$.

    c. Find $g(4)$.

    d. For what value(s) of x is $g(x) = 3$?

    e. For what value(s) of x is $g(x) = 0$?

    f. Write the domain of g.

    g. Write the range of g.

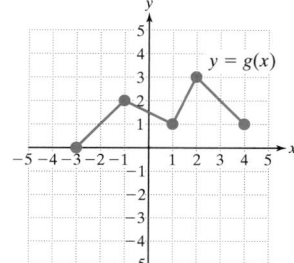

59. The graph of $y = H(x)$ is given.

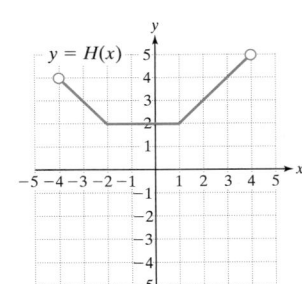

    a. Find $H(-3)$.

    b. Find $H(4)$.

    c. Find $H(3)$.

    d. For what value(s) of x is $H(x) = 3$?

    e. For what value(s) of x is $H(x) = 2$?

    f. Write the domain of H.

    g. Write the range of H.

60. The graph of $y = K(x)$ is given.

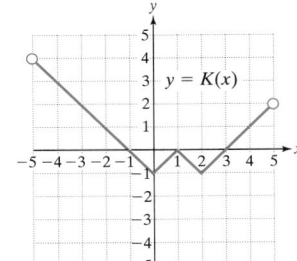

    a. Find $K(0)$.

    b. Find $K(-5)$.

    c. Find $K(1)$.

    d. For what value(s) of x is $K(x) = 0$?

    e. For what value(s) of x is $K(x) = 3$?

    f. Write the domain of K.

    g. Write the range of K.

Concept 5: Domain of a Function

61. Explain how to determine the domain of the function defined by $f(x) = \dfrac{x + 6}{x - 2}$.

62. Explain how to determine the domain of the function defined by $g(x) = \sqrt{x - 3}$.

For Exercises 63–78, find the domain. Write the answers in interval notation.

63. $k(x) = \dfrac{x - 3}{x + 6}$
    64. $m(x) = \dfrac{x - 1}{x - 4}$
    65. $f(t) = \dfrac{5}{t}$
    66. $g(t) = \dfrac{t - 7}{t}$

67. $h(p) = \dfrac{p - 4}{p^2 + 1}$
    68. $n(p) = \dfrac{p + 8}{p^2 + 2}$
    69. $h(t) = \sqrt{t + 7}$
    70. $k(t) = \sqrt{t - 5}$

71. $f(a) = \sqrt{a - 3}$
    72. $g(a) = \sqrt{a + 2}$
    73. $m(x) = \sqrt{1 - 2x}$
    74. $n(x) = \sqrt{12 - 6x}$

75. $p(t) = 2t^2 + t - 1$
    76. $q(t) = t^3 + t - 1$
    77. $f(x) = x + 6$
    78. $g(x) = 8x - \pi$

Mixed Exercises

79. The height (in feet) of a ball that is dropped from an 80-ft building is given by $h(t) = -16t^2 + 80$, where t is time in seconds after the ball is dropped.

    a. Find $h(1)$ and $h(1.5)$

    b. Interpret the meaning of the function values found in part (a).

80. A ball is dropped from a 50-m building. The height (in meters) after t sec is given by $h(t) = -4.9t^2 + 50$.

     a. Find $h(1)$ and $h(1.5)$.

     b. Interpret the meaning of the function values found in part (a).

81. If Alicia rides a bike at an average of 11.5 mph, the distance that she rides can be represented by $d(t) = 11.5t$, where t is the time in hours.

     a. Find $d(1)$ and $d(1.5)$.

     b. Interpret the meaning of the function values found in part (a).

82. If Miguel walks at an average of 5.9 km/hr, the distance that he walks can be represented by $d(t) = 5.9t$, where t is the time in hours.

     a. Find $d(1)$ and $d(2)$.

     b. Interpret the meaning of the function values found in part (a).

83. Brian's score on an exam is a function of the number of hours he spends studying. The function defined by $P(x) = \dfrac{100x^2}{50 + x^2}$ $(x \geq 0)$ indicates that he will achieve a score of $P\%$ if he studies for x hours.

     a. Evaluate $P(0)$, $P(5)$, $P(10)$, $P(15)$, $P(20)$, and $P(25)$. (Round to 1 decimal place.) Interpret $P(25)$ in the context of this problem.

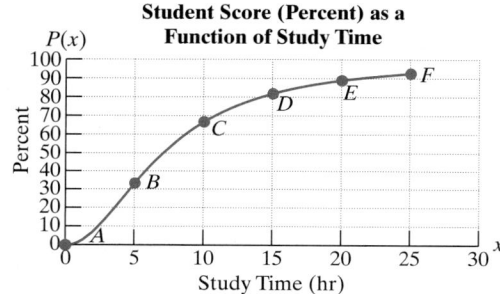

     b. Match the function values found in part (a) with the points A, B, C, D, E, and F on the graph.

Expanding Your Skills

For Exercises 84–85, find the domain. Write the answers in interval notation.

84. $q(x) = \dfrac{2}{\sqrt{x + 2}}$ **85.** $p(x) = \dfrac{8}{\sqrt{x - 4}}$

For Exercises 86–95, refer to the functions $y = f(x)$ and $y = g(x)$, defined as follows:

$$f = \{(-3, 5), (-7, -3), (-\tfrac{3}{2}, 4), (1.2, 5)\}$$
$$g = \{(0, 6), (2, 6), (6, 0), (1, 0)\}$$

86. Identify the domain of f.   **87.** Identify the range of f.

88. Identify the range of g.   **89.** Identify the domain of g.

90. For what value(s) of x is $f(x) = 5$?   **91.** For what value(s) of x is $f(x) = -3$?

92. For what value(s) of x is $g(x) = 0$?   **93.** For what value(s) of x is $g(x) = 6$?

94. Find $f(-7)$.   **95.** Find $g(0)$.

Graphing Calculator Exercises

96. Graph $k(t) = \sqrt{t} - 5$. Use the graph to support your answer to Exercise 70.

97. Graph $h(t) = \sqrt{t} + 7$. Use the graph to support your answer to Exercise 69.

98. a. Graph $h(t) = -4.9t^2 + 50$ on a viewing window defined by $0 \le t \le 3$ and $0 \le y \le 60$.

    b. Use the graph to approximate the function at $t = 1$. Use these values to support your answer to Exercise 80.

99. a. Graph $h(t) = -16t^2 + 80$ on a viewing window defined by $0 \le t \le 2$ and $0 \le y \le 100$.

    b. Use the graph to approximate the function at $t = 1$. Use these values to support your answer to Exercise 79.

Graphs of Functions

Section 8.3

1. Linear and Constant Functions

A function may be expressed as a mathematical equation that relates two or more variables. In this section, we will look at several elementary functions.

    We know from Section 3.2 that an equation in the form $y = k$, where k is a constant, is a horizontal line. In function notation, this can be written as $f(x) = k$. For example, the function defined by $f(x) = 3$ is a horizontal line, as shown in Figure 8-7.

    We say that a function defined by $f(x) = k$ is a constant function because for any value of x, the function value is constant.

    An equation of the form $y = mx + b$ is represented graphically by a line with slope m and y-intercept $(0, b)$. In function notation, this can be written as $f(x) = mx + b$. A function in this form is called a linear function. For example, the function defined by $f(x) = 2x - 3$ is a linear function with slope $m = 2$ and y-intercept $(0, -3)$ (Figure 8-8).

Concepts

1. Linear and Constant Functions
2. Graphs of Basic Functions
3. Definition of a Quadratic Function
4. Finding the x- and y-Intercepts of a Function Defined by $y = f(x)$
5. Determining Intervals of Increasing, Decreasing, or Constant Behavior

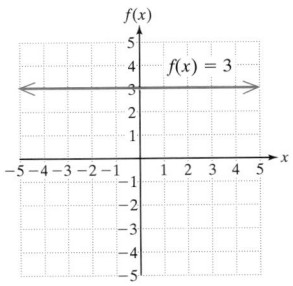

Figure 8-7

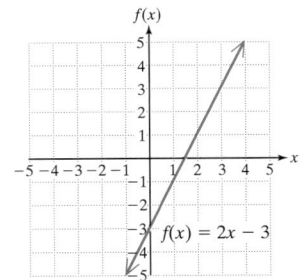

Figure 8-8

Definition of a Linear Function and a Constant Function

Let m and b represent real numbers such that $m \ne 0$. Then

A function that can be written in the form $f(x) = mx + b$ is a **linear function**.
A function that can be written in the form $f(x) = b$ is a **constant function**.

Note: The graphs of linear and constant functions are lines.

2. Graphs of Basic Functions

At this point, we are able to recognize the equations and graphs of linear and constant functions. In addition to linear and constant functions, the following equations define six basic functions that will be encountered in the study of algebra:

Equation	**Function Notation**
$y = x$	$f(x) = x$
$y = x^2$	$f(x) = x^2$
$y = x^3$	$f(x) = x^3$
$y = |x|$	$f(x) = |x|$
$y = \sqrt{x}$	$f(x) = \sqrt{x}$
$y = \dfrac{1}{x}$	$f(x) = \dfrac{1}{x}$

equivalent function notation

The graph of the function defined by $f(x) = x$ is linear, with slope $m = 1$ and y-intercept $(0, 0)$ (Figure 8-9).

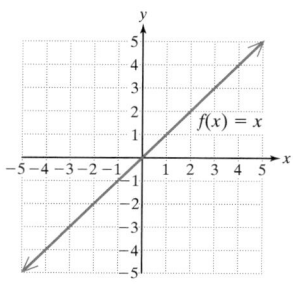

Figure 8-9

To determine the shapes of the other basic functions, we can plot several points to establish the pattern of the graph. Analyzing the equation itself may also provide insight to the domain, range, and shape of the function. To demonstrate this, we will graph $f(x) = x^2$ and $g(x) = \frac{1}{x}$.

Example 1 **Graphing Basic Functions**

Graph the functions defined by

a. $f(x) = x^2$ **b.** $g(x) = \dfrac{1}{x}$

Solution:

a. The domain of the function given by $f(x) = x^2$ (or equivalently $y = x^2$) is all real numbers.

To graph the function, choose arbitrary values of x within the domain of the function. Be sure to choose values of x that are positive and values that are negative to determine the behavior of the function to the right and left of the origin (Table 8-4). The graph of $f(x) = x^2$ is shown in Figure 8-10.

The function values are equated to the square of x, so $f(x)$ will always be greater than or equal to zero. Hence, the y-coordinates on the graph will never be negative. The range of the function is $\{y \mid y$ is a real number and $y \geq 0\}$. The arrows on each branch of the graph imply that the pattern continues indefinitely.

Table 8-4

x	$f(x) = x^2$
0	0
1	1
2	4
3	9
−1	1
−2	4
−3	9

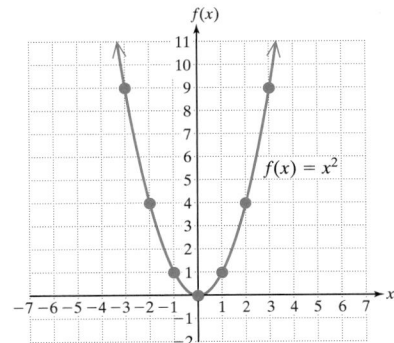

Figure 8-10

b. $g(x) = \dfrac{1}{x}$ Notice that $x = 0$ is not in the domain of the function. From the equation $y = \frac{1}{x}$, the y-values will be the reciprocal of the x-values. The graph defined by $g(x) = \frac{1}{x}$ is shown in Figure 8-11.

x	$g(x) = \dfrac{1}{x}$
1	1
2	$\frac{1}{2}$
3	$\frac{1}{3}$
−1	−1
−2	$-\frac{1}{2}$
−3	$-\frac{1}{3}$

x	$g(x) = \dfrac{1}{x}$
$\frac{1}{2}$	2
$\frac{1}{3}$	3
$\frac{1}{4}$	4
$-\frac{1}{2}$	−2
$-\frac{1}{3}$	−3
$-\frac{1}{4}$	−4

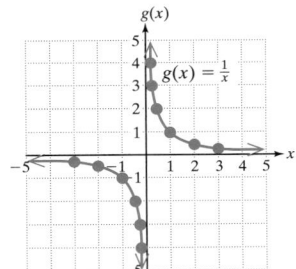

Figure 8-11

Notice that as x increases, the y-value decreases and gets closer to zero. In fact, as x approaches $-\infty$ or ∞, the graph gets closer to the x-axis. In this case, the x-axis is called a *horizontal asymptote*. Similarly, the graph of the function approaches the y-axis as x gets close to zero. In this case, the y-axis is called a *vertical asymptote*.

Calculator Connections

The graphs of the functions defined by $f(x) = x^2$ and $g(x) = \frac{1}{x}$ are shown in the following calculator displays.

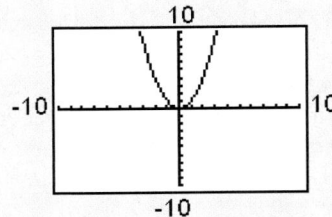

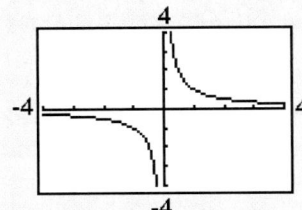

1. Graph $f(x) = -x^2$ by first making a table of points.

2. Graph $h(x) = |x| - 1$ by first making a table of points.

For your reference, we have provided the graphs of six basic functions in the following table.

Summary of Six Basic Functions and Their Graphs

Function	Graph	Domain and Range
1. $f(x) = x$	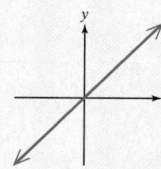	Domain $(-\infty, \infty)$ Range $(-\infty, \infty)$
2. $y = x^2$	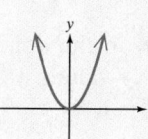	Domain $(-\infty, \infty)$ Range $[0, \infty)$
3. $y = x^3$	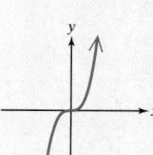	Domain $(-\infty, \infty)$ Range $(-\infty, \infty)$
4. $f(x) = |x|$	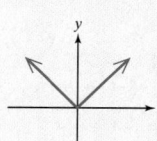	Domain $(-\infty, \infty)$ Range $[0, \infty)$
5. $y = \sqrt{x}$	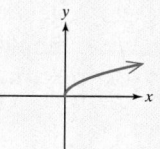	Domain $[0, \infty)$ Range $[0, \infty)$
6. $y = \dfrac{1}{x}$	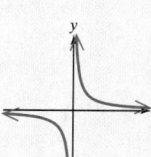	Domain $(-\infty, 0) \cup (0, \infty)$ Range $(-\infty, 0) \cup (0, \infty)$

Skill Practice Answers

1.

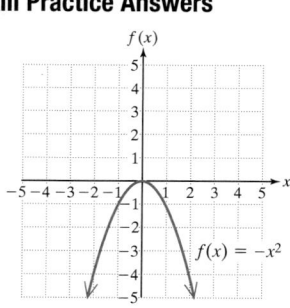

2.

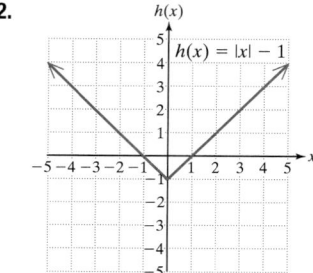

The shapes of these six graphs will be developed in the homework exercises. These functions are used often in the study of algebra. Therefore, we recommend that you associate an equation with its graph and commit each to memory.

3. Definition of a Quadratic Function

In Example 1 we graphed the function defined by $f(x) = x^2$ by plotting points. This function belongs to a special category called **quadratic functions**. A quadratic function can be written in the form $f(x) = ax^2 + bx + c$, where a, b, and c are real numbers and $a \neq 0$. The graph of a quadratic function is in the shape of a **parabola**. The leading coefficient, a, determines the direction of the parabola.

If $a > 0$, then the parabola opens upward, for example, $f(x) = x^2$. The minimum point on a parabola opening upward is called the vertex (Figure 8-12).

If $a < 0$, then the parabola opens downward, for example, $f(x) = -x^2$. The maximum point on a parabola opening downward is called the vertex (Figure 8-13).

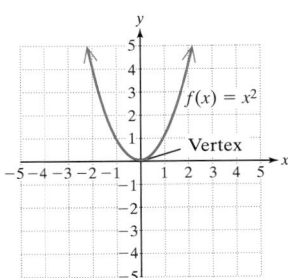

Figure 8-12

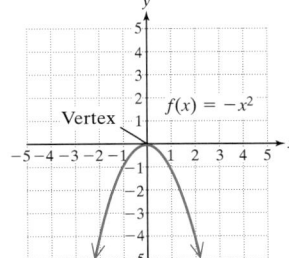

Figure 8-13

Example 2   Identifying Linear, Constant, and Quadratic Functions

Identify each function as linear, constant, quadratic, or none of these.

a. $f(x) = -4$ **b.** $f(x) = x^2 + 3x + 2$

c. $f(x) = 7 - 2x$ **d.** $f(x) = \dfrac{4x + 8}{8}$

Solution:

a. $f(x) = -4$ is a constant function. It is in the form $f(x) = b$, where $b = -4$.

b. $f(x) = x^2 + 3x + 2$ is a quadratic function. It is in the form $f(x) = ax^2 + bx + c$, where $a \neq 0$.

c. $f(x) = 7 - 2x$ is linear. Writing it in the form $f(x) = mx + b$, we get $f(x) = -2x + 7$, where $m = -2$ and $b = 7$.

d. $f(x) = \dfrac{4x + 8}{8}$ is linear. Writing it in the form $f(x) = mx + b$, we get

$$f(x) = \frac{4x}{8} + \frac{8}{8}$$

$$= \frac{1}{2}x + 1, \text{ where } m = \frac{1}{2} \text{ and } b = 1.$$

Skill Practice Identify whether the function is constant, linear, quadratic, or none of these.

3. $m(x) = -2x^2 - 3x + 7$ **4.** $n(x) = -6$

5. $W(x) = \dfrac{4}{3}x - \dfrac{1}{2}$ **6.** $R(x) = \dfrac{4}{3x} - \dfrac{1}{2}$

4. Finding the *x*- and *y*-Intercepts of a Function Defined by $y = f(x)$

In Section 3.2, we learned that to find an *x*-intercept, we substitute $y = 0$ and solve the equation for *x*. Using function notation, this is equivalent to finding the real solutions of the equation $f(x) = 0$. To find a *y*-intercept, substitute $x = 0$ and solve the equation for *y*. In function notation, this is equivalent to finding $f(0)$.

> **Finding the *x*- and *y*-Intercepts of a Function**
>
> Given a function defined by $y = f(x)$,
>
> **1.** The *x*-intercepts are the real solutions to the equation $f(x) = 0$.
> **2.** The *y*-intercept is given by $f(0)$.

Example 3 Finding the *x*- and *y*-Intercepts of a Function

Given the function defined by $f(x) = 2x - 4$:

a. Find the *x*-intercept(s).

b. Find the *y*-intercept.

c. Graph the function.

Solution:

a. To find the *x*-intercept(s), find the real solutions to the equation $f(x) = 0$.

$$f(x) = 2x - 4$$
$$0 = 2x - 4 \qquad \text{Substitute } f(x) = 0.$$
$$4 = 2x$$
$$2 = x \qquad \text{The } x\text{-intercept is } (2, 0).$$

b. To find the *y*-intercept, evaluate $f(0)$.

$$f(0) = 2(0) - 4 \qquad \text{Substitute } x = 0.$$
$$f(0) = -4 \qquad \text{The } y\text{-intercept is } (0, -4).$$

c. This function is linear, with a *y*-intercept of $(0, -4)$, an *x*-intercept of $(2, 0)$, and a slope of 2 (Figure 8-14).

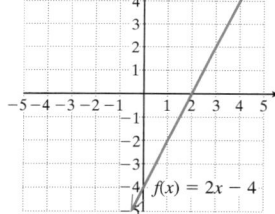

Figure 8-14

Skill Practice

7. Consider $f(x) = -5x + 1$.

a. Find the *x*-intercept.

b. Find the *y*-intercept.

c. Graph the function.

Skill Practice Answers

3. Quadratic   **4.** Constant
5. Linear   **6.** None of these
7a. $\left(\tfrac{1}{5}, 0\right)$ **b.** $(0, 1)$
c.

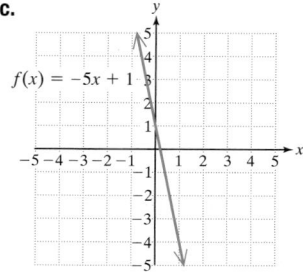

For the function pictured in Figure 8-15, estimate

a. The real values of *x* for which $f(x) = 0$.

b. The value of $f(0)$.

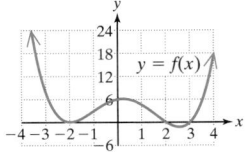

Figure 8-15

Solution:

a. The real values of *x* for which $f(x) = 0$ are the *x*-intercepts of the function. For this graph, the *x*-intercepts are located at $x = -2$, $x = 2$, and $x = 3$.

b. The value of $f(0)$ is the value of *y* at $x = 0$. That is, $f(0)$ is the *y*-intercept, $f(0) = 6$.

| Skill Practice |

8. Use the function pictured below.

   a. Estimate the real value(s) of *x* for which $f(x) = 0$.

   b. Estimate the value of $f(0)$.

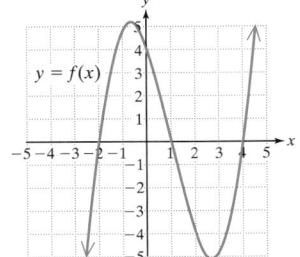

5. Determining Intervals of Increasing, Decreasing, or Constant Behavior

The function shown in Figure 8-16 represents monthly household cost for electricity based on average temperature for that month. At lower temperatures, people probably run their heat, and at higher temperatures, people probably run their air-conditioners. This leads to greater energy cost. However, when the average daily temperature is pleasant, such as between 65°F and 70°F, the heater and air-conditioner are turned off and the monthly electric bill is less.

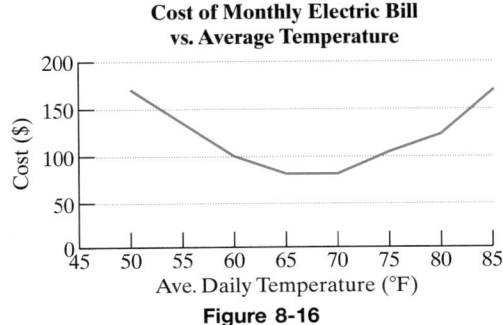

Figure 8-16

- Notice that for temperatures between 50°F and 65°F, the monthly cost decreases. We say that the function is *decreasing* on the interval $(50, 65)$.

- For daily temperatures between 70°F and 85°F, the monthly cost increases. We say that the function is *increasing* on the interval $(70, 85)$.

- For temperatures between 65°F and 70°F, the cost remained the same (or constant). We say that the function is *constant* on the interval $(65, 70)$.

Skill Practice Answers

8a. $x = -2$, $x = 1$, and $x = 4$
  b. $f(0) = 4$

In many applications, it is important to note the open intervals where a function is increasing, decreasing, or constant. An open interval, denoted by (a, b), consists of numbers strictly greater than a and strictly less than b.

> ### Intervals Over Which a Function is Increasing, Decreasing, or Constant
>
> Let I be an open interval in the domain of a function, f. Then,
>
> **1.** f is *increasing* on I if $f(a) < f(b)$ for all $a < b$ on I.
>
> **2.** f is *decreasing* on I if $f(a) > f(b)$ for all $a < b$ on I.
>
> **3.** f is *constant* on I if $f(a) = f(b)$ for all a and b on I.

TIP:
- A function is *increasing* on an interval if it goes "uphill" from left to right.
- A function is *decreasing* on an interval if it goes "downhill" from left to right.
- A function is *constant* on an interval if it is "level" or "flat."

Example 5 | ### Determining Where a Function is Increasing, Decreasing, or Constant

For the function pictured, determine the open interval(s) for which the function is

a. increasing

b. decreasing

c. constant

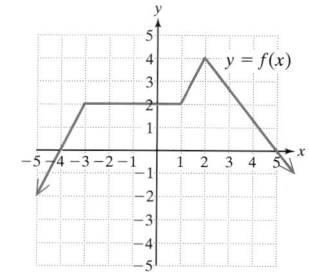

Solution:

a. As we trace the function from left to right, the y-values increase on the intervals $(-\infty, -3)$ and $(1, 2)$.

b. As we trace the function from left to right, the y-values decrease on the interval $(2, \infty)$.

c. A function is constant on an interval if the y-values remain unchanged (where the graph appears level). This function is constant over the interval $(-3, 1)$.

TIP: The intervals over which a function given by $y = f(x)$ is increasing, decreasing, or constant are always expressed in terms of x.

Skill Practice | Refer to the function pictured. Find the intervals for which the function is

9a. increasing

b. decreasing

c. constant

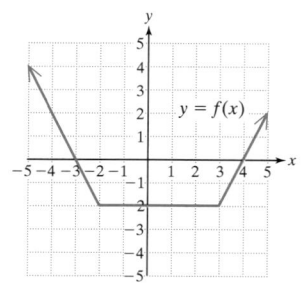

Skill Practice Answers

9a. $(3, \infty)$ **b.** $(-\infty, -2)$
 c. $(-2, 3)$

Section 8.3   Practice Exercises

Study Skills Exercise

1. Define the key terms.

   a. linear function   **b. constant function**   **c. quadratic function**   **d. parabola**

Review Exercises

2. Given: $g = \{(6, 1), (5, 2), (4, 3), (3, 4)\}$

   a. Is this relation a function?

   b. List the elements in the domain.

   c. List the elements in the range.

3. Given: $f = \{(7, 3), (2, 3), (-5, 3)\}$

   a. Is this relation a function?

   b. List the elements in the domain.

   c. List the elements in the range.

4. Given: $f(x) = \sqrt{x + 4}$

   a. Evaluate $f(0), f(-3), f(-4)$, and $f(-5)$, if possible.

   b. Write the domain of this function in interval notation.

5. Given: $g(x) = \dfrac{2}{x - 3}$

   a. Evaluate $g(2), g(4), g(5)$, and $g(3)$, if possible.

   b. Write the domain of this function in interval notation.

6. The force (measured in pounds) to stretch a certain spring x inches is given by $f(x) = 3x$. Evaluate $f(3)$ and $f(10)$, and interpret the results in the context of this problem.

7. The velocity in feet per second of a falling object is given by $V(t) = -32t$, where t is the time in seconds after the object was released. Evaluate $V(2)$ and $V(5)$, and interpret the results in the context of this problem.

Concept 1: Linear and Constant Functions

8. Fill in the blank with the word *vertical* or *horizontal*. The graph of a constant function is a _____ line.

9. For the linear function $f(x) = mx + b$, identify the slope and y-intercept.

10. Graph the constant function $f(x) = 2$. Then use the graph to identify the domain and range of f.

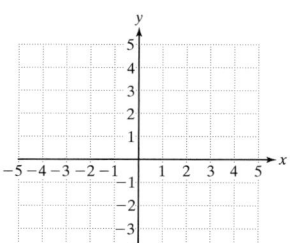

11. Graph the linear function $g(x) = -2x + 1$. Then use the graph to identify the domain and range of g.

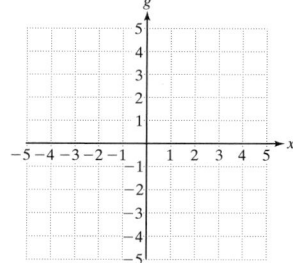

Concept 2: Graphs of Basic Functions

For Exercises 12–17, sketch a graph by completing the table and plotting the points.

12. $f(x) = \dfrac{1}{x}$

x	f(x)	x	f(x)
-2		$\frac{1}{4}$	
-1		$\frac{1}{2}$	
$-\frac{1}{2}$		1	
$-\frac{1}{4}$		2	

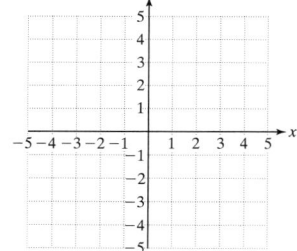

13. $g(x) = |x|$

x	g(x)
-2	
-1	
0	
1	
2	

14. $h(x) = x^3$

x	h(x)
-2	
-1	
0	
1	
2	

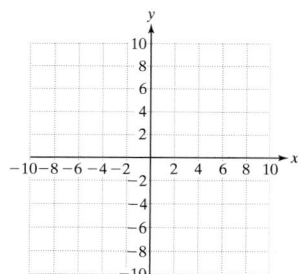

15. $k(x) = x$

x	k(x)
-2	
-1	
0	
1	
2	

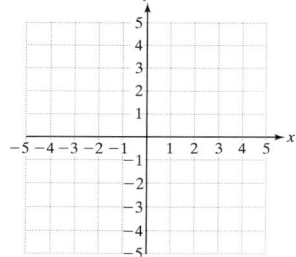

16. $q(x) = x^2$

x	q(x)
-2	
-1	
0	
1	
2	

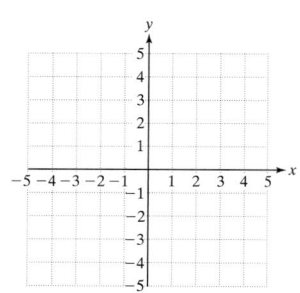

17. $p(x) = \sqrt{x}$

x	p(x)
0	
1	
4	
9	
16	

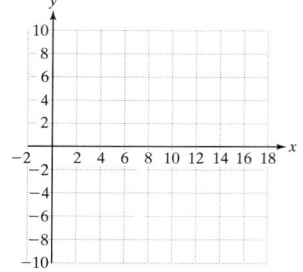

Concept 3: Definition of a Quadratic Function

For Exercises 18–29, determine if the function is constant, linear, quadratic, or none of these.

18. $f(x) = 2x^2 + 3x + 1$ **19.** $g(x) = -x^2 + 4x + 12$ **20.** $k(x) = -3x - 7$ **21.** $h(x) = -x - 3$

22. $m(x) = \dfrac{4}{3}$ **23.** $n(x) = 0.8$ **24.** $p(x) = \dfrac{2}{3x} + \dfrac{1}{4}$ **25.** $Q(x) = \dfrac{1}{5x} - 3$

26. $t(x) = \dfrac{2}{3}x + \dfrac{1}{4}$ **27.** $r(x) = \dfrac{1}{5}x - 3$ **28.** $w(x) = \sqrt{4 - x}$ **29.** $T(x) = -|x + 10|$

Concept 4: Finding the *x*- and *y*-Intercepts of a Function Defined by *y = f(x)*

For Exercises 30–37, find the *x*- and *y*-intercepts, and graph the function.

30. $f(x) = 5x - 10$ **31.** $f(x) = -3x + 12$ **32.** $g(x) = -6x + 5$

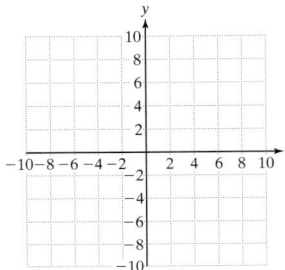   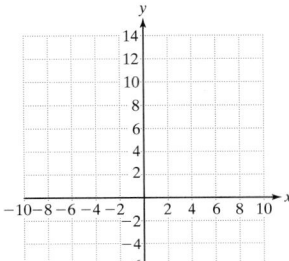   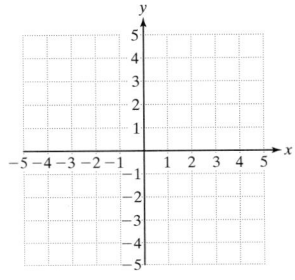

33. $h(x) = 2x + 9$ **34.** $f(x) = 18$ **35.** $g(x) = -7$

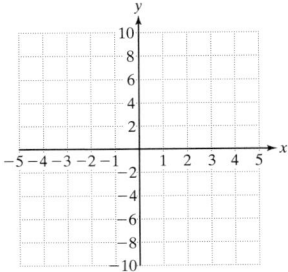   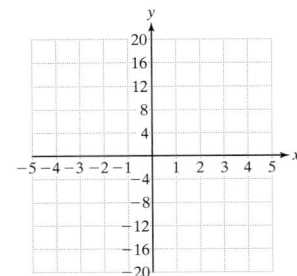   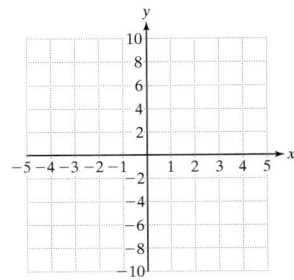

36. $g(x) = \dfrac{2}{3}x + \dfrac{1}{4}$ **37.** $h(x) = -\dfrac{5}{6}x + \dfrac{1}{2}$

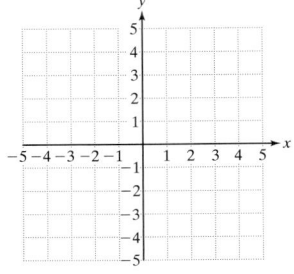   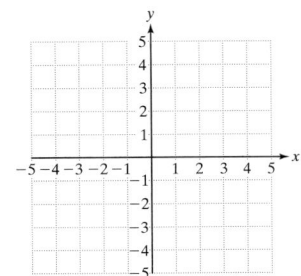

For Exercises 38–43, use the function pictured to estimate

    a. The real values of x for which $f(x) = 0$.   **b.** The value of $f(0)$.

38.

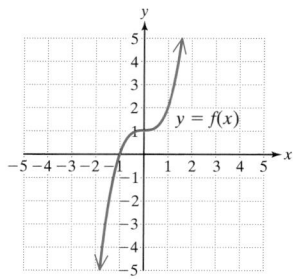

39.

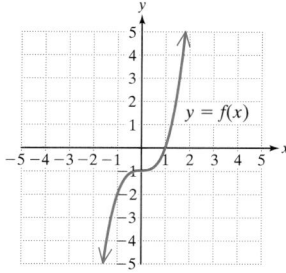

40.

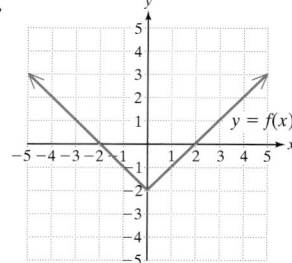

41.

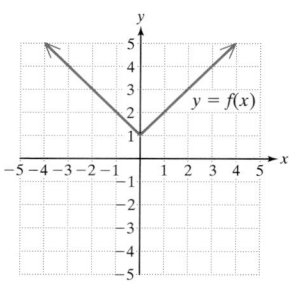

42.

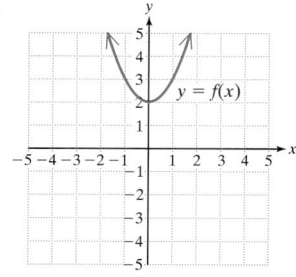

43.

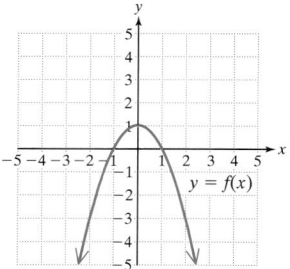

For Exercises 44–53,

    a. Identify the domain of the function.

    b. Identify the y-intercept of the function.

    c. Match the function with its graph by recognizing the basic shape of the function and using the results from parts (a) and (b). Plot additional points if necessary.

44. $q(x) = 2x^2$

45. $p(x) = -2x^2 + 1$

46. $h(x) = x^3 + 1$

47. $k(x) = x^3 - 2$

48. $r(x) = \sqrt{x + 1}$

49. $s(x) = \sqrt{x + 4}$

50. $f(x) = \dfrac{1}{x - 3}$

51. $g(x) = \dfrac{1}{x + 1}$

52. $k(x) = |x + 2|$

53. $h(x) = |x - 1| + 2$

i.

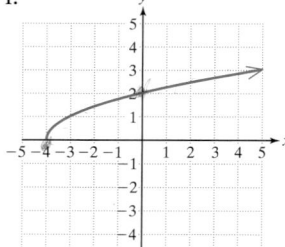

ii.

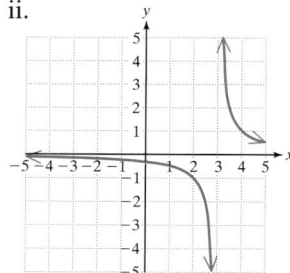

iii.

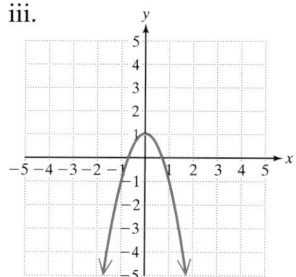

iv.

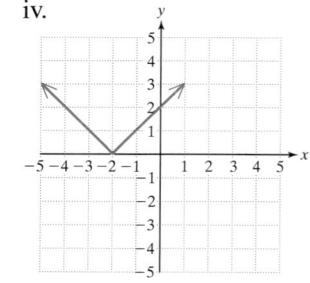

v.
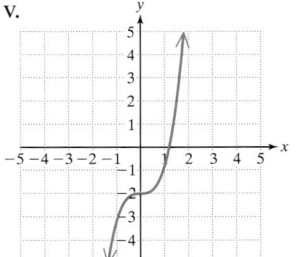

vi.
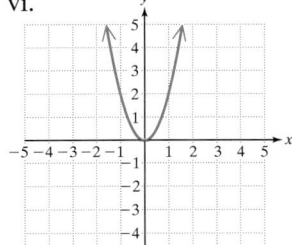

vii.
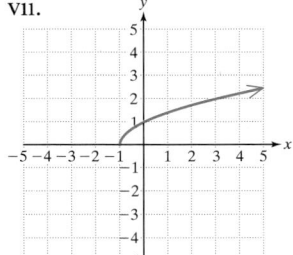

viii.
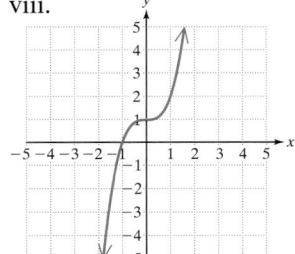

ix.
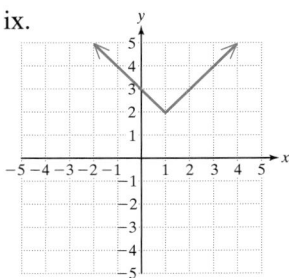

x.
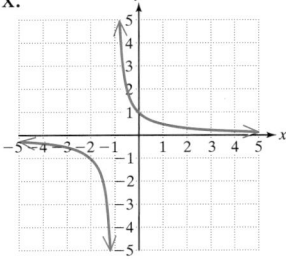

Concept 5: Determining Intervals of Increasing, Decreasing, or Constant Behavior

For Exercises 54–57, give the open interval(s) over which the function is

    a. increasing   **b.** decreasing   **c.** constant

54.

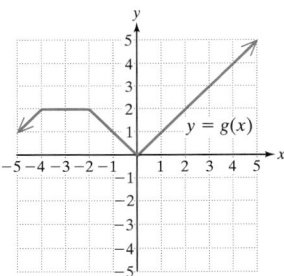

55.

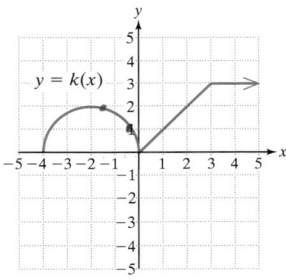

56.

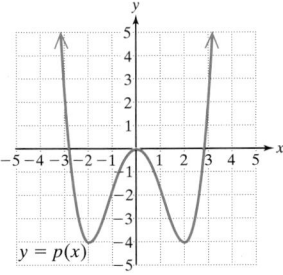

57.

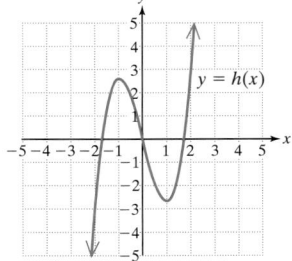

For Exercises 58–63, refer to the graphs of the six basic functions (see page 596). For each function, give the open interval(s) over which the function is

    a. increasing   **b.** decreasing   **c.** constant

58. $f(x) = x$ **59.** $g(x) = x^2$ **60.** $h(x) = x^3$

61. $m(x) = |x|$ **62.** $n(x) = \sqrt{x}$ **63.** $p(x) = \dfrac{1}{x}$

64. Refer back to the graph in Exercise 54. The function is increasing on the interval $(0, \infty)$. Suppose we arbitrarily select two values of x on this interval such as $x = 1$ and $x = 3$. Is it true that $g(1) < g(3)$? How does this relate to the definition of a function increasing on an interval?

65. Refer back to the graph in Exercise 55. The function is decreasing on the interval $(-2, 0)$. Suppose we arbitrarily select two values of x on this interval, such as $x = -1.5$ and $x = -0.5$. Is it true that $k(-1.5) > k(-0.5)$? How does this relate to the definition of a function decreasing on an interval?

Graphing Calculator Exercises

For Exercises 66–71, use a graphing calculator to graph the basic functions. Verify your answers from the table on page 596.

66. $f(x) = x$

67. $f(x) = x^2$

68. $f(x) = x^3$

69. $f(x) = |x|$

70. $f(x) = \sqrt{x}$

71. $f(x) = \dfrac{1}{x}$

Section 8.4   Variation

Concepts

1. Definition of Direct and Inverse Variation
2. Translations Involving Variation
3. Applications of Variation

1. Definition of Direct and Inverse Variation

In this section, we introduce the concept of variation. Direct and inverse variation models can show how one quantity varies in proportion to another.

> ### Definition of Direct and Inverse Variation
>
> Let k be a nonzero constant real number. Then the following statements are equivalent:
>
> **1.** y varies **directly** as x.
> y is directly proportional to x. $\left.\right\}$ $y = kx$
>
> **2.** y varies **inversely** as x.
> y is inversely proportional to x. $\left.\right\}$ $y = \dfrac{k}{x}$
>
> *Note:* The value of k is called the constant of variation.

For a car traveling 30 mph, the equation $d = 30t$ indicates that the distance traveled is *directly proportional* to the time of travel. For positive values of k, when two variables are directly related, as one variable increases, the other variable will also increase. Likewise, if one variable decreases, the other will decrease. In the equation $d = 30t$, the longer the time of the trip, the greater the distance traveled. The shorter the time of the trip, the shorter the distance traveled.

For positive values of k, when two variables are *inversely related*, as one variable increases, the other will decrease, and vice versa. Consider a car traveling between Toronto and Montreal, a distance of 500 km. The time required to make the trip is inversely proportional to the speed of travel: $t = 500/r$. As the rate of speed r increases, the quotient $500/r$ will decrease. Hence, the time will decrease. Similarly, as the rate of speed decreases, the trip will take longer.

2. Translations Involving Variation

The first step in using a variation model is to translate an English phrase into an equivalent mathematical equation.

Example 1   Translating to a Variation Model

Translate each expression into an equivalent mathematical model.

a. The circumference of a circle varies directly as the radius.

b. At a constant temperature, the volume of a gas varies inversely as the pressure.

c. The length of time of a meeting is directly proportional to the *square* of the number of people present.

Solution:

a. Let C represent circumference and r represent radius. The variables are directly related, so use the model $C = kr$.

b. Let V represent volume and P represent pressure. Because the variables are inversely related, use the model $V = \dfrac{k}{P}$.

c. Let t represent time and let N be the number of people present at a meeting. Because t is directly related to N^2, use the model $t = kN^2$.

Skill Practice   Translate to a variation model.

1. The time t it takes to drive a particular distance is inversely proportional to the speed s.

2. The amount of your paycheck P varies directly with the number of hours h that you work.

3. q varies inversely as the square of t.

Sometimes a variable varies directly as the product of two or more other variables. In this case, we have joint variation.

Definition of Joint Variation

Let k be a nonzero constant real number. Then the following statements are equivalent:

$$\left.\begin{array}{l} y \text{ varies } \textbf{jointly} \text{ as } w \text{ and } z. \\ y \text{ is jointly proportional to } w \text{ and } z. \end{array}\right\} \quad y = kwz$$

Skill Practice Answers

1. $t = \dfrac{k}{s}$ **2.** $P = kh$

3. $q = \dfrac{k}{t^2}$

Example 2 **Translating to a Variation Model**

Translate each expression into an equivalent mathematical model.

   a. y varies jointly as u and the square root of v.

   b. The gravitational force of attraction between two planets varies jointly as the product of their masses and inversely as the square of the distance between them.

Solution:

   a. $y = ku\sqrt{v}$

   b. Let m_1 and m_2 represent the masses of the two planets. Let F represent the gravitational force of attraction and d represent the distance between the planets. The variation model is $F = \dfrac{km_1m_2}{d^2}$.

Skill Practice Translate to a variation model.

   4. a varies jointly as b and c.
   5. x varies directly as the square root of y and inversely as z.

3. Applications of Variation

Consider the variation models $y = kx$ and $y = k/x$. In either case, if values for x and y are known, we can solve for k. Once k is known, we can use the variation equation to find y if x is known, or to find x if y is known. This concept is the basis for solving many problems involving variation.

> **Steps to Find a Variation Model**
>
>   **1.** Write a general variation model that relates the variables given in the problem. Let k represent the constant of variation.
>   **2.** Solve for k by substituting known values of the variables into the model from step 1.
>   **3.** Substitute the value of k into the original variation model from step 1.

Example 3 **Solving an Application Involving Direct Variation**

The variable z varies directly as w. When w is 16, z is 56.

   a. Write a variation model for this situation. Use k as the constant of variation.
   b. Solve for the constant of variation.
   c. Find the value of z when w is 84.

Solution:

   a. $z = kw$

Skill Practice Answers

4. $a = kbc$ **5.** $x = \dfrac{k\sqrt{y}}{z}$

b. $z = kw$

$56 = k(16)$ — Substitute known values for z and w. Then solve for the unknown value of k.

$\dfrac{56}{16} = \dfrac{k(16)}{16}$ — To isolate k, divide both sides by 16.

$\dfrac{7}{2} = k$ — Simplify $\dfrac{56}{16}$ to $\dfrac{7}{2}$.

c. With the value of k known, the variation model can now be written as $z = \frac{7}{2}w$.

$z = \dfrac{7}{2}(84)$ — To find z when $w = 84$, substitute $w = 84$ into the equation.

$z = 294$

Skill Practice

6. The variable q varies directly as the square of v. When v is 2, q is 40.
   a. Write a variation model for this relationship.
   b. Solve for the constant of variation.
   c. Find q when $v = 7$.

Example 4 | **Solving an Application Involving Direct Variation**

The speed of a racing canoe in still water varies directly as the square root of the length of the canoe.

a. If a 16-ft canoe can travel 6.2 mph in still water, find a variation model that relates the speed of a canoe to its length.

b. Find the speed of a 25-ft canoe.

Solution:

a. Let s represent the speed of the canoe and L represent the length. The general variation model is $s = k\sqrt{L}$. To solve for k, substitute the known values for s and L.

$s = k\sqrt{L}$

$6.2 = k\sqrt{16}$ — Substitute $s = 6.2$ mph and $L = 16$ ft.

$6.2 = k \cdot 4$

$\dfrac{6.2}{4} = \dfrac{4k}{4}$ — Solve for k.

$k = 1.55$

$s = 1.55\sqrt{L}$ — Substitute $k = 1.55$ into the model $s = k\sqrt{L}$.

b. $s = 1.55\sqrt{L}$

$= 1.55\sqrt{25}$ — Find the speed when $L = 25$ ft.

$= 7.75$ mph

Skill Practice Answers

6a. $q = kv^2$ **b.** $k = 10$
c. $q = 490$

Skill Practice

7. The amount of water needed by a mountain hiker varies directly as the time spent hiking. The hiker needs 2.4 L for a 3-hr hike.

   a. Write a model that relates the amount of water needed to the time of the hike.

   b. How much water will be needed for a 5-hr hike?

Example 5 Solving an Application Involving Inverse Variation

The loudness of sound measured in decibels (dB) varies inversely as the square of the distance between the listener and the source of the sound. If the loudness of sound is 17.92 dB at a distance of 10 ft from a stereo speaker, what is the decibel level 20 ft from the speaker?

Solution:

Let L represent the loudness of sound in decibels and d represent the distance in feet. The inverse relationship between decibel level and the square of the distance is modeled by

$$L = \frac{k}{d^2}$$

$$17.92 = \frac{k}{(10)^2} \qquad \text{Substitute } L = 17.92 \text{ dB and } d = 10 \text{ ft.}$$

$$17.92 = \frac{k}{100}$$

$$(17.92)100 = \frac{k}{100} \cdot 100 \qquad \text{Solve for } k \text{ (clear fractions).}$$

$$k = 1792$$

$$L = \frac{1792}{d^2} \qquad \text{Substitute } k = 1792 \text{ into the original model}$$
$$\qquad\qquad\qquad L = \frac{k}{d^2}.$$

With the value of k known, we can find L for any value of d.

$$L = \frac{1792}{(20)^2} \qquad \text{Find the loudness when } d = 20 \text{ ft.}$$

$$= 4.48 \text{ dB}$$

Notice that the loudness of sound is 17.92 dB at a distance 10 ft from the speaker. When the distance from the speaker is increased to 20 ft, the decibel level decreases to 4.48 dB. This is consistent with an inverse relationship. For $k > 0$, as one variable is increased, the other is decreased. It also seems reasonable that the farther one moves away from the source of a sound, the softer the sound becomes.

Skill Practice

8. The yield on a bond varies inversely as the price. The yield on a particular bond is 4% when the price is $100. Find the yield when the price is $80.

Skill Practice Answers

7a. $w = 0.8t$ **b.** 4 L   **8.** 5%

| Example 6 | Solving an Application Involving Joint Variation |

In the early morning hours of August 29, 2005, Hurricane Katrina plowed into the Gulf Coast of the United States, bringing unprecedented destruction to southern Louisiana, Mississippi, and Alabama.

The kinetic energy of an object varies jointly as the weight of the object at sea level and as the square of its velocity. During a hurricane, a $\frac{1}{2}$-lb stone traveling at 60 mph has 81 joules (J) of kinetic energy. Suppose the wind speed doubles to 120 mph. Find the kinetic energy.

Solution:

Let E represent the kinetic energy, let w represent the weight, and let v represent the velocity of the stone. The variation model is

$E = kwv^2$

$81 = k(0.5)(60)^2$ Substitute $E = 81$ J, $w = 0.5$ lb, and $v = 60$ mph.

$81 = k(0.5)(3600)$ Simplify exponents.

$81 = k(1800)$

$\dfrac{81}{1800} = \dfrac{k(\cancel{1800})}{\cancel{1800}}$ Divide by 1800.

$0.045 = k$ Solve for k.

With the value of k known, the model $E = kwv^2$ can be written as $E = 0.045wv^2$. We now find the kinetic energy of a $\frac{1}{2}$-lb stone traveling at 120 mph.

$$E = 0.045(0.5)(120)^2$$

$$= 324$$

The kinetic energy of a $\frac{1}{2}$-lb stone traveling at 120 mph is 324 J.

Skill Practice

9. The amount of simple interest earned in an account varies jointly as the interest rate and time of the investment. An account earns $40 in 2 years at 4% interest. How much interest would be earned in 3 years at a rate of 5%?

In Example 6, when the velocity increased by 2 times, the kinetic energy increased by 4 times (note that 324 J = 4·81 J). This factor of 4 occurs because the kinetic energy is proportional to the *square* of the velocity. When the velocity increased by 2 times, the kinetic energy increased by 2^2 times.

Skill Practice Answers
9. $75

Section 8.4   Practice Exercises

Study Skills Exercise

1. Define the key terms.

    a. direct variation   **b.** inverse variation   **c.** joint variation

Review Exercises

For Exercises 2–8, refer to the graph.

2. Find $f(-3)$.   **3.** Find $f(-4)$.   **4.** Find $f(0)$.

5. Find the value(s) of x for which $f(x) = 1$.

6. Find the value(s) of x for which $f(x) = 2$.

7. Write the domain of f.   **8.** Write the range of f.

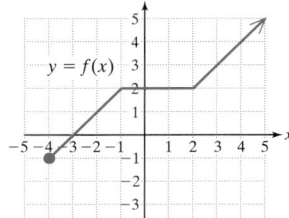

Concept 1: Definition of Direct and Inverse Variation

9. Suppose y varies directly as x, and $k > 0$.

    a. If x increases, then will y increase
    or decrease?

    b. If x decreases, will y increase
    or decrease?

10. Suppose y varies inversely as x, and $k > 0$.

    a. If x increases, then will y increase
    or decrease?

    b. If x decreases, then will y increase
    or decrease?

Concept 2: Translations Involving Variation

For Exercises 11–18, write a variation model. Use k as the constant of variation.

11. T varies directly as q.

12. P varies inversely as r.

13. W varies inversely as the square of p.

14. Y varies directly as the square root of z.

15. Q is directly proportional to x and inversely proportional to the cube of y.

16. M is directly proportional to the square of p and inversely proportional to the cube of n.

17. L varies jointly as w and the square root of v.

18. X varies jointly as w and the square of y.

Concept 3: Applications of Variation

For Exercises 19–24, find the constant of variation k.

19. y varies directly as x, and when x is 4, y is 18.

20. m varies directly as x and when x is 8, m is 22.

21. p is inversely proportional to q and when q is 16, p is 32.

22. T is inversely proportional to x and when x is 40, T is 200.

23. y varies jointly as w and v. When w is 50 and v is 0.1, y is 8.75.

24. N varies jointly as t and p. When t is 1 and p is 7.5, N is 330.

Solve Exercises 25–30 by using the steps found on page 608.

25. Z varies directly as the square of w, and $Z = 14$ when $w = 4$. Find Z when $w = 8$.

26. Q varies inversely as the square of p, and $Q = 4$ when $p = 3$. Find Q when $p = 2$.

27. L varies jointly as a and the square root of b, and $L = 72$ when $a = 8$ and $b = 9$. Find L when $a = \frac{1}{2}$ and $b = 36$.

28. Y varies jointly as the cube of x and the square root of w, and $Y = 128$ when $x = 2$ and $w = 16$. Find Y when $x = \frac{1}{2}$ and $w = 64$.

29. B varies directly as m and inversely as n, and $B = 20$ when $m = 10$ and $n = 3$. Find B when $m = 15$ and $n = 12$.

30. R varies directly as s and inversely as t, and $R = 14$ when $s = 2$ and $t = 9$. Find R when $s = 4$ and $t = 3$.

For Exercises 31–42, use a variation model to solve for the unknown value.

31. The amount of pollution entering the atmosphere varies directly as the number of people living in an area. If 80,000 people cause 56,800 tons of pollutants, how many tons enter the atmosphere in a city with a population of 500,000?

32. The area of a picture projected on a wall varies directly as the square of the distance from the projector to the wall. If a 10-ft distance produces a 16-ft² picture, what is the area of a picture produced when the projection unit is moved to a distance 20 ft from the wall?

33. The stopping distance of a car is directly proportional to the square of the speed of the car. If a car traveling at 40 mph has a stopping distance of 109 ft, find the stopping distance of a car that is traveling at 25 mph. (Round your answer to 1 decimal place.)

34. The intensity of a light source varies inversely as the square of the distance from the source. If the intensity is 48 lumens (lm) at a distance of 5 ft, what is the intensity when the distance is 8 ft?

35. The current in a wire varies directly as the voltage and inversely as the resistance. If the current is 9 amperes (A) when the voltage is 90 volts (V) and the resistance is 10 ohms (Ω), find the current when the voltage is 185 V and the resistance is 10 Ω.

36. The power in an electric circuit varies jointly as the current and the square of the resistance. If the power is 144 watts (W) when the current is 4 A and the resistance is 6 Ω, find the power when the current is 3 A and the resistance is 10 Ω.

37. The resistance of a wire varies directly as its length and inversely as the square of its diameter. A 40-ft wire with 0.1-in. diameter has a resistance of 4 Ω. What is the resistance of a 50-ft wire with a diameter of 0.20 in.?

38. The frequency of a vibrating string is inversely proportional to its length. A 24-in. piano string vibrates at 252 cycles/sec. What is the frequency of an 18-in. piano string?

39. The weight of a medicine ball varies directly as the cube of its radius. A ball with a radius of 3 in. weighs 4.32 lb. How much would a medicine ball weigh if its radius were 5 in.?

40. The surface area of a cube varies directly as the square of the length of an edge. The surface area is 24 ft^2 when the length of an edge is 2 ft. Find the surface area of a cube with an edge that is 5 ft.

41. The strength of a wooden beam varies jointly as the width of the beam and the square of the thickness of the beam and inversely as the length of the beam. A beam that is 48 in. long, 6 in. wide, and 2 in. thick can support a load of 417 lb. Find the maximum load that can be safely supported by a board that is 12 in. wide, 72 in. long, and 4 in. thick.

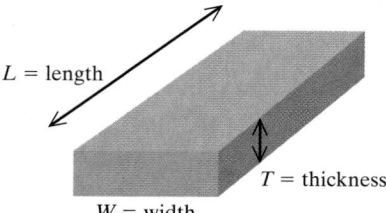

$L = $ length

$T = $ thickness

$W = $ width

42. The period of a pendulum is the length of time required to complete one swing back and forth. The period varies directly as the square root of the length of the pendulum. If it takes 1.8 sec for a 0.81-m pendulum to complete one period, what is the period of a 1-m pendulum?

Expanding Your Skills

43. The area A of a square varies directly as the square of the length l of its sides.

  a. Write a general variation model with k as the constant of variation.

  b. If the length of the sides is doubled, what effect will that have on the area?

  c. If the length of the sides is tripled, what effect will that have on the area?

44. In a physics laboratory, a spring is fixed to the ceiling. With no weight attached to the end of the spring, the spring is said to be in its equilibrium position. As weights are applied to the end of the spring, the force stretches the spring a distance d from its equilibrium position. A student in the laboratory collects the following data:

Force F (lb)	2	4	6	8	10
Distance d (cm)	2.5	5.0	7.5	10.0	12.5

  a. Based on the data, do you suspect a direct relationship between force and distance or an inverse relationship?

  b. Find a variation model that describes the relationship between force and distance.

Chapter 8 · SUMMARY

Section 8.1 · Introduction to Relations

Key Concepts

Any set of ordered pairs (x, y) is called a **relation in x and y**.

The **domain** of a relation is the set of first components in the ordered pairs in the relation. The **range** of a relation is the set of second components in the ordered pairs.

Examples

Example 1

Let $A = \{(0, 0), (1, 1), (2, 4), (3, 9), (-1, 1), (-2, 4), (-3, 9)\}$.

Domain of A: $\{0, 1, 2, 3, -1, -2, -3\}$

Range of A: $\{0, 1, 4, 9\}$

Example 2

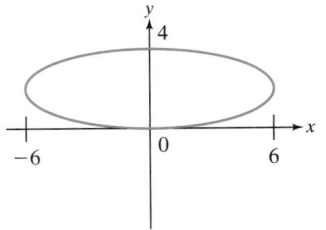

Domain: $[-6, 6]$

Range: $[0, 4]$

Section 8.2 · Introduction to Functions

Key Concepts

Given a relation in x and y, we say "**y is a function of x**" if for every element x in the domain, there corresponds exactly one element y in the range.

The Vertical Line Test for Functions

Consider a relation defined by a set of points (x, y) in a rectangular coordinate system. Then the graph defines y as a function of x if no vertical line intersects the graph in more than one point.

Examples

Example 1

Function $\{(1, 3), (2, 5), (6, 3)\}$

Nonfunction $\{(1, 3), (2, 5), (1, 4)\}$

Example 2

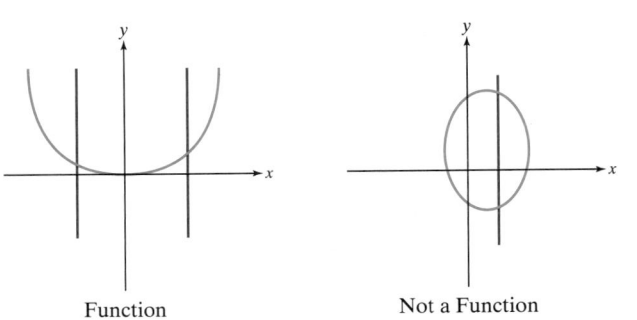

Function   Not a Function

Function Notation

$f(x)$ is the value of the function f at x.

The domain of a function defined by $y = f(x)$ is the set of x-values that when substituted into the function produces a real number. In particular,

- Exclude values of x that make the denominator of a fraction zero.
- Exclude values of x that make a negative value within a square root.

Example 3

Given $f(x) = -3x^2 + 5x$, find $f(-2)$.

$$f(-2) = -3(-2)^2 + 5(-2)$$
$$= -12 - 10$$
$$= -22$$

Example 4

Find the domain.

1. $f(x) = \dfrac{x + 4}{x - 5}; (-\infty, 5) \cup (5, \infty)$

2. $f(x) = \sqrt{x - 3}; [3, \infty)$

3. $f(x) = 3x^2 - 5; (-\infty, \infty)$

Section 8.3 Graphs of Functions

Key Concepts

A function of the form $f(x) = mx + b\ (m \neq 0)$ is a **linear function**. Its graph is a line with slope m and y-intercept $(0, b)$.

A function of the form $f(x) = k$ is a **constant function**. Its graph is a horizontal line.

Examples

Example 1

$f(x) = 2x - 3$

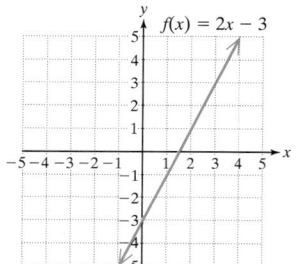

Example 2

$f(x) = 3$

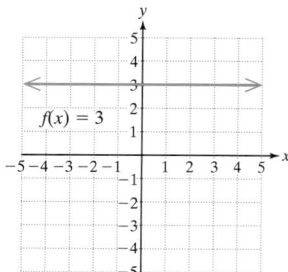

A function of the form $f(x) = ax^2 + bx + c \ (a \neq 0)$ is a **quadratic function**. Its graph is a **parabola**.

Graphs of basic functions:

$f(x) = x$

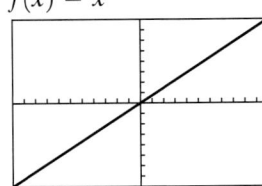

$f(x) = x^2$

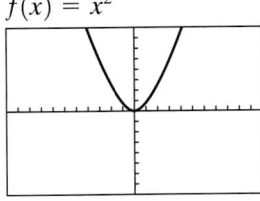

$f(x) = x^3$

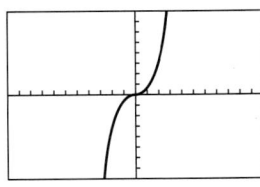

$f(x) = |x|$

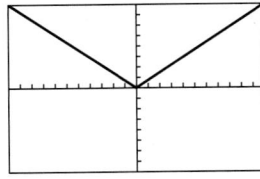

$f(x) = \sqrt{x}$

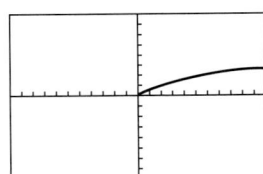

$f(x) = \dfrac{1}{x}$

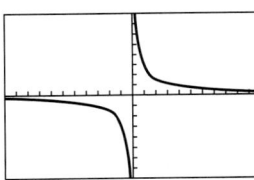

The x-intercepts of a function are determined by finding the real solutions to the equation $f(x) = 0$.

The y-intercept of a function is at $f(0)$.

Let I be an open interval in the domain of a function, f. Then,

1. f is *increasing* on I if $f(a) < f(b)$ for all $a < b$ on I.
2. f is *decreasing* on I if $f(a) > f(b)$ for all $a < b$ on I.
3. f is *constant* on I if $f(a) = f(b)$ for all a and b on I.

Example 3

$f(x) = x^2 - 2x - 1$

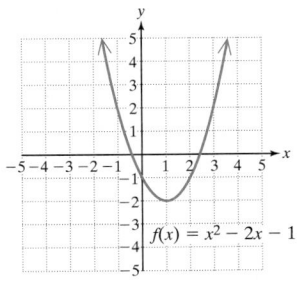

Example 4

Find the x- and y-intercepts for the function pictured.

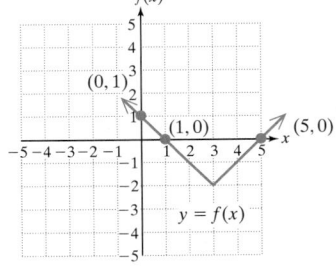

$f(x) = 0$, when $x = 1$ and $x = 5$.

The x-intercepts are $(1, 0)$ and $(5, 0)$.

$f(0) = 1$. The y-intercept is $(0, 1)$.

Example 5

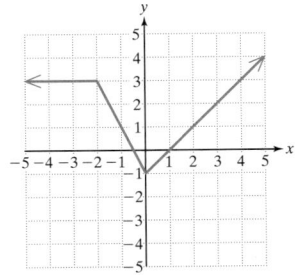

f is increasing on $(0, \infty)$

f is decreasing on $(-2, 0)$

f is constant on $(-\infty, -2)$

Section 8.4   Variation

Key Concepts	Examples

Key Concepts

Direct Variation

y varies directly as x.
y is directly proportional to x.
$$\left.\right\} \quad y = kx$$

Inverse Variation

y varies inversely as x.
y is inversely proportional to x.
$$\left.\right\} \quad y = \frac{k}{x}$$

Joint Variation

y varies jointly as w and z.
y is jointly proportional to w and z.
$$\left.\right\} \quad y = kwz$$

Steps to Find a Variation Model

1. Write a general variation model that relates the variables given in the problem. Let k represent the constant of variation.

2. Solve for k by substituting known values of the variables into the model from step 1.

3. Substitute the value of k into the original variation model from step 1.

Examples

Example 1

t varies directly as the square root of x.

$$t = k\sqrt{x}$$

Example 2

W is inversely proportional to the cube of x.

$$W = \frac{k}{x^3}$$

Example 3

y is jointly proportional to x and to the square of z.

$$y = kxz^2$$

Example 4

C varies directly as the square root of d and inversely as t. If $C = 12$ when d is 9 and t is 6, find C if d is 16 and t is 12.

Step 1: $C = \dfrac{k\sqrt{d}}{t}$

Step 2: $12 = \dfrac{k\sqrt{9}}{6} \Rightarrow 12 = \dfrac{k \cdot 3}{6} \Rightarrow k = 24$

Step 3: $C = \dfrac{24\sqrt{d}}{t} \Rightarrow C = \dfrac{24\sqrt{16}}{12} \Rightarrow C = 8$

Chapter 8   Review Exercises

Section 8.1

1. Write a relation with four ordered pairs for which the first element is the name of a parent and the second element is the name of the parent's child.

For Exercises 2–5, find the domain and range.

2. $\left\{ \left(\dfrac{1}{3}, 10 \right), \left(6, -\dfrac{1}{2} \right), \left(\dfrac{1}{4}, 4 \right), \left(7, \dfrac{2}{5} \right) \right\}$

3.

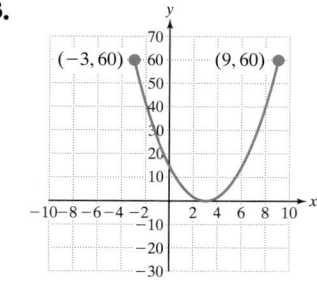

4.

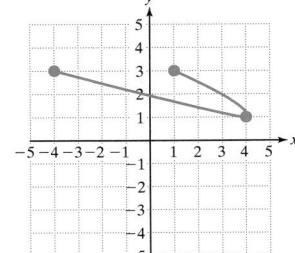

5.

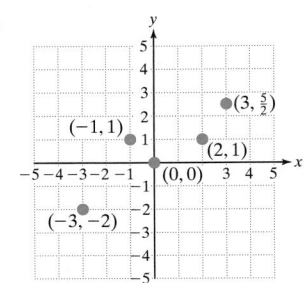

Section 8.2

6. Sketch a relation that is *not* a function. (Answers may vary.)

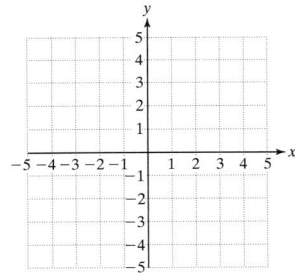

7. Sketch a relation that *is* a function. (Answers may vary.)

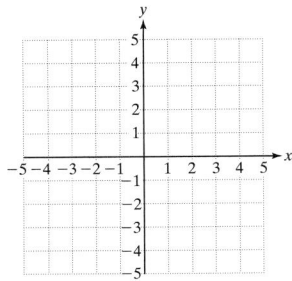

For Exercises 8–13:

a. Determine whether the relation defines y as a function of x.

b. Find the domain.

c. Find the range.

8.

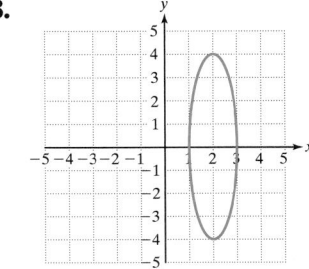

9.

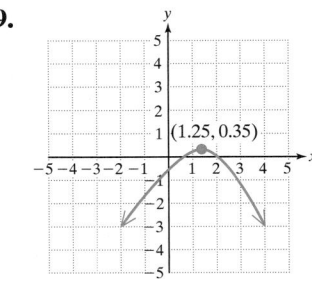

10. $\{(1, 3), (2, 3), (3, 3), (4, 3)\}$

11. $\{(0, 2), (0, 3), (4, 4), (0, 5)\}$

12.

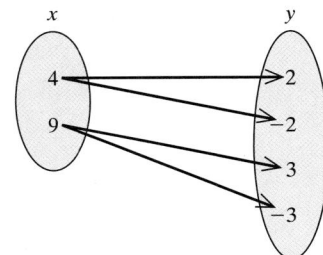

13.

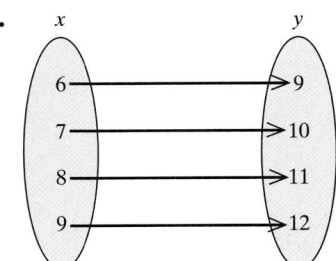

For Exercises 14–21, find the function values given $f(x) = 6x^2 - 4$.

14. $f(0)$ **15.** $f(1)$

16. $f(-1)$ **17.** $f(t)$

18. $f(b)$ **19.** $f(\pi)$

20. $f(\square)$ **21.** $f(-2)$

For Exercises 22–25, write the domain of each function in interval notation.

22. $g(x) = 7x^3 + 1$

23. $h(x) = \dfrac{x + 10}{x - 11}$

24. $k(x) = \sqrt{x - 8}$

25. $w(x) = \sqrt{x + 2}$

26. Anita is a waitress and makes \$6 per hour plus tips. Her tips average \$5 per table. In one 8-hr shift, Anita's pay can be described by $p(x) = 48 + 5x$, where x represents the number of tables she waits on. Find out how much Anita will earn if she waits on

   a. 10 tables   **b.** 15 tables   **c.** 20 tables

Section 8.3

For Exercises 27–32, sketch the functions from memory.

27. $h(x) = x$

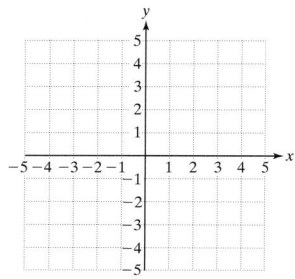

28. $f(x) = x^2$

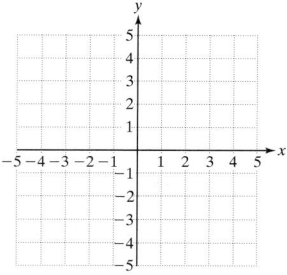

29. $g(x) = x^3$

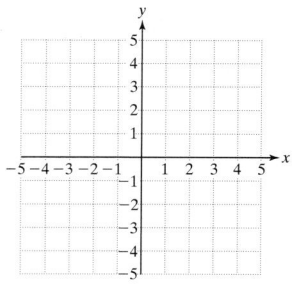

30. $w(x) = |x|$

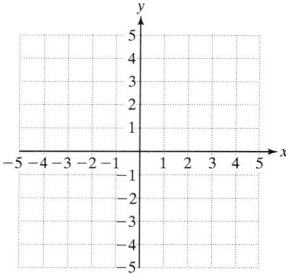

31. $s(x) = \sqrt{x}$

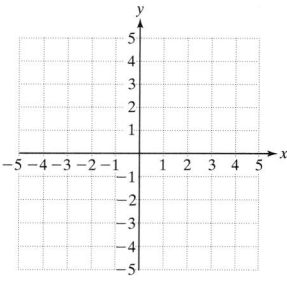

32. $r(x) = \dfrac{1}{x}$

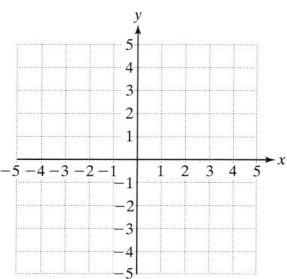

For Exercises 33–34, sketch the function and determine the open intervals for which the function is increasing, decreasing, and constant.

33. $q(x) = 3$

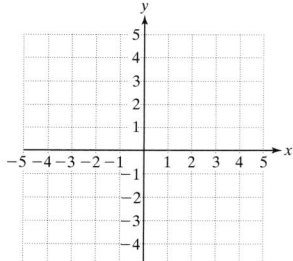

34. $k(x) = 2x + 1$

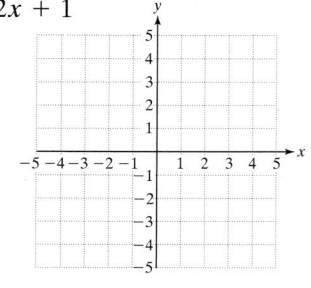

For Exercises 35–36, find the *x*- and *y*-intercepts.

35. $p(x) = 4x - 7$

36. $q(x) = -2x + 9$

37. The function defined by $b(t) = 0.7t + 4.5$ represents the per capita consumption of bottled water in the United States between 1985 and 2005. The values of $b(t)$ are measured in gallons, and $t = 0$ corresponds to the year 1985. (Source: U.S. Department of Agriculture.)

   a. Evaluate $b(0)$ and $b(7)$ and interpret the results in the context of this problem.

   b. What is the slope of this function? Interpret the slope in the context of this problem.

For Exercises 38–45, refer to the graph.

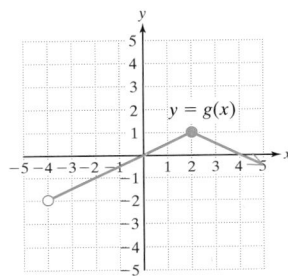

38. Find $g(-2)$.

39. Find $g(2)$.

40. For what value(s) of *x* is $g(x) = 0$?

41. For what value(s) of *x* is $g(x) = -4$?

42. Write the domain of *g*.

43. Write the range of *g*.

44. For what open interval(s) is *g* increasing?

45. For what open interval(s) is *g* decreasing?

46. Given: $r(x) = 2\sqrt{x - 4}$

   a. Find $r(4)$, $r(5)$, and $r(8)$.

   b. What is the domain of *r*?

47. Given: $h(x) = \dfrac{3}{x - 3}$

   a. Find $h(-3)$, $h(-1)$, $h(0)$, $h(2)$, $h(3)$, and $h(4)$.

   b. What is the domain of *h*?

Section 8.4

48. The force applied to a spring varies directly with the distance that the spring is stretched. When 6 lb of force is applied, the spring stretches 2 ft.

   a. Write a variation model using *k* as the constant of variation.

   b. Find *k*.

   c. How many feet will the spring stretch when 5 lb of pressure is applied?

49. Suppose *y* varies directly with the cube of *x* and $y = 32$ when $x = 2$. Find *y* when $x = 4$.

50. Suppose *y* varies jointly with *x* and the square root of *z*, and $y = 3$ when $x = 3$ and $z = 4$. Find *y* when $x = 8$ and $z = 9$.

51. The distance *d* that one can see to the horizon varies directly as the square root of the height above sea level. If a person 25 m above sea level can see 30 km, how far can a person see if she is 64 m above sea level?

Chapter 8   Test

For Exercises 1–3, **a.** determine if the relation defines y as a function of x, **b.** identify the domain, and **c.** identify the range.

1.

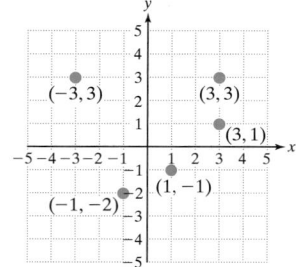

2.

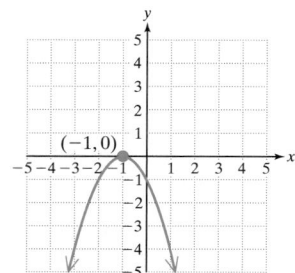

3. Explain how to find the x- and y-intercepts of a function defined by $y = f(x)$.

Graph the functions.

4. $f(x) = -3x - 1$

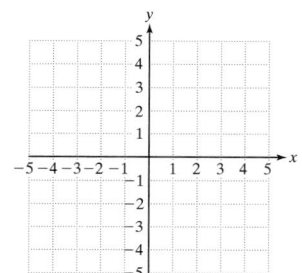

5. $k(x) = -2$

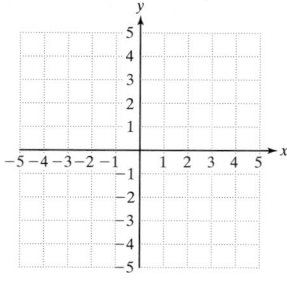

6. $p(x) = x^2$

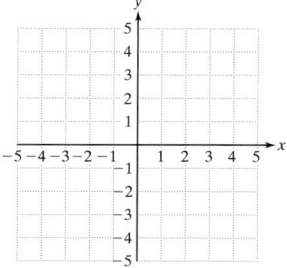

7. $w(x) = |x|$

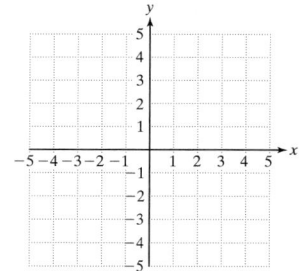

For Exercises 8–10, write the domain in interval notation.

8. $f(x) = \dfrac{x - 5}{x + 7}$

9. $f(x) = \sqrt{x + 7}$

10. $h(x) = (x + 7)(x - 5)$

11. Given: $r(x) = x^2 - 2x + 1$

   a. Find $r(-2)$, $r(-1)$, $r(0)$, $r(2)$, and $r(3)$.

   b. What is the domain of r?

12. The function defined by $s(t) = 1.6t + 36$ approximates the per capita consumption of soft drinks in the United States between 1985 and 2006. The values of $s(t)$ are measured in gallons, and $t = 0$ corresponds to the year 1985. (Source: U.S. Department of Agriculture.)

   a. Evaluate $s(0)$ and $s(7)$ and interpret the results in the context of this problem.

   b. What is the slope of the function? Interpret the slope in the context of this problem.

For Exercises 13–23, refer to the graph.

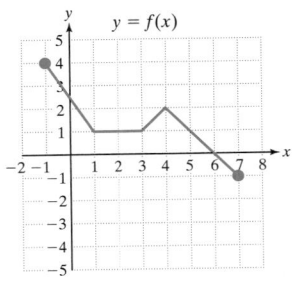

13. Find $f(1)$.

14. Find $f(4)$.

15. Write the domain of f.

16. Write the range of f.

17. True or false? The value $y = 5$ is in the range of f.

18. Find the x-intercept of the function.

19. For what value(s) of x is $f(x) = 0$?

20. For what value(s) of x is $f(x) = 1$?

21. For what open interval(s) is f increasing?

22. For what open intervals is f decreasing?

23. For what open intervals is f constant?

For Exercises 24–27, determine if the function is constant, linear, quadratic, or none of these.

24. $f(x) = -3x^2$

25. $g(x) = -3x$

26. $h(x) = -3$

27. $k(x) = -\dfrac{3}{x}$

28. Find the x- and y-intercepts for $f(x) = \dfrac{3}{4}x + 9$.

29. Write a variation model using k as the constant of variation. The variable x varies directly as y and inversely as the square of t.

30. The period of a pendulum varies directly as the square root of the length of the pendulum. If the period of the pendulum is 2.2 sec when the length is 4 ft, find the period when the length is 9 ft.

Chapters 1–8   Cumulative Review Exercises

1. Solve the equation.

$$\frac{1}{3}t + \frac{1}{5} = \frac{1}{10}(t - 2)$$

2. Simplify. $5 - 3(2 - \sqrt{25}) + 2 - 10 \div 5$

3. Solve the inequality. Write the solution set in interval notation.

$$4 \le -6y + 5$$

4. Determine the volume of the cone pictured here. Round your answer to the nearest whole unit.

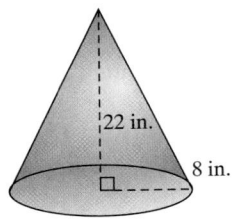

5. Find the pitch (slope) of the roof.

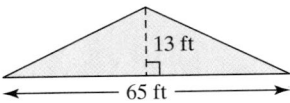

6. a. Explain how to find the x- and y-intercepts of a function $y = f(x)$.

   b. Find the y-intercept of the function defined by $f(x) = 3x + 2$.

   c. Find the x-intercept(s) of the function defined by $f(x) = 3x + 2$.

7. Solve the system.

$$-\frac{1}{4}x + \frac{1}{3}y = -1$$

$$\frac{1}{2}x - \frac{3}{10}y = 2$$

8. One positive number is two-thirds of another positive number. The larger number is 12 more than the smaller. Find the numbers.

9. Simplify: $\dfrac{x^2(y^3)^2 z}{x^0 y^4 z^6}$

10. Multiply: $(p + 6)(p^2 - 3p + 1)$

11. Add: $(p + 6) + (p^2 - 3p + 1)$

12. Factor: $2k^2 + k - 1$

13. Solve: $2k^2 + k - 1 = 0$

14. Write the domain of the function $f(x) = \dfrac{1}{x - 15}$ in interval notation.

15. Solve: $\dfrac{1}{x - 15} + \dfrac{x}{x - 1} = \dfrac{14}{x^2 - 16x + 15}$.

16. Simplify: $\dfrac{\dfrac{x + 1}{x - 1} - 1}{\dfrac{3}{x - 1} + 2}$

17. The linear function defined by $N(x) = 420x + 5260$ provides a model for the number of full-time-equivalent (FTE) students attending a community college from 1988 to 2006. Assume that $x = 0$ corresponds to the year 1988.

   a. Use this model to find the number of FTE students who attended the college in 1996.

   b. If this linear trend continues, predict the year in which the number of FTE students will reach 14,920.

18. State the domain and range of the relation. Is the relation a function?

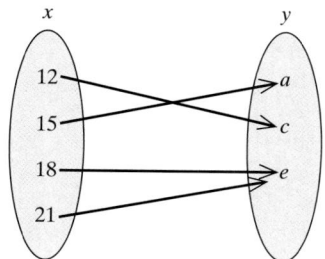

19. Given: $f(x) = \frac{1}{2}x - 1$ and $g(x) = 3x^2 - 2x$.

   a. Find $f(4)$.   b. Find $g(-3)$.

20. Simple interest varies jointly as the interest rate and as the time the money is invested. If an investment yields $1120 interest at 8% for 2 years, how much interest will the investment yield at 10% for 5 years?

Systems of Linear Equations in Three Variables

9.1 Systems of Linear Equations in Three Variables

9.2 Applications of Systems of Linear Equations in Three Variables

9.3 Solving Systems of Linear Equations by Using Matrices

9.4 Determinants and Cramer's Rule

In this chapter we extend our study of systems of linear equations to those involving three variables. We will also learn two new techniques for solving systems of linear equations using matrices and determinants.

Match each question on the left with an answer on the right. Then fill in the boxes below with the word or number to the right of the answer.

Given: $2x - 3y = 1$
$5x + 7y = -2$

1. Identify the coefficient matrix.
2. Identify the augmented matrix.
3. Identify the solution.

Given: $A = \begin{bmatrix} -3 & 1 \\ 8 & -10 \end{bmatrix}$

4. Compute det A.
5. Identify the order of A.
6. Identify the element in the second row, first column.

General:

7. Give an example of a row matrix.
8. Give an example of a column matrix.
9. Give an example of a square matrix.

$\left(\frac{1}{29}, -\frac{9}{29}\right)$	got
22	Math
$\begin{bmatrix} 2 & -3 \\ 5 & 7 \end{bmatrix}$	2
$[\pi \quad 1.5 \quad -\frac{3}{2}]$	10
$\begin{bmatrix} 2 & -3 & 1 \\ 5 & 7 & -2 \end{bmatrix}$	4
2 by 2	good
$\begin{bmatrix} -6.1 \\ 3.7 \end{bmatrix}$	is
8	B

4	8	1	5	9	6	2	3	7

Section 9.1

Systems of Linear Equations in Three Variables

1. Solutions to Systems of Linear Equations in Three Variables

In Sections 4.1–4.3, we solved systems of linear equations in two variables. In this section, we will expand the discussion to solving systems involving three variables.

A **linear equation in three variables** can be written in the form $Ax + By + Cz = D$, where A, B, and C are not all zero. For example, the equation $2x + 3y + z = 6$ is a linear equation in three variables. Solutions to this equation are **ordered triples** of the form (x, y, z) that satisfy the equation. Some solutions to the equation $2x + 3y + z = 6$ are

Solution:	Check:
$(1, 1, 1) \longrightarrow$	$2(1) + 3(1) + (1) = 6$ ✔ True
$(2, 0, 2) \longrightarrow$	$2(2) + 3(0) + (2) = 6$ ✔ True
$(0, 1, 3) \longrightarrow$	$2(0) + 3(1) + (3) = 6$ ✔ True

Infinitely many ordered triples serve as solutions to the equation $2x + 3y + z = 6$.

The set of all ordered triples that are solutions to a linear equation in three variables may be represented graphically by a plane in space. Figure 9-1 shows a portion of the plane $2x + 3y + z = 6$ in a 3-dimensional coordinate system.

A solution to a system of linear equations in three variables is an ordered triple that satisfies *each* equation. Geometrically, a solution is a point of intersection of the planes represented by the equations in the system.

A system of linear equations in three variables may have *one unique solution, infinitely many solutions,* or *no solution.*

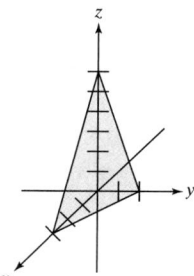

Figure 9-1

One unique solution (planes intersect at one point)
- The system is consistent.
- The system is independent.

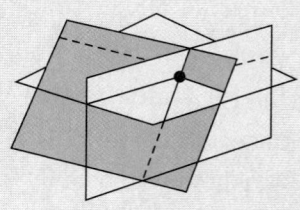

No solution (the three planes do not all intersect)
- The system is inconsistent.
- The system is independent.

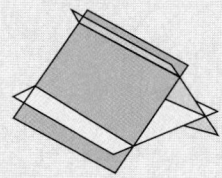 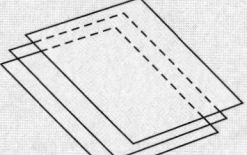 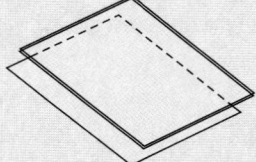

Infinitely many solutions (planes intersect at infinitely many points)
- The system is consistent.
- The system is dependent.

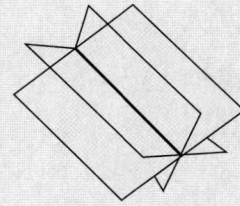 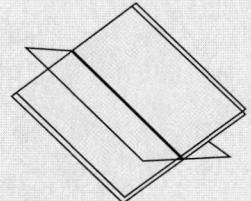 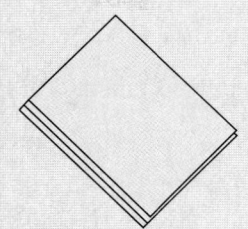

2. Solving Systems of Linear Equations in Three Variables

To solve a system involving three variables, the goal is to eliminate one variable. This reduces the system to two equations in two variables. One strategy for eliminating a variable is to pair up the original equations two at a time.

Solving a System of Three Linear Equations in Three Variables

1. Write each equation in standard form $Ax + By + Cz = D$.

2. Choose a pair of equations, and eliminate one of the variables by using the addition method.

3. Choose a different pair of equations and eliminate the *same* variable.

4. Once steps 2 and 3 are complete, you should have two equations in two variables. Solve this system by using the methods from Sections 4.2 and 4.3.

5. Substitute the values of the variables found in step 4 into any of the three original equations that contain the third variable. Solve for the third variable.

6. Check the ordered triple in each of the original equations.

Example 1 | **Solving a System of Linear Equations in Three Variables**

Solve the system.

$$2x + y - 3z = -7$$
$$3x - 2y + z = 11$$
$$-2x - 3y - 2z = 3$$

Solution:

$\boxed{A}$ $\quad 2x + y - 3z = -7$

$\boxed{B}$ $\quad 3x - 2y + z = 11$

$\boxed{C}$ $\quad -2x - 3y - 2z = 3$

Step 1: The equations are already in standard form.

- It is often helpful to label the equations.
- The y-variable can be easily eliminated from equations $\boxed{A}$ and $\boxed{B}$ and from equations $\boxed{A}$ and $\boxed{C}$. This is accomplished by creating opposite coefficients for the y-terms and then adding the equations.

Step 2:   Eliminate the y-variable from equations $\boxed{A}$ and $\boxed{B}$.

$$\boxed{A}\ \ 2x + \ y - 3z = -7 \xrightarrow{\text{Multiply by 2.}} 4x + 2y - 6z = -14$$

$$\boxed{B}\ \ 3x - 2y + \ z = \ 11 \xrightarrow{} \underline{3x - 2y + \ z = \ \ 11}$$

$$7x - 5z = \ -3\ \boxed{D}$$

TIP: It is important to note that in steps 2 and 3, the *same* variable is eliminated.

Step 3:   Eliminate the y-variable again, this time from equations $\boxed{A}$ and $\boxed{C}$.

$$\boxed{A}\ \ \ 2x + \ y - 3z = -7 \xrightarrow{\text{Multiply by 3.}} 6x + 3y - \ 9z = -21$$

$$\boxed{C}\ -2x - 3y - 2z = \ \ 3 \xrightarrow{} \underline{-2x - 3y - \ 2z = \ \ 3}$$

$$4x - 11z = -18\ \boxed{E}$$

Step 4:   Now equations $\boxed{D}$ and $\boxed{E}$ can be paired up to form a linear system in two variables. Solve this system.

$$\boxed{D}\ \ 7x - \ \ 5z = \ \ -3 \xrightarrow{\text{Multiply by -4.}} -28x + 20z = \ \ \ 12$$

$$\boxed{E}\ \ 4x - 11z = -18 \xrightarrow[\text{Multiply by 7.}]{} \underline{\ 28x - 77z = -126}$$

$$-57z = -114$$

$$z = 2$$

Once one variable has been found, substitute this value into either equation in the two-variable system, that is, either equation $\boxed{D}$ or $\boxed{E}$.

$$\boxed{D}\ \ 7x - \ \ 5z = -3$$

$$7x - 5(2) = -3 \qquad \text{Substitute } z = 2 \text{ into equation } \boxed{D}.$$

$$7x - \ \ 10 = -3$$

$$7x = \ \ 7$$

$$x = \ \ 1$$

$$\boxed{A}\ \ \ 2x + y - \ \ 3z = -7$$

$$2(1) + y - 3(2) = -7$$

$$2 + y - 6 = -7$$

$$y - 4 = -7$$

$$y = -3$$

Step 5:   Now that two variables are known, substitute these values for x and z into any of the original three equations to find the remaining variable y. Substitute $x = 1$ and $z = 2$ into equation $\boxed{A}$.

The solution is $(1, -3, 2)$.

Step 6:   Check the ordered triple in the three original equations.

Check:

$$2x + \ y - 3z = -7 \rightarrow \ \ 2(1) + \ (-3) - 3(2) = -7 \ ✔ \text{ True}$$

$$3x - 2y + \ z = \ 11 \rightarrow \ \ 3(1) - 2(-3) + \ (2) = \ 11 \ ✔ \text{ True}$$

$$-2x - 3y - 2z = \ \ 3 \rightarrow -2(1) - 3(-3) - 2(2) = \ \ 3 \ ✔ \text{ True}$$

Skill Practice

1. Solve the system.

$$x + 2y + \ z = \ \ 1$$

$$3x - \ y + 2z = \ 13$$

$$2x + 3y - \ z = -8$$

Skill Practice Answers

1. $(1, -2, 4)$

| Example 2 | **Solving a System of Linear Equations in Three Variables** |

Solve the system:
$$3x + y - 5 = 0$$
$$z = 7 + 2y$$
$$x + 3z = 17$$

Solution:

A	$3x + y = 5$
B	$-2y + z = 7$
C	$x + 3z = 17$

Step 1: Write equations in standard form, $Ax + By + Cz = D$.

Steps 2 and 3: Note equation $\boxed{C}$ is missing a y-variable. Therefore, we can eliminate the y-variable by pairing up equations $\boxed{A}$ and $\boxed{B}$.

A	$3x + y = 5$	$\xrightarrow{\text{Multiply by 2.}}$	$6x + 2y = 10$
B	$-2y + z = 7$	$\longrightarrow$	$-2y + z = 7$
			$6x + z = 17$ $\boxed{D}$

Step 4: Now we pair up equations $\boxed{C}$ and $\boxed{D}$ to eliminate the z-variable.

C	$x + 3z = 17$	$\longrightarrow$	$x + 3z = 17$
D	$6x + z = 17$	$\xrightarrow[\text{Multiply by } -3.]{}$	$-18x - 3z = -51$
			$-17x = -34$
			$x = 2$

Step 5: Now substitute $x = 2$ into equations $\boxed{A}$ and $\boxed{C}$ to find the remaining variables.

A	$3x + y = 5$		C	$x + 3z = 17$
	$3(2) + y = 5$			$(2) + 3z = 17$
	$6 + y = 5$			$3z = 15$
	$y = -1$			$z = 5$

The solution is $(2, -1, 5)$.

Step 6: Check the ordered triple in the three original equations.

Check: $3x + y - 5 = 0 \longrightarrow 3(2) + (-1) - 5 = 0$ ✔ True
$z = 7 + 2y \longrightarrow 5 = 7 + 2(-1)$ ✔ True
$x + 3z = 17 \longrightarrow (2) + 3(5) = 17$ ✔ True

| Skill Practice |

2. Solve the system.

$$4y = x + 4$$
$$2x + z = 5$$
$$y - 14 = 4z$$

Skill Practice Answers

2. $(4, 2, -3)$

| **Example 3** | **Solving a Dependent System of Linear Equations** |

Solve the system. If there is not a unique solution, label the system as either dependent or inconsistent.

$$\boxed{A}\quad 3x + y - z = 8$$
$$\boxed{B}\quad 2x - y + 2z = 3$$
$$\boxed{C}\quad x + 2y - 3z = 5$$

Solution:

The first step is to make a decision regarding the variable to eliminate. The y-variable is particularly easy to eliminate because the coefficients of y in equations $\boxed{A}$ and $\boxed{B}$ are already opposites. The y-variable can be eliminated from equations $\boxed{B}$ and $\boxed{C}$ by multiplying equation $\boxed{B}$ by 2.

$$\boxed{A}\quad 3x + y - z = 8$$
$$\boxed{B}\quad 2x - y + 2z = 3$$
$$\overline{5x + z = 11}\quad \boxed{D}$$

Pair up equations $\boxed{A}$ and $\boxed{B}$ to eliminate y.

$$\boxed{B}\quad 2x - y + 2z = 3 \xrightarrow{\text{Multiply by 2.}} 4x - 2y + 4z = 6$$
$$\boxed{C}\quad x + 2y - 3z = 5 \xrightarrow{} \underline{x + 2y - 3z = 5}$$
$$ 5x + z = 11 \quad \boxed{E}$$

Pair up equations $\boxed{B}$ and $\boxed{C}$ to eliminate y.

Because equations $\boxed{D}$ and $\boxed{E}$ are equivalent equations, it appears that this is a dependent system. By eliminating variables we obtain the identity $0 = 0$.

$$\boxed{D}\quad 5x + z = 11 \xrightarrow{\text{Multiply by } -1.} -5x - z = -11$$
$$\boxed{E}\quad 5x + z = 11 \xrightarrow{} \underline{5x + z = 11}$$
$$ 0 = 0$$

The result $0 = 0$ indicates that there are infinitely many solutions and that the system is dependent.

| **Skill Practice** |

3. Solve the system. If the system does not have a unique solution, identify the system as dependent or inconsistent.

$$x + y + z = 8$$
$$2x - y + z = 6$$
$$-5x - 2y - 4z = -30$$

Skill Practice Answers

3. Dependent system

Example 4 Solving an Inconsistent System of Linear Equations

Solve the system. If there is not a unique solution, identify the system as either dependent or inconsistent.

$$2x + 3y - 7z = 4$$
$$-4x - 6y + 14z = 1$$
$$5x + y - 3z = 6$$

Solution:

We will eliminate the x-variable.

A $2x + 3y - 7z = 4$ $\xrightarrow{\text{Multiply by 2.}}$ $4x + 6y - 14z = 8$

B $-4x - 6y + 14z = 1$ $\xrightarrow{\phantom{\text{Multiply by 2.}}}$ $\underline{-4x - 6y + 14z = 1}$

C $5x + y - 3z = 6$ $0 = 9$ (contradiction)

The result $0 = 9$ is a contradiction, indicating that the system has no solution. The system is inconsistent.

Skill Practice

4. Solve the system. If the system does not have a unique solution, identify the system as dependent or inconsistent.

$$x - 2y + z = 5$$
$$x - 3y + 2z = -7$$
$$-2x + 4y - 2z = 6$$

Skill Practice Answers

4. The system is inconsistent.

Section 9.1   Practice Exercises

Study Skills Exercise

1. Define the key term.

   a. linear equation in three variables   **b. ordered triple**

Review Exercises

For Exercises 2–4, solve the systems by using two methods: (**a**) the substitution method and (**b**) the addition method.

2. $3x + y = 4$
   $4x + y = 5$

3. $2x - 5y = 3$
   $-4x + 10y = 3$

4. $4x - 6y = 5$

   $x = \dfrac{3}{2}y + \dfrac{5}{4}$

5. Two cars leave Kansas City at the same time. One travels east and one travels west. After 3 hr the cars are 369 mi apart. If one car travels 7 mph slower than the other, find the speed of each car.

Concept 1: Solutions to Systems of Linear Equations in Three Variables

6. How many solutions are possible when solving a system of three equations with three variables?

7. Which of the following points are solutions to the system?

$$(2, 1, 7), (3, -10, -6), (4, 0, 2)$$

$$\begin{aligned} 2x - y + z &= 10 \\ 4x + 2y - 3z &= 10 \\ x - 3y + 2z &= 8 \end{aligned}$$

8. Which of the following points are solutions to the system?

$$(1, 1, 3), (0, 0, 4), (4, 2, 1)$$

$$\begin{aligned} -3x - 3y - 6z &= -24 \\ -9x - 6y + 3z &= -45 \\ 9x + 3y - 9z &= 33 \end{aligned}$$

9. Which of the following points are solutions to the system?

$$(12, 2, -2), (4, 2, 1), (1, 1, 1)$$

$$\begin{aligned} -x - y - 4z &= -6 \\ x - 3y + z &= -1 \\ 4x + y - z &= 4 \end{aligned}$$

10. Which of the following points are solutions to the system?

$$(0, 4, 3), (3, 6, 10), (3, 3, 1)$$

$$\begin{aligned} x + 2y - z &= 5 \\ x - 3y + z &= -5 \\ -2x + y - z &= -4 \end{aligned}$$

Concept 2: Solving Systems of Linear Equations in Three Variables

For Exercises 11–34, solve the system of equations.

11. $\begin{aligned} 2x + y - 3z &= -12 \\ 3x - 2y - z &= 3 \\ -x + 5y + 2z &= -3 \end{aligned}$

12. $\begin{aligned} -3x - 2y + 4z &= -15 \\ 2x + 5y - 3z &= 3 \\ 4x - y + 7z &= 15 \end{aligned}$

13. $\begin{aligned} x - 3y - 4z &= -7 \\ 5x + 2y + 2z &= -1 \\ 4x - y - 5z &= -6 \end{aligned}$

14. $\begin{aligned} 6x - 5y + z &= 7 \\ 5x + 3y + 2z &= 0 \\ -2x + y - 3z &= 11 \end{aligned}$

15. $\begin{aligned} -3x + y - z &= 8 \\ -4x + 2y + 3z &= -3 \\ 2x + 3y - 2z &= -1 \end{aligned}$

16. $\begin{aligned} 2x + 3y + 3z &= 15 \\ 3x - 6y - 6z &= -23 \\ -9x - 3y + 6z &= 8 \end{aligned}$

17. $\begin{aligned} 4x + 2z &= 12 + 3y \\ 2y &= 3x + 3z - 5 \\ y &= 2x + 7z + 8 \end{aligned}$

18. $\begin{aligned} y &= 2x + z + 1 \\ -3x - 1 &= -2y + 2z \\ 5x + 3z &= 16 - 3y \end{aligned}$

19. $\begin{aligned} x + y + z &= 6 \\ -x + y - z &= -2 \\ 2x + 3y + z &= 11 \end{aligned}$

20. $\begin{aligned} x - y - z &= -11 \\ x + y - z &= 15 \\ 2x - y + z &= -9 \end{aligned}$

21. $\begin{aligned} 2x - 3y + 2z &= -1 \\ x + 2y &= -4 \\ x + z &= 1 \end{aligned}$

22. $\begin{aligned} x + y + z &= 2 \\ 2x - z &= 5 \\ 3y + z &= 2 \end{aligned}$

23. $\begin{aligned} 4x + 9y &= 8 \\ 8x + 6z &= -1 \\ 6y + 6z &= -1 \end{aligned}$

24. $\begin{aligned} 3x + 2z &= 11 \\ y - 7z &= 4 \\ x - 6y &= 1 \end{aligned}$

25. $\begin{aligned} 2x + 3y - 2z &= 8 \\ x - 4y + z &= -8 \\ -4x + 3y + 2z &= 3 \end{aligned}$

26. $\begin{aligned} 5x - 3y + z &= 4 \\ -x + 6y + 4z &= 1 \\ 2x + 3y - z &= 3 \end{aligned}$

27. $\begin{aligned} 5y &= -7 + 4z \\ x + 3y + 3 &= 0 \\ 2x &= 14 + z \end{aligned}$

28. $\begin{aligned} 4(x - z) &= -1 \\ 3y + z &= 1 \\ 4x &= 3 + y \end{aligned}$

29. $\quad x + z = -16 + 3y$
$2(x + z) = \quad 12 - 5y$
$3x + 8 = \quad y - z$

30. $\quad 4(x - 2y) = \quad 8 - 3z$
$4y + 13 = \quad 3x + z$
$x - \quad 3y = -2(z + 2)$

31. $\frac{1}{3}x + \frac{1}{2}y + \frac{1}{6}z = \frac{1}{6}$
$\frac{1}{5}x + y - \frac{3}{5}z = \frac{4}{5}$
$x - \frac{4}{9}y + \frac{2}{3}z = \frac{1}{3}$

32. $\quad \frac{1}{2}x \qquad + \frac{5}{2}z = -1$
$\frac{1}{4}x - \frac{1}{6}y - \frac{2}{3}z = \quad 3$
$-\frac{1}{2}x + \frac{5}{8}y + \frac{1}{8}z = -3$

33. $0.1x - 0.6y + \quad z = \quad -6.8$
$0.7x \qquad - 0.3z = -14$
$y + 0.5z = \quad 8$

34. $\quad 0.5x + \quad y \qquad = 10$
$-0.2y + 0.5z = \quad 0.5$
$-0.3x \qquad - \quad z = -6$

Mixed Exercises

For Exercises 35–46, solve the system. If there is not a unique solution, label the system as either dependent or inconsistent.

35. $2x + \quad y + 3z = \quad 2$
$x - \quad y + 2z = -4$
$x + 3y - \quad z = \quad 1$

36. $\quad x + y = z$
$3(x - y) + 6z = 1 - y$
$7x + 3(y + 1) = 7 - z$

37. $\quad 6x - 2y + 2z = \quad 2$
$4x + 8y - 2z = \quad 5$
$-2x - 4y + \quad z = -2$

38. $\quad 3x + 2y + \quad z = 3$
$x - 3y + \quad z = 4$
$-6x - 4y - 2z = 1$

39. $\frac{1}{2}x + \frac{2}{3}y \qquad = \quad \frac{5}{2}$
$\frac{1}{5}x \qquad - \frac{1}{2}z = -\frac{3}{10}$
$\frac{1}{3}y - \frac{1}{4}z = \quad \frac{3}{4}$

40. $\frac{1}{2}x + \frac{1}{4}y + \quad z = 3$
$\frac{1}{8}x + \frac{1}{4}y + \frac{1}{4}z = \frac{9}{8}$
$x - \quad y - \frac{2}{3}z = \frac{1}{3}$

41. $3(2x + y) = 4(3z - 1)$
$y + 4z = \quad 2(3 + 5x)$
$2(x + z) = \qquad y - 1$

42. $\quad 5(y - 2z) = \quad 9x - 11$
$2(3x + \quad z) = \quad 5 + \quad y$
$3x = \quad z + 2(y + 1)$

43. $\quad 2x + \quad y = 3(z - 1)$
$3x - 2(y - 2z) = 1$
$2(2x - 3z) = -6 - 2y$

44. $\quad 2x + \quad y \qquad = -3$
$2y + 16z = -10$
$-7x - 3y + \quad 4z = \quad 8$

45. $\qquad -0.1y + 0.2z = 0.2$
$0.1x + 0.1y + 0.1z = 0.2$
$-0.1x \qquad + 0.3z = 0.2$

46. $0.1x - 0.2y \qquad = \quad 0$
$0.3y + 0.1z = -0.1$
$0.4x \qquad - 0.1z = \quad 1.2$

Expanding Your Skills

The systems in Exercises 47–50 are called homogeneous systems because each system has $(0, 0, 0)$ as a solution. However, if a system is dependent, it will have infinitely many more solutions. For each system determine whether $(0, 0, 0)$ is the only solution or if the system is dependent.

47. $\quad 2x - 4y + 8z = 0$
$-x - 3y + \quad z = 0$
$x - 2y + 5z = 0$

48. $\quad 2x - 4y + \quad z = 0$
$x - 3y - \quad z = 0$
$3x - \quad y + 2z = 0$

49. $\quad 4x - 2y - 3z = 0$
$-8x - \quad y + \quad z = 0$
$2x - \quad y - \frac{3}{2}z = 0$

50. $5x + \quad y \qquad = 0$
$4y - z = 0$
$5x + 5y - z = 0$

Applications of Systems of Linear Equations in Three Variables

1. Applications Involving Geometry

In this section we solve several applications of linear equations in three variables.

Example 1   Applying Systems of Linear Equations in Three Variables

In a triangle, the smallest angle measures 10° more than one-half of the largest angle. The middle angle measures 12° more than the smallest angle. Find the measure of each angle.

Solution:

Let x represent the measure of the smallest angle.

Let y represent the measure of the middle angle.

Let z represent the measure of the largest angle.

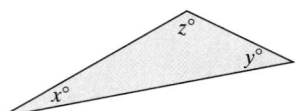

To solve for three variables, we need to establish three independent relationships among x, y, and z.

$\boxed{A}$ $x = \dfrac{z}{2} + 10$ The smallest angle measures 10° more than one-half the measure of the largest angle.

$\boxed{B}$ $y = x + 12$ The middle angle measures 12° more than the measure of the smallest angle.

$\boxed{C}$ $x + y + z = 180$ The sum of the interior angles of a triangle measures 180°.

Clear fractions and write each equation in standard form.

Standard Form

$\boxed{A}$ $x \qquad = \dfrac{z}{2} + 10 \xrightarrow{\text{Multiply by 2.}} 2x = z + 20 \longrightarrow 2x \quad - z = 20$

$\boxed{B}$ $\quad y \quad = x + 12 \xrightarrow{\hspace{4cm}} -x + y \quad = 12$

$\boxed{C}$ $x + y + z = \quad 180 \xrightarrow{\hspace{4cm}} x + y + z = 180$

Notice equation $\boxed{B}$ is missing the z-variable. Therefore, we can eliminate z again by pairing up equations $\boxed{A}$ and $\boxed{C}$.

$\boxed{A}$ $2x \qquad - z = \quad 20$

$\boxed{C}$ $\underline{x + y + z = 180}$

$\qquad 3x + y \qquad = 200$ $\boxed{D}$

$\boxed{B}$ $-x + y = 12$ $\xrightarrow{\text{Multiply by } -1.}$ $x - y = -12$ Pair up equations $\boxed{B}$ and

$\boxed{D}$ $3x + y = 200$ $\longrightarrow$ $\underline{3x + y = 200}$ $\boxed{D}$ to form a system of two variables.

$\qquad\qquad\qquad\qquad\qquad\qquad 4x \quad\; = 188$

$\qquad\qquad\qquad\qquad\qquad\qquad\; x = 47$ $\qquad$ Solve for x.

From equation $\boxed{B}$ we have $-x + y = 12$ $\longrightarrow$ $-47 + y = 12 \rightarrow y = 59$

From equation $\boxed{C}$ we have $x + y + z = 180 \rightarrow 47 + 59 + z = 180 \rightarrow z = 74$

The smallest angle measures $47°$, the middle angle measures $59°$, and the largest angle measures $74°$.

Skill Practice

1. The perimeter of a triangle is 30 in. The shortest side is 4 in. shorter than the longest side. The longest side is 6 in. less than the sum of the other two sides. Find the length of each side.

---

2. Applications Involving Mixtures

Example 2 Applying Systems of Linear Equations to Nutrition

Doctors have become increasingly concerned about the sodium intake in the U.S. diet. Recommendations by the American Medical Association indicate that most individuals should not exceed 2400 mg of sodium per day.

Liz ate 1 slice of pizza, 1 serving of ice cream, and 1 glass of soda for a total of 1030 mg of sodium. David ate 3 slices of pizza, no ice cream, and 2 glasses of soda for a total of 2420 mg of sodium. Melinda ate 2 slices of pizza, 1 serving of ice cream, and 2 glasses of soda for a total of 1910 mg of sodium. How much sodium is in one serving of each item?

Solution:

Let x represent the sodium content of 1 slice of pizza.

Let y represent the sodium content of 1 serving of ice cream.

Let z represent the sodium content of 1 glass of soda.

From Liz's meal we have: $\qquad$ $\boxed{A}$ $x + y + z = 1030$

From David's meal we have: $\qquad$ $\boxed{B}$ $3x \qquad + 2z = 2420$

From Melinda's meal we have: $\qquad$ $\boxed{C}$ $2x + y + 2z = 1910$

Equation $\boxed{B}$ is missing the y-variable. Eliminating y from equations $\boxed{A}$ and $\boxed{C}$, we have

$\boxed{A}$ $x + y + z = 1030$ $\xrightarrow{\text{Multiply by } -1.}$ $-x - y - z = -1030$

$\boxed{C}$ $2x + y + 2z = 1910$ $\longrightarrow$ $\underline{2x + y + 2z = \quad 1910}$

$\qquad\qquad\qquad\qquad\qquad\qquad\quad$ $\boxed{D}$ $x \qquad + z = \qquad 880$

Solve the system formed by equations $\boxed{B}$ and $\boxed{D}$.

$\boxed{B}$ $3x + 2z = 2420$ $\longrightarrow$ $3x + 2z = \quad 2420$

$\boxed{D}$ $x + z = 880$ $\xrightarrow{\text{Multiply by } -2.}$ $\underline{-2x - 2z = -1760}$

$\qquad\qquad\qquad\qquad\qquad\qquad\quad x \qquad\quad = \qquad 660$

Skill Practice Answers

1. 8 in., 10 in., and 12 in.

From equation $\boxed{D}$ we have $x + z = 880 \longrightarrow 660 + z = 880 \longrightarrow z = 220$

From equation $\boxed{A}$ we have $x + y + z = 1030 \longrightarrow 660 + y + 220 = 1030 \longrightarrow y = 150$

Therefore, 1 slice of pizza has 660 mg of sodium, 1 serving of ice cream has 150 mg of sodium, and 1 glass of soda has 220 mg of sodium.

Skill Practice

2. Annette, Barb, and Carlita work in a clothing shop. One day the three had combined sales of $1480. Annette sold $120 more than Barb. Barb and Carlita combined sold $280 more than Annette. How much did each person sell?

---

3. Finding an Equation of a Parabola

In Section 3.5 we wrote an equation of a line given two points using the point-slope formula: $y - y_1 = m(x - x_1)$. Now we learn to write an equation of a parabola given three points using systems of equations and the standard form of a parabola: $y = ax^2 + bx + c$.

This process involves substituting the x- and y-coordinates of three distinct, non-collinear points and solving the resulting system of three equations.

Example 3   **Writing an Equation of a Parabola**

Write an equation of a parabola that passes through the points $(1, -1)$, $(-1, -5)$, and $(2, 4)$.

Solution:

Substitute $(1, -1)$ into the equation $y = ax^2 + bx + c$

$$(-1) = a(1)^2 + b(1) + c$$
$$-1 = a + b + c$$
$$\text{(standard form)} \quad a + b + c = -1$$

Substitute $(-1, -5)$ into the equation $y = ax^2 + bx + c$

$$(-5) = a(-1)^2 + b(-1) + c$$
$$-5 = a - b + c$$
$$\text{(standard form)} \quad a - b + c = -5$$

Substitute $(2, 4)$ into the equation $y = ax^2 + bx + c$

$$(4) = a(2)^2 + b(2) + c$$
$$4 = 4a + 2b + c$$
$$\text{(standard form)} \quad 4a + 2b + c = 4$$

Solve the system: $\boxed{A}$ $\quad a + b + c = -1$

$\boxed{B}$ $\quad a - b + c = -5$

$\boxed{C}$ $\quad 4a + 2b + c = 4$

Notice that the c-variables all have a coefficient of 1. Therefore we choose to eliminate the c-variable.

$\boxed{A}$ $\quad a + b + c = -1 \longrightarrow \qquad a + b + c = -1$

$\boxed{B}$ $\quad a - b + c = -5 \xrightarrow{\text{Multiply by } -1.} \underline{-a + b - c = 5}$

$$2b = 4$$

$$b = 2 \quad \boxed{D}$$

$$\boxed{B} \quad a - b + c = -5 \longrightarrow \quad a - b + c = -5$$

$$\boxed{C} \quad 4a + 2b + c = \quad 4 \xrightarrow[\text{Multiply by } -1.]{} \underline{-4a - 2b - c = -4}$$

$$-3a - 3b \quad\quad = -9 \quad \boxed{E}$$

Substitute b with 2 in equation $\boxed{E}$:

$$-3a - 3(2) = -9$$
$$-3a \quad - 6 = -9$$
$$-3a \quad\quad = -3$$
$$a \quad\quad = 1$$

Substitute a and b in equation $\boxed{A}$ to solve for c: $(1) + (2) + c = -1$

$$3 + c = -1$$
$$c = -4$$

Substitute $a = 1$, $b = 2$, and $c = -4$ in the standard form of the parabola for the final answer.

$$y = (1)x^2 + (2)x + (-4) \longrightarrow y = x^2 + 2x - 4$$

Skill Practice

3. Write an equation of the parabola that passes through the points $(1, 1)$, $(-2, 1)$, and $(3, -9)$.

Skill Practice Answers

3. $a = -1$, $b = -1$, $c = 3$;
$y = -x^2 - x + 3$

Section 9.2 Practice Exercises

Review Exercises

For Exercises 1–4, solve the system. If there is not a unique solution label the system as either dependent or inconsistent.

1.
$$-5y - z = -8$$
$$x + 10y + 2z = 9$$
$$-3x + y = 21$$

2.
$$8x - y + 2z = 18$$
$$x + y - 5z = -21$$
$$4x - 0.5y + z = 9$$

3.
$$\frac{1}{2}x - y + \frac{5}{6}z = 4$$
$$\frac{3}{2}x - 3y + \frac{5}{2}z = 6$$
$$-\frac{1}{4}x + y - \frac{1}{3}z = -2$$

4.
$$12x - y + z = -6$$
$$-6x + 2y - 3z = 5$$
$$6x + y - z = 3$$

Concept 1: Applications Involving Geometry

5. A triangle has one angle that measures $5°$ more than twice the smallest angle, and the largest angle measures $11°$ less than 3 times the measure of the smallest angle. Find the measures of the three angles.

6. The largest angle of a triangle measures 4° less than 5 times the measure of the smallest angle. The middle angle measures twice that of the smallest angle. Find the measures of the three angles.

7. One angle of a triangle measures 6° more than twice the measure of the smallest angle. The third angle measures 1° less than four times the smallest. Find the measures of the three angles.

8. In a triangle the smallest angle measures 12° less than the middle angle. The measure of the largest angle is equal to the sum of the other two. Find the measures of the three angles.

9. The perimeter of a triangle is 55 cm. The measure of the shortest side is 8 cm less than the middle side. The measure of the longest side is 1 cm less than the sum of the other two sides. Find the lengths of the sides.

10. The perimeter of a triangle is 5 ft. The longest side of the triangle measures 20 in. more than the shortest side. The middle side is 3 times the measure of the shortest side. Find the lengths of the three sides in *inches*.

11. The perimeter of a triangle is 4.5 ft. The shortest side measures 2 in. less than half the longest side. The longest side measures 2 in. less than the sum of the other two sides. Find the lengths of the sides in *inches*.

12. The perimeter of a triangle is 7 m. The longest side measures twice the shortest side and the sum of the measures of the smallest side and the middle side is 1 m more than the measure of the longest side. Find the lengths of the sides.

Concept 2: Applications Involving Mixtures

13. Sean kept track of his fiber intake from three sources for 3 weeks. The first week he had 3 servings of a fiber supplement, 1 serving of oatmeal, and 4 servings of cereal, which totaled 19 g of fiber. The second week he had 2 servings of the fiber supplement, 4 servings of oatmeal, and 2 servings of cereal totaling 25 g. The third week he had 5 servings of the fiber supplement, 3 servings of oatmeal, and 2 servings of cereal for a total of 30 g. Find the amount of fiber in one serving of each of the following: the fiber supplement, the oatmeal, and the cereal.

14. Natalie kept track of her calcium intake from three sources for 3 days. The first day she had 1 glass of milk, 1 serving of ice cream, and 1 calcium supplement in pill form which totaled 1180 mg of calcium. The second day she had 2 glasses of milk, 1 serving of ice cream, and 1 calcium supplement totaling 1680 mg. The third day she had 1 glass of milk, 2 servings of ice cream, and 1 calcium supplement for a total of 1260 mg. Find the amount of calcium in one glass of milk, in one serving of ice cream, and in one calcium supplement.

15. Kyoki invested a total of $10,000 into bonds, mutual funds, and a money market account. He put the same amount of money in the money market account as he did in bonds. For 1 year, the bonds paid 5% interest, the mutual funds paid 8%, and the money market paid 4%. At the end of the year, Kyoki earned $660 in interest. How much did he invest in each account?

16. Walter had $25,000 to invest. He spit the money into three types of investment: small caps earning 6%, global market investments earning 10%, and a balanced fund earning 9%. He put twice as much money in the global account as he did in the balanced fund. If his earnings for the first year totaled $2160, how much did he invest in each account?

17. Winston deposited $4500 into three certificates of deposit: a 24-month CD paying 5% interest, a 30-month CD paying 5.5% interest, and an 18-month CD paying 4% interest. He put $1000 more in the CD with the 5.5% rate than he put in the 4% rate. After the first year, he earned a total of $225 interest. How much did he deposit into each CD?

18. Raeann deposited $8000 into three accounts at her credit union: a checking account that pays 1.2% interest, a savings account that pays 2.5% interest, and a money market account that pays 3% interest. If she put three times more money in the 3% account than she did in the 1.2% account, and her total interest for 1 year was $202, how much did she deposit into each account?

19. A movie theater charges $7 for adults, $5 for children under age 17, and $4 for seniors over age 60. For one showing of *Batman* the theater sold 222 tickets and took in $1383. If twice as many adult tickets were sold as the total of children and senior tickets, how many tickets of each kind were sold?

20. Goofie Golf has 18 holes that are par 3, par 4, or par 5. Most of the holes are par 4. In fact, there are 3 times as many par 4s as par 3s. There are 3 more par 5s than par 3s. How many of each type are there?

21. Combining peanuts, pecans, and cashews makes a party mixture of nuts. If the amount of peanuts equals the amount of pecans and cashews combined, and if there are twice as many cashews as pecans, how many ounces of each nut is used to make 48 oz of party mixture?

22. Souvenir hats, T-shirts, and jackets are sold at a rock concert. Three hats, two T-shirts, and one jacket cost $140. Two hats, two T-shirts, and two jackets cost $170. One hat, three T-shirts, and two jackets cost $180. Find the prices of the individual items.

23. In 2002, Baylor University in Waco, Texas, had twice as many students as Vanderbilt University in Nashville, Tennessee. Pace University in New York City had 2800 more students than Vanderbilt University. If the enrollment for all three schools totaled 27,200, find the enrollment for each school.

24. Annie and Maria traveled overseas for seven days and stayed in three different hotels in three different cities: Stockholm, Sweden; Oslo, Norway; and Paris, France.

    The total bill for all seven nights (not including tax) was $1040. The total tax was $106. The nightly cost (excluding tax) to stay at the hotel in Paris was $80 more than the nightly cost (excluding tax) to stay in Oslo. Find the cost per night for each hotel excluding tax.

City	Number of Nights	Cost/Night ($)	Tax Rate
Paris, France	1	x	8%
Stockholm, Sweden	4	y	11%
Oslo, Norway	2	z	10%

Concept 3: Finding an Equation of a Parabola

For Exercises 25–32, use the standard form of a parabola given by $y = ax^2 + bx + c$.

25. Write an equation of a parabola that passes through the points $(0, 4)$, $(1, 0)$, and $(-1, -10)$.

26. Write an equation of a parabola that passes through the points $(0, 3)$, $(3, 0)$, and $(-1, 8)$.

27. Write an equation of a parabola that passes through the points $(2, 1)$, $(-2, 5)$, and $(1, -4)$.

28. Write an equation of a parabola that passes through the points $(1, 2)$, $(-1, -6)$, and $(2, -3)$.

29. Write an equation of a parabola that passes through the points $(2, -4)$, $(1, 1)$, and $(-1, -7)$.

30. Write an equation of a parabola that passes through the points $(1, 4)$, $(-1, 6)$, and $(2, 18)$.

31. Write an equation of a parabola that passes through the points $(-3, -4)$, $(-2, -5)$, and $(1, 4)$.

32. Write an equation of a parabola that passes through the points $(4, 18)$, $(-2, 12)$, and $(-1, 8)$.

Section 9.3 — Solving Systems of Linear Equations by Using Matrices

Concepts

1. Introduction to Matrices
2. Solving Systems of Linear Equations by Using the Gauss-Jordan Method

1. Introduction to Matrices

In Sections 4.2, 4.3, and 9.1, we solved systems of linear equations by using the substitution method and the addition method. We now present a third method called the Gauss-Jordan method that uses matrices to solve a linear system.

A **matrix** is a rectangular array of numbers (the plural of *matrix* is *matrices*). The rows of a matrix are read horizontally, and the columns of a matrix are read vertically. Every number or entry within a matrix is called an element of the matrix.

The **order of a matrix** is determined by the number of rows and number of columns. A matrix with m rows and n columns is an $m \times n$ (read as "m by n") matrix. Notice that with the order of a matrix, the number of rows is given first, followed by the number of columns.

Example 1 — Determining the Order of a Matrix

Determine the order of each matrix.

a. $\begin{bmatrix} 2 & -4 & 1 \\ 5 & \pi & \sqrt{7} \end{bmatrix}$
b. $\begin{bmatrix} 1.9 \\ 0 \\ 7.2 \\ -6.1 \end{bmatrix}$
c. $\begin{bmatrix} 1 & 0 & 0 \\ 0 & 1 & 0 \\ 0 & 0 & 1 \end{bmatrix}$
d. $\begin{bmatrix} a & b & c \end{bmatrix}$

Solution:

a. This matrix has two rows and three columns. Therefore, it is a 2×3 matrix.

b. This matrix has four rows and one column. Therefore, it is a 4×1 matrix. A matrix with one column is called a **column matrix**.

c. This matrix has three rows and three columns. Therefore, it is a 3×3 matrix. A matrix with the same number of rows and columns is called a **square matrix**.

d. This matrix has one row and three columns. Therefore, it is a 1×3 matrix. A matrix with one row is called a **row matrix**.

Skill Practice Determine the order of the matrix.

1. $\begin{bmatrix} -5 & 2 \\ 1 & 3 \\ 8 & 9 \end{bmatrix}$ **2.** $[4 \;\; -8]$ **3.** $\begin{bmatrix} 5 \\ 10 \\ 15 \end{bmatrix}$ **4.** $\begin{bmatrix} 2 & -0.5 \\ -1 & 6 \end{bmatrix}$

A matrix can be used to represent a system of linear equations written in standard form. To do so, we extract the coefficients of the variable terms and the constants within the equation. For example, consider the system

$$2x - y = 5$$
$$x + 2y = -5$$

The matrix **A** is called the **coefficient matrix**.

$$\mathbf{A} = \begin{bmatrix} 2 & -1 \\ 1 & 2 \end{bmatrix}$$

If we extract both the coefficients and the constants from the equations, we can construct the **augmented matrix** of the system:

$$\left[\begin{array}{cc|c} 2 & -1 & 5 \\ 1 & 2 & -5 \end{array} \right]$$

A vertical bar is inserted into an augmented matrix to designate the position of the equal signs.

Example 2 **Writing the Augmented Matrix of a System of Linear Equations**

Write the augmented matrix for each linear system.

a. $-3x - 4y = 3$
$\quad 2x + 4y = 2$

b. $2x - 3z = 14$
$ \quad 2y + z = 2$
$ \quad x + y = 4$

Solution:

a. $\left[\begin{array}{cc|c} -3 & -4 & 3 \\ 2 & 4 & 2 \end{array} \right]$

b. $\left[\begin{array}{ccc|c} 2 & 0 & -3 & 14 \\ 0 & 2 & 1 & 2 \\ 1 & 1 & 0 & 4 \end{array} \right]$

TIP: Notice that zeros are inserted to denote the coefficient of each missing term.

Skill Practice Write the augmented matrix for the system.

5. $-x + y = 4$
$\quad 2x - y = 1$

6. $2x - y + z = 14$
$-3x + 4y = 8$
$ \quad x - y + 5z = 0$

Skill Practice Answers

1. 3×2 **2.** 1×2
3. 3×1 **4.** 2×2

5. $\left[\begin{array}{cc|c} -1 & 1 & 4 \\ 2 & -1 & 1 \end{array} \right]$

6. $\left[\begin{array}{ccc|c} 2 & -1 & 1 & 14 \\ -3 & 4 & 0 & 8 \\ 1 & -1 & 5 & 0 \end{array} \right]$

| **Example 3** | **Writing a Linear System from an Augmented Matrix** |

Write a system of linear equations represented by each augmented matrix.

a. $\begin{bmatrix} 2 & -5 & | & -8 \\ 4 & 1 & | & 6 \end{bmatrix}$　　**b.** $\begin{bmatrix} 2 & -1 & 3 & | & 14 \\ 1 & 1 & -2 & | & -5 \\ 3 & 1 & -1 & | & 2 \end{bmatrix}$

c. $\begin{bmatrix} 1 & 0 & 0 & | & 4 \\ 0 & 1 & 0 & | & -1 \\ 0 & 0 & 1 & | & 0 \end{bmatrix}$

Solution:

a. $2x - 5y = -8$
$4x + y = 6$

b. $2x - y + 3z = 14$
$x + y - 2z = -5$
$3x + y - z = 2$

c. $x + 0y + 0z = 4$　　　　　$x = 4$
$0x + y + 0z = -1$　or　$y = -1$
$0x + 0y + z = 0$　　　　　$z = 0$

| **Skill Practice** | Write a system of linear equations represented by each augmented matrix. |

7. $\begin{bmatrix} 2 & 3 & | & 5 \\ -1 & 8 & | & 1 \end{bmatrix}$　　**8.** $\begin{bmatrix} -3 & 2 & 1 & | & 4 \\ 14 & 1 & 0 & | & 20 \\ -8 & 3 & 5 & | & 6 \end{bmatrix}$　　**9.** $\begin{bmatrix} 1 & 0 & 0 & | & -5 \\ 0 & 1 & 0 & | & 2 \\ 0 & 0 & 1 & | & 0 \end{bmatrix}$

2. Solving Systems of Linear Equations by Using the Gauss-Jordan Method

We know that interchanging two equations results in an equivalent system of linear equations. Interchanging two rows in an augmented matrix results in an equivalent augmented matrix. Similarly, because each row in an augmented matrix represents a linear equation, we can perform the following elementary row operations that result in an equivalent augmented matrix.

Elementary Row Operations

The following *elementary row operations* performed on an augmented matrix produce an equivalent augmented matrix:

1. Interchange two rows.
2. Multiply every element in a row by a nonzero real number.
3. Add a multiple of one row to another row.

Skill Practice Answers

7. $2x + 3y = 5$
$-x + 8y = 1$
8. $-3x + 2y + z = 4$
$14x + y = 20$
$-8x + 3y + 5z = 6$
9. $x = -5, y = 2, z = 0$

When we are solving a system of linear equations by any method, the goal is to write a series of simpler but equivalent systems of equations until the solution is obvious. The *Gauss-Jordan method* uses a series of elementary row operations performed on the augmented matrix to produce a simpler augmented matrix. In particular, we want to produce an augmented matrix that has 1s along the diagonal of the matrix of coefficients and 0s for the remaining entries in the matrix of coefficients. A matrix written in this way is said to be written in **reduced row echelon form**. For example, the augmented matrix from Example 3(c) is written in reduced row echelon form.

$$\begin{bmatrix} 1 & 0 & 0 & | & 4 \\ 0 & 1 & 0 & | & -1 \\ 0 & 0 & 1 & | & 0 \end{bmatrix}$$

The solution to the corresponding system of equations is easily recognized as $x = 4$, $y = -1$, and $z = 0$.

Similarly, matrix **B** represents a solution of $x = a$ and $y = b$.

$$\mathbf{B} = \begin{bmatrix} 1 & 0 & | & a \\ 0 & 1 & | & b \end{bmatrix}$$

Example 4 | **Solving a System of Linear Equations by Using the Gauss-Jordan Method**

Solve by using the Gauss-Jordan method.

$$2x - y = 5$$
$$x + 2y = -5$$

Solution:

$$\begin{bmatrix} 2 & -1 & | & 5 \\ 1 & 2 & | & -5 \end{bmatrix}$$ Set up the augmented matrix.

$\xrightarrow{R_1 \Leftrightarrow R_2}$ $\begin{bmatrix} 1 & 2 & | & -5 \\ 2 & -1 & | & 5 \end{bmatrix}$ Switch row 1 and row 2 to get a 1 in the upper left position.

$\xrightarrow{-2R_1 + R_2 \Rightarrow R_2}$ $\begin{bmatrix} 1 & 2 & | & -5 \\ 0 & -5 & | & 15 \end{bmatrix}$ Multiply row 1 by -2 and add the result to row 2. This produces an entry of 0 below the upper left position.

$\xrightarrow{-\frac{1}{5}R_2 \Rightarrow R_2}$ $\begin{bmatrix} 1 & 2 & | & -5 \\ 0 & 1 & | & -3 \end{bmatrix}$ Multiply row 2 by $-\frac{1}{5}$ to produce a 1 along the diagonal in the second row.

$\xrightarrow{-2R_2 + R_1 \Rightarrow R_1}$ $\begin{bmatrix} 1 & 0 & | & 1 \\ 0 & 1 & | & -3 \end{bmatrix}$ Multiply row 2 by -2 and add the result to row 1. This produces a 0 in the first row, second column.

The matrix **C** is in reduced row echelon form. From the augmented matrix, we have $x = 1$ and $y = -3$. The solution to the system is $(1, -3)$.

$$\mathbf{C} = \begin{bmatrix} 1 & 0 & | & 1 \\ 0 & 1 & | & -3 \end{bmatrix}$$

Skill Practice

10. Solve by using the Gauss-Jordan method.

$$x - 2y = -21$$
$$2x + y = -2$$

Skill Practice Answers

10. $(-5, 8)$

The order in which we manipulate the elements of an augmented matrix to produce reduced row echelon form was demonstrated in Example 4. In general, the order is as follows.

- First produce a 1 in the first row, first column. Then use the first row to obtain 0s in the first column below this element.
- Next, if possible, produce a 1 in the second row, second column. Use the second row to obtain 0s above and below this element.
- Next, if possible, produce a 1 in the third row, third column. Use the third row to obtain 0s above and below this element.
- The process continues until reduced row echelon form is obtained.

Example 5   **Solving a System of Linear Equations by Using the Gauss-Jordan Method**

Solve by using the Gauss-Jordan method.

$$
\begin{aligned}
x &= -y + 5 \\
-2x + 2z &= y - 10 \\
3x + 6y + 7z &= 14
\end{aligned}
$$

Solution:

First write each equation in the system in standard form.

$$
\begin{aligned}
x &= -y + 5 &\longrightarrow& \quad x + y &= 5 \\
-2x + 2z &= y - 10 &\longrightarrow& \quad -2x - y + 2z &= -10 \\
3x + 6y + 7z &= 14 &\longrightarrow& \quad 3x + 6y + 7z &= 14
\end{aligned}
$$

$$
\begin{bmatrix}
1 & 1 & 0 & 5 \\
-2 & -1 & 2 & -10 \\
3 & 6 & 7 & 14
\end{bmatrix}
$$
Set up the augmented matrix.

$$
\begin{aligned}
2R_1 + R_2 &\Rightarrow R_2 \\
-3R_1 + R_3 &\Rightarrow R_3
\end{aligned}
\longrightarrow
\begin{bmatrix}
1 & 1 & 0 & 5 \\
0 & 1 & 2 & 0 \\
0 & 3 & 7 & -1
\end{bmatrix}
$$
Multiply row 1 by 2 and add the result to row 2. Multiply row 1 by -3 and add the result to row 3.

$$
\begin{aligned}
-1R_2 + R_1 &\Rightarrow R_1 \\
-3R_2 + R_3 &\Rightarrow R_3
\end{aligned}
\longrightarrow
\begin{bmatrix}
1 & 0 & -2 & 5 \\
0 & 1 & 2 & 0 \\
0 & 0 & 1 & -1
\end{bmatrix}
$$
Multiply row 2 by -1 and add the result to row 1. Multiply row 2 by -3 and add the result to row 3.

$$
\begin{aligned}
2R_3 + R_1 &\Rightarrow R_1 \\
-2R_3 + R_2 &\Rightarrow R_2
\end{aligned}
\longrightarrow
\begin{bmatrix}
1 & 0 & 0 & 3 \\
0 & 1 & 0 & 2 \\
0 & 0 & 1 & -1
\end{bmatrix}
$$
Multiply row 3 by 2 and add the result to row 1. Multiply row 3 by -2 and add the result to row 2.

From the reduced row echelon form of the matrix, we have $x = 3$, $y = 2$, and $z = -1$. The solution to the system is $(3, 2, -1)$.

Solve by using the Gauss-Jordan method.

11. $x + y + z = 2$
$x - y + z = 4$
$x + 4y + 2z = 1$

It is particularly easy to recognize a dependent or inconsistent system of equations from the reduced row echelon form of an augmented matrix. This is demonstrated in Examples 6 and 7.

Example 6 **Solving a Dependent System of Equations by Using the Gauss-Jordan Method**

Solve by using the Gauss-Jordan method.

$$x - 3y = 4$$

$$\frac{1}{2}x - \frac{3}{2}y = 2$$

Solution:

$$\begin{bmatrix} 1 & -3 & | & 4 \\ \frac{1}{2} & -\frac{3}{2} & | & 2 \end{bmatrix}$$ Set up the augmented matrix.

$$-\frac{1}{2}R_1 + R_2 \Rightarrow R_2 \quad \begin{bmatrix} 1 & -3 & | & 4 \\ 0 & 0 & | & 0 \end{bmatrix}$$ Multiply row 1 by $-\frac{1}{2}$ and add the result to row 2.

The second row of the augmented matrix represents the equation $0 = 0$; hence, the system is dependent. The solution is $\{(x, y) \,|\, x - 3y = 4\}$.

Skill Practice Solve by using the Gauss-Jordan method.

12. $4x - 6y = 16$
$6x - 9y = 24$

Example 7 **Solving an Inconsistent System of Equations by Using the Gauss-Jordan Method**

Solve by using the Gauss-Jordan method.

$$x + 3y = 2$$

$$-3x - 9y = 1$$

Solution:

$$\begin{bmatrix} 1 & 3 & | & 2 \\ -3 & -9 & | & 1 \end{bmatrix}$$ Set up the augmented matrix.

$$3R_1 + R_2 \Rightarrow R_2 \quad \begin{bmatrix} 1 & 3 & | & 2 \\ 0 & 0 & | & 7 \end{bmatrix}$$ Multiply row 1 by 3 and add the result to row 2.

The second row of the augmented matrix represents the contradiction $0 = 7$; hence, the system is inconsistent. There is no solution.

Skill Practice Answers

11. $(1, -1, 2)$
12. Infinitely many solutions;
$\{(x, y) \,|\, 4x - 6y = 16\}$;
dependent system

Skill Practice

13. Solve by using the Gauss-Jordan method.

$$6x + 10y = 1$$
$$15x + 25y = 3$$

Calculator Connections

Many graphing calculators have a matrix editor in which the user defines the order of the matrix and then enters the elements of the matrix. For example, the 2×3 matrix

$$\mathbf{D} = \begin{bmatrix} 2 & -3 & -13 \\ 3 & 1 & 8 \end{bmatrix}$$

is entered as shown.

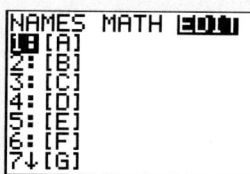 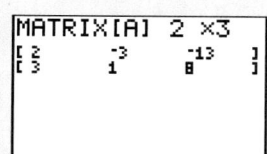

Once an augmented matrix has been entered into a graphing calculator, a *rref* function can be used to transform the matrix into reduced row echelon form.

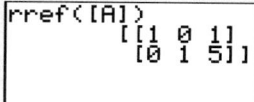

Section 9.3   Practice Exercises

Boost *your* GRADE at
mathzone.com!

- Practice Problems
- Self-Tests
- NetTutor
- e-Professors
- Videos

Study Skills Exercise

1. Define the key terms.

   a. matrix   **b. order of a matrix**   **c. column matrix**   **d. square matrix**

   e. row matrix   **f. coefficient matrix**   **g. augmented matrix**   **h. reduced row echelon form**

Review Exercises

For Exercises 2–5, solve the system by using any method.

2. $-\dfrac{1}{2}x = 3y + 1$
$\phantom{-\dfrac{1}{2}}x + 6y = 4$

3. $x - 6y = 9$
$x + 2y = 13$

4. $x + y - z = 8$
$x - 2y + z = 3$
$x + 3y + 2z = 7$

5. $2x - y + z = -4$
$-x + y + 3z = -7$
$x + 3y - 4z = 22$

Concept 1: Introduction to Matrices

For Exercises 6–14, (**a**) determine the order of each matrix and (**b**) determine if the matrix is a row matrix, a column matrix, a square matrix, or none of these.

6. $\begin{bmatrix} 4 \\ 5 \\ -3 \\ 0 \end{bmatrix}$

7. $\begin{bmatrix} 5 \\ -1 \\ 2 \end{bmatrix}$

8. $\begin{bmatrix} -9 & 4 & 3 \\ -1 & -8 & 4 \\ 5 & 8 & 7 \end{bmatrix}$

9. $\begin{bmatrix} 3 & -9 \\ -1 & -3 \end{bmatrix}$

10. $\begin{bmatrix} 4 & -7 \end{bmatrix}$

11. $\begin{bmatrix} 0 & -8 & 11 & 5 \end{bmatrix}$

12. $\begin{bmatrix} 5 & -8.1 & 4.2 & 0 \\ 4.3 & -9 & 18 & 3 \end{bmatrix}$

13. $\begin{bmatrix} \frac{1}{3} & \frac{3}{4} & 6 \\ -2 & 1 & -\frac{7}{8} \end{bmatrix}$

14. $\begin{bmatrix} 5 & 1 \\ -1 & 2 \\ 0 & 7 \end{bmatrix}$

For Exercises 15–19, set up the augmented matrix.

15. $x - 2y = -1$
    $2x + y = -7$

16. $x - 3y = 3$
    $2x - 5y = 4$

17. $y = 2x - 1$
    $y = -3x + 7$

18. $x - 2y = 5 - z$
    $2x + 6y + 3z = -2$
    $3x - y - 2z = 1$

19. $5x - 17 = -2z$
    $8x + 6z = 26 + y$
    $8x + 3y - 12z = 24$

For Exercises 20–23, write a system of linear equations represented by the augmented matrix.

20. $\left[\begin{array}{cc|c} 4 & 3 & 6 \\ 12 & 5 & -6 \end{array}\right]$

21. $\left[\begin{array}{cc|c} -2 & 5 & -15 \\ -7 & 15 & -45 \end{array}\right]$

22. $\left[\begin{array}{ccc|c} 1 & 0 & 0 & 4 \\ 0 & 1 & 0 & -1 \\ 0 & 0 & 1 & 7 \end{array}\right]$

23. $\left[\begin{array}{ccc|c} 1 & 0 & 0 & 0.5 \\ 0 & 1 & 0 & 6.1 \\ 0 & 0 & 1 & 3.9 \end{array}\right]$

Concept 2: Solving Systems of Linear Equations by Using the Gauss-Jordan Method

24. Given the matrix **E**

$$\mathbf{E} = \left[\begin{array}{cc|c} 3 & -2 & 8 \\ 9 & -1 & 7 \end{array}\right]$$

   a. What is the element in the second row and third column?

   b. What is the element in the first row and second column?

25. Given the matrix **F**

$$\mathbf{F} = \left[\begin{array}{cc|c} 1 & 8 & 0 \\ 12 & -13 & -2 \end{array}\right]$$

   a. What is the element in the second row and second column?

   b. What is the element in the first row and third column?

26. Given the matrix **Z**

$$\mathbf{Z} = \left[\begin{array}{cc|c} 2 & 1 & 11 \\ 2 & -1 & 1 \end{array}\right]$$

   write the matrix obtained by multiplying the elements in the first row by $\frac{1}{2}$.

27. Given the matrix **J**

$$\mathbf{J} = \left[\begin{array}{cc|c} 1 & 1 & 7 \\ 0 & 3 & -6 \end{array}\right]$$

   write the matrix obtained by multiplying the elements in the second row by $\frac{1}{3}$.

28. Given the matrix **K**

$$\mathbf{K} = \left[\begin{array}{cc|c} 5 & 2 & 1 \\ 1 & -4 & 3 \end{array}\right]$$

write the matrix obtained by interchanging rows 1 and 2.

29. Given the matrix **L**

$$\mathbf{L} = \left[\begin{array}{cc|c} 9 & 6 & 13 \\ -7 & 2 & 19 \end{array}\right]$$

write the matrix obtained by interchanging rows 1 and 2.

30. Given the matrix **M**

$$\mathbf{M} = \left[\begin{array}{cc|c} 1 & 5 & 2 \\ -3 & -4 & -1 \end{array}\right]$$

write the matrix obtained by multiplying the first row by 3 and adding the result to row 2.

31. Given the matrix **N**

$$\mathbf{N} = \left[\begin{array}{cc|c} 1 & 3 & -5 \\ -2 & 2 & 12 \end{array}\right]$$

write the matrix obtained by multiplying the first row by 2 and adding the result to row 2.

32. Given the matrix **R**

$$\mathbf{R} = \left[\begin{array}{ccc|c} 1 & 3 & 0 & -1 \\ 4 & 1 & -5 & 6 \\ -2 & 0 & -3 & 10 \end{array}\right]$$

a. Write the matrix obtained by multiplying the first row by -4 and adding the result to row 2.

b. Using the matrix obtained from part (a), write the matrix obtained by multiplying the first row by 2 and adding the result to row 3.

33. Given the matrix **S**

$$\mathbf{S} = \left[\begin{array}{ccc|c} 1 & 2 & 0 & 10 \\ 5 & 1 & -4 & 3 \\ -3 & 4 & 5 & 2 \end{array}\right]$$

a. Write the matrix obtained by multiplying the first row by -5 and adding the result to row 2.

b. Using the matrix obtained from part (a), write the matrix obtained by multiplying the first row by 3 and adding the result to row 3.

For Exercises 34–49, solve the systems by using the Gauss-Jordan method.

34. $\begin{aligned} x - 2y &= -1 \\ 2x + y &= -7 \end{aligned}$

35. $\begin{aligned} x - 3y &= 3 \\ 2x - 5y &= 4 \end{aligned}$

36. $\begin{aligned} x + 3y &= 6 \\ -4x - 9y &= 3 \end{aligned}$

37. $\begin{aligned} 2x - 3y &= -2 \\ x + 2y &= 13 \end{aligned}$

38. $\begin{aligned} x + 3y &= 3 \\ 4x + 12y &= 12 \end{aligned}$

39. $\begin{aligned} 2x + 5y &= 1 \\ -4x - 10y &= -2 \end{aligned}$

40. $\begin{aligned} x - y &= 4 \\ 2x + y &= 5 \end{aligned}$

41. $\begin{aligned} 2x - y &= 0 \\ x + y &= 3 \end{aligned}$

42. $\begin{aligned} x + 3y &= -1 \\ -3x - 6y &= 12 \end{aligned}$

43. $\begin{aligned} x + y &= 4 \\ 2x - 4y &= -4 \end{aligned}$

44. $\begin{aligned} 3x + y &= -4 \\ -6x - 2y &= 3 \end{aligned}$

45. $\begin{aligned} 2x + y &= 4 \\ 6x + 3y &= -1 \end{aligned}$

46. $\begin{aligned} x + y + z &= 6 \\ x - y + z &= 2 \\ x + y - z &= 0 \end{aligned}$

47. $\begin{aligned} 2x - 3y - 2z &= 11 \\ x + 3y + 8z &= 1 \\ 3x - y + 14z &= -2 \end{aligned}$

48. $\begin{aligned} x - 2y &= 5 - z \\ 2x + 6y + 3z &= -10 \\ 3x - y - 2z &= 5 \end{aligned}$

49. $\begin{aligned} 5x - 10z &= 15 \\ x - y + 6z &= 23 \\ x + 3y - 12z &= 13 \end{aligned}$

For Exercises 50–53, use the augmented matrices **A**, **B**, **C**, and **D** to answer true or false.

$$\mathbf{A} = \begin{bmatrix} 6 & -4 & | & 2 \\ 5 & -2 & | & 7 \end{bmatrix} \quad \mathbf{B} = \begin{bmatrix} 5 & -2 & | & 7 \\ 6 & -4 & | & 2 \end{bmatrix} \quad \mathbf{C} = \begin{bmatrix} 1 & -\frac{2}{3} & | & \frac{1}{3} \\ 5 & -2 & | & 7 \end{bmatrix} \quad \mathbf{D} = \begin{bmatrix} 5 & -2 & | & 7 \\ -12 & 8 & | & -4 \end{bmatrix}$$

50. The matrix **A** is a 2×3 matrix.

51. Matrix **B** is equivalent to matrix **A**.

52. Matrix **A** is equivalent to matrix **C**.

53. Matrix **B** is equivalent to matrix **D**.

54. What does the notation $R_2 \Leftrightarrow R_1$ mean when one is performing the Gauss-Jordan method?

55. What does the notation $2R_3 \Rightarrow R_3$ mean when one is performing the Gauss-Jordan method?

56. What does the notation $-3R_1 + R_2 \Rightarrow R_2$ mean when one is performing the Gauss-Jordan method?

57. What does the notation $4R_2 + R_3 \Rightarrow R_3$ mean when one is performing the Gauss-Jordan method?

Graphing Calculator Exercises

For Exercises 58–63, use the matrix features on a graphing calculator to express each augmented matrix in reduced row echelon form. Compare your results to the solution you obtained in the indicated exercise.

58. $\begin{bmatrix} 1 & -2 & | & -1 \\ 2 & 1 & | & -7 \end{bmatrix}$

Compare with Exercise 34.

59. $\begin{bmatrix} 1 & -3 & | & 3 \\ 2 & -5 & | & 4 \end{bmatrix}$

Compare with Exercise 35.

60. $\begin{bmatrix} 1 & 3 & | & 6 \\ -4 & -9 & | & 3 \end{bmatrix}$

Compare with Exercise 36.

61. $\begin{bmatrix} 2 & -3 & | & -2 \\ 1 & 2 & | & 13 \end{bmatrix}$

Compare with Exercise 37.

62. $\begin{bmatrix} 1 & 1 & 1 & | & 6 \\ 1 & -1 & 1 & | & 2 \\ 1 & 1 & -1 & | & 0 \end{bmatrix}$

Compare with Exercise 46.

63. $\begin{bmatrix} 2 & -3 & -2 & | & 11 \\ 1 & 3 & 8 & | & 1 \\ 3 & -1 & 14 & | & -2 \end{bmatrix}$

Compare with Exercise 47.

Section 9.4 Determinants and Cramer's Rule

1. Introduction to Determinants

Associated with every square matrix is a real number called the **determinant** of the matrix. A determinant of a square matrix $\mathbf{A}$, denoted $\det \mathbf{A}$, is written by enclosing the elements of the matrix within two vertical bars. For example,

$$\text{If} \quad \mathbf{A} = \begin{bmatrix} 2 & -1 \\ 6 & 0 \end{bmatrix} \quad \text{then} \quad \det \mathbf{A} = \begin{vmatrix} 2 & -1 \\ 6 & 0 \end{vmatrix}$$

$$\text{If} \quad \mathbf{B} = \begin{bmatrix} 0 & -5 & 1 \\ 4 & 0 & \frac{1}{2} \\ -2 & 10 & 1 \end{bmatrix} \quad \text{then} \quad \det \mathbf{B} = \begin{vmatrix} 0 & -5 & 1 \\ 4 & 0 & \frac{1}{2} \\ -2 & 10 & 1 \end{vmatrix}$$

Determinants have many applications in mathematics, including solving systems of linear equations, finding the area of a triangle, determining whether three points are collinear, and finding an equation of a line between two points.

The determinant of a 2 × 2 matrix is defined as follows:

Determinant of a 2 × 2 Matrix

The determinant of the matrix $\begin{bmatrix} a & b \\ c & d \end{bmatrix}$ is the real number $ad - bc$. It is written as

$$\begin{vmatrix} a & b \\ c & d \end{vmatrix} = ad - bc$$

Example 1 Evaluating a 2 × 2 Determinant

Evaluate the determinants.

a. $\begin{vmatrix} 6 & -2 \\ 5 & \frac{1}{3} \end{vmatrix}$ **b.** $\begin{vmatrix} 2 & -11 \\ 0 & 0 \end{vmatrix}$

Solution:

a. $\begin{vmatrix} 6 & -2 \\ 5 & \frac{1}{3} \end{vmatrix}$ For this determinant, $a = 6$, $b = -2$, $c = 5$, and $d = \frac{1}{3}$.

$$ad - bc = (6)\left(\frac{1}{3}\right) - (-2)(5)$$

$$= 2 + 10$$

$$= 12$$

b. $\begin{vmatrix} 2 & -11 \\ 0 & 0 \end{vmatrix}$ For this determinant, $a = 2$, $b = -11$, $c = 0$, $d = 0$.

$$ad - bc = (2)(0) - (-11)(0)$$

$$= 0 - 0$$

$$= 0$$

TIP: Example 1(b) illustrates that the value of a determinant having a row of all zeros is 0. The same is true for a determinant having a column of all zeros.

Evaluate the determinants.

1a. $\begin{vmatrix} 2 & 8 \\ -1 & 5 \end{vmatrix}$ **b.** $\begin{vmatrix} -6 & 0 \\ 4 & 0 \end{vmatrix}$

2. Determinant of a 3 × 3 Matrix

To find the determinant of a 3 × 3 matrix, we first need to define the **minor** of an element of the matrix. For any element of a 3 × 3 matrix, the minor of that element is the determinant of the 2 × 2 matrix obtained by deleting the row and column in which the element resides. For example, consider the matrix

$$\begin{bmatrix} 5 & -1 & 6 \\ 0 & -7 & 1 \\ 4 & 2 & 6 \end{bmatrix}$$

The minor of the element 5 is found by deleting the first row and first column and then evaluating the determinant of the remaining 2 × 2 matrix:

$$\begin{bmatrix} 5 & -1 & 6 \\ 0 & -7 & 1 \\ 4 & 2 & 6 \end{bmatrix}$$ Now evaluate the determinant: $\begin{vmatrix} -7 & 1 \\ 2 & 6 \end{vmatrix} = (-7)(6) - (1)(2)$

$$= -44$$

For this matrix, the minor for the element 5 is −44.

To find the minor of the element −7, delete the second row and second column, and then evaluate the determinant of the remaining 2 × 2 matrix.

$$\begin{bmatrix} 5 & -1 & 6 \\ 0 & -7 & 1 \\ 4 & 2 & 6 \end{bmatrix}$$ Now evaluate the determinant: $\begin{vmatrix} 5 & 6 \\ 4 & 6 \end{vmatrix} = (5)(6) - (6)(4) = 6$

For this matrix, the minor for the element −7 is 6.

Example 2 Determining the Minor for Elements in a 3 × 3 Matrix

Find the minor for each element in the first column of the matrix.

$$\begin{bmatrix} 3 & 4 & -1 \\ 2 & -4 & 5 \\ 0 & 1 & -6 \end{bmatrix}$$

Solution:

For 3: $\begin{bmatrix} 3 & 4 & -1 \\ 2 & -4 & 5 \\ 0 & 1 & -6 \end{bmatrix}$ The minor is: $\begin{vmatrix} -4 & 5 \\ 1 & -6 \end{vmatrix} = (-4)(-6) - (5)(1) = 19$

Skill Practice Answers
1a. 18   **b.** 0

For 2: $\begin{bmatrix} 3 & 4 & -1 \\ 2 & -4 & 5 \\ 0 & 1 & -6 \end{bmatrix}$ The minor is: $\begin{vmatrix} 4 & -1 \\ 1 & -6 \end{vmatrix} = (4)(-6) - (-1)(1) = -23$

For 0: $\begin{bmatrix} 3 & 4 & -1 \\ 2 & -4 & 5 \\ 0 & 1 & -6 \end{bmatrix}$ The minor is: $\begin{vmatrix} 4 & -1 \\ -4 & 5 \end{vmatrix} = (4)(5) - (-1)(-4) = 16$

Skill Practice

2. Find the minor for the element 3.

$$\begin{bmatrix} -1 & 8 & -6 \\ \frac{1}{2} & 3 & 2 \\ 5 & 7 & 4 \end{bmatrix}$$

The determinant of a 3 × 3 matrix is defined as follows.

Definition of a Determinant of a 3 × 3 Matrix

$$\begin{vmatrix} a_1 & b_1 & c_1 \\ a_2 & b_2 & c_2 \\ a_3 & b_3 & c_3 \end{vmatrix} = a_1 \cdot \begin{vmatrix} b_2 & c_2 \\ b_3 & c_3 \end{vmatrix} - a_2 \cdot \begin{vmatrix} b_1 & c_1 \\ b_3 & c_3 \end{vmatrix} + a_3 \cdot \begin{vmatrix} b_1 & c_1 \\ b_2 & c_2 \end{vmatrix}$$

From this definition, we see that the determinant of a 3 × 3 matrix can be written as

$$a_1 \cdot (\text{minor of } a_1) - a_2 \cdot (\text{minor of } a_2) + a_3 \cdot (\text{minor of } a_3)$$

Evaluating determinants in this way is called *expanding minors*.

Example 3 Evaluating a 3 × 3 Determinant

Evaluate the determinant. $\begin{vmatrix} 2 & 4 & 2 \\ 1 & -3 & 0 \\ -5 & 5 & -1 \end{vmatrix}$

Solution:

$$\begin{vmatrix} 2 & 4 & 2 \\ 1 & -3 & 0 \\ -5 & 5 & -1 \end{vmatrix} = 2 \cdot \begin{vmatrix} -3 & 0 \\ 5 & -1 \end{vmatrix} - (1) \cdot \begin{vmatrix} 4 & 2 \\ 5 & -1 \end{vmatrix} + (-5) \cdot \begin{vmatrix} 4 & 2 \\ -3 & 0 \end{vmatrix}$$

$$= 2[(-3)(-1) - (0)(5)] - 1[(4)(-1) - (2)(5)] - 5[(4)(0) - (2)(-3)]$$

$$= 2(3) - 1(-14) - 5(6)$$

$$= 6 + 14 - 30$$

$$= -10$$

Skill Practice Answers

2. $\begin{vmatrix} -1 & -6 \\ 5 & 4 \end{vmatrix} = 26$

3. Evaluate the determinant.

$$\begin{vmatrix} -2 & 4 & 9 \\ 5 & -1 & 2 \\ 1 & 1 & 6 \end{vmatrix}$$

Although we defined the determinant of a matrix by expanding the minors of the elements in the first column, *any row or column may be used.* However, we must choose the correct sign to apply to each term in the expansion. The following array of signs is helpful.

$$\begin{array}{ccc} + & - & + \\ - & + & - \\ + & - & + \end{array}$$

The signs alternate for each row and column, beginning with $+$ in the first row, first column.

TIP: There is another method to determine the signs for each term of the expansion. For the a_{ij} element, multiply the term by $(-1)^{i+j}$.

Example 4   **Evaluating a 3 × 3 Determinant**

Evaluate the determinant, by expanding minors about the elements in the second row.

$$\begin{vmatrix} 2 & 4 & 2 \\ 1 & -3 & 0 \\ -5 & 5 & -1 \end{vmatrix}$$

Solution:

Signs obtained from the array of signs

$$\begin{vmatrix} 2 & 4 & 2 \\ 1 & -3 & 0 \\ -5 & 5 & -1 \end{vmatrix} = -(1) \cdot \begin{vmatrix} 4 & 2 \\ 5 & -1 \end{vmatrix} + (-3) \cdot \begin{vmatrix} 2 & 2 \\ -5 & -1 \end{vmatrix} - (0) \cdot \begin{vmatrix} 2 & 4 \\ -5 & 5 \end{vmatrix}$$

$$= -1[(4)(-1) - (2)(5)] - 3[(2)(-1) - (2)(-5)] - 0$$

$$= -1(-14) - 3(8)$$

$$= 14 - 24$$

$$= -10$$

TIP: Notice that the value of the determinant obtained in Examples 3 and 4 is the same.

4. Evaluate the determinant.

$$\begin{vmatrix} 4 & -1 & 2 \\ 3 & 6 & -8 \\ 0 & \frac{1}{2} & 5 \end{vmatrix}$$

In Example 4, the third term in the expansion of minors was zero because the element 0 when multiplied by its minor is zero. To simplify the arithmetic in evaluating a determinant of a 3×3 matrix, expand about the row or column that has the most 0 elements.

Skill Practice Answers

3. -42
4. 154

Calculator Connections

The determinant of a matrix can be evaluated on a graphing calculator. First use the matrix editor to enter the elements of the matrix. Then use a *det* function to evaluate the determinant. The determinant from Examples 3 and 4 is evaluated below.

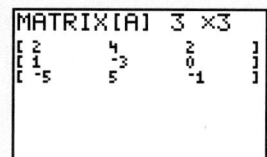 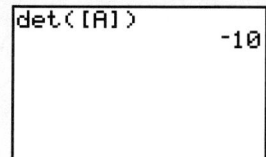

3. Cramer's Rule

In this section, we will learn another method called **Cramer's rule** to solve a system of linear equations.

Cramer's Rule for a 2 × 2 System of Linear Equations

The solution to the system

$$a_1 x + b_1 y = c_1$$
$$a_2 x + b_2 y = c_2$$

is given by $x = \dfrac{\mathbf{D}_x}{\mathbf{D}}$ and $y = \dfrac{\mathbf{D}_y}{\mathbf{D}}$

where $\mathbf{D} = \begin{vmatrix} a_1 & b_1 \\ a_2 & b_2 \end{vmatrix}$ (and $\mathbf{D} \neq 0$) $\mathbf{D}_x = \begin{vmatrix} c_1 & b_1 \\ c_2 & b_2 \end{vmatrix}$ $\mathbf{D}_y = \begin{vmatrix} a_1 & c_1 \\ a_2 & c_2 \end{vmatrix}$

Example 5 **Using Cramer's Rule to Solve a 2 × 2 System of Linear Equations**

Solve the system by using Cramer's rule.

$$3x - 5y = 11$$
$$-x + 3y = -5$$

Solution:

For this system: $a_1 = 3$ $b_1 = -5$ $c_1 = 11$

$a_2 = -1$ $b_2 = 3$ $c_2 = -5$

$$\mathbf{D} = \begin{vmatrix} 3 & -5 \\ -1 & 3 \end{vmatrix} = (3)(3) - (-5)(-1) = 9 - 5 = 4$$

$$\mathbf{D}_x = \begin{vmatrix} 11 & -5 \\ -5 & 3 \end{vmatrix} = (11)(3) - (-5)(-5) = 33 - 25 = 8$$

$$\mathbf{D}_y = \begin{vmatrix} 3 & 11 \\ -1 & -5 \end{vmatrix} = (3)(-5) - (11)(-1) = -15 + 11 = -4$$

Therefore, $\quad x = \dfrac{\mathbf{D}_x}{\mathbf{D}} = \dfrac{8}{4} = 2 \qquad y = \dfrac{\mathbf{D}_y}{\mathbf{D}} = \dfrac{-4}{4} = -1$

The solution is $(2, -1)$. <u>Check:</u> $3x - 5y = 11 \longrightarrow 3(2) - 5(-1) = 11$ ✔

$\qquad\qquad\qquad\qquad\qquad\qquad -x + 3y = -5 \longrightarrow -(2) + 3(-1) = -5$ ✔

Skill Practice

5. Solve using Cramer's rule.

$$2x + y = 5$$
$$-x - 3y = 5$$

TIP: Here are some memory tips to help you remember Cramer's rule.

Coefficients of
x-terms   *y*-terms

1. The determinant **D** is the determinant of the coefficients of *x* and *y*.

$$\mathbf{D} = \begin{vmatrix} a_1 & b_1 \\ a_2 & b_2 \end{vmatrix}$$

x-coefficients
replaced by c_1 and c_2

2. The determinant $\mathbf{D}_x$ has the column of *x*-term coefficients replaced by c_1 and c_2.

$$\mathbf{D}_x = \begin{vmatrix} c_1 & b_1 \\ c_2 & b_2 \end{vmatrix}$$

y-coefficients
replaced by c_1 and c_2

3. The determinant $\mathbf{D}_y$ has the column of *y*-term coefficients replaced by c_1 and c_2.

$$\mathbf{D}_y = \begin{vmatrix} a_1 & c_1 \\ a_2 & c_2 \end{vmatrix}$$

It is important to note that the linear equations must be written in standard form to apply Cramer's rule.

Example 6 **Using Cramer's Rule to Solve a 2 × 2 System of Linear Equations**

Solve the system by using Cramer's rule.

$$-16y = -40x - 7$$
$$40y = 24x + 27$$

Solution:

$$-16y = -40x - 7 \longrightarrow 40x - 16y = -7 \qquad \text{Rewrite each equation}$$
$$40y = 24x + 27 \longrightarrow -24x + 40y = 27 \qquad \text{in standard form.}$$

For this system:

$$a_1 = 40 \qquad b_1 = -16 \qquad c_1 = -7$$
$$a_2 = -24 \qquad b_2 = 40 \qquad c_2 = 27$$

$$\mathbf{D} = \begin{vmatrix} 40 & -16 \\ -24 & 40 \end{vmatrix} = (40)(40) - (-16)(-24) = 1216$$

$$\mathbf{D}_x = \begin{vmatrix} -7 & -16 \\ 27 & 40 \end{vmatrix} = (-7)(40) - (-16)(27) = 152$$

$$\mathbf{D}_y = \begin{vmatrix} 40 & -7 \\ -24 & 27 \end{vmatrix} = (40)(27) - (-7)(-24) = 912$$

Therefore, $\qquad x = \dfrac{\mathbf{D}_x}{\mathbf{D}} = \dfrac{152}{1216} = \dfrac{1}{8} \qquad y = \dfrac{\mathbf{D}_y}{\mathbf{D}} = \dfrac{912}{1216} = \dfrac{3}{4}$

The solution $\left(\frac{1}{8}, \frac{3}{4}\right)$ checks in the original equations.

Skill Practice

6. Solve using Cramer's rule.

$$9x - 12y = -8$$
$$18x + 30y = -7$$

Cramer's rule can be used to solve a 3×3 system of linear equations by using a similar pattern of determinants.

Cramer's Rule for a 3 × 3 System of Linear Equations

The solution to the system

$$a_1x + b_1y + c_1z = d_1$$
$$a_2x + b_2y + c_2z = d_2$$
$$a_3x + b_3y + c_3z = d_3$$

is given by

$$x = \frac{\mathbf{D}_x}{\mathbf{D}} \qquad y = \frac{\mathbf{D}_y}{\mathbf{D}} \qquad \text{and} \qquad z = \frac{\mathbf{D}_z}{\mathbf{D}}$$

where $\mathbf{D} = \begin{vmatrix} a_1 & b_1 & c_1 \\ a_2 & b_2 & c_2 \\ a_3 & b_3 & c_3 \end{vmatrix}$ (and $\mathbf{D} \neq 0$) $\qquad \mathbf{D}_x = \begin{vmatrix} d_1 & b_1 & c_1 \\ d_2 & b_2 & c_2 \\ d_3 & b_3 & c_3 \end{vmatrix}$

$$\mathbf{D}_y = \begin{vmatrix} a_1 & d_1 & c_1 \\ a_2 & d_2 & c_2 \\ a_3 & d_3 & c_3 \end{vmatrix} \qquad\qquad \mathbf{D}_z = \begin{vmatrix} a_1 & b_1 & d_1 \\ a_2 & b_2 & d_2 \\ a_3 & b_3 & d_3 \end{vmatrix}$$

Skill Practice Answers

6. $\left(-\dfrac{2}{3}, \dfrac{1}{6}\right)$

Example 7 | Using Cramer's Rule to Solve a 3 × 3 System of Linear Equations

Solve the system by using Cramer's rule.

$$x - 2y + 4z = 3$$
$$x - 4y + 3z = -5$$
$$x + 3y - 2z = 6$$

Solution:

$$D = \begin{vmatrix} 1 & -2 & 4 \\ 1 & -4 & 3 \\ 1 & 3 & -2 \end{vmatrix} = 1 \cdot \begin{vmatrix} -4 & 3 \\ 3 & -2 \end{vmatrix} - 1 \cdot \begin{vmatrix} -2 & 4 \\ 3 & -2 \end{vmatrix} + 1 \cdot \begin{vmatrix} -2 & 4 \\ -4 & 3 \end{vmatrix}$$

$$= 1(-1) - 1(-8) + 1(10)$$

$$= 17$$

$$D_x = \begin{vmatrix} 3 & -2 & 4 \\ -5 & -4 & 3 \\ 6 & 3 & -2 \end{vmatrix} = 3 \cdot \begin{vmatrix} -4 & 3 \\ 3 & -2 \end{vmatrix} - (-5) \cdot \begin{vmatrix} -2 & 4 \\ 3 & -2 \end{vmatrix} + 6 \cdot \begin{vmatrix} -2 & 4 \\ -4 & 3 \end{vmatrix}$$

$$= 3(-1) + 5(-8) + 6(10)$$

$$= 17$$

$$D_y = \begin{vmatrix} 1 & 3 & 4 \\ 1 & -5 & 3 \\ 1 & 6 & -2 \end{vmatrix} = 1 \cdot \begin{vmatrix} -5 & 3 \\ 6 & -2 \end{vmatrix} - 1 \cdot \begin{vmatrix} 3 & 4 \\ 6 & -2 \end{vmatrix} + 1 \cdot \begin{vmatrix} 3 & 4 \\ -5 & 3 \end{vmatrix}$$

$$= 1(-8) - 1(-30) + 1(29)$$

$$= 51$$

$$D_z = \begin{vmatrix} 1 & -2 & 3 \\ 1 & -4 & -5 \\ 1 & 3 & 6 \end{vmatrix} = 1 \cdot \begin{vmatrix} -4 & -5 \\ 3 & 6 \end{vmatrix} - 1 \cdot \begin{vmatrix} -2 & 3 \\ 3 & 6 \end{vmatrix} + 1 \cdot \begin{vmatrix} -2 & 3 \\ -4 & -5 \end{vmatrix}$$

$$= 1(-9) - 1(-21) + 1(22)$$

$$= 34$$

TIP: In Example 7, we expanded the determinants about the first column.

Hence

$$x = \frac{D_x}{D} = \frac{17}{17} = 1 \qquad y = \frac{D_y}{D} = \frac{51}{17} = 3 \qquad \text{and} \qquad z = \frac{D_z}{D} = \frac{34}{17} = 2$$

The solution is $(1, 3, 2)$.

Check: $x - 2y + 4z = 3$ $(1) - 2(3) + 4(2) = 3$ ✔

$x - 4y + 3z = -5$ $(1) - 4(3) + 3(2) = -5$ ✔

$x + 3y - 2z = 6$ $(1) + 3(3) - 2(2) = 6$ ✔

Skill Practice

7. Solve using Cramer's rule.

$$x + 3y - 3z = -14$$
$$x - 4y + z = 2$$
$$x + y + 2z = 6$$

Skill Practice Answers

7. $(-2, 0, 4)$

Cramer's rule may seem cumbersome for solving a 3×3 system of linear equations. However, it provides convenient formulas that can be programmed into a computer or calculator to solve for x, y, and z. Cramer's rule can also be extended to solve a 4×4 system of linear equations, a 5×5 system of linear equations, and in general an $n \times n$ system of linear equations.

It is important to remember that Cramer's rule does not apply if $\mathbf{D} = 0$. In such a case, the system of equations is either dependent or inconsistent, and another method must be used to analyze the system.

Example 8   **Analyzing a Dependent System of Equations**

Solve the system. Use Cramer's rule if possible.

$$2x - 3y = 6$$
$$-6x + 9y = -18$$

Solution:

$$\mathbf{D} = \begin{vmatrix} 2 & -3 \\ -6 & 9 \end{vmatrix} = (2)(9) - (-3)(-6) = 18 - 18 = 0$$

Because $\mathbf{D} = 0$, Cramer's rule does not apply. Using the addition method to solve the system, we have

$$\begin{array}{lcl} & \text{Multiply by 3.} & \\ 2x - 3y = 6 & \longrightarrow & 6x - 9y = 18 \\ -6x + 9y = -18 & \longrightarrow & \underline{-6x + 9y = -18} \\ & & 0 = 0 \quad \text{The system is dependent.} \end{array}$$

The solution is $\{(x, y) \mid 2x - 3y = 6\}$.

Skill Practice

8. Solve. Use Cramer's rule if possible.

$$x - 6y = 2$$
$$2x - 12y = -2$$

Skill Practice Answers

8. No solution; Inconsistent system

Study Skills Exercise

1. Define the key terms.

   a. determinant   b. minor   c. Cramer's rule

Concept 1: Introduction to Determinants

For Exercises 2–7, evaluate the determinant of the 2×2 matrix.

2. $\begin{vmatrix} -3 & 1 \\ 5 & 2 \end{vmatrix}$

3. $\begin{vmatrix} 5 & 6 \\ 4 & 8 \end{vmatrix}$

4. $\begin{vmatrix} -2 & 2 \\ -3 & -5 \end{vmatrix}$

5. $\begin{vmatrix} 5 & -1 \\ 1 & 0 \end{vmatrix}$

6. $\begin{vmatrix} \frac{1}{2} & 3 \\ -2 & 4 \end{vmatrix}$

7. $\begin{vmatrix} -3 & \frac{1}{4} \\ 8 & -2 \end{vmatrix}$

Concept 2: Determinant of a 3 × 3 Matrix

For Exercises 8–11, evaluate the minor corresponding to the given element from matrix **A**.

$$\mathbf{A} = \begin{bmatrix} 4 & -1 & 8 \\ 2 & 6 & 0 \\ -7 & 5 & 3 \end{bmatrix}$$

8. 4

9. −1

10. 2

11. 3

For Exercises 12–15, evaluate the minor corresponding to the given element from matrix **B**.

$$\mathbf{B} = \begin{bmatrix} -2 & 6 & 0 \\ 4 & -2 & 1 \\ 5 & 9 & -1 \end{bmatrix}$$

12. 6

13. 5

14. 1

15. 0

16. Construct the sign array for a 3×3 matrix.

17. Evaluate the determinant of matrix **B**, using expansion by minors.

$$\mathbf{B} = \begin{bmatrix} 0 & 1 & 2 \\ 3 & -1 & 2 \\ 3 & 2 & -2 \end{bmatrix}$$

a. About the first column

b. About the second row

18. Evaluate the determinant of matrix **C**, using expansion by minors.

$$\mathbf{C} = \begin{bmatrix} 4 & 1 & 3 \\ 2 & -2 & 1 \\ 3 & 1 & 2 \end{bmatrix}$$

a. About the first row

b. About the second column

19. When evaluating the determinant of a 3×3 matrix, explain the advantage of being able to choose any row or column about which to expand minors.

For Exercises 20–25, evaluate the determinants.

20. $\begin{vmatrix} 8 & 2 & -4 \\ 4 & 0 & 2 \\ 3 & 0 & -1 \end{vmatrix}$

21. $\begin{vmatrix} 5 & 2 & 1 \\ 3 & -6 & 0 \\ -2 & 8 & 0 \end{vmatrix}$

22. $\begin{vmatrix} -2 & 1 & 3 \\ 1 & 4 & 4 \\ 1 & 0 & 2 \end{vmatrix}$

23. $\begin{vmatrix} 3 & 2 & 1 \\ 1 & -1 & 2 \\ 1 & 0 & 4 \end{vmatrix}$

24. $\begin{vmatrix} -5 & 4 & 2 \\ 0 & 0 & 0 \\ 3 & -1 & 5 \end{vmatrix}$

25. $\begin{vmatrix} 0 & 5 & -8 \\ 0 & -4 & 1 \\ 0 & 3 & 6 \end{vmatrix}$

For Exercises 26–31, evaluate the determinants.

26. $\begin{vmatrix} x & 3 \\ y & -2 \end{vmatrix}$

27. $\begin{vmatrix} a & 2 \\ b & 8 \end{vmatrix}$

28. $\begin{vmatrix} a & 5 & -1 \\ b & -3 & 0 \\ c & 3 & 4 \end{vmatrix}$

29. $\begin{vmatrix} x & 0 & 3 \\ y & -2 & 6 \\ z & -1 & 1 \end{vmatrix}$

30. $\begin{vmatrix} p & 0 & q \\ r & 0 & s \\ t & 0 & u \end{vmatrix}$

31. $\begin{vmatrix} f & e & 0 \\ d & c & 0 \\ b & a & 0 \end{vmatrix}$

Concept 3: Cramer's Rule

For Exercises 32–34, evaluate the determinants represented by $\mathbf{D}$, $\mathbf{D}_x$, and $\mathbf{D}_y$.

32. $x - 4y = 2$
$3x + 2y = 1$

33. $4x + 6y = 9$
$-2x + y = 12$

34. $-3x + 8y = -10$
$5x + 5y = -13$

For Exercises 35–40, solve the system by using Cramer's rule.

35. $2x + y = 3$
$x - 4y = 6$

36. $2x - y = -1$
$3x + y = 6$

37. $x - 4y = 8$
$3x + 7y = 5$

38. $7x + 3y = 4$
$5x - 4y = 9$

39. $4x - 3y = 5$
$2x + 5y = 7$

40. $2x + 3y = 4$
$6x - 12y = -5$

41. When does Cramer's rule not apply in solving a system of equations?

42. How can a system be solved if Cramer's rule does not apply?

For Exercises 43–48, solve the system of equations by using Cramer's rule, if possible. If not possible, use another method.

43. $4x - 2y = 3$
$-2x + y = 1$

44. $6x - 6y = 5$
$x - y = 8$

45. $4x + y = 0$
$x - 7y = 0$

46. $-3x - 2y = 0$
$-x + 5y = 0$

47. $x + 5y = 3$
$2x + 10y = 6$

48. $-2x - 10y = -4$
$x + 5y = 2$

For Exercises 49–54, solve for the indicated variable by using Cramer's rule.

49. $2x - y + 3z = 9$
$x + 4y + 4z = 5$ for x
$3x + 2y + 2z = 5$

50. $x + 2y + 3z = 8$
$2x - 3y + z = 5$ for y
$3x - 4y + 2z = 9$

51. $3x - 2y + 2z = 5$
$6x + 3y - 4z = -1$ for z
$3x - y + 2z = 4$

52. $4x + 4y - 3z = 3$
$8x + 2y + 3y = 0$ for x
$4x - 4y + 6z = -3$

53. $5x + 6z = 5$
$-2x + y = -6$ for y
$3y - z = 3$

54. $8x + y = 1$
$7y + z = 0$ for y
$x - 3z = -2$

For Exercises 55–58, solve the system by using Cramer's rule, if possible.

55.

$$x \qquad\qquad = 3$$
$$-x + 3y \qquad = 3$$
$$y + 2z = 4$$

56. $4x \qquad + z = 7$
$$y \qquad = 2$$
$$x \qquad + z = 4$$

57. $x + y + \ 8z = 3$
$$2x + y + 11z = 4$$
$$x \qquad + \ 3z = 0$$

58. $-8x + y + z = 6$
$$2x - y + z = 3$$
$$3x \qquad - z = 0$$

Expanding Your Skills

For Exercises 59–62, solve the equation.

59. $\begin{vmatrix} 6 & x \\ 2 & -4 \end{vmatrix} = 14$

60. $\begin{vmatrix} y & -2 \\ 8 & 7 \end{vmatrix} = 30$

61. $\begin{vmatrix} 3 & 1 & 0 \\ 0 & 4 & -2 \\ 1 & 0 & w \end{vmatrix} = 10$

62. $\begin{vmatrix} -1 & 0 & 2 \\ 4 & t & 0 \\ 0 & -5 & 3 \end{vmatrix} = -4$

For Exercises 63–64, evaluate the determinant by using expansion by minors about the first column.

63. $\begin{vmatrix} 1 & 0 & 3 & 0 \\ 0 & 1 & 2 & 4 \\ -2 & 0 & 0 & 1 \\ 4 & -1 & -2 & 0 \end{vmatrix}$

64. $\begin{vmatrix} 5 & 2 & 0 & 0 \\ 0 & 4 & -1 & 1 \\ -1 & 0 & 3 & 0 \\ 0 & -2 & 1 & 0 \end{vmatrix}$

For Exercises 65–66, refer to the following system of four variables.

$$x + y + z + w = 0$$
$$2x \qquad - z + w = 5$$
$$2x + y \qquad - w = 0$$
$$y + z \qquad = -1$$

65. a. Evaluate the determinant **D**.

    b. Evaluate the determinant $\mathbf{D}_x$.

    c. Solve for x by computing $\dfrac{\mathbf{D}_x}{\mathbf{D}}$.

66. a. Evaluate the determinant $\mathbf{D}_y$.

    b. Solve for y by computing $\dfrac{\mathbf{D}_y}{\mathbf{D}}$.

Chapter 9   SUMMARY

Section 9.1   Systems of Linear Equations in Three Variables

Key Concepts

A **linear equation in three variables** can be written in the form $Ax + By + Cz = D$, where A, B, and C are not all zero. The graph of a linear equation in three variables is a plane in space.

A solution to a system of linear equations in three variables is an **ordered triple** that satisfies each equation. Graphically, a solution is a point of intersection among three planes.

A system of linear equations in three variables may have one unique solution, infinitely many solutions (dependent system), or no solution (inconsistent system).

Examples

Example 1

$\boxed{A}$ $x + 2y - z = 4$

$\boxed{B}$ $3x - y + z = 5$

$\boxed{C}$ $2x + 3y + 2z = 7$

$\boxed{A}$ and $\boxed{B}$
$$\begin{aligned} x + 2y - z &= 4 \\ 3x - y + z &= 5 \\ \hline 4x + y &= 9 \quad \boxed{D} \end{aligned}$$

$2 \cdot \boxed{A}$ and $\boxed{C}$
$$\begin{aligned} 2x + 4y - 2z &= 8 \\ 2x + 3y + 2z &= 7 \\ \hline 4x + 7y &= 15 \quad \boxed{E} \end{aligned}$$

$\boxed{D}$ $4x + y = 9 \;\rightarrow\; -4x - y = -9$

$\boxed{E}$
$$\begin{aligned} 4x + 7y &= 15 \;\rightarrow\; \underline{4x + 7y = 15} \\ 6y &= 6 \\ y &= 1 \end{aligned}$$

Substitute $y = 1$ into either equation $\boxed{D}$ or $\boxed{E}$.

$\boxed{D}$
$$\begin{aligned} 4x + (1) &= 9 \\ 4x &= 8 \\ x &= 2 \end{aligned}$$

Substitute $x = 2$ and $y = 1$ into equation $\boxed{A}$, $\boxed{B}$, or $\boxed{C}$.

$\boxed{A}$
$$\begin{aligned} (2) + 2(1) - z &= 4 \\ z &= 0 \end{aligned}$$

The solution is $(2, 1, 0)$.

| Section 9.2 | Applications of Systems of Linear Equations in Three Variables |

Key Concepts

In application problems where there are three unknowns, it is often useful to set up a system of linear equations in three variables.

With three variables, three independent relationships among x, y, and z are required.

Examples

Example 1

In a triangle, the middle angle measures 55° more than the smallest angle. The largest angle measures 10° more than the sum of the other two angles. Find the measure of the angles.

Let x = measure of the smallest angle

Let y = measure of the middle angle

Let z = measure of the largest angle

First recall that the sum of the angles in a triangle is 180°.

standard form

$$x + y + z = 180 \longrightarrow x + y + z = 180$$
$$y = x + 55 \longrightarrow -x + y \quad = 55$$
$$z = x + y + 10 \longrightarrow -x - y + z = 10$$

Solving the system yields, $x = 15$, $y = 70$, and $z = 95$.

The measures of the three angles are 15°, 70°, and 95°.

Section 9.3   Solving Systems of Linear Equations by Using Matrices

Key Concepts

A **matrix** is a rectangular array of numbers displayed in rows and columns. Every number or entry within a matrix is called an element of the matrix.

The **order of a matrix** is determined by the number of rows and number of columns. A matrix with m rows and n columns is an $m \times n$ matrix.

A system of equations written in standard form can be represented by an **augmented matrix** consisting of the coefficients of the terms of each equation in the system.

The Gauss-Jordan method can be used to solve a system of equations by using the following elementary row operations on an augmented matrix.

1. Interchange two rows.
2. Multiply every element in a row by a nonzero real number.
3. Add a multiple of one row to another row.

These operations are used to write the matrix in **reduced row echelon form**.

$$\begin{bmatrix} 1 & 0 & | & a \\ 0 & 1 & | & b \end{bmatrix}$$

which represents the solution, $x = a$ and $y = b$.

Examples

Example 1

$[1 \quad 2 \quad 5]$ is a 1×3 matrix (called a row matrix).

$\begin{bmatrix} -1 & 8 \\ 1 & 5 \end{bmatrix}$ is a 2×2 matrix (called a square matrix).

$\begin{bmatrix} 4 \\ 1 \end{bmatrix}$ is a 2×1 matrix (called a column matrix).

Example 2

The augmented matrix for

$$\begin{aligned} 4x + y &= -12 \\ x - 2y &= 6 \end{aligned}$$

is $\begin{bmatrix} 4 & 1 & | & -12 \\ 1 & -2 & | & 6 \end{bmatrix}$

Example 3

Solve the system from Example 2 by using the Gauss-Jordan method.

$R_1 \Leftrightarrow R_2$ $\begin{bmatrix} 1 & -2 & | & 6 \\ 4 & 1 & | & -12 \end{bmatrix}$

$\xrightarrow{-4R_1 + R_2 \Rightarrow R_2}$ $\begin{bmatrix} 1 & -2 & | & 6 \\ 0 & 9 & | & -36 \end{bmatrix}$

$\xrightarrow{\frac{1}{9}R_2 \Rightarrow R_2}$ $\begin{bmatrix} 1 & -2 & | & 6 \\ 0 & 1 & | & -4 \end{bmatrix}$

$\xrightarrow{2R_2 + R_1 \Rightarrow R_1}$ $\begin{bmatrix} 1 & 0 & | & -2 \\ 0 & 1 & | & -4 \end{bmatrix}$

Solution:

$x = -2$ and $y = -4$

Section 9.4 Determinants and Cramer's Rule

Key Concepts

The **determinant** of matrix $\mathbf{A} = \begin{bmatrix} a & b \\ c & d \end{bmatrix}$

is denoted $\det\mathbf{A} = \begin{vmatrix} a & b \\ c & d \end{vmatrix}$.

The determinant of a 2×2 matrix is

defined as $\begin{vmatrix} a & b \\ c & d \end{vmatrix} = ad - bc$.

The determinant of a 3×3 matrix is defined by

$$\begin{vmatrix} a_1 & b_1 & c_1 \\ a_2 & b_2 & c_2 \\ a_3 & b_3 & c_3 \end{vmatrix} = a_1 \cdot \begin{vmatrix} b_2 & c_2 \\ b_3 & c_3 \end{vmatrix} - a_2 \cdot \begin{vmatrix} b_1 & c_1 \\ b_3 & c_3 \end{vmatrix}$$
$$+ a_3 \cdot \begin{vmatrix} b_1 & c_1 \\ b_2 & c_2 \end{vmatrix}$$

Cramer's rule can be used to solve a 2×2 system of linear equations.

$\begin{matrix} a_1x + b_1y = c_1 \\ a_2x + b_2y = c_2 \end{matrix}$ is given by $x = \dfrac{\mathbf{D}_x}{\mathbf{D}}$

and $y = \dfrac{\mathbf{D}_y}{\mathbf{D}}$

where $\mathbf{D} = \begin{vmatrix} a_1 & b_1 \\ a_2 & b_2 \end{vmatrix}$ (and $\mathbf{D} \neq 0$),

$\mathbf{D}_x = \begin{vmatrix} c_1 & b_1 \\ c_2 & b_2 \end{vmatrix}$ and $\mathbf{D}_y = \begin{vmatrix} a_1 & c_1 \\ a_2 & c_2 \end{vmatrix}$.

Examples

Example 1

For $\mathbf{A} = \begin{bmatrix} 7 & -2 \\ 3 & 2 \end{bmatrix}$,

$\det\mathbf{A} = \begin{vmatrix} 7 & -2 \\ 3 & 2 \end{vmatrix} = 7(2) - (-2)(3)$
$= 14 - (-6)$
$= 20$

Example 2

For $\mathbf{B} = \begin{bmatrix} 3 & -1 & 4 \\ 0 & 5 & 1 \\ 6 & -2 & -3 \end{bmatrix}$,

$\det\mathbf{B} = \begin{vmatrix} 3 & -1 & 4 \\ 0 & 5 & 1 \\ 6 & -2 & -3 \end{vmatrix}$

$= 3 \cdot \begin{vmatrix} 5 & 1 \\ -2 & -3 \end{vmatrix} - 0 \cdot \begin{vmatrix} -1 & 4 \\ -2 & -3 \end{vmatrix} + 6 \cdot \begin{vmatrix} -1 & 4 \\ 5 & 1 \end{vmatrix}$

$= 3(-15 + 2) - 0(3 + 8) + 6(-1 - 20)$
$= -39 - 0 + (-126)$
$= -165$

Example 3

Solve $\begin{matrix} 2x + 8y = 0 \\ x + 3y = 1 \end{matrix}$

$\mathbf{D} = \begin{vmatrix} 2 & 8 \\ 1 & 3 \end{vmatrix} = -2, \quad \mathbf{D}_x = \begin{vmatrix} 0 & 8 \\ 1 & 3 \end{vmatrix} = -8,$

$\mathbf{D}_y = \begin{vmatrix} 2 & 0 \\ 1 & 1 \end{vmatrix} = 2$

Therefore, $x = \dfrac{-8}{-2} = 4, \; y = \dfrac{2}{-2} = -1$

Chapter 9 Review Exercises

Section 9.1

For Exercises 1–6, solve the systems of equations. If a system does not have a unique solution, label the system as either dependent or inconsistent.

1.
$$5x + 5y + 5z = 30$$
$$-x + y + z = 2$$
$$10x + 6y - 2z = 4$$

2.
$$5x + 3y - z = 5$$
$$x + 2y + z = 6$$
$$-x - 2y - z = 8$$

3.
$$x + y + z = 4$$
$$-x - 2y - 3z = -6$$
$$2x + 4y + 6z = 12$$

4.
$$3x + 4z = 5$$
$$2y + 3z = 2$$
$$2x - 5y = 8$$

5.
$$3(x - y) = -1 - 5z - y$$
$$3y + 6 = 2z + x$$
$$5x + 1 = 3(z - y) - y$$

6.
$$2a = -3c - 2$$
$$2b - 8 = 5c$$
$$7a + 3b = 5$$

Section 9.2

7. The perimeter of a right triangle is 30 ft. One leg is 2 ft longer than twice the shortest leg. The hypotenuse is 2 ft less than 3 times the shortest leg. Find the lengths of the sides of this triangle.

8. Three pumps are working to drain a construction site. Working together, the pumps can pump 950 gal/hr of water. The slowest pump pumps 150 gal/hr less than the fastest pump. The fastest pump pumps 150 gal/hr less than the sum of the other two pumps. How many gallons can each pump drain per hour?

9. Theresa had $12,000 to invest among three mutual funds. Fund A had a 9% yield for the year. Fund B had a 14% yield, and Fund C lost 4% for the year. She invested three times as much in Fund A as in Fund C. If she gained $1020, in 1 year, how much was invested in each fund?

10. In a triangle, the largest angle measures 76° more than the sum of the other two angles. The middle angle measures 3 times the smallest angle. Find the measure of each angle.

For Exercises 11–12, write an equation of a parabola in the form $y = ax^2 + bx + c$ for a parabola passing through the three given points.

11. $(1, 8), (3, 0), (0, 15)$

12. $(2, -5), (-1, 4), (-3, 0)$

Section 9.3

For Exercises 13–16, determine the order of each matrix.

13.
$$\begin{bmatrix} 2 & 4 & -1 \\ 5 & 0 & -3 \\ -1 & 6 & 10 \end{bmatrix}$$

14.
$$\begin{bmatrix} -5 & 6 \\ 9 & 2 \\ 0 & -3 \end{bmatrix}$$

15. $\begin{bmatrix} 0 & 13 & -4 & 16 \end{bmatrix}$

16.
$$\begin{bmatrix} 7 \\ 12 \\ -4 \end{bmatrix}$$

For Exercises 17–18, set up the augmented matrix.

17.
$$x + y = 3$$
$$x - y = -1$$

18.
$$x - y + z = 4$$
$$2x - y + 3z = 8$$
$$-2x + 2y - z = -9$$

For Exercises 19–20, write a corresponding system of equations from the augmented matrix.

19. $\begin{bmatrix} 1 & 0 & | & 9 \\ 0 & 1 & | & -3 \end{bmatrix}$

20. $\begin{bmatrix} 1 & 0 & 0 & | & -5 \\ 0 & 1 & 0 & | & 2 \\ 0 & 0 & 1 & | & -8 \end{bmatrix}$

21. Given the matrix $\mathbf{C}$

$$\mathbf{C} = \begin{bmatrix} 1 & 3 & | & 1 \\ 4 & -1 & | & 6 \end{bmatrix}$$

    a. What is the element in the second row and first column?

    b. Write the matrix obtained by multiplying the first row by -4 and adding the result to row 2.

22. Given the matrix **D**

$$\mathbf{D} = \begin{bmatrix} 1 & 2 & 0 & | & -3 \\ 4 & -1 & 1 & | & 0 \\ -3 & 2 & 2 & | & 5 \end{bmatrix}$$

  a. Write the matrix obtained by multiplying the first row by -4 and adding the result to row 2.

  b. Using the matrix obtained in part (a), write the matrix obtained by multiplying the first row by 3 and adding the result to row 3.

For Exercises 23–26, solve the system by using the Gauss-Jordan method.

23. $x + y = 3$
  $x - y = -1$

24. $4x + 3y = 6$
  $12x + 5y = -6$

25. $x - y + z = -4$
  $2x + y - 2z = 9$
  $x + 2y + z = 5$

26. $x - y + z = 4$
  $2x - y + 3z = 8$
  $-2x + 2y - z = -9$

Section 9.4

For Exercises 27–30, evaluate the determinant.

27. $\begin{vmatrix} 5 & -2 \\ 2 & -3 \end{vmatrix}$

28. $\begin{vmatrix} -6 & 1 \\ 0 & 10 \end{vmatrix}$

29. $\begin{vmatrix} \frac{1}{2} & 3 \\ 1 & 8 \end{vmatrix}$

30. $\begin{vmatrix} 9 & 3 \\ -2 & \frac{2}{3} \end{vmatrix}$

For Exercises 31–34, evaluate the minor corresponding to the given element from matrix **A**.

$$\mathbf{A} = \begin{bmatrix} 8 & 2 & 0 \\ -1 & 4 & -2 \\ 3 & -3 & 6 \end{bmatrix}$$

31. 8   **32.** 2   **33.** -2 **34.** 4

For Exercises 35–38, evaluate the determinant.

35. $\begin{vmatrix} 2 & 1 & 0 \\ -4 & 3 & -1 \\ 3 & 0 & 1 \end{vmatrix}$

36. $\begin{vmatrix} 1 & 0 & 2 \\ 3 & -2 & 4 \\ 0 & 1 & 1 \end{vmatrix}$

37. $\begin{vmatrix} 4 & -2 & 0 \\ 9 & 5 & 4 \\ 1 & 2 & 0 \end{vmatrix}$

38. $\begin{vmatrix} -1 & 0 & 2 \\ 5 & -2 & 6 \\ 3 & 0 & -4 \end{vmatrix}$

For Exercises 39–44, solve the system using Cramer's rule.

39. $3x - 4y = 1$
  $2x + 3y = 12$

40. $3x + 2y = 11$
  $-x + y = 3$

41. $3x + 2y = 9$
  $x - 3y = 1$

42. $-x + 3y = -8$
  $2x - y = -2$

43. $2x + 3y - z = -7$
  $x + 3z = 10$
  $2y + z = -1$

44. $6y + 4z = -12$
  $3x + 3y = 9$
  $2x - 3z = 10$

For Exercises 45–46, solve the system of equations using Cramer's rule if possible. If not possible, use another method.

45. $2x - y = 1$
  $4x - 2y = 2$

46. $x + y - 3z = -1$
  $y - z = 6$
  $-x + 2y = 1$

Chapter 9   Test

For Exercises 1–3, solve the system of equations.

1. $2x + 2y + 4z = -6$
  $3x + y + 2z = 29$
  $x - y - z = 44$

2. $2(x + z) = 6 + x - 3y$
  $2x = 11 + y - z$
  $x + 2(y + z) = 8$

3. $3x - 4y = 6 - 2z$

$x - 3y + 2z = 9$

$2x = y - 1$

4. The perimeter of a triangle is 43 ft. The shortest side is one-third the length of the middle side. The longest side is 3 feet more than four times the shortest side. Find the lengths of the three sides.

5. Bennet invested a total of $11,500 among three mutual funds. At the end of one year, Fund A had a 6% yield, Fund B had an 8% yield, and Fund C lost 5%. Twice as much money was invested in Fund B as in Fund C. At the end of one year, the total gain was $550. How much was invested in each fund?

6. Working together, Joanne, Kent, and Geoff can process 504 orders per day for their business. Kent can process 20 more orders per day than Joanne can process. Geoff can process 104 fewer orders per day than Kent and Joanne combined. Find the number of orders that each person can process per day.

7. Find an equation of a parabola, $y = ax^2 + bx + c$, that passes through the points $(0, -8)$, $(-2, 0)$, and $(1, -15)$.

8. Write an example of a 3×2 matrix.

9. Given the matrix **A**

$$\mathbf{A} = \left[\begin{array}{ccc|c} 1 & 2 & 1 & -3 \\ 4 & 0 & 1 & -2 \\ -5 & -6 & 3 & 0 \end{array}\right]$$

a. Write the matrix obtained by multiplying the first row by -4 and adding the result to row 2.

b. Using the matrix obtained in part (a), write the matrix obtained by multiplying the first row by 5 and adding the result to row 3.

For Exercises 10–11, solve by using the Gauss-Jordan method.

10. $5x - 4y = 34$

$x - 2y = 8$

11. $x + y + z = 1$

$2x + y = 0$

$-2y - z = 5$

For Exercises 12–13, find the determinant of the matrix.

12. $\begin{bmatrix} 2 & -3 \\ 1 & 2 \end{bmatrix}$

13. $\begin{bmatrix} 0 & 5 & -2 \\ 0 & 0 & 2 \\ 2 & 3 & 1 \end{bmatrix}$

For Exercises 14–16, use Cramer's rule to solve for y.

14. $6x - 5y = 13$

$-2x - 2y = 9$

15. $2x + 2z = 2$

$5x + 3y = 4$

$3y - 4z = 4$

16. Solve the system: $6x - 2y = 0$

$3x - y = 0$

For Exercises 17–18, solve using any method. If a system does not have a unique solution, state whether the system is inconsistent or dependent.

17. $\frac{1}{3}x + y + \frac{1}{3}z = \frac{5}{3}$

$\frac{1}{4}x + \frac{1}{2}y - \frac{1}{8}z = -\frac{1}{4}$

$x + \frac{1}{3}y + \frac{5}{6}z = \frac{13}{3}$

18. $3x + 2y + 6z = 10$

$x + 3y + 2z = 4$

$2x - y + 4z = 6$

Chapters 1–9 Cumulative Review Exercises

For Exercises 1–2, solve the equation.

1. $-5(2x - 1) - 2(3x + 1) = 7 - 2(8x + 1)$

2. $\frac{1}{2}(a - 2) - \frac{3}{4}(2a + 1) = -\frac{1}{6}$

3. Simplify the expression. $\dfrac{(6a^2 b)^{-2}}{a^{-5} b^0}$

4. Solve the inequality. Write the answer in interval notation.

$$-3y - 2(y + 1) < 5$$

5. Identify the slope and the x- and y-intercepts of the line $5x - 2y = 15$.

6. Graph: $y = -\dfrac{1}{3}x - 4$

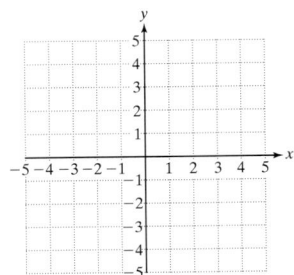

7. Find an equation for the line that passes through the points $(3, -8)$ and $(2, -4)$. Write the answer in slope-intercept form.

8. Solve the system by using the addition method.

$$2x - 3y = 6$$

$$\frac{1}{2}x - \frac{3}{4}y = 1$$

9. Factor: $a^2 - 64$

10. Factor: $a^3 - 64$

11. Divide: $(4x^3 - 13x^2 + 8x - 20) \div (x - 3)$

12. Determine the LCM of $10x^2y^3$ and $15xy^4$.

13. Simplify: $\dfrac{x + 15}{x^2 + 3x} + \dfrac{x}{x + 3} - \dfrac{5}{x}$

14. Given $g(x) = \dfrac{x + 6}{x - 1}$, find

 a. $g(0)$ **b.** $g(1)$ **c.** $g(-6)$

15. Solve the system.

$$3x + 2y + 3z = 3$$

$$4x - 5y + 7z = 1$$

$$2x + 3y - 2z = 6$$

16. Determine the order of the matrix.

$$\begin{bmatrix} 4 & 5 & 1 \\ -2 & 6 & 0 \end{bmatrix}$$

17. Given $y = f(x)$ as shown,

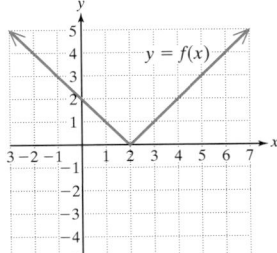

 a. Determine $f(-1)$.

 b. Determine $f(6)$.

 c. Find the values of x for which $f(x) = 2$.

 d. Write the domain in interval notation.

 e. Write the range in interval notation.

18. Solve the system by using the Gauss-Jordan method.

$$2x - 4y = -2$$

$$4x + y = 5$$

19. Find the determinant of matrix $\mathbf{C} = \begin{bmatrix} 8 & -2 \\ 3 & 1 \end{bmatrix}$.

20. Solve the system using Cramer's rule.

$$2x - 4y = -2$$

$$4x + y = 5$$

More Equations and Inequalities

As you study Chapter 10 you will be able to recognize and solve a variety of equations and inequalities. As you work through the chapter, write the solution set for each inequality here. Then use the letter next to each answer to complete the puzzle.

Solve each inequality.

_____ 1. $x < 3$ and $x \geq -2$ i. $(-\infty, \infty)$

_____ 2. $x < 3$ or $x \geq -2$ s. $(-2, 3)$

_____ 3. $x > 3$ and $x \leq -2$ d. $[-2, 3)$

_____ 4. $x > 3$ or $x \leq -2$ v. $(-\infty, -2] \cup (3, \infty)$

_____ 5. $\dfrac{x-3}{x+2} \leq 0$ i. $\{\,\}$

_____ 6. $\dfrac{x-3}{x+2} \geq 0$ o. $(-\infty, -2) \cup (3, \infty)$

_____ 7. $(x-3)(x+2) < 0$ n. $(-2, 3]$

_____ 8. $(x-3)(x+2) > 0$ i. $(-\infty, -2) \cup [3, \infty)$

He wears glasses during his math class because it improves . . .

$$\overline{}\ \overline{}\ \overline{}\ \overline{}\ \overline{}\ \overline{}\ \overline{}\ \overline{}$$
$$1\quad 6\quad 4\quad 2\quad 7\quad 3\quad 8\quad 5$$

Compound Inequalities

1. Union and Intersection

In Chapter 2 we graphed linear inequalities and expressed the solution set in interval notation and in set-builder notation. In this chapter, we will solve **compound inequalities** that involve the union or intersection of two or more inequalities.

> ### A Union B and A Intersection B
>
> The **union** of sets A and B, denoted $A \cup B$, is the set of elements that belong to set A or to set B or to both sets A and B.
>
> The **intersection** of two sets A and B, denoted $A \cap B$, is the set of elements common to both A and B.

The concepts of the union and intersection of two sets are illustrated in Figures 10-1 and 10-2.

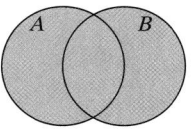

$A \cup B$
A union B
The elements in A *or* B *or* both
Figure 10-1

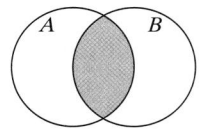

$A \cap B$
A intersection B
The elements in A *and* B
Figure 10-2

Example 1   **Finding the Union and Intersection of Two Intervals**

Find the union or intersection as indicated.

a. $\left(-\infty, \dfrac{1}{2}\right) \cap [-3, 4)$ **b.** $(-\infty, -2) \cup [-4, 3)$

Solution:

a. $\left(-\infty, \dfrac{1}{2}\right) \cap [-3, 4)$ To find the intersection, graph each interval separately. Then find the real numbers common to both intervals.

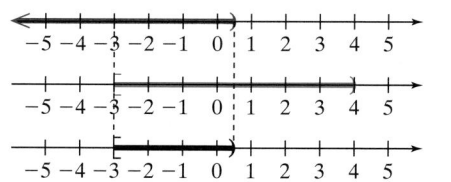

$\left(-\infty, \dfrac{1}{2}\right)$

$[-3, 4)$

The intersection is the "overlap" of the two intervals: $[-3, \frac{1}{2})$

The intersection is $[-3, \frac{1}{2})$.

b. $(-\infty, -2) \cup [-4, 3)$ To find the union, graph each interval separately. The union is the collection of real numbers that lie in the first interval, the second interval, or both intervals.

$(-\infty, -2)$

$[-4, 3)$

The union consists of all real numbers in the red interval along with the real numbers in the blue interval: $(-\infty, 3)$

The union is $(-\infty, 3)$.

Skill Practice

1. Find the intersection $(-\infty, 2) \cap [-5, \infty)$. Write the answer in interval notation.
2. Find the union $(-\infty, -3) \cup (-\infty, 0)$. Write the answer in interval notation.

---

2. Solving Compound Inequalities: And

The solution to two inequalities joined by the word *and* is the intersection of their solution sets. The solution to two inequalities joined by the word *or* is the union of their solution sets.

Steps to Solve a Compound Inequality

1. Solve and graph each inequality separately.
2. • If the inequalities are joined by the word *and*, find the intersection of the two solution sets.
   • If the inequalities are joined by the word *or*, find the union of the two solution sets.
3. Express the solution set in interval notation or in set-builder notation.

As you work through the examples in this section, remember that multiplying or dividing an inequality by a negative factor reverses the direction of the inequality sign.

Example 2 Solving Compound Inequalities: And

Solve the compound inequalities.

a. $-2x < 6$ and $x + 5 \le 7$

b. $4.4a + 3.1 < -12.3$ and $-2.8a + 9.1 < -6.3$

c. $-\dfrac{2}{3}x \le 6$ and $-\dfrac{1}{2}x < 1$

Solution:

a. $-2x < 6$ and $x + 5 \le 7$ — Solve each inequality separately.

$\dfrac{-2x}{-2} > \dfrac{6}{-2}$ and $x \le 2$ — Reverse the first inequality sign.

$x > -3$ and $x \le 2$

Skill Practice Answers

1. $[-5, 2)$ **2.** $(-\infty, 0)$

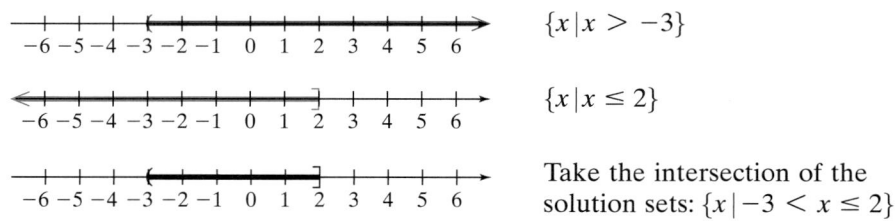

$\{x \mid x > -3\}$

$\{x \mid x \le 2\}$

Take the intersection of the solution sets: $\{x \mid -3 < x \le 2\}$

The solution is $\{x \mid -3 < x \le 2\}$ or equivalently, in interval notation, $(-3, 2]$.

b. $4.4a + 3.1 < -12.3$ and $-2.8a + 9.1 < -6.3$

$4.4a < -15.4$ and $-2.8a < -15.4$

$\dfrac{4.4a}{4.4} < \dfrac{-15.4}{4.4}$ and $\dfrac{-2.8a}{-2.8} > \dfrac{-15.4}{-2.8}$ Reverse the second inequality sign.

$a < -3.5$ and $a > 5.5$

$\{a \mid a < -3.5\}$

$\{a \mid a > 5.5\}$

The intersection of the solution sets is the empty set: $\{\ \ \}$

There are no real numbers that are simultaneously less than -3.5 and greater than 5.5. Hence, there is no solution.

c. $-\dfrac{2}{3}x \le 6$ and $-\dfrac{1}{2}x < 1$

$-\dfrac{3}{2}\left(-\dfrac{2}{3}x\right) \ge -\dfrac{3}{2}(6)$ and $-2\left(-\dfrac{1}{2}x\right) > -2(1)$ Solve each inequality separately.

$x \ge -9$ and $x > -2$

$\{x \mid x \ge -9\}$

$\{x \mid x > -2\}$

Take the intersection of the solution sets: $\{x \mid x > -2\}$

The solution is $\{x \mid x > -2\}$ or, in interval notation, $(-2, \infty)$.

Skill Practice   Solve the compound inequalities.

3. $5x + 2 \ge -8$ and $-4x > -24$ **4.** $-\dfrac{1}{3}y - \dfrac{1}{2} > 2$ and $6y + 2 > 2$

5. $-2.1x > 4.2$ and $3.5x < -10.5$

In Section 2.7, we learned that the inequality $a < x < b$ is the intersection of two simultaneous conditions implied on x.

$a < x < b$ is equivalent to $a < x$ and $x < b$

Skill Practice Answers

3. $\{x \mid -2 \le x < 6\}$; $[-2, 6)$
4. No solution
5. $\{x \mid x < -3\}$; $(-\infty, -3)$

Example 3 Solving Compound Inequalities: And

Solve the inequality $-2 \leq -3x + 1 < 5$.

Solution:

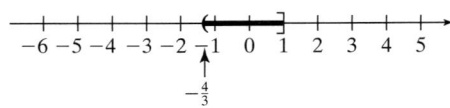

$-2 \leq -3x + 1$	and	$-3x + 1 < 5$	Set up the intersection of two inequalities.
$-3 \leq -3x$	and	$-3x < 4$	Solve each inequality.
$\dfrac{-3}{-3} \geq \dfrac{-3x}{-3}$	and	$\dfrac{-3x}{-3} > \dfrac{4}{-3}$	Reverse the direction of the inequality signs.
$1 \geq x$	and	$x > -\dfrac{4}{3}$	
$x \leq 1$	and	$x > -\dfrac{4}{3}$	Rewrite the inequalities.
$-\dfrac{4}{3} < x \leq 1$			Take the intersection of the solution sets.

The solution is $\{x \mid -\frac{4}{3} < x \leq 1\}$ or, equivalently in interval notation, $(-\frac{4}{3}, 1]$.

Skill Practice Solve the inequality.

6. $-6 < 5 - 2x \leq 1$

TIP: As an alternative approach to Example 3, we can isolate the variable x in the "middle" portion of the inequality. Recall that the operations performed on the middle part of the inequality must also be performed on the left- and right-hand sides.

$-2 \leq -3x + 1 < 5$

$-2 - 1 \leq -3x + 1 - 1 < 5 - 1$ Subtract 1 from all three parts of the inequality.

$-3 \leq -3x < 4$ Simplify.

$\dfrac{-3}{-3} \geq \dfrac{-3x}{-3} > \dfrac{4}{-3}$ Divide by -3 in all three parts of the inequality. (Remember to reverse inequality signs.)

$1 \geq x > -\dfrac{4}{3}$ Simplify.

$-\dfrac{4}{3} < x \leq 1$ Rewrite the inequality.

Skill Practice Answers

6. $\left\{x \mid 2 \leq x < \dfrac{11}{2}\right\}$; $\left[2, \dfrac{11}{2}\right)$

3. Solving Compound Inequalities: Or

> **Example 4**   Solving Compound Inequalities: Or

Solve the compound inequalities.

    a. $-3y - 5 > 4$ or $4 - y \leq 6$

    b. $4x + 3 < 16$ or $-2x < 3$

    c. $\dfrac{1}{3}x < 2$ or $-\dfrac{1}{2}x + 1 > 0$

Solution:

    a. $-3y - 5 > 4$ or $4 - y \leq 6$

       $-3y > 9$ or $-y \leq 2$ Solve each inequality separately.

       $\dfrac{-3y}{-3} < \dfrac{9}{-3}$ or $\dfrac{-y}{-1} \geq \dfrac{2}{-1}$ Reverse the inequality signs.

       $y < -3$ or $y \geq -2$

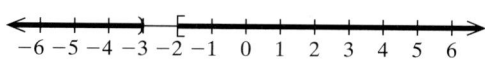

$\{y \mid y < -3\}$

$\{y \mid y \geq -2\}$

Take the union of the solution sets $\{y \mid y < -3$ or $y \geq -2\}$.

The solution is $\{y \mid y < -3$ or $y \geq -2\}$ or, equivalently in interval notation, $(-\infty, -3) \cup [-2, \infty)$.

    b. $4x + 3 < 16$ or $-2x < 3$

       $4x < 13$ or $x > -\dfrac{3}{2}$ Solve each inequality separately.

       $x < \dfrac{13}{4}$ or $x > -\dfrac{3}{2}$

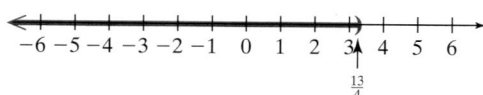

$\{x \mid x < \frac{13}{4}\}$

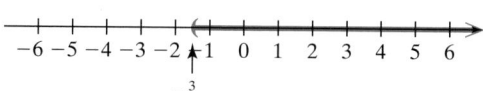

$\{x \mid x > -\frac{3}{2}\}$

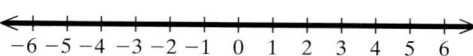

Take the union of the solution sets.

The union of the solution sets is $\{x \mid x$ is any real number$\}$ or equivalently $(-\infty, \infty)$.

c. $\quad \dfrac{1}{3}x < 2 \qquad$ or $\qquad -\dfrac{1}{2}x + 1 > 0$

$\quad 3\left(\dfrac{1}{3}x\right) < 3(2) \qquad$ or $\qquad -\dfrac{1}{2}x > -1 \qquad$ Solve each inequality
separately.

$\qquad x < 6 \qquad$ or $\qquad -2\left(-\dfrac{1}{2}x\right) < -2(-1)$

$\qquad x < 6 \qquad$ or $\qquad\qquad x < 2$

$\qquad \{x \mid x < 6\}$

$\qquad \{x \mid x < 2\}$

$\qquad$ Take the union of the solution
sets: $\{x \mid x < 6\}$.

The union of the solution sets is $\{x \mid x < 6\}$ or, in interval notation, $(-\infty, 6)$.

Skill Practice Solve the compound inequalities.

7. $-10t - 8 \geq 12$ or $3t - 6 > 3$ $\qquad$ **8.** $x - 7 > -2$ or $-6x > -48$

9. $2.5x > 10$ or $-0.75x < 3$

4. Applications of Compound Inequalities

Compound inequalities are used in many applications, as shown in Examples 5
and 6.

Example 5 Translating Compound Inequalities

The normal level of thyroid-stimulating hormone (TSH) for adults ranges from
0.4 to 4.8 microunits per milliliter (μU/mL). Let x represent the amount of TSH
measured in microunits per milliliter.

a. Write an inequality representing the normal range of TSH.

b. Write a compound inequality representing abnormal TSH levels.

Solution:

a. $0.4 \leq x \leq 4.8$ $\qquad$ **b.** $x < 0.4$ $\quad$ or $\quad x > 4.8$

Skill Practice

10. The length of a normal human pregnancy is from 37 to 41 weeks, inclusive.
   a. Write an inequality representing the normal length of a pregnancy.
   b. Write a compound inequality representing an abnormal length for a
   pregnancy.

Skill Practice Answers

7. $\{t \mid t \leq -2 \text{ or } t > 3\}$;
$\quad (-\infty, -2] \cup (3, \infty)$
8. All real numbers; $(-\infty, \infty)$
9. $\{x \mid x > -4\}$; $(-4, \infty)$
10a. $37 \leq w \leq 41$
$\quad$ **b.** $w < 37$ or $w > 41$

| Example 6 | Translating and Solving a Compound Inequality |

The sum of a number and 4 is between -5 and 12. Find all such numbers.

Solution:

Let x represent a number.

$-5 < x + 4 < 12$	Translate the inequality.
$-5 - 4 < x + 4 - 4 < 12 - 4$	Subtract 4 from all three parts of the inequality.
$-9 < x < 8$	

The number may be any real number between -9 and 8: $\{x \mid -9 < x < 8\}$.

| Skill Practice |

Skill Practice Answers

11. Any real number between 5 and 10: $\{n \mid 5 < n < 10\}$

11. The sum of twice a number and 11 is between 21 and 31. Find all such numbers.

| Section 10.1 | **Practice Exercises** |

Study Skills Exercise

1. Define the key terms.

   a. compound inequality   **b.** intersection   **c.** union

Review Exercises

For Exercises 2–8, review solving linear inequalities from Section 2.7. Write the answers in interval notation.

2. $6u + 5 > 2$ **3.** $-2 + 3z \le 4$ **4.** $-\dfrac{3}{4}p \le 12$ **5.** $-6q > -\dfrac{1}{3}$

6. $-1.5 < 0.1x \le 8.1$ **7.** $-0.2 < 2.6 + 7t < 4$ **8.** $0 \le 4x - 1 < 5$

Concept 1: Union and Intersection

For Exercises 9–14, find the intersection and union of sets as indicated. Write the answers in interval notation.

9. **a.** $(-2, 5) \cap [-1, \infty)$

   b. $(-2, 5) \cup [-1, \infty)$

10. **a.** $(-\infty, 4) \cap [-1, 5)$

   b. $(-\infty, 4) \cup [-1, 5)$

11. **a.** $\left(-\dfrac{5}{2}, 3\right) \cap \left(-1, \dfrac{9}{2}\right)$

   b. $\left(-\dfrac{5}{2}, 3\right) \cup \left(-1, \dfrac{9}{2}\right)$

12. **a.** $(-3.4, 1.6) \cap (-2.2, 4.1)$

   b. $(-3.4, 1.6) \cup (-2.2, 4.1)$

13. **a.** $(-4, 5] \cap (0, 2]$

   b. $(-4, 5] \cup (0, 2]$

14. **a.** $[-1, 5) \cap (0, 3)$

   b. $[-1, 5) \cup (0, 3)$

Concept 2: Solving Compound Inequalities: And

For Exercises 15–24, solve the inequality and graph the solution. Write the answer in interval notation.

15. $y - 7 \geq -9$ and $y + 2 \leq 5$

16. $a + 6 > -2$ and $5a < 30$

17. $2t + 7 < 19$ and $5t + 13 > 28$

18. $5p + 2p \geq -21$ and $-9p + 3p \geq -24$

19. $21k - 11 \leq 6k + 19$ and
$3k - 11 < -k + 7$

20. $6w - 1 > 3w - 11$ and
$-3w + 7 \leq 8w - 13$

21. $\frac{2}{3}(2p - 1) \geq 10$ and $\frac{4}{5}(3p + 4) \geq 20$

22. $5(a + 3) + 9 < 2$ and $3(a - 2) + 6 < 10$

23. $-2 < -x - 12$ and $-14 < 5(x - 3) + 6x$

24. $-8 \geq -3y - 2$ and $3(y - 7) + 16 > 4y$

25. Write $-4 \leq t < \frac{3}{4}$ as two separate inequalities.

26. Write $-2.8 < y \leq 15$ as two separate inequalities.

27. Explain why $6 < x < 2$ has no solution.

28. Explain why $4 < t < 1$ has no solution.

29. Explain why $-5 > y > -2$ has no solution.

30. Explain why $-3 > w > -1$ has no solution.

For Exercises 31–40, solve the inequality and graph the solution set. Write the answer in interval notation.

31. $0 \leq 2b - 5 < 9$

32. $-6 < 3k - 9 \leq 0$

33. $-1 < \frac{a}{6} \leq 1$

34. $-3 \leq \frac{1}{2}x < 0$

35. $-\frac{2}{3} < \frac{y - 4}{-6} < \frac{1}{3}$

36. $\frac{1}{3} > \frac{t - 4}{-3} > -2$

37. $5 \leq -3x - 2 \leq 8$

38. $-1 < -2x + 4 \leq 5$

39. $12 > 6x + 3 \geq 0$

40. $-4 \geq 2x - 5 > -7$

Concept 3: Solving Compound Inequalities: Or

For Exercises 41–54, solve the inequality and graph the solution set. Write the answer in interval notation.

41. $h + 4 < 0$ or $6h > -12$

42. $5y > 12$ or $y - 3 < -2$

43. $2y - 1 \geq 3$ or $y < -2$

44. $x < 0$ or $3x + 1 \geq 7$

45. $1.2 > 7.2z - 9.6$ or $3.1 \leq 6.3 - 1.6z$

46. $9.5 > 3.1 + 0.8z$ or $-2.8 > 6.1 + 0.89z$

47. $5(x - 1) \geq -5$ or $5 - x \leq 11$

48. $-p + 7 \geq 10$ or $3(p - 1) \leq 12$

49. $\dfrac{5}{3}v \leq 5$ or $-v - 6 < 1$

50. $\dfrac{3}{8}u + 1 > 0$ or $-2u \geq -4$

51. $\dfrac{3t - 1}{10} > \dfrac{1}{2}$ or $\dfrac{3t - 1}{10} < -\dfrac{1}{2}$

52. $\dfrac{6 - x}{12} > \dfrac{1}{4}$ or $\dfrac{6 - x}{12} < -\dfrac{1}{6}$

53. $0.5w + 5 < 2.5w - 4$ or $0.3w \leq -0.1w - 1.6$

54. $1.25a + 3 \leq 0.5a - 6$ or $2.5a - 1 \geq 9 - 1.5a$

Mixed Exercises

For Exercises 55–66, solve the inequality. Write the answer in interval notation.

55. a. $3x - 5 < 19$ and $-2x + 3 < 23$

    b. $3x - 5 < 19$ or $-2x + 3 < 23$

56. a. $0.5(6x + 8) > 0.8x - 7$ and $4(x + 1) < 7.2$

    b. $0.5(6x + 8) > 0.8x - 7$ or $4(x + 1) < 7.2$

57. a. $8x - 4 \geq 6.4$ or $0.3(x + 6) \leq -0.6$

    b. $8x - 4 \geq 6.4$ and $0.3(x + 6) \leq -0.6$

58. a. $-2r + 4 \leq -8$ or $3r + 5 \leq 8$

    b. $-2r + 4 \leq -8$ and $3r + 5 \leq 8$

59. $-4 \leq \dfrac{2 - 4x}{3} < 8$

60. $-1 < \dfrac{3 - x}{2} \leq 0$

61. $5 \geq -4(t - 3) + 3t$ or

    $6 < 12t + 8(4 - t)$

62. $3 > -(w - 3) + 4w$ or

    $-5 \geq -3(w - 5) + 6w$

63. $-7(3 - x) < 9[x - 3(x + 1)]$ and

    $6x - (4 - x) + 3 < -3[8 - 2(x + 1)]$

64. $-5[x + 3(2 - x)] \leq 4(3 - 5x) + 7$ and

    $4 - [-4 - (2x - 3)] \geq 9 - (10 - 6x)$

65. $\dfrac{-x + 3}{2} > \dfrac{4 + x}{5}$ or $\dfrac{1 - x}{4} > \dfrac{2 - x}{3}$

66. $\dfrac{y - 7}{-3} < \dfrac{1}{4}$ or $\dfrac{y + 1}{-2} > -\dfrac{1}{3}$

Concept 4: Applications of Compound Inequalities

67. The normal number of white blood cells for human blood is between 4800 and 10,800 cells per cubic millimeter, inclusive. Let x represent the number of white blood cells per cubic millimeter.

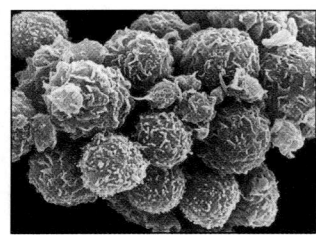

    a. Write an inequality representing the normal range of white blood cells per cubic millimeter.

    b. Write a compound inequality representing abnormal levels of white blood cells per cubic millimeter.

68. Normal hemoglobin levels in human blood for adult males are between 13 and 16 grams per deciliter (g/dL), inclusive. Let x represent the level of hemoglobin measured in grams per deciliter.

a. Write an inequality representing normal hemoglobin levels for adult males.

b. Write a compound inequality representing abnormal levels of hemoglobin for adult males.

69. The normal number of platelets in human blood is between 2.0×10^5 and 3.5×10^5 platelets per cubic millimeter, inclusive. Let x represent the number of platelets per cubic millimeter.

a. Write an inequality representing a normal platelet count per cubic millimeter.

b. Write a compound inequality representing abnormal platelet counts per cubic millimeter.

70. Normal hemoglobin levels in human blood for adult females are between 12 and 15 g/dL, inclusive. Let x represent the level of hemoglobin measured in grams per deciliter.

a. Write an inequality representing normal hemoglobin levels for adult females.

b. Write a compound inequality representing abnormal levels of hemoglobin for adult females.

71. Twice a number is between -3 and 12. Find all such numbers.

72. The difference of a number and 6 is between 0 and 8. Find all such numbers.

73. One plus twice a number is either greater than 5 or less than -1. Find all such numbers.

74. One-third of a number is either less than -2 or greater than 5. Find all such numbers.

Polynomial and Rational Inequalities

Section 10.2

1. Solving Inequalities Graphically

Concepts

1. Solving Inequalities Graphically
2. Solving Polynomial Inequalities by Using the Test Point Method
3. Solving Rational Inequalities by Using the Test Point Method
4. Inequalities with "Special Case" Solution Sets

In Sections 2.7 and 10.1, we solved simple and compound linear inequalities. In this section we will solve polynomial and rational inequalities. We begin by defining a quadratic inequality.

Quadratic inequalities are inequalities that can be written in any of the following forms:

$$ax^2 + bx + c \geq 0 \qquad ax^2 + bx + c \leq 0$$
$$ax^2 + bx + c > 0 \qquad ax^2 + bx + c < 0 \qquad \text{where } a \neq 0$$

Recall from Section 8.3 that the graph of a quadratic function defined by $f(x) = ax^2 + bx + c \, (a \neq 0)$ is a parabola that opens upward or downward. The quadratic inequality $f(x) > 0$ or equivalently $ax^2 + bx + c > 0$ is asking the question, "For what values of x is the value of the function positive (above the x-axis)?" The inequality $f(x) < 0$ or equivalently $ax^2 + bx + c < 0$ is asking, "For what values of x is the value of the function negative (below the x-axis)?" The graph of a quadratic function can be used to answer these questions.

Example 1 Using a Graph to Solve a Quadratic Inequality

Use the graph of $f(x) = x^2 - 6x + 8$ in Figure 10-3 to solve the inequalities.

a. $x^2 - 6x + 8 < 0$ **b.** $x^2 - 6x + 8 > 0$

Solution:

From Figure 10-3, we see that the graph of $f(x) = x^2 - 6x + 8$ is a parabola opening upward. The function factors as $f(x) = (x - 2)(x - 4)$. The x-intercepts are at $x = 2$ and $x = 4$, and the y-intercept is $(0, 8)$.

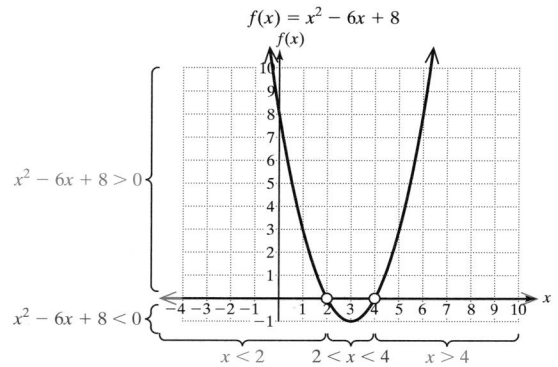

Figure 10-3

a. The solution to $x^2 - 6x + 8 < 0$ is the set of real numbers x for which $f(x) < 0$. Graphically, this is the set of all x-values corresponding to the points where the parabola is below the x-axis (shown in red). Hence

$$x^2 - 6x + 8 < 0 \qquad \text{for } \{x \,|\, 2 < x < 4\} \text{ or equivalently, } (2, 4)$$

b. The solution to $x^2 - 6x + 8 > 0$ is the set of x-values for which $f(x) > 0$. This is the set of x-values where the parabola is above the x-axis (shown in blue). Hence

$$x^2 - 6x + 8 > 0 \qquad \text{for } \{x \,|\, x < 2 \quad \text{or} \quad x > 4\} \text{ or } (-\infty, 2) \cup (4, \infty)$$

> **Skill Practice**

1. Refer to the graph of $f(x) = x^2 + 3x - 4$ to solve the inequalities.
 a. $x^2 + 3x - 4 > 0$
 b. $x^2 + 3x - 4 < 0$

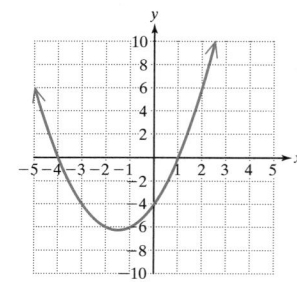

Notice that $x = 2$ and $x = 4$ define the boundaries of the solution sets to the inequalities in Example 1. These values are the solutions to the related equation $x^2 - 6x + 8 = 0$.

> **TIP:** The inequalities in Example 1 are strict inequalities. Therefore, $x = 2$ and $x = 4$ (where $f(x) = 0$) are not included in the solution set. However, the corresponding inequalities using the symbols $\leq$ and $\geq$ do include the values where $f(x) = 0$. Hence,
>
> The solution to $\quad x^2 - 6x + 8 \leq 0 \quad$ is $\quad \{x \,|\, 2 \leq x \leq 4\}$ or $[2, 4]$
>
> The solution to $\quad x^2 - 6x + 8 \geq 0 \quad$ is $\quad \{x \,|\, x \leq 2 \text{ or } x \geq 4\}$ or $(-\infty, -2] \cup [4, \infty)$

Skill Practice Answers

1a. $\{x \,|\, x < -4 \text{ or } x > 1\};$
$(-\infty, -4) \cup (1, \infty)$
b. $\{x \,|\, -4 < x < 1\}; (-4, 1)$

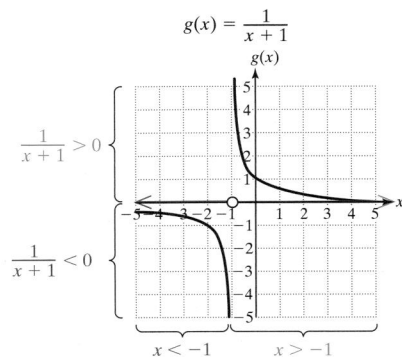

Example 2 Using a Graph to Solve a Rational Inequality

Use the graph of $g(x) = \dfrac{1}{x + 1}$ in Figure 10-4 to solve the inequalities.

a. $\dfrac{1}{x + 1} < 0$ **b.** $\dfrac{1}{x + 1} > 0$

Solution:

Figure 10-4

a. Figure 10-4 indicates that $g(x)$ is below the x-axis for $x < -1$ (shown in red). Therefore, the solution to $\dfrac{1}{x + 1} < 0$ is $\{x \mid x < -1\}$ or, equivalently, $(-\infty, -1)$.

b. Figure 10-4 indicates that $g(x)$ is above the x-axis for $x > -1$ (shown in blue). Therefore, the solution to $\dfrac{1}{x + 1} > 0$ is $\{x \mid x > -1\}$ or, equivalently, $(-1, \infty)$.

Skill Practice

2. Refer to the graph of $g(x) = \dfrac{1}{x - 2}$ to solve the inequalities.

     a. $\dfrac{1}{x - 2} > 0$ **b.** $\dfrac{1}{x - 2} < 0$

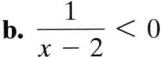

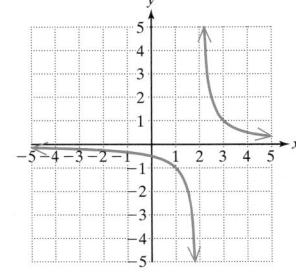

TIP: Notice that $x = -1$ defines the boundary of the solution sets to the inequalities in Example 2. At $x = -1$ the inequality is undefined.

Skill Practice Answers

2a. $\{x \mid x > 2\}; (2, \infty)$
    b. $\{x \mid x < 2\}; (-\infty, 2)$

2. Solving Polynomial Inequalities by Using the Test Point Method

Examples 1 and 2 demonstrate that the boundary points of an inequality provide the boundaries of the solution set.

> ### Boundary Points
>
> The **boundary points** of an inequality consist of the real solutions to the related equation and the points where the inequality is undefined.

Testing points in regions bounded by these points is the basis of the **test point method** to solve inequalities.

> ### Solving Inequalities by Using the Test Point Method
>
> **1.** Find the boundary points of the inequality.
>
> **2.** Plot the boundary points on the number line. This divides the number line into regions.
>
> **3.** Select a test point from each region and substitute it into the original inequality.
> - If a test point makes the original inequality true, then that region is part of the solution set.
>
> **4.** Test the boundary points in the original inequality.
> - If a boundary point makes the original inequality true, then that point is part of the solution set.

Example 3 Solving Polynomial Inequalities by Using the Test Point Method

Solve the inequalities by using the test point method.

a. $2x^2 + 5x < 12$ **b.** $x(x - 2)(x + 4)^2(x - 4) > 0$

Solution:

a. $2x^2 + 5x < 12$ **Step 1:** Find the boundary points. Because polynomials are defined for all values of x, the only boundary points are the real solutions to the related equation.

$2x^2 + 5x = 12$ Solve the related equation.

$2x^2 + 5x - 12 = 0$

$(2x - 3)(x + 4) = 0$

$x = \frac{3}{2} \qquad x = -4$ The boundary points are $\frac{3}{2}$ and -4.

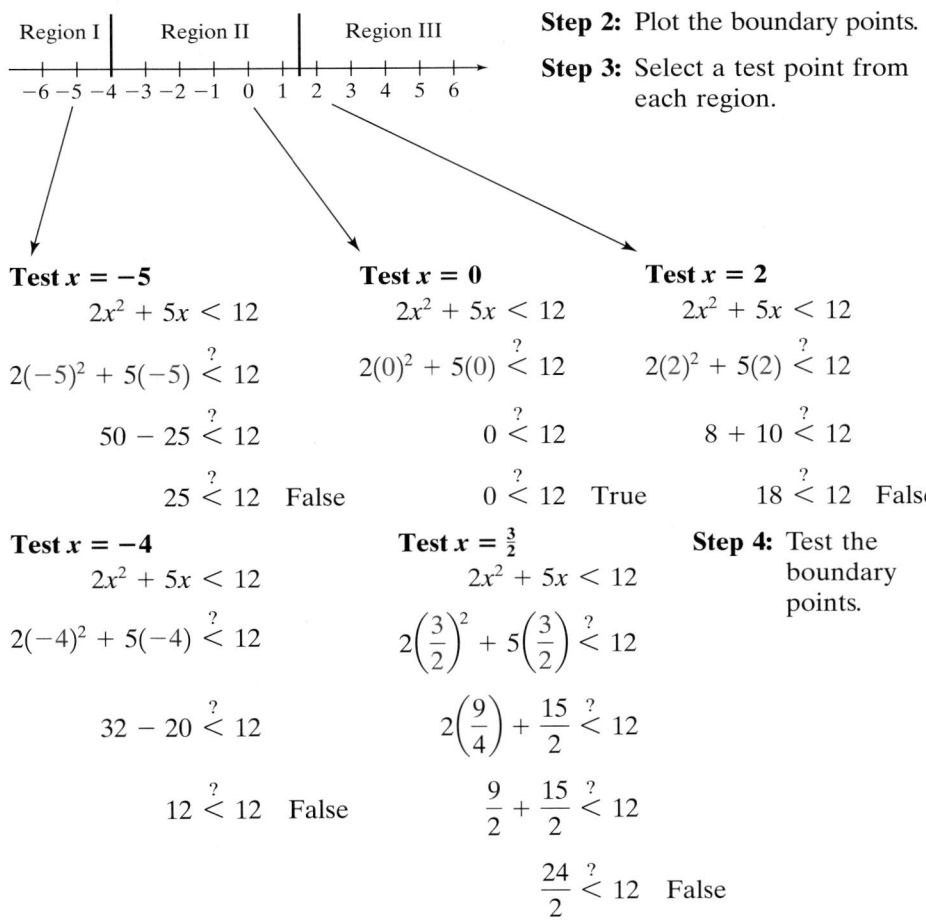

Step 2: Plot the boundary points.

Step 3: Select a test point from each region.

Test $x = -5$
$$2x^2 + 5x < 12$$
$$2(-5)^2 + 5(-5) \overset{?}{<} 12$$
$$50 - 25 \overset{?}{<} 12$$
$$25 \overset{?}{<} 12 \quad \text{False}$$

Test $x = 0$
$$2x^2 + 5x < 12$$
$$2(0)^2 + 5(0) \overset{?}{<} 12$$
$$0 \overset{?}{<} 12$$
$$0 \overset{?}{<} 12 \quad \text{True}$$

Test $x = 2$
$$2x^2 + 5x < 12$$
$$2(2)^2 + 5(2) \overset{?}{<} 12$$
$$8 + 10 \overset{?}{<} 12$$
$$18 \overset{?}{<} 12 \quad \text{False}$$

Test $x = -4$
$$2x^2 + 5x < 12$$
$$2(-4)^2 + 5(-4) \overset{?}{<} 12$$
$$32 - 20 \overset{?}{<} 12$$
$$12 \overset{?}{<} 12 \quad \text{False}$$

Test $x = \frac{3}{2}$
$$2x^2 + 5x < 12$$
$$2\left(\frac{3}{2}\right)^2 + 5\left(\frac{3}{2}\right) \overset{?}{<} 12$$
$$2\left(\frac{9}{4}\right) + \frac{15}{2} \overset{?}{<} 12$$
$$\frac{9}{2} + \frac{15}{2} \overset{?}{<} 12$$
$$\frac{24}{2} \overset{?}{<} 12 \quad \text{False}$$

Step 4: Test the boundary points.

Neither boundary point makes the inequality true. Therefore, the boundary points are not included in the solution set.

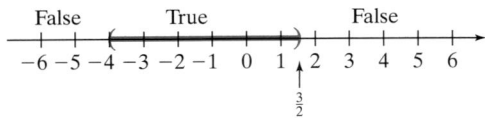

The solution is $\{x \mid -4 < x < \frac{3}{2}\}$ or equivalently in interval notation $(-4, \frac{3}{2})$.

TIP: The strict inequality, $<$, excludes values of x for which $2x^2 + 5x = 12$. This implies that the boundary points are not included in the solution set.

Calculator Connections

Graph $Y_1 = 2x^2 + 5x$ and $Y_2 = 12$.
Notice that $Y_1 < Y_2$ for $\{x \mid -4 < x < \frac{3}{2}\}$.

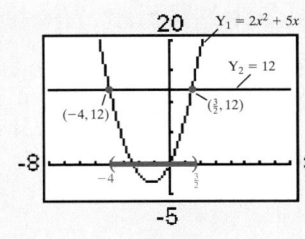

b. $x(x - 2)(x + 4)^2(x - 4) > 0$

$$x(x - 2)(x + 4)^2(x - 4) = 0 \qquad \textbf{Step 1:} \text{ Find the boundary points.}$$
$$x = 0 \qquad x = 2 \qquad x = -4 \qquad x = 4$$

I   II   III   IV   V

$$-6 \ -5 \ -4 \ -3 \ -2 \ -1 \quad 0 \quad 1 \quad 2 \quad 3 \quad 4 \quad 5 \quad 6$$

Step 2: Plot the boundary points.

Step 3: Select a test point from each region.

Test $x = -5$: $-5(-5 - 2)(-5 + 4)^2(-5 - 4) \overset{?}{>} 0$ $-315 \overset{?}{>} 0$ False

Test $x = -1$: $-1(-1 - 2)(-1 + 4)^2(-1 - 4) \overset{?}{>} 0$ $-135 \overset{?}{>} 0$ False

Test $x = 1$: $1(1 - 2)(1 + 4)^2(1 - 4) \overset{?}{>} 0$ $75 \overset{?}{>} 0$ True

Test $x = 3$: $3(3 - 2)(3 + 4)^2(3 - 4) \overset{?}{>} 0$ $-147 \overset{?}{>} 0$ False

Test $x = 5$: $5(5 - 2)(5 + 4)^2(5 - 4) \overset{?}{>} 0$ $1215 \overset{?}{>} 0$ True

False   False   True False   True

$$-6 \ -5 \ -4 \ -3 \ -2 \ -1 \quad 0 \quad 1 \quad 2 \quad 3 \quad 4 \quad 5 \quad 6$$

Step 4: The boundary points are not included because the inequality, $>$, is strict.

The solution is $\{x \mid 0 < x < 2 \text{ or } x > 4\}$ or, equivalently in interval notation, $(0, 2) \cup (4, \infty)$.

Calculator Connections

Graph $Y_1 = x(x - 2)(x + 4)^2(x - 4)$. Y_1 is positive (above the x-axis) for $\{x \mid 0 < x < 2 \text{ or } x > 4\}$ or equivalently $(0, 2) \cup (4, \infty)$.

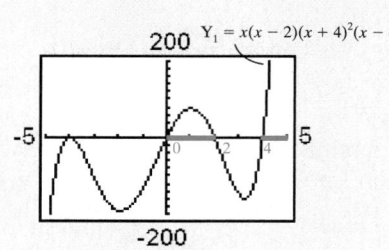

Skill Practice Solve the inequalities by using the test point method. Write the answers in interval notation.

3. $x^2 + x > 6$ **4.** $t(t - 5)(t + 2)^2 > 0$

3. Solving Rational Inequalities by Using the Test Point Method

The test point method can be used to solve rational inequalities. A **rational inequality** is an inequality in which one or more terms is a rational expression. The solution set to a rational inequality must exclude all values of the variable that make the inequality undefined. That is, exclude all values that make the denominator equal to zero for any rational expression in the inequality.

Skill Practice Answers
3. $(-\infty, -3) \cup (2, \infty)$
4. $(-\infty, -2) \cup (-2, 0) \cup (5, \infty)$

Example 4   Solving a Rational Inequality by Using the Test Point Method

Solve the inequality by using the test point method. $\dfrac{x + 2}{x - 4} \le 3$

Solution:

$$\frac{x+2}{x-4} \leq 3$$

Step 1: Find the boundary points. Note that the inequality is undefined for $x = 4$. Hence $x = 4$ is automatically a boundary point. To find any other boundary points, solve the related equation.

$$\frac{x+2}{x-4} = 3$$

$$(x-4)\left(\frac{x+2}{x-4}\right) = (x-4)(3)$$ 　　　　　Clear fractions.

$$x + 2 = 3(x - 4)$$ 　　　　　Solve for x.

$$x + 2 = 3x - 12$$

$$-2x = -14$$

$$x = 7$$

The solution to the related equation is $x = 7$, and the inequality is undefined for $x = 4$. Therefore, the boundary points are $x = 4$ and $x = 7$.

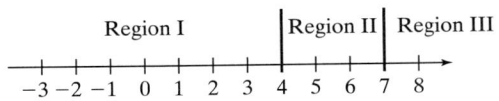

Region I 　　 Region II 　 Region III

Step 2: Plot boundary points.

Step 3: Select test points.

Test $x = 0$

$$\frac{x+2}{x-4} \leq 3$$

$$\frac{0+2}{0-4} \overset{?}{\leq} 3$$

$$-\frac{1}{2} \overset{?}{\leq} 3 \quad \text{True}$$

Test $x = 5$

$$\frac{x+2}{x-4} \leq 3$$

$$\frac{5+2}{5-4} \overset{?}{\leq} 3$$

$$\frac{7}{1} \overset{?}{\leq} 3 \quad \text{False}$$

Test $x = 8$

$$\frac{x+2}{x-4} \leq 3$$

$$\frac{8+2}{8-4} \overset{?}{\leq} 3$$

$$\frac{10}{4} \overset{?}{\leq} 3$$

$$\frac{5}{2} \overset{?}{\leq} 3 \quad \text{True}$$

Test $x = 4$:

$$\frac{x+2}{x-4} \leq 3$$

$$\frac{4+2}{4-4} \overset{?}{\leq} 3$$

$$\frac{6}{0} \overset{?}{\leq} 3 \quad \text{Undefined}$$

Test $x = 7$:

$$\frac{x+2}{x-4} \leq 3$$

$$\frac{7+2}{7-4} \overset{?}{\leq} 3$$

$$\frac{9}{3} \overset{?}{\leq} 3 \quad \text{True}$$

Step 4: Test the boundary points.

The boundary point $x = 4$ cannot be included in the solution set, because it is undefined in the inequality. The boundary point $x = 7$ makes the original inequality true and must be included in the solution set.

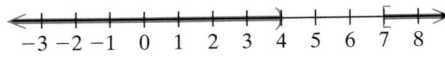

The solution is $\{x \mid x < 4 \text{ or } x \geq 7\}$ or, equivalently in interval notation, $(-\infty, 4) \cup [7, \infty)$.

Calculator Connections

Graph $Y_1 = \dfrac{x+2}{x-4}$ and $Y_2 = 3$.

Y_1 has a vertical asymptote at $x = 4$. Furthermore, $Y_1 = Y_2$ at $x = 7$. $Y_1 \leq Y_2$ (that is, Y_1 is below Y_2) for $x < 4$ and for $x \geq 7$.

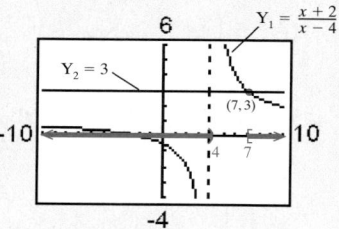

Skill Practice Solve the inequality by using the test point method. Write the answer in interval notation.

5. $\dfrac{x-5}{x+4} \leq -1$

Example 5 **Solving a Rational Inequality by Using the Test Point Method**

Solve the inequality. $\quad \dfrac{3}{x-1} > 0$

Solution:

$$\frac{3}{x-1} > 0$$

$$\frac{3}{x-1} = 0$$

$$\cancel{(x-1)} \cdot \left(\frac{3}{\cancel{x-1}}\right) = (x-1) \cdot 0$$

$$3 = 0$$

Step 1: Find the boundary points. Note that the inequality is undefined for $x = 1$, so $x = 1$ is a boundary point. To find any other boundary points, solve the related equation.

Clear fractions.

There is no solution to the related equation.

The only boundary point is $x = 1$.

Region I | Region II

$-5\ -4\ -3\ -2\ -1\quad 0\quad 1\quad 2\quad 3\quad 4\quad 5$

Step 2: Plot boundary points.

Test $x = 0$:   **Test $x = 2$:**   **Step 3:** Select test points.

$$\frac{3}{(0)-1} \overset{?}{>} 0 \qquad\qquad \frac{3}{(2)-1} \overset{?}{>} 0$$

$$\frac{3}{-1} \overset{?}{>} 0 \text{ False} \qquad \frac{3}{1} > 0 \text{ True}$$

Step 4: The boundary point $x = 1$ cannot be included in the solution set because it is undefined in the original inequality.

False   True

$-5\ -4\ -3\ -2\ -1\quad 0\quad 1\quad 2\quad 3\quad 4\quad 5$

Skill Practice Answers

5. $\left(-4, \dfrac{1}{2}\right]$

The solution is $\{x \mid x > 1\}$ or equivalently in interval notation, $(1, \infty)$.

Skill Practice Solve the inequality.

6. $\dfrac{-5}{y + 2} < 0$

4. Inequalities with "Special Case" Solution Sets

The solution to an inequality is often one or more regions on the real number line. Sometimes, however, the solution to an inequality may be a single point on the number line, the empty set, or the set of all real numbers.

Example 6 Solving Inequalities

Solve the inequalities.

a. $x^2 + 6x + 9 \geq 0$ b. $x^2 + 6x + 9 > 0$

c. $x^2 + 6x + 9 \leq 0$ d. $x^2 + 6x + 9 < 0$

Solution:

a. $x^2 + 6x + 9 \geq 0$ Notice that $x^2 + 6x + 9$ is a perfect square trinomial.

  $(x + 3)^2 \geq 0$ Factor $x^2 + 6x + 9 = (x + 3)^2$.

The quantity $(x + 3)^2$ is a perfect square and is greater than or equal to zero for all real numbers, x. The solution is all real numbers, $(-\infty, \infty)$.

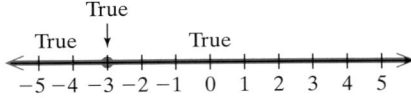

b. $x^2 + 6x + 9 > 0$

  $(x + 3)^2 > 0$ This is the same inequality as in part (a) with the exception that the inequality is strict. The solution set does not include the point where $x^2 + 6x + 9 = 0$. Therefore, the boundary point $x = -3$ is *not* included in the solution set.

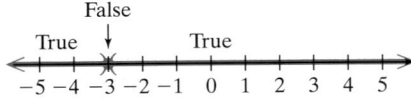

The solution set is $\{x \mid x < -3 \text{ or } x > -3\}$ or equivalently $(-\infty, -3) \cup (-3, \infty)$.

c. $x^2 + 6x + 9 \leq 0$

  $(x + 3)^2 \leq 0$

A perfect square cannot be less than zero. However, $(x + 3)^2$ is equal to zero at $x = -3$. Therefore, the solution set is $\{-3\}$.

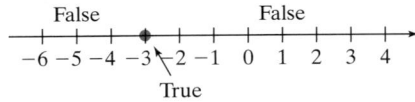

TIP: The graph of $f(x) = x^2 + 6x + 9$, or equivalently $f(x) = (x + 3)^2$, is equal to zero at $x = -3$ and positive (above the x-axis) for all other values of x in its domain.

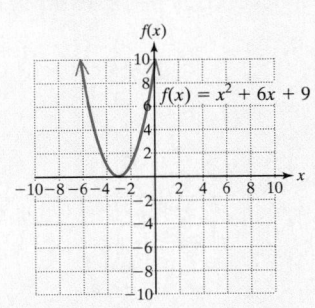

Skill Practice Answers

6. $(-2, \infty)$

d. $x^2 + 6x + 9 < 0$

$(x + 3)^2 < 0$

A perfect square cannot be negative; therefore, there are no real numbers x such that $(x + 3)^2 < 0$. There is no solution.

Skill Practice Answers

7. All real numbers; $(-\infty, \infty)$
8. $(-\infty, 2) \cup (2, \infty)$
9. $\{2\}$
10. No solution

Skill Practice Solve the inequalities.

7. $x^2 - 4x + 4 \geq 0$ **8.** $x^2 - 4x + 4 > 0$

9. $x^2 - 4x + 4 \leq 0$ **10.** $x^2 - 4x + 4 < 0$

Section 10.2   Practice Exercises

Boost *your* GRADE at mathzone.com!

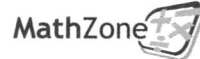

- Practice Problems
- Self-Tests
- NetTutor
- e-Professors
- Videos

Study Skills Exercise

1. Define the key terms.

    a. quadratic inequality   **b. boundary points**

    c. test point method   **d. rational inequality**

Review Exercises

For Exercises 2–8, solve the compound inequalities. Write the solutions in interval notation.

2. $6x - 10 > 8$ or $8x + 2 < 5$ **3.** $3(a - 1) + 2 > 0$ or $2a > 5a + 12$

4. $5(k - 2) > -25$ and $7(1 - k) > 7$ **5.** $2y + 4 \geq 10$ and $5y - 3 \leq 13$

6. $0 < 3(x + 1) \leq 4$ **7.** $6 \geq 4 - 2x \geq -2$

8. $-4 > 5 - x > -6$

Concept 1: Solving Inequalities Graphically

For Exercises 9–12, estimate from the graph the intervals for which the inequality is true.

9.

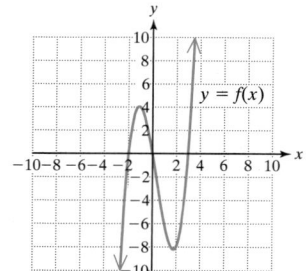

   a. $f(x) > 0$ **b.** $f(x) < 0$

   c. $f(x) \leq 0$ **d.** $f(x) \geq 0$

10.

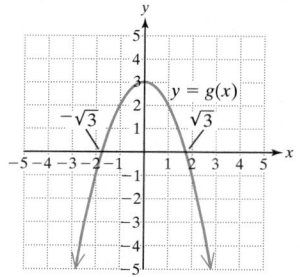

   a. $g(x) < 0$ **b.** $g(x) > 0$

   c. $g(x) \geq 0$ **d.** $g(x) \leq 0$

11.

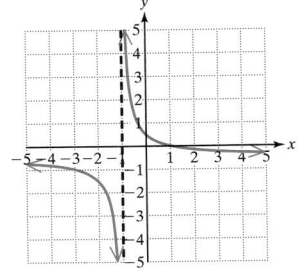

a. $h(x) \geq 0$ **b.** $h(x) \leq 0$

c. $h(x) < 0$ **d.** $h(x) > 0$

12.

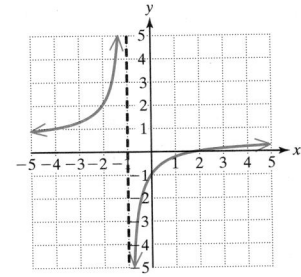

a. $k(x) \leq 0$ **b.** $k(x) \geq 0$

c. $k(x) > 0$ **d.** $k(x) < 0$

Concept 2: Solving Polynomial Inequalities by Using the Test Point Method

For Exercises 13–18, solve the equation and related inequalities.

13. a. $3(4 - x)(2x + 1) = 0$

 b. $3(4 - x)(2x + 1) < 0$

 c. $3(4 - x)(2x + 1) > 0$

14. a. $5(y + 6)(3 - 5y) = 0$

 b. $5(y + 6)(3 - 5y) < 0$

 c. $5(y + 6)(3 - 5y) > 0$

15. a. $x^2 + 7x = 30$

 b. $x^2 + 7x < 30$

 c. $x^2 + 7x > 30$

16. a. $q^2 - 4q = 5$

 b. $q^2 - 4q \leq 5$

 c. $q^2 - 4q \geq 5$

17. a. $2p(p - 2) = p + 3$

 b. $2p(p - 2) \leq p + 3$

 c. $2p(p - 2) \geq p + 3$

18. a. $3w(w + 4) = 10 - w$

 b. $3w(w + 4) < 10 - w$

 c. $3w(w + 4) > 10 - w$

For Exercises 19–38, solve the polynomial inequality. Write the answer in interval notation.

19. $(t - 7)(t + 1) < 0$

20. $(p - 4)(p + 2) > 0$

21. $-6(4 + 2x)(5 - x) > 0$

22. $-8(2t + 5)(6 - t) < 0$

23. $m(m + 1)^2(m + 5) \leq 0$

24. $w^2(3 - w)(w + 2) \geq 0$

25. $a^2 - 12a \leq -32$

26. $w^2 + 20w \geq -64$

27. $5x^2 - 4x - 1 > 0$

28. $3x^2 + x - 4 > 0$

29. $x^2 + x \leq 6$

30. $x^2 - 8 \leq 2x$

31. $b^2 - 121 < 0$

32. $c^2 - 25 < 0$

33. $3p(p - 2) - 3 \geq 2p$

34. $2t(t + 3) - t \leq 12$

35. $x^3 - x^2 \leq 12x$

36. $x^3 + 36 > 4x^2 + 9x$

37. $w^3 + w^2 > 4w + 4$

38. $2p^3 - 5p^2 \leq 3p$

Concept 3: Solving Rational Inequalities by Using the Test Point Method

For Exercises 39–42, solve the equation and related inequalities.

39. a. $\dfrac{10}{x-5} = 5$

b. $\dfrac{10}{x-5} < 5$

c. $\dfrac{10}{x-5} > 5$

40. a. $\dfrac{8}{a+1} = 4$

b. $\dfrac{8}{a+1} > 4$

c. $\dfrac{8}{a+1} < 4$

41. a. $\dfrac{z+2}{z-6} = -3$

b. $\dfrac{z+2}{z-6} \le -3$

c. $\dfrac{z+2}{z-6} \ge -3$

42. a. $\dfrac{w-8}{w+6} = 2$

b. $\dfrac{w-8}{w+6} \le 2$

c. $\dfrac{w-8}{w+6} \ge 2$

For Exercises 43–54, solve the rational inequalities. Write the answer in interval notation.

43. $\dfrac{2}{x-1} \ge 0$

44. $\dfrac{-3}{x+2} \le 0$

45. $\dfrac{b+4}{b-4} > 0$

46. $\dfrac{a+1}{a-3} < 0$

47. $\dfrac{3}{2x-7} < -1$

48. $\dfrac{8}{4x+9} > 1$

49. $\dfrac{x+1}{x-5} \ge 4$

50. $\dfrac{x-2}{x+6} \le 5$

51. $\dfrac{1}{x} \le 2$

52. $\dfrac{1}{x} \ge 3$

53. $\dfrac{(x+2)^2}{x} > 0$

54. $\dfrac{(x-3)^2}{x} < 0$

Concept 4: Inequalities with "Special Case" Solution Sets

For Exercises 55–70, solve the inequalities.

55. $x^2 + 10x + 25 \ge 0$

56. $x^2 + 6x + 9 < 0$

57. $x^2 + 2x + 1 < 0$

58. $x^2 + 8x + 16 \ge 0$

59. $\dfrac{x^2}{x^2+4} < 0$

60. $\dfrac{x^2}{x^2+4} \ge 0$

61. $x^4 + 3x^2 \le 0$

62. $x^4 + 2x^2 \le 0$

63. $x^2 + 24x + 144 > 0$

64. $x^2 + 12x + 36 < 0$

65. $x^2 + 24x + 144 \le 0$

66. $x^2 + 12x + 36 \ge 0$

67. $x^2 - 14x + 49 < 0$

68. $x^2 + 16x + 64 \ge 0$

69. $-4x^2 + 4x \le 1$

70. $-9x^2 + 6x > 1$

Mixed Exercises

For Exercises 71–94, identify the inequality as one of the following types: linear, quadratic, rational, or polynomial (degree > 2). Then solve the inequality and write the answer in interval notation.

71. $2y^2 - 8 \le 24$

72. $8p^2 - 18 > 0$

73. $(5x+2)^2 > -4$

74. $(3-7x)^2 < 0$

75. $4(x-2) < 6x - 3$

76. $-7(3-y) > 4 + 2y$

77. $\dfrac{2x+3}{x+1} \le 2$

78. $\dfrac{5x-1}{x+3} \ge 5$

79. $4x^3 - 40x^2 + 100x > 0$

80. $2y^3 - 12y^2 + 18y < 0$

81. $2p^3 > 4p^2$

82. $w^3 \le 5w^2$

83. $\dfrac{1}{x^2+3} < -4$

84. $\dfrac{1}{4t^2+5} > -2$

85. $x^2 - 4 < 0$

86. $y^2 - 9 > 0$

87. $x^2 + x - 2 \ge 0$

88. $2t^2 + 7t + 3 \le 0$

89. $\dfrac{a + 2}{a - 5} \geq 0$ **90.** $\dfrac{t + 1}{t - 2} \leq 0$ **91.** $2 \geq t - 3$

92. $-5p + 8 < p$ **93.** $4x^2 + 9 \geq 12x$ **94.** $20x - 100 > x^2$

Graphing Calculator Exercises

95. To solve the inequality $\dfrac{x}{x - 2} > 0$

enter Y_1 as $x/(x - 2)$ and determine where the graph is above the x-axis. Write the solution in interval notation.

96. To solve the inequality $\dfrac{x}{x - 2} < 0$

enter Y_1 as $x/(x - 2)$ and determine where the graph is below the x-axis. Write the solution in interval notation.

97. To solve the inequality $x^2 - 1 < 0$, enter Y_1 as $x^2 - 1$ and determine where the graph is below the x-axis. Write the solution in interval notation.

98. To solve the inequality $x^2 - 1 > 0$, enter Y_1 as $x^2 - 1$ and determine where the graph is above the x-axis. Write the solution in interval notation.

For Exercises 99–102, determine the solution by graphing the inequalities.

99. $x^2 + 10x + 25 \leq 0$ **100.** $-x^2 + 10x - 25 \geq 0$

101. $\dfrac{8}{x^2 + 2} < 0$ **102.** $\dfrac{-6}{x^2 + 3} > 0$

Absolute Value Equations

Section 10.3

1. Solving Absolute Value Equations

An equation of the form $|x| = a$ is called an **absolute value equation**. The solution includes all real numbers whose absolute value equals a. For example, the solutions to the equation $|x| = 4$ are $x = 4$ as well as $x = -4$, because $|4| = 4$ and $|-4| = 4$. In Chapter 1, we introduced a geometric interpretation of $|x|$. The absolute value of a number is its distance from zero on the number line (Figure 10-5). Therefore, the solutions to the equation $|x| = 4$ are the values of x that are 4 units away from zero.

Concepts

1. Solving Absolute Value Equations

2. Solving Equations Having Two Absolute Values

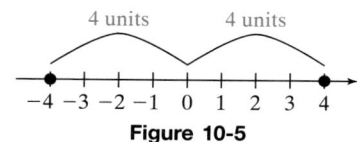

Figure 10-5

Absolute Value Equations of the Form $|x| = a$

If a is a real number, then

1. If $a \geq 0$, the equation $|x| = a$ is equivalent to $x = a$ or $x = -a$.
2. If $a < 0$, there is no solution to the equation $|x| = a$.

Example 1   Solving Absolute Value Equations

Solve the absolute value equations.

   a. $|x| = 5$ **b.** $|w| - 2 = 12$ **c.** $|p| = 0$ **d.** $|x| = -6$

Solution:

a.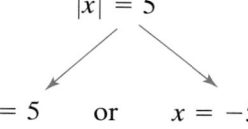

   The equation is in the form $|x| = a$, where $a = 5$.

   $x = 5$ or $x = -5$

   Rewrite the equation as $x = a$ or $x = -a$.

b. $|w| - 2 = 12$

> **TIP:** The absolute value must be isolated on one side of the equal sign before $|x| = a$ can be rewritten as $x = a$ or $x = -a$.

   $|w| = 14$

   Isolate the absolute value to write the equation in the form $|w| = a$.

   $w = 14$ or $w = -14$

   Rewrite the equation as $w = a$ or $w = -a$.

c.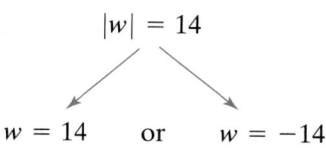

   $p = 0$ or $p = -0$

   Rewrite as two equations. Notice that the second equation $p = -0$ is the same as the first equation. Intuitively, $p = 0$ is the only number whose absolute value equals 0.

d. $|x| = -6$

   No solution

   This equation is of the form $|x| = a$, but a is negative. There is no number whose absolute value is negative.

Skill Practice   Solve the absolute value equations.

   1. $|y| = 7$ **2.** $|v| + 6 = 10$ **3.** $|w| = 0$ **4.** $|z| = -12$

Skill Practice Answers

1. $y = 7$ or $y = -7$
2. $v = 4$ or $v = -4$
3. $w = 0$ **4.** No solution

We have solved absolute value equations of the form $|x| = a$. Notice that x can represent any algebraic quantity. For example, to solve the equation $|2w - 3| = 5$, we still rewrite the absolute value equation as two equations. In this case, we set the quantity $2w - 3$ equal to 5 and to -5, respectively.

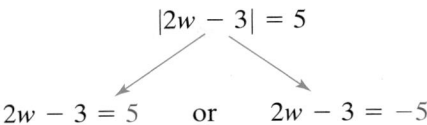

$$|2w - 3| = 5$$

$$2w - 3 = 5 \quad \text{or} \quad 2w - 3 = -5$$

Steps to Solve an Absolute Value Equation

1. Isolate the absolute value. That is, write the equation in the form $|x| = a$, where a is a constant real number.

2. If $a < 0$, there is no solution.

3. Otherwise, if $a \geq 0$, rewrite the absolute value equation as $x = a$ or $x = -a$.

4. Solve the individual equations from step 3.

5. Check the answers in the original absolute value equation.

Example 2 **Solving Absolute Value Equations**

Solve the absolute value equations.

a. $|2w - 3| = 5$ **b.** $|2c - 5| + 6 = 2$

Solution:

a. $|2w - 3| = 5$ The equation is already in the form $|x| = a$, where $x = 2w - 3$.

$2w - 3 = 5 \quad \text{or} \quad 2w - 3 = -5$ Rewrite as two equations.

$2w = 8 \quad \text{or} \quad 2w = -2$ Solve each equation.

$w = 4 \quad \text{or} \quad w = -1$

Check: $w = 4$ Check: $w = -1$ Check the solutions in the original equation.

$|2w - 3| = 5 \quad\quad |2w - 3| = 5$

$|2(4) - 3| \stackrel{?}{=} 5 \quad |2(-1) - 3| \stackrel{?}{=} 5$

$|8 - 3| \stackrel{?}{=} 5 \quad |-2 - 3| \stackrel{?}{=} 5$

$|5| \stackrel{?}{=} 5 \checkmark \quad |-5| \stackrel{?}{=} 5 \checkmark$

Calculator Connections

To confirm the answers to Example 2(a), graph $Y_1 = \text{abs}(2x - 3)$ and $Y_2 = 5$. The solutions to the equation $|2w - 3| = 5$ are the x-coordinates of the points of intersection $(4, 5)$ and $(-1, 5)$.

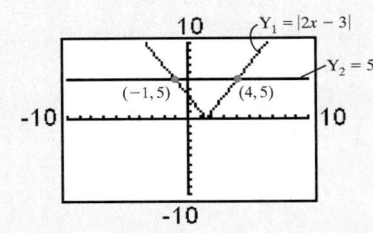

b. $|2c - 5| + 6 = 2$

$\qquad |2c - 5| = -4$ — Isolate the absolute value. The equation is in the form $|x| = a$, where $x = 2c - 5$ and $a = -4$. Because $a < 0$, there is no solution.

$\qquad$ No solution — There are no numbers c that will make an absolute value equal to a negative number.

Calculator Connections

The graphs of $Y_1 = \text{abs}(2x - 5) + 6$ and $Y_2 = 2$ do not intersect.
$\qquad$ Therefore, there is no solution to the equation $|2c - 5| + 6 = 2$.

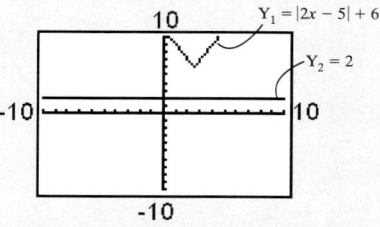

Skill Practice Solve the absolute value equations.

5. $|4x + 1| = 9$ $\qquad$ **6.** $|3z + 10| + 3 = 1$

Example 3 Solving Absolute Value Equations

Solve the absolute value equations.

a. $-2\left|\dfrac{2}{5}p + 3\right| - 7 = -19$ $\qquad$ **b.** $|4.1 - p| + 6.9 = 6.9$

Solution:

a. $-2\left|\dfrac{2}{5}p + 3\right| - 7 = -19$

$\qquad -2\left|\dfrac{2}{5}p + 3\right| = -12$ $\qquad$ Isolate the absolute value.

$\qquad \dfrac{-2\left|\dfrac{2}{5}p + 3\right|}{-2} = \dfrac{-12}{-2}$

$\qquad \left|\dfrac{2}{5}p + 3\right| = 6$

$\dfrac{2}{5}p + 3 = 6 \qquad \text{or} \qquad \dfrac{2}{5}p + 3 = -6$ $\qquad$ Rewrite as two equations.

$2p + 15 = 30 \qquad \text{or} \qquad 2p + 15 = -30$ $\qquad$ Multiply all terms by 5 to clear fractions.

$2p = 15 \qquad \text{or} \qquad 2p = -45$

$p = \dfrac{15}{2} \qquad \text{or} \qquad p = -\dfrac{45}{2}$ $\qquad$ Both solutions check in the original equation.

b. $|4.1 - p| + 6.9 = 6.9$

$\qquad\quad |4.1 - p| = 0$ Isolate the absolute value.

$\quad 4.1 - p = 0 \quad$ or $\quad 4.1 - p = -0$ Rewrite as two equations. Notice that the equations are the same.

$\qquad\qquad -p = -4.1$ Subtract 4.1 from both sides.

$\qquad\qquad\quad p = 4.1$ <u>Check:</u> $p = 4.1$

$\qquad\qquad\qquad\qquad\qquad\qquad |4.1 - p| + 6.9 = 6.9$

$\qquad\qquad\qquad\qquad\qquad\quad |4.1 - 4.1| + 6.9 \overset{?}{=} 6.9$

$\qquad\qquad\qquad\qquad\qquad\qquad\qquad |0| + 6.9 \overset{?}{=} 6.9$

The solution is 4.1. $6.9 = 6.9 \checkmark$

Skill Practice Solve the absolute value equations.

7. $3\left|\dfrac{3}{2}a + 1\right| + 2 = 14$ **8.** $|1.2 + x| - 3.5 = -3.5$

---

2. Solving Equations Having Two Absolute Values

Some equations have two absolute values. The solutions to the equation $|x| = |y|$ are $x = y$ or $x = -y$. That is, if two quantities have the same absolute value, then the quantities are equal or the quantities are opposites.

> ### Equality of Absolute Values
>
> $$|x| = |y| \text{ implies that } x = y \text{ or } x = -y.$$

Example 4 **Solving an Equation Having Two Absolute Values**

Solve the equations.

a. $|2w - 3| = |5w + 1|$ **b.** $|x - 4| = |x + 8|$

Solution:

a.
$$|2w - 3| = |5w + 1|$$

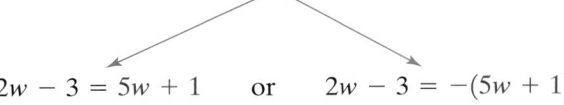

$2w - 3 = 5w + 1 \quad$ or $\quad 2w - 3 = -(5w + 1)$ Rewrite as two equations, $x = y$ or $x = -y$.

$2w - 3 = 5w + 1 \quad$ or $\quad 2w - 3 = -5w - 1$ Solve for w.

$-3w - 3 = 1 \qquad$ or $\qquad 7w - 3 = -1$

$\quad -3w = 4 \qquad$ or $\qquad\quad 7w = 2$

$\qquad\; w = -\dfrac{4}{3} \qquad$ or $\qquad\quad\; w = \dfrac{2}{7}$

The solutions are $-\dfrac{4}{3}$ and $\dfrac{2}{7}$.   Both values check in the original equation.

Avoiding Mistakes:

To take the opposite of the quantity $5w + 1$, use parentheses and apply the distributive property.

Skill Practice Answers

7. $a = 2$ or $a = -\dfrac{10}{3}$

8. $x = -1.2$

b. $|x - 4| = |x + 8|$

$x - 4 = x + 8$ or $x - 4 = -(x + 8)$ Rewrite as two equations, $x = y$ or $x = -y$.

$-4 = 8$ or $x - 4 = -x - 8$ Solve for x.

⤴ contradiction

$2x - 4 = -8$

$2x = -4$

$x = -2$

The only solution is -2. $x = -2$ checks in the original equation.

Skill Practice Answers

9. $x = \dfrac{4}{5}$ or $x = -2$

10. $t = \dfrac{1}{4}$

Skill Practice Solve the equations.

9. $|3 - 2x| = |3x - 1|$ **10.** $|4t + 3| = |4t - 5|$

Section 10.3   Practice Exercises

Study Skills Exercise

1. Define the key term **absolute value equation**.

Review Exercises

For Exercises 2–7, solve the inequalities. Write the answers in interval notation.

2. $3(a + 2) - 6 > 2$ and $-2(a - 3) + 14 > -3$ **3.** $3x - 5 \geq 7x + 3$ or $2x - 1 \leq 4x - 5$

4. $\dfrac{4}{y - 4} \geq 3$ **5.** $\dfrac{3}{t + 1} \leq 2$

6. $3(x - 2)(x + 4)(2x - 1) < 0$ **7.** $x^3 - 7x^2 - 8x > 0$

Concept 1: Solving Absolute Value Equations

For Exercises 8–39, solve the absolute value equations.

8. $|p| = 7$ **9.** $|q| = 10$  **10.** $|x| + 5 = 11$ **11.** $|x| - 3 = 20$

12. $|y| = \sqrt{2}$ **13.** $|y| = \dfrac{5}{8}$ **14.** $|w| - 3 = -5$ **15.** $|w| + 4 = -8$

16. $|3q| = 0$ **17.** $|4p| = 0$ **18.** $\left|3x - \dfrac{1}{2}\right| = \dfrac{1}{2}$ **19.** $|4x + 1| = 6$

20. $\left|\dfrac{7z}{3} - \dfrac{1}{3}\right| + 3 = 6$ **21.** $\left|\dfrac{w}{2} + \dfrac{3}{2}\right| - 2 = 7$ **22.** $\left|\dfrac{5y + 2}{2}\right| = 6$ **23.** $\left|\dfrac{2t - 1}{3}\right| = 5$

24. $|0.2x - 3.5| = -5.6$ **25.** $|1.81 + 2x| = -2.2$ **26.** $1 = -4 + \left|2 - \dfrac{1}{4}w\right|$ **27.** $-12 = -6 - |6 - 2x|$

28. $10 = 4 + |2y + 1|$ **29.** $-1 = -|5x + 7|$ **30.** $-2|3b - 7| - 9 = -9$

31. $-3|5x + 1| + 4 = 4$ **32.** $-2|x + 3| = 5$ **33.** $-3|x - 5| = 7$

34. $0 = |6x - 9|$ **35.** $7 = |4k - 6| + 7$ **36.** $\left|-\dfrac{1}{5} - \dfrac{1}{2}k\right| = \dfrac{9}{5}$

37. $\left|-\dfrac{1}{6} - \dfrac{2}{9}h\right| = \dfrac{1}{2}$ **38.** $-3|2 - 6x| + 5 = -10$ **39.** $5|1 - 2x| - 7 = 3$

Concept 2: Solving Equations Having Two Absolute Values

For Exercises 40–53, solve the absolute value equations.

40. $|4x - 2| = |-8|$ **41.** $|3x + 5| = |-5|$ **42.** $|4w + 3| = |2w - 5|$

43. $|3y + 1| = |2y - 7|$ **44.** $|2y + 5| = |7 - 2y|$ **45.** $|9a + 5| = |9a - 1|$

46. $\left|\dfrac{4w - 1}{6}\right| = \left|\dfrac{2w}{3} + \dfrac{1}{4}\right|$ **47.** $\left|\dfrac{3p + 2}{4}\right| = \left|\dfrac{1}{2}p - 2\right|$ **48.** $|2h - 6| = |2h + 5|$

49. $|6n - 7| = |4 - 6n|$ **50.** $|3.5m - 1.2| = |8.5m + 6|$ **51.** $|11.2n + 9| = |7.2n - 2.1|$

52. $|4x - 3| = -|2x - 1|$ **53.** $-|3 - 6y| = |8 - 2y|$

Expanding Your Skills

54. Write an absolute value equation whose solution is the set of real numbers 6 units from zero on the number line.

55. Write an absolute value equation whose solution is the set of real numbers $\frac{7}{2}$ units from zero on the number line.

56. Write an absolute value equation whose solution is the set of real numbers $\frac{4}{3}$ units from zero on the number line.

57. Write an absolute value equation whose solution is the set of real numbers 9 units from zero on the number line.

For Exercises 58–63, solve the absolute value equations.

58. $|5y - 3| + \sqrt{5} = 1 + \sqrt{5}$ **59.** $|2x - \sqrt{3}| + 4 = 4 + \sqrt{3}$ **60.** $|\sqrt{3} + x| = 7$

61. $\sqrt{2} + |w - 8| = 3 + 4\sqrt{2}$ **62.** $|w - \sqrt{6}| = |3w + \sqrt{6}|$ **63.** $\left|\dfrac{\sqrt{5}}{2}x - 4\right| = 6$

Graphing Calculator Exercises

For Exercises 64–71, enter the left side of the equation as Y_1 and enter the right side of the equation as Y_2. Then use the *Intersect* feature or *Zoom* and *Trace* to approximate the x-values where the two graphs intersect (if they intersect).

64. $|4x - 3| = 5$ **65.** $|x - 4| = 3$ **66.** $|8x + 1| + 8 = 1$

67. $|3x - 2| + 4 = 2$ **68.** $|x - 3| = |x + 2|$ **69.** $|x + 4| = |x - 2|$

70. $|2x - 1| = |-x + 3|$ **71.** $|3x| = |2x - 5|$

Section 10.4   Absolute Value Inequalities

Concepts

1. Solving Absolute Value Inequalities by Definition
2. Solving Absolute Value Inequalities by the Test Point Method
3. Translating to an Absolute Value Expression

1. Solving Absolute Value Inequalities by Definition

In Section 10.3, we studied absolute value equations in the form $|x| = a$. In this section we will solve absolute value *inequalities*. An inequality in any of the forms $|x| < a$, $|x| \le a$, $|x| > a$, or $|x| \ge a$ is called an **absolute value inequality**.

Recall that an absolute value represents distance from zero on the real number line. Consider the following absolute value equation and inequalities.

1. $|x| = 3$ **Solution:**

  $x = 3$ or $x = -3$ The set of all points 3 units from zero on the number line

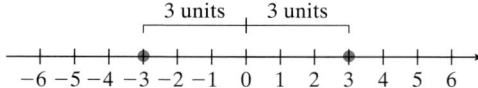

2. $|x| > 3$

Solution:

$x < -3$ or $x > 3$ The set of all points more than 3 units from zero

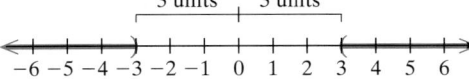

3. $|x| < 3$

Solution:

$-3 < x < 3$ The set of all points less than 3 units from zero

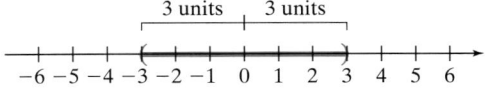

Absolute Value Equations and Inequalities

Let a be a real number such that $a > 0$. Then

Equation/ Inequality	Solution (Equivalent Form)	Graph
$|x| = a$	$x = -a$ or $x = a$	
$|x| > a$	$x < -a$ or $x > a$	
$|x| < a$	$-a < x < a$	

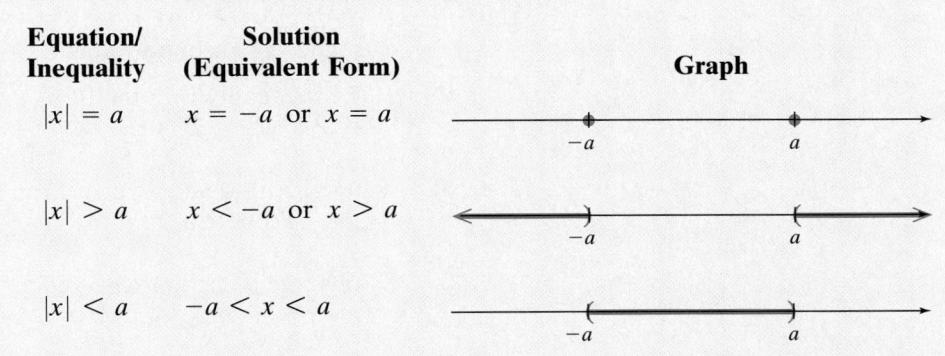

To solve an absolute value inequality, first isolate the absolute value and then rewrite the absolute value inequality in its equivalent form.

Example 1   Solving Absolute Value Inequalities

Solve the inequalities.

a. $|3w + 1| - 4 < 7$ **b.** $3 \le 1 + \left|\dfrac{1}{2}t - 5\right|$

Solution:

a. $|3w + 1| - 4 < 7$

        $|3w + 1| < 11$ ◄─── Isolate the absolute value first.

        The inequality is in the form $|x| < a$, where $x = 3w + 1$.

$-11 < 3w + 1 < 11$ Rewrite in the equivalent form $-a < x < a$.

   $-12 < 3w < 10$ Solve for w.

     $-4 < w < \dfrac{10}{3}$

The solution is $\{w \mid -4 < w < \frac{10}{3}\}$ or, equivalently in interval notation, $(-4, \frac{10}{3})$.

Graph $Y_1 = \text{abs}(3x + 1) - 4$ and $Y_2 = 7$.
On the given display window, $Y_1 < Y_2$
(Y_1 is below Y_2) for $-4 < x < \frac{10}{3}$.

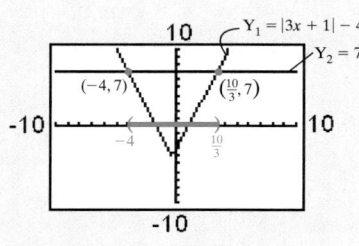

b. $3 \leq 1 + \left|\dfrac{1}{2}t - 5\right|$

$\qquad 1 + \left|\dfrac{1}{2}t - 5\right| \geq 3$ Write the inequality with the absolute value on the left.

$\qquad \left|\dfrac{1}{2}t - 5\right| \geq 2$ Isolate the absolute value.

     The inequality is in the form $|x| \geq a$, where $x = \frac{1}{2}t - 5$.

$\dfrac{1}{2}t - 5 \leq -2$ or $\dfrac{1}{2}t - 5 \geq 2$ Rewrite in the equivalent form $x \leq -a$ or $x \geq a$.

$\dfrac{1}{2}t \leq 3$ or $\dfrac{1}{2}t \geq 7$ Solve the compound inequality.

$2\left(\dfrac{1}{2}t\right) \leq 2(3)$ or $2\left(\dfrac{1}{2}t\right) \geq 2(7)$ Clear fractions.

$t \leq 6$ or $t \geq 14$ 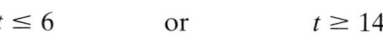

The solution is $\{t \mid t \leq 6 \text{ or } t \geq 14\}$ or, equivalently in interval notation, $(-\infty, 6] \cup [14, \infty)$.

Graph $Y_1 = \text{abs}((1/2)x - 5) + 1$ and $Y_2 = 3$.
On the given display window, $Y_1 \geq Y_2$ for $x \leq 6$ or $x \geq 14$.

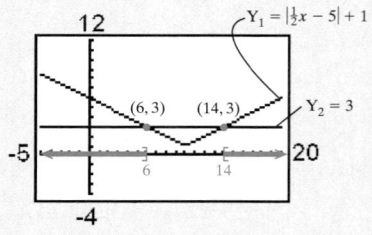

Skill Practice   Solve the inequalities. Write the solutions in interval notation.

1. $|2t + 5| + 2 \leq 11$ **2.** $5 < 1 + \left|\dfrac{1}{3}c - 1\right|$

Skill Practice Answers
1. $[-7, 2]$
2. $(-\infty, -9) \cup (15, \infty)$

By definition, the absolute value of a real number will always be nonnegative. Therefore, the absolute value of any expression will always be greater than

a negative number. Similarly, an absolute value can never be less than a negative number. Let a represent a positive real number. Then

- The solution to the inequality $|x| > -a$ is all real numbers, $(-\infty, \infty)$.
- There is no solution to the inequality $|x| < -a$.

Example 2 **Solving Absolute Value Inequalities**

Solve the inequalities.

a. $|3d - 5| + 7 < 4$ **b.** $|3d - 5| + 7 > 4$

Solution:

a. $|3d - 5| + 7 < 4$ Isolate the absolute value. An absolute value
    $|3d - 5| < -3$ expression cannot be less than a negative
                                number. Therefore, there is no solution.

    No solution

b. $|3d - 5| + 7 > 4$ Isolate the absolute value. The inequality is in
    $|3d - 5| > -3$ the form $|x| > a$, where a is negative. An
                                absolute value of any real number is greater
                                than a negative number. Therefore, the
    All real numbers, $(-\infty, \infty)$ solution is all real numbers.

Calculator Connections

By graphing $Y_1 = \text{abs}(3x - 5) + 7$ and $Y_2 = 4$, we see that $Y_1 > Y_2$ (Y_1 is above Y_2) for all real numbers x on the given display window.

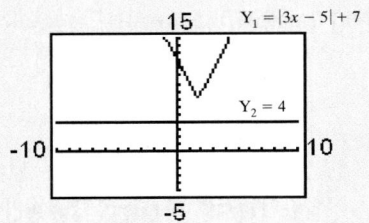

Skill Practice Solve the inequalities.

3. $|4p + 2| + 6 < 2$ **4.** $|4p + 2| + 6 > 2$

Example 3 **Solving Absolute Value Inequalities**

Solve the inequalities.

a. $|4x + 2| \geq 0$ **b.** $|4x + 2| > 0$

Solution:

a. $|4x + 2| \geq 0$ ◄── The absolute value is already isolated.

The absolute value of any real number is nonnegative. Therefore, the solution is all real numbers, $(-\infty, \infty)$.

b. $|4x + 2| > 0$

An absolute value will be greater than zero at all points *except where it is equal to zero*. That is, the point(s) for which $|4x + 2| = 0$ must be excluded from the solution set.

Skill Practice Answers

3. No solution
4. All real numbers; $(-\infty, \infty)$

$$|4x + 2| = 0$$

$$4x + 2 = 0 \qquad \text{or} \qquad 4x + 2 = -0 \qquad \text{The second equation is the same as the first.}$$

$$4x = -2$$

$$x = -\frac{1}{2} \qquad\qquad\qquad \text{Therefore, exclude } x = -\tfrac{1}{2} \text{ from the solution.}$$

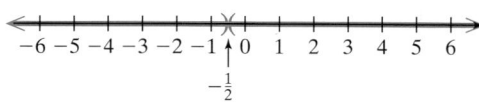

The solution is $\{x \mid x \neq -\frac{1}{2}\}$ or equivalently in interval notation, $(-\infty, -\frac{1}{2}) \cup (-\frac{1}{2}, \infty)$.

Calculator Connections

Graph $Y_1 = \text{abs}(4x + 2)$. From the graph, $Y_1 = 0$ at $x = -\frac{1}{2}$ (the x-intercept). On the given display window, $Y_1 > 0$ for $x < -\frac{1}{2}$ or $x > -\frac{1}{2}$.

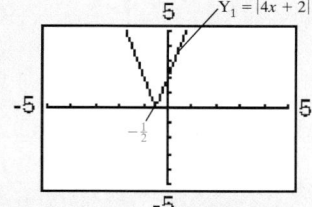

Skill Practice Solve the inequalities.

5. $|3x - 1| \geq 0$ **6.** $|3x - 1| > 0$

2. Solving Absolute Value Inequalities by the Test Point Method

For each problem in Example 1, the absolute value inequality was converted to an equivalent compound inequality. However, sometimes students have difficulty setting up the appropriate compound inequality. To avoid this problem, you may want to use the test point method to solve absolute value inequalities.

Solving Inequalities by Using the Test Point Method

1. Find the boundary points of the inequality. (Boundary points are the real solutions to the related equation and points where the inequality is undefined.)

2. Plot the boundary points on the number line. This divides the number line into regions.

3. Select a test point from each region and substitute it into the original inequality.
 - If a test point makes the original inequality true, then that region is part of the solution set.

4. Test the boundary points in the original inequality.
 - If a boundary point makes the original inequality true, then that point is part of the solution set.

Skill Practice Answers

5. $(-\infty, \infty)$

6. $\left\{x \mid x \neq \frac{1}{3}\right\}; \left(-\infty, \frac{1}{3}\right) \cup \left(\frac{1}{3}, \infty\right)$

To demonstrate the use of the test point method, we will repeat the absolute value inequalities from Example 1. Notice that regardless of the method used, the absolute value is always isolated *first* before any further action is taken.

| Example 4 | Solving Absolute Value Inequalities by the Test Point Method |

Solve the inequalities by using the test point method.

a. $|3w + 1| - 4 < 7$ **b.** $3 \leq 1 + \left|\frac{1}{2}t - 5\right|$

Solution:

a. $|3w + 1| - 4 < 7$

$|3w + 1| < 11$ ←————————————— Isolate the absolute value.

$|3w + 1| = 11$ **Step 1:** Solve the related equation.

$3w + 1 = 11$ or $3w + 1 = -11$ Write as an equivalent system of two equations.

$3w = 10$ or $3w = -12$

$w = \dfrac{10}{3}$ or $w = -4$ These are the only boundary points.

Step 2: Plot the boundary points.

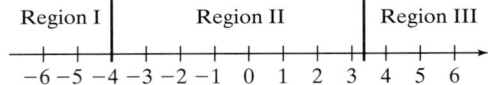

Step 3: Select a test point from each region.

Test $w = -5$:   **Test $w = 0$:**   **Test $w = 4$:**

$|3(-5) + 1| - 4 \overset{?}{<} 7$ $|3(0) + 1| - 4 \overset{?}{<} 7$ $|3(4) + 1| - 4 \overset{?}{<} 7$

$|-14| - 4 \overset{?}{<} 7$ $|1| - 4 \overset{?}{<} 7$ $|13| - 4 \overset{?}{<} 7$

$14 - 4 \overset{?}{<} 7$ $-3 \overset{?}{<} 7$ True $13 - 4 \overset{?}{<} 7$

$10 \overset{?}{<} 7$ False $9 \overset{?}{<} 7$ False

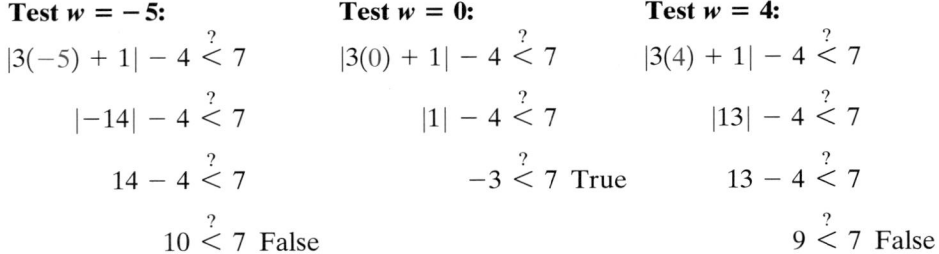

Step 4: Because the original inequality is a strict inequality, the boundary points (where equality occurs) are not included.

The solution is $\{w | -4 < w < \frac{10}{3}\}$ or, equivalently in interval notation, $(-4, \frac{10}{3})$.

b. $3 \le 1 + \left|\dfrac{1}{2}t - 5\right|$

$1 + \left|\dfrac{1}{2}t - 5\right| \ge 3$ Write the inequality with the absolute value on the left.

$\left|\dfrac{1}{2}t - 5\right| \ge 2$ ⟵   Isolate the absolute value.

$\left|\dfrac{1}{2}t - 5\right| = 2$ **Step 1:** Solve the related equation.

$\dfrac{1}{2}t - 5 = 2 \qquad \text{or} \qquad \dfrac{1}{2}t - 5 = -2$ Write the union of two equations.

$\qquad \dfrac{1}{2}t = 7 \qquad \text{or} \qquad \dfrac{1}{2}t = 3$

$\qquad\quad t = 14 \qquad \text{or} \qquad\quad t = 6$ These are the boundary points.

Step 2: Plot the boundary points.

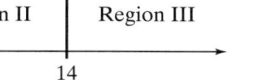

Region I | Region II | Region III

6   14

Step 3: Select a test point from each region.

Test $t = 0$:

$3 \overset{?}{\le} 1 + \left|\dfrac{1}{2}(0) - 5\right|$

$3 \overset{?}{\le} 1 + |0 - 5|$

$3 \overset{?}{\le} 1 + |-5|$

$3 \le 1 + 5$ True

Test $t = 10$:

$3 \overset{?}{\le} 1 + \left|\dfrac{1}{2}(10) - 5\right|$

$3 \overset{?}{\le} 1 + |5 - 5|$

$3 \overset{?}{\le} 1 + |0|$

$3 \overset{?}{\le} 1$ False

Test $t = 16$:

$3 \overset{?}{\le} 1 + \left|\dfrac{1}{2}(16) - 5\right|$

$3 \overset{?}{\le} 1 + |8 - 5|$

$3 \overset{?}{\le} 1 + |3|$

$3 \le 4$ True

Step 4: The original inequality uses the sign $\ge$. Therefore, the boundary points (where equality occurs) must be part of the solution set.

True   False   True

6   14

The solution is $\{x \mid x \le 6 \text{ or } x \ge 14\}$ or, equivalently in interval notation, $(-\infty, 6] \cup [14, \infty)$.

Skill Practice   Solve the inequalities by using the test point method.

7. $10 \ge 6 + |3t - 4|$ **8.** $\left|\dfrac{1}{2}c + 4\right| + 1 > 6$

Skill Practice Answers

7. $\left[0, \dfrac{8}{3}\right]$

8. $(-\infty, -18) \cup (2, \infty)$

Example 5 Solving Absolute Value Inequalities

Solve the inequalities.

a. $\left| \dfrac{1}{3}x + 4 \right| < 0$ **b.** $\left| \dfrac{1}{3}x + 4 \right| \leq 0$

Solution:

a. $\left| \dfrac{1}{3}x + 4 \right| < 0$ ◄—— The absolute value is already isolated.

No solution   Because the absolute value of any real number is nonnegative, an absolute value cannot be strictly less than zero. Therefore, there is no solution to this inequality.

b. $\left| \dfrac{1}{3}x + 4 \right| \leq 0$ ◄—— The absolute value is already isolated.

An absolute value will never be less than zero. However, an absolute value may be equal to zero. Therefore, the only solutions to this inequality are the solutions to the related equation.

$\left| \dfrac{1}{3}x + 4 \right| = 0$ Set up the related equation.

$\dfrac{1}{3}x + 4 = 0$

$\dfrac{1}{3}x = -4$

$x = -12$ This is the only boundary point.

The solution set is $\{-12\}$.

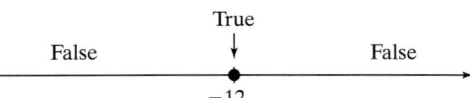

The solution set is $\{-12\}$.

Calculator Connections

Graph $Y_1 = \text{abs}((1/3)x + 4)$. Notice that on the given viewing window the graph of Y_1 does not extend below the x-axis. Therefore, there is no solution to the inequality $Y_1 < 0$.
    Because $Y_1 = 0$ at $x = -12$, the inequality $Y_1 \leq 0$ has a solution at $x = -12$.

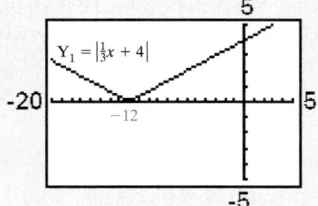

Skill Practice Solve the inequalities.

9. $\left| \dfrac{3}{5}x + 3 \right| < 0$ **10.** $\left| \dfrac{3}{5}x + 3 \right| \leq 0$

Skill Practice Answers

9. No solution   **10.** $\{-5\}$

3. Translating to an Absolute Value Expression

Absolute value expressions can be used to describe distances. The distance between c and d is given by $|c - d|$. For example, the distance between -2 and 3 on the number line is $|(-2) - 3| = 5$ as expected.

Example 6   Expressing Distances with Absolute Value

Write an absolute value inequality to represent the following phrases.

a. All real numbers x, whose distance from zero is greater than 5 units

b. All real numbers x, whose distance from -7 is less than 3 units

Solution:

a. All real numbers x, whose distance from zero is greater than 5 units

$|x - 0| > 5$ or simply $|x| > 5$

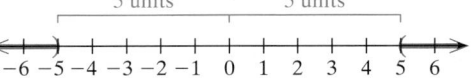

b. All real numbers x, whose distance from -7 is less than 3 units

$|x - (-7)| < 3$ or simply $|x + 7| < 3$

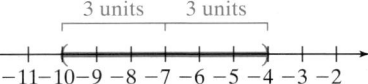

Skill Practice   Write an absolute value inequality to represent the following phrases.

11. All real numbers whose distance from zero is greater than 10 units
12. All real numbers whose distance from 4 is less than 6

Absolute value expressions can also be used to describe boundaries for measurement error.

Example 7   Expressing Measurement Error with Absolute Value

Latoya measured a certain compound on a scale in the chemistry lab at school. She measured 8 g of the compound, but the scale is only accurate to ± 0.1 g. Write an absolute value inequality to express an interval for the true mass, x, of the compound she measured.

Solution:

Because the scale is only accurate to ± 0.1 g, the true mass, x, of the compound may deviate by as much as 0.1 g above or below 8 g. This may be expressed as an absolute value inequality:

$|x - 8.0| \leq 0.1$ or equivalently $7.9 \leq x \leq 8.1$

Skill Practice Answers

11. $|x| > 10$
12. $|x - 4| < 6$

Skill Practice

13. Vonzell molded a piece of metal in her machine shop. She measured the thickness at 12 mm. Her machine is accurate to ±0.05 mm. Write an absolute value inequality to express an interval for the true measurement of the thickness, t, of the metal.

Skill Practice Answers

13. $|t - 12| \le 0.05$

Section 10.4   Practice Exercises

Boost *your* GRADE at mathzone.com!

• Practice Problems   • e-Professors
• Self-Tests   • Videos
• NetTutor

Study Skills Exercise

1. Define the key term **absolute value inequality**.

Review Exercises

For Exercises 2–3, solve the equations.

2. $|10x - 6| = -5$

3. $2 = |5 - 7x| + 1$

For Exercises 4–7, solve the inequalities and graph the solution set. Write the solution in interval notation.

4. $-15 < 3w - 6 \le -9$

5. $5 - 2y \le 1$ and $3y + 2 \ge 14$

6. $m - 7 \le -5$ or $m - 7 \ge -10$

7. $3b - 2 < 7$ or $b - 2 > 4$

Concepts 1 and 2: Solving Absolute Value Inequalities

For Exercises 8–18, solve the equations and inequalities. For each inequality, graph the solution set and express the solution in interval notation.

8. **a.** $|x| = 5$

   b. $|x| > 5$

   c. $|x| < 5$

9. **a.** $|a| = 4$

   b. $|a| > 4$

   c. $|a| < 4$

10. **a.** $|x - 3| = 7$

    b. $|x - 3| > 7$

    c. $|x - 3| < 7$

11. **a.** $|w + 2| = 6$

    b. $|w + 2| > 6$

    c. $|w + 2| < 6$

12. **a.** $|p| = -2$

    b. $|p| > -2$

    c. $|p| < -2$

13. **a.** $|x| = -14$

    b. $|x| > -14$

    c. $|x| < -14$

14. **a.** $|y + 1| = -6$

    b. $|y + 1| > -6$

    c. $|y + 1| < -6$

15. **a.** $|z - 4| = -3$

    b. $|z - 4| > -3$

    c. $|z - 4| < -3$

16. **a.** $|x| = 0$

    b. $|x| > 0$

    c. $|x| < 0$

17. a. $|p + 3| = 0$

   b. $|p + 3| > 0$

           ————————————→

   c. $|p + 3| < 0$

           ————————————→

18. a. $|k - 7| = 0$

   b. $|k - 7| > 0$

           ————————————→

   c. $|k - 7| < 0$

           ————————————→

For Exercises 19–48, solve the absolute value inequalities by using either the definition or the test point method. Graph the solution set and write the solution in interval notation.

19. $|x| > 6$

20. $|x| \leq 6$

21. $|t| \leq 3$

22. $|p| > 3$

23. $|y + 2| \geq 0$

24. $0 \leq |7n + 2|$

25. $5 \leq |2x - 1|$

26. $|x - 2| \geq 7$

27. $|k - 7| < -3$

28. $|h + 2| < -9$

29. $\left|\dfrac{w - 2}{3}\right| - 3 \leq 1$

30. $\left|\dfrac{x + 3}{2}\right| - 2 \geq 4$

31. $14 \leq |9 - 4y|$

32. $1 > |2m - 7|$

33. $\left|\dfrac{2x + 1}{4}\right| < 5$

34. $\left|\dfrac{x - 4}{5}\right| \leq 7$

35. $8 < |4 - 3x| + 12$

36. $-16 < |5x - 1| - 1$

37. $5 - |2m + 1| > 5$

38. $3 - |5x + 3| > 3$

39. $|p + 5| \leq 0$

40. $|y + 1| - 4 \leq -4$

41. $|z - 6| + 5 > 5$

42. $|2c - 1| - 4 > -4$

43. $5|2y - 6| + 3 \geq 13$

44. $7|y + 1| - 3 \geq 11$

45. $-3|6 - t| + 1 > -5$

46. $-4|8 - x| + 2 > -14$

47. $|0.02x + 0.06| - 0.1 < 0.05$

48. $|0.05x - 0.04| - 0.01 < 0.11$

Concept 3: Translating to an Absolute Value Expression

For Exercises 49–52, write an absolute value inequality equivalent to the expression given.

49. All real numbers whose distance from 0 is greater than 7

50. All real numbers whose distance from −3 is less than 4

51. All real numbers whose distance from 2 is at most 13

52. All real numbers whose distance from 0 is at least 6

53. A 32-oz jug of orange juice may not contain exactly 32 oz of juice. The possibility of measurement error exists when the jug is filled in the factory. If the maximum measurement error is ±0.05 oz, write an absolute value inequality representing the range of volumes, x, in which the orange juice jug may be filled.

54. The length of a board is measured to be 32.3 in. The maximum measurement error is ±0.2 in. Write an absolute value inequality that represents the range for the length of the board, x.

55. A bag of potato chips states that its weight is $6\frac{3}{4}$ oz. The maximum measurement error is ±$\frac{1}{8}$ oz. Write an absolute value inequality that represents the range for the weight, x, of the bag of chips.

56. A $\frac{7}{8}$-in. bolt varies in length by at most ±$\frac{1}{16}$ in. Write an absolute value inequality that represents the range for the length, x, of the bolt.

57. The width, w, of a bolt is supposed to be 2 cm but may have a 0.01-cm margin of error. Solve $|w - 2| \le 0.01$, and interpret the solution to the inequality in the context of this problem.

58. In the 2004 election, Senator Barak Obama was projected to receive 70% of the votes with a margin of error of 3%. Solve $|p - 0.70| \le 0.03$, and interpret the solution to the inequality in the context of this problem.

Expanding Your Skills

For Exercises 59–62, match the graph with the inequality:

59.

60.

61.

62.

a. $|x - 2| < 4$ **b.** $|x - 1| > 4$ **c.** $|x - 3| < 2$ **d.** $|x - 5| > 1$

Graphing Calculator Exercises

To solve an absolute value inequality by using a graphing calculator, let Y_1 equal the left side of the inequality and let Y_2 equal the right side of the inequality. Graph both Y_1 and Y_2 on a standard viewing window and use an *Intersect* feature or *Zoom* and *Trace* to approximate the intersection of the graphs. To solve $Y_1 > Y_2$, determine all x-values where the graph of Y_1 is above the graph of Y_2. To solve $Y_1 < Y_2$, determine all x-values where the graph of Y_1 is below the graph of Y_2.

For Exercises 63–72, solve the inequalities using a graphing calculator.

63. $|x + 2| > 4$ **64.** $|3 - x| > 6$

65. $\left| \dfrac{x + 1}{3} \right| < 2$

66. $\left| \dfrac{x - 1}{4} \right| < 1$

67. $|x - 5| < -3$

68. $|x + 2| < -2$

69. $|2x + 5| > -4$

70. $|1 - 2x| > -4$

71. $|6x + 1| \leq 0$

72. $|3x - 4| \leq 0$

Chapter 10   Problem Recognition Exercises— Equations and Inequalities

For Exercises 1–24, identify the category for each equation or inequality (choose from the list here). Then solve the equation or inequality.

- Linear
- Quadratic
- Polynomial (degree greater than 2)
- Rational
- Absolute value

1. $z^2 + 10z + 9 > 0$

2. $5a - 2 = 6(a + 4)$

3. $\dfrac{x - 4}{2x + 4} \geq 1$

4. $4x^2 - 7x = 2$

5. $|3x - 1| + 4 < 6$

6. $9t^2 + 6t + 1 < 0$

7. $\dfrac{1}{2}p - \dfrac{2}{3} < \dfrac{1}{6}p - 4$

8. $p^2 + 3p \leq 4$

9. $3y^2 - 5y - 2 \geq 0$

10. $\dfrac{x + 6}{x + 4} = 7$

11. $3(2x - 4) \geq 1 - (x - 3)$

12. $|6x + 5| + 3 = 2$

13. $(x - 3)(2x + 1)(x + 5) \geq 0$

14. $-x^2 - 18x - 81 \leq 0$

15. $\dfrac{-6}{y - 2} < 2$

16. $x^2 + 4 = 5x$

17. $\left| \dfrac{x}{4} + 2 \right| + 6 > 6$

18. $x^2 - 36 < 0$

19. $3y^3 + 5y^2 - 12y - 20 = 0$

20. $|8 - 2x| = 16$

21. $5b - 10 - 3b = 3(b - 2) - 4$

22. $\dfrac{x}{x - 5} \geq 0$

23. $|6x + 1| = |5 + 6x|$

24. $4x(x - 5)^2(2x + 3) < 0$

Linear Inequalities in Two Variables

Section 10.5

1. Graphing Linear Inequalities in Two Variables

A **linear inequality in two variables** x and y is an inequality that can be written in one of the following forms: $ax + by < c$, $ax + by > c$, $ax + by \leq c$, or $ax + by \geq c$, provided a and b are not both zero.

    A solution to a linear inequality in two variables is an ordered pair that makes the inequality true. For example, solutions to the inequality $x + y < 6$ are ordered pairs (x, y) such that the sum of the x- and y-coordinates is less than 6. This inequality has an infinite number of solutions, and therefore it is convenient to express the solution set as a graph.

    To graph a linear inequality in two variables, we will follow these steps.

Concepts

1. Graphing Linear Inequalities in Two Variables
2. Compound Linear Inequalities in Two Variables
3. Graphing a Feasible Region

> **Graphing a Linear Inequality in Two Variables**
>
> 1. Solve for y, if possible.
> 2. Graph the related equation. Draw a dashed line if the inequality is strict, $<$ or $>$. Otherwise, draw a solid line.
> 3. Shade above or below the line as follows:
>   - Shade *above* the line if the inequality is of the form $y > ax + b$ or $y \geq ax + b$.
>   - Shade *below* the line if the inequality is of the form $y < ax + b$ or $y \leq ax + b$.

This process is demonstrated in Example 1.

Example 1   Graphing a Linear Inequality in Two Variables

Graph the solution set. $-3x + y \leq 1$

Solution:

$$-3x + y \leq 1$$

$$y \leq 3x + 1 \qquad \text{Solve for } y.$$

Next graph the line defined by the related equation $y = 3x + 1$.

Because the inequality is of the form $y \leq ax + b$, the solution to the inequality is the region *below* the line $y = 3x + 1$. See Figure 10-6.

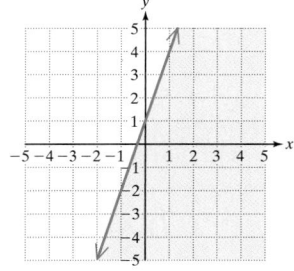

Figure 10-6

Skill Practice   Graph the solution set.

1. $2x + y \geq -4$

After graphing the solution to a linear inequality, we can verify that we have shaded the correct side of the line by using test points. In Example 1, we can pick an arbitrary ordered pair within the shaded region. Then substitute the x- and y-coordinates in the original inequality. If the result is a true statement, then that ordered pair is a solution to the inequality and suggests that other points from the same region are also solutions.

For example, the point $(0, 0)$ lies within the shaded region (Figure 10-7).

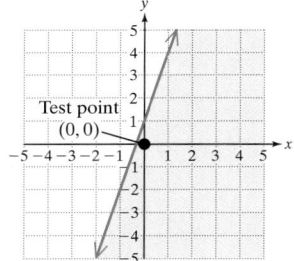

Figure 10-7

$$-3x + y \leq 1 \qquad \text{Substitute } (0, 0) \text{ in the original inequality.}$$

$$-3(0) + (0) \overset{?}{\leq} 1$$

$$0 + 0 \leq 1 \quad ✔ \text{ True} \qquad \text{The point } (0, 0) \text{ from the shaded region is a solution.}$$

In Example 2, we will graph the solution set to a strict inequality. A strict inequality uses the symbol $<$ or $>$. In such a case, the boundary line will be drawn as a dashed line. This indicates that the boundary itself is *not* part of the solution set.

Example 2   **Graphing a Linear Inequality in Two Variables**

Graph the solution set. $-4y < 5x$

Solution:

$$-4y < 5x$$

$$\frac{-4y}{-4} > \frac{5x}{-4} \qquad \text{Solve for } y. \text{ Reverse the inequality sign.}$$

$$y > -\frac{5}{4}x$$

Skill Practice Answers

1.

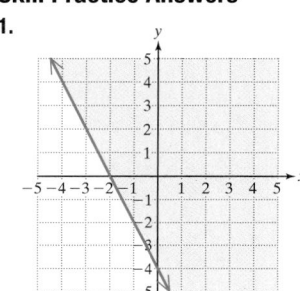

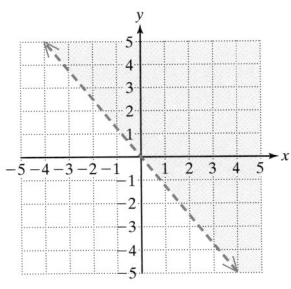

Figure 10-8

Graph the line defined by the related equation, $y = -\frac{5}{4}x$. The boundary line is drawn as a dashed line because the inequality is strict. Also note that the line passes through the origin.

Because the inequality is of the form $y > ax + b$, the solution to the inequality is the region *above* the line. See Figure 10-8.

Skill Practice   Graph the solution set.

2. $-3y < x$

In Example 2, we cannot use the origin as a test point, because the point $(0, 0)$ is on the boundary line. Be sure to select a test point strictly within the shaded region. In this case, we choose $(2, 1)$. See Figure 10-9.

$$-4y < 5x$$
$$-4(1) \overset{?}{<} 5(2) \qquad \text{Substitute } (2, 1) \text{ in the original inequality.}$$
$$-4 < 10 \ \checkmark \ \text{True} \qquad \text{The point } (2, 1) \text{ from the shaded region is a solution to the original inequality.}$$

In Example 3, we encounter a situation in which we cannot solve for the y-variable.

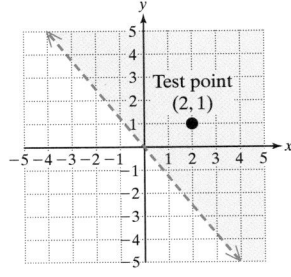

Figure 10-9

Example 3   **Graphing a Linear Inequality in Two Variables**

Graph the solution set. $4x \geq -12$

Solution:

$4x \geq -12$

$x \geq -3$

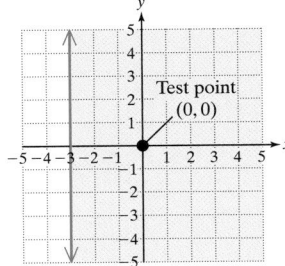

Figure 10-10

In this inequality, there is no y-variable. However, we can simplify the inequality by solving for x.

Graph the related equation $x = -3$. This is a vertical line. The boundary is drawn as a solid line, because the inequality is not strict, $\geq$.

To shade the appropriate region, refer to the inequality, $x \geq -3$. The points for which x is greater than -3 are to the right of $x = -3$. Therefore, shade the region to the *right* of the line (Figure 10-10).

Selecting a test point such as $(0, 0)$ from the shaded region indicates that we have shaded the correct side of the line.

$$4x \geq -12 \qquad \text{Substitute } x = 0.$$
$$4(0) \geq -12 \ \checkmark \qquad \text{True}$$

Skill Practice   Graph the solution set.

3. $-2x \geq 2$

Skill Practice Answers

2.

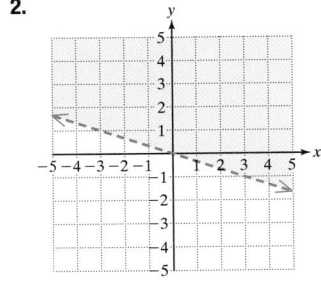

3.

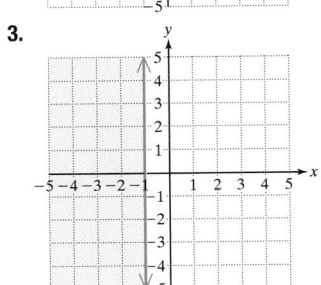

2. Compound Linear Inequalities in Two Variables

Some applications require us to find the union or intersection of two or more linear inequalities.

> **Example 4** **Graphing a Compound Linear Inequality**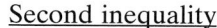

Graph the solution set of the compound inequality.

$$y > \tfrac{1}{2}x + 1 \quad \text{and} \quad x + y < 1$$

Solution:

Solve each inequality for y.

<u>First inequality</u>

$y > \tfrac{1}{2}x + 1$

The inequality is of the form $y > ax + b$. Graph *above* the boundary line. (See Figure 10-11).

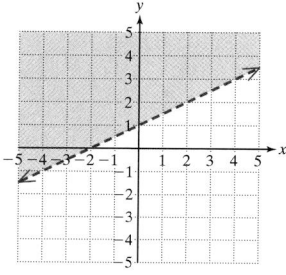

Figure 10-11

<u>Second inequality</u>

$x + y < 1$

$\qquad y < -x + 1$

The inequality is of the form $y < ax + b$. Graph *below* the boundary line. (See Figure 10-12).

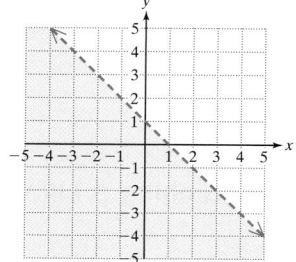

Figure 10-12

The region bounded by the inequalities is the region above the line $y = \tfrac{1}{2}x + 1$ and below the line $y = -x + 1$. This is the intersection or "overlap" of the two regions (shown in purple in Figure 10-13).

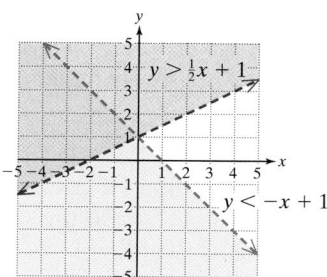

Figure 10-13

The intersection is the solution set to the system of inequalities. See Figure 10-14.

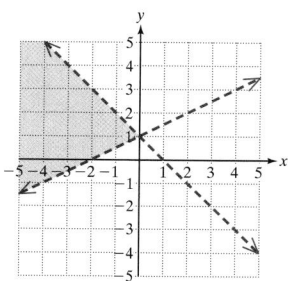

Figure 10-14

Skill Practice Graph the solution set.

4. $x - 3y > 3$

$y < -2x + 4$

Example 5 demonstrates the union of the solution sets of two linear inequalities.

Example 5 **Graphing a Compound Linear Inequality**

Graph the solution set of the compound inequality.

$$3y \le 6 \quad \text{or} \quad y - x \le 0$$

Solution:

<u>First inequality</u>

$3y \le 6$

$y \le 2$

The graph of $y \le 2$ is the region on and below the horizontal line $y = 2$. (See Figure 10-15.)

<u>Second inequality</u>

$y - x \le 0$

$y \le x$

The inequality is of the form $y \le ax + b$. Graph a solid line and the region below the line. (See Figure 10-16.)

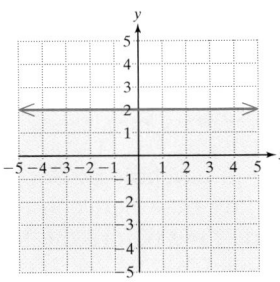

Figure 10-15

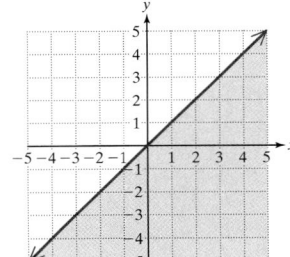

Figure 10-16

The solution to the compound inequality $3y \le 6$ or $y - x \le 0$ is the union of these regions, Figure 10-17.

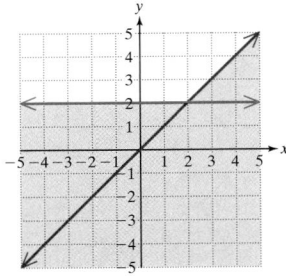

Figure 10-17

Skill Practice Graph the solution set.

5. $2y \le 4 \quad \text{or} \quad y \le x + 1$

Skill Practice Answers

4.

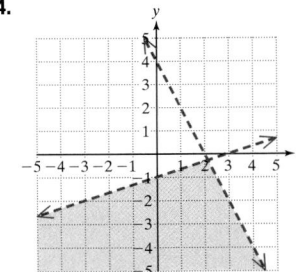

5.

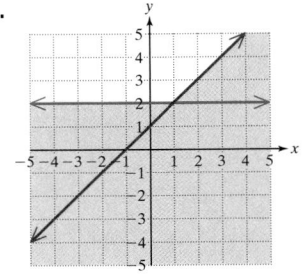

Example 6   Graphing Compound Linear Inequalities

Describe the region of the plane defined by the following systems of inequalities.

$$x \leq 0 \quad \text{and} \quad y \geq 0$$

Solution:

$x \leq 0$ $x \leq 0$ on the y-axis and in the second and third quadrants.

$y \geq 0$ $y \geq 0$ on the x-axis and in the first and second quadrants.

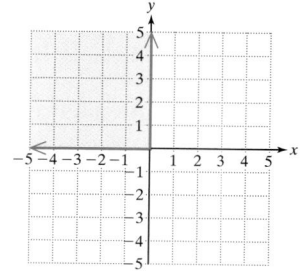

The intersection of these regions is the set of points in the second quadrant (with the boundary included).

Skill Practice   Graph the region defined by the system of inequalities.

6. $x \leq 0 \quad$ and $\quad y \leq 0$

---

3. Graphing a Feasible Region

When two variables are related under certain constraints, a system of linear inequalities can be used to show a region of feasible values for the variables.

Example 7   Graphing a Feasible Region

Susan has two tests on Friday: one in chemistry and one in psychology. Because the two classes meet in consecutive hours, she has no study time between tests. Susan estimates that she has a maximum of 12 hr of study time before the tests, and she must divide her time between chemistry and psychology.

Let x represent the number of hours Susan spends studying chemistry.

Let y represent the number of hours Susan spends studying psychology.

a. Find a set of inequalities to describe the constraints on Susan's study time.

b. Graph the constraints to find the feasible region defining Susan's study time.

Solution:

a. Because Susan cannot study chemistry or psychology for a negative period of time, we have $x \geq 0$ and $y \geq 0$.
Furthermore, her total time studying cannot exceed 12 hr: $x + y \leq 12$.

A system of inequalities that defines the constraints on Susan's study time is

$$x \geq 0$$
$$y \geq 0$$
$$x + y \leq 12$$

Skill Practice Answers

6.

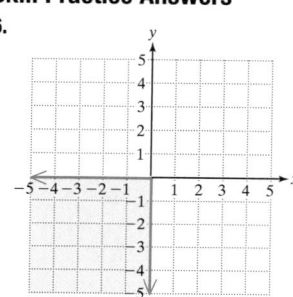

b. The first two conditions $x \geq 0$ and $y \geq 0$ represent the set of points in the first quadrant. The third condition $x + y \leq 12$ represents the set of points below and including the line $x + y = 12$ (Figure 10-18).

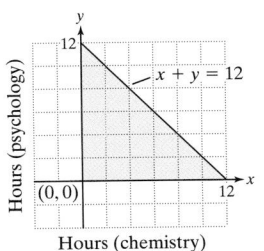

Figure 10-18

Discussion:

1. Refer to the feasible region drawn in Example 7(b). Is the ordered pair $(8, 5)$ part of the feasible region?

No. The ordered pair $(8, 5)$ indicates that Susan spent 8 hr studying chemistry and 5 hr studying psychology. This is a total of 13 hr, which exceeds the constraint that Susan only had 12 hr to study. The point $(8, 5)$ lies outside the feasible region, above the line $x + y = 12$ (Figure 10-19).

2. Is the ordered pair $(7, 3)$ part of the feasible region?

Yes. The ordered pair $(7, 3)$ indicates that Susan spent 7 hr studying chemistry and 3 hr studying psychology.

This point lies within the feasible region and satisfies all three constraints.

$$x \geq 0 \longrightarrow \qquad 7 \geq 0 \qquad \text{True}$$

$$y \geq 0 \longrightarrow \qquad 3 \geq 0 \qquad \text{True}$$

$$x + y \leq 12 \longrightarrow (7) + (3) \leq 12 \qquad \text{True}$$

Notice that the ordered pair $(7, 3)$ corresponds to a point where Susan is not making full use of the 12 hr of study time.

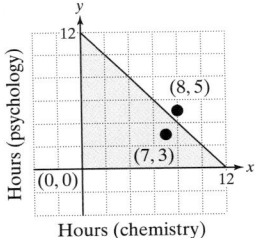

Figure 10-19

3. Suppose there was one additional constraint imposed on Susan's study time. She knows she needs to spend at least twice as much time studying chemistry as she does studying psychology. Graph the feasible region with this additional constraint.

Because the time studying chemistry must be at least twice the time studying psychology, we have $x \geq 2y$.

This inequality may also be written as $y \leq \dfrac{x}{2}$

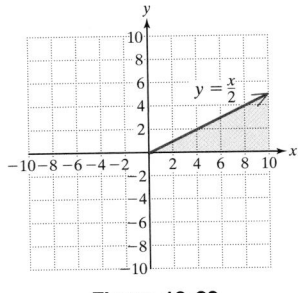

Figure 10-20

Figure 10-20 shows the first quadrant with the constraint $y \leq \dfrac{x}{2}$.

4. At what point in the feasible region is Susan making the most efficient use of her time for both classes?

First and foremost, Susan must make use of *all* 12 hr. This occurs for points along the line $x + y = 12$. Susan will also want to study for both classes with approximately twice as much time devoted to chemistry. Therefore, Susan will be deriving the maximum benefit at the point of intersection of the line $x + y = 12$ and the line $y = \dfrac{x}{2}$.

Using the substitution method, replace $y = \dfrac{x}{2}$ into the equation $x + y = 12$.

$$x + \frac{x}{2} = 12$$

$$2x + x = 24 \qquad \text{Clear fractions.}$$

$$3x = 24$$

$$x = 8 \qquad \text{Solve for } x.$$

$$y = \frac{(8)}{2} \qquad \text{To solve for } y, \text{ substitute } x = 8$$

$$\text{into the equation } y = \frac{x}{2}.$$

$$y = 4$$

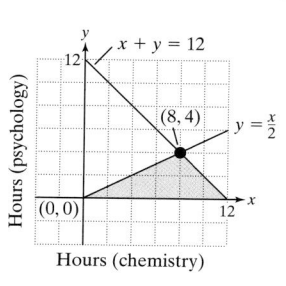

Therefore Susan should spend 8 hr studying chemistry and 4 hr studying psychology.

Skill Practice Answers

7a. $x \geq 0$ and $y \geq 0$
b. $x + y \leq 30$
c.

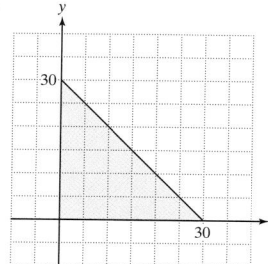

Skill Practice

7. A local pet rescue group has a total of 30 cages that can be used to hold cats and dogs. Let x represent the number of cages used for cats, and let y represent the number used for dogs.
   a. Write a set of inequalities to express the fact that the number of cat and dog cages cannot be negative.
   b. Write an inequality to describe the constraint on the total number of cages for cats and dogs.
   c. Graph the system of inequalities to find the feasible region describing the available cages.

Section 10.5   Practice Exercises

Boost *your* GRADE at mathzone.com!

MathZone

- Practice Problems
- Self-Tests
- NetTutor
- e-Professors
- Videos

Study Skills Exercise

1. Define the key term **linear inequality in two variables**.

Review Exercises

For Exercises 2–5, solve the inequalities.

2. $5 < x + 1$ and $-2x + 6 \geq -6$

3. $5 - x \leq 4$ and $6 > 3x - 3$

4. $4 - y < 3y + 12$ or $-2(y + 3) \geq 12$

5. $-2x < 4$ or $3x - 1 \leq -13$

Concept 1: Graphing Linear Inequalities in Two Variables

For Exercises 6–9, decide if the following points are solutions to the inequality.

6. $2x - y > 8$
   a. $(3, -5)$
   b. $(-1, -10)$
   c. $(4, -2)$
   d. $(0, 0)$

7. $3y + x < 5$
   a. $(-1, 7)$
   b. $(5, 0)$
   c. $(0, 0)$
   d. $(2, -3)$

8. $y \leq -2$

    a. $(5, -3)$ **b.** $(-4, -2)$

    c. $(0, 0)$ **d.** $(3, 2)$

9. $x \geq 5$

    a. $(4, 5)$ **b.** $(5, -1)$

    c. $(8, 8)$ **d.** $(0, 0)$

For Exercises 10–15, decide which inequality symbol should be used $(<, >, \geq, \leq)$ by looking at the graph.

10.

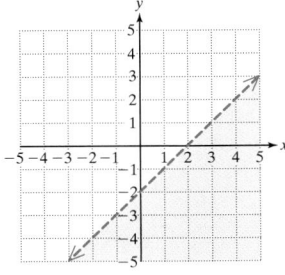

$x - y$ _____ 2

11.

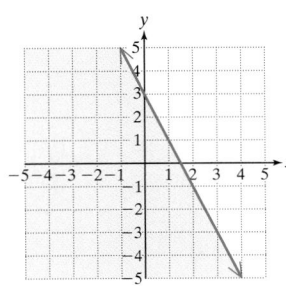

y _____ $-2x + 3$

12.

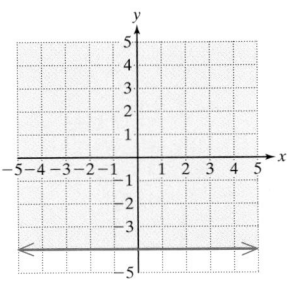

y _____ -4

13.

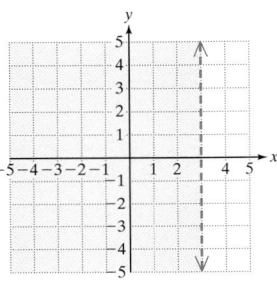

x _____ 3

14.

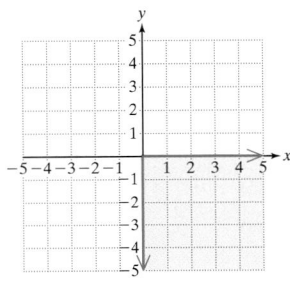

x _____ 0 and y _____ 0

15.

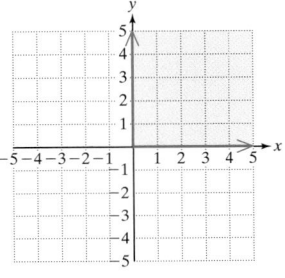

x _____ 0 and y _____ 0

For Exercises 16–39, graph the solution set.

16. $x - 2y > 4$

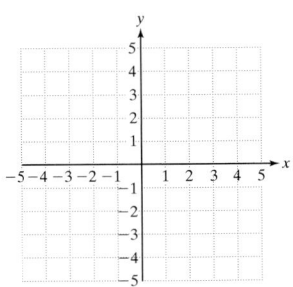

17. $x - 3y > 6$

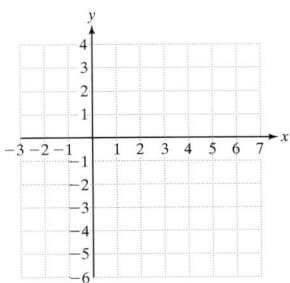

18. $5x - 2y < 10$

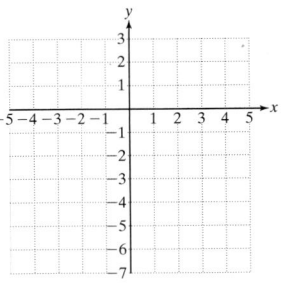

19. $x - 3y < 8$

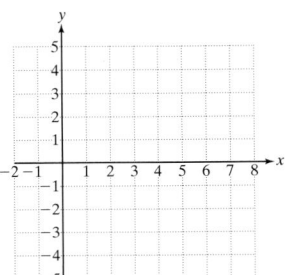

20. $2x \leq -6y + 12$

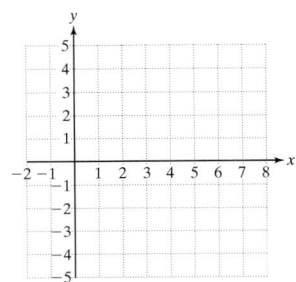

21. $4x < 3y + 12$

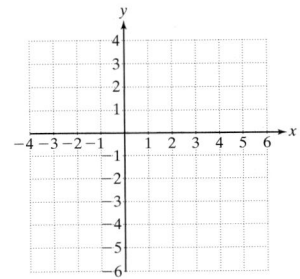

22. $2y \le 4x$

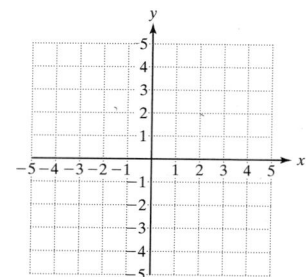

23. $-6x < 2y$

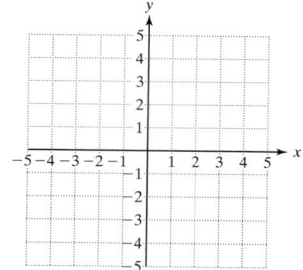

24. $y \ge -2$

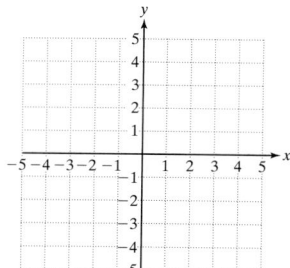

25. $y \ge 5$

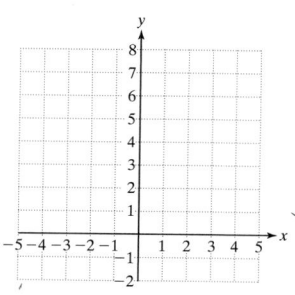

26. $4x < 5$

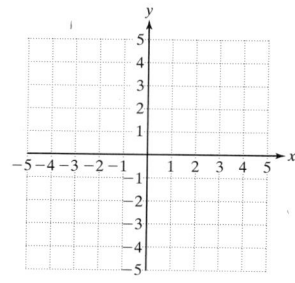

27. $x + 6 < 7$

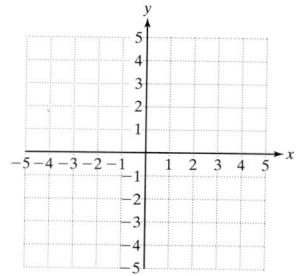

28. $y \ge \dfrac{2}{5}x - 4$

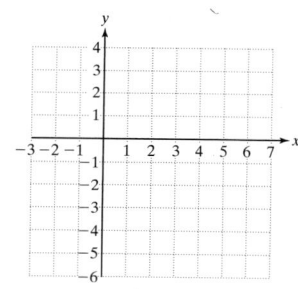

29. $y \ge -\dfrac{5}{2}x - 4$

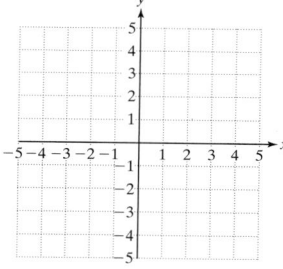

30. $y \le \dfrac{1}{3}x + 6$

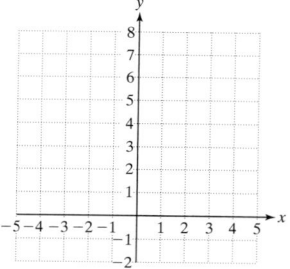

31. $y \le -\dfrac{1}{4}x + 2$

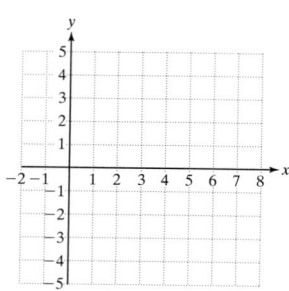

32. $y - 5x > 0$

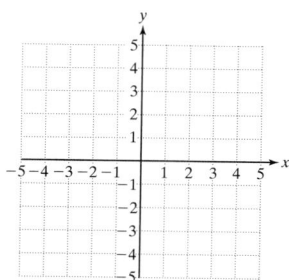

33. $y - \dfrac{1}{2}x > 0$

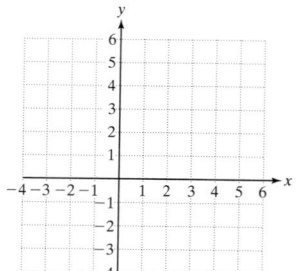

34. $\dfrac{x}{5} + \dfrac{y}{4} < 1$

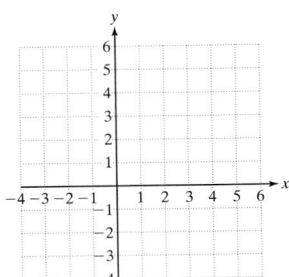

35. $x + \dfrac{y}{2} \geq 2$

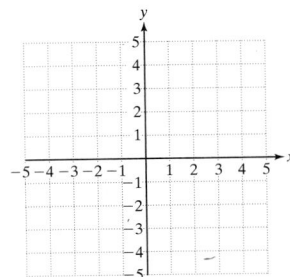

36. $0.1x + 0.2y \leq 0.6$

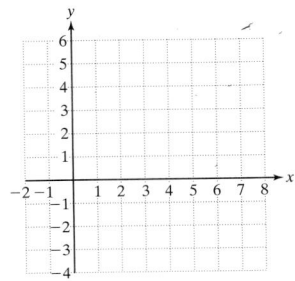

37. $0.3x - 0.2y < 0.6$

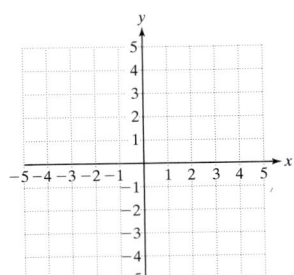

38. $x \leq -\dfrac{2}{3}y$

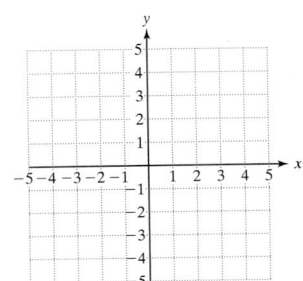

39. $x \geq -\dfrac{5}{4}y$

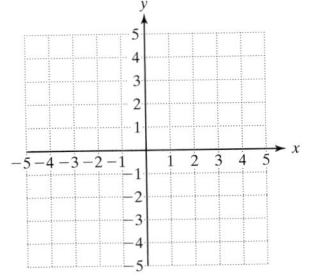

Concept 2: Compound Linear Inequalities in Two Variables

For Exercises 40–55, graph the solution set of each compound inequality.

40. $y < 4$ and $y > -x + 2$

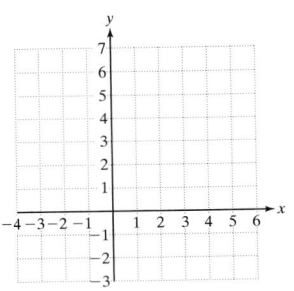

41. $y < 3$ and $x + 2y < 6$

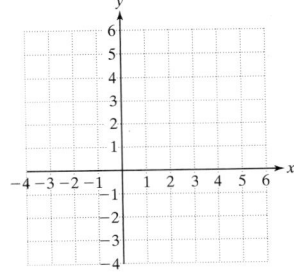

42. $2x + y \leq 5$ or $x \geq 3$

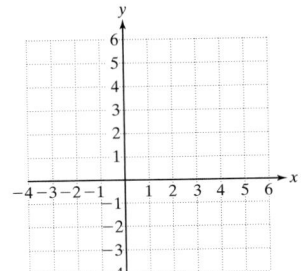

43. $x + 3y \geq 3$ or $x \leq -2$

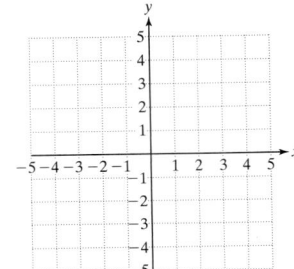

44. $x + y < 3$ and $4x + y < 6$

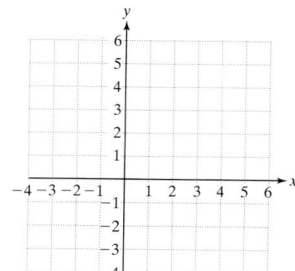

45. $x + y < 4$ and $3x + y < 9$

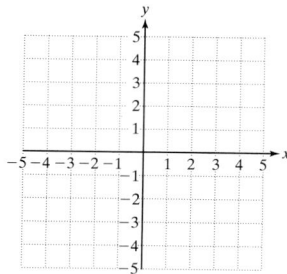

46. $2x - y \leq 2$ or $2x + 3y \geq 6$

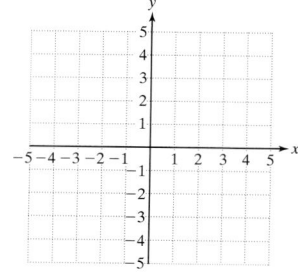

47. $3x + 2y \geq 4$ or $x - y \leq 3$

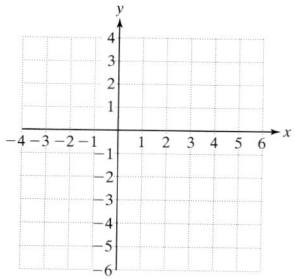

48. $x > 4$ and $y < 2$

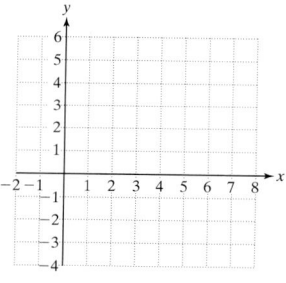

49. $x < 3$ and $y > 4$

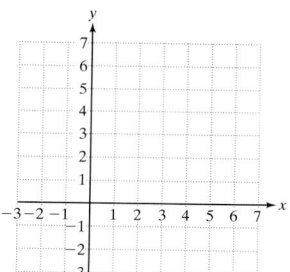

50. $x \leq -2$ or $y \leq 0$

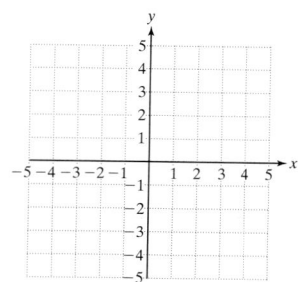

51. $x \geq 0$ or $y \geq -3$

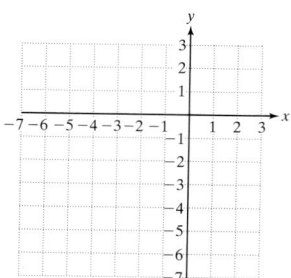

52. $x > 0$ and $x + y < 6$

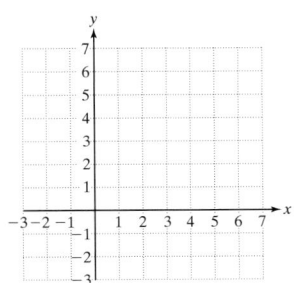

53. $x < 0$ and $x + y < 2$

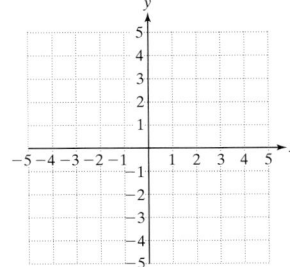

54. $y \le 0$ or $x - y \le -4$

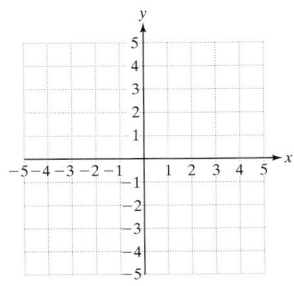

55. $y \ge 0$ or $x - y \ge -3$

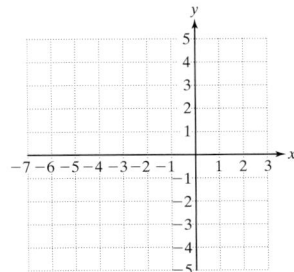

Concept 3: Graphing a Feasible Region

For Exercises 56–61, graph the feasible regions.

56. $x + y \le 3$ and

$x \ge 0, y \ge 0$

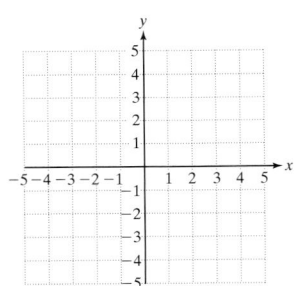

57. $x - y \le 2$ and

$x \ge 0, y \ge 0$

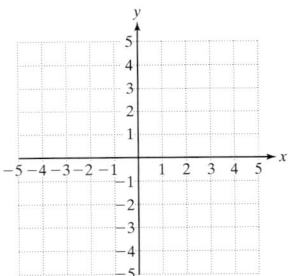

58. $y < \dfrac{1}{2}x - 3$ and

$x \le 0, y \ge -5$

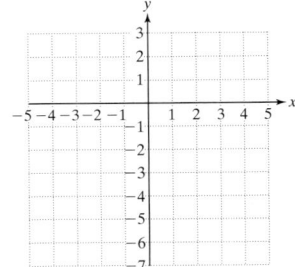

59. $y > \dfrac{1}{2}x - 3$ and

$x \ge -2, y \le 0$

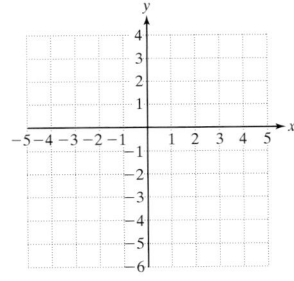

60. $x \ge 0, y \ge 0$

$x + y \le 8$ and

$3x + 5y \le 30$

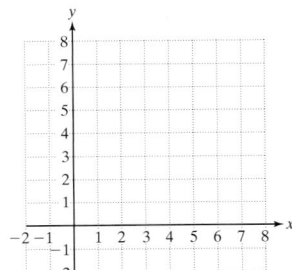

61. $x \ge 0, y \ge 0$

$x + y \le 5$ and

$x + 2y \le 6$

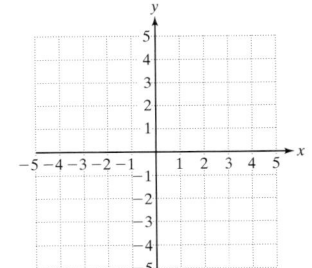

62. A manufacturer produces two models of desks. Model A requires $1\frac{1}{2}$ hr to stain and finish and $1\frac{1}{4}$ hr to assemble. Model B requires 2 hr to stain and finish and $\frac{3}{4}$ hr to assemble. The total amount of time available for staining and finishing is 12 hr and for assembling is 6 hr. Let x represent the number of Model A desks, and let y represent the number of Model B desks.

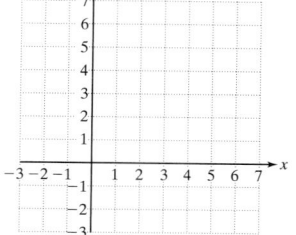

   a. Write two inequalities that express the fact that the number of desks to be produced cannot be negative.

   b. Write an inequality in terms of the number of Model A and Model B desks that can be produced if the total time for staining and finishing is at most 12 hr.

   c. Write an inequality in terms of the number of Model A and Model B desks that can be produced if the total time for assembly is no more than 6 hr.

   d. Identify the feasible region formed by graphing the preceding inequalities.

   e. Is the point $(4, 1)$ in the feasible region? What does the point $(4, 1)$ represent in the context of this problem?

   f. Is the point $(5, 4)$ in the feasible region? What does the point $(5, 4)$ represent in the context of this problem?

63. In scheduling two drivers for delivering pizza, James needs to have at least 65 hr scheduled this week. His two drivers, Karen and Todd, are not allowed to get overtime, so each one can work at most 40 hr. Let x represent the number of hours that Karen can be scheduled, and let y represent the number of hours Todd can be scheduled.

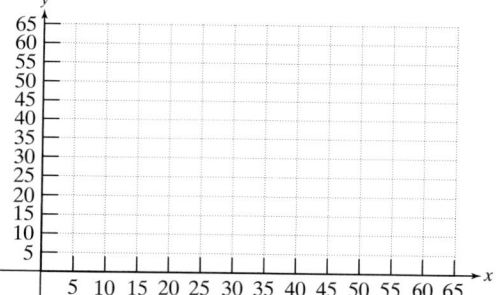

   a. Write two inequalities that express the fact that Karen and Todd cannot work a negative number of hours.

   b. Write two inequalities that express the fact that neither Karen nor Todd is allowed overtime (i.e., each driver can have at most 40 hr).

   c. Write an inequality that expresses the fact that the total number of hours from both Karen and Todd needs to be at least 65 hr.

   d. Graph the feasible region formed by graphing the inequalities.

   e. Is the point $(35, 40)$ in the feasible region? What does the point $(35, 40)$ represent in the context of this problem?

   f. Is the point $(20, 40)$ in the feasible region? What does the point $(20, 40)$ represent in the context of this problem?

Chapter 10 SUMMARY

Section 10.1 Compound Inequalities

Key Concepts

Solve two or more inequalities joined by *and* by finding the intersection of their solution sets. Solve two or more inequalities joined by *or* by finding the union of the solution sets.

Examples

Example 1

$$-7x + 3 \geq -11 \quad \text{and} \quad 1 - x < 4.5$$
$$-7x \geq -14 \quad \text{and} \quad -x < 3.5$$
$$x \leq 2 \quad \text{and} \quad x > -3.5$$

$x \leq 2$

$x > -3.5$

The solution is $\{x \mid -3.5 < x \leq 2\}$ or equivalently $(-3.5, 2]$.

Example 2

$$5y + 1 \geq 6 \quad \text{or} \quad 2y - 5 \leq -11$$
$$5y \geq 5 \quad \text{or} \quad 2y \leq -6$$
$$y \geq 1 \quad \text{or} \quad y \leq -3$$

$y \geq 1$

$y \leq -3$

The solution is $\{y \mid y \geq 1 \text{ or } y \leq -3\}$ or equivalently $(-\infty, -3] \cup [1, \infty)$.

Section 10.2   Polynomial and Rational Inequalities

Key Concepts

The Test Point Method to Solve Polynomial and Rational Inequalities

1. Find the boundary points of the inequality. (Boundary points are the real solutions to the related equation and points where the inequality is undefined.)
2. Plot the boundary points on the number line. This divides the number line into regions.
3. Select a test point from each region and substitute it into the original inequality.
   - If a test point makes the original inequality true, then that region is part of the solution set.
4. Test the boundary points in the original inequality.
   - If a boundary point makes the original inequality true, then that point is part of the solution set.

Examples

Example 1

$$\frac{28}{2x - 3} \le 4$$ The inequality is undefined for $x = \frac{3}{2}$. Find other possible boundary points by solving the related equation.

$$\frac{28}{2x - 3} = 4$$ Related equation

$$(2x - 3) \cdot \frac{28}{2x - 3} = (2x - 3) \cdot 4$$

$$28 = 8x - 12$$

$$40 = 8x$$

$$x = 5$$

The boundaries are $x = \dfrac{3}{2}$ and $x = 5$

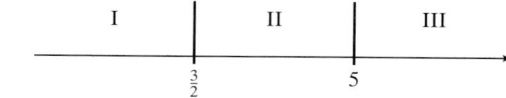

Region I: Test $x = 1$: $\dfrac{28}{2(1) - 3} \overset{?}{\le} 4$ True

Region II: Test $x = 2$: $\dfrac{28}{2(2) - 3} \overset{?}{\le} 4$ False

Region III: Test $x = 6$: $\dfrac{28}{2(6) - 3} \overset{?}{\le} 4$ True

The boundary point $x = \frac{3}{2}$ is not included because $\dfrac{28}{2x - 3}$ is undefined there. The boundary $x = 5$ does check in the original inequality.

The solution is $\left(-\infty, \frac{3}{2}\right) \cup [5, \infty)$.

Section 10.3 — Absolute Value Equations

Key Concepts

The equation $|x| = a$ is an absolute value equation. For $a \geq 0$, the solution to the equation $|x| = a$ is $x = a$ or $x = -a$.

Steps to Solve an Absolute Value Equation

1. Isolate the absolute value to write the equation in the form $|x| = a$.
2. If $a < 0$, there is no solution.
3. Otherwise, rewrite the equation $|x| = a$ as $x = a$ or $x = -a$.
4. Solve the equations from step 3.
5. Check answers in the original equation.

The solution to the equation $|x| = |y|$ is $x = y$ or $x = -y$.

Examples

Example 1

$$|2x - 3| + 5 = 10$$
$$|2x - 3| = 5 \qquad \text{Isolate the absolute value.}$$

$2x - 3 = 5$	or	$2x - 3 = -5$
$2x = 8$	or	$2x = -2$
$x = 4$	or	$x = -1$

Example 2

$$|x + 2| + 5 = 1$$
$$|x + 2| = -4 \qquad \text{No solution}$$

Example 3

$$|2x - 1| = |x + 4|$$

$2x - 1 = x + 4$	or	$2x - 1 = -(x + 4)$
$x = 5$	or	$2x - 1 = -x - 4$
	or	$3x = -3$
	or	$x = -1$

Section 10.4 — Absolute Value Inequalities

Key Concepts

Solutions to Absolute Value Inequalities

$$|x| > a \Rightarrow x < -a \quad \text{or} \quad x > a$$
$$|x| < a \Rightarrow -a < x < a$$

Examples

Example 1

$$|5x - 2| < 12$$
$$-12 < 5x - 2 < 12$$
$$-10 < 5x < 14$$
$$-2 < x < \frac{14}{5}$$

The solution is $\left(-2, \frac{14}{5}\right)$.

Test Point Method to Solve Inequalities

1. Find the boundary points of the inequality. (Boundary points are the real solutions to the related equation and points where the inequality is undefined.)
2. Plot the boundary points on the number line. This divides the number line into regions.
3. Select a test point from each region and substitute it into the original inequality.
   - If a test point makes the original inequality true, then that region is part of the solution set.
4. Test the boundary points in the original inequality.
   - If a boundary point makes the original inequality true, then that point is part of the solution set.

If *a* is *negative* (*a* < 0), then

1. $|x| < a$ has no solution.
2. $|x| > a$ is true for all real numbers.

Example 2

$$|x - 3| + 2 \geq 7$$

$$|x - 3| \geq 5 \qquad \text{Isolate the absolute value.}$$

$$|x - 3| = 5 \qquad \text{Related equation}$$

$$x - 3 = 5 \quad \text{or} \quad x - 3 = -5$$

$$x = 8 \quad \text{or} \quad x = -2 \quad \text{Boundary points}$$

Region I:

Test $x = -3$: $\quad |(-3) - 3| + 2 \overset{?}{\geq} 7 \quad$ True

Region II:

Test $x = 0$: $\quad |(0) - 3| + 2 \overset{?}{\geq} 7 \quad$ False

Region III:

Test $x = 9$: $\quad |(9) - 3| + 2 \overset{?}{\geq} 7 \quad$ True

The solution is $(-\infty, -2] \cup [8, \infty)$.

Example 3

$$|x + 5| > -2$$

The solution is all real numbers because an absolute value will always be greater than a negative number.

$$(-\infty, \infty)$$

Section 10.5   Linear Inequalities in Two Variables

Key Concepts

A **linear inequality in two variables** is an inequality of the form $ax + by < c$, $ax + by > c$, $ax + by \leq c$, or $ax + by \geq c$.

Graphing a Linear Inequality in Two Variables

1. Solve for *y*, if possible.
2. Graph the related equation. Draw a dashed line if the inequality is strict, < or >. Otherwise, draw a solid line.
3. Shade above or below the line according to the following convention.
   - Shade *above* the line if the inequality is of the form $y > ax + b$ or $y \geq ax + b$.
   - Shade *below* the line if the inequality is of the form $y < ax + b$ or $y \leq ax + b$.

Examples

Example 1

Graph the solution to the inequality $2x - y < 4$. Graph the related equation, $y = 2x - 4$, with a dashed line.

Solve for *y*: $2x - y < 4$

$$-y < -2x + 4$$

$$y > 2x - 4$$

Shade above the line.

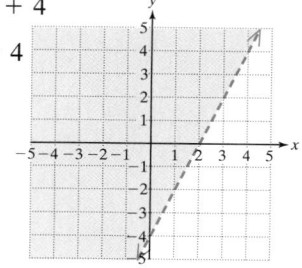

The union or intersection of two or more linear inequalities is the union or intersection of the solution sets.

Example 2

Graph.

$x < 0$ and $y > 2$

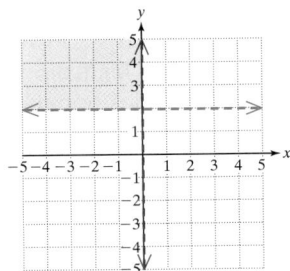

Example 3

Graph.

$x \leq 0$ or $y \geq 2$

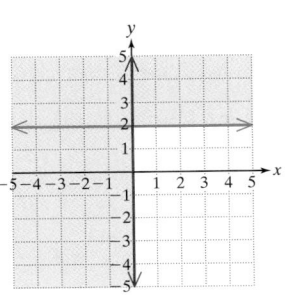

Chapter 10   Review Exercises

Section 10.1

For Exercises 1–10, solve the compound inequalities. Write the solutions in interval notation.

1. $4m > -11$ and $4m - 3 \leq 13$

2. $4n - 7 < 1$ and $7 + 3n \geq -8$

3. $-3y + 1 \geq 10$ and $-2y - 5 \leq -15$

4. $\dfrac{1}{2} - \dfrac{h}{12} \leq \dfrac{-7}{12}$ and $\dfrac{1}{2} - \dfrac{h}{10} > -\dfrac{1}{5}$

5. $\dfrac{2}{3}t - 3 \leq 1$ or $\dfrac{3}{4}t - 2 > 7$

6. $2(3x + 1) < -10$ or $3(2x - 4) \geq 0$

7. $-7 < -7(2w + 3)$ or $-2 < -4(3w - 1)$

8. $5(p + 3) + 4 > p - 1$ or $4(p - 1) + 2 > p + 8$

9. $2 \geq -(b - 2) - 5b \geq -6$

10. $-4 \leq \dfrac{1}{2}(x - 1) < -\dfrac{3}{2}$

11. The product of $\frac{1}{3}$ and the sum of a number and 3 is between -1 and 5. Find all such numbers.

12. Normal levels of total cholesterol vary according to age. For adults between 25 and 40 years old, the normal range is generally accepted to be between 140 and 225 mg/dL (milligrams per deciliter), inclusive.

   a. Write an inequality representing the normal range for total cholesterol for adults between 25 and 40 years old.

   b. Write a compound inequality representing abnormal ranges for total cholesterol for adults between 25 and 40 years old.

13. Normal levels of total cholesterol vary according to age. For adults younger than 25 years old, the normal range is generally accepted to be between 125 and 200 mg/dL, inclusive.

   a. Write an inequality representing the normal range for total cholesterol for adults younger than 25 years old.

   b. Write a compound inequality representing abnormal ranges for total cholesterol for adults younger than 25 years old.

14. In certain applications in statistics, a data value that is more than 3 standard deviations from the mean is said to be an *outlier* (a value unusually far from the average). If μ represents the mean of population and σ represents the population standard deviation, then the inequality $|x - \mu| > 3\sigma$ can be used to test whether a data value, x, is an outlier.

   The mean height, μ, of adult men is 69.0 in. (5′9″) and the standard deviation, σ, of the height of adult men is 3.0. Determine whether the heights of the following men are outliers.

   a. Shaquille O'Neal, 7′1″ = 85 in.

   b. Charlie Ward, 6′3″ = 75 in.

   c. Elmer Fudd, 4′5″ = 53 in.

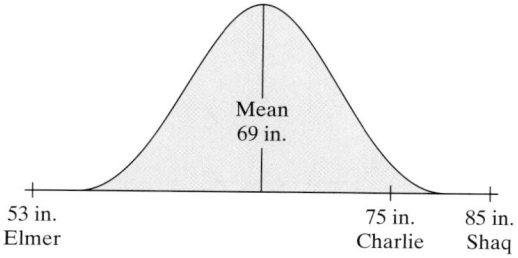

15. Explain the difference between the solution sets of the following compound inequalities.

   a. $x \le 5$ and $x \ge -2$

   b. $x \le 5$ or $x \ge -2$

Section 10.2

16. Solve the equation and inequalities. How do your answers to parts (a), (b), and (c) relate to the graph of $g(x) = x^2 - 4$?

   a. $x^2 - 4 = 0$

   b. $x^2 - 4 < 0$

   c. $x^2 - 4 > 0$

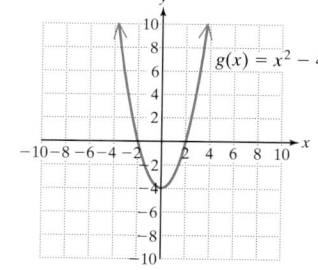

17. Solve the equation and inequalities. How do your answers to parts (a), (b), (c), and (d) relate to the graph of $k(x) = \dfrac{4x}{(x - 2)}$?

   a. $\dfrac{4x}{x - 2} = 0$

b. For which values is $k(x)$ undefined?

   c. $\dfrac{4x}{x - 2} \ge 0$

   d. $\dfrac{4x}{x - 2} \le 0$

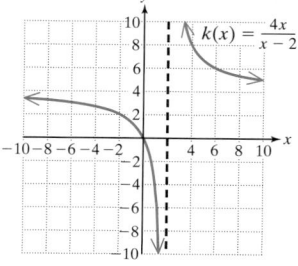

For Exercises 18–29, solve the inequalities. Write the answers in interval notation.

18. $w^2 - 4w - 12 < 0$

19. $t^2 + 6t + 9 \ge 0$

20. $\dfrac{12}{x + 2} \le 6$

21. $\dfrac{8}{p - 1} \ge -4$

22. $3y(y - 5)(y + 2) > 0$

23. $-3c(c + 2)(2c - 5) < 0$

24. $-x^2 - 4x \ge 3$

25. $y^2 + 4y > 5$

26. $\dfrac{w + 1}{w - 3} > 1$

27. $\dfrac{2a}{a + 3} \le 2$

28. $t^2 + 10t + 25 \le 0$

29. $-x^2 - 4x < 4$

Section 10.3

For Exercises 30–41, solve the absolute value equations.

30. $|x| = 10$

31. $|x| = 17$

32. $|8.7 - 2x| = 6.1$

33. $|5.25 - 5x| = 7.45$

34. $16 = |x + 2| + 9$

35. $5 = |x - 2| + 4$

36. $|4x - 1| + 6 = 4$

37. $|3x - 1| + 7 = 3$

38. $\left|\dfrac{7x - 3}{5}\right| + 4 = 4$

39. $\left|\dfrac{4x + 5}{-2}\right| - 3 = -3$

40. $|3x - 5| = |2x + 1|$

41. $|8x + 9| = |8x - 1|$

42. Which absolute value expression represents the distance between 3 and -2 on the number line?

$$|3 - (-2)| \qquad |-2 - 3|$$

Section 10.4

43. Write the compound inequality $x < -5$ or $x > 5$ as an absolute value inequality.

44. Write the compound inequality $-4 < x < 4$ as an absolute value inequality.

For Exercises 45–46, write an absolute value inequality that represents the solution sets graphed here.

45.
$-6 \qquad 6$

46.
$-\frac{2}{3} \qquad \frac{2}{3}$

For Exercises 47–60, solve the absolute value inequalities. Graph the solution set and write the solution in interval notation.

47. $|x + 6| \geq 8$

48. $|x + 8| \leq 3$

49. $2|7x - 1| + 4 > 4$

50. $4|5x + 1| - 3 > -3$

51. $|3x + 4| - 6 \leq -4$

52. $|5x - 3| + 3 \leq 6$

53. $\left|\dfrac{x}{2} - 6\right| < 5$

54. $\left|\dfrac{x}{3} + 2\right| < 2$

55. $|4 - 2x| + 8 \geq 8$

56. $|9 + 3x| + 1 \geq 1$

57. $-2|5.2x - 7.8| < 13$

58. $-|2.5x + 15| < 7$

59. $|3x - 8| < -1$

60. $|x + 5| < -4$

61. State one possible situation in which an absolute value inequality will have no solution.

62. State one possible situation in which an absolute value inequality will have a solution of all real numbers.

63. The Neilsen ratings estimated that the percent, p, of the television viewing audience watching *American Idol* was 20% with a 3% margin of error. Solve the inequality $|p - 0.20| \leq 0.03$ and interpret the answer in the context of this problem.

64. The length, L, of a screw is supposed to be $3\frac{3}{8}$ in. Due to variation in the production equipment, there is a $\frac{1}{4}$-in. margin of error. Solve the inequality $|L - 3\frac{3}{8}| \leq \frac{1}{4}$ and interpret the answer in the context of this problem.

Section 10.5

For Exercises 65–72, solve the inequalities by graphing.

65. $2x > -y + 5$

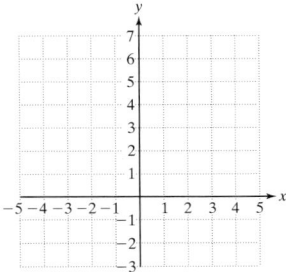

66. $2x \leq -8 - 3y$

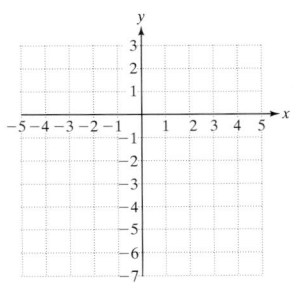

67. $y \geq -\dfrac{2}{3}x + 3$

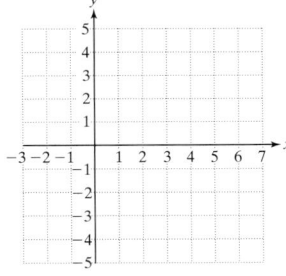

68. $y > \dfrac{3}{4}x - 2$

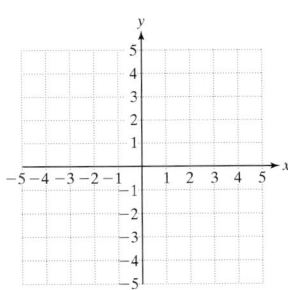

69. $x > -3$

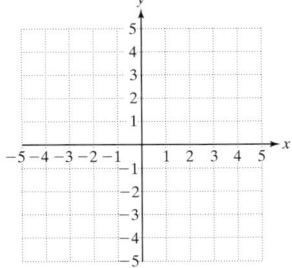

70. $x \leq 2$

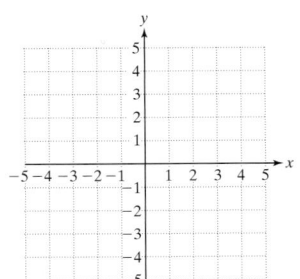

71. $x \geq \dfrac{1}{2}y$

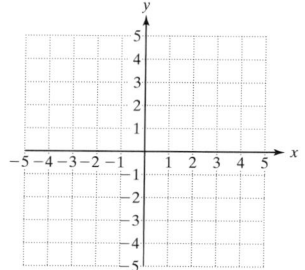

72. $x < \dfrac{2}{5}y$

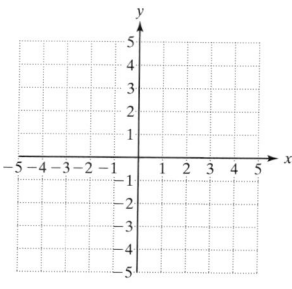

76. $x \geq 0$, $y \geq 0$,   and $y \leq -\dfrac{2}{3}x + 4$

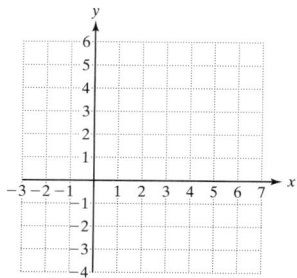

For Exercises 73–76, graph the system of inequalities.

73. $2x - y > -2$ and $2x - y \leq 2$

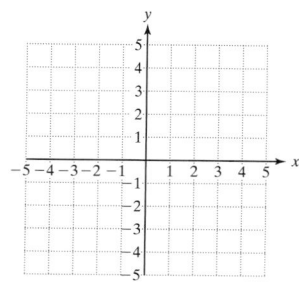

74. $3x + y \geq 6$ or $3x + y < -6$

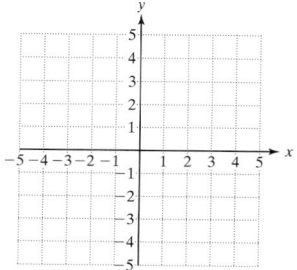

75. $x \geq 0$, $y \geq 0$,   and $y \geq -\dfrac{3}{2}x + 4$

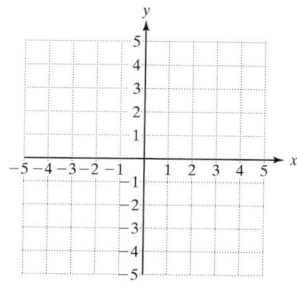

77. A pirate's treasure is buried on a small, uninhabited island in the eastern Caribbean. A shipwrecked sailor finds a treasure map at the base of a coconut palm tree. The treasure is buried within the intersection of three linear inequalities. The palm tree is located at the origin, and the positive y-axis is oriented due north. The scaling on the map is in 1-yd increments. Find the region where the sailor should dig for the treasure.

$$-2x + y \leq 4$$
$$y \leq -x + 6$$
$$y \geq 0$$

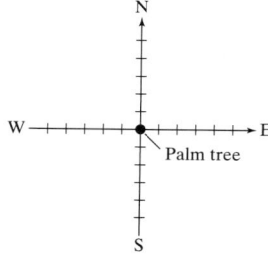

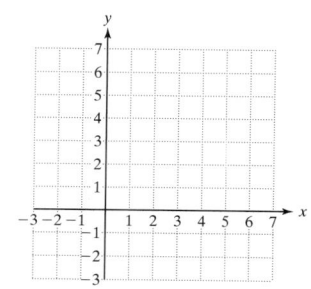

78. Suppose a farmer has 100 acres of land on which to grow oranges and grapefruit. Furthermore, because of demand from his customers, he wants to plant at least 4 times as many acres of orange trees as grapefruit trees.

Let x represent the number of acres of orange trees.

Let y represent the number of acres of grapefruit trees.

  a. Write two inequalities that express the fact that the farmer cannot use a negative number of acres to plant orange and grapefruit trees.

  b. Write an inequality that expresses the fact that the total number of acres used for growing orange and grapefruit trees is at most 100.

  c. Write an inequality that expresses the fact that the farmer wants to plant at least 4 times as many orange trees as grapefruit trees.

  d. Sketch the inequalities in parts (a)–(c) to find the feasible region for the farmer's distribution of orange and grapefruit trees.

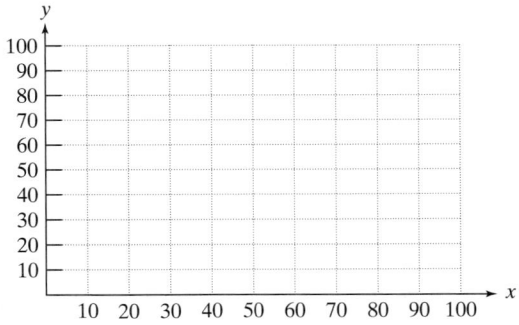

Chapter 10   Test

For Exercises 1–5, solve the compound inequalities. Write the answers in interval notation.

1. $-2 \le 3x - 1 \le 5$

2. $-\dfrac{3}{5}x - 1 \le 8$ or $-\dfrac{2}{3}x \ge 16$

3. $-2x - 3 > -3$ and $x + 3 \ge 0$

4. $5x + 1 \le 6$ or $2x + 4 > -6$

5. $2x - 3 > 1$ and $x + 4 < -1$

6. The normal range in humans of the enzyme adenosine deaminase (ADA), is between 9 and 33 IU (international units), inclusive. Let x represent the ADA level in international units.

  a. Write an inequality representing the normal range for ADA.

  b. Write a compound inequality representing abnormal ranges for ADA.

For Exercises 7–12, solve the polynomial and rational inequalities.

7. $\dfrac{2x - 1}{x - 6} \le 0$ **8.** $50 - 2a^2 > 0$

9. $y^3 + 3y^2 - 4y - 12 < 0$ **10.** $\dfrac{3}{w + 3} > 2$

11. $25x^2 - 10x + 1 < 0$

12. $t^2 + 22t + 121 \le 0$

For Exercises 13–14, solve the absolute value equations.

13. $\left| \dfrac{1}{2}x + 3 \right| - 4 = 4$

14. $|3x + 4| = |x - 12|$

15. Solve the following equation and inequalities. How do your answers to parts (a)–(c) relate to the graph of $f(x) = |x - 3| - 4$?

  a. $|x - 3| - 4 = 0$

  b. $|x - 3| - 4 < 0$

  c. $|x - 3| - 4 > 0$

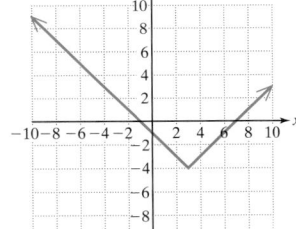

For Exercises 16–19, solve the absolute value inequalities. Write the answers in interval notation.

16. $|3 - 2x| + 6 < 2$ **17.** $|3x - 8| > 9$

18. $|0.4x + 0.3| - 0.2 < 7$

19. $|7 - 3x| + 1 > -3$

20. The mass of a small piece of metal is measured to be 15.41 g. If the measurement error is at most ± 0.01 g, write an absolute value inequality that represents the possible mass, x, of the piece of metal.

21. Graph the solution to the inequality $2x - 5y \geq 10$.

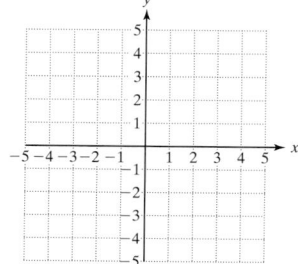

For Exercises 22–23, graph the solution to the compound inequality.

22. $x + y < 3$ and
$3x - 2y > -6$

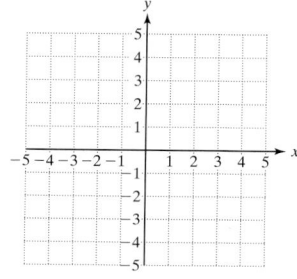

23. $5x \leq 5$ or $x + y \leq 0$

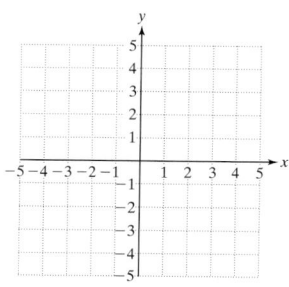

24. After menopause, women are at higher risk for hip fractures as a result of low calcium. As early as their teen years, women need at least 1200 mg of calcium per day (the USDA recommended daily allowance). One 8-oz glass of skim milk contains 300 mg of calcium, and one Antacid (regular strength) contains 400 mg of calcium. Let x represent the number of 8-oz glasses of milk that a woman drinks per day. Let y represent the number of Antacid tablets (regular strength) that a woman takes per day.

  a. Write two inequalities that express the fact that the number of glasses of milk and the number of Antacid taken each day cannot be negative.

  b. Write a linear inequality in terms of x and y for which the daily calcium intake is at least 1200 mg.

  c. Graph the inequalities.

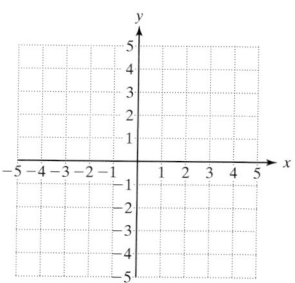

Chapters 1–10   Cumulative Review Exercises

1. Perform the indicated operations.
$$(2x - 3)(x - 4) - (x - 5)^2$$

2. Solve the equation. $-9m + 3 = 2m(m - 4)$

For Exercises 3–4, solve the equation and inequalities. Write the solution to the inequalities in interval notation.

3. a. $2|3 - p| - 4 = 2$

  b. $2|3 - p| - 4 < 2$

  c. $2|3 - p| - 4 > 2$

4. a. $\left|\dfrac{y - 2}{4}\right| - 6 = -3$

  b. $\left|\dfrac{y - 2}{4}\right| - 6 < -3$

  c. $\left|\dfrac{y - 2}{4}\right| - 6 > -3$

5. Graph the inequality. $4x - y > 12$

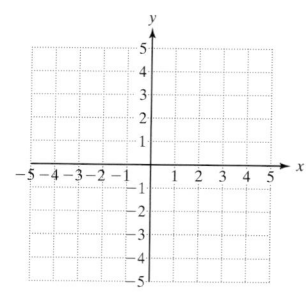

6. The time in minutes required for a rat to run through a maze depends on the number of trials, n, that the rat has practiced.

$$t(n) = \frac{3n + 15}{n + 1} \qquad n \ge 1$$

   a. Find $t(1)$, $t(50)$, and $t(500)$, and interpret the results in the context of this problem. Round to 2 decimal places, if necessary.

   b. Does there appear to be a limiting time in which the rat can complete the maze?

   c. How many trials are required so that the rat is able to finish the maze in under 5 min?

7. **a.** Solve the inequality. $\quad 2x^2 + x - 10 \ge 0$

   b. How does the answer in part (a) relate to the graph of the function $f(x) = 2x^2 + x - 10$?

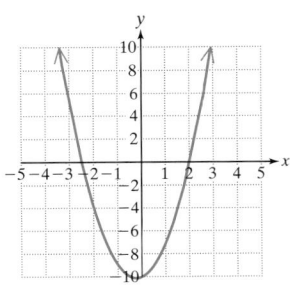

8. Shade the region defined by the compound inequality.

$$3x + y \le -2 \qquad \text{or} \qquad y \ge 1$$

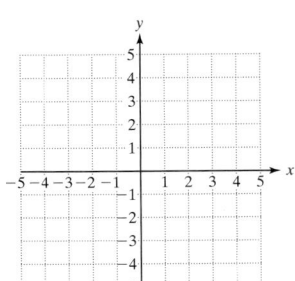

9. Simplify the expression.

$$2 - 3(x - 5) + 2[4 - (2x + 6)]$$

10. McDonald's corporation is the world's largest food service retailer. At the end of 1996, McDonald's operated 2.1×10^4 restaurants

in over 100 countries. Worldwide sales in 1996 were nearly 3.18×10^{10}. Find the average sales per restaurant in 1996. Write the answer in scientific notation.

11. **a.** Divide the polynomials.

$$\frac{2x^4 - x^3 + 5x - 7}{x^2 + 2x - 1}$$

   Identify the quotient and remainder.

   b. Check your answer by multiplying.

12. The area of a trapezoid is given by $A = \frac{1}{2}h(b_1 + b_2)$.

   a. Solve for b_1.

   b. Find b_1 when $h = 4$ cm, $b_2 = 6$ cm, and $A = 32$ cm^2.

13. The speed of a car varies inversely as the time to travel a fixed distance. A car traveling the speed limit of 60 mph travels between two points in 10 sec. How fast is a car moving if it takes only 8 sec to cover the same distance?

14. Two angles are supplementary. One angle measures 9° more than twice the other angle. Find the measures of the angles.

15. Chemique invests $3000 less in an account earning 5% simple interest than she does in an account bearing 6.5% simple interest. At the end of one year, she earns a total $770 in interest. Find the amount invested in each account.

16. Determine algebraically whether the lines are parallel, perpendicular, or neither.

$$4x - 2y = 5$$
$$-3x + 6y = 10$$

17. Solve the inequality.

$$\frac{-2}{x - 5} \ge 3$$

For Exercises 18–19, find the x- and y-intercepts and slope (if they exist) of the lines. Then graph the lines.

18. $3x + 5 = 8$

19. $\dfrac{1}{2}x + y = 4$

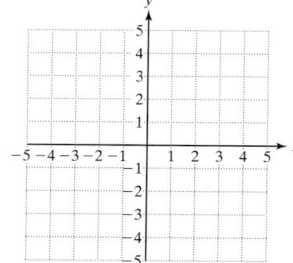
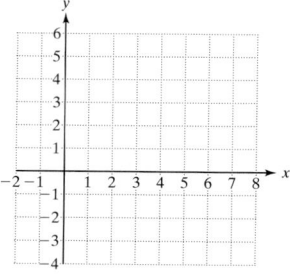

20. Find an equation of the line with slope $-\frac{2}{3}$ passing through the point $(4, -7)$. Write the final answer in slope-intercept form.

21. Solve the system of equations.

$$3x + y = z + 2$$
$$y = 1 - 2x$$
$$3z = -2y$$

22. Identify the order of the matrices.

a.
$$\begin{bmatrix} 2 & 4 & 5 \\ -1 & 0 & 1 \\ 9 & 2 & 3 \\ 3 & 0 & 1 \end{bmatrix}$$

b.
$$\begin{bmatrix} 5 & 6 & 3 \\ 6 & 0 & -1 \\ 0 & 1 & -2 \end{bmatrix}$$

23. Against a headwind, a plane can travel 6240 mi in 13 hr. On the return trip flying with the same wind, the plane can fly 6240 mi in 12 hr. Find the wind speed and the speed of the plane in still air.

24. The profit that a company makes manufacturing computer desks is given by

$$P(x) = -\tfrac{1}{5}(x - 20)(x - 650) \qquad x \geq 0$$

where x is the number of desks produced and $P(x)$ is the profit in dollars.

a. Is this function constant, linear, or quadratic?

b. Find $P(0)$ and interpret the result in the context of this problem.

c. Find the values of x where $P(x) = 0$. Interpret the results in the context of this problem.

25. Given $h(x) = \sqrt{50 - x}$, find the domain of h.

26. Simplify completely. $\dfrac{x^{-1} - y^{-1}}{y^{-2} - x^{-2}}$

27. Divide. $\dfrac{a^3 + 64}{16 - a^2} \div \dfrac{a^3 - 4a^2 + 16a}{a^2 - 3a - 4}$

28. Perform the indicated operations.

$$\dfrac{1}{x^2 - 7x + 10} + \dfrac{1}{x^2 + 8x - 20}$$

29. Solve the system of equations using Cramer's rule.

$$x + 5y = 6$$
$$-2x + y = -12$$

30. Factor the expression.

$$3x^3 - 6x^2 - 12x + 24$$

Radicals and Complex Numbers

11.1 Definition of an *n*th Root

11.2 Rational Exponents

11.3 Simplifying Radical Expressions

11.4 Addition and Subtraction of Radicals

11.5 Multiplication of Radicals

11.6 Rationalization

Problem Recognition Exercises—Operations on Radicals

11.7 Radical Equations

11.8 Complex Numbers

In this chapter we study radical expressions. This includes operations on square roots, cube roots, fourth roots, and so on. We also revisit the Pythagorean theorem and its applications.

As you work through this chapter, try to simplify the expressions or solve the equations. The following is a Sudoku puzzle. Use the clues given to fill in the boxes labeled a through n. Then fill in the remaining part of the grid so that every row, every column, and every 2 × 3 box contains the digits 1 through 6.

Clues

a. 2^2

b. $\sqrt[3]{27}$

c. $\sqrt[5]{32}$

d. $\sqrt[100]{1}$

e. Solution to $\sqrt{2x + 10} - 1 = 3$

f. Solution to $\sqrt{31 + x} - 1 = 5$

g. $\left| -\sqrt[4]{16} \right|$

h. Length of side *a*.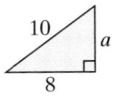

i. Solution to $\sqrt[3]{x + 123} = 5$

j. $125^{1/3}$

k. $\left(\sqrt{5} + 2 \right)\left(\sqrt{5} - 2 \right)$

l. $\left(\dfrac{1}{64} \right)^{-2/3} - \left(\dfrac{1}{169} \right)^{-1/2}$

m. $(3 + 2i) - (-4 - 3i) + (-2 - 5i)$

n. $\sqrt{10^2 - 8^2}$

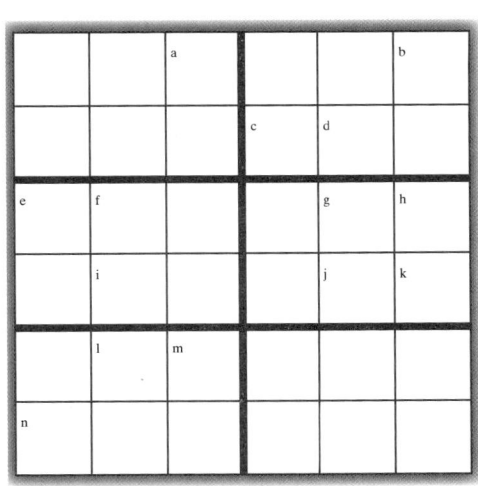

Section 11.1

Definition of an *n*th Root

1. Definition of a Square Root

The inverse operation to squaring a number is to find its square roots. For example, finding a square root of 36 is equivalent to asking, "what number when squared equals 36?"

One obvious answer to this question is 6 because $(6)^2 = 36$, but -6 will also work, because $(-6)^2 = 36$.

Definition of a Square Root

b is a **square root** of a if $b^2 = a$.

TIP: All positive real numbers have two real-valued square roots: one positive and one negative. Zero has only one square root, which is 0 itself. Finally, for any negative real number, there are no real-valued square roots.

Example 1   Identifying Square Roots

Identify the square roots of the real numbers.

a. 25   **b.** 49   **c.** 0   **d.** -9

Solution:

a. 5 is a square root of 25 because $(5)^2 = 25$.

-5 is a square root of 25 because $(-5)^2 = 25$.

b. 7 is a square root of 49 because $(7)^2 = 49$.

-7 is a square root of 49 because $(-7)^2 = 49$.

c. 0 is a square root of 0 because $(0)^2 = 0$.

d. There are no real numbers that when squared will equal a negative number; therefore, there are no real-valued square roots of -9.

Skill Practice   Identify the square roots of the real numbers.

1. 64   **2.** 16   **3.** 1   **4.** -100

Recall from Section 1.2 that the positive square root of a real number can be denoted with a **radical sign** $\sqrt{\ }$.

Notation for Positive and Negative Square Roots

Let a represent a positive real number. Then

1. $\sqrt{a}$ is the *positive* square root of a. The positive square root is also called the **principal square root**.

2. $-\sqrt{a}$ is the *negative* square root of a.

3. $\sqrt{0} = 0$

Skill Practice Answers

1. -8 and 8
2. -4 and 4
3. -1 and 1
4. No real-valued square roots

> **Example 2**   **Simplifying a Square Root**
>
> Simplify the square roots.
>
> **a.** $\sqrt{36}$ **b.** $\sqrt{\dfrac{4}{9}}$ **c.** $\sqrt{0.04}$
>
> **Solution:**
>
> **a.** $\sqrt{36}$ denotes the positive square root of 36.
>
> $\sqrt{36} = 6$
>
> **b.** $\sqrt{\dfrac{4}{9}}$ denotes the positive square root of $\dfrac{4}{9}$.
>
> $\sqrt{\dfrac{4}{9}} = \dfrac{2}{3}$
>
> **c.** $\sqrt{0.04}$ denotes the positive square root of 0.04.
>
> $\sqrt{0.04} = 0.2$
>
> **Skill Practice**   Simplify the square roots.
>
> **5.** $\sqrt{81}$ **6.** $\sqrt{\dfrac{36}{49}}$ **7.** $\sqrt{0.09}$

The numbers 36, $\frac{4}{9}$, and 0.04 are **perfect squares** because their square roots are rational numbers.

Radicals that cannot be simplified to rational numbers are irrational numbers. Recall that an irrational number cannot be written as a terminating or repeating decimal. For example, the symbol $\sqrt{13}$ is used to represent the *exact* value of the square root of 13. The symbol $\sqrt{42}$ is used to represent the *exact* value of the square root of 42. These values can be approximated by a rational number by using a calculator.

$$\sqrt{13} \approx 3.605551275 \qquad \sqrt{42} \approx 6.480740698$$

Calculator Connections

Use a calculator to approximate the values of $\sqrt{13}$ and $\sqrt{42}$.

```
√(13)
          3.605551275
√(42)
          6.480740698
```

TIP: Before using a calculator to evaluate a square root, try estimating the value first.

$\sqrt{13}$ must be a number between 3 and 4 because $\sqrt{9} < \sqrt{13} < \sqrt{16}$.

$\sqrt{42}$ must be a number between 6 and 7 because $\sqrt{36} < \sqrt{42} < \sqrt{49}$.

A negative number cannot have a real number as a square root because no real number when squared is negative. For example, $\sqrt{-25}$ is *not* a real number because there is no real number b for which $(b)^2 = -25$.

Skill Practice Answers

5. 9   **6.** $\dfrac{6}{7}$ **7.** 0.3

| **Example 3** | **Evaluating Square Roots** |

Simplify the square roots, if possible.

a. $\sqrt{-144}$ b. $-\sqrt{144}$ c. $\sqrt{-0.01}$ d. $-\sqrt{\dfrac{1}{9}}$

Solution:

a. $\sqrt{-144}$ is *not* a real number.

b. $-\sqrt{144}$

$$= -1 \cdot \sqrt{144}$$
$$\qquad\quad\downarrow \qquad \downarrow$$
$$= -1 \cdot \quad 12$$
$$= -12$$

> **TIP:** For the expression $-\sqrt{144}$, the factor of -1 is *outside* the radical.

c. $\sqrt{-0.01}$ is *not* a real number.

d. $-\sqrt{\dfrac{1}{9}}$

$$= -1 \cdot \sqrt{\dfrac{1}{9}}$$
$$= -1 \cdot \dfrac{1}{3}$$
$$= -\dfrac{1}{3}$$

| **Skill Practice** | Simplify the square roots, if possible. |

8. $-\sqrt{64}$ **9.** $\sqrt{-81}$ **10.** $\sqrt{-\dfrac{1}{4}}$ **11.** $-\sqrt{0.25}$

2. Definition of an *n*th Root

Finding a square root of a number is the inverse process of squaring a number. This concept can be extended to finding a third root (called a cube root), a fourth root, and in general an **nth root**.

> **Definition of an *n*th Root**
>
> b is an *n*th root of a if $b^n = a$.
>
> For example, 2 is a fourth root of 16 because $2^4 = 16$.

The radical sign $\sqrt{}$ is used to denote the principal square root of a number. The symbol $\sqrt[n]{}$ is used to denote the principal *n*th root of a number. In the expression $\sqrt[n]{a}$, n is called the **index** of the radical, and a is called the **radicand**. For a square root, the index is 2, but it is usually not written ($\sqrt[2]{a}$ is denoted simply as $\sqrt{a}$). A radical with an index of 3 is called a **cube root**, denoted by $\sqrt[3]{a}$.

Definition of $\sqrt[n]{a}$

1. If *n* is a positive *even* integer and $a > 0$, then $\sqrt[n]{a}$ is the principal (positive) *n*th root of *a*.
2. If $n > 1$ is an *odd* integer, then $\sqrt[n]{a}$ is the principal *n*th root of *a*.
3. If $n > 1$ is an integer, then $\sqrt[n]{0} = 0$.

For the purpose of simplifying radicals, it is helpful to know the following powers:

Perfect Cubes	Perfect Fourth Powers	Perfect Fifth Powers
$1^3 = 1$	$1^4 = 1$	$1^5 = 1$
$2^3 = 8$	$2^4 = 16$	$2^5 = 32$
$3^3 = 27$	$3^4 = 81$	$3^5 = 243$
$4^3 = 64$	$4^4 = 256$	$4^5 = 1024$
$5^3 = 125$	$5^4 = 625$	$5^5 = 3125$

Example 4 **Identifying the Principal *n*th Root of a Real Number**

Simplify the expressions, if possible.

a. $\sqrt{4}$ b. $\sqrt[3]{64}$ c. $\sqrt[5]{-32}$ d. $\sqrt[4]{81}$
e. $\sqrt[6]{1,000,000}$ f. $\sqrt{-100}$ g. $\sqrt[4]{-16}$

Solution:

a. $\sqrt{4} = 2$ because $(2)^2 = 4$

b. $\sqrt[3]{64} = 4$ because $(4)^3 = 64$

c. $\sqrt[5]{-32} = -2$ because $(-2)^5 = -32$

d. $\sqrt[4]{81} = 3$ because $(3)^4 = 81$

e. $\sqrt[6]{1,000,000} = 10$ because $(10)^6 = 1,000,000$

f. $\sqrt{-100}$ is not a real number.   No real number when squared equals -100.

g. $\sqrt[4]{-16}$ is not a real number.   No real number when raised to the fourth power equals -16.

Calculator Connections

A calculator can be used to approximate *n*th roots by using the $\boxed{\sqrt{}}$ function. On most calculators, the index is entered first.

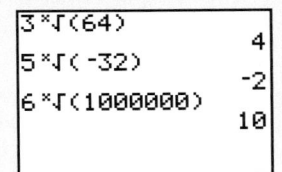

Skill Practice Simplify, if possible.

12. $\sqrt[4]{16}$ 13. $\sqrt[3]{1000}$ 14. $\sqrt[5]{-1}$ 15. $\sqrt[5]{32}$

16. $\sqrt[5]{100,000}$ 17. $\sqrt{-36}$ 18. $\sqrt[3]{-27}$

Examples 4(f) and 4(g) illustrate that an *n*th root of a negative quantity is not a real number if the index is even. This is because no real number raised to an even power is negative.

3. Roots of Variable Expressions

Finding an *n*th root of a variable expression is similar to finding an *n*th root of a numerical expression. For roots with an even index, however, particular care must be taken to obtain a nonnegative result.

Skill Practice Answers

12. 2   **13.** 10   **14.** −1
15. 2   **16.** 10
17. Not a real number
18. −3

> **Definition of $\sqrt[n]{a^n}$**
>
> 1. If n is a positive *odd* integer, then $\sqrt[n]{a^n} = a$.
> 2. If n is a positive *even* integer, then $\sqrt[n]{a^n} = |a|$.

The absolute value bars are necessary for roots with an even index because the variable a may represent a positive quantity or a negative quantity. By using absolute value bars, $\sqrt[n]{a^n} = |a|$ is nonnegative and represents the principal nth root of a.

Example 5   Simplifying Expressions of the Form $\sqrt[n]{a^n}$

Simplify the expressions.

a. $\sqrt[4]{(-3)^4}$ **b.** $\sqrt[5]{(-3)^5}$ **c.** $\sqrt{(x + 2)^2}$ **d.** $\sqrt[3]{(a + b)^3}$ **e.** $\sqrt{y^4}$

Solution:

a. $\sqrt[4]{(-3)^4} = |-3| = 3$

Because this is an *even*-indexed root, absolute value bars are necessary to make the answer positive.

b. $\sqrt[5]{(-3)^5} = -3$

This is an *odd*-indexed root, so absolute value bars are not used.

c. $\sqrt{(x + 2)^2} = |x + 2|$

Because this is an *even*-indexed root, absolute value bars are necessary. The sign of the quantity $x + 2$ is unknown; however, $|x + 2| \geq 0$ regardless of the value of x.

d. $\sqrt[3]{(a + b)^3} = a + b$

This is an *odd*-indexed root, so absolute value bars are not used.

e. $\sqrt{y^4} = \sqrt{(y^2)^2}$

$\quad\quad = |y^2|$

Because this is an even-indexed root, use absolute value bars.

$\quad\quad = y^2$

However, because y^2 is nonnegative, the absolute value bars are not necessary.

Skill Practice   Simplify the expressions.

19. $\sqrt{(-4)^2}$ **20.** $\sqrt[3]{(-4)^3}$ **21.** $\sqrt{y^2}$

22. $\sqrt[4]{(a + 1)^4}$ **23.** $\sqrt[4]{v^8}$

If n is an even integer, then $\sqrt[n]{a^n} = |a|$; however, if the variable a is assumed to be *nonnegative,* then the absolute value bars may be dropped. That is, $\sqrt[n]{a^n} = a$ provided $a \geq 0$. In many examples and exercises, we will make the assumption that the variables within a radical expression are positive real numbers. In such a case, the absolute value bars are not needed to evaluate $\sqrt[n]{a^n}$.

It is helpful to become familiar with the patterns associated with perfect squares and perfect cubes involving variable expressions.

Skill Practice Answers

19. 4   **20.** -4
21. $|y|$ **22.** $|a + 1|$
23. v^2

The following powers of x are perfect squares:

Perfect Squares

$(x^1)^2 = x^2$

$(x^2)^2 = x^4$

$(x^3)^2 = x^6$

$(x^4)^2 = x^8$

TIP: In general, any expression raised to an even power (a multiple of 2) is a perfect square.

The following powers of x are perfect cubes:

Perfect Cubes

$(x^1)^3 = x^3$

$(x^2)^3 = x^6$

$(x^3)^3 = x^9$

$(x^4)^3 = x^{12}$

TIP: In general, any expression raised to a power that is a multiple of 3 is a perfect cube.

These patterns may be extended to higher powers.

Example 6 **Simplifying *n*th Roots**

Simplify the expressions. Assume that all variables are positive real numbers.

a. $\sqrt{y^8}$ **b.** $\sqrt[3]{27a^3}$ **c.** $\sqrt[5]{\dfrac{a^5}{b^5}}$ **d.** $-\sqrt[4]{\dfrac{81x^4y^8}{16}}$

Solution:

a. $\sqrt{y^8} = \sqrt{(y^4)^2} = y^4$

b. $\sqrt[3]{27a^3} = \sqrt[3]{(3a)^3} = 3a$

c. $\sqrt[5]{\dfrac{a^5}{b^5}} = \sqrt[5]{\left(\dfrac{a}{b}\right)^5} = \dfrac{a}{b}$

d. $-\sqrt[4]{\dfrac{81x^4y^8}{16}} = -\sqrt[4]{\left(\dfrac{3xy^2}{2}\right)^4} = -\dfrac{3xy^2}{2}$

Skill Practice Simplify the expressions. Assume all variables represent positive real numbers.

24. $\sqrt{t^6}$ **25.** $\sqrt[3]{y^{12}}$ **26.** $\sqrt[4]{x^{12}y^4}$ **27.** $\sqrt[5]{\dfrac{32}{b^{10}}}$

---

4. Pythagorean Theorem

In Section 6.7, we used the Pythagorean theorem in several applications. For the triangle shown in Figure 11-1, the **Pythagorean theorem** may be stated as $a^2 + b^2 = c^2$. In this formula, a and b are the legs of the right triangle, and c is the hypotenuse. Notice that the hypotenuse is the longest side of the right triangle and is opposite the 90° angle.

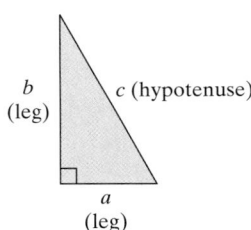

b
(leg)

c (hypotenuse)

a
(leg)

Figure 11-1

Skill Practice Answers

24. t^3 **25.** y^4

26. x^3y **27.** $\dfrac{2}{b^2}$

Example 7   Applying the Pythagorean Theorem

Use the Pythagorean theorem and the definition of the principal square root of a positive real number to find the length of the unknown side.

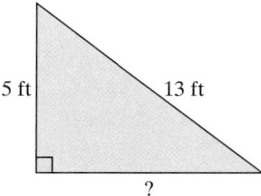

Solution:

Label the sides of the triangle.

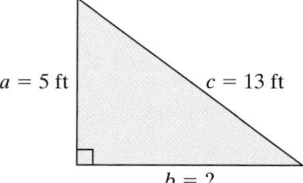

$$a^2 + b^2 = c^2$$ Apply the Pythagorean theorem.

$$(5)^2 + b^2 = (13)^2$$

$$25 + b^2 = 169$$ Simplify.

$$b^2 = 169 - 25$$ Isolate b^2.

$$b^2 = 144$$ By definition, b must be one of the square roots of 144. Because b represents the length of a side of a triangle, choose the positive square root of 144.

$$b = 12$$

The third side is 12 ft long.

Skill Practice

28. Use the Pythagorean theorem and the definition of the principal square root to find the length of the unknown side of the right triangle.

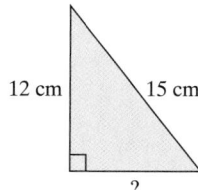

Example 8   Applying the Pythagorean Theorem

Two boats leave a dock at 12:00 noon. One travels due north at 6 mph, and the other travels due east at 8 mph (Figure 11-2). How far apart are the two boats after 2 hr?

Solution:

The boat traveling north travels a distance of $(6 \text{ mph})(2 \text{ hr}) = 12 \text{ mi}$. The boat traveling east travels a distance of $(8 \text{ mph})(2 \text{ hr}) = 16 \text{ mi}$. The

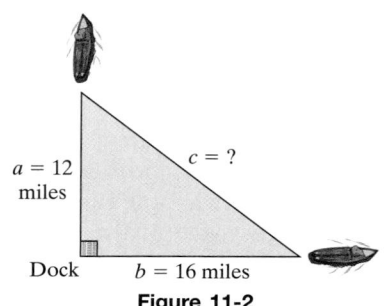

Figure 11-2

Skill Practice Answers

28. 9 cm

course of the boats forms a right triangle where the hypotenuse represents the distance between them.

$$a^2 + b^2 = c^2$$

$$(12)^2 + (16)^2 = c^2 \qquad \text{Apply the Pythagorean theorem.}$$

$$144 + 256 = c^2 \qquad \text{Simplify.}$$

$$400 = c^2$$

$$20 = c \qquad \text{By definition, } c \text{ must be one of the square roots of } 400. \text{ Choose the positive square root of } 400 \text{ to represent distance between the two boats.}$$

The boats are 20 mi apart.

Skill Practice

29. Two cars leave from the same place at the same time. One travels west at 40 mph, and the other travels north at 30 mph. How far apart are they after 2 hr?

5. Radical Functions

If n is an integer greater than 1, then a function written in the form $f(x) = \sqrt[n]{x}$ is called a **radical function**. Note that if n is an even integer, then the function will be a real number only if the radicand is nonnegative. Therefore, the domain is restricted to nonnegative real numbers, or equivalently, $[0, \infty)$. If n is an odd integer, then the domain is all real numbers.

Example 9   **Determining the Domain of Radical Functions**

For each function, write the domain in interval notation.

a. $g(t) = \sqrt[4]{t - 2}$ **b.** $h(a) = \sqrt[3]{a - 4}$ **c.** $k(x) = \sqrt{3 - 5x}$

Solution:

a. $g(t) = \sqrt[4]{t - 2}$ The index is even. The radicand must be nonnegative.

$t - 2 \geq 0$ Set the radicand greater than or equal to zero.

$t \geq 2$ Solve for t.

The domain is $[2, \infty)$.

Calculator Connections

The domain of $g(t) = \sqrt[4]{t - 2}$ can be confirmed from its graph.

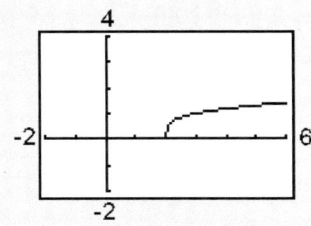

The domain of $h(a) = \sqrt[3]{a - 4}$ can be confirmed from its graph.

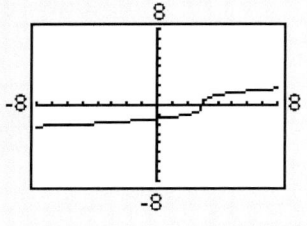

b. $h(a) = \sqrt[3]{a - 4}$ The index is odd; therefore, the domain is all real numbers.

The domain is $(-\infty, \infty)$.

c. $k(x) = \sqrt{3 - 5x}$ The index is even; therefore, the radicand must be nonnegative.

$3 - 5x \geq 0$ Set the radicand greater than or equal to zero.

$-5x \geq -3$ Solve for x.

$\dfrac{-5x}{-5} \leq \dfrac{-3}{-5}$ Reverse the inequality sign, because we are dividing by a negative number.

$x \leq \dfrac{3}{5}$

The domain is $\left(-\infty, \frac{3}{5}\right]$.

The domain of the function defined by $k(x) = \sqrt{3 - 5x}$ can be confirmed from its graph.

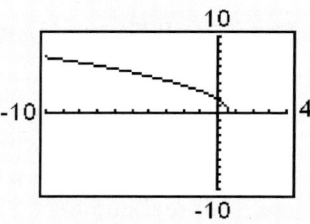

Skill Practice Answers

30. $[-5, \infty)$ **31.** $(-\infty, \infty)$
32. $\left(-\infty, \frac{1}{2}\right]$

Skill Practice   For each function, write the domain in interval notation.

30. $f(x) = \sqrt{x + 5}$ **31.** $g(t) = \sqrt[3]{t - 9}$ **32.** $h(a) = \sqrt{1 - 2a}$

Section 11.1   **Practice Exercises**

Boost *your* GRADE at mathzone.com!    MathZone
- Practice Problems
- Self-Tests
- NetTutor
- e-Professors
- Videos

Study Skills Exercise

1. Define the key terms.

a. **square root**   b. **radical sign**   c. **principal square root**   d. **perfect square**

e. **nth root**   f. **index**   g. **radicand**   h. **cube root**

i. **Pythagorean theorem**   j. **radical function**

Concept 1: Definition of a Square Root

2. Simplify the expression $\sqrt[3]{8}$. Explain how you can check your answer.

3. a. Find the square roots of 64.

   b. Find $\sqrt{64}$.

   c. Explain the difference between the answers in part (a) and part (b).

4. a. Find the square roots of 121.

   b. Find $\sqrt{121}$.

   c. Explain the difference between the answers in part (a) and part (b).

5. a. What is the principal square root of 81?

   b. What is the negative square root of 81?

6. a. What is the principal square root of 100?

   b. What is the negative square root of 100?

7. Using the definition of a square root, explain why $\sqrt{-36}$ is not a real number.

For Exercises 8–19, evaluate the roots without using a calculator. Identify those that are not real numbers.

8. $\sqrt{25}$

9. $\sqrt{49}$

10. $-\sqrt{25}$

11. $-\sqrt{49}$

12. $\sqrt{-25}$

13. $\sqrt{-49}$

14. $\sqrt{\dfrac{100}{121}}$

15. $\sqrt{\dfrac{64}{9}}$

16. $\sqrt{0.64}$

17. $\sqrt{0.81}$

18. $-\sqrt{0.0144}$

19. $-\sqrt{0.16}$

Concept 2: Definition of an *n*th Root

20. Using the definition of an *n*th root, explain why $\sqrt[4]{-16}$ is not a real number.

For Exercises 21–22, evaluate the roots without using a calculator. Identify those that are not real numbers.

21. a. $\sqrt{64}$ **b.** $\sqrt[3]{64}$ **c.** $-\sqrt{64}$

    d. $-\sqrt[3]{64}$ **e.** $\sqrt{-64}$ **f.** $\sqrt[3]{-64}$

22. a. $\sqrt{16}$ **b.** $\sqrt[4]{16}$ **c.** $-\sqrt{16}$

    d. $-\sqrt[4]{16}$ **e.** $\sqrt{-16}$ **f.** $\sqrt[4]{-16}$

For Exercises 23–36, evaluate the roots without using a calculator. Identify those that are not real numbers.

23. $\sqrt[3]{-27}$

24. $\sqrt[3]{-125}$

25. $\sqrt[3]{\dfrac{1}{8}}$

26. $\sqrt[5]{\dfrac{1}{32}}$

27. $\sqrt[5]{32}$

28. $\sqrt[4]{1}$

29. $\sqrt[3]{-\dfrac{125}{64}}$

30. $\sqrt[3]{-\dfrac{8}{27}}$

31. $\sqrt[4]{-1}$

32. $\sqrt[6]{-1}$

33. $\sqrt[6]{1{,}000{,}000}$

34. $\sqrt[4]{10{,}000}$

35. $-\sqrt[3]{0.008}$

36. $-\sqrt[4]{0.0016}$

Concept 3: Roots of Variable Expressions

For Exercises 37–54, simplify the radical expressions.

37. $\sqrt{a^2}$

38. $\sqrt[4]{a^4}$

39. $\sqrt[3]{a^3}$

40. $\sqrt[5]{a^5}$

41. $\sqrt[6]{a^6}$

42. $\sqrt[7]{a^7}$

43. $\sqrt{x^4}$

44. $\sqrt[3]{y^{12}}$

45. $\sqrt{x^2 y^4}$

46. $\sqrt[3]{(u+v)^3}$

47. $-\sqrt[3]{\dfrac{x^3}{y^3}},\ \ y \neq 0$

48. $\sqrt[4]{\dfrac{a^4}{b^8}},\ \ b \neq 0$

49. $\dfrac{2}{\sqrt[4]{x^4}}, \quad x \neq 0$ **50.** $\sqrt{(-5)^2}$ **51.** $\sqrt[3]{(-92)^3}$ **52.** $\sqrt[6]{(50)^6}$

53. $\sqrt[10]{(-2)^{10}}$ **54.** $\sqrt[5]{(-2)^5}$

For Exercises 55–70, simplify the expressions. Assume all variables are positive real numbers.

55. $\sqrt{x^2y^4}$ **56.** $\sqrt{16p^2}$ **57.** $\sqrt{\dfrac{a^6}{b^2}}$ **58.** $\sqrt{\dfrac{w^2}{z^4}}$

59. $-\sqrt{\dfrac{25}{q^2}}$ **60.** $-\sqrt{\dfrac{p^6}{81}}$ **61.** $\sqrt{9x^2y^4z^2}$ **62.** $\sqrt{4a^4b^2c^6}$

63. $\sqrt{\dfrac{h^2k^4}{16}}$ **64.** $\sqrt{\dfrac{4x^2}{y^8}}$ **65.** $-\sqrt[3]{\dfrac{t^3}{27}}$ **66.** $\sqrt[4]{\dfrac{16}{w^4}}$

67. $\sqrt[5]{32y^{10}}$ **68.** $\sqrt[3]{64x^6y^3}$ **69.** $\sqrt[6]{64p^{12}q^{18}}$ **70.** $\sqrt[4]{16r^{12}s^8}$

Concept 4: Pythagorean Theorem

For Exercises 71–74, find the length of the third side of each triangle by using the Pythagorean theorem.

71.
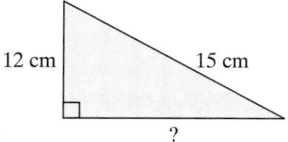

72.
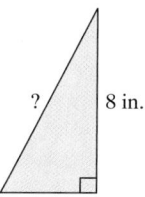

73.
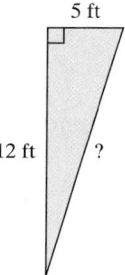

74.
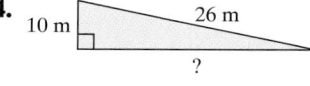

For Exercises 75–78, use the Pythagorean theorem.

75. Roberto and Sherona began running from the same place at the same time. They ran along two different paths that formed right angles with each other. Roberto ran 4 mi and stopped, while Sherona ran 3 mi and stopped. How far apart were they when they stopped?

76. Leine and Laura began hiking from their campground. Laura headed south while Leine headed east. Laura walked 12 mi and Leine walked 5 mi. How far apart were they when they stopped walking?

77. Two mountain bikers take off from the same place at the same time. One travels north at 4 mph, and the other travels east at 3 mph. How far apart are they after 5 hr?

78. Professor Ortiz leaves campus on her bike, heading west at 6 mph. Professor Wilson leaves campus at the same time and walks south at 2.5 mph. How far apart are they after 4 hr?

Concept 5: Radical Functions

For Exercises 79–82, evaluate the function for the given values of *x*. Then write the domain of the function in interval notation.

79. $h(x) = \sqrt{x - 2}$

   a. $h(0)$

   b. $h(1)$

   c. $h(2)$

   d. $h(3)$

   e. $h(6)$

80. $k(x) = \sqrt{x + 1}$

   a. $k(-3)$

   b. $k(-2)$

   c. $k(-1)$

   d. $k(0)$

   e. $k(3)$

81. $g(x) = \sqrt[3]{x - 2}$

   a. $g(-6)$

   b. $g(1)$

   c. $g(2)$

   d. $g(3)$

82. $f(x) = \sqrt[3]{x + 1}$

   a. $f(-9)$

   b. $f(-2)$

   c. $f(0)$

   d. $f(7)$

For each function defined in Exercises 83–86, write the domain in interval notation.

83. $q(x) = \sqrt{x + 5}$

84. $p(x) = \sqrt{1 - x}$

85. $R(x) = \sqrt[3]{x + 1}$

86. $T(x) = \sqrt{x - 10}$

For Exercises 87–90, match the function with the graph. Use the domain information from Exercises 83–86.

87. $p(x) = \sqrt{1 - x}$

88. $q(x) = \sqrt{x + 5}$

89. $T(x) = \sqrt{x - 10}$

90. $R(x) = \sqrt[3]{x + 1}$

a.

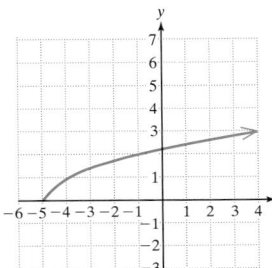

b.

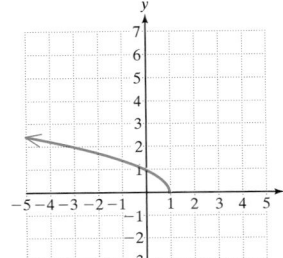

c.

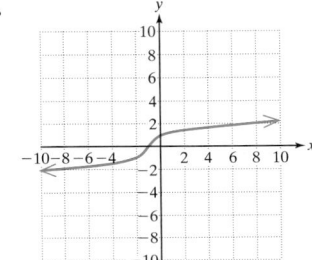

d.

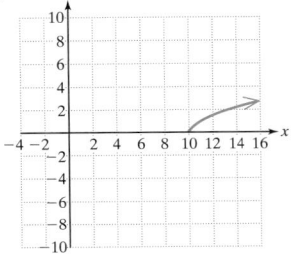

Mixed Exercises

For Exercises 91–94, translate the English phrase to an algebraic expression.

91. The sum of *q* and the square of *p*

92. The product of 11 and the cube root of *x*

93. The quotient of 6 and the cube root of *x*

94. The difference of *y* and the principal square root of *x*

95. If a square has an area of 64 in.², then what are the lengths of the sides?

$s = ?$

$A = 64\ \text{in.}^2 \quad s = ?$

96. If a square has an area of 121 m², then what are the lengths of the sides?

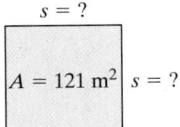

$s = ?$

$A = 121\ \text{m}^2 \quad s = ?$

Graphing Calculator Exercises

For Exercises 97–104, use a calculator to evaluate the expressions to four decimal places.

97. $\sqrt{69}$

98. $\sqrt{5798}$

99. $2 + \sqrt[3]{5}$

100. $3 - 2\sqrt[4]{10}$

101. $7\sqrt[4]{25}$

102. $-3\sqrt[3]{9}$

103. $\dfrac{3 - \sqrt{19}}{11}$

104. $\dfrac{5 + 2\sqrt{15}}{12}$

105. Graph $h(x) = \sqrt{x - 2}$ on the standard viewing window. Use the graph to confirm the domain found in Exercise 79.

106. Graph $k(x) = \sqrt{x + 1}$ on the standard viewing window. Use the graph to confirm the domain found in Exercise 80.

107. Graph $g(x) = \sqrt[3]{x - 2}$ on the standard viewing window. Use the graph to confirm the domain found in Exercise 81.

108. Graph $f(x) = \sqrt[3]{x + 1}$ on the standard viewing window. Use the graph to confirm the domain found in Exercise 82.

Section 11.2   Rational Exponents

Concepts

1. Definition of $a^{1/n}$ and $a^{m/n}$
2. Converting Between Rational Exponents and Radical Notation
3. Properties of Rational Exponents
4. Applications Involving Rational Exponents

1. Definition of $a^{1/n}$ and $a^{m/n}$

In Sections 5.1–5.3, the properties for simplifying expressions with integer exponents were presented. In this section, the properties are expanded to include expressions with rational exponents. We begin by defining expressions of the form $a^{1/n}$.

Definition of $a^{1/n}$

Let a be a real number, and let n be an integer such that $n > 1$. If $\sqrt[n]{a}$ is a real number, then

$$a^{1/n} = \sqrt[n]{a}$$

Example 1   **Evaluating Expressions of the Form $a^{1/n}$**

Convert the expression to radical form and simplify, if possible.

  a. $(-8)^{1/3}$ **b.** $81^{1/4}$ **c.** $-100^{1/2}$ **d.** $(-100)^{1/2}$ **e.** $16^{-1/2}$

Solution:

  a. $(-8)^{1/3} = \sqrt[3]{-8} = -2$

  b. $81^{1/4} = \sqrt[4]{81} = 3$

  c. $-100^{1/2} = -1 \cdot 100^{1/2}$ The exponent applies only to the base of 100.

           $= -1\sqrt{100}$

           $= -10$

  d. $(-100)^{1/2}$ is not a real number because $\sqrt{-100}$ is not a real number.

e. $16^{-1/2} = \dfrac{1}{16^{1/2}}$

Write the expression with a positive exponent.

Recall that $b^{-n} = \dfrac{1}{b^n}$.

$\quad = \dfrac{1}{\sqrt{16}}$

$\quad = \dfrac{1}{4}$

Skill Practice Simplify.

1. $(-64)^{1/3}$ **2.** $16^{1/4}$ **3.** $-36^{1/2}$ **4.** $(-36)^{1/2}$ **5.** $64^{-1/3}$

If $\sqrt[n]{a}$ is a real number, then we can define an expression of the form $a^{m/n}$ in such a way that the multiplication property of exponents still holds true. For example,

$$16^{3/4} \nearrow (16^{1/4})^3 = (\sqrt[4]{16})^3 = (2)^3 = 8$$
$$\searrow (16^3)^{1/4} = \sqrt[4]{16^3} = \sqrt[4]{4096} = 8$$

Definition of $a^{m/n}$

Let a be a real number, and let m and n be positive integers such that m and n share no common factors and $n > 1$. If $\sqrt[n]{a}$ is a real number, then

$$a^{m/n} = (a^{1/n})^m = (\sqrt[n]{a})^m \qquad \text{and} \qquad a^{m/n} = (a^m)^{1/n} = \sqrt[n]{a^m}$$

The rational exponent in the expression $a^{m/n}$ is essentially performing two operations. The numerator of the exponent raises the base to the mth power. The denominator takes the nth root.

Example 2   **Evaluating Expressions of the Form $a^{m/n}$**

Convert each expression to radical form and simplify.

a. $8^{2/3}$ **b.** $100^{5/2}$ **c.** $\left(\dfrac{1}{25}\right)^{3/2}$ **d.** $4^{-3/2}$ **e.** $(-81)^{3/4}$

Solution:

a. $8^{2/3} = (\sqrt[3]{8})^2$ Take the cube root of 8 and square the result.

$\quad\quad = (2)^2$ Simplify.

$\quad\quad = 4$

b. $100^{5/2} = (\sqrt{100})^5$ Take the square root of 100 and raise the result to the fifth power.

$\quad\quad = (10)^5$ Simplify.

$\quad\quad = 100,000$

Calculator Connections

A calculator can be used to confirm the results of Example 2(a)–2(c).

```
(8)^(2/3)
              4
(100)^(5/2)
         100000
(1/25)^(3/2)
           .008
```

Skill Practice Answers

1. -4 **2.** 2 **3.** -6
4. Not a real number   **5.** $\frac{1}{4}$

c. $\left(\dfrac{1}{25}\right)^{3/2} = \left(\sqrt{\dfrac{1}{25}}\right)^3$ Take the square root of $\frac{1}{25}$ and cube the result.

$\qquad\qquad = \left(\dfrac{1}{5}\right)^3$ Simplify.

$\qquad\qquad = \dfrac{1}{125}$

d. $4^{-3/2} = \dfrac{1}{4^{3/2}}$ Write the expression with positive exponents.

$\qquad\quad = \dfrac{1}{(\sqrt{4})^3}$ Take the square root of 4 and cube the result.

$\qquad\quad = \dfrac{1}{2^3}$ Simplify.

$\qquad\quad = \dfrac{1}{8}$

e. $(-81)^{3/4}$ is not a real number because $\sqrt[4]{-81}$ is not a real number.

Skill Practice   Convert each expression to radical form and simplify.

6. $9^{3/2}$ **7.** $8^{5/3}$ **8.** $32^{-4/5}$ **9.** $\left(\dfrac{1}{27}\right)^{4/3}$ **10.** $(-4)^{3/2}$

---

2. Converting Between Rational Exponents and Radical Notation

Example 3   **Using Radical Notation and Rational Exponents**

Convert each expression to radical notation. Assume all variables represent positive real numbers.

a. $a^{3/5}$ **b.** $(5x^2)^{1/3}$ **c.** $3y^{1/4}$ **d.** $z^{-3/4}$

Solution:

a. $a^{3/5} = \sqrt[5]{a^3}$

b. $(5x^2)^{1/3} = \sqrt[3]{5x^2}$

c. $3y^{1/4} = 3\sqrt[4]{y}$ Note that the coefficient 3 is not raised to the $\frac{1}{4}$ power.

d. $z^{-3/4} = \dfrac{1}{z^{3/4}} = \dfrac{1}{\sqrt[4]{z^3}}$

Skill Practice Answers

6. 27   **7.** 32   **8.** $\dfrac{1}{16}$

9. $\dfrac{1}{81}$ **10.** Not a real number

11. $\sqrt[5]{t^4}$ **12.** $\sqrt[4]{2y^3}$

13. $10\sqrt{p}$ **14.** $\dfrac{1}{\sqrt[3]{q^2}}$

Skill Practice   Convert each expression to radical notation. Assume all variables represent positive real numbers.

11. $t^{4/5}$ **12.** $(2y^3)^{1/4}$ **13.** $10p^{1/2}$ **14.** $q^{-2/3}$

Example 4 Using Radical Notation and Rational Exponents

Convert each expression to an equivalent expression by using rational exponents. Assume that all variables represent positive real numbers.

a. $\sqrt[4]{b^3}$ **b.** $\sqrt{7a}$ **c.** $7\sqrt{a}$

Solution:

a. $\sqrt[4]{b^3} = b^{3/4}$

b. $\sqrt{7a} = (7a)^{1/2}$

c. $7\sqrt{a} = 7a^{1/2}$

Skill Practice Convert to an equivalent expression using rational exponents. Assume all variables represent positive real numbers.

15. $\sqrt[3]{x^2}$ **16.** $\sqrt{5y}$ **17.** $5\sqrt{y}$

3. Properties of Rational Exponents

In Sections 5.1–5.3, several properties and definitions were introduced to simplify expressions with integer exponents. These properties also apply to rational exponents.

Properties of Exponents and Definitions

Let a and b be nonzero real numbers. Let m and n be rational numbers such that a^m, a^n, b^n, and b^m are real numbers.

Description	Property	Example
1. Multiplying like bases	$a^m a^n = a^{m+n}$	$x^{1/3} x^{4/3} = x^{5/3}$
2. Dividing like bases	$\dfrac{a^m}{a^n} = a^{m-n}$	$\dfrac{x^{3/5}}{x^{1/5}} \neq x^{2/5}$
3. The power rule	$(a^m)^n = a^{mn}$	$(2^{1/3})^{1/2} = 2^{1/6}$
4. Power of a product	$(ab)^m = a^m b^m$	$(xy)^{1/2} = x^{1/2} y^{1/2}$
5. Power of a quotient	$\left(\dfrac{a}{b}\right)^m = \dfrac{a^m}{b^m}$	$\left(\dfrac{4}{25}\right)^{1/2} = \dfrac{4^{1/2}}{25^{1/2}} = \dfrac{2}{5}$

Description	Definition	Example
1. Negative exponents	$a^{-m} = \left(\dfrac{1}{a}\right)^m = \dfrac{1}{a^m}$	$(8)^{-1/3} = \left(\dfrac{1}{8}\right)^{1/3} = \dfrac{1}{2}$
2. Zero exponent	$a^0 = 1$	$5^0 = 1$

Example 5 Simplifying Expressions with Rational Exponents

Use the properties of exponents to simplify the expressions. Assume all variables represent positive real numbers.

a. $y^{2/5} y^{3/5}$ **b.** $\left(\dfrac{s^{1/2} t^{1/3}}{w^{3/4}}\right)^4$ **c.** $\left(\dfrac{81cd^{-2}}{3c^{-2}d^4}\right)^{1/3}$

Skill Practice Answers

15. $x^{2/3}$ **16.** $(5y)^{1/2}$
17. $5y^{1/2}$

Solution:

a. $y^{2/5}y^{3/5} = y^{(2/5)+(3/5)}$ Multiply like bases by adding exponents.

   $= y^{5/5}$ Simplify.

   $= y$

b. $\left(\dfrac{s^{1/2}t^{1/3}}{w^{3/4}}\right)^4 = \dfrac{s^{(1/2)\cdot 4}t^{(1/3)\cdot 4}}{w^{(3/4)\cdot 4}}$ Apply the power rule. Multiply exponents.

   $= \dfrac{s^2 t^{4/3}}{w^3}$ Simplify.

c. $\left(\dfrac{81cd^{-2}}{3c^{-2}d^4}\right)^{1/3} = (27c^{1-(-2)}d^{-2-4})^{1/3}$ Simplify inside parentheses. Subtract exponents.

   $= (27c^3 d^{-6})^{1/3}$

   $= \left(\dfrac{27c^3}{d^6}\right)^{1/3}$ Rewrite using positive exponents.

   $= \dfrac{27^{1/3}c^{3/3}}{d^{6/3}}$ Apply the power rule. Multiply exponents.

   $= \dfrac{3c}{d^2}$ Simplify.

Skill Practice   Use the properties of exponents to simplify the expressions. Assume all variables represent positive real numbers.

18. $x^{1/2} \cdot x^{3/4}$ **19.** $\left(\dfrac{a^{1/3}b^{1/2}}{c^{5/8}}\right)^6$ **20.** $\left(\dfrac{32y^2 z^{-3}}{2y^{-2} z^5}\right)^{1/4}$

---

4. Applications Involving Rational Exponents

The expression

$$r = \left(\dfrac{12{,}500}{5000}\right)^{1/6} - 1$$

is easily evaluated on a graphing calculator.

```
(12500/5000)^(1/
6)-1
      .1649930508
```

Example 6   Applying Rational Exponents

Suppose P dollars in principal is invested in an account that earns interest annually. If after t years the investment grows to A dollars, then the annual rate of return r on the investment is given by

$$r = \left(\dfrac{A}{P}\right)^{1/t} - 1$$

Find the annual rate of return on $5000 which grew to $12,500 after 6 years.

Solution:

$r = \left(\dfrac{A}{P}\right)^{1/t} - 1$

$= \left(\dfrac{12{,}500}{5000}\right)^{1/6} - 1$ Substitute $A = \$12{,}500$, $P = \$5000$, and $t = 6$.

$= (2.5)^{1/6} - 1$

$\approx 1.165 - 1$

≈ 0.165 or 16.5% The annual rate of return is 16.5%.

Skill Practice Answers

18. $x^{5/4}$ **19.** $\dfrac{a^2 b^3}{c^{15/4}}$ **20.** $\dfrac{2y}{z^2}$

21. The radius r of a sphere of volume V is given by $r = \left(\dfrac{3V}{4\pi}\right)^{1/3}$. Find the radius of a sphere whose volume is 113.04 in.^{3} (Use 3.14 for π.)

Skill Practice Answers

21. Approximately 3 in.

Section 11.2 Practice Exercises

Boost *your* GRADE at mathzone.com!

MathZone

• Practice Problems
• Self-Tests
• NetTutor

• e-Professors
• Videos

For the exercises in this set, assume that all variables represent positive real numbers unless otherwise stated.

Study Skills Exercises

1. Before you do your homework for this section, go back to Sections 5.1–5.3 and review the properties of exponents. Do several problems from each of the Practice Exercise sets. This will help you with the concepts in Section 11.2.

2. Define the key terms.

   a. $a^{1/n}$ **b.** $a^{m/n}$

Review Exercises

3. Given: $\sqrt[3]{27}$

   a. Identify the index.

   b. Identify the radicand.

4. Given: $\sqrt{18}$

   a. Identify the index.

   b. Identify the radicand.

For Exercises 5–8, evaluate the radicals (if possible).

5. $\sqrt{25}$ **6.** $\sqrt[3]{8}$ **7.** $\sqrt[4]{81}$ **8.** $\left(\sqrt[4]{16}\right)^3$

Concept 1: Definition of $a^{1/n}$ and $a^{m/n}$

For Exercises 9–18, convert the expressions to radical form and simplify.

9. $144^{1/2}$ **10.** $16^{1/4}$  **11.** $-144^{1/2}$ **12.** $-16^{1/4}$

13. $(-144)^{1/2}$ **14.** $(-16)^{1/4}$ **15.** $(-64)^{1/3}$ **16.** $(-32)^{1/5}$

17. $(25)^{-1/2}$ **18.** $(27)^{-1/3}$

19. Explain how to interpret the expression $a^{m/n}$ as a radical.

20. Explain why $(\sqrt[3]{8})^4$ is easier to evaluate than $\sqrt[3]{8^4}$.

For Exercises 21–24, simplify the expression, if possible.

21. a. $16^{3/4}$ **b.** $-16^{3/4}$ **c.** $(-16)^{3/4}$ **d.** $16^{-3/4}$ **e.** $-16^{-3/4}$ **f.** $(-16)^{-3/4}$

22. a. $81^{3/4}$ **b.** $-81^{3/4}$ **c.** $(-81)^{3/4}$ **d.** $81^{-3/4}$ **e.** $-81^{-3/4}$ **f.** $(-81)^{-3/4}$

23. a. $25^{3/2}$ **b.** $-25^{3/2}$ **c.** $(-25)^{3/2}$ **d.** $25^{-3/2}$ **e.** $-25^{-3/2}$ **f.** $(-25)^{-3/2}$

24. a. $4^{3/2}$ **b.** $-4^{3/2}$ **c.** $(-4)^{3/2}$ **d.** $4^{-3/2}$ **e.** $-4^{-3/2}$ **f.** $(-4)^{-3/2}$

For Exercises 25–50, simplify the expression.

25. $64^{-3/2}$ **26.** $81^{-3/2}$ **27.** $243^{3/5}$ **28.** $1^{5/3}$

29. $-27^{-4/3}$ **30.** $-16^{-5/4}$ **31.** $\left(\dfrac{100}{9}\right)^{-3/2}$ **32.** $\left(\dfrac{49}{100}\right)^{-1/2}$

33. $(-4)^{-3/2}$ **34.** $(-49)^{-3/2}$ **35.** $(-8)^{1/3}$ **36.** $(-9)^{1/2}$

37. $-8^{1/3}$ **38.** $-9^{1/2}$ **39.** $27^{-2/3}$ **40.** $125^{-1/3}$

41. $\dfrac{1}{36^{-1/2}}$ **42.** $\dfrac{1}{16^{-1/2}}$ **43.** $\dfrac{1}{1000^{-1/3}}$ **44.** $\dfrac{1}{81^{-3/4}}$

45. $\left(\dfrac{1}{8}\right)^{2/3} + \left(\dfrac{1}{4}\right)^{1/2}$ **46.** $\left(\dfrac{1}{8}\right)^{-2/3} + \left(\dfrac{1}{4}\right)^{-1/2}$ **47.** $\left(\dfrac{1}{16}\right)^{-3/4} - \left(\dfrac{1}{49}\right)^{-1/2}$ **48.** $\left(\dfrac{1}{16}\right)^{1/4} - \left(\dfrac{1}{49}\right)^{1/2}$

49. $\left(\dfrac{1}{4}\right)^{1/2} + \left(\dfrac{1}{64}\right)^{-1/3}$ **50.** $\left(\dfrac{1}{36}\right)^{1/2} + \left(\dfrac{1}{64}\right)^{-5/6}$

Concept 2: Converting Between Rational Exponents and Radical Notation

For Exercises 51–58, convert each expression to radical notation.

51. $q^{2/3}$ **52.** $t^{3/5}$ **53.** $6y^{3/4}$ **54.** $8b^{4/9}$

55. $(x^2y)^{1/3}$ **56.** $(c^2d)^{1/6}$ **57.** $(qr)^{-1/5}$ **58.** $(7x)^{-1/4}$

For Exercises 59–66, write each expression by using rational exponents rather than radical notation.

59. $\sqrt[3]{x}$ **60.** $\sqrt[4]{a}$ **61.** $10\sqrt{b}$ **62.** $-2\sqrt[3]{t}$

63. $\sqrt[3]{y^2}$ **64.** $\sqrt[6]{z^5}$ **65.** $\sqrt[4]{a^2b^3}$ **66.** $\sqrt{abc}$

Concept 3: Properties of Rational Exponents

For Exercises 67–90, simplify the expressions by using the properties of rational exponents. Write the final answer using positive exponents only.

67. $x^{1/4}x^{-5/4}$ **68.** $2^{2/3}2^{-5/3}$ **69.** $\dfrac{p^{5/3}}{p^{2/3}}$ **70.** $\dfrac{q^{5/4}}{q^{1/4}}$

71. $(y^{1/5})^{10}$ **72.** $(x^{1/2})^8$ **73.** $6^{-1/5}6^{3/5}$ **74.** $a^{-1/3}a^{2/3}$

75. $\dfrac{4t^{-1/3}}{t^{4/3}}$ **76.** $\dfrac{5s^{-1/3}}{s^{5/3}}$ **77.** $(a^{1/3}a^{1/4})^{12}$ **78.** $(x^{2/3}x^{1/2})^6$

79. $(5a^2c^{-1/2}d^{1/2})^2$ **80.** $(2x^{-1/3}y^2z^{5/3})^3$ **81.** $\left(\dfrac{x^{-2/3}}{y^{-3/4}}\right)^{12}$ **82.** $\left(\dfrac{m^{-1/4}}{n^{-1/2}}\right)^{-4}$

83. $\left(\dfrac{16w^{-2}z}{2wz^{-8}}\right)^{1/3}$

84. $\left(\dfrac{50p^{-1}q}{2pq^{-3}}\right)^{1/2}$

85. $(25x^2y^4z^6)^{1/2}$

86. $(8a^6b^3c^9)^{2/3}$

87. $(x^2y^{-1/3})^6(x^{1/2}yz^{2/3})^2$

88. $(a^{-1/3}b^{1/2})^4(a^{-1/2}b^{3/5})^{10}$

89. $\left(\dfrac{x^{3m}y^{2m}}{z^{5m}}\right)^{1/m}$

90. $\left(\dfrac{a^{4n}b^{3n}}{c^n}\right)^{1/n}$

Concept 4: Applications Involving Rational Exponents

91. If the area A of a square is known, then the length of its sides, s, can be computed by the formula $s = A^{1/2}$.

    a. Compute the length of the sides of a square having an area of 100 in.^{2}

    b. Compute the length of the sides of a square having an area of 72 in.^{2} Round your answer to the nearest 0.1 in.

92. The radius r of a sphere of volume V is given by $r = \left(\dfrac{3V}{4\pi}\right)^{1/3}$. Find the radius of a sphere having a volume of 85 in.^{3} Round your answer to the nearest 0.1 in.

93. If P dollars in principal grows to A dollars after t years with annual interest, then the interest rate is given by $r = \left(\dfrac{A}{P}\right)^{1/t} - 1$.

    a. In one account, \$10,000 grows to \$16,802 after 5 years. Compute the interest rate. Round your answer to a tenth of a percent.

    b. In another account \$10,000 grows to \$18,000 after 7 years. Compute the interest rate. Round your answer to a tenth of a percent.

    c. Which account produced a higher average yearly return?

94. Is $(a + b)^{1/2}$ the same as $a^{1/2} + b^{1/2}$? If not, give a counterexample.

Expanding Your Skills

For Exercises 95–100, write the expression as a single radical.

95. $\sqrt{\sqrt[3]{x}}$

96. $\sqrt[3]{\sqrt{x}}$

97. $\sqrt[4]{\sqrt{y}}$

98. $\sqrt{\sqrt[4]{y}}$

99. $\sqrt[5]{\sqrt[3]{w}}$

100. $\sqrt[3]{\sqrt[4]{w}}$

For Exercises 101–108, use a calculator to approximate the expressions and round to 4 decimal places, if necessary.

101. $9^{1/2}$

102. $125^{-1/3}$

103. $50^{-1/4}$

104. $(172)^{3/5}$

105. $\sqrt[3]{5^2}$

106. $\sqrt[4]{6^3}$

107. $\sqrt{10^3}$

108. $\sqrt[3]{16}$

Simplifying Radical Expressions

1. Multiplication and Division Properties of Radicals

You may have already noticed certain properties of radicals involving a product or quotient.

Multiplication and Division Properties of Radicals

Let a and b represent real numbers such that $\sqrt[n]{a}$ and $\sqrt[n]{b}$ are both real. Then

1. $\sqrt[n]{ab} = \sqrt[n]{a} \cdot \sqrt[n]{b}$ *Multiplication property of radicals*

2. $\sqrt[n]{\dfrac{a}{b}} = \dfrac{\sqrt[n]{a}}{\sqrt[n]{b}}$ $b \neq 0$ *Division property of radicals*

Properties 1 and 2 follow from the properties of rational exponents.

$$\sqrt[n]{ab} = (ab)^{1/n} = a^{1/n}b^{1/n} = \sqrt[n]{a} \cdot \sqrt[n]{b}$$

$$\sqrt[n]{\frac{a}{b}} = \left(\frac{a}{b}\right)^{1/n} = \frac{a^{1/n}}{b^{1/n}} = \frac{\sqrt[n]{a}}{\sqrt[n]{b}}$$

The multiplication and division properties of radicals indicate that a product or quotient within a radicand can be written as a product or quotient of radicals, provided the roots are real numbers. For example:

$$\sqrt{144} = \sqrt{16} \cdot \sqrt{9}$$

$$\sqrt{\frac{25}{36}} = \frac{\sqrt{25}}{\sqrt{36}}$$

The reverse process is also true. A product or quotient of radicals can be written as a single radical provided the roots are real numbers and they have the same indices.

$$\sqrt{3} \cdot \sqrt{12} = \sqrt{36}$$

$$\frac{\sqrt[3]{8}}{\sqrt[3]{125}} = \sqrt[3]{\frac{8}{125}}$$

In algebra it is customary to simplify radical expressions as much as possible.

Simplified Form of a Radical

Consider any radical expression where the radicand is written as a product of prime factors. The expression is in *simplified form* if all the following conditions are met:

1. The radicand has no factor raised to a power greater than or equal to the index.
2. The radicand does not contain a fraction.
3. There are no radicals in the denominator of a fraction.

For example, the following radicals are not simplified.

1. The expression $\sqrt[3]{x^5}$ fails rule 1.

2. The expression $\sqrt{\dfrac{1}{4}}$ fails rule 2.

3. The expression $\dfrac{1}{\sqrt[3]{8}}$ fails rule 3.

2. Simplifying Radicals by Using the Multiplication Property of Radicals

The expression $\sqrt{x^2}$ is not simplified because it fails condition 1. Because x^2 is a perfect square, $\sqrt{x^2}$ is easily simplified:

$$\sqrt{x^2} = x \qquad \text{for } x \geq 0$$

However, how is an expression such as $\sqrt{x^9}$ simplified? This and many other radical expressions are simplified by using the multiplication property of radicals. The following examples illustrate how nth powers can be removed from the radicands of nth roots.

Example 1 Using the Multiplication Property to Simplify a Radical Expression

Use the multiplication property of radicals to simplify the expression $\sqrt{x^9}$. Assume $x \geq 0$.

Solution:

The expression $\sqrt{x^9}$ is equivalent to $\sqrt{x^8 \cdot x}$. Applying the multiplication property of radicals, we have

$$\sqrt{x^9} = \sqrt{x^8 \cdot x}$$

$$= \sqrt{x^8} \cdot \sqrt{x} \qquad \text{Apply the multiplication property of radicals.}$$

Note that x^8 is a perfect square because $x^8 = (x^4)^2$.

$$= x^4\sqrt{x} \qquad \text{Simplify.}$$

Skill Practice Simplify the expression. Assume that $a > 0$.

1. $\sqrt{a^{11}}$

In Example 1, the expression x^9 is not a perfect square. Therefore, to simplify $\sqrt{x^9}$, it was necessary to write the expression as the product of the largest perfect square and a remaining or "left-over" factor: $\sqrt{x^9} = \sqrt{x^8 \cdot x}$. This process also applies to simplifying nth roots, as shown in Example 2.

Example 2 Using the Multiplication Property to Simplify a Radical Expression

Use the multiplication property of radicals to simplify each expression. Assume all variables represent positive real numbers.

a. $\sqrt[4]{b^7}$ **b.** $\sqrt[3]{w^7 z^9}$

Solution:

The goal is to rewrite each radicand as the product of the largest perfect square (perfect cube, perfect fourth power, and so on) and a left-over factor.

a. $\sqrt[4]{b^7} = \sqrt[4]{b^4 \cdot b^3}$ b^4 is the largest perfect fourth power in the radicand.

$$= \sqrt[4]{b^4} \cdot \sqrt[4]{b^3} \qquad \text{Apply the multiplication property of radicals.}$$

$$= b\sqrt[4]{b^3} \qquad \text{Simplify.}$$

Skill Practice Answers
1. $a^5\sqrt{a}$

b. $\sqrt[3]{w^7z^9} = \sqrt[3]{(w^6z^9)\cdot(w)}$ w^6z^9 is the largest perfect cube in the radicand.

$\qquad\quad = \sqrt[3]{w^6z^9}\cdot\sqrt[3]{w}$ Apply the multiplication property of radicals.

$\qquad\quad = w^2z^3\sqrt[3]{w}$ Simplify.

Skill Practice   Simplify the expressions. Assume all variables represent positive real numbers.

2. $\sqrt[4]{v^{25}}$ **3.** $\sqrt[3]{p^8q^{12}}$

Each expression in Example 2 involves a radicand that is a product of variable factors. If a numerical factor is present, sometimes it is necessary to factor the coefficient before simplifying the radical.

Example 3   **Using the Multiplication Property to Simplify Radicals**

Use the multiplication property of radicals to simplify the expressions. Assume all variables represent positive real numbers.

a. $\sqrt{56}$ **b.** $6\sqrt{50}$ **c.** $\sqrt[3]{40x^3y^5z^7}$

Solution:

a. $\sqrt{56} = \sqrt{2^3\cdot7}$ Factor the radicand. $\begin{array}{r}2\,\lfloor 56 \\ 2\,\lfloor 28 \\ 2\,\lfloor 14 \\ 7\end{array}$

$\qquad\quad = \sqrt{(2^2)\cdot(2\cdot7)}$ 2^2 is the largest perfect square in the radicand.

$\qquad\quad = \sqrt{2^2}\cdot\sqrt{2\cdot7}$ Apply the multiplication property of radicals.

$\qquad\quad = 2\sqrt{14}$ Simplify.

Avoiding Mistakes:

The multiplication property of radicals allows us to simplify a product of *factors* within a radical. For example:

$\sqrt{x^2y^2} = \sqrt{x^2}\cdot\sqrt{y^2} = xy$

However, this rule does not apply to *terms* that are added or subtracted within the radical. For example:

$\sqrt{x^2+y^2}$ and $\sqrt{x^2-y^2}$

cannot be simplified.

Calculator Connections

A calculator can be used to support the solution to Example 3(a). The decimal approximation for $\sqrt{56}$ and $2\sqrt{14}$ agree for the first 10 digits. This in itself does not make $\sqrt{56} = 2\sqrt{14}$. It is the multiplication property of radicals that guarantees that the expressions are equal.

```
√(56)
          7.483314774
2*√(14)
          7.483314774
```

---

b. $6\sqrt{50} = 6\sqrt{2\cdot5^2}$ Factor the radicand. $\begin{array}{r}2\,\lfloor 50 \\ 5\,\lfloor 25 \\ 5\end{array}$

$\qquad\quad = 6\cdot\sqrt{5^2}\cdot\sqrt{2}$ Apply the multiplication property of radicals.

$\qquad\quad = 6\cdot5\cdot\sqrt{2}$ Simplify.

$\qquad\quad = 30\sqrt{2}$ Simplify.

Skill Practice Answers

2. $v^6\sqrt[4]{v}$ **3.** $p^2q^4\sqrt[3]{p^2}$

c. $\sqrt[3]{40x^3y^5z^7}$

$= \sqrt[3]{2^3 5 x^3 y^5 z^7}$ Factor the radicand.

$= \sqrt[3]{(2^3 x^3 y^3 z^6) \cdot (5y^2 z)}$ $2^3 x^3 y^3 z^6$ is the largest perfect cube.

$= \sqrt[3]{2^3 x^3 y^3 z^6} \cdot \sqrt[3]{5y^2 z}$ Apply the multiplication property of radicals.

$= 2xyz^2 \sqrt[3]{5y^2 z}$ Simplify.

$\begin{array}{r} 2\,|\,\underline{40} \\ 2\,|\,\underline{20} \\ 2\,|\,\underline{10} \\ 5 \end{array}$

Skill Practice   Simplify the radicals. Assume all variables represent positive real numbers.

4. $\sqrt{24}$ **5.** $5\sqrt{18}$ **6.** $\sqrt[4]{32a^{10}b^{19}}$

3. Simplifying Radicals by Using the Division Property of Radicals

The division property of radicals indicates that a radical of a quotient can be written as the quotient of the radicals and vice versa, provided all roots are real numbers.

Example 4   **Using the Division Property to Simplify Radicals**

Simplify the expressions. Assume all variables represent positive real numbers.

a. $\sqrt{\dfrac{a^7}{a^3}}$ **b.** $\dfrac{\sqrt[3]{3}}{\sqrt[3]{81}}$ **c.** $\dfrac{7\sqrt{50}}{15}$ **d.** $\sqrt[4]{\dfrac{2c^5}{32cd^8}}$

Solution:

a. $\sqrt{\dfrac{a^7}{a^3}}$ The radicand contains a fraction. However, the fraction can be reduced to lowest terms.

$= \sqrt{a^4}$

$= a^2$ Simplify the radical.

b. $\dfrac{\sqrt[3]{3}}{\sqrt[3]{81}}$ The expression has a radical in the denominator.

$= \sqrt[3]{\dfrac{3}{81}}$ Because the radicands have a common factor, write the expression as a single radical (division property of radicals).

$= \sqrt[3]{\dfrac{1}{27}}$ Reduce to lowest terms.

$= \dfrac{1}{3}$ Simplify.

Skill Practice Answers

4. $2\sqrt{6}$ **5.** $15\sqrt{2}$
6. $2a^2 b^4 \sqrt[4]{2a^2 b^3}$

c. $\dfrac{7\sqrt{50}}{15}$ Simplify $\sqrt{50}$.

$= \dfrac{7\sqrt{5^2 \cdot 2}}{15}$ 5^2 is the largest perfect square in the radicand.

$= \dfrac{7\sqrt{5^2} \cdot \sqrt{2}}{15}$ Multiplication property of radicals

$= \dfrac{7 \cdot 5\sqrt{2}}{15}$ Simplify the radicals.

$= \dfrac{7 \cdot \overset{1}{5}\sqrt{2}}{\underset{3}{15}}$ Reduce to lowest terms.

$= \dfrac{7\sqrt{2}}{3}$

d. $\sqrt[4]{\dfrac{2c^5}{32cd^8}}$ The radicand contains a fraction.

$= \sqrt[4]{\dfrac{c^4}{16d^8}}$ Simplify the factors in the radicand.

$= \dfrac{\sqrt[4]{c^4}}{\sqrt[4]{16d^8}}$ Apply the division property of radicals.

$= \dfrac{c}{2d^2}$ Simplify.

Skill Practice Simplify the expressions. Assume all variables represent positive real numbers.

Skill Practice Answers

7. v^8 **8.** 2
9. $\dfrac{2\sqrt{3}}{3}$ **10.** $\dfrac{3x^5}{y^6}$

7. $\sqrt{\dfrac{v^{21}}{v^5}}$ **8.** $\dfrac{\sqrt[5]{64}}{\sqrt[5]{2}}$ **9.** $\dfrac{2\sqrt{300}}{30}$ **10.** $\sqrt[3]{\dfrac{54x^{17}y}{2x^2y^{19}}}$

Section 11.3   Practice Exercises

Boost *your* GRADE at mathzone.com!

MathZone

• Practice Problems   • e-Professors
• Self-Tests   • Videos
• NetTutor

For the exercises in this set, assume that all variables represent positive real numbers unless otherwise stated.

Review Exercises

For Exercises 1–4, simplify the expression. Write the answer with positive exponents only.

1. $(a^2b^{-4})^{1/2}\left(\dfrac{a}{b^{-3}}\right)$ **2.** $\left(\dfrac{p^4}{q^{-6}}\right)^{-1/2}(p^3q^{-2})$ **3.** $(x^{1/3}y^{5/6})^{-6}$ **4.** $\left(\dfrac{m^{\frac{3}{4}}}{n^{-\frac{1}{2}}}\right)^4$

5. Write $x^{4/7}$ in radical notation.

6. Write $y^{2/5}$ in radical notation.

7. Write $\sqrt{y^9}$ by using rational exponents.

8. Write $\sqrt[3]{x^2}$ by using rational exponents.

Concept 2: Simplifying Radicals by Using the Multiplication Property of Radicals

For Exercises 9–30, simplify the radicals.

9. $\sqrt{x^{11}}$

10. $\sqrt{p^{15}}$

11. $\sqrt[3]{q^7}$

12. $\sqrt[3]{r^{17}}$

13. $\sqrt{a^5 b^4}$

14. $\sqrt{c^9 d^6}$

15. $-\sqrt[4]{x^8 y^{13}}$

16. $-\sqrt[4]{p^{16} q^{17}}$

17. $\sqrt{28}$

18. $\sqrt{63}$

19. $\sqrt{80}$

20. $\sqrt{108}$

21. $\sqrt[3]{54}$

22. $\sqrt[3]{250}$

23. $\sqrt{25ab^3}$

24. $\sqrt{64m^5 n^{20}}$

25. $\sqrt{18a^6 b^3}$

26. $\sqrt{72m^5 n^2}$

27. $\sqrt[3]{-16x^6 yz^3}$

28. $\sqrt[3]{-192a^6 bc^2}$

29. $\sqrt[4]{80w^4 z^7}$

30. $\sqrt[4]{32p^8 qr^5}$

Concept 3: Simplifying Radicals by Using the Division Property of Radicals

For Exercises 31–42, simplify the radicals.

31. $\sqrt{\dfrac{x^3}{x}}$

32. $\sqrt{\dfrac{y^5}{y}}$

33. $\dfrac{\sqrt{p^7}}{\sqrt{p^3}}$

34. $\dfrac{\sqrt{q^{11}}}{\sqrt{q^5}}$

35. $\sqrt{\dfrac{50}{2}}$

36. $\sqrt{\dfrac{98}{2}}$

37. $\dfrac{\sqrt[3]{3}}{\sqrt[3]{24}}$

38. $\dfrac{\sqrt[3]{3}}{\sqrt[3]{81}}$

39. $\dfrac{5\sqrt[3]{16}}{6}$

40. $\dfrac{7\sqrt{18}}{9}$

41. $\dfrac{5\sqrt[3]{72}}{12}$

42. $\dfrac{3\sqrt[3]{250}}{10}$

Mixed Exercises

For Exercises 43–58, simplify the radicals.

43. $5\sqrt{18}$

44. $2\sqrt{24}$

45. $-6\sqrt{75}$

46. $-8\sqrt{8}$

47. $\sqrt{25x^4 y^3}$

48. $\sqrt{125p^3 q^2}$

49. $\sqrt[3]{27x^2 y^3 z^4}$

50. $\sqrt[3]{108a^3 bc^2}$

51. $\sqrt[3]{\dfrac{16a^2 b}{2a^2 b^4}}$

52. $\sqrt[4]{\dfrac{3s^2 t^4}{10,000}}$

53. $\sqrt[5]{\dfrac{32x}{y^{10}}}$

54. $\sqrt[3]{\dfrac{-16j^3}{k^3}}$

55. $\dfrac{\sqrt{50x^3 y}}{\sqrt{9y^4}}$

56. $\dfrac{\sqrt[3]{-27a^4}}{\sqrt[3]{8a}}$

57. $\sqrt{2^3 a^{14} b^8 c^{31} d^{22}}$

58. $\sqrt{7^5 u^{12} v^{20} w^{65} x^{80}}$

For Exercises 59–62, write a mathematical expression for the English phrase and simplify.

59. The quotient of 1 and the cube root of w^6

60. The principal square root of the quotient of h and 49

61. The principal square root of the quantity k raised to the third power

62. The cube root of $2x^4$

For Exercises 63–66, find the third side of the right triangle. Write your answer as a radical and simplify.

63.

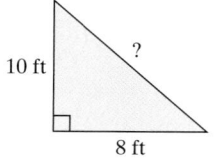

10 ft   ?

8 ft

64.

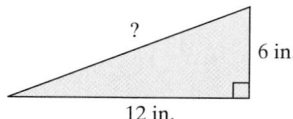

?   6 in.

12 in.

65.

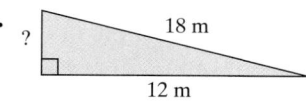

?   18 m

12 m

66.

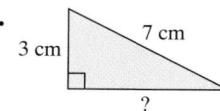

3 cm   7 cm

?

67. On a baseball diamond, the bases are 90 ft apart. Find the exact distance from home plate to second base. Then round to the nearest tenth of a foot.

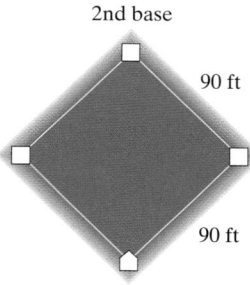

2nd base

90 ft

90 ft

Home plate

68. Linda is at the beach flying a kite. The kite is directly over a sand castle 60 ft away from Linda. If 100 ft of kite string is out (ignoring any sag in the string), how high is the kite? (Assume that Linda is 5 ft tall.) See figure.

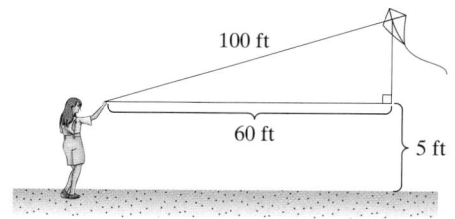

100 ft

60 ft

5 ft

Expanding Your Skills

69. Tom has to travel from town A to town C across a small mountain range. He can travel one of two routes. He can travel on a four-lane highway from A to B and then from B to C at an average speed of 55 mph. Or he can travel on a two-lane road directly from town A to town C, but his average speed will be only 35 mph. If Tom is in a hurry, which route will take him to town C faster?

70. One side of a rectangular pasture is 80 ft in length. The diagonal distance is 110 yd. If fencing costs \$3.29 per foot, how much will it cost to fence the pasture?

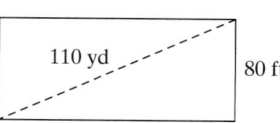

110 yd   80 ft

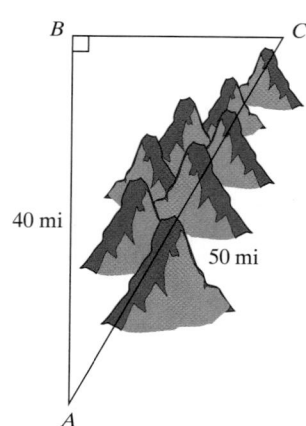

B   C

40 mi

50 mi

A

Addition and Subtraction of Radicals	Section 11.4

1. Definition of *Like* Radicals

Definition of *Like* Radicals

Two radical terms are said to be ***like* radicals** if they have the same index and the same radicand.

The following are pairs of *like* radicals:

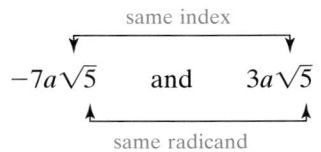

                                 Indices and radicands are the same.

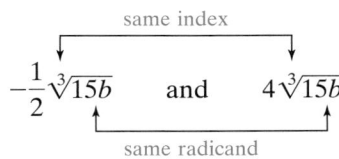

                                 Indices and radicands are the same.

These pairs are not *like* radicals:

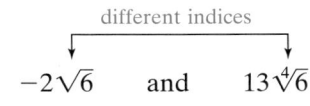

        $-2\sqrt{6}$ and $13\sqrt[4]{6}$ Radicals have different indices.

   $1.3cd\sqrt{3}$ and $-3.7cd\sqrt{10}$ Radicals have different radicands.

                 different radicands

2. Addition and Subtraction of Radicals

To add or subtract *like* radicals, use the distributive property. For example:

$$2\sqrt{5} + 6\sqrt{5} = (2 + 6)\sqrt{5}$$
$$= 8\sqrt{5}$$
$$9\sqrt[3]{2y} - 4\sqrt[3]{2y} = (9 - 4)\sqrt[3]{2y}$$
$$= 5\sqrt[3]{2y}$$

Example 1	Adding and Subtracting Radicals

Add or subtract as indicated.

a. $6\sqrt{11} + 2\sqrt{11}$ **b.** $\sqrt{3} + \sqrt{3}$

c. $-2\sqrt[3]{ab} + 7\sqrt[3]{ab} - \sqrt[3]{ab}$ **d.** $\dfrac{1}{4}x\sqrt{3y} - \dfrac{3}{2}x\sqrt{3y}$

Solution:

a. $6\sqrt{11} + 2\sqrt{11}$

$\qquad = (6 + 2)\sqrt{11}$ $\qquad\qquad$ Apply the distributive property.

$\qquad = 8\sqrt{11}$ $\qquad\qquad\qquad$ Simplify.

b. $\sqrt{3} + \sqrt{3}$

$\qquad = 1\sqrt{3} + 1\sqrt{3}$ $\qquad\qquad$ Note that $\sqrt{3} = 1\sqrt{3}$.

$\qquad = (1 + 1)\sqrt{3}$ $\qquad\qquad$ Apply the distributive property.

$\qquad = 2\sqrt{3}$ $\qquad\qquad\qquad$ Simplify.

Avoiding Mistakes:

The process of adding *like* radicals with the distributive property is similar to adding *like* terms. The end result is that the numerical coefficients are added and the radical factor is unchanged.

$$\sqrt{3} + \sqrt{3} = 1\sqrt{3} + 1\sqrt{3} = 2\sqrt{3}$$

Be careful: $\sqrt{3} + \sqrt{3} \neq \sqrt{6}$

In general: $\sqrt{x} + \sqrt{y} \neq \sqrt{x + y}$

---

c. $-2\sqrt[3]{ab} + 7\sqrt[3]{ab} - \sqrt[3]{ab}$

$\qquad = (-2 + 7 - 1)\sqrt[3]{ab}$ $\qquad\qquad$ Apply the distributive property.

$\qquad = 4\sqrt[3]{ab}$ $\qquad\qquad\qquad\qquad$ Simplify.

d. $\dfrac{1}{4}x\sqrt{3y} - \dfrac{3}{2}x\sqrt{3y}$

$\qquad = \left(\dfrac{1}{4} - \dfrac{3}{2}\right)x\sqrt{3y}$ $\qquad\qquad$ Apply the distributive property.

$\qquad = \left(\dfrac{1}{4} - \dfrac{6}{4}\right)x\sqrt{3y}$ $\qquad\qquad$ Get a common denominator.

$\qquad = -\dfrac{5}{4}x\sqrt{3y}$ $\qquad\qquad\qquad$ Simplify.

Skill Practice   Add or subtract as indicated.

1. $5\sqrt{6} - 8\sqrt{6}$ $\qquad\qquad$ **2.** $\sqrt{10} + \sqrt{10}$ $\qquad\qquad$ **3.** $5\sqrt[3]{xy} - 3\sqrt[3]{xy} + 7\sqrt[3]{xy}$

4. $\dfrac{5}{6}y\sqrt{2} + \dfrac{1}{4}y\sqrt{2}$

Example 2 shows that it is often necessary to simplify radicals before adding or subtracting.

Skill Practice Answers

1. $-3\sqrt{6}$ $\qquad$ **2.** $2\sqrt{10}$

3. $9\sqrt[3]{xy}$ $\qquad$ **4.** $\dfrac{13}{12}y\sqrt{2}$

Example 2 Adding and Subtracting Radicals

Simplify the radicals and add or subtract as indicated. Assume all variables represent positive real numbers.

a. $3\sqrt{8} + \sqrt{2}$

b. $8\sqrt{x^3y^2} - 3y\sqrt{x^3}$

c. $\sqrt{50x^2y^5} - 13y\sqrt{2x^2y^3} + xy\sqrt{98y^3}$

Solution:

a. $3\sqrt{8} + \sqrt{2}$ The radicands are different. Try simplifying the radicals first.

$\quad = 3 \cdot 2\sqrt{2} + \sqrt{2}$ Simplify: $\sqrt{8} = \sqrt{2^3} = 2\sqrt{2}$

$\quad = 6\sqrt{2} + \sqrt{2}$

$\quad = (6 + 1)\sqrt{2}$ Apply the distributive property.

$\quad = 7\sqrt{2}$ Simplify.

b. $8\sqrt{x^3y^2} - 3y\sqrt{x^3}$ The radicands are different. Simplify the radicals first.

$\quad = 8xy\sqrt{x} - 3xy\sqrt{x}$ Simplify $\sqrt{x^3y^2} = xy\sqrt{x}$ and $\sqrt{x^3} = x\sqrt{x}$.

$\quad = (8 - 3)xy\sqrt{x}$ Apply the distributive property.

$\quad = 5xy\sqrt{x}$ Simplify.

c. $\sqrt{50x^2y^5} - 13y\sqrt{2x^2y^3} + xy\sqrt{98y^3}$ Simplify each radical.

$\quad = 5xy^2\sqrt{2y} - 13xy^2\sqrt{2y} + 7xy^2\sqrt{2y}$

$$\begin{cases} \sqrt{50x^2y^5} = \sqrt{5^2 2x^2y^5} \\ \qquad = 5xy^2\sqrt{2y} \\ -13y\sqrt{2x^2y^3} = -13xy^2\sqrt{2y} \\ xy\sqrt{98y^3} = xy\sqrt{7^2 2y^3} \\ \qquad = 7xy^2\sqrt{2y} \end{cases}$$

$\quad = (5 - 13 + 7)xy^2\sqrt{2y}$ Apply the distributive property.

$\quad = -xy^2\sqrt{2y}$

Skill Practice Simplify the radicals and add or subtract as indicated. Assume all variables represent positive real numbers.

5. $\sqrt{75} + 2\sqrt{3}$ **6.** $4\sqrt{a^2b} - 6a\sqrt{b}$

7. $-3\sqrt{2y^3} + 5y\sqrt{18y} - 2\sqrt{50y^3}$

Skill Practice Answers

5. $7\sqrt{3}$ **6.** $-2a\sqrt{b}$

7. $2y\sqrt{2y}$

Section 11.4   Practice Exercises

For the exercises in this set, assume that all variables represent positive real numbers, unless otherwise stated.

Study Skills Exercise

1. Define the key term *like* **radicals**.

Review Exercises

For Exercises 2–5, simplify the radicals.

2. $\sqrt[3]{-16s^4t^9}$ **3.** $-\sqrt[4]{x^7y^4}$ **4.** $\sqrt{36a^2b^3}$ **5.** $\dfrac{\sqrt[3]{7b^8}}{\sqrt[3]{56b^2}}$

6. Write the expression $(4x^2)^{3/2}$ as a radical and simplify.

7. Convert to rational exponents and simplify. $\sqrt[5]{3^5x^{15}y^{10}}$

For Exercises 8–10, simplify the expressions. Write the answer with positive exponents only.

8. $\dfrac{a^{1/6}}{a^{2/3}}$ **9.** $y^{2/3}y^{1/4}$ **10.** $(x^{1/2}y^{-3/4})^{-4}$

Concept 1: Definition of *Like* Radicals

For Exercises 11–12, determine if the radical terms are *like*.

11. a. $\sqrt{2}$ and $\sqrt[3]{2}$ **12. a.** $7\sqrt[3]{x}$ and $\sqrt[3]{x}$

    b. $\sqrt{2}$ and $3\sqrt{2}$ **b.** $\sqrt[3]{x}$ and $\sqrt[4]{x}$

    c. $\sqrt{2}$ and $\sqrt{5}$ **c.** $2\sqrt[4]{x}$ and $x\sqrt[4]{2}$

13. Explain the similarities between the following pairs of expressions.

    a. $7\sqrt{5} + 4\sqrt{5}$ and $7x + 4x$

    b. $-2\sqrt{6} - 9\sqrt{3}$ and $-2x - 9y$

14. Explain the similarities between the following pairs of expressions.

    a. $-4\sqrt{3} + 5\sqrt{3}$ and $-4z + 5z$

    b. $13\sqrt{7} - 18$ and $13a - 18$

Concept 2: Addition and Subtraction of Radicals

For Exercises 15–32, add or subtract the radical expressions, if possible.

15. $3\sqrt{5} + 6\sqrt{5}$

16. $5\sqrt{a} + 3\sqrt{a}$

17. $3\sqrt[3]{t} - 2\sqrt[3]{t}$

18. $6\sqrt[3]{7} - 2\sqrt[3]{7}$

19. $6\sqrt{10} - \sqrt{10}$

20. $13\sqrt{11} - \sqrt{11}$

21. $\sqrt[4]{3} + 7\sqrt[4]{3} - \sqrt[4]{14}$

22. $2\sqrt{11} + 3\sqrt{13} + 5\sqrt{11}$

23. $8\sqrt{x} + 2\sqrt{y} - 6\sqrt{x}$

24. $10\sqrt{10} - 8\sqrt{10} + \sqrt{2}$

25. $\sqrt[3]{ab} + a\sqrt[3]{b}$

26. $x\sqrt[4]{y} - y\sqrt[4]{x}$

27. $\sqrt{2t} + \sqrt[3]{2t}$

28. $\sqrt[4]{5c} + \sqrt[3]{5c}$

29. $\frac{5}{6}z\sqrt[3]{6} + \frac{7}{9}z\sqrt[3]{6}$

30. $\frac{3}{4}a\sqrt[4]{b} + \frac{1}{6}a\sqrt[4]{b}$

31. $0.81x\sqrt{y} - 0.11x\sqrt{y}$

32. $7.5\sqrt{pq} - 6.3\sqrt{pq}$

33. Explain the process for adding the two radicals. Then find the sum. $3\sqrt{2} + 7\sqrt{50}$

34. Explain the process for adding the two radicals. Then find the sum. $\sqrt{8} + \sqrt{32}$

For Exercises 35–60, add or subtract the radical expressions as indicated.

35. $\sqrt{36} + \sqrt{81}$

36. $3\sqrt{80} - 5\sqrt{45}$

37. $2\sqrt{12} + \sqrt{48}$

38. $5\sqrt{32} + 2\sqrt{50}$

39. $4\sqrt{7} + \sqrt{63} - 2\sqrt{28}$

40. $8\sqrt{3} - 2\sqrt{27} + \sqrt{75}$

41. $5\sqrt{18} + \sqrt{32} - 4\sqrt{50}$

42. $7\sqrt{72} - \sqrt{8} + 4\sqrt{50}$

43. $\sqrt[3]{81} - \sqrt[3]{24}$

44. $17\sqrt[3]{81} - 2\sqrt[3]{24}$

45. $3\sqrt{2a} - \sqrt{8a} - \sqrt{72a}$

46. $\sqrt{12t} - \sqrt{27t} + 5\sqrt{3t}$

47. $2s^2\sqrt[3]{s^2t^6} + 3t^2\sqrt[3]{8s^8}$

48. $4\sqrt[3]{x^4} - 2x\sqrt[3]{x}$

49. $7\sqrt[3]{x^4} - x\sqrt[3]{x}$

50. $6\sqrt[3]{y^{10}} - 3y^2\sqrt[3]{y^4}$

51. $5p\sqrt{20p^2} + p^2\sqrt{80}$

52. $2q\sqrt{48q^2} - \sqrt{27q^4}$

53. $\sqrt[3]{a^2b} - \sqrt[3]{8a^2b}$

54. $w\sqrt{80} - 3\sqrt{125w^2}$

55. $11\sqrt[3]{54cd^3} - 2\sqrt[3]{2cd^3} + d\sqrt[3]{16c}$

56. $x\sqrt[3]{64x^5y^2} - x^2\sqrt[3]{x^2y^2} + 5\sqrt[3]{x^8y^2}$

57. $\frac{3}{2}ab\sqrt{24a^3} + \frac{4}{3}\sqrt{54a^5b^2} - a^2b\sqrt{150a}$

58. $mn\sqrt{72n} + \frac{2}{3}n\sqrt{8m^2n} - \frac{5}{6}\sqrt{50m^2n^3}$

59. $x\sqrt[3]{16} - 2\sqrt[3]{27x} + \sqrt[3]{54x^3}$

60. $5\sqrt[4]{y^5} - 2y\sqrt[4]{y} + \sqrt[4]{16y^7}$

Mixed Exercises

For Exercises 61–66, answer true or false. If an answer is false, explain why or give a counterexample.

61. $\sqrt{x} + \sqrt{y} = \sqrt{x + y}$

62. $\sqrt{x} + \sqrt{x} = 2\sqrt{x}$

63. $5\sqrt[3]{x} + 2\sqrt[3]{x} = 7\sqrt[3]{x}$

64. $6\sqrt{x} + 5\sqrt[3]{x} = 11\sqrt{x}$

65. $\sqrt{y} + \sqrt{y} = \sqrt{2y}$

66. $\sqrt{c^2 + d^2} = c + d$

For Exercises 67–70, translate the English phrase to an algebraic expression. Simplify each expression, if possible.

67. The sum of the principal square root of 48 and the principal square root of 12

68. The sum of the cube root of 16 and the cube root of 2

69. The difference of 5 times the cube root of x^6 and the square of x

70. The sum of the cube of y and the principal fourth root of y^{12}

For Exercises 71–74, write an English phrase that translates the mathematical expression. (Answers may vary.)

71. $\sqrt{18} - 5^2$ **72.** $4^3 - \sqrt[3]{4}$ **73.** $\sqrt[4]{x} + y^3$ **74.** $a^4 + \sqrt{a}$

For Exercises 75–76, find the exact value of the perimeter, and then approximate the value to 1 decimal place.

75.

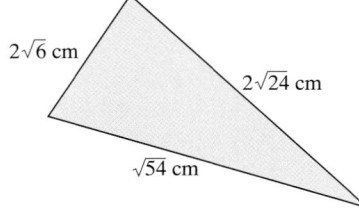

76.

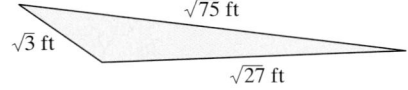

77. The figure has perimeter $14\sqrt{2}$ ft. Find the value of x.

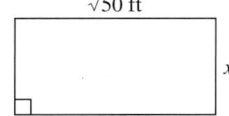

78. The figure has perimeter $12\sqrt{7}$. Find the value of x.

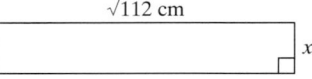

Expanding Your Skills

79. a. An irregularly shaped garden is shown in the figure. All distances are expressed in yards. Find the perimeter. *Hint:* Use the Pythagorean theorem to find the length of each side. Write the final answer in radical form.

  b. Approximate your answer to 2 decimal places.

  c. If edging costs $1.49 per foot and sales tax is 6%, find the total cost of edging the garden.

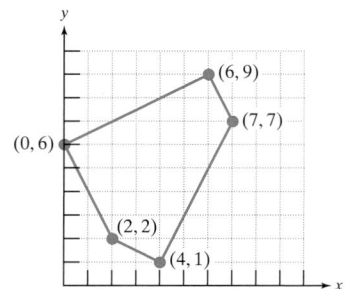

Multiplication of Radicals

1. Multiplication Property of Radicals

In this section we will learn how to multiply radicals by using the multiplication property of radicals first introduced in Section 11.3.

1. Multiplication Property of Radicals
2. Expressions of the Form $(\sqrt[n]{a})^n$
3. Special Case Products
4. Multiplying Radicals with Different Indices

The Multiplication Property of Radicals

Let a and b represent real numbers such that $\sqrt[n]{a}$ and $\sqrt[n]{b}$ are both real. Then

$$\sqrt[n]{a} \cdot \sqrt[n]{b} = \sqrt[n]{ab}$$

To multiply two radical expressions, we use the multiplication property of radicals along with the commutative and associative properties of multiplication.

Example 1 Multiplying Radical Expressions

Multiply the expressions and simplify the result. Assume all variables represent positive real numbers.

a. $(3\sqrt{2})(5\sqrt{6})$ **b.** $(2x\sqrt{y})(-7\sqrt{xy})$ **c.** $(15c\sqrt[3]{cd})\left(\frac{1}{3}\sqrt[3]{cd^2}\right)$

Solution:

a. $(3\sqrt{2})(5\sqrt{6})$

$\quad = (3 \cdot 5)(\sqrt{2} \cdot \sqrt{6})$ Commutative and associative properties of multiplication

$\quad = 15\sqrt{12}$ Multiplication property of radicals

$\quad = 15\sqrt{2^2 3}$

$\quad = 15 \cdot 2\sqrt{3}$ Simplify the radical.

$\quad = 30\sqrt{3}$

b. $(2x\sqrt{y})(-7\sqrt{xy})$

$\quad = (2x)(-7)(\sqrt{y} \cdot \sqrt{xy})$ Commutative and associative properties of multiplication

$\quad = -14x\sqrt{xy^2}$ Multiplication property of radicals

$\quad = -14xy\sqrt{x}$ Simplify the radical.

c. $(15c\sqrt[3]{cd})\left(\frac{1}{3}\sqrt[3]{cd^2}\right)$

$\quad = \left(15c \cdot \frac{1}{3}\right)(\sqrt[3]{cd} \cdot \sqrt[3]{cd^2})$ Commutative and associative properties of multiplication

$\quad = 5c\sqrt[3]{c^2 d^3}$ Multiplication property of radicals

$\quad = 5cd\sqrt[3]{c^2}$ Simplify the radical.

Skill Practice Multiply the expressions and simplify the results. Assume all variables represent positive real numbers.

1. $\sqrt{5} \cdot \sqrt{3}$

2. $(4\sqrt{6})(-2\sqrt{6})$

3. $(3ab\sqrt{b})(-2\sqrt{ab})$

4. $(2\sqrt[3]{4ab})(5\sqrt[3]{2a^2b})$

When multiplying radical expressions with more than one term, we use the distributive property.

Example 2 Multiplying Radical Expressions

Multiply the radical expressions. Assume all variables represent positive real numbers.

a. $3\sqrt{11}(2 + \sqrt{11})$

b. $(\sqrt{5} + 3\sqrt{2})(2\sqrt{5} - \sqrt{2})$

c. $(2\sqrt{14} + \sqrt{7})(6 - \sqrt{14} + 8\sqrt{7})$

d. $(-10a\sqrt{b} + 7b)(a\sqrt{b} + 2b)$

Solution:

a. $3\sqrt{11}(2 + \sqrt{11})$

$= 3\sqrt{11} \cdot (2) + 3\sqrt{11} \cdot \sqrt{11}$	Apply the distributive property.
$= 6\sqrt{11} + 3\sqrt{11^2}$	Multiplication property of radicals
$= 6\sqrt{11} + 3 \cdot 11$	Simplify the radical.
$= 6\sqrt{11} + 33$	

b. $(\sqrt{5} + 3\sqrt{2})(2\sqrt{5} - \sqrt{2})$

$= 2\sqrt{5^2} - \sqrt{10} + 6\sqrt{10} - 3\sqrt{2^2}$	Apply the distributive property.
$= 2 \cdot 5 + 5\sqrt{10} - 3 \cdot 2$	Simplify radicals and combine *like* radicals.
$= 10 + 5\sqrt{10} - 6$	
$= 4 + 5\sqrt{10}$	Combine *like* terms.

c. $(2\sqrt{14} + \sqrt{7})(6 - \sqrt{14} + 8\sqrt{7})$

$= 12\sqrt{14} - 2\sqrt{14^2} + 16\sqrt{7^2 \cdot 2} + 6\sqrt{7} - \sqrt{7^2 \cdot 2} + 8\sqrt{7^2}$	Apply the distributive property.
$= 12\sqrt{14} - 2 \cdot 14 + 16 \cdot 7\sqrt{2} + 6\sqrt{7} - 7\sqrt{2} + 8 \cdot 7$	Simplify the radicals.
$= 12\sqrt{14} - 28 + 112\sqrt{2} + 6\sqrt{7} - 7\sqrt{2} + 56$	Simplify.
$= 12\sqrt{14} + 105\sqrt{2} + 6\sqrt{7} + 28$	Combine *like* terms.

Skill Practice Answers

1. $\sqrt{15}$

2. -48

3. $-6ab^2\sqrt{a}$

4. $20a\sqrt[3]{b^2}$

d. $(-10a\sqrt{b} + 7b)(a\sqrt{b} + 2b)$

$= -10a^2\sqrt{b^2} - 20ab\sqrt{b} + 7ab\sqrt{b} + 14b^2$ Apply the distributive property.

$= -10a^2b - 13ab\sqrt{b} + 14b^2$ Simplify and combine *like* terms.

Skill Practice Multiply the radical expressions. Assume all variables represent positive real numbers.

5. $5\sqrt{5}(2\sqrt{5} - 2)$

6. $(2\sqrt{3} - 3\sqrt{10})(\sqrt{3} + 2\sqrt{10})$

7. $(\sqrt{6} + 3\sqrt{2})(-4\sqrt{6} + 2\sqrt{2})$

8. $(x\sqrt{y} + y)(3x\sqrt{y} - 2y)$

---

2. Expressions of the Form $(\sqrt[n]{a})^n$

The multiplication property of radicals can be used to simplify an expression of the form $(\sqrt{a})^2$, where $a \geq 0$.

$$(\sqrt{a})^2 = \sqrt{a} \cdot \sqrt{a} = \sqrt{a^2} = a \qquad \text{where } a \geq 0$$

This logic can be applied to nth roots. If $\sqrt[n]{a}$ is a real number, then $(\sqrt[n]{a})^n = a$.

Example 3 | **Simplifying Radical Expressions**

Simplify the expressions. Assume all variables represent positive real numbers.

a. $(\sqrt{11})^2$ **b.** $(\sqrt[5]{z})^5$ **c.** $(\sqrt[3]{pq})^3$

Solution:

a. $(\sqrt{11})^2 = 11$ **b.** $(\sqrt[5]{z})^5 = z$ **c.** $(\sqrt[3]{pq})^3 = pq$

Skill Practice Simplify.

9. $(\sqrt{14})^2$ **10.** $(\sqrt[7]{q})^7$ **11.** $(\sqrt[5]{3z})^5$

---

3. Special Case Products

From Example 2, you may have noticed a similarity between multiplying radical expressions and multiplying polynomials.

In Section 5.6 we learned that the square of a binomial results in a perfect square trinomial:

$$(a + b)^2 = a^2 + 2ab + b^2$$

$$(a - b)^2 = a^2 - 2ab + b^2$$

The same patterns occur when squaring a radical expression with two terms.

Example 4 | **Squaring a Two-Term Radical Expression**

Square the radical expressions. Assume all variables represent positive real numbers.

a. $(\sqrt{d} + 3)^2$ **b.** $(5\sqrt{y} - \sqrt{2})^2$

Skill Practice Answers

5. $50 - 10\sqrt{5}$
6. $-54 + \sqrt{30}$
7. $-12 - 20\sqrt{3}$
8. $3x^2y + xy\sqrt{y} - 2y^2$
9. 14 **10.** q **11.** $3z$

Solution:

a. $(\sqrt{d} + 3)^2$

This expression is in the form $(a + b)^2$, where $a = \sqrt{d}$ and $b = 3$.

$$a^2 + 2ab + b^2$$

$$= (\sqrt{d})^2 + 2(\sqrt{d})(3) + (3)^2 \qquad \text{Apply the formula}$$
$$(a + b)^2 = a^2 + 2ab + b^2.$$

$$= d + 6\sqrt{d} + 9 \qquad \text{Simplify.}$$

TIP: The product $(\sqrt{d} + 3)^2$ can also be found by using the distributive property:

$$(\sqrt{d} + 3)^2 = (\sqrt{d} + 3)(\sqrt{d} + 3) = \sqrt{d} \cdot \sqrt{d} + \sqrt{d} \cdot 3 + 3 \cdot \sqrt{d} + 3 \cdot 3$$

$$= \sqrt{d^2} + 3\sqrt{d} + 3\sqrt{d} + 9$$

$$= d + 6\sqrt{d} + 9$$

b. $(5\sqrt{y} - \sqrt{2})^2$

This expression is in the form $(a - b)^2$, where $a = 5\sqrt{y}$ and $b = \sqrt{2}$.

$$a^2 - 2ab + b^2$$

$$= (5\sqrt{y})^2 - 2(5\sqrt{y})(\sqrt{2}) + (\sqrt{2})^2 \qquad \text{Apply the formula}$$
$$(a - b)^2 = a^2 - 2ab + b^2.$$

$$= 25y - 10\sqrt{2y} + 2 \qquad \text{Simplify.}$$

Skill Practice Square the radical expressions. Assume all variables represent positive real numbers.

12. $(\sqrt{a} - 5)^2$ **13.** $(4\sqrt{x} + 3)^2$

Recall from Section 5.6 that the product of two conjugate binomials results in a difference of squares.

$$(a + b)(a - b) = a^2 - b^2$$

The same pattern occurs when multiplying two conjugate radical expressions.

Example 5 Multiplying Conjugate Radical Expressions

Multiply the radical expressions. Assume all variables represent positive real numbers.

a. $(\sqrt{3} + 2)(\sqrt{3} - 2)$ **b.** $\left(\dfrac{1}{3}\sqrt{s} - \dfrac{3}{4}\sqrt{t}\right)\left(\dfrac{1}{3}\sqrt{s} + \dfrac{3}{4}\sqrt{t}\right)$

Solution:

a. $(\sqrt{3} + 2)(\sqrt{3} - 2)$ The expression is in the form $(a + b)(a - b)$, where $a = \sqrt{3}$ and $b = 2$.

$$\overset{a^2 - b^2}{\Big\downarrow\ \ \Big\downarrow}$$

$= (\sqrt{3})^2 - (2)^2$ Apply the formula $(a + b)(a - b) = a^2 - b^2$.

$= 3 - 4$ Simplify.

$= -1$

TIP: The product $(\sqrt{3} + 2)(\sqrt{3} - 2)$ can also be found by using the distributive property.

$$(\sqrt{3} + 2)(\sqrt{3} - 2) = \sqrt{3} \cdot \sqrt{3} + \sqrt{3} \cdot (-2) + 2 \cdot \sqrt{3} + 2 \cdot (-2)$$

$$= 3 - 2\sqrt{3} + 2\sqrt{3} - 4$$

$$= 3 - 4$$

$$= -1$$

b. $\left(\dfrac{1}{3}\sqrt{s} - \dfrac{3}{4}\sqrt{t}\right)\left(\dfrac{1}{3}\sqrt{s} + \dfrac{3}{4}\sqrt{t}\right)$ This expression is in the form $(a - b)(a + b)$, where $a = \frac{1}{3}\sqrt{s}$ and $b = \frac{3}{4}\sqrt{t}$.

$$\overset{a^2 - b^2}{\Big\downarrow\ \ \Big\downarrow}$$

$= \left(\dfrac{1}{3}\sqrt{s}\right)^2 - \left(\dfrac{3}{4}\sqrt{t}\right)^2$ Apply the formula $(a + b)(a - b) = a^2 - b^2$.

$= \dfrac{1}{9}s - \dfrac{9}{16}t$ Simplify.

Skill Practice Multiply the conjugates. Assume all variables represent positive real numbers.

14. $(\sqrt{5} + 3)(\sqrt{5} - 3)$ **15.** $\left(\dfrac{1}{2}\sqrt{a} + \dfrac{2}{5}\sqrt{b}\right)\left(\dfrac{1}{2}\sqrt{a} - \dfrac{2}{5}\sqrt{b}\right)$

4. Multiplying Radicals with Different Indices

The product of two radicals can be simplified provided the radicals have the same index. If the radicals have different indices, then we can use the properties of rational exponents to obtain a common index.

Example 6   **Multiplying Radicals with Different Indices**

Multiply the expressions. Write the answers in radical form.

a. $\sqrt[3]{5} \cdot \sqrt[4]{5}$ **b.** $\sqrt[3]{7} \cdot \sqrt{2}$

Skill Practice Answers

14. -4 **15.** $\dfrac{1}{4}a - \dfrac{4}{25}b$

To approximate expressions such as $\sqrt[3]{7} \cdot \sqrt{2}$ on a calculator, you can enter the expressions using rational exponents. The original problem and simplified form show agreement to 10 digits.

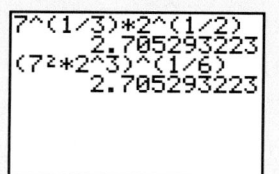

Solution:

a. $\sqrt[3]{5} \cdot \sqrt[4]{5}$

$= 5^{1/3}5^{1/4}$ Rewrite each expression with rational exponents.

$= 5^{(1/3)+(1/4)}$ Because the bases are equal, we can add the exponents.

$= 5^{(4/12)+(3/12)}$ Write the fractions with a common denominator.

$= 5^{7/12}$ Simplify the exponent.

$= \sqrt[12]{5^7}$ Rewrite the expression as a radical.

b. $\sqrt[3]{7} \cdot \sqrt{2}$

$= 7^{1/3}2^{1/2}$ Rewrite each expression with rational exponents.

$= 7^{2/6}2^{3/6}$ Write the rational exponents with a common denominator.

$= (7^2 2^3)^{1/6}$ Apply the power rule of exponents.

$= \sqrt[6]{7^2 2^3}$ Rewrite the expression as a single radical.

> **Skill Practice** Multiply the expressions. Write the answers in radical form. Assume all variables represent positive real numbers.

16. $\sqrt{x} \cdot \sqrt[3]{x}$ **17.** $\sqrt[4]{a^3} \cdot \sqrt[3]{b}$

Skill Practice Answers
16. $\sqrt[6]{x^5}$ **17.** $\sqrt[12]{a^9 b^4}$

Section 11.5   Practice Exercises

- Practice Problems
- Self-Tests
- NetTutor
- e-Professors
- Videos

For the exercises in this set, assume that all variables represent positive real numbers, unless otherwise stated.

Review Exercises

For Exercises 1–3, simplify the radicals.

1. $\sqrt{54p^3 q^6}$ **2.** $\sqrt[3]{-16x^5 y^6 z^7}$ **3.** $-\sqrt{20a^2 b^3 c}$

For Exercises 4–6, simplify the expressions. Write the answer with positive exponents only.

4. $x^{1/3}y^{1/4}x^{-1/6}y^{1/3}$ **5.** $p^{1/8}q^{1/2}p^{-1/4}q^{3/2}$ **6.** $\dfrac{b^{1/4}}{b^{3/2}}$

For Exercises 7–8, add or subtract as indicated.

7. $-2\sqrt[3]{7} + 4\sqrt[3]{7}$ **8.** $4\sqrt{8x^3} - x\sqrt{50x}$

Concept 1: Multiplication Property of Radicals

For Exercises 9–44, multiply the radical expressions.

9. $\sqrt[3]{7} \cdot \sqrt[3]{3}$

10. $\sqrt[4]{6} \cdot \sqrt[4]{2}$

11. $\sqrt{2} \cdot \sqrt{10}$

12. $\sqrt[3]{4} \cdot \sqrt[3]{12}$

13. $\sqrt[4]{16} \cdot \sqrt[4]{64}$

14. $\sqrt{5x^3} \cdot \sqrt{10x^4}$

15. $(4\sqrt[3]{4})(2\sqrt[3]{5})$

16. $(2\sqrt{5})(3\sqrt{7})$

17. $(8a\sqrt{b})(-3\sqrt{ab})$

18. $(p\sqrt[4]{q^3})(\sqrt[4]{pq})$

19. $\sqrt{30} \cdot \sqrt{12}$

20. $\sqrt{20} \cdot \sqrt{54}$

21. $\sqrt{6x}\sqrt{12x}$

22. $(\sqrt{3ab^2})(\sqrt{21a^2b})$

23. $\sqrt{5a^3b^2}\sqrt{20a^3b^3}$

24. $\sqrt[3]{m^2n^2} \cdot \sqrt[3]{48m^4n^2}$

25. $(4\sqrt{3xy^3})(-2\sqrt{6x^3y^2})$

26. $(2\sqrt[4]{3x})(4\sqrt[4]{27x^6})$

27. $(\sqrt[3]{4a^2b})(\sqrt[3]{2ab^3})(\sqrt[3]{54a^2b})$

28. $(\sqrt[3]{9x^3y})(\sqrt[3]{6xy})(\sqrt[3]{8x^2y^5})$

29. $\sqrt{3}(4\sqrt{3} - 6)$

30. $3\sqrt{5}(2\sqrt{5} + 4)$

31. $\sqrt{2}(\sqrt{6} - \sqrt{3})$

32. $\sqrt{5}(\sqrt{3} + \sqrt{7})$

33. $-3\sqrt{x}(\sqrt{x} + 7)$

34. $-2\sqrt{y}(8 - \sqrt{y})$

35. $(\sqrt{3} + 2\sqrt{10})(4\sqrt{3} - \sqrt{10})$

36. $(8\sqrt{7} - \sqrt{5})(\sqrt{7} + 3\sqrt{5})$

37. $(\sqrt{x} + 4)(\sqrt{x} - 9)$

38. $(\sqrt{w} - 2)(\sqrt{w} - 9)$

39. $(\sqrt[3]{y} + 2)(\sqrt[3]{y} - 3)$

40. $(4 + \sqrt[5]{p})(5 + \sqrt[5]{p})$

41. $(\sqrt{a} - 3\sqrt{b})(9\sqrt{a} - \sqrt{b})$

42. $(11\sqrt{m} + 4\sqrt{n})(\sqrt{m} + \sqrt{n})$

43. $(\sqrt{p} + 2\sqrt{q})(8 + 3\sqrt{p} - \sqrt{q})$

44. $(5\sqrt{s} - \sqrt{t})(\sqrt{s} + 5 + 6\sqrt{t})$

Concept 2: Expressions of the Form $\left(\sqrt[n]{a}\right)^n$

For Exercises 45–52, simplify the expressions. Assume all variables represent positive real numbers.

45. $(\sqrt{15})^2$

46. $(\sqrt{58})^2$

47. $(\sqrt{3y})^2$

48. $(\sqrt{19yz})^2$

49. $(\sqrt[3]{6})^3$

50. $(\sqrt[5]{24})^5$

51. $\sqrt{709} \cdot \sqrt{709}$

52. $\sqrt{401} \cdot \sqrt{401}$

Concept 3: Special Case Products

53. a. Write the formula for the product of two conjugates. $(x + y)(x - y) = $

   b. Multiply $(x + 5)(x - 5)$.

54. a. Write the formula for squaring a binomial. $(x + y)^2 = $

   b. Multiply $(x + 5)^2$.

For Exercises 55–66, multiply the special products.

55. $(\sqrt{3} + x)(\sqrt{3} - x)$

56. $(y + \sqrt{6})(y - \sqrt{6})$

57. $(\sqrt{6} + \sqrt{2})(\sqrt{6} - \sqrt{2})$

58. $(\sqrt{15} + \sqrt{5})(\sqrt{15} - \sqrt{5})$

59. $(8\sqrt{x} + 2\sqrt{y})(8\sqrt{x} - 2\sqrt{y})$

60. $(4\sqrt{s} + 11\sqrt{t})(4\sqrt{s} - 11\sqrt{t})$

61. $(\sqrt{13} + 4)^2$

62. $(6 - \sqrt{11})^2$

63. $(\sqrt{p} - \sqrt{7})^2$

64. $(\sqrt{q} + \sqrt{2})^2$

65. $(\sqrt{2a} - 3\sqrt{b})^2$

66. $(\sqrt{3w} + 4\sqrt{z})^2$

Mixed Exercises

For Exercises 67–74, identify each statement as true or false. If an answer is false, explain why.

67. $\sqrt{3} \cdot \sqrt{2} = \sqrt{6}$

68. $\sqrt{5} \cdot \sqrt[3]{2} = \sqrt{10}$

69. $(x - \sqrt{5})^2 = x - 5$

70. $3(2\sqrt{5x}) = 6\sqrt{5x}$

71. $5(3\sqrt{4x}) = 15\sqrt{20x}$

72. $\dfrac{\sqrt{5x}}{5} = \sqrt{x}$

73. $\dfrac{3\sqrt{x}}{3} = \sqrt{x}$

74. $(\sqrt{t} - 1)(\sqrt{t} + 1) = t - 1$

For Exercises 75–82, square the expressions.

75. $\sqrt{39}$

76. $\sqrt{21}$

77. $-\sqrt{6x}$

78. $-\sqrt{8a}$

79. $\sqrt{3x + 1}$

80. $\sqrt{x - 1}$

81. $\sqrt{x + 3} - 4$

82. $\sqrt{x + 1} + 3$

For Exercises 83–86, find the exact area.

83.
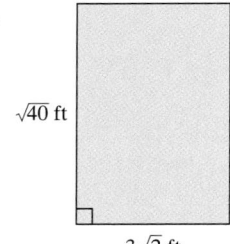
$\sqrt{40}$ ft

$3\sqrt{2}$ ft

84.
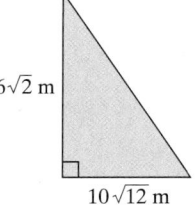
$6\sqrt{2}$ m

$10\sqrt{12}$ m

85.
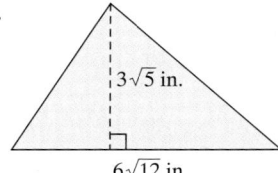
$3\sqrt{5}$ in.

$6\sqrt{12}$ in.

86.
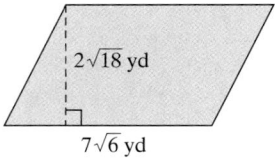
$2\sqrt{18}$ yd

$7\sqrt{6}$ yd

Concept 4: Multiplying Radicals with Different Indices

For Exercises 87–96, multiply or divide the radicals with different indices. Write the answers in radical form and simplify.

87. $\sqrt{x} \cdot \sqrt[4]{x}$

88. $\sqrt[3]{y} \cdot \sqrt{y}$

89. $\sqrt[5]{2z} \cdot \sqrt[3]{2z}$

90. $\sqrt[3]{5w} \cdot \sqrt[4]{5w}$

91. $\sqrt[3]{p^2} \cdot \sqrt{p^3}$

92. $\sqrt[4]{q^3} \cdot \sqrt[3]{q^2}$

93. $\dfrac{\sqrt{u^3}}{\sqrt[3]{u}}$

94. $\dfrac{\sqrt{v^5}}{\sqrt[4]{v}}$

95. $\dfrac{\sqrt{(a + b)}}{\sqrt[3]{(a + b)}}$

96. $\dfrac{\sqrt[3]{(q - 1)}}{\sqrt[4]{(q - 1)}}$

For Exercises 97–100, multiply the radicals with different indices.

97. $\sqrt[3]{x} \cdot \sqrt[6]{y}$

98. $\sqrt{a} \cdot \sqrt[6]{b}$

99. $\sqrt[4]{8} \cdot \sqrt{3}$

100. $\sqrt{11} \cdot \sqrt[6]{2}$

Expanding Your Skills

101. Multiply $(\sqrt[3]{a} + \sqrt[3]{b})(\sqrt[3]{a^2} - \sqrt[3]{ab} + \sqrt[3]{b^2})$.

102. Multiply $(\sqrt[3]{a} - \sqrt[3]{b})(\sqrt[3]{a^2} + \sqrt[3]{ab} + \sqrt[3]{b^2})$.

Rationalization

1. Simplified Form of a Radical

In this section we will learn additional techniques to simplify a radical.

> **Simplified Form of a Radical**
>
> Consider any radical expression in which the radicand is written as a product of prime factors. The expression is in simplified form if all the following conditions are met:
>
> 1. The radicand has no factor raised to a power greater than or equal to the index.
> 2. The radicand does not contain a fraction.
> 3. No radicals are in the denominator of a fraction.

The third condition restricts radicals from the denominator of an expression. The process to remove a radical from the denominator is called **rationalizing the denominator**. In many cases, rationalizing the denominator creates an expression that is computationally simpler. In Examples 3 and 4 we will show that

$$\frac{6}{\sqrt{3}} = 2\sqrt{3} \quad \text{and} \quad \frac{-2}{2 + \sqrt{6}} = 2 - \sqrt{6}$$

We will demonstrate the process to rationalize the denominator as two separate cases:

- Rationalizing the denominator (one term)
- Rationalizing the denominator (two terms involving square roots)

2. Rationalizing the Denominator—One Term

To begin the first case, recall that the nth root of a perfect nth power simplifies completely.

$$\sqrt{x^2} = x \qquad x \geq 0$$
$$\sqrt[3]{x^3} = x$$
$$\sqrt[4]{x^4} = x \qquad x \geq 0$$
$$\sqrt[5]{x^5} = x$$

$$\cdots$$

Therefore, to rationalize a radical expression, use the multiplication property of radicals to create an nth root of an nth power.

Example 1 **Rationalizing Radical Expressions**

Fill in the missing radicand to rationalize the radical expressions. Assume all variables represent positive real numbers.

a. $\sqrt{a} \cdot \sqrt{?} = \sqrt{a^2} = a$
b. $\sqrt[3]{y} \cdot \sqrt[3]{?} = \sqrt[3]{y^3} = y$

c. $\sqrt[4]{2z^3} \cdot \sqrt[4]{?} = \sqrt[4]{2^4 z^4} = 2z$

Solution:

a. $\sqrt{a} \cdot \sqrt{?} = \sqrt{a^2} = a$ What multiplied by $\sqrt{a}$ will equal $\sqrt{a^2}$?

$\sqrt{a} \cdot \sqrt{a} = \sqrt{a^2} = a$

b. $\sqrt[3]{y} \cdot \sqrt[3]{?} = \sqrt[3]{y^3} = y$ What multiplied by $\sqrt[3]{y}$ will equal $\sqrt[3]{y^3}$?

$\sqrt[3]{y} \cdot \sqrt[3]{y^2} = \sqrt[3]{y^3} = y$

c. $\sqrt[4]{2z^3} \cdot \sqrt[4]{?} = \sqrt[4]{2^4 z^4} = 2z$ What multiplied by $\sqrt[4]{2z^3}$ will equal $\sqrt[4]{2^4 z^4}$?

$\sqrt[4]{2z^3} \cdot \sqrt[4]{2^3 z} = \sqrt[4]{2^4 z^4} = 2z$

| **Skill Practice** | Fill in the missing radicand to rationalize the radical expression. |

1. $\sqrt{7} \cdot \sqrt{?}$ **2.** $\sqrt[5]{t^2} \cdot \sqrt[5]{?}$ **3.** $\sqrt[3]{5x^2} \cdot \sqrt[3]{?}$

To rationalize the denominator of a radical expression, multiply the numerator and denominator by an appropriate expression to create an nth root of an nth power in the denominator.

| **Example 2** | **Rationalizing the Denominator—One Term** |

Simplify the expression

$$\frac{5}{\sqrt[3]{a}} \qquad a \neq 0$$

Solution:

To remove the radical from the denominator, a cube root of a perfect cube is needed in the denominator. Multiply numerator and denominator by $\sqrt[3]{a^2}$ because $\sqrt[3]{a} \cdot \sqrt[3]{a^2} = \sqrt[3]{a^3} = a$.

$$\frac{5}{\sqrt[3]{a}} = \frac{5}{\sqrt[3]{a}} \cdot \frac{\sqrt[3]{a^2}}{\sqrt[3]{a^2}}$$

$$= \frac{5\sqrt[3]{a^2}}{\sqrt[3]{a^3}} \qquad \text{Multiply the radicals.}$$

$$= \frac{5\sqrt[3]{a^2}}{a} \qquad \text{Simplify.}$$

| **Skill Practice** | Simplify the expression. Assume $y > 0$. |

4. $\dfrac{2}{\sqrt[4]{y}}$

Note that for $a \neq 0$, the expression $\dfrac{\sqrt[3]{a^2}}{\sqrt[3]{a^2}} = 1$. In Example 2, multiplying the expression $\dfrac{5}{\sqrt[3]{a}}$ by this ratio does not change its value.

Skill Practice Answers

1. 7   **2.** t^3 **3.** $5^2 x$

4. $\dfrac{2\sqrt[4]{y^3}}{y}$

Example 3 **Rationalizing the Denominator—One Term**

Simplify the expressions. Assume all variables represent positive real numbers.

a. $\dfrac{6}{\sqrt{3}}$ **b.** $\sqrt{\dfrac{y^5}{7}}$ **c.** $\dfrac{15}{\sqrt[3]{25s}}$ **d.** $\dfrac{\sqrt{125p^3}}{\sqrt{5p}}$

Solution:

a. To rationalize the denominator, a square root of a perfect square is needed. Multiply numerator and denominator by $\sqrt{3}$ because $\sqrt{3} \cdot \sqrt{3} = \sqrt{3^2} = 3$.

$\dfrac{6}{\sqrt{3}} = \dfrac{6}{\sqrt{3}} \cdot \dfrac{\sqrt{3}}{\sqrt{3}}$ Rationalize the denominator.

$= \dfrac{6\sqrt{3}}{\sqrt{3^2}}$ Multiply the radicals.

$= \dfrac{6\sqrt{3}}{3}$ Simplify.

$= 2\sqrt{3}$ Reduce to lowest terms.

b. $\sqrt{\dfrac{y^5}{7}}$ The radical contains an irreducible fraction.

$= \dfrac{\sqrt{y^5}}{\sqrt{7}}$ Apply the division property of radicals.

$= \dfrac{y^2\sqrt{y}}{\sqrt{7}}$ Remove factors from the radical in the numerator.

$= \dfrac{y^2\sqrt{y}}{\sqrt{7}} \cdot \dfrac{\sqrt{7}}{\sqrt{7}}$ Rationalize the denominator.
Note: $\sqrt{7} \cdot \sqrt{7} = \sqrt{7^2} = 7$.

$= \dfrac{y^2\sqrt{7y}}{\sqrt{7^2}}$

$= \dfrac{y^2\sqrt{7y}}{7}$ Simplify.

c. $\dfrac{15}{\sqrt[3]{25s}}$

$= \dfrac{15}{\sqrt[3]{5^2 s}} \cdot \dfrac{\sqrt[3]{5s^2}}{\sqrt[3]{5s^2}}$ Because $25 = 5^2$, one additional factor of 5 is needed to form a perfect cube. Two additional factors of s are needed to make a perfect cube. Multiply numerator and denominator by $\sqrt[3]{5s^2}$.

$= \dfrac{15\sqrt[3]{5s^2}}{\sqrt[3]{5^3 s^3}}$

$= \dfrac{15\sqrt[3]{5s^2}}{5s}$ Simplify the perfect cube.

$= \dfrac{\overset{3}{\cancel{15}}\sqrt[3]{5s^2}}{\underset{1}{\cancel{5}s}}$ Reduce to lowest terms.

$= \dfrac{3\sqrt[3]{5s^2}}{s}$

Calculator Connections

A calculator can be used to support the solution to a simplified radical. The calculator approximations of the expressions $6/\sqrt{3}$ and $2\sqrt{3}$ agree to 10 digits.

```
6/√(3)
        3.464101615
2√(3)
        3.464101615
```

Avoiding Mistakes:

A factor within a radicand cannot be simplified with a factor outside the radicand. For example, $\dfrac{\sqrt{7y}}{7}$ cannot be simplified.

> **TIP:** In the expression $\dfrac{15\sqrt[3]{5s^2}}{5s}$, the factor of 15 and the factor of 5 may be reduced because both are outside the radical.
>
> $$\frac{15\sqrt[3]{5s^2}}{5s} = \frac{15}{5} \cdot \frac{\sqrt[3]{5s^2}}{s} = \frac{3\sqrt[3]{5s^2}}{s}$$

d. $\dfrac{\sqrt{125p^3}}{\sqrt{5p}}$ Notice that the radicands in the numerator and denominator share common factors.

$= \sqrt{\dfrac{125p^3}{5p}}$ Rewrite the expression by using the division property of radicals.

$= \sqrt{25p^2}$ Simplify the fraction within the radicand.

$= 5p$ Simplify the radical.

Skill Practice Simplify the expressions.

5. $\dfrac{12}{\sqrt{2}}$ **6.** $\sqrt{\dfrac{8}{3}}$ **7.** $\dfrac{18}{\sqrt[3]{3y^2}}$ **8.** $\dfrac{\sqrt[3]{16x^4}}{\sqrt[3]{2x}}$

3. Rationalizing the Denominator—Two Terms

Example 4 demonstrates how to rationalize a two-term denominator involving square roots.

First, recall from the multiplication of polynomials that the product of two conjugates results in a difference of squares.

$$(a + b)(a - b) = a^2 - b^2$$

If either a or b has a square root factor, the expression will simplify without a radical. That is, the expression is *rationalized.* For example:

$$(2 + \sqrt{6})(2 - \sqrt{6}) = (2)^2 - (\sqrt{6})^2$$
$$= 4 - 6$$
$$= -2$$

Example 4   Rationalizing the Denominator—Two Terms

Simplify the expression by rationalizing the denominator.

$$\frac{-2}{2 + \sqrt{6}}$$

Solution:

$$\frac{-2}{2 + \sqrt{6}}$$

$= \dfrac{(-2)}{(2 + \sqrt{6})} \cdot \dfrac{(2 - \sqrt{6})}{(2 - \sqrt{6})}$ Multiply the numerator and denominator by the conjugate of the denominator.

conjugates

Skill Practice Answers

5. $6\sqrt{2}$ **6.** $\dfrac{2\sqrt{6}}{3}$

7. $\dfrac{6\sqrt[3]{9y}}{y}$ **8.** $2x$

$$= \frac{-2(2 - \sqrt{6})}{(2)^2 - (\sqrt{6})^2}$$

In the denominator, apply the formula
$(a + b)(a - b) = a^2 - b^2$.

$$= \frac{-2(2 - \sqrt{6})}{4 - 6}$$

Simplify.

$$= \frac{-2(2 - \sqrt{6})}{-2}$$

$$= \frac{\cancel{-2}(2 - \sqrt{6})}{\cancel{-2}}$$

$$= 2 - \sqrt{6}$$

TIP: Recall that only factors within a fraction may be simplified (not terms).

$$\frac{ab}{a} = b$$

but

$$\frac{a + b}{a} \text{ cannot be simplified.}$$

> **Skill Practice** Simplify by rationalizing the denominator.
>
> **9.** $\dfrac{4}{3 + \sqrt{5}}$

Example 5 **Rationalizing the Denominator—Two Terms**

Rationalize the denominator of the expression. Assume all variables represent positive real numbers and $c \neq d$.

$$\frac{\sqrt{c} + \sqrt{d}}{\sqrt{c} - \sqrt{d}}$$

Solution:

$$\frac{\sqrt{c} + \sqrt{d}}{\sqrt{c} - \sqrt{d}}$$

$$= \frac{(\sqrt{c} + \sqrt{d})}{(\sqrt{c} - \sqrt{d})} \cdot \frac{(\sqrt{c} + \sqrt{d})}{(\sqrt{c} + \sqrt{d})}$$

Multiply numerator and denominator by the conjugate of the denominator.

conjugates

$$= \frac{(\sqrt{c} + \sqrt{d})^2}{(\sqrt{c})^2 - (\sqrt{d})^2}$$

In the denominator apply the formula
$(a + b)(a - b) = a^2 - b^2$.

$$= \frac{(\sqrt{c} + \sqrt{d})^2}{c - d}$$

Simplify.

$$= \frac{(\sqrt{c})^2 + 2\sqrt{c}\sqrt{d} + (\sqrt{d})^2}{c - d}$$

In the numerator apply the formula
$(a + b)^2 = a^2 + 2ab + b^2$.

$$= \frac{c + 2\sqrt{cd} + d}{c - d}$$

> **Skill Practice** Simplify by rationalizing the denominator. Assume that $y \geq 0$.
>
> **10.** $\dfrac{\sqrt{3} - \sqrt{y}}{\sqrt{3} + \sqrt{y}}$

Skill Practice Answers

9. $3 - \sqrt{5}$

10. $\dfrac{3 - 2\sqrt{3y} + y}{3 - y}$

Section 11.6   Practice Exercises

For the exercises in this set, assume that all variables represent positive real numbers unless otherwise stated.

Study Skills Exercise

1. Define the key term **rationalizing the denominator**.

Review Exercises

2. Multiply and simplify. $\sqrt{x} \cdot \sqrt{x}$

For Exercises 3–10, perform the indicated operations.

3. $2y\sqrt{45} + 3\sqrt{20y^2}$ **4.** $3x\sqrt{72x} - 9\sqrt{50x^3}$ **5.** $(-6\sqrt{y} + 3)(3\sqrt{y} + 1)$ **6.** $(\sqrt{w} + 12)(2\sqrt{w} - 4)$

7. $(8 - \sqrt{t})^2$ **8.** $(\sqrt{p} + 4)^2$ **9.** $(\sqrt{2} + \sqrt{7})(\sqrt{2} - \sqrt{7})$ **10.** $(\sqrt{3} + 5)(\sqrt{3} - 5)$

Concept 2: Rationalizing the Denominator—One Term

The radical expressions in Exercises 11–18 have radicals in the denominator. Fill in the missing radicands to rationalize the radical expressions in the denominators.

11. $\dfrac{x}{\sqrt{5}} = \dfrac{x}{\sqrt{5}} \cdot \dfrac{\sqrt{?}}{\sqrt{?}}$

12. $\dfrac{2}{\sqrt{x}} = \dfrac{2}{\sqrt{x}} \cdot \dfrac{\sqrt{?}}{\sqrt{?}}$

13. $\dfrac{7}{\sqrt[3]{x}} = \dfrac{7}{\sqrt[3]{x}} \cdot \dfrac{\sqrt[3]{?}}{\sqrt[3]{?}}$

14. $\dfrac{5}{\sqrt[4]{y}} = \dfrac{5}{\sqrt[4]{y}} \cdot \dfrac{\sqrt[4]{?}}{\sqrt[4]{?}}$

15. $\dfrac{8}{\sqrt{3z}} = \dfrac{8}{\sqrt{3z}} \cdot \dfrac{\sqrt{??}}{\sqrt{??}}$

16. $\dfrac{10}{\sqrt{7w}} = \dfrac{10}{\sqrt{7w}} \cdot \dfrac{\sqrt{??}}{\sqrt{??}}$

17. $\dfrac{1}{\sqrt[4]{2a^2}} = \dfrac{1}{\sqrt[4]{2a^2}} \cdot \dfrac{\sqrt[4]{??}}{\sqrt[4]{??}}$

18. $\dfrac{1}{\sqrt[3]{6b^2}} = \dfrac{1}{\sqrt[3]{6b^2}} \cdot \dfrac{\sqrt[3]{??}}{\sqrt[3]{??}}$

For Exercises 19–50, rationalize the denominator.

19. $\dfrac{1}{\sqrt{3}}$

20. $\dfrac{1}{\sqrt{7}}$

21. $\sqrt{\dfrac{1}{x}}$

22. $\sqrt{\dfrac{1}{z}}$

23. $\dfrac{6}{\sqrt{2y}}$

24. $\dfrac{9}{\sqrt{3t}}$

25. $\dfrac{-2a}{\sqrt{a}}$

26. $\dfrac{-7b}{\sqrt{b}}$

27. $\dfrac{6}{\sqrt{8}}$

28. $\dfrac{2}{\sqrt{48}}$

29. $\dfrac{3}{\sqrt[3]{2}}$

30. $\dfrac{2}{\sqrt[3]{7}}$

31. $\dfrac{-6}{\sqrt[4]{x}}$

32. $\dfrac{-2}{\sqrt[5]{y}}$

33. $\dfrac{7}{\sqrt[3]{4}}$

34. $\dfrac{1}{\sqrt[3]{9}}$

35. $\sqrt[3]{\dfrac{4}{w^2}}$

36. $\sqrt[3]{\dfrac{5}{z^2}}$

37. $\sqrt[4]{\dfrac{16}{3}}$

38. $\sqrt[4]{\dfrac{81}{8}}$

39. $\dfrac{2}{\sqrt[3]{4x^2}}$

40. $\dfrac{6}{\sqrt[3]{3y^2}}$

41. $\sqrt[3]{\dfrac{16x^3}{y}}$

42. $\sqrt{\dfrac{5}{9x}}$

43. $\dfrac{\sqrt{x^4y^5}}{\sqrt{10x}}$

44. $\sqrt[4]{\dfrac{10x^2}{15xy^3}}$

45. $\dfrac{8}{7\sqrt{24}}$

46. $\dfrac{5}{3\sqrt{50}}$

47. $\dfrac{1}{\sqrt{x^7}}$

48. $\dfrac{1}{\sqrt{y^5}}$

49. $\dfrac{2}{\sqrt{8x^5}}$

50. $\dfrac{6}{\sqrt{27t^7}}$

Concept 3: Rationalizing the Denominator—Two Terms

51. What is the conjugate of $\sqrt{2} - \sqrt{6}$?

52. What is the conjugate of $\sqrt{11} + \sqrt{5}$?

53. What is the conjugate of $\sqrt{x} + 23$?

54. What is the conjugate of $17 - \sqrt{y}$?

For Exercises 55–74, rationalize the denominators.

55. $\dfrac{4}{\sqrt{2} + 3}$

56. $\dfrac{6}{4 - \sqrt{3}}$

57. $\dfrac{8}{\sqrt{6} - 2}$

58. $\dfrac{-12}{\sqrt{5} - 3}$

59. $\dfrac{\sqrt{7}}{\sqrt{3} + 2}$

60. $\dfrac{\sqrt{8}}{\sqrt{3} + 1}$

61. $\dfrac{-1}{\sqrt{p} + \sqrt{q}}$

62. $\dfrac{6}{\sqrt{a} - \sqrt{b}}$

63. $\dfrac{x - 5}{\sqrt{x} + \sqrt{5}}$

64. $\dfrac{y - 2}{\sqrt{y} - \sqrt{2}}$

65. $\dfrac{-7}{2\sqrt{a} - 5\sqrt{b}}$

66. $\dfrac{-4}{3\sqrt{w} - 2\sqrt{z}}$

67. $\dfrac{3\sqrt{5} - \sqrt{3}}{\sqrt{3} + \sqrt{5}}$

68. $\dfrac{2\sqrt{5} + \sqrt{6}}{\sqrt{6} - \sqrt{5}}$

69. $\dfrac{3\sqrt{10}}{2 + \sqrt{10}}$

70. $\dfrac{4\sqrt{7}}{3 + \sqrt{7}}$

71. $\dfrac{2\sqrt{3} + \sqrt{7}}{3\sqrt{3} - \sqrt{7}}$

72. $\dfrac{5\sqrt{2} - \sqrt{5}}{5\sqrt{2} + \sqrt{5}}$

73. $\dfrac{\sqrt{5} + 4}{2 - \sqrt{5}}$

74. $\dfrac{3 + \sqrt{2}}{\sqrt{2} - 5}$

Mixed Exercises

For Exercises 75–78, translate the English phrase to an algebraic expression. Then simplify the expression.

75. 16 divided by the cube root of 4

76. 21 divided by the principal fourth root of 27

77. 4 divided by the difference of x and the principal square root of 2

78. 8 divided by the sum of y and the principal square root of 3

79. The time T (in seconds) for a pendulum to make one complete swing back and forth is approximated by

$$T(x) = 2\pi\sqrt{\dfrac{x}{32}}$$

where x is the length of the pendulum in feet.

Determine the exact time required for one swing for a pendulum that is 1 ft long. Then approximate the time to the nearest hundredth of a second.

80. An object is dropped off a building x meters tall. The time T (in seconds) required for the object to hit the ground is given by

$$T(x) = \sqrt{\frac{10x}{49}}$$

Find the exact time required for the object to hit the ground if it is dropped off the First National Plaza Building in Chicago, a height of 230 m. Then round the time to the nearest hundredth of a second.

Expanding Your Skills

For Exercises 81–86, simplify each term of the expression. Then add or subtract as indicated.

81. $\dfrac{\sqrt{6}}{2} + \dfrac{1}{\sqrt{6}}$

82. $\dfrac{1}{\sqrt{7}} + \sqrt{7}$

83. $\sqrt{15} - \sqrt{\dfrac{3}{5}} + \sqrt{\dfrac{5}{3}}$

84. $\sqrt{\dfrac{6}{2}} - \sqrt{12} + \sqrt{\dfrac{2}{6}}$

85. $\sqrt[3]{25} + \dfrac{3}{\sqrt[3]{5}}$

86. $\dfrac{1}{\sqrt[3]{4}} + \sqrt[3]{54}$

For Exercises 87–90, rationalize the numerator by multiplying both numerator and denominator by the conjugate of the numerator.

87. $\dfrac{\sqrt{3}+6}{2}$

88. $\dfrac{\sqrt{7}-2}{5}$

89. $\dfrac{\sqrt{a}-\sqrt{b}}{\sqrt{a}+\sqrt{b}}$

90. $\dfrac{\sqrt{p}+\sqrt{q}}{\sqrt{p}-\sqrt{q}}$

Chapter 11 Problem Recognition Exercises— Operations on Radicals

Perform the indicated operations and simplify if possible. Assume that all variables represent positive real numbers.

1. $(\sqrt{3})(\sqrt{6})$

2. $(\sqrt{2})(\sqrt{14})$

3. $\sqrt{3} + \sqrt{6}$

4. $\sqrt{2} + \sqrt{14}$

5. $\dfrac{\sqrt{6}}{\sqrt{3}}$

6. $\dfrac{\sqrt{14}}{\sqrt{2}}$

7. $(3+\sqrt{z})(3-\sqrt{z})$

8. $(4-\sqrt{y})(4+\sqrt{y})$

9. $(2\sqrt{5}+1)(\sqrt{5}-2)$

10. $(4\sqrt{3}-5)(\sqrt{3}+4)$

11. $2\sqrt{x^2y} - 3x\sqrt{y}$

12. $8\sqrt{a^3b^2} + 3a\sqrt{ab^2}$

13. $-3\sqrt{2}(4\sqrt{2}+2\sqrt{3}+1)$

14. $-8\sqrt{5}(2\sqrt{5}-\sqrt{3}-2)$

15. $\dfrac{2}{\sqrt{x}-7}$

16. $\dfrac{5}{\sqrt{y}+4}$

17. $\dfrac{9}{\sqrt{3}}$

18. $\dfrac{15}{\sqrt{5}}$

19. $\sqrt{\dfrac{7}{x}}$

20. $\sqrt{\dfrac{11}{y}}$

21. $\sqrt{y^4z^{11}}$

22. $\sqrt{8q^6}$

23. $\sqrt[3]{27p^8}$

24. $\sqrt[3]{125u^{11}v^{12}}$

25. $\dfrac{\sqrt{10x^3}}{\sqrt{x}}$

26. $\dfrac{\sqrt{15y^3}}{\sqrt{5y}}$

27. $6\sqrt{75} - 5\sqrt{12}$

28. $\sqrt{90} - \sqrt{40}$

29. $\sqrt[4]{\dfrac{1}{81}}$

30. $\sqrt[3]{\dfrac{125}{27}}$

31. $\dfrac{x-5}{\sqrt{x}+\sqrt{5}}$

32. $\dfrac{y-7}{\sqrt{y}+\sqrt{7}}$

33. $(4\sqrt{x} + \sqrt{y})(\sqrt{x} - 3\sqrt{y})$

34. $(\sqrt{2} + 7)^2$

35. $(\sqrt{3} + \sqrt{5})^2$

36. $(\sqrt{5} - \sqrt{11})^2$

37. $(\sqrt{x} - 6)^2$

38. $(2\sqrt{3} - 10)(2\sqrt{3} + 10)$

39. $(\sqrt{u} - 3\sqrt{v})(\sqrt{u} + 3\sqrt{v})$

40. $2\sqrt{6} - 5\sqrt{6}$

41. $5\sqrt{a} + 7\sqrt{a} - \sqrt{a}$

42. $x\sqrt{18} + \sqrt{2x^2}$

43. $4\sqrt{75} - 20\sqrt{3}$

44. $\sqrt{5}(\sqrt{5} + \sqrt{7})$

45. $\sqrt{a}(\sqrt{a} + 2)$

46. $(3\sqrt{2} - 4)(5\sqrt{2} + 1)$

Radical Equations

Section 11.7

1. Solutions to Radical Equations

An equation with one or more radicals containing a variable is called a **radical equation**. For example, $\sqrt[3]{x} = 5$ is a radical equation. Recall that $(\sqrt[n]{a})^n = a$, provided $\sqrt[n]{a}$ is a real number. The basis of solving a radical equation is to eliminate the radical by raising both sides of the equation to a power equal to the index of the radical.

To solve the equation $\sqrt[3]{x} = 5$, cube both sides of the equation.

$$\sqrt[3]{x} = 5$$
$$(\sqrt[3]{x})^3 = (5)^3$$
$$x = 125$$

Concepts

1. Solutions to Radical Equations
2. Solving Radical Equations Involving One Radical
3. Solving Radical Equations Involving More than One Radical
4. Applications of Radical Equations and Functions

By raising each side of a radical equation to a power equal to the index of the radical, a new equation is produced. Note, however, that some of or all the solutions to the new equation may *not* be solutions to the original radical equation. For this reason, it is necessary to *check all potential solutions* in the original equation. For example, consider the equation $x = 4$. By squaring both sides we produce a quadratic equation.

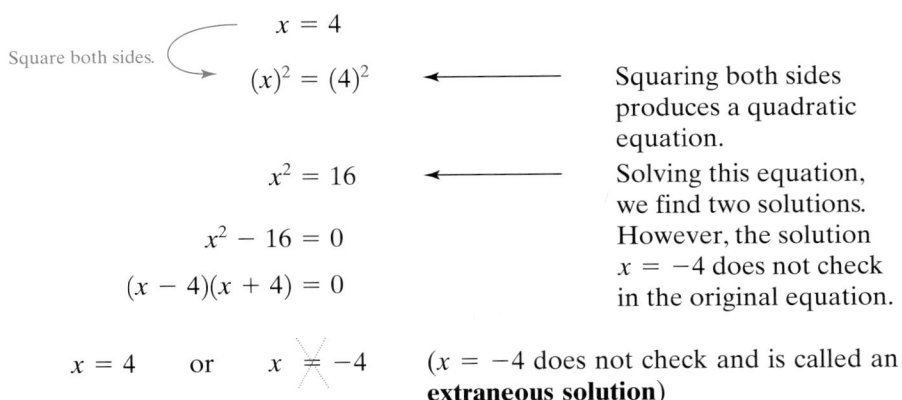

Square both sides.

$$x = 4$$
$$(x)^2 = (4)^2 \quad \longleftarrow \quad \text{Squaring both sides produces a quadratic equation.}$$
$$x^2 = 16 \quad \longleftarrow \quad \text{Solving this equation, we find two solutions. However, the solution } x = -4 \text{ does not check in the original equation.}$$
$$x^2 - 16 = 0$$
$$(x - 4)(x + 4) = 0$$

$$x = 4 \quad \text{or} \quad x \ne -4 \qquad (x = -4 \text{ does not check and is called an } \textbf{extraneous solution})$$

Steps to Solve a Radical Equation

1. Isolate the radical. If an equation has more than one radical, choose one of the radicals to isolate.

2. Raise each side of the equation to a power equal to the index of the radical.

3. Solve the resulting equation. If the equation still has a radical, repeat steps 1 and 2.

*4. Check the potential solutions in the original equation.

*Extraneous solutions can arise only when both sides of the equation are raised to an *even power.* Therefore, an equation with odd-index roots will not have an extraneous solution. However, it is still recommended that you check *all* potential solutions regardless of the type of root.

2. Solving Radical Equations Involving One Radical

> **Example 1**   Solving Equations Containing One Radical

Solve the equations.

a. $\sqrt{p} + 5 = 9$ **b.** $\sqrt{3x - 2} + 4 = 5$ **c.** $(w - 1)^{1/3} - 2 = 2$

d. $7 = \sqrt[4]{x + 3} + 9$ **e.** $y + \sqrt{y - 2} = 8$

Solution:

a. $\sqrt{p} + 5 = 9$

$\qquad \sqrt{p} = 4$ Isolate the radical.

$\qquad (\sqrt{p})^2 = 4^2$ Because the index is 2, square both sides.

$\qquad p = 16$

$\qquad$ Check: $p = 16$ Check $p = 16$ as a potential solution.

$\qquad \sqrt{p} + 5 = 9$

$\qquad \sqrt{16} + 5 \overset{?}{=} 9$

$\qquad 4 + 5 = 9$ ✔   True, $p = 16$ is a solution to the original equation.

b. $\sqrt{3x - 2} + 4 = 5$

$\qquad \sqrt{3x - 2} = 1$ Isolate the radical.

$\qquad (\sqrt{3x - 2})^2 = (1)^2$ Because the index is 2, square both sides.

$\qquad 3x - 2 = 1$ Simplify.

$\qquad 3x = 3$ Solve the resulting equation.

$\qquad x = 1$

$\qquad$ Check: $x = 1$ Check $x = 1$ as a potential solution.

$\qquad \sqrt{3x - 2} + 4 = 5$

$\qquad \sqrt{3(1) - 2} + 4 \overset{?}{=} 5$

$\qquad \sqrt{1} + 4 \overset{?}{=} 5$

$\qquad 5 = 5$ ✔   True, $x = 1$ is a solution to the original equation.

c. $(w - 1)^{1/3} - 2 = 2$ Note that $(w - 1)^{1/3} = \sqrt[3]{w - 1}$.

$\sqrt[3]{w - 1} - 2 = 2$

$\sqrt[3]{w - 1} = 4$ Isolate the radical.

$(\sqrt[3]{w - 1})^3 = (4)^3$ Because the index is 3, cube both sides.

$w - 1 = 64$ Simplify.

$w = 65$

Check: $w = 65$

$(w - 1)^{1/3} - 2 = 2$ Check $w = 65$ as a potential solution.

$\sqrt[3]{65 - 1} - 2 \overset{?}{=} 2$

$\sqrt[3]{64} - 2 \overset{?}{=} 2$

$4 - 2 \overset{?}{=} 2$

$2 = 2$ ✔   True, $w = 65$ is a solution to the original equation.

d. $7 = \sqrt[4]{x + 3} + 9$

$-2 = \sqrt[4]{x + 3}$ Isolate the radical.

$(-2)^4 = (\sqrt[4]{x + 3})^4$ Because the index is 4, raise both sides to the fourth power.

$16 = x + 3$

$x = 13$ Solve for x.

Check: $x = 13$

$7 = \sqrt[4]{x + 3} + 9$

$7 \overset{?}{=} \sqrt[4]{(13) + 3} + 9$

$7 \overset{?}{=} \sqrt[4]{16} + 9$

$7 \neq 2 + 9$ $x = 13$ is *not* a solution to the original equation.

The equation $7 = \sqrt[4]{x + 3} + 9$ has no solution.

e. $y + \sqrt{y - 2} = 8$

$\sqrt{y - 2} = 8 - y$ Isolate the radical.

$(\sqrt{y - 2})^2 = (8 - y)^2$ Because the index is 2, square both sides.

Note that $(8 - y)^2 = (8 - y)(8 - y) = 64 - 16y + y^2$.

$y - 2 = 64 - 16y + y^2$ Simplify.

$0 = y^2 - 17y + 66$ The equation is quadratic. Set one side equal to zero. Write the other side in descending order.

$0 = (y - 11)(y - 6)$ Factor.

$y - 11 = 0$ or $y - 6 = 0$ Set each factor equal to zero.

$y = 11$ or $y = 6$ Solve.

TIP: After isolating the radical in Example 1(d), the equation shows a fourth root equated to a negative number:

$$-2 = \sqrt[4]{x + 3}$$

By definition, a principal fourth root of any real number must be nonnegative. Therefore, there can be no real solution to this equation.

Avoiding Mistakes:

Be sure to square both sides of the equation, not the individual terms.

Check: $y = 11$

$$y + \sqrt{y - 2} = 8$$

$$11 + \sqrt{11 - 2} \stackrel{?}{=} 8$$

$$11 + \sqrt{9} \stackrel{?}{=} 8$$

$$11 + 3 \stackrel{?}{=} 8$$

$$14 \neq 8$$

$y = 11$ is not a solution to the original equation.

Check: $y = 6$

$$y + \sqrt{y - 2} = 8$$

$$6 + \sqrt{6 - 2} \stackrel{?}{=} 8$$

$$6 + \sqrt{4} \stackrel{?}{=} 8$$

$$6 + 2 \stackrel{?}{=} 8$$

$$8 = 8 ✔$$

True, $y = 6$ is a solution to the original equation.

Skill Practice Solve the equations.

1. $\sqrt{x} - 3 = 2$ **2.** $\sqrt{5y + 1} - 2 = 4$ **3.** $(t + 2)^{1/3} + 5 = 3$

4. $\sqrt[4]{b - 1} + 6 = 3$ **5.** $\sqrt{x + 1} + 5 = x$

3. Solving Radical Equations Involving More than One Radical

Example 2 Solving Equations with Two Radicals

Solve the radical equation.

$$\sqrt[3]{2x - 4} = \sqrt[3]{1 - 8x}$$

Solution:

$$\sqrt[3]{2x - 4} = \sqrt[3]{1 - 8x}$$

$$(\sqrt[3]{2x - 4})^3 = (\sqrt[3]{1 - 8x})^3 \qquad \text{Because the index is 3, cube both sides.}$$

$$2x - 4 = 1 - 8x \qquad \text{Simplify.}$$

$$10x - 4 = 1 \qquad \text{Solve the resulting equation.}$$

$$10x = 5$$

$$x = \frac{1}{2} \qquad \text{Solve for } x.$$

Check: $x = \frac{1}{2}$

$$\sqrt[3]{2x - 4} = \sqrt[3]{1 - 8x}$$

$$\sqrt[3]{2\left(\frac{1}{2}\right) - 4} \stackrel{?}{=} \sqrt[3]{1 - 8\left(\frac{1}{2}\right)}$$

$$\sqrt[3]{1 - 4} \stackrel{?}{=} \sqrt[3]{1 - 4}$$

$$\sqrt[3]{-3} = \sqrt[3]{-3} ✔$$

Therefore, $x = \frac{1}{2}$ is a solution to the original equation.

Skill Practice Answers

1. $x = 25$ **2.** $y = 7$

3. $t = -10$

4. No solution ($b = 82$ does not check)

5. $x = 8$ ($x = 3$ does not check.)

Calculator Connections

The expressions on the right- and left-hand sides of the equation $\sqrt[3]{2x - 4} = \sqrt[3]{1 - 8x}$ are each functions of x. Consider the graphs of the functions:

$$Y_1 = \sqrt[3]{2x - 4} \quad \text{and} \quad Y_2 = \sqrt[3]{1 - 8x}$$

The x-coordinate of the point of intersection of the two functions is the solution to the equation $\sqrt[3]{2x - 4} = \sqrt[3]{1 - 8x}$. The point of intersection can be approximated by using *Zoom* and *Trace* or by using an *Intersect* function.

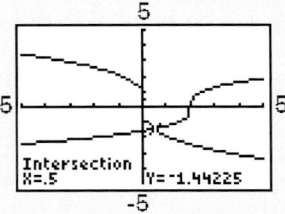

Skill Practice Solve the equation.

6. $\sqrt[5]{2y - 1} = \sqrt[5]{10y + 3}$

Example 3 **Solving Equations with Two Radicals**

Solve the equation. $\sqrt{3m + 1} - \sqrt{m + 4} = 1$

Solution:

$$\sqrt{3m + 1} - \sqrt{m + 4} = 1$$

$$\sqrt{3m + 1} = \sqrt{m + 4} + 1 \qquad \text{Isolate one of the radicals.}$$

$$(\sqrt{3m + 1})^2 = (\sqrt{m + 4} + 1)^2 \qquad \text{Square both sides.}$$

$$3m + 1 = m + 4 + 2\sqrt{m + 4} + 1$$

$$\textit{Note: } (\sqrt{m + 4} + 1)^2$$
$$= (\sqrt{m + 4})^2 + 2(1)\sqrt{m + 4} + (1)^2$$
$$= m + 4 + 2\sqrt{m + 4} + 1$$

$$3m + 1 = m + 5 + 2\sqrt{m + 4} \qquad \text{Combine } \textit{like} \text{ terms.}$$

$$2m - 4 = 2\sqrt{m + 4} \qquad \text{Isolate the radical again.}$$

$$m - 2 = \sqrt{m + 4} \qquad \text{Divide both sides by 2.}$$

$$(m - 2)^2 = (\sqrt{m + 4})^2 \qquad \text{Square both sides again.}$$

$$m^2 - 4m + 4 = m + 4 \qquad \text{The resulting equation is quadratic.}$$

$$m^2 - 5m = 0 \qquad \text{Set the quadratic equation equal to zero.}$$

$$m(m - 5) = 0 \qquad \text{Factor.}$$

$$m = 0 \quad \text{or} \quad m = 5$$

Skill Practice Answers

6. $y = -\dfrac{1}{2}$

<u>Check:</u> $m = 0$

$$\sqrt{3(0) + 1} - \sqrt{(0) + 4} \overset{?}{=} 1$$

$$\sqrt{1} - \sqrt{4} \overset{?}{=} 1$$

$$1 - 2 \neq 1 \qquad \text{Does not check}$$

<u>Check:</u> $m = 5$

$$\sqrt{3(5) + 1} - \sqrt{(5) + 4} = 1$$

$$\sqrt{16} - \sqrt{9} \overset{?}{=} 1$$

$$4 - 3 = 1 \ \checkmark$$

The solution is $m = 5$ (the value $m = 0$ does not check).

> **Skill Practice**   Solve the equation.
>
> **7.** $\sqrt{3c + 1} - \sqrt{c - 1} = 2$

4. Applications of Radical Equations and Functions

Example 4   Applying a Radical Equation in Geometry

For a pyramid with a square base, the length of a side of the base b is given by

$$b = \sqrt{\frac{3V}{h}}$$

where V is the volume and h is the height.

The Pyramid of the Pharoah Khufu (known as the Great Pyramid) at Giza has a square base (Figure 11-3). If the distance around the bottom of the pyramid is 921.6 m and the height is 146.6 m, what is the volume of the pyramid?

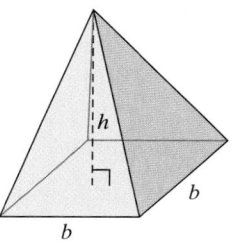

Figure 11-3

Solution:

The length of a side b (in meters) is given by $\dfrac{921.6}{4} = 230.4$ m.

$$b = \sqrt{\frac{3V}{h}}$$

$$230.4 = \sqrt{\frac{3V}{146.6}} \qquad \text{Substitute } b = 230.4 \text{ and } h = 146.6.$$

$$(230.4)^2 = \left(\sqrt{\frac{3V}{146.6}}\right)^2 \qquad \text{Because the index is 2, square both sides.}$$

Skill Practice Answers

7. $c = 1; c = 5$

$$53,084.16 = \frac{3V}{146.6}$$ Simplify.

$$(53,084.16)(146.6) = \frac{3V}{\cancel{146.6}}(\cancel{146.6})$$ Multiply both sides by 146.6.

$$(53,084.16)(146.6) = 3V$$
Divide both sides by 3.
$$\frac{(53,084.16)(146.6)}{3} = \frac{3V}{3}$$

$$2,594,046 \approx V$$

The volume of the Great Pyramid at Giza is approximately 2,594,046 m³.

Skill Practice

8. The length of the legs, s, of an isosceles right triangle is $s = \sqrt{2A}$, where A is the area. If the lengths of the legs are 9 in., find the area.

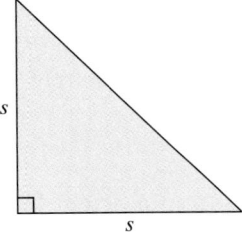

Example 5   **Applying a Radical Function**

On a certain surface, the speed $s(x)$ (in miles per hour) of a car before the brakes were applied can be approximated from the length of its skid marks x (in feet) by

$$s(x) = 3.8\sqrt{x} \qquad x \geq 0$$

See Figure 11-4.

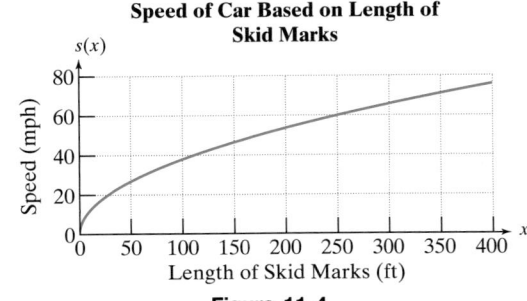

Speed of Car Based on Length of Skid Marks

Figure 11-4

a. Find the speed of a car before the brakes were applied if its skid marks are 361 ft long.

b. How long would you expect the skid marks to be if the car had been traveling the speed limit of 50 mph? (Round to the nearest foot.)

Skill Practice Answers
8. 40.5 in.²

Solution:

a. $s(x) = 3.8\sqrt{x}$

$\quad\quad s(361) = 3.8\sqrt{361}$ Substitute $x = 361$.

$\quad\quad\quad\quad\quad = 3.8(19)$

$\quad\quad\quad\quad\quad = 72.2$

If the skid marks are 361 ft, the car was traveling approximately 72.2 mph before the brakes were applied.

b. $s(x) = 3.8\sqrt{x}$

$\quad\quad 50 = 3.8\sqrt{x}$ Substitute $s(x) = 50$ and solve for x.

$\quad\quad \dfrac{50}{3.8} = \sqrt{x}$ Isolate the radical.

$\quad\quad \left(\dfrac{50}{3.8}\right)^2 = x$

$\quad\quad\quad x \approx 173$

If the car had been going the speed limit (50 mph), then the length of the skid marks would have been approximately 173 ft.

[**Skill Practice**]

9. When an object is dropped from a height of 64 ft, the time $t(h)$ in seconds it takes to reach a height h in feet is given by

$$t(h) = \frac{1}{4}\sqrt{64 - h}$$

a. Find the time to reach a height of 28 ft from the ground.

b. What is the height after 1 sec?

Skill Practice Answers

9a. 1.5 sec   **b.** 48 ft

| Section 11.7 | **Practice Exercises**

Boost *your* GRADE at mathzone.com!

MathZone

• Practice Problems
• Self-Tests
• NetTutor

• e-Professors
• Videos

Study Skills Exercise

1. Define the key terms.

    a. radical equation   **b. extraneous solution**

Review Exercises

2. Identify the equation as linear or quadratic. Then solve the equation.

    a. $2x + 3 = 23$ **b.** $2x^2 - 9x = 5$

For Exercises 3–6, simplify the radical expressions, if possible. Assume all variables represent positive real numbers.

3. $\sqrt{\dfrac{9w^3}{16}}$ **4.** $\sqrt{\dfrac{a^2}{3}}$ **5.** $\sqrt[3]{54c^4}$ **6.** $\sqrt{\dfrac{49}{5t^3}}$

For Exercises 7–10, simplify each expression. Assume all radicands represent positive real numbers.

7. $(\sqrt{4x-6})^2$ **8.** $(\sqrt{5y+2})^2$ **9.** $(\sqrt[3]{9p+7})^3$ **10.** $(\sqrt[3]{4t+13})^3$

Concept 2: Solving Radical Equations Involving One Radical

For Exercises 11–26, solve the equations.

11. $\sqrt{4x}=6$ **12.** $\sqrt{2x}=8$ **13.** $\left(\sqrt{5y+1}=4\right.$ **14.** $\sqrt{9z-5}=11$

15. $(2z-3)^{1/2}=9$ **16.** $(8+3a)^{1/2}=5$ **17.** $\sqrt[3]{x-2}-3=0$ **18.** $\sqrt[3]{2x-5}-2=0$

19. $(15-w)^{1/3}=-5$ **20.** $(k+18)^{1/3}=-2$ **21.** $3+\sqrt{x-16}=0$ **22.** $12+\sqrt{2x+1}=0$

23. $2\sqrt{6a+7}-2a=0$ **24.** $2\sqrt{3-w}-w=0$ **25.** $\sqrt[4]{2x-5}=-1$ **26.** $\sqrt[4]{x+16}=-4$

For Exercises 27–30, assume all variables represent positive real numbers.

27. Solve for V: $r=\sqrt[3]{\dfrac{3V}{4\pi}}$ **28.** Solve for V: $r=\sqrt{\dfrac{V}{h\pi}}$

29. Solve for h^2: $r=\pi\sqrt{r^2+h^2}$ **30.** Solve for d: $s=1.3\sqrt{d}$

For Exercises 31–36, square the expression as indicated.

31. $(a+5)^2$ **32.** $(b+7)^2$ **33.** $(\sqrt{5a}-3)^2$

34. $(2+\sqrt{b})^2$ **35.** $(\sqrt{r-3}+5)^2$ **36.** $(2-\sqrt{2t-4})^2$

For Exercises 37–42, solve the radical equations, if possible.

37. $\sqrt{a^2+2a+1}=a+5$ **38.** $\sqrt{b^2-5b-8}=b+7$ **39.** $\sqrt{25w^2-2w-3}=5w-4$

40. $\sqrt{4p^2-2p+1}=2p-3$ **41.** $\sqrt{5y+1}+2=y+3$ **42.** $\sqrt{2x-2}+3=x+2$

Concept 3: Solving Radical Equations Involving More than One Radical

For Exercises 43–64, solve the radical equations, if possible.

43. $\sqrt[4]{h+4}=\sqrt[4]{2h-5}$ **44.** $\sqrt[4]{3b+6}=\sqrt[4]{7b-6}$ **45.** $\sqrt[3]{5a+3}-\sqrt[3]{a-13}=0$

46. $\sqrt[3]{k-8}-\sqrt[3]{4k+1}=0$ **47.** $\sqrt{5a-9}=\sqrt{5a}-3$ **48.** $\sqrt{8+b}=2+\sqrt{b}$

49. $\sqrt{2h+5}-\sqrt{2h}=1$ **50.** $\sqrt{3k-5}-\sqrt{3k}=-1$ **51.** $\sqrt{t-9}-\sqrt{t}=3$

52. $\sqrt{y-16}-\sqrt{y}=4$ **53.** $6=\sqrt{x^2+3}-x$ **54.** $2=\sqrt{y^2+5}-y$

55. $\sqrt{3t - 7} = 2 - \sqrt{3t + 1}$ **56.** $\sqrt{p - 6} = \sqrt{p + 2} - 4$ ● **57.** $\sqrt{z + 1} + \sqrt{2z + 3} = 1$

58. $\sqrt{2y + 6} = \sqrt{7 - 2y} + 1$ **59.** $\sqrt{6m + 7} - \sqrt{3m + 3} = 1$ **60.** $\sqrt{5w + 1} - \sqrt{3w} = 1$

61. $2 + 2\sqrt{2t + 3} + 2\sqrt{3t - 5} = 0$ **62.** $6 + 3\sqrt{3x + 1} + 3\sqrt{x - 1} = 0$ **63.** $4\sqrt{y} + 6 = 13$

64. $\sqrt{5x - 8} = 2\sqrt{x - 1}$

Concept 4: Applications of Radical Equations and Functions

65. If an object is dropped from an initial height h, its velocity at impact with the ground is given by

$$v = \sqrt{2gh}$$

where g is the acceleration due to gravity and h is the initial height.

a. Find the initial height (in feet) of an object if its velocity at impact is 44 ft/sec. (Assume that the acceleration due to gravity is $g = 32$ ft/sec^2.)

b. Find the initial height (in meters) of an object if its velocity at impact is 26 m/sec. (Assume that the acceleration due to gravity is $g = 9.8$ m/sec^2.) Round to the nearest tenth of a meter.

66. The time T (in seconds) required for a pendulum to make one complete swing back and forth is approximated by

$$T = 2\pi\sqrt{\frac{L}{g}}$$

where g is the acceleration due to gravity and L is the length of the pendulum (in feet).

a. Find the length of a pendulum that requires 1.36 sec to make one complete swing back and forth. (Assume that the acceleration due to gravity is $g = 32$ ft/sec^2.) Round to the nearest tenth of a foot.

b. Find the time required for a pendulum to complete one swing back and forth if the length of the pendulum is 4 ft. (Assume that the acceleration due to gravity is $g = 32$ ft/sec^2.) Round to the nearest tenth of a second.

67. The time $t(d)$ in seconds it takes an object to drop d meters is given by

$$t(d) = \sqrt{\frac{d}{4.9}}$$

a. Approximate the height of the Texas Commerce Tower in Houston if it takes an object 7.89 sec to drop from the top. Round to the nearest meter.

b. Approximate the height of the Shanghai World Financial Center if it takes an object 9.69 sec to drop from the top. Round to the nearest meter.

68. The airline cost for x thousand passengers to travel round trip from New York to Atlanta is given by

$$C(x) = \sqrt{0.3x + 1}$$

where $C(x)$ is measured in millions of dollars and $x \geq 0$.

a. Find the airline's cost for 10,000 passengers ($x = 10$) to travel from New York to Atlanta.

b. If the airline charges $320 per passenger, find the profit made by the airline for flying 10,000 passengers from New York to Atlanta.

c. Approximate the number of passengers who traveled from New York to Atlanta if the total cost for the airline was $4 million.

Expanding Your Skills

69. The number of hours needed to cook a turkey that weighs x lb can be approximated by

$$t(x) = 0.90\sqrt[5]{x^3}$$

where $t(x)$ is the time in hours and x is the weight of the turkey in pounds.

a. Find the weight of a turkey that cooked for 4 hr. Round to the nearest pound.

b. Find $t(18)$ and interpret the result. Round to the nearest tenth of an hour.

For Exercises 70–73, use the Pythagorean theorem to find a, b, or c.

$$a^2 + b^2 = c^2$$

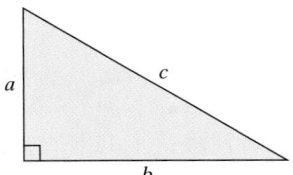

70. Find b when $a = 2$ and $c = y$.

71. Find b when $a = h$ and $c = 5$.

72. Find a when $b = x$ and $c = 8$.

73. Find a when $b = 14$ and $c = k$.

Graphing Calculator Exercises

74. Refer to Exercise 12. Graph Y_1 and Y_2 on a viewing window defined by $-10 \leq x \leq 40$ and $-5 \leq y \leq 10$.

$$Y_1 = \sqrt{2x} \qquad \text{and} \qquad Y_2 = 8$$

Use an *Intersect* feature to approximate the x-coordinate of the point of intersection of the two graphs to support your solution to Exercise 12.

75. Refer to Exercise 11. Graph Y_1 and Y_2 on a viewing window defined by $-10 \leq x \leq 20$ and $-5 \leq y \leq 10$.

$$Y_1 = \sqrt{4x} \qquad \text{and} \qquad Y_2 = 6$$

Use an *Intersect* feature to approximate the x-coordinate of the point of intersection of the two graphs to support your solution to Exercise 11.

76. Refer to Exercise 44. Graph Y_1 and Y_2 on a viewing window defined by $-5 \le x \le 20$ and $-1 \le y \le 4$.

$$Y_1 = \sqrt[4]{3x + 6} \qquad \text{and} \qquad Y_2 = \sqrt[4]{7x - 6}$$

Use an *Intersect* feature to approximate the *x*-coordinate of the point of intersection of the two graphs to support your solution to Exercise 44.

77. Refer to Exercise 43. Graph Y_1 and Y_2 on a viewing window defined by $-5 \le x \le 20$ and $-1 \le y \le 4$.

$$Y_1 = \sqrt[4]{x + 4} \qquad \text{and} \qquad Y_2 = \sqrt[4]{2x - 5}$$

Use an *Intersect* feature to approximate the *x*-coordinate of the point of intersection of the two graphs to support your solution to Exercise 43.

Section 11.8 — Complex Numbers

Concepts

1. Definition of *i*
2. Powers of *i*
3. Definition of a Complex Number
4. Addition, Subtraction, and Multiplication of Complex Numbers
5. Division and Simplification of Complex Numbers

1. Definition of *i*

In Section 11.1, we learned that there are no real-valued square roots of a negative number. For example, $\sqrt{-9}$ is not a real number because no real number when squared equals -9. However, the square roots of a negative number are defined over another set of numbers called the **imaginary numbers**. The foundation of the set of imaginary numbers is the definition of the imaginary number *i* as $i = \sqrt{-1}$.

Definition of *i*

$$i = \sqrt{-1}$$

Note: From the definition of *i*, it follows that $i^2 = -1$.

Using the imaginary number *i*, we can define the square root of any negative real number.

Definition of $\sqrt{-b}$ for $b > 0$

Let *b* be a real number such that $b > 0$. Then $\sqrt{-b} = i\sqrt{b}$.

| **Example 1** | **Simplifying Expressions in Terms of i** |

Simplify the expressions in terms of i.

a. $\sqrt{-64}$ **b.** $\sqrt{-50}$ **c.** $-\sqrt{-4}$ **d.** $\sqrt{-29}$

Solution:

a. $\sqrt{-64} = i\sqrt{64}$

$\qquad = 8i$

b. $\sqrt{-50} = i\sqrt{50}$

$\qquad = i\sqrt{5^2 \cdot 2}$

$\qquad = 5i\sqrt{2}$

c. $-\sqrt{-4} = -1 \cdot \sqrt{-4}$

$\qquad = -1 \cdot 2i$

$\qquad = -2i$

d. $\sqrt{-29} = i\sqrt{29}$

| **Skill Practice** | Simplify the expressions in terms of i.

1. $\sqrt{-81}$ **2.** $\sqrt{-20}$ **3.** $\sqrt{-7}$ **4.** $-\sqrt{-36}$

Avoiding Mistakes:

In an expression such as $i\sqrt{29}$, the i is usually written in front of the square root. The expression $\sqrt{29}\,i$ is also correct, but may be misinterpreted as $\sqrt{29i}$ (with i incorrectly placed under the radical).

The multiplication and division properties of radicals were presented in Sections 11.3 and 11.5 as follows:

If a and b represent real numbers such that $\sqrt[n]{a}$ and $\sqrt[n]{b}$ are both real, then

$$\sqrt[n]{ab} = \sqrt[n]{a} \cdot \sqrt[n]{b} \quad \text{and} \quad \sqrt[n]{\frac{a}{b}} = \frac{\sqrt[n]{a}}{\sqrt[n]{b}} \quad b \neq 0$$

The conditions that $\sqrt[n]{a}$ and $\sqrt[n]{b}$ must both be real numbers prevent us from applying the multiplication and division properties of radicals for square roots with a negative radicand. Therefore, to multiply or divide radicals with a negative radicand, write the radical in terms of the imaginary number i first. This is demonstrated in Example 2.

| **Example 2** | **Simplifying a Product or Quotient in Terms of i** |

Simplify the expressions.

a. $\dfrac{\sqrt{-100}}{\sqrt{-25}}$ **b.** $\sqrt{-25} \cdot \sqrt{-9}$ **c.** $\sqrt{-5} \cdot \sqrt{-5}$

Solution:

a. $\dfrac{\sqrt{-100}}{\sqrt{-25}}$

$= \dfrac{10i}{5i}$ Simplify each radical in terms of i *before* dividing.

$= 2$ Simplify.

Skill Practice Answers

1. $9i$ **2.** $2i\sqrt{5}$ **3.** $i\sqrt{7}$
4. $-6i$

b. $\sqrt{-25} \cdot \sqrt{-9}$

$= 5i \cdot 3i$ Simplify each radical in terms of i first *before* multiplying.

$= 15i^2$ Multiply.

$= 15(-1)$ Recall that $i^2 = -1$.

$= -15$ Simplify.

c. $\sqrt{-5} \cdot \sqrt{-5}$

$= i\sqrt{5} \cdot i\sqrt{5}$

$= i^2 \cdot (\sqrt{5})^2$

$= -1 \cdot 5$

$= -5$

Skill Practice   Simplify the expressions.

5. $\dfrac{\sqrt{-36}}{\sqrt{-9}}$ **6.** $\sqrt{-16} \cdot \sqrt{-49}$ **7.** $\sqrt{-2} \cdot \sqrt{-2}$

Avoiding Mistakes:

In Example 2, we wrote the radical expressions in terms of *i* first, before multiplying or dividing. If we had mistakenly applied the multiplication or division property first, we would have obtained an incorrect answer.

Correct: $\sqrt{-25} \cdot \sqrt{-9}$

$= (5i)(3i) = 15i^2$

$= 15(-1) = -15$
       ↑ correct

Be careful: $\sqrt{-25} \cdot \sqrt{-9}$ $\sqrt{-25}$ and $\sqrt{-9}$ are not real numbers. Therefore, the multiplication property of radicals cannot be applied.

$\neq \sqrt{225} = 15$
↑ (incorrect answer)

---

2. Powers of *i*

From the definition of $i = \sqrt{-1}$, it follows that

$i = i$

$i^2 = -1$

$i^3 = -i$ because $i^3 = i^2 \cdot i = (-1)i = -i$

$i^4 = 1$ because $i^4 = i^2 \cdot i^2 = (-1)(-1) = 1$

$i^5 = i$ because $i^5 = i^4 \cdot i = (1)i = i$

$i^6 = -1$ because $i^6 = i^4 \cdot i^2 = (1)(-1) = -1$

This pattern of values $i, -1, -i, 1, i, -1, -i, 1, \ldots$ continues for all subsequent powers of i. Here is a list of several powers of i.

Skill Practice Answers

5. 2   **6.** -28 **7.** -2

Powers of _i_

$$i^1 = i \qquad i^5 = i \qquad i^9 = i$$
$$i^2 = -1 \qquad i^6 = -1 \qquad i^{10} = -1$$
$$i^3 = -i \qquad i^7 = -i \qquad i^{11} = -i$$
$$i^4 = 1 \qquad i^8 = 1 \qquad i^{12} = 1$$

To simplify higher powers of _i_, we can decompose the expression into multiples of i^4 ($i^4 = 1$) and write the remaining factors as i, i^2, or i^3.

Example 3 **Simplifying Powers of _i_**

Simplify the powers of _i_.

a. i^{13} **b.** i^{18} **c.** i^{107} **d.** i^{32}

Solution:

a. $i^{13} = (i^{12}) \cdot (i)$
$$= (i^4)^3 \cdot (i)$$
$$= (1)^3(i) \qquad \text{Recall that } i^4 = 1.$$
$$= i \qquad \text{Simplify.}$$

b. $i^{18} = (i^{16}) \cdot (i^2)$
$$= (i^4)^4 \cdot (i^2)$$
$$= (1)^4 \cdot (-1) \qquad i^4 = 1 \text{ and } i^2 = -1$$
$$= -1 \qquad \text{Simplify.}$$

c. $i^{107} = (i^{104}) \cdot (i^3)$
$$= (i^4)^{26}(i^3)$$
$$= (1)^{26}(-i) \qquad i^4 = 1 \text{ and } i^3 = -i$$
$$= -i \qquad \text{Simplify.}$$

d. $i^{32} = (i^4)^8$
$$= (1)^8 \qquad i^4 = 1$$
$$= 1 \qquad \text{Simplify.}$$

Skill Practice Simplify the powers of _i_.

8. i^8 **9.** i^{22} **10.** i^{45} **11.** i^{31}

3. Definition of a Complex Number

We have already learned the definitions of the integers, rational numbers, irrational numbers, and real numbers. In this section, we define the complex numbers.

Definition of a Complex Number

A **complex number** is a number of the form $a + bi$, where a and b are real numbers and $i = \sqrt{-1}$.

Notes:

- If $b = 0$, then the complex number $a + bi$ is a real number.
- If $b \neq 0$, then we say that $a + bi$ is an imaginary number.
- The complex number $a + bi$ is said to be written in standard form. The quantities a and b are called the real and imaginary parts (respectively) of the complex number.
- The complex numbers $a - bi$ and $a + bi$ are called **conjugates**.

From the definition of a complex number, it follows that all real numbers are complex numbers and all imaginary numbers are complex numbers. Figure 11-5 illustrates the relationship among the sets of numbers we have learned so far.

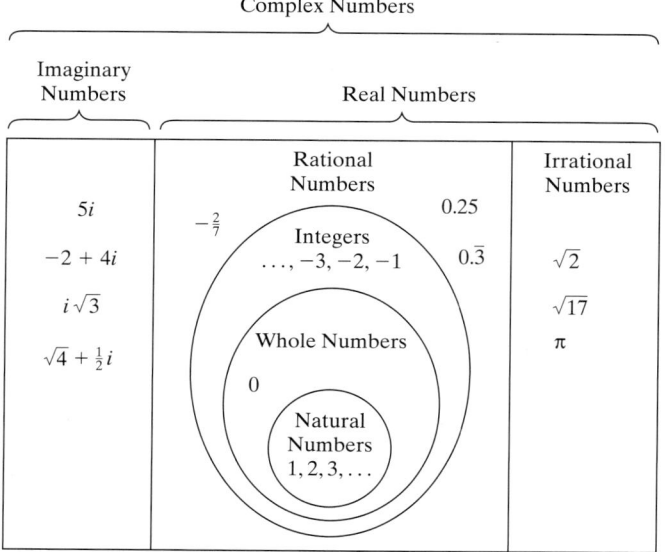

Figure 11-5

Example 4   **Identifying the Real and Imaginary Parts of a Complex Number**

Identify the real and imaginary parts of the complex numbers.

a. $-8 + 2i$ **b.** $\dfrac{3}{2}$ **c.** $-1.75i$

Solution:

a. $-8 + 2i$ -8 is the real part, and 2 is the imaginary part.

b. $\dfrac{3}{2} = \dfrac{3}{2} + 0i$ Rewrite $\frac{3}{2}$ in the form $a + bi$.
$\frac{3}{2}$ is the real part, and 0 is the imaginary part.

c. $-1.75i$

     $= 0 + -1.75i$ Rewrite $-1.75i$ in the form $a + bi$.
0 is the real part, and -1.75 is the imaginary part.

Skill Practice Identify the real and imaginary parts of the complex numbers.

12. $22 - 14i$ **13.** -50 **14.** $15i$

TIP: Example 4(b) illustrates that a real number is also a complex number.

$$\frac{3}{2} = \frac{3}{2} + 0i$$

Example 4(c) illustrates that an imaginary number is also a complex number.

$$-1.75i = 0 + -1.75i$$

4. Addition, Subtraction, and Multiplication of Complex Numbers

The operations of addition, subtraction, and multiplication of real numbers also apply to imaginary numbers. To add or subtract complex numbers, combine the real parts and combine the imaginary parts. The commutative, associative, and distributive properties that apply to real numbers also apply to complex numbers.

| **Example 5** | **Adding, Subtracting, and Multiplying Complex Numbers** |

a. Add: $(1 - 5i) + (-3 + 7i)$

b. Subtract: $\left(-\frac{1}{4} + \frac{3}{5}i\right) - \left(\frac{1}{2} - \frac{1}{10}i\right)$

c. Multiply: $(10 - 5i)(2 + 3i)$

d. Multiply: $(1.2 + 0.5i)(1.2 - 0.5i)$

Solution:

real parts

a. $(1 - 5i) + (-3 + 7i) = (1 + -3) + (-5 + 7)i$ Add real parts. Add imaginary parts.

imaginary parts

$= -2 + 2i$ Simplify.

b. $\left(-\dfrac{1}{4} + \dfrac{3}{5}i\right) - \left(\dfrac{1}{2} - \dfrac{1}{10}i\right) = -\dfrac{1}{4} + \dfrac{3}{5}i - \dfrac{1}{2} + \dfrac{1}{10}i$ Apply the distributive property.

$= \left(-\dfrac{1}{4} - \dfrac{1}{2}\right) + \left(\dfrac{3}{5} + \dfrac{1}{10}\right)i$ Add real parts. Add imaginary parts.

$= \left(-\dfrac{1}{4} - \dfrac{2}{4}\right) + \left(\dfrac{6}{10} + \dfrac{1}{10}\right)i$ Get common denominators.

$= -\dfrac{3}{4} + \dfrac{7}{10}i$ Simplify.

c. $(10 - 5i)(2 + 3i)$

$$= (10)(2) + (10)(3i) + (-5i)(2) + (-5i)(3i)$$ Apply the distributive property.

$$= 20 + 30i - 10i - 15i^2$$

$$= 20 + 20i - (15)(-1)$$ Recall $i^2 = -1$.

$$= 20 + 20i + 15$$

$$= 35 + 20i$$ Write in the form $a + bi$.

d. $(1.2 + 0.5i)(1.2 - 0.5i)$

The expressions $(1.2 + 0.5i)$ and $(1.2 - 0.5i)$ are conjugates. The product is a difference of squares.

$$(a + b)(a - b) = a^2 - b^2$$

$$(1.2 + 0.5i)(1.2 - 0.5i) = (1.2)^2 - (0.5i)^2$$ Apply the formula, where $a = 1.2$ and $b = 0.5i$.

$$= 1.44 - 0.25i^2$$

$$= 1.44 - 0.25(-1)$$ Recall $i^2 = -1$.

$$= 1.44 + 0.25$$

$$= 1.69$$

> **TIP:** The complex numbers $(1.2 + 0.5i)$ and $(1.2 - 0.5i)$ can also be multiplied by using the distributive property:
>
> $$(1.2 + 0.5i)(1.2 - 0.5i) = 1.44 - 0.6i + 0.6i - 0.25i^2$$
>
> $$= 1.44 - 0.25(-1)$$
>
> $$= 1.69$$

Skill Practice   Perform the indicated operations.

15. $\left(\dfrac{1}{2} - \dfrac{1}{4}i\right) + \left(\dfrac{3}{5} + \dfrac{2}{3}i\right)$ **16.** $(-6 + 11i) - (-9 - 12i)$

17. $(4 - 6i)(2 - 3i)$ **18.** $(1.5 + 0.8i)(1.5 - 0.8i)$

5. Division and Simplification of Complex Numbers

The product of a complex number and its conjugate is a real number. For example:

$$(5 + 3i)(5 - 3i) = 25 - 9i^2$$

$$= 25 - 9(-1)$$

$$= 25 + 9$$

$$= 34$$

To divide by a complex number, multiply the numerator and denominator by the conjugate of the denominator. This produces a real number in the denominator so that the resulting expression can be written in the form $a + bi$.

Skill Practice Answers

15. $\dfrac{11}{10} + \dfrac{5}{12}i$ **16.** $3 + 23i$

17. $-10 - 24i$ **18.** 2.89

> **Example 6** Dividing by a Complex Number

Divide the complex numbers and write the answer in the form $a + bi$.

$$\frac{4 - 3i}{5 + 2i}$$

Solution:

$\dfrac{4 - 3i}{5 + 2i}$ Multiply the numerator and denominator by the conjugate of the denominator:

$$\frac{(4 - 3i)}{(5 + 2i)} \cdot \frac{(5 - 2i)}{(5 - 2i)} = \frac{(4)(5) + (4)(-2i) + (-3i)(5) + (-3i)(-2i)}{(5)^2 - (2i)^2}$$

$$= \frac{20 - 8i - 15i + 6i^2}{25 - 4i^2}$$ Simplify numerator and denominator.

$$= \frac{20 - 23i + 6(-1)}{25 - 4(-1)}$$ Recall $i^2 = -1$.

$$= \frac{20 - 23i - 6}{25 + 4}$$

$$= \frac{14 - 23i}{29}$$ Simplify.

$$= \frac{14}{29} - \frac{23i}{29}$$ Write in the form $a + bi$.

> **Skill Practice** Divide the complex numbers. Write the answer in the form $a + bi$.

19. $\dfrac{2 + i}{3 - 2i}$

> **Example 7** Simplifying Complex Numbers

Simplify the complex numbers.

a. $\dfrac{6 + \sqrt{-18}}{9}$ **b.** $\dfrac{4 - \sqrt{-36}}{2}$

Solution:

a. $\dfrac{6 + \sqrt{-18}}{9} = \dfrac{6 + i\sqrt{18}}{9}$ Write the radical in terms of i.

$$= \frac{6 + 3i\sqrt{2}}{9}$$ Simplify $\sqrt{18} = 3\sqrt{2}$.

$$= \frac{3(2 + i\sqrt{2})}{9}$$ Factor the numerator.

$$= \frac{\overset{1}{3}(2 + i\sqrt{2})}{\underset{3}{9}}$$ Simplify.

$$= \frac{2 + i\sqrt{2}}{3} \text{ or } \frac{2}{3} + \frac{\sqrt{2}}{3}i$$

Skill Practice Answers

19. $\dfrac{4}{13} + \dfrac{7}{13}i$

b. $\dfrac{4 - \sqrt{-36}}{2} = \dfrac{4 - i\sqrt{36}}{2}$ Write the radical in terms of i.

$= \dfrac{4 - 6i}{2}$ Simplify $\sqrt{36} = 6$.

$= \dfrac{2(2 - 3i)}{2}$ Factor the numerator.

$= \dfrac{\overset{1}{2}(2 - 3i)}{\underset{1}{2}}$ Simplify.

$= 2 - 3i$

Skill Practice Answers

20. $\dfrac{4 - i\sqrt{6}}{3}$ or $\dfrac{4}{3} - \dfrac{\sqrt{6}}{3}i$

21. $-3 + 2i$

Skill Practice   Simplify the complex numbers.

20. $\dfrac{8 - \sqrt{-24}}{6}$ **21.** $\dfrac{-12 + \sqrt{-64}}{4}$

Section 11.8   Practice Exercises

Study Skills Exercises

1. Compare the process for dividing two imaginary numbers such as $\dfrac{1 + 2i}{3 - i}$ with the process of rationalizing the radical expression $\dfrac{1 + \sqrt{2}}{\sqrt{3} - 1}$. What do the processes have in common?

2. Define the key terms.

   a. imaginary numbers   **b.** i **c. complex number**   **d. conjugate**

Review Exercises

For Exercises 3–6, perform the indicated operations.

3. $-2\sqrt{5} - 3\sqrt{50} + \sqrt{125}$ **4.** $\sqrt[3]{2x}(\sqrt[3]{2x} - \sqrt[3]{4x^2})$ **5.** $(3 - \sqrt{x})(3 + \sqrt{x})$

6. $(\sqrt{5} + \sqrt{2})^2$

For Exercises 7–10, solve the equations.

7. $\sqrt[3]{3p + 7} - \sqrt[3]{2p - 1} = 0$ **8.** $\sqrt[3]{t - 5} - \sqrt[3]{2t + 1} = 0$ **9.** $\sqrt{36c + 15} = 6\sqrt{c} + 1$

10. $\sqrt{4a + 29} = 2\sqrt{a} + 5$

Concept 1: Definition of i

11. Simplify the expressions $\sqrt{-1}$ and $-\sqrt{1}$.   **12.** Simplify i^2.

For Exercises 13–34, simplify the expressions.

13. $\sqrt{-144}$ **14.** $\sqrt{-81}$ **15.** $\sqrt{-3}$ **16.** $\sqrt{-17}$

17. $\sqrt{-20}$ **18.** $\sqrt{-75}$ **19.** $2\sqrt{-25} \cdot 3\sqrt{-4}$ **20.** $(-4\sqrt{-9})(-3\sqrt{-1})$

21. $3\sqrt{-18} + 5\sqrt{-32}$ **22.** $5\sqrt{-45} + 3\sqrt{-80}$ **23.** $7\sqrt{-63} - 4\sqrt{-28}$ **24.** $7\sqrt{-3} - 4\sqrt{-27}$

25. $\sqrt{-7} \cdot \sqrt{-7}$ **26.** $\sqrt{-11} \cdot \sqrt{-11}$ **27.** $\sqrt{-9} \cdot \sqrt{-16}$ **28.** $\sqrt{-25} \cdot \sqrt{-36}$

29. $\sqrt{-15} \cdot \sqrt{-6}$ **30.** $\sqrt{-12} \cdot \sqrt{-50}$ **31.** $\dfrac{\sqrt{-50}}{\sqrt{25}}$ **32.** $\dfrac{\sqrt{-27}}{\sqrt{9}}$

33. $\dfrac{\sqrt{-90}}{\sqrt{10}}$ **34.** $\dfrac{\sqrt{-125}}{\sqrt{45}}$

Concept 2: Powers of i

For Exercises 35–46, simplify the powers of i.

35. i^7 **36.** i^{38} **37.** i^{64} **38.** i^{75}

39. i^{41} **40.** i^{25} **41.** i^{52} **42.** i^0

43. i^{23} **44.** i^{103} **45.** i^6 **46.** i^{82}

Concept 3: Definition of a Complex Number

47. What is the conjugate of a complex number $a + bi$?

48. True or false?

     a. Every real number is a complex number.

     b. Every complex number is a real number.

For Exercises 49–56, identify the real and imaginary parts of the complex number.

49. $-5 + 12i$ **50.** $22 - 16i$ **51.** $-6i$ **52.** $10i$

53. 35 **54.** -1 **55.** $\dfrac{3}{5} + i$ **56.** $-\dfrac{1}{2} - \dfrac{1}{4}i$

Concept 4: Addition, Subtraction, and Multiplication of Complex Numbers

For Exercises 57–80, perform the indicated operations. Write the answer in the form $a + bi$.

57. $(2 - i) + (5 + 7i)$ **58.** $(5 - 2i) + (3 + 4i)$ **59.** $\left(\dfrac{1}{2} + \dfrac{2}{3}i\right) - \left(\dfrac{1}{5} - \dfrac{5}{6}i\right)$

60. $\left(\dfrac{11}{10} - \dfrac{7}{5}i\right) - \left(-\dfrac{2}{5} + \dfrac{3}{5}i\right)$ **61.** $(1 + 3i) + (4 - 3i)$ **62.** $(-2 + i) + (1 - i)$

63. $(2 + 3i) - (1 - 4i) + (-2 + 3i)$ **64.** $(2 + 5i) - (7 - 2i) + (-3 + 4i)$

65. $(8i)(3i)$ **66.** $(2i)(4i)$ **67.** $6i(1 - 3i)$ **68.** $-i(3 + 4i)$

69. $(2 - 10i)(3 + 2i)$ **70.** $(4 + 7i)(1 - i)$ **71.** $(-5 + 2i)(5 + 2i)$ **72.** $(4 - 11i)(-4 - 11i)$

73. $(4 + 5i)^2$ **74.** $(3 - 2i)^2$ **75.** $(2 + i)(3 - 2i)(4 + 3i)$ **76.** $(3 - i)(3 + i)(4 - i)$

77. $(-4 - 6i)^2$ **78.** $(-3 - 5i)^2$ **79.** $\left(-\dfrac{1}{2} - \dfrac{3}{4}i\right)\left(-\dfrac{1}{2} + \dfrac{3}{4}i\right)$ **80.** $\left(-\dfrac{2}{3} + \dfrac{1}{6}i\right)\left(-\dfrac{2}{3} - \dfrac{1}{6}i\right)$

Concept 5: Division and Simplification of Complex Numbers

For Exercises 81–94, divide the complex numbers. Write the answer in the form $a + bi$.

81. $\dfrac{2}{1 + 3i}$ **82.** $\dfrac{-2}{3 + i}$ **83.** $\dfrac{-i}{4 - 3i}$ **84.** $\dfrac{3 - 3i}{1 - i}$

85. $\dfrac{5 + 2i}{5 - 2i}$ **86.** $\dfrac{7 + 3i}{4 - 2i}$ **87.** $\dfrac{3 + 7i}{-2 - 4i}$ **88.** $\dfrac{-2 + 9i}{-1 - 4i}$

89. $\dfrac{13i}{-5 - i}$ **90.** $\dfrac{15i}{-2 - i}$ **91.** $\dfrac{2 + 3i}{6i}$ (*Hint:* The denominator can be written as $0 + 6i$. Therefore, the conjugate is $0 - 6i$, or simply $-6i$.)

92. $\dfrac{4 - i}{2i}$ **93.** $\dfrac{-10 + i}{i}$ **94.** $\dfrac{-6 - i}{-i}$

For Exercises 95–100, simplify the complex numbers.

95. $\dfrac{2 + \sqrt{-16}}{8}$ **96.** $\dfrac{6 - \sqrt{-4}}{4}$ **97.** $\dfrac{-6 + \sqrt{-72}}{6}$ **98.** $\dfrac{-20 + \sqrt{-500}}{10}$

99. $\dfrac{-8 - \sqrt{-48}}{4}$ **100.** $\dfrac{-18 - \sqrt{-72}}{3}$

Chapter 11 SUMMARY

Section 11.1 Definition of an *n*th root

Key Concepts

b is an ***n*th root** of *a* if $b^n = a$.

The expression $\sqrt{a}$ represents the principal square root of *a*.

The expression $\sqrt[n]{a}$ represents the principal *n*th root of *a*.

$\sqrt[n]{a^n} = |a|$ if *n* is even.

$\sqrt[n]{a^n} = a$ if *n* is odd.

$\sqrt[n]{a}$ is not a real number if $a < 0$ and *n* is even.

$f(x) = \sqrt[n]{x}$ defines a **radical function**.

The Pythagorean Theorem

$a^2 + b^2 = c^2$

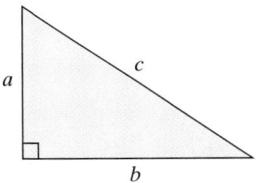

Examples

Example 1

2 is a square root of 4.

−2 is a square root of 4.

−3 is a cube root of −27.

Example 2

$\sqrt{36} = 6$ $\sqrt[3]{-64} = -4$

Example 3

$\sqrt[4]{(x + 3)^4} = |x + 3|$ $\sqrt[5]{(x + 3)^5} = x + 3$

Example 4

$\sqrt[4]{-16}$ is not a real number.

Example 5

For $g(x) = \sqrt{x}$ the domain is $[0, \infty)$.

For $h(x) = \sqrt[3]{x}$ the domain is $(-\infty, \infty)$.

Example 6

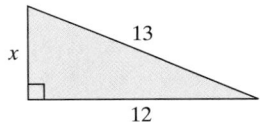

$x^2 + 12^2 = 13^2$

$x^2 + 144 = 169$

$x^2 = 25$

$x = \sqrt{25}$

$x = 5$

Section 11.2 — Rational Exponents

Key Concepts

Let a be a real number and n be an integer such that $n > 1$. If $\sqrt[n]{a}$ exists, then

$$a^{1/n} = \sqrt[n]{a}$$

$$a^{m/n} = \left(\sqrt[n]{a}\right)^m = \sqrt[n]{a^m}$$

All properties of integer exponents hold for rational exponents, provided the roots are real-valued.

Examples

Example 1

$$121^{1/2} = \sqrt{121} = 11$$

Example 2

$$27^{2/3} = \left(\sqrt[3]{27}\right)^2 = (3)^2 = 9$$

Example 3

$$p^{1/3} \cdot p^{1/4}$$
$$= p^{1/3 + 1/4}$$
$$= p^{4/12 + 3/12}$$
$$= p^{7/12}$$
$$= \sqrt[12]{p^7}$$

Section 11.3 — Simplifying Radical Expressions

Key Concepts

Let a and b represent real numbers such that $\sqrt[n]{a}$ and $\sqrt[n]{b}$ are both real. Then

$$\sqrt[n]{ab} = \sqrt[n]{a} \cdot \sqrt[n]{b} \qquad \text{Multiplication property}$$

$$\sqrt[n]{\frac{a}{b}} = \frac{\sqrt[n]{a}}{\sqrt[n]{b}} \qquad \text{Division property}$$

A radical expression whose radicand is written as a product of prime factors is in simplified form if all the following conditions are met:

1. The radicand has no factor raised to a power greater than or equal to the index.
2. The radicand does not contain a fraction.
3. No radicals are in the denominator of a fraction.

Examples

Example 1

$$\sqrt{3} \cdot \sqrt{5} = \sqrt{15}$$

Example 2

$$\sqrt{\frac{x}{9}} = \frac{\sqrt{x}}{\sqrt{9}} = \frac{\sqrt{x}}{3}$$

Example 3

$$\sqrt[3]{16x^5y^7}$$
$$= \sqrt[3]{2^4x^5y^7}$$
$$= \sqrt[3]{2^3x^3y^6 \cdot 2x^2y}$$
$$= \sqrt[3]{2^3x^3y^6} \cdot \sqrt[3]{2x^2y}$$
$$= 2xy^2\sqrt[3]{2x^2y}$$

Section 11.4   Addition and Subtraction of Radicals

Key Concepts

Like **radicals** have radical factors with the same index and the same radicand.

    Use the distributive property to add and subtract *like* radicals.

Examples

Example 1

$$3x\sqrt{7} - 5x\sqrt{7} + x\sqrt{7}$$
$$= (3 - 5 + 1) \cdot x\sqrt{7}$$
$$= -x\sqrt{7}$$

Example 2

$$x\sqrt[4]{16x} - 3\sqrt[4]{x^5}$$
$$= 2x\sqrt[4]{x} - 3x\sqrt[4]{x}$$
$$= (2 - 3)x\sqrt[4]{x}$$
$$= -x\sqrt[4]{x}$$

Section 11.5   Multiplication of Radicals

Key Concepts

The Multiplication Property of Radicals

If $\sqrt[n]{a}$ and $\sqrt[n]{b}$ are real numbers, then

$$\sqrt[n]{a} \cdot \sqrt[n]{b} = \sqrt[n]{ab}$$

    To multiply or divide radicals with different indices, convert to rational exponents and use the properties of exponents.

Examples

Example 1

$$3\sqrt{2}(\sqrt{2} + 5\sqrt{7} - \sqrt{6})$$
$$= 3\sqrt{4} + 15\sqrt{14} - 3\sqrt{12}$$
$$= 3 \cdot 2 + 15\sqrt{14} - 3 \cdot 2\sqrt{3}$$
$$= 6 + 15\sqrt{14} - 6\sqrt{3}$$

Example 2

$$\sqrt{p} \cdot \sqrt[5]{p^2}$$
$$= p^{1/2} \cdot p^{2/5}$$
$$= p^{5/10} \cdot p^{4/10}$$
$$= p^{9/10}$$
$$= \sqrt[10]{p^9}$$

Section 11.6   **Rationalization**

Key Concepts

The process of removing a radical from the denominator of an expression is called **rationalizing the denominator**.

Rationalizing a denominator with one term

Examples

Example 1

Rationalize:

$$\frac{4}{\sqrt[4]{2y^3}}$$

$$= \frac{4}{\sqrt[4]{2y^3}} \cdot \frac{\sqrt[4]{2^3 y}}{\sqrt[4]{2^3 y}}$$

$$= \frac{4\sqrt[4]{8y}}{\sqrt[4]{2^4 y^4}}$$

$$= \frac{4\sqrt[4]{8y}}{2y}$$

$$= \frac{2\sqrt[4]{8y}}{y}$$

Rationalizing a denominator with two terms involving square roots

Example 2

Rationalize the denominator:

$$\frac{\sqrt{2}}{\sqrt{x} - \sqrt{3}}$$

$$= \frac{\sqrt{2}}{(\sqrt{x} - \sqrt{3})} \cdot \frac{(\sqrt{x} + \sqrt{3})}{(\sqrt{x} + \sqrt{3})}$$

$$= \frac{\sqrt{2x} + \sqrt{6}}{x - 3}$$

Section 11.7 Radical Equations

Key Concepts

Steps to Solve a Radical Equation

1. Isolate the radical. If an equation has more than one radical, choose one of the radicals to isolate.
2. Raise each side of the equation to a power equal to the index of the radical.
3. Solve the resulting equation. If the equation still has a radical, repeat steps 1 and 2.
4. Check the potential solutions in the original equation.

Examples

Example 1

Solve:

$$\sqrt{b-5} - \sqrt{b+3} = 2$$
$$\sqrt{b-5} = \sqrt{b+3} + 2$$
$$(\sqrt{b-5})^2 = (\sqrt{b+3} + 2)^2$$
$$b - 5 = b + 3 + 4\sqrt{b+3} + 4$$
$$b - 5 = b + 7 + 4\sqrt{b+3}$$
$$-12 = 4\sqrt{b+3}$$
$$-3 = \sqrt{b+3}$$
$$(-3)^2 = (\sqrt{b+3})^2$$
$$9 = b + 3$$
$$6 = b$$

Check:

$$\sqrt{6-5} - \sqrt{6+3} \stackrel{?}{=} 2$$
$$\sqrt{1} - \sqrt{9} \stackrel{?}{=} 2$$
$$1 - 3 \neq 2 \qquad \text{Does not check.}$$

No solution.

Section 11.8 Complex Numbers

Key Concepts

$i = \sqrt{-1}$ and $i^2 = -1$

For a real number $b > 0$, $\sqrt{-b} = i\sqrt{b}$

A **complex number** is in the form $a + bi$, where a and b are real numbers. The a is called the real part, and the b is called the imaginary part.

To add or subtract complex numbers, combine the real parts and combine the imaginary parts.

Examples

Example 1

$$\sqrt{-4} \cdot \sqrt{-9}$$
$$= (2i)(3i)$$
$$= 6i^2$$
$$= -6$$

Example 2

$$(3 - 5i) - (2 + i) + (3 - 2i)$$
$$= 3 - 5i - 2 - i + 3 - 2i$$
$$= 4 - 8i$$

Multiply complex numbers by using the distributive property.

Example 3

$(1 + 6i)(2 + 4i)$

$= 2 + 4i + 12i + 24i^2$

$= 2 + 16i + 24(-1)$

$= -22 + 16i$

Divide complex numbers by multiplying the numerator and denominator by the **conjugate** of the denominator.

Example 4

$\dfrac{3}{2 - 5i}$

$= \dfrac{3}{2 - 5i} \cdot \dfrac{(2 + 5i)}{(2 + 5i)} = \dfrac{6 + 15i}{4 - 25i^2}$

$= \dfrac{6 + 15i}{29}$ or $\dfrac{6}{29} + \dfrac{15}{29}i$

Chapter 11   Review Exercises

For the exercises in this set, assume that all variables represent positive real numbers unless otherwise stated.

Section 11.1

1. True or false?

   a. The principal nth root of an even-indexed root is always positive.

   b. The principal nth root of an odd-indexed root is always positive.

2. Explain why $\sqrt{(-3)^2} \neq -3$.

3. Are the following statements true or false?

   a. $\sqrt{a^2 + b^2} = a + b$

   b. $\sqrt{(a + b)^2} = a + b$

For Exercises 4–6, simplify the radicals.

4. $\sqrt{\dfrac{50}{32}}$ **5.** $\sqrt[4]{625}$ **6.** $\sqrt{(-6)^2}$

7. Evaluate the function values for $f(x) = \sqrt{x - 1}$.

   a. $f(10)$ **b.** $f(1)$ **c.** $f(8)$

   d. Write the domain of f in interval notation.

8. Evaluate the function values for $g(t) = \sqrt{5 + t}$.

   a. $g(-5)$ **b.** $g(-4)$ **c.** $g(4)$

   d. Write the domain of g in interval notation.

9. Translate the English expression to an algebraic expression: Four more than the quotient of the cube root of $2x$ and the principal fourth root of $2x$.

For Exercises 10–11, simplify the expression. Assume that x and y represent *any* real number.

10. a. $\sqrt{x^2}$ **b.** $\sqrt[3]{x^3}$

   c. $\sqrt[4]{x^4}$ **d.** $\sqrt[5]{(x + 1)^5}$

11. a. $\sqrt{4y^2}$ **b.** $\sqrt[3]{27y^3}$

   c. $\sqrt[100]{y^{100}}$ **d.** $\sqrt[101]{y^{101}}$

12. Use the Pythagorean theorem to find the length of the third side of the triangle.

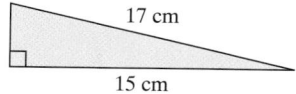

Section 11.2

13. Are the properties of exponents the same for rational exponents and integer exponents? Give an example. (Answers may vary.)

14. In the expression $x^{m/n}$ what does n represent?

15. Explain the process of eliminating a negative exponent from an algebraic expression.

For Exercises 16–20, simplify the expressions. Write the answer with positive exponents only.

16. $(-125)^{1/3}$

17. $16^{-1/4}$

18. $\left(\dfrac{1}{16}\right)^{-3/4} - \left(\dfrac{1}{8}\right)^{-2/3}$

19. $(b^{1/2} \cdot b^{1/3})^{12}$

20. $\left(\dfrac{x^{-1/4}y^{-1/3}z^{3/4}}{2^{1/3}x^{-1/3}y^{2/3}}\right)^{-12}$

For Exercises 21–22, rewrite the expressions by using rational exponents.

21. $\sqrt[4]{x^3}$

22. $\sqrt[3]{2y^2}$

For Exercises 23–25, use a calculator to approximate the expressions to 4 decimal places.

23. $10^{1/3}$ **24.** $17.8^{2/3}$ **25.** $147^{4/5}$

26. An initial investment of P dollars is made in an account in which the return is compounded quarterly. The amount in the account can be determined by

$$A = P\left(1 + \dfrac{r}{4}\right)^{t/3}$$

where r is the annual rate of return and t is the time in months.

When she is 20 years old, Jenna invests $5000 in a mutual fund that grows by an average of 11% per year compounded quarterly. How much money does she have

a. After 6 months?   **b.** After 1 year?

c. At age 40?   **d.** At age 50?

e. At age 65?

Section 11.3

27. List the criteria for a radical expression to be simplified.

For Exercises 28–31, simplify the radicals.

28. $\sqrt{108}$

29. $\sqrt[4]{x^5yz^4}$

30. $\sqrt{5x} \cdot \sqrt{20x}$

31. $\sqrt[3]{\dfrac{-16x^7y^6}{z^9}}$

32. Write an English phrase that describes the following mathematical expressions: (Answers may vary.)

a. $\sqrt{\dfrac{2}{x}}$

b. $(x + 1)^3$

33. An engineering firm made a mistake when building a $\frac{1}{4}$-mi bridge in the Florida Keys. The bridge was made without adequate expansion joints to prevent buckling during the heat of summer. During mid-June, the bridge expanded 1.5 ft, causing a vertical bulge in the middle. Calculate the height of the bulge h in feet. (*Note:* 1 mi = 5280 ft.) Round to the nearest foot.

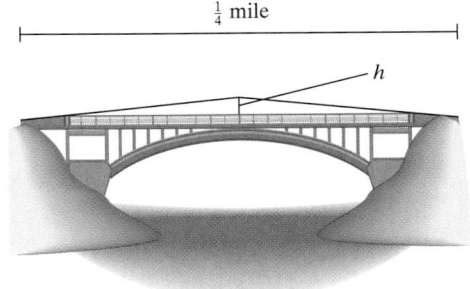

Section 11.4

34. Complete the following statement: Radicals may be added or subtracted if . . .

For Exercises 35–38, determine whether the radicals may be combined, and explain your answer.

35. $\sqrt[3]{2x} - 2\sqrt{2x}$

36. $2 + \sqrt{x}$

37. $\sqrt[4]{3xy} + 2\sqrt[4]{3xy}$

38. $-4\sqrt{32} + 7\sqrt{50}$

For Exercises 39–42, add or subtract as indicated.

39. $4\sqrt{7} - 2\sqrt{7} + 3\sqrt{7}$

40. $2\sqrt[3]{64} + 3\sqrt[3]{54} - 16$

41. $\sqrt{50} + 7\sqrt{2} - \sqrt{8}$

42. $x\sqrt[3]{16x^2} - 4\sqrt[3]{2x^5} + 5x\sqrt[3]{54x^2}$

For Exercises 43–44, answer true or false. If an answer is false, explain why. Assume all variables represent positive real numbers.

43. $5 + 3\sqrt{x} = 8\sqrt{x}$

44. $\sqrt{y} + \sqrt{y} = \sqrt{2y}$

Section 11.5

For Exercises 45–56, multiply the radicals and simplify the answer.

45. $\sqrt{3} \cdot \sqrt{12}$ **46.** $\sqrt[4]{4} \cdot \sqrt[4]{8}$

47. $-2\sqrt{3}(\sqrt{3} - 3\sqrt{3})$ **48.** $-3\sqrt{5}(2\sqrt{3} - \sqrt{5})$

49. $(2\sqrt{x} - 3)(2\sqrt{x} + 3)$ **50.** $(\sqrt{y} + 4)(\sqrt{y} - 4)$

51. $(\sqrt{7y} - \sqrt{3x})^2$ **52.** $(2\sqrt{3w} + 5)^2$

53. $(-\sqrt{z} - \sqrt{6})(2\sqrt{z} + 7\sqrt{6})$

54. $(3\sqrt{a} - \sqrt{5})(\sqrt{a} + 2\sqrt{5})$

55. $\sqrt[3]{u} \cdot \sqrt{u^5}$ **56.** $\sqrt{2} \cdot \sqrt[4]{w^3}$

Section 11.6

For Exercises 57–64, rationalize the denominator.

57. $\sqrt{\dfrac{7}{2y}}$ **58.** $\sqrt{\dfrac{5}{3w}}$

59. $\dfrac{4}{\sqrt[3]{9p^2}}$ **60.** $\dfrac{-2}{\sqrt[3]{2x}}$

61. $\dfrac{-5}{\sqrt{15} - \sqrt{10}}$ **62.** $\dfrac{-6}{\sqrt{7} - \sqrt{5}}$

63. $\dfrac{t - 3}{\sqrt{t} - \sqrt{3}}$ **64.** $\dfrac{w - 7}{\sqrt{w} - \sqrt{7}}$

65. Translate the mathematical expression to an English phrase. (Answers may vary.)
$$\frac{\sqrt{2}}{x^2}$$

Section 11.7

Solve the radical equations in Exercises 66–73, if possible.

66. $\sqrt{2y} = 7$ **67.** $\sqrt{a - 6} - 5 = 0$

68. $\sqrt[3]{2w - 3} + 5 = 2$

69. $\sqrt[4]{p + 12} - \sqrt[4]{5p - 16} = 0$

70. $\sqrt{t} + \sqrt{t - 5} = 5$

71. $\sqrt{8x + 1} = -\sqrt{x - 13}$

72. $\sqrt{2m^2 + 4} - \sqrt{9m} = 0$

73. $\sqrt{x + 2} = 1 - \sqrt{2x + 5}$

74. A tower is supported by stabilizing wires. Find the exact length of each wire, and then round to the nearest tenth of a meter.

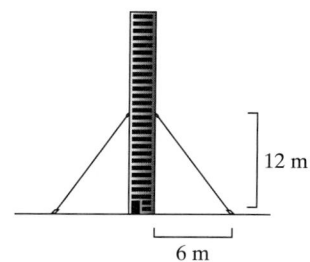

75. The velocity, $v(d)$, of an ocean wave depends on the water depth d as the wave approaches land.
$$v(d) = \sqrt{32d}$$
where $v(d)$ is in feet per second and d is in feet.

    a. Find $v(20)$ and interpret its value. Round to 1 decimal place.

    b. Find the depth of the water at a point where a wave is traveling at 16 ft/sec.

Section 11.8

76. Define a complex number.

77. Define an imaginary number.

78. Consider the following expressions.
$$\frac{3}{4 + 6i} \quad \text{and} \quad \frac{3}{\sqrt{4} + \sqrt{6}}$$
Compare the process of dividing by a complex number to the process of rationalizing the denominator.

For Exercises 79–82, rewrite the expressions in terms of i.

79. $\sqrt{-16}$

80. $-\sqrt{-5}$

81. $\sqrt{-75} \cdot \sqrt{-3}$

82. $\dfrac{-\sqrt{-24}}{\sqrt{6}}$

For Exercises 83–86, simplify the powers of i.

83. i^{38}

84. i^{101}

85. i^{19}

86. $i^{1000} + i^{1002}$

For Exercises 87–90, perform the indicated operations. Write the final answer in the form $a + bi$.

87. $(-3 + i) - (2 - 4i)$

88. $(1 + 6i)(3 - i)$

89. $(4 - 3i)(4 + 3i)$

90. $(5 - i)^2$

For Exercises 91–92, write the expressions in the form $a + bi$, and determine the real and imaginary parts.

91. $\dfrac{17 - 4i}{-4}$

92. $\dfrac{-16 - 8i}{8}$

For Exercises 93–94, divide and simplify. Write the final answer in the form $a + bi$.

93. $\dfrac{2 - i}{3 + 2i}$

94. $\dfrac{10 + 5i}{2 - i}$

For Exercises 95–96, simplify the expression.

95. $\dfrac{-8 + \sqrt{-40}}{12}$

96. $\dfrac{6 - \sqrt{-144}}{3}$

Chapter 11   Test

1. a. What is the principal square root of 36?

b. What is the negative square root of 36?

2. Which of the following are real numbers?

a. $-\sqrt{100}$

b. $\sqrt{-100}$

c. $-\sqrt[3]{1000}$

d. $\sqrt[3]{-1000}$

3. Simplify.

a. $\sqrt[3]{y^3}$

b. $\sqrt[4]{y^4}$

For Exercises 4–11, simplify the radicals. Assume that all variables represent positive numbers.

4. $\sqrt[4]{81}$

5. $\sqrt{\dfrac{16}{9}}$

6. $\sqrt[3]{32}$

7. $\sqrt{a^4 b^3 c^5}$

8. $\sqrt{3x} \cdot \sqrt{6x^3}$

9. $\sqrt{\dfrac{32w^6}{3w}}$

10. $\sqrt[6]{7} \cdot \sqrt{y}$

11. $\dfrac{\sqrt[3]{10}}{\sqrt[4]{10}}$

12. a. Evaluate the function values $f(-8), f(-6)$, $f(-4)$, and $f(-2)$ for $f(x) = \sqrt{-2x - 4}$.

b. Write the domain of f in interval notation.

13. Use a calculator to evaluate $\dfrac{-3 - \sqrt{5}}{17}$ to 4 decimal places.

For Exercises 14–15, simplify the expressions. Assume that all variables represent positive numbers.

14. $-27^{1/3}$

15. $\dfrac{t^{-1} \cdot t^{1/2}}{t^{1/4}}$

16. Add or subtract as indicated

$$3\sqrt{5} + 4\sqrt{5} - 2\sqrt{20}$$

17. Multiply the radicals.

a. $3\sqrt{x}(\sqrt{2} - \sqrt{5})$

b. $(2\sqrt{5} - 3\sqrt{x})(4\sqrt{5} + \sqrt{x})$

18. Rationalize the denominator. Assume $x > 0$.

a. $\dfrac{-2}{\sqrt[3]{x}}$

b. $\dfrac{\sqrt{x} + 2}{3 - \sqrt{x}}$

19. Rewrite the expressions in terms of i.

   a. $\sqrt{-8}$ **b.** $2\sqrt{-16}$ **c.** $\dfrac{2 + \sqrt{-8}}{4}$

For Exercises 20–26, perform the indicated operation and simplify completely. Write the final answer in the form $a + bi$.

20. $(3 - 5i) - (2 + 6i)$ **21.** $(4 + i)(8 + 2i)$

22. $\sqrt{-16} \cdot \sqrt{-49}$ **23.** $(4 - 7i)^2$

24. $(2 - 10i)(2 + 10i)$ **25.** $\dfrac{3 - 2i}{3 - 4i}$

26. $(10 + 3i)[(-5i + 8) - (5 - 3i)]$

27. If the volume V of a sphere is known, the radius of the sphere can be computed by

$$r(V) = \sqrt[3]{\dfrac{3V}{4\pi}}.$$

   Find $r(10)$ to 2 decimal places. Interpret the meaning in the context of the problem.

28. A patio 20 ft wide has a slanted roof, as shown in the picture. Find the length of the roof if there is an 8-in. overhang. Round the answer to the nearest foot.

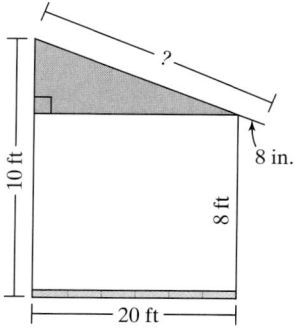

For Exercises 29–31, solve the radical equation.

29. $\sqrt[3]{2x + 5} = -3$

30. $\sqrt{5x + 8} = \sqrt{5x - 1} + 1$

31. $\sqrt{t + 7} - \sqrt{2t - 3} = 2$

Chapters 1–11 Cumulative Review Exercises

1. Simplify the expression.
$6^2 - 2[5 - 8(3 - 1) + 4 \div 2]$

2. Simplify the expression.
$3x - 3(-2x + 5) - 4y + 2(3x + 5) - y$

3. Solve the equation: $9(2y + 8) = 20 - (y + 5)$

4. Solve the inequality. Write the answer in interval notation.

$$2a - 4 < -14$$

5. Write an equation of the line that is parallel to the line $2x + y = 9$ and passes through the point $(3, -1)$. Write the answer in slope-intercept form.

6. On the same coordinate system, graph the line $2x + y = 9$ and the line that you derived in Exercise 5. Verify that these two lines are indeed parallel.

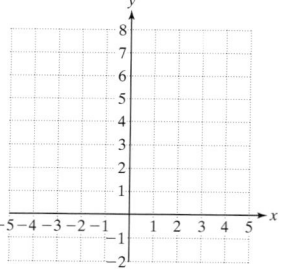

7. Solve the system of equations by using the addition method.

$$2x - 3y = 0$$
$$-4x + 3y = -1$$

8. Determine if $(2, -2, \frac{1}{2})$ is a solution to the system.

$$2x + y - 4z = 0$$

$$x - y + 2z = 5$$

$$3x + 2y + 2z = 4$$

9. Write a system of linear equations from the augmented matrix. Use x, y, and z for the variables.

$$\begin{bmatrix} 1 & 0 & 0 & | & 6 \\ 0 & 1 & 0 & | & 3 \\ 0 & 0 & 1 & | & 8 \end{bmatrix}$$

10. Given the function defined by $f(x) = 4x - 2$.

  a. Find $f(-2), f(0), f(4)$, and $f(\frac{1}{2})$.

  b. Write the ordered pairs that correspond to the function values in part (a).

  c. Graph $y = f(x)$.

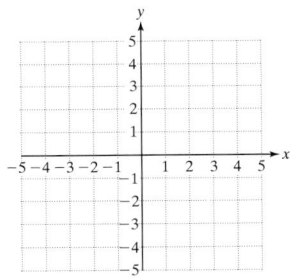

11. Determine if the graph defines y as a function of x.

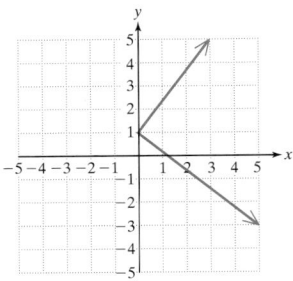

12. Simplify the expression. Write the final answer with positive exponents only.

$$\left(\frac{a^3 b^{-1} c^3}{ab^{-5}c^2} \right)^2$$

13. Simplify the expression. Write the final answer with positive exponents only.

$$\left(\frac{a^{3/2} b^{-1/4} c^{1/3}}{ab^{-5/4}c^0} \right)^{12}$$

14. Multiply or divide as indicated, and write the answer in scientific notation.

  a. $(3.5 \times 10^7)(4 \times 10^{-12})$

  b. $\dfrac{6.28 \times 10^5}{2.0 \times 10^{-4}}$

15. Multiply the polynomials $(2x + 5)(x - 3)$. What is the degree of the product?

16. Perform the indicated operations and simplify.
$$\sqrt{3}(\sqrt{5} + \sqrt{6} + \sqrt{3})$$

17. Divide $(x^2 - x - 12) \div (x + 3)$.

18. Simplify and subtract: $\sqrt[4]{\dfrac{1}{16}} - \sqrt[3]{\dfrac{8}{27}}$

19. Simplify: $\sqrt[3]{\dfrac{54c^4}{cd^3}}$

20. Add: $4\sqrt{45b^3} + 5b\sqrt{80b}$

21. Divide: $\dfrac{13i}{3 + 2i}$ Write the answer in the form $a + bi$.

22. Solve the equation.

$$\frac{5}{y - 2} - \frac{3}{y - 4} = \frac{6}{y^2 - 6y + 8}$$

23. Add: $\dfrac{3}{x^2 + 5x} + \dfrac{-2}{x^2 - 25}$

24. Divide: $\dfrac{a + 10}{2a^2 - 11a - 6} \div \dfrac{a^2 + 12a + 20}{6 - a}$

25. Perform the indicated operations.
$$(-5x^2 - 4x + 8) - (3x - 5)^2$$

26. Simplify. $\dfrac{-4}{\sqrt{3} - \sqrt{5}}$

27. Divide. $\dfrac{4}{3 - 5i}$

28. Solve. $12x^2 + 4x - 21 = 0$

29. Factor. $x^2 + 6x + 9 - y^2$

30. Factor. $x^6 + 8$

Quadratic Equations and Functions

12

12.1 Square Root Property and Completing the Square

12.2 Quadratic Formula

12.3 Equations in Quadratic Form

12.4 Graphs of Quadratic Functions

12.5 Vertex of a Parabola and Applications

In Chapter 12 we revisit quadratic equations. In earlier chapters, we solved quadratic equations by factoring and applying the zero product rule. In this chapter, we present two additional techniques to solve quadratic equations. However, these techniques can be used to solve quadratic equations even if the equations are not factorable.

Complete the word scramble to familiarize yourself with the key terms for this chapter. As a hint, there is a clue for each word.

1. r a e q s u __ __ __ __ __ __   (_____ root property)

2. e i t g c o m p l n __ __ __ __ __ __ __ __ __ __ (_____ the square)

3. d q a t r i u a c __ __ __ __ __ __ __ __ __   (_____ formula)

4. t a i i c i d s r m n n __ __ __ __ __ __ __ __ __ __ __ __ $(b^2 - 4ac)$

5. a o b r p l a a __ __ __ __ __ __ __ __   (shape of a quadratic function)

6. x t e e r v __ __ __ __ __ __   (maximum or minimum point on the graph of a parabola)

7. s x i a __ __ __ __   (_____ of symmetry)

Section 12.1	# Square Root Property and Completing the Square

1. Solving Quadratic Equations by Using the Square Root Property

In Section 6.7 we learned how to solve a quadratic equation by factoring and applying the zero product rule. For example,

$$x^2 = 81$$

$x^2 - 81 = 0$	Set one side equal to zero.
$(x - 9)(x + 9) = 0$	Factor.
$x - 9 = 0$ or $x + 9 = 0$	Set each factor equal to zero.
$x = 9$ or $x = -9$	

The solutions are $x = 9$ and $x = -9$.

It is important to note that the zero product rule can only be used if the equation is factorable. In this section and Section 12.2, we will learn how to solve quadratic equations, factorable and nonfactorable.

The first technique utilizes the **square root property**.

The Square Root Property

For any real number, k, if $x^2 = k$, then $x = \sqrt{k}$ or $x = -\sqrt{k}$.

Note: The solution may also be written as $x = \pm\sqrt{k}$, read x equals "plus or minus the square root of k."

Example 1	Solving Quadratic Equations by Using the Square Root Property

Use the square root property to solve the equations.

a. $x^2 = 81$ **b.** $3x^2 + 75 = 0$ **c.** $(w + 3)^2 = 20$

Solution:

a. $x^2 = 81$	The equation is in the form $x^2 = k$.
$x = \pm\sqrt{81}$	Apply the square root property.
$x = \pm 9$	

The solutions are $x = 9$ and $x = -9$. Notice that this is the same solution obtained by factoring and applying the zero product rule.

Avoiding Mistakes:

A common mistake is to forget the $\pm$ symbol when solving the equation $x^2 = k$:

$$x = \pm\sqrt{k}$$

b. $3x^2 + 75 = 0$	Rewrite the equation to fit the form $x^2 = k$.
$3x^2 = -75$	
$x^2 = -25$	The equation is now in the form $x^2 = k$.
$x = \pm\sqrt{-25}$	Apply the square root property.
$= \pm 5i$	

The solutions are $x = 5i$ and $x = -5i$.

Check: $x = 5i$

$3x^2 + 75 = 0$

$3(5i)^2 + 75 \stackrel{?}{=} 0$

$3(25i^2) + 75 \stackrel{?}{=} 0$

$3(-25) + 75 \stackrel{?}{=} 0$

$-75 + 75 = 0$ ✔

Check: $x = -5i$

$3x^2 + 75 = 0$

$3(-5i)^2 + 75 \stackrel{?}{=} 0$

$3(25i^2) + 75 \stackrel{?}{=} 0$

$3(-25) + 75 \stackrel{?}{=} 0$

$-75 + 75 = 0$ ✔

c. $(w + 3)^2 = 20$ The equation is in the form $x^2 = k$, where $x = (w + 3)$.

$w + 3 = \pm\sqrt{20}$ Apply the square root property.

$w + 3 = \pm\sqrt{2^2 \cdot 5}$ Simplify the radical.

$w + 3 = \pm 2\sqrt{5}$

$w = -3 \pm 2\sqrt{5}$ Solve for w.

The solutions are $w = -3 + 2\sqrt{5}$ and $w = -3 - 2\sqrt{5}$.

| **Skill Practice** | Use the square root property to solve the equations. |

1. $a^2 = 100$ **2.** $8x^2 + 72 = 0$ **3.** $(t - 5)^2 = 18$

2. Solving Quadratic Equations by Completing the Square

In Example 1(c) we used the square root property to solve an equation where the square of a binomial was equal to a constant.

$$\underbrace{(w + 3)^2}_{\text{square of a binomial}} = \overset{\uparrow}{\underset{\text{constant}}{20}}$$

The square of a binomial is the factored form of a perfect square trinomial. For example:

Perfect Square Trinomial   Factored Form

$x^2 + 10x + 25 \longrightarrow (x + 5)^2$

$t^2 - 6t + 9 \longrightarrow (t - 3)^2$

$p^2 - 14p + 49 \longrightarrow (p - 7)^2$

For a perfect square trinomial with a leading coefficient of 1, the constant term is the square of one-half the linear term coefficient. For example:

$$x^2 + 10x + 25$$
$$[\tfrac{1}{2}(10)]^2$$

In general an expression of the form $x^2 + bx + n$ is a perfect square trinomial if $n = (\tfrac{1}{2}b)^2$. The process to create a perfect square trinomial is called **completing the square**.

Skill Practice Answers

1. $a = \pm 10$ **2.** $x = \pm 3i$
3. $t = 5 \pm 3\sqrt{2}$

Example 2   Completing the Square

Determine the value of n that makes the polynomial a perfect square trinomial. Then factor the expression as the square of a binomial.

a. $x^2 + 12x + n$ **b.** $x^2 - 26x + n$

c. $x^2 + 11x + n$ **d.** $x^2 - \dfrac{4}{7}x + n$

Solution:

The expressions are in the form $x^2 + bx + n$. The value of n equals the square of one-half the linear term coefficient $\left(\frac{1}{2}b\right)^2$.

a. $x^2 + 12x + n$

$\quad x^2 + 12x + 36 \qquad n = \left[\frac{1}{2}(12)\right]^2 = (6)^2 = 36.$

$\quad (x + 6)^2 \qquad\qquad$ Factored form

b. $x^2 - 26x + n$

$\quad x^2 - 26x + 169 \qquad n = \left[\frac{1}{2}(-26)\right]^2 = (-13)^2 = 169.$

$\quad (x - 13)^2 \qquad\qquad$ Factored form

c. $x^2 + 11x + n$

$\quad x^2 + 11x + \dfrac{121}{4} \qquad n = \left[\frac{1}{2}(11)\right]^2 = \left(\frac{11}{2}\right)^2 = \frac{121}{4}.$

$\quad \left(x + \dfrac{11}{2}\right)^2 \qquad\qquad$ Factored form

d. $x^2 - \dfrac{4}{7}x + n$

$\quad x^2 - \dfrac{4}{7}x + \dfrac{4}{49} \qquad n = \left[\frac{1}{2}\left(-\frac{4}{7}\right)\right]^2 = \left(-\frac{2}{7}\right)^2 = \frac{4}{49}$

$\quad \left(x - \dfrac{2}{7}\right)^2 \qquad\qquad$ Factored form

Skill Practice   Determine the value of n that makes the polynomial a perfect square trinomial. Then factor.

4. $x^2 + 20x + n$ **5.** $y^2 - 16y + n$

6. $a^2 - 5a + n$ **7.** $w^2 + \dfrac{7}{3}w + n$

The process of completing the square can be used to write a quadratic equation $ax^2 + bx + c = 0 \ (a \neq 0)$ in the form $(x - h)^2 = k$. Then the square root property can be used to solve the equation. The following steps outline the procedure.

Skill Practice Answers

4. $n = 100; (x + 10)^2$

5. $n = 64; (y - 8)^2$

6. $n = \dfrac{25}{4}; \left(a - \dfrac{5}{2}\right)^2$

7. $n = \dfrac{49}{36}; \left(w + \dfrac{7}{6}\right)^2$

Solving a Quadratic Equation in the Form $ax^2 + bx + c = 0$ $(a \neq 0)$ by Completing the Square and Applying the Square Root Property

1. Divide both sides by a to make the leading coefficient 1.

2. Isolate the variable terms on one side of the equation.

3. Complete the square. (Add the square of one-half the linear term coefficient to both sides of the equation. Then factor the resulting perfect square trinomial.)

4. Apply the square root property and solve for x.

Example 3 **Solving Quadratic Equations by Completing the Square and Applying the Square Root Property**

Solve the quadratic equations by completing the square and applying the square root property.

a. $x^2 - 6x + 13 = 0$ **b.** $2x(2x - 10) = -30 + 6x$

Solution:

a. $x^2 - 6x + 13 = 0$

$x^2 - 6x + 13 = 0$	**Step 1:** Since the leading coefficient a is equal to 1, we do not have to divide by a. We can proceed to step 2.
$x^2 - 6x = -13$	**Step 2:** Isolate the variable terms on one side.
$x^2 - 6x + 9 = -13 + 9$	**Step 3:** To complete the square, add $[\frac{1}{2}(-6)]^2 = 9$ to both sides of the equation.
$(x - 3)^2 = -4$	Factor the perfect square trinomial.
$x - 3 = \pm\sqrt{-4}$	**Step 4:** Apply the square root property.
$x - 3 = \pm 2i$	Simplify the radical.
$x = 3 \pm 2i$	Solve for x.

The solutions are imaginary numbers and can be written as $x = 3 + 2i$ and $x = 3 - 2i$.

b.

$2x(2x - 10) = -30 + 6x$	
$4x^2 - 20x = -30 + 6x$	Clear parentheses.
$4x^2 - 26x + 30 = 0$	Write the equation in the form $ax^2 + bx + c = 0$.
$\dfrac{4x^2}{4} - \dfrac{26x}{4} + \dfrac{30}{4} = \dfrac{0}{4}$	**Step 1:** Divide both sides by the leading coefficient 4.
$x^2 - \dfrac{13}{2}x + \dfrac{15}{2} = 0$	

$$x^2 - \frac{13}{2}x = -\frac{15}{2}$$

Step 2: Isolate the variable terms on one side.

$$x^2 - \frac{13}{2}x + \frac{169}{16} = -\frac{15}{2} + \frac{169}{16}$$

Step 3: Add $[\frac{1}{2}(-\frac{13}{2})]^2 = (-\frac{13}{4})^2 = \frac{169}{16}$ to both sides.

$$\left(x - \frac{13}{4}\right)^2 = -\frac{120}{16} + \frac{169}{16}$$

Factor the perfect square trinomial. Rewrite the right-hand side with a common denominator.

$$\left(x - \frac{13}{4}\right)^2 = \frac{49}{16}$$

$$x - \frac{13}{4} = \pm\sqrt{\frac{49}{16}}$$

Step 4: Apply the square root property.

$$x - \frac{13}{4} = \pm\frac{7}{4}$$

Simplify the radical.

$$x = \frac{13}{4} + \frac{7}{4} = \frac{20}{4} = 5$$

$$x = \frac{13}{4} \pm \frac{7}{4}$$

$$x = \frac{13}{4} - \frac{7}{4} = \frac{6}{4} = \frac{3}{2}$$

The solutions are rational numbers: $x = \frac{3}{2}$ and $x = 5$. The check is left to the reader.

Skill Practice   Solve by completing the square and applying the square root property.

8. $z^2 - 4z - 2 = 0$ **9.** $2y(y - 1) = 3 - y$

TIP: In general, if the solutions to a quadratic equation are rational numbers, the equation can be solved by factoring and using the zero product rule. Consider the equation from Example 3(b).

$$2x(2x - 10) = -30 + 6x$$
$$4x^2 - 20x = -30 + 6x$$
$$4x^2 - 26x + 30 = 0$$
$$2(2x^2 - 13x + 15) = 0$$
$$2(x - 5)(2x - 3) = 0$$
$$x = 5 \quad \text{or} \quad x = \frac{3}{2}$$

Skill Practice Answers

8. $z = 2 \pm \sqrt{6}$

9. $y = \frac{3}{2}$ and $y = -1$

3. Literal Equations

| **Example 4**   Solving a Literal Equation |

Ignoring air resistance, the distance d (in meters) that an object falls in t sec is given by the equation

$$d = 4.9t^2 \qquad \text{where } t \geq 0$$

a. Solve the equation for t. Do not rationalize the denominator.

b. Using the equation from part (a), determine the amount of time required for an object to fall 500 m. Round to the nearest second.

Solution:

a. $d = 4.9t^2$

$\dfrac{d}{4.9} = t^2$ Isolate the quadratic term. The equation is in the form $t^2 = k$.

$t = \pm\sqrt{\dfrac{d}{4.9}}$ Apply the square root property.

$ = \sqrt{\dfrac{d}{4.9}}$ Because $t \geq 0$, reject the negative solution.

b. $t = \sqrt{\dfrac{d}{4.9}}$

$ = \sqrt{\dfrac{500}{4.9}}$ Substitute $d = 500$.

$t \approx 10.1$

The object will require approximately 10.1 sec to fall 500 m.

| **Skill Practice** |

10. The formula for the area of a circle is $A = \pi r^2$, where r is the radius.

    a. Solve for r. (Do not rationalize the denominator.)

    b. Use the equation from part (a) to find the radius when the area is 15.7 cm^2. (Use 3.14 for π and round to 2 decimal places.)

Skill Practice Answers

10a. $r = \sqrt{\dfrac{A}{\pi}}$

    b. The radius is $\sqrt{5} \approx 2.24$ cm.

| **Section 12.1** | **Practice Exercises** |

Boost *your* GRADE at mathzone.com!

 MathZone

- Practice Problems
- Self-Tests
- NetTutor
- e-Professors
- Videos

Study Skills Exercise

  1. Define the key terms.

     a. square root property   **b. completing the square**

Concept 1: Solving Quadratic Equations by Using the Square Root Property

For Exercises 2–17, solve the equations by using the square root property.

2. $x^2 = 100$ **3.** $y^2 = 4$ **4.** $a^2 = 5$ **5.** $k^2 - 7 = 0$

6. $3v^2 + 33 = 0$ **7.** $-2m^2 = 50$ **8.** $(p - 5)^2 = 9$ **9.** $(q + 3)^2 = 4$

10. $(3x - 2)^2 - 5 = 0$ **11.** $(2y + 3)^2 - 7 = 0$ **12.** $(h - 4)^2 = -8$ **13.** $(t + 5)^2 = -18$

14. $6p^2 - 3 = 2$ **15.** $15 = 4 + 3w^2$ **16.** $\left(x - \dfrac{3}{2}\right)^2 + \dfrac{7}{4} = 0$ **17.** $\left(m + \dfrac{4}{5}\right)^2 - \dfrac{3}{25} = 0$

18. Given the equation $x^2 = k$, match the following statements.

    a. If $k > 0$, then _____   **i.** there will be one real solution.

    b. If $k < 0$, then _____   **ii.** there will be two real solutions.

    c. If $k = 0$, then _____   **iii.** there will be two imaginary solutions.

19. State two methods that can be used to solve the equation $x^2 - 36 = 0$. Then solve the equation by using both methods.

20. State two methods that can be used to solve the equation $x^2 - 9 = 0$. Then solve the equation by using both methods.

Concept 2: Solving Quadratic Equations by Completing the Square

For Exercises 21–30, find the value of n so that the expression is a perfect square trinomial. Then factor the trinomial.

21. $x^2 - 6x + n$ **22.** $x^2 + 12x + n$ **23.** $t^2 + 8t + n$ **24.** $v^2 - 18v + n$

25. $c^2 - c + n$ **26.** $x^2 + 9x + n$ **27.** $y^2 + 5y + n$ **28.** $a^2 - 7a + n$

29. $b^2 + \dfrac{2}{5}b + n$ **30.** $m^2 - \dfrac{2}{7}m + n$

31. Summarize the steps used in solving a quadratic equation by completing the square and applying the square root property.

32. What types of quadratic equations can be solved by completing the square and applying the square root property?

For Exercises 33–52, solve the quadratic equation by completing the square and applying the square root property.

33. $t^2 + 8t + 15 = 0$ **34.** $m^2 + 6m + 8 = 0$ **35.** $x^2 + 6x = -16$ **36.** $x^2 - 4x = 3$

37. $p^2 + 4p + 6 = 0$ **38.** $q^2 + 2q + 2 = 0$ **39.** $y^2 - 3y - 10 = 0$ **40.** $-24 = -2y^2 + 2y$

41. $2a^2 + 4a + 5 = 0$ **42.** $3a^2 + 6a - 7 = 0$ **43.** $9x^2 - 36x + 40 = 0$ **44.** $9y^2 - 12y + 5 = 0$

45. $p^2 - \dfrac{2}{5}p = \dfrac{2}{25}$ **46.** $n^2 - \dfrac{2}{3}n = \dfrac{1}{9}$ **47.** $(2w + 5)(w - 1) = 2$ **48.** $(3p - 5)(p + 1) = -3$

49. $n(n - 4) = 7$ **50.** $m(m + 10) = 2$ **51.** $2x(x + 6) = 14$ **52.** $3x(x - 2) = 24$

Concept 3: Literal Equations

53. The distance (in feet) that an object falls in t sec is given by the equation $d = 16t^2$, where $t \geq 0$.

    a. Solve the equation for t.

    b. Using the equation from part (a), determine the amount of time required for an object to fall 1024 ft.

54. The volume of a can that is 4 in. tall is given by the equation $V = 4\pi r^2$, where r is the radius of the can, measured in inches.

    a. Solve the equation for r. Do not rationalize the denominator.

    b. Using the equation from part (a), determine the radius of a can with volume of 12.56 in.³ Use 3.14 for π.

For Exercises 55–60, solve for the indicated variable.

55. $A = \pi r^2$ for r $(r > 0)$

56. $E = mc^2$ for c $(c > 0)$

57. $a^2 + b^2 + c^2 = d^2$ for a $(a > 0)$

58. $a^2 + b^2 = c^2$ for b $(b > 0)$

59. $V = \dfrac{1}{3}\pi r^2 h$ for r $(r > 0)$

60. $V = \dfrac{1}{3}s^2 h$ for s

61. A corner shelf is to be made from a triangular piece of plywood, as shown in the diagram. Find the distance x that the shelf will extend along the walls. Assume that the walls are at right angles. Round the answer to a tenth of a foot.

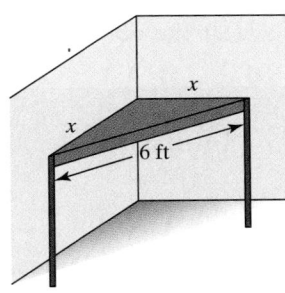

62. A square has an area of 50 in.² What are the lengths of the sides? (Round to 1 decimal place.)

63. The amount of money A in an account with an interest rate r compounded annually is given by

$$A = P(1 + r)^t$$

where P is the initial principal and t is the number of years the money is invested.

    a. If a \$10,000 investment grows to \$11,664 after 2 years, find the interest rate.

    b. If a \$6000 investment grows to \$7392.60 after 2 years, find the interest rate.

    c. Jamal wants to invest \$5000. He wants the money to grow to at least \$6500 in 2 years to cover the cost of his son's first year at college. What interest rate does Jamal need for his investment to grow to \$6500 in 2 years? Round to the nearest hundredth of a percent.

64. The volume of a box with a square bottom and a height of 4 in. is given by $V(x) = 4x^2$, where x is the length (in inches) of the sides of the bottom of the box.

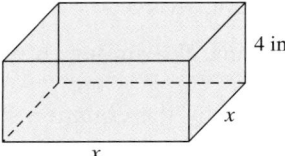

a. If the volume of the box is 289 in.^{3}, find the dimensions of the box.

b. Are there two possible answers to part (a)? Why or why not?

65. A textbook company has discovered that the profit for selling its books is given by

$$P(x) = -\frac{1}{8}x^2 + 5x$$

where x is the number of textbooks produced (in thousands) and $P(x)$ is the corresponding profit (in thousands of dollars). The graph of the function is shown at right.

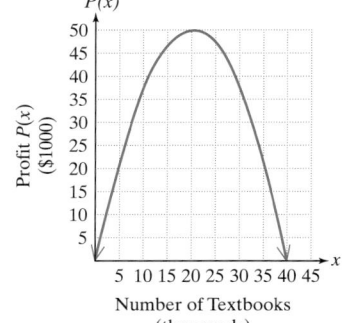

a. Approximate the number of books required to make a profit of $20,000. [*Hint:* Let $P(x) = 20$. Then complete the square to solve for x.] Round to one decimal place.

b. Why are there two answers to part (a)?

66. If we ignore air resistance, the distance (in feet) that an object travels in free fall can be approximated by $d(t) = 16t^2$, where t is the time in seconds after the object was dropped.

a. If the CN Tower in Toronto is 1815 ft high, how long will it take an object to fall from the top of the building? Round to one decimal place.

b. If the Renaissance Tower in Dallas is 886 ft high, how long will it take an object to fall from the top of the building? Round to one decimal place.

Quadratic Formula

1. Derivation of the Quadratic Formula

If we solve a general quadratic equation $ax^2 + bx + c = 0 \, (a \neq 0)$ by completing the square and using the square root property, the result is a formula that gives the solutions for x in terms of a, b, and c.

$ax^2 + bx + c = 0$ — Begin with a quadratic equation in standard form.

$\dfrac{ax^2}{a} + \dfrac{bx}{a} + \dfrac{c}{a} = \dfrac{0}{a}$ — Divide by the leading coefficient.

$x^2 + \dfrac{b}{a}x + \dfrac{c}{a} = 0$

$x^2 + \dfrac{b}{a}x = -\dfrac{c}{a}$ — Isolate the terms containing x.

$x^2 + \dfrac{b}{a}x + \left(\dfrac{1}{2} \cdot \dfrac{b}{a}\right)^2 = \left(\dfrac{1}{2} \cdot \dfrac{b}{a}\right)^2 - \dfrac{c}{a}$ — Add the square of $\frac{1}{2}$ the linear term coefficient to both sides of the equation.

$\left(x + \dfrac{b}{2a}\right)^2 = \dfrac{b^2}{4a^2} - \dfrac{c}{a}$ — Factor the left side as a perfect square.

$\left(x + \dfrac{b}{2a}\right)^2 = \dfrac{b^2 - 4ac}{4a^2}$ — Combine fractions on the right side by getting a common denominator.

$x + \dfrac{b}{2a} = \pm\sqrt{\dfrac{b^2 - 4ac}{4a^2}}$ — Apply the square root property.

$x + \dfrac{b}{2a} = \dfrac{\pm\sqrt{b^2 - 4ac}}{2a}$ — Simplify the denominator.

$x = -\dfrac{b}{2a} \pm \dfrac{\sqrt{b^2 - 4ac}}{2a}$ — Subtract $\dfrac{b}{2a}$ from both sides.

$= \dfrac{-b \pm \sqrt{b^2 - 4ac}}{2a}$ — Combine fractions.

The solution to the equation $ax^2 + bx + c = 0$ for x in terms of the coefficients a, b, and c is given by the **quadratic formula**.

The Quadratic Formula

For any quadratic equation of the form $ax^2 + bx + c = 0 \, (a \neq 0)$ the solutions are

$$x = \frac{-b \pm \sqrt{b^2 - 4ac}}{2a}$$

The quadratic formula gives us another technique to solve a quadratic equation. This method will work regardless of whether the equation is factorable or not factorable.

2. Solving Quadratic Equations by Using the Quadratic Formula

| Example 1 | Solving a Quadratic Equation by Using the Quadratic Formula |

Solve the quadratic equation by using the quadratic formula.

$$2x^2 - 3x = 5$$

Solution:

$$2x^2 - 3x = 5$$

$$2x^2 - 3x - 5 = 0$$ Write the equation in the form $ax^2 + bx + c = 0$.

$$a = 2, \qquad b = -3, \qquad c = -5$$ Identify a, b, and c.

> **TIP:** Always remember to write the *entire* numerator over 2a.

$$x = \frac{-b \pm \sqrt{b^2 - 4ac}}{2a}$$ Apply the quadratic formula.

$$x = \frac{-(-3) \pm \sqrt{(-3)^2 - 4(2)(-5)}}{2(2)}$$ Substitute $a = 2$, $b = -3$, and $c = -5$.

$$= \frac{3 \pm \sqrt{9 + 40}}{4}$$ Simplify.

$$= \frac{3 \pm \sqrt{49}}{4}$$

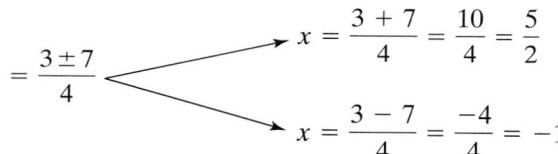

$$= \frac{3 \pm 7}{4}$$

$$x = \frac{3 + 7}{4} = \frac{10}{4} = \frac{5}{2}$$

$$x = \frac{3 - 7}{4} = \frac{-4}{4} = -1$$

There are two rational solutions, $x = \frac{5}{2}$ and $x = -1$. Both solutions check in the original equation.

| Skill Practice | Solve the equation by using the quadratic formula.

1. $6x^2 - 5x = 4$

| Example 2 | Solving a Quadratic Equation by Using the Quadratic Formula |

Solve the quadratic equation by using the quadratic formula.

$$-x(x - 6) = 11$$

Solution:

$$-x(x - 6) = 11$$

$$-x^2 + 6x - 11 = 0$$ Write the equation in the form $ax^2 + bx + c = 0$.

Skill Practice Answers

1. $x = -\frac{1}{2}; x = \frac{4}{3}$

$$-1(-x^2 + 6x - 11) = -1(0)$$

$$x^2 - 6x + 11 = 0$$

If the leading coefficient of the quadratic polynomial is negative, we suggest multiplying both sides of the equation by -1. Although this is not mandatory, it is generally easier to simplify the quadratic formula when the value of a is positive.

$a = 1, b = -6,$ and $c = 11$ Identify a, b, and c.

$$x = \frac{-b \pm \sqrt{b^2 - 4ac}}{2a}$$ Apply the quadratic formula.

$$x = \frac{-(-6) \pm \sqrt{(-6)^2 - 4(1)(11)}}{2(1)}$$ Substitute $a = 1$, $b = -6$, and $c = 11$.

$$= \frac{6 \pm \sqrt{36 - 44}}{2}$$ Simplify.

$$= \frac{6 \pm \sqrt{-8}}{2}$$

$$= \frac{6 \pm 2i\sqrt{2}}{2}$$ Simplify the radical.

$$= \frac{2(3 \pm i\sqrt{2})}{2}$$ Factor the numerator.

$$= \frac{2(3 \pm i\sqrt{2})}{2}$$ Simplify the fraction to lowest terms.

$$= 3 \pm i\sqrt{2} \qquad \begin{matrix} x = 3 + i\sqrt{2} \\ x = 3 - i\sqrt{2} \end{matrix}$$

There are two imaginary solutions, $x = 3 + i\sqrt{2}$ and $x = 3 - i\sqrt{2}$.

Skill Practice   Solve the equation by using the quadratic formula.

2. $y(y + 3) = 2$

3. Using the Quadratic Formula in Applications

Example 3   Using the Quadratic Formula in an Application

A delivery truck travels south from Hartselle, Alabama, to Birmingham, Alabama, along Interstate 65. The truck then heads east to Atlanta, Georgia, along Interstate 20. The distance from Birmingham to Atlanta is 8 mi less than twice the distance from Hartselle to Birmingham. If the straight-line distance from Hartselle to Atlanta is 165 mi, find the distance from Hartselle to Birmingham and from Birmingham to Atlanta. (Round the answers to the nearest mile.)

Skill Practice Answers

2. $y = \dfrac{-3 \pm \sqrt{17}}{2}$

Solution:

The motorist travels due south and then due east. Therefore, the three cities form the vertices of a right triangle (Figure 12-1).

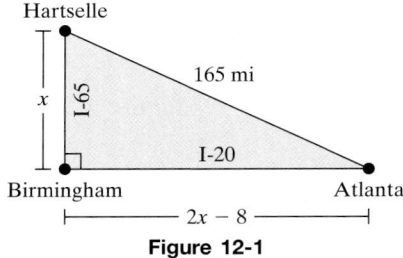

Figure 12-1

Let x represent the distance between Hartselle and Birmingham.

Then $2x - 8$ represents the distance between Birmingham and Atlanta.

Use the Pythagorean theorem to establish a relationship among the three sides of the triangle.

$$(x)^2 + (2x - 8)^2 = (165)^2$$

$$x^2 + 4x^2 - 32x + 64 = 27{,}225$$

$$5x^2 - 32x - 27{,}161 = 0 \qquad \text{Write the equation in the form } ax^2 + bx + c = 0.$$

$$a = 5 \qquad b = -32 \qquad c = -27{,}161 \qquad \text{Identify } a, b, \text{ and } c.$$

$$x = \frac{-(-32) \pm \sqrt{(-32)^2 - 4(5)(-27{,}161)}}{2(5)} \qquad \text{Apply the quadratic formula.}$$

$$x = \frac{32 \pm \sqrt{1024 + 543{,}220}}{10} \qquad \text{Simplify.}$$

$$x = \frac{32 \pm \sqrt{544{,}244}}{10}$$

$$x = \frac{32 + \sqrt{544{,}244}}{10} \approx 76.97 \text{ mi} \qquad \text{or}$$

$$x = \frac{32 - \sqrt{544{,}244}}{10} \approx -70.57 \text{ mi} \qquad \text{We reject the negative solution because distance cannot be negative.}$$

Recall that x represents the distance from Hartselle to Birmingham; therefore, to the nearest mile, the distance between Hartselle and Birmingham is 77 mi.

The distance between Birmingham and Atlanta is $2x - 8 = 2(77) - 8 = 146$ mi.

Skill Practice

3. Steve and Tammy leave a campground, hiking on two different trails. Steve heads south and Tammy heads east. By lunchtime they are 9 mi apart. Steve walked 3 mi more than twice as many miles as Tammy. Find the distance each person hiked. (Round to the nearest tenth of a mile.)

Skill Practice Answers

3. Tammy hiked 2.8 mi, and Steve hiked 8.6 mi.

| Example 4 | **Analyzing a Quadratic Function**

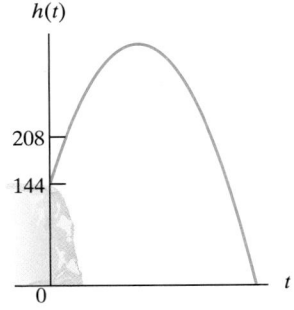

Figure 12-2

A model rocket is launched straight up from the side of a 144-ft cliff (Figure 12-2). The initial velocity is 112 ft/sec. The height of the rocket $h(t)$ is given by

$$h(t) = -16t^2 + 112t + 144$$

where $h(t)$ is measured in feet and t is the time in seconds. Find the time(s) at which the rocket is 208 ft above the ground.

Solution:

$$h(t) = -16t^2 + 112t + 144$$

$$208 = -16t^2 + 112t + 144 \qquad \text{Substitute 208 for } h(t).$$

$$16t^2 - 112t + 64 = 0 \qquad \text{Write the equation in the form } at^2 + bt + c = 0.$$

$$\frac{16t^2}{16} - \frac{112t}{16} + \frac{64}{16} = \frac{0}{16} \qquad \text{Divide by 16. This makes the coefficients smaller, and it is less cumbersome to apply the quadratic formula.}$$

$$t^2 - 7t + 4 = 0 \qquad \text{The equation is not factorable. Apply the quadratic formula.}$$

$$t = \frac{-(-7) \pm \sqrt{(-7)^2 - 4(1)(4)}}{2(1)} \qquad \text{Let } a = 1, b = -7, \text{ and } c = 4.$$

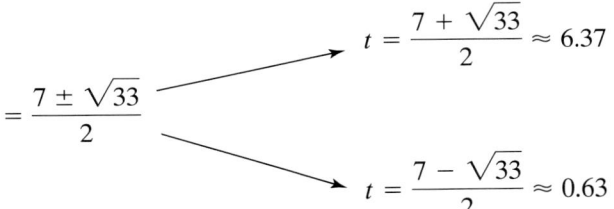

$$= \frac{7 \pm \sqrt{33}}{2}$$

$$t = \frac{7 + \sqrt{33}}{2} \approx 6.37$$

$$t = \frac{7 - \sqrt{33}}{2} \approx 0.63$$

The rocket will reach a height of 208 ft after approximately 0.63 sec (on the way up) and after 6.37 sec (on the way down).

| Skill Practice |

4. An object is launched from the top of a 96-ft building with an initial velocity of 64 ft/sec. The height $h(t)$ of the rocket is given by

$$h(t) = -16t^2 + 64t + 96$$

Find the time it takes for the object to hit the ground. (*Hint:* $h(t) = 0$ when the object hits the ground.)

4. Discriminant

The radicand within the quadratic formula is the expression $b^2 - 4ac$. This is called the **discriminant**. The discriminant can be used to determine the number of solutions to a quadratic equation as well as whether the solutions are rational, irrational, or imaginary numbers.

Skill Practice Answers

4. $2 + \sqrt{10} \approx 5.16$ sec

Using the Discriminant to Determine the Number and Type of Solutions to a Quadratic Equation

Consider the equation $ax^2 + bx + c = 0$, where a, b, and c are rational numbers and $a \neq 0$. The expression $b^2 - 4ac$ is called the *discriminant*. Furthermore,

1. If $b^2 - 4ac > 0$, then there will be two real solutions.
    a. If $b^2 - 4ac$ is a perfect square, the solutions will be rational numbers.
    b. If $b^2 - 4ac$ is not a perfect square, the solutions will be irrational numbers.
2. If $b^2 - 4ac < 0$, then there will be two imaginary solutions.
3. If $b^2 - 4ac = 0$, then there will be one rational solution.

Example 5   **Using the Discriminant**

Use the discriminant to determine the type and number of solutions for each equation.

  a. $2x^2 - 5x + 9 = 0$ **b.** $3x^2 = -x + 2$

  c. $-2x(2x - 3) = -1$ **d.** $3.6x^2 = -1.2x - 0.1$

Solution:

For each equation, first write the equation in standard form $ax^2 + bx + c = 0$. Then determine the discriminant.

Equation	Discriminant	Solution Type and Number
a. $2x^2 - 5x + 9 = 0$	$b^2 - 4ac$ $= (-5)^2 - 4(2)(9)$ $= 25 - 72$ $= -47$	Because $-47 < 0$, there will be two imaginary solutions.
b. $3x^2 = -x + 2$ $3x^2 + x - 2 = 0$	$b^2 - 4ac$ $= (1)^2 - 4(3)(-2)$ $= 1 - (-24)$ $= 25$	Because $25 > 0$ and 25 is a perfect square, there will be two rational solutions.
c. $-2x(2x - 3) = -1$ $-4x^2 + 6x = -1$ $-4x^2 + 6x + 1 = 0$	$b^2 - 4ac$ $= (6)^2 - 4(-4)(1)$ $= 36 - (-16)$ $= 52$	Because $52 > 0$, but 52 is *not* a perfect square, there will be two irrational solutions.

d. $3.6x^2 = -1.2x - 0.1$

$3.6x^2 + 1.2x + 0.1 = 0$

$b^2 - 4ac$

$= (1.2)^2 - 4(3.6)(0.1)$

$= 1.44 - 1.44$

$= 0$

Because the discriminant equals 0, there will be only one rational solution.

Skill Practice Determine the discriminant. Use the discriminant to determine the type and number of solutions for the equation.

5. $3y^2 + y + 3 = 0$

6. $4t^2 = 6t - 2$

7. $3t(t + 1) = 9$

8. $\frac{2}{3}x^2 - \frac{2}{3}x + \frac{1}{6} = 0$

With the discriminant we can determine the number of real-valued solutions to the equation $ax^2 + bx + c = 0$, and thus the number of x-intercepts to the function $f(x) = ax^2 + bx + c$. The following illustrations show the graphical interpretation of the three cases of the discriminant.

$f(x) = x^2 - 4x + 3$

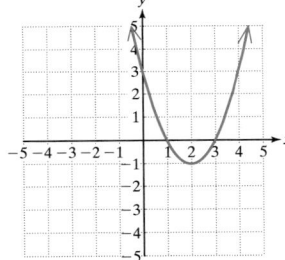

Use $x^2 - 4x + 3 = 0$ to find the value of the discriminant.

$$b^2 - 4ac = (-4)^2 - 4(1)(3)$$
$$= 4$$

Since the discriminant is positive, there are two real solutions to the quadratic equation. Therefore, there are two x-intercepts to the corresponding quadratic function, $(1, 0)$ and $(3, 0)$.

$f(x) = x^2 - x + 1$

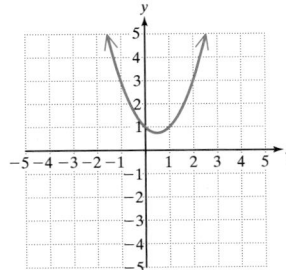

Use $x^2 - x + 1 = 0$ to find the value of the discriminant.

$$b^2 - 4ac = (-1)^2 - 4(1)(1)$$
$$= -3$$

Since the discriminant is negative, there are no real solutions to the quadratic equation. Therefore, there are no x-intercepts to the corresponding quadratic function.

$f(x) = x^2 - 2x + 1$

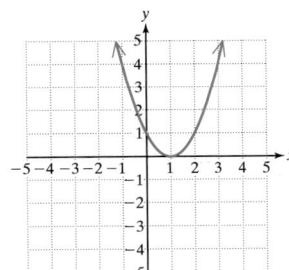

Use $x^2 - 2x + 1 = 0$ to find the value of the discriminant.

$$b^2 - 4ac = (-2)^2 - 4(1)(1)$$
$$= 0$$

Since the discriminant is zero, there is one real solution to the quadratic equation. Therefore, there is one x-intercept to the corresponding quadratic function, $(1, 0)$.

Skill Practice Answers

5. -35; two imaginary solutions
6. 4; two rational solutions
7. 117; two irrational solutions
8. 0; one rational solution

> **Example 6** Finding the x- and y-Intercepts of a Quadratic Function

Given: $f(x) = \dfrac{1}{4}x^2 + \dfrac{1}{4}x + \dfrac{1}{2}$

a. Find the y-intercept.   **b.** Find the x-intercept(s).

Solution:

a. The y-intercept is given by $f(0) = \dfrac{1}{4}(0)^2 + \dfrac{1}{4}(0) + \dfrac{1}{2}$

$$= \dfrac{1}{2}$$

The y-intercept is located at $(0, \frac{1}{2})$.

b. The x-intercepts are given by the real solutions to the equation $f(x) = 0$. In this case, we have

$$f(x) = \dfrac{1}{4}x^2 + \dfrac{1}{4}x + \dfrac{1}{2} = 0$$

$$4\left(\dfrac{1}{4}x^2 + \dfrac{1}{4}x + \dfrac{1}{2}\right) = 4(0) \qquad \text{Multiply by 4 to clear fractions.}$$

$$x^2 + x + 2 = 0 \qquad \text{The equation is in the form } ax^2 + bx + c = 0, \text{ where } a = 1, b = 1, \text{ and } c = 2.$$

$$x = \dfrac{-(1) \pm \sqrt{(1)^2 - 4(1)(2)}}{2(1)} \qquad \text{Apply the quadratic formula.}$$

$$= \dfrac{-1 \pm \sqrt{-7}}{2} \qquad \text{Simplify.}$$

$$= -\dfrac{1}{2} \pm \dfrac{\sqrt{7}}{2}i$$

These solutions are *imaginary numbers.* Because there are no real solutions to the equation $f(x) = 0$, the function has no x-intercepts. The function is graphed in Figure 12-3.

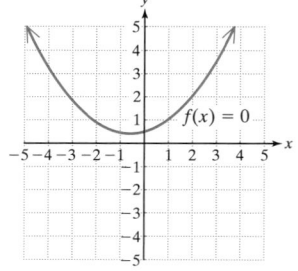

Figure 12-3

> **Skill Practice**

9. Find the x- and y-intercepts of the function given by

$$f(x) = \dfrac{1}{2}x^2 - x - \dfrac{5}{2}$$

Skill Practice Answers

9. x-intercepts:
$(1 + \sqrt{6}, 0) \approx (3.45, 0),$
$(1 - \sqrt{6}, 0) \approx (-1.45, 0)$;
y-intercept: $\left(0, -\dfrac{5}{2}\right)$

5. Mixed Review: Methods to Solve a Quadratic Equation

Three methods have been presented to solve quadratic equations.

> **Methods to Solve a Quadratic Equation**
> - Factor and use the zero product rule (Section 6.7).
> - Use the square root property. Complete the square, if necessary (Section 12.1).
> - Use the quadratic formula (Section 12.2).

Using the zero product rule is often the simplest method, but it works only if you can factor the equation. The square root property and the quadratic formula can be used to solve any quadratic equation. Before solving a quadratic equation, take a minute to analyze it first. Each problem must be evaluated individually before choosing the most efficient method to find its solutions.

Example 7 | Solving Quadratic Equations by Using Any Method

Solve the quadratic equations by using any method.

a. $(x + 3)^2 + x^2 - 9x = 8$ **b.** $\dfrac{x^2}{2} + \dfrac{5}{2} = -x$ **c.** $(p - 2)^2 - 11 = 0$

Solution:

a.
$$(x + 3)^2 + x^2 - 9x = 8$$

$$x^2 + 6x + 9 + x^2 - 9x - 8 = 0$$ 　 Clear parentheses and write the equation in the form $ax^2 + bx + c = 0$.

$$2x^2 - 3x + 1 = 0$$ 　 This equation is factorable.

$$(2x - 1)(x - 1) = 0$$ 　 Factor.

$2x - 1 = 0$ 　 or 　 $x - 1 = 0$ 　 Apply the zero product rule.

$x = \frac{1}{2}$ 　 or 　 $x = 1$ 　 Solve for x.

This equation could have been solved by using any of the three methods, but factoring was the most efficient method.

b. 　 $\dfrac{x^2}{2} + \dfrac{5}{2} = -x$ 　 Clear fractions and write the equation in the form $ax^2 + bx + c = 0$.

$$x^2 + 5 = -2x$$

$$x^2 + 2x + 5 = 0$$ 　 This equation does not factor. Use the quadratic formula.

$a = 1$ 　 $b = 2$ 　 $c = 5$ 　 Identify a, b, and c.

$$x = \frac{-(2) \pm \sqrt{(2)^2 - 4(1)(5)}}{2(1)}$$ 　 Apply the quadratic formula.

$$= \frac{-2 \pm \sqrt{-16}}{2}$$ 　 Simplify.

$$= \frac{-2 \pm 4i}{2} \qquad \text{Simplify the radical.}$$

$$= -1 \pm 2i \qquad \text{Reduce to lowest terms.}$$

$$\frac{-2 \pm 4i}{2} = \frac{\cancel{2}(-1 \pm 2i)}{\cancel{2}}$$

This equation could also have been solved by completing the square and applying the square root property.

c. $(p - 2)^2 - 11 = 0$

$$(p - 2)^2 = 11 \qquad \text{The equation is in the form } x^2 = k, \text{ where } x = (p - 2).$$

$$p - 2 = \pm\sqrt{11} \qquad \text{Apply the square root property.}$$

$$p = 2 \pm \sqrt{11} \qquad \text{Solve for } p.$$

This problem could have been solved by the quadratic formula, but that would have involved clearing parentheses and collecting *like* terms first.

Skill Practice Answers

10. $t = -1; t = \dfrac{5}{3}$

11. $x = \dfrac{-11 \pm \sqrt{97}}{4}$

12. $y = 3 \pm i\sqrt{3}$

Skill Practice　Solve the quadratic equations by using any method.

10. $2t(t - 1) + t^2 = 5$ 　　　**11.** $0.2x^2 + 1.1x = -0.3$ 　　　**12.** $(2y - 6)^2 = -12$

Section 12.2　　Practice Exercises

- Practice Problems
- Self-Tests
- NetTutor
- e-Professors
- Videos

Study Skills Exercise

1. Define the key terms.

　　a. quadratic formula　　　　**b. discriminant**

Review Exercises

2. Show by substitution that $x = -3 + \sqrt{5}$ is a solution to $x^2 + 6x + 4 = 0$.

For Exercises 3–6, solve by completing the square and using the square root property.

3. $(x + 5)^2 = 49$ 　　　**4.** $16 = (2x - 3)^2$ 　　　**5.** $x^2 - 2x + 10 = 0$ 　　　**6.** $x^2 - 4x + 15 = 0$

For Exercises 7–10, simplify the expressions.

7. $\dfrac{16 - \sqrt{320}}{4}$ 　　　**8.** $\dfrac{18 + \sqrt{180}}{3}$ 　　　**9.** $\dfrac{14 - \sqrt{-147}}{7}$ 　　　**10.** $\dfrac{10 - \sqrt{-175}}{5}$

Concept 2: Solving Quadratic Equations by Using the Quadratic Formula

11. What form should a quadratic equation be in so that the quadratic formula can be applied?

12. Describe the circumstances in which the square root property can be used as a method for solving a quadratic equation.

13. Write the quadratic formula from memory.

For Exercises 14–39, solve the equation by using the quadratic formula.

14. $a^2 + 11a - 12 = 0$

15. $5b^2 - 14b - 3 = 0$

16. $9y^2 - 2y + 5 = 0$

17. $2t^2 + 3t - 7 = 0$

18. $12p^2 - 4p + 5 = 0$

19. $-5n^2 + 4n - 6 = 0$

20. $-z^2 = -2z - 35$

21. $12x^2 - 5x = 2$

 **22.** $a^2 + 3a = 8$

23. $k^2 + 4 = 6k$

24. $25x^2 - 20x + 4 = 0$

25. $9y^2 = -12y - 4$

26. $w^2 - 6w + 14 = 0$

27. $2m^2 + 3m = -2$

28. $(x + 2)(x - 3) = 1$

29. $3y(y + 1) - 7y(y + 2) = 6$

30. $-4a^2 - 2a + 3 = 0$

31. $-2m^2 - 5m + 3 = 0$

32. $\frac{1}{2}y^2 + \frac{2}{3} = -\frac{2}{3}y$
(*Hint:* Clear the fractions first.)

33. $\frac{2}{3}p^2 - \frac{1}{6}p + \frac{1}{2} = 0$

34. $\frac{1}{5}h^2 + h + \frac{3}{5} = 0$

35. $\frac{1}{4}w^2 + \frac{7}{4}w + 1 = 0$

36. $0.01x^2 + 0.06x + 0.08 = 0$
(*Hint:* Clear the decimals first.)

37. $0.5y^2 - 0.7y + 0.2 = 0$

38. $0.3t^2 + 0.7t - 0.5 = 0$

39. $0.01x^2 + 0.04x - 0.07 = 0$

For Exercises 40–43, factor the expression. Then use the zero product rule and the quadratic formula to solve the equation. There should be three solutions to each equation.

40. a. Factor. $x^3 - 27$

b. Solve. $x^3 - 27 = 0$

41. a. Factor. $64x^3 + 1$

b. Solve. $64x^3 + 1 = 0$

42. a. Factor. $3x^3 - 6x^2 + 6x$

b. Solve. $3x^3 - 6x^2 + 6x = 0$

43. a. Factor. $5x^3 + 5x^2 + 10x$

b. Solve. $5x^3 + 5x^2 + 10x = 0$

44. The volume of a cube is 27 ft³. Find the lengths of the sides.

45. The volume of a rectangular box is 64 ft³. If the width is 3 times longer than the height, and the length is 9 times longer than the height, find the dimensions of the box.

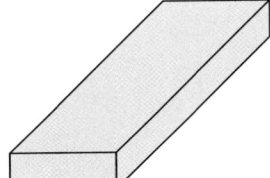

Concept 3: Using the Quadratic Formula in Applications

46. The hypotenuse of a right triangle measures 4 in. One leg of the triangle is 2 in. longer than the other leg. Find the lengths of the legs of the triangle. Round to one decimal place.

47. The length of one leg of a right triangle is 1 cm more than twice the length of the other leg. The hypotenuse measures 6 cm. Find the lengths of the legs. Round to 1 decimal place.

48. The hypotenuse of a right triangle is 10.2 m long. One leg is 2.1 m shorter than the other leg. Find the lengths of the legs. Round to 1 decimal place.

49. The hypotenuse of a right triangle is 17 ft long. One leg is 3.4 ft longer than the other leg. Find the lengths of the legs.

50. The fatality rate (in fatalities per 100 million vehicle miles driven) can be approximated for drivers x years old according to the function, $F(x) = 0.0036x^2 - 0.35x + 9.2$. Source: U.S. Department of Transportation

     a. Approximate the fatality rate for drivers 16 years old.

     b. Approximate the fatality rate for drivers 40 years old.

     c. Approximate the fatality rate for drivers 80 years old.

     d. For what age(s) is the fatality rate approximately 2.5?

51. Mitch throws a baseball straight up in the air from a cliff that is 48 ft high. The initial velocity is 48 ft/sec. The height (in feet) of the object after t sec is given by $h(t) = -16t^2 + 48t + 48$. Find the time at which the height of the object is 64 ft.

52. An astronaut on the moon throws a rock into the air from the deck of a spacecraft that is 8 m high. The initial velocity of the rock is 2.4 m/sec. The height (in meters) of the rock after t sec is given by $h(t) = -0.8t^2 + 2.4t + 8$. Find the time at which the height of the rock is 6 m.

53. The braking distance (in feet) of a car going v mph is given by

$$d(v) = \frac{v^2}{20} + v \qquad v \geq 0$$

     a. How fast would a car be traveling if its braking distance were 150 ft? Round to the nearest mile per hour.

     b. How fast would a car be traveling if its braking distance were 100 ft? Round to the nearest mile per hour.

Concept 4: Discriminant

For Exercises 54–61:

     a. Write the equation in the form $ax^2 + bx + c = 0, a > 0$.

     b. Find the value of the discriminant.

     c. Use the discriminant to determine the number and type of solutions.

54. $x^2 + 2x = -1$ **55.** $12y - 9 = 4y^2$ **56.** $19m^2 = 8m$

57. $2n - 5n^2 = 0$ **58.** $5p^2 - 21 = 0$ **59.** $3k^2 = 7$

60. $4n(n - 2) - 5n(n - 1) = 4$ **61.** $(2x + 1)(x - 3) = -9$

For Exercises 62–65, find the x- and y-intercepts of the quadratic function.

62. $f(x) = x^2 - 5x + 3$ **63.** $g(x) = -x^2 + x - 1$

64. $f(x) = 2x^2 + x + 5$ **65.** $h(x) = -3x^2 + 2x + 2$

Concept 5: Mixed Review: Methods to Solve a Quadratic Equation

For Exercises 66–83, solve the quadratic equations by using any method.

66. $a^2 + 3a + 4 = 0$

67. $4z^2 + 7z = 0$

68. $x^2 - 2 = 0$

69. $b^2 + 7 = 0$

70. $4y^2 + 8y - 5 = 0$

71. $k^2 - k + 8 = 0$

72. $\left(x + \dfrac{1}{2}\right)^2 + 4 = 0$

73. $(2y + 3)^2 = 9$

74. $2y(y - 3) = -1$

75. $w(w - 5) = 4$

76. $(2t + 5)(t - 1) = (t - 3)(t + 8)$

77. $(b - 1)(b + 4) = (3b + 2)(b + 1)$

78. $a^2 - 2a = 3$

79. $x^2 - x = 30$

80. $32z^2 - 20z - 3 = 0$

81. $8k^2 - 14k + 3 = 0$

82. $3p^2 - 27 = 0$

83. $5h^2 - 125 = 0$

Sometimes students shy away from completing the square and using the square root property to solve a quadratic equation. However, sometimes this process leads to a simple solution. For Exercises 84–85, solve the equations two ways.

    a. Solve the equation by completing the square and applying the square root property.

    b. Solve the equation by applying the quadratic formula.

    c. Which technique was easier for you?

84. $x^2 + 6x = 5$

85. $x^2 - 10x = -27$

Expanding Your Skills

86. An artist has been commissioned to make a stained glass window in the shape of a regular octagon. The octagon must fit inside an 18-in. square space. See the figure.

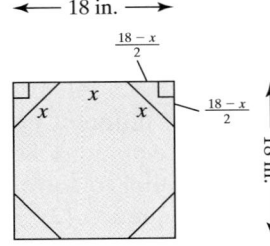

    a. Let x represent the length of each side of the octagon. Verify that the legs of the small triangles formed by the corners of the square can be expressed as $\dfrac{18 - x}{2}$.

    b. Use the Pythagorean theorem to set up an equation in terms of x that represents the relationship between the legs of the triangle and the hypotenuse.

    c. Simplify the equation by clearing parentheses and clearing fractions.

    d. Solve the resulting quadratic equation by using the quadratic formula. Use a calculator and round your answers to the nearest tenth of an inch.

    e. There are two solutions for x. Which one is appropriate and why?

Graphing Calculator Exercises

87. Graph $Y_1 = 64x^3 + 1$. Compare the x-intercepts with the solutions to the equation $64x^3 + 1 = 0$ found in Exercise 41.

88. Graph $Y_1 = x^3 - 27$. Compare the x-intercepts with the solutions to the equation $x^3 - 27 = 0$ found in Exercise 40.

89. Graph $Y_1 = 5x^3 + 5x^2 + 10x$. Compare the x-intercepts with the solutions to the equation $5x^3 + 5x^2 + 10x = 0$ found in Exercise 43.

90. Graph $Y_1 = 3x^3 - 6x^2 + 6x$. Compare the x-intercepts with the solutions to the equation $3x^3 - 6x^2 + 6x = 0$ found in Exercise 42.

91. The recent population (in thousands) of Ecuador can be approximated by $P(t) = 1.12t^2 + 204.4t + 6697$, where $t = 0$ corresponds to the year 1974.

    a. Approximate the number of people in Ecuador in the year 2000. (Round to the nearest thousand.)

    b. If this trend continues, approximate the number of people in Ecuador in the year 2010. (Round to the nearest thousand.)

    c. In what year after 1974 did the population of Ecuador reach 10 million? Round to the nearest year. (*Hint:* 10 million equals 10,000 thousands.)

    d. Use a graphing calculator to graph the function P on the window $0 \leq x \leq 20$, $4000 \leq y \leq 12{,}000$. Use a *Trace* feature to approximate the year in which the population in Ecuador first exceeded 10 million (10,000 thousands).

Section 12.3   Equations in Quadratic Form

Concepts

1. Solving Equations by Using Substitution
2. Solving Equations Reducible to a Quadratic

1. Solving Equations by Using Substitution

We have learned to solve a variety of different types of equations, including linear, radical, rational, and polynomial equations. Sometimes, however, it is necessary to use a quadratic equation as a tool to solve other types of equations.

In Example 1, we will solve the equation $(2x^2 - 5)^2 - 16(2x^2 - 5) + 39 = 0$.

Notice that the terms in the equation are written in descending order by degree. Furthermore, the first two terms have the same base, $2x^2 - 5$, and the exponent on the first term is exactly double the exponent on the second term. The third term is a constant. An equation in this pattern is called **quadratic in form**.

exponent is double 3rd term is constant

$$(2x^2 - 5)^2 - 16(2x^2 - 5)^1 + 39 = 0.$$

To solve this equation we will use substitution as demonstrated in Example 1.

Example 1   Solving an Equation in Quadratic Form

Solve the equation.

$$(2x^2 - 5)^2 - 16(2x^2 - 5) + 39 = 0$$

Solution:

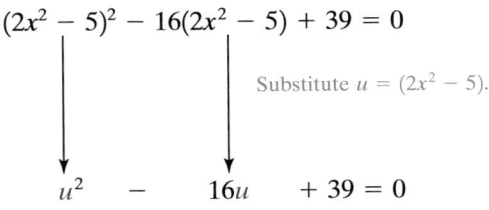

$(2x^2 - 5)^2 - 16(2x^2 - 5) + 39 = 0$ Substitute $u = (2x^2 - 5)$.   If the substitution $u = (2x^2 - 5)$ is made, the equation becomes quadratic in the variable u.

$u^2 \quad - \quad 16u \quad + 39 = 0$ The equation is in the form $au^2 + bu + c = 0$.

$(u - 13)(u - 3) = 0$ The equation is factorable.

$u = 13 \quad$ or $\quad u = 3$ Apply the zero product rule.

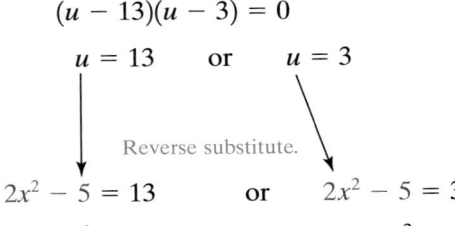

Reverse substitute.

$2x^2 - 5 = 13 \qquad$ or $\qquad 2x^2 - 5 = 3$

$2x^2 = 18 \qquad$ or $\qquad 2x^2 = 8$

$x^2 = 9 \qquad$ or $\qquad x^2 = 4$ Write the equations in the form $x^2 = k$.

$x = \pm\sqrt{9} \qquad$ or $\qquad x = \pm\sqrt{4}$ Apply the square root property.

$= \pm 3 \qquad$ or $\qquad = \pm 2$

The solutions are $x = 3, x = -3, x = 2, x = -2$.   Substituting these values in the original equation verifies that these are all valid solutions.

> **Avoiding Mistakes:**
> When using substitution, it is critical to reverse substitute to solve the equation in terms of the original variable.

Skill Practice   Solve the equation.

1. $(3t^2 - 10)^2 + 5(3t^2 - 10) - 14 = 0$

For an equation written in descending order, notice that u was set equal to the variable factor on the middle term. This is generally the case.

Example 2   Solving an Equation in Quadratic Form

Solve the equation.

$$p^{2/3} - 2p^{1/3} = 8$$

Solution:

$p^{2/3} - 2p^{1/3} = 8$

$p^{2/3} - 2p^{1/3} - 8 = 0$ Set the equation equal to zero.

Skill Practice Answers
1. $t = 1, t = -1, t = 2, t = -2$

$$(p^{1/3})^2 - 2(p^{1/3})^1 - 8 = 0 \qquad \text{Make the substitution } u = p^{1/3}.$$

Then the equation is in the form $au^2 + bu + c = 0$.

Substitute $u = p^{1/3}$.

$$u^2 - 2u - 8 = 0$$

$$(u - 4)(u + 2) = 0 \qquad \text{The equation is factorable.}$$

$$u = 4 \qquad \text{or} \qquad u = -2 \qquad \text{Apply the zero product rule.}$$

Reverse substitute.

$$p^{1/3} = 4 \qquad \text{or} \qquad p^{1/3} = -2$$

$$\sqrt[3]{p} = 4 \qquad \text{or} \qquad \sqrt[3]{p} = -2 \qquad \text{The equations are radical equations.}$$

$$(\sqrt[3]{p})^3 = (4)^3 \qquad \text{or} \qquad (\sqrt[3]{p})^3 = (-2)^3 \qquad \text{Cube both sides.}$$

$$p = 64 \qquad \text{or} \qquad p = -8$$

Check: $p = 64$ Check: $p = -8$

$$p^{2/3} - 2p^{1/3} = 8 \qquad\qquad p^{2/3} - 2p^{1/3} = 8$$

$$(64)^{2/3} - 2(64)^{1/3} \stackrel{?}{=} 8 \qquad (-8)^{2/3} - 2(-8)^{1/3} \stackrel{?}{=} 8$$

$$16 - 2(4) \stackrel{?}{=} 8 \qquad\qquad 4 - 2(-2) \stackrel{?}{=} 8$$

$$8 = 8 \checkmark \qquad\qquad 4 + 4 = 8 \checkmark$$

The solutions are $p = 64$ and $p = -8$.

| Skill Practice | Solve the equation.

2. $y^{2/3} - y^{1/3} = 12$

| Example 3 | **Solving a Quadratic Equation in Quadratic Form**

Solve the equation.

$$x - \sqrt{x} - 12 = 0$$

Solution:

The equation can be solved by first isolating the radical and then squaring both sides (this is left as an exercise—see Exercise 24). However, this equation is also quadratic in form. By writing $\sqrt{x}$ as $x^{1/2}$, we see that the exponent on the first term is exactly double the exponent on the middle term.

$$x^1 - x^{1/2} - 12 = 0$$

$$u^2 - u - 12 = 0 \qquad \text{Let } u = x^{1/2}.$$

$$(u - 4)(u + 3) = 0 \qquad \text{Factor.}$$

$$u = 4 \quad \text{or} \quad u = -3 \qquad \text{Solve for } u.$$

$$x^{1/2} = 4 \quad \text{or} \quad x^{1/2} = -3 \qquad \text{Back substitute.}$$

$$\sqrt{x} = 4 \quad \text{or} \quad \sqrt{x} \cancel{=} -3 \qquad \text{Solve each equation for } x. \text{ Notice that the principal square root of a number cannot equal } -3.$$

$$x = 16$$

The solution $x = 16$ checks in the original equation.

Skill Practice Answers

2. $y = 64, y = -27$

3. $z - \sqrt{z} - 2 = 0$

2. Solving Equations Reducible to a Quadratic

Some equations are reducible to a quadratic equation. In Example 4, we solve a polynomial equation by factoring. The resulting factors are quadratic.

Example 4 Solving a Polynomial Equation

Solve the equation.

$$4x^4 + 7x^2 - 2 = 0$$

Solution:

$4x^4 + 7x^2 - 2 = 0$ This is a polynomial equation.

$(4x^2 - 1)(x^2 + 2) = 0$ Factor.

$4x^2 - 1 = 0$ or $x^2 + 2 = 0$ Set each factor equal to zero. Notice that the two equations are quadratic. Each can be solved by the square root property.

$x^2 = \dfrac{1}{4}$ or $x^2 = -2$

$x = \pm\sqrt{\dfrac{1}{4}}$ or $x = \pm\sqrt{-2}$ Apply the square root property.

$x = \pm\dfrac{1}{2}$ or $x = \pm i\sqrt{2}$ Simplify the radicals.

The equation has four solutions: $x = \frac{1}{2}$, $x = -\frac{1}{2}$, $x = i\sqrt{2}$, $x = -i\sqrt{2}$.

Skill Practice

4. $9x^4 + 35x^2 - 4 = 0$

Example 5 Solving a Rational Equation

Solve the equation.

$$\frac{3y}{y + 2} - \frac{2}{y - 1} = 1$$

Solution:

$\dfrac{3y}{y + 2} - \dfrac{2}{y - 1} = 1$ This is a rational equation. The LCD is $(y + 2)(y - 1)$.

$\left(\dfrac{3y}{y + 2} - \dfrac{2}{y - 1}\right) \cdot (y + 2)(y - 1) = 1 \cdot (y + 2)(y - 1)$ Multiply both sides by the LCD.

$\dfrac{3y}{y+2} \cdot (y+2)(y - 1) - \dfrac{2}{y-1} \cdot (y + 2)(y - 1) = 1 \cdot (y + 2)(y - 1)$

$3y(y - 1) - 2(y + 2) = (y + 2)(y - 1)$ Clear fractions.

Skill Practice Answers

3. $z = 4$

4. $x = \pm\dfrac{1}{3}, x = \pm 2i$

$$3y^2 - 3y - 2y - 4 = y^2 - y + 2y - 2 \qquad \text{Apply the distributive property.}$$

$$3y^2 - 5y - 4 = y^2 + y - 2 \qquad \text{The equation is quadratic.}$$

$$2y^2 - 6y - 2 = 0 \qquad \text{Write the equation in descending order.}$$

$$\frac{2y^2}{2} - \frac{6y}{2} - \frac{2}{2} = \frac{0}{2} \qquad \text{Each coefficient in the equation is divisible by 2. Therefore, if we divide both sides by 2, the coefficients in the equation are smaller. This will make it easier to apply the quadratic formula.}$$

$$y^2 - 3y - 1 = 0$$

$$y = \frac{-(-3) \pm \sqrt{(-3)^2 - 4(1)(-1)}}{2(1)} \qquad \text{Apply the quadratic formula.}$$

$$= \frac{3 \pm \sqrt{9 + 4}}{2}$$

$$= \frac{3 \pm \sqrt{13}}{2} \quad \begin{cases} y = \dfrac{3 + \sqrt{13}}{2} \\[2mm] y = \dfrac{3 - \sqrt{13}}{2} \end{cases}$$

The values $y = \dfrac{3 + \sqrt{13}}{2}$ and $y = \dfrac{3 - \sqrt{13}}{2}$ are both defined in the original equation.

Skill Practice

Skill Practice Answers

5. $t = \dfrac{-5 \pm 3\sqrt{5}}{2}$

5. $\dfrac{t}{2t - 1} - \dfrac{1}{t + 4} = 1$

Section 12.3   Practice Exercises

Study Skills Exercise

1. What is meant by an equation in **quadratic form**?

Review Exercises

For Exercises 2–7, solve the quadratic equations.

2. $16 = (2x - 3)^2$

3. $\left(x - \dfrac{3}{2}\right)^2 = \dfrac{7}{4}$

4. $n(n - 6) = -13$

5. $x(x + 8) = -16$

6. $6k^2 + 7k = 6$

7. $2x^2 - 8x - 44 = 0$

Concept 1: Solving Equations by Using Substitution

8. a. Solve the quadratic equation by factoring. $u^2 + 10u + 24 = 0$

   b. Solve the equation by using substitution. $(y^2 + 5y)^2 + 10(y^2 + 5y) + 24 = 0$

9. **a.** Solve the quadratic equation by factoring. $u^2 - 2u - 35 = 0$

   b. Solve the equation by using substitution. $(w^2 - 6w)^2 - 2(w^2 - 6w) - 35 = 0$

10. **a.** Solve the quadratic equation by factoring. $u^2 - 2u - 24 = 0$

    b. Solve the equation by using substitution. $(x^2 - 5x)^2 - 2(x^2 - 5x) - 24 = 0$

11. **a.** Solve the quadratic equation by factoring. $u^2 - 4u + 3 = 0$

    b. Solve the equation by using substitution. $(2p^2 + p)^2 - 4(2p^2 + p) + 3 = 0$

For Exercises 12–23, solve the equation by using substitution.

12. $(x^2 - 2x)^2 + 2(x^2 - 2x) = 3$

13. $(x^2 + x)^2 - 8(x^2 + x) = -12$

14. $m^{2/3} - m^{1/3} - 6 = 0$

15. $2n^{2/3} + 7n^{1/3} - 15 = 0$

16. $2t^{2/5} + 7t^{1/5} + 3 = 0$

17. $p^{2/5} + p^{1/5} - 2 = 0$

18. $y + 6\sqrt{y} = 16$

19. $p - 8\sqrt{p} = -15$

20. $2x + 3\sqrt{x} - 2 = 0$

21. $3t + 5\sqrt{t} - 2 = 0$

22. $16\left(\dfrac{x+6}{4}\right)^2 + 8\left(\dfrac{x+6}{4}\right) + 1 = 0$

23. $9\left(\dfrac{x+3}{2}\right)^2 - 6\left(\dfrac{x+3}{2}\right) + 1 = 0$

24. In Example 3, we solved the equation $x - \sqrt{x} - 12 = 0$ by using substitution. Now solve this equation by first isolating the radical and then squaring both sides. Don't forget to check the potential solutions in the original equation. Do you obtain the same solution as in Example 3?

Concept 2: Solving Equations Reducible to a Quadratic

For Exercises 25–34, solve the equations.

25. $t^4 + t^2 - 12 = 0$

26. $w^4 + 4w^2 - 45 = 0$

27. $x^2(9x^2 + 7) = 2$

28. $y^2(4y^2 + 17) = 15$

29. $\dfrac{y}{10} - 1 = -\dfrac{12}{5y}$

30. $\dfrac{x+5}{x} + \dfrac{x}{2} = \dfrac{x+19}{4x}$

31. $\dfrac{3x}{x+1} - \dfrac{2}{x-3} = 1$

32. $\dfrac{2t}{t-3} - \dfrac{1}{t+4} = 1$

33. $\dfrac{x}{2x-1} = \dfrac{1}{x-2}$

34. $\dfrac{z}{3z+2} = \dfrac{2}{z+1}$

Mixed Exercises

For Exercises 35–54, solve the equations.

35. $x^4 - 16 = 0$

36. $t^4 - 625 = 0$

37. $(4x + 5)^2 + 3(4x + 5) + 2 = 0$

38. $2(5x + 3)^2 - (5x + 3) - 28 = 0$

39. $4m^4 - 9m^2 + 2 = 0$

40. $x^4 - 7x^2 + 12 = 0$

41. $x^6 - 9x^3 + 8 = 0$

42. $x^6 - 26x^3 - 27 = 0$

43. $\sqrt{x^2 + 20} = 3\sqrt{x}$

44. $\sqrt{4t + 1} = t + 1$

45. $2\left(\dfrac{t - 4}{3}\right)^2 - \left(\dfrac{t - 4}{3}\right) - 3 = 0$

46. $\left(\dfrac{x + 1}{5}\right)^2 - 3\left(\dfrac{x + 1}{5}\right) - 10 = 0$

47. $x^{2/3} + x^{1/3} = 20$

48. $x^{2/5} - 3x^{1/5} = -2$

49. $m^4 + 2m^2 - 8 = 0$

50. $2c^4 + c^2 - 1 = 0$

51. $a^3 + 16a - a^2 - 16 = 0$ *Hint:* Factor by grouping first.

52. $b^3 + 9b - b^2 - 9 = 0$

53. $x^3 + 5x - 4x^2 - 20 = 0$

54. $y^3 + 8y - 3y^2 - 24 = 0$

Graphing Calculator Exercises

55. a. Solve the equation $x^4 + 4x^2 + 4 = 0$.

b. How many solutions are real and how many solutions are imaginary?

c. How many x-intercepts do you anticipate for the function defined by $y = x^4 + 4x^2 + 4$?

d. Graph $Y_1 = x^4 + 4x^2 + 4$ on a standard viewing window.

56. a. Solve the equation $x^4 - 2x^2 + 1 = 0$.

b. How many solutions are real and how many solutions are imaginary?

c. How many x-intercepts do you anticipate for the function defined by $y = x^4 - 2x^2 + 1$?

d. Graph $Y_1 = x^4 - 2x^2 + 1$ on a standard viewing window.

57. a. Solve the equation $x^4 - x^3 - 6x^2 = 0$.

b. How many solutions are real and how many solutions are imaginary?

c. How many x-intercepts do you anticipate for the function defined by $y = x^4 - x^3 - 6x^2$?

d. Graph $Y_1 = x^4 - x^3 - 6x^2$ on a standard viewing window.

58. a. Solve the equation $x^4 - 10x^2 + 9 = 0$.

b. How many solutions are real and how many solutions are imaginary?

c. How many x-intercepts do you anticipate for the function defined by $y = x^4 - 10x^2 + 9$?

d. Graph $Y_1 = x^4 - 10x^2 + 9$ on a standard viewing window.

Graphs of Quadratic Functions

In Section 6.7, we defined a quadratic function as a function of the form $f(x) = ax^2 + bx + c$ $(a \neq 0)$. We also learned that the graph of a quadratic function is a **parabola**. In this section we will learn how to graph parabolas.

A parabola opens upward if $a > 0$ (Figure 12-4) and opens downward if $a < 0$ (Figure 12-5). If a parabola opens upward, the **vertex** is the lowest point on the graph. If a parabola opens downward, the **vertex** is the highest point on the graph. The **axis of symmetry** is the vertical line that passes through the vertex.

Concepts

1. Quadratic Functions of the Form $f(x) = x^2 + k$
2. Quadratic Functions of the Form $f(x) = (x - h)^2$
3. Quadratic Functions of the Form $f(x) = ax^2$
4. Quadratic Functions of the Form $f(x) = a(x - h)^2 + k$

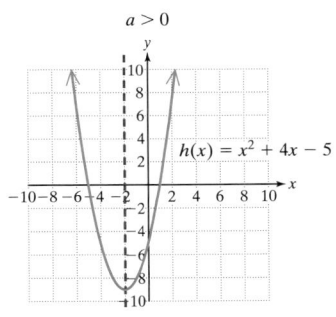

Figure 12-4

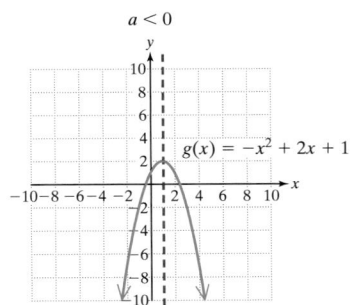

Figure 12-5

1. Quadratic Functions of the Form $f(x) = x^2 + k$

One technique for graphing a function is to plot a sufficient number of points on the function until the general shape and defining characteristics can be determined. Then sketch a curve through the points.

Example 1 Graphing Quadratic Functions of the Form $f(x) = x^2 + k$

Graph the functions f, g, and h on the same coordinate system.

$$f(x) = x^2 \qquad g(x) = x^2 + 1 \qquad h(x) = x^2 - 2$$

Solution:

Several function values for $f, g,$ and h are shown in Table 12-1 for selected values of x. The corresponding graphs are pictured in Figure 12-6.

Table 12-1

x	$f(x) = x^2$	$g(x) = x^2 + 1$	$h(x) = x^2 - 2$
-3	9	10	7
-2	4	5	2
-1	1	2	-1
0	0	1	-2
1	1	2	-1
2	4	5	2
3	9	10	7

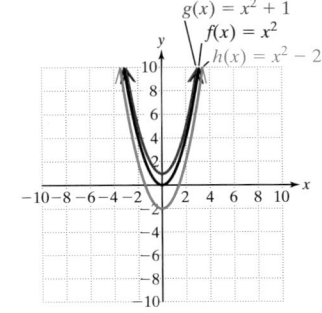

Figure 12-6

Notice that the graphs of $g(x) = x^2 + 1$ and $h(x) = x^2 - 2$ take on the same shape as $f(x) = x^2$. However, the y-values of g are 1 greater than the y-values of f. Hence the graph of $g(x) = x^2 + 1$ is the same as the graph of $f(x) = x^2$ shifted *up* 1 unit. Likewise the y-values of h are 2 less than those of f. The graph of $h(x) = x^2 - 2$ is the same as the graph of $f(x) = x^2$ shifted *down* 2 units.

Skill Practice

1. Refer to the graph of $f(x) = x^2 + k$ to determine the value of k.

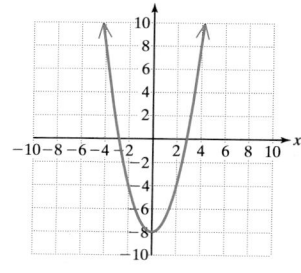

The functions in Example 1 illustrate the following properties of quadratic functions of the form $f(x) = x^2 + k$.

Graphs of $f(x) = x^2 + k$

If $k > 0$, then the graph of $f(x) = x^2 + k$ is the same as the graph of $y = x^2$ shifted *up* k units.

If $k < 0$, then the graph of $f(x) = x^2 + k$ is the same as the graph of $y = x^2$ shifted *down* $|k|$ units.

Calculator Connections

Try experimenting with a graphing calculator by graphing functions of the form $y = x^2 + k$ for several values of k.

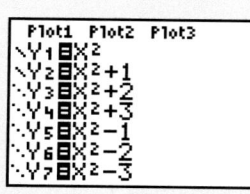 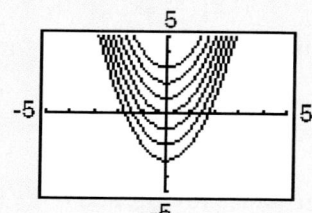

Example 2   **Graphing Quadratic Functions of the Form $f(x) = x^2 + k$**

Sketch the functions defined by

a. $m(x) = x^2 - 4$ **b.** $n(x) = x^2 + \dfrac{7}{2}$

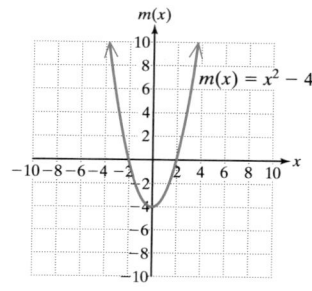

Solution:

a. $m(x) = x^2 - 4$

$m(x) = x^2 + (-4)$

Because $k = -4$, the graph is obtained by shifting the graph of $y = x^2$ down $|-4|$ units (Figure 12-7).

Figure 12-7

Skill Practice Answers

1. $k = -8$

b. $n(x) = x^2 + \dfrac{7}{2}$

Because $k = \frac{7}{2}$, the graph is obtained by shifting the graph of $y = x^2$ up $\frac{7}{2}$ units (Figure 12-8).

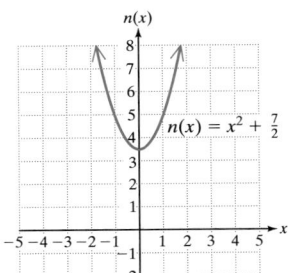

Figure 12-8

Skill Practice Graph the functions.

2. $g(x) = x^2 + 3$ **3.** $h(x) = x^2 - 5$

2. Quadratic Functions of the Form $f(x) = (x - h)^2$

The graph of $f(x) = x^2 + k$ represents a vertical shift (up or down) of the function $y = x^2$. Example 3 shows that functions of the form $f(x) = (x - h)^2$ represent a horizontal shift (left or right) of the function $y = x^2$.

Example 3 **Graphing Quadratic Functions of the Form** $f(x) = (x - h)^2$

Graph the functions f, g, and h on the same coordinate system.

$$f(x) = x^2 \qquad g(x) = (x + 1)^2 \qquad h(x) = (x - 2)^2$$

Solution:

Several function values for f, g, and h are shown in Table 12-2 for selected values of x. The corresponding graphs are pictured in Figure 12-9.

Table 12-2

x	$f(x) = x^2$	$g(x) = (x + 1)^2$	$h(x) = (x - 2)^2$
-4	16	9	36
-3	9	4	25
-2	4	1	16
-1	1	0	9
0	0	1	4
1	1	4	1
2	4	9	0
3	9	16	1
4	16	25	4
5	25	36	9

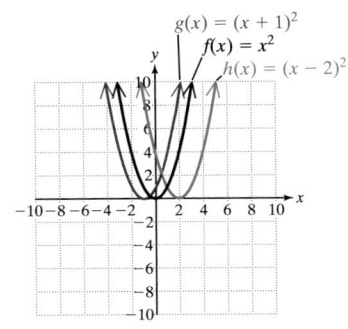

Figure 12-9

Skill Practice Answers

2–3.

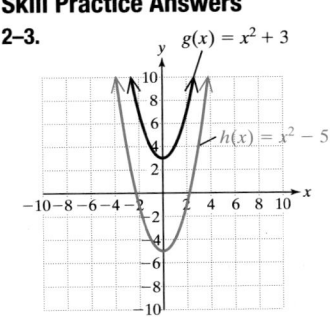

4. Refer to the graph of $f(x) = (x - h)^2$ to determine the value of h.

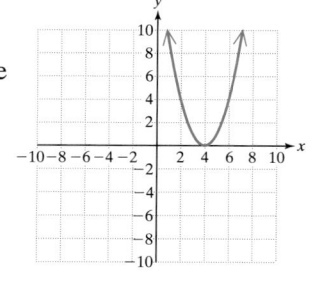

Example 3 illustrates the following properties of quadratic functions of the form $f(x) = (x - h)^2$.

Graphs of $f(x) = (x - h)^2$

If $h > 0$, then the graph of $f(x) = (x - h)^2$ is the same as the graph of $y = x^2$ shifted h units to the *right*.

If $h < 0$, then the graph of $f(x) = (x - h)^2$ is the same as the graph of $y = x^2$ shifted $|h|$ units to the *left*.

From Example 3 we have

$$h(x) = (x - 2)^2 \qquad \text{and} \qquad g(x) = (x + 1)^2$$

$$g(x) = [x - (-1)]^2$$

$y = x^2$ shifted 2 units to the right $\qquad$ $y = x^2$ shifted $|-1|$ unit to the left

---

Example 4 | **Graphing Functions of the Form $f(x) = (x - h)^2$**

Sketch the functions p and q.

a. $p(x) = (x - 7)^2$ **b.** $q(x) = (x + 1.6)^2$

Solution:

a. $p(x) = (x - 7)^2$

Because $h = 7 > 0$, shift the graph of $y = x^2$ to the *right* 7 units (Figure 12-10).

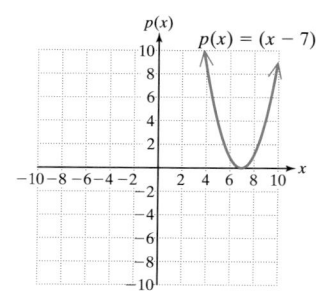

$p(x) = (x - 7)^2$

Figure 12-10

Skill Practice Answers

4. $h = 4$

b. $q(x) = (x + 1.6)^2$

$q(x) = [x - (-1.6)]^2$

Because $h = -1.6 < 0$, shift the graph of $y = x^2$ to the *left* 1.6 units (Figure 12-11).

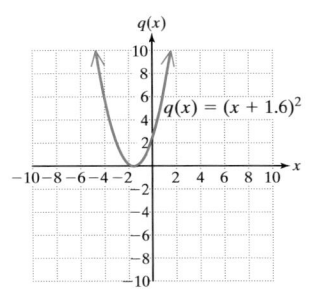

Figure 12-11

Skill Practice Graph the functions.

5. $g(x) = (x + 3)^2$ **6.** $h(x) = (x - 6)^2$

---

3. Quadratic Functions of the Form $f(x) = ax^2$

Examples 5 and 6 investigate functions of the form $f(x) = ax^2$ $(a \neq 0)$.

Example 5 Graphing Functions of the Form $f(x) = ax^2$

Graph the functions f, g, and h on the same coordinate system.

$$f(x) = x^2 \qquad g(x) = 2x^2 \qquad h(x) = \frac{1}{2}x^2$$

Solution:

Several function values for f, g, and h are shown in Table 12-3 for selected values of x. The corresponding graphs are pictured in Figure 12-12.

Table 12-3

x	$f(x) = x^2$	$g(x) = 2x^2$	$h(x) = \frac{1}{2}x^2$
-3	9	18	$\frac{9}{2}$
-2	4	8	2
-1	1	2	$\frac{1}{2}$
0	0	0	0
1	1	2	$\frac{1}{2}$
2	4	8	2
3	9	18	$\frac{9}{2}$

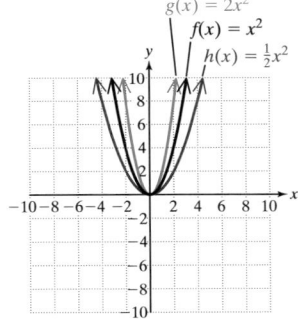

Figure 12-12

Skill Practice

7. Graph the functions f, g, and h on the same coordinate system.

$$f(x) = x^2$$

$$g(x) = 3x^2$$

$$h(x) = \frac{1}{3}x^2$$

Skill Practice Answers

5–6. $g(x) = (x + 3)^2$ y $h(x) = (x - 6)^2$

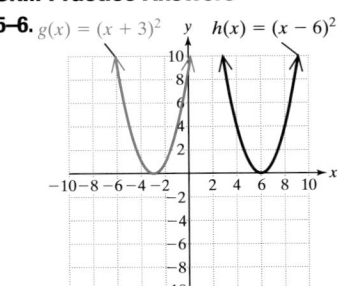

7.

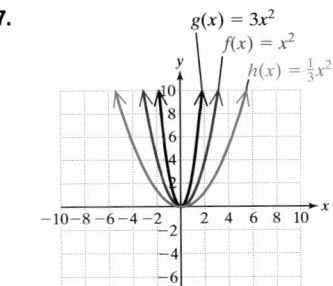

In Example 5, the function values defined by $g(x) = 2x^2$ are twice those of $f(x) = x^2$. The graph of $g(x) = 2x^2$ is the same as the graph of $f(x) = x^2$ *stretched vertically* by a factor of 2 (the graph appears narrower than $f(x) = x^2$).

In Example 5, the function values defined by $h(x) = \frac{1}{2}x^2$ are one-half those of $f(x) = x^2$. The graph of $h(x) = \frac{1}{2}x^2$ is the same as the graph of $f(x) = x^2$ *shrunk vertically* by a factor of $\frac{1}{2}$ (the graph appears wider than $f(x) = x^2$).

Example 6   **Graphing Functions of the Form $f(x) = ax^2$**

Graph the functions f, g, and h on the same coordinate system.

$$f(x) = -x^2 \qquad g(x) = -3x^2 \qquad h(x) = -\frac{1}{3}x^2$$

Solution:

Several function values for f, g, and h are shown in Table 12-4 for selected values of x. The corresponding graphs are pictured in Figure 12-13.

Table 12-4

x	$f(x) = -x^2$	$g(x) = -3x^2$	$h(x) = -\frac{1}{3}x^2$
-3	-9	-27	-3
-2	-4	-12	$-\frac{4}{3}$
-1	-1	-3	$-\frac{1}{3}$
0	0	0	0
1	-1	-3	$-\frac{1}{3}$
2	-4	-12	$-\frac{4}{3}$
3	-9	-27	-3

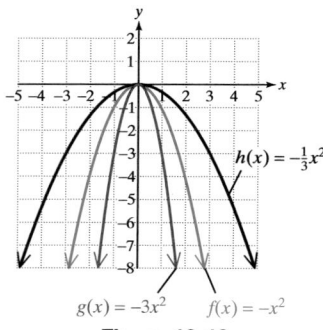

Figure 12-13

Skill Practice

8. Graph the functions f, g, and h on the same coordinate system.

$$f(x) = -x^2$$
$$g(x) = -2x^2$$
$$h(x) = -\frac{1}{2}x^2$$

Skill Practice Answers

8.

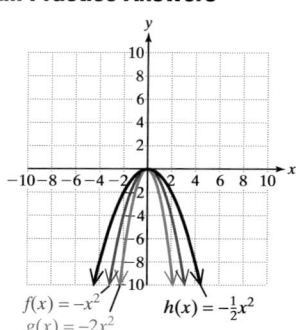

Example 6 illustrates that if the coefficient of the square term is negative, the parabola opens down. The graph of $g(x) = -3x^2$ is the same as the graph of $f(x) = -x^2$ with a *vertical stretch* by a factor of $|-3|$. The graph of $h(x) = -\frac{1}{3}x^2$ is the same as the graph of $f(x) = -x^2$ with a *vertical shrink* by a factor of $|-\frac{1}{3}|$.

Graphs of $f(x) = ax^2$

1. If $a > 0$, the parabola opens upward. Furthermore,
   - If $0 < a < 1$, then the graph of $f(x) = ax^2$ is the same as the graph of $y = x^2$ with a *vertical shrink* by a factor of a.
   - If $a > 1$, then the graph of $f(x) = ax^2$ is the same as the graph of $y = x^2$ with a *vertical stretch* by a factor of a.
2. If $a < 0$, the parabola opens downward. Furthermore,
   - If $0 < |a| < 1$, then the graph of $f(x) = ax^2$ is the same as the graph of $y = -x^2$ with a *vertical shrink* by a factor of $|a|$.
   - If $|a| > 1$, then the graph of $f(x) = ax^2$ is the same as the graph of $y = -x^2$ with a *vertical stretch* by a factor of $|a|$.

4. Quadratic Functions of the Form $f(x) = a(x - h)^2 + k$

We can summarize our findings from Examples 1–6 by graphing functions of the form $f(x) = a(x - h)^2 + k$ $(a \neq 0)$.

The graph of $y = x^2$ has its vertex at the origin $(0, 0)$. The graph of $f(x) = a(x - h)^2 + k$ is the same as the graph of $y = x^2$ shifted to the right or left h units and shifted up or down k units. Therefore, the vertex is shifted from $(0, 0)$ to (h, k). The axis of symmetry is the vertical line through the vertex, that is, the line $x = h$.

Graphs of $f(x) = a(x - h)^2 + k$

1. The vertex is located at (h, k).

2. The axis of symmetry is the line $x = h$.

3. If $a > 0$, the parabola opens upward, and k is the **minimum value** of the function.

4. If $a < 0$, the parabola opens downward, and k is the **maximum value** of the function.

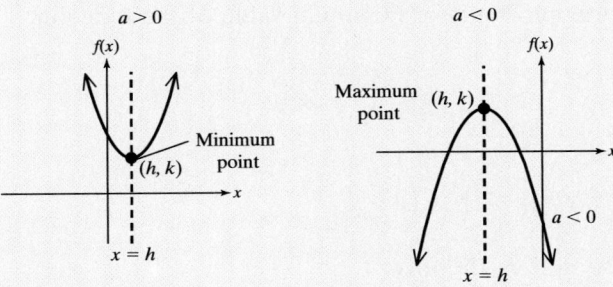

| Example 7 | **Graphing a Function of the Form** $f(x) = a(x - h)^2 + k$ |

Given the function defined by

$$f(x) = 2(x - 3)^2 + 4$$

a. Identify the vertex.
b. Sketch the function.
c. Identify the axis of symmetry.
d. Identify the maximum or minimum value of the function.

Solution:

a. The function is in the form $f(x) = a(x - h)^2 + k$, where $a = 2$, $h = 3$, and $k = 4$. Therefore, the vertex is $(3, 4)$.

b. The graph of f is the same as the graph of $y = x^2$ shifted to the right 3 units, shifted up 4 units, and stretched vertically by a factor of 2 (Figure 12-14).

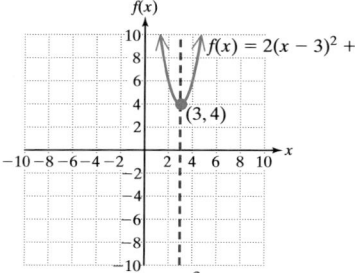

c. The axis of symmetry is the line $x = 3$.

d. Because $a > 0$, the function opens upward. Therefore, the minimum function value is 4. Notice that the minimum value is the minimum y-value on the graph.

Figure 12-14

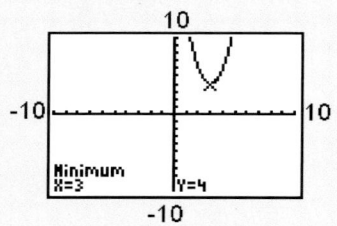

Skill Practice

9. Given the function defined by $g(x) = 3(x + 1)^2 - 3$
   a. Identify the vertex.
   b. Sketch the graph.
   c. Identify the axis of symmetry.
   d. Identify the maximum or minimum value of the function.

| **Example 8** | **Graphing a Function of the Form** $f(x) = a(x - h)^2 + k$ |

Given the function defined by

$$g(x) = -(x + 2)^2 - \frac{7}{4}$$

a. Identify the vertex.

b. Sketch the function.

c. Identify the axis of symmetry.

d. Identify the maximum or minimum value of the function.

Solution:

a. $g(x) = -(x + 2)^2 - \dfrac{7}{4}$

$$= -1[x - (-2)]^2 + \left(-\frac{7}{4}\right)$$

The function is in the form $g(x) = a(x - h)^2 + k$, where $a = -1$, $h = -2$, and $k = -\frac{7}{4}$. Therefore, the vertex is $(-2, -\frac{7}{4})$.

b. The graph of g is the same as the graph of $y = x^2$ shifted to the left 2 units, shifted down $\frac{7}{4}$ units, and opening downward (Figure 12-15).

c. The axis of symmetry is the line $x = -2$.

d. The parabola opens downward, so the maximum function value is $-\frac{7}{4}$.

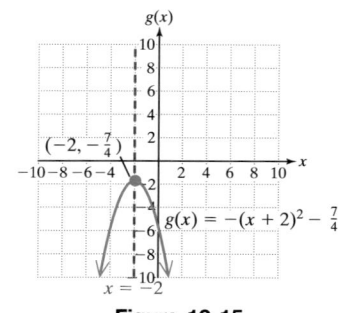

Figure 12-15

Skill Practice Answers

9a. Vertex: $(-1, -3)$
b.

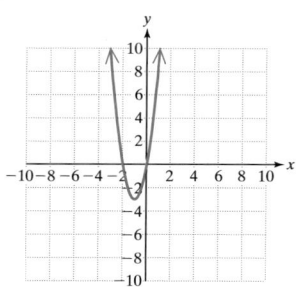

c. Axis of symmetry: $x = -1$
d. Minimum value: -3

Skill Practice

10. Given the function defined by $h(x) = -\dfrac{1}{2}(x - 4)^2 + 2$

    a. Identify the vertex.

    b. Sketch the graph.

    c. Identify the axis of symmetry.

    d. Identify the maximum or minimum value of the function.

Skill Practice Answers

10a. Vertex: (4, 2)

b.

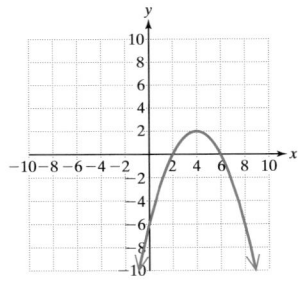

c. Axis of symmetry: $x = 4$

d. Maximum value: 2

Section 12.4   Practice Exercises

**Boost *your* GRADE at
mathzone.com!**

- Practice Problems
- Self-Tests
- NetTutor

- e-Professors
- Videos

Study Skills Exercise

1. Define the key terms.

    a. parabola   **b. vertex**   **c. axis of symmetry**

    d. maximum value   **e. minimum value**

Review Exercises

For Exercises 2–8, solve the equations.

    2. $x^2 + x - 5 = 0$ **3.** $(y - 3)^2 = -4$ **4.** $\sqrt{2a + 7} = a + 1$ **5.** $5t(t - 2) = -3$

    6. $2z^2 - 3z - 9 = 0$ **7.** $x^{2/3} + 5x^{1/3} + 6 = 0$ **8.** $m^2(m^2 + 6) = 27$

Concept 1: Quadratic Functions of the Form $f(x) = x^2 + k$

    9. Describe how the value of k affects the graphs of functions of the form $f(x) = x^2 + k$.

For Exercises 10–19, graph the functions.

    10. $g(x) = x^2 + 1$ **11.** $f(x) = x^2 + 2$ **12.** $p(x) = x^2 - 3$

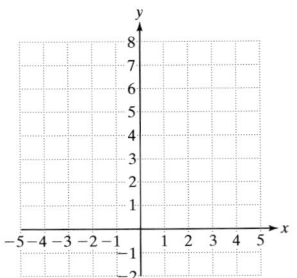 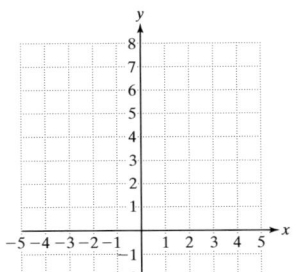 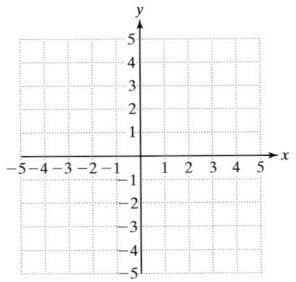

13. $q(x) = x^2 - 4$

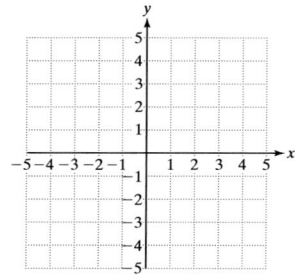

14. $T(x) = x^2 + \dfrac{3}{4}$

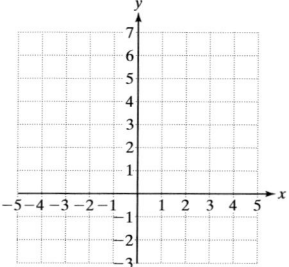

15. $S(x) = x^2 + \dfrac{3}{2}$

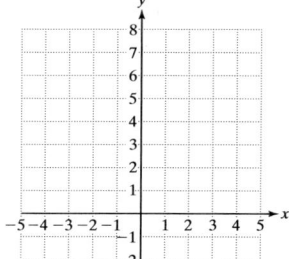

16. $M(x) = x^2 - \dfrac{5}{4}$

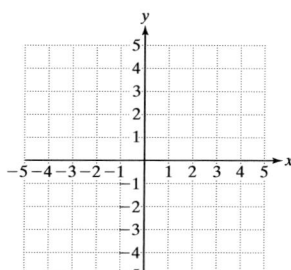

17. $n(x) = x^2 - \dfrac{1}{3}$

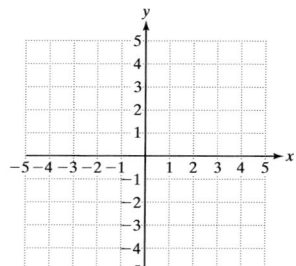

18. $P(x) = x^2 + \dfrac{1}{2}$

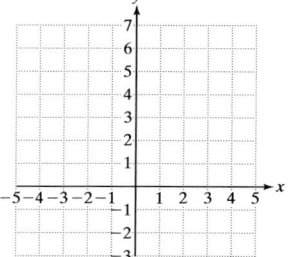

19. $Q(x) = x^2 + \dfrac{1}{4}$

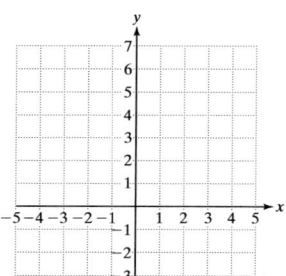

Concept 2: Quadratic Functions of the Form $f(x) = (x - h)^2$

20. Describe how the value of h affects the graphs of functions of the form $f(x) = (x - h)^2$.

For Exercises 21–30, graph the functions.

21. $r(x) = (x + 1)^2$

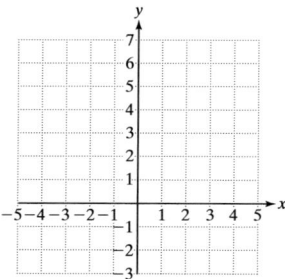

22. $h(x) = (x + 2)^2$

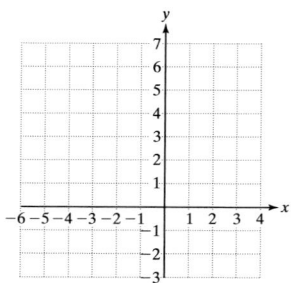

23. $k(x) = (x - 3)^2$

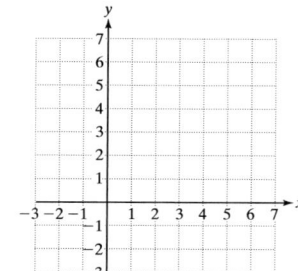

24. $L(x) = (x - 4)^2$

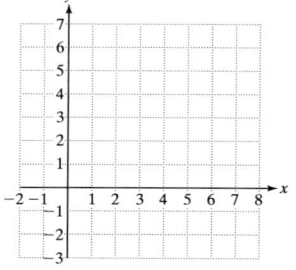

25. $A(x) = \left(x + \dfrac{3}{4}\right)^2$ **26.** $r(x) = \left(x + \dfrac{3}{2}\right)^2$ **27.** $W(x) = \left(x - \dfrac{5}{4}\right)^2$ **28.** $V(x) = \left(x - \dfrac{1}{3}\right)^2$

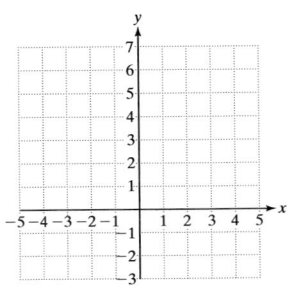 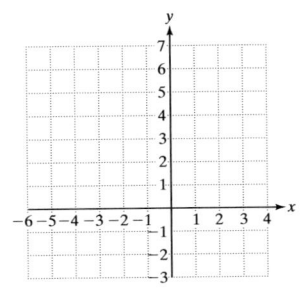 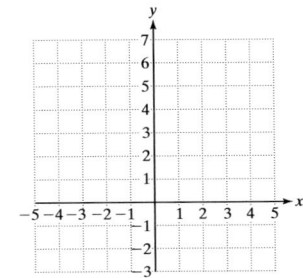 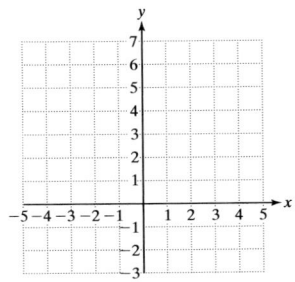

29. $M(x) = (x + 4.5)^2$ **30.** $N(x) = (x - 3.5)^2$

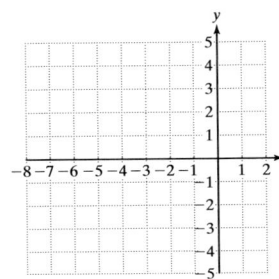 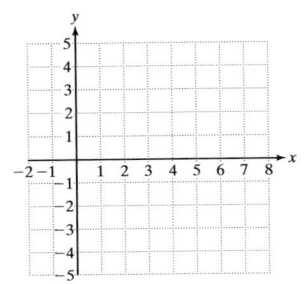

Concept 3: Quadratic Functions of the Form $f(x) = ax^2$

31. Describe how the value of a affects the graph of a function of the form $f(x) = ax^2$, where $a \neq 0$.

32. How do you determine whether the graph of a function defined by $h(x) = ax^2 + bx + c$ $(a \neq 0)$ opens up or down?

For Exercises 33–40, graph the functions.

33. $f(x) = 2x^2$ **34.** $g(x) = 3x^2$ **35.** $h(x) = \dfrac{1}{2}x^2$ **36.** $f(x) = \dfrac{1}{3}x^2$

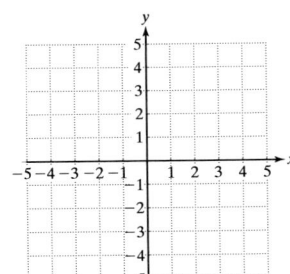 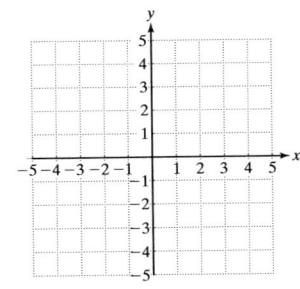 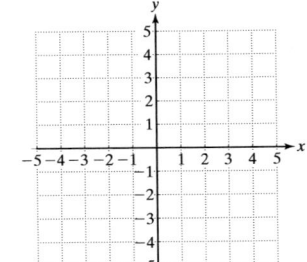 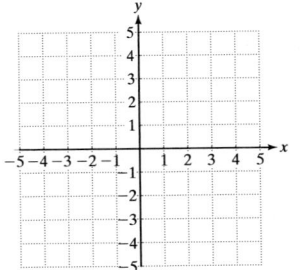

37. $c(x) = -x^2$ **38.** $g(x) = -2x^2$ **39.** $v(x) = -\dfrac{1}{3}x^2$ **40.** $f(x) = -\dfrac{1}{2}x^2$

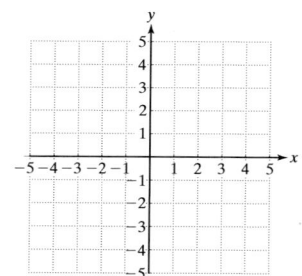 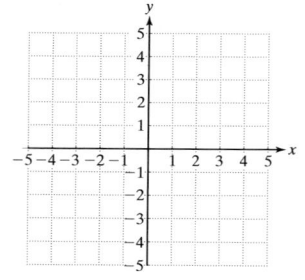 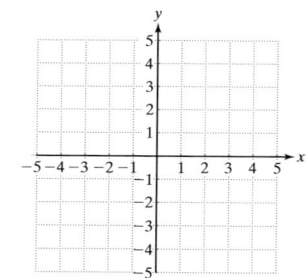 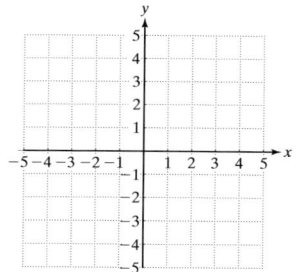

Concept 4: Quadratic Functions of the Form $f(x) = a(x - h)^2 + k$

For Exercises 41–48, match the function with its graph.

41. $f(x) = -\dfrac{1}{4}x^2$ **42.** $g(x) = (x + 3)^2$ **43.** $k(x) = (x - 3)^2$ **44.** $h(x) = \dfrac{1}{4}x^2$

45. $t(x) = x^2 + 2$ **46.** $m(x) = x^2 - 4$ **47.** $p(x) = (x + 1)^2 - 3$ **48.** $n(x) = -(x - 2)^2 + 3$

a.
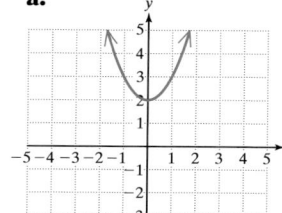

b.
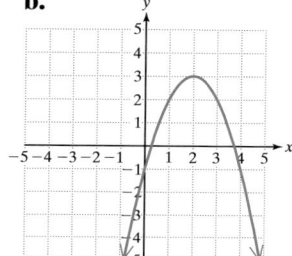

c.
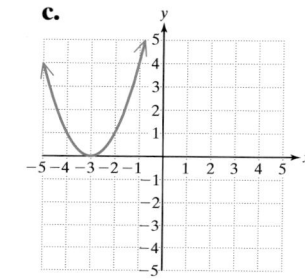

d.
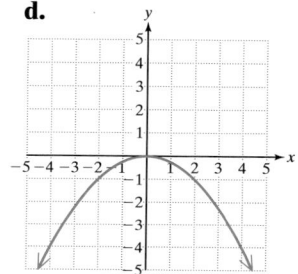

e.
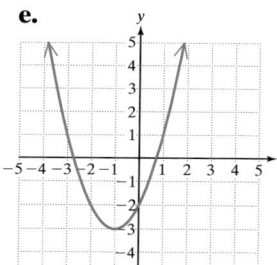

f.
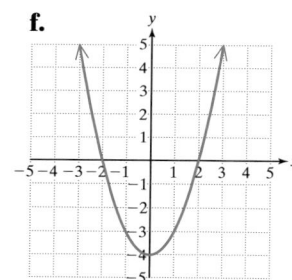

g.
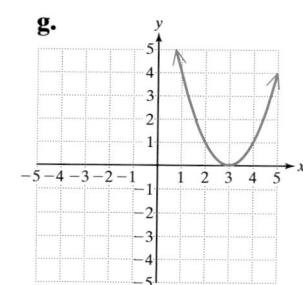

h.
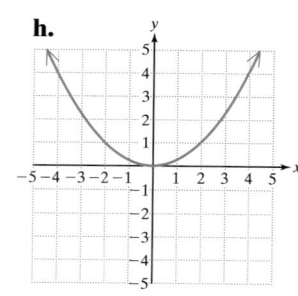

For Exercises 49–66, graph the parabola and the axis of symmetry. Label the coordinates of the vertex, and write the equation of the axis of symmetry.

49. $y = (x - 3)^2 + 2$ **50.** $y = (x - 2)^2 + 3$ **51.** $y = (x + 1)^2 - 3$ **52.** $y = (x + 3)^2 - 1$

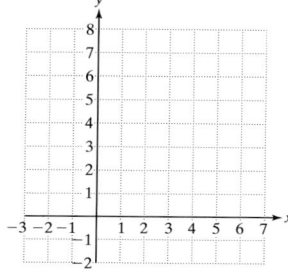 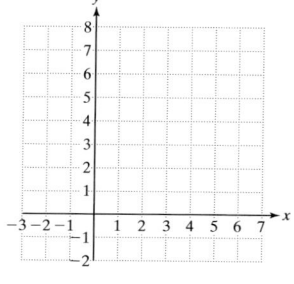 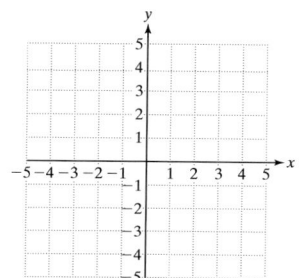 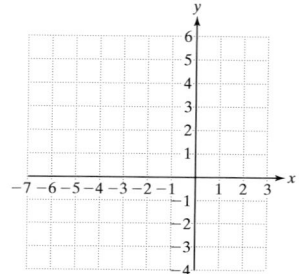

53. $y = -(x - 4)^2 - 2$ **54.** $y = -(x - 2)^2 - 4$ **55.** $y = -(x + 3)^2 + 3$ **56.** $y = -(x + 2)^2 + 2$

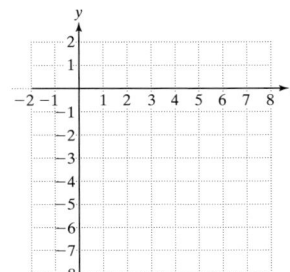 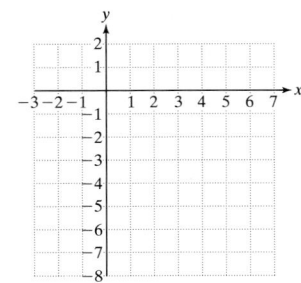 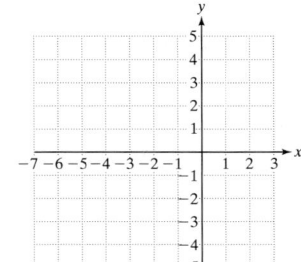 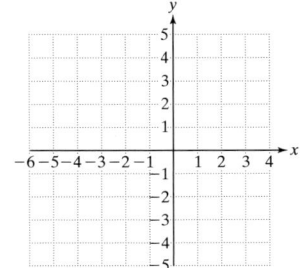

57. $y = (x + 1)^2 + 1$ **58.** $y = (x - 4)^2 - 4$ **59.** $y = 3(x - 1)^2$ **60.** $y = -3(x - 1)^2$

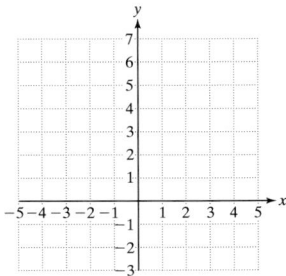 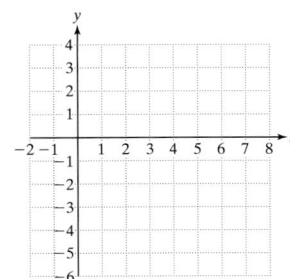 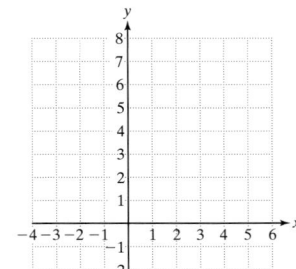 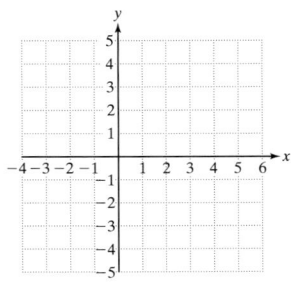

61. $y = -4x^2 + 3$ **62.** $y = 4x^2 + 3$ **63.** $y = 2(x + 3)^2 - 1$

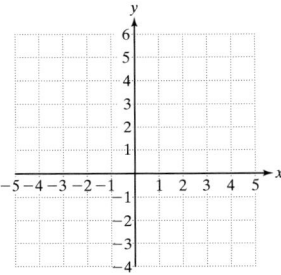 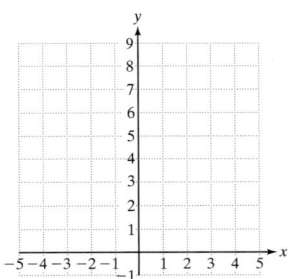 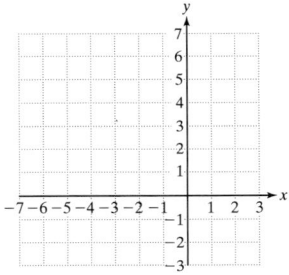

64. $y = -2(x + 3)^2 - 1$ **65.** $y = -\dfrac{1}{4}(x - 1)^2 + 2$ **66.** $y = \dfrac{1}{4}(x - 1)^2 + 2$

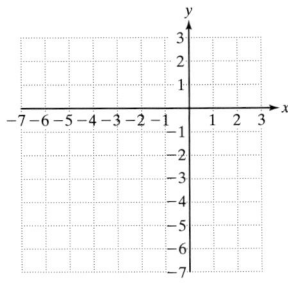 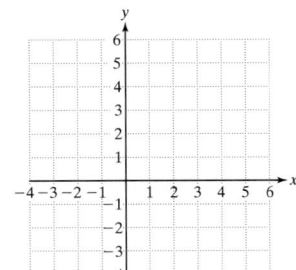 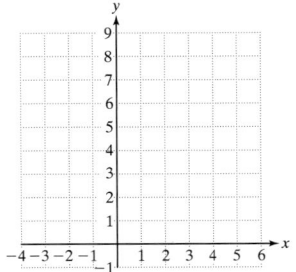

For Exercises 67–78, write the coordinates of the vertex and determine if the vertex is a maximum point or a minimum point. Then write the maximum or minimum value.

67. $f(x) = 4(x - 6)^2 - 9$

68. $g(x) = 3(x - 4)^2 - 7$

69. $p(x) = -\dfrac{2}{5}(x - 2)^2 + 5$

70. $h(x) = -\dfrac{3}{7}(x - 5)^2 + 10$

71. $k(x) = \dfrac{1}{2}(x + 8)^2$

72. $m(x) = \dfrac{2}{9}(x + 11)^2$

73. $n(x) = -6x^2 + \dfrac{21}{4}$

74. $q(x) = -4x^2 + \dfrac{1}{6}$

75. $A(x) = 2(x - 7)^2 - \dfrac{3}{2}$

76. $B(x) = 5(x - 3)^2 - \dfrac{1}{4}$

77. $F(x) = 7x^2$

78. $G(x) = -7x^2$

79. True or false: The function defined by $g(x) = -5x^2$ has a maximum value but no minimum value.

80. True or false: The function defined by $f(x) = 2(x - 5)^2$ has a maximum value but no minimum value.

81. True or false: If the vertex $(-2, 8)$ represents a minimum point, then the minimum value is -2.

82. True or false: If the vertex $(-2, 8)$ represents a maximum point, then the maximum value is 8.

83. A suspension bridge is 120 ft long. Its supporting cable hangs in a shape that resembles a parabola. The function defined by $H(x) = \frac{1}{90}(x - 60)^2 + 30$ (where $0 \le x \le 120$) approximates the height of the supporting cable a distance of x ft from the end of the bridge (see figure).

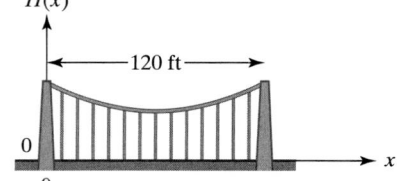

a. What is the location of the vertex of the parabolic cable?

b. What is the minimum height of the cable?

c. How high are the towers at either end of the supporting cable?

84. A 50-m bridge over a crevasse is supported by a parabolic arch. The function defined by $f(x) = -0.16(x - 25)^2 + 100$ (where $0 \le x \le 50$) approximates the height of the supporting arch x meters from the end of the bridge (see figure).

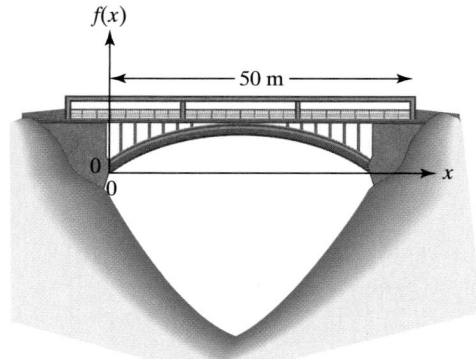

a. What is the location of the vertex of the arch?

b. What is the maximum height of the arch (relative to the origin)?

Graphing Calculator Exercises

For Exercises 85–88, verify the maximum and minimum points found in Exercises 67–70, by graphing each function on the calculator.

85. $Y_1 = 4(x - 6)^2 - 9$ (Exercise 67)

86. $Y_1 = 3(x - 4)^2 - 7$ (Exercise 68)

87. $Y_1 = -\dfrac{2}{5}(x - 2)^2 + 5$ (Exercise 69)

88. $Y_1 = -\dfrac{3}{7}(x - 5)^2 + 10$ (Exercise 70)

Vertex of a Parabola and Applications

1. Writing a Quadratic Function in the Form $f(x) = a(x - h)^2 + k$

A quadratic function can be expressed as $f(x) = ax^2 + bx + c$ $(a \neq 0)$. However, by completing the square, we can write the function in the form $f(x) = a(x - h)^2 + k$. In this form, the vertex is easily recognized as (h, k).

The process to complete the square in this context is similar to the steps outlined in Section 12.1. However, all algebraic steps are performed on the right-hand side of the equation.

Concepts

1. Writing a Quadratic Function in the Form $f(x) = a(x - h)^2 + k$
2. Vertex Formula
3. Determining the Vertex and Intercepts of a Quadratic Function
4. Vertex of a Parabola: Applications

Example 1 — Writing a Quadratic Function in the Form $f(x) = a(x - h)^2 + k$ $(a \neq 0)$

Given: $f(x) = x^2 + 8x + 13$

a. Write the function in the form $f(x) = a(x - h)^2 + k$.

b. Identify the vertex, axis of symmetry, and minimum function value.

Solution:

a. $f(x) = x^2 + 8x + 13$

Rather than dividing by the leading coefficient on both sides, we will factor out the leading coefficient from the variable terms on the right-hand side.

$= 1(x^2 + 8x) + 13$

$= 1(x^2 + 8x \quad\quad) + 13$

Next, complete the square on the expression within the parentheses: $[\frac{1}{2}(8)]^2 = 16$.

$= 1(x^2 + 8x + 16 - 16) + 13$

Rather than add 16 to both sides of the function, we *add and subtract 16* within the parentheses on the right-hand side. This has the effect of adding 0 to the right-hand side.

Avoiding Mistakes:

Do not factor out the leading coefficient from the constant term.

$$= 1(x^2 + 8x + 16) - 16 + 13$$

Use the associative property of addition to regroup terms and isolate the perfect square trinomial within the parentheses.

$$= (x + 4)^2 - 3$$

Factor and simplify.

b. $f(x) = (x + 4)^2 - 3$

The vertex is $(-4, -3)$.

The axis of symmetry is $x = -4$.

Because $a > 0$, the parabola opens upward.

The minimum value is -3 (Figure 12-16).

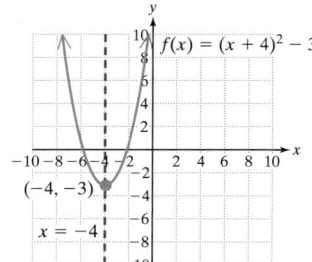

Figure 12-16

> **Skill Practice**
>
> **1.** Given: $f(x) = x^2 + 8x - 1$
>   **a.** Write the equation in the form $f(x) = a(x - h)^2 + k$.
>   **b.** Identify the vertex, axis of symmetry, and minimum value of the function.

Example 2 **Analyzing a Quadratic Function**

Given: $f(x) = -2x^2 + 12x - 16$

a. Write the function in the form $f(x) = a(x - h)^2 + k$.

b. Find the vertex, axis of symmetry, and maximum function value.

c. Find the x- and y-intercepts.

d. Graph the function.

Solution:

a. $f(x) = -2x^2 + 12x - 16$

To find the vertex, write the function in the form $f(x) = a(x - h)^2 + k$.

$$= -2(x^2 - 6x \quad\quad) - 16$$

Factor the leading coefficient from the variable terms.

$$= -2(x^2 - 6x + 9 - 9) - 16$$

Add and subtract the quantity $[\frac{1}{2}(-6)]^2 = 9$ within the parentheses.

$$= -2(x^2 - 6x + 9) + (-2)(-9) - 16$$

To remove the term -9 from the parentheses, we must first apply the distributive property. When -9 is removed from the parentheses, it carries with it a factor of -2.

$$= -2(x - 3)^2 + 18 - 16$$

Factor and simplify.

$$= -2(x - 3)^2 + 2$$

b. $f(x) = -2(x - 3)^2 + 2$

The vertex is $(3, 2)$. The axis of symmetry is $x = 3$. Because $a < 0$, the parabola opens downward and the maximum value is 2.

c. The y-intercept is given by $f(0) = -2(0)^2 + 12(0) - 16 = -16$. The y-intercept is $(0, -16)$.

To find the x-intercept(s), find the real solutions to the equation $f(x) = 0$.

$$f(x) = -2x^2 + 12x - 16$$

$$0 = -2x^2 + 12x - 16 \qquad \text{Substitute } f(x) = 0.$$

$$0 = -2(x^2 - 6x + 8) \qquad \text{Factor.}$$

$$0 = -2(x - 4)(x - 2)$$

$$x = 4 \qquad \text{or} \qquad x = 2$$

The x-intercepts are $(4, 0)$ and $(2, 0)$.

d. Using the information from parts (a)–(c), sketch the graph (Figure 12-17).

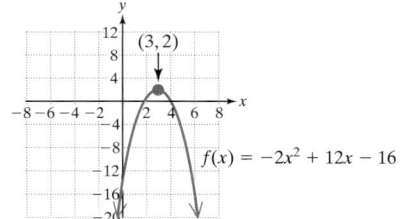

Figure 12-17

2. Given: $g(x) = -x^2 + 6x - 5$

   a. Write the equation in the form $g(x) = a(x - h)^2 + k$.

   b. Identify the vertex, axis of symmetry, and maximum value of the function.

   c. Determine the x- and y-intercepts.

   d. Graph the function.

2. Vertex Formula

Completing the square and writing a quadratic function in the form $f(x) = a(x - h)^2 + k$ $(a \neq 0)$ is one method to find the vertex of a parabola. Another method is to use the vertex formula. The **vertex formula** can be derived by completing the square on the function defined by $f(x) = ax^2 + bx + c$ $(a \neq 0)$.

$$f(x) = ax^2 + bx + c \qquad (a \neq 0)$$

$$= a\left(x^2 + \frac{b}{a}x \qquad\qquad\right) + c \qquad \text{Factor } a \text{ from the variable terms.}$$

$$= a\left(x^2 + \frac{b}{a}x + \frac{b^2}{4a^2} - \frac{b^2}{4a^2}\right) + c \qquad \text{Add and subtract } [\tfrac{1}{2}(b/a)]^2 = b^2/(4a^2) \text{ within the parentheses.}$$

$$= a\left(x^2 + \frac{b}{a}x + \frac{b^2}{4a^2}\right) + (a)\left(-\frac{b^2}{4a^2}\right) + c \qquad \text{Apply the distributive property and remove the term } -b^2/(4a^2) \text{ from the parentheses.}$$

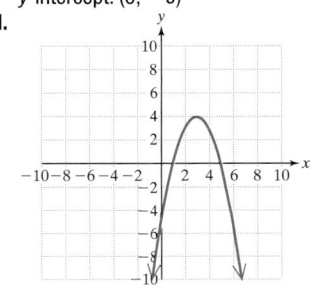

$$= a\left(x + \frac{b}{2a}\right)^2 - \frac{b^2}{4a} + c$$ Factor the trinomial and simplify.

$$= a\left(x + \frac{b}{2a}\right)^2 + c - \frac{b^2}{4a}$$ Apply the commutative property of addition to reverse the last two terms.

$$= a\left(x + \frac{b}{2a}\right)^2 + \frac{4ac}{4a} - \frac{b^2}{4a}$$ Obtain a common denominator.

$$= a\left(x + \frac{b}{2a}\right)^2 + \frac{4ac - b^2}{4a}$$

$$= a\left[x - \left(-\frac{b}{2a}\right)\right]^2 + \frac{4ac - b^2}{4a}$$

$$f(x) = a(x \quad - \quad h)^2 \quad + \quad k$$

The function is in the form $f(x) = a(x - h)^2 + k$, where

$$h = \frac{-b}{2a} \quad \text{and} \quad k = \frac{4ac - b^2}{4a}$$

Hence, the vertex is

$$\left(\frac{-b}{2a}, \frac{4ac - b^2}{4a}\right)$$

Although the y-coordinate of the vertex is given as $(4ac - b^2)/(4a)$, it is usually easier to determine the x-coordinate of the vertex first and then find y by evaluating the function at $x = -b/(2a)$.

The Vertex Formula

For $f(x) = ax^2 + bx + c$ $(a \neq 0)$, the vertex is given by

$$\left(\frac{-b}{2a}, \frac{4ac - b^2}{4a}\right) \quad \text{or} \quad \left(\frac{-b}{2a}, f\left(-\frac{b}{2a}\right)\right)$$

---

3. Determining the Vertex and Intercepts of a Quadratic Function

Example 3   Determining the Vertex and Intercepts of a Quadratic Function

Given: $h(x) = x^2 - 2x + 5$

a. Use the vertex formula to find the vertex.

b. Find the x- and y-intercepts.

c. Sketch the function.

Solution:

a. $h(x) = x^2 - 2x + 5$

$a = 1 \qquad b = -2 \qquad c = 5 \qquad$ Identify a, b, and c.

The x-coordinate of the vertex is $\dfrac{-b}{2a} = \dfrac{-(-2)}{2(1)} = 1$.

The y-coordinate of the vertex is $h(1) = (1)^2 - 2(1) + 5 = 4$.

The vertex is $(1, 4)$.

b. The y-intercept is given by $h(0) = (0)^2 - 2(0) + 5 = 5$.

The y-intercept is $(0, 5)$.

To find the x-intercept(s), find the real solutions to the equation $h(x) = 0$.

$h(x) = x^2 - 2x + 5$

$0 = x^2 - 2x + 5 \qquad$ This quadratic equation is not factorable. Apply the quadratic formula: $a = 1$, $b = -2$, $c = 5$.

$x = \dfrac{-(-2) \pm \sqrt{(-2)^2 - 4(1)(5)}}{2(1)}$

$ = \dfrac{2 \pm \sqrt{4 - 20}}{2(1)}$

$ = \dfrac{2 \pm \sqrt{-16}}{2}$

$ = \dfrac{2 \pm 4i}{2}$

$ = 1 \pm 2i$

The solutions to the equation $h(x) = 0$ are not real numbers. Therefore, there are no x-intercepts.

c.

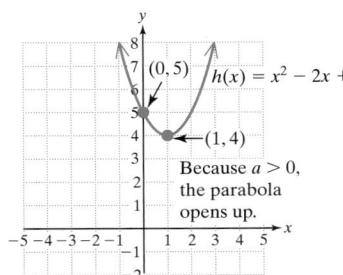

Figure 12-18

TIP: The location of the vertex and the direction that the parabola opens can be used to determine whether the function has any x-intercepts.

Given $h(x) = x^2 - 2x + 5$, the vertex $(1, 4)$ is *above* the x-axis. Furthermore, because $a > 0$, the parabola opens upward. Therefore, it is not possible for the function h to cross the x-axis (Figure 12-18).

Skill Practice Answers

3a. Vertex: $(-2, 2)$
b. x-intercepts: none; y-intercepts: $(0, 6)$
c.

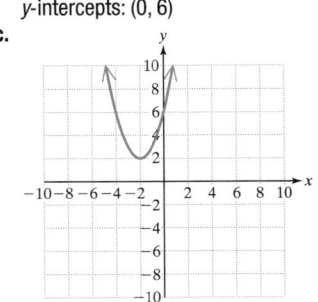

Skill Practice

3. Given: $f(x) = x^2 + 4x + 6$

a. Use the vertex formula to find the vertex of the parabola.

b. Determine the x- and y-intercepts.

c. Sketch the graph.

4. Vertex of a Parabola: Applications

> ### Example 4   Applying a Quadratic Function

The crew from Extravaganza Entertainment launches fireworks at an angle of 60° from the horizontal. The height of one particular type of display can be approximated by the following function:

$$h(t) = -16t^2 + 128\sqrt{3}t$$

where $h(t)$ is measured in feet and t is measured in seconds.

  a. How long will it take the fireworks to reach their maximum height? Round to the nearest second.

  b. Find the maximum height. Round to the nearest foot.

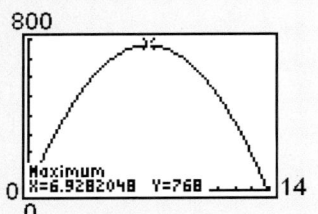
Solution:

$$h(t) = -16t^2 + 128\sqrt{3}t$$

This parabola opens downward; therefore, the maximum height of the fireworks will occur at the vertex of the parabola.

$$a = -16 \qquad b = 128\sqrt{3} \qquad c = 0$$

Identify a, b, and c, and apply the vertex formula.

The x-coordinate of the vertex is

$$\frac{-b}{2a} = \frac{-128\sqrt{3}}{2(-16)} = \frac{-128\sqrt{3}}{-32} \approx 6.9$$

The y-coordinate of the vertex is approximately

$$h(6.9) = -16(6.9)^2 + 128\sqrt{3}(6.9) \approx 768$$

The vertex is $(6.9, 768)$.

  a. The fireworks will reach their maximum height in 6.9 sec.

  b. The maximum height is 768 ft.

> **Skill Practice**

4. An object is launched into the air with an initial velocity of 48 ft/sec from the top of a building 288 ft high. The height $h(t)$ of the object after t seconds is given by

$$h(t) = -16t^2 + 48t + 288$$

  a. Find the time it takes for the object to reach its maximum height.

  b. Find the maximum height.

> ### Example 5   Applying a Quadratic Function

A group of students start a small company that produces CDs. The weekly profit (in dollars) is given by the function

$$P(x) = -2x^2 + 80x - 600$$

where x represents the number of CDs produced.

  a. Find the x-intercepts of the profit function, and interpret the meaning of the x-intercepts in the context of this problem.

  b. Find the y-intercept of the profit function, and interpret its meaning in the context of this problem.

Skill Practice Answers

4a. 1.5 sec   **b.** 324 ft

c. Find the vertex of the profit function, and interpret its meaning in the context of this problem.

d. Sketch the profit function.

Solution:

a. $P(x) = -2x^2 + 80x - 600$ The x-intercepts are the real solutions of the equation $P(x) = 0$.

$$0 = -2x^2 + 80x - 600$$
$$0 = -2(x^2 - 40x + 300)$$
$$0 = -2(x - 10)(x - 30)$$ Solve by factoring.
$$x = 10 \quad \text{or} \quad x = 30$$

The x-intercepts are $(10, 0)$ and $(30, 0)$. The x-intercepts represent points where the profit is zero. These are called break-even points. The break-even points occur when 10 CDs are produced and also when 30 CDs are produced.

b. The y-intercept is $P(0) = -2(0)^2 + 80(0) - 600 = -600$. The y-intercept is $(0, -600)$. The y-intercept indicates that if no CDs are produced, the company has a $600 loss.

c. The x-coordinate of the vertex is

$$\frac{-b}{2a} = \frac{-80}{2(-2)} = 20$$

The y-coordinate is $P(20) = -2(20)^2 + 80(20) - 600 = 200$.

The vertex is $(20, 200)$.
A maximum weekly profit of $200 is obtained when 20 CDs are produced.

d. Using the information from parts (a)–(c), sketch the profit function (Figure 12-19).

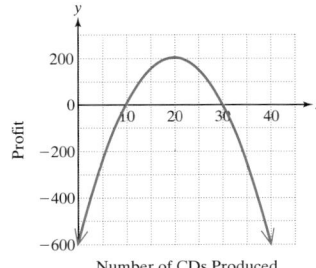

Number of CDs Produced

Figure 12-19

Skill Practice

5. The weekly profit function for a tutoring service is given by $P(x) = -4x^2 + 120x - 500$, where x is the number of hours that the tutors are scheduled to work and $P(x)$ represents profit in dollars.

a. Find the x-intercepts of the profit function, and interpret their meaning in the context of this problem.

b. Find the y-intercept of the profit function, and interpret its meaning in the context of this problem.

c. Find the vertex of the profit function, and interpret its meaning in the context of this problem.

d. Sketch the profit function.

Skill Practice Answers

5a. x-intercepts: $(5, 0)$ and $(25, 0)$. The x-intercepts represent the break-even points, where the profit is 0.

b. y-intercept: $(0, -500)$. If the service does no tutoring, it will have a $500 loss.

c. Vertex: $(15, 400)$. A maximum weekly profit of $400 is obtained when 15 hr of tutoring is scheduled.

d.

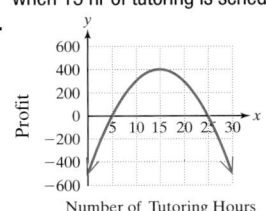

Number of Tutoring Hours

> **Example 6**   Applying a Quadratic Function

The average number of visits to office-based physicians is a function of the age of the patient.

$$N(x) = 0.0014x^2 - 0.0658x + 2.65$$

where x is a patient's age in years and $N(x)$ is the average number of doctor visits per year (Figure 12-20). Find the age for which the number of visits to office-based physicians is a minimum.

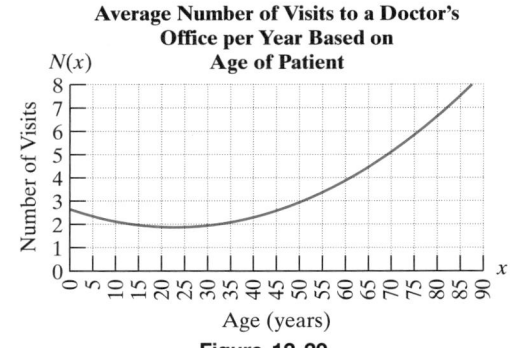

Average Number of Visits to a Doctor's Office per Year Based on Age of Patient

Figure 12-20

(Source: U.S. National Center for Health Statistics.)

Solution:

$$N(x) = 0.0014x^2 - 0.0658x + 2.65$$

$$\frac{-b}{2a} = \frac{-(-0.0658)}{2(0.0014)} = 23.5 \qquad \text{Find the } x\text{-coordinate of the vertex.}$$

The average number of visits to office-based physicians is lowest for people approximately 23.5 years old.

> **Skill Practice**

6. In a recent year, an Internet company started up, became successful, and then quickly went bankrupt. The price of a share of the company's stock is given by the function

$$S(x) = -2.25x^2 + 40.5x + 42.75$$

where x is the number of months that the stock was on the market. Find the number of months at which the stock price was a maximum.

Skill Practice Answers

6. The maximum price for the stock occurred after 9 months.

Section 12.5   Practice Exercises

- Practice Problems
- Self-Tests
- NetTutor
- e-Professors
- Videos

Study Skills Exercise

1. Write the **vertex formula**.

Review Exercises

2. How does the graph of $f(x) = -2x^2$ compare with the graph of $y = x^2$?

3. How does the graph of $p(x) = \frac{1}{4}x^2$ compare with the graph of $y = x^2$?

4. How does the graph of $Q(x) = x^2 - \frac{8}{3}$ compare with the graph of $y = x^2$?

5. How does the graph of $r(x) = x^2 + 7$ compare with the graph of $y = x^2$?

6. How does the graph of $s(x) = (x - 4)^2$ compare with the graph of $y = x^2$?

7. How does the graph of $t(x) = (x + 10)^2$ compare with the graph of $y = x^2$?

8. Find the coordinates of the vertex of the graph of $g(x) = 2(x + 3)^2 - 4$.

For Exercises 9–16, find the value of n to complete the square.

9. $x^2 - 8x + n$ **10.** $x^2 + 4x + n$ **11.** $y^2 + 7y + n$ **12.** $a^2 - a + n$

13. $b^2 + \frac{2}{9}b + n$ **14.** $m^2 - \frac{2}{7}m + n$ **15.** $t^2 - \frac{1}{3}t + n$ **16.** $p^2 + \frac{1}{4}p + n$

Concept 1: Writing a Quadratic Function in the Form $f(x) = a(x - h)^2 + k$

For Exercises 17–30, write the function in the form $f(x) = a(x - h)^2 + k$ by completing the square. Then identify the vertex.

17. $g(x) = x^2 - 8x + 5$ **18.** $h(x) = x^2 + 4x + 5$ **19.** $n(x) = 2x^2 + 12x + 13$

20. $f(x) = 4x^2 + 16x + 19$ **21.** $p(x) = -3x^2 + 6x - 5$ **22.** $q(x) = -2x^2 + 12x - 11$

23. $k(x) = x^2 + 7x - 10$ **24.** $m(x) = x^2 - x - 8$ **25.** $f(x) = x^2 + 8x + 1$

26. $g(x) = x^2 + 5x - 2$ **27.** $F(x) = 5x^2 + 10x + 1$ **28.** $G(x) = 4x^2 + 4x - 7$

29. $P(x) = -2x^2 + x$ **30.** $Q(x) = 3x^2 - 12x$

Concept 2: Vertex Formula

For Exercises 31–44, find the vertex by using the vertex formula.

31. $Q(x) = x^2 - 4x + 7$ **32.** $T(x) = x^2 - 8x + 17$ **33.** $r(x) = -3x^2 - 6x - 5$

34. $s(x) = -2x^2 - 12x - 19$ **35.** $N(x) = x^2 + 8x + 1$ **36.** $M(x) = x^2 + 6x - 5$

37. $m(x) = \frac{1}{2}x^2 + x + \frac{5}{2}$ **38.** $n(x) = \frac{1}{2}x^2 + 2x + 3$ **39.** $k(x) = -x^2 + 2x + 2$

40. $h(x) = -x^2 + 4x - 3$ **41.** $f(x) = 2x^2 + 4x + 6$ **42.** $g(x) = 3x^2 + 12x + 9$

43. $A(x) = -\frac{1}{3}x^2 + x$ **44.** $B(x) = -\frac{2}{3}x^2 - 2x$

Concept 3: Determining the Vertex and Intercepts of a Quadratic Function

For Exercises 45–50

    a. Find the vertex.

    b. Find the y-intercept.

    c. Find the x-intercept(s), if they exist.

    d. Use this information to graph the function.

45. $y = x^2 + 9x + 8$ **46.** $y = x^2 + 7x + 10$ **47.** $y = 2x^2 - 2x + 4$

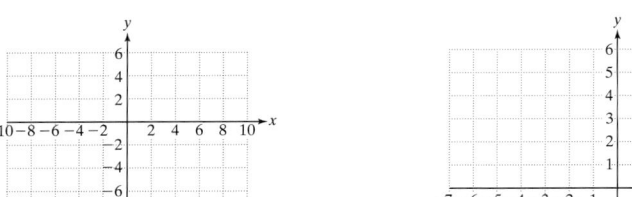
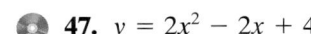
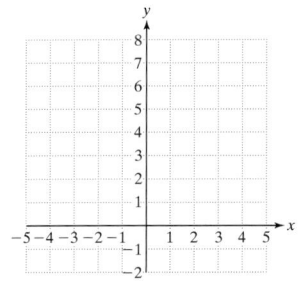

48. $y = 2x^2 - 12x + 19$ **49.** $y = -x^2 + 3x - \dfrac{9}{4}$ **50.** $y = -x^2 - \dfrac{3}{2}x - \dfrac{9}{16}$

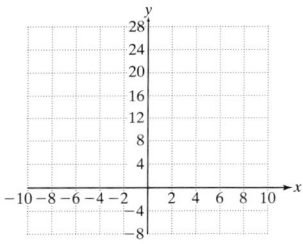
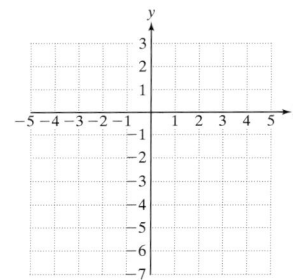
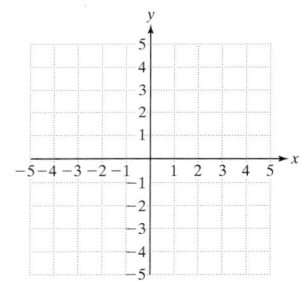

Concept 4: Vertex of a Parabola: Applications

51. Mia sells used MP3 players. The average cost to package MP3 players is given by the equation $C(x) = 2x^2 - 40x + 2200$, where x is the number of MP3 players she packages in a week. How many players must she package to minimize her average cost?

52. Ben sells used iPods. The average cost to package iPods is given by the equation $C(x) = 3x^2 - 120x + 1300$, where x is the number of iPods packaged per month. Determine the number of iPods that Ben needs to package to minimize the average cost.

53. The pressure x in an automobile tire can affect its wear. Both overinflated and underinflated tires can lead to poor performance and poor mileage. For one particular tire, the function P represents the number of miles that a tire lasts (in thousands) for a given pressure x.

$$P(x) = -0.857x^2 + 56.1x - 880$$

where x is the tire pressure in pounds per square inch (psi).

    a. Find the tire pressure that will yield the maximum mileage. Round to the nearest pound per square inch.

    b. What is the maximum number of miles that a tire can last? Round to the nearest thousand.

54. A baseball player throws a ball, and the height of the ball (in feet) can be approximated by

$$y(x) = -0.011x^2 + 0.577x + 5$$

where x is the horizontal position of the ball measured in feet from the origin.

a. For what value of x will the ball reach its highest point? Round to the nearest foot.

b. What is the maximum height of the ball? Round to the nearest tenth of a foot.

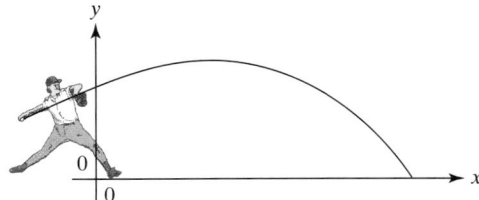

55. For a fund-raising activity, a charitable organization produces cookbooks to sell in the community. The profit (in dollars) depends on the number of cookbooks produced, x, according to

$$p(x) = -\frac{1}{50}x^2 + 12x - 550, \text{ where } x \geq 0.$$

a. How much profit is made when 100 cookbooks are produced?

b. Find the y-intercept of the profit function, and interpret its meaning in the context of this problem.

c. How many cookbooks must be produced for the organization to break even? (*Hint:* Find the x-intercepts.)

d. Find the vertex.

e. Sketch the function.

f. How many cookbooks must be produced to maximize profit? What is the maximum profit?

56. A jewelry maker sells bracelets at art shows. The profit (in dollars) depends on the number of bracelets produced, x, according to

$$p(x) = -\frac{1}{10}x^2 + 42x - 1260 \text{ where } x \geq 0.$$

a. How much profit does the jeweler make when 10 bracelets are produced?

b. Find the y-intercept of the profit function, and interpret its meaning in the context of this problem.

c. How many bracelets must be produced for the jeweler to break even? Round to the nearest whole unit.

d. Find the vertex.

e. Sketch the function.

f. How many bracelets must be produced to maximize profit? What is the maximum profit?

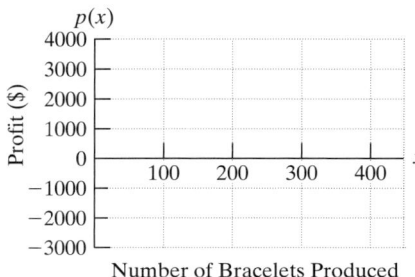

57. Gas mileage depends in part on the speed of the car. The gas mileage of a subcompact car is given by the function $m(x) = -0.04x^2 + 3.6x - 49$, where $20 \leq x \leq 70$ represents the speed in miles per hour and $m(x)$ is given in miles per gallon. At what speed will the car get the maximum gas mileage?

58. Gas mileage depends in part on the speed of the car. The gas mileage of a luxury car is given by the function $L(x) = -0.015x^2 + 1.44x - 21$, where $25 \leq x \leq 70$ represents the speed in miles per hour and $L(x)$ is given in miles per gallon. At what speed will the car get the maximum gas mileage?

59. Tetanus bacillus bacterium is cultured to produce tetanus toxin used in an inactive form for the tetanus vaccine. The amount of toxin produced per batch increases with time and then becomes unstable. The amount of toxin (in grams) as a function of time t (in hours) can be approximated by the following function.

$$b(t) = -\frac{1}{1152}t^2 + \frac{1}{12}t$$

How many hours will it take to produce the maximum yield?

60. The bacterium *Pseudomonas aeruginosa* is cultured with an initial population of 10^4 active organisms. The population of active bacteria increases up to a point, and then due to a limited food supply and an increase of waste products, the population of living organisms decreases. Over the first 48 hr, the population can be approximated by the following function.

$$P(t) = -1718.75t^2 + 82{,}500t + 10{,}000 \qquad \text{where } 0 \leq t \leq 48$$

Find the time required for the population to reach its maximum value.

Expanding Your Skills

61. A farmer wants to fence a rectangular corral adjacent to the side of a barn; however, she has only 200 ft of fencing and wants to enclose the largest possible area. See the figure.

a. If x represents the length of the corral and y represents the width, explain why the dimensions of the corral are subject to the constraint $2x + y = 200$.

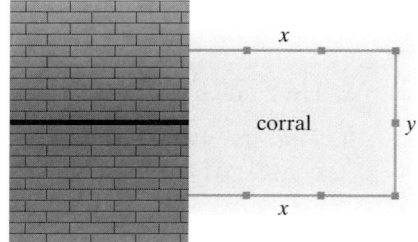

b. The area of the corral is given by $A = xy$. Use the constraint equation from part (a) to express A as a function of x, where $0 < x < 100$.

c. Use the function from part (b) to find the dimensions of the corral that will yield the maximum area. [*Hint:* Find the vertex of the function from part (b).]

62. A veterinarian wants to construct two equal-sized pens of maximum area out of 240 ft of fencing. See the figure.

    a. If x represents the length of each pen and y represents the width of each pen, explain why the dimensions of the pens are subject to the constraint $3x + 4y = 240$.

    b. The area of each individual pen is given by $A = xy$. Use the constraint equation from part (a) to express A as a function of x, where $0 < x < 80$.

    c. Use the function from part (b) to find the dimensions of an individual pen that will yield the maximum area. [*Hint:* Find the vertex of the function from part (b).]

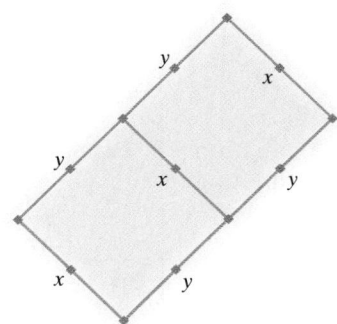

Graphing Calculator Exercises

For Exercises 63–68, graph the functions in Exercises 45–50 on a graphing calculator. Use the *Max* or *Min* feature or *Zoom* and *Trace* to approximate the vertex.

63. $Y_1 = x^2 + 9x + 8$ (Exercise 45)

64. $Y_1 = x^2 + 7x + 10$ (Exercise 46)

65. $Y_1 = 2x^2 - 2x + 4$ (Exercise 47)

66. $Y_1 = 2x^2 - 12x + 19$ (Exercise 48)

67. $Y_1 = -x^2 + 3x - \dfrac{9}{4}$ (Exercise 49)

68. $Y_1 = -x^2 - \dfrac{3}{2}x - \dfrac{9}{16}$ (Exercise 50)

Chapter 12   SUMMARY

Section 12.1   Square Root Property and Completing the Square

Key Concepts

The **square root property** states that

If $x^2 = k$ then $x = \pm\sqrt{k}$

Follow these steps to solve a quadratic equation in the form $ax^2 + bx + c = 0 \,(a \neq 0)$ by completing the square and applying the square root property:

1. Divide both sides by a to make the leading coefficient 1.
2. Isolate the variable terms on one side of the equation.
3. Complete the square: Add the square of one-half the linear term coefficient to both sides of the equation. Then factor the resulting perfect square trinomial.
4. Apply the square root property and solve for x.

Examples

Example 1

$$(x - 5)^2 = -13$$

$$x - 5 = \pm\sqrt{-13} \qquad \text{(square root property)}$$

$$x = 5 \pm i\sqrt{13}$$

Example 2

$$2x^2 - 6x - 5 = 0$$

$$\frac{2x^2}{2} - \frac{6x}{2} - \frac{5}{2} = \frac{0}{2}$$

$$x^2 - 3x = \frac{5}{2}$$

$$\textit{Note: } \left[\frac{1}{2} \cdot (-3)\right]^2 = \frac{9}{4}$$

$$x^2 - 3x + \frac{9}{4} = \frac{5}{2} + \frac{9}{4}$$

$$x^2 - 3x + \frac{9}{4} = \frac{10}{4} + \frac{9}{4}$$

$$\left(x - \frac{3}{2}\right)^2 = \frac{19}{4}$$

$$x - \frac{3}{2} = \pm\sqrt{\frac{19}{4}}$$

$$x = \frac{3}{2} \pm \frac{\sqrt{19}}{2} \qquad \text{or} \qquad x = \frac{3 \pm \sqrt{19}}{2}$$

Section 12.2 Quadratic Formula

Key Concepts

The solutions to a quadratic equation $ax^2 + bx + c = 0$ $(a \neq 0)$ are given by the **quadratic formula**

$$x = \frac{-b \pm \sqrt{b^2 - 4ac}}{2a}$$

The **discriminant** of a quadratic equation $ax^2 + bx + c = 0$ is $b^2 - 4ac$. If a, b, and c are rational numbers, then

1. If $b^2 - 4ac > 0$, then there will be two real solutions. Moreover,
   a. If $b^2 - 4ac$ is a perfect square, the solutions will be rational numbers.
   b. If $b^2 - 4ac$ is not a perfect square, the solutions will be irrational numbers.
2. If $b^2 - 4ac < 0$, then there will be two imaginary solutions.
3. If $b^2 - 4ac = 0$, then there will be one rational solution.

Three methods to solve a quadratic equation are

1. Factoring and applying the zero product rule.
2. Completing the square and applying the square root property.
3. Using the quadratic formula.

Example

Example 1

$$3x^2 - 2x + 4 = 0$$

$$a = 3 \qquad b = -2 \qquad c = 4$$

$$x = \frac{-(-2) \pm \sqrt{(-2)^2 - 4(3)(4)}}{2(3)}$$

$$= \frac{2 \pm \sqrt{4 - 48}}{6}$$

$$= \frac{2 \pm \sqrt{-44}}{6} \qquad$$ The discriminant is -44. Therefore, there will be two imaginary solutions.

$$= \frac{2 \pm 2i\sqrt{11}}{6}$$

$$= \frac{1 \pm i\sqrt{11}}{3}$$

Section 12.3 Equations in Quadratic Form

Key Concepts

Substitution can be used to solve equations that are in quadratic form.

Examples

Example 1

$$x^{2/3} - x^{1/3} - 12 = 0$$

Let $u = x^{1/3}$. Therefore, $u^2 = (x^{1/3})^2 = x^{2/3}$

$$u^2 - u - 12 = 0$$

$$(u - 4)(u + 3) = 0$$

$$u = 4 \qquad \text{or} \qquad u = -3$$

$$x^{1/3} = 4 \qquad \text{or} \qquad x^{1/3} = -3$$

$$x = 64 \qquad \text{or} \qquad x = -27 \quad \text{Cube both sides.}$$

Section 12.4   **Graphs of Quadratic Functions**

Key Concepts

A quadratic function of the form $f(x) = x^2 + k$ shifts the graph of $y = x^2$ up k units if $k > 0$ and down $|k|$ units if $k < 0$.

Examples

Example 1

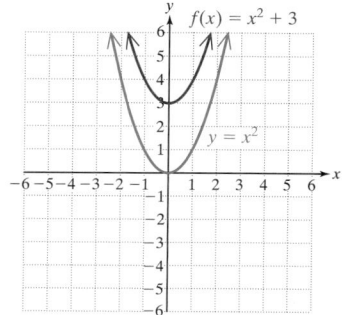

A quadratic function of the form $f(x) = (x - h)^2$ shifts the graph of $y = x^2$ right h units if $h > 0$ and left $|h|$ units if $h < 0$.

Example 2

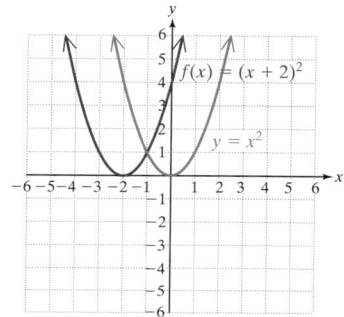

The graph of a quadratic function of the form $f(x) = ax^2$ is a parabola that opens up when $a > 0$ and opens down when $a < 0$. If $|a| > 1$, the graph of $y = x^2$ is stretched vertically by a factor of $|a|$. If $0 < |a| < 1$, the graph of $y = x^2$ is shrunk vertically by a factor of $|a|$.

Example 3

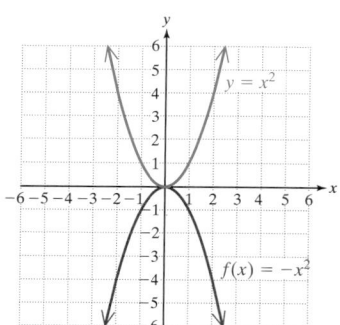

A quadratic function of the form $f(x) = a(x - h)^2 + k$ has vertex (h, k). If $a > 0$, the vertex represents the minimum point. If $a < 0$, the vertex represents the maximum point.

Example 4

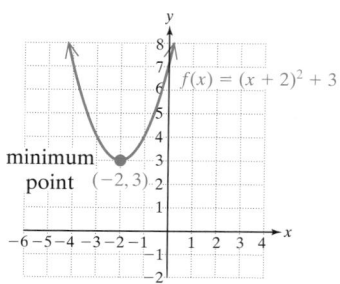

Section 12.5   Vertex of a Parabola and Applications

Key Concepts

Completing the square is a technique used to write a quadratic function $f(x) = ax^2 + bx + c$ $(a \neq 0)$ in the form $f(x) = a(x - h)^2 + k$ for the purpose of identifying the vertex (h, k).

Examples

Example 1

$$f(x) = 3x^2 + 6x + 11$$
$$= 3(x^2 + 2x \qquad) + 11$$
$$= 3(x^2 + 2x + 1 - 1) + 11$$
$$= 3(x^2 + 2x + 1) - 3 + 11$$
$$= 3(x + 1)^2 + 8$$
$$= 3[x - (-1)]^2 + 8$$

The vertex is $(-1, 8)$. Because $a = 3 > 0$, the parabola opens upward and the vertex $(-1, 8)$ is a minimum point.

The **vertex formula** finds the vertex of a quadratic function $f(x) = ax^2 + bx + c$ $(a \neq 0)$.

The vertex is

$$\left(\frac{-b}{2a}, \frac{4ac - b^2}{4a}\right) \quad \text{or} \quad \left(\frac{-b}{2a}, f\left(\frac{-b}{2a}\right)\right)$$

Example 2

$$f(x) = -5x^2 + 4x - 1$$
$$a = -5 \qquad b = 4 \qquad c = -1$$
$$x = \frac{-4}{2(-5)} = \frac{2}{5}$$
$$f\left(\frac{2}{5}\right) = -5\left(\frac{2}{5}\right)^2 + 4\left(\frac{2}{5}\right) - 1 = -\frac{1}{5}$$

The vertex is $(\frac{2}{5}, -\frac{1}{5})$. Because $a = -5 < 0$, the parabola opens downward and the vertex $(\frac{2}{5}, -\frac{1}{5})$ is a maximum point.

Chapter 12   Review Exercises

Section 12.1

For Exercises 1–8, solve the equations by using the square root property.

1. $x^2 = 5$

2. $2y^2 = -8$

3. $a^2 = 81$

4. $3b^2 = -19$

5. $(x - 2)^2 = 72$

6. $(2x - 5)^2 = -9$

7. $(3y - 1)^2 = 3$

8. $3(m - 4)^2 = 15$

9. The length of each side of an equilateral triangle is 10 in. Find the height of the triangle. Round the answer to the nearest tenth of an inch.

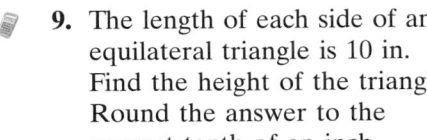

10. Use the square root property to find the length of the sides of a square whose area is 81 in.^{2}

11. Use the square root property to find the length of the sides of a square whose area is 150 in.^{2} Round the answer to the nearest tenth of an inch.

For Exercises 12–15, find the value of n so that the expression is a perfect square trinomial. Then factor the trinomial.

12. $x^2 + 16x + n$

13. $x^2 - 9x + n$

14. $y^2 + \frac{1}{2}y + n$

15. $z^2 - \frac{2}{5}z + n$

For Exercises 16–21, solve the equation by completing the square and applying the square root property.

16. $w^2 + 4w + 13 = 0$ **17.** $4y^2 - 12y + 13 = 0$

18. $3x^2 + 2x = 1$ **19.** $b^2 + \frac{7}{2}b = 2$

20. $2x^2 = 12x + 6$ **21.** $-t^2 + 8t - 25 = 0$

Section 12.2

22. Explain how the discriminant can determine the type and number of solutions to a quadratic equation with rational coefficients.

For Exercises 23–28, determine the type (rational, irrational, or imaginary) and number of solutions for the equations by using the discriminant.

23. $x^2 - 5x = -6$ **24.** $2y^2 = -3y$

25. $z^2 + 23 = 17z$ **26.** $a^2 + a + 1 = 0$

27. $10b + 1 = -25b^2$ **28.** $3x^2 + 15 = 0$

For Exercises 29–36, solve the equations by using the quadratic formula.

29. $y^2 - 4y + 1 = 0$ **30.** $m^2 - 5m + 25 = 0$

31. $6a(a - 1) = 10 + a$ **32.** $3x(x - 3) = x - 8$

33. $b^2 - \frac{4}{25} = \frac{3}{5}b$ **34.** $k^2 + 0.4k = 0.05$

35. $-32 + 4x - x^2 = 0$ **36.** $8y - y^2 = 10$

For Exercises 37–40, solve using any method.

37. $3x^2 - 4x = 6$ **38.** $\frac{b}{8} - \frac{2}{b} = \frac{3}{4}$

39. $y^2 + 14y = -46$ **40.** $(a + 1)^2 = 11$

41. The landing distance that a certain plane will travel on a runway is determined by the initial landing speed at the instant the plane touches down. The function D relates landing distance in feet to initial landing speed s:

$$D(s) = \frac{1}{10}s^2 - 3s + 22 \text{ for } s \geq 50$$

where s is in feet per second.

a. Find the landing distance for a plane traveling 150 ft/sec at touchdown.

b. If the landing speed is too fast, the pilot may run out of runway. If the speed is too slow, the plane may stall. Find the maximum initial landing speed of a plane for a runway that is 1000 ft long. Round to 1 decimal place.

42. The recent population (in thousands) of Kenya can be approximated by $P(t) = 4.62t^2 + 564.6t + 13{,}128$, where t is the number of years since 1974.

a. If this trend continues, approximate the number of people in Kenya in the year 2025.

b. In what year after 1974 will the population of Kenya reach 50 million? (*Hint:* 50 million equals 50,000 thousand.)

Section 12.3

For Exercises 43–52, solve the equations by using substitution, if necessary.

43. $x - 4\sqrt{x} - 21 = 0$

44. $n - 6\sqrt{n} + 8 = 0$

45. $y^4 - 11y^2 + 18 = 0$

46. $2m^4 - m^2 - 3 = 0$

47. $t^{2/5} + t^{1/5} - 6 = 0$

48. $p^{2/5} - 3p^{1/5} + 2 = 0$

49. $\frac{2t}{t + 1} + \frac{-3}{t - 2} = 1$

50. $\frac{1}{m - 2} - \frac{m}{m + 3} = 2$

51. $(x^2 + 5)^2 + 2(x^2 + 5) - 8 = 0$

52. $(x^2 - 3)^2 - 5(x^2 - 3) + 4 = 0$

Section 12.4

For Exercises 53–60, graph the functions.

53. $g(x) = x^2 - 5$

54. $f(x) = x^2 + 3$

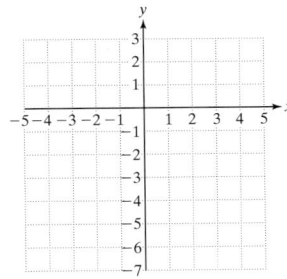

55. $h(x) = (x - 5)^2$

56. $k(x) = (x + 3)^2$

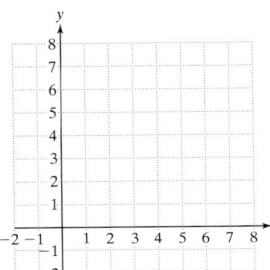
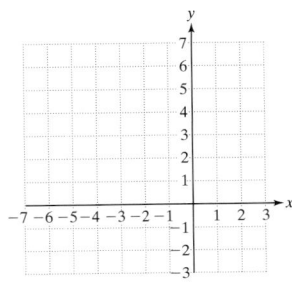

57. $m(x) = -2x^2$

58. $n(x) = -4x^2$

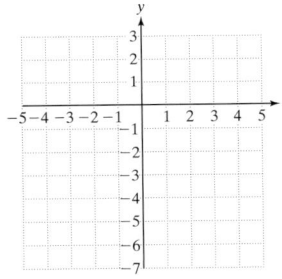
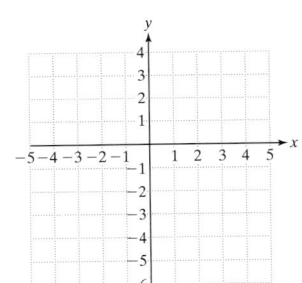

59. $p(x) = -2(x - 5)^2 - 5$

60. $q(x) = -4(x + 3)^2 + 3$

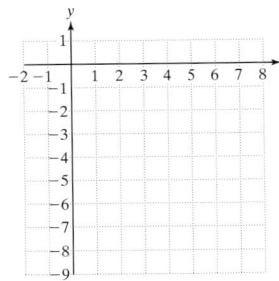
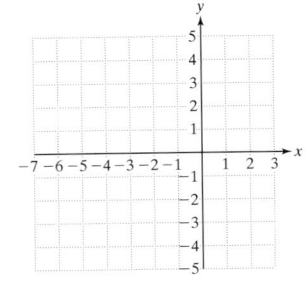

For Exercises 61–62, write the coordinates of the vertex of the parabola and determine if the vertex is a maximum point or a minimum point. Then write the maximum or the minimum value.

61. $t(x) = \dfrac{1}{3}(x - 4)^2 + \dfrac{5}{3}$

62. $s(x) = -\dfrac{5}{7}(x - 1)^2 - \dfrac{1}{7}$

For Exercises 63–64, write the equation of the axis of symmetry of the parabola.

63. $a(x) = -\dfrac{3}{2}\left(x + \dfrac{2}{11}\right)^2 - \dfrac{4}{13}$

64. $w(x) = -\dfrac{4}{3}\left(x - \dfrac{3}{16}\right)^2 + \dfrac{2}{9}$

Section 12.5

For Exercises 65–68, write the function in the form $f(x) = a(x - h)^2 + k$ by completing the square. Then write the coordinates of the vertex.

65. $z(x) = x^2 - 6x + 7$

66. $b(x) = x^2 - 4x - 44$

67. $p(x) = -5x^2 - 10x - 13$

68. $q(x) = -3x^2 - 24x - 54$

For Exercises 69–72, find the coordinates of the vertex of each function by using the vertex formula.

69. $f(x) = -2x^2 + 4x - 17$

70. $g(x) = -4x^2 - 8x + 3$

71. $m(x) = 3x^2 - 3x + 11$

72. $n(x) = 3x^2 + 2x - 7$

73. For the quadratic equation $y = -(x + 2)^2 + 4$

    a. Write the coordinates of the vertex.

    b. Find the x- and y-intercepts.

c. Use this information to sketch a graph of the parabola.

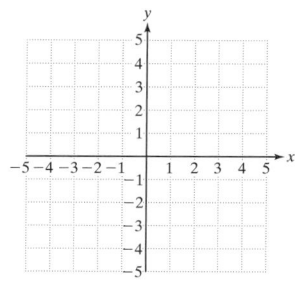

74. The height of a projectile fired vertically into the air from the ground is given by the equation $h(t) = -16t^2 + 96t$, where t represents the number of seconds that the projectile has been in the air. How long will it take the projectile to reach its maximum height?

75. The weekly profit for a small catering service is given by $P(x) = -4x^2 + 1200x$, where x is the number of meals prepared. Find the number of meals that should be prepared to obtain the maximum profit.

Chapter 12   Test

For Exercises 1–3, solve the equation by using the square root property.

1. $(x + 3)^2 = 25$

2. $(p - 2)^2 = 12$

3. $(m + 1)^2 = -1$

4. Find the value of n so that the expression is a perfect square trinomial. Then factor the trinomial $d^2 + 7d + n$.

For Exercises 5–6, solve the equation by completing the square and applying the square root property.

5. $2x^2 + 12x - 36 = 0$

6. $2x^2 = 3x - 7$

For Exercises 7–8

  a. Write the equation in standard form $ax^2 + bx + c = 0$.

  b. Identify a, b, and c.

  c. Find the discriminant.

  d. Determine the number and type (rational, irrational, or imaginary) of solutions.

7. $x^2 - 3x = -12$

8. $y(y - 2) = -1$

For Exercises 9–10, solve the equation by using the quadratic formula.

9. $3x^2 - 4x + 1 = 0$

10. $x(x + 6) = -11 - x$

11. The base of a triangle is 3 ft less than twice the height. The area of the triangle is 14 ft². Find the base and the height. Round the answers to the nearest tenth of a foot.

12. A circular garden has an area of approximately 450 ft². Find the radius. Round the answer to the nearest tenth of a foot.

For Exercises 13–15, solve the equation by using substitution, if necessary.

13. $x - \sqrt{x} - 6 = 0$

14. $y^{2/3} + 2y^{1/3} = 8$

15. $(3y - 8)^2 - 13(3y - 8) + 30 = 0$

16. $p^4 - 15p^2 = -54$

17. $3 = \dfrac{y}{2} - \dfrac{1}{y + 1}$

For Exercises 18–21, find the x- and y-intercepts of the function. Then match the function with its graph.

18. $f(x) = x^2 - 6x + 8$

19. $k(x) = x^3 + 4x^2 - 9x - 36$

20. $p(x) = -2x^2 - 8x - 6$

21. $q(x) = x^3 - x^2 - 12x$

a.

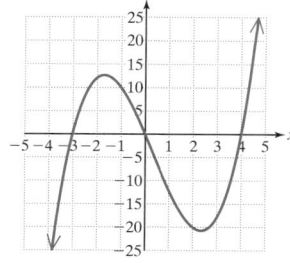

b.

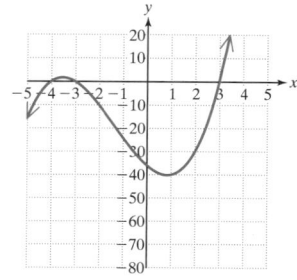

c.

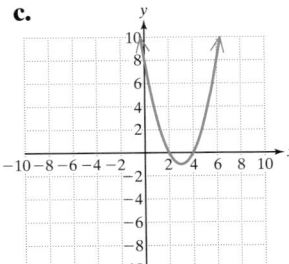

d.

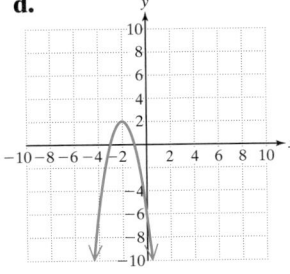

22. A child launches a toy rocket from the ground. The height of the rocket can be determined by its horizontal distance from the launch pad x by

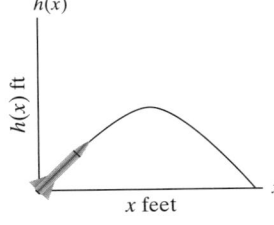

$$h(x) = -\frac{x^2}{256} + x$$

where x and $h(x)$ are in feet.

How many feet from the launch pad will the rocket hit the ground?

23. The recent population (in millions) of India can be approximated by $P(t) = 0.135t^2 + 12.6t + 600$, where $t = 0$ corresponds to the year 1974.

    a. If this trend continues, approximate the number of people in India in the year 2014.

    b. Approximate the year in which the population of India reached 1 billion (1000 million). (Round to the nearest year.)

24. Explain the relationship between the graphs of $y = x^2$ and $y = x^2 - 2$.

25. Explain the relationship between the graphs of $y = x^2$ and $y = (x + 3)^2$.

26. Explain the relationship between the graphs of $y = 4x^2$ and $y = -4x^2$.

27. Given the function defined by

$$f(x) = -(x - 4)^2 + 2$$

    a. Identify the vertex of the parabola.

    b. Does this parabola open upward or downward?

    c. Does the vertex represent the maximum or minimum point of the function?

    d. What is the maximum or minimum value of the function f?

    e. What is the axis of symmetry for this parabola?

28. For the function defined by $g(x) = 2x^2 - 20x + 51$, find the vertex by using two methods.

    a. Complete the square to write $g(x)$ in the form $g(x) = a(x - h)^2 + k$. Identify the vertex.

    b. Use the vertex formula to find the vertex.

29. A farmer has 400 ft of fencing with which to enclose a rectangular field. The field is situated such that one of its sides is adjacent to a river and requires no fencing. The area of the field (in square feet) can be modeled by

$$A(x) = -\frac{x^2}{2} + 200x$$

where x is the length of the side parallel to the river (measured in feet).

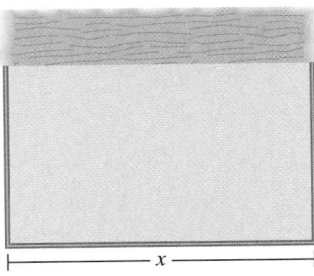

Use the function to determine the maximum area that can be enclosed.

Chapters 1–12   Cumulative Review Exercises

1. Given: $A = \{2, 4, 6, 8, 10\}$ and $B = \{2, 8, 12, 16\}$

   a. Find $A \cup B$.   b. Find $A \cap B$.

2. Perform the indicated operations and simplify.

   $$(2x^2 - 5) - (x + 3)(5x - 2)$$

3. Simplify completely. $4^0 - \left(\dfrac{1}{2}\right)^{-3} - 81^{1/2}$

4. Perform the indicated operations. Write the answer in scientific notation:

   $$(3.0 \times 10^{12})(6.0 \times 10^{-3})$$

5. a. Factor completely. $x^3 + 2x^2 - 9x - 18$

   b. Divide by using long division. Identify the quotient and remainder.

   $$(x^3 + 2x^2 - 9x - 18) \div (x - 3)$$

6. Multiply. $(\sqrt{x} - \sqrt{2})(\sqrt{x} + \sqrt{2})$

7. Simplify. $\dfrac{4}{\sqrt{2x}}$

8. Jacques invests a total of $10,000 in two mutual funds. After 1 year, one fund produced 12% growth, and the other lost 3%. Find the amount invested in each fund if the total investment grew by $900.

9. Solve the system of equations.

   $$\dfrac{1}{9}x - \dfrac{1}{3}y = -\dfrac{13}{9}$$

   $$x - \dfrac{1}{2}y = \dfrac{9}{2}$$

10. An object is fired straight up into the air from an initial height of 384 ft with an initial velocity of 160 ft/sec. The height in feet is given by

    $$h(t) = -16t^2 + 160t + 384$$

    where t is the time in seconds after launch.

    a. Find the height of the object after 3 sec.

    b. Find the height of the object after 7 sec.

    c. Find the time required for the object to hit the ground.

11. Solve the equation. $(x - 3)^2 + 16 = 0$

12. Solve the equation. $2x^2 + 5x - 1 = 0$

13. What number would have to be added to the quantity $x^2 + 10x$ to make it a perfect square trinomial?

14. Factor completely. $2x^3 + 250$

15. Graph the line.
    $3x - 5y = 10$

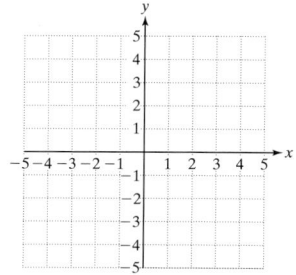

16. a. Find the x-intercepts of the function defined by $g(x) = 2x^2 - 9x + 10$.

    b. What is the y-intercept of $y = g(x)$?

17. Michael Jordan was the NBA leading scorer for 10 of 12 seasons between 1987 and 1998. In his 1998 season, he scored a total of 2357 points consisting of 1-point free throws, 2-point field goals, and 3-point field goals. He scored 286 more 2-point shots than he did free throws. The number of 3-point shots was 821 less than the number of 2-point shots. Determine the number of free throws, 2-point shots, and 3-point shots scored by Michael Jordan during his 1998 season.

18. Explain why this relation is *not* a function.

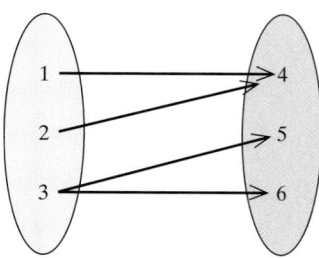

19. Graph the function defined by $f(x) = \dfrac{1}{x}$.

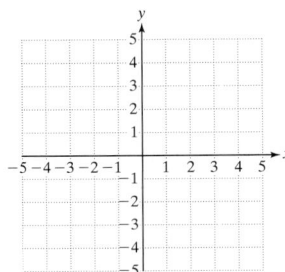

20. The quantity y varies directly as x and inversely as z. If $y = 15$ when $x = 50$ and $z = 10$, find y when $x = 65$ and $z = 5$.

21. Given the function defined by $g(x) = \sqrt{2 - x}$, find the function values (if they exist) over the set of real numbers.

   a. $g(-7)$ **b.** $g(0)$ **c.** $g(3)$

22. Let $m(x) = \sqrt{x + 4}$ and $n(x) = x^2 + 2$. Find

   a. The domain of m **b.** The domain of n

23. Consider the function $y = f(x)$ graphed here. Find

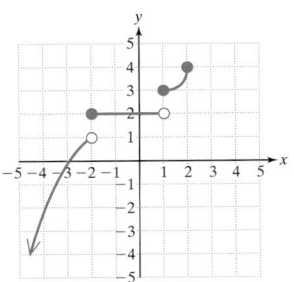

   a. The domain   **b.** The range

   c. $f(-2)$ **d.** $f(1)$ **e.** $f(0)$

   f. For what value(s) of x is $f(x) = 0$?

24. a. Solve the equation: $|3x - 4| = 2$

   b. Solve the inequality: $|3x - 4| > 2$

25. Solve for f. $\dfrac{1}{p} + \dfrac{1}{q} = \dfrac{1}{f}$

26. Solve. $\dfrac{15}{t^2 - 2t - 8} = \dfrac{1}{t - 4} + \dfrac{2}{t + 2}$

27. Simplify. $\dfrac{y - \dfrac{4}{y - 3}}{y - 4}$

28. Graph the inequality: $4x + y \le 8$

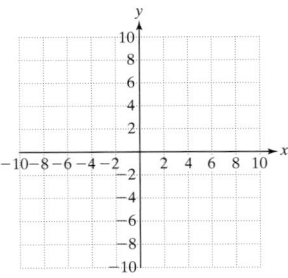

29. Given: the function defined by $f(x) = 2(x - 3)^2 + 1$

   a. Write the coordinates of the vertex.

   b. Does the graph of the function open upward or downward?

   c. Write the coordinates of the y-intercept.

   d. Find the x-intercepts, if possible.

   e. Sketch the function.

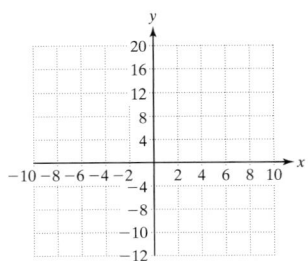

30. Find the vertex of the parabola.

$$f(x) = x^2 - 16x + 2$$

Exponential and Logarithmic Functions

13

13.1 Algebra and Composition of Functions

13.2 Inverse Functions

13.3 Exponential Functions

13.4 Logarithmic Functions

13.5 Properties of Logarithms

13.6 The Irrational Number *e*

   Problem Recognition Exercises—Logarithmic and Exponential Forms

13.7 Logarithmic and Exponential Equations

Chapter 13 is devoted to the study exponential and logarithmic functions. These functions are used to study many naturally occurring phenomena such as population growth, exponential decay of radioactive matter, and growth of investments.

   The following is a Sudoku puzzle. As you work through this chapter, try to simplify the expressions or solve the equations in the clues given below. Use the clues to fill in the boxes labeled a–n. Then fill in the remaining part of the grid so that every row, every column, and every 2 × 3 box contains the digits 1 through 6.

Clues

a. $\log_2 8$

b. $\ln e$

c. Solution to $3^{2x-10} = 9$

d. Solution to $\ln(x + 3) = \ln 8$

e. $1 + \log 1$

f. $\log_2 16x - \log_2 x$

g. Solution to $\log_{12}(x - 4) = 1 - \log_{12} x$

h. $\dfrac{\ln 2^2}{\ln 2}$

i. $f(f^{-1}(6))$

j. $\ln e^4$

k. $\log_7 7^5$

l. e^0

m. $\left(\dfrac{1}{2}\right)^{-x} = 16$

n. $\log 100$

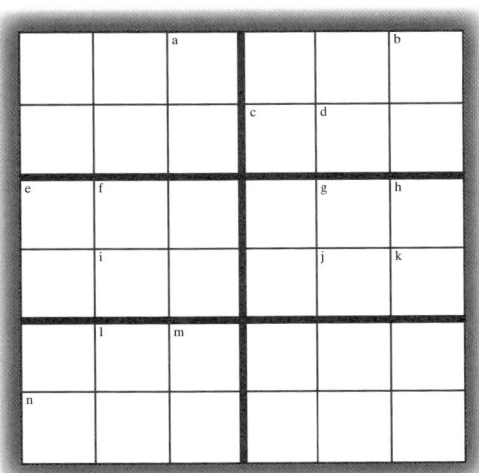

Section 13.1 Algebra and Composition of Functions

1. Algebra of Functions

Addition, subtraction, multiplication, and division can be used to create a new function from two or more functions. The domain of the new function will be the intersection of the domains of the original functions. Finding domains of functions was first introduced in Section 8.2.

Sum, Difference, Product, and Quotient of Functions

Given two functions f and g, the functions $f + g$, $f - g$, $f \cdot g$, and $\frac{f}{g}$ are defined as

$$(f + g)(x) = f(x) + g(x)$$
$$(f - g)(x) = f(x) - g(x)$$
$$(f \cdot g)(x) = f(x) \cdot g(x)$$
$$\left(\frac{f}{g}\right)(x) = \frac{f(x)}{g(x)} \quad \text{provided } g(x) \neq 0$$

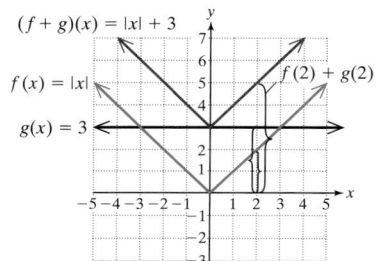

Figure 13-1

For example, suppose $f(x) = |x|$ and $g(x) = 3$. Taking the sum of the functions produces a new function denoted by $(f + g)$. In this case, $(f + g)(x) = |x| + 3$. Graphically, the y-values of the function $(f + g)$ are given by the sum of the corresponding y-values of f and g. This is depicted in Figure 13-1. The function $(f + g)$ appears in red. In particular, notice that $(f + g)(2) = f(2) + g(2) = 2 + 3 = 5$.

Example 1 Adding, Subtracting, and Multiplying Functions

Given: $g(x) = 4x$ $h(x) = x^2 - 3x$ $k(x) = \sqrt{x - 2}$

a. Find $(g + h)(x)$ and write the domain of $(g + h)$ in interval notation.

b. Find $(h - g)(x)$ and write the domain of $(h - g)$ in interval notation.

c. Find $(g \cdot k)(x)$ and write the domain of $(g \cdot k)$ in interval notation.

Solution:

a. $(g + h)(x) = g(x) + h(x)$
$$= (4x) + (x^2 - 3x)$$
$$= 4x + x^2 - 3x$$
$$= x^2 + x \qquad \text{The domain is all real numbers } (-\infty, \infty).$$

b. $(h - g)(x) = h(x) - g(x)$
$$= (x^2 - 3x) - (4x)$$
$$= x^2 - 3x - 4x$$
$$= x^2 - 7x \qquad \text{The domain is all real numbers } (-\infty, \infty).$$

c. $(g \cdot k)(x) = g(x) \cdot k(x)$
$$= (4x)(\sqrt{x - 2})$$
$$= 4x\sqrt{x - 2} \qquad \text{The domain is } [2, \infty) \text{ because } x - 2 \geq 0$$
$$\text{for } x \geq 2.$$

Skill Practice Given:

$$f(x) = x - 1$$
$$g(x) = 5x^2 + x$$
$$h(x) = \sqrt{5 - x}$$

Perform the indicated operations. Write the domain of the resulting function in interval notation.

1. $(f + g)(x)$ **2.** $(g - f)(x)$ **3.** $(f \cdot h)(x)$

Example 2 Dividing Functions

Given the functions defined by $h(x) = x^2 - 3x$ and $k(x) = \sqrt{x - 2}$, find $\left(\frac{k}{h}\right)(x)$ and write the domain of $\left(\frac{k}{h}\right)$ in interval notation.

Solution:

$$\left(\frac{k}{h}\right)(x) = \frac{\sqrt{x - 2}}{x^2 - 3x}$$

To find the domain, we must consider the restrictions on x imposed by the square root and by the fraction.

- From the numerator we have $x - 2 \geq 0$ or, equivalently, $x \geq 2$.
- From the denominator we have $x^2 - 3x \neq 0$ or, equivalently, $x(x - 3) \neq 0$. Hence, $x \neq 3$ and $x \neq 0$.

Thus, the domain of $\frac{k}{h}$ is the set of real numbers greater than or equal to 2, but not equal to 3 or 0. This is shown graphically in Figure 13-2.

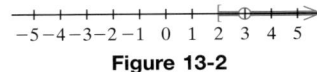

Figure 13-2

The domain is $[2, 3) \cup (3, \infty)$.

Skill Practice Given:

$$f(x) = \sqrt{x + 1}$$
$$g(x) = x^2 + 2x$$

4. Find $\left(\frac{f}{g}\right)(x)$ and write the domain interval notation.

---

2. Composition of Functions

Composition of Functions

The **composition** of f and g, denoted $f \circ g$, is defined by the rule

$$(f \circ g)(x) = f(g(x)) \quad \text{provided that } g(x) \text{ is in the domain of } f$$

The composition of g and f, denoted $g \circ f$, is defined by the rule

$$(g \circ f)(x) = g(f(x)) \quad \text{provided that } f(x) \text{ is in the domain of } g$$

Note: $f \circ g$ is also read as "f compose g," and $g \circ f$ is also read as "g compose f."

Skill Practice Answers

1. $5x^2 + 2x - 1$; domain: $(-\infty, \infty)$
2. $5x^2 + 1$; domain: $(-\infty, \infty)$
3. $(x - 1)\sqrt{5 - x}$; domain: $(-\infty, 5]$
4. $\dfrac{\sqrt{x + 1}}{x^2 + 2x}$; domain: $[-1, 0) \cup (0, \infty)$

For example, given $f(x) = 2x - 3$ and $g(x) = x + 5$, we have

$$(f \circ g)(x) = f(g(x))$$

$$= f(x + 5) \qquad \text{Substitute } g(x) = x + 5 \text{ into the function } f.$$

$$= 2(x + 5) - 3$$

$$= 2x + 10 - 3$$

$$= 2x + 7$$

In this composition, the function g is the innermost operation and acts on x first. Then the output value of function g becomes the domain element of the function f, as shown in the figure.

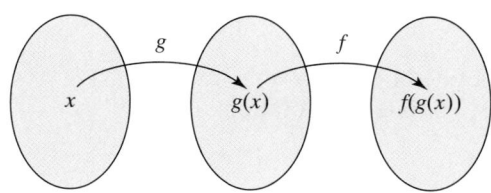

Example 3   Composing Functions

Given: $f(x) = x - 5$, $g(x) = x^2$, and $n(x) = \sqrt{x + 2}$

a. Find $(f \circ g)(x)$ and write the domain of $(f \circ g)$ in interval notation.

b. Find $(g \circ f)(x)$ and write the domain of $(g \circ f)$ in interval notation.

c. Find $(n \circ f)(x)$ and write the domain of $(n \circ f)$ in interval notation.

Solution:

a. $(f \circ g)(x) = f(g(x))$

$$= f(x^2) \qquad \text{Evaluate the function } f \text{ at } x^2.$$

$$= (x^2) - 5 \qquad \text{Replace } x \text{ by } x^2 \text{ in the function } f.$$

$$= x^2 - 5 \qquad \text{The domain is all real numbers } (-\infty, \infty).$$

b. $(g \circ f)(x) = g(f(x))$

$$= g(x - 5) \qquad \text{Evaluate the function } g \text{ at } (x - 5).$$

$$= (x - 5)^2 \qquad \text{Replace } x \text{ by } (x - 5) \text{ in function } g.$$

$$= x^2 - 10x + 25 \qquad \text{The domain is all real numbers } (-\infty, \infty).$$

c. $(n \circ f)(x) = n(f(x))$

$$= n(x - 5) \qquad \text{Evaluate the function } n \text{ at } x - 5.$$

$$= \sqrt{(x - 5) + 2} \qquad \text{Replace } x \text{ by the quantity } (x - 5) \text{ in function } n.$$

$$= \sqrt{x - 3} \qquad \text{The domain is } [3, \infty).$$

TIP: Examples 3(a) and 3(b) illustrate that the order in which two functions are composed may result in different functions. That is, $f \circ g$ does not necessarily equal $g \circ f$.

Skill Practice   Given $f(x) = 2x^2$, $g(x) = x + 3$, and $h(x) = \sqrt{x - 1}$,

5. Find $(f \circ g)(x)$. Write the domain of $(f \circ g)$ in interval notation.

6. Find $(g \circ f)(x)$. Write the domain of $(g \circ f)$ in interval notation.

7. Find $(h \circ g)(x)$. Write the domain of $(h \circ g)$ in interval notation.

Skill Practice Answers

5. $2x^2 + 12x + 18$; domain: $(-\infty, \infty)$
6. $2x^2 + 3$; domain: $(-\infty, \infty)$
7. $\sqrt{x + 2}$; domain: $[-2, \infty)$

3. Multiple Operations on Functions

> **Example 4** Combining Functions

Given the functions defined by $f(x) = x - 7$ and $h(x) = 2x^3$, find the function values, if possible.

a. $(f \cdot h)(3)$ **b.** $\left(\dfrac{h}{f}\right)(7)$ **c.** $(h \circ f)(2)$

Solution:

a. $(f \cdot h)(3) = f(3) \cdot h(3)$ $\qquad$ $(f \cdot h)(3)$ is a product (not a composition).

$\qquad = (3 - 7) \cdot 2(3)^3$

$\qquad = (-4) \cdot 2(27)$

$\qquad = -216$

b. The function $\dfrac{h}{f}$ has restrictions on its domain.

$$\left(\frac{h}{f}\right)(x) = \frac{h(x)}{f(x)} = \frac{2x^3}{x - 7}$$

Therefore, $x = 7$ is not in the domain, and $\left(\dfrac{h}{f}\right)(7) = \dfrac{h(7)}{f(7)}$ is undefined.

Avoiding Mistakes:

If you had tried evaluating the function $\frac{h}{f}$ at $x = 7$, the denominator would be zero and the function undefined.

$$\frac{h(7)}{f(7)} = \frac{2(7)^3}{7 - 7}$$

---

c. $(h \circ f)(2) = h(f(2))$ $\qquad$ Evaluate $f(2)$ first. $f(2) = 2 - 7 = -5$.

$\qquad = h(-5)$ $\qquad$ Substitute the result into function h.

$\qquad = 2(-5)^3$

$\qquad = 2(-125)$

$\qquad = -250$

> **Skill Practice** Given:

$$h(x) = x + 4$$
$$k(x) = x^2 - 3$$

8. Find $(h \cdot k)(-2)$. $\qquad$ **9.** Find $\left(\dfrac{h}{k}\right)(4)$. $\qquad$ **10.** Find $(k \circ h)(1)$.

Skill Practice Answers

8. 2 $\qquad$ **9.** $\dfrac{8}{13}$ $\qquad$ **10.** 22

Example 5 **Finding Function Values from a Graph**

For the functions f and g pictured, find the function values if possible.

a. $g(2)$

b. $(f - g)(-3)$

c. $\left(\dfrac{g}{f}\right)(5)$

d. $(f \circ g)(4)$

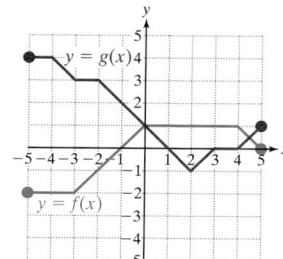

Solution:

a. $g(2) = -1$

The value $g(2)$ represents the y-value of $y = g(x)$ (the red graph) when $x = 2$. Because the point $(2, -1)$ lies on the graph, $g(2) = -1$.

b. $(f - g)(-3) = f(-3) - g(-3)$ Evaluate the difference of $f(-3)$ and $g(-3)$.

$\qquad = -2 - (3)$ Estimate function values from the graph.

$\qquad = -5$

c. $\left(\dfrac{g}{f}\right)(5) = \dfrac{g(5)}{f(5)}$ Evaluate the quotient of $g(5)$ and $f(5)$.

$\qquad = \dfrac{1}{0}$ (undefined)

The function $\frac{g}{f}$ is undefined at 5 because the denominator is zero.

d. $(f \circ g)(4) = f(g(4))$ From the red graph, find the value of $g(4)$ first.

$\qquad = f(0)$ From the blue graph, find the value of f at $x = 0$.

$\qquad = 1$

Skill Practice Find the values from the graph.

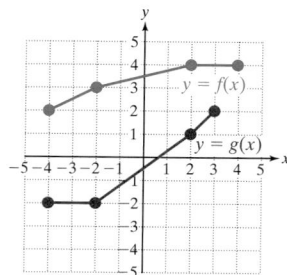

11. $g(3)$ **12.** $(f + g)(-4)$ **13.** $\left(\dfrac{f}{g}\right)(2)$ **14.** $(g \circ f)(-2)$

Skill Practice Answers

11. 2 **12.** 0 **13.** 4 **14.** 2

Section 13.1   Practice Exercises

Study Skills Exercise

1. Define the key term **composition of functions**.

Concept 1: Algebra of Functions

For Exercises 2–13, refer to the functions defined below.

$$f(x) = x + 4 \qquad g(x) = 2x^2 + 4x$$

$$h(x) = \sqrt{x - 1} \qquad k(x) = \frac{1}{x}$$

Find the indicated functions. Write the domain in interval notation.

2. $(f + g)(x)$

3. $(f - g)(x)$

4. $(g - f)(x)$

5. $(f + h)(x)$

6. $(f \cdot h)(x)$

7. $(h \cdot k)(x)$

8. $(g \cdot f)(x)$

9. $(f \cdot k)(x)$

10. $\left(\dfrac{h}{f}\right)(x)$

11. $\left(\dfrac{g}{f}\right)(x)$

12. $\left(\dfrac{f}{g}\right)(x)$

13. $\left(\dfrac{f}{h}\right)(x)$

Concept 2: Composition of Functions

For Exercises 14–22, find the indicated functions and their domains. Use $f, g, h,$ and k as defined in Exercises 2–13.

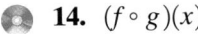

14. $(f \circ g)(x)$

15. $(f \circ k)(x)$

16. $(g \circ f)(x)$

17. $(k \circ f)(x)$

18. $(k \circ h)(x)$

19. $(h \circ k)(x)$

20. $(k \circ g)(x)$

21. $(g \circ k)(x)$

22. $(f \circ h)(x)$

23. Based on your answers to Exercises 15 and 17 is it true in general that $(f \circ k)(x) = (k \circ f)(x)$?

24. Based on your answers to Exercises 14 and 16 is it true in general that $(f \circ g)(x) = (g \circ f)(x)$?

For Exercises 25–30, find $(f \circ g)(x)$ and $(g \circ f)(x)$.

25. $f(x) = x^2 - 3x + 1, g(x) = 5x$

26. $f(x) = 3x^2 + 8, g(x) = 2x - 4$

27. $f(x) = |x|, g(x) = x^3 - 1$

28. $f(x) = \dfrac{1}{x + 2}, g(x) = |x + 2|$

29. For $h(x) = 5x - 4$, find $(h \circ h)(x)$.

30. For $k(x) = -x^2 + 1$, find $(k \circ k)(x)$.

Concept 3: Multiple Operations on Functions

For Exercises 31–44, refer to the functions defined below.

$$m(x) = x^3 \qquad n(x) = x - 3$$

$$r(x) = \sqrt{x + 4} \qquad p(x) = \frac{1}{x + 2}$$

Find the function values if possible.

31. $(m \cdot r)(0)$ **32.** $(n \cdot p)(0)$ **33.** $(m + r)(-4)$ **34.** $(n - m)(4)$

 **35.** $(r \circ n)(3)$ **36.** $(n \circ r)(5)$ **37.** $(p \circ m)(-1)$ **38.** $(m \circ n)(5)$

39. $(m \circ p)(2)$ **40.** $(r \circ m)(2)$ **41.** $(r + p)(-3)$ **42.** $(n + p)(-2)$

43. $(m \circ p)(-2)$ **44.** $(r \circ m)(-2)$

For Exercises 45–62, approximate the function values from the graph, if possible.

45. $f(-4)$ **46.** $f(1)$ **47.** $g(-2)$

48. $g(3)$  **49.** $(f + g)(2)$ **50.** $(g - f)(3)$

51. $(f \cdot g)(-1)$ **52.** $(g \cdot f)(-4)$ **53.** $\left(\dfrac{g}{f}\right)(0)$

54. $\left(\dfrac{f}{g}\right)(-2)$ **55.** $\left(\dfrac{f}{g}\right)(0)$ **56.** $\left(\dfrac{g}{f}\right)(-2)$

57. $(g \circ f)(-1)$ **58.** $(f \circ g)(0)$ **59.** $(f \circ g)(-4)$

60. $(g \circ f)(-4)$ **61.** $(g \circ g)(2)$ **62.** $(f \circ f)(-2)$

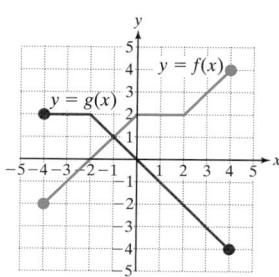

For Exercises 63–78, approximate the function values from the graph, if possible.

63. $a(-3)$ **64.** $a(1)$ **65.** $b(-1)$

66. $b(3)$ **67.** $(a - b)(-1)$ **68.** $(a + b)(0)$

69. $(b \cdot a)(1)$ **70.** $(a \cdot b)(2)$ **71.** $(b \circ a)(0)$

72. $(a \circ b)(-2)$ **73.** $(a \circ b)(-4)$ **74.** $(b \circ a)(-3)$

75. $\left(\dfrac{b}{a}\right)(3)$ **76.** $\left(\dfrac{a}{b}\right)(4)$ **77.** $(a \circ a)(-2)$ **78.** $(b \circ b)(1)$

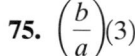 **79.** The cost in dollars of producing x toy cars is $C(x) = 2.2x + 1$. The revenue received is $R(x) = 5.98x$. To calculate profit, subtract the cost from the revenue.

    a. Write and simplify a function P that represents profit in terms of x.

    b. Find the profit of producing 50 toy cars.

80. The cost in dollars of producing x lawn chairs is $C(x) = 2.5x + 10.1$. The revenue for selling x chairs is $R(x) = 6.99x$. To calculate profit, subtract the cost from the revenue.

a. Write and simplify a function P that represents profit in terms of x.

b. Find the profit in producing 100 lawn chairs.

81. The functions defined by $D(t) = 0.925t + 26.958$ and $R(t) = 0.725t + 20.558$ approximate the amount of child support (in billions of dollars) that was due and the amount of child support actually received in the United States between the years 2000 and 2006. In each case, $t = 0$ corresponds to the year 2000.

a. Find the function F defined by $F(t) = D(t) - R(t)$. What does F represent in the context of this problem?

b. Find $F(0)$, $F(2)$, and $F(4)$. What do these function values represent in the context of this problem?

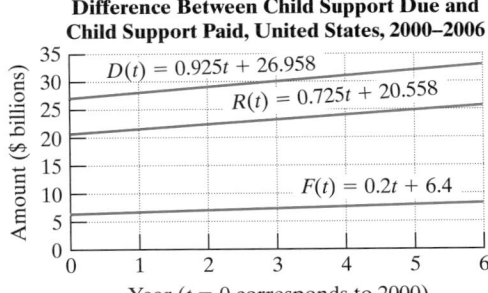

Difference Between Child Support Due and Child Support Paid, United States, 2000–2006

(Source: U.S. Bureau of the Census)

82. If t represents the number of years after 1900, then the rural and urban populations in the South (United States) between the years 1900 and 1970 can be approximated by

$$r(t) = -3.497t^2 + 266.2t + 20,220$$

where $t = 0$ corresponds to the year 1900 and $r(t)$ represents the rural population in thousands.

$$u(t) = 0.0566t^3 + 0.952t^2 + 177.8t + 4593$$

where $t = 0$ corresponds to the year 1900 and $u(t)$ represents the urban population in thousands.

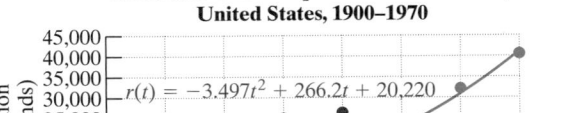

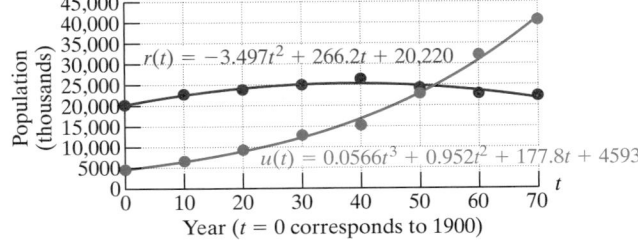

Rural and Urban Populations in the South, United States, 1900–1970

(Source: Historical Abstract of the United States)

a. Find the function T defined by $T(t) = r(t) + u(t)$. What does the function T represent in the context of this problem?

b. Use the function T to approximate the total population in the South for the year 1940.

83. Joe rides a bicycle and his wheels revolve at 80 revolutions per minute (rpm). Therefore, the total number of revolutions, r, is given by $r(t) = 80t$, where t is the time in minutes. For each revolution of the wheels of the bike, he travels approximately 7 ft. Therefore, the total distance he travels, D, depends on the total number of revolutions, r, according to the function $D(r) = 7r$.

a. Find $(D \circ r)(t)$ and interpret its meaning in the context of this problem.

b. Find Joe's total distance in feet after 10 min.

84. The area, A, of a square is given by the function $a(x) = x^2$, where x is the length of the sides of the square. If carpeting costs \$9.95 per square yard, then the cost, C, to carpet a square room is given by $C(a) = 9.95a$, where a is the area of the room in square yards.

a. Find $(C \circ a)(x)$ and interpret its meaning in the context of this problem.

b. Find the cost to carpet a square room if its floor dimensions are 15 yd by 15 yd.

| Section 13.2 | Inverse Functions |

Concepts

1. Introduction to Inverse Functions
2. Definition of a One-to-One Function
3. Finding an Equation of the Inverse of a Function
4. Definition of the Inverse of a Function

1. Introduction to Inverse Functions

In Section 8.2, we defined a function as a set of ordered pairs (x, y) such that for every element x in the domain, there corresponds exactly one element y in the range. For example, the function f relates the price, x (in dollars), of a USB Flash drive to the amount of memory that it holds, y (in megabytes).

$$\underset{\$}{\text{price}, x} \quad \underset{\text{megabytes}, y}{}$$

$$\begin{array}{cc} \downarrow & \downarrow \end{array}$$

$$f = \{(52, 512), (29, 256), (25, 128)\}$$

That is, the amount of memory depends on how much money a person has to spend. Now suppose we create a new function in which the values of x and y are interchanged. The new function is called the inverse of f and is denoted by f^{-1}. This relates the amount of memory, x, to the cost, y.

$$\underset{\text{megabytes}, x}{} \quad \underset{\$}{\text{price}, y}$$

$$\begin{array}{cc} \downarrow & \downarrow \end{array}$$

$$f^{-1} = \{(521, 52), (256, 29), (128, 25)\}$$

Avoiding Mistakes:

f^{-1} denotes the inverse of a function. The -1 does not represent an exponent.

Notice that interchanging the x- and y-values has the following outcome. The domain of f is the same as the range of f^{-1}, and the range of f is the domain of f^{-1}.

2. Definition of a One-to-One Function

A necessary condition for a function f to have an inverse function is that no two ordered pairs in f have different x-coordinates and the same y-coordinate. A function that satisfies this condition is called a **one-to-one function**. The function relating the price of a USB Flash drive to its memory is a one-to-one function. However, consider the function g defined by

$$g = \{(1, 4), (2, 3), (-2, 4)\}$$

same y

different x

This function is not one-to-one because the range element 4 has two different x-coordinates, 1 and -2. Interchanging the x- and y-values produces a relation that violates the definition of a function.

$$\{(4, 1), (3, 2), (4, -2)\}$$

same x

different y

This relation is not a function because for $x = 4$ there are two different y-values, $y = 1$ and $y = -2$.

In Section 8.2, you learned the vertical line test to determine visually if a graph represents a function. We use a **horizontal line test** to determine whether a function is one-to-one.

Horizontal Line Test

Consider a function defined by a set of points (x, y) in a rectangular coordinate system. The graph of the ordered pairs defines y as a *one-to-one* function of x if no horizontal line intersects the graph in more than one point.

To understand the horizontal line test, consider the functions f and g.

$$f = \{(1, 2.99), (1.5, 4.49), (4, 11.96)\} \qquad g = \{(1, 4), (2, 3), (-2, 4)\}$$

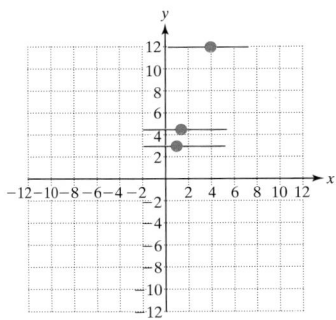   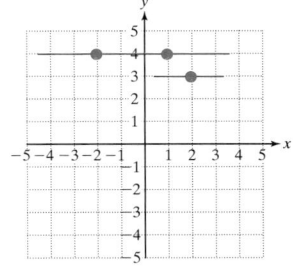

This function is one-to-one.
No horizontal line intersects more than once.

This function is *not* one-to-one.
A horizontal line intersects more than once.

Example 1 | **Identifying One-to-One Functions**

Determine whether the function is one-to-one.

a.

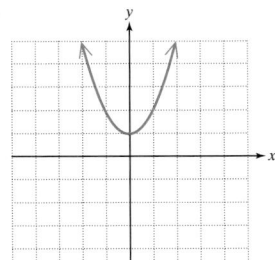

b.

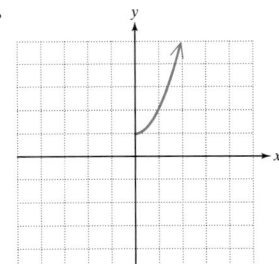

Solution:

a. Function is not one-to-one.
A horizontal line intersects in more than one point.

b. Function is one-to-one.
No horizontal line intersects more than once.

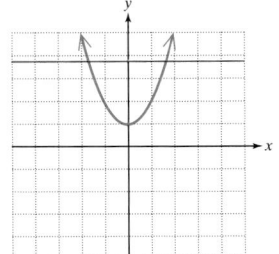   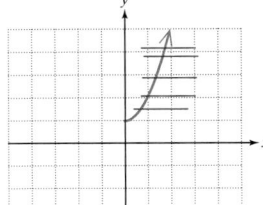

Skill Practice   Use the horizontal line test to determine if the functions are one-to-one.

1.

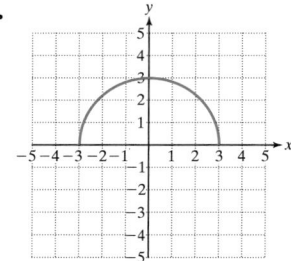

2.

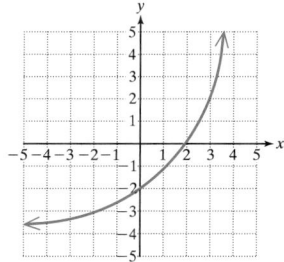

3. Finding an Equation of the Inverse of a Function

Another way to view the construction of the inverse of a function is to find a function that performs the inverse operations in the reverse order. For example, the function defined by $f(x) = 2x + 1$ multiplies x by 2 and then adds 1. Therefore, the inverse function must *subtract* 1 from x and *divide* by 2. We have

$$f^{-1}(x) = \frac{x - 1}{2}$$

To facilitate the process of finding an equation of the inverse of a one-to-one function, we offer the following steps.

Finding an Equation of an Inverse of a Function

For a one-to-one function defined by $y = f(x)$, the equation of the inverse can be found as follows:

1. Replace $f(x)$ by y.

2. Interchange x and y.

3. Solve for y.

4. Replace y by $f^{-1}(x)$.

Example 2   Finding an Equation of the Inverse of a Function

Find the inverse. $f(x) = 2x + 1$

Solution:

We know the graph of f is a nonvertical line. Therefore, $f(x) = 2x + 1$ defines a one-to-one function. To find the inverse we have

$$y = 2x + 1 \qquad \textbf{Step 1:} \text{ Replace } f(x) \text{ by } y.$$

$$x = 2y + 1 \qquad \textbf{Step 2:} \text{ Interchange } x \text{ and } y.$$

Skill Practice Answers

1. Not one-to-one
2. One-to-one

$$x - 1 = 2y \qquad \textbf{Step 3: } \text{Solve for } y. \text{ Subtract 1 from both sides.}$$

$$\frac{x - 1}{2} = y \qquad \text{Divide both sides by 2.}$$

$$f^{-1}(x) = \frac{x - 1}{2} \qquad \textbf{Step 4: } \text{Replace } y \text{ by } f^{-1}(x).$$

Skill Practice

3. Find the inverse of $f(x) = 4x + 6$.

The key step in determining the equation of the inverse of a function is to interchange x and y. By so doing, a point (a, b) on f corresponds to a point (b, a) on f^{-1}. For this reason, the graphs of f and f^{-1} are symmetric with respect to the line $y = x$ (Figure 13-3). Notice that the point $(-3, -5)$ of the function f corresponds to the point $(-5, -3)$ of f^{-1}. Likewise, $(1, 3)$ of f corresponds to $(3, 1)$ of f^{-1}.

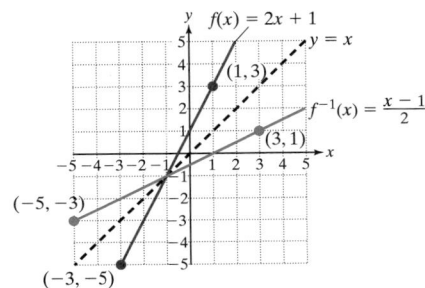

Figure 13-3

Example 3 **Finding an Equation of the Inverse of a Function**

Find the inverse of the one-to-one function. $\qquad g(x) = \sqrt[3]{5x} - 4$

Solution:

$$y = \sqrt[3]{5x} - 4 \qquad \textbf{Step 1: } \text{Replace } g(x) \text{ by } y.$$

$$x = \sqrt[3]{5y} - 4 \qquad \textbf{Step 2: } \text{Interchange } x \text{ and } y.$$

$$x + 4 = \sqrt[3]{5y} \qquad \textbf{Step 3: } \text{Solve for } y. \text{ Add 4 to both sides.}$$

$$(x + 4)^3 = (\sqrt[3]{5y})^3 \qquad \text{To eliminate the cube root, cube both sides.}$$

$$(x + 4)^3 = 5y \qquad \text{Simplify the right side.}$$

$$\frac{(x + 4)^3}{5} = y \qquad \text{Divide both sides by 5.}$$

$$g^{-1}(x) = \frac{(x + 4)^3}{5} \qquad \textbf{Step 4: } \text{Replace } y \text{ by } g^{-1}(x).$$

Skill Practice

4. Find the inverse of $h(x) = \sqrt[3]{2x} - 1$.

Skill Practice Answers

3. $f^{-1}(x) = \dfrac{x - 6}{4}$

4. $h^{-1}(x) = \dfrac{x^3 + 1}{2}$

The graphs of g and g^{-1} from Example 3 are shown in Figure 13-4. Once again we see that the graphs of a function and its inverse are symmetric with respect to the line $y = x$.

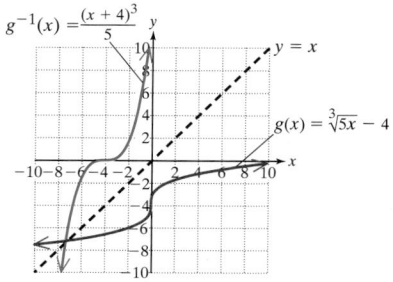

$g^{-1}(x) = \dfrac{(x+4)^3}{5}$

$y = x$

$g(x) = \sqrt[3]{5x} - 4$

Figure 13-4

For a function that is not one-to-one, sometimes we can restrict its domain to create a new function that is one-to-one. This is demonstrated in Example 4.

Example 4   **Finding the Equation of an Inverse of a Function with a Restricted Domain**

Given the function defined by $m(x) = x^2 + 4$ for $x \geq 0$, find an equation defining m^{-1}.

Solution:

From Section 12.4, we know that $y = x^2 + 4$ is a parabola with vertex at $(0, 4)$ (Figure 13-5). The graph represents a function that is not one-to-one. However, with the restriction on the domain $x \geq 0$, the graph of $m(x) = x^2 + 4$, $x \geq 0$, consists of only the "right" branch of the parabola (Figure 13-6). This *is* a one-to-one function.

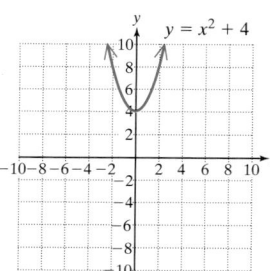

$y = x^2 + 4$

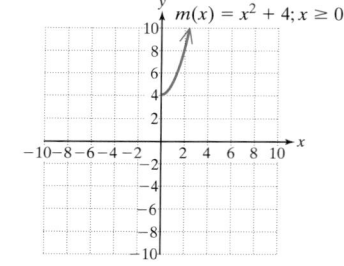

$m(x) = x^2 + 4;\, x \geq 0$

Figure 13-5   **Figure 13-6**

To find the inverse, we have

$y = x^2 + 4 \qquad x \geq 0$ **Step 1:** Replace $m(x)$ by y.

$x = y^2 + 4 \qquad y \geq 0$ **Step 2:** Interchange x and y. Notice that the restriction $x \geq 0$ becomes $y \geq 0$.

$x - 4 = y^2 \qquad y \geq 0$ **Step 3:** Solve for y. Subtract 4 from both sides.

$\sqrt{x - 4} = y \qquad y \geq 0$ Apply the square root property. Notice that we obtain the *positive* square root of $x - 4$ because of the restriction $y \geq 0$.

$m^{-1}(x) = \sqrt{x - 4}$

Step 4: Replace y by $m^{-1}(x)$. Notice that the domain of m^{-1} has the same values as the range of m.

Figure 13-7 shows the graphs of m and m^{-1}. Compare the domain and range of m and m^{-1}.

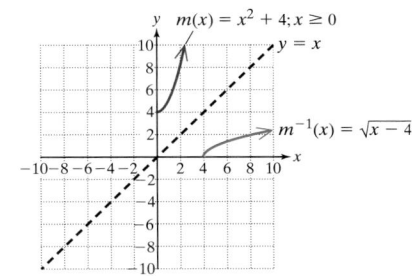

Domain of m: $[0, \infty)$
Range of m: $[4, \infty)$

Domain of m^{-1}: $[4, \infty)$
Range of m^{-1}: $[0, \infty)$

Figure 13-7

Skill Practice

5. Find the inverse. $g(x) = x^2 - 2 \qquad x \geq 0$

---

4. Definition of the Inverse of a Function

Definition of an Inverse Function

If f is a one-to-one function represented by ordered pairs of the form (x, y), then the **inverse function**, denoted f^{-1}, is the set of ordered pairs denoted by ordered pairs of the form (y, x).

An important relationship between a function and its inverse is shown in Figure 13-8.

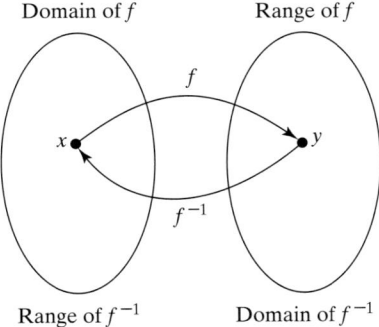

Figure 13-8

Recall that the domain of f is the range of f^{-1} and the range of f is the domain of f^{-1}. The operations performed by f are reversed by f^{-1}. This leads to the inverse function property.

Inverse Function Property

If f is a one-to-one function, then g is the inverse of f if and only if

$$(f \circ g)(x) = x \qquad \text{for all } x \text{ in the domain of } g$$

and

$$(g \circ f)(x) = x \qquad \text{for all } x \text{ in the domain of } f$$

Example 5 **Composing a Function with Its Inverse**

Show that the functions are inverses.

$$h(x) = 2x + 1 \qquad \text{and} \qquad k(x) = \frac{x-1}{2}$$

Solution:

To show that the functions h and k are inverses, we need to confirm that $(h \circ k)(x) = x$ and $(k \circ h)(x) = x$.

$$(h \circ k)(x) = h(k(x)) = h\left(\frac{x-1}{2}\right)$$

$$= 2\left(\frac{x-1}{2}\right) + 1$$

$$= x - 1 + 1$$

$$= x \checkmark \qquad (h \circ k)(x) = x \text{ as desired.}$$

$$(k \circ h)(x) = k(h(x)) = k(2x+1)$$

$$= \frac{(2x+1) - 1}{2}$$

$$= \frac{2x + 1 - 1}{2}$$

$$= \frac{2x}{2}$$

$$= x \checkmark \qquad (k \circ h)(x) = x \text{ as desired.}$$

The functions h and k are inverses because $(h \circ k)(x) = x$ and $(k \circ h)(x) = x$ for all real numbers x.

Skill Practice

6. Show that the functions are inverses.

$$f(x) = 3x - 2 \qquad \text{and} \qquad g(x) = \frac{x+2}{3}$$

Skill Practice Answers

6. $(f \circ g)(x) = f(g(x))$
$$= 3\left(\frac{x+2}{3}\right) - 2 = x$$
$(g \circ f)(x) = g(f(x))$
$$= \frac{3x - 2 + 2}{3} = x$$

Section 13.2   **Practice Exercises**

Study Skills Exercise

1. Define the key terms.

    a. inverse function   **b. one-to-one function**   **c. horizontal line test**

Review Exercises

2. Write the domain and range of the relation $\{(3, 4), (5, -2), (6, 1), (3, 0)\}$.

For Exercises 3–8, determine if the relation is a function by using the vertical line test. (See Section 8.2.)

3.
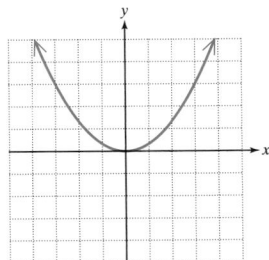

4.
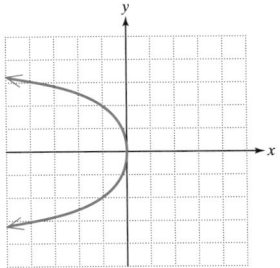

5.
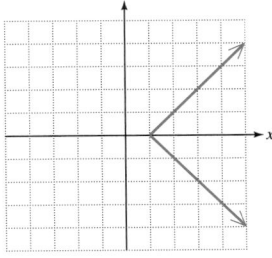

6.
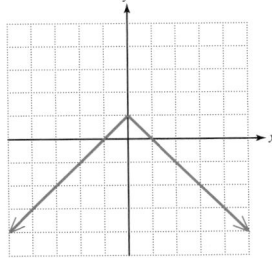

7.
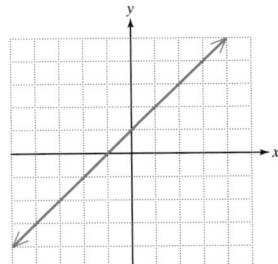

8.
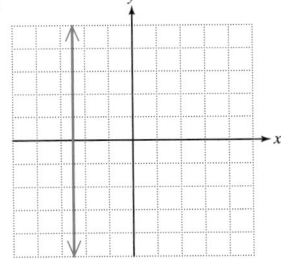

Concept 1: Introduction to Inverse Functions

For Exercises 9–12, write the inverse function.

9. $g = \{(3, 5), (8, 1), (-3, 9), (0, 2)\}$

10. $f = \{(-6, 2), (-9, 0), (-2, -1), (3, 4)\}$

11. $r = \{(a, 3), (b, 6), (c, 9)\}$

12. $s = \{(-1, x), (-2, y), (-3, z)\}$

Concept 2: Definition of a One-to-One Function

13. The table relates a state, x, to the number of representatives in the House of Representatives, y, in the year 2006. Does this relation define a one-to-one function? If so, write a function defining the inverse as a set of ordered pairs.

State x	Number of Representatives y
Colorado	7
California	53
Texas	32
Louisiana	7
Pennsylvania	19

14. The table relates a city x to its average January temperature y. Does this relation define a one-to-one function? If so, write a function defining the inverse as a set of ordered pairs.

City x	Temperature y (°C)
Gainesville, Florida	13.6
Keene, New Hampshire	−6.0
Wooster, Ohio	−4.0
Rock Springs, Wyoming	−6.0
Lafayette, Louisiana	10.9

For Exercises 15–20, determine if the function is one-to-one by using the horizontal line test.

15.

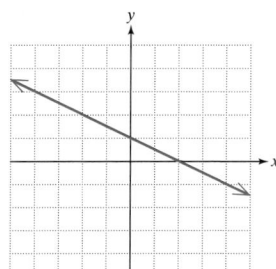

16.

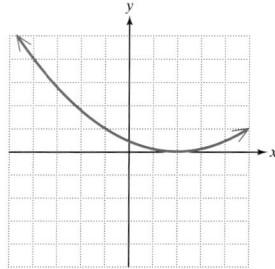

17.

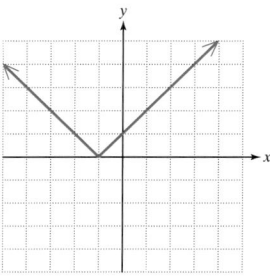

18.

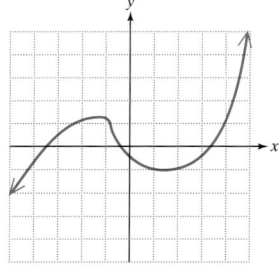

19.

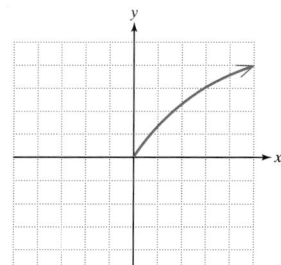

20.

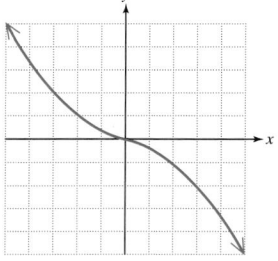

Concept 3: Finding an Equation of the Inverse of a Function

For Exercises 21–30, write an equation of the inverse for each one-to-one function as defined.

21. $h(x) = x + 4$

22. $k(x) = x - 3$

23. $m(x) = \dfrac{1}{3}x - 2$

24. $n(x) = 4x + 2$

25. $p(x) = -x + 10$

26. $q(x) = -x - \dfrac{2}{3}$

27. $f(x) = x^3$

28. $g(x) = \sqrt[3]{x}$

29. $g(x) = \sqrt[3]{2x - 1}$

30. $f(x) = x^3 - 4$

31. The function defined by $f(x) = 0.3048x$ converts a length of x feet into $f(x)$ meters.

    a. Find the equivalent length in meters for a 4-ft board and a 50-ft wire.

    b. Find an equation defining $y = f^{-1}(x)$.

    c. Use the inverse function from part (b) to find the equivalent length in feet for a 1500-m race in track and field. Round to the nearest tenth of a foot.

32. The function defined by $s(x) = 1.47x$ converts a speed of x mph to $s(x)$ ft/sec.

   a. Find the equivalent speed in feet per second for a car traveling 60 mph.

   b. Find an equation defining $y = s^{-1}(x)$.

   c. Use the inverse function from part (b) to find the equivalent speed in miles per hour for a train traveling 132 ft/sec. Round to the nearest tenth.

For Exercises 33–39, answer true or false.

33. The function defined by $y = 2$ has an inverse function defined by $x = 2$.

34. The domain of any one-to-one function is the same as the domain of its inverse.

35. All linear functions with a nonzero slope have an inverse function.

36. The function defined by $g(x) = |x|$ is one-to-one.

37. The function defined by $k(x) = x^2$ is one-to-one.

38. The function defined by $h(x) = |x|$ for $x \geq 0$ is one-to-one.

39. The function defined by $L(x) = x^2$ for $x \geq 0$ is one-to-one.

40. Explain how the domain and range of a one-to-one function and its inverse are related.

41. If $(0, b)$ is the y-intercept of a one-to-one function, what is the x-intercept of its inverse?

42. If $(a, 0)$ is the x-intercept of a one-to-one function, what is the y-intercept of its inverse?

43. Can you think of any function that is its own inverse?

44. a. Find the domain and range of the function defined by $f(x) = \sqrt{x - 1}$.

   b. Find the domain and range of the function defined by $f^{-1}(x) = x^2 + 1, x \geq 0$.

45. a. Find the domain and range of the function defined by $g(x) = x^2 - 4, x \leq 0$.

   b. Find the domain and range of the function defined by $g^{-1}(x) = -\sqrt{x + 4}$.

For Exercises 46–49, the graph of $y = f(x)$ is given.

   a. State the domain of f.   b. State the range of f.

   c. State the domain of f^{-1}.   d. State the range of f^{-1}.

   e. Graph the function defined by $y = f^{-1}(x)$. The line $y = x$ is provided for your reference.

46.

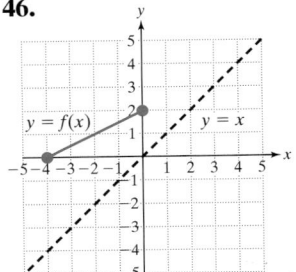

47.

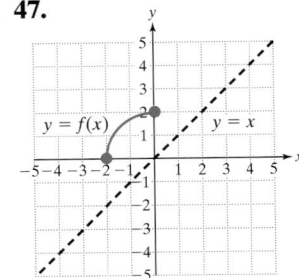

48.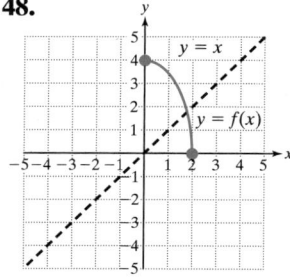

49.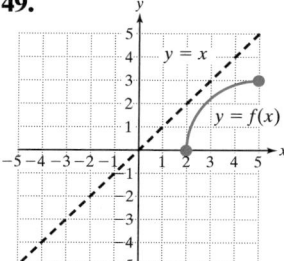

Concept 4: Definition of the Inverse of a Function

For Exercises 50–55, verify that f and g are inverse functions by showing that

    a. $(f \circ g)(x) = x$ **b.** $(g \circ f)(x) = x$

50. $f(x) = 6x + 1$ and $g(x) = \dfrac{x - 1}{6}$

51. $f(x) = 5x - 2$ and $g(x) = \dfrac{x + 2}{5}$

52. $f(x) = \dfrac{\sqrt[3]{x}}{2}$ and $g(x) = 8x^3$

53. $f(x) = \dfrac{\sqrt[3]{x}}{3}$ and $g(x) = 27x^3$

54. $f(x) = x^2 + 1$, $x \geq 0$, and $g(x) = \sqrt{x - 1}$, $x \geq 1$

55. $f(x) = x^2 - 3$, $x \geq 0$, and $g(x) = \sqrt{x + 3}$, $x \geq -3$

Expanding Your Skills

For Exercises 56–67, write an equation of the inverse of the one-to-one function.

56. $f(x) = \dfrac{x - 1}{x + 1}$

57. $p(x) = \dfrac{3 - x}{x + 3}$

58. $t(x) = \dfrac{2}{x - 1}$

59. $w(x) = \dfrac{4}{x + 2}$

60. $g(x) = x^2 + 9$ $x \geq 0$

61. $m(x) = x^2 - 1$ $x \geq 0$

62. $n(x) = x^2 + 9$ $x \leq 0$

63. $g(x) = x^2 - 1$ $x \leq 0$

64. $q(x) = \sqrt{x + 4}$

65. $v(x) = \sqrt{x + 16}$

66. $z(x) = -\sqrt{x + 4}$

67. $u(x) = -\sqrt{x + 16}$

Graphing Calculator Exercises

For Exercises 68–71, use a graphing calculator to graph each function on the standard viewing window defined by $-10 \leq x \leq 10$ and $-10 \leq y \leq 10$. Use the graph of the function to determine if the function is one-to-one on the interval $-10 \leq x \leq 10$. If the function is one-to-one, find its inverse and graph both functions on the standard viewing window.

68. $f(x) = \sqrt[3]{x + 5}$

69. $k(x) = x^3 - 4$

70. $g(x) = 0.5x^3 - 2$

71. $m(x) = 3x - 4$

Exponential Functions

1. Definition of an Exponential Function

The concept of a function was first introduced in Section 8.2. Since then we have learned to recognize several categories of functions, including constant, linear, rational, and quadratic functions. In this section and the next, we will define two new types of functions called exponential and logarithmic functions.

To introduce the concept of an exponential function, consider the following salary plans for a new job. Plan A pays $1 million for a month's work. Plan B starts with 2¢ on Day 1, and every day thereafter the salary is doubled.

At first glance, the million-dollar plan appears to be more favorable. Look, however, at Table 13-1, which shows the daily payments for 30 days under plan B.

Table 13-1

Day	Payment	Day	Payment	Day	Payment
1	2¢	11	$20.48	21	$20,971.52
2	4¢	12	$40.96	22	$41,943.04
3	8¢	13	$81.92	23	$83,886.08
4	16¢	14	$163.84	24	$167,772.16
5	32¢	15	$327.68	25	$335,544.32
6	64¢	16	$655.36	26	$671,088.64
7	$1.28	17	$1310.72	27	$1,342,177.28
8	$2.56	18	$2621.44	28	$2,684,354.56
9	$5.12	19	$5242.88	29	$5,368,709.12
10	$10.24	20	$10,485.76	30	$10,737,418.24

Notice that the salary on the 30th day for plan B is over $10 million. Taking the sum of the payments, we see the total salary for the 30-day period is $21,474,836.46.

The daily salary for plan B can be represented by the function $y = 2^x$, where x is the number of days on the job and y is the salary for that day. An interesting feature of this function is that for every positive 1-unit change in x, the y-value doubles. The function $y = 2^x$ is called an exponential function.

Definition of an Exponential Function

Let b be any real number such that $b > 0$ and $b \neq 1$. Then for any real number x, a function of the form $y = b^x$ is called an **exponential function**.

An exponential function is easily recognized as a function with a constant base and variable exponent. Notice that the base of an exponential function must be a positive real number not equal to 1.

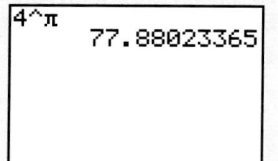
2. Approximating Exponential Expressions with a Calculator

Up to this point, we have evaluated exponential expressions with integer exponents and with rational exponents, for example, $4^3 = 64$ and $4^{1/2} = \sqrt{4} = 2$. However, how do we evaluate an exponential expression with an irrational exponent such as 4^π? In such a case, the exponent is a nonterminating and nonrepeating decimal. The value of 4^π can be thought of as the limiting value of a sequence of approximations using rational exponents:

$$4^{3.14} \approx 77.7084726$$

$$4^{3.141} \approx 77.81627412$$

$$4^{3.1415} \approx 77.87023095$$

$$\vdots$$

$$4^\pi \approx 77.88023365$$

An exponential expression can be evaluated at all rational numbers and at all irrational numbers. Hence, the domain of an exponential function is all real numbers.

Example 1   Approximating Exponential Expressions with a Calculator

Approximate the expressions. Round the answers to four decimal places.

a. $8^{\sqrt{3}}$ **b.** $5^{-\sqrt{17}}$ **c.** $\sqrt{10}^{\sqrt{2}}$

Solution:

a. $8^{\sqrt{3}} \approx 36.6604$ **b.** $5^{-\sqrt{17}} \approx 0.0013$ **c.** $\sqrt{10}^{\sqrt{2}} \approx 5.0946$

Skill Practice   Approximate the value of the expressions. Round the answers to four decimal places.

1. 9^π **2.** $15^{\sqrt{5}}$ **3.** $\sqrt{7}^{\sqrt{3}}$

3. Graphs of Exponential Functions

The functions defined by $f(x) = 2^x$, $g(x) = 3^x$, $h(x) = 5^x$, and $k(x) = (\frac{1}{2})^x$ are all examples of exponential functions. Example 2 illustrates the two general shapes of exponential functions.

Example 2   Graphing an Exponential Function

Graph the functions f and g.

a. $f(x) = 2^x$ **b.** $g(x) = (\frac{1}{2})^x$

Skill Practice Answers

1. 995.0416   **2.** 426.4028
3. 5.3936

Solution:

Table 13-2 shows several function values $f(x)$ and $g(x)$ for both positive and negative values of x. The graph is shown in Figure 13-9.

Table 13-2

x	$f(x) = 2^x$	$g(x) = (\frac{1}{2})^x$
-4	$\frac{1}{16}$	16
-3	$\frac{1}{8}$	8
-2	$\frac{1}{4}$	4
-1	$\frac{1}{2}$	2
0	1	1
1	2	$\frac{1}{2}$
2	4	$\frac{1}{4}$
3	8	$\frac{1}{8}$
4	16	$\frac{1}{16}$

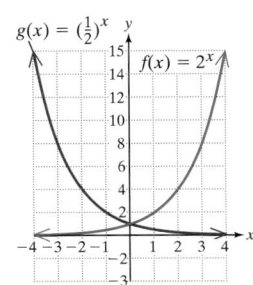

Figure 13-9

Skill Practice Graph the functions.

4. $f(x) = 3^x$ **5.** $f(x) = \left(\frac{1}{3}\right)^x$

The graphs in Figure 13-9 illustrate several important features of exponential functions.

Graphs of $f(x) = b^x$

The graph of an exponential function defined by $f(x) = b^x$ ($b > 0$ and $b \neq 1$) has the following properties.

1. If $b > 1$, f is an *increasing* exponential function, sometimes called an **exponential growth function**.

   If $0 < b < 1$, f is a *decreasing* exponential function, sometimes called an **exponential decay function**.

2. The domain is the set of all real numbers, $(-\infty, \infty)$.

3. The range is $(0, \infty)$.

4. The x-axis is a horizontal asymptote.

5. The function passes through the point $(0, 1)$ because $f(0) = b^0 = 1$.

These properties indicate that the graph of an exponential function is an *increasing function* if the base is greater than 1. Furthermore, the base affects its "steepness." Consider the graphs of $f(x) = 2^x$, $h(x) = 3^x$, and $k(x) = 5^x$ (Figure 13-10). For every positive 1-unit change in x, $f(x) = 2^x$ increases by 2 times, $h(x) = 3^x$ increases by 3 times, and $k(x) = 5^x$ increases by 5 times (Table 13-3).

Skill Practice Answers

4.

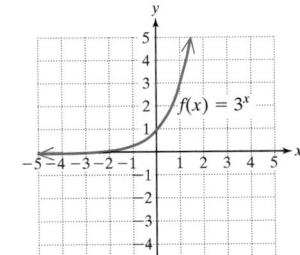

5.

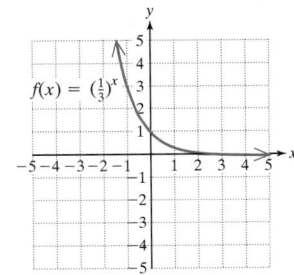

Table 13-3

x	$f(x) = 2^x$	$h(x) = 3^x$	$k(x) = 5^x$
-3	$\frac{1}{8}$	$\frac{1}{27}$	$\frac{1}{125}$
-2	$\frac{1}{4}$	$\frac{1}{9}$	$\frac{1}{25}$
-1	$\frac{1}{2}$	$\frac{1}{3}$	$\frac{1}{5}$
0	1	1	1
1	2	3	5
2	4	9	25
3	8	27	125

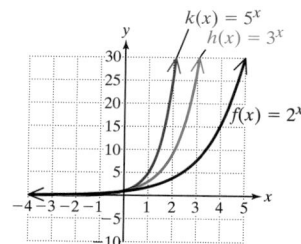

Figure 13-10

The graph of an exponential function is a *decreasing function* if the base is between 0 and 1. Consider the graphs of $g(x) = (\frac{1}{2})^x$, $m(x) = (\frac{1}{3})^x$, and $n(x) = (\frac{1}{5})^x$ (Table 13-4 and Figure 13-11).

Table 13-4

x	$g(x) = (\frac{1}{2})^x$	$m(x) = (\frac{1}{3})^x$	$n(x) = (\frac{1}{5})^x$
-3	8	27	125
-2	4	9	25
-1	2	3	5
0	1	1	1
1	$\frac{1}{2}$	$\frac{1}{3}$	$\frac{1}{5}$
2	$\frac{1}{4}$	$\frac{1}{9}$	$\frac{1}{25}$
3	$\frac{1}{8}$	$\frac{1}{27}$	$\frac{1}{125}$

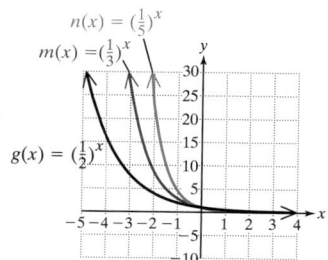

Figure 13-11

4. Applications of Exponential Functions

Exponential growth and decay can be found in a variety of real-world phenomena; for example,

- Population growth can often be modeled by an exponential function.
- The growth of an investment under compound interest increases exponentially.
- The mass of a radioactive substance decreases exponentially with time.
- The temperature of a cup of coffee decreases exponentially as it approaches room temperature.

A substance that undergoes radioactive decay is said to be radioactive. The *half-life* of a radioactive substance is the amount of time it takes for one-half of the original amount of the substance to change into something else. That is, after each half-life the amount of the original substance decreases by one-half.

In 1898, Marie Curie discovered the highly radioactive element radium. She shared the 1903 Nobel Prize in physics for her research on radioactivity and was awarded the 1911 Nobel Prize in chemistry for her discovery of radium and polonium. Radium 226 (an isotope of radium) has a half-life of 1620 years and decays into radon 222 (a radioactive gas).

Marie and Pierre Curie

| Example 3 | **Applying an Exponential Decay Function**

In a sample originally having 1 g of radium 226, the amount of radium 226 present after t years is given by

$$A(t) = (\tfrac{1}{2})^{t/1620}$$

where A is the amount of radium in grams and t is the time in years.

a. How much radium 226 will be present after 1620 years?

b. How much radium 226 will be present after 3240 years?

c. How much radium 226 will be present after 4860 years?

Solution:

$$A(t) = \left(\frac{1}{2}\right)^{t/1620}$$

a. $A(1620) = \left(\dfrac{1}{2}\right)^{1620/1620}$ Substitute $t = 1620$.

$$= \left(\frac{1}{2}\right)^{1}$$

$$= 0.5$$

After 1620 years (1 half-life), 0.5 g remains.

b. $A(3240) = \left(\dfrac{1}{2}\right)^{3240/1620}$ Substitute $t = 3240$.

$$= \left(\frac{1}{2}\right)^{2}$$

$$= 0.25$$

After 3240 years (2 half-lives), the amount of the original substance is reduced by one-half, 2 times: 0.25 g remains.

c. $A(4860) = \left(\dfrac{1}{2}\right)^{4860/1620}$ Substitute $t = 4860$.

$$= \left(\frac{1}{2}\right)^{3}$$

$$= 0.125$$

After 4860 years (3 half-lives), the amount of the original substance is reduced by one-half, 3 times: 0.125 g remains.

| Skill Practice |

6. Cesium 137 is a radioactive metal with a short half-life of 30 years. In a sample originally having 1 g of cesium 137, the amount of cesium 137 present after t years is given by

$$A(t) = \left(\frac{1}{2}\right)^{t/30}$$

a. How much cesium 137 will be present after 30 years?

b. How much cesium 137 will be present after 90 years?

Skill Practice Answers

6a. 0.5 g   **b.** 0.125 g

Exponential functions are often used to model population growth. Suppose the initial value of a population at some time $t = 0$ is P_0. If the rate of increase of a population is r, then after 1, 2, and 3 years, the new population can be found as follows:

After 1 year: $\left(\begin{array}{c}\text{Total}\\\text{population}\end{array}\right) = \left(\begin{array}{c}\text{initial}\\\text{population}\end{array}\right) + \left(\begin{array}{c}\text{increase in}\\\text{population}\end{array}\right)$

$\qquad\qquad = P_0 + P_0 r$

$\qquad\qquad = P_0(1 + r)$ Factor out P_0.

After 2 years: $\left(\begin{array}{c}\text{Total}\\\text{population}\end{array}\right) = \left(\begin{array}{c}\text{population}\\\text{after 1 year}\end{array}\right) + \left(\begin{array}{c}\text{increase in}\\\text{population}\end{array}\right)$

$\qquad\qquad = P_0(1 + r) + P_0(1 + r)r$

$\qquad\qquad = P_0(1 + r)1 + P_0(1 + r)r$

$\qquad\qquad = P_0(1 + r)(1 + r)$ Factor out $P_0(1 + r)$.

$\qquad\qquad = P_0(1 + r)^2$

After 3 years: $\left(\begin{array}{c}\text{Total}\\\text{population}\end{array}\right) = \left(\begin{array}{c}\text{population}\\\text{after 2 years}\end{array}\right) + \left(\begin{array}{c}\text{increase in}\\\text{population}\end{array}\right)$

$\qquad\qquad = P_0(1 + r)^2 + P_0(1 + r)^2 r$

$\qquad\qquad = P_0(1 + r)^2 1 + P_0(1 + r)^2 r$

$\qquad\qquad = P_0(1 + r)^2(1 + r)$ Factor out $P_0(1 + r)^2$.

$\qquad\qquad = P_0(1 + r)^3$

This pattern continues, and after t years, the population $P(t)$ is given by

$$P(t) = P_0(1 + r)^t$$

Example 4 **Applying an Exponential Growth Function**

The population of the Bahamas in 2000 was estimated at 300,000 with an annual rate of increase of 2%.

a. Find a mathematical model that relates the population of the Bahamas as a function of the number of years after 2000.

b. If the annual rate of increase remains the same, use this model to predict the population of the Bahamas in the year 2010. Round to the nearest thousand.

Solution:

a. The initial population is $P_0 = 300{,}000$, and the rate of increase is $r = 2\%$.

$P(t) = P_0(1 + r)^t$ Substitute $P_0 = 300{,}000$ and $r = 0.02$

$\quad = 300{,}000(1 + 0.02)^t$

$\quad = 300{,}000(1.02)^t$ Here $t = 0$ corresponds to the year 2000.

b. Because the initial population ($t = 0$) corresponds to the year 2000, we use $t = 10$ to find the population in the year 2010.

$$P(10) = 300{,}000(1.02)^{10}$$

$$\approx 366{,}000$$

Skill Practice

7. The population of Colorado in 2000 was approximately 4,300,000 with an annual increase of 2.6%.

　a. Find a mathematical model that relates the population of Colorado as function of the number of years after 2000.

　b. If the annual increase remains the same, use this model to predict the population of Colorado in 2010.

Skill Practice Answers

7a. $P(t) = 4{,}300{,}000(1.026)^t$
　b. Approximately 5,558,000

Section 13.3　Practice Exercises

Boost *your* GRADE at mathzone.com!　MathZone+

• Practice Problems　• e-Professors
• Self-Tests　• Videos
• NetTutor

Study Skills Exercise

1. Define the key terms.

　a. exponential function　**b.** exponential growth function　**c.** exponential decay function

Review Exercises

For Exercises 2–7, find the functions, using f and g as given. $f(x) = 2x^2 + x + 2$ 　$g(x) = 3x - 1$

2. $(f + g)(x)$

3. $(g - f)(x)$

4. $(f \cdot g)(x)$

5. $\left(\dfrac{g}{f}\right)(x)$

6. $(f \circ g)(x)$

7. $(g \circ f)(x)$

For Exercises 8–10, find the inverse function.

8. $\{(2, 3), (0, 0), (-8, 4), (10, 12)\}$

9. $\left\{(-13, 14), \left(\dfrac{1}{2}, -\dfrac{1}{2}\right), (6, 30), \left(-\dfrac{5}{3}, 0\right), (0, 1)\right\}$

10. $\{(a, b), (c, d), (e, f)\}$

For Exercises 11–18, evaluate the expression without the use of a calculator.

11. 5^2

12. 2^{-3}

13. 10^{-3}

14. 3^4

15. $36^{1/2}$

16. $27^{1/3}$

17. $16^{3/4}$

18. $8^{2/3}$

Concept 2: Approximating Exponential Expressions with a Calculator

For Exercises 19–26, evaluate the expression by using a calculator. Round to 4 decimal places.

19. $5^{1.1}$

20. $2^{\sqrt{3}}$

21. 10^{π}

22. $3^{4.8}$

23. $36^{-\sqrt{2}}$

24. $27^{-0.5126}$

25. $16^{-0.04}$

26. $8^{-0.61}$

27. Solve for x.

    a. $3^x = 9$ **b.** $3^x = 27$ **c.** Between what two consecutive integers must the solution to $3^x = 11$ lie?

28. Solve for x.

    a. $5^x = 125$ **b.** $5^x = 625$ **c.** Between what two consecutive integers must the solution to $5^x = 130$ lie?

29. Solve for x.

    a. $2^x = 16$ **b.** $2^x = 32$ **c.** Between what two consecutive integers must the solution to $2^x = 30$ lie?

30. Solve for x.

    a. $4^x = 16$ **b.** $4^x = 64$ **c.** Between what two consecutive integers must the solution to $4^x = 20$ lie?

31. For $f(x) = \left(\frac{1}{5}\right)^x$ find $f(0), f(1), f(2), f(-1),$ and $f(-2)$.

32. For $g(x) = \left(\frac{2}{3}\right)^x$ find $g(0), g(1), g(2), g(-1),$ and $g(-2)$.

33. For $h(x) = \pi^x$ use a calculator to find $h(0), h(1), h(-1), h\left(\sqrt{2}\right),$ and $h(\pi)$. Round to two decimal places.

34. For $k(x) = \left(\sqrt{5}\right)^x$ use a calculator to find $k(0), k(1), k(-1), k(\pi),$ and $k\left(\sqrt{2}\right)$. Round to two decimal places.

35. For $r(x) = 3^{x+2}$ find $r(0), r(1), r(2), r(-1), r(-2),$ and $r(-3)$.

36. For $s(x) = 2^{2x-1}$ find $s(0), s(1), s(2), s(-1),$ and $s(-2)$.

Concept 3: Graphs of Exponential Functions

37. How do you determine whether the graph of $f(x) = b^x$ $(b > 0, b \neq 1)$ is increasing or decreasing?

38. For $f(x) = b^x$ $(b > 0, b \neq 1)$, find $f(0)$.

Graph the functions defined in Exercises 39–46. Plot at least three points for each function.

39. $f(x) = 4^x$ **40.** $g(x) = 6^x$ **41.** $m(x) = \left(\frac{1}{8}\right)^x$

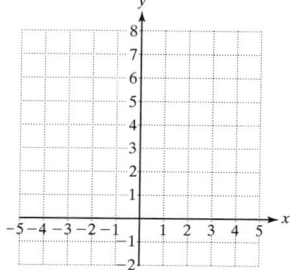   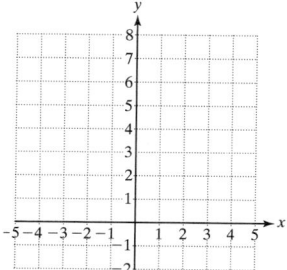   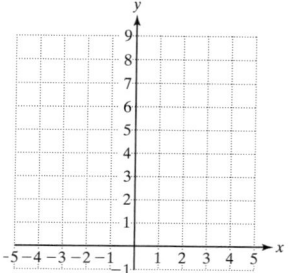

42. $n(x) = \left(\dfrac{1}{3}\right)^x$

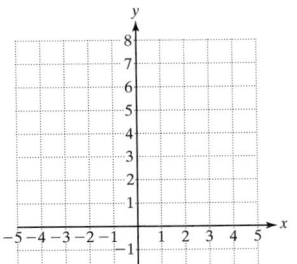

43. $h(x) = 2^{x+1}$

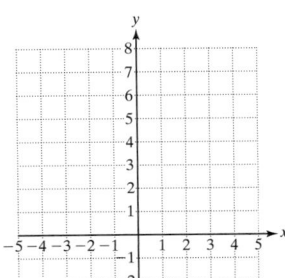

44. $k(x) = 5^{x-1}$

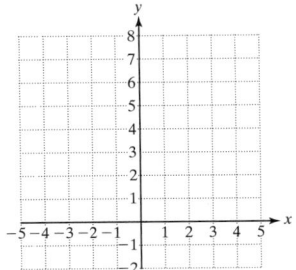

45. $g(x) = 5^{-x}$

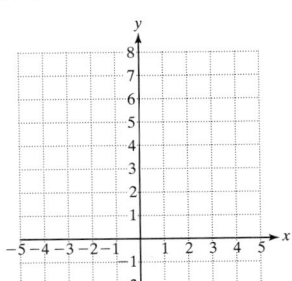

46. $f(x) = 2^{-x}$

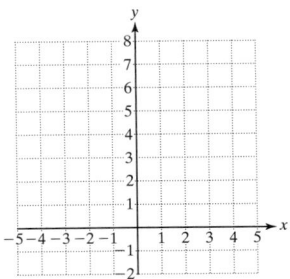

Concept 4: Applications of Exponential Functions

47. The half-life of the element radon (Rn 86) is 3.8 days. In a sample originally containing 1 g of radon, the amount left after t days is given by $A(t) = (0.5)^{t/3.8}$. (Round to two decimal places, if necessary.)

  a. How much radon will be present after 7.6 days?

  b. How much radon will be present after 10 days?

48. Nobelium, an element discovered in 1958, has a half-life of 10 min under certain conditions. In a sample containing 1 g of nobelium, the amount left after t min is given by $A(t) = (0.5)^{t/10}$. (Round to three decimal places.)

  a. How much nobelium is left after 5 min?

  b. How much nobelium is left after 1 hr?

49. Once an antibiotic is introduced to bacteria, the number of bacteria decreases exponentially. For example, beginning with 1 million bacteria, the amount present t days from the time penicillin is introduced is given by the function $A(t) = 1,000,000(2)^{-t/5}$. Rounding to the nearest thousand, determine how many bacteria are present after

  a. 2 days   **b.** 1 week   **c.** 2 weeks

50. Once an antibiotic is introduced to bacteria, the number of bacteria decreases exponentially. For example, beginning with 1 million bacteria, the amount present t days from the time streptomycin is introduced is given by the function $A(t) = 1,000,000(2)^{-t/10}$. Rounding to the nearest thousand, determine how many bacteria are present after

  a. 5 days   **b.** 1 week   **c.** 2 weeks

51. The population of Bangladesh was 141,340,000 in 2004 with an annual growth rate of 1.5%.

    a. Find a mathematical model that relates the population of Bangladesh as a function of the number of years after 2004.

    b. If the annual rate of increase remains the same, use this model to predict the population of Bangladesh in the year 2050. Round to the nearest million.

52. The population of Figi was 886,000 in 2004 with an annual growth rate of 1.07%.

    a. Find a mathematical model that relates the population of Figi as a function of the number of years after 2004.

    b. If the annual rate of increase remains the same, use this model to predict the population of Figi in the year 2050. Round to the nearest thousand.

53. Suppose $1000 is initially invested in an account and the value of the account grows exponentially. If the investment doubles in 7 years, then the amount in the account t years after the initial investment is given by

$$A(t) = 1000(2)^{t/7}$$

where t is expressed in years and $A(t)$ is the amount in the account.

    a. Find the amount in the account after 5 years.

    b. Find the amount in the account after 10 years.

    c. Find $A(0)$ and $A(7)$ and interpret the answers in the context of this problem.

54. Suppose $1500 is initially invested in an account and the value of the account grows exponentially. If the investment doubles in 8 years, then the amount in the account t years after the initial investment is given by

$$A(t) = 1500(2)^{t/8}$$

where t is expressed in years and $A(t)$ is the amount in the account.

    a. Find the amount in the account after 5 years.

    b. Find the amount in the account after 10 years.

    c. Find $A(0)$ and $A(8)$ and interpret the answers in the context of this problem.

Graphing Calculator Exercises

For Exercises 55–62, graph the functions on your calculator to support your solutions to the indicated exercises.

55. $f(x) = 4^x$
(see Exercise 39)

56. $g(x) = 6^x$
(see Exercise 40)

57. $m(x) = (\frac{1}{8})^x$
(see Exercise 41)

58. $n(x) = (\frac{1}{3})^x$
(see Exercise 42)

59. $h(x) = 2^{x+1}$
(see Exercise 43)

60. $k(x) = 5^{x-1}$
(see Exercise 44)

61. $g(x) = 5^{-x}$ (see Exercise 45)

62. $f(x) = 2^{-x}$ (see Exercise 46)

63. The function defined by $A(x) = 1000(2)^{x/7}$ represents the total amount A in an account x years after an initial investment of $1000.

    a. Graph $y = A(x)$ on the window where $0 \leq x \leq 25$ and $0 \leq y \leq 10{,}000$.

    b. Use *Zoom* and *Trace* to approximate the times required for the account to reach $2000, $4000, and $8000.

64. The function defined by $A(x) = 1500(2)^{x/8}$ represents the total amount A in an account x years after an initial investment of $1500.

    a. Graph $y = A(x)$ on the window where $0 \leq x \leq 40$ and $0 \leq y \leq 25{,}000$.

    b. Use *Zoom* and *Trace* to approximate the times required for the account to reach $3000, $6000, and $12,000.

Logarithmic Functions

1. Definition of a Logarithmic Function

Consider the following equations in which the variable is located in the exponent of an expression. In some cases the solution can be found by inspection because the constant on the right-hand side of the equation is a perfect power of the base.

Equation	Solution
$5^x = 5$ ⟶	$x = 1$
$5^x = 20$ ⟶	$x = ?$
$5^x = 25$ ⟶	$x = 2$
$5^x = 60$ ⟶	$x = ?$
$5^x = 125$ ⟶	$x = 3$

The equation $5^x = 20$ cannot be solved by inspection. However, we might suspect that x is between 1 and 2. Similarly, the solution to the equation $5^x = 60$ is between 2 and 3. To solve an exponential equation for an unknown exponent, we must use a new type of function called a logarithmic function.

Concepts

1. Definition of a Logarithmic Function
2. Evaluating Logarithmic Expressions
3. The Common Logarithmic Function
4. Graphs of Logarithmic Functions
5. Applications of the Common Logarithmic Function

> ### Definition of a Logarithm Function
>
> If x and b are positive real numbers such that $b \neq 1$, then $y = \log_b x$ is called the **logarithmic function** with base b and
>
> $$y = \log_b x \quad \text{is equivalent to} \quad b^y = x$$
>
> *Note:* In the expression $y = \log_b x$, y is called the **logarithm**, b is called the **base**, and x is called the **argument**.

The expression $y = \log_b x$ is equivalent to $b^y = x$ and indicates that *the logarithm y is the exponent to which b must be raised to obtain x.* The expression $y = \log_b x$ is called the logarithmic form of the equation, and the expression $b^y = x$ is called the exponential form of the equation.

The definition of a logarithmic function suggests a close relationship with an exponential function of the same base. In fact, a logarithmic function is the inverse of the corresponding exponential function. For example, the following steps find the inverse of the exponential function defined by $f(x) = b^x$.

$$f(x) = b^x$$

$$y = b^x \qquad \text{Replace } f(x) \text{ by } y.$$

$$x = b^y \qquad \text{Interchange } x \text{ and } y.$$

$$y = \log_b x \qquad \text{Solve for } y \text{ using the definition of a logarithmic function.}$$

$$f^{-1}(x) = \log_b x \qquad \text{Replace } y \text{ by } f^{-1}(x).$$

Example 1   **Converting from Logarithmic Form to Exponential Form**

Rewrite the logarithmic equations in exponential form.

a. $\log_2 32 = 5$ **b.** $\log_{10}\left(\dfrac{1}{1000}\right) = -3$ **c.** $\log_5 1 = 0$

Solution:

	Logarithmic Form		**Exponential Form**
a.	$\log_2 32 = 5$	$\Leftrightarrow$	$2^5 = 32$
b.	$\log_{10}\left(\dfrac{1}{1000}\right) = -3$	$\Leftrightarrow$	$10^{-3} = \dfrac{1}{1000}$
c.	$\log_5 1 = 0$	$\Leftrightarrow$	$5^0 = 1$

Skill Practice   Rewrite the logarithmic equations in exponential form.

1. $\log_3 9 = 2$ **2.** $\log_{10}\left(\dfrac{1}{100}\right) = -2$ **3.** $\log_8 1 = 0$

Skill Practice Answers

1. $3^2 = 9$ **2.** $10^{-2} = \dfrac{1}{100}$

3. $8^0 = 1$

2. Evaluating Logarithmic Expressions

| Example 2 | Evaluating Logarithmic Expressions

Evaluate the logarithmic expressions.

a. $\log_{10} 10{,}000$ **b.** $\log_5 \left(\dfrac{1}{125} \right)$ **c.** $\log_{1/2} \left(\dfrac{1}{8} \right)$

d. $\log_b b$ **e.** $\log_c \left(c^7 \right)$ **f.** $\log_3 \left(\sqrt[4]{3} \right)$

Solution:

a. $\log_{10} 10{,}000$ is the exponent to which 10 must be raised to obtain 10,000.

     $y = \log_{10} 10{,}000$ Let y represent the value of the logarithm.

     $10^y = 10{,}000$ Rewrite the expression in exponential form.

     $y = 4$

Therefore, $\log_{10} 10{,}000 = 4$.

b. $\log_5 \left(\frac{1}{125} \right)$ is the exponent to which 5 must be raised to obtain $\frac{1}{125}$.

     $y = \log_5 \left(\dfrac{1}{125} \right)$ Let y represent the value of the logarithm.

     $5^y = \dfrac{1}{125}$ Rewrite the expression in exponential form.

     $y = -3$

Therefore, $\log_5 \left(\frac{1}{125} \right) = -3$.

c. $\log_{1/2} \left(\frac{1}{8} \right)$ is the exponent to which $\frac{1}{2}$ must be raised to obtain $\frac{1}{8}$.

     $y = \log_{1/2} \left(\dfrac{1}{8} \right)$ Let y represent the value of the logarithm.

     $\left(\dfrac{1}{2} \right)^y = \dfrac{1}{8}$ Rewrite the expression in exponential form.

     $y = 3$

Therefore, $\log_{1/2} \left(\frac{1}{8} \right) = 3$.

d. $\log_b b$ is the exponent to which b must be raised to obtain b.

     $y = \log_b b$ Let y represent the value of the logarithm.

     $b^y = b$ Rewrite the expression in exponential form.

     $y = 1$

Therefore, $\log_b b = 1$.

e. $\log_c \left(c^7 \right)$ is the exponent to which c must be raised to obtain c^7.

     $y = \log_c \left(c^7 \right)$ Let y represent the value of the logarithm.

     $c^y = c^7$ Rewrite the expression in exponential form.

     $y = 7$

Therefore, $\log_c \left(c^7 \right) = 7$.

f. $\log_3\left(\sqrt[4]{3}\right) = \log_3\left(3^{1/4}\right)$ is the exponent to which 3 must be raised to obtain $3^{1/4}$.

$$y = \log_3\left(3^{1/4}\right) \qquad \text{Let } y \text{ represent the value of the logarithm.}$$

$$3^y = 3^{1/4} \qquad \text{Rewrite the expression in exponential form.}$$

$$y = \frac{1}{4}$$

Therefore, $\log_3\left(\sqrt[4]{3}\right) = \frac{1}{4}$

Skill Practice Evaluate the logarithmic expressions.

4. $\log_{10} 1000$ **5.** $\log_4\left(\dfrac{1}{16}\right)$ **6.** $\log_{1/3} 3$

7. $\log_x x$ **8.** $\log_b\left(b^{10}\right)$ **9.** $\log_5\left(\sqrt[3]{5}\right)$

3. The Common Logarithmic Function

The logarithmic function with base 10 is called the **common logarithmic function** and is denoted by $y = \log x$. Notice that the base is not explicitly written but is understood to be 10. That is, $y = \log_{10} x$ is written simply as $y = \log x$.

Calculator Connections

On most calculators, the $\boxed{\log}$ key is used to compute logarithms with base 10. For example, we know the expression $\log(1{,}000{,}000) = 6$ because $10^6 = 1{,}000{,}000$. Use the $\boxed{\log}$ key to show this result on a calculator.

```
log(1000000)
                6
```

Example 3 Evaluating Common Logarithms on a Calculator

Evaluate the common logarithms. Round the answers to four decimal places.

a. $\log 420$ **b.** $\log\left(8.2 \times 10^9\right)$ **c.** $\log(0.0002)$

Solution:

a. $\log 420 \approx 2.6232$

b. $\log\left(8.2 \times 10^9\right) \approx 9.9138$

c. $\log(0.0002) \approx -3.6990$

Skill Practice Evaluate the common logarithms. Round answers to 4 decimal places.

10. $\log 1200$ **11.** $\log\left(6.3 \times 10^5\right)$ **12.** $\log(0.00025)$

Skill Practice Answers

4. 3   **5.** -2 **6.** -1
7. 1   **8.** 10   **9.** $\dfrac{1}{3}$
10. 3.0792   **11.** 5.7993
12. -3.6021

4. Graphs of Logarithmic Functions

In Section 13.3 we studied the graphs of exponential functions. In this section, we will graph logarithmic functions. First, recall that $f(x) = \log_b x$ is the inverse of $g(x) = b^x$. Therefore, the graph of $y = f(x)$ is symmetric to the graph of $y = g(x)$ about the line $y = x$, as shown in Figures 13-12 and 13-13.

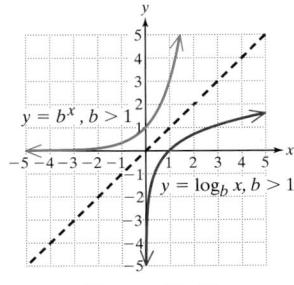

Figure 13-12	**Figure 13-13**

From Figures 13-12 and 13-13, we see that the range of $y = b^x$ is the set of positive real numbers. As expected, the domain of its inverse, the logarithmic function $y = \log_b x$, is also the set of positive real numbers. Therefore, the **domain of the logarithmic function** $y = \log_b x$ is the set of positive real numbers.

Example 4 | **Graphing Logarithmic Functions**

Graph the functions and compare the graphs to examine the effect of the base on the shape of the graph.

a. $y = \log_2 x$ **b.** $y = \log x$

Solution:

We can write each equation in its equivalent exponential form and create a table of values (Table 13-5). To simplify the calculations, choose integer values of y and then solve for x.

$$y = \log_2 x \quad \text{or} \quad 2^y = x \qquad y = \log x \quad \text{or} \quad 10^y = x$$

Choose values for y.

Table 13-5

$x = 2^y$	$x = 10^y$	y
$\frac{1}{8}$	$\frac{1}{1000}$	-3
$\frac{1}{4}$	$\frac{1}{100}$	-2
$\frac{1}{2}$	$\frac{1}{10}$	-1
1	1	0
2	10	1
4	100	2
8	1000	3

Solve for x.

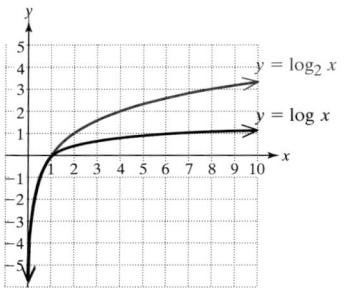

Figure 13-14

The graphs of $y = \log_2 x$ and $y = \log x$ are shown in Figure 13-14. Both graphs exhibit the same general behavior, and the steepness of the curve is affected by the base. The function $y = \log x$ requires a 10-fold increase in x to increase the y-value by 1 unit. The function $y = \log_2 x$ requires a 2-fold increase in x to increase the y-value by 1 unit. In addition, the following characteristics are true for both graphs.

- The domain is the set of real numbers x such that $x > 0$.
- The range is the set of real numbers.
- The y-axis is a vertical asymptote.
- Both graphs pass through the point $(1, 0)$.

Skill Practice   Graph the functions

13. $y = \log_3 x$.

Example 4 illustrates that a logarithmic function with base $b > 1$ is an increasing function. In Example 5, we see that if the base b is between 0 and 1, the function decreases over its entire domain.

Example 5   Graphing a Logarithmic Function

Graph $y = \log_{1/4} x$.

Solution:

The equation $y = \log_{1/4} x$ can be written in exponential form as $(\frac{1}{4})^y = x$. By choosing several values for y, we can solve for x and plot the corresponding points (Table 13-6).

The expression $y = \log_{1/4} x$ defines a decreasing logarithmic function (Figure 13-15). Notice that the vertical asymptote, domain, and range are the same for both increasing and decreasing logarithmic functions.

Table 13-6

$x = (\frac{1}{4})^y$	y
64	-3
16	-2
4	-1
1	0
$\frac{1}{4}$	1
$\frac{1}{16}$	2
$\frac{1}{64}$	3

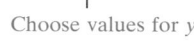
Solve for x.

Choose values for y.

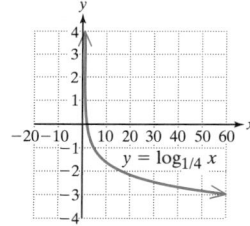

Figure 13-15

Skill Practice

14. Graph $y = \log_{1/2} x$.

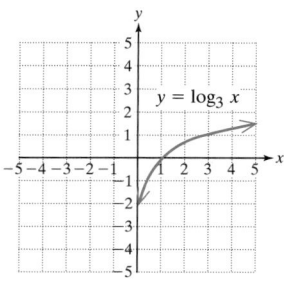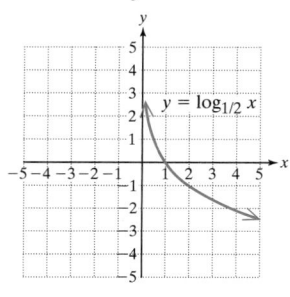

When graphing a logarithmic equation, it is helpful to know its domain.

Example 6 **Identifying the Domain of a Logarithmic Function**

Find the domain of the functions.

a. $f(x) = \log(4 - x)$ **b.** $g(x) = \log(2x + 6)$

Solution:

The domain of the function $y = \log_b(x)$ is the set of positive real numbers. That is, the argument x must be greater than zero: $x > 0$.

a. $f(x) = \log(4 - x)$ The argument is $4 - x$.

     $4 - x > 0$ The argument of the logarithm must be greater than zero.

        $-x > -4$ Solve for x.

        $x < 4$ Divide by -1 and reverse the inequality sign.

The domain is $(-\infty, 4)$.

b. $g(x) = \log(2x + 6)$ The argument is $2x + 6$.

     $2x + 6 > 0$ The argument of the logarithm must be greater than zero.

        $2x > -6$ Solve for x.

        $x > -3$

The domain is $(-3, \infty)$.

Skill Practice Find the domain of the functions.

15. $f(x) = \log_3(x + 7)$ **16.** $g(x) = \log(4 - 8x)$

Calculator Connections

The graphs of $Y_1 = \log(4 - x)$ and $Y_2 = \log(2x + 6)$ are shown here and can be used to confirm the solutions to Example 6. Notice that each function has a vertical asymptote at the value of x where the argument equals zero.

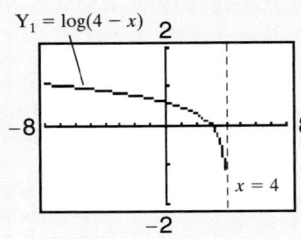

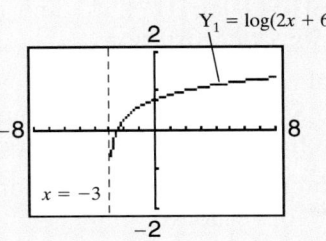

The general shape and important features of exponential and logarithmic functions are summarized as follows.

Skill Practice Answers

15. Domain: $(-7, \infty)$

16. Domain: $\left(-\infty, \dfrac{1}{2}\right)$

Graphs of Exponential and Logarithmic Functions—A Summary

Exponential Functions

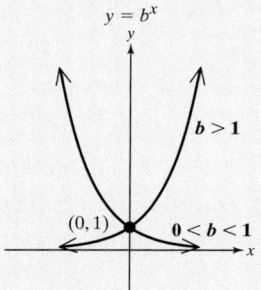

$y = b^x$

Logarithmic Functions

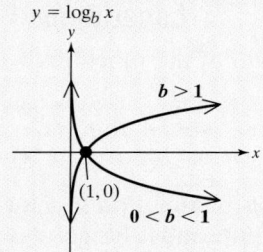

$y = \log_b x$

Domain: $(-\infty, \infty)$
Range: $(0, \infty)$
Horizontal asymptote: $y = 0$
Passes through $(0, 1)$
If $b > 1$, the function is increasing.
If $0 < b < 1$, the function is decreasing.

Domain: $(0, \infty)$
Range: $(-\infty, \infty)$
Vertical asymptote: $x = 0$
Passes through $(1, 0)$
If $b > 1$, the function is increasing.
If $0 < b < 1$, the function is decreasing.

Notice that the roles of x and y are interchanged for the functions $y = b^x$ and $b^y = x$. Therefore, it is not surprising that the domain and range are reversed between exponential and logarithmic functions. Moreover, an exponential function passes through $(0, 1)$, whereas a logarithmic function passes through $(1, 0)$. An exponential function has a horizontal asymptote at $y = 0$, whereas a logarithmic function has a vertical asymptote at $x = 0$.

5. Applications of the Common Logarithmic Function

Example 7   Applying a Common Logarithm to Compute pH

The pH (hydrogen potential) of a solution is defined as

$$pH = -\log [H^+]$$

where $[H^+]$ represents the concentration of hydrogen ions in moles per liter (mol/L). The pH scale ranges from 0 to 14. The midpoint of this range, 7, represents a neutral solution. Values below 7 are progressively more acidic, and values above 7 are progressively more alkaline. Based on the equation $pH = -\log [H^+]$, a 1-unit change in pH means a 10-fold change in hydrogen ion concentration.

a. Normal rain has a pH of 5.6. However, in some areas of the northeastern United States the rainwater is more acidic. What is the pH of a rain sample for which the concentration of hydrogen ions is 0.0002 mol/L?

b. Find the pH of household ammonia if the concentration of hydrogen ions is 1.0×10^{-11} mol/L.

Solution:

a. $pH = -\log [H^+]$

$= -\log (0.0002)$ Substitute $[H^+] = 0.0002$.

≈ 3.7 (To compare this value with a familiar substance, note that the pH of orange juice is roughly 3.5.)

b. $pH = -\log[H^+]$

$\quad = -\log(1.0 \times 10^{-11})$ Substitute $[H^+] = 1.0 \times 10^{-11}$.

$\quad = -\log(10^{-11})$

$\quad = -(-11)$

$\quad = 11$

The pH of household ammonia is 11.

Skill Practice

17. A new all-natural shampoo on the market claims its hydrogen ion concentration is 5.88×10^{-7} mol/L. Use the formula $pH = -\log[H^+]$ to calculate the pH level of the shampoo.

Example 8 **Applying Logarithmic Functions to a Memory Model**

One method of measuring a student's retention of material after taking a course is to retest the student at specified time intervals after the course has been completed. A student's score on a calculus test t months after completing a course in calculus is approximated by

$$S(t) = 85 - 25\log(t + 1)$$

where t is the time in months after completing the course and $S(t)$ is the student's score as a percent.

a. What was the student's score at $t = 0$?

b. What was the student's score after 2 months?

c. What was the student's score after 1 year?

Solution:

a. $S(t) = 85 - 25\log(t + 1)$

$S(0) = 85 - 25\log(0 + 1)$ Substitute $t = 0$.

$\quad = 85 - 25\log 1$ $\log 1 = 0$ because $10^0 = 1$.

$\quad = 85 - 25(0)$

$\quad = 85 - 0$

$\quad = 85$ The student's score at the time the course was completed was 85%.

b. $S(t) = 85 - 25\log(t + 1)$

$S(2) = 85 - 25\log(2 + 1)$

$\quad = 85 - 25\log 3$ Use a calculator to approximate $\log 3$.

$\quad \approx 73.1$ The student's score dropped to 73.1%.

c. $S(t) = 85 - 25\log(t + 1)$

$S(12) = 85 - 25\log(12 + 1)$

$\quad = 85 - 25\log 13$ Use a calculator to approximate $\log 13$.

$\quad \approx 57.2$ The student's score dropped to 57.2%.

Skill Practice Answers

17. $pH \approx 6.23$

Skill Practice

18. The memory model for a student's score on a statistics test t months after completion of the course in statistics is approximated by

$$S(t) = 92 - 28 \log (t + 1)$$

    a. What was the student's score at the time the course was completed ($t = 0$)?

    b. What was her score after 1 month?

Skill Practice Answers

    c. What was the score after 2 months?

18a. 92   **b.** 83.6   **c.** 78.6

Section 13.4   Practice Exercises

Boost *your* GRADE at mathzone.com!

 MathZone

- Practice Problems
- Self-Tests
- NetTutor
- e-Professors
- Videos

Study Skills Exercise

1. Define the key terms.

    a. logarithmic function   **b. logarithm**   **c. base of a logarithm**

    d. argument   **e. common logarithmic function**   **f. domain of a logarithmic function**

Review Exercises

2. For which graph of $y = b^x$ is $0 < b < 1$?

3. For which graph of $y = b^x$ is $b > 1$?

i.
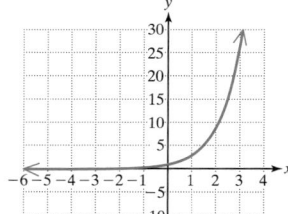

ii.
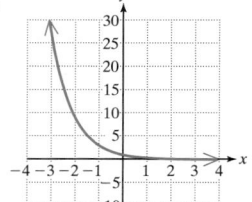

4. Let $f(x) = 6^x$.

    a. Find $f(-2), f(-1), f(0), f(1),$ and $f(2)$.

    b. Graph $y = f(x)$.

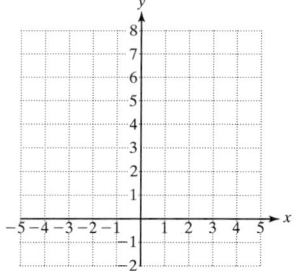

5. Let $g(x) = 3^x$.

    a. Find $g(-2), g(-1), g(0), g(1),$ and $g(2)$.

    b. Graph $y = g(x)$.

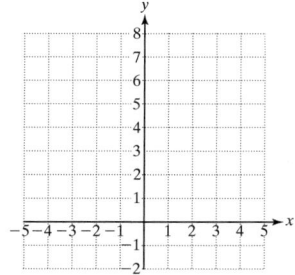

6. Let $r(x) = \left(\frac{3}{4}\right)^x$.

   a. Find $r(-2), r(-1), r(0), r(1),$ and $r(2)$.

   b. Graph $y = r(x)$.

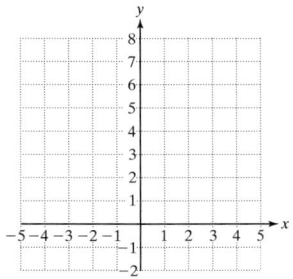

7. Let $s(x) = \left(\frac{2}{5}\right)^x$.

   a. Find $s(-2), s(-1), s(0), s(1),$ and $s(2)$.

   b. Graph $y = s(x)$.

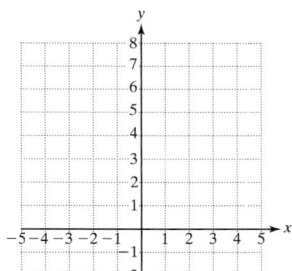

Concept 1: Definition of a Logarithmic Function

8. For the equation $y = \log_b x$, identify the base, the argument, and the logarithm.

9. Rewrite the equation in exponential form. $y = \log_b x$

For Exercises 10–21, write the equation in logarithmic form.

10. $3^x = 81$

11. $10^3 = 1000$

12. $5^2 = 25$

13. $8^{1/3} = 2$

14. $7^{-1} = \dfrac{1}{7}$

15. $8^{-2} = \dfrac{1}{64}$

16. $b^x = y$

17. $b^y = x$

18. $e^x = y$

19. $e^y = x$

20. $\left(\dfrac{1}{3}\right)^{-2} = 9$

21. $\left(\dfrac{5}{2}\right)^{-1} = \dfrac{2}{5}$

For Exercises 22–33, write the equation in exponential form.

22. $\log_5 625 = 4$

23. $\log_{125} 25 = \dfrac{2}{3}$

24. $\log_{10}(0.0001) = -4$

25. $\log_{25}\left(\dfrac{1}{5}\right) = -\dfrac{1}{2}$

26. $\log_6 36 = 2$

27. $\log_2 128 = 7$

28. $\log_b 15 = x$

29. $\log_b 82 = y$

30. $\log_3 5 = x$

31. $\log_2 7 = x$

32. $\log_{1/4} x = 10$

33. $\log_{1/2} x = 6$

Concept 2: Evaluating Logarithmic Expressions

For Exercises 34–49, find the logarithms without the use of a calculator.

34. $\log_7 49$

35. $\log_3 27$

36. $\log_{10} 0.1$

37. $\log_2\left(\dfrac{1}{16}\right)$

38. $\log_{16} 4$

39. $\log_8 2$

40. $\log_{7/2} 1$

41. $\log_{1/2} 2$

42. $\log_3 3^5$

43. $\log_9 9^3$

44. $\log_{10} 10$

45. $\log_7 1$

46. $\log_a a^3$

47. $\log_r r^4$

48. $\log_x \sqrt{x}$

49. $\log_y \sqrt[3]{y}$

Concept 3: The Common Logarithmic Function

For Exercises 50–58, find the common logarithm without the use of a calculator.

50. $\log 10$

51. $\log 100$

52. $\log 1000$

53. $\log 10{,}000$

54. $\log (1.0 \times 10^6)$

55. $\log 0.1$

56. $\log (0.01)$

57. $\log (0.001)$

58. $\log (1.0 \times 10^{-6})$

For Exercises 59–70, use a calculator to approximate the logarithms. Round to four decimal places.

59. $\log 6$

60. $\log 18$

61. $\log \pi$

62. $\log \left(\dfrac{1}{8}\right)$

63. $\log \left(\dfrac{1}{32}\right)$

64. $\log \sqrt{5}$

65. $\log (0.0054)$

66. $\log (0.0000062)$

67. $\log (3.4 \times 10^5)$

68. $\log (4.78 \times 10^9)$

69. $\log (3.8 \times 10^{-8})$

70. $\log (2.77 \times 10^{-4})$

71. Given: $\log 10 = 1$ and $\log 100 = 2$

    a. Estimate $\log 93$.

    b. Estimate $\log 12$.

    c. Evaluate the logarithms in parts (a) and (b) on a calculator and compare to your estimates.

72. Given: $\log \left(\tfrac{1}{10}\right) = -1$ and $\log 1 = 0$

    a. Estimate $\log \left(\tfrac{9}{10}\right)$.

    b. Estimate $\log \left(\tfrac{1}{5}\right)$.

    c. Evaluate the logarithms in parts (a) and (b) on a calculator and compare to your estimates.

Concept 4: Graphs of Logarithmic Functions

73. Let $f(x) = \log_4 x$.

    a. Find the values of $f(\tfrac{1}{64}), f(\tfrac{1}{16}), f(\tfrac{1}{4}), f(1), f(4)$, $f(16)$, and $f(64)$.

    b. Graph $y = f(x)$.

74. Let $g(x) = \log_2 x$.

    a. Find the values of $g(\tfrac{1}{8}), g(\tfrac{1}{4}), g(\tfrac{1}{2}), g(1), g(2)$, $g(4)$, and $g(8)$.

    b. Graph $y = g(x)$.

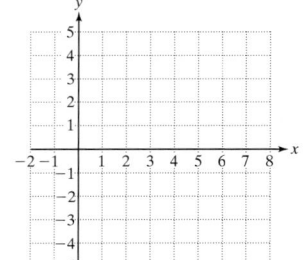

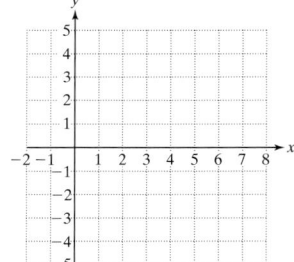

Graph the logarithmic functions in Exercises 75–78 by writing the function in exponential form and making a table of points.

75. $y = \log_3 x$

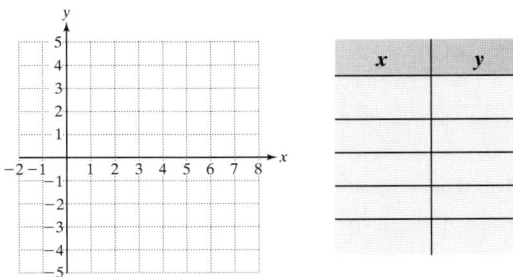

76. $y = \log_5 x$

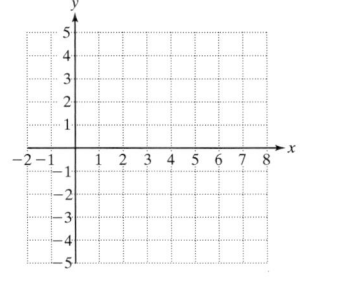
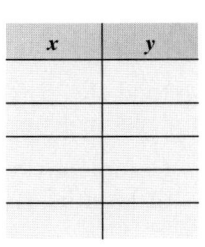

77. $y = \log_{1/2} x$

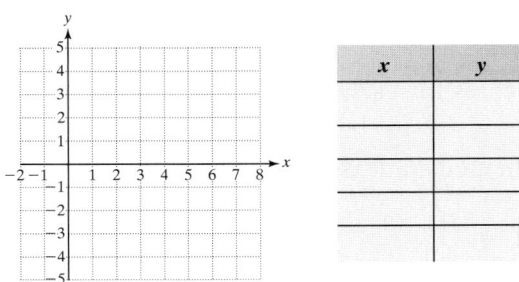

78. $y = \log_{1/3} x$

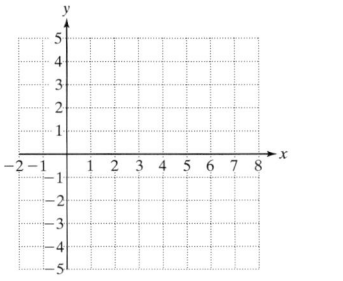

For Exercises 79–84, find the domain of the function and express the domain in interval notation.

79. $y = \log_7 (x - 5)$

80. $y = \log_3 (2x + 1)$

81. $y = \log_3 (x + 1.2)$

82. $y = \log \left(x - \dfrac{1}{2} \right)$

83. $y = \log x^2$

84. $y = \log (x^2 + 1)$

Concept 5: Applications of the Common Logarithmic Function

For Exercises 85–86, use the formula $pH = -\log[H^+]$, where $[H^+]$ represents the concentration of hydrogen ions in moles per liter. Round to the hundredths place.

85. Normally, the level of hydrogen ions in the blood is approximately 4.47×10^{-8} mol/L. Find the pH level of blood.

86. The normal pH level for streams and rivers is between 6.5 and 8.5. A high level of bacteria in a particular stream caused environmentalists to test the water. The level of hydrogen ions was found to be 0.006 mol/L. What is the pH level of the stream?

87. A graduate student in education is doing research to compare the effectiveness of two different teaching techniques designed to teach vocabulary to sixth-graders. The first group of students (group 1) was taught with method I, in which the students worked individually to complete the assignments in a workbook. The second group (group 2) was taught with method II, in which the students worked cooperatively in groups of four to complete the assignments in the same workbook.

None of the students knew any of the vocabulary words before the study began. After completing the assignments in the workbook, the students were then tested on the vocabulary at 1-month intervals to assess how much material they had retained over time. The students' average score t months after completing the assignments are given by the following functions:

Method I: $S_1(t) = 91 - 30 \log (t + 1)$, where t is the time in months and $S_1(t)$ is the average score of students in group 1.

Method II: $S_2(t) = 88 - 15 \log (t + 1)$, where t is the time in months and $S_2(t)$ is the average score of students in group 2.

a. Complete the table to find the average scores for each group of students after the indicated number of months. Round to one decimal place.

t (months)	0	1	2	6	12	24
$S_1(t)$						
$S_2(t)$						

b. Based on the table of values, what were the average scores for each group immediately after completion of the assigned material ($t = 0$)?

c. Based on the table of values, which teaching method helped students retain the material better for a long time?

88. Generally, the more money a company spends on advertising, the higher the sales. Let a represent the amount of money spent on advertising (in $100s). Then the amount of money in sales $S(a)$ (in $1000s) is given by

$$S(a) = 10 + 20 \log (a + 1) \text{ where } a \geq 0.$$

a. The value of $S(1) \approx 16.0$ means that if $100 is spent on advertising, $16,000 is returned in sales. Find the values of $S(11)$, $S(21)$, and $S(31)$. Round to 1 decimal place. Interpret the meaning of each function value in the context of this problem.

b. The graph of $y = S(a)$ is shown here. Use the graph and your answers from part (a) to explain why the money spent in advertising becomes less "efficient" as it is used in larger amounts.

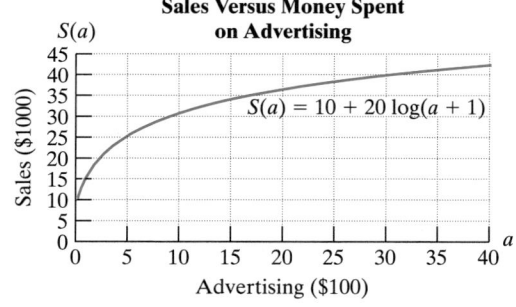

Sales Versus Money Spent on Advertising

Graphing Calculator Exercises

For Exercises 89–94, graph the function on an appropriate viewing window. From the graph, identify the domain of the function and the location of the vertical asymptote.

89. $y = \log(x + 6)$

90. $y = \log(2x + 4)$

91. $y = \log(0.5x - 1)$

92. $y = \log(x + 8)$

93. $y = \log(2 - x)$

94. $y = \log(3 - x)$

Properties of Logarithms   Section 13.5

1. Properties of Logarithms

You have already been exposed to certain properties of logarithms that follow directly from the definition. Recall

$\qquad y = \log_b x \qquad$ is equivalent to $\qquad b^y = x \qquad$ for $x > 0, b > 0$, and $b \neq 1$

The following properties follow directly from the definition.

$$\log_b 1 = 0 \qquad \text{Property 1}$$
$$\log_b b = 1 \qquad \text{Property 2}$$
$$\log_b b^p = p \qquad \text{Property 3}$$
$$b^{\log_b x} = x \qquad \text{Property 4}$$

Concepts

1. Properties of Logarithms
2. Expanded Logarithmic Expressions
3. Single Logarithmic Expressions

Example 1 **Applying the Properties of Logarithms to Simplify Expressions**

Use the properties of logarithms to simplify the expressions. Assume that all variable expressions within the logarithms represent positive real numbers.

a. $\log_8 8 + \log_8 1$ **b.** $10^{\log(x+2)}$ **c.** $\log_{1/2}\left(\dfrac{1}{2}\right)^x$

Solution:

a. $\log_8 8 + \log_8 1 = 1 + 0 = 1$ Properties 2 and 1

b. $10^{\log(x+2)} = x + 2$ Property 4

c. $\log_{1/2}\left(\dfrac{1}{2}\right)^{x} = x$ Property 3

Skill Practice Use the properties of logarithms to simplify the expressions.

1. $\log_5 1 + \log_5 5$ **2.** $15^{\log_{15} 7}$ **3.** $\log_{1/3}\left(\dfrac{1}{3}\right)^{c}$

Three additional properties are useful when simplifying logarithmic expressions. The first is the product property for logarithms.

> ### Product Property for Logarithms
> Let b, x, and y be positive real numbers where $b \neq 1$. Then
> $$\log_b (xy) = \log_b x + \log_b y$$
> The logarithm of a product equals the sum of the logarithms of the factors.

Proof:

Let $M = \log_b x$, which implies $b^M = x$.

Let $N = \log_b y$, which implies $b^N = y$.

Then $xy = b^M b^N = b^{M+N}$.

Writing the expression $xy = b^{M+N}$ in logarithmic form, we have,

$$\log_b (xy) = M + N$$
$$\log_b (xy) = \log_b x + \log_b y \ \checkmark$$

To demonstrate the product property for logarithms, simplify the following expressions by using the order of operations.

$$\log_3 (3 \cdot 9) \overset{?}{=} \log_3 3 + \log_3 9$$
$$\log_3 27 \overset{?}{=} 1 + 2$$
$$3 = 3 \ \checkmark$$

> ### Quotient Property for Logarithms
> Let b, x, and y be positive real numbers where $b \neq 1$. Then
> $$\log_b \left(\dfrac{x}{y}\right) = \log_b x - \log_b y$$
> The logarithm of a quotient equals the difference of the logarithms of the numerator and denominator.

Skill Practice Answers

1. 1 **2.** 7 **3.** c

The proof of the quotient property for logarithms is similar to the proof of the product property and is omitted here. To demonstrate the quotient property for logarithms, simplify the following expressions by using the order of operations.

$$\log\left(\frac{1,000,000}{100}\right) \overset{?}{=} \log 1,000,000 - \log 100$$

$$\log 10,000 \overset{?}{=} 6 - 2$$

$$4 = 4 \checkmark$$

Power Property for Logarithms

Let b and x be positive real numbers where $b \neq 1$. Let p be any real number. Then

$$\log_b x^p = p \log_b x$$

Proof:

Let $M = \log_b x$, which implies $b^M = x$.

Raise both sides to the power p: $(b^M)^p = x^p$ or equivalently $b^{Mp} = x^p$

Write the expression $b^{Mp} = x^p$ in logarithmic form: $\log_b x^p = Mp = pM$ or equivalently $\log_b x^p = p \log_b x$. $\checkmark$

To demonstrate the power property for logarithms, simplify the following expressions by using the order of operations.

$$\log_4 4^2 \overset{?}{=} 2 \log_4 4$$

$$2 \overset{?}{=} 2 \cdot 1$$

$$2 = 2 \checkmark$$

The properties of logarithms are summarized in the box.

Properties of Logarithms

Let b, x, and y be positive real numbers where $b \neq 1$, and let p be a real number. Then the following properties of logarithms are true.

1. $\log_b 1 = 0$	**5.** $\log_b (xy) = \log_b x + \log_b y$	**Product property for logarithms**
2. $\log_b b = 1$	**6.** $\log_b\left(\dfrac{x}{y}\right) = \log_b x - \log_b y$	**Quotient property for logarithms**
3. $\log_b b^p = p$	**7.** $\log_b x^p = p \log_b x$	**Power property for logarithms**
4. $b^{\log_b x} = x$		

2. Expanded Logarithmic Expressions

In many applications it is advantageous to expand a logarithm into a sum or difference of simpler logarithms.

> **Example 2**   **Writing a Logarithmic Expression in Expanded Form**
>
> Write the expressions as the sum or difference of logarithms of x, y, and z. Assume all variables represent positive real numbers.
>
> **a.** $\log_3\left(\dfrac{xy^3}{z^2}\right)$ **b.** $\log\left(\dfrac{\sqrt{x+y}}{10}\right)$ **c.** $\log_b\sqrt[5]{\dfrac{x^4}{yz^3}}$

Solution:

a. $\log_3\left(\dfrac{xy^3}{z^2}\right)$

$= \log_3 xy^3 - \log_3 z^2$ Quotient property for logarithms (property 6)

$= [\log_3 x + \log_3 y^3] - \log_3 z^2$ Product property for logarithms (property 5)

$= \log_3 x + 3\log_3 y - 2\log_3 z$ Power property for logarithms (property 7)

b. $\log\left(\dfrac{\sqrt{x+y}}{10}\right)$

$= \log(\sqrt{x+y}) - \log(10)$ Quotient property for logarithms (property 6)

$= \log(x+y)^{1/2} - 1$ Write $\sqrt{x+y}$ as $(x+y)^{1/2}$ and simplify $\log 10 = 1$.

$= \dfrac{1}{2}\log(x+y) - 1$ Power property for logarithms (property 7)

c. $\log_b\sqrt[5]{\dfrac{x^4}{yz^3}}$

$= \log_b\left(\dfrac{x^4}{yz^3}\right)^{1/5}$ Write $\sqrt[5]{\dfrac{x^4}{yz^3}}$ as $\left(\dfrac{x^4}{yz^3}\right)^{1/5}$.

$= \dfrac{1}{5}\log_b\left(\dfrac{x^4}{yz^3}\right)$ Power property for logarithms (property 7)

$= \dfrac{1}{5}[\log_b x^4 - \log_b(yz^3)]$ Quotient property for logarithms (property 6)

$= \dfrac{1}{5}[\log_b x^4 - (\log_b y + \log_b z^3)]$ Product property for logarithms (property 5)

$= \dfrac{1}{5}[\log_b x^4 - \log_b y - \log_b z^3]$ Distributive property

$$= \frac{1}{5}[4 \log_b x - \log_b y - 3 \log_b z]$$

Power property for logarithms
(property 7)

or $\quad \dfrac{4}{5} \log_b x - \dfrac{1}{5} \log_b y - \dfrac{3}{5} \log_b z$

Skill Practice Write the expression as the sum or difference of logarithms of a, b, and c. Assume all variables represent positive real numbers.

4. $\log_5 \left(\dfrac{a^2 b^3}{c} \right)$ **5.** $\log \dfrac{(a-b)^2}{\sqrt{10}}$ **6.** $\log_3 \sqrt[4]{\dfrac{a}{bc^5}}$

3. Single Logarithmic Expressions

In some applications it is necessary to write a sum or difference of logarithms as a single logarithm.

Example 3 **Writing a Sum or Difference of Logarithms as a Single Logarithm**

Rewrite the expressions as a single logarithm, and simplify the result, if possible. Assume all variable expressions within the logarithms represent positive real numbers.

a. $\log_2 560 - \log_2 7 - \log_2 5$ **b.** $2 \log x - \dfrac{1}{2} \log y + 3 \log z$

c. $\dfrac{1}{2}[\log_5 (x^2 - y^2) - \log_5 (x + y)]$

Solution:

a. $\log_2 560 - \log_2 7 - \log_2 5$

$= \log_2 560 - (\log_2 7 + \log_2 5)$ Factor out -1 from the last two terms.

$= \log_2 560 - \log_2 (7 \cdot 5)$ Product property for logarithms (property 5)

$= \log_2 \left(\dfrac{560}{7 \cdot 5} \right)$ Quotient property for logarithms (property 6)

$= \log_2 16$ Simplify inside parentheses.

$= 4$

b. $2 \log x - \dfrac{1}{2} \log y + 3 \log z$

$= \log x^2 - \log y^{1/2} + \log z^3$ Power property for logarithms (property 7)

$= \log x^2 + \log z^3 - \log y^{1/2}$ Group terms with positive coefficients.

$= \log (x^2 z^3) - \log y^{1/2}$ Product property for logarithms (property 5)

$= \log \left(\dfrac{x^2 z^3}{y^{1/2}} \right)$ or $\log \left(\dfrac{x^2 z^3}{\sqrt{y}} \right)$ Quotient property for logarithms (property 6)

Skill Practice Answers

4. $2 \log_5 a + 3 \log_5 b - \log_5 c$

5. $2 \log (a - b) - \dfrac{1}{2}$

6. $\dfrac{1}{4} \log_3 a - \dfrac{1}{4} \log_3 b - \dfrac{5}{4} \log_3 c$

c. $\frac{1}{2}\left[\log_5(x^2 - y^2) - \log_5(x + y)\right]$

$= \frac{1}{2}\log_5\left(\frac{x^2 - y^2}{x + y}\right)$ Quotient property for logarithms (property 6)

$= \frac{1}{2}\log_5\left[\frac{(x + y)(x - y)}{x + y}\right]$ Factor and simplify within the parentheses.

$= \frac{1}{2}\log_5(x - y)$

$= \log_5(x - y)^{1/2}$ or $\log_5\sqrt{x - y}$ Power property for logarithms (property 7)

Skill Practice Write the expression as a single logarithm, and simplify the result, if possible.

7. $\log_3 54 + \log_3 10 - \log_3 20$ **8.** $3\log x + \frac{1}{3}\log y - 2\log z$

9. $\frac{1}{2}\left[\log_2(a^2 + ab) - \log_2 a\right]$

It is important to note that the properties of logarithms may be used to write a single logarithm as a sum or difference of logarithms. Furthermore, the properties may be used to write a sum or difference of logarithms as a single logarithm. In either case, these operations may change the domain.

For example, consider the function $y = \log_b x^2$. Using the power property for logarithms, we have $y = 2\log_b x$. Consider the domain of each function:

$y = \log_b x^2$ Domain: $(-\infty, 0) \cup (0, \infty)$

$y = 2\log_b x$ Domain: $(0, \infty)$

These two functions are equivalent only for values of x in the intersection of the two domains, that is, for $(0, \infty)$.

Skill Practice Answers

7. 3   **8.** $\log\left(\dfrac{x^3\sqrt[3]{y}}{z^2}\right)$

9. $\log_2\sqrt{a + b}$

Section 13.5   Practice Exercises

Study Skills Exercise

1. Define the key terms.

    a. product property of logarithms   **b. quotient property of logarithms**

    c. power property of logarithms

Review Exercises

For Exercises 2–5, find the values of the logarithmic and exponential expressions without using a calculator.

2. 8^{-2}

3. $\log 10{,}000$

4. $\log_2 32$

5. 6^{-1}

For Exercises 6–9, approximate the values of the logarithmic and exponential expressions by using a calculator.

6. $(\sqrt{2})^{\pi}$

7. $\log 8$

8. $\log 27$

9. $\pi^{\sqrt{2}}$

For Exercises 10–13, match the function with the appropriate graph.

10. $f(x) = 4^x$

11. $q(x) = \left(\dfrac{1}{5}\right)^x$

12. $h(x) = \log_5 x$

13. $k(x) = \log_{1/3} x$

a.
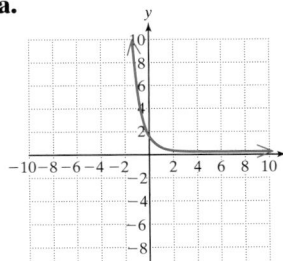

b.
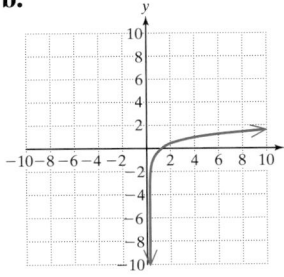

c.
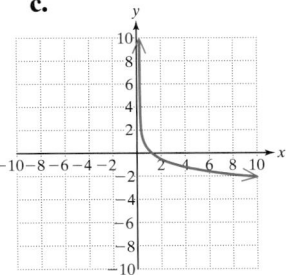

d.
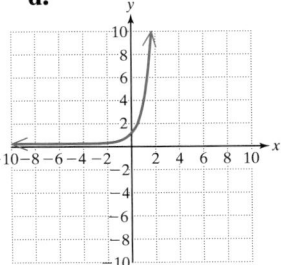

Concept 1: Properties of Logarithms

14. Property 1 of logarithms states that $\log_b 1 = 0$. Write an example of this property.

15. Property 2 of logarithms states that $\log_b b = 1$. Write an example of this property.

16. Property 3 of logarithms states that $\log_b b^p = p$. An example is $\log_6 6^3 = 3$. Write another example of this property.

17. Property 4 of logarithms states that $b^{\log_b x} = x$. An example is $2^{\log_2 5} = 5$. Write another example of this property.

For Exercises 18–39, evaluate each expression.

18. $\log_3 3$

19. $\log 10$

20. $\log_5 5^4$

21. $\log_4 4^5$

22. $6^{\log_6 11}$

23. $7^{\log_7 2}$

24. $\log 10^3$

25. $\log_6 6^3$

26. $\log_3 1$

27. $\log_8 1$

28. $10^{\log 9}$

29. $8^{\log_8 5}$

30. $\log_{1/2} 1$

31. $\log_{1/3}\left(\dfrac{1}{3}\right)$

32. $\log_2 1 + \log_2 2^3$

33. $\log 10^4 + \log 10$

34. $\log_4 4 + \log_2 1$

35. $\log_7 7 + \log_4 4^2$

36. $\log_{1/4}\left(\dfrac{1}{4}\right)^{2x}$

37. $\log_{2/3}\left(\dfrac{2}{3}\right)^p$

38. $\log_a a^4$

39. $\log_y y^2$

40. Compare the expressions by approximating their values on a calculator. Which two expressions appear to be equivalent?

    a. $\log (3 \cdot 5)$ **b.** $\log 3 \cdot \log 5$ **c.** $\log 3 + \log 5$

41. Compare the expressions by approximating their values on a calculator. Which two expressions appear to be equivalent?

a. $\log\left(\dfrac{6}{5}\right)$ **b.** $\dfrac{\log 6}{\log 5}$ **c.** $\log 6 - \log 5$

42. Compare the expressions by approximating their values on a calculator. Which two expressions appear to be equivalent?

a. $\log 20^2$ **b.** $[\log 20]^2$ **c.** $2 \log 20$

43. Compare the expressions by approximating their values on a calculator. Which two expressions appear to be equivalent?

a. $\log \sqrt{4}$ **b.** $\dfrac{1}{2} \log 4$ **c.** $\sqrt{\log 4}$

Concept 2: Expanded Logarithmic Expressions

For Exercises 44–61, expand into sums and/or differences of logarithms. Assume all variables represent positive real numbers.

44. $\log_3\left(\dfrac{x}{5}\right)$ **45.** $\log_2\left(\dfrac{y}{z}\right)$ **46.** $\log(2x)$

47. $\log_6(xyz)$ **48.** $\log_5 x^4$ **49.** $\log_7 z^{1/3}$

50. $\log_4\left(\dfrac{ab}{c}\right)$ **51.** $\log_2\left(\dfrac{x}{yz}\right)$ **52.** $\log_b\left(\dfrac{\sqrt{xy}}{z^3 w}\right)$

53. $\log\left(\dfrac{a\sqrt[3]{b}}{cd^2}\right)$ **54.** $\log_2\left(\dfrac{x+1}{y^2\sqrt{z}}\right)$ **55.** $\log\left(\dfrac{a+1}{b\sqrt[3]{c}}\right)$

56. $\log\left(\sqrt[3]{\dfrac{ab^2}{c}}\right)$ **57.** $\log_5\left(\sqrt[4]{\dfrac{w^3 z}{x^2}}\right)$ **58.** $\log\left(\dfrac{1}{w^5}\right)$

59. $\log_3\left(\dfrac{1}{z^4}\right)$ **60.** $\log_b\left(\dfrac{\sqrt{a}}{b^3 c}\right)$ **61.** $\log_x\left(\dfrac{x}{y\sqrt{z}}\right)$

Concept 3: Single Logarithmic Expressions

For Exercises 62–75, write the expressions as a single logarithm. Assume all variables represent positive real numbers.

62. $\log x + \log y - \log z$

63. $\log C + \log A + \log B + \log I + \log N$

64. $2 \log_3 x - 3 \log_3 y + \log_3 z$

65. $\dfrac{1}{3} \log_2 x - 5 \log_2 y - 3 \log_2 z$

66. $2 \log_3 a - \dfrac{1}{4} \log_3 b + \log_3 c$

67. $\log_5 a - \dfrac{1}{2} \log_5 b - 3 \log_5 c$

68. $\log_b x - 3 \log_b x + 4 \log_b x$

69. $2 \log_3 z + \log_3 z - \dfrac{1}{2} \log_3 z$

70. $5 \log_8 a - \log_8 1 + \log_8 8$

71. $\log_2 2 + 2 \log_2 b - \log_2 1$

72. $2 \log (x + 6) + \dfrac{1}{3} \log y - 5 \log z$

73. $\dfrac{1}{4} \log (a + 1) - 2 \log b - 4 \log c$

74. $\log_b (x + 1) - \log_b (x^2 - 1)$

75. $\log_x (p^2 - 4) - \log_x (p - 2)$

Expanding Your Skills

For Exercises 76–85, find the values of the logarithms given that $\log_b 2 \approx 0.693$, $\log_b 3 \approx 1.099$, and $\log_b 5 \approx 1.609$.

76. $\log_b 6$

77. $\log_b 4$

78. $\log_b 12$

79. $\log_b 25$

80. $\log_b 81$

81. $\log_b 30$

82. $\log_b \left(\dfrac{5}{2} \right)$

83. $\log_b \left(\dfrac{25}{3} \right)$

84. $\log_b 10^6$

85. $\log_b 2^{12}$

86. The intensity of sound waves is measured in decibels and is calculated by the formula

$$B = 10 \log \left(\dfrac{I}{I_0} \right)$$

where I_0 is the minimum detectable decibel level.

a. Expand this formula by using the properties of logarithms.

b. Let $I_0 = 10^{-16}$ W/cm^2 and simplify.

87. The Richter scale is used to measure the intensity of an earthquake and is calculated by the formula

$$R = \log \left(\dfrac{I}{I_0} \right)$$

where I_0 is the minimum level detectable by a seismograph.

a. Expand this formula by using the properties of logarithms.

b. Let $I_0 = 1$ and simplify.

Graphing Calculator Exercises

88. a. Graph $Y_1 = \log x^2$ and state its domain.

b. Graph $Y_2 = 2 \log x$ and state its domain.

c. For what values of x are the expressions $\log x^2$ and $2 \log x$ equivalent?

89. a. Graph $Y_1 = \log (x - 1)^2$ and state its domain.

b. Graph $Y_2 = 2 \log (x - 1)$ and state its domain.

c. For what values of x are the expressions $\log (x - 1)^2$ and $2 \log (x - 1)$ equivalent?

Section 13.6   The Irrational Number *e*

1. The Irrational Number *e*

The exponential function base 10 is particularly easy to work with because integral powers of 10 represent different place positions in the base-10 numbering system. In this section, we introduce another important exponential function whose base is an irrational number called *e*.

Consider the expression $(1 + \frac{1}{x})^x$. The value of the expression for increasingly large values of *x* approaches a constant (Table 13-7).

Table 13-7

x	$\left(1 + \dfrac{1}{x}\right)^x$
100	2.70481382942
1000	2.71692393224
10,000	2.71814592683
100,000	2.71826823717
1,000,000	2.71828046932
1,000,000,000	2.71828182710

As *x* approaches infinity, the expression $(1 + \frac{1}{x})^x$ approaches a constant value that we call *e*. From Table 13-7, this value is approximately 2.718281828.

$$e \approx 2.718281828$$

The value of *e* is an irrational number (a nonterminating, nonrepeating decimal) and is a universal constant, as is π.

Example 1   Graphing $f(x) = e^x$

Graph the function defined by $f(x) = e^x$.

Solution:

Because the base of the function is greater than 1 ($e \approx 2.718281828$), the graph is an increasing exponential function. We can use a calculator to evaluate $f(x) = e^x$ at several values of *x*.

If you are using your calculator correctly, your answers should match those found in Table 13-8. Values are rounded to 3 decimal places. The corresponding graph of $f(x) = e^x$ is shown in Figure 13-16.

Table 13-8

x	$f(x) = e^x$
−3	0.050
−2	0.135
−1	0.368
0	1.000
1	2.718
2	7.389
3	20.086

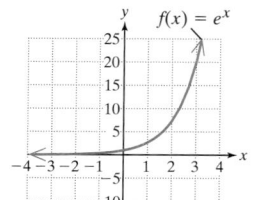

Figure 13-16

1. Graph $f(x) = e^x + 1$.

2. Computing Compound Interest

One particularly interesting application of exponential functions is in computing compound interest.

1. If the number of compounding periods per year is finite, then the amount in an account is given by

$$A(t) = P\left(1 + \frac{r}{n}\right)^{nt}$$

where P is the initial principal, r is the annual interest rate, n is the number of times compounded per year, and t is the time in years that the money is invested.

2. If the number of compound periods per year is infinite, then interest is said to be **compounded continuously**. In such a case, the amount in an account is given by

$$A(t) = Pe^{rt}$$

where P is the initial principal, r is the annual interest rate, and t is the time in years that the money is invested.

Example 2   Computing the Balance on an Account

Suppose $5000 is invested in an account earning 6.5% interest. Find the balance in the account after 10 years under the following compounding options.

a. Compounded annually

b. Compounded quarterly

c. Compounded monthly

d. Compounded daily

e. Compounded continuously

Solution:

Compounding Option	n Value	Formula	Result
Annually	$n = 1$	$A(10) = 5000\left(1 + \frac{0.065}{1}\right)^{(1)(10)}$	$9385.69
Quarterly	$n = 4$	$A(10) = 5000\left(1 + \frac{0.065}{4}\right)^{(4)(10)}$	$9527.79
Monthly	$n = 12$	$A(10) = 5000\left(1 + \frac{0.065}{12}\right)^{(12)(10)}$	$9560.92
Daily	$n = 365$	$A(10) = 5000\left(1 + \frac{0.065}{365}\right)^{(365)(10)}$	$9577.15
Continuously	Not applicable	$A(10) = 5000e^{(0.065)(10)}$	$9577.70

Skill Practice Answers

1.

Notice that there is a $191.46 difference in the account balance between annual compounding and daily compounding. However, the difference between compounding daily and compounding continuously is small—$0.55. As n gets infinitely large, the function defined by

$$A(t) = P\left(1 + \frac{r}{n}\right)^{nt}$$

converges to $A(t) = Pe^{rt}$.

Skill Practice

2. Suppose $1000 is invested at 5%. Find the balance after 8 years under the following options.
   a. Compounded annually
   b. Compounded quarterly
   c. Compounded monthly
   d. Compounded daily
   e. Compounded continuously

3. The Natural Logarithmic Function

Recall that the common logarithmic function $y = \log x$ has a base of 10. Another important logarithmic function is called the **natural logarithmic function**. The natural logarithmic function has a base of e and is written as $y = \ln x$. That is,

$$y = \ln x = \log_e x$$

Calculator Connections

```
ln(1)
               0
ln(2)
      .6931471806
ln(3)
      1.098612289
```

Example 3   Graphing $y = \ln x$

Graph $y = \ln x$.

Solution:

Because the base of the function $y = \ln x$ is e and $e > 1$, the graph is an increasing logarithmic function. We can use a calculator to find specific points on the graph of $y = \ln x$ by pressing the $\boxed{\ln}$ key.

Practice using your calculator by evaluating $\ln x$ for the following values of x. If you are using your calculator correctly, your answers should match those found in Table 13-9. Values are rounded to 3 decimal places. The corresponding graph of $y = \ln x$ is shown in Figure 13-17.

Table 13-9

x	$\ln x$
1	0.000
2	0.693
3	1.099
4	1.386
5	1.609
6	1.792
7	1.946

Note that the natural logarithmic function $f(x) = \ln x$ is the inverse of $g(x) = e^x$. That means that the graphs are symmetric to each other about the line $y = x$.

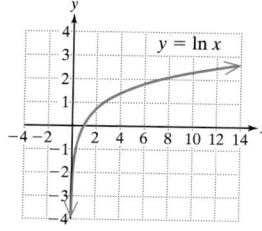

Figure 13-17

3. Graph $y = \ln x + 1$.

The properties of logarithms stated in Section 13.5 are also true for natural logarithms.

Properties of the Natural Logarithmic Function

Let x and y be positive real numbers, and let p be a real number. Then the following properties are true.

1. $\ln 1 = 0$ **5.** $\ln (xy) = \ln x + \ln y$ Product property for logarithms

2. $\ln e = 1$ **6.** $\ln \left(\dfrac{x}{y}\right) = \ln x - \ln y$ Quotient property for logarithms

3. $\ln e^p = p$ **7.** $\ln x^p = p \ln x$ Power property for logarithms

4. $e^{\ln x} = x$

Example 4   **Simplifying Expressions with Natural Logarithms**

Simplify the expressions. Assume that all variable expressions within the logarithms represent positive real numbers.

a. $\ln e$ **b.** $\ln 1$ **c.** $\ln (e^{x+1})$ **d.** $e^{\ln (x+1)}$

Solution:

a. $\ln e = 1$ Property 2

b. $\ln 1 = 0$ Property 1

c. $\ln (e^{x+1}) = x + 1$ Property 3

d. $e^{\ln (x+1)} = x + 1$ Property 4

Simplify.

4. $\ln e^2$ **5.** $-3 \ln 1$ **6.** $\ln e^{(x+y)}$ **7.** $e^{\ln (3x)}$

Example 5   **Writing a Sum or Difference of Natural Logarithms as a Single Logarithm**

Write the expression as a single logarithm. Assume that all variable expressions within the logarithms represent positive real numbers.

$$\ln (x^2 - 9) - \ln (x - 3) - 2 \ln x$$

Solution:

$\ln (x^2 - 9) - \ln (x - 3) - 2 \ln x$

$= \ln (x^2 - 9) - \ln (x - 3) - \ln x^2$ Power property for logarithms (property 7)

$= \ln (x^2 - 9) - [\ln (x - 3) + \ln x^2]$ Factor out -1 from the last two terms.

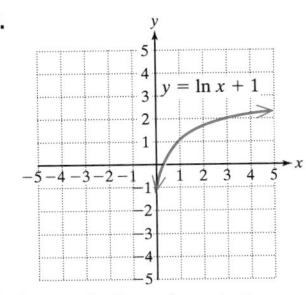

$$= \ln (x^2 - 9) - \ln [(x - 3)x^2] \qquad \text{Product property for logarithms (property 5)}$$

$$= \ln \left[\frac{x^2 - 9}{(x - 3)x^2} \right] \qquad \text{Quotient property for logarithms (property 6)}$$

$$= \ln \left[\frac{(x - 3)(x + 3)}{(x - 3)x^2} \right] \qquad \text{Factor.}$$

$$= \ln \left(\frac{x + 3}{x^2} \right) \qquad \text{Simplify.}$$

Skill Practice

8. Write as a single logarithm.

$$\ln (x + 4) - \ln (x^2 - 16) + \ln x$$

Example 6   **Writing a Logarithmic Expression in Expanded Form**

Write the expression

$$\ln \left(\frac{e}{x^2 \sqrt{y}} \right)$$

as a sum or difference of logarithms of x and y. Assume all variable expressions within the logarithm represent positive real numbers.

Solution:

$$\ln \left(\frac{e}{x^2 \sqrt{y}} \right)$$

$$= \ln e - \ln (x^2 \sqrt{y}) \qquad \text{Quotient property for logarithms (property 6)}$$

$$= \ln e - [\ln x^2 + \ln \sqrt{y}] \qquad \text{Product property for logarithms (property 5)}$$

$$= 1 - \ln x^2 - \ln y^{1/2} \qquad \text{Distributive property. Also simplify } \ln e = 1.$$

$$= 1 - 2 \ln x - \frac{1}{2} \ln y \qquad \text{Power property for logarithms (property 7)}$$

Skill Practice

9. Write as a sum or difference of logarithms of x and y. Assume all variables represent positive real numbers.

$$\ln \left(\frac{x^3 \sqrt{y}}{e^2} \right)$$

---

4. Change-of-Base Formula

Skill Practice Answers

8. $\ln \left(\dfrac{x}{x - 4} \right)$

9. $3 \ln x + \dfrac{1}{2} \ln y - 2$

A calculator can be used to approximate the value of a logarithm with a base of 10 or a base of e by using the $\boxed{\log}$ key or the $\boxed{\ln}$ key, respectively. However, to use a calculator to evaluate a logarithmic expression with a base other than 10 or e, we must use the **change-of-base formula**.

Change-of-Base Formula

Let a and b be positive real numbers such that $a \neq 1$ and $b \neq 1$. Then for any positive real number x,

$$\log_b x = \frac{\log_a x}{\log_a b}$$

Proof:

Let $M = \log_b x$, which implies that $b^M = x$.

Take the logarithm, base a, on both sides: $\log_a b^M = \log_a x$

Apply the power property for logarithms: $M \cdot \log_a b = \log_a x$

Divide both sides by $\log_a b$: $\dfrac{M \cdot \log_a b}{\log_a b} = \dfrac{\log_a x}{\log_a b}$

$$M = \frac{\log_a x}{\log_a b}$$

Because $M = \log_b x$, we have $\log_b x = \dfrac{\log_a x}{\log_a b}$ ✓

The change-of-base formula converts a logarithm of one base to a ratio of logarithms of a different base. For the sake of using a calculator, we often apply the change-of-base formula with base 10 or base e.

| **Example 7** | **Using the Change-of-Base Formula** |

a. Use the change-of-base formula to evaluate $\log_4 80$ by using base 10. (Round to three decimal places.)

b. Use the change-of-base formula to evaluate $\log_4 80$ by using base e. (Round to three decimal places.)

Solution:

a. $\log_4 80 = \dfrac{\log_{10} 80}{\log_{10} 4} = \dfrac{\log 80}{\log 4} \approx \dfrac{1.903089987}{0.6020599913} \approx 3.161$

b. $\log_4 80 = \dfrac{\log_e 80}{\log_e 4} = \dfrac{\ln 80}{\ln 4} \approx \dfrac{4.382026635}{1.386294361} \approx 3.161$

To check the result, we see that $4^{3.161} \approx 80$.

Skill Practice

10. Use the change-of-base formula to evaluate $\log_5 95$ by using base 10. Round to three decimal places.

11. Use the change-of-base formula to evaluate $\log_5 95$ by using base e. Round to three decimal places.

Skill Practice Answers

10. 2.829   **11.** 2.829

The change-of-base formula can be used to graph logarithmic functions with bases other than 10 or e. For example, to graph $Y_1 = \log_2 x$, we can enter the function as either

$$Y_1 = \frac{\log x}{\log 2} \quad \text{or} \quad Y_1 = \frac{\ln x}{\ln 2}$$

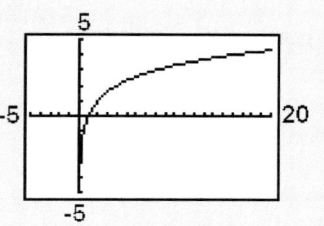

5. Applications of the Natural Logarithmic Function

Plant and animal tissue contains both carbon 12 and carbon 14. Carbon 12 is a stable form of carbon, whereas carbon 14 is a radioactive isotope with a half-life of approximately 5730 years. While a plant or animal is living, it takes in carbon from the atmosphere either through photosynthesis or through its food. The ratio of carbon 14 to carbon 12 in a living organism is constant and is the same as the ratio found in the atmosphere.

When a plant or animal dies, it no longer ingests carbon from the atmosphere. The amount of stable carbon 12 remains unchanged from the time of death, but the carbon 14 begins to decay. Because the rate of decay is constant, a tissue sample can be dated by comparing the percentage of carbon 14 still present to the percentage of carbon 14 assumed to be in its original living state.

The age of a tissue sample is a function of the percent of carbon 14 still present in the organism according to the following model:

$$A(p) = \frac{\ln p}{-0.000121}$$

where $A(p)$ is the age in years and p is the percentage (in decimal form) of carbon 14 still present.

Example 8   **Applying the Natural Logarithmic Function to Radioactive Decay**

Using the formula

$$A(p) = \frac{\ln p}{-0.000121}$$

a. Find the age of a bone that has 72% of its initial carbon 14.

b. Find the age of the Iceman, a body uncovered in the mountains of northern Italy in 1991. Samples of his hair revealed that 51.4% of the original carbon 14 was present after his death.

Solution:

a. $A(p) = \dfrac{\ln p}{-0.000121}$

$A(0.72) = \dfrac{\ln (0.72)}{-0.000121}$ Substitute 0.72 for p.

≈ 2715 years

b. $A(p) = \dfrac{\ln p}{-0.000121}$

$A(0.514) = \dfrac{\ln (0.514)}{-0.000121}$ Substitute 0.514 for *p*.

≈ 5500 years   The body of the Iceman is approximately 5500 years old.

Skill Practice

12. Use the formula $A(p) = \dfrac{\ln p}{-0.000121}$ (where $A(p)$ is the age in years and *p* is the percent of carbon 14 still present) to determine the age of a human skull that has 90% of its initial carbon 14.

Skill Practice Answers

12. ≈ 871 years old

Section 13.6   Practice Exercises

Boost *your* GRADE at mathzone.com!

MathZone

• Practice Problems
• Self-Tests
• NetTutor
• e-Professors
• Videos

Study Skills Exercise

1. Define the key terms.

a. *e*

b. continuously compounded interest

c. natural logarithmic function

d. change-of-base formula

Review Exercises

For Exercises 2–5, fill out the tables and graph the functions. For Exercises 4 and 5 round to two decimal places.

2. $f(x) = \left(\dfrac{3}{2}\right)^x$

x	*f(x)*
−3	
−2	
−1	
0	
1	
2	
3	

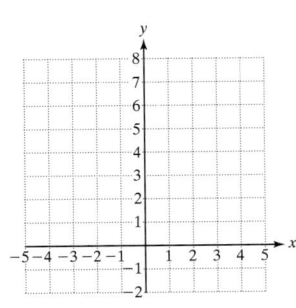

3. $g(x) = \left(\dfrac{1}{5}\right)^x$

x	*g(x)*
−3	
−2	
−1	
0	
1	
2	
3	

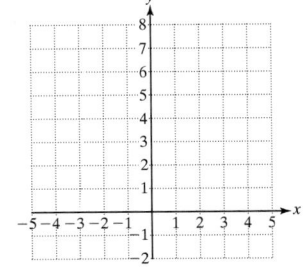

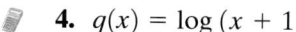 **4.** $q(x) = \log(x + 1)$

x	$q(x)$
-0.5	
0	
4	
9	

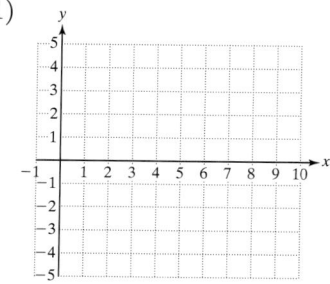

5. $r(x) = \log x$

x	$r(x)$
0.5	
1	
5	
10	

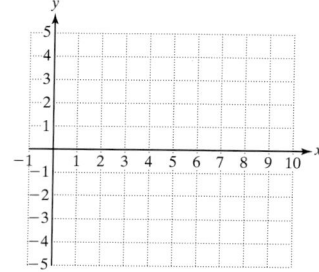

Concept 1: The Irrational Number *e*

For Exercises 6–9, graph the equation by completing the table and plotting points. Identify the domain. Round to two decimal places when necessary.

 6. $y = e^{x+1}$

x	y
-4	
-3	
-2	
-1	
0	
1	

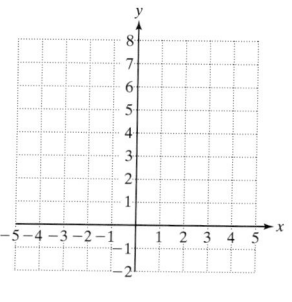

7. $y = e^{x+2}$

x	y
-5	
-4	
-3	
-2	
-1	
0	

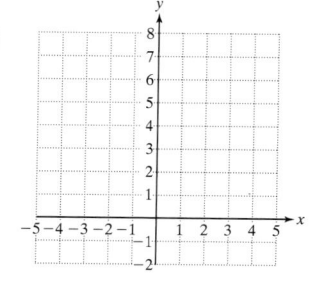

8. $y = e^x + 2$

x	y
-2	
-1	
0	
1	
2	
3	

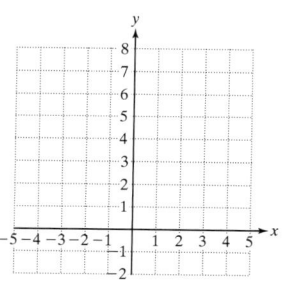

9. $y = e^x - 1$

x	y
-4	
-3	
-2	
-1	
0	
1	

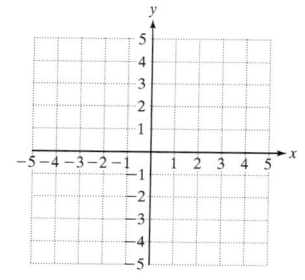

Concept 2: Computing Compound Interest

In Exercises 10–15, use the model

$$A(t) = P\left(1 + \frac{r}{n}\right)^{nt}$$

for interest compounded n times per year. Use the model $A(t) = Pe^{rt}$ for interest compounded continuously.

10. Suppose an investor deposits $10,000 in a certificate of deposit for 5 years for which the interest is compounded monthly. Find the total amount of money in the account for the following interest rates. Compare your answers and comment on the effect of interest rate on an investment.

   a. $r = 4.0\%$ **b.** $r = 6.0\%$ **c.** $r = 8.0\%$ **d.** $r = 9.5\%$

11. Suppose an investor deposits $5000 in a certificate of deposit for 8 years for which the interest is compounded quarterly. Find the total amount of money in the account for the following interest rates. Compare your answers and comment on the effect of interest rate on an investment.

   a. $r = 4.5\%$ **b.** $r = 5.5\%$ **c.** $r = 7.0\%$ **d.** $r = 9.0\%$

12. Suppose an investor deposits $8000 in a savings account for 10 years at 4.5% interest. Find the total amount of money in the account for the following compounding options. Compare your answers. How does the number of compound periods per year affect the total investment?

a. Compounded annually

b. Compounded quarterly

c. Compounded monthly

d. Compounded daily

e. Compounded continuously

13. Suppose an investor deposits $15,000 in a savings account for 8 years at 5.0% interest. Find the total amount of money in the account for the following compounding options. Compare your answers. How does the number of compound periods per year affect the total investment?

a. Compounded annually

b. Compounded quarterly

c. Compounded monthly

d. Compounded daily

e. Compounded continuously

14. Suppose an investor deposits $5000 in an account bearing 6.5% interest compounded continuously. Find the total amount in the account for the following time periods. How does the length of time affect the amount of interest earned?

a. 5 years   **b.** 10 years   **c.** 15 years   **d.** 20 years   **e.** 30 years

15. Suppose an investor deposits $10,000 in an account bearing 6.0% interest compounded continuously. Find the total amount in the account for the following time periods. How does the length of time affect the amount of interest earned?

a. 5 years   **b.** 10 years   **c.** 15 years   **d.** 20 years   **e.** 30 years

Concept 3: The Natural Logarithmic Function

For Exercises 16–19, graph the equation by completing the table and plotting the points. Identify the domain. Round to two decimal places when necessary.

16. $y = \ln(x - 2)$

x	y
2.25	
2.50	
2.75	
3	
4	
5	
6	

17. $y = \ln(x - 1)$

x	y
1.25	
1.50	
1.75	
2	
3	
4	
5	

18. $y = \ln x - 1$

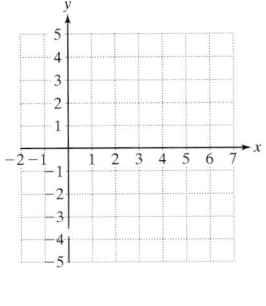

x	y
0.25	
0.5	
0.75	
1	
2	
3	
4	

19. $y = \ln x + 2$

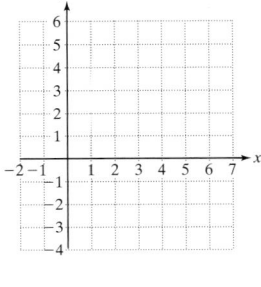

x	y
0.25	
0.5	
0.75	
1	
2	
3	
4	

20. a. Graph $f(x) = 10^x$

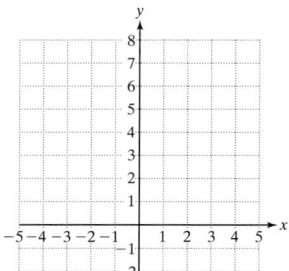

b. Identify the domain and range of f.

c. Graph $g(x) = \log x$.

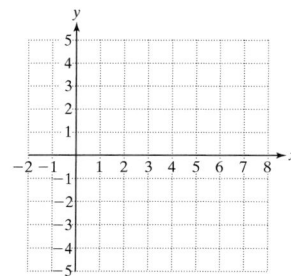

d. Identify the domain and range of g.

21. a. Graph $f(x) = e^x$.

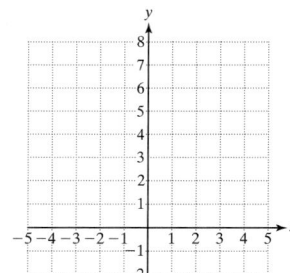

b. Identify the domain and range of f.

c. Graph $g(x) = \ln x$.

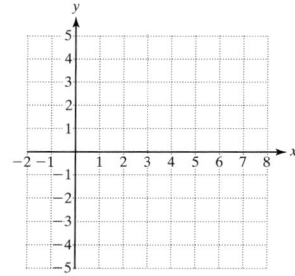

d. Identify the domain and range of g.

For Exercises 22–29, simplify the expressions. Assume all variables represent positive real numbers.

22. $\ln e$

23. $\ln e^2$

24. $\ln 1$

25. $e^{\ln x}$

26. $\ln e^{-6}$

27. $\ln e^{(5-3x)}$

28. $e^{\ln (2x + 3)}$

29. $e^{\ln 4}$

For Exercises 30–37, write the expression as a single logarithm. Assume all variables represent positive real numbers.

30. $6 \ln p + \dfrac{1}{3} \ln q$

31. $2 \ln w + \ln z$

32. $\dfrac{1}{2} (\ln x - 3 \ln y)$

33. $\dfrac{1}{3} (4 \ln a - \ln b)$

34. $2 \ln a - \ln b - \dfrac{1}{3} \ln c$

35. $-\ln x + 3 \ln y - \ln z$

36. $4 \ln x - 3 \ln y - \ln z$

37. $\dfrac{1}{2} \ln c + \ln a - 2 \ln b$

For Exercises 38–45, write the expression as a sum and/or difference of $\ln a$, $\ln b$, and $\ln c$. Assume all variables represent positive real numbers.

38. $\ln \left(\dfrac{a}{b} \right)^2$

39. $\ln \sqrt[3]{\dfrac{a}{b}}$

40. $\ln (b^2 \cdot e)$

41. $\ln (\sqrt{c} \cdot e)$

42. $\ln \left(\dfrac{a^4 \sqrt{b}}{c} \right)$

43. $\ln \left(\dfrac{\sqrt{ab}}{c^3} \right)$

44. $\ln \left(\dfrac{ab}{c^2} \right)^{1/5}$

45. $\ln \sqrt{2ab}$

Concept 4: Change-of-Base Formula

46. a. Evaluate $\log_6 200$ by computing $\dfrac{\log 200}{\log 6}$ to four decimal places.

b. Evaluate $\log_6 200$ by computing $\dfrac{\ln 200}{\ln 6}$ to four decimal places.

c. How do your answers to parts (a) and (b) compare?

47. a. Evaluate $\log_8 120$ by computing $\dfrac{\log 120}{\log 8}$ to four decimal places.

b. Evaluate $\log_8 120$ by computing $\dfrac{\ln 120}{\ln 8}$ to four decimal places.

c. How do your answers to parts (a) and (b) compare?

For Exercises 48–59, use the change-of-base formula to approximate the logarithms to 4 decimal places. Check your answers by using the exponential key on your calculator.

48. $\log_2 7$

49. $\log_3 5$

50. $\log_8 24$

51. $\log_4 17$

52. $\log_8 (0.012)$

53. $\log_7 (0.251)$

54. $\log_9 1$

55. $\log_2 \left(\dfrac{1}{5}\right)$

56. $\log_4 \left(\dfrac{1}{100}\right)$

57. $\log_5 (0.0025)$

58. $\log_7 (0.0006)$

59. $\log_2 (0.24)$

Concept 5: Applications of the Natural Logarithmic Function

Under continuous compounding, the amount of time *t* in years required for an investment to double is a function of the interest rate *r*:

$$t = \frac{\ln 2}{r}$$

Use the formula for Exercises 60–62.

60. a. If you invest $5000, how long will it take the investment to reach $10,000 if the interest rate is 4.5%? Round to one decimal place.

b. If you invest $5000, how long will it take the investment to reach $10,000 if the interest rate is 10%? Round to one decimal place.

c. Using the doubling time found in part (b), how long would it take a $5000 investment to reach $20,000 if the interest rate is 10%?

61. a. If you invest $3000, how long will it take the investment to reach $6000 if the interest rate is 5.5%? Round to one decimal place.

b. If you invest $3000, how long will it take the investment to reach $6000 if the interest rate is 8%? Round to one decimal place.

c. Using the doubling time found in part (b), how long would it take a $3000 investment to reach $12,000 if the interest rate is 8%?

62. a. If you invest $4000, how long will it take the investment to reach $8000 if the interest rate is 3.5%? Round to one decimal place.

b. If you invest $4000, how long will it take the investment to reach $8000 if the interest rate is 5%? Round to one decimal place.

c. Using the doubling time found in part (b), how long would it take a $4000 investment to reach $16,000 if the interest rate is 5%?

63. On August 31, 1854, an epidemic of cholera was discovered in London, England, resulting from a contaminated community water pump at Broad Street. By the end of September more than 600 citizens who drank water from the pump had died.

The cumulative number of deaths from cholera in the 1854 London epidemic can be approximated by

$$D(t) = 91 + 160 \ln (t + 1)$$

where t is the number of days after the start of the epidemic ($t = 0$ corresponds to September 1, 1854).

a. Approximate the total number of deaths as of September 1 ($t = 0$).

b. Approximate the total number of deaths as of September 5, September 10, and September 20.

Graphing Calculator Connections

64. a. Graph the function defined by $f(x) = \log_3 x$ by graphing $Y_1 = \dfrac{\log x}{\log 3}$.

b. Graph the function defined by $f(x) = \log_3 x$ by graphing $Y_2 = \dfrac{\ln x}{\ln 3}$.

c. Does it appear that $Y_1 = Y_2$ on the standard viewing window?

65. a. Graph the function defined by $f(x) = \log_7 x$ by graphing $Y_1 = \dfrac{\log x}{\log 7}$.

b. Graph the function defined by $f(x) = \log_7 x$ by graphing $Y_2 = \dfrac{\ln x}{\ln 7}$.

c. Does it appear that $Y_1 = Y_2$ on the standard viewing window?

66. a. Graph the function defined by $f(x) = \log_{1/3} x$ by graphing $Y_1 = \dfrac{\log x}{\log \left(\frac{1}{3}\right)}$.

b. Graph the function defined by $f(x) = \log_{1/3} x$ by graphing $Y_2 = \dfrac{\ln x}{\ln \left(\frac{1}{3}\right)}$.

c. Does it appear that $Y_1 = Y_2$ on the standard viewing window?

67. a. Graph the function defined by $f(x) = \log_{1/7} x$ by graphing $Y_1 = \dfrac{\log x}{\log \left(\frac{1}{7}\right)}$.

b. Graph the function defined by $f(x) = \log_{1/7} x$ by graphing $Y_2 = \dfrac{\ln x}{\ln \left(\frac{1}{7}\right)}$.

c. Does it appear that $Y_1 = Y_2$ on the standard viewing window?

For Exercises 68–70, graph the functions on your calculator.

68. Graph $s(x) = \log_{1/2} x$

69. Graph $y = e^{x-2}$

70. Graph $y = e^{x-1}$

Chapter 13   Problem Recognition Exercises—Logarithmic and Exponential Forms

Fill out the table by writing the exponential expressions in logarithmic form, and the logarithmic expressions in exponential form. Use the fact that:

$$y = \log_b x \quad \text{is equivalent to} \quad b^y = x$$

	Exponential Form	Logarithmic Form
1.	$2^5 = 32$	
2.		$\log_3 81 = 4$
3.	$z^y = x$	
4.		$\log_b a = c$
5.	$10^3 = 1000$	
6.		$\log 10 = 1$
7.	$e^a = b$	
8.		$\ln p = q$
9.	$\left(\frac{1}{2}\right)^2 = \frac{1}{4}$	
10.		$\log_{1/3} 9 = -2$
11.	$10^{-2} = 0.01$	
12.		$\log 4 = x$
13.	$e^0 = 1$	
14.		$\ln e = 1$
15.	$25^{1/2} = 5$	
16.		$\log_{16} 2 = \frac{1}{4}$
17.	$e^t = s$	
18.		$\ln w = r$
19.	$15^{-2} = \frac{1}{225}$	
20.		$\log_3 p = -1$

Section 13.7   Logarithmic and Exponential Equations

1. Solving Logarithmic Equations

Equations containing one or more logarithms are called **logarithmic equations**. For example,

$$\ln(2x + 5) = 1 \qquad \text{and} \qquad \log_4 = 1 - \log_4(x - 3)$$

are logarithmic equations. To solve a basic logarithmic equation such as $\ln(2x + 5) = 1$, write the expression in exponential form:

$$\ln(2x + 5) = 1 \Rightarrow e^1 = 2x + 5. \text{ Now solve for } x.$$

$$e - 5 = 2x$$

$$\frac{e - 5}{2} = x \quad \text{or} \quad x \approx -1.14$$

To solve equations containing more than one logarithm of first degree, use the following guidelines.

Guidelines to Solve Logarithmic Equations

1. Isolate the logarithms on one side of the equation.
2. Write a sum or difference of logarithms as a single logarithm.
3. Rewrite the equation in step 2 in exponential form.
4. Solve the resulting equation from step 3.
5. Check all solutions to verify that they are within the domain of the logarithmic expressions in the original equation.

Example 1   Solving a Logarithmic Equation

Solve the equation.

$$\log_4 x = 1 - \log_4(x - 3)$$

Solution:

$$\log_4 x = 1 - \log_4(x - 3)$$

$$\log_4 x + \log_4(x - 3) = 1 \qquad \text{Isolate the logarithms on one side of the equation.}$$

$$\log_4[x(x - 3)] = 1 \qquad \text{Write as a single logarithm.}$$

$$\log_4(x^2 - 3x) = 1 \qquad \text{Simplify inside the parentheses.}$$

$$x^2 - 3x = 4^1 \qquad \text{Write the equation in exponential form.}$$

$$x^2 - 3x - 4 = 0 \qquad \text{The resulting equation is quadratic.}$$

$$(x - 4)(x + 1) = 0 \qquad \text{Factor.}$$

$$x = 4 \quad \text{or} \quad x = -1 \qquad \text{Apply the zero product rule.}$$

Notice that $x = -1$ is *not* a solution because $\log_4 x$ is not defined at $x = -1$. However, $x = 4$ *is* defined in both expressions $\log_4 x$ and $\log_4 (x - 3)$. We can substitute $x = 4$ into the original equation to show that it checks.

<u>Check:</u> $x = 4$

$$\log_4 x = 1 - \log_4 (x - 3)$$

$$\log_4 4 \overset{?}{=} 1 - \log_4 (4 - 3)$$

$$1 \overset{?}{=} 1 - \log_4 1$$

$$1 \overset{?}{=} 1 - 0 \checkmark \text{ True}$$

The solution is $x = 4$.

Skill Practice

1. Solve the equation.

$$\log_3 (x - 8) = 2 - \log_3 x$$

TIP: The equation from Example 1 involved the logarithmic functions $y = \log_4 x$ and $y = \log_4 (x - 3)$. The domains of these functions are $\{x \mid x > 0\}$ and $\{x \mid x > 3\}$, respectively. Therefore, the solutions to the equation are restricted to x-values in the intersection of these two sets, that is, $\{x \mid x > 3\}$. The solution $x = 4$ satisfies this requirement, whereas $x = -1$ does not.

Example 2 Solving Logarithmic Equations

Solve the equations.

a. $\log (x + 300) = 3.7$

b. $\ln (x + 2) + \ln (x - 1) = \ln (9x - 17)$

Solution:

a. $\log (x + 300) = 3.7$ The equation has a single logarithm that is already isolated.

$$10^{3.7} = x + 300$$ Write the equation in exponential form.

$$10^{3.7} - 300 = x$$ Solve for x.

$$x = 10^{3.7} - 300 \approx 4711.87$$

Check the exact value of x in the original equation.

<u>Check:</u> $x = 10^{3.7} - 300$

$$\log (x + 300) = 3.7$$

$$\log [(10^{3.7} - 300) + 300] \overset{?}{=} 3.7$$

$$\log (10^{3.7} - 300 + 300) \overset{?}{=} 3.7$$

$$\log 10^{3.7} \overset{?}{=} 3.7$$ Property 3 of logarithms: $\log_b b^p = p$

$$3.7 = 3.7 \checkmark \text{ True}$$

The solution $x = 10^{3.7} - 300$ checks.

Skill Practice Answers

1. $x = 9$ ($x = -1$ does not check)

b. $\ln(x + 2) + \ln(x - 1) = \ln(9x - 17)$

$\ln(x + 2) + \ln(x - 1) - \ln(9x - 17) = 0$ Isolate the logarithms on one side.

$$\ln\left[\frac{(x + 2)(x - 1)}{9x - 17}\right] = 0$$ Write as a single logarithm.

$$e^0 = \frac{(x + 2)(x - 1)}{9x - 17}$$ Write the equation in exponential form.

$$1 = \frac{(x + 2)(x - 1)}{9x - 17}$$ Simplify.

$$(1) \cdot (9x - 17) = \left[\frac{(x + 2)(x - 1)}{9x - 17}\right] \cdot (9x - 17)$$ Multiply by the LCD.

$$9x - 17 = (x + 2)(x - 1)$$ The equation is quadratic.

$$9x - 17 = x^2 + x - 2$$

$$0 = x^2 - 8x + 15$$

$$0 = (x - 5)(x - 3)$$

$$x = 5 \quad \text{or} \quad x = 3$$

The solutions $x = 5$ and $x = 3$ are both within the domain of the logarithmic functions in the original equation.

Check: $x = 5$

$\ln(x + 2) + \ln(x - 1) = \ln(9x - 17)$

$\ln(5 + 2) + \ln(5 - 1) \overset{?}{=} \ln[9(5) - 17]$

$\ln 7 + \ln 4 \overset{?}{=} \ln(45 - 17)$

$\ln(7 \cdot 4) \overset{?}{=} \ln(28)$

$\ln 28 = \ln 28$ ✓ True

Check: $x = 3$

$\ln(x + 2) + \ln(x - 1) = \ln(9x - 17)$

$\ln(3 + 2) + \ln(3 - 1) \overset{?}{=} \ln[9(3) - 17]$

$\ln 5 + \ln 2 \overset{?}{=} \ln(27 - 17)$

$\ln(5 \cdot 2) \overset{?}{=} \ln(10)$

$\ln 10 = \ln 10$ ✓ True

Both solutions check.

Skill Practice   Solve the equations.

2. $\log(2p + 6) = 1$ **3.** $\ln(t - 3) + \ln(t - 1) = \ln(2t - 5)$

2. Applications of Logarithmic Equations

Example 3   Applying a Logarithmic Equation to Earthquake Intensity

The magnitude of an earthquake (the amount of seismic energy released at the hypocenter of the earthquake) is measured on the Richter scale. The Richter scale value R is determined by the formula

$$R = \log\left(\frac{I}{I_0}\right)$$

where I is the intensity of the earthquake and I_0 is the minimum measurable intensity of an earthquake. (I_0 is a "zero-level" quake—one that is barely detected by a seismograph.)

Skill Practice Answers

2. $p = 2$
3. $t = 4$ ($t = 2$ does not check)

a. Compare the Richter scale values of earthquakes that are (i) 100,000 times (10^5 times) more intense than I_0 and (ii) 1,000,000 times (10^6 times) more intense than I_0.

b. On October 17, 1989, an earthquake measuring 7.1 on the Richter scale occurred in the Loma Prieta area in the Santa Cruz Mountains. The quake devastated parts of San Francisco and Oakland, California, bringing 63 deaths and over 3700 injuries. Determine how many times more intense this earthquake was than a zero-level quake.

Solution:

a. $R = \log\left(\dfrac{I}{I_0}\right)$

i. Earthquake 100,000 times I_0

$R = \log\left(\dfrac{10^5 \cdot I_0}{I_0}\right)$　　Substitute $I = 10^5 I_0$.

$= \log 10^5$

$= 5$

ii. Earthquake 1,000,000 times I_0

$R = \log\left(\dfrac{10^6 \cdot I_0}{I_0}\right)$　　Substitute $I = 10^6 I_0$.

$= \log 10^6$

$= 6$

Notice that the value on the Richter scale corresponds to the magnitude (power of 10) of the energy released. That is, a 1-unit increase on the Richter scale represents a 10-fold increase in the intensity of an earthquake.

b. $R = \log\left(\dfrac{I}{I_0}\right)$

$7.1 = \log\left(\dfrac{I}{I_0}\right)$　　Substitute $R = 7.1$.

$\dfrac{I}{I_0} = 10^{7.1}$　　Write the equation in exponential form.

$I = 10^{7.1} \cdot I_0$　　Solve for I.

The Loma Prieta earthquake in 1989 was $10^{7.1}$ times ($\approx 12,590,000$ times) more intense than a zero-level earthquake.

Skill Practice

4. In December 2004, an earthquake in the Indian Ocean measuring 9.0 on the Richter scale caused a tsunami that killed more than 300,000 people. Determine how many times more intense this earthquake was than a zero-level quake. Use the formula

$$R = \log\left(\dfrac{I}{I_0}\right)$$

where I is the intensity of the quake, I_0 is the measure of a zero-level quake, and R is the Richter scale value.

Skill Practice Answers

4. The earthquake was 10^9 times more intense than a zero-level quake.

3. Solving Exponential Equations

An equation with one or more exponential expressions is called an **exponential equation**. The following property is often useful in solving exponential equations.

Equivalence of Exponential Expressions

Let x, y, and b be real numbers such that $b > 0$ and $b \neq 1$. Then

$$b^x = b^y \quad \text{implies} \quad x = y$$

The equivalence of exponential expressions indicates that if two exponential expressions of the same base are equal, their exponents must be equal.

Example 4   **Solving Exponential Equations**

Solve the equations.

a. $4^{2x-9} = 64$ **b.** $(2^x)^{x+3} = \dfrac{1}{4}$

Solution:

a. $4^{2x-9} = 64$

$\qquad 4^{2x-9} = 4^3$ Write both sides with a common base.

$\qquad 2x - 9 = 3$ If $b^x = b^y$, then $x = y$.

$\qquad\qquad 2x = 12$ Solve for x.

$\qquad\qquad\ x = 6$

To check, substitute $x = 6$ into the original equation.

$$4^{2(6)-9} \stackrel{?}{=} 64$$

$$4^{12-9} \stackrel{?}{=} 64$$

$$4^3 = 64 \ \checkmark \ \text{True}$$

b. $(2^x)^{x+3} = \dfrac{1}{4}$

$\qquad 2^{x^2+3x} = 2^{-2}$ Apply the multiplication property of exponents. Write both sides of the equation with a common base.

$\qquad x^2 + 3x = -2$ If $b^x = b^y$, then $x = y$.

$\qquad x^2 + 3x + 2 = 0$ The resulting equation is quadratic.

$\qquad (x+2)(x+1) = 0$ Solve for x.

$\qquad x = -2 \quad \text{or} \quad x = -1$

The check is left to the reader.

Skill Practice   Solve the equations.

Skill Practice Answers

5. $x = \dfrac{2}{3}$ **6.** $x = 1$ or $x = 4$

5. $2^{3x+1} = 8$ **6.** $(3^x)^{x-5} = \dfrac{1}{81}$

Example 5 Solving an Exponential Equation

Solve the equation. $4^x = 25$

Solution:

Because 25 cannot be written as an integral power of 4, we cannot use the property that if $b^x = b^y$, then $x = y$. Instead we can rewrite the equation in its corresponding logarithmic form to solve for x.

$$4^x = 25$$

$$x = \log_4 25 \qquad \text{Write the equation in logarithmic form}$$

$$= \frac{\ln 25}{\ln 4} \approx 2.322 \qquad \text{Change-of-base formula}$$

The same result can be reached by taking a logarithm of any base on both sides of the equation. Then by applying the power property of logarithms, the unknown exponent can be written as a factor.

$$4^x = 25$$

$$\log 4^x = \log 25 \qquad \text{Take the common logarithm of both sides.}$$

$$x \log 4 = \log 25 \qquad \text{Apply the power property of logarithms to express the exponent as a factor. This is now a linear equation in } x.$$

$$\frac{x \log 4}{\log 4} = \frac{\log 25}{\log 4} \qquad \text{Solve for } x.$$

$$x = \frac{\log 25}{\log 4} \approx 2.322$$

Skill Practice

7. Solve the equation.

$$5^x = 32$$

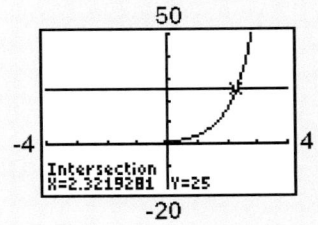

Guidelines to Solve Exponential Equations

1. Isolate one of the exponential expressions in the equation.

2. Take a logarithm on both sides of the equation. (The natural logarithmic function or the common logarithmic function is often used so that the final answer can be approximated with a calculator.)

3. Use the power property of logarithms (property 7) to write exponents as factors. Recall: $\log_b M^p = p \log_b M$.

4. Solve the resulting equation from step 3.

Example 6 Solving Exponential Equations by Taking a Logarithm on Both Sides

Solve the equations.

a. $2^{x+3} = 7^x$ **b.** $e^{-3.6x} = 9.74$

Skill Practice Answers

7. $x = \dfrac{\ln 32}{\ln 5} \approx 2.153$

Solution:

a. $$2^{x+3} = 7^x$$

$$\ln 2^{(x+3)} = \ln 7^x$$ Take the natural logarithm of both sides.

$$(x + 3)\ln 2 = x \ln 7$$ Use the power property of logarithms.

$$x(\ln 2) + 3(\ln 2) = x \ln 7$$ Apply the distributive property.

$$x(\ln 2) - x(\ln 7) = -3 \ln 2$$ Collect x-terms on one side.

$$x(\ln 2 - \ln 7) = -3 \ln 2$$ Factor out x.

$$\frac{x(\ln 2 - \ln 7)}{(\ln 2 - \ln 7)} = \frac{-3 \ln 2}{(\ln 2 - \ln 7)}$$ Solve for x.

$$x = \frac{-3 \ln 2}{(\ln 2 - \ln 7)} \approx 1.660$$

TIP: The exponential equation $2^{x+3} = 7^x$ could have been solved by taking a logarithm of *any* base on both sides of the equation. For example, using base 10 yields

$$\log 2^{x+3} = \log 7^x$$

$$(x + 3)\log 2 = x \log 7$$ Apply the power property for logarithms.

$$x \log 2 + 3 \log 2 = x \log 7$$ Apply the distributive property.

$$x \log 2 - x \log 7 = -3 \log 2$$ Collect x-terms on one side of the equation.

$$x(\log 2 - \log 7) = -3 \log 2$$ Factor out x.

$$x = \frac{-3 \log 2}{\log 2 - \log 7} \approx 1.660$$

b. $$e^{-3.6x} = 9.74$$

$$\ln e^{-3.6x} = \ln 9.74$$ The exponential expression has a base of e, so it is convenient to take the natural logarithm of both sides.

$$(-3.6x)\ln e = \ln 9.74$$ Use the power property of logarithms.

$$-3.6x = \ln 9.74$$ Simplify (recall that $\ln e = 1$).

$$x = \frac{\ln 9.74}{-3.6} \approx -0.632$$

Skill Practice   Solve the equations.

8. $3^x = 8^{x+2}$ 9. $e^{-0.2t} = 7.52$

Skill Practice Answers

8. $x = \dfrac{2 \ln 8}{\ln 3 - \ln 8} \approx -4.240$

9. $t = \dfrac{\ln 7.52}{-0.2} \approx -10.088$

4. Applications of Exponential Equations

> **Example 7** Applying an Exponential Function to World Population

The population of the world was estimated to have reached 6.5 billion in April 2006. The population growth rate for the world is estimated to be 1.4%. (Source: U.S. Census Bureau)

$$P(t) = 6.5(1.014)^t$$

represents the world population in billions as a function of the number of years after April 2006. ($t = 0$ represents April 2006).

a. Use the function to estimate the world population in April 2010.

b. Use the function to estimate the amount of time after April 2006 required for the world population to reach 13 billion.

Solution:

a.
$$P(t) = 6.5(1.014)^t$$

$P(4) = 6.5(1.014)^4$ The year 2010 corresponds to $t = 4$.

≈ 6.87 In 2010, the world's population will be approximately 6.87 billion.

b.
$$P(t) = 6.5(1.014)^t$$

$13 = 6.5(1.014)^t$ Substitute $P(t) = 13$ and solve for t.

$\dfrac{13}{6.5} = \dfrac{6.5(1.014)^t}{6.5}$ Isolate the exponential expression on one side of the equation.

$2 = 1.014^t$

$\ln 2 = \ln 1.014^t$ Take the natural logarithm of both sides.

$\ln 2 = t \ln 1.014$ Use the power property of logarithms.

$\dfrac{\ln 2}{\ln 1.014} = \dfrac{t \ln 1.014}{\ln 1.014}$ Solve for t.

$t = \dfrac{\ln 2}{\ln 1.014} \approx 50$ The population will reach 13 billion (double the April 2006 value) approximately 50 years after 2006.

Note: It has taken thousands of years for the world's population to reach 6.5 billion. However, with a growth rate of 1.4%, it will take only 50 years to gain an additional 6.5 billion.

> **Skill Practice** Use the population function from Example 7.

10. Estimate the world population in April 2020.

11. Estimate the year in which the world population will reach 9 billion.

Skill Practice Answers

10. Approximately 7.9 billion
11. The year 2029

On Friday, April 25, 1986, a nuclear accident occurred at the Chernobyl nuclear reactor, resulting in radioactive contaminates being released into the atmosphere. The most hazardous isotopes released in this accident were ^{137}Cs (cesium 137), ^{131}I (iodine 131), and ^{90}Sr (strontium 90). People living close to Chernobyl (in Ukraine) were at risk of radiation exposure from inhalation, from absorption through the skin, and from food contamination. Years after the incident, scientists have seen an increase in the incidence of thyroid disease among children living in the contaminated areas. Because iodine is readily absorbed in the thyroid gland, scientists suspect that radiation from iodine 131 is the cause.

Example 8 **Applying an Exponential Equation to Radioactive Decay**

The half-life of radioactive iodine ^{131}I is 8.04 days. If 10 g of iodine 131 is initially present, then the amount of radioactive iodine still present after t days is approximated by

$$A(t) = 10e^{-0.0862t}$$

where t is the time in days.

a. Use the model to approximate the amount of ^{131}I still present after 2 weeks. Round to the nearest 0.1 g.

b. How long will it take for the amount of ^{131}I to decay to 0.5 g? Round to the nearest 0.1 year.

Solution:

a. $A(t) = 10e^{-0.0862t}$

$A(14) = 10e^{-0.0862(14)}$ Substitute $t = 14$ (2 weeks).

≈ 3.0 g

b. $A(t) = 10e^{-0.0862t}$

$0.5 = 10e^{-0.0862t}$ Substitute $A = 0.5$.

$\dfrac{0.5}{10} = \dfrac{\cancel{10}e^{-0.0862t}}{\cancel{10}}$ Isolate the exponential expression.

$0.05 = e^{-0.0862t}$

$\ln(0.05) = \ln(e^{-0.0862t})$ Take the natural logarithm of both sides.

$\ln(0.05) = -0.0862t$ The resulting equation is linear.

$\dfrac{\ln(0.05)}{-0.0862} = \dfrac{-0.0862t}{-0.0862}$ Solve for t.

$t = \dfrac{\ln(0.05)}{-0.0862} \approx 34.8$ years

Note: Radioactive iodine (^{131}I) is used in medicine in appropriate dosages to treat patients with hyperactive (overactive) thyroids. Because iodine is readily absorbed in the thyroid gland, the radiation is localized and will reduce the size of the thyroid while minimizing damage to surrounding tissues.

Skill Practice Radioactive strontium 90 (^{90}Sr) has a half-life of 28 years. If 100 g of strontium 90 is initially present, the amount left after t years is approximated by

$$A(t) = 100e^{-0.0248t}$$

12. Find the amount of ^{90}Sr present after 85 years.

13. How long will it take for the amount of ^{90}Sr to decay to 40 g?

Skill Practice Answers

12. 12.1 g   **13.** 36.9 years

Section 13.7 Practice Exercises

Boost *your* GRADE at
mathzone.com!

 MathZone

• Practice Problems   • e-Professors
• Self-Tests   • Videos
• NetTutor

Study Skills Exercise

1. Define the key terms.

   a. logarithmic equation   **b.** exponential equation

Review Exercises

2. a. Graph $f(x) = e^x$.

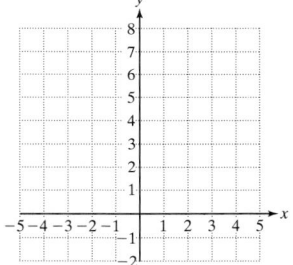

   b. Write the domain and range in interval notation.

3. a. Graph $g(x) = 3^x$.

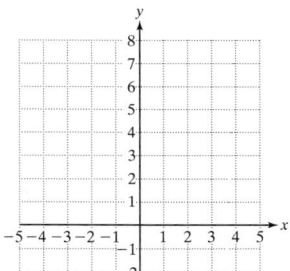

   b. Write the domain and range in interval notation.

4. a. Graph $h(x) = \ln x$.

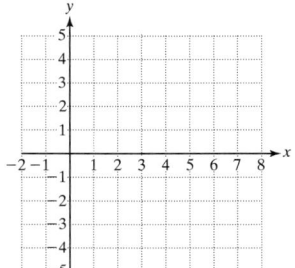

   b. Write the equation of the vertical asymptote.

   c. Write the domain and range in interval notation.

5. a. Graph $k(x) = \log x$.

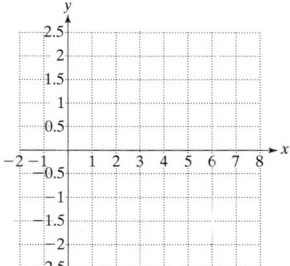

   b. Write the equation of the vertical asymptote.

   c. Write the domain and range in interval notation.

For Exercises 6–9, write the expression as a single logarithm. Assume all variables represent positive real numbers.

6. $\log_b (x - 1) + \log_b (x + 2)$

7. $\log_b x + \log_b (2x + 3)$

8. $\log_b x - \log_b (1 - x)$

9. $\log_b (x + 2) - \log_b (3x - 5)$

Concept 1: Solving Logarithmic Equations

Solve the logarithmic equations in Exercises 10–41.

10. $\log_3 x = 2$

11. $\log_4 x = 9$

12. $\log p = 42$

13. $\log q = \dfrac{1}{2}$

14. $\ln x = 0.08$

15. $\ln x = 19$

16. $\log_2 x = -4$

17. $\log x = -1$

18. $\log_x 25 = 2 \qquad (x > 0)$

19. $\log_x 100 = 2 \qquad (x > 0)$

20. $\log_b 10{,}000 = 4 \qquad (b > 0)$

21. $\log_b e^3 = 3 \qquad (b > 0)$

22. $\log_y 5 = \dfrac{1}{2} \qquad (y > 0)$

23. $\log_b 8 = \dfrac{1}{2} \qquad (b > 0)$

24. $\log_4 (c + 5) = 3$

25. $\log_5 (a - 4) = 2$

26. $\log_5 (4y + 1) = 1$

27. $\log_6 (5t - 2) = 1$

28. $\ln (1 - x) = 0$

29. $\log_4 (2 - x) = 1$

30. $\log_3 8 - \log_3 (x + 5) = 2$

31. $\log_2 (x + 3) - \log_2 (x + 2) = 1$

32. $\log_3 k + \log_3 (2k + 3) = 2$

33. $\log_2 (h - 1) + \log_2 (h + 1) = 3$

34. $\log (x + 2) = \log (3x - 6)$

35. $\log x = \log (1 - x)$

36. $\ln x - \ln (4x - 9) = 0$

37. $\ln (x + 5) - \ln x = \ln (4x)$

38. $\log_5 (3t + 2) - \log_5 t = \log_5 4$

39. $\log (6y - 7) + \log y = \log 5$

40. $\log (4m) = \log 2 + \log (m - 3)$

41. $\log (-h) + \log 3 = \log (2h - 15)$

Concept 2: Applications of Logarithmic Equations

42. On May 28, 2004, an earthquake that measured 6.3 on the Richter scale occurred in northern Iran. Use the formula $R = \log \left(\dfrac{I}{I_0} \right)$ to determine how many times more intense this earthquake was than a zero-level quake.

43. Papua, Indonesia, had a 7.0 earthquake on February 4, 2004. Determine how many times more intense this earthquake was than a zero-level quake.

The decibel level of sound can be found by the equation $D = 10 \log \left(\dfrac{I}{I_0} \right)$, where I is the intensity of the sound and I_0 is the intensity of the least audible sound that an average healthy person can hear. Generally I_0 is found as 10^{-12} watt per square meter (W/m^2). Use this information to answer Exercises 44–45.

44. Given that heavy traffic has a decibel level of 89.3 and that $I_0 = 10^{-12}$, find the intensity of the sound of heavy traffic.

45. Given that normal conversation has a decibel level of 65 and $I_0 = 10^{-12}$, find the intensity of the sound of normal conversation.

Concept 3: Solving Exponential Equations

For Exercises 46–61, solve the exponential equation by using the property that $b^x = b^y$ implies $x = y$, for $b > 0$ and $b \neq 1$.

46. $5^x = 625$

47. $3^x = 81$

48. $2^{-x} = 64$

49. $6^{-x} = 216$

50. $36^x = 6$

51. $343^x = 7$

52. $4^{2x-1} = 64$

53. $5^{3x-1} = 125$

54. $81^{3x-4} = \dfrac{1}{243}$

55. $4^{2x-7} = \dfrac{1}{128}$

56. $\left(\dfrac{2}{3}\right)^{-x+4} = \dfrac{8}{27}$

57. $\left(\dfrac{1}{4}\right)^{3x+2} = \dfrac{1}{64}$

58. $16^{-x+1} = 8^{5x}$

59. $27^{(1/3)x} = \left(\dfrac{1}{9}\right)^{2x-1}$

60. $(4^x)^{x+1} = 16$

61. $(3^x)^{x+2} = \dfrac{1}{3}$

For Exercises 62–73, solve the exponential equations by taking the logarithm of both sides. (Round the answers to three decimal places.)

62. $8^a = 21$

63. $6^y = 39$

64. $e^x = 8.1254$

65. $e^x = 0.3151$

66. $10^t = 0.0138$

67. $10^p = 16.8125$

68. $e^{0.07h} = 15$

69. $e^{0.03k} = 4$

70. $32e^{0.04m} = 128$

71. $8e^{0.05n} = 160$

72. $3^{x+1} = 5^x$

73. $2^{x-1} = 7^x$

Concept 4: Applications of Exponential Equations

74. The population of China can be modeled by the function

$$P(t) = 1237(1.0095)^t$$

where $P(t)$ is in millions and t is the number of years since 1998.

   a. Using this model, what was the population in the year 2002?

   b. Predict the population in the year 2012.

   c. If this growth rate continues, in what year will the population reach 2 billion people (2 billion is 2000 million)?

75. The population of Delhi, India, can be modeled by the function

$$P(t) = 9817(1.031)^t$$

where $P(t)$ is in thousands and t is the number of years since 2001.

   a. Using this model, predict the population in the year 2010.

   b. If this growth rate continues, in what year will the population reach 15 million (15 million is 15,000 thousand)?

76. The growth of a certain bacteria in a culture is given by the model

$$A(t) = 500e^{0.0277t}$$

where $A(t)$ is the number of bacteria and t is time in minutes.

   a. What is the initial number of bacteria?

   b. What is the population after 10 min?

   c. How long will it take for the population to double (that is, reach 1000)?

77. The population of the bacteria *Salmonella typhimurium* is given by the model

$$A(t) = 300e^{0.01733t}$$

where $A(t)$ is the number of bacteria and t is time in minutes.

   a. What is the initial number of bacteria?

   b. What is the population after 10 min?

   c. How long will it take for the population to double?

78. Suppose $5000 is invested at 7% interest compounded continuously. How long will it take for the investment to grow to $10,000? Use the model $A(t) = Pe^{rt}$ and round to the nearest tenth of a year.

79. Suppose $2000 is invested at 10% interest compounded continuously. How long will it take for the investment to triple? Use the model $A(t) = Pe^{rt}$ and round to the nearest year.

80. Phosphorus 32 (^{32}P) has a half-life of approximately 14 days. If 10 g of ^{32}P is present initially, then the amount of phosphorus 32 still present after t days is given by $A(t) = 10(0.5)^{t/14}$.

   a. Find the amount of phosphorus 32 still present after 5 days. Round to the nearest tenth of a gram.

   b. Find the amount of time necessary for the amount of ^{32}P to decay to 4 g. Round to the nearest tenth of a day.

81. Polonium 210 (^{210}Po) has a half-life of approximately 138.6 days. If 4 g of ^{210}Po is present initially, then the amount of polonium 210 still present after t days is given by $A(t) = 4e^{-0.005t}$.

   a. Find the amount of polonium 138 still present after 50 days. Round to the nearest tenth of a gram.

   b. Find the amount of time necessary for the amount of ^{32}P to decay to 0.5 g. Round to the nearest tenth of a day.

82. Suppose you save $10,000 from working an extra job. Rather than spending the money, you decide to save the money for retirement by investing in a mutual fund that averages 12% per year. How long will it take for this money to grow to $1,000,000? Use the model $A(t) = Pe^{rt}$ and round to the nearest tenth of a year.

83. The model $A = Pe^{rt}$ is used to compute the total amount of money in an account after t years at an interest rate r, compounded continuously. The value P is the initial principal. Find the amount of time required for the investment to double as a function of the interest rate. (*Hint:* Substitute $A = 2P$ and solve for t.)

Expanding Your Skills

84. The isotope of plutonium of mass 238 (written ^{238}Pu) is used to make thermoelectric power sources for spacecraft. The heat and electric power derived from such units have made the Voyager, Gallileo, and Cassini missions to the outer reaches of our solar system possible. The half-life of ^{238}Pu is 87.7 years.

   Suppose a hypothetical space probe is launched in the year 2002 with 2.0 kg of ^{238}Pu. Then the amount of ^{238}Pu available to power the spacecraft decays over time according to

$$P(t) = 2e^{-0.0079t}$$

where $t \geq 0$ is the time in years and $P(t)$ is the amount of plutonium still present (in kilograms).

   a. Suppose the space probe is due to arrive at Pluto in the year 2045. How much plutonium will remain when the spacecraft reaches Pluto? Round to 2 decimal places.

   b. If 1.5 kg of ^{238}Pu is required to power the spacecraft's data transmitter, will there be enough power in the year 2045 for us to receive close-up images of Pluto?

85. ^{99m}Tc is a radionuclide of technetium that is widely used in nuclear medicine. Although its half-life is only 6 hr, the isotope is continuously produced via the decay of its longer-lived parent ^{99}Mo (molybdenum 99) whose half-life is approximately 3 days. The ^{99}Mo generators (or "cows") are sold to hospitals in which the ^{99m}Tc can be "milked" as needed over a period of a few weeks. Once separated from its parent, the ^{99m}Tc may be chemically incorporated into a variety of imaging agents, each of which is designed to be taken up by a specific target organ within the body. Special cameras, sensitive to the gamma rays emitted by the technetium, are then used to record a "picture" (similar in appearance to an X-ray film) of the selected organ.

Suppose a technician prepares a sample of ^{99m}Tc-pyrophosphate to image the heart of a patient suspected of having had a mild heart attack. If the injection contains 10 millicuries (mCi) of ^{99m}Tc at 1:00 P.M., then the amount of technetium still present is given by

$$T(t) = 10e^{-0.1155t}$$

where $t > 0$ represents the time in hours after 1:00 P.M. and $T(t)$ represents the amount of ^{99m}Tc (in millicuries) still present.

a. How many millicuries of ^{99m}Tc will remain at 4:20 P.M. when the image is recorded? Round to the nearest tenth of a millicurie.

b. How long will it take for the radioactive level of the ^{99m}Tc to reach 2 mCi? Round to the nearest tenth of an hour.

For Exercises 86–89, solve the equations.

86. $(\log x)^2 - 2 \log x - 15 = 0$
(*Hint:* Let $u = \log x$.)

87. $(\log_2 z)^2 - 3 \log_2 z - 4 = 0$

88. $(\log_3 w)^2 + 5 \log_3 w + 6 = 0$

89. $(\ln x)^2 - 2 \ln x = 0$

Graphing Calculator Exercises

90. The amount of money a company receives from sales is related to the money spent on advertising according to

$$S(x) = 400 + 250 \log x, \quad x \geq 1$$

where $S(x)$ is the amount in sales (in \$1000s) and x is the amount spent on advertising (in \$1000s).

a. The value of $S(1) = 400$ means that if \$1000 is spent on advertising, the total sales will be \$400,000.

i. Find the total sales for this company if \$11,000 is spent on advertising.

ii. Find the total sales for this company if \$21,000 is spent on advertising.

iii. Find the total sales for this company if \$31,000 is spent on advertising.

b. Graph the function $y = S(x)$ on a window where $0 \leq x \leq 40$ and $0 \leq y \leq 1000$. Using the graph and your answers from part (a), describe the relationship between total sales and the money spent on advertising. As the money spent on advertising is increased, what happens to the rate of increase of total sales?

c. How many advertising dollars are required for the total sales to reach \$1,000,000? That is, for what value of x will $S(x) = 1000$?

91. Graph $Y_1 = 8^\wedge x$ and $Y_2 = 21$ on a window where $0 \le x \le 5$ and $0 \le y \le 40$. Use the graph and an *Intersect* feature or *Zoom* and *Trace* to support your answer to Exercise 62.

92. Graph $Y_1 = 6^\wedge x$ and $Y_2 = 39$ on a window where $0 \le x \le 5$ and $0 \le y \le 50$. Use the graph and an *Intersect* feature or *Zoom* and *Trace* to support your answer to Exercise 63.

Chapter 13   SUMMARY

Section 13.1   Algebra and Composition of Functions

Key Concepts

The Algebra of Functions

Given two functions f and g, the functions $f + g, f - g, f \cdot g$, and $\dfrac{f}{g}$ are defined as

$(f + g)(x) = f(x) + g(x)$

$(f - g)(x) = f(x) - g(x)$

$(f \cdot g)(x) = f(x) \cdot g(x)$

$\left(\dfrac{f}{g}\right)(x) = \dfrac{f(x)}{g(x)}$ provided $g(x) \ne 0$

Examples

Example 1

Let $g(x) = 5x + 1$ and $h(x) = x^3$. Find:

1. $(g + h)(3) = g(3) + h(3) = 16 + 27 = 43$

2. $(g \cdot h)(-1) = g(-1) \cdot h(-1) = (-4) \cdot (-1) = 4$

3. $(g - h)(x) = 5x + 1 - x^3$

4. $\left(\dfrac{g}{h}\right)(x) = \dfrac{5x + 1}{x^3}$

Composition of Functions

The **composition** of f and g, denoted $f \circ g$, is defined by the rule

$(f \circ g)(x) = f(g(x))$ provided that $g(x)$ is in the domain of f

Example 2

Find $(f \circ g)(x)$ and $(g \circ f)(x)$ given the functions defined by $f(x) = 4x + 3$ and $g(x) = 7x$.

$(f \circ g)(x) = f(g(x))$

$\qquad\qquad = f(7x)$

$\qquad\qquad = 4(7x) + 3$

$\qquad\qquad = 28x + 3$

The **composition** of g and f, denoted $g \circ f$, is defined by the rule

$(g \circ f)(x) = g(f(x))$ provided that $f(x)$ is in the domain of g

$(g \circ f)(x) = g(f(x))$

$\qquad\qquad = g(4x + 3)$

$\qquad\qquad = 7(4x + 3)$

$\qquad\qquad = 28x + 21$

Section 13.2 Inverse Functions

Key Concepts

Horizontal Line Test

Consider a function defined by a set of points (x, y) in a rectangular coordinate system. Then the graph defines y as a one-to-one function of x if no horizontal line intersects the graph in more than one point.

Examples

Example 1

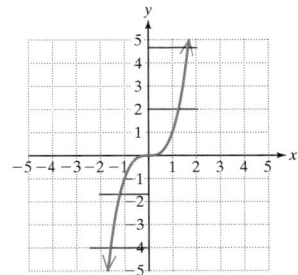

The function is one-to-one because it passes the horizontal line test.

Finding an Equation of the Inverse of a Function

For a one-to-one function defined by $y = f(x)$, the equation of the inverse can be found as follows:

1. Replace $f(x)$ by y.
2. Interchange x and y.
3. Solve for y.
4. Replace y by $f^{-1}(x)$.

Example 2

Find the inverse of the one-to-one function defined by $f(x) = 3 - x^3$.

1. $y = 3 - x^3$

2. $x = 3 - y^3$

3. $\quad x - 3 = -y^3$

$\quad -x + 3 = y^3$

$\quad \sqrt[3]{-x + 3} = y$

4. $f^{-1}(x) = \sqrt[3]{-x + 3}$

The graphs defined by $y = f(x)$ and $y = f^{-1}(x)$ are symmetric with respect to the line $y = x$.

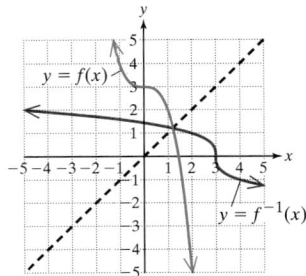

Definition of an Inverse Function

If f is a one-to-one function represented by ordered pairs of the form (x, y), then the inverse function, denoted f^{-1}, is the set of ordered pairs defined by ordered pairs of the form (y, x).

Inverse Function Property

If f is a one-to-one function, then g is the inverse of f if and only if $(f \circ g)(x) = x$ for all x in the domain of g, and $(g \circ f)(x) = x$ for all x in the domain of f.

Example 3

Verify that the functions defined by $f(x) = x - 1$ and $g(x) = x + 1$ are inverses.

$(f \circ g)(x) = f(x + 1) = (x + 1) - 1 = x$

$(g \circ f)(x) = g(x - 1) = (x - 1) + 1 = x$

Section 13.3   Exponential Functions

Key Concepts

A function $y = b^x$ $(b > 0, b \neq 1)$ is an **exponential function**.

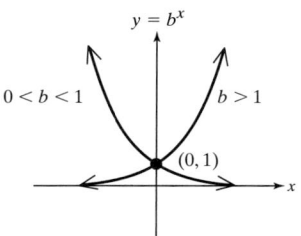

The domain is $(-\infty, \infty)$.

The range is $(0, \infty)$.

The line $y = 0$ is a horizontal asymptote.

The y-intercept is $(0, 1)$.

Examples

Example 1

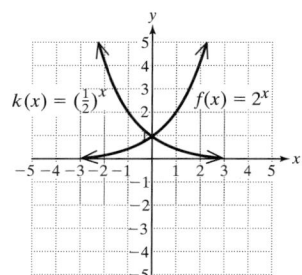

Section 13.4   Logarithmic Functions

Key Concepts

The function $y = \log_b x$ is a **logarithmic function**.

$$y = \log_b x \quad \Leftrightarrow \quad b^y = x \quad (x > 0, b > 0, b \neq 1)$$

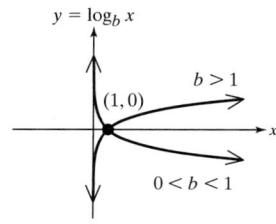

For $y = \log_b x$, the domain is $(0, \infty)$.

The range is $(-\infty, \infty)$.

The line $x = 0$ is a vertical asymptote.

The x-intercept is $(1, 0)$.

The function $y = \log x$ is the **common logarithmic function** (base 10).

Examples

Example 1

$\log_4 64 = 3$ because $4^3 = 64$

Example 2

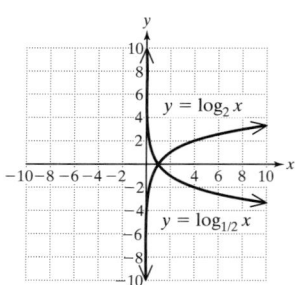

Example 3

$\log 10{,}000 = 4$ because $10^4 = 10{,}000$

Section 13.5   Properties of Logarithms

Key Concepts

Let b, x, and y be positive real numbers where $b \neq 1$, and let p be a real number. Then the following properties are true.

1. $\log_b 1 = 0$

2. $\log_b b = 1$

3. $\log_b b^p = p$

4. $b^{\log_b(x)} = x$

5. $\log_b(xy) = \log_b x + \log_b y$

6. $\log_b\left(\dfrac{x}{y}\right) = \log_b x - \log_b y$

7. $\log_b x^p = p\log_b x$

The **properties of logarithms** can be used to write multiple logarithms as a single logarithm.

The **properties of logarithms** can be used to write a single logarithm as a sum or difference of logarithms.

Examples

Example 1

1. $\log_5 1 = 0$

2. $\log_6 6 = 1$

3. $\log_4 4^7 = 7$

4. $2^{\log_2(5)} = 5$

5. $\log(5x) = \log 5 + \log x$

6. $\log_7\left(\dfrac{z}{10}\right) = \log_7 z - \log_7 10$

7. $\log x^5 = 5\log x$

Example 2

$$\log x - \frac{1}{2}\log y - 3\log z$$

$$= \log x - (\log y^{1/2} + \log z^3)$$

$$= \log x - \log(\sqrt{y}z^3)$$

$$= \log\left(\frac{x}{\sqrt{y}z^3}\right)$$

Example 3

$$\log \sqrt[3]{\frac{x}{y^2}}$$

$$= \frac{1}{3}\log\left(\frac{x}{y^2}\right)$$

$$= \frac{1}{3}(\log x - \log y^2)$$

$$= \frac{1}{3}(\log x - 2\log y)$$

$$= \frac{1}{3}\log x - \frac{2}{3}\log y$$

Section 13.6   The Irrational Number e

Key Concepts

As x becomes infinitely large, the expression $\left(1 + \dfrac{1}{x}\right)^x$ approaches the irrational number e, where $e \approx 2.718281$.

The balance of an account earning compound interest n times per year is given by

$$A(t) = P\left(1 + \frac{r}{n}\right)^{nt}$$

where P = principal, r = interest rate, t = time in years, and n = number of compound periods per year.

The balance of an account earning interest continuously is given by

$$A(t) = Pe^{rt}$$

The function $y = e^x$ is the exponential function with base e.

The **natural logarithm function** $y = \ln x$ is the logarithm function with base e.

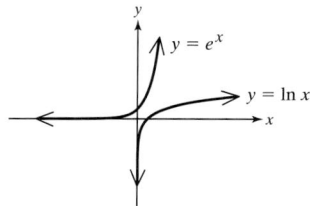

Change-of-Base Formula

$$\log_b x = \frac{\log_a x}{\log_a b} \qquad a > 0, a \neq 1, b > 0, b \neq 1$$

Examples

Example 1

Find the account balance for $8000 invested for 10 years at 7% compounded quarterly.

$$P = 8000 \qquad t = 10 \qquad r = 0.07 \qquad n = 4$$

$$A(10) = 8000\left(1 + \frac{0.07}{4}\right)^{(4)(10)}$$

$$= \$16{,}012.78$$

Example 2

Find the account balance for the same investment compounded continuously.

$$P = 8000 \qquad t = 10 \qquad r = 0.07$$

$$A(t) = 8000e^{(0.07)(10)}$$

$$= \$16{,}110.02$$

Example 3

Use a calculator to approximate the value of the expressions.

$$e^{7.5} \approx 1808.04$$

$$e^{-\pi} \approx 0.0432$$

$$\ln 107 \approx 4.6728$$

$$\ln\left(\frac{1}{\sqrt{2}}\right) \approx -0.3466$$

Example 4

$$\log_3 59 = \frac{\log 59}{\log 3} \approx 3.7115$$

Section 13.7 Logarithmic and Exponential Equations

Key Concepts

Guidelines to Solve Logarithmic Equations

1. Isolate the logarithms on one side of the equation.
2. Write a sum or difference of logarithms as a single logarithm.
3. Rewrite the equation in step 2 in exponential form.
4. Solve the resulting equation from step 3.
5. Check all solutions to verify that they are within the domain of the logarithmic expressions in the equation.

The equivalence of exponential expressions can be used to solve **exponential equations**.

If $b^x = b^y$ then $x = y$

Guidelines to Solve Exponential Equations

1. Isolate one of the exponential expressions in the equation.
2. Take a logarithm of both sides of the equation.
3. Use the power property of logarithms to write exponents as factors.
4. Solve the resulting equation from step 3.

Examples

Example 1

$\log(3x - 1) + 1 = \log(2x + 1)$

Step 1: $\log(3x - 1) - \log(2x + 1) = -1$

Step 2: $\log\left(\dfrac{3x - 1}{2x + 1}\right) = -1$

Step 3: $10^{-1} = \dfrac{3x - 1}{2x + 1}$

Step 4: $\dfrac{1}{10} = \dfrac{3x - 1}{2x + 1}$

$2x + 1 = 10(3x - 1)$

$2x + 1 = 30x - 10$

$-28x = -11$

$x = \dfrac{11}{28}$

Step 5: $x = \dfrac{11}{28}$ Checks in original equation

Example 2

$5^{2x} = 125$

$5^{2x} = 5^3$ implies that $2x = 3$

$x = \dfrac{3}{2}$

Example 3

$4^{x+1} - 2 = 1055$

Step 1: $4^{x+1} = 1057$

Step 2: $\ln(4^{x+1}) = \ln 1057$

Step 3: $(x + 1)\ln 4 = \ln 1057$

Step 4: $x + 1 = \dfrac{\ln 1057}{\ln 4}$

$x = \dfrac{\ln 1057}{\ln 4} - 1 \approx 4.023$

Chapter 13   Review Exercises

Section 13.1

For Exercises 1–8, refer to the functions defined here.

$$f(x) = x - 7 \qquad g(x) = -2x^3 - 8x$$

$$m(x) = \sqrt{x} \qquad n(x) = \frac{1}{x - 2}$$

Find the indicated function values. Write the domain in interval notation.

1. $(f - g)(x)$

2. $(f + g)(x)$

3. $(f \cdot n)(x)$

4. $(f \cdot m)(x)$

5. $\left(\dfrac{f}{g}\right)(x)$

6. $\left(\dfrac{g}{f}\right)(x)$

7. $(m \circ f)(x)$

8. $(n \circ f)(x)$

For Exercises 9–12, refer to the functions defined for Exercises 1–8. Find the function values, if possible.

9. $(m \circ g)(-2)$

10. $(n \circ g)(-1)$

11. $(f \circ g)(4)$

12. $(g \circ f)(8)$

13. Given $f(x) = 2x + 1$ and $g(x) = x^2$

    a. Find $(g \circ f)(x)$.

    b. Find $(f \circ g)(x)$.

    c. Based on your answers to parts (a) and (b), is $f \circ g$ equal to $g \circ f$?

For Exercises 14–19, refer to the graph. Approximate the function values, if possible.

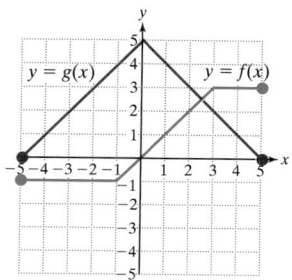

14. $\left(\dfrac{f}{g}\right)(1)$

15. $(f \cdot g)(-2)$

16. $(f + g)(-4)$

17. $(f - g)(2)$

18. $(g \circ f)(-3)$

19. $(f \circ g)(4)$

Section 13.2

For Exercises 20–21, determine if the function is one-to-one by using the horizontal line test.

20.

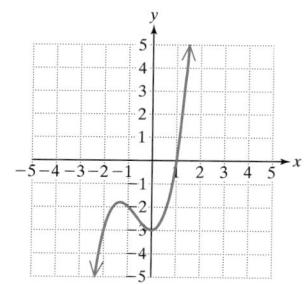

21.

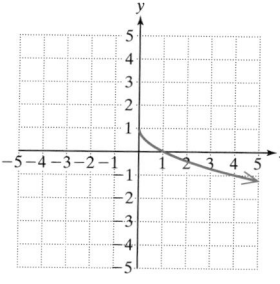

For Exercises 22–26, write the inverse for each one-to-one function.

22. $\{(3, 5), (2, 9), (0, -1), (4, 1)\}$

23. $q(x) = \dfrac{3}{4}x - 2$

24. $g(x) = \sqrt[5]{x} + 3$

25. $f(x) = (x - 1)^3$

26. $n(x) = \dfrac{4}{x - 2}$

27. Verify that the functions defined by $f(x) = 5x - 2$ and $g(x) = \frac{1}{5}x + \frac{2}{5}$ are inverses by showing that $(f \circ g)(x) = x$ and $(g \circ f)(x) = x$.

28. Graph the functions q and q^{-1} from Exercise 23 on the same coordinate axes. What can you say about the relationship between these two graphs?

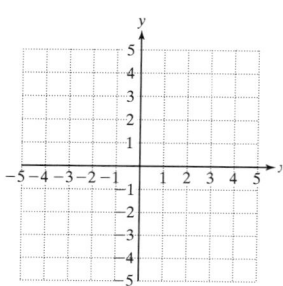

29. a. Find the domain and range of the function defined by $h(x) = \sqrt{x + 1}$.

    b. Find the domain and range of the function defined by $k(x) = x^2 - 1, x \geq 0$.

30. Determine the inverse of the function $p(x) = \sqrt{x} + 2$.

Section 13.3

For Exercises 31–38, evaluate the exponential expressions. Use a calculator and round to 3 decimal places, if necessary.

31. 4^5

32. 6^{-2}

33. $8^{1/3}$

34. $\left(\dfrac{1}{100}\right)^{-1/2}$

35. 2^{π}

36. $5^{\sqrt{3}}$

37. $(\sqrt{7})^{1/2}$

38. $\left(\dfrac{3}{4}\right)^{4/3}$

For Exercises 39–42, graph the functions.

39. $f(x) = 3^x$

40. $g(x) = \left(\dfrac{1}{4}\right)^x$

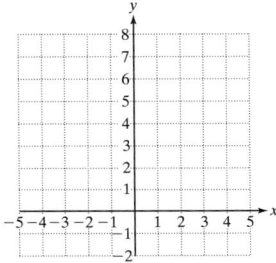

41. $h(x) = 5^{-x}$

42. $k(x) = \left(\dfrac{2}{5}\right)^{-x}$

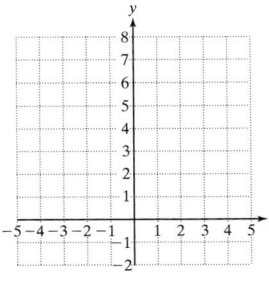
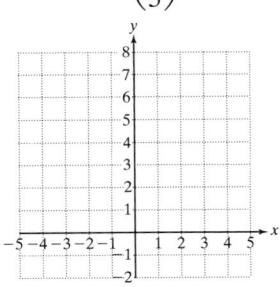

43. a. Does the graph of $y = b^x, b > 0, b \neq 1$, have a vertical or a horizontal asymptote?

   b. Write an equation of the asymptote.

44. Background radiation is radiation that we are exposed to from naturally occurring sources including the soil, the foods we eat, and the sun. Background radiation varies depending on where we live. A typical background radiation level is 150 millirems (mrem) per year. (A rem is a measure of energy produced from radiation.) Suppose a substance emits 30,000 mrem per year and has a half-life of 5 years. The function defined by

$$A(t) = 30,000\left(\dfrac{1}{2}\right)^{t/5}$$

gives the radiation level (in millirems) of this substance after t years.

   a. What is the radiation level after 5 years?

   b. What is the radiation level after 15 years?

   c. Will the radiation level of this substance be below the background level of 150 mrem after 50 years?

Section 13.4

For Exercises 45–52, evaluate the logarithms without using a calculator.

45. $\log_3\left(\dfrac{1}{27}\right)$

46. $\log_5 1$

47. $\log_7 7$

48. $\log_2 2^8$

49. $\log_2 16$

50. $\log_3 81$

51. $\log 100,000$

52. $\log_8\left(\dfrac{1}{8}\right)$

For Exercises 53–54, graph the logarithmic functions.

53. $q(x) = \log_3 x$

54. $r(x) = \log_{1/2} x$

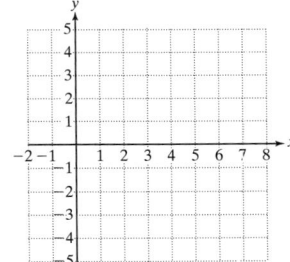
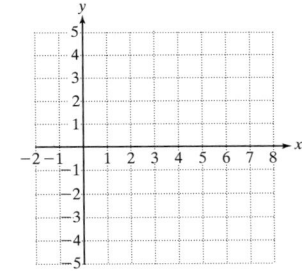

55. a. Does the graph of $y = \log_b x$ have a vertical or a horizontal asymptote?

   b. Write an equation of the asymptote.

56. Acidity of a substance is measured by its pH. The pH can be calculated by the formula $pH = -\log[H^+]$, where $[H^+]$ is the hydrogen ion concentration.

   a. What is the pH of a fruit with a hydrogen ion concentration of 0.00316 mol/L? Round to one decimal place.

   b. What is the pH of an antacid tablet with $[H^+] = 3.16 \times 10^{-10}$? Round to one decimal place.

Section 13.5

For Exercises 57–60, evaluate the logarithms without using a calculator.

57. $\log_8 8$

58. $\log_{11} 11^6$

59. $\log_{1/2} 1$

60. $12^{\log_{12} 7}$

61. Complete the properties. Assume x, y, and b are positive real numbers such that $b \neq 1$.

    a. $\log_b (xy) =$

    b. $\log_b x - \log_b y =$

    c. $\log_b x^p =$

For Exercises 62–65, write the logarithmic expressions as a single logarithm.

62. $\dfrac{1}{4}(\log_b y - 4\log_b z + 3\log_b x)$

63. $\dfrac{1}{2}\log_3 a + \dfrac{1}{2}\log_3 b - 2\log_3 c - 4\log_3 d$

64. $\log 540 - 3\log 3 - 2\log 2$

65. $-\log_4 18 + \log_4 6 + \log_4 3 - \log_4 1$

66. Which of the following is equivalent to $\dfrac{2\log 7}{\log 7 + \log 6}$?

    a. $\dfrac{\log 7}{\log 6}$ **b.** $\dfrac{\log 49}{\log 42}$ **c.** $\dfrac{2}{\log 6}$

67. Which of the following is equivalent to $\dfrac{\log 8^{-3}}{\log 2 + \log 4}$?

    a. -3 **b.** $-3\log\left(\dfrac{4}{3}\right)$ **c.** $\dfrac{-3\log 4}{\log 3}$

Section 13.6

For Exercises 68–75, use a calculator to approximate the expressions to 4 decimal places.

68. e^5

69. $e^{\sqrt{7}}$

70. $32e^{0.008}$

71. $58e^{-0.0125}$

72. $\ln 6$

73. $\ln\left(\dfrac{1}{9}\right)$

74. $\log 22$

75. $\log e^3$

For Exercises 76–79, use the change-of-base formula to approximate the logarithms to 4 decimal places.

76. $\log_2 10$

77. $\log_9 80$

78. $\log_5 (0.26)$

79. $\log_4 (0.0062)$

80. An investor wants to deposit $20,000 in an account for 10 years at 5.25% interest. Compare the amount she would have if her money were invested with the following different compounding options. Use

$$A(t) = P\left(1 + \frac{r}{n}\right)^{nt}$$

for interest compounded n times per year and $A(t) = Pe^{rt}$ for interest compounded continuously.

    a. Compounded annually

    b. Compounded quarterly

    c. Compounded monthly

    d. Compounded continuously

81. To measure a student's retention of material at the end of a course, researchers give the student a test on the material every month for 24 months after the course is over. The student's average score t months after completing the course is given by

$$S(t) = 75e^{-0.5t} + 20$$

where t is the time in months and $S(t)$ is the test score.

    a. Find $S(0)$ and interpret the result.

    b. Find $S(6)$ and interpret the result.

    c. Find $S(12)$ and interpret the result.

    d. The graph of $y = S(t)$ is shown here. Does it appear that the student's average score is approaching a limiting value? Explain.

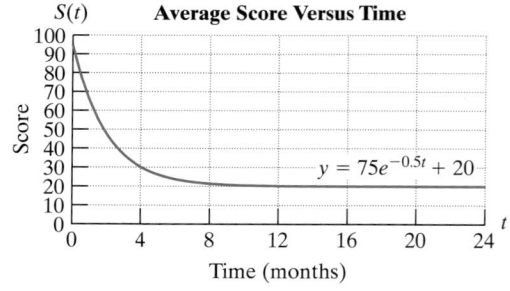

Average Score Versus Time

$y = 75e^{-0.5t} + 20$

Section 13.7

For Exercises 82–89, identify the domain. Write the answer in interval notation.

82. $f(x) = e^x$

83. $g(x) = e^{x+6}$

84. $h(x) = e^{x-3}$

85. $k(x) = \ln x$

86. $q(x) = \ln (x + 5)$

87. $p(x) = \ln (x - 7)$

88. $r(x) = \ln (4 - 3x)$

89. $w(x) = \ln (5 - x)$

Solve the logarithmic equations in Exercises 90–97. If necessary, round to two decimal places.

90. $\log_5 x = 3$

91. $\log_7 x = -2$

92. $\log_6 y = 3$

93. $\log_3 y = \dfrac{1}{12}$

94. $\log (2w - 1) = 3$

95. $\log_2 (3w + 5) = 5$

96. $\log p - 1 = -\log (p - 3)$

97. $\log_4 (2 + t) - 3 = \log_4 (3 - 5t)$

Solve the exponential equations in Exercises 98–105. If necessary, round to four decimal places.

98. $4^{3x+5} = 16$

99. $5^{7x} = 625$

100. $4^a = 21$

101. $5^a = 18$

102. $e^{-x} = 0.1$

103. $e^{-2x} = 0.06$

104. $10^{2n} = 1512$

105. $10^{-3m} = \dfrac{1}{821}$

106. $2^{x+3} = 7^x$

107. $14^{x-5} = 6^x$

108. Radioactive iodine (^{131}I) is used to treat patients with a hyperactive (overactive) thyroid. Patients with this condition may have symptoms that include rapid weight loss, heart palpitations, and high blood pressure. The half-life of radioactive iodine is 8.04 days. If a patient is given an initial dose of 2 μg, then the amount of iodine in the body after t days is approximated by

$$A(t) = 2e^{-0.0862t}$$

where t is the time in days and $A(t)$ is the amount (in micrograms) of ^{131}I remaining.

a. How much radioactive iodine is present after a week? Round to two decimal places.

b. How much radioactive iodine is present after 30 days? Round to two decimal places.

c. How long will it take for the level of radioactive iodine to reach 0.5 μg?

109. The growth of a certain bacterium in a culture is given by the model $A(t) = 150e^{0.007t}$, where $A(t)$ is the number of bacteria and t is time in minutes. Let $t = 0$ correspond to the initial number of bacteria.

a. What is the initial number of bacteria?

b. What is the population after $\frac{1}{2}$ hr?

c. How long will it take for the population to double?

110. The value of a car is depreciated with time according to

$$V(t) = 15{,}000e^{-0.15t}$$

where $V(t)$ is the value in dollars and t is the time in years.

a. Find $V(0)$ and interpret the result in the context of this problem.

b. Find $V(10)$ and interpret the result in the context of this problem. Round to the nearest dollar.

c. Find the time required for the value of the car to drop to $5000. Round to the nearest tenth of a year.

d. The graph of $y = V(t)$ is shown here. Does it appear that the value of the car is approaching a limiting value? If so what does the limiting value appear to be?

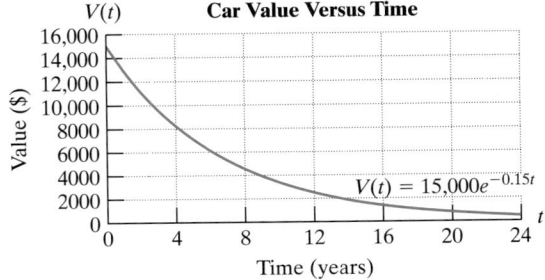

Car Value Versus Time

$V(t) = 15{,}000e^{-0.15t}$

Chapter 13   Test

For Exercises 1–8, refer to these functions.

$$f(x) = x - 4 \qquad g(x) = \sqrt{x + 2} \qquad h(x) = \frac{1}{x}$$

Find the function values if possible.

1. $\left(\dfrac{f}{g}\right)(x)$

2. $(h \cdot g)(x)$

3. $(g \circ f)(x)$

4. $(h \circ f)(x)$

5. $(f - g)(7)$

6. $(h + f)(2)$

7. $(h \circ g)(14)$

8. $(g \circ f)(0)$

9. If $f(x) = x - 4$ and $g(x) = \sqrt{x + 2}$, write the domain of the function $\frac{g}{f}$.

10. Explain how to determine graphically if a function is one-to-one.

11. Which of the functions is one-to-one?

a.

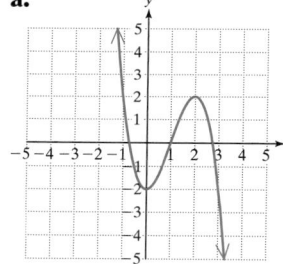

b.

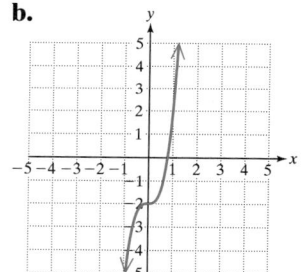

12. Write an equation of the inverse of the one-to-one function defined by $f(x) = \frac{1}{4}x + 3$.

13. Write an equation of the inverse of the function defined by $g(x) = (x - 1)^2$, $x \geq 1$.

14. Given the graph of the function $y = p(x)$, graph its inverse $p^{-1}(x)$.

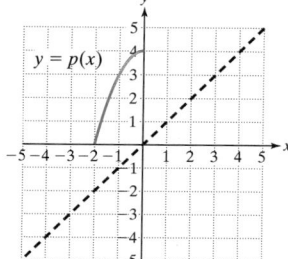

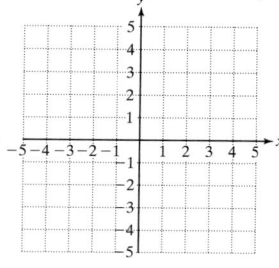

15. Use a calculator to approximate the expression to 4 decimal places.

a. $10^{2/3}$ **b.** $3^{\sqrt{10}}$ **c.** 8^{π}

16. Graph $f(x) = 4^{x-1}$.

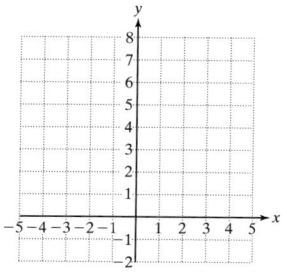

17. a. Write in logarithmic form. $16^{3/4} = 8$

b. Write in exponential form. $\log_x 31 = 5$

18. Graph $g(x) = \log_3 x$.

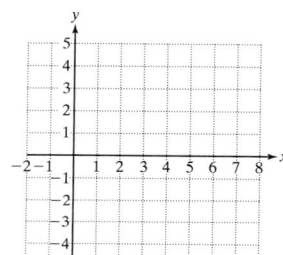

19. Complete the change-of-base formula:
$\log_b n = \underline{\hspace{1cm}}$

20. Use a calculator to approximate the expression to four decimal places:

a. $\log 21$ **b.** $\log_4 13$ **c.** $\log_{1/2} 6$

21. Using the properties of logarithms, expand and simplify. Assume all variables represent positive real numbers.

a. $-\log_3\left(\dfrac{3}{9x}\right)$ **b.** $\log\left(\dfrac{1}{10^5}\right)$

22. Write as a single logarithm. Assume all variables represent positive real numbers.

a. $\dfrac{1}{2}\log_b x + 3\log_b y$ **b.** $\log a - 4\log a$

23. Use a calculator to approximate the expression to four decimal places, if necessary.

a. $e^{1/2}$

b. e^{-3}

c. $\ln\left(\dfrac{1}{3}\right)$

d. $\ln e$

24. Identify the graphs as $y = e^x$ or $y = \ln(x)$.

a.

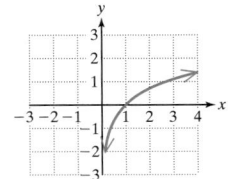

b.

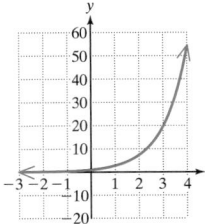

25. Researchers found that t months after taking a course, students remembered $p\%$ of the material according to

$$p(t) = 92 - 20\ln(t + 1)$$

where $0 \le t \le 24$ is the time in months.

a. Find $p(4)$ and interpret the results.

b. Find $p(12)$ and interpret the results.

c. Find $p(0)$ and interpret the results.

26. The population of New York City has a 2% growth rate and can be modeled by the function $P(t) = 8008(1.02)^t$, where $P(t)$ is in thousands and t is in years ($t = 0$ corresponds to the year 2000).

a. Using this model, predict the population in the year 2010.

b. If this growth rate continues, in what year will the population reach 12 million (12 million is 12,000 thousand)?

27. A certain bacterial culture grows according to

$$P(t) = \frac{1,500,000}{1 + 5000e^{-0.8t}}$$

where P is the population of the bacteria and t is the time in hours.

a. Find $P(0)$ and interpret the result. Round to the nearest whole number.

b. How many bacteria will be present after 6 hr?

c. How many bacteria will be present after 12 hr?

d. How many bacteria will be present after 18 hr?

e. From the graph does it appear that the population of bacteria is reaching a limiting value? If so, what does the limiting value appear to be?

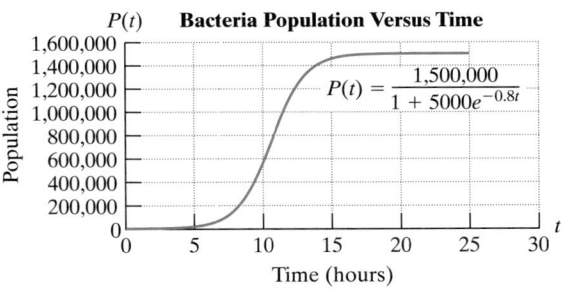

For Exercises 28–33, solve the exponential and logarithmic equations. If necessary, round to 3 decimal places.

28. $\log x + \log(x - 21) = 2$

29. $\log_{1/2} x = -5$

30. $\ln(x + 7) = 2.4$

31. $3^{x+4} = \dfrac{1}{27}$

32. $4^x = 50$

33. $e^{2.4x} = 250$

34. Atmospheric pressure P decreases exponentially with altitude x according to

$$P(x) = 760e^{-0.000122x}$$

where $P(x)$ is the pressure measured in millimeters of mercury (mm Hg) and x is the altitude measured in meters.

a. Find $P(2500)$ and interpret the result. Round to one decimal place.

b. Find the pressure at sea level.

c. Find the altitude at which the pressure is 633 mm Hg.

35. Use the formula $A(t) = Pe^{rt}$ to compute the value of an investment under continuous compounding.

a. If $2000 is invested at 7.5% compounded continuously, find the value of the investment after 5 years.

b. How long will it take the investment to double? Round to two decimal places.

Chapters 1–13   Cumulative Review Exercises

1. Simplify completely.

$$\frac{8 - 4 \cdot 2^2 + 15 \div 5}{|-3 + 7|}$$

2. Divide.

$$\frac{-8p^2 + 4p^3 + 6p^5}{8p^2}$$

3. Divide $(t^4 - 13t^2 + 36) \div (t - 2)$. Identify the quotient and remainder.

4. Simplify. $\sqrt{x^2 - 6x + 9}$

5. Simplify. $\dfrac{4}{\sqrt[3]{40}}$

6. Simplify. Write the answer with positive exponents only.

$$\frac{2^{2/5}c^{-1/4}d^{1/5}}{2^{-8/5}c^{3/4}d^{1/10}}$$

7. Find the area of the rectangle.

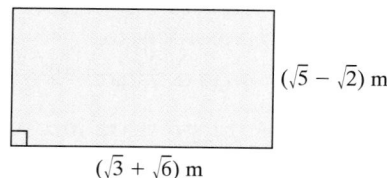

$(\sqrt{5} - \sqrt{2})$ m

$(\sqrt{3} + \sqrt{6})$ m

8. Perform the indicated operation.

$$\frac{4 - 3i}{2 + 5i}$$

9. Find the measure of each angle in the right triangle.

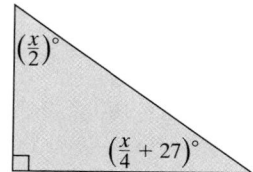

$\left(\frac{x}{2}\right)^\circ$

$\left(\frac{x}{4} + 27\right)^\circ$

10. Find the positive slope of the sides of a pyramid with a square base 66 ft on a side and height of 22 ft.

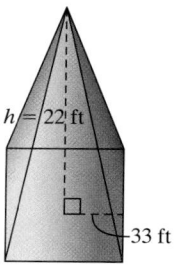

$h = 22$ ft

33 ft

11. How many liters of pure alcohol must be mixed with 8 L of 20% alcohol to bring the concentration up to 50% alcohol?

12. Bank robbers leave the scene of a crime and head north through winding mountain roads to their hideaway. Their average rate of speed is 40 mph. The police leave 6 min later in pursuit. If the police car averages 50 mph traveling the same route, how long will it take the police to catch the bank robbers?

13. Solve the system by using the Gauss-Jordan method.

$$5x + 10y = 25$$
$$-2x + 6y = -20$$

14. Solve for w. $-2[w - 3(w + 1)] = 4 - 7(w + 3)$

15. Solve for x. $ax - c = bx + d$

16. Solve for t. $s = \frac{1}{2}gt^2$ $t \geq 0$

17. Solve for T. $\sqrt{1 - kT} = \dfrac{V_0}{V}$

18. Find the x-intercepts of the function defined by $f(x) = |x - 5| - 2$.

19. Let $f(t) = 6$, $g(t) = -5t$, and $h(t) = 2t^2$. Find

   a. $(f \cdot g)(t)$ b. $(g \circ h)(t)$ c. $(h - g)(t)$

20. Solve for q. $|2q - 5| = |2q + 5|$

21. a. Find an equation of the line parallel to the y-axis and passing through the point (2, 6).

b. Find an equation of the line perpendicular to the y-axis and passing through the point (2, 6).

c. Find an equation of the line perpendicular to the line $2x + y = 4$ and passing through the point (2, 6). Write the answer in slope-intercept form.

22. The smallest angle in a triangle measures one-half the largest angle. The smallest angle measures 20° less than the middle angle. Find the measures of all three angles.

23. Solve the system.

$$\frac{1}{2}x - \frac{1}{4}y = 1$$

$$-2x + y = -4$$

24. Match the function with the appropriate graph.

  i. $f(x) = \ln(x)$ ii. $g(x) = 3^x$

  iii. $h(x) = x^2$ iv. $k(x) = -2x - 3$

  v. $L(x) = |x|$ vi. $m(x) = \sqrt{x}$

  vii. $n(x) = \sqrt[3]{x}$ viii. $p(x) = x^3$

  ix. $q(x) = \dfrac{1}{x}$ x. $r(x) = x$

a.
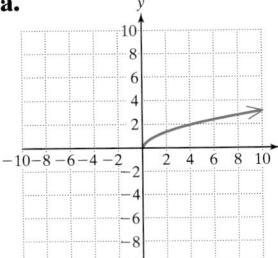

b.
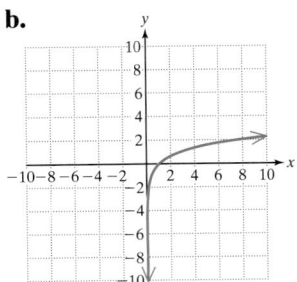

c.
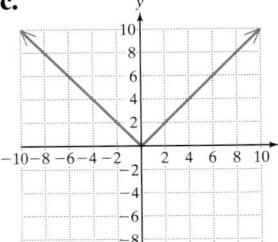

d.
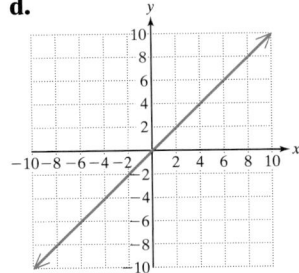

e.
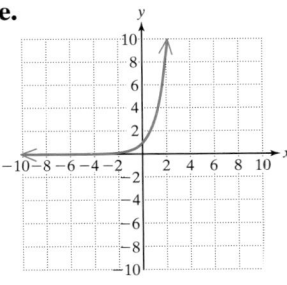

f.
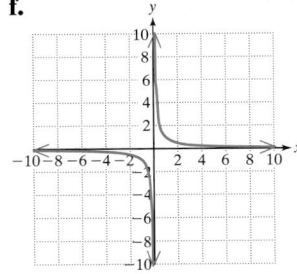

g.
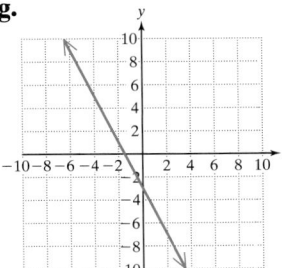

h.
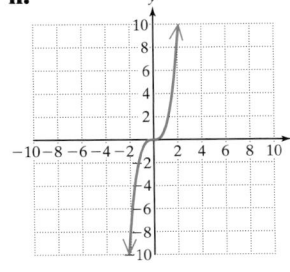

i.
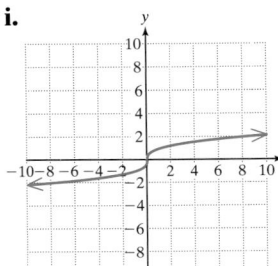

j.
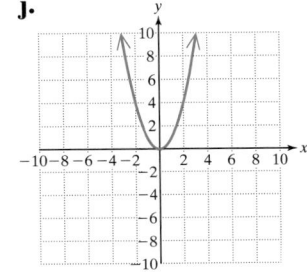

25. Find the domain. $f(x) = \sqrt{2x - 1}$.

26. Find $f^{-1}(x)$, given $f(x) = 5x - \frac{2}{3}$.

 27. The volume of a gas varies directly as its temperature and inversely with pressure. At a temperature of 100 kelvins (K) and a pressure of 10 newtons per square meter (N/m²), the gas occupies a volume of 30 m³. Find the volume at a temperature of 200 K and pressure of 15 N/m².

28. Perform the indicated operations.

$$\frac{5x - 10}{x^2 - 4x + 4} \div \frac{5x^2 - 125}{25 - 5x} \cdot \frac{x^3 + 125}{10x + 5}$$

29. Perform the indicated operations.

$$\frac{x}{x - y} + \frac{y}{y - x} + x$$

30. Given the equation

$$\frac{2}{x-4} = \frac{5}{x+2}$$

a. Are there any restrictions on x for which the rational expressions are undefined?

b. Solve the equation.

c. Solve the related inequality.

$$\frac{2}{x-4} \geq \frac{5}{x+2}$$

Write the answer in interval notation.

31. Two more than 3 times the reciprocal of a number is $\frac{5}{4}$ less than the number. Find all such numbers.

32. Solve the equation. $\sqrt{-x} = x + 6$

33. Solve the inequality $2|x-3| + 1 > -7$. Write the answer in interval notation.

34. Four million *Escherichia coli (E. coli)* bacteria are present in a laboratory culture. An antibacterial agent is introduced and the population of bacteria decreases by one-half every 6 hr according to

$$P(t) = 4,000,000\left(\frac{1}{2}\right)^{t/6} \qquad t \geq 0$$

where t is the time in hours.

a. Find the population of bacteria after 6, 12, 18, 24, and 30 hr.

b. Sketch a graph of $y = P(t)$ based on the function values found in part (a).

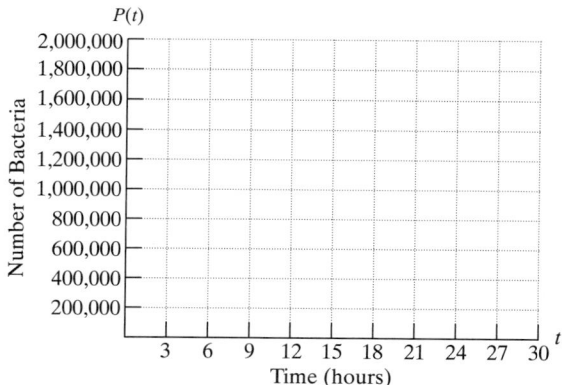

c. Predict the time required for the population to decrease to 15,625 bacteria.

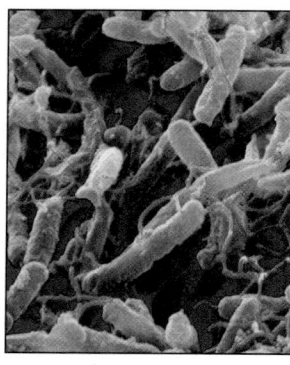

35. Evaluate the expressions without a calculator.

a. $\log_7 49$

b. $\log_4\left(\frac{1}{64}\right)$

c. $\log(1,000,000)$

d. $\ln e^3$

36. Use a calculator to approximate the expressions to four decimal places.

a. $\pi^{4.7}$

b. e^{π}

c. $(\sqrt{2})^{-5}$

d. $\log 5362$

e. $\ln(0.67)$

f. $\log_4 37$

37. Solve the equation. $e^x = 100$

38. Solve the equation. $\log_3(x+6) - 3 = -\log_3 x$

39. Write the following expression as a single logarithm.

$$\frac{1}{2}\log z - 2\log x - 3\log y$$

40. Write the following expression as a sum or difference of logarithms

$$\ln\left(\sqrt[3]{\frac{x^2}{y}}\right)$$

Conic Sections and Nonlinear Systems

14.1 Distance Formula and Circles

14.2 More on the Parabola

14.3 The Ellipse and Hyperbola

14.4 Nonlinear Systems of Equations in Two Variables

14.5 Nonlinear Inequalities and Systems of Inequalities

In Chapter 14 we present several new types of graphs, called conic sections. These include circles, parabolas, ellipses, and hyperbolas. These shapes are found in a variety of applications. For example, a reflecting telescope has a mirror whose cross section is in the shape of a parabola, and planetary orbits are modeled by ellipses.

As you work through the chapter, you will encounter a variety of equations associated with the conic sections. Match the equation with its description, and then use the letter next to each answer to complete the puzzle.

_____ 1. $\dfrac{x^2}{a^2} + \dfrac{y^2}{b^2} = 1$

_____ 2. $x^2 + y^2 = r^2$

_____ 3. $x = a(y - k)^2 + h$

_____ 4. $d = \sqrt{(x_2 - x_1)^2 + (y_2 - y_1)^2}$

_____ 5. $y = a(x - h)^2 + k$

_____ 6. $(x - h)^2 + (y - k)^2 = r^2$

_____ 7. $\dfrac{x^2}{a^2} - \dfrac{y^2}{b^2} = 1$

a. standard form of a parabola with horizontal axis of symmetry

o. standard form of an ellipse centered at the origin

f. standard form of a circle centered at (h, k)

t. standard form of a hyperbola with horizontal transverse axis

s. standard form of a circle centered at the origin

c. standard form of a parabola with vertical axis of symmetry

r. distance formula

He became a math teacher due to some prime $\overline{}\ \overline{}\ \overline{}\ \overline{}\ \overline{}\ \overline{}\ \overline{}$
$\ 6\ \ 3\ \ 5\ \ 7\ \ 1\ \ 4\ \ 2$

Section 14.1   Distance Formula and Circles

1. Distance Formula

Suppose we are given two points (x_1, y_1) and (x_2, y_2) in a rectangular coordinate system. The distance between the two points can be found by using the Pythagorean theorem (Figure 14-1).

First draw a right triangle with the distance d as the hypotenuse. The length of the horizontal leg a is $|x_2 - x_1|$, and the length of the vertical leg b is $|y_2 - y_1|$. From the Pythagorean theorem we have

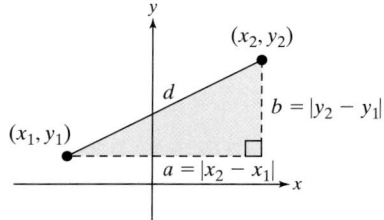

Figure 14-1

$$d^2 = a^2 + b^2 \qquad \text{Pythagorean theorem}$$
$$= (x_2 - x_1)^2 + (y_2 - y_1)^2$$

$$d = \pm\sqrt{(x_2 - x_1)^2 + (y_2 - y_1)^2}$$
$$= \sqrt{(x_2 - x_1)^2 + (y_2 - y_1)^2} \qquad \text{Because distance is positive, reject the negative value.}$$

The Distance Formula

The distance d between the points (x_1, y_1) and (x_2, y_2) is
$$d = \sqrt{(x_2 - x_1)^2 + (y_2 - y_1)^2}$$

Example 1   Finding the Distance Between Two Points

Find the distance between the points $(-2, 3)$ and $(4, -1)$ (Figure 14-2).

Solution:

$$(-2, 3) \qquad \text{and} \qquad (4, -1)$$
$$(x_1, y_1) \qquad \text{and} \qquad (x_2, y_2) \qquad \text{Label the points.}$$
$$d = \sqrt{(x_2 - x_1)^2 + (y_2 - y_1)^2}$$
$$= \sqrt{[(4) - (-2)]^2 + [(-1) - (3)]^2} \qquad \text{Apply the distance formula.}$$
$$= \sqrt{(6)^2 + (-4)^2}$$
$$= \sqrt{36 + 16}$$
$$= \sqrt{52}$$
$$= \sqrt{2^2 \cdot 13}$$
$$= 2\sqrt{13}$$

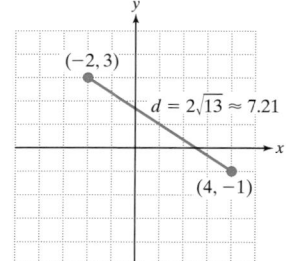

Figure 14-2

Skill Practice

1. Find the distance between the points $(-4, -2)$ and $(2, -5)$.

Skill Practice Answers

1. $3\sqrt{5}$

TIP: The order in which the points are labeled does not affect the result of the distance formula. For example, if the points in Example 1 had been labeled in reverse, the distance formula would still yield the same result:

$(-2, 3)$ and $(4, -1)$ $\qquad d = \sqrt{(x_2 - x_1)^2 + (y_2 - y_1)^2}$

(x_2, y_2) and (x_1, y_1) $\qquad\quad = \sqrt{[(-2) - (4)]^2 + [3 - (-1)]^2}$

$\qquad\qquad\qquad\qquad\qquad\quad = \sqrt{(-6)^2 + (4)^2}$

$\qquad\qquad\qquad\qquad\qquad\quad = \sqrt{36 + 16}$

$\qquad\qquad\qquad\qquad\qquad\quad = \sqrt{52}$

$\qquad\qquad\qquad\qquad\qquad\quad = 2\sqrt{13}$

2. Circles

A **circle** is defined as the set of all points in a plane that are equidistant from a fixed point called the **center**. The fixed distance from the center is called the **radius** and is denoted by r, where $r > 0$.

Suppose a circle is centered at the point (h, k) and has radius, r (Figure 14-3). The distance formula can be used to derive an equation of the circle.

Let (x, y) be any arbitrary point on the circle. Then, by definition, the distance between (h, k) and (x, y) must be r.

$$\sqrt{(x - h)^2 + (y - k)^2} = r$$

$$(x - h)^2 + (y - k)^2 = r^2 \qquad \text{Square both sides.}$$

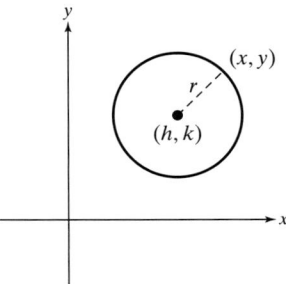

Figure 14-3

Standard Equation of a Circle

The **standard equation of a circle**, centered at (h, k) with radius r, is given by

$$(x - h)^2 + (y - k)^2 = r^2 \qquad \text{where } r > 0$$

Note: If a circle is centered at the origin $(0, 0)$, then $h = 0$ and $k = 0$, and the equation simplifies to $x^2 + y^2 = r^2$.

Example 2 **Graphing a Circle**

Find the center and radius of each circle. Then graph the circle.

a. $(x - 3)^2 + (y + 4)^2 = 36$ $\qquad$ **b.** $x^2 + \left(y - \dfrac{10}{3}\right)^2 = \dfrac{25}{9}$

c. $x^2 + y^2 = 10$

Solution:

a. $\qquad (x - 3)^2 + (y + 4)^2 = 36$

$\qquad (x - 3)^2 + [y - (-4)]^2 = (6)^2$

$\qquad\quad h = 3, k = -4, \text{and } r = 6$

The center is $(3, -4)$ and the radius is 6 (Figure 14-4).

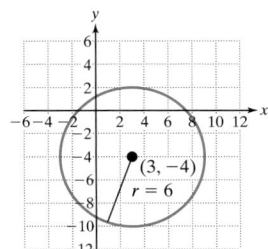

Figure 14-4

Calculator Connections

Graphing calculators are designed to graph *functions*, in which y is written in terms of x. A circle is not a function. However, it can be graphed as the union of two functions—one representing the top semicircle and the other representing the bottom semicircle.

Solving for y in Example 2(a), we have

$$(x - 3)^2 + (y + 4)^2 = 36$$

Graph these functions as Y_1 and Y_2, using a square viewing window.

$$(y + 4)^2 = 36 - (x - 3)^2$$

$$y + 4 = \pm\sqrt{36 - (x - 3)^2}$$

$$y = -4 \pm \sqrt{36 - (x - 3)^2}$$

$$Y_1 = -4 + \sqrt{36 - (x - 3)^2}$$

$$Y_2 = -4 - \sqrt{36 - (x - 3)^2}$$

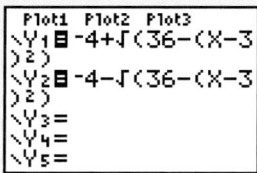
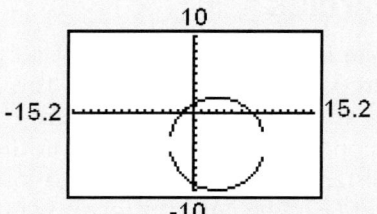

Notice that the image from the calculator does not show the upper and lower semicircles connecting at their endpoints, when in fact the semicircles should "hook up." This is due to the calculator's limited resolution.

b. $x^2 + \left(y - \dfrac{10}{3}\right)^2 = \dfrac{25}{9}$

$(x - 0)^2 + \left(y - \dfrac{10}{3}\right)^2 = \left(\dfrac{5}{3}\right)^2$

The center is $\left(0, \frac{10}{3}\right)$ and the radius is $\frac{5}{3}$ (Figure 14-5).

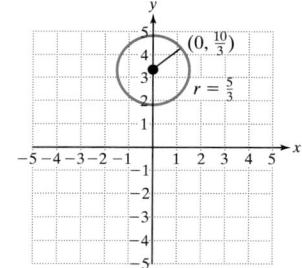

Figure 14-5

c. $x^2 + y^2 = 10$

$(x - 0)^2 + (y - 0)^2 = (\sqrt{10})^2$

The center is $(0, 0)$ and the radius is $\sqrt{10} \approx 3.16$ (Figure 14-6).

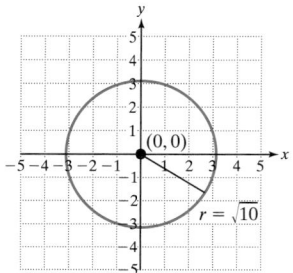

Figure 14-6

Skill Practice Find the center and radius of each circle. Then graph the circle.

2. $(x + 1)^2 + (y - 2)^2 = 9$ **3.** $\left(x + \dfrac{7}{2}\right)^2 + y^2 = \dfrac{9}{4}$

4. $x^2 + y^2 = 15$

Sometimes it is necessary to complete the square to write an equation of a circle in standard form.

Example 3 Writing the Equation of a Circle in the Form $(x - h)^2 + (y - k)^2 = r^2$

Identify the center and radius of the circle given by the equation $x^2 + y^2 + 2x - 16y + 61 = 0$.

Solution:

$$x^2 + y^2 + 2x - 16y + 61 = 0$$

To identify the center and radius, write the equation in the form $(x - h)^2 + (y - k)^2 = r^2$.

$$(x^2 + 2x \quad) + (y^2 - 16y \quad) = -61$$

Group the x-terms and group the y-terms. Move the constant to the right-hand side.

$$(x^2 + 2x + 1) + (y^2 - 16y + 64) = -61 + 1 + 64$$

- Complete the square on x. Add $\left[\frac{1}{2}(2)\right]^2 = 1$ to both sides of the equation.
- Complete the square on y. Add $\left[\frac{1}{2}(-16)\right]^2 = 64$ to both sides of the equation.

$$(x + 1)^2 + (y - 8)^2 = 4$$

Factor and simplify.

$$[x - (-1)]^2 + (y - 8)^2 = 2^2$$

The center is $(-1, 8)$ and the radius is 2.

Skill Practice

5. Identify the center and radius of the circle given by the equation

$$x^2 + y^2 - 10x + 4y - 7 = 0$$

Skill Practice Answers

2. Center; $(-1, 2)$; radius: 3

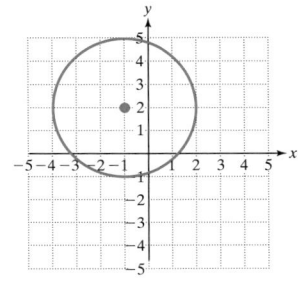

3. Center: $\left(-\dfrac{7}{2}, 0\right)$; radius: $\dfrac{3}{2}$

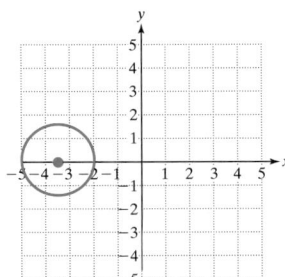

4. Center: $(0, 0)$; radius: $\sqrt{15}$

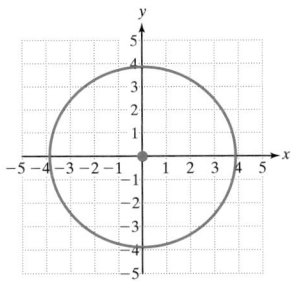

5. Center: $(5, -2)$; radius: 6

3. Writing an Equation of a Circle

Example 4   Writing an Equation of a Circle

Write an equation of the circle shown in Figure 14-7.

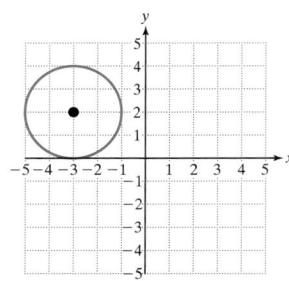

Figure 14-7

Solution:

The center is $(-3, 2)$; therefore, $h = -3$ and $k = 2$.
From the graph, $r = 2$.

$$(x - h)^2 + (y - k)^2 = r^2$$
$$[x - (-3)]^2 + (y - 2)^2 = (2)^2$$
$$(x + 3)^2 + (y - 2)^2 = 4$$

Skill Practice

6. Write an equation for a circle whose center is the point $(6, -1)$ and whose radius is 8.

Skill Practice Answers

6. $(x - 6)^2 + (y + 1)^2 = 64$

Section 14.1   **Practice Exercises**

Study Skills Exercise

1. Define the key terms.

   a. distance formula   **b. circle**   **c. center**   **d. radius**   **e. standard equation of a circle**

Concept 1: Distance Formula

For Exercises 2–16, use the distance formula to find the distance between the two points.

2. $(-2, 7)$ and $(3, -9)$

3. $(1, 10)$ and $(-2, 4)$

4. $(0, 5)$ and $(-3, 8)$

5. $(6, 7)$ and $(3, 2)$

6. $\left(\dfrac{2}{3}, \dfrac{1}{5}\right)$ and $\left(-\dfrac{5}{6}, \dfrac{3}{10}\right)$

7. $\left(-\dfrac{1}{2}, \dfrac{5}{8}\right)$ and $\left(-\dfrac{3}{2}, \dfrac{1}{4}\right)$

8. $(4, 13)$ and $(4, -6)$

9. $(-2, 5)$ and $(-2, 9)$

10. $(8, -6)$ and $(-2, -6)$

11. $(7, 2)$ and $(15, 2)$

12. $(-6, -2)$ and $(-1, -4)$

13. $(-1, -5)$ and $(-3, -2)$

14. $(3\sqrt{5}, 2\sqrt{7})$ and $(-\sqrt{5}, -3\sqrt{7})$ **15.** $(4\sqrt{6}, -2\sqrt{2})$ and $(2\sqrt{6}, \sqrt{2})$ **16.** $(6, 0)$ and $(0, -1)$

17. Explain how to find the distance between 5 and -7 on the y-axis.

18. Explain how to find the distance between 15 and -37 on the x-axis.

19. Find the values of y such that the distance between the points $(4, 7)$ and $(-4, y)$ is 10 units.

20. Find the values of x such that the distance between the points $(-4, -2)$ and $(x, 3)$ is 13 units.

21. Find the values of x such that the distance between the points $(x, 2)$ and $(4, -1)$ is 5 units.

22. Find the values of y such that the distance between the points $(-5, 2)$ and $(3, y)$ is 10 units.

For Exercises 23–26, determine if the three points define the vertices of a right triangle.

23. $(-3, 2), (-2, -4)$, and $(3, 3)$

24. $(1, -2), (-2, 4)$, and $(7, 1)$

25. $(-3, -2), (4, -3)$, and $(1, 5)$

26. $(1, 4), (5, 3)$, and $(2, 0)$

Concept 2: Circles

For Exercises 27–48, identify the center and radius of the circle and then graph the circle. Complete the square, if necessary.

27. $(x - 4)^2 + (y + 2)^2 = 9$ **28.** $(x - 3)^2 + (y + 1)^2 = 16$ **29.** $(x + 1)^2 + (y + 1)^2 = 1$

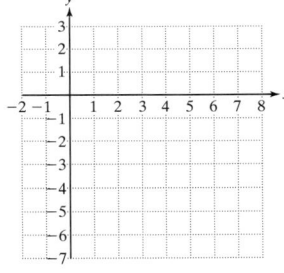   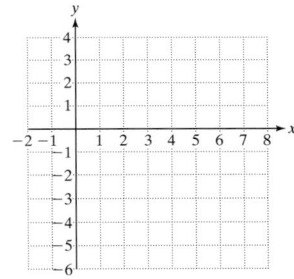   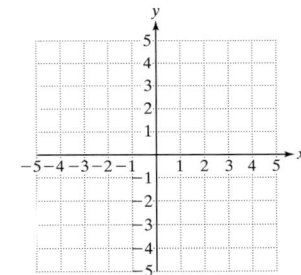

30. $(x - 4)^2 + (y - 4)^2 = 4$ **31.** $x^2 + (y - 5)^2 = 25$ **32.** $(x + 1)^2 + y^2 = 1$

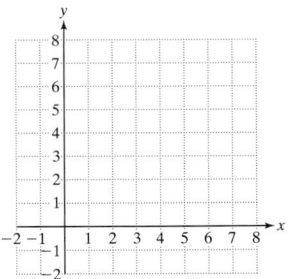   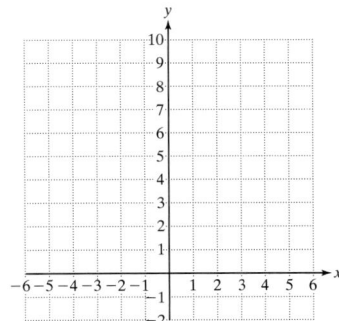   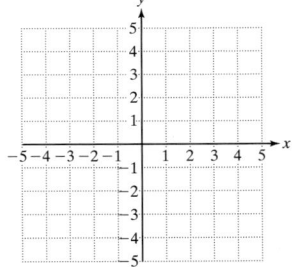

33. $(x - 3)^2 + y^2 = 8$

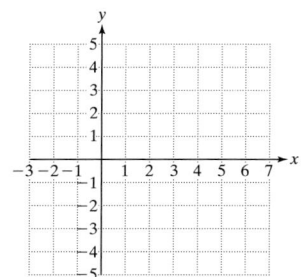

34. $x^2 + (y + 2)^2 = 20$

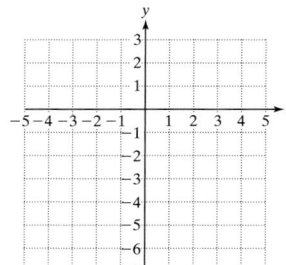

35. $x^2 + y^2 = 6$

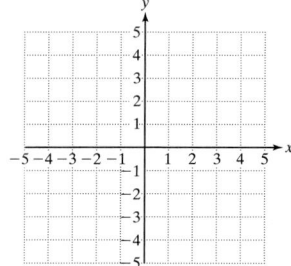

36. $x^2 + y^2 = 15$

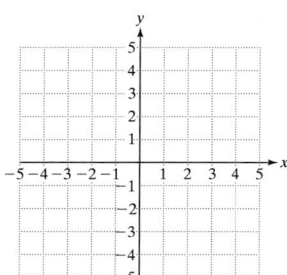

37. $\left(x + \dfrac{4}{5}\right)^2 + y^2 = \dfrac{64}{25}$

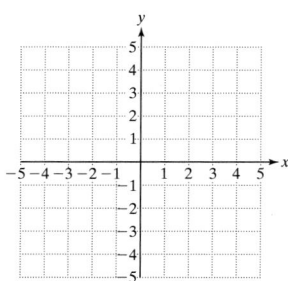

38. $x^2 + \left(y - \dfrac{5}{2}\right)^2 = \dfrac{9}{4}$

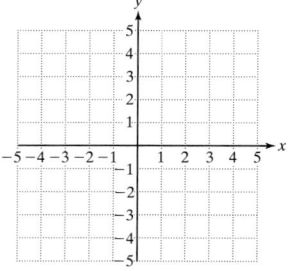

39. $x^2 + y^2 - 2x - 6y - 26 = 0$

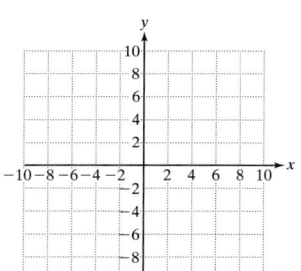

40. $x^2 + y^2 + 4x - 8y + 16 = 0$

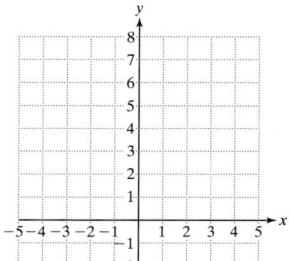

41. $x^2 + y^2 - 6y + 5 = 0$

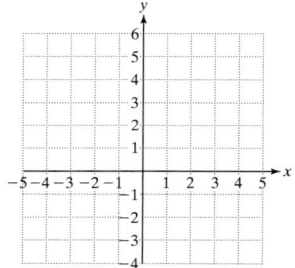

42. $x^2 + 2x + y^2 - 24 = 0$

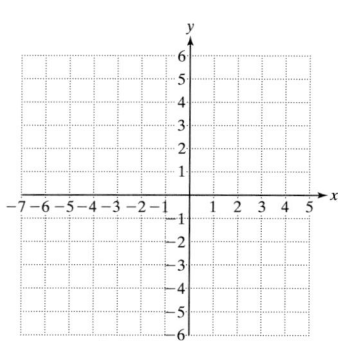

43. $x^2 + y^2 + 6y + \dfrac{65}{9} = 0$

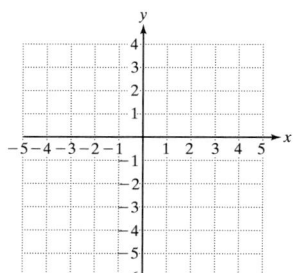

44. $x^2 + y^2 + 12x + \dfrac{143}{4} = 0$

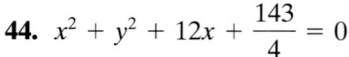
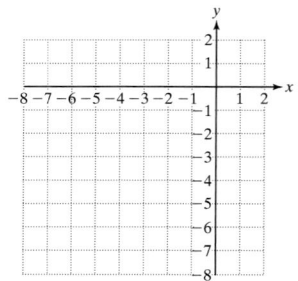

45. $x^2 + y^2 - 12x + 12y + 71 = 0$ **46.** $x^2 + y^2 + 2x + 4y - 4 = 0$ **47.** $2x^2 + 2y^2 = 32$

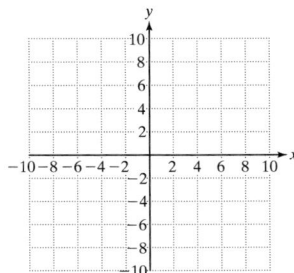   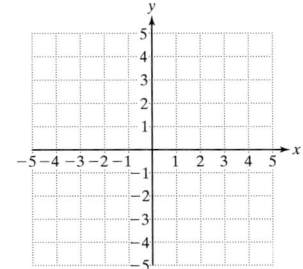   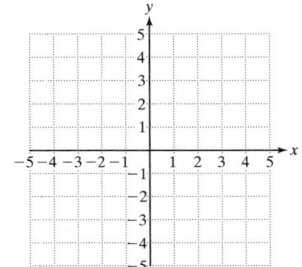

48. $3x^2 + 3y^2 = 3$

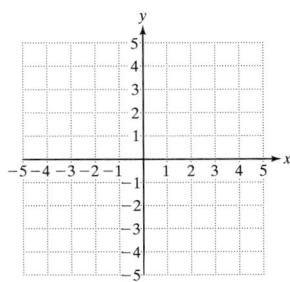

Concept 3: Writing an Equation of a Circle

For Exercises 49–54, write an equation that represents the graph of the circle.

49.

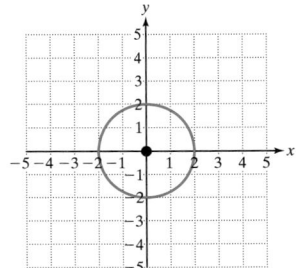

50.

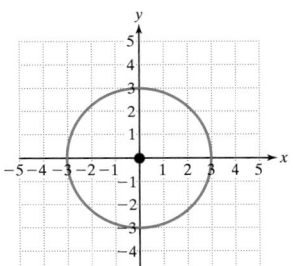

51.

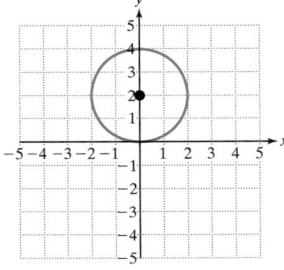

52.

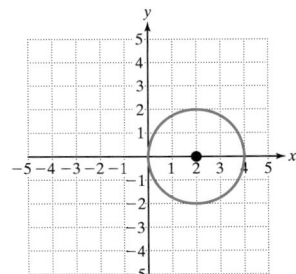

53.

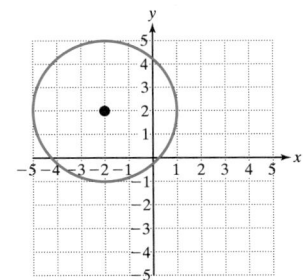

54.

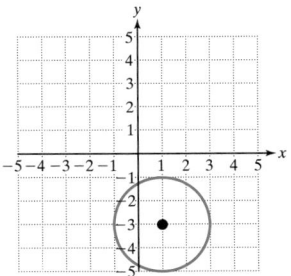

55. Write an equation of a circle centered at the origin with a radius of 7 m.

56. Write an equation of a circle centered at the origin with a radius of 12 m.

57. Write an equation of a circle centered at $(-3, -4)$ with a diameter of 12 ft.

58. Write an equation of a circle centered at $(5, -1)$ with a diameter of 8 ft.

Expanding Your Skills

59. Write an equation of a circle that has the points $(-2, 3)$ and $(2, 3)$ as endpoints of a diameter.

60. Write an equation of a circle that has the points $(-1, 3)$ and $(-1, -3)$ as endpoints of a diameter.

61. Write an equation of a circle whose center is at $(4, 4)$ and is tangent to the x- and y-axes. (*Hint:* Sketch the circle first.)

62. Write an equation of a circle whose center is at $(-3, 3)$ and is tangent to the x- and y-axes. (*Hint:* Sketch the circle first.)

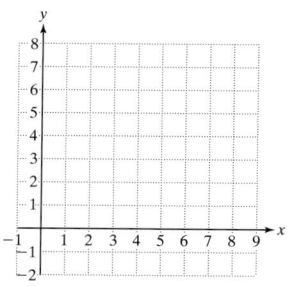

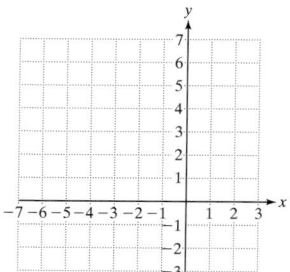

63. Write an equation of a circle whose center is at $(1, 1)$ and that passes through the point $(-4, 3)$.

64. Write an equation of a circle whose center is at $(-3, -1)$ and that passes through the point $(5, -2)$.

Graphing Calculator Exercises

For Exercises 65–70, graph the circles from the indicated exercise on a square viewing window, and approximate the center and the radius from the graph.

65. $(x - 4)^2 + (y + 2)^2 = 9$ (Exercise 27)

66. $(x - 3)^2 + (y + 1)^2 = 16$ (Exercise 28)

67. $x^2 + (y - 5)^2 = 25$ (Exercise 31)

68. $(x + 1)^2 + y^2 = 1$ (Exercise 32)

69. $x^2 + y^2 = 6$ (Exercise 35)

70. $x^2 + y^2 = 15$ (Exercise 36)

More on the Parabola

1. Introduction to Conic Sections

Recall that the graph of a second-degree equation of the form $y = ax^2 + bx + c$ ($a \neq 0$) is a parabola. In Section 14.1 we learned that the graph of $(x - h)^2 + (y - k)^2 = r^2$ is a circle. These and two other types of figures called ellipses and hyperbolas are called **conic sections**. Conic sections derive their names because each is the intersection of a plane and a double-napped cone (Figure 14-8).

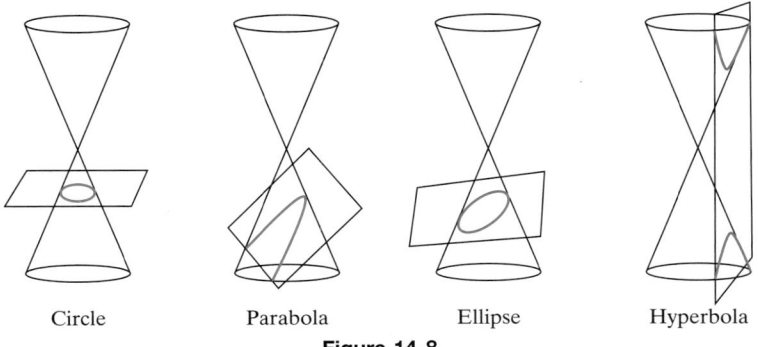

Circle   Parabola   Ellipse   Hyperbola

Figure 14-8

2. Parabola—Vertical Axis of Symmetry

A **parabola** is defined by a set of points in a plane that are equidistant from a fixed line (called the directrix) and a fixed point (called the focus) not on the directrix. Parabolas have numerous real-world applications. For example, a reflecting telescope has a mirror with the cross section in the shape of a parabola. A parabolic mirror has the property that incoming rays of light are reflected from the surface of the mirror to the focus.

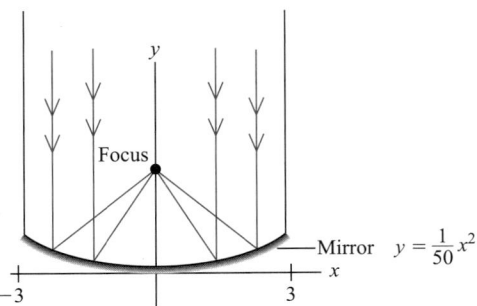

The graph of the solutions to the quadratic equation $y = ax^2 + bx + c$ is a parabola. We graphed parabolas of this type in Section 12.4 by writing the equation in the form $y = a(x - h)^2 + k$. Recall that the **vertex** (h, k) is the highest or

the lowest point of a parabola. The **axis of symmetry** of the parabola is a line that passes through the vertex and is perpendicular to the directrix (Figure 14-9).

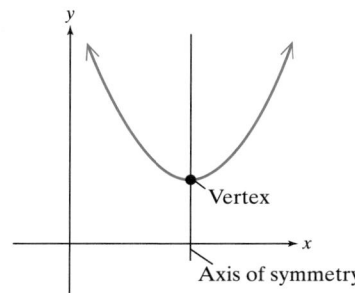

Vertex

Axis of symmetry

Figure 14-9

Standard Form of the Equation of a Parabola—Vertical Axis of Symmetry

The standard form of the equation of a parabola with vertex (h, k) and vertical axis of symmetry is

$$y = a(x - h)^2 + k \qquad \text{where } a \neq 0$$

If $a > 0$, then the parabola opens upward; and if $a < 0$, the parabola opens downward.

The axis of symmetry is given by $x = h$.

In Section 12.5, we learned that it is sometimes necessary to complete the square to write the equation of a parabola in standard form.

Example 1   Graphing a Parabola by First Completing the Square

Given: the equation of the parabola $y = -2x^2 + 4x + 1$

a. Write the equation in standard form $y = a(x - h)^2 + k$.

b. Identify the vertex and axis of symmetry. Determine if the parabola opens upward or downward.

c. Graph the parabola.

Solution:

a. Complete the square to write the equation in the form $y = a(x - h)^2 + k$.

$y = -2x^2 + 4x + 1$

$y = -2(x^2 - 2x) + 1$ \hfill Factor out -2 from the variable terms.

$y = -2(x^2 - 2x + 1 - 1) + 1$ \hfill Add and subtract the quantity $\left[\frac{1}{2}(-2)\right]^2 = 1$.

$y = -2(x^2 - 2x + 1) + (-2)(-1) + 1$ \hfill Remove the -1 term from within the parentheses by first applying the distributive property. When -1 is removed from the parentheses, it carries with it the factor of -2 from outside the parentheses.

$$y = -2(x - 1)^2 + 2 + 1$$
$$y = -2(x - 1)^2 + 3$$

The equation is in the form $y = a(x - h)^2 + k$ where $a = -2$, $h = 1$, and $k = 3$.

b. The vertex is $(1, 3)$. Because a is negative, $(a = -2 < 0)$, the parabola opens downward. The axis of symmetry is $x = 1$.

c. To graph the parabola, we know that its orientation is downward. Furthermore, we know the vertex is $(1, 3)$. To find other points on the parabola, select several values of x and solve for y. Recall that the y-intercept is found by substituting $x = 0$ and solving for y.

x	y
1	3	← Vertex
-1	-5
0	1	← y-intercept
2	1
3	-5

Choose x.   Solve for y.

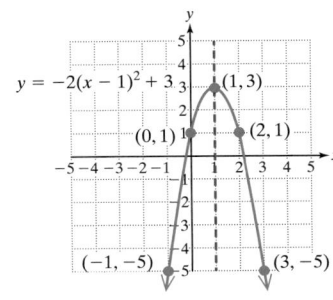

Skill Practice

1. Given: the equation of the parabola $y = 2x^2 - 4x + 5$
   a. Write the equation in standard form.
   b. Identify the vertex and axis of symmetry.
   c. Graph the parabola.

3. Parabola—Horizontal Axis of Symmetry

We have seen that the graph of a parabola $y = ax^2 + bx + c$ opens upward if $a > 0$ and downward if $a < 0$. A parabola can also open to the left or right. In such a case, the "roles" of x and y are essentially interchanged in the equation. Thus, the graph of $x = ay^2 + by + c$ opens to the right if $a > 0$ (Figure 14-10) and to the left if $a < 0$ (Figure 14-11).

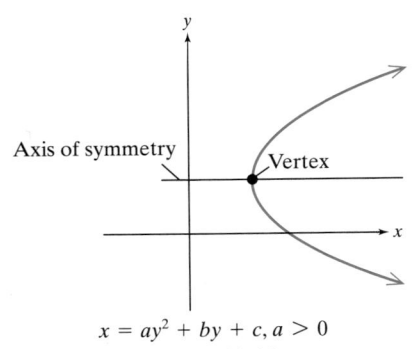

$x = ay^2 + by + c, a > 0$

Figure 14-10

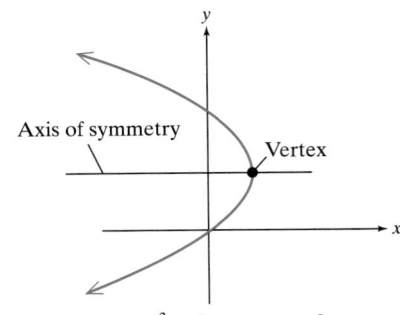

$x = ay^2 + by + c, a < 0$

Figure 14-11

Skill Practice Answers

1a. $y = 2(x - 1)^2 + 3$
b. Vertex: $(1, 3)$; axis of symmetry: $x = 1$
c.

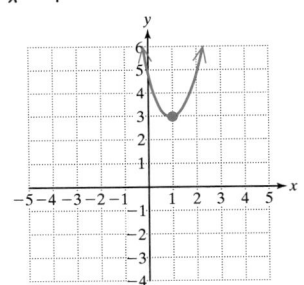

Standard Form of the Equation of a Parabola—Horizontal Axis of Symmetry

The standard form of the equation of a parabola with vertex (h, k) and horizontal axis of symmetry is

$$x = a(y - k)^2 + h \qquad \text{where } a \neq 0$$

If $a > 0$, then the parabola opens to the right and if $a < 0$, the parabola opens to the left.

The axis of symmetry is given by $y = k$.

Example 2   **Graphing a Parabola with a Horizontal Axis of Symmetry**

Given the equation of the parabola $x = 4y^2$,

a. Determine the coordinates of the vertex and the equation of the axis of symmetry.

b. Use the value of a to determine if the parabola opens to the right or left.

c. Plot several points and graph the parabola.

Solution:

a. The equation can be written in the form $x = a(y - k)^2 + h$:

$$x = 4(y - 0)^2 + 0$$

Therefore, $h = 0$ and $k = 0$.

The vertex is $(0, 0)$. The axis of symmetry is $y = 0$ (the x-axis).

b. Because a is positive ($a = 4 > 0$), the parabola opens to the right.

c. The vertex of the parabola is $(0, 0)$. To find other points on the graph, select values for y and solve for x.

x	y
0	0	←—Vertex
4	1
4	−1
16	2
16	−2

Solve for x.   Choose y.

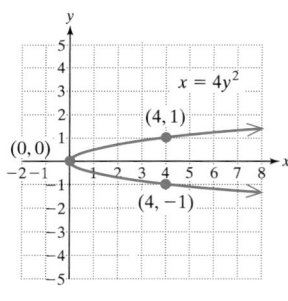

Skill Practice Answers

2a. Vertex: $(1, 0)$; axis of symmetry: $y = 0$
b. Parabola opens left.
c.

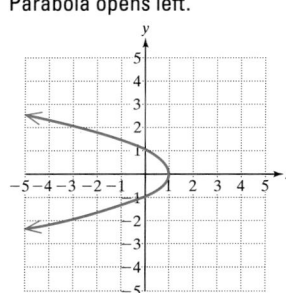

Skill Practice

2. Given: the equation $x = -y^2 + 1$

a. Identify the vertex and the axis of symmetry.

b. Determine if the parabola opens to the right or to the left.

c. Graph the parabola.

Calculator Connections

Graphing calculators are designed to graph functions, where y is written as a function of x. The parabola from Example 2 is not a function. It can be graphed, however, as the union of two functions, one representing the top branch and the other representing the bottom branch.

Solving for y in Example 2, we have: $x = 4y^2$

$$y^2 = \frac{x}{4}$$

$$y = \pm\sqrt{\frac{x}{4}}$$

$$y = \pm\frac{\sqrt{x}}{2}$$

Graph these functions as Y_1 and Y_2.

$$Y_1 = \frac{\sqrt{x}}{2} \quad \text{and} \quad Y_2 = -\frac{\sqrt{x}}{2}$$

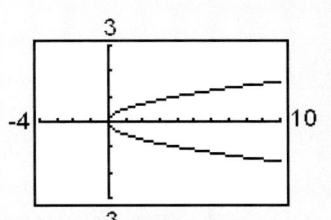

Example 3 Graphing a Parabola by First Completing the Square

Given the equation of the parabola $x = -y^2 + 8y - 14$,

a. Write the equation in standard form $x = a(y - k)^2 + h$.

b. Identify the vertex and axis of symmetry. Determine if the parabola opens to the right or left.

c. Graph the parabola.

Solution:

a. Complete the square to write the equation in the form $x = a(y - k)^2 + h$.

$x = -y^2 + 8y - 14$

$x = -1(y^2 - 8y) - 14$ Factor out -1 from the variable terms.

$x = -1(y^2 - 8y + 16 - 16) - 14$ Add and subtract the quantity $\left[\frac{1}{2}(-8)\right]^2 = 16$.

$x = -1(y^2 - 4y + 16) + (-1)(-16) - 14$ Remove the -16 term from within the parentheses by first applying the distributive property. When -16 is removed from the parentheses, it carries with it the factor of -1 from outside the parentheses.

$x = -1(y - 4)^2 + 16 - 14$

$x = -1(y - 4)^2 + 2$

The equation is in the form $x = a(y - k)^2 + h$, where $a = -1$, $h = 2$, and $k = 4$.

b. The vertex is $(2, 4)$. Because a is negative $(a = -1 < 0)$, the parabola opens to the left. The axis of symmetry is $y = 4$.

c.

x	y
2	4	←——Vertex
1	5
1	3
−2	6
−2	2

↑ Solve for x.   ↑ Choose y.

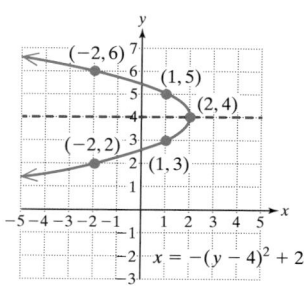

$x = -(y - 4)^2 + 2$

Skill Practice

3. Given the equation of the parabola $x = \dfrac{1}{8}(y + 2)^2 + 3$,

   a. Determine the coordinates of the vertex.
   b. Determine if the parabola opens to the right or to the left.
   c. Graph the parabola.

4. Vertex Formula

From Section 12.5, we learned that the vertex formula can also be used to find the vertex of a parabola.

For a parabola defined by $y = ax^2 + bx + c$,

- The x-coordinate of the vertex is given by $x = \frac{-b}{2a}$.
- The y-coordinate of the vertex is found by substituting this value for x into the original equation and solving for y.

For a parabola defined by $x = ay^2 + by + c$,

- The y-coordinate of the vertex is given by $y = \frac{-b}{2a}$.
- The x-coordinate of the vertex is found by substituting this value for y into the original equation and solving for x.

Example 4 **Finding the Vertex of a Parabola by Using the Vertex Formula**

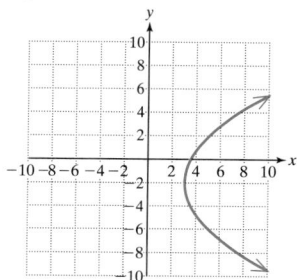

Find the vertex by using the vertex formula.

a. $x = y^2 + 4y + 5$ **b.** $y = \dfrac{1}{2}x^2 - 3x + \dfrac{5}{2}$

Solution:

a. $x = y^2 + 4y + 5$ The parabola is in the form $x = ay^2 + by + c$.

   $a = 1$ $b = 4$ $c = 5$ Identify a, b, and c.

The y-coordinate of the vertex is given by $y = \dfrac{-b}{2a} = \dfrac{-(4)}{2(1)} = -2$.

The x-coordinate of the vertex is found by substitution:
$x = (-2)^2 + 4(-2) + 5 = 1$.

The vertex is $(1, -2)$.

b. $y = \dfrac{1}{2}x^2 - 3x + \dfrac{5}{2}$
The parabola is in the form
$y = ax^2 + bx + c$.

$a = \dfrac{1}{2} \qquad b = -3 \qquad c = \dfrac{5}{2}$
Identify a, b, and c.

The x-coordinate of the vertex is given by $x = \dfrac{-b}{2a} = \dfrac{-(-3)}{2(\frac{1}{2})} = 3$.

The y-coordinate of the vertex is found by substitution.

$$y = \frac{1}{2}(3)^2 - 3(3) + \frac{5}{2}$$

$$= \frac{9}{2} - 9 + \frac{5}{2}$$

$$= \frac{14}{2} - 9$$

$$= 7 - 9$$

$$= -2$$

The vertex is $(3, -2)$.

Skill Practice Find the vertex by using the vertex formula.

4. $x = -4y^2 + 12y$

5. $y = \dfrac{3}{4}x^2 + 3x + 5$

Skill Practice Answers

4. Vertex: $\left(9, \dfrac{3}{2}\right)$

5. Vertex: $(-2, 2)$

Section 14.2 Practice Exercises

Boost *your* GRADE at
mathzone.com!

MathZone
- Practice Problems
- Self-Tests
- NetTutor
- e-Professors
- Videos

Study Skills Exercise

1. Define the key terms.

    a. conic sections   **b. parabola**   **c. axis of symmetry**   **d. vertex**

Review Exercises

2. Determine the distance between the points $(1, 1)$ and $(2, -2)$.

3. Determine the distance from the origin to the point $(4, -3)$.

4. Determine the distance between the points $(11, 3)$ and $(5, 3)$.

For Exercises 5–8, identify the center and radius of the circle and then graph the circle.

5. $x^2 + (y + 1)^2 = 16$

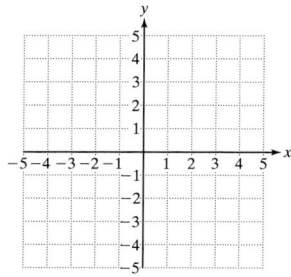

6. $(x - 3)^2 + y^2 = 4$

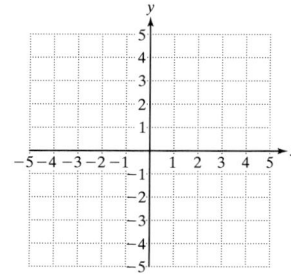

7. $(x - 3)^2 + (y + 3)^2 = 1$

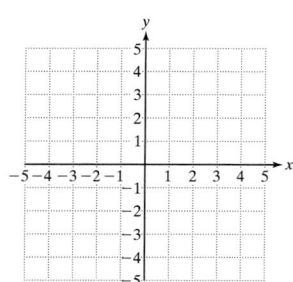

8. $(x + 2)^2 + (y + 4)^2 = 9$

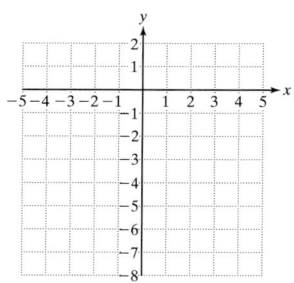

Concept 2: Parabola—Vertical Axis of Symmetry

For Exercises 9–16, use the equation of the parabola in standard form $y = a(x - h)^2 + k$ to determine the coordinates of the vertex and the equation of the axis of symmetry. Then graph the parabola.

9. $y = (x + 2)^2 + 1$

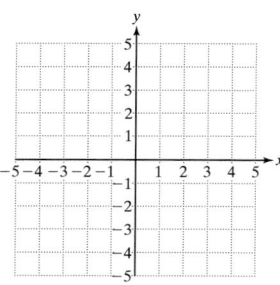

10. $y = (x - 1)^2 + 1$

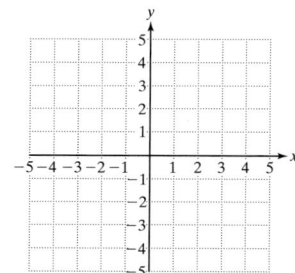

11. $y = x^2 - 4x + 3$

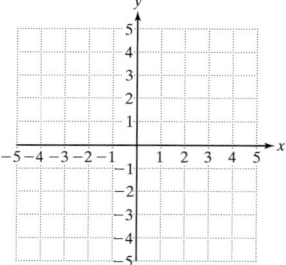

12. $y = x^2 + 6x + 5$

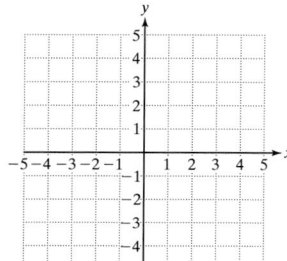

13. $y = -2x^2 + 8x$

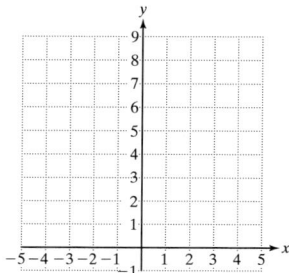

14. $y = -3x^2 - 6x$

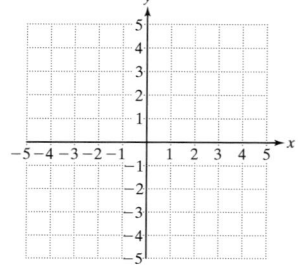

15. $y = -x^2 - 3x + 2$

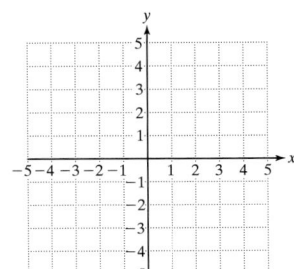

16. $y = -x^2 + x - 4$

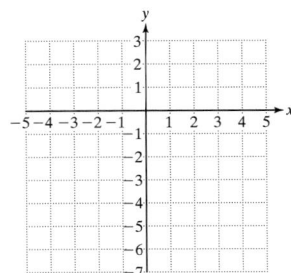

Concept 3: Parabola—Horizontal Axis of Symmetry

For Exercises 17–24, use the equation of the parabola in standard form $x = a(y - k)^2 + h$ to determine the coordinates of the vertex and the axis of symmetry. Then graph the parabola.

17. $x = y^2 - 3$

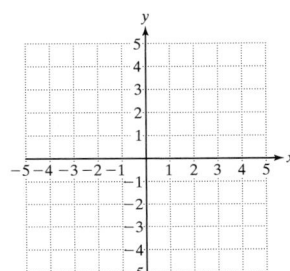

18. $x = y^2 + 1$

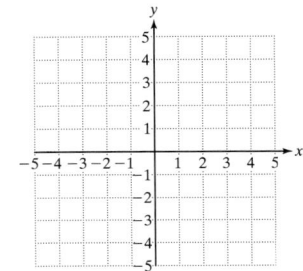

19. $x = -(y - 3)^2 - 3$

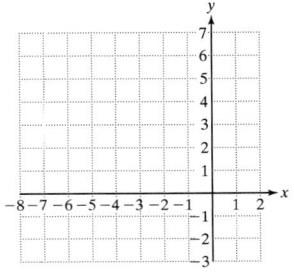

20. $x = -2(y + 2)^2 + 1$

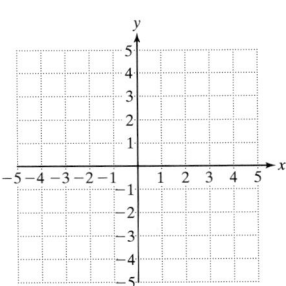

21. $x = -y^2 + 4y - 4$

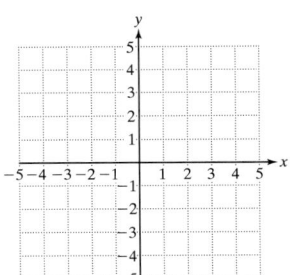

22. $x = -4y^2 - 4y - 2$

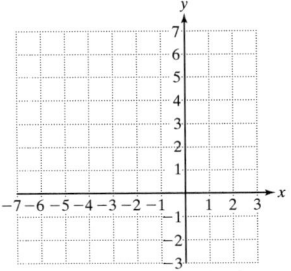

23. $x = y^2 - 2y + 2$

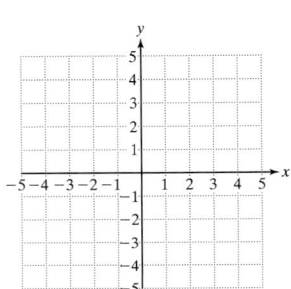

24. $x = y^2 + 4y + 1$

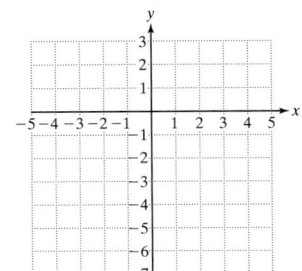

Concept 4: Vertex Formula

For Exercises 25–34, determine the vertex by using the vertex formula.

25. $y = x^2 - 4x + 3$

26. $y = x^2 + 6x - 2$

27. $x = y^2 + 2y + 6$

28. $x = y^2 - 8y + 3$

29. $y = -2x^2 + 8x$

30. $y = -3x^2 - 6x$

31. $y = x^2 - 3x + 2$

32. $y = x^2 + x - 4$

33. $x = -2y^2 + 16y + 1$

34. $x = -3y^2 - 6y + 7$

Mixed Exercises

35. Explain how to determine whether a parabola opens upward, downward, left, or right.

36. Explain how to determine whether a parabola has a vertical or horizontal axis of symmetry.

For Exercises 37–48, use the equation of the parabola first to determine whether the axis of symmetry is vertical or horizontal. Then determine if the parabola opens upward, downward, left, or right.

37. $y = (x - 2)^2 + 3$

38. $y = (x - 4)^2 + 2$

39. $y = -2(x + 1)^2 - 4$

40. $y = -3(x + 2)^2 - 1$

41. $x = y^2 + 4$

42. $x = y^2 - 2$

43. $x = -(y + 3)^2 + 2$

44. $x = -2(y - 1)^2 - 3$

45. $y = -2x^2 - 5$

46. $y = -x^2 + 3$

47. $x = 2y^2 + 3y - 2$

48. $x = y^2 - 5y + 1$

Section 14.3   The Ellipse and Hyperbola

1. Standard Form of an Equation of an Ellipse

In this section we will study the two remaining conic sections: the ellipse and the hyperbola. An **ellipse** is the set of all points (x, y) such that the sum of the distance between (x, y) and two distinct points is a constant. The fixed points are called the foci (plural of *focus*) of the ellipse.

To visualize an ellipse, consider the following application. Suppose Sonya wants to cut an elliptical rug from a rectangular rug to avoid a stain made by the family dog. She places two tacks along the center horizontal line. Then she ties the ends of a slack piece of rope to each tack. With the rope pulled tight, she traces out a curve. This curve is an ellipse, and the tacks are located at the foci of the ellipse (Figure 14-12).

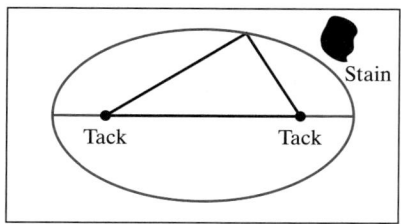

Figure 14-12

We will first graph ellipses that are centered at the origin.

Standard Form of an Equation of an Ellipse Centered at the Origin

An ellipse with the center at the origin has the equation $\dfrac{x^2}{a^2} + \dfrac{y^2}{b^2} = 1$ where

a and b are positive real numbers. In the standard form of the equation, the right side must equal 1.

To graph an ellipse centered at the origin, find the x- and y-intercepts.

To find the x-intercepts, let $y = 0$.

$$\frac{x^2}{a^2} + \frac{0}{b^2} = 1$$

$$\frac{x^2}{a^2} = 1$$

$$x^2 = a^2$$

$$x = \pm\sqrt{a^2}$$

$$x = \pm a$$

The x-intercepts are $(a, 0)$ and $(-a, 0)$.

To find the y-intercepts, let $x = 0$.

$$\frac{0}{a^2} + \frac{y^2}{b^2} = 1$$

$$\frac{y^2}{b^2} = 1$$

$$y^2 = b^2$$

$$y = \pm\sqrt{b^2}$$

$$y = \pm b$$

The y-intercepts are $(0, b)$ and $(0, -b)$.

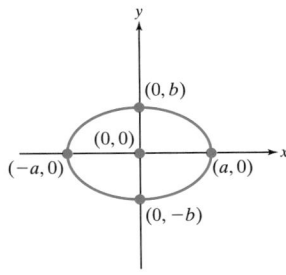

Example 1 Graphing an Ellipse

Graph the ellipse given by the equation. $\dfrac{x^2}{9} + \dfrac{y^2}{4} = 1$

Solution:

The equation can be written as $\dfrac{x^2}{3^2} + \dfrac{y^2}{2^2} = 1$; therefore, $a = 3$ and $b = 2$.

Graph the intercepts $(3, 0)$, $(-3, 0)$, $(0, 2)$, $(0, -2)$ and sketch the ellipse.

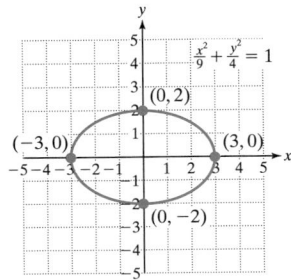

Skill Practice

1. Graph the ellipse given by the equation $x^2 + \dfrac{y^2}{16} = 1$.

Skill Practice Answers

1.

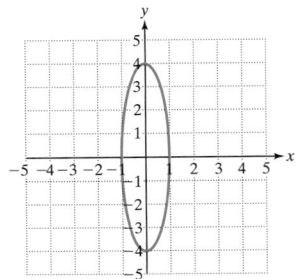

Example 2 **Graphing an Ellipse**

Graph the ellipse given by the equation. $25x^2 + y^2 = 25$

Solution:

First, to write the equation in standard form, divide both sides by 25:

$$x^2 + \frac{y^2}{25} = 1$$

The equation can then be written as $\frac{x^2}{1^2} + \frac{y^2}{5^2} = 1$; therefore, $a = 1$ and $b = 5$.

Graph the intercepts $(1, 0)$, $(-1, 0)$, $(0, 5)$, and $(0, -5)$ and sketch the ellipse.

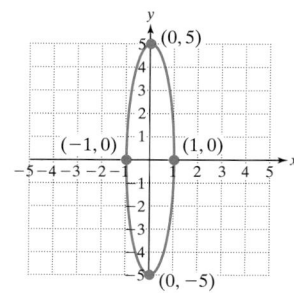

Skill Practice

2. Graph the ellipse given by the equation $25x^2 + 16y^2 = 400$.

A circle is a special case of an ellipse where $a = b$. Therefore, it is not surprising that we graph an ellipse centered at the point (h, k) in much the same way we graph a circle.

Example 3 **Graphing an Ellipse Whose Center is Not at the Origin**

Skill Practice Answers

2.

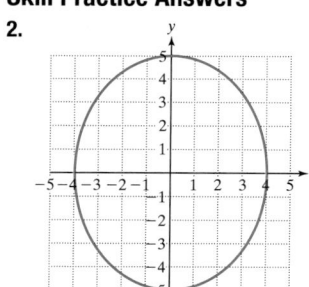

3.

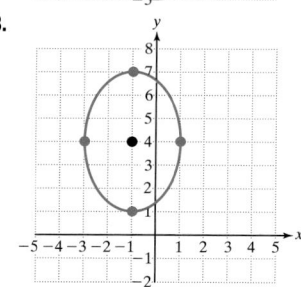

Graph the ellipse give by the equation. $\dfrac{(x - 1)^2}{16} + \dfrac{(y + 3)^2}{4} = 1$

Solution:

Just as we would find the center of a circle, we see that the center of the ellipse is $(1, -3)$. Now we can use the values of a and b to help us plot four strategic points to define the curve.

The equation can be written as $\dfrac{(x - 1)^2}{(4)^2} + \dfrac{(y + 3)^2}{(2)^2} = 1.$

From this, we have $a = 4$ and $b = 2$. To sketch the curve, locate the center at $(1, -3)$. Then move $a = 4$ units to the left and right of the center and plot two points. Similarly, move $b = 2$ units up and down from the center and plot two points.

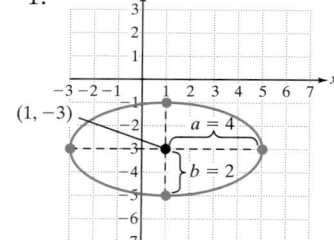

Skill Practice

3. Graph the ellipse. $\dfrac{(x + 1)^2}{4} + \dfrac{(y - 4)^2}{9} = 1$

2. Standard Forms of an Equation of a Hyperbola

A **hyperbola** is the set of all points (x, y) such that the *difference* of the distances between (x, y) and two distinct points is a constant. The fixed points are called the foci of the hyperbola. The graph of a hyperbola has two parts, called branches. Each part resembles a parabola but is a slightly different shape. A hyperbola has two vertices that lie on an axis of symmetry called the **transverse axis**. For the hyperbolas studied here, the transverse axis is either horizontal or vertical.

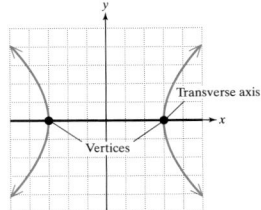 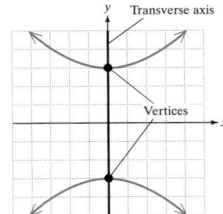

Standard Forms of an Equation of a Hyperbola with Center at the Origin

Let a and b represent positive real numbers.

Horizontal transverse axis:

The standard form of an equation of a hyperbola with a *horizontal transverse axis* and center at the origin is given by $\dfrac{x^2}{a^2} - \dfrac{y^2}{b^2} = 1$.

Note: The x-term is positive. The branches of the hyperbola open left and right.

Vertical transverse axis:

The standard form of an equation of a hyperbola with a *vertical transverse axis* and center at the origin is given by $\dfrac{y^2}{b^2} - \dfrac{x^2}{a^2} = 1$.

Note: The y-term is positive. The branches of the hyperbola open up and down.

In the standard forms of an equation of a hyperbola, the right side must equal 1.

To graph a hyperbola centered at the origin, first construct a reference rectangle. Draw this rectangle by using the points (a, b), $(-a, b)$, $(a, -b)$, and $(-a, -b)$. Asymptotes lie on the diagonals of the rectangle. The branches of the hyperbola are drawn to approach the asymptotes.

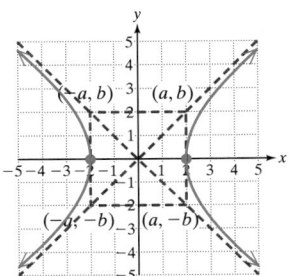 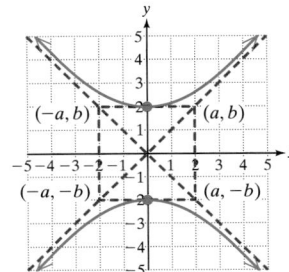

| **Example 4** | **Graphing a Hyperbola** |

Graph the hyperbola given by the equation. $\dfrac{x^2}{36} - \dfrac{y^2}{9} = 1$

    a. Determine whether the transverse axis is horizontal or vertical.

    b. Draw the reference rectangle and asymptotes.

    c. Graph the hyperbola and label the vertices.

Solution:

    a. Since the x-term is positive, the transverse axis is horizontal.

> **TIP:** In the equation
> $\dfrac{x^2}{36} - \dfrac{y^2}{9} = 1$, the
> x-term is positive.
> Therefore, the hyperbola
> opens in the x-direction
> (left/right).

    b. The equation can be written $\dfrac{x^2}{6^2} - \dfrac{y^2}{3^2} = 1$; therefore, $a = 6$ and $b = 3$. Graph the reference rectangle from the points $(6, 3)$, $(6, -3)$, $(-6, 3)$, $(-6, -3)$.

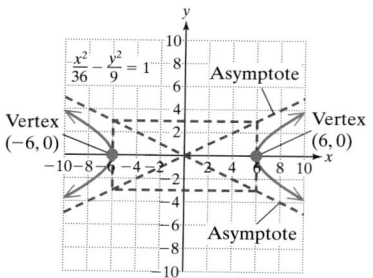

    c. The coordinates of the vertices are $(-6, 0)$ and $(6, 0)$.

| **Skill Practice** |

    4. Graph the hyperbola $\dfrac{x^2}{1} - \dfrac{y^2}{9} = 1$.

| **Example 5** | **Graphing a Hyperbola** |

Graph the hyperbola given by the equation. $y^2 - 4x^2 - 16 = 0$

    a. Write the equation in standard form to determine whether the transverse axis is horizontal or vertical.

    b. Draw the reference rectangle and asymptotes.

    c. Graph the hyperbola and label the vertices.

Solution:

    a. Isolate the variable terms and divide by 16:

$$\dfrac{y^2}{16} - \dfrac{x^2}{4} = 1.$$

Since the y-term is positive, the transverse axis is vertical.

    b. The equation can be written $\dfrac{y^2}{4^2} - \dfrac{x^2}{2^2} = 1$;

therefore, $a = 2$ and $b = 4$. Graph the reference rectangle from the points $(2, 4)$, $(2, -4)$, $(-2, 4)$, $(-2, -4)$.

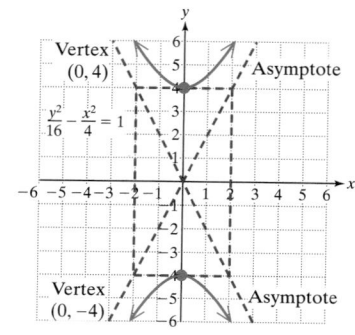

Skill Practice Answers

4.

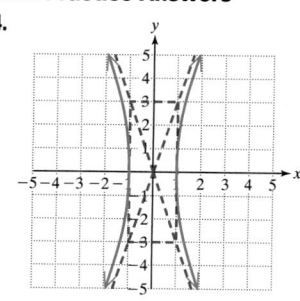

c. The coordinates of the vertices are $(0, 4)$ and $(0, -4)$.

Skill Practice

5. Graph the hyperbola $\dfrac{y^2}{4} - \dfrac{x^2}{9} = 1$.

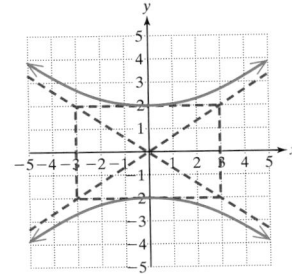

Section 14.3 Practice Exercises

Boost *your* GRADE at
mathzone.com!

MathZone

- Practice Problems
- Self-Tests
- NetTutor
- e-Professors
- Videos

Study Skills Exercise

1. Define the key terms.

  a. ellipse   **b. hyperbola**   **c. transverse axis of a hyperbola**

Review Exercises

2. Write the standard form of a circle with center at (h, k) and radius, r.

For Exercises 3–4, identify the center and radius of the circle.

3. $x^2 + y^2 - 16x + 12y = 0$

4. $x^2 + y^2 + 4x + 4y - 1 = 0$

For Exercises 5–6, identify the vertex and the axis of symmetry.

5. $y = 3(x + 3)^2 - 1$

6. $x = -\dfrac{1}{4}(y - 1)^2 - 6$

7. Write an equation for a circle whose center has coordinates $(\frac{1}{2}, \frac{5}{2})$ with radius equal to $\frac{1}{2}$.

8. Write the equation for the circle centered at the origin and with radius equal to $\frac{1}{8}$.

Concept 1: Standard Form of an Equation of an Ellipse

For Exercises 9–16, find the x- and y-intercepts and graph the ellipse.

 9. $\dfrac{x^2}{4} + \dfrac{y^2}{9} = 1$

10. $\dfrac{x^2}{16} + \dfrac{y^2}{25} = 1$

11. $\dfrac{x^2}{16} + \dfrac{y^2}{9} = 1$

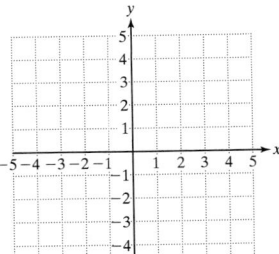

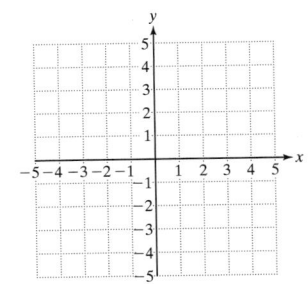

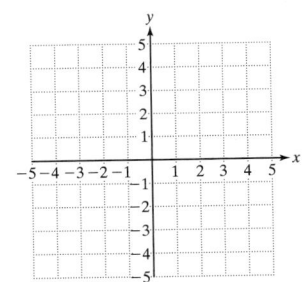

12. $\dfrac{x^2}{36} + \dfrac{y^2}{4} = 1$

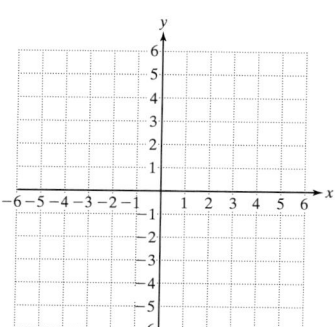

13. $4x^2 + y^2 = 4$

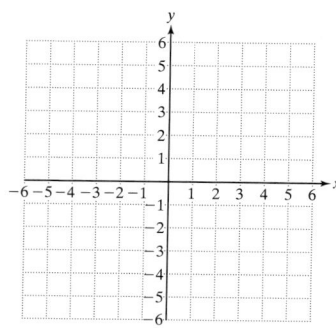

14. $9x^2 + y^2 = 36$

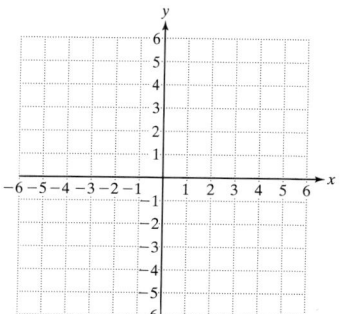

15. $x^2 + 25y^2 - 25 = 0$

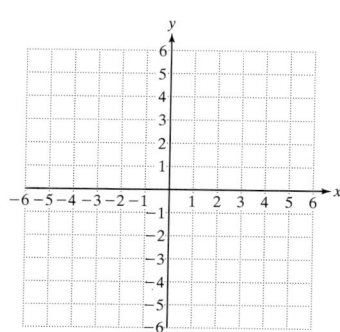

16. $4x^2 + 9y^2 = 144$

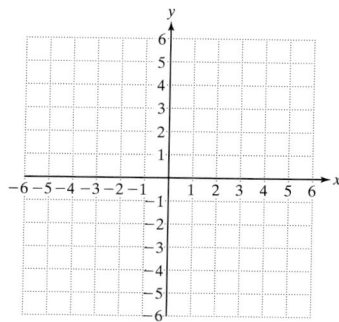

For Exercises 17–24, identify the center (h, k) of the ellipse. Then graph the ellipse.

17. $\dfrac{(x-4)^2}{4} + \dfrac{(y-5)^2}{9} = 1$

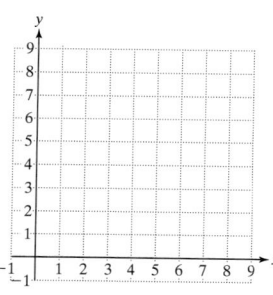

18. $\dfrac{(x+2)^2}{9} + \dfrac{(y-1)^2}{16} = 1$

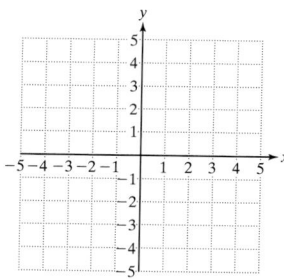

19. $\dfrac{(x+1)^2}{25} + \dfrac{(y-2)^2}{9} = 1$

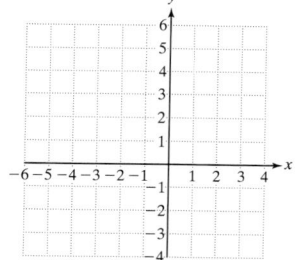

20. $\dfrac{(x-4)^2}{25} + \dfrac{(y+1)^2}{16} = 1$

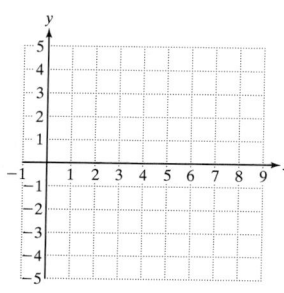

21. $\dfrac{(x-2)^2}{9} + (y+3)^2 = 1$

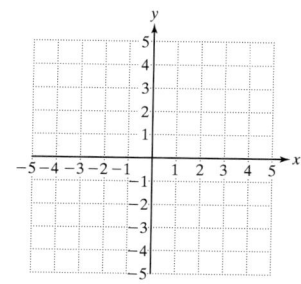

22. $(x-5)^2 + \dfrac{(y-3)^2}{4} = 1$

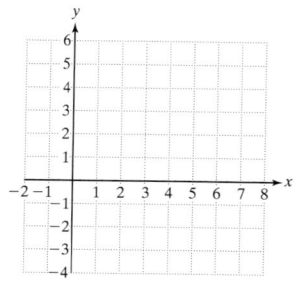

23. $\dfrac{x^2}{36} + \dfrac{(y-1)^2}{25} = 1$

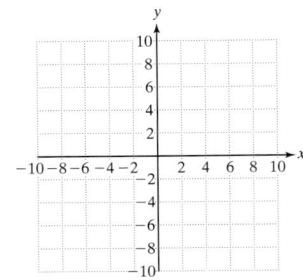

24. $\dfrac{(x+5)^2}{4} + \dfrac{y^2}{16} = 1$

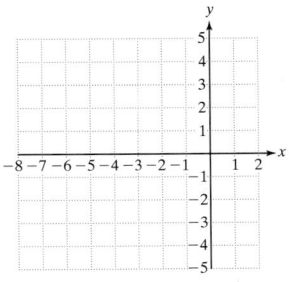

Concept 2: Standard Forms of an Equation of a Hyperbola

For Exercises 25–32, determine whether the transverse axis is horizontal or vertical.

25. $\dfrac{y^2}{6} - \dfrac{x^2}{18} = 1$

26. $\dfrac{y^2}{10} - \dfrac{x^2}{4} = 1$

27. $\dfrac{x^2}{20} - \dfrac{y^2}{15} = 1$

28. $\dfrac{x^2}{12} - \dfrac{y^2}{9} = 1$

29. $x^2 - y^2 = 12$

30. $x^2 - y^2 = 15$

31. $x^2 - 3y^2 = -9$

32. $2x^2 - y^2 = -10$

For Exercises 33–40, use the equation in standard form to graph the hyperbola. Label the vertices of the hyperbola.

33. $\dfrac{x^2}{25} - \dfrac{y^2}{16} = 1$

34. $\dfrac{x^2}{9} - \dfrac{y^2}{36} = 1$

35. $\dfrac{y^2}{4} - \dfrac{x^2}{4} = 1$

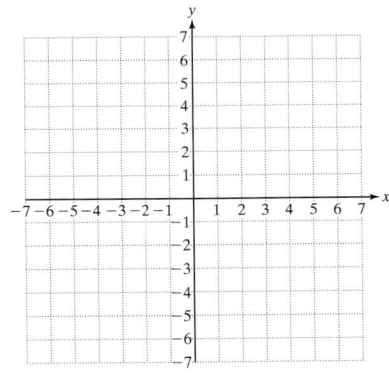

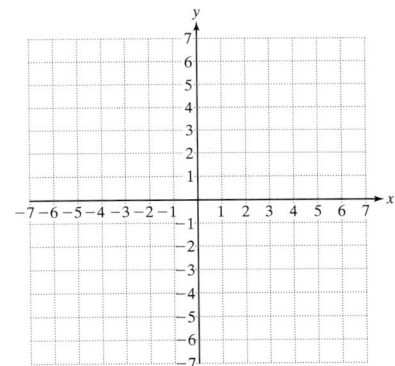

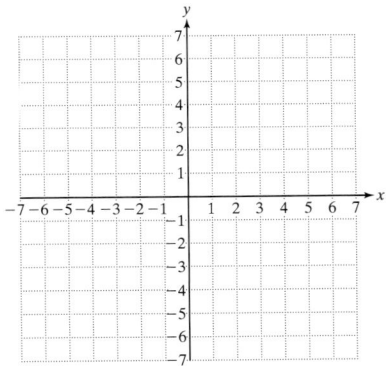

36. $\dfrac{y^2}{9} - \dfrac{x^2}{9} = 1$

37. $36x^2 - y^2 = 36$

38. $x^2 - 25y^2 = 25$

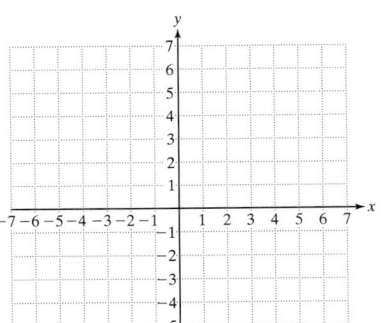

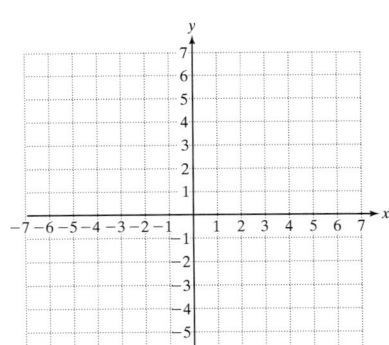

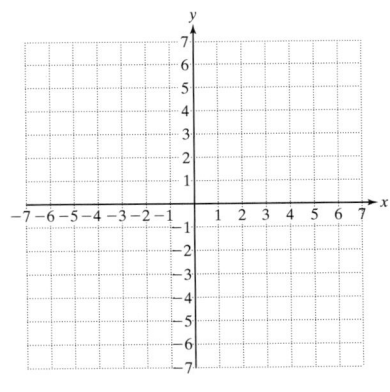

39. $y^2 - 4x^2 - 16 = 0$

40. $y^2 - 4x^2 - 36 = 0$

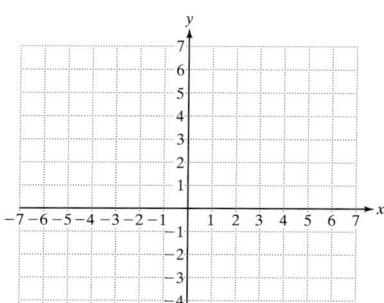

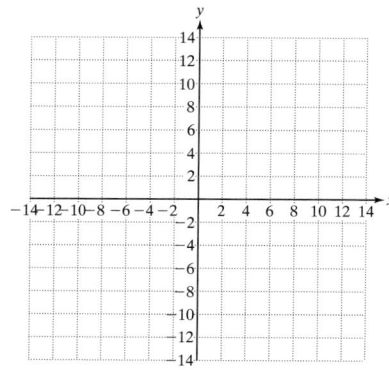

Mixed Exercises

For Exercises 41–52, determine if the equation represents an ellipse or a hyperbola.

41. $\dfrac{x^2}{6} - \dfrac{y^2}{10} = 1$

42. $\dfrac{x^2}{14} + \dfrac{y^2}{2} = 1$

43. $\dfrac{y^2}{4} + \dfrac{x^2}{16} = 1$

44. $\dfrac{x^2}{5} + \dfrac{y^2}{10} = 1$

45. $4x^2 + y^2 = 16$

46. $-3x^2 - 4y^2 = -36$

47. $-y^2 + 2x^2 = -10$

48. $x^2 - y^2 = -1$

49. $5x^2 + y^2 - 10 = 0$

50. $5x^2 - 3y^2 = 15$

51. $y^2 - 6x^2 = 6$

52. $3x^2 + 5y^2 = 15$

Expanding Your Skills

53. An arch for a tunnel is in the shape of a semiellipse. The distance between vertices is 120 ft, and the height to the top of the arch is 50 ft. Find the height of the arch 10 ft from the center. Round to the nearest foot.

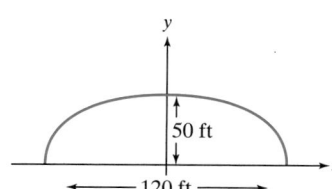

54. A bridge over a gorge is supported by an arch in the shape of a semiellipse. The length of the bridge is 400 ft, and the height is 100 ft. Find the height of the arch 50 ft from the center. Round to the nearest foot.

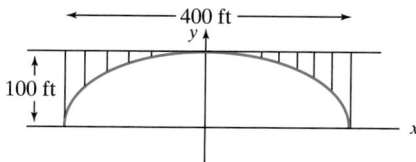

In Exercises 55–58, graph the hyperbola centered at (h, k) by following these guidelines. First locate the center. Then draw the reference rectangle relative to the center (h, k). Using the reference rectangle as a guide, sketch the hyperbola. Label the center and vertices.

55. $\dfrac{(x-1)^2}{9} - \dfrac{(y+2)^2}{4} = 1$

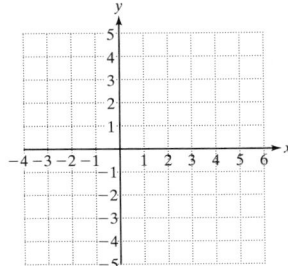

56. $\dfrac{(x+2)^2}{16} - \dfrac{(y-3)^2}{4} = 1$

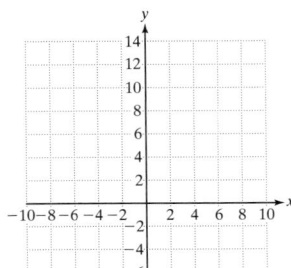

57. $\dfrac{(y-1)^2}{4} - (x+3)^2 = 1$

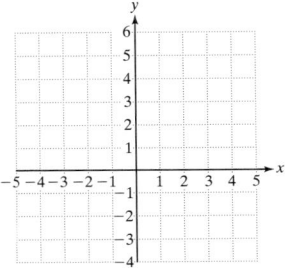

58. $(y+2)^2 - \dfrac{(x+3)^2}{4} = 1$

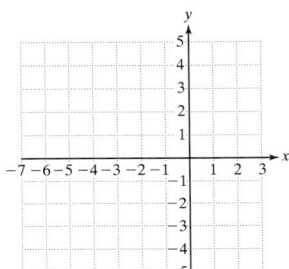

Nonlinear Systems of Equations in Two Variables

1. Solving Nonlinear Systems of Equations by the Substitution Method

Recall that a linear equation in two variables x and y is an equation that can be written in the form $ax + by = c$, where a and b are not both zero. In Sections 4.1–4.3, we solved systems of linear equations in two variables by using the graphing method, the substitution method, and the addition method. In this section, we will solve *nonlinear* systems of equations by using the same methods. A **nonlinear system of equations** is a system in which at least one of the equations is nonlinear.

Graphing the equations in a nonlinear system helps to determine the number of solutions and to approximate the coordinates of the solutions. The substitution method is used most often to solve a nonlinear system of equations analytically.

Concepts

1. Solving Nonlinear Systems of Equations by the Substitution Method

2. Solving Nonlinear Systems of Equations by the Addition Method

> **Example 1**   Solving a Nonlinear System of Equations

Given the system

$$x - 7y = -25$$

$$x^2 + y^2 = 25$$

a. Solve the system by graphing.

b. Solve the system by the substitution method.

Solution:

a. $x - 7y = -25$ is a line (the slope-intercept form is $y = \frac{1}{7}x + \frac{25}{7}$).

$x^2 + y^2 = 25$ is a circle centered at the origin with radius 5.

From Figure 14-13, we appear to have two solutions $(-4, 3)$ and $(3, 4)$.

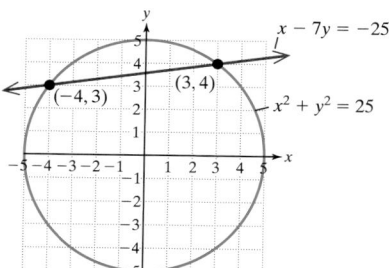

Figure 14-13

b. To use the substitution method, isolate one of the variables from one of the equations. We will solve for x in the first equation.

$$\boxed{A} \qquad x - 7y = -25 \xrightarrow{\text{Solve for } x} x = 7y - 25$$

$$\boxed{B} \qquad x^2 + y^2 = 25$$

$$\boxed{B} \quad (7y - 25)^2 + y^2 = 25 \qquad \text{Substitute } (7y - 25) \text{ for } x \text{ in the second equation.}$$

$$49y^2 - 350y + 625 + y^2 = 25 \qquad \text{The resulting equation is quadratic in } y.$$

$$50y^2 - 350y + 600 = 0 \qquad \text{Set the equation equal to zero.}$$

$$50(y^2 - 7y + 12) = 0 \qquad \text{Factor.}$$

$$50(y - 3)(y - 4) = 0$$

$$y = 3 \quad \text{or} \quad y = 4$$

For each value of y, find the corresponding x value from the equation $x = 7y - 25$.

$y = 3$: $x = 7(3) - 25 = -4$ The solution is $(-4, 3)$.

$y = 4$: $x = 7(4) - 25 = 3$ The solution is $(3, 4)$.
(See Figure 14-13.)

1. Given the system

$$2x + y = 5$$
$$x^2 + y^2 = 50$$

a. Solve the system by graphing.

b. Solve the system by the substitution method.

Example 2 **Solving a Nonlinear System by the Substitution Method**

Given the system

$$y = \sqrt{x}$$
$$x^2 + y^2 = 20$$

a. Sketch the graphs.

b. Solve the system by the substitution method.

Solution:

a. $y = \sqrt{x}$ is one of the six basic functions graphed in Section 8.3.

$x^2 + y^2 = 20$ is a circle centered at the origin with radius $\sqrt{20} \approx 4.5$.

From Figure 14-14, we see that there is one solution.

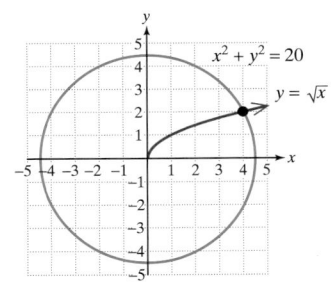

Figure 14-14

b. To use the substitution method, we will substitute $y = \sqrt{x}$ into equation $\boxed{B}$.

$\boxed{A}$ $\qquad y = \sqrt{x}$

$\boxed{B}$ $\qquad x^2 + y^2 = 20$

$\boxed{B}$ $\quad x^2 + (\sqrt{x})^2 = 20$ $\qquad$ Substitute $y = \sqrt{x}$ into the second equation.

$\qquad\qquad x^2 + x = 20$

$\qquad\quad x^2 + x - 20 = 0$ $\qquad$ Set the second equation equal to zero.

$\qquad (x + 5)(x - 4) = 0$ $\qquad$ Factor.

$x \cancel{=} -5 \quad$ or $\quad x = 4$ $\qquad$ Reject $x = -5$ because it is not in the domain of $y = \sqrt{x}$.

Substitute $x = 4$ into the equation $y = \sqrt{x}$.

If $x = 4$, then $y = \sqrt{4} = 2$. $\qquad$ The solution is $(4, 2)$.

Skill Practice Answers

1a.

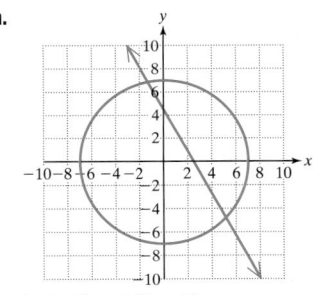

b. $(-1, 7)$ and $(5, -5)$

Skill Practice

2. Given the system

$$y = \sqrt{2x}$$
$$x^2 + y^2 = 8$$

a. Sketch the graphs.

b. Solve the system by using the substitution method.

Calculator Connections

Graph the equations from Example 2 to confirm your solution to the system of equations. Use an *Intersect* feature or *Zoom* and *Trace* to approximate the point of intersection. Recall that the circle must be entered into the calculator as two functions:

$$Y_1 = \sqrt{20 - x^2}$$
$$Y_2 = -\sqrt{20 - x^2}$$
$$Y_3 = \sqrt{x}$$

Example 3　**Solving a Nonlinear System by Using the Substitution Method**

Solve the system by using the substitution method.

$$y = \sqrt[3]{x}$$
$$y = x$$

Calculator Connections

Graph the equations $y = \sqrt[3]{x}$ and $y = x$ to support the solutions to Example 3.

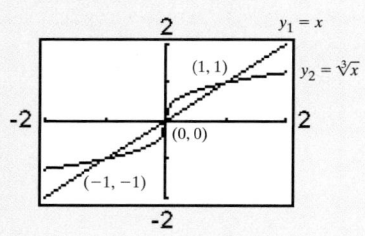

Skill Practice Answers

2a.

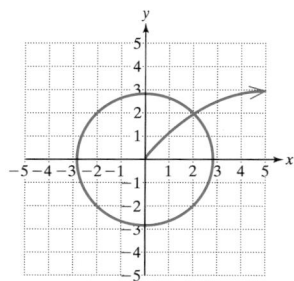

b. $(2, 2)$

Solution:

$\boxed{A}$　$y = \sqrt[3]{x}$

$\boxed{B}$　$y = x$

$\sqrt[3]{x} = x$　　Because y is isolated in both equations, we can equate the expressions for y.

$(\sqrt[3]{x})^3 = (x)^3$　　To solve the radical equation, raise both sides to the third power.

$x = x^3$　　This is a third-degree polynomial equation.

$0 = x^3 - x$　　Set the equation equal to zero.

$0 = x(x^2 - 1)$　　Factor out the GCF.

$0 = x(x + 1)(x - 1)$　　Factor completely.

$x = 0$　　or　　$x = -1$　　or　　$x = 1$

For each value of x, find the corresponding y-value from either original equation. We will use equation $\boxed{B}$: $y = x$.

If $x = 0$, then $y = 0$.　　The solution is $(0, 0)$.

If $x = -1$, then $y = -1$.　　The solution is $(-1, -1)$.

If $x = 1$, then $y = 1$.　　The solution is $(1, 1)$.

3. Solve the system by using the substitution method.

$$y = \sqrt[3]{9x}$$

$$y = x$$

2. Solving Nonlinear Systems of Equations by the Addition Method

The substitution method is used most often to solve a system of nonlinear equations. In some situations, however, the addition method offers an efficient means of finding a solution. Example 4 demonstrates that we can eliminate a variable from both equations provided the terms containing that variable are *like* terms.

Example 4 Solving a Nonlinear System of Equations by the Addition Method

Solve the system. $2x^2 + y^2 = 17$

$$x^2 + 2y^2 = 22$$

Solution:

$\boxed{A}$ $2x^2 + y^2 = 17$ Notice that the y^2-terms are *like* in each equation.

$\boxed{B}$ $x^2 + 2y^2 = 22$ To eliminate the y^2-terms, multiply the first equation by -2.

$\boxed{A}$ $2x^2 + y^2 = 17$ $\xrightarrow{\text{Multiply by } -2.}$ $-4x^2 - 2y^2 = -34$

$\boxed{B}$ $x^2 + 2y^2 = 22$ $\longrightarrow$ $\underline{\quad x^2 + 2y^2 = \quad 22}$

$\qquad\qquad\qquad\qquad\qquad\qquad -3x^2 \qquad\quad = -12$ Eliminate the y^2 term.

$$\frac{-3x^2}{-3} = \frac{-12}{-3}$$

$$x^2 = 4$$

$$x = \pm 2$$

TIP: In Example 4, the x^2-terms are also *like* in both equations. We could have eliminated the x^2-terms by multiplying equation $\boxed{B}$ by -2.

Substitute each value of x into one of the original equations to solve for y. We will use equation $\boxed{A}$: $2x^2 + y^2 = 17$.

$x = 2$: $\qquad \boxed{A} \quad 2(2)^2 + y^2 = 17$

$\qquad\qquad\qquad\qquad 8 + y^2 = 17$

$\qquad\qquad\qquad\qquad\quad y^2 = 9$

$\qquad\qquad\qquad\qquad\quad y = \pm 3$ $\qquad$ The solutions are $(2, 3)$ and $(2, -3)$.

$x = -2$: $\qquad \boxed{A} \quad 2(-2)^2 + y^2 = 17$

$\qquad\qquad\qquad\qquad 8 + y^2 = 17$

$\qquad\qquad\qquad\qquad\quad y^2 = 9$

$\qquad\qquad\qquad\qquad\quad y = \pm 3$ $\qquad$ The solutions are $(-2, 3)$ and $(-2, -3)$.

Skill Practice

4. Solve the system by using the addition method.

$$x^2 - y^2 = 24$$
$$3x^2 + y^2 = 76$$

Calculator Connections

The solutions to Example 4 can be checked from the graphs of the equations.

For the equation $2x^2 + y^2 = 17$, we have $y = \pm\sqrt{17 - 2x^2}$

For the equation $x^2 + 2y^2 = 22$, we have $y = \pm\sqrt{\dfrac{22 - x^2}{2}}$

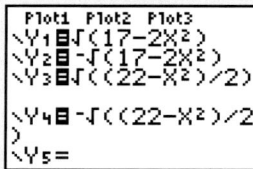

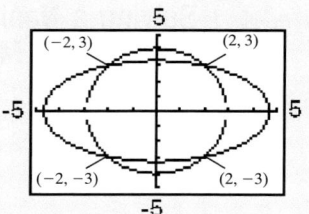

TIP: It is important to note that the addition method can be used only if two equations share a pair of *like* terms. The substitution method is effective in solving a wider range of systems of equations. The system in Example 4 could also have been solved by using substitution.

| A | $2x^2 + y^2 = 17$ $\xrightarrow{\text{Solve for } y^2}$ $y^2 = 17 - 2x^2$ |

| B | $x^2 + 2y^2 = 22$ |

| B | $x^2 + 2(17 - 2x^2) = 22$ $x = 2$: $y^2 = 17 - 2(2)^2$

$x^2 + 34 - 4x^2 = 22$ $y^2 = 9$

$-3x^2 = -12$ $y = \pm 3$ The solutions are (2, 3) and (2, −3).

$x^2 = 4$

$x = \pm 2$ ⟶ $x = -2$: $y^2 = 17 - 2(-2)^2$

$y^2 = 9$

$y = \pm 3$ The solutions are (−2, 3) and (−2, −3).

Skill Practice Answers

4. (5, 1); (5, −1); (−5, 1); (−5, −1)

Section 14.4	**Practice Exercises**

Study Skills Exercise

1. Define the key term **nonlinear system of equations**.

Review Exercises

2. Write the distance formula between two points (x_1, y_1) and (x_2, y_2) from memory.

3. Find the distance between the two points $(8, -1)$ and $(1, -8)$.

4. Write an equation representing the set of all points 2 units from the point $(-1, 1)$.

5. Write an equation representing the set of all points 8 units from the point $(-5, 3)$.

For Exercises 6–13, determine if the equation represents a parabola, circle, ellipse, or hyperbola.

6. $x^2 + y^2 = 15$

7. $\dfrac{x^2}{4} - \dfrac{y^2}{2} = 1$

8. $y = (x - 6)^2 + 4$

9. $\dfrac{(x + 1)^2}{2} + \dfrac{(y + 1)^2}{5} = 1$

10. $\dfrac{y^2}{3} - \dfrac{x^2}{3} = 1$

11. $3x^2 + 3y^2 = 1$

12. $\dfrac{x^2}{9} + \dfrac{y^2}{12} = 1$

13. $x = (y + 2)^2 - 5$

Concept 1: Solving Nonlinear Systems of Equations by the Substitution Method

For Exercises 14–21, use sketches to explain.

14. How many points of intersection are possible between a line and a parabola?

15. How many points of intersection are possible between a line and a circle?

16. How many points of intersection are possible between two distinct circles?

17. How many points of intersection are possible between two distinct parabolas of the form $y = ax^2 + bx + c$, $a \neq 0$?

18. How many points of intersection are possible between a circle and a parabola?

19. How many points of intersection are possible between two distinct lines?

20. How many points of intersection are possible with an ellipse and a hyperbola?

21. How many points of intersection are possible with an ellipse and a parabola?

For Exercises 22–27, sketch each system of equations. Then solve the system by the substitution method.

22. $\quad y = x + 3$

$\quad\quad x^2 + y = 9$

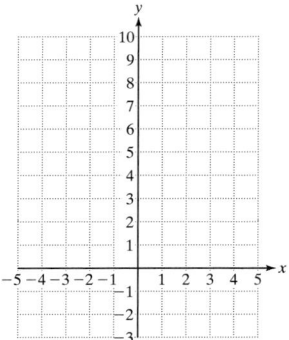

23. $\quad y = x - 2$

$\quad\quad x^2 + y = 4$

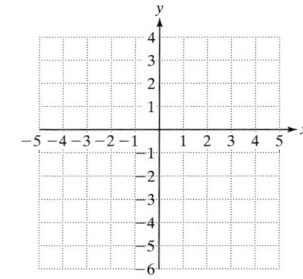

24. $x^2 + y^2 = 1$

$\quad\quad y = x + 1$

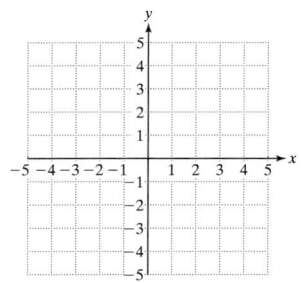

25. $x^2 + y^2 = 25$

$\quad\quad y = 2x$

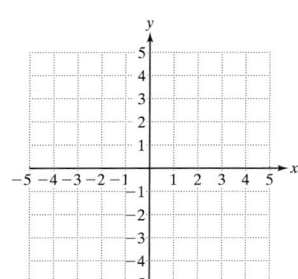

26. $x^2 + y^2 = 6$

$\quad\quad y = x^2$

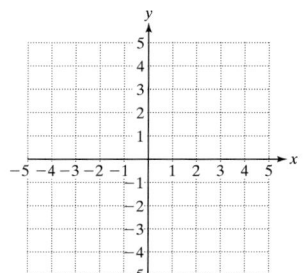

27. $x^2 + y^2 = 12$

$\quad\quad y = x^2$

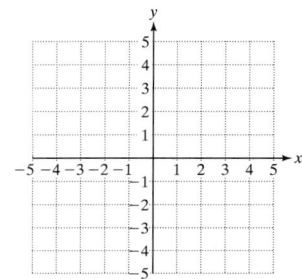

For Exercises 28–36, solve the system by the substitution method.

28. $x^2 + y^2 = 20$

$\quad\quad y = \sqrt{x}$

29. $x^2 + y^2 = 30$

$\quad\quad y = \sqrt{x}$

30. $y = x^2$

$\quad\quad y = -\sqrt{x}$

31. $y = -x^2$

$\quad\quad y = -\sqrt{x}$

32. $y = x^2$

$\quad\quad y = (x - 3)^2$

33. $y = (x + 4)^2$

$\quad\quad y = x^2$

34. $y = x^2 + 6x$

$\quad\quad y = 4x$

35. $y = 3x^2 - 6x$

$\quad\quad y = 3x$

36. $x^2 - 5x + y = 0$

$\quad\quad y = 3x + 1$

Concept 2: Solving Nonlinear Systems of Equations by the Addition Method

For Exercises 37–50, solve the system of nonlinear equations by the addition method.

37. $x^2 + y^2 = 13$

$\quad\quad x^2 - y^2 = 5$

38. $4x^2 - y^2 = 4$

$\quad\quad 4x^2 + y^2 = 4$

39. $9x^2 + 4y^2 = 36$

$\quad\quad x^2 + y^2 = 9$

40. $x^2 + y^2 = 4$

$\quad\quad 2x^2 + y^2 = 8$

41. $3x^2 + 4y^2 = 16$

$\quad\quad 2x^2 - 3y^2 = 5$

42. $2x^2 - 5y^2 = -2$

$\quad\quad 3x^2 + 2y^2 = 35$

43. $y = x^2 - 2$

$y = -x^2 + 2$

44. $y = x^2$

$y = -x^2 + 8$

45. $\dfrac{x^2}{4} + \dfrac{y^2}{9} = 1$

$x^2 + y^2 = 4$

46. $\dfrac{x^2}{16} + \dfrac{y^2}{4} = 1$

$x^2 + y^2 = 4$

47. $x^2 + 6y^2 = 9$

$\dfrac{x^2}{9} + \dfrac{y^2}{12} = 1$

48. $\dfrac{x^2}{10} + \dfrac{y^2}{10} = 1$

$2x^2 + y^2 = 11$

49. $x^2 - xy = -4$

$2x^2 - xy = 12$

50. $x^2 - xy = 3$

$2x^2 + xy = 6$

Expanding Your Skills

51. The sum of two numbers is 7. The sum of the squares of the numbers is 25. Find the numbers.

52. The sum of the squares of two numbers is 100. The sum of the numbers is 2. Find the numbers.

53. The sum of the squares of two numbers is 32. The difference of the squares of the numbers is 18. Find the numbers.

54. The sum of the squares of two numbers is 24. The difference of the squares of the numbers is 8. Find the numbers.

Graphing Calculator Exercises

For Exercises 55–58, use the *Intersect* feature or *Zoom* and *Trace* to approximate the solutions to the system.

55. $y = x + 3$ (Exercise 22)

$x^2 + y = 9$

56. $y = x - 2$ (Exercise 23)

$x^2 + y = 4$

57. $y = x^2$ (Exercise 30)

$y = -\sqrt{x}$

58. $y = -x^2$ (Exercise 31)

$y = -\sqrt{x}$

For Exercises 59–60, graph the system on a square viewing window. What can be said about the solution to the system?

59. $x^2 + y^2 = 4$

$y = x^2 + 3$

60. $x^2 + y^2 = 16$

$y = -x^2 - 5$

Section 14.5

Nonlinear Inequalities and Systems of Inequalities

Concepts

1. Nonlinear Inequalities in Two Variables
2. Systems of Nonlinear Inequalities in Two Variables

1. Nonlinear Inequalities in Two Variables

In Section 10.5 we graphed the solution sets to linear inequalities in two variables, such as $y \le 2x + 1$. This involved graphing the related equation (a line in the xy-plane) and then shading the appropriate region above or below the line. See Figure 14-15. In this section, we will employ the same technique to solve nonlinear inequalities in two variables.

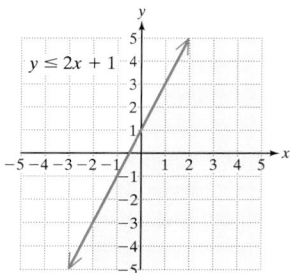

Figure 14-15

> **Example 1**   Graphing a Nonlinear Inequality in Two Variables

Graph the solution set of the inequality $x^2 + y^2 < 16$.

Solution:

The related equation $x^2 + y^2 = 16$ is a circle of radius 4, centered at the origin. Graph the related equation by using a dashed curve because the points satisfying the equation $x^2 + y^2 = 16$ are not part of the solution to the strict inequality $x^2 + y^2 < 16$. See Figure 14-16.

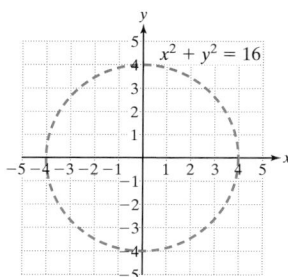

Figure 14-16

Notice that the dashed curve separates the xy-plane into two regions, one "inside" the circle, the other "outside" the circle. Select a test point from each region and test the point in the original inequality.

Test Point "Inside": (0, 0)

$$x^2 + y^2 < 16$$
$$(0)^2 + (0)^2 \overset{?}{<} 16$$
$$0 \overset{?}{<} 16 \quad \text{True}$$

Test Point "Outside": (4, 4)

$$x^2 + y^2 < 16$$
$$(4)^2 + (4)^2 \overset{?}{<} 16$$
$$32 \overset{?}{<} 16 \quad \text{False}$$

The inequality $x^2 + y^2 < 16$ is true at the test point $(0, 0)$. Therefore, the solution set is the region "inside" the circle. See Figure 14-17.

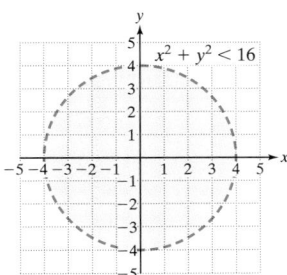

Figure 14-17

Skill Practice

1. Graph the solution set of the inequality $x^2 + y^2 \geq 9$.

Example 2 **Graphing a Nonlinear Inequality in Two Variables**

Graph the solution set of the inequality $9y^2 \geq 36 + 4x^2$.

Solution:

First graph the related equation $9y^2 = 36 + 4x^2$. Notice that the equation can be written in the standard form of a hyperbola.

$$9y^2 = 36 + 4x^2$$

$$9y^2 - 4x^2 = 36 \qquad \text{Subtract } 4x^2 \text{ from both sides.}$$

$$\frac{y^2}{4} - \frac{x^2}{9} = 1 \qquad \text{Divide both sides by 36.}$$

Graph the hyperbola as a solid curve, because the original inequality includes equality. See Figure 14-18.

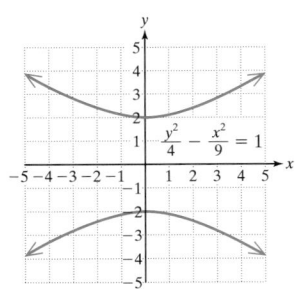

Figure 14-18

The hyperbola divides the xy-plane into three regions: a region above the upper branch, a region between the branches, and a region below the lower branch. Select a test point from each region.

$$9y^2 \geq 36 + 4x^2$$

Skill Practice Answers

1.

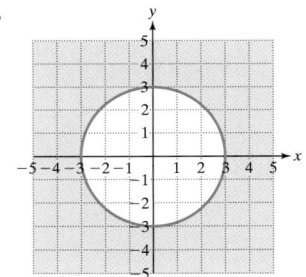

Test: (0, 3)

$$9(3)^2 \overset{?}{\geq} 36 + 4(0)^2$$

$$81 \overset{?}{\geq} 36 \quad \text{True}$$

Test: (0, 0)

$$9(0)^2 \overset{?}{\geq} 36 + 4(0)^2$$

$$0 \overset{?}{\geq} 36 \quad \text{False}$$

Test: (0, −3)

$$9(-3)^2 \overset{?}{\geq} 36 + 4(0)^2$$

$$81 \overset{?}{\geq} 36 \quad \text{True}$$

Shade the regions above the top branch and below the bottom branch of the hyperbola. See Figure 14-19.

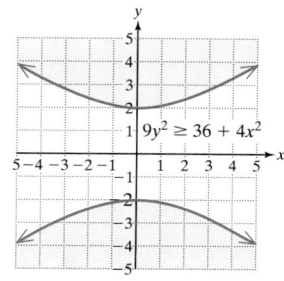

Figure 14-19

> **Skill Practice**
>
> **2.** Graph the solution set of the inequality $9x^2 < 144 - 16y^2$.

2. Systems of Nonlinear Inequalities in Two Variables

In Section 14.4 we solved systems of nonlinear equations in two variables. The solution set for such a system is the set of ordered pairs that satisfy both equations simultaneously. We will now solve systems of nonlinear inequalities in two variables. Similarly, the solution set is the set of all ordered pairs that simultaneously satisfy each inequality. To solve a system of inequalities, we will graph the solution to each individual inequality and then take the intersection of the solution sets.

> **Example 3**   Graphing a System of Nonlinear Inequalities in Two Variables

Graph the solution set.

$$y > e^x$$

$$y < -x^2 + 4$$

Solution:

The solution to $y > e^x$ is the set of points above the curve $y = e^x$. See Figure 14-20. The solution to $y < -x^2 + 4$ is the set of points below the parabola $y = -x^2 + 4$. See Figure 14-21.

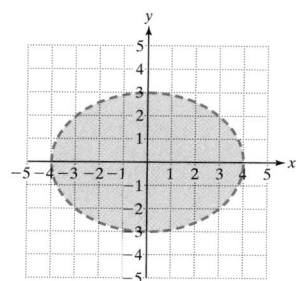
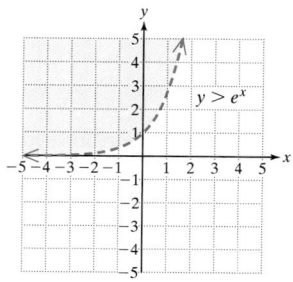

Figure 14-20

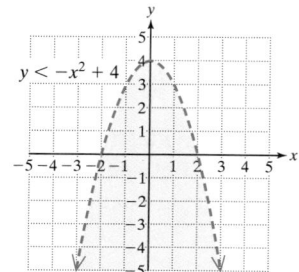

Figure 14-21

The solution to the system of inequalities is the intersection of the solution sets of the individual inequalities. See Figure 14-22.

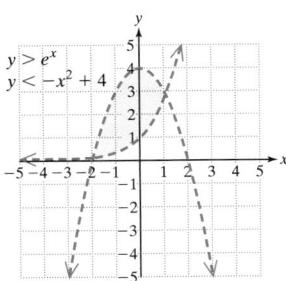

$y > e^x$
$y < -x^2 + 4$

Figure 14-22

Skill Practice Answers

3.

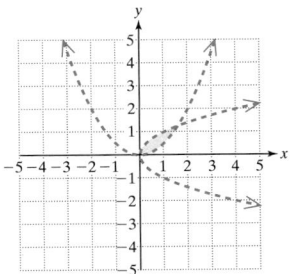

Skill Practice

3. Graph the solution set.

$$y > \frac{1}{2}x^2$$

$$x > y^2$$

Section 14.5   Practice Exercises

Boost *your* GRADE at mathzone.com!

MathZone

- Practice Problems
- Self-Tests
- NetTutor
- e-Professors
- Videos

Review Exercises

For Exercises 1–11, match the equation with its graph.

1. $y = \left(\dfrac{1}{3}\right)^x$

2. $y = 4x - 1$

3. $y = -4x^2$

4. $y = e^x$

5. $y = x^3$

6. $\dfrac{x^2}{4} + \dfrac{y^2}{9} = 1$

7. $\dfrac{x^2}{4} - \dfrac{y^2}{9} = 1$

8. $y = \dfrac{1}{x}$

9. $y = \log_2(x)$

10. $y = \sqrt{x}$

11. $(x + 2)^2 + (y - 1)^2 = 4$

a.

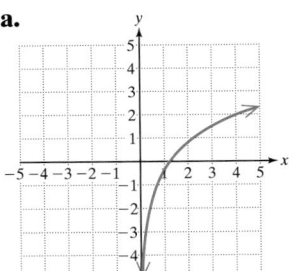

b.

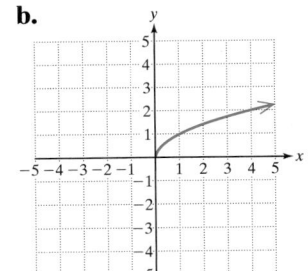

c.

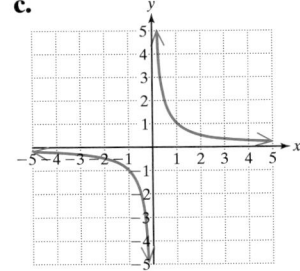

d.

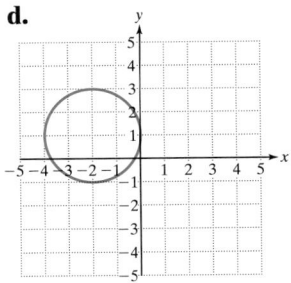

e. 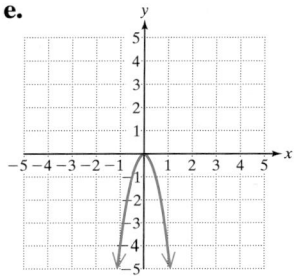   **f.** 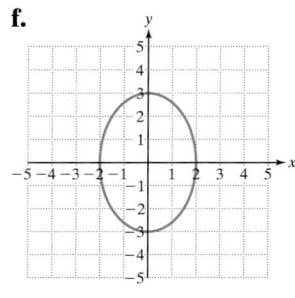   **g.** 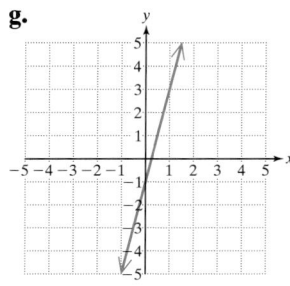   **h.**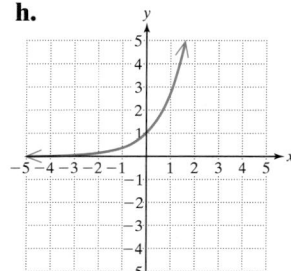

i. 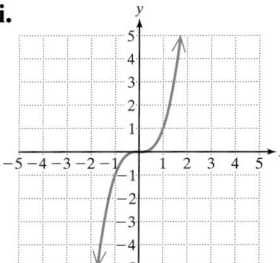   **j.** 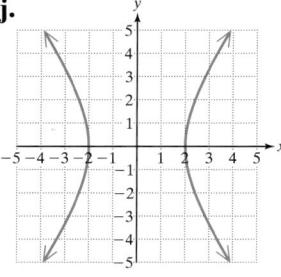   **k.**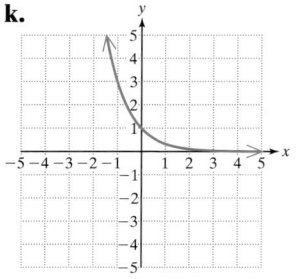

Concept 1: Nonlinear Inequalities in Two Variables

12. True or false? The point $(1, 1)$ satisfies the inequality $-x^2 + y^3 > 1$.

13. True or false? The point $(4, -2)$ satisfies the inequality $4x^2 - 2x + 1 + y^2 < 3$.

14. True or false? The point $(5, 4)$ satisfies the system of inequalities.

$$\frac{x^2}{36} + \frac{y^2}{25} < 1$$
$$x^2 + y^2 \geq 4$$

15. True or false? The point $(1, -2)$ satisfies the system of inequalities.

$$y < x^2$$
$$y > x^2 - 4$$

16. True or false? The point $(-3, 5)$ satisfies the system of inequalities.

$$(x + 3)^2 + (y - 5)^2 \leq 2$$
$$y > x^2$$

17. a. Graph the solution set of $x^2 + y^2 \leq 9$.

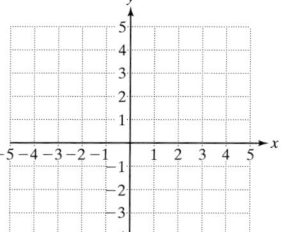

18. a. Graph the solution set of $\dfrac{x^2}{4} + \dfrac{y^2}{9} \geq 1$.

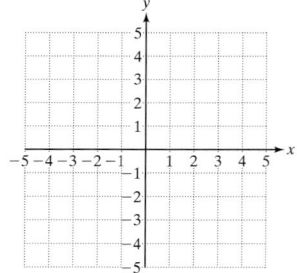

b. Describe the solution set for the inequality $x^2 + y^2 \geq 9$.

c. Describe the solution set of the equation $x^2 + y^2 = 9$

b. Describe the solution set for the inequality $\dfrac{x^2}{4} + \dfrac{y^2}{9} \leq 1$.

c. Describe the solution set for the equation $\dfrac{x^2}{4} + \dfrac{y^2}{9} = 1$

19. a. Graph the solution set of $y \geq x^2 + 1$.

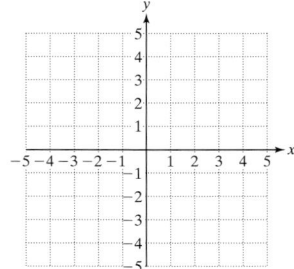

b. How would the solution change for the strict inequality $y > x^2 + 1$?

20. a. Graph the solution set of $\dfrac{x^2}{4} - \dfrac{y^2}{9} \leq 1$.

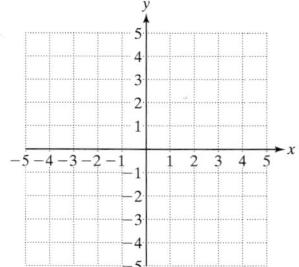

b. How would the solution change for the strict inequality $\dfrac{x^2}{4} - \dfrac{y^2}{9} < 1$?

For Exercises 21–36, graph the solution set.

21. $2x + y \geq 1$

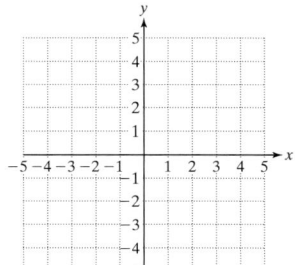

22. $3x + 2y \geq 6$

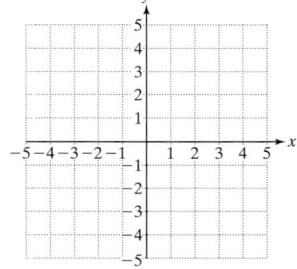

23. $x \leq y^2$

24. $y \leq -x^2$

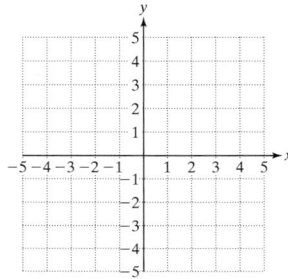

25. $(x - 1)^2 + (y + 2)^2 > 9$

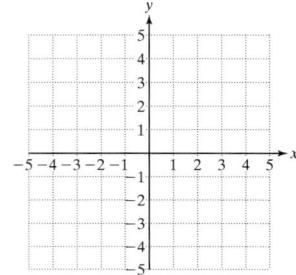

26. $(x + 1)^2 + (y - 4)^2 > 1$

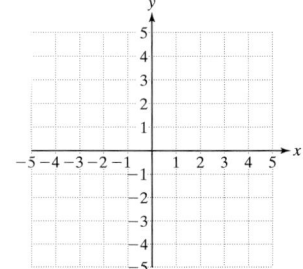

27. $x + y^2 \geq 4$

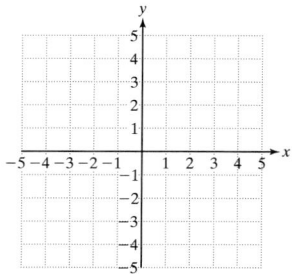

28. $x^2 + 2x + y - 1 \leq 0$

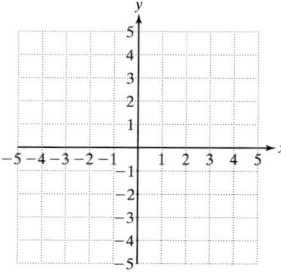

29. $9x^2 - y^2 > 9$

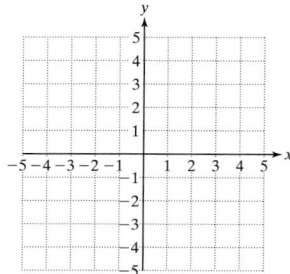

30. $y^2 - 4x^2 \leq 4$

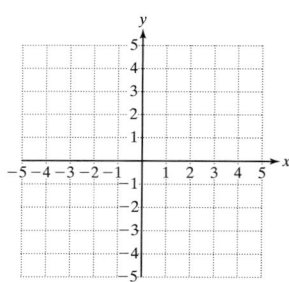

31. $x^2 + 16y^2 \leq 16$

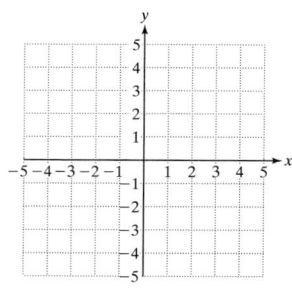

32. $4x^2 + y^2 \leq 4$

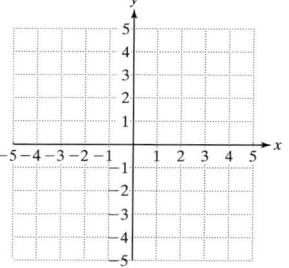

33. $y \leq \ln x$

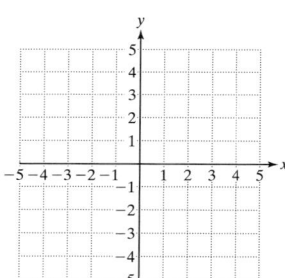

34. $y \leq \log x$

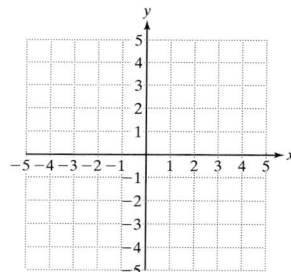

35. $y > 5^x$

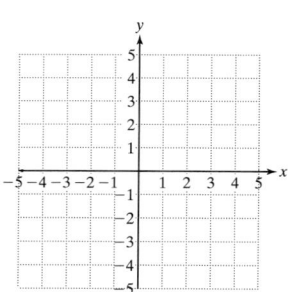

36. $y \geq \left(\dfrac{1}{3}\right)^x$

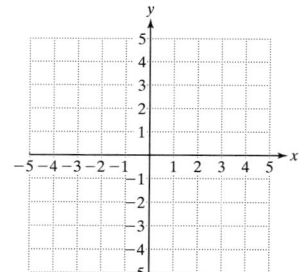

Concept 2: Systems of Nonlinear Inequalities in Two Variables

For Exercises 37–50, graph the solution set to the system of inequalities.

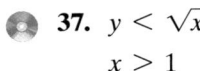 **37.** $y < \sqrt{x}$
$x > 1$

38. $y \geq \sqrt{x}$
$x \geq 0$

39. $\dfrac{x^2}{36} + \dfrac{y^2}{25} < 1$
$x^2 + y^2 \geq 4$

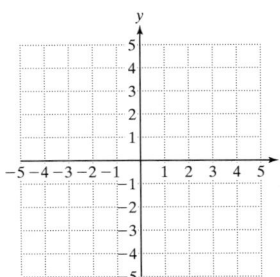

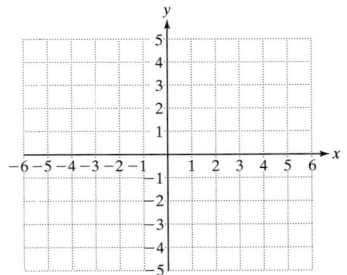

40. $x^2 - y^2 \geq 1$
$x \leq 0$

41. $y < x^2$
$y > x^2 - 4$

42. $y^2 - x^2 \geq 1$
$y \geq 0$

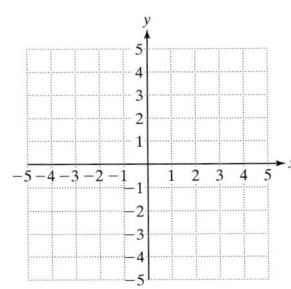

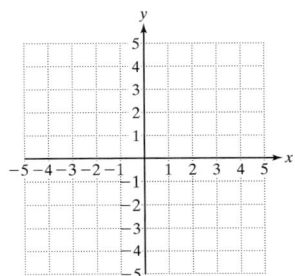

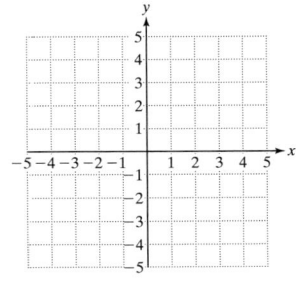

43. $y < \dfrac{1}{x}$
$y > 0$
$y < x$

44. $y > x^3$
$y < 8$
$x > 0$

 45. $x^2 + y^2 \geq 25$
$x^2 + y^2 \leq 9$

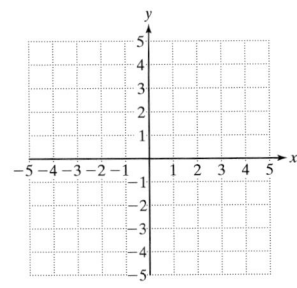

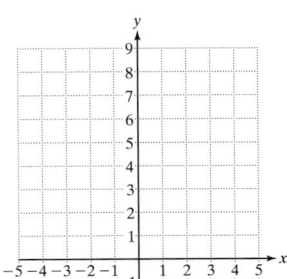

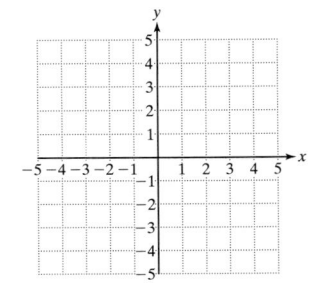

46. $\dfrac{x^2}{4} + \dfrac{y^2}{25} \geq 1$

$x^2 + \dfrac{y^2}{4} \leq 1$

47. $x < -(y - 1)^2 + 3$

$x + y > 2$

48. $x > (y - 2)^2 + 1$

$x - y < 1$

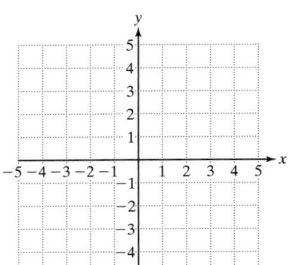 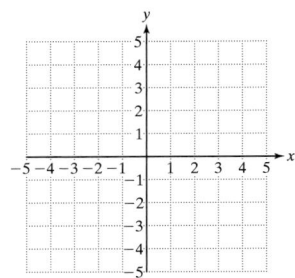 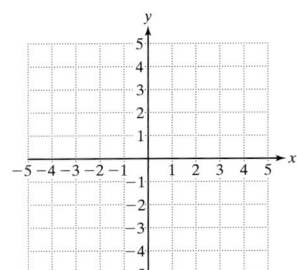

49. $x^2 + y^2 < 25$

$y < \dfrac{4}{3}x$

$y > -\dfrac{4}{3}x$

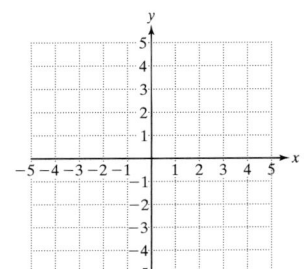

50. $y < e^x$

$y > 1$

$x < 2$

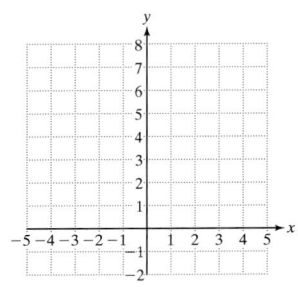

Expanding Your Skills

For Exercises 51–54, graph the compound inequalities.

51. $x^2 + y^2 \leq 36$ or $x + y \geq 0$

52. $y \leq -x^2 + 4$ or $y \geq x^2 - 4$

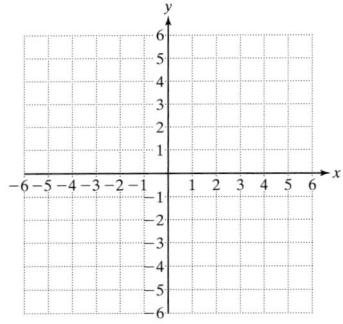 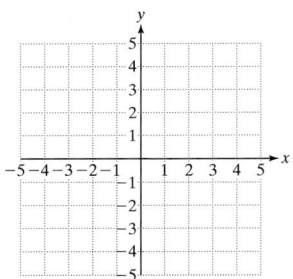

53. $y + 1 \geq x^2$ or $y + 1 \leq -x^2$

54. $(x + 2)^2 + (y + 3)^2 \leq 4$ or $x \geq y^2$

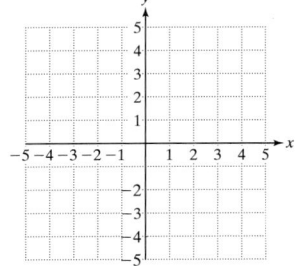 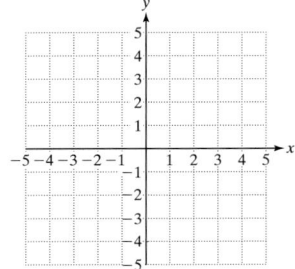

Chapter 14 SUMMARY

Section 14.1 Distance Formula and Circles

Key Concepts

The **distance between two points** (x_1, y_1) and (x_2, y_2) is

$$d = \sqrt{(x_2 - x_1)^2 + (y_2 - y_1)^2}$$

The standard equation of a **circle** with center (h, k) and radius r is

$$(x - h)^2 + (y - k)^2 = r^2$$

Examples

Example 1

Find the distance between two points.

$(5, -2)$ and $(-1, -6)$

$$d = \sqrt{(-1 - 5)^2 + [-6 - (-2)]^2}$$
$$= \sqrt{(-6)^2 + (-4)^2}$$
$$= \sqrt{36 + 16}$$
$$= \sqrt{52} = 2\sqrt{13}$$

Example 2

Find the center and radius of the circle.

$$x^2 + y^2 - 8x + 6y = 0$$
$$(x^2 - 8x + 16) + (y^2 + 6y + 9) = 16 + 9$$
$$(x - 4)^2 + (y + 3)^2 = 25$$

The center is $(4, -3)$ and the radius is 5.

Section 14.2 More on the Parabola

Key Concepts

A **parabola** is the set of points in a plane that are equidistant from a fixed line (called the directrix) and a fixed point (called the focus) not on the directrix.

The standard form of an equation of a parabola with **vertex** (h, k) and vertical **axis of symmetry** is

$$y = a(x - h)^2 + k$$

- The equation of the axis of symmetry is $x = h$.
- If $a > 0$, then the parabola opens upward.
- If $a < 0$, the parabola opens downward.

Examples

Example 1

Given the parabola, $y = (x - 2)^2 + 1$

The vertex is $(2, 1)$.

The axis of symmetry is $x = 2$.

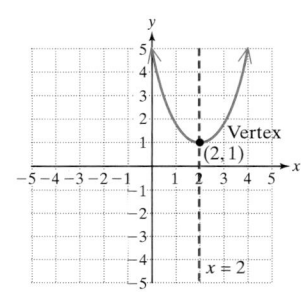

The standard form of an equation of a parabola with vertex (h, k) and horizontal axis of symmetry is

$$x = a(y - k)^2 + h$$

- The equation of the axis of symmetry is $y = k$.
- If $a > 0$, then the parabola opens to the right.
- If $a < 0$, the parabola opens to the left.

Example 2

Given the parabola $x = -\dfrac{1}{4}y^2 + 1$,

determine the coordinates of the vertex and the equation of the axis of symmetry.

$$x = -\frac{1}{4}(y - 0)^2 + 1$$

The vertex is $(1, 0)$.

The axis of symmetry is $y = 0$.

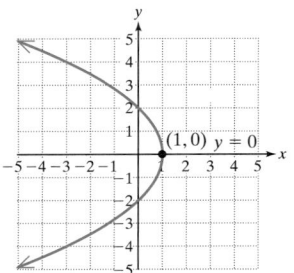

Section 14.3   The Ellipse and Hyperbola

Key Concepts

An **ellipse** is the set of all points (x, y) such that the sum of the distances between (x, y) and two distinct points (called foci) is constant.

Standard Form of an Ellipse with Center at the Origin

An ellipse with the center at the origin has the equation

$$\frac{x^2}{a^2} + \frac{y^2}{b^2} = 1$$

where a and b are positive real numbers.

For an ellipse centered at the origin, the x-intercepts are given by $(a, 0)$ and $(-a, 0)$, and the y-intercepts are given by $(0, b)$ and $(0, -b)$.

A **hyperbola** is the set of all points (x, y) such that the difference of the distances between (x, y) and two distinct points is a constant. The fixed points are called the foci of the hyperbola.

Examples

Example 1

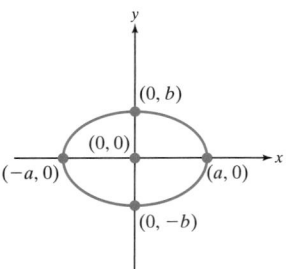

Example 2

$$\frac{x^2}{25} + \frac{y^2}{9} = 1$$

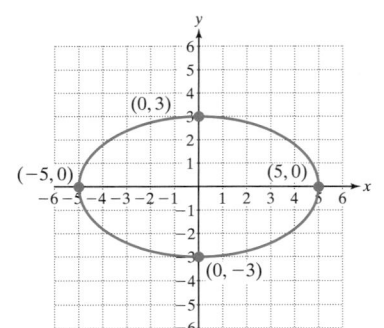

Standard Forms of an Equation of a Hyperbola

Let a and b represent positive real numbers.

Horizontal Transverse Axis. The standard form of an equation of a hyperbola with a horizontal transverse axis and center at the origin is given by

$$\frac{x^2}{a^2} - \frac{y^2}{b^2} = 1$$

Vertical Transverse Axis. The standard form of an equation of a hyperbola with a vertical transverse axis and center at the origin is given by

$$\frac{y^2}{a^2} - \frac{x^2}{b^2} = 1$$

Example 3

$$\frac{x^2}{4} - \frac{y^2}{16} = 1$$

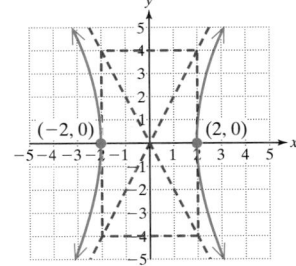

Example 4

$$\frac{y^2}{4} - \frac{x^2}{16} = 1$$

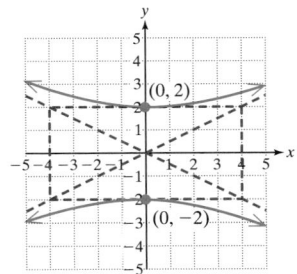

Section 14.4 Nonlinear Systems of Equations in Two Variables

Key Concepts

A **nonlinear system of equations** can be solved by graphing or by the substitution method.

$2x^2 + y^2 = 15$

$x^2 - y = 0$

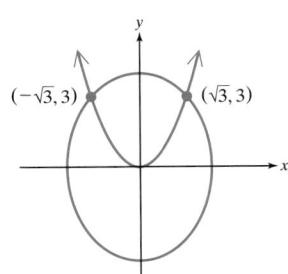

Examples

Example 1

$\boxed{A}$ $2x^2 + y^2 = 15$

$\boxed{B}$ $x^2 - y = 0$ Solve for y: $y = x^2$

$\boxed{A}$ $2x^2 + (x^2)^2 = 15$ Substitute in first equation.

$2x^2 + x^4 = 15$

$x^4 + 2x^2 - 15 = 0$

$(x^2 + 5)(x^2 - 3) = 0$

$x^2 + 5 = 0$ or $x^2 - 3 = 0$

$\cancel{x^2 = -5}$ or $x^2 = 3$

$x = \pm\sqrt{3}$

If $x = \sqrt{3}$ then $y = (\sqrt{3})^2 = 3$.

If $x = -\sqrt{3}$ then $y = (-\sqrt{3})^2 = 3$.

Points of intersection are $(\sqrt{3}, 3)$ and $(-\sqrt{3}, 3)$.

A nonlinear system may also be solved by using the *addition method* when the equations share *like* terms.

Example 2

$$2x^2 + y^2 = 4 \xrightarrow{\text{Mult. by } -5.} -10x^2 - 5y^2 = -20$$

$$3x^2 + 5y^2 = 13 \longrightarrow \underline{\quad 3x^2 + 5y^2 = 13 \quad}$$

$$-7x^2 \qquad\quad = -7$$

$$\frac{-7x^2}{-7} = \frac{-7}{-7}$$

$$x^2 = 1 \longrightarrow x = \pm 1$$

If $x = 1$, $\qquad\qquad 2(1)^2 + y^2 = 4$

$$y^2 = 2$$

$$y = \pm\sqrt{2}$$

If $x = -1$, $\qquad\quad 2(-1)^2 + y^2 = 4$

$$y^2 = 2$$

$$y = \pm\sqrt{2}$$

The points of intersection are $(1, \sqrt{2})$, $(1, -\sqrt{2})$, $(-1, \sqrt{2})$, and $(-1, -\sqrt{2})$.

Section 14.5 Nonlinear Inequalities and Systems of Inequalities

Key Concepts

Graph a nonlinear inequality by using the test point method. That is, graph the related equation. Then choose test points in each region to determine where the inequality is true.

Examples

Example 1

$y \geq x^2$

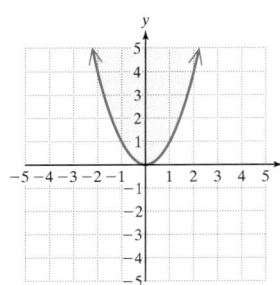

Example 2

$x^2 + y^2 < 4$

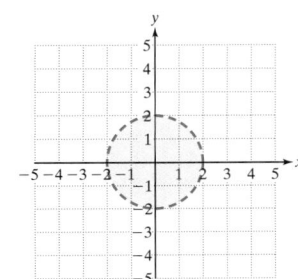

Graph a system of nonlinear inequalities by finding the intersection of the solution sets. That is, graph the solution set for each individual inequality, then take the intersection.

Example 3

$$x \geq 0, \qquad y > x^2, \qquad \text{and} \qquad x^2 + y^2 < 4$$

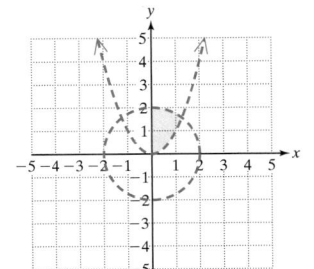

Chapter 14   Review Exercises

Section 14.1

For Exercises 1–4, find the distance between the two points by using the distance formula.

1. $(-6, 3)$ and $(0, 1)$ **2.** $(4, 13)$ and $(-1, 5)$

3. Find x such that $(x, 5)$ is 5 units from $(2, 9)$.

4. Find x such that $(-3, 4)$ is 3 units from $(x, 1)$.

Points are said to be collinear if they all lie on the same line. If three points are collinear, then the distance between the outermost points will equal the sum of the distances between the middle point and each of the outermost points. For Exercises 5–6, determine if the three points are collinear.

5. $(-2, -3), (1, 3)$, and $(5, 11)$

6. $(-2, 11), (0, 5)$, and $(4, -7)$

For Exercises 7–10, find the center and the radius of the circle.

7. $(x - 12)^2 + (y - 3)^2 = 16$

8. $(x + 7)^2 + (y - 5)^2 = 81$

9. $(x + 3)^2 + (y + 8)^2 = 20$

10. $(x - 1)^2 + (y + 6)^2 = 32$

11. A stained glass window is in the shape of a circle with a 16-in. diameter. Find an equation of the circle relative to the origin for each of the following graphs.

a.

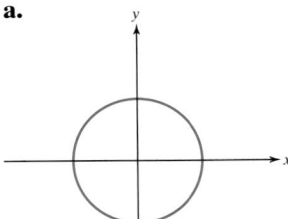

b.

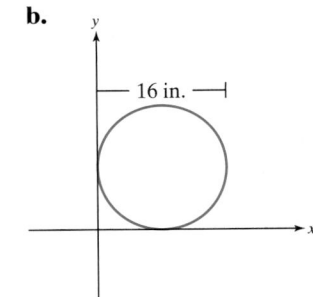

For Exercises 12–15, write the equation of the circle in standard form by completing the square.

12. $x^2 + y^2 + 12x - 10y + 51 = 0$

13. $x^2 + y^2 + 4x + 16y + 60 = 0$

14. $x^2 + y^2 - x - 4y + \dfrac{1}{4} = 0$

15. $x^2 + y^2 - 6x - \dfrac{2}{3}y + \dfrac{1}{9} = 0$

16. Write an equation of a circle with center at the origin and a diameter of 7 m.

17. Write an equation of a circle with center at $(0, 2)$ and a diameter of 6 m.

Section 14.2

For Exercises 18–21, determine whether the axis of symmetry is vertical or horizontal and if the parabola opens upward, downward, left, or right.

18. $y = -2(x - 3)^2 + 2$

19. $x = 3(y - 9)^2 + 1$

20. $x = -(y + 4)^2 - 8$

21. $y = (x + 3)^2 - 10$

For Exercises 22–25, determine the coordinates of the vertex and the equation of the axis of symmetry. Then use this information to graph the parabola.

22. $x = -(y - 1)^2$

23. $y = (x + 2)^2$

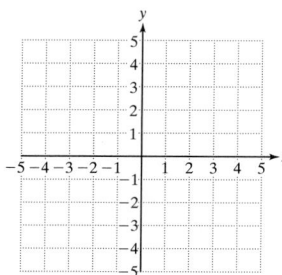
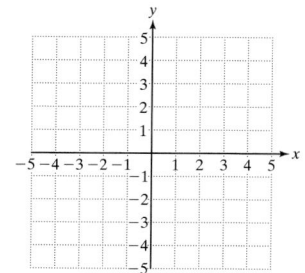

24. $y = -\dfrac{1}{4}x^2$

25. $x = 2y^2 - 1$

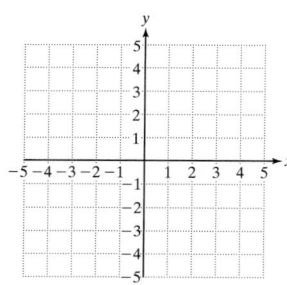
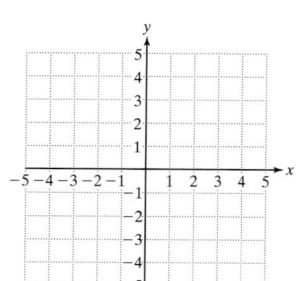

For Exercises 26–29, write the equation in standard form $y = a(x - h)^2 + k$ or $x = a(y - k)^2 + h$. Then identify the vertex, and axis of symmetry.

26. $y = x^2 - 6x + 5$

27. $x = y^2 + 4y + 2$

28. $x = -4y^2 + 4y$

29. $y = -2x^2 - 2x$

Section 14.3

For Exercises 30–31, identify the x- and y-intercepts. Then graph the ellipse.

30. $\dfrac{x^2}{9} + \dfrac{y^2}{25} = 1$

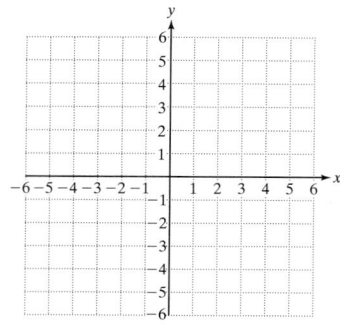

31. $x^2 + 4y^2 = 36$

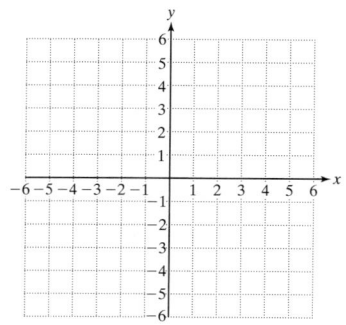

For Exercises 32–33, identify the center of the ellipse and graph the ellipse.

32. $\dfrac{(x - 5)^2}{4} + \dfrac{(y + 3)^2}{16} = 1$

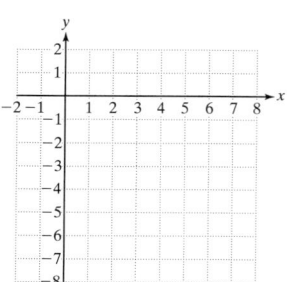

33. $\dfrac{x^2}{25} + \dfrac{(y - 2)^2}{9} = 1$

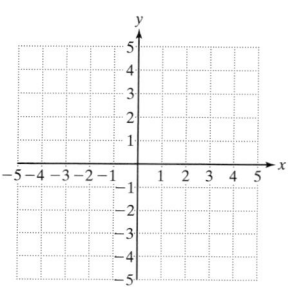

For Exercises 34–37, determine whether the transverse axis is horizontal or vertical.

34. $\dfrac{x^2}{12} - \dfrac{y^2}{16} = 1$

35. $\dfrac{y^2}{9} - \dfrac{x^2}{9} = 1$

36. $\dfrac{y^2}{24} - \dfrac{x^2}{10} = 1$

37. $\dfrac{x^2}{6} - \dfrac{y^2}{16} = 1$

For Exercises 38–39, graph the hyperbola by first drawing the reference rectangle and the asymptotes. Label the vertices.

38. $\dfrac{x^2}{4} - y^2 = 1$

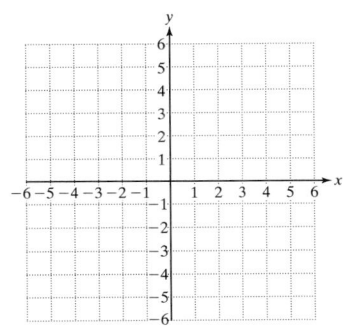

39. $y^2 - x^2 = 16$

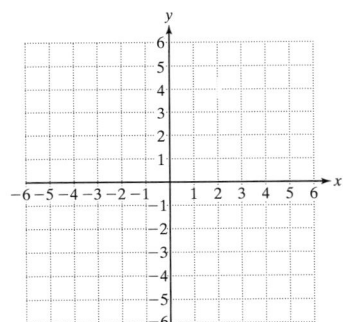

For Exercises 40–43, identify the equations as representing an ellipse or a hyperbola.

40. $\dfrac{x^2}{4} - \dfrac{y^2}{9} = 1$ **41.** $\dfrac{x^2}{16} + \dfrac{y^2}{9} = 1$

42. $\dfrac{x^2}{4} + \dfrac{y^2}{1} = 1$ **43.** $\dfrac{y^2}{1} - \dfrac{x^2}{16} = 1$

Section 14.4

For Exercises 44–47,

    a. Identify each equation as representing a line, a parabola, a circle, an ellipse, or a hyperbola.

    b. Graph both equations on the same coordinate system.

    c. Solve the system analytically and verify the answers from the graph.

44. $3x + 2y = 10$
       $y = x^2 - 5$

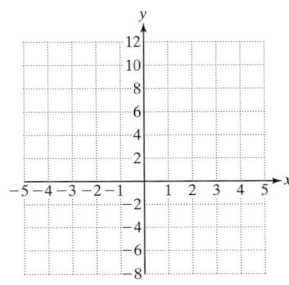

45. $4x + 2y = 10$
       $y = x^2 - 10$

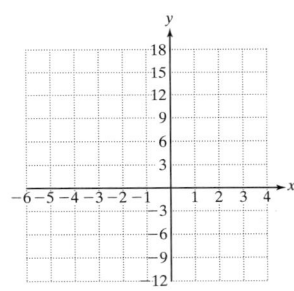

46. $x^2 + y^2 = 9$
       $2x + y = 3$

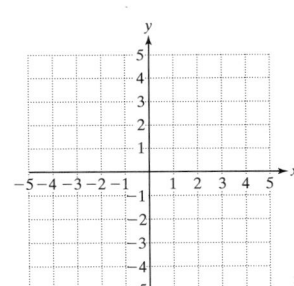

47. $x^2 + y^2 = 16$
       $x - 2y = 8$

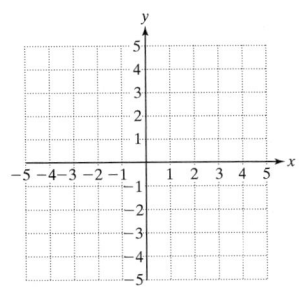

For Exercises 48–53, solve the system of nonlinear equations by using either the substitution method or the addition method.

48. $x^2 + 2y^2 = 8$

$2x - y = 2$

49. $x^2 + 4y^2 = 29$

$x - y = -4$

50. $x - y = 4$

$y^2 = 2x$

51. $y = x^2$

$6x^2 - y^2 = 8$

52. $x^2 + y^2 = 10$

$x^2 + 9y^2 = 18$

53. $x^2 + y^2 = 61$

$x^2 - y^2 = 11$

Section 14.5

For Exercises 54–61, graph the solution set to the inequality.

54. $3x + y \leq 4$

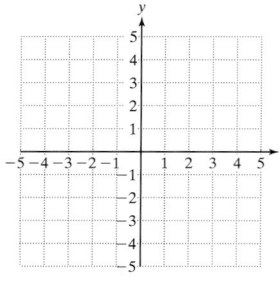

55. $x - 2y \geq -2$

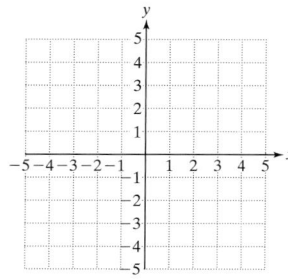

56. $\dfrac{x^2}{16} + \dfrac{y^2}{81} < 1$

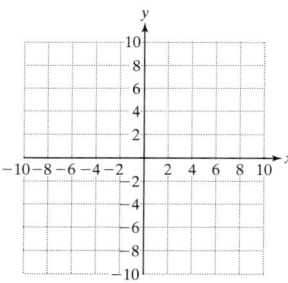

57. $\dfrac{x^2}{25} + \dfrac{y^2}{4} > 1$

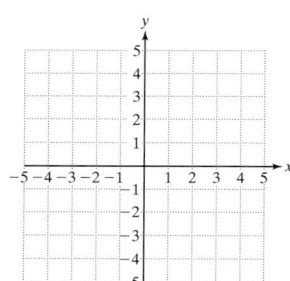

58. $(x - 3)^2 + (y + 1)^2 \geq 9$

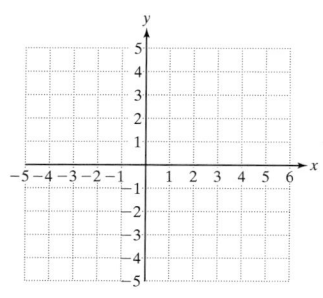

59. $(x + 2)^2 + (y + 1)^2 \leq 4$

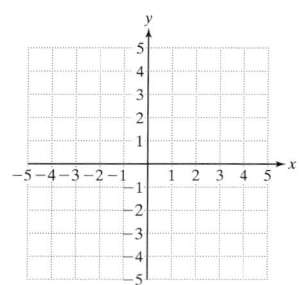

60. $y > (x - 1)^2$

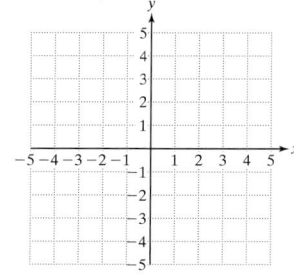

61. $y > x^2 - 1$

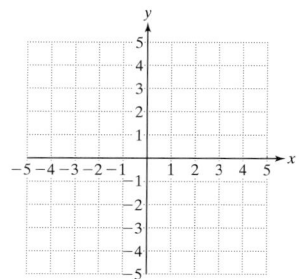

For Exercises 62–64, graph the solution set to the system of nonlinear inequalities.

62. $y > 2^x$

$x^2 + y^2 < 4$

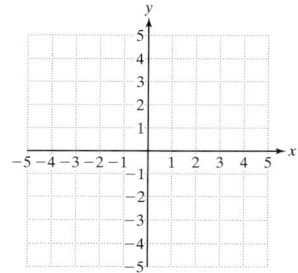

63. $y < 2^x$

$x^2 + y^2 < 9$

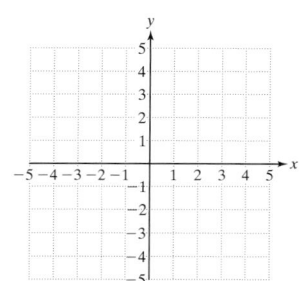

64. $\dfrac{x^2}{4} - y^2 < 1$

$x^2 + y^2 < 9$

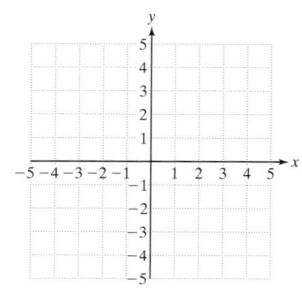

Chapter 14   Test

1. Determine the coordinates of the vertex and the equation of the axis of symmetry. Then graph the parabola.

$$x = -(y - 2)^2 + 3$$

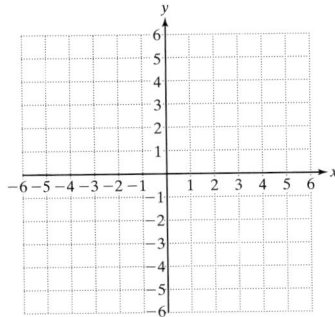

2. Determine the center and radius of the circle.

$$\left(x - \frac{5}{6}\right)^2 + \left(y + \frac{1}{3}\right)^2 = \frac{25}{49}$$

3. Write the equation in standard form $y = a(x - h)^2 + k$, and graph the parabola.

$$y = x^2 + 4x + 5$$

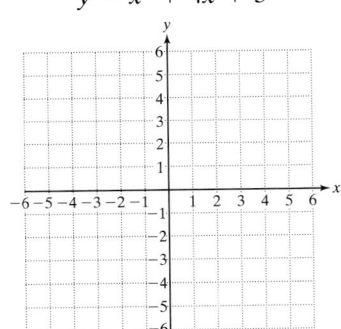

4. Use the distance formula to find the distance between the two points $(-5, 19)$ and $(-3, 13)$.

5. Determine the center and radius of the circle.

$$x^2 + y^2 - 4y - 5 = 0$$

6. Let $(0, 4)$ be the center of a circle that passes through the point $(-2, 5)$.

   a. What is the radius of the circle?

   b. Write the equation of the circle in standard form.

7. Graph the ellipse.

$$\frac{x^2}{16} + \frac{y^2}{49} = 1$$

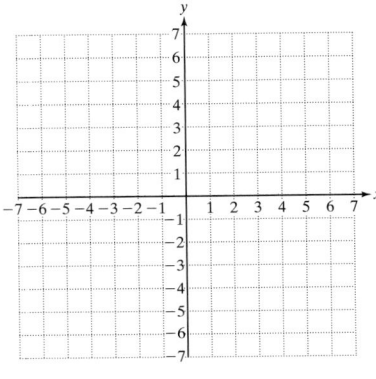

8. Graph the ellipse.

$$\frac{(x + 4)^2}{25} + (y - 3)^2 = 1$$

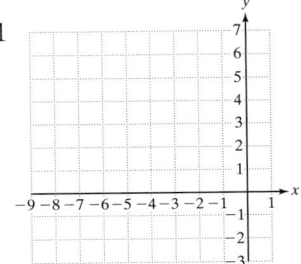

9. Graph the hyperbola.

$$y^2 - \frac{x^2}{4} = 1$$

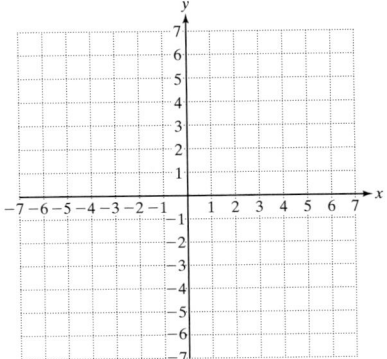

10. Solve the systems and identify the correct graph of the equations.

   a. $16x^2 + 9y^2 = 144$ **b.** $x^2 + 4y^2 = 4$

       $4x - 3y = -12$ $4x - 3y = -12$

i. 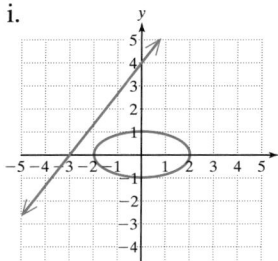   ii.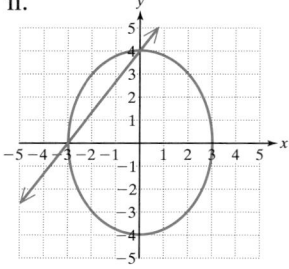

11. Describe the circumstances in which a nonlinear system of equations can be solved by using the addition method.

12. Solve the system by using either the substitution method or the addition method.

$$25x^2 + 4y^2 = 100$$
$$25x^2 - 4y^2 = 100$$

For Exercises 13–16, graph the solution set.

13. $x \le y^2 + 1$

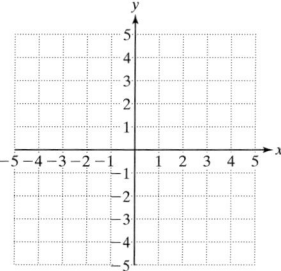

14. $y \ge -\dfrac{1}{3}x + 1$

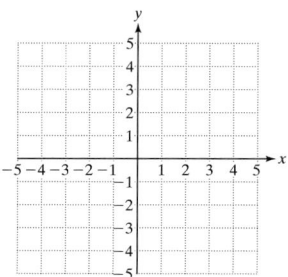

15. $x < y^2 + 1$

$\quad y > -\dfrac{1}{3}x + 1$

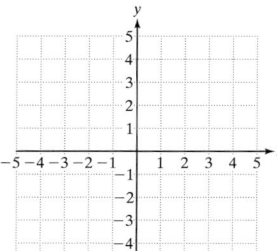

16. $y < \sqrt{x}$

$\quad y > x - 2$

$\quad x > 0$

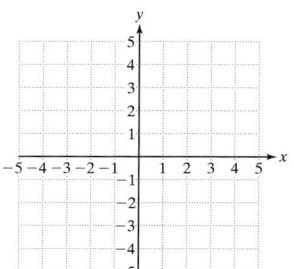

Chapters 1–14   Cumulative Review Exercises

1. Solve the equation.

$$5(2y - 1) = 2y - 4 + 8y - 1$$

2. Solve the inequality. Graph the solution and write the solution in interval notation.

$$4(x - 1) + 2 > 3x + 8 - 2x$$

3. The product of two integers is 150. If one integer is 5 less than twice the other, find the integers.

4. For $5y - 3x - 15 = 0$:

   a. Find the x- and y-intercepts.

   b. Find the slope.

   c. Graph the line.

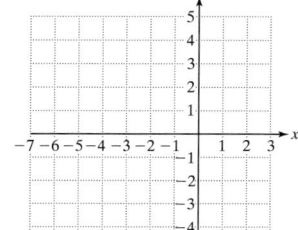

5. The amount of money spent on motor vehicles and parts each year since the year 2000 is shown in the graph. Let $x = 0$ correspond to the year 2000. Let y represent the amount of money spent on motor vehicles and parts (in billions of dollars).

   a. Use any two data points to find the slope of the line.

   b. Find an equation of the line through the points. Write the answer in slope-intercept form.

   c. Use the linear model found in part (b) to predict the amount spent on motor vehicles and parts in the year 2010.

Amount Spent (in $ billions) on Motor Vehicles and Parts in United States

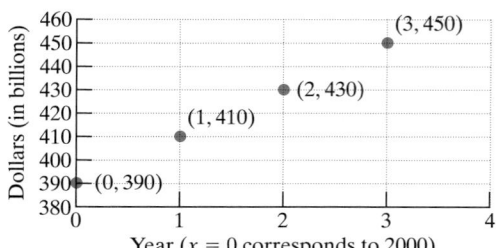

Source: Bureau of Economic Analysis, U.S. Department of Commerce

6. Find the slope and y-intercept of $3x - 4y = 6$ by first writing the equation in slope-intercept form.

7. A collection of dimes and quarters has a total value of \$2.45. If there are 17 coins, how many of each type are there?

8. Solve the system.
$$x + y \qquad = -1$$
$$2x \qquad - z = \quad 3$$
$$y + 2z = -1$$

9. a. Given the matrix $\mathbf{A} = \begin{bmatrix} 1 & -2 & | & -8 \\ 0 & 3 & | & 6 \end{bmatrix}$, write the matrix obtained by multiplying the elements in the second row by $\frac{1}{3}$.

  b. Using the matrix obtained from part (a), write the matrix obtained by multiplying the second row by 2 and adding the result to the first row.

10. Solve the following system.
$$4x - 2y = 7$$
$$-3x + 5y = 0$$

11. Solve using Cramer's rule:
$$3x - 4y = \quad 6$$
$$x + 2y = 12$$

12. For $f(x) = 3x - x^2 - 12$, find the function values $f(0), f(-1), f(2),$ and $f(4)$.

13. For $g = \{(2, 5), (8, -1), (3, 0), (-5, 5)\}$ find the function values $g(2), g(8), g(3),$ and $g(-5)$.

14. The quantity z varies jointly as y and as the square of x. If z is 80 when x is 5 and y is 2, find z when $x = 2$ and $y = 5$.

15. For $f(x) = \sqrt{x + 1}$ and $g(x) = x^2 + 6$, find $(g \circ f)(x)$.

16. a. Find the value of the expression $x^3 + x^2 + x + 1$ for $x = -2$.

  b. Factor the expression $x^3 + x^2 + x + 1$ and find the value when x is -2.

  c. Compare the values for parts (a) and (b).

17. Factor completely.
$$x^2 - y^2 - 6x - 6y$$

18. Multiply: $(x - 4)(x^2 + 2x + 1)$

19. Solve for x: $2x(x - 7) = x - 18$

20. Reduce the expression: $\dfrac{3a^2 - a - 2}{3a^2 + 8a + 4}$

21. Subtract: $\dfrac{2}{x + 3} - \dfrac{x}{x - 2}$

22. Solve: $\dfrac{2}{x + 3} - \dfrac{x}{x - 2} = \dfrac{-4}{x^2 + x - 6}$

23. Solve the radical equations.
  a. $\sqrt{2x - 5} = -3$
  b. $\sqrt[3]{2x - 5} = -3$

24. Perform the indicated operations with complex numbers.
  a. $6i(4 + 5i)$ **b.** $\dfrac{3}{4 - 5i}$

25. Find the length of the missing side.

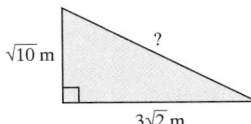

26. An automobile starts from rest and accelerates at a constant rate for 10 sec. The distance, $d(t)$, in feet traveled by the car is given by
$$d(t) = 4.4t^2$$
where $0 \le t \le 10$ is the time in seconds.
  a. How far has the car traveled after 2, 3, and 4 sec, respectively?
  b. How long will it take for the car to travel 281.6 ft?

27. Solve the equation $125w^3 + 1 = 0$ by factoring and using the quadratic formula. (*Hint:* You will find one real solution and two imaginary solutions.)

28. Solve the rational equation.

$$\frac{x}{x+2} - \frac{3}{x-1} = \frac{1}{x^2+x-2}$$

29. Find the coordinates of the vertex of the parabola defined by $f(x) = x^2 + 10x - 11$ by completing the square.

30. Graph the quadratic function defined by $g(x) = -x^2 - 2x + 3$.

   a. Label the x-intercepts.

   b. Label the y-intercept.

   c. Label the vertex.

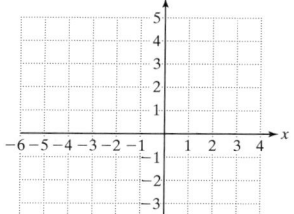

31. Solve the inequality and write the answer in interval notation.

$$|x - 9| - 3 < 7$$

32. Solve the inequality: $|2x - 5| \geq 4$

33. Write the expression in logarithmic form.
$8^{5/3} = 32$

34. Solve the equation: $5^2 = 125^x$

35. For $h(x) = x^3 - 1$, find $h^{-1}(x)$.

36. Write an equation representing the set of all points 4 units from the point $(0, 5)$.

37. Can a circle and a parabola intersect in only one point? Explain.

38. Solve the system of nonlinear equations.

$$x^2 + y^2 = 16$$
$$y = -x^2 - 4$$

39. Graph the solution set.

$$y^2 - x^2 < 1$$

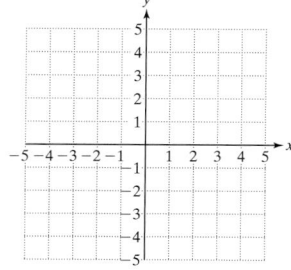

40. Graph the solution set to this system.

$$y > \left(\frac{1}{2}\right)^x$$
$$x < 0$$

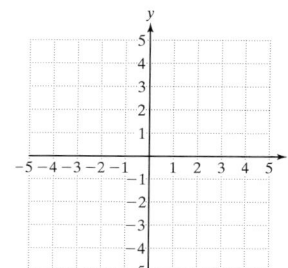

Beginning Algebra Review

Set of Real Numbers

Review A

1. Sets of Real Numbers

The numbers we are familiar with in everyday life comprise the **set of real numbers**. Every real number corresponds to a unique point on a number line.

For example:

Several subsets of the real numbers are given below. We use the symbols { } to enclose the elements of a set.

Set	Definition
Natural numbers	$\{1, 2, 3, \ldots\}$
Whole numbers	$\{0, 1, 2, 3, \ldots\}$
Integers	$\{\ldots, -3, -2, -1, 0, 1, 2, 3, \ldots\}$
Rational numbers	the set of numbers of the form $\frac{p}{q}$ where p and q are integers and $q \neq 0$.
Irrational numbers	the set of real numbers that are not rational

Rational numbers can be written as a ratio of two integers such as $\frac{1}{3}$ or $\frac{3}{4}$. When expressed in decimal form a rational number is either a repeating decimal or a terminating decimal such as $0.\overline{3}$ or 0.75. The decimal form of an irrational number is non-repeating and non-terminating. Examples of irrational numbers are π and $\sqrt{2}$.

2. Symbols and Mathematical Language

In mathematics, the symbol $<$ (meaning "is less than") and the symbol $>$ (meaning "is greater than") are used to express inequalities. These and other inequality symbols are summarized in Table A-1.

Table A-1

Inequality	In Words	Example
$a < b$	a is less than b	$5 < 7$
$a > b$	a is greater than b	$8 > 3$
$a \leq b$	a is less than or equal to b	$8 \leq 9$
$a \geq b$	a is greater than or equal to b	$3 \geq 3$
$a < x < b$	x is between a and b	$3 < 4 < 5$
$a \neq b$	a is not equal to b	$4 \neq 6$

Concepts

1. Sets of Real Numbers
2. Symbols and Mathematical Language
3. Operations on Real Numbers
4. Exponents and Square Roots
5. Order of Operations
6. Properties of Real Numbers
7. Simplifying Expressions

the real number line, p. 42

subsets of real numbers, p. 43

identifying rational numbers, p. 44

inequality symbols, p. 46

The operations of addition, subtraction, multiplication, and division can be denoted several ways. These along with other common operations on real numbers are summarized in (Table A-2).

Table A-2

common mathematical translations, p. 53

Operation	Symbols	Translation/Example
Addition	$a + b$	"the **sum** of a and b" "the sum of 8 and -4" $\Rightarrow 8 + (-4)$
Subtraction	$a - b$	"the **difference** of a and b" "the difference of -8 and 2" $\Rightarrow -8 - 2$
Multiplication	$a \times b, a \cdot b, a(b),$ $(a)b, (a)(b), ab$	"the **product** of a and b" "the product of 3 and -2" $\Rightarrow 3(-2)$
Division	$a \div b, \dfrac{a}{b}, a/b, b\overline{)a}$	"the **quotient** of a and b" "the quotient of 10 and 5" $\Rightarrow \dfrac{10}{5}$
Absolute value	$\lvert a \rvert$	"the **absolute value** of a" "the absolute value of -3" $\Rightarrow \lvert -3 \rvert$
Opposite	$-a$	"the **opposite** of a" "the opposite of 7" $\Rightarrow -(7)$
Reciprocal	$\dfrac{1}{a} \, (a \neq 0)$	"the **reciprocal** of a" "the reciprocal of 5" $\Rightarrow \dfrac{1}{5}$
Square root	$\sqrt{a}$	"the **square root** of a" "the square root of 64" $\Rightarrow \sqrt{64}$

3. Operations on Real Numbers

absolute value and opposite, pp. 47

The **absolute value** of a real number, a, denoted $\lvert a \rvert$, is its distance from 0 on the number line. Two numbers that are the same distance from 0 but on opposite sides of the number line are called **opposites**. The opposite of a is denoted $-a$.

Example 1　Finding Opposites and Absolute Value

a. Find the opposite of 5.

b. Find the opposite of $-\dfrac{1}{2}$.

c. Evaluate $\lvert -6 \rvert$.

d. Evaluate $\lvert 2.7 \rvert$.

e. Evaluate $-\lvert -8 \rvert$.

Solution:

a. The opposite of 5 is -5.

b. The opposite of $-\dfrac{1}{2}$ is $\dfrac{1}{2}$.

c. $\lvert -6 \rvert = 6$

d. $\lvert 2.7 \rvert = 2.7$

e. $-\lvert -8 \rvert = -8$　　Take the opposite of the absolute value of -8.

The rules for adding, subtracting, multiplying, and dividing real numbers are given as follows.

Addition of Two Real Numbers

• To add two numbers with the *same sign:* Add the absolute values of the numbers and apply the common sign to the sum.	example: $-3 + (-7) = -(3 + 7) = -10$
• To add two numbers with *unlike signs:* Subtract the smaller absolute value from the larger absolute value. Then apply the sign of the number having the larger absolute value.	examples: $10 + (-4) = 10 - 4 = 6$ $-8 + 5 = -(8 - 5) = -3$

adding real numbers, p. 63

Subtraction of Two Real Numbers

Add the opposite of the second number to the first number. In symbols: $a - b = a + (-b)$	examples: $8 - (-4) = 8 + (4) = 12$ $-3 - 6 = -3 + (-6) = -9$

subtracting real numbers, p. 72

Multiplication and Division of Two Real Numbers

• The product or quotient of two real numbers with the *same* sign is positive.	example: $-20 \div (-4) = 5$
• The product or quotient of two real numbers with *unlike* signs is negative.	example: $\dfrac{30}{-2} = -15$
<u>Notes:</u> • The product of any real number and 0 is 0. • The quotient of 0 and any nonzero real number is 0. • The quotient of a real number and 0 is undefined.	example: $5 \cdot 0 = 0$ example: $0 \div 6 = 0$ example: $6 \div 0$ is undefined

multiplying and dividing real numbers, pp. 81

4. Exponents and Square Roots

exponents, p. 54

We can use **exponents** to represent repeated multiplication. For example, the product

$$3 \cdot 3 \cdot 3 \cdot 3 \cdot 3 \qquad \text{can be written as} \qquad 3^5.$$

with labels: exponent → 5, base → 3

The expression 3^5 is written in exponential form. The exponent (or **power**) is 5 and represents the number of times the **base**, 3, is multiplied. The expression 3^5 is read as "three to the fifth power." In general,

$$x^n = \underbrace{x \cdot x \cdot x \cdot \cdots \cdot x}_{\text{the factor } x \text{ occurs } n \text{ times}}$$

square roots, p. 55

The number b is a **square root** of a if $b^2 = a$. A **radical sign**, $\sqrt{}$, is used to denote the principle (or positive) square root of a nonnegative real number.

Example 2 **Evaluating Expressions Containing Exponents and Square Roots**

Evaluate.   **a.** 2^4 **b.** $\sqrt{36}$

Solution:

a. $2^4 = 2 \cdot 2 \cdot 2 \cdot 2 = 16$

b. $\sqrt{36} = 6$ because $6^2 = 36$.

---

5. Order of Operations

Algebraic expressions often have more than one operation. In such a case, it is important to follow the order of operations.

order of operations, p. 56

Order of Operations

1. Simplify expressions within parentheses and other grouping symbols first. These include absolute value bars, fraction bars, and radicals. If imbedded parentheses are present, start with the innermost parenthesis.
2. Evaluate expressions involving exponents and radicals.
3. Perform multiplication or division in the order that they occur from left to right.
4. Perform addition or subtraction in the order that they occur from left to right.

Example 3	Applying the Order of Operations

Simplify. $50 \div [15 - 2(4 + 6) + 15] \cdot 2^3$

Solution:

$50 \div [15 - 2(4 + 6) + 15] \cdot 2^3$

$= 50 \div [15 - 2(10) + 15] \cdot 2^3$	Simplify inside the inner parentheses.
$= 50 \div [15 - 20 + 15] \cdot 2^3$	Simplify inside [ ]. Within [ ], we perform multiplication before addition or subtraction.
$= 50 \div [-5 + 15] \cdot 2^3$	Within [ ], subtract and add from left to right.
$= 50 \div [10] \cdot 2^3$	Simplify within [ ].
$= 50 \div [10] \cdot 8$	Evaluate the exponential expressions.
$= 5 \cdot 8$	Divide and multiply from left to right.
$= 40$	Multiply.

Example 4	Applying the Order of Operations within a Formula

Given $x = -3$, evaluate the expressions.   **a.** x^2 **b.** $-x^2$

Solution:

a. x^2

$= (\ )^2$	Use parentheses in place of the variable.
$= (-3)^2$	Substitute -3 for x.
$= 9$	Evaluate expressions with exponents.

b. $-x^2$

$= -(\ )^2$	Use parentheses in place of the variable.
$= -(-3)^2$	Substitute -3 for x.
$= -(9)$	Simplify $(-3)^2$ first. Note that $(-3)^2 = 9$.
$= -9$	Now take the opposite of 9.

TIP: The expression $-(-3)^2$ is equivalent to $-1(-3)^2$. The order of operations indicates that we should square -3 first, and then multiply the result by -1.

6. Properties of Real Numbers

Several properties of real numbers are reviewed in Table A-3.

Table A-3

Property Name	Algebraic Representation	Example	Description/Notes
Commutative property of addition	$a + b = b + a$	$5 + 3 = 3 + 5$	The order in which two real numbers are added or multiplied does not affect the result.
Commutative property of multiplication	$a \cdot b = b \cdot a$	$(5)(3) = (3)(5)$	
Associative property of addition	$(a + b) + c = a + (b + c)$	$(2 + 3) + 7 = 2 + (3 + 7)$	The manner in which two real numbers are grouped under addition or multiplication does not affect the result.
Associative property of multiplication	$(a \cdot b)c = a(b \cdot c)$	$(2 \cdot 3)7 = 2(3 \cdot 7)$	
Distributive property of multiplication over addition	$a(b + c) = ab + ac$	$3(5 + 2) = 3 \cdot 5 + 3 \cdot 2$	A factor outside the parentheses is multiplied by each term inside the parentheses.
Identity property of addition	0 is the identity element for addition because $a + 0 = 0 + a = a$	$5 + 0 = 0 + 5 = 5$	Any number added to the identity element, 0, will remain unchanged.
Identity property of multiplication	1 is the identity element for multiplication because $a \cdot 1 = 1 \cdot a = a$	$5 \cdot 1 = 1 \cdot 5 = 5$	Any number multiplied by the identity element, 1, will remain unchanged.
Inverse property of addition	a and $(-a)$ are additive inverses because $a + (-a) = 0$ and $(-a) + a = 0$	$3 + (-3) = 0$	The sum of a number and its additive inverse (**opposite**) is the identity element, 0.
Inverse property of multiplication	a and $\frac{1}{a}$ are multiplicative inverses because $a \cdot \dfrac{1}{a} = 1$ and $\dfrac{1}{a} \cdot a = 1$ (provided $a \neq 0$)	$5 \cdot \frac{1}{5} = 1$	The product of a number and its multiplicative inverse (**reciprocal**) is the identity element, 1.

7. Simplifying Expressions

We can simplify expressions containing variables by adding or subtracting *like* terms. Recall that *like* terms have the same variables and the corresponding variables are raised to the same powers. To combine *like* terms, we use the distributive property. The same result may also be obtained by adding or subtracting the coefficients of each term and leaving the variable factors unchanged.

$$5x + 2x - 3x = 4x$$

Example 5 | Simplifying Expressions

Simplify. $4x - 3(2x - 8) - 1 + 5x$

Solution:

$4x - 3(2x - 8) - 1 + 5x$

$= 4x - 6x + 24 - 1 + 5x$ Apply the distributive property.

$= 4x - 6x + 5x + 24 - 1$ Arrange *like* terms together.

$= 3x + 23$ Combine *like* terms.

Review A   Practice Exercises

Boost *your* GRADE at
mathzone.com!

 MathZone

- Practice Problems
- Self-Tests
- NetTutor
- e-Professors
- Videos

For Exercises 1–8, refer to sets

$A = \{-6, -\sqrt{5}, -\frac{7}{8}, -0.\overline{3}, 0, \frac{4}{9}, \pi, 7, 10.4\}$ and $B = \{-2, -\pi, -\frac{1}{2}, 0, 1, \sqrt{3}, \frac{5}{4}, 8\}$

1. List all rational numbers in set A.

2. List all irrational numbers in set A.

3. List all integers in set A.

4. List all whole numbers in set A.

5. List all natural numbers in set B.

6. List all integers in set B.

7. List all irrational numbers in set B.

8. List all rational numbers in set B.

For Exercises 9–16, write the expression in words.

9. $5 > -2$ **10.** $8 \leq 9$ **11.** $-3 \geq -4$ **12.** $\frac{1}{2} > -\frac{1}{2}$

13. $6 < 10 < 12$ **14.** $-1 < 0 < 1$ **15.** $2 \neq -5$ **16.** $\frac{2}{3} \neq -\frac{2}{3}$

For Exercises 17–26, translate the English phrase into an algebraic expression.

17. The product of 6 and 3

18. The difference of 7 and 4

19. The quotient of 20 and 5

20. The sum of 2 and 12

21. The absolute value of -5

22. The absolute value of -13

23. The sum of 2 and the absolute value of -8.

24. The difference of 10 and the absolute value of -1

25. The product of 4 and the reciprocal of 3

26. The quotient of 8 and the square root of 16

For Exercises 27–46, add or subtract as indicated.

27. $8 + (-10)$ **28.** $23 + (-9)$ **29.** $51 - 32$ **30.** $12 - 15$

31. $0 + 16$ **32.** $26 + 0$ **33.** $-13 - 20$ **34.** $-5 - 30$

35. $-2.1 + (-2.2)$ **36.** $-4.0 + (-0.5)$ **37.** $-19 - (-3)$ **38.** $-7 - (-28)$

39. $0 - 19$ **40.** $0 - 41$ **41.** $5.2 - (-0.4)$ **42.** $0.03 - (-1.2)$

43. $-\dfrac{5}{6} - \dfrac{3}{8}$ **44.** $-\dfrac{5}{2} - \dfrac{1}{6}$ **45.** $-\dfrac{7}{3} - \left(-\dfrac{1}{9}\right)$ **46.** $-\dfrac{7}{10} - \left(-\dfrac{4}{5}\right)$

For Exercises 47–66, multiply or divide as indicated.

47. $3(-9)$ **48.** $(-8)(2)$ **49.** $(-5)(-7)$ **50.** $(-11)(-3)$

51. $0 \cdot 32$ **52.** $0 \cdot (-12)$ **53.** $(2.05)(-4.2)$ **54.** $(-10.7)(3.4)$

55. $\left(-\dfrac{5}{9}\right) \cdot \left(-\dfrac{3}{2}\right)$ **56.** $\left(-\dfrac{1}{6}\right) \cdot \left(-\dfrac{12}{7}\right)$ **57.** $-56 \div (-8)$ **58.** $-51 \div (-3)$

59. $-78 \div 13$ **60.** $112 \div (-16)$ **61.** $0 \div (-5)$ **62.** $0 \div 12$

63. $3.1 \div 0$ **64.** $-1.4 \div 0$ **65.** $\left(-\dfrac{7}{8}\right) \div \left(-\dfrac{7}{4}\right)$ **66.** $\left(-\dfrac{3}{11}\right) \div \left(\dfrac{21}{22}\right)$

For Exercises 67–80, simplify using the order of operations.

67. $|-14| - |-5|$ **68.** $\sqrt{81} - |-4|$ **69.** $10 \div 5 \cdot 4$ **70.** $12 \div 2 - 3$

71. $6 + 42 \div 7 - 10$ **72.** $21 - 3 \cdot 5 + 6$ **73.** $(8 - 5)^2 - (2 + 3)^2$ **74.** $16 - (10 - 3)^2 + 2 \cdot 7$

75. $3^3 - 2 + 5(3 - 7)$ **76.** $6 - 4^2 + (8 \div 4 + 5)$ **77.** $12 - \sqrt{25} + 4^2 \div 2$ **78.** $\sqrt{81} - \sqrt{16} \div 4 + 18$

79. $\dfrac{6 + 8 - (-2)}{-40 \div 8 - 1}$ **80.** $\dfrac{\sqrt{32 \div 2} + 7}{-52 + (7 \cdot 4 + 2)}$

For Exercises 81–84, evaluate the expressions given $a = 3$, $b = -5$, $c = 9$, and $d = -2$.

81. $a^2 - b^2$ **82.** $c \div a - b$ **83.** $b + \sqrt{c} \cdot d$ **84.** $-a^2 + c$

85. The area of a trapezoid is given by the formula $A = \frac{1}{2}(b_1 + b_2)h$. Find the area of the given trapezoid.

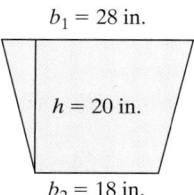

$b_1 = 28$ in.

$h = 20$ in.

$b_2 = 18$ in.

86. The surface area of a rectangular solid with a square base is given by $S = 4ab + 2a^2$. Find the surface area of the given rectangular solid.

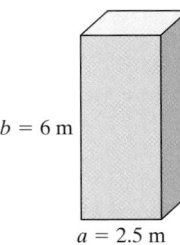

$b = 6$ m

$a = 2.5$ m

For Exercises 87–92, identify the property that is shown. Choose from the commutative property of addition, commutative property of multiplication, associative property of addition, associative property of multiplication, or distributive property of multiplication over addition.

87. $4(8 \cdot 14) = (4 \cdot 8)(14)$

88. $(7 + 4) + 6 = 7 + (4 + 6)$

89. $-6(3 + 9) = -6(3) + -6(9)$

90. $9 \cdot 2 = 2 \cdot 9$

91. $5 + 15 = 15 + 5$

92. $9(11 + 4) = 9(11) + 9(4)$

93. What is the identity element for multiplication? Use it in an example.

94. What is the identity element for addition? Use it in an example.

95. a. What is another name for multiplicative inverse?

   b. Write the multiplicative inverse of 4.

96. a. What is another name for the additive inverse?

   b. Write the additive inverse of 6.

For Exercises 97–104, simplify the expression by applying the distributive property.

97. $5(a + 9)$

98. $3(x + 12)$

99. $-7(y - z)$

100. $-2(a - b)$

101. $\frac{1}{3}(6b + 9c - 15)$

102. $\frac{1}{5}(10h - 20k - 5)$

103. $-(2x - 3y + 8)$

104. $-(-3p - q + 9)$

For Exercises 105–112, simplify the expression by clearing parentheses and combining like terms.

105. $5a + 2(a + 7)$

106. $6b - 3(b + 4)$

107. $8y - 3(y + 2) - 19$

108. $9k - 3(k - 4) + 8$

109. $4w + 9u - (5w - 3u)$

110. $-5m + 2(m + 4n) - 7n$

111. $2g - 4[2 - 3(g - 4h) - 2]$

112. $3p + 8[q - (p - 4) - 3p]$

Linear Equations in One Variable

1. Solving Linear Equations in One Variable

An equation that can be written in the form $ax + b = 0$, where $a \neq 0$, is called a **linear equation in one variable**. The steps for solving a linear equation in one variable are given next.

Steps for Solving a Linear Equation in One Variable

1. Consider clearing fractions or decimals (if any are present) by multiplying both sides of the equation by a common denominator of all terms.
2. Simplify both sides of the equation by clearing parentheses and combining *like* terms.
3. Use the addition and subtraction properties of equality to collect the variable terms on one side of the equation.
4. Use the addition and subtraction properties of equality to collect the constant terms on the other side of the equation.
5. Use the multiplication and division properties of equality to make the coefficient of the variable term equal to 1.
6. Check your answer.

Example 1 Solving Linear Equations in One Variable

Solve. $\dfrac{1}{3}x + \dfrac{1}{4}x = x + 5$

Solution:

$$\frac{1}{3}x + \frac{1}{4}x = x + 5 \qquad \text{The LCD (least common denominator) is 12.}$$

$$12\left(\frac{1}{3}x + \frac{1}{4}x\right) = 12(x + 5) \qquad \text{Multiply both sides by 12.}$$

$$\frac{12}{1}\left(\frac{1}{3}x\right) + \frac{12}{1}\left(\frac{1}{4}x\right) = \frac{12}{1}(x) + \frac{12}{1}(5) \qquad \text{Apply the distributive property.}$$

$$4x + 3x = 12x + 60$$

$$7x = 12x + 60 \qquad \text{Combine } like \text{ terms.}$$

$$-5x = 60 \qquad \text{Subtract } 12x \text{ from both sides.}$$

$$x = -12 \qquad \text{Divide both sides by } -5.$$

The answer can be checked by substituting $x = -12$ back into the original equation and verifying that the left-hand side equals the right-hand side.

2. Problem Solving

Solving word problems in mathematics takes practice and often a little patience. We also offer the following guidelines for problem solving.

Problem-Solving Guidelines

Step 1: Read the problem carefully.

Step 2: Assign labels to unknown quantities.

Step 3: Develop an equation in words.

Step 4: Write a mathematical equation.

Step 5: Solve the equation.

Step 6: Interpret the results and write the answer in words.

problem-solving flowchart, p. 142

Example 2   **Solving an Application Involving Principal and Interest**

Kadriana took out two student loans for a total of $8500. One loan was for 5% simple interest and the other was for 8% simple interest. If the total interest after one year was $530, find the amount borrowed at each rate.

Solution:

We have two unknown values. We arbitrarily let x represent the amount borrowed at 5%. Then the remaining principal, $8500 - x$, is the amount borrowed at 8%.

Step 1: Read the problem.

interest problems, p. 158

	5% Loan	**8% Loan**	**Total**
Principal	x	$8500 - x$	
Interest	$0.05x$	$0.08(8500 - x)$	530

Step 2: Label the variables. A chart may help to organize the information given in the problem.

$$\begin{pmatrix} \text{Interest from} \\ \text{5\% loan} \end{pmatrix} + \begin{pmatrix} \text{Interest from} \\ \text{8\% loan} \end{pmatrix} = \begin{pmatrix} \text{Total} \\ \text{interest} \end{pmatrix}$$

Step 3: Verbal equation.

$$0.05x + 0.08(8500 - x) = 530$$

Step 4: Mathematical equation.

$$0.05x + 680 - 0.08x = 530$$

$$-0.03x + 680 = 530$$

Step 5: Solve the equation. Combine *like* terms.

$$-0.03x = -150$$

Subtract 680 from both sides.

$$\frac{-0.03x}{-0.03} = \frac{-150}{-0.03}$$

Divide both sides by -0.03.

$$x = 5000$$

Step 6: Write the answer in words.

The amount borrowed at 5% is $5000. The amount borrowed at 8% is given by $8500 - x = 8500 - 5000 = 3500$. Thus, $3500 was borrowed at 8%.

distance, rate, time problems, p. 148

| **Example 3** | **Solving an Application Involving Distance, Rate, and Time** |

Two cars leave a rest area at 12:00 P.M. One car travels north on Interstate 25 through Colorado and the other travels south. The southbound car travels 10 mph slower than the northbound car. After 2 hours, the cars are 260 miles apart. How fast is the northbound car traveling?

Solution: **Step 1:** Read the problem.

Let x represent the speed of the northbound car.   **Step 2:** Label the variables.

	Distance	Rate	Time
Northbound Car	$2(x)$	x	2
Southbound Car	$2(x - 10)$	$x - 10$	2

$d = rt$ distance = (rate)(time)

$$\begin{pmatrix} \text{Distance by} \\ \text{northbound car} \end{pmatrix} + \begin{pmatrix} \text{distance by} \\ \text{southbound car} \end{pmatrix} = \begin{pmatrix} \text{total} \\ \text{distance} \end{pmatrix}$$ **Step 3:** Verbal equation.

$\qquad 2x \qquad + \qquad 2(x - 10) \qquad = 260$ **Step 4:** Mathematical equation.

$$2x + 2x - 20 = 260$$ **Step 5:** Solve the equation.

$$4x - 20 = 260$$

$$4x = 280$$

$$x = 70$$

The northbound car travels at 70 mph.   **Step 6:** Interpret the answer.

3. Linear Inequalities

A linear inequality is a mathematical sentence written with one of the symbols, $<, >, \leq$, or $\geq$. The solution to an inequality can be visualized graphically on a number line or can be written in interval notation. For any real numbers a and b, Table A-4 summarizes the solution sets for five general inequalities.

Table A-4

interval notation, p. 177

Inequality	Graph	Interval Notation
$x > a$		(a, ∞)
$x \geq a$		$[a, \infty)$
$x < a$		$(-\infty, a)$
$x \leq a$		$(-\infty, a]$
$a < x < b$		(a, b)

The use of a parenthesis (or) indicates that an endpoint is not included in the solution. The use of a square bracket [or] indicates that an endpoint *is* included in the solution. Note that a parenthesis is always used for infinity.

The steps to solve a linear inequality mirror those to solve a linear equation. **However, if an inequality is multiplied or divided by a negative number, then the direction of the inequality sign must be reversed.** This is demonstrated in Example 4.

Example 4 Solving a Linear Inequality

Solve the inequality. Graph the solution and write the solution in interval notation.

solving linear inequalities, p. 180

$$-4(x - 3) + 1 \leq x + 23$$

Solution:

$-4(x - 3) + 1 \leq x + 23$	
$-4x + 12 + 1 \leq x + 23$	Apply the distributive property.
$-4x + 13 \leq x + 23$	Combine *like* terms.
$-4x - x + 13 \leq x - x + 23$	Subtract x from both sides.
$-5x + 13 \leq 23$	
$-5x + 13 - 13 \leq 23 - 13$	Subtract 13 from both sides.
$-5x \leq 10$	
$\dfrac{-5x}{-5} \geq \dfrac{10}{-5}$	Divide both sides by -5. Reverse the direction of the inequality sign.
$x \geq -2$	

Graph:

Interval notation: $[-2, \infty)$

Review B   Practice Exercises

For Exercises 1–30, solve the equation.

1. $4x = -84$

2. $-3y = 72$

3. $\dfrac{t}{5} = 12$

4. $\dfrac{r}{3} = 13$

5. $6 + p = -14$

6. $-5 + q = -23$

7. $7x + 12 = 33$

8. $-3x - 4 = -1$

9. $4(x + 5) = 14$

10. $5(y - 2) = -22$

11. $3b + 25 = 5b - 21$

12. $4c - 9 = 6c + 17$

13. $3w + 2(w - 7) = 7(w - 1)$

14. $x + 5(x - 4) = 3(x - 8) - 5$

15. $\frac{1}{5}y - \frac{3}{10}y = y - \frac{2}{5}y + 1$

16. $\frac{2}{3}q + 2 - \frac{1}{9}q = \frac{5}{9}q + \frac{4}{3}q$

17. $\frac{5}{6}(x + 2) = \frac{1}{3}x - \frac{3}{2}$

18. $\frac{1}{4}(6 - p) = \frac{3}{8}(2 + 3p)$

19. $0.2a + 6 = 1.8a - 2.8$

20. $0.3(v - 3) = 1.4v - 3.1$

21. $0.72t - 1.18 = 0.28(4 - t)$

22. $0.08(100 - 73x) = 1.2(1 - 5x)$

23. $-4(x + 1) - 2x = -2(3x + 2) - x$

24. $5(z + 3) - 2z + 6(1 - z) = 4 - 2(1 + z)$

25. $11 - 5(y + 3) = -2[2y - 3(1 - y)]$

26. $-4p + 2[p - 4(2 - p)] = 4(p - 5)$

27. $-8m - 2[4 - 3(m + 1)] = 4(3 - m) + 4$

28. $10x - 5[1 + 3(1 - 2x)] = [4(x - 3) + 27] + x$

29. $4 - 6[2y - 2(y + 3)] = 7 - y$

30. $-3(x + 4) + 2x + 12 = -6(x - 5)$

For Exercises 31–40, solve.

31. Five times the product of 8 and a number is 20. What is the number?

32. One half of the total of 6 times a number and 20 is 4. What is the number?

33. The sum of three consecutive integers is -57. Find the integers.

34. The sum of three consecutive odd integers is 111. Find the integers.

35. If Casey invested $2500 in an account and earned $112.50 in interest the first year, what was the simple interest rate?

36. Roberto deposited $4050 in an account that made $202.50 simple interest the first year. Find the simple interest rate.

37. Ketul made a total investment of $3500 into two accounts earning 4.2% simple interest and 3.8% simple interest. If his total interest after one year was $141.40, how much was invested in each account?

38. Katlin borrowed $1800, part from her parents and part from a bank. The bank charged 8% interest but her parents only charged 2%. If she paid off the loan in one year and paid a total of $102 interest, how much did she borrow from her parents?

39. A car and a truck leave home at the same time traveling in opposite directions. The truck travels 4 mph slower than the car. If the distance between the two vehicles is 372 miles after 3 hours, find the speed of each vehicle.

40. Juan and Peter both leave a shopping center at the same time going in opposite directions. Juan is on his bike and travels 3 mph faster than Peter who is on his skateboard. After 1.5 hours they are 19.5 miles apart. How fast does Peter travel.

For Exercises 41–56, solve the inequality. Graph the solution set and write the set in interval notation.

41. $3 - x > 5$

$-7\ -6\ -5\ -4\ -3\ -2\ -1$

42. $4 - 2d < 6$

$-2\ -1\ \ 0\ \ 1\ \ 2\ \ 3\ \ 4\ \ 5$

43. $4y + 6 \geq -18$

$-7\ -6\ -5\ -4\ -3\ -2\ -1\ \ 0\ \ 1$

44. $3r - 13 \leq 27$

$8\ \ 9\ \ 10\ 11\ 12\ \ \ \ 14$

45. $8m - 15 < 9m - 13$

$-3\ -2\ -1\ \ 0\ \ 1\ \ 2\ \ 3\ \ 4$

46. $6n + 23 > -3n - 13$

$-5\ -4\ -3\ -2\ -1\ \ 0\ \ 1\ \ 2$

47. $1 - 5(2t - 7) < 38t$

$0\ \ 1\ \ 2\ \ 3\ \ 4\ \ 5\ \ 6$

48. $9r + 3(r - 14) \geq 6$

$3\ \ 4\ \ 5\ \ 6\ \ 7\ \ 8\ \ 9\ \ 10$

49. $-5 < 4p + 2 \leq 6$

$-3\ -2\ -1\ \ 0\ \ 1\ \ 2$

50. $-8 \leq 7 + 5k \leq 17$

$-4\ -3\ -2\ -1\ \ 0\ \ 1\ \ 2\ \ 3$

51. $2 > -y + 1 > -6$

$-2\ -1\ \ 0\ \ 1\ \ 2\ \ 3\ \ 4\ \ 5\ \ 6\ \ 7\ \ 8$

52. $0 \geq -3x \geq -10$

$-1\ \ 0\ \ 1\ \ 2\ \ 3\ \ 4$

53. $\dfrac{1}{7}h + 2 > \dfrac{5}{14}h - \dfrac{1}{2}$

$5\ \ 6\ \ 7\ \ 8\ \ 9\ \ 10\ 11\ 12\ 13$

54. $\dfrac{2}{3}b - \dfrac{1}{6} \leq 3b + \dfrac{5}{6}$

$-1\ \ 0\ \ 1\ \ 2\ \ 3\ \ 4\ \ 5\ \ 6$

55. $0.4w + 3.1(w - 1) \leq -3.8$

$-5\ -4\ -3\ -2\ -1\ \ 0\ \ 1$

56. $2.1x - 2(x + 5.1) > -8.2$

$19\ 20\ 21\ 22\ 23\ 24\ 25\ 26$

Linear Equations in Two Variables

1. Plotting Ordered Pairs

In Figure A-1, we review some important features of a rectangular coordinate system. The x- and y-axes are perpendicular lines that cross at a common point called the **origin**. The arrows indicate the positive direction for the x- and y-axes. The four regions in the graph are called **quadrants**. Each point in a rectangular coordinate system is represented by an **ordered pair**. Figure A-2 shows the location of the following ordered pairs: $(4, 1), (-2, 5), (5, -2), (-2, 1), (0, 4)$, and $\left(-\frac{7}{2}, -2\right)$. For example, to plot the point represented by $(-2, 5)$ move -2 units in the x-direction (left) and 5 units in the y-direction (up) and draw a dot.

Concepts

1. Plotting Ordered Pairs
2. Graphing Linear Equations in Two Variables
3. x- and y-Intercepts
4. Horizontal and Vertical Lines
5. Slope of a Line
6. Slope-Intercept Form
7. Point-Slope Formula

rectangular coordinate system, p. 203

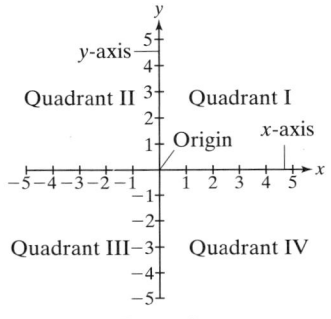

Figure A-1

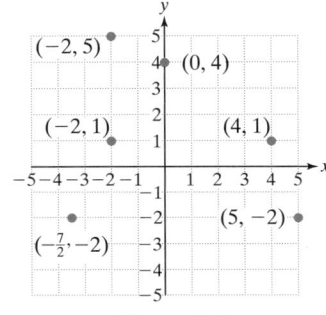

Figure A-2

2. Graphing Linear Equations in Two Variables

An equation that can be written in the form $Ax + By = C$ (where A and B are not both 0) is called a linear equation in two variables. The equation $3x + y = 5$ is a linear equation in two variables. A solution to such an equation is an ordered pair (x, y) that when substituted into the equation makes the equation a true statement. For example, the ordered pair $(-1, 8)$ is a solution to the equation $3x + y = 5$ because $3(-1) + 8 = 5$. A linear equation in two variables has infinitely many solutions that when plotted form a line in a rectangular coordinate system. To find a solution to a linear equation in two variables, we substitute any real number for one of the variables in the equation and then solve for the other variable.

Example 1 · Graphing a Linear Equation in Two Variables

Complete the ordered pairs to form solutions to the equation $3x + y = 5$. Then graph the equation.

$$(1, \quad) \quad (\quad, -4) \quad (0, \quad) \quad (\quad, 0)$$

Solution:

The ordered pairs can also be represented in tabular form as follows.

x	y
1	
	-4
0	
	0

$3x + y = 5$

$3(1) + y = 5$

$3x + (-4) = 5$

$3(0) + y = 5$

$3x + (0) = 5$

x	y	Ordered pair
1	2	$(1, 2)$
3	-4	$(3, -4)$
0	5	$(0, 5)$
$\frac{5}{3}$	0	$(\frac{5}{3}, 0)$

The ordered pairs are graphed in Figure A-3. Notice that the points all "line up." The line drawn through the points represents all solutions to the equation. Therefore, we say the line represents a graph of the equation.

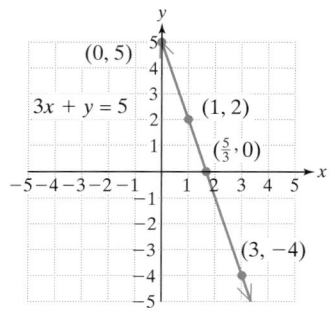

Figure A-3

---

3. x- and y-Intercepts

In Example 1, the points $(\frac{5}{3}, 0)$ and $(0, 5)$ lie on the x-axis and y-axis, respectively. The point $(\frac{5}{3}, 0)$ is called an x-intercept and the point $(0, 5)$ is called a y-intercept. In general, the x- and y-intercepts of a graph are points where the graph intersects the x- and y-axes, respectively. Notice that the x-intercept $(\frac{5}{3}, 0)$ has a y-coordinate of 0. The y-intercept $(0, 5)$ has an x-coordinate of 0.

Finding the *x*- and *y*-Intercepts of a Linear Equation in Two Variables

Consider an equation in the variables x and y.

1. To find the x-intercept substitute $y = 0$ into the equation and solve for x.
2. To find the y-intercept substitute $x = 0$ into the equation and solve for y.

Example 2   **Finding *x*- and *y*-Intercepts**

Given $7x - 3y = 6$, find the x- and y-intercepts.

Solution:

To find the x-intercept,
substitute zero for y:

$$7x - 3(0) = 6$$

$$7x = 6$$

$$x = \frac{6}{7}$$

To find the y-intercept,
substitute zero for x:

$$7(0) - 3y = 6$$

$$-3y = 6$$

$$y = -2$$

The x-intercept is $\left(\frac{6}{7}, 0\right)$ and the y-intercept is $(0, -2)$.

---

4. Horizontal and Vertical Lines

In Section 3.2, we learned that an equation of the form $x = k$ is a vertical line in a rectangular coordinate system. An equation of the form $y = k$ is a horizontal line.

horizontal and vertical lines, p. 219

Example 3   **Graphing Horizontal and Vertical Lines**

Graph the lines.   **a.** $x = 4$ **b.** $-2y = 5$

graphing horizontal and vertical lines, p. 219

Solution:

a. The equation $x = 4$ is in the form $x = k$. Therefore, the graph of the equation is a vertical line that passes through the x-axis at 4.

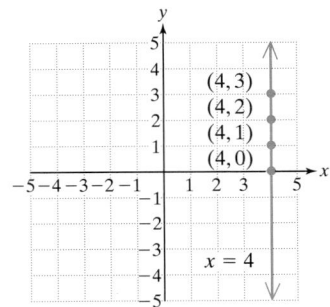

The solutions to the equation $x = 4$ are ordered pairs whose x-coordinate is 4 and whose y-coordinate has no restriction. Several such solutions are $(4, 0)$, $(4, 1)$, $(4, 2)$, $(4, 3)$, and so on.

b. The equation $-2y = 5$ is equivalent to $y = -\frac{5}{2}$. The equation is in the form $y = k$. Therefore, the graph of the equation is a horizontal line passing through the y-axis at $-\frac{5}{2}$.

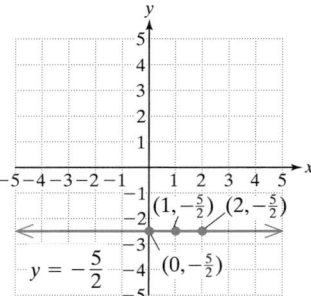

The solutions to the equation $y = -\frac{5}{2}$ are ordered pairs whose y-coordinate is $-\frac{5}{2}$ and whose x-coordinate has no restriction. Several such solutions are $(0, -\frac{5}{2})$, $(1, -\frac{5}{2})$, $(2, -\frac{5}{2})$, and so on.

5. Slope of a Line

The slope of a line (often denoted by m) is a measure of the line's "steepness." It measures the ratio of an incremental change in y to an incremental change in x between two points on the line.

The slope is $\frac{3}{2}$.

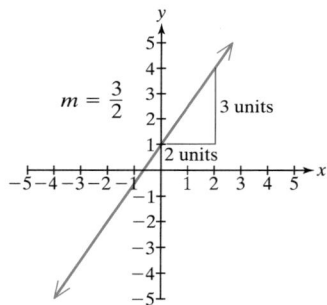

The slope is 4.

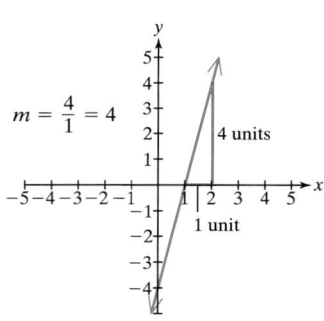

The slope of a line may be positive, negative, zero, or undefined.

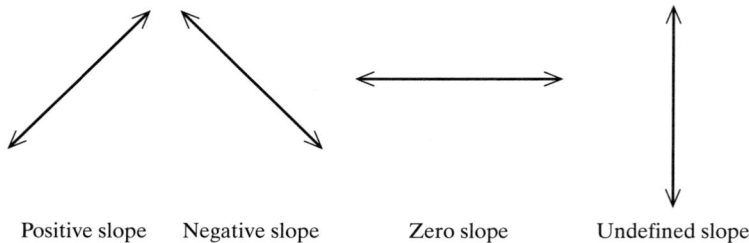

Positive slope   Negative slope   Zero slope   Undefined slope

To find the slope of a line given two points, (x_1, y_1) and (x_2, y_2), we use the slope formula:

$$m = \frac{y_2 - y_1}{x_2 - x_1}.$$

| Example 4 | Finding the Slope Given Two Points on a Line |

Find the slope of the line between $(-3, -4)$ and $(7, -1)$.

slope formula, p. 232

Solution:

(x_1, y_1) (x_2, y_2)
$(-3, -4)$ $(7, -1)$ Label the points.

$m = \dfrac{y_2 - y_1}{x_2 - x_1} = \dfrac{-1 - (-4)}{7 - (-3)} = \dfrac{3}{10}$ The slope is $\dfrac{3}{10}$.

The slopes of two lines can be used to determine whether the lines are parallel or perpendicular.

parallel and perpendicular lines, p. 235

- Two lines that are parallel have the same slope. That is, $m_1 = m_2$.
- The slopes of perpendicular lines are related such that one slope is the opposite of the reciprocal of the other.

  That is, $m_1 = -\dfrac{1}{m_2}$.

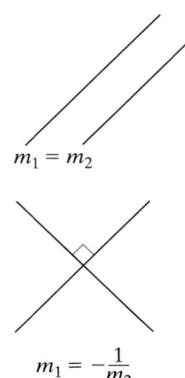

$m_1 = m_2$

$m_1 = -\dfrac{1}{m_2}$

6. Slope-Intercept Form

A linear equation in two variables written in the form $y = mx + b$ is said to be in slope-intercept form. In such a case, m is the slope, and $(0, b)$ is the y-intercept. To write a linear equation in slope-intercept form, solve the equation for y.

| Example 5 | Writing an Equation in Slope-Intercept Form |

Given $2x + 5y = 20$,

slope-intercept form of a line, p. 245

  a. Write the equation in slope-intercept form.

  b. Identify the slope and y-intercept.

  c. Graph the line using the slope and y-intercept.

Solution:

  a. $2x + 5y = 20$ To write in the form $y = mx + b$, solve for y.

$5y = -2x + 20$ Subtract $2x$ from both sides.

$\dfrac{5y}{5} = \dfrac{-2x}{5} + \dfrac{20}{5}$ Divide by 5.

$y = -\dfrac{2}{5}x + 4$ The equation is in slope-intercept form.

graphing a line using the slope and y-intercept, p. 247

b. The slope is $-\frac{2}{5}$ and the y-intercept is $(0, 4)$.

c. To graph the line using the slope and y-intercept, first plot the y-intercept. Then we can use the slope to find a second point on the line. For example,

The slope can be interpreted as $\frac{-2}{5}$, which indicates that we move two units in the negative y-direction (down) and 5 units in the positive x-direction (right).

The slope can also be interpreted as $\frac{2}{-5}$ which indicates that we move two units in the positive y direction (up) and 5 units in the negative x-direction (left). See Figure A-4.

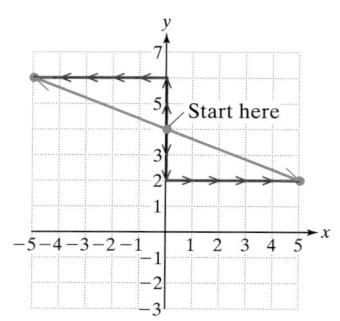

Figure A-4

Example 6 | **Determining Whether Lines are Parallel, Perpendicular, or Neither**

comparing slopes of lines, p. 248

Determine whether the lines l_1 and l_2 are parallel, perpendicular, or neither.

$l_1: 2x - 3y = 12$ $l_2: 6x + 4y = 4$

Solution:

Write each line in slope-intercept form and compare the slopes.

$l_1: 2x - 3y = 12$ $l_2: 6x + 4y = 4$

$$-3y = -2x + 12$$ $$4y = -6x + 4$$

$$\frac{-3y}{-3} = \frac{-2x}{-3} + \frac{12}{-3}$$ $$\frac{4y}{4} = \frac{-6x}{4} + \frac{4}{4}$$

$$y = \frac{2}{3}x - 4$$ $$y = -\frac{3}{2}x + 1$$

The slope of l_1 is $\frac{2}{3}$.   The slope of l_2 is $-\frac{3}{2}$.

The slope of l_1 is the opposite of the reciprocal of the slope of l_2. Therefore, the lines are perpendicular.

---

7. Point-Slope Formula

The point-slope formula is a useful tool to find an equation of a line given a point on the line and the slope of the line.

point-slope formula, p. 256

Point-Slope Formula

The **point-slope formula** is given by

$$y - y_1 = m(x - x_1)$$

where m is the slope of the line and (x_1, y_1) is a known point on the line.

Example 7	**Writing an Equation of a Line Given Two Points**

Write an equation of the line passing through the points $(-2, -1)$ and $(-4, 5)$. Write the final answer in slope-intercept form and in standard form.

Solution:

We can find the slope of the line by using the slope formula. Once the slope is known, we can apply the point-slope formula using the slope and either given point.

$(-2, -1)$ and $(-4, 5)$
(x_1, y_1) and (x_2, y_2) Label the points.

$m = \dfrac{y_2 - y_1}{x_2 - x_1} = \dfrac{5 - (-1)}{-4 - (-2)} = \dfrac{6}{-2} = -3$ Apply the slope formula.

$y - y_1 = m(x - x_1)$

$y - (-1) = -3[(x - (-2))]$ Apply the point-slope formula.

$y + 1 = -3(x + 2)$

$y + 1 = -3x - 6$

$y = -3x - 7$ (slope-intercept form)   Solve for y.

or

$3x + y = -7$ (standard form)   Write in the form $Ax + By = C$.

Review C	**Practice Exercises**

For Exercises 1–8, plot the points on a rectangular coordinate system.

1. $(2, 4)$ **2.** $(4, 2)$ **3.** $(4, -3)$

4. $(0, 2)$ **5.** $(-3, 0)$ **6.** $(0, 0)$

7. $\left(-\dfrac{3}{2}, -2\right)$ **8.** $(-1, 3)$

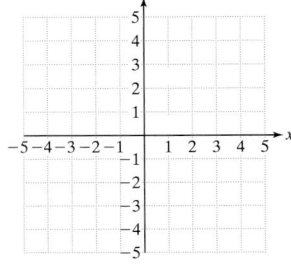

For Exercises 9–16, identify the quadrant in which each point is located.

9. $(20, -6)$ **10.** $(-3, -3)$ **11.** $\left(-13, -\dfrac{7}{8}\right)$ **12.** $\left(\dfrac{12}{7}, 5\right)$

13. $(\pi, 14)$ **14.** $(7.8, -42)$ **15.** $(-3.8, 6.2)$ **16.** $(-7.8, 42)$

For Exercises 17–22, complete the table and graph the line.

17. $x + y = 4$

x	y
3	
	-1
0	

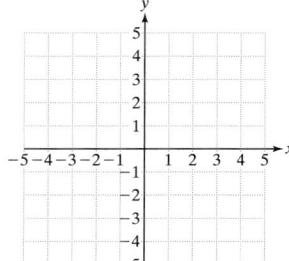

18. $3x - y = 6$

x	y
2	
	1
1	

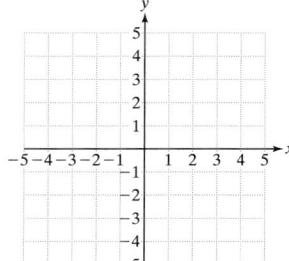

19. $y = 5x$

x	y
0	
$\frac{2}{5}$	
-1	

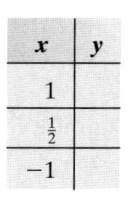

20. $y = -2x + 3$

x	y
1	
$\frac{1}{2}$	
-1	

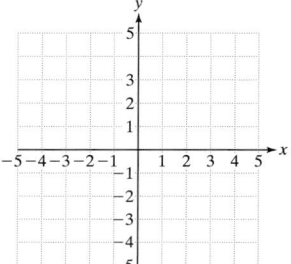

21. $y = \frac{2}{3}x + 1$

x	y
3	
-3	
0	

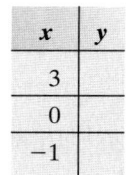

22. $y = -\frac{4}{3}x + 2$

x	y
3	
0	
-1	

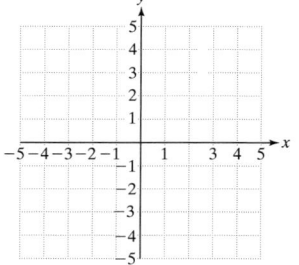

For Exercises 23–34, find the x- and y-intercepts.

23. $3x - 2y = 18$

24. $x - 4y = 16$

25. $5x + y = -15$

26. $-6x - y = 6$

27. $\frac{1}{2}x + y = 2$

28. $x - \frac{1}{3}y = -4$

29. $-3x + 4y = 0$

30. $6x - 5y = 0$

31. $y = -\frac{2}{3}x - 3$

32. $y = \frac{3}{5}x + 1$

33. $y = 6x - 5$

34. $y = -2x - 3$

For Exercises 35–40, write the equation in the form $x = k$ or $y = k$ and identify the line as horizontal or vertical.

35. $2x = 6$

36. $5y = 10$

37. $-3y + 1 = 2$

38. $7x - 1 = 3$

39. $5x = 0$

40. $\frac{2}{3}y = 0$

For Exercises 41–50, write the equation in slope-intercept form (if possible). Then identify the slope and the *y*-intercept (if they exist).

41. $4x - 5y = 6$

42. $2x - 5y = 11$

43. $-6x + 2y = 3$

44. $-x + 3y = 4$

45. $5x - 1 = 4$

46. $2y - 3 = 0$

47. $x + 4y = 0$

48. $8x + 2y = 0$

49. $y + 4 = 8$

50. $x - 3 = 5$

For Exercises 51–56, graph the line given the slope and *y*-intercept.

51. Slope is $\dfrac{1}{2}$, *y*-intercept is $(0, -2)$

52. Slope is -3, *y*-intercept is $(0, 1)$

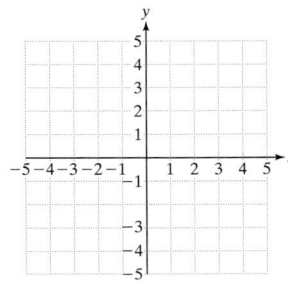 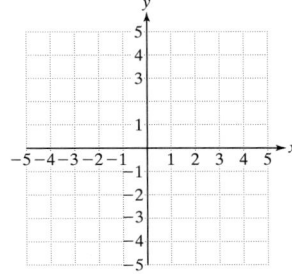

53. Slope is $-\dfrac{3}{2}$, *y*-intercept is $(0, 3)$

54. Slope is $\dfrac{5}{3}$, *y*-intercept is $(0, -1)$

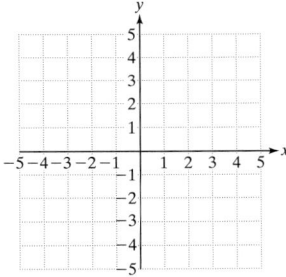 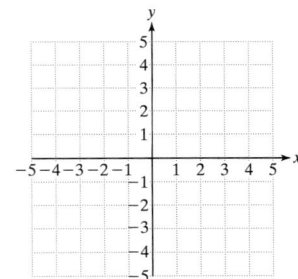

55. Slope is 4, *y*-intercept is $(0, -2)$

56. Slope is 2, *y*-intercept is $(0, -3)$

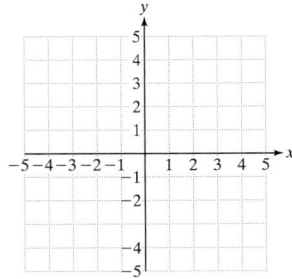 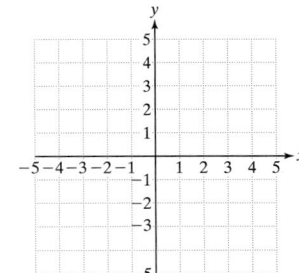

For Exercises 57–68, graph the line.

57. $2x - 3y = 12$

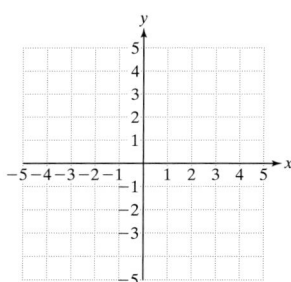

58. $2x + 4y = -8$

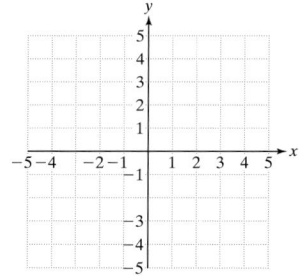

59. $x - 2y = -4$

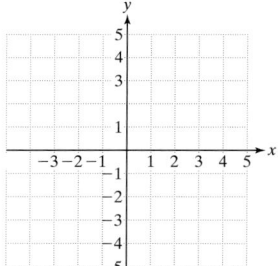

60. $6x - y = 4$

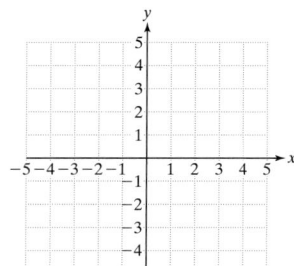

61. $4x + y = 0$

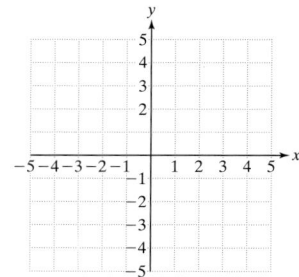

62. $-x - 3y = 0$

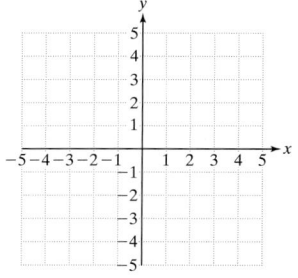

63. $y = -\dfrac{2}{3}x + 1$

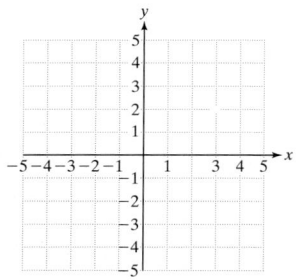

64. $y = \dfrac{1}{4}x - 1$

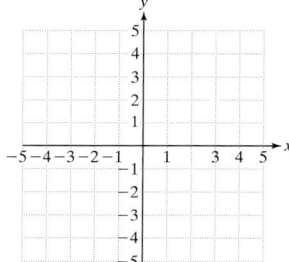

65. $-2x + 1 = 5$

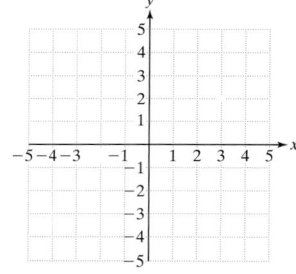

66. $4y - 2 = 6$

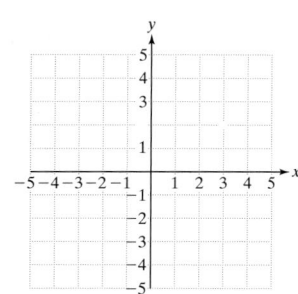

67. $y = 3x - 3$

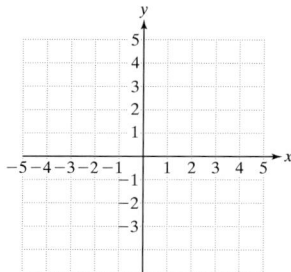

68. $y = -4x - 4$

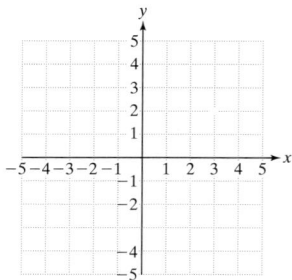

For Exercises 69–76, determine if the lines are parallel, perpendicular, or neither.

69. l_1: $-2x + y = 1$
     l_2: $2x - y = 3$

70. l_1: $y = 3x + 6$
     l_2: $-x + 3y = -3$

71. l_1: $-2x + 5y = 5$
     l_2: $5x - 2y = -2$

72. l_1: $\quad -7x + y = 1$
    l_2: $\quad -7x = 0$

73. l_1: $\quad x + y = 7$
    l_2: $\quad x - y = -2$

74. l_1: $\quad y = -\dfrac{1}{4}x + 2$
    l_2: $\quad 2x + 8y = 24$

75. l_1: $\quad y = 6$
    l_2: $\quad y = -2$

76. l_1: $\quad x = 3$
    l_2: $\quad y = 0$

For Exercises 77–86, find an equation of the line given the following information. Write the final answer in slope-intercept form and in standard form.

77. The slope is -3 and the line passes through $(-5, -2)$.

78. The slope is $\frac{1}{2}$ and the line passes through $(5, 0)$.

79. The line passes through $(-3, -4)$ and $(-1, 4)$.

80. The line passes through $(-1, 7)$ and $(4, -3)$.

81. The line passes through $(4, 2)$ and is parallel to the line defined by $3y = 6x - 1$.

82. The line passes through $(-5, 6)$ and is parallel to the line defined by $4x = y - 3$.

83. The line passes through $(1, -2)$ and is perpendicular to the line defined by $-2x + 4y = 5$.

84. The line passes through $(3, 0)$ and is perpendicular to the line defined by $3x + 9y = 1$.

85. The line passes through $(4, -9)$ and is parallel to the line $x = 3$.

86. The line passes through $(-7, 2)$ and is perpendicular to the line $2y = 8$.

Systems of Linear Equations in Two Variables | Review D

1. Solving Systems of Equations by Graphing

Two or more linear equations can form a system of linear equations. For example, the following is a system of linear equations.

$$2x + y = 5$$
$$-x + y = 2$$

A solution to a system of equations is an ordered pair that satisfies both equations. Graphically, a solution to a system of linear equations is a point of intersection of the two lines.

Concepts

1. Solving Systems of Equations by Graphing
2. Solving Systems of Equations by the Substitution Method
3. Solving Systems of Equations by the Addition Method
4. Applications of Systems of Equations

solving systems by graphing, p. 290

Example 1 Solving a System of Linear Equations by Graphing

Solve by graphing. $\quad 2x + y = 5$
                 $-x + y = 2$

Solution:

First, graph the line defined by each equation. To graph the lines, we will use the slope-intercept form of each equation.

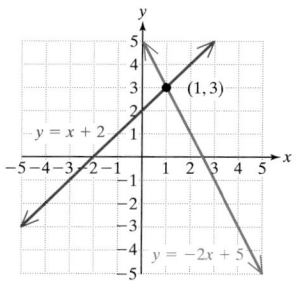

$$2x + y = 5 \xrightarrow{\text{slope-intercept form}} y = -2x + 5$$

$$-x + y = 2 \xrightarrow{\hspace{2cm}} y = x + 2$$

From the graph, we estimate the point of intersection to be (1, 3). The solution must be checked in each original equation.

$$2x + y = 5 \longrightarrow 2(1) + (3) \overset{?}{=} 5 \longrightarrow 5 = 5 \ ✔ \ \text{True}$$

$$-x + y = 2 \longrightarrow -(1) + (3) \overset{?}{=} 2 \longrightarrow 2 = 2 \ ✔ \ \text{True}$$

The solution is (1, 3).

2. Solving Systems of Equations by the Substitution Method

solving systems by substitution, p. 302

The graphing method to solve a system of linear equations can be cumbersome because of the limited accuracy of the graph. We will review two algebraic methods to solve systems of linear equations. The first is called the substitution method.

Example 2   Solving a System of Linear Equations Using the Substitution Method

Solve by using the substitution method: $3x - y = 6$
$2x + 5y = 4$

Solution:

The first step in the substitution method is to isolate one of the variables in one of the equations. It is generally easiest to solve for a variable whose coefficient is 1 or −1. Therefore, we will isolate y in the first equation.

First equation: $3x - y = 6$

$$-y = -3x + 6$$

$$y = \underbrace{3x - 6}$$

Second equation: $2x + 5y = 4$ Now substitute the quantity $(3x - 6)$ for y in

$$2x + 5(3x - 6) = 4 \qquad \text{the other equation.}$$

$$2x + 15x - 30 = 4$$

$$17x - 30 = 4$$

$$17x = 34$$

$$x = 2$$

We can find the value of y by substituting $x = 2$ into either of the two equations. However, we will use the equation $y = 3x - 6$ because y is already isolated.

$y = 3x - 6$ First equation

$y = 3(2) - 6$ Substitute $x = 2$.

$y = 0$

The solution to the system is (2, 0).   The check is left to the reader.

3. Solving Systems of Equations by the Addition Method

A second method to solve a system of linear equations is called the addition method (also called the elimination method). The basis of this method is to eliminate one of the variables by adding two equations that have opposite coefficients for that variable.

| Example 3 | Solving a System of Linear Equations by the Addition Method |

Solve by using the addition method.
$$2x = 5y - 4$$
$$3(x + y) = y + 13$$

solving systems by the addition method, p. 312

Solution:

To use the addition method, first write each equation in standard form $Ax + By = C$.

$$2x = 5y - 4 \xrightarrow{\text{subtract } 5y} 2x - 5y = -4 \longrightarrow 2x - 5y = -4$$

$$3(x + y) = y + 13 \xrightarrow[\text{clear parentheses}]{} 3x + 3y = y + 13 \xrightarrow[\text{subtract } y]{} 3x + 2y = 13$$

We will now eliminate one of the variables by multiplying one or both equations by appropriate constants to create opposite coefficients on that variable. Suppose we want to eliminate the x-variable. To do so, we want the coefficients on x to be 6 and -6 (the number 6 is the LCM of the original coefficients 2 and 3).

$$2x - 5y = -4 \xrightarrow{\text{To create an } x\text{-coefficient of 6, multiply both sides by 3.}} 6x - 15y = -12$$

$$3x + 2y = 13 \xrightarrow[\text{To create an } x\text{-coefficient of } -6, \text{ multiply both sides by } -2.]{} -6x - 4y = -26$$

$$-19y = -38$$

Add the equations.
With the x-variable
eliminated, solve for y.

$$y = 2$$

Substitute $y = 2$ back into either of the original equations. We will use the first.

$$2x = 5y - 4$$

$$2x = 5(2) - 4 \qquad \text{Substitute } y = 2 \text{ into the first equation.}$$

$$2x = 10 - 4 \qquad \text{Solve for } x.$$

$$2x = 6$$

$$x = 3$$

The solution is $(3, 2)$.   The check is left for the reader.

Two linear equations in a system may represent parallel lines. In such a case, the system has no solution because there is no point of intersection. A system of equations that has no solution is called an **inconsistent system**.

definition of an inconsistent system, p. 291

| Example 4 | Identifying an Inconsistent System |

Solve the system of linear equations.
$$2x + y = 4$$
$$4x + 2y = 0$$

Solution:

Using the addition method we have

$$2x + y = 4 \xrightarrow{\quad \text{multiply by } -2 \quad} -4x - 2y = -8$$

$$4x + 2y = 0 \xrightarrow{\hspace{4cm}} \underline{4x + 2y = 0}$$

$$0 = -8 \text{ (contradiction)}$$

This system has no solution. The system is inconsistent.

Graphing the linear equations in Example 4, we see that the lines are indeed parallel and that there is no point of intersection (Figure A-5). In general if a system of equations results in a contradiction (such as $0 = -8$), then there is no solution.

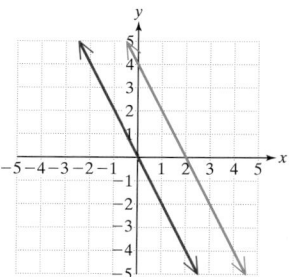

Figure A-5

Two linear equations may represent the same line. In such a case, all points on the common line are solutions to the system. That is, there are infinitely many solutions, and the system is called a **dependent system**.

Example 5 | **Solving a Dependent System of Linear Equations**

Solve the system. $6x - 9y = 18$

$$\frac{1}{2}x - \frac{3}{4}y = \frac{3}{2}$$

definition of a dependent system, p. 291

Solution:

First note that we can clear fractions in the second equation by multiplying by the LCD of all fractions in the equation.

$$6x - 9y = 18 \xrightarrow{\hspace{2cm}} 6x - 9y = 18 \xrightarrow{\hspace{2cm}} 6x - 9y = 18$$

$$\frac{1}{2}x - \frac{3}{4}y = \frac{3}{2} \xrightarrow{\text{multiply by } 4} 2x - 3y = 6 \xrightarrow{\text{multiply by } -3} \underline{-6x + 9y = -18}$$

$$0 = 0$$

The system reduces to the identity, $0 = 0$. This tells us that the equations have the same (identical) solution set, and that the system of equations is dependent.

Both equations can be written in slope-intercept form as $y = \frac{2}{3}x - 2$. All points on this line are solutions to the system. Therefore, we can write the solution set as $\{(x, y) \mid y = \frac{2}{3}x - 2\}$.

4. Applications of Systems of Equations

A system of equations is often useful to solve application problems where there are two unknown quantities.

| Example 6 | Solving an Application Using a System of Linear Equations |

Ashlee invested a total of $6000 in two accounts, a money market account and a certificate of deposit (CD). The money market pays 3.5% simple interest, and the CD pays 5% simple interest. At the end of one year, she earned $276 in interest. Find the amount invested in each account.

applications of linear equations in two variables. p. 323

Solution:

Let x represent the amount invested at 3.5%.
Let y represent the amount invested at 5%.

We have two variables, and therefore, we need two independent equations relating the variables. One equation relates the amount invested (principal), the second equation relates the amount of interest earned.

$$\begin{pmatrix} \text{Principal} \\ \text{invested at 3.5\%} \end{pmatrix} + \begin{pmatrix} \text{Principal} \\ \text{invested at 5\%} \end{pmatrix} = \begin{pmatrix} \text{Total} \\ \text{invested} \end{pmatrix} \qquad x + y = 6000$$

$$\begin{pmatrix} \text{Interest} \\ \text{earned at 3.5\%} \end{pmatrix} + \begin{pmatrix} \text{Interest} \\ \text{earned at 5\%} \end{pmatrix} = \begin{pmatrix} \text{Total} \\ \text{interest} \end{pmatrix} \qquad 0.035x + 0.05y = 276$$

$x + y = 6000$ $\xrightarrow{\text{multiply by } -35}$ $-35x - 35y = -210{,}000$

$0.035x + 0.05y = 276$ $\xrightarrow{\text{multiply by } 1000}$ $\underline{35x + 50y = 276{,}000}$

$$15y = 66{,}000$$

$$y = 4400$$

$x + y = 6000$

$x + 4400 = 6000$ Substitute $y = 4400$ back into the first equation.

$x = 1600$

Interpret the results: $x = 1600$ means that $1600 was invested at 3.5%.
$y = 4400$ means that $4400 was invested at 5%.

| Review D | **Practice Exercises** |

Boost *your* GRADE at mathzone.com!

MathZone

- Practice Problems
- Self-Tests
- NetTutor
- e-Professors
- Videos

For Exercises 1–4, solve the system by graphing.

1. $3x + y = 1$

$-2x + 2y = -6$

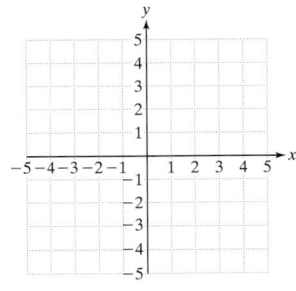

2. $3x - 2y = -6$

$4x + y = 3$

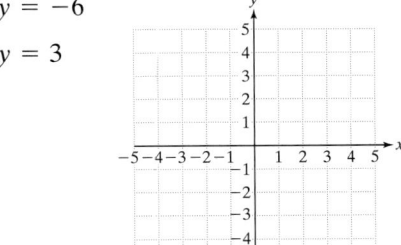

3. $2x = 8$

$-3x + 5y = -2$

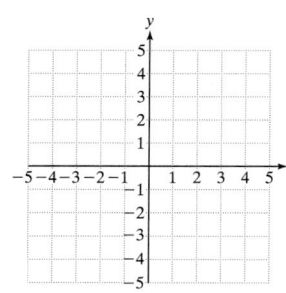

4. $y + 3 = 1$

$x - 2y = 3$

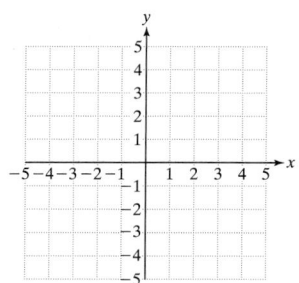

For Exercises 5–8, solve the systems by using the substitution method.

5. $y = 3x - 4$

$-2x + 5y = -7$

6. $x = 4y + 1$

$-2x + 5y = -20$

7. $9x - 4y = 9$

$x + 6y = 1$

8. $-4x - 2y = -14$

$3x + y = 9$

For Exercises 9–12, solve the systems by using the addition method.

9. $2x + y = -5$

$-3x + 4y = -31$

10. $-6x + 5y = 1$

$x - 3y = -11$

11. $5x - 6y = 16$

$3x - 4y = 10$

12. $7x + 2y = 19$

$5x + 3y = 12$

For Exercises 13–24, solve the systems using any method.

13. $5x + 2y = 10$

$-10x - 4y = 20$

14. $-3x + y = 6$

$9x - 3y = 0$

15. $3x + y - 7 = x - 4$

$3x - 4y + 4 = -6y + 5$

16. $7y - 8y - 3 = -3x + 4$

$5(2x - y) - 12 = 13$

17. $-2x = -y - 14$

$4x - 2y = 28$

18. $y = 0.4x - 0.3$

$\dfrac{2}{5}x - y = \dfrac{3}{10}$

19. $x + y = 3200$

$0.06x + 0.04y = 172$

20. $x + y = 4500$

$0.07x + 0.05y = 291$

21. $\dfrac{1}{3}x + \dfrac{1}{2}y = \dfrac{2}{3}$

$y = \dfrac{2}{3}x - \dfrac{4}{3}$

22. $-y = \dfrac{1}{8}x - \dfrac{1}{8}$

$-\dfrac{1}{4}x + \dfrac{1}{2}y = \dfrac{13}{8}$

23. $u - v = 3$

$6u - 4v = 13$

24. $5v + 6w = -11$

$3v + w = -4$

25. Lynn has $10,000 to invest. In one account he earns 6% simple interest and in the other he earns 4.5% simple interest. How much did he invest in each account if he makes $570 in interest after 1 year?

26. Greg has $8500 to invest. In a CD (certificate of deposit) he earns 5.5% simple interest. In a money market account he earns 4% simple interest. How much did he invest in each account if he made $415 in interest in one year?

27. Two angles are complementary (the sum of their measures is 90°). The difference in the measures of the larger angle and the smaller angle is 14°. Find the measure of each angle.

28. Two angles are supplementary (the sum of their measures is 180°). The measure of one angle is 11 times the measure of the other angle. Find the measure of each angle.

29. The sum of Denisha's age and her sister's age is 48. If Denisha is 4 years older than her sister, find each of their ages.

30. In a right triangle, one acute angle measures $10°$ less than 4 times the measure of the other acute angle. Find the measure of each acute angle.

Polynomials and Properties of Exponents

Review E

1. Properties of Exponents

Recall that expressions with exponents can be simplified using the following definitions and properties.

Concepts

1. Properties of Exponents
2. Scientific Notation
3. Addition and Subtraction of Polynomials
4. Multiplication of Polynomials
5. Division of Polynomials

Properties and Definitions Involving Exponents

Assume that a and b are real numbers such that $a \neq 0$, $b \neq 0$ and m and n are integers.

Property/Definition		Example
Multiplication of like bases:	$b^m b^n = b^{m+n}$	$x^3 x^5 = x^{3+5} = x^8$
Division of like bases:	$\dfrac{b^m}{b^n} = b^{m-n}$	$\dfrac{y^7}{y^3} = y^{7-3} = y^4$
The power rule:	$(b^m)^n = b^{m \cdot n}$	$(5^4)^2 = 5^{4 \cdot 2} = 5^8$
Power of a product:	$(ab)^m = a^m b^m$	$(5x)^2 = 5^2 x^2 = 25x^2$
Power of a quotient:	$\left(\dfrac{a}{b}\right)^m = \dfrac{a^m}{b^m}$	$\left(\dfrac{x}{4}\right)^3 = \dfrac{x^3}{4^3} = \dfrac{x^3}{64}$
Definition of b^0:	$b^0 = 1$	$5^0 = 1$
Definition of b^{-n}:	$b^{-n} = \left(\dfrac{1}{b}\right)^n = \dfrac{1}{b^n}$	$x^{-6} = \left(\dfrac{1}{x}\right)^6 = \dfrac{1}{x^6}$

multiplying and dividing common bases, p. 346

power rule for exponents, p. 354

definition of b^0 and b^{-n}, p. 359

These properties and definitions may be combined to simplify exponential expressions.

Example 1 Simplifying Exponential Expressions

Simplify. **a.** $\dfrac{x^4 x^6}{x^3}$ **b.** $(2y^3)^4$ **c.** $\left(\dfrac{pw^2}{2z^5}\right)^3$

d. $\dfrac{4x^6 z^{-3}}{y^{-2}}$ **e.** $\left(\dfrac{1}{2}\right)^{-3} + \left(\dfrac{1}{4}\right)^0$

Solution:

a. $\dfrac{x^4 x^6}{x^3} = \dfrac{x^{4+6}}{x^3} = \dfrac{x^{10}}{x^3} = x^{10-3} = x^7$

b. $(2y^3)^4 = 2^4(y^3)^4 = 2^4 y^{3 \cdot 4} = 16y^{12}$

c. $\left(\dfrac{pw^2}{2z^5}\right)^3 = \dfrac{p^3w^6}{2^3\,z^{15}} = \dfrac{p^3w^6}{8z^{15}}$

d. $\dfrac{4x^6z^{-3}}{y^{-2}} = \dfrac{4x^6y^2}{z^3}$

e. $\left(\dfrac{1}{2}\right)^{-3} + \left(\dfrac{1}{4}\right)^0 = \left(\dfrac{2}{1}\right)^3 + 1 = 8 + 1 = 9$

2. Scientific Notation

Recall that scientific notation is a convenient method to write very large or very small numbers. A number is in scientific notation if it is written in the form $a \times 10^n$, where $1 \le |a| < 10$ and n is an integer. For example,

Number	Equivalent Form	Scientific Notation
4200	4.2×1000	4.2×10^3
0.008	$8.0 \times 0.001 = 8.0 \times \dfrac{1}{1000} = 8.0 \times \dfrac{1}{10^3}$	8.0×10^{-3}

scientific notation, p. 367

Writing a number in scientific notation can be done easily by moving the decimal point to the left or right and then multiplying the resulting number by the appropriate power of 10.

$$730{,}000 = 7.3 \times 10^5 \qquad\qquad 0.000046 = 4.6 \times 10^{-5}$$

multiplying and dividing numbers in scientific notation, p. 370

> **Example 2** | **Multiplying and Dividing Numbers in Scientific Notation**

Perform the indicated operations and write the answer in scientific notation.
$$\frac{(6.0 \times 10^3)(9.0 \times 10^5)}{2.0 \times 10^{-2}}$$

Solution:

$\dfrac{(6.0 \times 10^3)(9.0 \times 10^5)}{2.0 \times 10^{-2}} = \dfrac{(6.0)(9.0) \times 10^3 10^5}{2.0 \times 10^{-2}}$ Multiply the numerators.

$= \dfrac{54.0 \times 10^8}{2.0 \times 10^{-2}}$ Simplify.

$= \left(\dfrac{54.0}{2.0}\right) \times \dfrac{10^8}{10^{-2}}$ Divide.

$= 27.0 \times 10^{8-(-2)}$ Simplify.

$= 27.0 \times 10^{10}$ This number is not in proper scientific notation.

$= (2.7 \times 10^1) \times 10^{10}$ Write 27.0 as 2.7×10^1.

$= 2.7 \times 10^{11}$

3. Addition and Subtraction of Polynomials

To add or subtract polynomials, we add or subtract *like* terms.

Example 3 Adding and Subtracting Polynomials

Perform the indicated operations.

$$\left(2x^4 - 5x^2 - \frac{3}{4}x\right) + (7x^2 + 13) - \left(8x^4 + x^3 + \frac{1}{4}x\right)$$

Solution:

$$\left(2x^4 - 5x^2 - \frac{3}{4}x\right) + (7x^2 + 13) - \left(8x^4 + x^3 + \frac{1}{4}x\right)$$

$$= 2x^4 - 5x^2 - \frac{3}{4}x + 7x^2 + 13 - 8x^4 - x^3 - \frac{1}{4}x \qquad \text{Apply the distributive property.}$$

$$= 2x^4 - 8x^4 - x^3 - 5x^2 + 7x^2 - \frac{3}{4}x - \frac{1}{4}x + 13 \qquad \text{Collect } like \text{ terms together.}$$

$$= -6x^4 - x^3 + 2x^2 - \frac{4}{4}x + 13 \qquad \text{Add } like \text{ terms.}$$

$$= -6x^4 - x^3 + 2x^2 - x + 13 \qquad \text{Simplify.}$$

adding and subtracting polynomials, pp. 379

---

4. Multiplication of Polynomials

To multiply polynomials, multiply each term in the first polynomial by each term in the second and combine *like* terms.

Example 4 Multiplying Polynomials

Multiply. **a.** $(2y + 5)(6y^2 - 4y + 3)$ **b.** $(2x - 3y)(5x + 7y)$

Solution:

a. $(2y + 5)(6y^2 - 4y + 3)$

$= 2y(6y^2) + 2y(-4y) + 2y(3) + 5(6y^2) + 5(-4y) + 5(3) \qquad$ Multiply.

$= 12y^3 - 8y^2 + 6y + 30y^2 - 20y + 15 \qquad$ Simplify.

$= 12y^3 + 22y^2 - 14y + 15 \qquad$ Combine *like* terms.

b. $(2x - 3y)(5x + 7y)$

$= 2x(5x) + 2x(7y) + (-3y)(5x) + (-3y)(7y) \qquad$ Multiply.

$= 10x^2 + 14xy - 15xy - 21y^2 \qquad$ Simplify.

$= 10x^2 - xy - 21y^2 \qquad$ Combine *like* terms.

multiplying polynomials, p. 388

When squaring a binomial such as $(a + b)^2$, we can write the expression as $(a + b)(a + b)$. Proceeding as in Example 4, we have $(a + b)(a + b) = a^2 + ab + ab + b^2 = a^2 + 2ab + b^2$. The resulting trinomial is a **perfect square trinomial**.

To multiply the product of conjugates such as $(a + b)(a - b)$, we have $a^2 - ab + ab - b^2 = a^2 - b^2$. This result is called a **difference of squares**.

These general results are summarized in Table A-5.

Table A-5

special case products, p. 391

Product	Example
The square of a binomial results in a perfect square trinomial.	
1. $(a + b)^2 = a^2 + 2ab + b^2$	1. $(2x + 5y)^2 = (2x)^2 + 2(2x)(5y) + (5y)^2$ $= 4x^2 + 20xy + 25y^2$
2. $(a - b)^2 = a^2 - 2ab + b^2$	2. $(4m - 3n)^2 = (4m)^2 - 2(4m)(3n) + (3n)^2$ $= 16m^2 - 24mn + 9n^2$
The product of conjugate factors results in a difference of squares.	
3. $(a + b)(a - b) = a^2 - b^2$	3. $(7w + 3)(7w - 3) = (7w)^2 - (3)^2$ $= 49w^2 - 9$

5. Division of Polynomials

Division of polynomials is presented in two separate cases.

1. To divide a polynomial by a monomial divisor (that is, when the divisor has only one term), we use the properties:

$$\frac{a + b}{c} = \frac{a}{c} + \frac{b}{c} \quad \text{and} \quad \frac{a - b}{c} = \frac{a}{c} - \frac{b}{c}.$$

2. If the divisor has more than one term, we use long division.

Example 5   **Dividing a Polynomial by a Monomial**

division by a monomial, p. 398

Divide. $(15xy^2 - 5x^2y^3 + 10x^3y^6) \div (5x^2y^2)$

Solution:

We can write the expression as $\dfrac{15xy^2 - 5x^2y^3 + 10x^3y^6}{5x^2y^2}$

$= \dfrac{15xy^2}{5x^2y^2} - \dfrac{5x^2y^3}{5x^2y^2} + \dfrac{10x^3y^6}{5x^2y^2}$ Divide each term in the numerator by $5x^2y^2$.

$= \dfrac{3}{x} - y + 2xy^4$ Simplify each term.

Example 6 | Using Long Division to Divide Polynomials

Divide. $(8x^3 + 14x^2 - 7) \div (x + 3)$

Solution:

Write the polynomials in long division format. The term $0x$ is inserted in the dividend as a place holder for the missing power of x.

long division, p 399

$$
\begin{array}{r}
8x^2 \\
x + 3 \overline{\smash{\big)}\, 8x^3 + 14x^2 + 0x - 7}
\end{array}
$$

Divide the leading term of the dividend by the leading term of the divisor $(8x^3/x)$. The result, $8x^2$, is the first term in the quotient.

$$
\begin{array}{r}
8x^2 \\
x + 3 \overline{\smash{\big)}\, 8x^3 + 14x^2 + 0x - 7} \\
-(8x^3 + 24x^2)
\end{array}
$$

Multiply $8x^2$ by the divisor and subtract the result.

$$
\begin{array}{r}
8x^2 \\
x + 3 \overline{\smash{\big)}\, 8x^3 + 14x^2 + 0x - 7} \\
-8x^3 - 24x^2 \\
\hline
-10x^2 + 0x
\end{array}
$$

Subtract the quantity $8x^3 + 24x^2$. To do this we can add the opposite.

Bring down the next column and repeat the process.

$$
\begin{array}{r}
8x^2 - 10x \\
x + 3 \overline{\smash{\big)}\, 8x^3 + 14x^2 + 0x - 7} \\
-8x^3 - 24x^2 \\
\hline
-10x^2 + 0x \\
-(-10x - 30x)
\end{array}
$$

$$
\begin{array}{r}
8x^2 - 10x + 30 \\
x + 3 \overline{\smash{\big)}\, 8x^3 + 14x^2 + 0x - 7} \\
-8x^3 - 24x^2 \\
\hline
-10x^2 + 0x \\
10x + 30x \\
\hline
30x - 7 \\
-(30x + 90)
\end{array}
$$

$$
\begin{array}{r}
8x^2 - 10x + 30 \\
x + 3 \overline{\smash{\big)}\, 8x^3 + 14x^2 + 0x - 7} \\
-8x^3 - 24x^2 \\
\hline
-10x^2 + 0x \\
10x + 30x \\
\hline
30x - 7 \\
-30x - 90 \\
\hline
-97
\end{array}
$$

Therefore, $(8x^3 + 14x^2 - 7) \div (x + 3) = 8x^2 - 10x + 30 + \dfrac{-97}{x + 3}$.

Review E   Practice Exercises

- Practice Problems   • e-Professors
- Self-Tests   • Videos
- NetTutor

Assume that all variables represent positive numbers.

For Exercises 1–8, simplify and state the property used. Choose from multiplication of like bases, division of like bases, power rule, power of a product, or power of a quotient.

1. $6^2 \cdot 6^5$ **2.** $(x^3)^4$ **3.** $\dfrac{y^{10}}{y^7}$ **4.** $(10p)^5$

5. $\left(\dfrac{w}{4}\right)^4$ **6.** $x^{10} \cdot x^3 \cdot x$ **7.** $(q^5)^6$ **8.** $\dfrac{z^4}{z^4}$

For Exercises 9–28, simplify. Write the answers with positive exponents.

9. 3^0 **10.** 5^{-1} **11.** $\left(\dfrac{1}{2}\right)^{-1}$ **12.** $(-6)^0$

13. $(-5)^{-2}$ **14.** $\left(\dfrac{2}{3}\right)^{-2}$ **15.** $6^{-1} + (-4)^0$ **16.** $(-2)^{-2} + 8^{-1}$

17. $\dfrac{r^2 \cdot r^6}{r^{10}}$ **18.** $\dfrac{t^3 \cdot t^7}{t^{13}}$ **19.** $\dfrac{b^3}{b^4 \cdot b^{-3}}$ **20.** $\dfrac{c^{-2}}{c^{-5} \cdot c^2}$

21. $(3x^2)^3$ **22.** $(-7y^4)^2$ **23.** $\dfrac{6w^2z}{2z^{-1}}$ **24.** $\dfrac{-8m^4n^{-2}}{4n^3}$

25. $\left(\dfrac{a^2 b^{-1}}{c^3 d^{-4}}\right)^{-2}$ **26.** $\left(\dfrac{p^{-3} q}{r^2 s^{-1}}\right)^{-3}$ **27.** $(5^0 w^2 z) \cdot (w^{-3} z^{-4})$ **28.** $(4p^{-3}q) \cdot (p^{-2}q^4)$

For Exercises 29–36, write the number in scientific notation.

29. 3,050,000   **30.** 82,500,000   **31.** 0.0000251   **32.** 0.0038

33. There were 89,600,000 people who watched the Super Bowl on Sunday, February 1, 2004.

34. Mont Blanc, a mountain in the Alps, is about 15,800 ft high.

35. A single blood cell is approximately 0.00039 in. long.

36. The shortest flash of light recorded was 0.000 000 000 000 000 69 seconds.

For Exercises 37–44, perform the operation. Write the answer in scientific notation.

37. $(2.4 \times 10^1)(1.5 \times 10^3)$ **38.** $(8 \times 10^{-5})(1.2 \times 10^9)$ **39.** $\dfrac{6.3 \times 10^{14}}{3.0 \times 10^{-2}}$

40. $\dfrac{9.6 \times 10^{-12}}{4.0 \times 10^3}$ **41.** $(3.0 \times 10^{-5})(7.3 \times 10^2)$ **42.** $(2.1 \times 10^{13})(5.0 \times 10^{-2})$

43. $\dfrac{(4.0 \times 10^3)(1.2 \times 10^{-4})}{(6.0 \times 10^{10})}$ **44.** $\dfrac{(2.8 \times 10^2)(4.0 \times 10^4)}{(7.0 \times 10^{-6})}$

For Exercises 45–50, add or subtract as indicated.

45. $(5p^3 + 2p - 3) + (8p^3 - 4p^2 + 14)$

46. $(2m^2 - 3m + 9) - (-3m^2 + 2m - 8)$

47. $\left(10n^2 - \dfrac{3}{8}n + 2\right) - \left(11n^2 + \dfrac{5}{8}n - 6\right)$

48. $(a^2 - 5a + 10) + \left(-3a^3 + \dfrac{1}{2}a^2 - 13\right)$

49. $(-u^2v + 6u^2 - 2uv^2) + (11u^2 + 7uv^2) - (9u^2v + u^2 + 3uv^2)$

50. $(8a^2b - 5ab + 12ab^2) - (4a^2b - 6ab - ab^2) + (10a^2b + 3ab^2)$

For Exercises 51–66, multiply the polynomials.

51. $4p(p^3 - 5p^2 + 2p + 8)$

52. $5mn(-3m^2 + 6mn + n^2)$

53. $(3x + y)(-2x - 5y)$

54. $(4u - 7v)(3u - v)$

55. $(8b + 4)(2b - 3)$

56. $(-5a + 2)(5a - 3)$

57. $(x - 3y)(x^2 - 3xy + 5y^2)$

58. $(3p - 5q)(p - 2pq + q)$

59. $(3h - 8)(3h + 8)$

60. $(k - 5)(k + 5)$

61. $(4x - 5)^2$

62. $(3y + 8)^2$

63. $\left(\dfrac{1}{3}t^2 - 9\right)\left(\dfrac{1}{3}t^2 + 9\right)$

64. $\left(ab - \dfrac{3}{4}\right)\left(ab + \dfrac{3}{4}\right)$

65. $(0.2x^2 - 3)^2$

66. $(1.2y^2 + 4)^2$

For Exercises 67–76, divide the polynomials.

67. $(6a^3b^2 - 18a^2b^2 + 3ab^2 - 9ab) \div (3ab)$

68. $(8x^4y^4 + 4x^3y^2 - 12x^2y^4 + 4x^2y^2) \div (4x^2y^2)$

69. $(x^3 - 2x^2 - 4x + 33) \div (x + 3)$

70. $(2y^3 - 7y^2 + y + 10) \div (y - 2)$

71. $(2p^3 - 17p^2 + 39p - 10) \div (p - 5)$

72. $(t^3 - t^2 - 10t + 4) \div (t + 3)$

73. $(10m^5 - 16m^4 + 8m^3 + 8m^2 - 2m) \div (-2m)$

74. $(-25b^5 + 10b^4 - 5b^3 - 35b^2) \div (-5b^2)$

75. $(8a^3 - 9) \div (2a - 3)$

76. $(32k^3 + 5) \div (2k - 1)$

Factoring Polynomials and Solving Quadratic Equations

1. Factoring Polynomials

The steps to factor a polynomial are outlined in detail in Section 6.6. Recall that the first step in factoring is to factor out the greatest common factor, GCF. Then we proceed with an appropriate technique of factoring based on whether the polynomial has two terms (a binomial), three terms (a trinomial), or four terms.

1. Factoring Polynomials
2. Solving Quadratic Equations Using the Zero Product Rule

> **Example 1** Factoring Binomials

Factor completely.   **a.** $81w^4 - 1$ **b.** $2y^3 + 128$

Solution:

factoring a difference of squares, p. 458

When factoring a binomial, determine if it fits any of the following patterns.

$$a^2 - b^2 = (a + b)(a - b) \qquad \text{Difference of squares (Section 6.5)}$$
$$a^3 - b^3 = (a - b)(a^2 + ab + b^2) \qquad \text{Difference of cubes (Section 6.5)}$$
$$a^3 + b^3 = (a + b)(a^2 - ab + b^2) \qquad \text{Sum of cubes (Section 6.5)}$$

a. $81w^4 - 1$

The GCF is 1.

$= (9x^2)^2 - (1)^2$

This is a difference of squares, $a^2 - b^2$, where $a = 9x^2$ and $b = 1$.

Avoiding Mistakes:

The binomial $9x^2 - 1$ is a difference of squares and may be factored as $(3x + 1)(3x - 1)$. However, the binomial $9x^2 + 1$ is a sum of squares and may not be factored over the real numbers.

$= (9x^2 + 1)(9x^2 - 1)$

Apply the formula $a^2 - b^2 = (a + b)(a - b)$.

$= (9x^2 + 1)(3x + 1)(3x - 1)$

The factor $9x^2 - 1$ factors further as a difference of squares. $9x^2 - 1 = (3x + 1)(3x - 1)$.

factoring a sum or difference of cubes, p. 460

b. $2y^3 + 128$

The GCF is 2.

$= 2(y^3 + 64)$

Factor out 2.

$= 2[(y)^3 + (4)^3]$

The resulting binomial is a sum of cubes, $a^3 + b^3$, where $a = y$ and $b = 4$.

$= 2(y + 4)(y^2 - 4y + 16)$

Apply the formula $a^3 + b^3 = (a + b)(a^2 - ab + b^2)$

> **TIP:** Recall that a factoring problem may be checked by multiplication.
>
> $2(y + 4)(y^2 - 4y + 16) = (2y + 8)(y^2 - 4y + 16)$
> $$= 2y(y^2) + 2y(-4y) + 2y(16) + 8(y^2) + 8(-4y) + 8(16)$$
> $$= 2y^3 - 8y^2 + 32y + 8y^2 - 32y + 128$$
> $$= 2y^3 + 128$$

> **Example 2** Factoring Trinomials

Factor.   **a.** $49w^2 - 28w + 4$ **b.** $5x^4y - 8x^3y + 3x^2y$

Solution:

Check first to see if the trinomial is a perfect square trinomial, in which case factor as the square of a binomial.

$$\text{Perfect square trinomials} \quad \begin{cases} a^2 + 2ab + b^2 = (a + b)^2 \\ a^2 - 2ab + b^2 = (a - b)^2 \end{cases}$$

Otherwise factor the trinomial using the trial-and-error method (Section 6.3) or the ac-method (Section 6.4).

a. $49w^2 - 28w + 4$ The GCF is 1.

$= (7w)^2 - 2(7w)(2) + (2)^2$ The polynomial is a perfect square trinomial, $a^2 - 2ab + b^2$ where $a = 7w$ and $b = 2$.

$= (7w - 2)^2$ Apply the formula $a^2 - 2ab + b^2 = (a - b)^2$.

factoring a perfect square trinomial, p. 443, 453

b. $5x^4y - 8x^3y + 3x^2y$ The GCF is x^2y.

$= x^2y(5x^2 - 8x + 3)$ Factor out x^2y. The factored form of $5x^2 - 8x + 3$ must be a product of binomials of the form:

factoring trinomials with the trial-and-error method, p. 438

factors of 5

$(\square x - \square)(\square x - \square)$

factors of 3

factoring trinomials with the ac-method, p. 449

The signs within parentheses must both be negative for a product of 3 and a sum of $-8x$.

Trying all possible combinations we have:

$(5x - 1)(x - 3) = 5x^2 - 15x - x + 3$

$= 5x^2 - 16x + 3$ Wrong middle term.

$(5x - 3)(x - 1) = 5x^2 - 5x - 3x + 3$

$= 5x^2 - 8x + 3$ Correct!

$= x^2y(5x - 3)(x - 1)$ Factored completely.

Example 3   **Factoring a Four-Term Polynomial**

Factor completely. $6ab + 8ax + 15bx + 20x^2$

factoring a four-term polynomial by grouping, p. 427

Solution:

This polynomial has 4 terms. Therefore, try factoring by grouping.

$6ab + 8ax + 15bx + 20x^2$ The GCF is 1.

$= 6ab + 8ax \mid + 15bx + 20x^2$ Group the first pair of terms and the second pair of terms.

$= 2a(3b + 4x) + 5x(3b + 4x)$ Factor out the GCF from each pair. Note that the two resulting terms share a common binomial factor.

$= (3b + 4x)(2a + 5x)$ Factor out the common binomial factor.

2. Solving Quadratic Equations Using the Zero Product Rule

definition of a quadratic equation, p. 471

A quadratic equation is an equation of the form $ax^2 + bx + c = 0$, where $a \neq 0$. To solve a quadratic equation in this form, we factor the expression $ax^2 + bx + c$ and set each factor equal to zero.

Example 4 Solving Quadratic Equations

solving a quadratic equation by factoring, p. 472

Solve.　　**a.** $(2x - 1)(x - 3) = 0$　　**b.** $10x^2 + 50x = 140$

Solution:

a. $(2x - 1)(x - 3) = 0$　　　　　The equation is already in factored form and set equal to zero.

　　$2x - 1 = 0$　or　$x - 3 = 0$　　　Set each factor equal to zero.

　　　　$2x = 1$　or　　　$x = 3$　　　Solve each equation.

　　　　$x = \dfrac{1}{2}$　or　　　$x = 3$　　　The solutions are $x = \dfrac{1}{2}$ and $x = 3$.

b.　　　　$10x^2 + 50x = 140$

　　$10x^2 + 50x - 140 = 0$　　　　　Write the equation in the form $ax^2 + bx + c = 0$.

　　$10(x^2 + 5x - 14) = 0$　　　　　Factor out the GCF.

　　$10(x - 2)(x + 7) = 0$　　　　　Factor the trinomial.

　　$10 \not= 0$　or　$x - 2 = 0$　or　$x + 7 = 0$　　Set each factor equal to zero.

　　　　　　　　$x = 2$　or　　　$x = -7$　　The solutions are $x = 2$ and $x = -7$.

Review F　　Practice Exercises

- Practice Problems
- Self-Tests
- NetTutor
- e-Professors
- Videos

1. Write the formula to factor the difference of squares, $a^2 - b^2$.

2. Write the formula to factor the difference of cubes, $a^3 - b^3$.

3. Write the formula to factor the sum of cubes, $a^3 + b^3$.

4. Is it possible to factor the sum of squares $a^2 + b^2$ over the real numbers?

For Exercises 5–50,

　a. Identify the category in which the polynomial best fits. Choose from

- difference of squares
- sum of squares

- difference of cubes
- sum of cubes
- trinomial (perfect square trinomial)
- trinomial (nonperfect square trinomial)
- four terms—grouping
- none of these

Hint: (You may have to factor out the greatest common factor first to categorize the polynomial.)

b. Factor the polynomial completely.

5. $t^2 - 100$

6. $4m^2 - 49n^2$

7. $y^3 + 27$

8. $x^3 + 1$

9. $d^2 + 3d - 28$

10. $c^2 + 5c - 24$

11. $x^2 - 12x + 36$

12. $p^2 + 16p + 64$

13. $2ax^2 - 5ax + 2bx - 5b$

14. $8x^2 - 4bx + 2ax - ab$

15. $10y^2 + 3y - 4$

16. $12z^2 + 11z + 2$

17. $10p^2 - 640$

18. $50a^2 - 72$

19. $z^4 - 64z$

20. $t^4 - 8t$

21. $b^3 - 4b^2 - 45b$

22. $y^3 - 14y^2 + 40y$

23. $9w^2 + 24wx + 16x^2$

24. $4k^2 - 20kp + 25p^2$

25. $60x^2 - 20x + 30ax - 10a$

26. $50x^2 - 200x + 10cx - 40c$

27. $x^3 + 4x^2 - 9x - 36$

28. $m^3 + 5m^2 - 4m - 20$

29. $w^4 - 16$

30. $k^4 - 81$

31. $t^6 - 8$

32. $p^6 + 27$

33. $8p^2 - 22p + 5$

34. $9m^2 - 3m - 20$

35. $36y^2 - 12y + 1$

36. $9a^2 + 42a + 49$

37. $x^2 + 4x - xy - 4y$

38. $m^2 - 6m + mn - 6n$

39. $2x^2 + 50$

40. $4y^2 + 64$

41. $12r^2s^2 + 7rs^2 - 10s^2$

42. $7z^2w^2 - 10zw^2 - 8w^2$

43. $x^2 + 8xy - 33y^2$

44. $s^2 - 9st - 36t^2$

45. $m^6 + n^3$

46. $a^3 - b^6$

47. $x^2(a + b) - x(a + b) - 12(a + b)$

48. $y^2(y + 2) + 10y(y + 2) + 9(y + 2)$

49. $x^2 - 4x$

50. $y^2 - 9y$

For Exercises 51–64, solve the equation.

51. $(2x + 5)(x - 3) = 0$

52. $(4x - 1)(x + 6) = 0$

53. $x(x - 5) = 0$

54. $y(y + 7) = 0$

55. $5(w + 3) = 0$

56. $-2(t - 10) = 0$

57. $z^2 - 2z - 8 = 0$

58. $p^2 - 15p - 16 = 0$

59. $6x^2 - 7x = 5$

60. $4x^2 = x + 14$

61. $2x(x - 10) = -9x - 12$

62. $x(x - 12) + 5 = -30$

63. $w^3 - w^2 - w + 1 = 0$

64. $z^3 - 4z^2 - 16z + 64 = 0$

definition of a rational expression, p. 494

simplifying rational expressions to lowest terms, p. 496

multiplying and dividing rational expressions, p. 504

Rational Expressions

1. Definition of a Rational Expression

A rational expression is an expression of the form $\frac{p}{q}$, where p and q are polynomials. The following are rational expressions.

$$\frac{x^2 + x - 4}{x - 2}, \quad \frac{1}{x^3}, \quad \frac{2}{5}$$

The domain of a rational expression is the set of all real numbers except those that make the denominator zero. For example, the domain of

$$\frac{x^2 + x - 4}{x - 2}$$ is $\{x \mid x \text{ is a real number and } x \neq 2\}$.

$$\frac{1}{x^3}$$ is $\{x \mid x \text{ is a real number and } x \neq 0\}$.

$$\frac{2}{5}$$ is $\{x \mid x \text{ is any real number}\}$.

2. Simplifying Rational Expressions to Lowest Terms

The expressions $\frac{6}{12}, \frac{2}{4}$, and $\frac{1}{2}$ are all equivalent. However, the value $\frac{1}{2}$ is in lowest terms. To simplify a rational expression to lowest terms, first factor the numerator and denominator. Then reduce the common factors that form a ratio of 1.

Example 1　Simplifying a Rational Expression to Lowest Terms

Simplify to lowest terms. $\dfrac{x^2 + 10x + 9}{x^2 - 1}$

Solution:

$$\frac{x^2 + 10x + 9}{x^2 - 1} = \frac{(x + 1)(x + 9)}{(x + 1)(x - 1)} \quad \text{Factor numerator and denominator.}$$

$$= \frac{\cancel{(x + 1)}(x + 9)}{\cancel{(x + 1)}(x - 1)} \quad \text{Simplify } (x + 1)/(x + 1) \text{ to 1.}$$

$$= \frac{x + 9}{x - 1} \quad (\text{where } x \neq 1 \text{ and } x \neq -1)$$

In Example 1, it is important to realize that the original restrictions on the variable $x \neq 1$ and $x \neq -1$, are still in effect even after the expression is reduced to lowest terms. That is,

$$\frac{x^2 + 10x + 9}{x^2 - 1} = \frac{x + 9}{x - 1} \quad \text{for } x \neq 1 \text{ and } x \neq -1.$$

3. Multiplying and Dividing Rational Expressions

To multiply two rational expressions, factor the numerator and denominator of both expressions. Then simplify common factors that form a ratio of 1.

Example 2 | **Multiplying Rational Expressions**

Multiply and simplify the result. $\dfrac{2x - 4}{4x + 4} \cdot \dfrac{2x^2 - 7x - 4}{x^2 - 6x + 8}$

Solution:

$\dfrac{2x - 4}{4x + 4} \cdot \dfrac{2x^2 - 7x - 4}{x^2 - 6x + 8} = \dfrac{2(x - 2)}{4(x + 1)} \cdot \dfrac{(2x + 1)(x - 4)}{(x - 4)(x - 2)}$ Factor.

$\qquad\qquad\qquad = \dfrac{\overset{1}{2}(\overset{1}{x - 2})}{\underset{}{2} \cdot 2(x + 1)} \cdot \dfrac{(2x + 1)(\overset{1}{x - 4})}{(x - 4)(x - 2)}$ Simplify ratios of 1.

$\qquad\qquad\qquad = \dfrac{2x + 1}{2(x + 1)}$

> **TIP:** In the expression $\dfrac{2x + 1}{2(x + 1)}$, do not try to "cancel" the 2's. The 2 in the numerator is not a factor of the numerator.

To divide two rational expressions, multiply the first expression by the reciprocal of the second expression.

Example 3 | **Dividing Rational Expressions**

Divide and simplify. $\dfrac{-x^2 + x}{-x^2} \div \dfrac{3x^2 + 2x - 5}{5x}$

Solution:

$\dfrac{-x^2 + x}{-x^2} \div \dfrac{3x^2 + 2x - 5}{5x} = \dfrac{-x^2 + x}{-x^2} \cdot \dfrac{5x}{3x^2 + 2x - 5}$ Multiply the first expression by the reciprocal of the second.

$\qquad\qquad\qquad = \dfrac{-x(x - 1)}{-x \cdot x} \cdot \dfrac{5x}{(3x + 5)(x - 1)}$ Factor.

$\qquad\qquad\qquad = \dfrac{(\overset{1}{-x})(\overset{1}{x - 1})}{(-x) \cdot x} \cdot \dfrac{5x}{(3x + 5)(\overset{1}{x - 1})}$ Simplify ratios of 1.

$\qquad\qquad\qquad = \dfrac{5}{3x + 5}$

---

4. Adding and Subtracting Rational Expressions

To add or subtract a rational expression it is necessary that the expressions have a common denominator. To add or subtract expressions with a common denominator add or subtract the terms in the numerators and write the result over the common denominator. Then simplify the expression to lowest terms if possible.

Example 4 | **Subtracting Rational Expressions with a Common Denominator**

Subtract. $\dfrac{x^2 - 2x}{x - 5} - \dfrac{8x - 25}{x - 5}$

adding and subtracting rational expressions with the same denominator, p. 518

Solution:

$$\frac{x^2 - 2x}{x - 5} - \frac{8x - 25}{x - 5} = \frac{x^2 - 2x - (8x - 25)}{x - 5}$$

Subtract terms in the numerator. Use parentheses to help you remember to subtract both terms in the numerator of the second expression.

$$= \frac{x^2 - 2x - 8x + 25}{x - 5}$$

Apply the distributive property.

$$= \frac{x^2 - 10x + 25}{x - 5}$$

Combine *like* terms.

$$= \frac{(x - 5)(x - 5)}{x - 5}$$

Factor and simplify to lowest terms.

$$= x - 5$$

To add or subtract two rational expressions that do not have a common denominator, it is necessary to convert each expression to an equivalent rational expression with a common denominator. The steps below outline the process to find the least common denominator of two or more rational expressions.

finding the LCD, p. 511

Steps to find the LCD of Two or More Rational Expressions

1. Factor all denominators completely.
2. The LCD is the product of unique factors from the denominators, where each factor is raised to the highest power to which it appears in any denominator.

For example, the LCD of the expressions $\frac{2}{5x^2y^4}$ and $\frac{1}{7x^3y}$ is given by the product:

$$5^1 \cdot 7^1 \cdot x^3 \cdot y^4 = 35x^3y^4.$$

To convert a rational expression to an equivalent expression with a common denominator, we multiply the numerator and denominator by the factors in the LCD that are missing from the denominator of the original expression. This is mathematically feasible because we are multiplying the original expression by 1. For example, we convert $\frac{2}{5x^2y^4}$ and $\frac{1}{7x^3y}$ to equivalent expressions with a denominator of $35x^3y^4$ as follows.

$$\frac{2}{5x^2y^4} = \frac{2 \cdot 7x}{5x^2y^4 \cdot 7x} = \frac{14x}{35x^3y^4} \qquad \text{Note that } \frac{7x}{7x} = 1.$$

$$\frac{1}{7x^3y} = \frac{1 \cdot 5y^3}{7x^3y \cdot 5y^3} = \frac{5y^3}{35x^3y^4} \qquad \text{Note that } \frac{5y^3}{5y^3} = 1.$$

adding and subtracting rational
expressions with different
denominators, p. 520

Example 5 **Subtracting Two Rational Expressions with Different Denominators**

Subtract the rational expressions. $\dfrac{3x}{x^2 + 7x + 10} - \dfrac{2x}{x^2 + 6x + 8}$

Solution:

$$\dfrac{3x}{x^2 + 7x + 10} - \dfrac{2x}{x^2 + 6x + 8}$$

$$= \dfrac{3x}{(x + 2)(x + 5)} - \dfrac{2x}{(x + 4)(x + 2)}$$

Factor the denominators.
The LCD is
$(x + 2)(x + 5)(x + 4)$.

$$= \dfrac{3x(x + 4)}{(x + 2)(x + 5)(x + 4)} - \dfrac{2x(x + 5)}{(x + 4)(x + 2)(x + 5)}$$

Convert each
expression to an
equivalent expression
with the LCD.

$$= \dfrac{3x(x + 4) - 2x(x + 5)}{(x + 2)(x + 5)(x + 4)}$$

Subtract the numerators.

$$= \dfrac{3x^2 + 12x - 2x^2 - 10x}{(x + 2)(x + 5)(x + 4)}$$

Apply the distributive
property.

$$= \dfrac{x^2 + 2x}{(x + 2)(x + 5)(x + 4)}$$

Combine *like* terms.

$$= \dfrac{x(x + 2)}{(x + 2)(x + 5)(x + 4)}$$

Factor and simplify to
lowest terms.

$$= \dfrac{x}{(x + 5)(x + 4)}$$

5. Simplifying Complex Fractions

A complex fraction is a fraction whose numerator or denominator contains one or more rational expressions. For example:

$$\dfrac{1 - \dfrac{4}{3y}}{y - \dfrac{16}{9y}}$$

One method to simplify such an expression is to multiply the numerator and denominator of the complex fraction by the LCD of all four individual fractions. This is demonstrated in Example 6.

Example 6   Simplifying a Complex Fraction

complex fractions, p. 532

Simplify. $\dfrac{1 - \dfrac{4}{3y}}{y - \dfrac{16}{9y}}$

Solution:

$\dfrac{1 - \dfrac{4}{3y}}{y - \dfrac{16}{9y}}$ The LCD of the expressions $\dfrac{1}{1}, \dfrac{4}{3y}, \dfrac{y}{1}$, and $\dfrac{16}{9y}$ is $9y$.

$\dfrac{1 - \dfrac{4}{3y}}{y - \dfrac{16}{9y}} = \dfrac{9y\left(1 - \dfrac{4}{3y}\right)}{9y\left(y - \dfrac{16}{9y}\right)} = \dfrac{(9y) \cdot 1 - (9y) \cdot \dfrac{4}{3y}}{(9y) \cdot y - (9y) \cdot \dfrac{16}{9y}}$ Apply the distributive property.

$= \dfrac{9y - 12}{9y^2 - 16}$ Simplify.

$= \dfrac{3(\overset{1}{\cancel{3y - 4}})}{\cancel{(3y - 4)}(3y + 4)}$ Factor and reduce to lowest terms.

$= \dfrac{3}{3y + 4}$

The expression in Example 6 could also have been simplified by using the order of operations to combine the expressions in the numerator and denominator separately and then dividing the resulting expressions.

6. Solving Rational Equations

solving rational equations, p. 537

An equation with one or more rational expressions is called a rational equation. To solve a rational equation, we offer the following steps.

Steps to Solve a Rational Equation

1. Factor the denominators of all rational expressions.
2. Identify the LCD of all expressions in the equation.
3. Multiply both sides of the equation by the LCD.
4. Solve the resulting equation.
5. Check potential solutions in the original equation.

Example 7   Solving a Rational Equation

Solve. $\dfrac{4}{y - 4} + \dfrac{y}{2} = \dfrac{y}{y - 4}$

Solution:

$$\frac{4}{y-4} + \frac{y}{2} = \frac{y}{y-4}$$

Step 1: The denominators are already factored.

Step 2: The LCD is $2(y-4)$.

$$2(y-4)\left(\frac{4}{y-4} + \frac{y}{2}\right) = 2(y-4)\left(\frac{y}{y-4}\right)$$

Step 3: Multiply by the LCD.

$$\frac{2(y-4)}{1} \cdot \left(\frac{4}{y-4}\right) + \frac{2(y-4)}{1} \cdot \frac{y}{2} = \frac{2(y-4)}{1} \cdot \left(\frac{y}{y-4}\right)$$

Apply the distributive property.

$$2 \cdot 4 + (y-4) \cdot y = 2 \cdot y$$

Step 4: Solve the resulting equation.

$$8 + y^2 - 4y = 2y$$

Apply the distributive property to clear parentheses.

$$y^2 - 4y - 2y + 8 = 0$$

The equation is quadratic. Set the equation equal to zero.

$$y^2 - 6y + 8 = 0$$

$$(y-4)(y-2) = 0$$

Factor.

$$y - 4 = 0 \quad \text{or} \quad y - 2 = 0$$

Set each factor equal to zero.

$$y = 4 \quad \text{or} \quad y = 2$$

Step 5: Check each potential solution.

Check $y = 4$:

$$\frac{4}{(4)-4} + \frac{(4)}{2} \stackrel{?}{=} \frac{(4)}{(4)-4}$$

$$\frac{4}{0} + \frac{4}{2} \stackrel{?}{=} \frac{4}{0} \text{ (Undefined)}$$

Check $y = 2$:

$$\frac{4}{(2)-4} + \frac{(2)}{2} \stackrel{?}{=} \frac{(2)}{(2)-4}$$

$$\frac{4}{-2} + 1 \stackrel{?}{=} \frac{2}{-2}$$

$$-2 + 1 = -1 ✔$$

The solution is $y = 2$. The value $y = 4$ does not check because it makes the denominator zero in one or more of the rational expressions.

TIP: Before solving a rational equation you might consider noting any values of the variable for which the equation is undefined. In Example 7, the equation
$$\frac{4}{y-4} + \frac{y}{2} = \frac{y}{y-4}$$ is not defined for $y = 4$ because this value makes the denominator zero in the first and third rational expressions. If one of your potential solutions matches a restricted value, then you automatically know that this value cannot be a solution to the original equation.

Review G   Practice Exercises

Boost *your* GRADE at mathzone.com!

MathZone

- Practice Problems
- Self-Tests
- NetTutor
- e-Professors
- Videos

For Exercises 1–6, determine the domain of the expression.

1. $\dfrac{x+2}{x+4}$

2. $\dfrac{x-5}{x-1}$

3. $\dfrac{2x}{25x^2-9}$

4. $\dfrac{5y}{4y^2-49}$

5. $\dfrac{t-7}{8}$

6. $\dfrac{x+2}{5}$

For Exercises 7–12, reduce the expression to lowest terms.

7. $\dfrac{3x^4y^7}{12xy^8}$

8. $\dfrac{21ab^5}{7a^3b^4}$

9. $\dfrac{t^2 - 4}{2t - 4}$

10. $\dfrac{m^2 - 9}{3m - 9}$

11. $\dfrac{2y^2 + 5y - 12}{2y^2 + y - 6}$

12. $\dfrac{2x^2 + 20x + 48}{4x^2 - 144}$

For Exercises 13–22, multiply or divide as indicated and reduce to lowest terms.

13. $\dfrac{2x}{15y^2} \cdot \dfrac{3y^5}{4x^2}$

14. $\dfrac{4s^2t}{8t^3} \cdot \dfrac{2t^5}{6s^4}$

15. $\dfrac{5y - 15}{10y + 40} \cdot \dfrac{2y^2 + y - 28}{2y^2 - 13y + 21}$

16. $\dfrac{x^2 - 36}{x^2 - 4x - 12} \cdot \dfrac{4x + 8}{4x - 24}$

17. $\dfrac{a^2b^3}{5c^2} \div \dfrac{ab}{15c^3}$

18. $\dfrac{m^3}{14n^5} \div \dfrac{m^7}{21n}$

19. $\dfrac{p^2 - 36}{2p - 4} \div \dfrac{2p + 12}{p^2 - 2p}$

20. $\dfrac{y^2 + y}{y^2 - y - 12} \div \dfrac{5y + 5}{y^2 - 5y + 4}$

21. $\dfrac{t^2 + 7t}{3 - t} \div \dfrac{t^2 - 49}{t^2 - 3t}$

22. $\dfrac{x^2 - 2x}{2 - x} \div \dfrac{x^2 + 5x}{x^2 - 25}$

For Exercises 23–26, find the LCD for each pair of expressions.

23. $\dfrac{1}{4x^3y^7}, \dfrac{1}{8xy^{10}}$

24. $\dfrac{1}{16ab^4}, \dfrac{1}{24a^2b^2}$

25. $\dfrac{1}{x^2 - x - 12}, \dfrac{1}{x^2 - 9}$

26. $\dfrac{1}{x^2 + 10x + 9}, \dfrac{1}{x^2 - 81}$

For Exercises 27–40, add or subtract as indicated.

27. $\dfrac{1}{2x} - \dfrac{5}{6x}$

28. $\dfrac{7}{15t} - \dfrac{4}{5t}$

29. $\dfrac{1}{2x^3y} + \dfrac{5}{4xy^2}$

30. $\dfrac{1}{3a^2b^4} + \dfrac{7}{9a^2b}$

31. $\dfrac{x^2}{x - 7} - \dfrac{14x - 49}{x - 7}$

32. $\dfrac{z^2}{z - 10} - \dfrac{20z - 100}{z - 10}$

33. $\dfrac{7x}{4x - 2} + \dfrac{5x}{2x - 1}$

34. $\dfrac{-8}{5x + 4} + \dfrac{7}{10x + 8}$

35. $\dfrac{x}{x^2 + x - 12} - \dfrac{2}{x^2 + 3x - 4}$

36. $\dfrac{4y}{y^2 - 4y + 4} + \dfrac{3}{y^2 - 7y + 10}$

37. $\dfrac{-3}{y - 2} + \dfrac{2y + 11}{y^2 + y - 6} - \dfrac{2}{y + 3}$

38. $\dfrac{-2}{x + 2} + \dfrac{x}{x - 1} + \dfrac{3x - 6}{x^2 + x - 2}$

39. $\dfrac{5}{x - 3} - \dfrac{x + 2}{x - 3}$

40. $\dfrac{3x}{x + 2} - \dfrac{4x + 2}{x + 2}$

For Exercises 41–48, simplify the complex fractions.

41. $\dfrac{\dfrac{3}{a} + \dfrac{4}{b}}{\dfrac{7}{a} - \dfrac{1}{b}}$

42. $\dfrac{\dfrac{2}{p} - \dfrac{4}{q}}{\dfrac{2}{p} + \dfrac{1}{q}}$

43. $\dfrac{8x - \dfrac{1}{2x}}{1 - \dfrac{1}{4x}}$

44. $\dfrac{\dfrac{1}{2} - \dfrac{1}{2t}}{1 - \dfrac{1}{t^2}}$

45. $\dfrac{\dfrac{u^2 - v^2}{uv}}{\dfrac{u + v}{uv}}$

46. $\dfrac{\dfrac{a^2 - b^2}{ab}}{\dfrac{2a - 2b}{ab}}$

47. $\dfrac{\dfrac{1}{y - 5}}{\dfrac{2}{y + 5} + \dfrac{1}{y^2 - 25}}$

48. $\dfrac{\dfrac{3}{x + 1}}{\dfrac{2}{x^2 - 1} + \dfrac{1}{x - 1}}$

For Exercises 49–56, solve the equation.

49. $\dfrac{1}{8} - \dfrac{3}{4x} = \dfrac{1}{2x}$

50. $\dfrac{3}{10} - \dfrac{4}{5x} = \dfrac{1}{x}$

51. $\dfrac{5}{x - 3} = \dfrac{2}{x - 3}$

52. $\dfrac{6}{x + 2} = \dfrac{4}{x + 2}$

53. $1 + \dfrac{6}{x} = -\dfrac{8}{x^2}$

54. $1 = \dfrac{8}{x} - \dfrac{15}{x^2}$

55. $\dfrac{4x + 11}{x^2 - 4} - \dfrac{5}{x + 2} = \dfrac{2x}{x^2 - 4}$

56. $\dfrac{3x}{x + 1} - 2 = \dfrac{12}{x^2 - 1}$

Student Answer Appendix

Chapter R

Chapter Opener Puzzles

$$\frac{p}{5}\ \frac{o}{3}\ \frac{i}{8}\ \frac{n}{1}\ \frac{t}{6}\ \frac{l}{9}\ \frac{e}{2}\ \frac{s}{7}\ \frac{s}{4}$$

Section R.2 Practice Exercises, pp. 18–22

1. Numerator: 7; denominator: 8; proper
3. Numerator: 9; denominator: 5; improper
5. Numerator: 6; denominator: 6; improper
7. Numerator: 12; denominator: 1; improper
9. $\frac{3}{4}$ 11. $\frac{4}{3}$ 13. $\frac{1}{6}$ 15. $\frac{2}{2}$ 17. $\frac{5}{2}$ or $2\frac{1}{2}$
19. $\frac{6}{2}$ or 3 21. The set of whole numbers includes the

number 0 and the set of natural numbers does not.

23. For example: $\frac{2}{4}$ 25. Prime 27. Composite
29. Composite 31. Prime 33. $2 \times 2 \times 3 \times 3$
35. $2 \times 3 \times 7$ 37. $2 \times 5 \times 11$ 39. $3 \times 3 \times 3 \times 5$
41. $\frac{1}{5}$ 43. $\frac{3}{8}$ 45. $\frac{7}{8}$ 47. $\frac{3}{4}$ 49. $\frac{5}{8}$ 51. $\frac{3}{4}$
53. False: When adding or subtracting fractions, it
is necessary to have a common denominator. 55. $\frac{4}{3}$
57. $\frac{2}{3}$ 59. $\frac{9}{2}$ 61. $\frac{3}{5}$ 63. $\frac{5}{3}$ 65. $\frac{90}{13}$
67. \$300 69. 4 graduated with honors 71. 8 aprons
73. 8 jars 75. $\frac{3}{7}$ 77. $\frac{1}{2}$ 79. 24 81. 40
83. 90 85. $\frac{7}{8}$ 87. $\frac{3}{40}$ 89. $\frac{3}{26}$ 91. $\frac{29}{36}$
93. $\frac{7}{10}$ 95. $\frac{35}{48}$ 97. $\frac{37}{24}$ or $1\frac{13}{24}$ 99. $\frac{51}{28}$ or $1\frac{23}{28}$
101. 46 103. $\frac{14}{5}$ or $2\frac{4}{5}$ 105. $\frac{11}{54}$ 107. $8\frac{19}{24}$ in.
109. $\frac{16}{7}$ or $2\frac{2}{7}$ 111. $\frac{3}{2}$ or $1\frac{1}{2}$ 113. $\frac{43}{6}$ or $7\frac{1}{6}$
115. $\frac{11}{7}$ or $1\frac{4}{7}$ 117. $1\frac{7}{12}$ hr 119. $2\frac{1}{4}$ lb
121. $\frac{3}{8}$ in. 123. $4\frac{3}{4}$ in.

Section R.3 Practice Exercises, pp. 32–39

1. b, e, i 3. 108 cm 5. 1 ft 7. $11\frac{1}{2}$ in.
9. 31.4 ft 11. a, f, g 13. 40 ft^2 15. 37.21 in.^{2}
17. 0.0004 m^2 19. 40 mi^2 21. 132.665 cm^2
23. 66 in.^{2} 25. 6 km^2 27. 212.00652 cm^3
29. 39 in.^{3} 31. 113.04 cm^3 33. 1695.6 cm^3
35. 3052.08 in.^{3} 37. 113.04 cm^3 39. 32.768 ft^3
41. a. \$0.31/ft^2 b. \$129 43. Perimeter
45. a. 50.24 in.^{2} b. 113.04 in.^{2} c. One 12-in. pizza

47. 289.3824 cm^3 49. True 51. True 53. True
55. True 57. 45° 59. Not possible 61. For
example, 100°, 80° 63. a. $\angle 1$ and $\angle 3$, $\angle 2$ and $\angle 4$
b. $\angle 1$ and $\angle 2$, $\angle 2$ and $\angle 3$, $\angle 3$ and $\angle 4$, $\angle 1$ and $\angle 4$
c. $m(\angle 1) = 100°, m(\angle 2) = 80°, m(\angle 3) = 100°$
65. 57° 67. 78° 69. 147° 71. 58° 73. 7
75. 1 77. 1 79. 5
81. $m(\angle a) = 45°, m(\angle b) = 135°, m(\angle c) = 45°$,
$m(\angle d) = 135°, m(\angle e) = 45°, m(\angle f) = 135°, m(\angle g) = 45°$
83. Scalene 85. Isosceles
87. No, a 90° angle plus an angle greater than 90° would
make the sum of the angles greater than 180°.
89. 40° 91. 37°
93. $m(\angle a) = 80°, m(\angle b) = 80°, m(\angle c) = 100°$,
$m(\angle d) = 100°, m(\angle e) = 65°, m(\angle f) = 115°, m(\angle g) = 115°$,
$m(\angle h) = 35°, m(\angle i) = 145°, m(\angle j) = 145°$
95. $m(\angle a) = 70°, m(\angle b) = 65°, m(\angle c) = 65°$,
$m(\angle d) = 110°, m(\angle e) = 70°, m(\angle f) = 110°$,
$m(\angle g) = 115°, m(\angle h) = 115°, m(\angle i) = 65°$,
$m(\angle j) = 70°, m(\angle k) = 65°$
97. 82 ft 99. 36 in.^{2} 101. 15.2464 cm^2

Chapter 1

Chapter Opener Puzzle

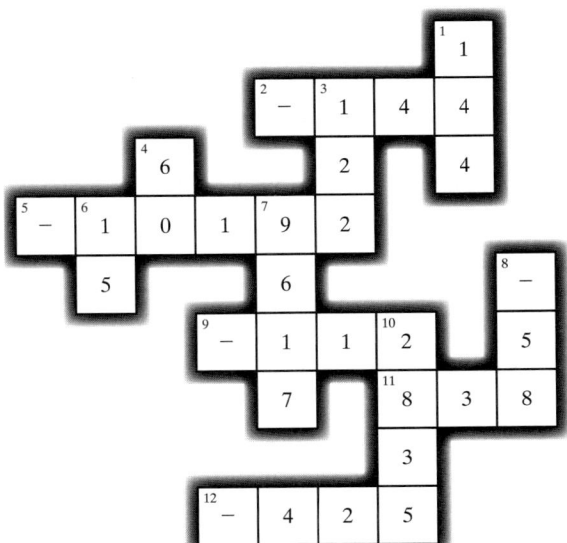

Section 1.1 Practice Exercises, pp. 50–53

3.
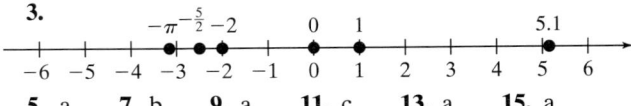

5. a 7. b 9. a 11. c 13. a 15. a
17. b 19. c
21. $0.29, 3.8, \frac{1}{9}, \frac{1}{3}, \frac{1}{8}, \frac{1}{5}, -0.125, -3.24, -3, -6, 0.\overline{2}, 0.\overline{6}$

23. For example: $\pi, -\sqrt{2}, \sqrt{3}$

25. For example: $-4, -1, 0$

27. For example: $-\dfrac{3}{4}, \dfrac{1}{2}, 0.206$ **29.** $-\dfrac{3}{2}, -4, 0.\bar{6}, 0, 1$

31. 1   **33.** $-4, 0, 1$

35. a. $>$ **b.** $>$ **c.** $<$ **d.** $>$

37. -18 **39.** 6.1   **41.** $\dfrac{5}{8}$ **43.** $-\dfrac{7}{3}$ **45.** 3

47. $-\dfrac{7}{3}$ **49.** 8   **51.** 2   **53.** 1.5   **55.** -1.5

57. $\dfrac{3}{2}$ **59.** False, $|n|$ is never negative.

61. True   **63.** False   **65.** True   **67.** False
69. False   **71.** False   **73.** True   **75.** True
77. False   **79.** True   **81.** True   **83.** True
85. For all $a < 0$

Section 1.2 Calculator Connections, pp. 59–60

1. 2   **2.** 91   **1.–3.**
3. 84

```
(4+6)/(8-3)
                    2
110-5*(2+1)-4
                   91
100-2*(5-3)^3
                   84
```

4. 12   **5.** 49   **6.** 18   **7.** 4   **8.** 27   **9.** 0.5

4.–6.
```
3+(4-1)²
                  12
(12-6+1)²
                  49
3*8-√(32+2²)
                  18
```

7.–9.
```
√(18-2)
                   4
(4*3-3*3)^3
                  27
(20-3²)/(26-2²)
                  .5
```

Section 1.2 Practice Exercises, pp. 60–63

3. $-4, 5.\bar{6}, 0, 4.02, \dfrac{7}{9}$ **5.** 9.2   **7.** 34.2

9. 15   **11.** 20   **13.** 11   **15.** 8   **17.** $\dfrac{2}{9}$

19. 4.8   **21.** 1040 ft   **23.** 1000 yd^3 **25.** 10^6

27. $7x^2y^2$ **29.** $3wz^4$ **31.** $\left(\dfrac{2}{3}t\right)^3$

33. a. y **b.** Yes, 1   **35.** $y \cdot y \cdot y \cdot y$ **37.** $8c \cdot 8c$

39. $x \cdot x \cdot y \cdot y \cdot y$ **41.** $3 \cdot a \cdot a \cdot a \cdot b$ **43.** 64

45. $\dfrac{1}{32}$ **47.** 0.64   **49.** 169   **51.** 8   **53.** 3

55. 7   **57.** 6   **59.** $\dfrac{1}{8}$ **61.** $\dfrac{7}{10}$ **63.** 19

65. 40   **67.** 7   **69.** $\dfrac{7}{6}$ **71.** $\dfrac{4}{3}$ **73.** 20

75. 21   **77.** 7   **79.** 17   **81.** 2   **83.** $\dfrac{7}{2}$

85. $\dfrac{9}{37}$ **87.** 225   **89.** 44   **91.** $b + 6$

93. $\dfrac{4}{k}$ or $4 \div k$ **95.** $t - 3$ **97.** $9 - 3p$

99. $2(x - 3)$ **101.** $t - 14$ **103.** 300   **105.** 1
107. 5   **109.** 2   **111.** 32
113. The difference of 18 and x

115. The sum of y and 12
117. One more than the product of 7 and x
119. 6 cubed   **121.** The square root of 10
123. 10 squared
125. Multiplication or division are performed in order from left to right. Addition or subtraction are performed in order from left to right.

127. 1   **129.** $\dfrac{1}{4}$

Section 1.3 Practice Exercises, pp. 69–71

3. $>$ **5.** $>$ **7.** $>$ **9.** -8 **11.** 7
13. 6   **15.** 4   **17.** -7 **19.** 6   **21.** -35
23. -14 **25.** -4 **27.** -8 **29.** 0   **31.** 0
33. -3 **35.** -7 **37.** 0   **39.** -23 **41.** 0

43. -27 **45.** 1.3   **47.** $\dfrac{3}{16}$ **49.** $-\dfrac{1}{4}$ **51.** $-\dfrac{13}{9}$

53. $-\dfrac{7}{24}$ **55.** -9.2 or $-\dfrac{46}{5}$ **57.** $-\dfrac{21}{20}$ or -1.05

59. 0   **61.** $-\dfrac{21}{22}$ **63.** $\dfrac{1}{4}$ **65.** -0.0124

67. -6.17 **69.** -5630.15
71. To add two numbers with the same sign, add their absolute values and apply the common sign.
73. 27   **75.** -6 **77.** 15   **79.** 15   **81.** $-3 + 5; 2$
83. $21 + 4; 25$ **85.** $-7 + 24; 17$ **87.** $2(-6 + 10); 8$
89. $(4 + (-1)) + (-6); -3$ **91.** $4 + (-9) + 2; -3°F$
93. $3 + (-5) + 14; 12$-yd gain
95. a. $40.02 + (-40.96)$ **b.** Yes
97. a. $-50,000 + (-32,000) + (-5000) + 13,000 + 26,000$
b. $-\$48,000$

Section 1.4 Calculator Connections, p. 77

1. -13 **2.** -2 **3.** 711   **1.–3.**
```
-8+(-5)
                 -13
4+(-5)+(-1)
                  -2
627-(-84)
                 711
```

4. -0.18 **5.** -17.7 **6.** -990 **7.** -17 **8.** 38

4.–6.
```
-0.06-0.12
                -.18
-3.2+(-14.5)
               -17.7
-472+(-518)
                -990
```

7.–8.
```
-12-9+4
                 -17
209-108+(-63)
                  38
```

Section 1.4 Practice Exercises, pp. 77–80

3. x^2 **5.** $-b + 2$ **7.** 9   **9.** 5   **11.** -7
13. -9 **15.** 4   **17.** 10   **19.** -3 **21.** 21
23. -21 **25.** -4 **27.** 12   **29.** 16   **31.** -16
33. -12 **35.** -21 **37.** -29 **39.** 25   **41.** -4

43. $-\dfrac{17}{9}$ **45.** $\dfrac{23}{15}$ **47.** $\dfrac{1}{14}$ **49.** $-\dfrac{17}{24}$ **51.** 9.1

53. -7.79 **55.** -34 **57.** -45 **59.** -196.37
61. -149.11 **63.** 0.00258   **65.** $18 - (-1); 19$
67. $8 - 21; -13$ **69.** $-2 - (-18); 16$
71. $-19 - (-31); 12$ **73.** $-3 - 7; -10$
75. $1200 - 500 + 800; \$1500$ **77.** $134°F$

79. 20,602 ft   **81.** −5   **83.** 8   **85.** 13
87. −18   **89.** −18   **91.** $\frac{4}{7}$ **93.** $-\frac{5}{12}$ **95.** 3
97. −7   **99.** −9   **101.** −9   **103.** −2
105. 9   **107.** 1

Chapter 1 Problem Recognition Exercises—Addition and Subtraction of Signed Numbers, p. 80

1. Add their absolute values and apply a negative sign.
2. Subtract the smaller absolute value from the larger absolute value. Apply the sign of the number with the larger absolute value.
3. 41   **4.** 13   **5.** 31   **6.** 46   **7.** −1.3
8. −3.6   **9.** −16   **10.** −7   **11.** $-\frac{1}{12}$ **12.** $\frac{7}{24}$
13. −36   **14.** −59   **15.** −12   **16.** −50   **17.** $-\frac{19}{6}$
18. $-\frac{8}{5}$ **19.** −5   **20.** −32   **21.** 0   **22.** 0
23. −7.7   **24.** −10.5   **25.** −114   **26.** −56
27. −32   **28.** −46   **29.** −60   **30.** −70   **31.** −30
32. −400   **33.** 8   **34.** 2   **35.** −8 + 20; 12
36. −11 − (−2); −9

Section 1.5 Calculator Connections, p. 88

1. −30   **2.** −2   **3.** 625 1.–3.

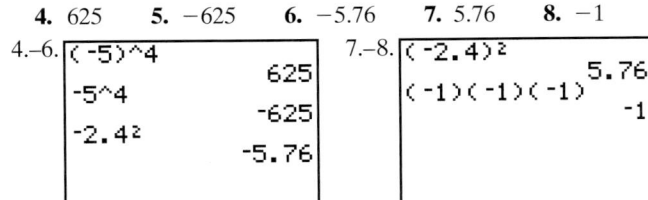

4. 625   **5.** −625   **6.** −5.76   **7.** 5.76   **8.** −1
4.–6.

9. 4   **10.** −36   9.–10.

Section 1.5 Practice Exercises, pp. 88–91

3. True   **5.** True   **7.** False   **9.** 6 + 6
11. (−6) + (−6) + (−6) + (−6) + (−6)
13. −12   **15.** −45   **17.** 130   **19.** 128   **21.** 100
23. −100   **25.** $-\frac{125}{8}$ **27.** 0.0001
29. (−6)(3) = −18   **31.** −4 · 0 = 0
33. No number multiplied by zero equals −4.
35. −9   **37.** 3   **39.** 7   **41.** $-\frac{1}{11}$ **43.** 48
45. −48   **47.** −48   **49.** 48   **51.** 26   **53.** −26
55. −26   **57.** 26   **59.** 0   **61.** Undefined

63. 0   **65.** 0   **67.** $-\frac{3}{2}$ **69.** $\frac{1}{4}$ **71.** −13
73. −5383.37   **75.** 0.129   **77.** −8   **79.** 0.3125, or $\frac{5}{16}$
81. 49   **83.** −49   **85.** $-\frac{1}{125}$ **87.** 0.000001
89. −0.0001   **91.** −16   **93.** 55   **95.** −2.163
97. 410   **99.** $-\frac{5}{36}$ **101.** $\frac{6}{23}$ **103.** −60
105. 90   **107.** 5   **109.** 3   **111.** $\frac{17}{16}$ or $1\frac{1}{16}$
113. $\frac{7}{52}$ **115.** 21   **117.** $\frac{4}{5}$ **119.** 13
121. 6   **123.** $\frac{1}{7}$ **125.** Undefined   **127.** 6
129. −16   **131.** 1   **133.** 42   **135.** −60
137. −8   **139.** $\frac{1}{4}$
141. Yes, the parentheses indicate that the divisor is the quantity 5x.
143. (−0.4)(−1.258); 0.5032   **145.** $-\frac{3}{14} \div \frac{1}{7}; -\frac{3}{2}$
147. 0.5 + (−2)(0.125); 0.25   **149.** $-5 - \left(-\frac{5}{6}\right)\frac{3}{8}; -\frac{75}{16}$
151. −2(6) + 5 = −7; loss of $7
153. a. −34   **b.** 5040   **c.** In part (a), we subtract; in part (b), we multiply.

Section 1.6 Practice Exercises, pp. 101–105

3. 8   **5.** −8   **7.** $-\frac{9}{2}$, or −4.5   **9.** 0   **11.** $\frac{7}{8}$
13. $\frac{1}{3}$ **15.** $-\frac{4}{45}$ **17.** $\frac{11}{15}$ **19.** −8 + 5
21. x + 8   **23.** 4(5)   **25.** −12x
27. x + (−3); −3 + x   **29.** 4p + (−9); −9 + 4p
31. 4(p · p); 4p²   **33.** (−5 · 3)x; −15x
35. $\left(\frac{6}{11} \cdot \frac{11}{6}\right)x; x$ **37.** $\left(-4 \cdot -\frac{1}{4}\right)t; t$ **39.** Reciprocal
41. 0   **43.** 30x + 6   **45.** −2a − 16
47. 15c − 3d   **49.** −7y + 14   **51.** $-\frac{2}{3}x + 4$
53. $\frac{1}{3}m - 1$ **55.** $\frac{3}{2} + 3s$ **57.** −2p − 10
59. 3w + 5z   **61.** 4x + 8y − 4z   **63.** 6w − x + 3y
65. 6 + 2x   **67.** 24z   **69.** b   **71.** i   **73.** g
75. d   **77.** h   **79.**

Term	Coefficient
2x	2
−y	−1
18xy	18
5	5

81.

Term	Coefficient
−x	−1
8y	8
−9x²y	−9
−3	−3

83. The variable factors are different.
85. The variables are the same and raised to the same power.
87. For example: $5y, -2x, 6$ **89.** $2p - 12$

91. $-6y^2 - 7y$ **93.** $7x^2 + x - 4$ **95.** $3t - \dfrac{7}{5}$

97. $-10.4w + 3.3$ **99.** $-8a - 20$ **101.** $10r$
103. $-6x - 12$ **105.** $-16y - 27$ **107.** 23
109. $-3q + 1$ **111.** $-314p + 107$ **113.** $-7b + 8$
115. $-\dfrac{9}{5}p + \dfrac{5}{2}$ **117.** $3k + 11$ **119.** $-y + 25$

121. $-6.12q + 29.72$ **123.** Equivalent
125. Not equivalent. The terms are not *like* terms and cannot be combined.
127. Not equivalent; subtraction is not commutative.
129. Equivalent
131. Not equivalent; coefficients for the variable terms and constants are different.
133. $5\frac{1}{8} + (18\frac{2}{5} + 1\frac{3}{5})$ is easier.

Chapter 1 Review Exercises, pp. 109–111

1. a. $1, 7$ **b.** $7, -4, 0, 1$ **c.** $7, 0, 1$ **d.** $7, \frac{1}{3}, -4, 0, -0.\overline{2}, 1$
e. $-\sqrt{3}, \pi$ **f.** $7, \frac{1}{3}, -4, 0, -\sqrt{3}, -0.\overline{2}, \pi, 1$

2. $\dfrac{1}{2}$ **3.** 6 **4.** $\sqrt{7}$ **5.** 0 **6.** False   **7.** False

8. True   **9.** True   **10.** True   **11.** True   **12.** False

13. True   **14.** $x \cdot \dfrac{2}{3}$, or $\dfrac{2}{3}x$ **15.** $\dfrac{7}{y}$, or $7 \div y$

16. $2 + 3b$ **17.** $a - 5$ **18.** $5k + 2$ **19.** $13z - 7$
20. $\dfrac{6}{x} - 18$ **21.** $3y + 12$ **22.** $z - \dfrac{3}{8}$ **23.** $2p - 5$

24. 216 **25.** 225 **26.** 6 **27.** $\dfrac{1}{16}$ **28.** $\dfrac{1}{10}$

29. $\dfrac{27}{8}$ **30.** 13 **31.** 11 **32.** 7 **33.** 10

34. 2 **35.** 4 **36.** 15 **37.** -17 **38.** $\dfrac{11}{63}$

39. $-\dfrac{5}{22}$ **40.** $-\dfrac{14}{15}$ **41.** $-\dfrac{27}{10}$ **42.** -2.15

43. -4.28 **44.** 3 **45.** 8 **46.** 7
47. When a and b are both negative or when a and b have different signs and the number with the larger absolute value is negative
48. $-1.9°C$ **49.** -12 **50.** 33 **51.** -1
52. -17 **53.** $-\dfrac{29}{18}$ **54.** $-\dfrac{19}{24}$ **55.** -1.2

56. -4.25 **57.** -10.2 **58.** -12.09 **59.** $\dfrac{10}{3}$

60. $-\dfrac{17}{20}$ **61.** -1 **62.** If $a < b$

63. $-7 - (-18); 11$ **64.** $-6 - 41; -47$
65. $7 - 13; -6$ **66.** $[20 - (-7)] - 5; 22$
67. $[6 + (-12)] - 21; -27$ **68.** $125 - (-50); 175°F$

69. -170 **70.** -91 **71.** -2 **72.** 3 **73.** $-\dfrac{1}{6}$

74. $-\dfrac{8}{11}$ **75.** 0 **76.** Undefined   **77.** 0 **78.** 0

79. 2.25 **80.** 30 **81.** $-\dfrac{3}{2}$ **82.** $\dfrac{1}{4}$ **83.** -30

84. 450 **85.** $\dfrac{1}{4}$ **86.** $-\dfrac{1}{7}$ **87.** -2 **88.** $\dfrac{18}{7}$

89. 17 **90.** 6 **91.** $-\dfrac{7}{120}$ **92.** 4.4 **93.** -2

94. 26 **95.** -23 **96.** 80 **97.** 70.6 **98.** True
99. False, any nonzero real number raised to an even power is positive.
100. True   **101.** True
102. False, the product of two negative numbers is positive.
103. True   **104.** True   **105.** For example: $2 + 3 = 3 + 2$
106. For example: $(2 + 3) + 4 = 2 + (3 + 4)$
107. For example: $5 + (-5) = 0$
108. For example: $7 + 0 = 7$
109. For example: $5 \cdot 2 = 2 \cdot 5$
110. For example: $(8 \cdot 2)10 = 8(2 \cdot 10)$
111. For example: $3 \cdot \dfrac{1}{3} = 1$ **112.** For example: $8 \cdot 1 = 8$

113. $5x - 2y = 5x + (-2y)$, then use the commutative property of addition.
114. $3a - 9y = 3a + (-9y)$, then use the commutative property of addition.
115. $3y, 10x, -12, xy$ **116.** $3, 10, -12, 1$
117. a. $8a - b - 10$ **b.** $-7p - 11q + 16$
118. a. $-8z - 18$ **b.** $20w - 40y + 5$
119. $p - 2$ **120.** $-h + 14$ **121.** $-14q - 1$
122. $-5.7b + 2.4$ **123.** $4x + 24$ **124.** $50y + 105$

Chapter 1 Test, p. 111–112

1. Rational, all repeating decimals are rational numbers.
2.

3. a. False   **b.** True   **c.** True   **d.** True
4. a. $(4x)(4x)(4x)$ **b.** $4 \cdot x \cdot x \cdot x$
5. a. Twice the difference of a and b **b.** The difference of twice a and b
6. $\dfrac{\sqrt{c}}{d^2}$ or $\sqrt{c} \div d^2$ **7.** 6 **8.** -19 **9.** -12

10. 28 **11.** $-\dfrac{7}{8}$ **12.** 4.66 **13.** -32

14. -12 **15.** Undefined   **16.** -28 **17.** 0

18. 96 **19.** $\dfrac{2}{3}$ **20.** -8 **21.** 9 **22.** $\dfrac{1}{3}$

23. $-\dfrac{3}{5}$

24. a. Commutative property of multiplication
b. Identity property of addition   **c.** Associative property of addition   **d.** Inverse property of multiplication
e. Associative property of multiplication
25. $-12x + 2y - 4$ **26.** $-12m - 24p + 21$
27. $40x - 49$ **28.** $-4p - 23$ **29.** $4p - \dfrac{4}{3}$

30. 5 **31.** 18 **32.** -6 **33.** -32
34. $12 - (-4); 16$ **35.** $6 - 8; -2$

Chapter 2
Chapter Opener Puzzle

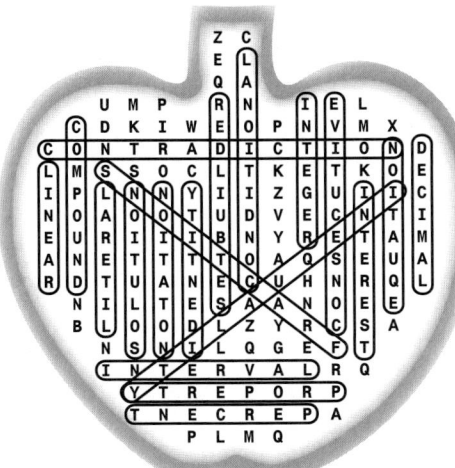

Section 2.1 Practice Exercises, pp. 122–124

3. Expression   **5.** Equation
7. Substitute the value into the equation and determine if the right-hand side is equal to the left-hand side.
9. No   **11.** Yes   **13.** Yes   **15.** $x = 12$
17. $w = -8$ **19.** $n = 16$ **21.** $b = 12$ **23.** $b = 0$
25. $y = 7.6$ **27.** $p = -\frac{23}{3}$ or $-7\frac{2}{3}$ **29.** $x = 3$
31. $p = -1.155$ **33.** $k = 4.0629$ **35.** $w = 4$
37. $q = -3$ **39.** $k = 0$ **41.** $z = -7$
43. $h = \frac{3}{7}$ **45.** $b = -24$ **47.** $w = -\frac{4}{3}$
49. $y = -32$ **51.** $q = -5.5$ **53.** $x = 0.014$
55. $31 + x = 13; x = -18$ **57.** $-12 + x = -15; x = -3$
59. $-3x = 24; x = -8$ **61.** $18 = \frac{x}{2}; x = 36$
63. $x - \frac{2}{3} = \frac{1}{3}; x = 1$ **65.** $b = -2$ **67.** $k = 2$
69. $p = 20$ **71.** $y = \frac{57}{4}$ or $14\frac{1}{4}$ **73.** $d = -20$
75. $t = -9$ **77.** $t = 2$ **79.** $s = -\frac{10}{3}$ or $-3\frac{1}{3}$
81. $s = 13$ **83.** $m = -25$ **85.** $p = \frac{11}{4}$ or $2\frac{3}{4}$
87. $w = 2$ **89.** $w = 12.9$ **91.** $m = -1.9834$
93. No   **95.** Yes   **97.** No   **99.** Yes
101. For example: $1 + x = 3$ **103.** For example: $-4t = 40$
105. For example: $6k + 1 = 7$ **107.** $y = 4$
109. $t = -2$

Section 2.2 Practice Exercises, pp. 131–133

3. $-5z + 2$ **5.** $10p - 10$
7. To simplify an expression, clear parentheses and combine *like* terms. To solve an equation, use the addition, subtraction, multiplication, and division properties of equality to isolate the variable.
9. $y = -3$ **11.** $z = -5$ **13.** $z = 2$ **15.** $y = 6$

17. $y = \frac{5}{2}$ or $2\frac{1}{2}$ **19.** $m = 1$ **21.** $x = -\frac{17}{8}$ or $-2\frac{1}{8}$
23. $x = -42$ **25.** $p = -\frac{3}{4}$ **27.** $w = 5$
29. $h = -4$ **31.** $a = -26$ **33.** $r = \frac{50}{3}$ or $16\frac{2}{3}$
35. $z = -8$ **37.** $x = -\frac{7}{3}$ or $-2\frac{1}{3}$ **39.** $y = 0$
41. $p = \frac{9}{2}$ or $4\frac{1}{2}$ **43.** $x = -\frac{1}{3}$ **45.** $y = 12$
47. $s = -6$ **49.** $t = 0$ **51.** $x = -2$
53. $p = \frac{10}{3}$ or $3\frac{1}{3}$ **55.** $w = 6$ **57.** $n = -0.25$
59. Contradiction; no solution
61. Conditional equation; $x = -15$
63. Identity; all real numbers
65. One solution   **67.** Infinitely many solutions
69. $t = 10$ **71.** $y = 2$ **73.** $m = 0.6$
75. $n = -\frac{3}{4}$ **77.** No solution   **79.** $x = 28$
81. $p = \frac{5}{4}$ **83.** $r = 7$ **85.** $y = -20$
87. All real numbers   **89.** $g = \frac{4}{3}$ **91.** $w = 0$
93. $a = -18$ **95.** $a = 4.5$
97. For example: $4x - 3 = 4x + 1$

Section 2.3 Practice Exercises, pp. 139–141

3. $x = -2$ **5.** $y = -5$ **7.** No solution   **9.** $18, 36$
11. $100; 1000; 10,000$ **13.** $30, 60$ **15.** $x = 4$
17. $y = -\frac{19}{2}$ **19.** $q = -\frac{15}{4}$ **21.** $w = 8$
23. $m = 3$ **25.** $s = 15$ **27.** No solution
29. All real numbers   **31.** $x = 5$ **33.** $w = 2$
35. $x = -15$ **37.** $y = 6$ **39.** $w = 3$
41. All real numbers   **43.** $x = 67$ **45.** $p = 90$
47. $y = -2$ **49.** $x = -3.8$ **51.** No solution
53. $x = -0.25$ **55.** $b = 7$ **57.** $y = -8$
59. $a = 20$ **61.** $x = 0$ **63.** $h = 3$
65. $x = \frac{8}{3}$ or $2\frac{2}{3}$ **67.** No solution   **69.** $w = \frac{25}{2}$ or $12\frac{1}{2}$
71. All real numbers   **73.** $a = -6$ **75.** $h = \frac{1}{3}$
77. $t = -6$ **79.** $x = \frac{23}{12}$ **81.** $k = \frac{1}{10}$
83. The number is $\frac{2}{3}$.   **85.** The number is $\frac{13}{24}$.
87. $c = -2$ **89.** $h = 2$

Section 2.4 Practice Exercises, pp. 151–155

3. $x + 16 = -31; x = -47$ **5.** $x - 6 = -3; x = 3$
7. $x - 16 = -1; x = 15$ **9.** The number is -4.
11. The number is -3.   **13.** The number is 5.
15. The number is -5.   **17.** The number is 9.
19. a. $x + 1, x + 2$ **b.** $x - 1, x - 2$
21. The integers are -34 and -33.
23. The integers are 13 and 15.

25. The page numbers are 470 and 471.

27. The sides are 14, 15, 16, 17, and 18 in.

29. Karen's age is 35, and Clarann's age is 23.

31. There were 165 Republicans and 269 Democrats.

33. The lengths of the pieces are 33 cm and 53 cm.

35. 4.698 million watch *The Dr. Phil Show.*

37. The Congo River is 4370 km long, and the Nile River is 6825 km.

39. The area of Africa is 30,065,000 km^2. The area of Asia is 44,579,000 km^2.

41. She hikes 3 mph to the lake.

43. The plane travels 600 mph in still air.

45. The slower car travels 48 mph and the faster car travels 52 mph.

47. The speeds of the vehicles are 40 mph and 50 mph.

49. The rates of the boats are 20 mph and 40 mph.

51. The number is −80.   **53.** 42, 43, and 44

55. The speed of the current is 4 mph.

57. The number is 11.

59. Jennifer Lopez made $37 million, and U2 made $69 million.

61. The boats will meet in $\frac{3}{4}$ hr (45 min).

63. The number is 10.

65. The deepest point in the Arctic Ocean is 5122 m.

Section 2.5 Practice Exercises, pp. 160–163

3. 1. Read the problem carefully.   2. Assign labels to unknown quantities.   3. Develop a verbal model.   4. Write a mathematical equation.   5. Solve the equation. 6. Interpret the results, and write the final answer in words.

5. The pyramid at Saqqara is 60 m high, and the Great Pyramid is 137 m high.

7. 65%   **9.** 72%   **11.** 58.52   **13.** 436.8

15. 840   **17.** 1320   **19.** Approximately 91,000 cases

21. 3%   **23. a.** $89.90   **b.** $809.10

25. The original price was $21.95.

27. The markup rate is 88%.

29. Patrick will have to pay $317.96.

31. The price was $24.00.

33. The tax rate is 7.5%.

35. The price of a ticket is $67.

37. The original cost of one CD is $18.

39. Roxanne will have to pay $160.

41. Mike borrowed $3250.   **43.** The rate is 8%.

45. Rafael will have $3300.   **47.** Sherica invested $5500.

49. Dan earned $26,000.

51. Anna sold $10,400 worth of appliances.

53. Jessica's rate is 8%.

Section 2.6 Calculator Connections, p. 169

1. 140.056   **2.** 31.831   **3.** −80   **4.** 2

1.–2.
```
880/(2π)
        140.0563499
1600/(π*(4)²)
        31.83098862
```
3.–4.
```
20/((-0.05)*(5))
             -80
10/(0.5*(6+4))
             2
```

Section 2.6 Practice Exercises, pp. 170–173

3. $y = -5$ **5.** $x = 0$ **7.** $y = -2$

9. $b = P - a - c$ **11.** $d = e - c$

13. $x = y - 35$ **15.** $r = \dfrac{d}{t}$ **17.** $V_1 = \dfrac{P_2 V_2}{P_1}$

19. $y = -2 - x$ **21.** $x = 6y - 10$

23. $y = \dfrac{-5x + 10}{2}$ or $y = -\dfrac{5}{2}x + 5$

25. $x = \dfrac{y - 13}{3}$ or $x = \dfrac{1}{3}y - \dfrac{13}{3}$

27. $y = \dfrac{-6x + 4}{-3}$ or $y = 2x - \dfrac{4}{3}$

29. $x = \dfrac{-by + c}{a}$ or $x = -\dfrac{b}{a}y + \dfrac{c}{a}$

31. $L = \dfrac{P - 2w}{2}$ or $L = \dfrac{P}{2} - w$

33. $x = \dfrac{z - 3y}{3}$ or $x = \dfrac{z}{3} - y$

35. $a = 2Q + b$ **37.** $c = 3A - a - b$ **39.** $m = \dfrac{Fd^2}{GM}$

41. The length is 12 cm, and the width is 5 cm.

43. The length is 40 yd, and the width is 30 yd.

45. The sides are 4 ft, 5 ft, and 7 ft.

47. The measures of the angles are 20°, 50°, and 110°.

49. The measures of the angles are 38°, 28°, and 114°.

51. $y = 26.5$; the measures of the angles are 28.5° and 61.5°.

53. The angles are 55° and 35°.

55. The angles are 23.5° and 66.5°.

57. The angles are 60° and 120°.

59. $x = 20$; the vertical angles measure 37°.

61. a. $A = lw$ **b.** $w = \dfrac{A}{l}$ **c.** The width is 29.5 ft.

63. a. $P = 2l + 2w$ **b.** $l = \dfrac{P - 2w}{2}$ **c.** The length is 103 m.

65. a. $A = \dfrac{1}{2}bh$ **b.** $h = \dfrac{2A}{b}$ **c.** The height is 4 km.

67. a. 415.48 m^2 **b.** 10,386.89 m^3

Section 2.7 Practice Exercises, pp. 184–188

3. $x = -3$

Set-Builder Notation	Graph	Interval Notation
5. $\{x \mid x \geq 6\}$	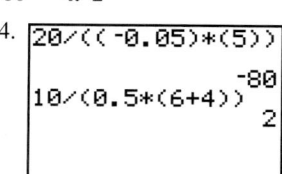	$[6, \infty)$
7. $\{x \mid x \leq 2.1\}$		$(-\infty, 2.1]$
9. $\{x \mid -2 < x \leq 7\}$		$(-2, 7]$
11. $\left\{x \mid x > \dfrac{3}{4}\right\}$		$\left(\dfrac{3}{4}, \infty\right)$
13. $\{x \mid -1 < x < 8\}$		$(-1, 8)$
15. $\{x \mid x < -14\}$		$(-\infty, -14)$
17. $\{x \mid x \geq 18\}$		$[18, \infty)$
19. $\{x \mid x < -0.6\}$		$(-\infty, -0.6)$
21. $\{x \mid -3.5 \leq x < 7.1\}$		$[-3.5, 7.1)$

23. a. $x = 3$ **b.** $x > 3$

25. a. $p = 13$ **b.** $p \le 13$

27. a. $c = -3$ **b.** $c < -3$

29. a. $z = -\dfrac{3}{2}$ **b.** $z \ge -\dfrac{3}{2}$

31.

33. $-3 < x < 5$

35. $2 \le x \le 6$

37. $(-\infty, 1]$

39. $(10, \infty)$

41. $(3, \infty)$

43. $(-\infty, 8]$

45. $(2, \infty)$

47. $(-\infty, -2)$

49. $[14, \infty)$

51. $[-24, \infty)$

53. $[-3, 3)$

55. $\left(0, \dfrac{5}{2}\right)$

57. $(10, 12)$

59. $[-1, 4)$

61. $[90, \infty)$

63. $(-9, \infty)$

65. $\left[-\dfrac{15}{2}, \infty\right)$

67. $(-\infty, -3)$

69. $\left[-\dfrac{1}{3}, \infty\right)$

71. $(-3, \infty)$

73. $(-\infty, 7)$

75. $(-\infty, -5]$

77. $\left(-\infty, \dfrac{15}{4}\right]$

79. $(-\infty, 9)$

81. $(-3, \infty)$

83. No   **85.** Yes

87. a. A $[93, 100]$; B+ $[89, 93)$; B $[84, 89)$; C+ $[80, 84)$;

C $[75, 80)$; F $[0, 75)$ **b.** B   **c.** C   **89.** $L \ge 10$

91. $w > 75$ **93.** $t \le 72$ **95.** $L \ge 8$ **97.** $a \le 2$

99. $300 < c < 400$ **101.** More than 10.1 in. of snow is
needed.   **103. a.** \$384   **b.** It costs \$1112.96 for 148 shirts
and \$1104.00 for 150 shirts. 150 shirts cost less than 148 shirts
because the discount is greater.

105. a. $14.95 + 0.22x < 18.95 + 0.18x$ **b.** $x < 100$;
Company A costs less than Company B if the mileage is less
than 100 miles.

107. $(-\infty, -6)$

109. $(-\infty, -7)$

111. $(-\infty, -1.\overline{3}]$

Chapter 2 Review Exercises, pp. 195–198

1. a. Equation   **b.** Expression   **c.** Equation   **d.** Equation
2. A linear equation can be written in the form $ax + b = 0$,
$a \ne 0$.   **3. a.** Nonlinear   **b.** Linear   **c.** Nonlinear
d. Linear   **4. a.** No   **b.** No   **c.** Yes   **d.** No

5. $a = -8$ **6.** $z = 15$ **7.** $k = \dfrac{21}{4}$ **8.** $r = 70$

9. $x = -\dfrac{21}{5}$ **10.** $t = -60$ **11.** $k = -\dfrac{10}{7}$

12. $m = 27$ **13.** The number is 60.

14. The number is $\dfrac{7}{24}$.   **15.** The number is -8.

16. The number is -2.   **17.** $d = 1$ **18.** $c = -\dfrac{3}{5}$

19. $c = \dfrac{9}{4}$ **20.** $w = -6$ **21.** $b = -3$ **22.** $h = 18$

23. $p = \dfrac{3}{4}$ **24.** $t = \dfrac{11}{8}$ **25.** $a = 0$ **26.** $c = \dfrac{1}{8}$

27. $b = \dfrac{3}{8}$ **28.** $x = \dfrac{17}{3}$ **29.** A contradiction has no
solution and an identity is true for all real numbers.
30. a. Identity   **b.** Conditional equation   **c.** Contradiction
d. Identity   **e.** Contradiction   **31.** $x = 6$ **32.** $y = 22$
33. $z = -3$ **34.** $y = -9$ **35.** $p = -10$

36. $y = -7$ **37.** $t = \dfrac{5}{3}$ **38.** $w = -\dfrac{9}{4}$ **39.** $q = 2.5$

40. $z = -4$ **41.** $a = -4.2$ **42.** $t = 2.5$
43. $x = -312$ **44.** $x = 200$ **45.** No solution
46. No solution   **47.** All real numbers
48. All real numbers   **49.** The number is 30.
50. The number is 11.   **51.** The number is -7.
52. The number is -10.
53. The integers are $66, 68$, and 70.
54. The integers are $27, 28$, and 29.
55. The sides are $25, 26$, and 27 in.
56. The sides are $36, 37, 38, 39$, and 40 cm.
57. The minimum salary was \$30,000 in 1980.
58. Indiana has 6.2 million people and Kentucky has
4.1 million.   **59.** The Mooney travels 190 mph, and the
Cessna travels 140 mph.   **60.** The jogger runs 6 mph and
the bicyclist travels 12 mph.   **61.** 23.8   **62.** 28.8
63. 12.5%   **64.** 95%   **65.** 160   **66.** 1750
67. The sale price is \$26.39.   **68.** The total price is
\$28.24.   **69. a.** She will earn \$840.   **b.** Her balance will
be \$3840.   **70.** He invested \$12,000.

71. $K = C + 273$ **72.** $C = K - 273$ **73.** $s = \dfrac{P}{4}$

74. $s = \dfrac{P}{3}$ **75.** $x = \dfrac{y - b}{m}$ **76.** $x = \dfrac{c - a}{b}$

77. $y = \dfrac{-2 - 2x}{5}$ **78.** $b = \dfrac{Q - 4a}{4}$ or $b = \dfrac{Q}{4} - a$

79. a. $h = \dfrac{3V}{\pi r^2}$ **b.** The height is 5.1 in.

80. The height is 7 m.   **81.** The measures of the angles are 22°, 78°, and 80°.   **82.** The measures of the angles are 32° and 58°.   **83.** The length is 5 ft, and the width is 4 ft.
84. $x = 20$. The angle measure is 65°.
85. The measure of angle y is 53°.

86. a. $(-2, \infty)$

b. $\left(-\infty, \dfrac{1}{2}\right]$

c. $(-1, 4]$

87. a. \$637   **b.** 300 plants cost \$1410, and 295 plants cost \$1416. It is cheaper to buy 300 plants because there is a greater discount.

88. $(-\infty, 17)$

89. $\left(-\dfrac{1}{3}, \infty\right)$

90. $(-\infty, -6]$

91. $\left(-\infty, -\dfrac{14}{5}\right]$

92. $[49, \infty)$

93. $(-\infty, 34.5)$

94. $(18, \infty)$

95. $\left(-\infty, \dfrac{5}{2}\right]$

96. $(-3, 7]$

97. $[-6, 5]$

98. More than 2.5 in. is required.
99. a. $1.50x > 33 + 0.4x$ **b.** $x > 30$; a profit is realized if more than 30 hot dogs are sold.

Chapter 2 Test, pp. 198–199

1. b, d   **2. a.** $5x + 7$ **b.** $x = 9$ **3.** $t = -16$

4. $p = 12$ **5.** $t = -\dfrac{16}{9}$ **6.** $x = \dfrac{7}{3}$ **7.** $p = 15$

8. $d = \dfrac{13}{4}$ **9.** $x = \dfrac{20}{21}$ **10.** No solution

11. $x = -3$ **12.** $c = -47$ **13.** All real numbers

14. $y = -3x - 4$ **15.** $r = \dfrac{C}{2\pi}$ **16.** 90

17. The numbers are 18 and 13.   **18.** The sides are 61, 62, 63, 64, and 65 in.   **19.** Each basketball ticket was \$36.32, and each hockey ticket was \$40.64.

20. The cost was \$82.00.   **21.** Clarita originally borrowed \$5000.   **22.** The field is 110 m long and 75 m wide.
23. $y = 30$; The measures of the angles are 30°, 39°, and 111°.
24. One family travels 55 mph and the other travels 50 mph.
25. The measures of the angles are 32° and 58°.
26. a. $(-\infty, 0)$

b. $[-2, 5)$

27. $(-2, \infty)$

28. $(-\infty, -4]$

29. $[-5, 1]$

30. More than 26.5 in. is required.

Chapters 1–2 Cumulative Review Exercises, pp. 199–200

1. $\dfrac{1}{2}$ **2.** -7 **3.** $-\dfrac{5}{12}$ **4.** 16   **5.** 4
6. $\sqrt{5^2 - 9}$; 4   **7.** $-14 + 12$; -2 **8.** $-7x^2y, 4xy, -6$
9. $9x + 13$ **10.** $t = 4$ **11.** $x = -7.2$ **12.** All real numbers   **13.** $x = \dfrac{24}{7}$ **14.** $x = -\dfrac{4}{7}$
15. $w = -80$ **16.** The numbers are 77 and 79.
17. The cost before tax was \$350.00.   **18.** The height is $\dfrac{41}{6}$ cm or $6\dfrac{5}{6}$ cm.
19. $(-2, \infty)$ **20.** $[-1, 9]$

Chapter 3

Chapter Opener Puzzle

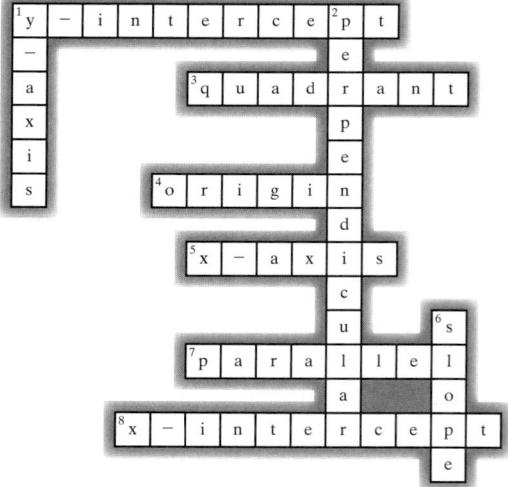

Section 3.1 Practice Exercises, pp. 207–212

3. a. Month 10   **b.** 30   **c.** Between months 3 and 5 and between months 10 and 12   **d.** Months 8 and 9
e. Month 3   **f.** 80   **5. a.** On day 1 the price per share was \$89.25.   **b.** \$1.75   **c.** $-\$2.75$

7.

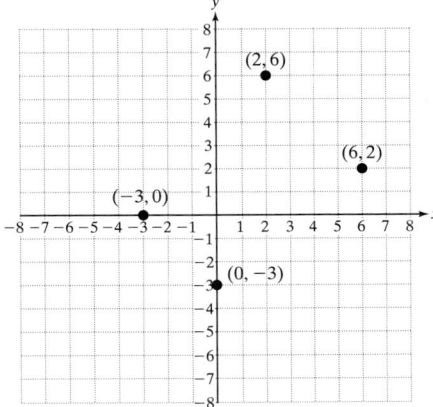

9.

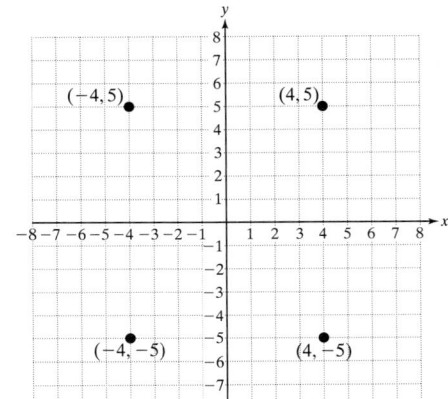

11.

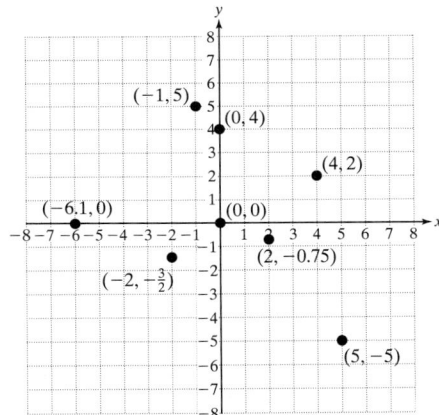

13. IV   **15.** II   **17.** III   **19.** I   **21.** 0

23. $(\frac{7}{8}, 0)$ is located on the x-axis.

25. $A(-4, 2)$, $B(\frac{1}{2}, 4)$, $C(3, -4)$, $D(-3, -4)$, $E(0, -3)$, $F(5, 0)$

27. a. $(250, 225)$, $(175, 193)$, $(315, 330)$, $(220, 209)$, $(450, 570)$, $(400, 480)$, $(190, 185)$; the ordered pair $(250, 225)$ means that 250 people produce $225 in popcorn sales.

b.

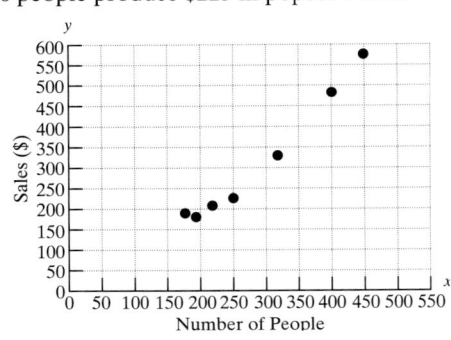

29. a. $(10, 6800)$ means that in 1990, the poverty level was $6800.

b.

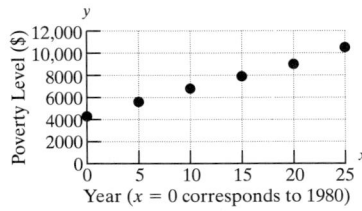

31. a. $(1, -10.2)$, $(2, -9.0)$, $(3, -2.5)$, $(4, 5.7)$, $(5, 13.0)$, $(6, 18.3)$, $(7, 20.9)$, $(8, 19.6)$, $(9, 14.8)$, $(10, 8.7)$, $(11, 2.0)$, $(12, -6.9)$.

b.

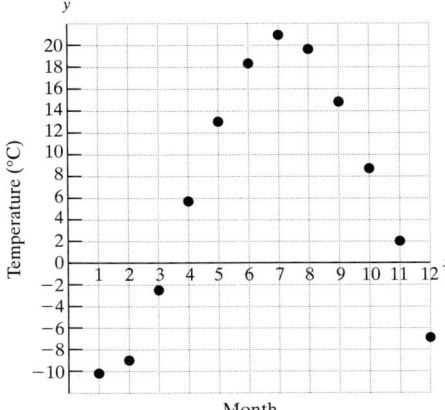

33. a. $A(400, 200)$, $B(200, -150)$, $C(-300, -200)$, $D(-300, 250)$, $E(0, 450)$ **b.** 450 m

Section 3.2 Calculator Connections, pp. 221–222

1.

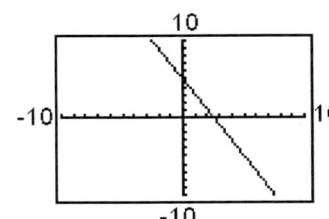

2.

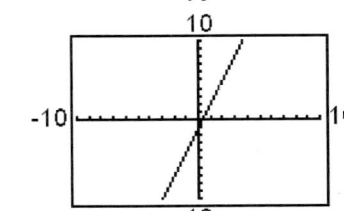

3.

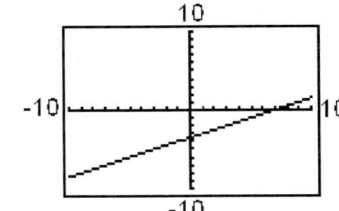

4.

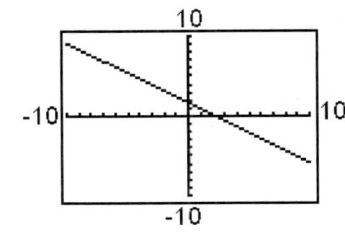

5.

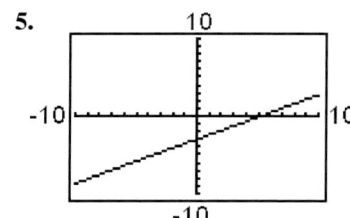

6.

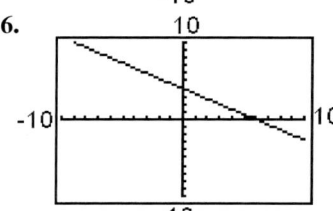

7.

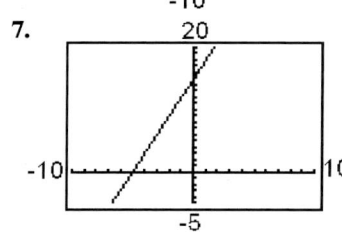

8.

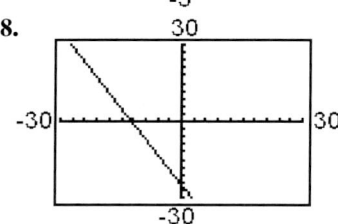

9.

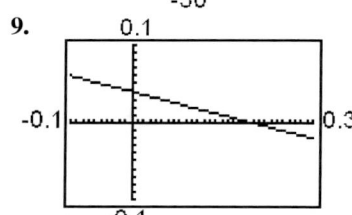

10.

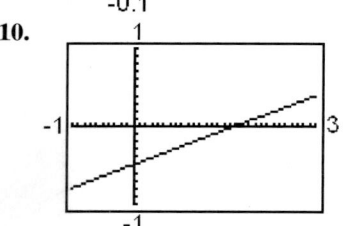

Section 3.2 Practice Exercises, pp. 223–230

3. $(2, 4)$; quadrant I   **5.** $(0, -1)$; y-axis
7. $(3, -4)$; quadrant IV   **9.** Yes   **11.** Yes
13. No   **15.** No   **17.** Yes

19.

x	y
1	−3
−2	0
−3	1
−4	2

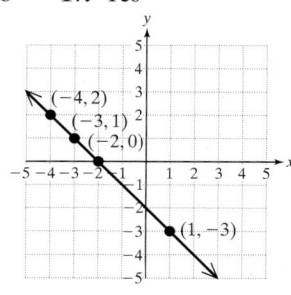

21.

x	y
−2	3
−1	0
−4	9

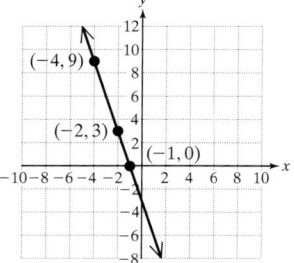

23.

x	y
0	4
2	0
3	−2

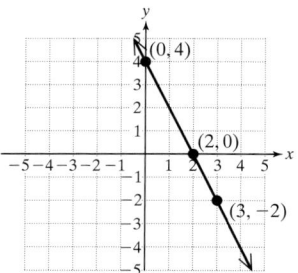

25.

x	y
0	−2
5	−5
10	−8

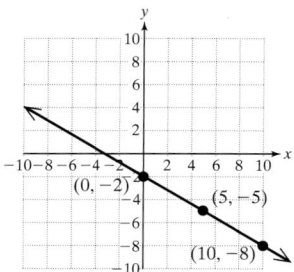

27.

x	y
−2	−14/3
9/2	4
0	−2

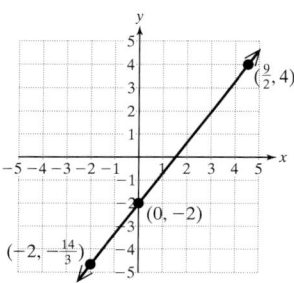

29.

x	y
0	−2
−3	−2
5	−2

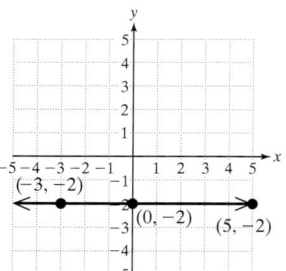

31.

x	y
3/2	−1
3/2	2
3/2	−3

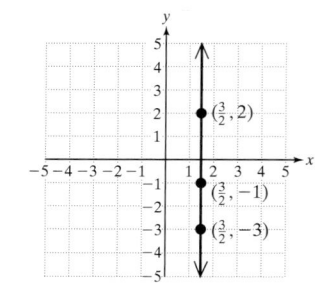

33.

x	y
0	4.6
1	3.4
2	2.2

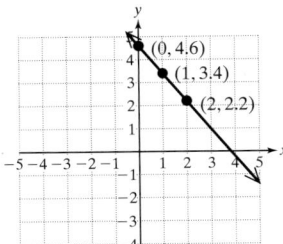

35. $x - y = 4$

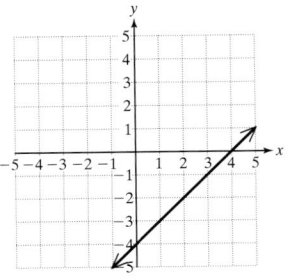

37. $2x - 5y = 10$ **39.** $y = -2x$

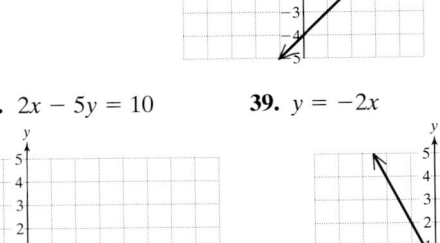

41. $y = \frac{1}{4}x - 2$ **43.** $-x + y = 0$

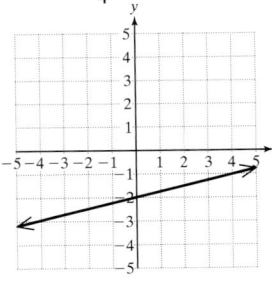 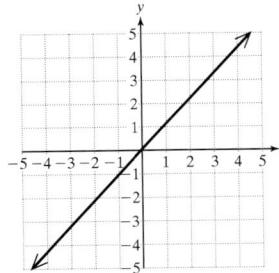

45. $4x - 5y = 15$ **47.** $-30x - 20y = 60$

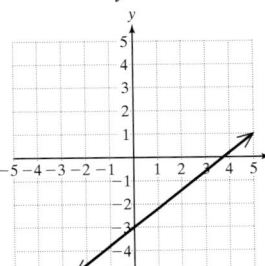 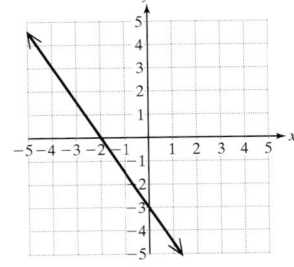

49. y-axis   **51.** x-intercept: $(-1, 0)$; y-intercept: $(0, -3)$
53. x-intercept: $(-4, 0)$; y-intercept: $(0, 1)$

55. x-intercept: $(-9, 0)$;
y-intercept: $(0, 3)$
$x - 3y = -9$

57. x-intercept: $\left(\frac{8}{3}, 0\right)$;
y-intercept: $(0, 2)$
$y = -\frac{3}{4}x + 2$

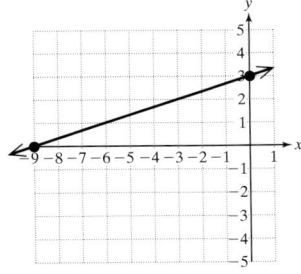 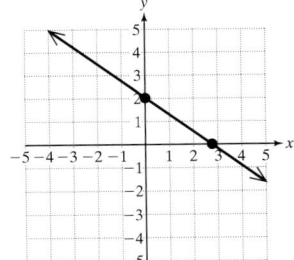

59. x-intercept: $(-4, 0)$;
y-intercept: $(0, 8)$
$2x + 8 = y$

61. x-intercept: $(0, 0)$;
y-intercept: $(0, 0)$
$2x - 2y = 0$

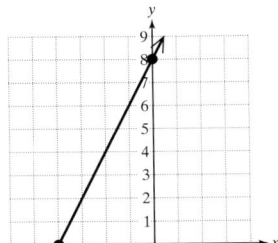 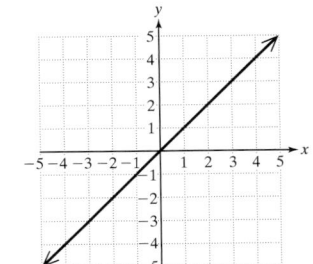

63. x-intercept: $(10, 0)$;
y-intercept: $(0, 5)$
$20x = -40y + 200$

65. x-intercept: $(-2, 0)$;
y-intercept: $(0, -1.5)$
$-8.1x - 10.8y = 16.2$

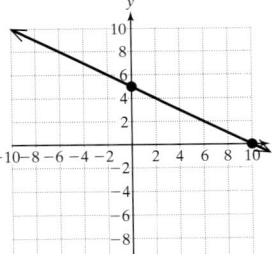 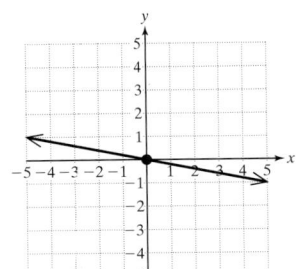

67. x-intercept: $(0, 0)$; y-intercept: $(0, 0)$ $x = -5y$

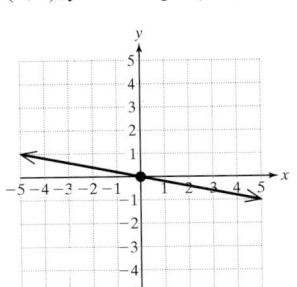

69. b, c, d   **71.** False, $x = 3$ is vertical   **73.** True

75. a. Vertical
c. x-intercept: $(3, 0)$;
no y-intercept

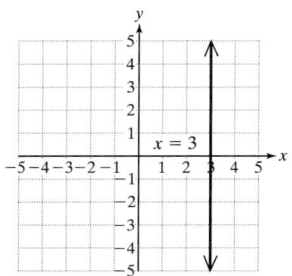

77. a. Horizontal
c. no x-intercept;
y-intercept: $(0, -4)$

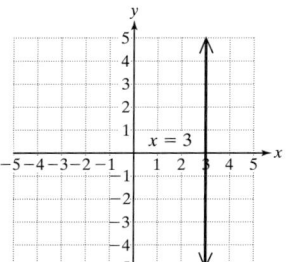

79. a. Vertical
c. x-intercept: $(4, 0)$;
no y-intercept

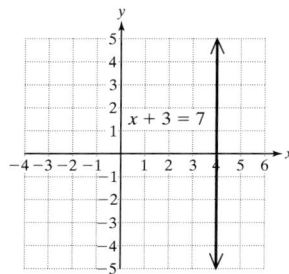

81. a. Horizontal
c. All points on the x-axis
are x-intercepts;
y-intercept: $(0, 0)$

83. a. Vertical
c. x-intercept: $(\frac{3}{2}, 0)$;
no y-intercept

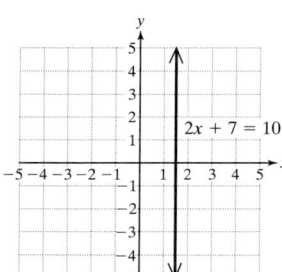

85. a. Horizontal
c. no x-intercept;
y-intercept: $(0, \frac{3}{2})$

87. a. $y = 17.95$ **b.** $x = 145$
c. $(55, 17.95)$ Collecting 55 lb of cans yields $17.95.
$(145, 80.05)$ Collecting 145 lb of cans yields $80.05.
d.

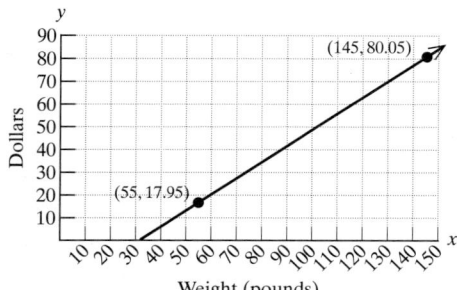

89. a. $y = 10,068$ **b.** $x = 3$ **c.** $(1, 10068)$ One year after
purchase the value of the car is $10,068. $(3, 7006)$ Three years
after purchase the value of the car is $7006.

Section 3.3 Practice Exercises, pp. 238–245

3. x-intercept: $(6, 0)$;
y-intercept: $(0, -2)$
$x - 3y = 6$

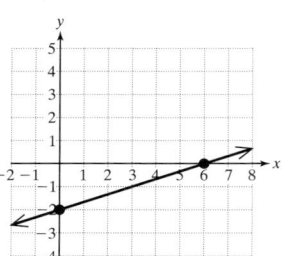

5. x-intercept: $(0, 0)$;
y-intercept: $(0, 0)$
$$y = \frac{2}{3}x$$

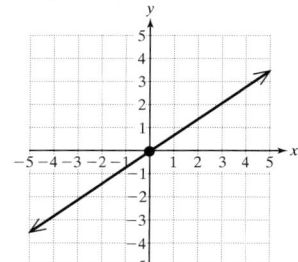

7. x-intercept: $(2, 0)$;
y-intercept: $(0, 8)$
$4x + y = 8$

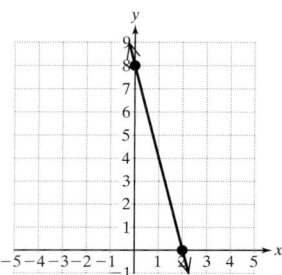

9. $m = \pm\frac{1}{3}$ **11.** $m = \frac{5}{12}$ **13.** Undefined

15. Positive **17.** Negative **19.** Zero **21.** Undefined

23. Positive **25.** $m = 2$ **27.** $m = -\frac{1}{3}$ **29.** $m = 0$

31. The slope is undefined. **33.** $\frac{1}{3}$ **35.** -3

37. $\frac{3}{5}$ **39.** Zero **41.** Undefined **43.** $\frac{28}{5}$

45. $-\frac{7}{8}$ **47.** -0.45 or $-\frac{9}{20}$ **49.** -0.15 or $-\frac{3}{20}$

51. a. -2 **b.** $\frac{1}{2}$ **53. a.** 0 **b.** undefined

55. a. $\frac{4}{5}$ **b.** $-\frac{5}{4}$ **57. a.** undefined **b.** 0

59. l_1: $m = 2$, l_2: $m = 2$; parallel

61. l_1: $m = 5$, l_2: $m = -\frac{1}{5}$; perpendicular

63. l_1: $m = \frac{1}{4}$, l_2: $m = 4$; neither

65. l_1: m is undefined, l_2: m is undefined; parallel
67. a. $m = 47$ **b.** The number of male prisoners increased
at a rate of 47 thousand per year during this time period.
69. a. $m = 420$. The median income for females in the
United States increased at a rate of approximately $420 per
year between 2000 and 2005. **b.** No, the rate of increase in
women's median income is less than the rate of increase in
men's income.
71. a. 1 mile **b.** 2 miles **c.** 3 miles **d.** $m = 0.2$; The
distance between a lightning strike and an observer increases
by 0.2 miles for every additional second between seeing
lightning and hearing thunder.

73. a. $m = -\frac{2}{5}$ **75.** $m = 1$ **77.** The slope is undefined.

79. For example:
(2, 0) and (−6, −6)

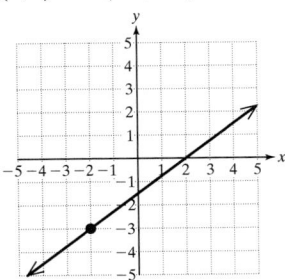

81. For example:
(0, 1) and (−2, 5)

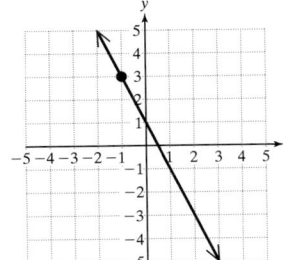

4. The lines may appear parallel; however, they are not parallel because the slopes are different.

5. The lines may appear to coincide on a graph; however, they are not the same line because the *y*-intercepts are different.

Section 3.4 Practice Exercises, pp. 251–255

3. *x*-intercept: (10, 0); *y*-intercept: (0, −2)
5. *x*-intercept: none; *y*-intercept: (0, −3)
7. *x*-intercept: (0, 0); *y*-intercept: (0, 0)
9. *x*-intercept: (4, 0); *y*-intercept: none

11. $m = \dfrac{2}{3}$; *y*-intercept: (0, 5)

13. $m = -1$; *y*-intercept: (0, 6)
15. $m = -4$; *y*-intercept: (0, 0)

17. $m = 1$; *y*-intercept: $\left(0, -\dfrac{5}{3}\right)$

19. $m = -\dfrac{3}{2}$; *y*-intercept: $\left(0, \dfrac{9}{2}\right)$

21. $m = \dfrac{7}{3}$; *y*-intercept: (0, 2)

23. $m = 1$; *y*-intercept: (0, −1)
25. Undefined slope; no *y*-intercept
27. $m = 0$; *y*-intercept: (0, −8)

29. $m = \dfrac{5}{6}$; *y*-intercept: (0, 0)

83.

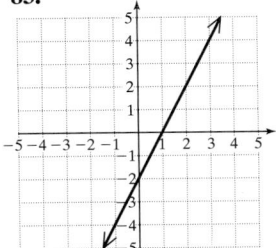

85.

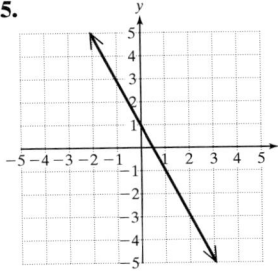

87.

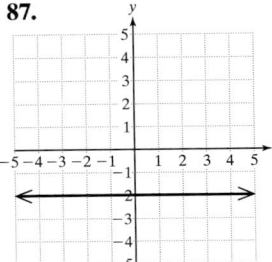

89.

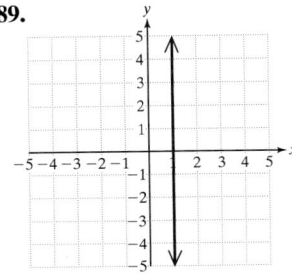

31.

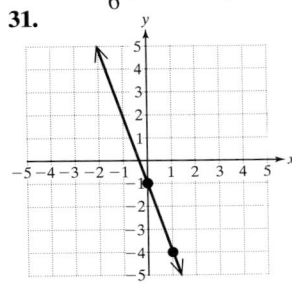

33.

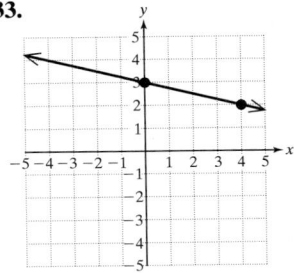

91. $\dfrac{2t}{2c + d}$ or $\dfrac{-2t}{-2c - d}$

93. $\left(0, \dfrac{c}{b}\right)$

95. For example: (1, 5)

35. d   **37.** a   **39.** f

41. $y = \dfrac{5}{2}x - 1$ **43.** $y = 6x + 8$

Section 3.4 Calculator Connections, p. 250

1. Perpendicular

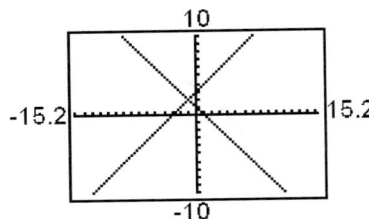

2. Parallel

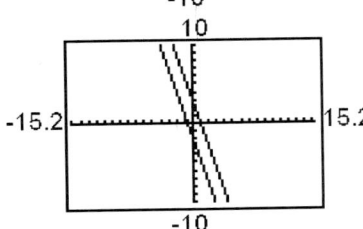

3. Neither

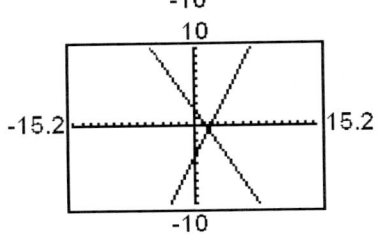

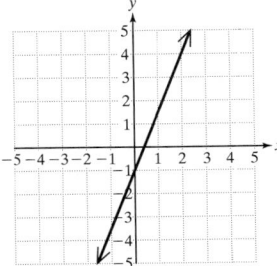

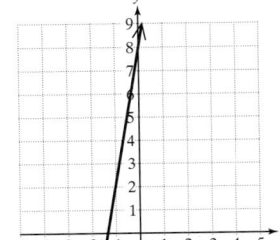

45. $y = 3x + 7$ **47.** $y = x$

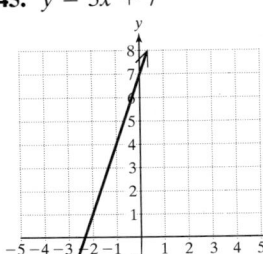

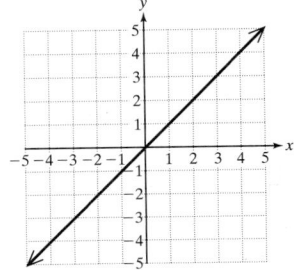

49. $y = -\dfrac{2}{5}x$ **51.** $y = 1$

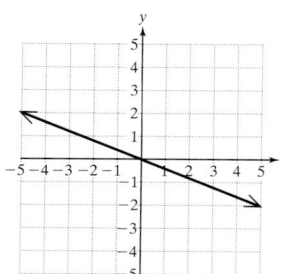

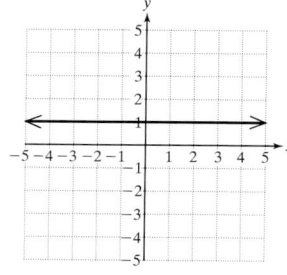

53. $x = -3$ **55.** $y = 2x - 3$

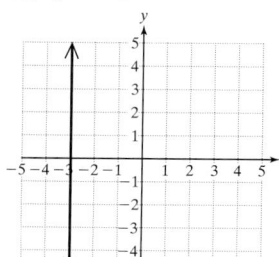

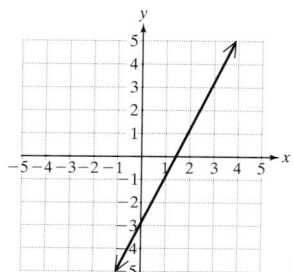

57. Neither   **59.** Perpendicular   **61.** Parallel
63. Perpendicular   **65.** Neither   **67.** Parallel
69. Perpendicular   **71.** Parallel   **73.** Perpendicular

75. Neither   **77.** Parallel   **79.** $y = \dfrac{2}{3}x - 1$

81. $y = -14x + 2$ **83.** $y = \dfrac{6}{7}$ **85.** $y = -3x$

87. $y = -4x - 3$

89. a. $m = 0.10$. The cost increases $0.10/min of long distance.   **b.** $(0, 16.95)$. The monthly bill is $16.95 if 0 minutes of long distance are used.   **c.** The cost is $40.35.

91. $m = -\dfrac{2}{5}$ **93.** $m = \dfrac{4}{3}$

Section 3.5 Practice Exercises, pp. 261–265

3. $2x - 3y = -3$ **5.** $3 - y = 9$

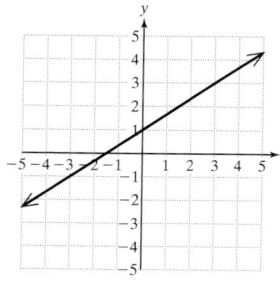

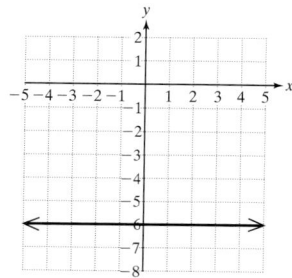

7. 9   **9.** 0   **11.** $y = 3x + 7$ or $3x - y = -7$

13. $y = -4x - 14$ or $4x + y = -14$

15. $y = -\dfrac{1}{2}x - \dfrac{1}{2}$ or $x + 2y = -1$

17. $y = \dfrac{1}{4}x + 8$ or $x - 4y = -32$

19. $y = 4.5x - 25.6$ or $45x - 10y = 256$
21. $y = -2$

23. $y = -2x + 1$ **25.** $y = \dfrac{1}{2}x - 1$

27. $y = 2x - 2$ or $2x - y = 2$

29. $y = -\dfrac{5}{8}x - \dfrac{19}{8}$ or $5x + 8y = -19$

31. $y = -0.2x - 2.86$ or $20x + 100y = -286$
33. $y = 4x + 13$ or $4x - y = -13$

35. $y = -\dfrac{3}{2}x + 6$ or $3x + 2y = 12$

37. $y = -2x - 8$ or $2x + y = -8$

39. $y = -\dfrac{1}{5}x - 6$ or $x + 5y = -30$

41. $y = 3x - 8$ or $3x - y = 8$

43. $y = -\dfrac{1}{4}x - \dfrac{3}{2}$ or $x + 4y = -6$

45. iv   **47.** vi   **49.** iii   **51.** $y = 1$ **53.** $x = 2$

55. $x = \dfrac{5}{2}$ **57.** $y = 2$ **59.** $x = -6$ **61.** $x = -4$

63. a. -0.35 **b.** $y = -0.35x + 33.9$ **c.** In 2000, approximately 21.65% of women smoked.

Section 3.6 Calculator Connections, pp. 270–271

1. 13.3

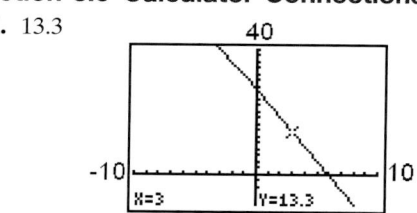

2. -42.3

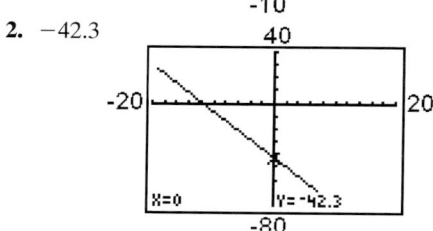

3. 345

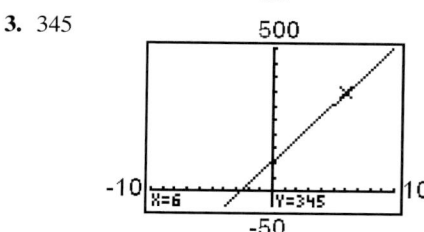

4. 95

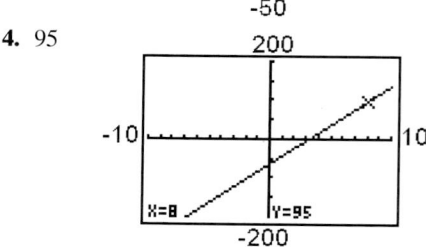

Section 3.6 Practice Exercises, pp. 271–275

3. x-intercept: $(6, 0)$; y-intercept: $(0, 5)$
5. x-intercept: $(-2, 0)$; y-intercept: $(0, -4)$

7. *x*-intercept: none; *y*-intercept: $(0, -9)$
9. a. *x*, year **b.** *y*, minimum hourly wage **c.** $3.32 per hour **d.** $5.82 per hour **e.** The *y*-intercept is $(0, 0.82)$ and indicates that the minimum hourly wage in the year 1960 $(x = 0)$ was approximately $0.82/hour. **f.** The slope is 0.10 and indicates that minimum hourly wage rises an average of $0.10 per year.
11. a. 0.75 or equivalently $\frac{3}{4}$. O'Neal would expect to score 0.75 point for each minute played. This is equivalent to scoring 3 points for 4 minutes played. **b.** 0.50 or equivalently $\frac{1}{2}$. Iverson would expect to score 0.5 point for each minute played. This is equivalent to scoring 1 point for 2 minutes played. **c.** O'Neal: 27 points; Iverson: 18 points
13. a. *y*, temperature **b.** *x*, latitude **c.** 30.7° **d.** 13.4°
e. $m = -2.333$. The average temperature in January decreases 2.333° per 1° of latitude. **f.** $(53.2, 0)$. At 53.2° latitude, the average temperature in January is 0°.
15. a. $y = 2.5x + 31$ **b.** 43.5 in.
17. a. $y = 2x + 2$ **b.** $m = 2$. For each additional mile, the time is increased by 2 min. **c.** 38 min
19. a. $y = 0.08x + 18.95$ **b.** $25.91
21. a. $y = 90x + 105$ **b.** $1185.00
23. a. $y = 25x + 5000$ **b.** $8750.00
25. a. $y = 0.5x + 35$ **b.** $210.00

Chapter 3 Review Exercises, pp. 280–284

1.

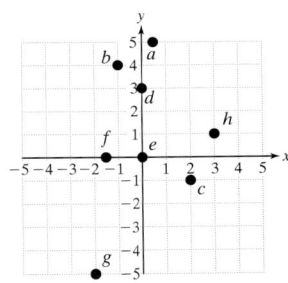

2. $A(4, -3); B(-3, -2); C(\frac{5}{2}, 5); D(-4, 1); E(-\frac{1}{2}, 0); F(0, -5)$
3. III **4.** II **5.** IV **6.** I **7.** IV **8.** III
9. *x*-axis **10.** *y*-axis
11. a. On day 1, the price was $26.25. **b.** Day 2 **c.** $2.25
12. a. In 2003 (8 years after 1995), there was only 1 space shuttle launch (this was the year that the Columbia and its crew were lost).
b.

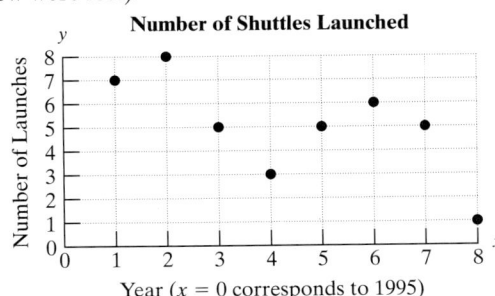

Number of Shuttles Launched

13. No   **14.** No   **15.** Yes   **16.** Yes
17.

x	*y*
2	1
3	4
1	-2

18.

x	*y*
0	2
-2	$\frac{7}{3}$
-6	3

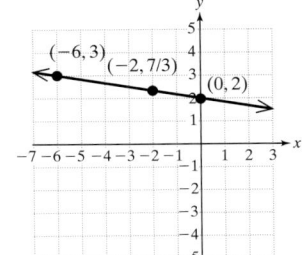

19.

x	*y*
0	-1
3	1
-6	-5

20.

x	*y*
0	-3
-3	3
1	-5

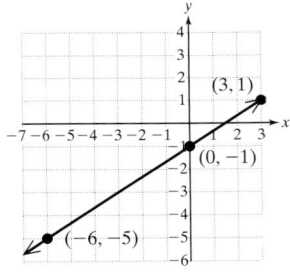

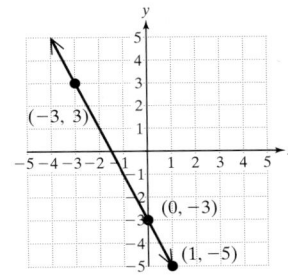

21.

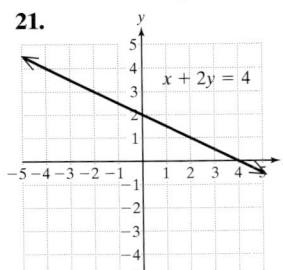

22.

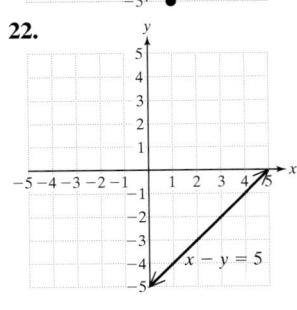

23.

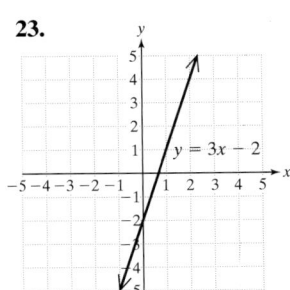

24.

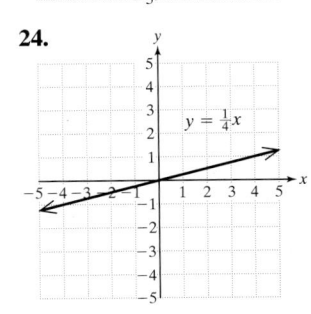

25. Vertical

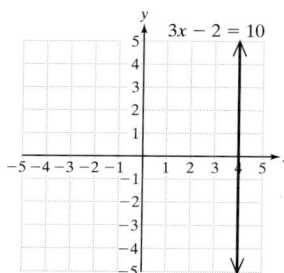

26. Vertical

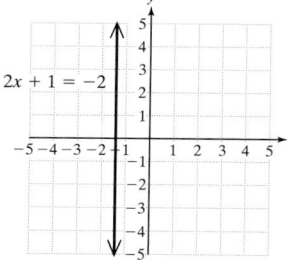

27. Horizontal

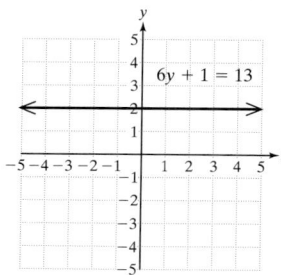

28. Horizontal

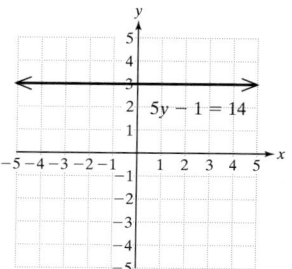

29. x-intercept: $(-3, 0)$; y-intercept: $\left(0, \frac{3}{2}\right)$

30. x-intercept: $(3, 0)$; y-intercept: $(0, 6)$

31. x-intercept: $(0, 0)$; y-intercept: $(0, 0)$

32. x-intercept: $(0, 0)$; y-intercept: $(0, 0)$

33. x-intercept: none; y-intercept: $(0, -4)$

34. x-intercept: none; y-intercept: $(0, 2)$

35. x-intercept: $\left(-\frac{5}{2}, 0\right)$; y-intercept: none

36. x-intercept: $\left(\frac{1}{3}, 0\right)$; y-intercept: none

37. $m = \frac{12}{5}$ **38.** -2 **39.** $-\frac{2}{3}$ **40.** 8

41. Undefined   **42.** 0   **43. a.** -5 **b.** $\frac{1}{5}$

44. a. 0   **b.** Undefined   **45.** $m_1 = \frac{2}{3}$ $m_2 = \frac{2}{3}$; parallel

46. $m_1 = 8$, $m_2 = 8$; parallel

47. $m_1 = -\frac{5}{12}$, $m_2 = \frac{12}{5}$; perpendicular

48. $m_1 =$ undefined, $m_2 = 0$; perpendicular

49. $y = \frac{5}{2}x - 5$; $m = \frac{5}{2}$; y-intercept: $(0, -5)$

50. $y = -\frac{3}{4}x + 3$; $m = -\frac{3}{4}$; y-intercept: $(0, 3)$

51. $y = \frac{1}{3}x$; $m = \frac{1}{3}$; y-intercept: $(0, 0)$

52. $y = \frac{12}{5}$; $m = 0$; y-intercept: $\left(0, \frac{12}{5}\right)$

53. $y = -\frac{5}{2}$; $m = 0$; y-intercept: $\left(0, -\frac{5}{2}\right)$

54. $y = x$; $m = 1$; y-intercept: $(0, 0)$ **55.** Neither

56. Perpendicular   **57.** Parallel   **58.** Parallel

59. Perpendicular   **60.** $y = -\frac{4}{3}x - 1$ or $4x + 3y = -3$

61. $y = 5x$ or $5x - y = 0$ **62.** For example: $y = 3x + 2$

63. For example: $5x + 2y = -4$ **64.** $m = \frac{y_2 - y_1}{x_2 - x_1}$

65. $y - y_1 = m(x - x_1)$ **66.** For example: $x = 6$

67. For example: $y = -5$

68. $y = -6x + 2$ or $6x + y = 2$

69. $y = \frac{2}{3}x + \frac{5}{3}$ or $2x - 3y = -5$

70. $y = \frac{1}{4}x - 4$ or $x - 4y = 16$ **71.** $y = -5$

72. $y = \frac{6}{5}x + 6$ or $6x - 5y = -30$

73. $y = 4x + 31$ or $4x - y = -31$

74. a. x, age   **b.** y, height   **c.** 47.8 in.   **d.** The slope is 2.4 and indicates that the average height for girls increases at a rate of 2.4 in. per year.   **75. a.** $m = 137$ **b.** The number of prescriptions increased by 137 million per year during this time period.   **c.** $y = 137x + 2140$ **d.** 3921 million

76. a. $y = 30.6x + 264$ **b.** \$814.8 billion

77. a. $y = 20x + 55$ **b.** \$235

78. a. $y = 8x + 700$ **b.** \$1340

Chapter 3 Test, pp. 284–286

1. a. II   **b.** IV   **c.** III   **2.** 0   **3.** 0

4. a. $(5, 46)$ At age 5 the boy's height was 46 in. $(7, 50)$ $(9, 55)$ $(11, 60)$ **b.**

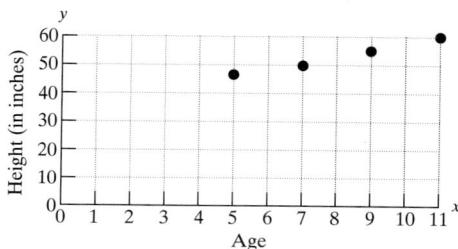

c. 57.5 in.   **d.** No, his height will maximize in his teen years.

5. a. No   **b.** Yes   **c.** Yes   **d.** Yes

6.

x	y
0	-2
4	-1
6	$-\frac{1}{2}$

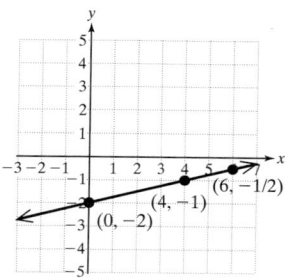

7. a. 202 beats per minute   **b.** $(20, 200)$ $(30, 190)$ $(40, 180)$ $(50, 170)$ $(60, 160)$

8. Horizontal

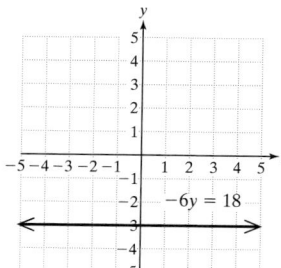

9. Vertical

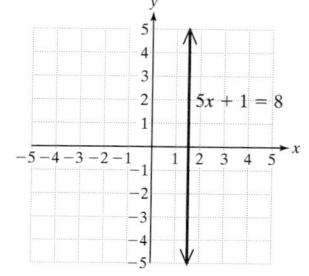

10. x-intercept: $\left(-\dfrac{3}{2}, 0\right)$; y-intercept: $(0, 2)$

11. $\dfrac{2}{5}$ **12. a.** $\dfrac{1}{3}$ **b.** $\dfrac{4}{3}$ **13. a.** $-\dfrac{1}{4}$ **b.** 4

14. a. Undefined **b.** 0

15. x-intercept: $\left(-\dfrac{1}{4}, 0\right)$;
y-intercept: $(0, 2)$
$y = 8x + 2$

16. x-intercept: $(0, 0)$;
y-intercept: $(0, 0)$
$2x + 9y = 0$

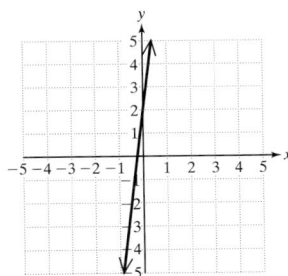

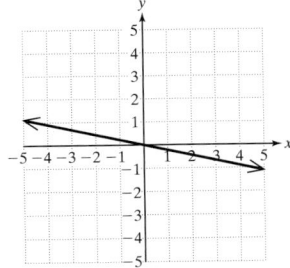

17. x-intercept: $(3, 0)$;
y-intercept: none
$x - 3 = 0$

18. x-intercept: none;
y-intercept: $(0, -3)$
$-4y = 12$

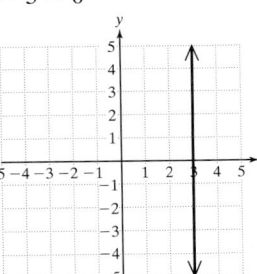

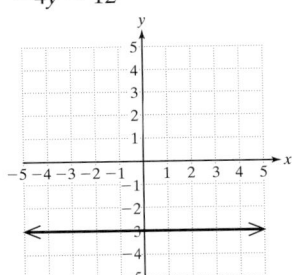

19. Perpendicular   **20.** $y = \dfrac{1}{4}x + \dfrac{1}{2}$ or $x - 4y = -2$

21. $y = -\dfrac{7}{2}x + 15$ or $7x + 2y = 30$

22. $y = -6$ **23.** $y = -\dfrac{1}{3}x + 1$ or $x + 3y = 3$

24. $y = 3x + 8$ or $3x - y = -8$ **25. a.** $\dfrac{3}{4}$ **b.** $\dfrac{1}{2}$

26. a. $y = 1.5x + 10$ **b.** \$25

27. a. $m = 20$; The slope indicates that there is an increase of 20 thousand medical doctors per year. **b.** $y = 20x + 414$
c. 1014 thousand or, equivalently, 1,014,000

Chapters 1–3 Cumulative Review Exercises, pp. 286–287

1. a. Rational **b.** Rational **c.** Irrational **d.** Rational

2. a. $-\dfrac{2}{3}; \dfrac{2}{3}$ **b.** $-5.3; 5.3$ **3.** 69 **4.** -13 **5.** 18

6. $\dfrac{3}{4} \div -\dfrac{7}{8}; -\dfrac{6}{7}$ **7.** $(-2.1)(-6); 12.6$

8. The associative property of addition   **9.** $x = 4$

10. $m = 5$ **11.** $y = -\dfrac{9}{2}$ **12.** $z = -2$ **13.** 9241 mi^2

14. $a = \dfrac{c - b}{3}$ **15.**

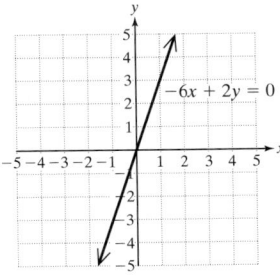

16. x-intercept: $(-2, 0)$; y-intercept: $(0, 1)$

17. $y = -\dfrac{3}{2}x - 6$; slope: $-\dfrac{3}{2}$; y-intercept: $(0, -6)$

18. $2x + 3 = 5$ can be written as $x = 1$, which represents a vertical line. A vertical line of the form $x = k$ $(k \neq 0)$ has an x-intercept of $(k, 0)$ and no y-intercept.   **19.** $y = -3x + 1$ or $3x + y = 1$ **20.** $y = \frac{2}{3}x + 6$ or $2x - 3y = -18$

Chapter 4

Chapter Opener Puzzle

One drink costs \$2.00 and one small popcorn costs \$3.50.

Section 4.1 Calculator Connections, pp. 295–296

1. $(2, 1)$

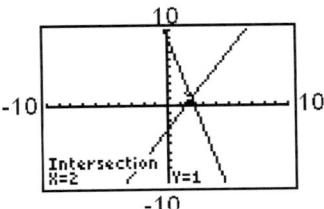

2. $(6, -1)$

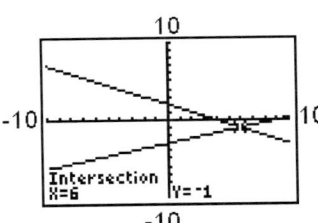

3. $(3, 1)$

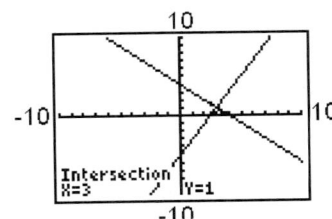

4. $(-2, 0)$

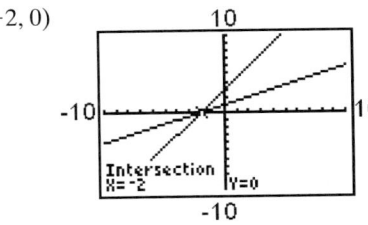

5. No solution, inconsistent system

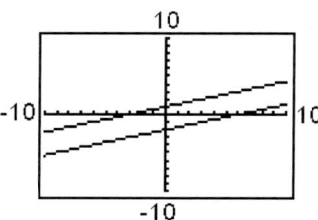

6. Dependent system

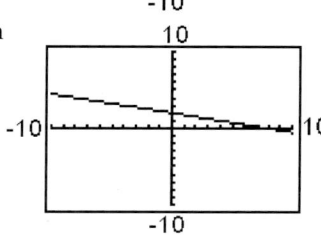

Section 4.1 Practice Exercises, pp. 296–301

3. Yes   **5.** No   **7.** Yes   **9.** No   **11.** b   **13.** d

15. a.

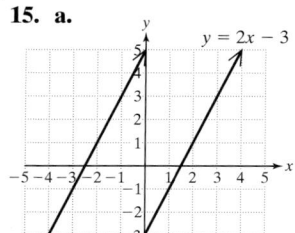

b.

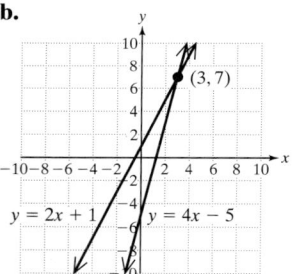

c.

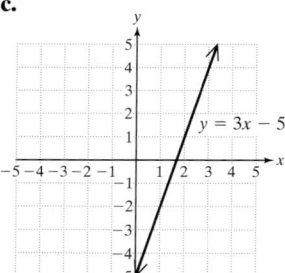

17. c   **19.** a   **21.** a   **23.** b   **25.** c

27.

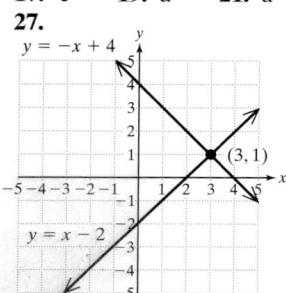

29.

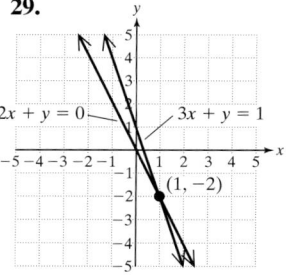

31.

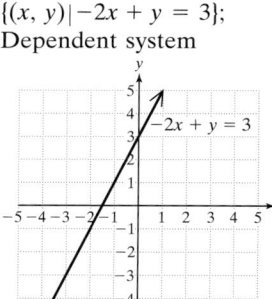

33. No solution; Inconsistent system

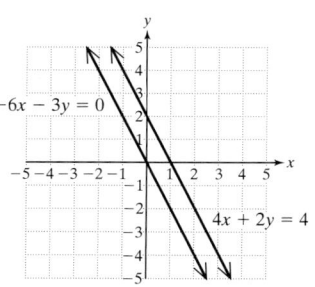

35. Infinitely many solutions $\{(x, y)\,|-2x + y = 3\}$; Dependent system

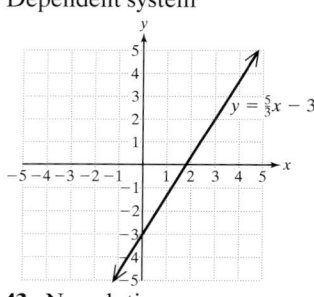

37.

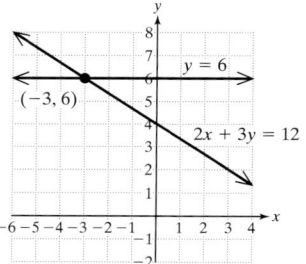

39. Infinitely many solutions $\{(x, y)\,|\,y = \frac{5}{3}x - 3\}$; Dependent system

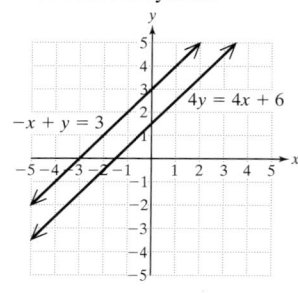

41.

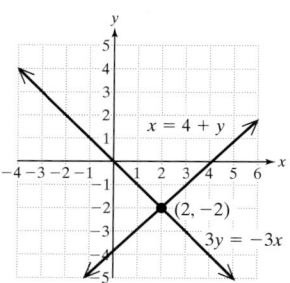

43. No solution; Inconsistent system

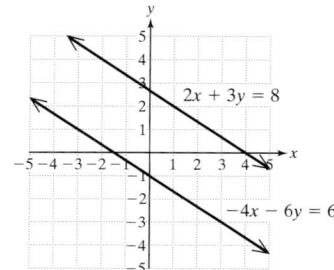

45.

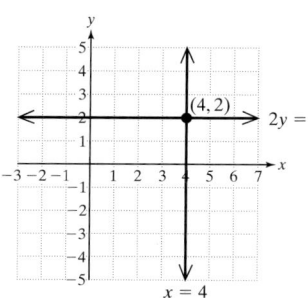

47. No solution; Inconsistent system

49.

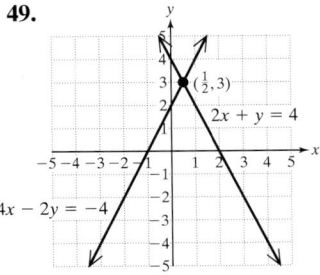

51. Infinitely many solutions
$\{(x, y) | y = 0.5x + 2\}$; Dependent system

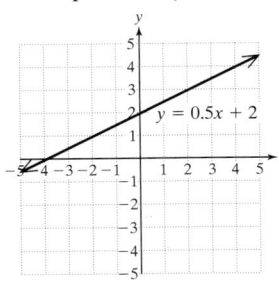

53. 4 lessons will cost $120 for each instructor.
55. The point of intersection is below the x-axis and cannot have a positive y-coordinate.
57. For example: $4x + y = 9$; $-2x - y = -5$
59. For example: $2x + 2y = 1$

Section 4.2 Practice Exercises, pp. 309–311

1. $y = 2x - 4$; $y = 2x - 4$; coinciding lines
3. $y = -\dfrac{2}{3}x + 2$; $y = x - 5$; intersecting lines
5. $y = 4x - 4$; $y = 4x - 13$; parallel lines
7. $(3, -6)$ **9.** $(0, 4)$ **11. a.** y in the second equation is easiest to solve for because its coefficient is 1.   **b.** $(1, 5)$
13. $(5, 2)$ **15.** $(10, 5)$ **17.** $\left(\dfrac{1}{2}, 3\right)$ **19.** $(5, 3)$
21. $(1, 0)$ **23.** $(1, 4)$ **25.** No solution; Inconsistent system   **27.** Infinitely many solutions $\{(x, y) | 2x - 6y = -2\}$; Dependent system   **29.** $(5, -7)$
31. $\left(-5, \dfrac{3}{2}\right)$ **33.** $(2, -5)$ **35.** $(-4, 6)$ **37.** $(0, 2)$
39. Infinitely many solutions $\{(x, y) | y = 0.25x + 1\}$; Dependent system   **41.** $(1, 1)$
43. No solution; Inconsistent system   **45.** $(-1, 5)$
47. $(-6, -4)$ **49.** The numbers are 48 and 58.
51. The numbers are 13 and 39.   **53.** The angles are $165°$ and $15°$.   **55.** The angles are $70°$ and $20°$.
57. The angles are $42°$ and $48°$.   **59.** For example, $(0, 3), (1, 5), (-1, 1)$

Section 4.3 Practice Exercises, pp. 319–322

3. No   **5.** Yes   **7. a.** True   **b.** False, multiply the second equation by 5.   **9. a.** x would be easier.
   b. $(0, -3)$ **11.** $(4, -1)$ **13.** $(4, 3)$ **15.** $(2, 3)$
17. $(1, -4)$ **19.** $(1, -1)$ **21.** $(-4, -6)$
23. There are infinitely many solutions. The lines coincide.
25. The system will have no solution. The lines are parallel.
27. No solution; Inconsistent system   **29.** Infinitely many solutions; $\{(x, y) | 4x - 3y = 6\}$; Dependent system
31. $(2, -2)$ **33.** $(0, 3)$ **35.** $(5, 2)$ **37.** No solution; Inconsistent system   **39.** $(-5, 0)$ **41.** $(2.5, -0.5)$
43. $\left(\dfrac{1}{2}, 0\right)$ **45.** $(-3, 2)$ **47.** $\left(\dfrac{7}{4}, 3\right)$ **49.** $(0, 1)$

51. No solution; Inconsistent system   **53.** $(0, -5)$
55. $(4, -2)$ **57.** Infinitely many solutions; $\{(a, b) | a = 5 + 2b\}$; Dependent system
59. The numbers are 17 and 19.
61. The numbers are -1 and 3.   **63.** $\left(\dfrac{7}{9}, \dfrac{5}{9}\right)$
65. $\left(\dfrac{7}{16}, -\dfrac{7}{8}\right)$ **67.** $(1, 3)$
69. One line within the system of equations would have to "bend" for the system to have exactly two points of intersection. This is not possible.   **71.** $A = -5, B = 2$

Section 4.4 Practice Exercises, pp. 330–333

1. $(-1, 4)$ **3.** $\left(\dfrac{5}{2}, 1\right)$ **5.** The numbers are 4 and 16.
7. The angles are $80°$ and $10°$.
9. Tapes are $10.50 each, and CDs are $15.50 each.
11. Technology stock costs $16 per share, and the mutual fund costs $11 per share.
13. Shanelle invested $3500 in the 10% account and $6500 in the 7% account.
15. $9000 is borrowed at 6%, and $3000 is borrowed at 9%.
17. 15 gal of the 50% mixture should be mixed with 10 gal of the 40% mixture.
19. 12 gal of the 45% disinfectant solution should be mixed with 8 gal of the 30% disinfectant solution.
21. The speed of the boat in still water is 6 mph, and speed of the current is 2 mph.
23. The speed of the plane in still air is 300 mph, and wind is 20 mph.   **25.** There are 17 dimes and 22 nickels.
27. a. 835 free throws and 1597 field goals
b. 4029 points   **c.** Approximately 50 points per game
29. The speed of the plane in still air is 160 mph, and wind is 40 mph.
31. 12 lb of candy should be mixed with 8 lb of nuts.
33. $15,000 is invested in the 5.5% account, and $45,000 is invested in the 6.5% account.
35. 20 oz of Miracle-Gro should be mixed with 40 oz of Green Light.
37. Dallas scored 30 points, and Buffalo scored 13 points.
39. There were 300 women and 200 men in the survey.
41. There are six 15-sec commercials and sixteen 30-sec commercials.

Chapter 4 Review Exercises, pp. 337–340

1. Yes   **2.** No   **3.** No   **4.** Yes
5. Intersecting lines (the lines have different slopes)
6. Intersecting lines (the lines have different slopes)
7. Parallel lines (the lines have the same slope but different y-intercepts)
8. Intersecting lines (the lines have different slopes)
9. Coinciding lines (the lines have the same slope and same y-intercept)
10. Intersecting lines (the lines have different slopes)

11. $(0, -2)$
Consistent; independent

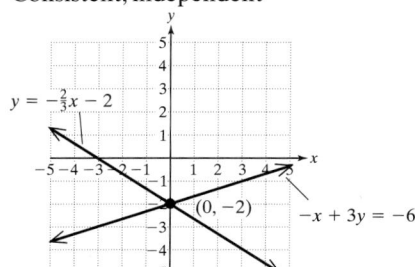

$y = -\frac{2}{3}x - 2$

$(0, -2)$

$-x + 3y = -6$

12. $(-2, 3)$

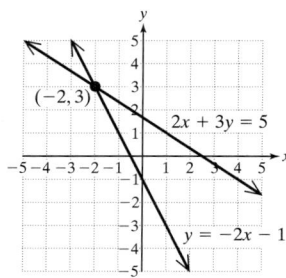

$(-2, 3)$

$2x + 3y = 5$

$y = -2x - 1$

13. Infinitely many solutions
$\{(x, y)\,|\,2x + y = 5\};$
Dependent system

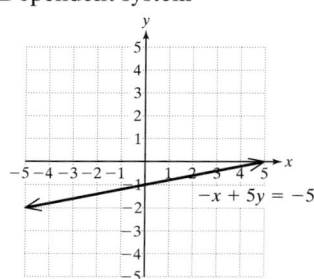

$2x + y = 5$

14. Infinitely many solutions
$\{(x, y)\,|\,y = \frac{1}{5}x - 1\};$
Dependent system

15. $(1, -1)$

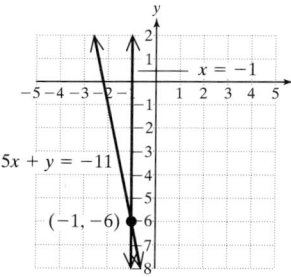

$-x + 5y = -5$

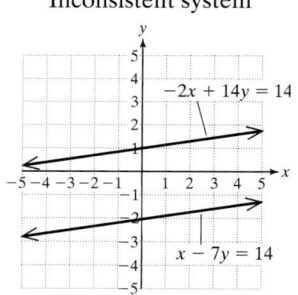

$6x - 3y = 9$

$y = -1$

$(1, -1)$

16. $(-1, -6)$

17. No solution;
Inconsistent system

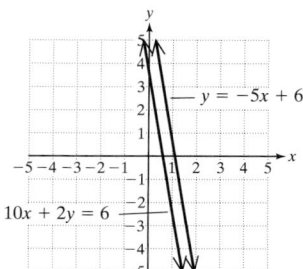

$x = -1$

$5x + y = -11$

$(-1, -6)$

$-2x + 14y = 14$

$x - 7y = 14$

18. No solution; Inconsistent system

$y = -5x + 6$

$10x + 2y = 6$

19. 200 miles   **20.** $\left(\frac{2}{3}, -2\right)$ **21.** $(-4, 1)$

22. No solution; Inconsistent system   **23.** Infinitely
many solutions; $\{(x, y)\,|\,y = -2x + 2\};$ Dependent system
24. a. x in the first equation is easiest to solve for because
its coefficient is 1.   **b.** $\left(6, \frac{5}{2}\right)$

25. a. y in the second equation is easiest to solve for
because its coefficient is 1.   **b.** $\left(\frac{9}{2}, 3\right)$ **26.** $(5, -4)$

27. $(0, 4)$ **28.** Infinitely many solutions;
$\{(x, y)\,|\,x - 3y = 9\};$ Dependent system
29. No solution; Inconsistent system
30. The numbers are 50 and 8.
31. The angles are $42°$ and $48°$.
32. The angles are $115\frac{1}{3}°$ and $64\frac{2}{3}°$.
33. See page 321.   **34. b.** $(-3, -2)$ **35. b.** $(2, 2)$
36. $(2, -1)$ **37.** $(-6, 2)$ **38.** $\left(-\frac{1}{2}, \frac{1}{3}\right)$ **39.** $\left(\frac{1}{4}, -\frac{2}{5}\right)$
40. Infinitely many solutions; $\{(x, y)\,|\,-4x - 6y = -2\};$
Dependent system   **41.** No solution; Inconsistent system
42. $(-4, -2)$ **43.** $(1, 0)$ **44. b.** $(5, -3)$
45. b. $(-2, -1)$
46. There were 8 adult tickets and 52 children's tickets sold.
47. Emillo invested $2500 in the 5% account and $17,500 in
the 8% account.   **48.** 5 liters of the 20% solution should
be mixed with 10 L of the 14% solution.   **49.** The speed of
the boat is 18 mph, and that of the current is 2 mph.
50. Suzanne has seven quarters and three dollar coins.
51. A hot dog costs $4.50 and a drink costs $3.50.
52. 3000 women and 2700 men   **53.** The score was 72 on
the first round and 82 on the second round.

Chapter 4 Test, pp. 340–341

1. $y = -\frac{5}{2}x - 3;$ $y = -\frac{5}{2}x + 3;$ Parallel lines

2.

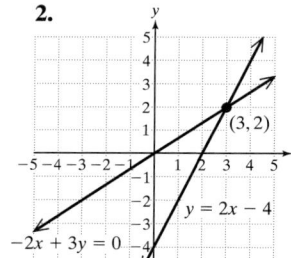

$(3, 2)$

$y = 2x - 4$

$-2x + 3y = 0$

3. Infinitely many solutions;
$\{(x, y)\,|\,2x + 4y = 6\};$
Dependent system

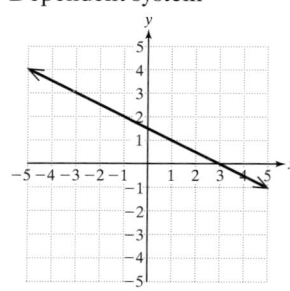

4. $(-2, 0)$
5. Swoopes scored 614 points and Jackson scored 597.
6. $\left(2, -\frac{1}{3}\right)$ **7.** 12 mL of the 50% acid solution should
be mixed with 24 mL of the 20% solution.
8. a. No solution   **b.** Infinitely many solutions
c. One solution   **9.** $(-5, 4)$ **10.** No solution
11. $(3, -5)$ **12.** $(-1, 2)$ **13.** Infinitely many
solutions; $\{(x, y)\,|\,10x + 2y = -8\}$ **14.** $(1, -2)$
15. CDs cost $8 each and DVDs cost $11 each.

16. a. $18 was required. **b.** They used 24 quarters and 12 $1 bills.   **17.** $1200 was borrowed at 10%, and $3800 was borrowed at 8%.   **18.** He scored 155 receiving touchdowns and 10 touchdowns rushing.   **19.** The plane travels 470 mph in still air, and the wind speed is 30 mph.   **20.** The cake has 340 calories, and the ice cream has 120 calories.   **21.** There are 178 men and 62 women on the force.

Chapters 1–4 Cumulative Review Exercises, pp. 341–342

1. $\dfrac{11}{6}$ **2.** $x = -\dfrac{21}{2}$ **3.** No solution

4. $y = \dfrac{3}{2}x - 3$ **5.** $\left[\dfrac{3}{11}, \infty\right)$

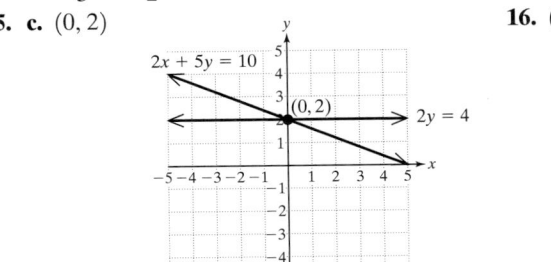

6. The angles are 37°, 33°, and 110°.   **7.** The rates of the hikers are 2 mph and 4 mph.   **8.** Jesse Ventura received approximately 762,200 votes.   **9.** 36% of the goal has been achieved.   **10.** The angles are 36.5° and 53.5°.

11. $x = 5z + m$ **12.** $y = \dfrac{2}{3}x - 2$

13. a. $-\dfrac{2}{3}$ **b.** $\dfrac{3}{2}$ **14.** $y = -3x + 3$

15. c. $(0, 2)$ **16.** $(0, 2)$

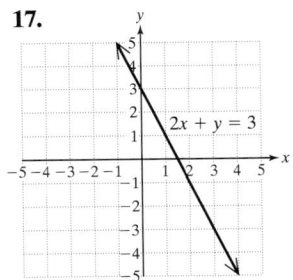

17.

18. 20 gal of the 15% solution should be mixed with 40 gal of the 60% solution.   **19.** x is 27°; y is 63°   **20. a.** 1.4 **b.** Between 1920 and 1990, the winning speed in the Indianapolis 500 increased on average by 1.4 mph per year.

Chapter 5

Chapter Opener Puzzle

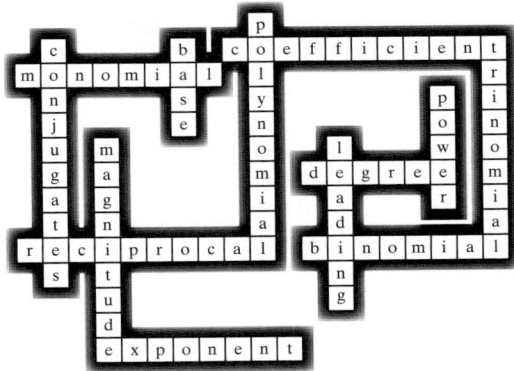

Section 5.1 Calculator Connections, p. 350

1.–3.
```
(1.06)^5
       1.338225578
(1.02)^40
       2.208039664
5000(1.06)^5
       6691.127888
```

4.–6.
```
2000(1.02)^40
       4416.079327
3000(1+.06)^2
       3370.8
1000(1+.05)^3
       1157.625
```

Section 5.1 Practice Exercises, pp. 351–354

3. Base: x; exponent: 4   **5.** Base: 3; exponent: 5
7. Base: -1; exponent: 4   **9.** Base: q; exponent: 1
11. v **13.** y **15.** $(-6b)^2$ **17.** $-6b^2$
19. $(y + 2)^4$ **21.** $\dfrac{-2}{t^3}$
23. No; $-3^4 = -81$ and $(-3)^4 = 81$
25. Yes; $-5^3 = -125$ and $(-5)^3 = -125$
27. Yes; $\left(\dfrac{1}{5}\right)^2 = \dfrac{1}{25}$ and $\dfrac{1}{5^2} = \dfrac{1}{25}$
29. Yes; $\left(\dfrac{7}{10}\right)^3 = \dfrac{343}{1000}$ and $(0.7)^3 = 0.343$ **31.** 20
33. 1   **35.** 0   **37.** $-\dfrac{1}{64}$ **39.** $-\dfrac{9}{25}$ **41.** 0
43. -3 **45.** 72   **47.** 33   **49.** 80   **51.** 400
53. 81   **55.** 81   **57.** 1920   **59.** -80
61. a. $(y \cdot y)(y \cdot y \cdot y \cdot y) = y^6$
b. $(3 \cdot 3)(3 \cdot 3 \cdot 3 \cdot 3) = 3^6$ **63.** w^{11} **65.** p^5
67. 6^{12} **69.** $\left(\dfrac{1}{x}\right)^3$ **71.** b^{17} **73.** z^{79}
75. a. $\dfrac{w \cdot w \cdot w \cdot w \cdot w}{w \cdot w} = w^3$ **b.** $\dfrac{4 \cdot 4 \cdot 4 \cdot 4 \cdot 4}{4 \cdot 4} = 4^3$
77. z **79.** b^{11} **81.** 2^2 **83.** 3^4 **85.** w **87.** n^7
89. 5^{10} **91.** 3^3 **93.** w^5 **95.** $30x^5y^4$ **97.** $42p^7q^{11}$

99. $7c^8d^{13}$ **101.** $4h^2k$ **103.** $\dfrac{w^6z^2}{2}$ **105.** $\dfrac{15m^4np^3}{4}$

107. $50a^6b^2c$ **109.** $-wx^9y^3$ **111.** \$2433.31

113. \$12,155.06 **115.** 1963 ft^2 **117.** 268 in.^{3}

119. y^{3a} **121.** q **123.** w^{3n+3} **125.** t^3

Section 5.2 Practice Exercises, pp. 357–358

1. 4^9 **3.** a^{20} **5.** d^9 **7.** 7^6

9. When multiplying expressions with the same base, add the exponents. When dividing expressions with the same base, subtract the exponents. **11.** 5^{12}

13. 12^6 **15.** y^{14} **17.** w^{25} **19.** a^{36} **21.** y^{14}

23. They are both equal to 2^6.

25. $2^{(2^4)} = 2^{16}$ $(2^2)^4 = 2^8$; the expression $2^{(2^4)}$ is greater than $(2^2)^4$. **27.** $25w^2$ **29.** $s^4r^4t^4$ **31.** $\dfrac{16}{r^4}$

33. $\dfrac{x^5}{y^5}$ **35.** $81a^4$ **37.** $-27a^3b^3c^3$ **39.** $-\dfrac{64}{x^3}$

41. $\dfrac{a^2}{b^2}$ **43.** $6^3u^6v^{12}$ or $216\,u^6v^{12}$ **45.** $5x^8y^4$

47. $-h^{28}$ **49.** m^{12} **51.** $\dfrac{4^5}{r^5s^{20}}$ or $\dfrac{1024}{r^5s^{20}}$

53. $\dfrac{3^5p^5}{q^{15}}$ or $\dfrac{243p^5}{q^{15}}$ **55.** y^{14} **57.** x^{31} **59.** $a^{26}b^{18}$

61. $16p^8q^{16}$ **63.** $-m^{35}n^{15}$ **65.** $25a^{18}b^6$ **67.** $3x^{12}y^4$

69. $\dfrac{4c^2d^6}{9}$ **71.** $\dfrac{c^{27}d^{31}}{2}$ **73.** $-\dfrac{27a^9b^3}{c^6}$

75. $16b^{26}$ **77.** x^{2m} **79.** $125a^{6n}$ **81.** $\dfrac{m^{2b}}{n^{3b}}$

83. $\dfrac{3^na^{3n}}{5^nb^{4n}}$

Section 5.3 Practice Exercises, pp. 365–367

3. c^9 **5.** y **7.** 3^6 or 729 **9.** $7^4w^{28}z^8$ or $2401w^{28}z^8$

11. $\dfrac{25k^6}{h^{14}}$ **13. a.** 1 **b.** 1 **15.** 1 **17.** 1 **19.** -1

21. 1 **23.** 1 **25.** 6 **27.** $-p$ **29. a.** $\dfrac{1}{4^3}$ **b.** $\dfrac{1}{4^3}$

31. $\dfrac{4}{5}$ **33.** -27 **35.** $\dfrac{1}{c^5}$ **37.** $\dfrac{1}{16}$ **39.** $\dfrac{1}{3z}$

41. $\dfrac{7}{y}$ **43.** $-\dfrac{1}{64}$ **45.** $\dfrac{1}{w^2}$ **47.** $-\dfrac{1}{r^5}$ **49.** b^6

51. $\dfrac{y^5}{y^{-3}} = y^{5-(-3)} = y^8$ **53.** $5b^{-2} = 5 \cdot \dfrac{1}{b^2} = \dfrac{5}{b^2}$

55. $\dfrac{1}{s}$ **57.** 1 **59.** b^6 **61.** $\dfrac{d^{10}}{c^8}$ **63.** $\dfrac{y^{33}}{6^3x^3}$ or $\dfrac{y^{33}}{216x^3}$

65. $\dfrac{1}{q^8}$ **67.** u^4 **69.** p^8 **71.** $\dfrac{1}{s^7}$ **73.** $\dfrac{1}{9}$ **75.** 1

77. $\dfrac{1}{k^{10}h^2}$ **79.** $\dfrac{1}{p^{11}}$ **81.** $\dfrac{1}{25}$ **83.** 1 **85.** $\dfrac{1}{27u^6}$

87. $\dfrac{-10q^4}{p^6}$ **89.** $\dfrac{5}{3x^6}$ **91.** $9s^4t^4$ **93.** $\dfrac{x^{12}}{25}$ **95.** $\dfrac{m^2}{2n^6}$

97. $\dfrac{3}{16}$ **99.** $\dfrac{8}{9}$ **101.** $\dfrac{3}{8}$ **103.** $\dfrac{5}{4}$ **105.** $\dfrac{5}{9}$

Section 5.4 Calculator Connections, p. 372

1.–2.

```
(5.2E6)*(4.6E-3)
            23920
(2.19E-8)*(7.84E
-4)
     1.71696E-11
```

3.–4.

```
(4.76E-5)/(2.38E
9)
           2E-14
(8.5E4)/(4.0E-1)
        212500
```

5.

```
((9.6E7)*(4.0E-3
))/(2.0E-2)
        19200000
```

6.

```
((5.0E-12)*(6.4E
-5))/((1.6E-8)*(
4.0E2))
          5E-11
```

Section 5.4 Practice Exercises, pp. 373–375

3. b^{13} **5.** 10^{13} **7.** $\dfrac{1}{y^5}$ **9.** $\dfrac{1}{10^5}$ **11.** w^4

13. 10^4 **15.** 100,000 **17.** $\dfrac{1}{10,000}$ or 0.0001

19. 10 **21.** 100

23. Move the decimal point between the 2 and 3 and multiply by 10^{-10}; 2.3×10^{-10}.

25. 5×10^4 **27.** 2.08×10^5 **29.** 6.01×10^6

31. 8×10^{-6} **33.** 1.25×10^{-4} **35.** 6.708×10^{-3}

37. 1.7×10^{-24} g **39.** \$2.7 \times 10^{10}$

41. 6.8×10^7 gal; 1.0×10^2 miles

43. Move the decimal point nine places to the left; 0.000 000 0031 **45.** 0.00005 **47.** 2800 **49.** 0.000603

51. 2,400,000 **53.** 0.019 **55.** 7032

57. 0.000 000 000 001 g **59.** 1600 calories and 2800 calories

61. 5.0×10^4 **63.** 3.6×10^{11} **65.** 2.2×10^4

67. 2.25×10^{-13} **69.** 3.2×10^{14} **71.** 2.432×10^{-10}

73. 3.0×10^{13} **75.** 6.0×10^5 **77.** 1.38×10^1

79. 5.0×10^{-14} **81.** 3.75 in. **83.** \$2.97 \times 10^{10}$

85. a. 6.5×10^7 **b.** 2.3725×10^{10} days **c.** 5.694×10^{11} hr **d.** 2.04984×10^{15} sec

87. Each person would have to pay \$25,000.

89. 1.25×10^{-2} ft or 0.0125 ft

Chapter 5 Problem Recognition Exercises— Properties of Exponents, p. 376

1. t^8 **2.** 2^8 or 256 **3.** y^5 **4.** p^6 **5.** r^4s^8

6. $a^3b^9c^6$ **7.** w^6 **8.** $\dfrac{1}{m^{16}}$ **9.** $\dfrac{x^4z^3}{y^7}$ **10.** $\dfrac{a^3c^8}{b^6}$

11. 1.25×10^3 **12.** 1.24×10^5 **13.** 8.0×10^8

14. 6.0×10^{-9} **15.** p^{15} **16.** p^{15} **17.** $\dfrac{1}{v^2}$

18. $c^{50}d^{40}$ **19.** 3   **20.** -4 **21.** $\dfrac{b^9}{2^{15}}$ **22.** $\dfrac{81}{y^6}$

23. $\dfrac{16y^4}{81x^4}$ **24.** $\dfrac{25d^6}{36c^2}$ **25.** $3a^7b^5$ **26.** $64x^7y^{11}$

27. $\dfrac{y^4}{x^8}$ **28.** $\dfrac{1}{a^{10}b^{10}}$ **29.** $\dfrac{1}{t^2}$ **30.** $\dfrac{1}{p^7}$ **31.** $\dfrac{8w^6x^9}{27}$

32. $\dfrac{25b^8}{16c^6}$ **33.** $\dfrac{q^3s}{r^2t^5}$ **34.** $\dfrac{m^2p^3q}{n^3}$ **35.** $\dfrac{1}{y^{13}}$

36. w^{10} **37.** $-\dfrac{1}{8a^{18}b^6}$ **38.** $\dfrac{4x^{18}}{9y^{10}}$ **39.** $\dfrac{k^8}{5h^6}$ **40.** $\dfrac{6n^{10}}{m^{12}}$

Section 5.5 Practice Exercises, pp. 383–387

3. $\dfrac{15}{x^3}$ **5.** $\dfrac{2}{t^4}$ **7.** $\dfrac{1}{3^{12}}$

9. 4.0×10^{-2} is scientific notation in which 10 is raised to the -2 power. 4^{-2} is not scientific notation and 4 is being raised to the -2 power.   **11.** $-7x^4 + 7x^2 + 9x + 6$

13. Binomial; 10; 2   **15.** Monomial; 6; 2
17. Binomial; -1; 4   **19.** Trinomial; 12; 4
21. Monomial; 23; 0   **23.** Monomial; -32; 3
25. a. 134 ft, 114 ft, and 86 ft   **b.** 150 ft
27. a. \$740, \$1190, and \$1140   **b.** $-\$60$; when no candles are produced, the business will lose \$60.
29. The exponents on the x-factors are different.
31. $35x^2y$ **33.** $10y$ **35.** $8b^2 - 9$ **37.** $4y^2 + y - 9$
39. $4a - 8c$ **41.** $a - \dfrac{1}{2}b - 2$ **43.** $\dfrac{4}{3}z^2 - \dfrac{5}{3}$
45. $7.9t^3 - 3.4t^2 + 6t - 4.2$ **47.** $-4h + 5$
49. $2m^2 - 3m + 15$ **51.** $-3v^3 - 5v^2 - 10v - 22$
53. $9t^4 + 8t + 39$ **55.** $-8a^3b^2$ **57.** $-53x^3$
59. $-5a - 3$ **61.** $16k + 9$ **63.** $2s + 14$
65. $3t^2 - 4t - 3$ **67.** $-2r - 3s + 3t$
69. $\dfrac{3}{4}x + \dfrac{1}{3}y - \dfrac{3}{10}$ **71.** $-\dfrac{2}{3}h^2 + \dfrac{3}{5}h - \dfrac{5}{2}$
73. $2.4x^4 - 3.1x^2 - 4.4x - 6.7$
75. $4b^3 + 12b^2 - 5b - 12$ **77.** $-3x^3 - 2x^2 + 11x - 31$
79. $4y^3 + 2y^2 + 2$ **81.** $3a^2 - 3a + 5$
83. $9ab^2 - 3ab + 16a^2b$ **85.** $4z^5 + z^4 + 9z^3 - 3z - 2$
87. $2x^4 + 11x^3 - 3x^2 + 8x - 4$ **89.** $-2w^2 - 7w + 18$
91. $-p^2q - 4pq^2 + 3pq$ **93.** 0   **95.** $-5ab + 6ab^2$
97. $11y^2 - 10y - 4$ **99.** For example, $x^3 + 6$
101. For example, $8x^5$ **103.** For example, $-6x^2 + 2x + 5$

Section 5.6 Practice Exercises, pp. 394–397

3. $-2y^2$ **5.** $-8y^4$ **7.** $8uvw^2$ **9.** $7u^2v^2w^4$
11. $3t^3 + 3t$ **13.** $9t^4$ **15.** $-12y$ **17.** $21p$
19. $12a^{14}b^8$ **21.** $-2c^{10}d^{12}$
23. $16p^2q^2 - 24p^2q + 40pq^2$ **25.** $-4k^3 + 52k^2 + 24k$
27. $-45p^3q - 15p^4q^3 + 30pq^2$ **29.** $y^2 - y - 90$
31. $m^2 - 14m + 24$ **33.** $2p^2 - p - 3$
35. $12w^2 + 29w + 15$ **37.** $-p^2 + 12p - 11$
39. $12x^2 + 28x - 5$ **41.** $8a^2 - 22a + 9$
43. $9t^2 - 18t - 7$ **45.** $3x^2 + 28x + 32$
47. $5s^3 + 8s^2 - 7s - 6$ **49.** $27w^3 - 8$
51. $p^4 + 5p^3 - 2p^2 - 21p + 5$ **53.** $9a^2 - 16b^2$
55. $81k^2 - 36$ **57.** $\dfrac{1}{4} - t^2$ **59.** $u^6 - 25v^2$

61. $4 - 9a^2$ **63.** $\dfrac{4}{9} - p^2$ **65.** $a^2 + 2ab + b^2$
67. $x^2 - 2xy + y^2$ **69.** $4c^2 + 20c + 25$
71. $9t^4 - 24st^2 + 16s^2$ **73.** $t^2 - 14t + 49$
75. $16q^2 + 24q + 9$
77. a. 36   **b.** 20   **c.** $(a+b)^2 \ne a^2 + b^2$ in general
79. a. $9x^2 + 6xy + y^2$ **b.** $9x^2y^2$ **c.** $(a+b)^2$ is the square of a binomial, $a^2 + 2ab + b^2$; $(ab)^2$ is the square of a monomial, a^2b^2.   **81.** $36 - y^2$
83. $49q^2 - 42q + 9$ **85.** $3t^3 - 12t$
87. $r^3 - 15r^2 + 63r - 49$ **89.** $81w^2 - 16z^2$
91. $25s^2 - 30st + 9t^2$ **93.** $10a^2 - 13ab + 4b^2$
95. $s^2 + \dfrac{147}{5}s - 18$ **97.** $4k^3 - 4k^2 - 5k - 25$
99. $u^3 + u^2 - 10u + 8$ **101.** $w^3 + w^2 + 4w + 12$
103. $\dfrac{4}{25}p^2 - \dfrac{4}{5}pq + q^2$ **105.** $8y^4 + 10y^3 - 7y^2 - 11y + 3$
107. $4h^2 - 7.29$ **109.** $5k^6 - 19k^3 + 18$
111. $6.25y^2 + 5.5y + 1.21$ **113.** $h^3 + 9h^2 + 27h + 27$
115. $8a^3 - 48a^2 + 96a - 64$
117. $6w^4 + w^3 - 15w^2 - 11w - 5$
119. $30x^3 + 55x^2 - 10x$ **121.** $2y^3 - y^2 - 15y + 18$
123. $x + 6$ **125.** $k = 6$ or -6

Section 5.7 Practice Exercises, pp. 407–410

1. $6z^5 - 10z^4 - 4z^3 - z^2 - 6$ **3.** $10x^2 - 29xy - 3y^2$
5. $11x - 2y$ **7.** $y^2 - \dfrac{3}{4}y + \dfrac{1}{2}$ **9.** $a^3 + 27$
11. Use long division when the divisor is a polynomial with two or more terms.
13. a. $5t^2 + 6t$ **15.** $3a^2 + 2a - 7$ **17.** $x^2 + 4x - 1$
19. $3p^2 - p$ **21.** $1 + \dfrac{2}{m}$ **23.** $-2y^2 + y - 3$
25. $x^2 - 6x - \dfrac{1}{4} + \dfrac{2}{x}$ **27.** $a - 1 + \dfrac{b}{a}$
29. $3t - 1 + \dfrac{3}{2t} - \dfrac{1}{2t^2} + \dfrac{2}{t^3}$ **31. a.** $z + 2 + \dfrac{1}{z+5}$
33. $t + 3$ **35.** $7b + 4$ **37.** $k - 6$
39. $2p^2 + 3p - 4$ **41.** $k - 2 + \dfrac{-4}{k+1}$
43. $2x^2 - x + 6 + \dfrac{2}{2x-3}$ **45.** $a - 3 + \dfrac{18}{a+3}$
47. $4x^2 + 8x + 13$ **49.** $w^2 + 5w - 2 + \dfrac{1}{w^2-3}$
51. $n^2 + n - 6$ **53.** $x - 1 + \dfrac{-8}{5x^2+5x+1}$
55. Multiply $(x-2)(x^2+4) = x^3 - 2x^2 + 4x - 8$, which does not equal $x^3 - 8$.   **57.** The divisor must be of the form $x - r$.   **59.** No, the divisor must be of the form $x - r$.   **61. a.** $x - 5$ **b.** $x^2 + 3x + 11$ **c.** 58
63. $x + 6$ **65.** $t - 4$ **67.** $5y + 10 + \dfrac{11}{y-1}$
69. $3y^2 - 2y + 2 + \dfrac{-3}{y+3}$ **71.** $x^2 - x - 2$
73. $4w^3 + 2w^2 + 6$
75. Monomial division; $3a^2 + 4a$
77. Long division; $p + 2$

79. Long division; $t^3 - 2t^2 + 5t - 10 + \dfrac{4}{t + 2}$

81. Long division; $w^2 + 3 + \dfrac{1}{w^2 - 2}$

83. Long division; $n^2 + 4n + 16$

85. Monomial division; $-3r + 4 - \dfrac{3}{r^2}$ **87. a.** -84

b. $4x^2 - 6x + 16 + \dfrac{-84}{x + 4}$ **c.** The values are the same.

89. $x + 1$ **91.** $x^3 + x^2 + x + 1$ **93.** $x + 1 + \dfrac{1}{x - 1}$

95. $x^3 + x^2 + x + 1 + \dfrac{1}{x - 1}$

Chapter 5 Problem Recognition Exercises—
Operations on Polynomials p. 410

1. $2x^3 - 8x^2 + 14x - 12$ **2.** $-3y^4 - 20y^2 - 32$
3. $x^2 - 1$ **4.** $4y^2 + 12$ **5.** $36y^2 - 84y + 49$
6. $9z^2 + 12z + 4$ **7.** $36y^2 - 49$ **8.** $9z^2 - 4$
9. $-x^2 - 3x + 4$ **10.** $5m^2 - 4m + 1$
11. $16x^2 + 8xy + y^2$ **12.** $4a^2 + 4ab + b^2$ **13.** $16x^2y^2$
14. $4a^2b^2$ **15.** $-7m^2 - 16m$
16. $-4n^5 + n^4 + 6n^2 - 7n + 2$

17. $8x^2 + 16x + 34 + \dfrac{74}{x - 2}$

18. $-4x^2 - 10x - 30 + \dfrac{-95}{x - 3}$

19. $6x^3 + 5x^2y - 6xy^2 + y^3$ **20.** $6a^3 - a^2b + 5ab^2 + 2b^3$
21. $x^3 + y^6$ **22.** $m^6 + 1$ **23.** $4b$ **24.** $-12z$
25. $a^4 - 4b^2$ **26.** $y^6 - 36z^2$ **27.** $64u^2 + 48uv + 9v^2$

28. $4p^2 - 4pt + t^2$ **29.** $4p + 4 + \dfrac{-2}{2p - 1}$

30. $2v - 7 + \dfrac{29}{2v + 3}$ **31.** $4x^2y^2$ **32.** $-9pq$

33. $10a^2 - 57a + 54$ **34.** $28a^2 - 17a - 3$

35. $\dfrac{9}{49}x^2 - \dfrac{1}{4}$ **36.** $\dfrac{4}{25}y^2 - \dfrac{16}{9}$

37. $-\dfrac{11}{9}x^3 + \dfrac{5}{9}x^2 - \dfrac{1}{2}x - 4$ **38.** $-\dfrac{13}{10}y^2 - \dfrac{9}{10}y + \dfrac{4}{15}$

39. $1.3x^2 - 0.3x - 0.5$ **40.** $5w^3 - 4.1w^2 + 2.8w - 1.2$

Chapter 5 Review Exercises, pp. 414–417

1. Base: 5; exponent: 3   **2.** Base: x; exponent: 4
3. Base: -2; exponent: 0   **4.** Base: y; exponent: 1
5. a. 36   **b.** 36   **c.** -36
6. a. 64   **b.** -64 **c.** -64
7. 5^{13} **8.** a^{11} **9.** x^9 **10.** 6^9 **11.** 10^3
12. y^6 **13.** b^8 **14.** 7^7 **15.** k **16.** 1
17. 2^8 **18.** q^6
19. Exponents are added only when multiplying factors with the same base. In such a case, the base does not change.
20. Exponents are subtracted only when dividing factors with the same base. In such a case, the base does not change.
21. \$7146.10   **22.** \$22,050   **23.** 7^{12} **24.** c^{12}

25. p^{18} **26.** 9^{28} **27.** $\dfrac{a^2}{b^2}$ **28.** $\dfrac{1}{3^4}$ **29.** $\dfrac{5^2}{c^4d^{10}}$

30. $-\dfrac{m^{10}}{4^5n^{30}}$ **31.** $2^4a^4b^8$ **32.** $x^{14}y^2$

33. $-\dfrac{3^3x^9}{5^3y^6z^3}$ **34.** $\dfrac{r^{15}}{s^{10}t^{30}}$ **35.** a^{11} **36.** 8^2

37. $4h^{14}$ **38.** $2p^{14}q^{13}$ **39.** $\dfrac{x^6y^2}{4}$ **40.** a^9b^6

41. 1   **42.** 1   **43.** 1   **44.** -1 **45.** 2   **46.** 1

47. $\dfrac{1}{z^5}$ **48.** $\dfrac{1}{10^4}$ **49.** $\dfrac{1}{36a^2}$ **50.** $\dfrac{6}{a^2}$ **51.** $\dfrac{17}{16}$

52. $\dfrac{10}{9}$ **53.** $\dfrac{1}{t^8}$ **54.** $\dfrac{1}{r}$ **55.** $\dfrac{2y^7}{x^6}$ **56.** $\dfrac{4a^6bc}{5}$

57. $\dfrac{n^{16}}{16m^8}$ **58.** $\dfrac{u^{15}}{27v^6}$ **59.** $\dfrac{k^{21}}{5}$ **60.** $\dfrac{h^9}{9}$ **61.** $\dfrac{1}{2}$

62. $\dfrac{5}{4}$ **63. a.** 9.7×10^7 **b.** 4.2×10^{-3} in.

c. 1.66241×10^8 km^2 **64. a.** $0.000\ 000\ 0001$
b. $257{,}300{,}000$ **c.** \$256,000   **65.** 9.43×10^5
66. 1.55×10^{10} **67.** 2.5×10^8 **68.** 1.638×10^3
69. $\approx 9.5367 \times 10^{13}$. This number is too big to fit on most calculator displays.
70. $\approx 1.1529 \times 10^{-12}$. This number is too small to fit on most calculator displays.
71. a. $\approx 5.84 \times 10^8$ miles   **b.** $\approx 6.67 \times 10^4$ mph
72. a. $\approx 2.26 \times 10^8$ miles   **b.** $\approx 1.07 \times 10^5$ mph
73. a. Trinomial   **b.** 4   **c.** 7
74. a. Binomial   **b.** 7   **c.** -5
75. $7x - 3$ **76.** $-y^2 - 14y - 2$ **77.** $14a^2 - 2a - 6$

78. $10w^4 + 2w^3 - 7w + 4$ **79.** $\dfrac{15}{2}x^3 + \dfrac{1}{4}x^2 + \dfrac{1}{2}x + 2$

80. $0.01b^5 + b^4 - 0.1b^3 + 0.3b + 0.33$
81. $-2x^2 - 9x - 6$ **82.** $-5x^2 - 9x - 12$
83. For example, $-5x^2 + 2x - 4$
84. For example, $6x^6 + 8$ **85.** $6w + 6$
86. $-75x^6y^4$ **87.** $18a^8b^4$ **88.** $15c^4 - 35c^2 + 25c$
89. $-2x^3 - 10x^2 + 6x$ **90.** $5k^2 + k - 4$
91. $20t^2 + 3t - 2$ **92.** $6q^2 + 47q - 8$
93. $2a^2 + 4a - 30$ **94.** $49a^2 + 7a + \dfrac{1}{4}$

95. $b^2 - 8b + 16$ **96.** $8p^3 - 27$
97. $-2w^3 - 5w^2 - 5w + 4$ **98.** $b^2 - 16$

99. $\dfrac{1}{9}r^8 - s^4$ **100.** $49z^4 - 84z^2 + 36$

101. $2h^5 + h^4 - h^3 + h^2 - h + 3$ **102.** $2x^2 + 3x - 20$
103. $4y^2 - 2y$ **104.** $2a^2b - a - 3b$

105. $-3x^2 + 2x - 1$ **106.** $-\dfrac{z^5w^3}{2} + \dfrac{3zw}{4} + \dfrac{1}{z}$

107. $m^2 + \dfrac{5}{6}m - 1$ **108.** $6n^2 - 2n + 4$

109. $x + 2$ **110.** $2t + 5$ **111.** $p - 3 + \dfrac{5}{2p + 7}$

112. $a + 6 + \dfrac{-4}{5a - 3}$ **113.** $b^2 + 5b + 25$

114. $z + 4$ **115.** $y^2 - 4y + 2 + \dfrac{9y - 4}{y^2 + 3}$

116. $t^2 - 3t + 1 + \dfrac{-2t - 6}{3t^2 + t + 1}$ **117.** $w^2 + w - 1$

Chapter 5 Test, pp. 417–418

1. $\dfrac{(3 \cdot 3 \cdot 3 \cdot 3) \cdot (3 \cdot 3 \cdot 3)}{3 \cdot 3 \cdot 3 \cdot 3 \cdot 3 \cdot 3} = 3$ **2.** 9^6 **3.** q^8

4. $27a^6b^3$ **5.** $\dfrac{16x^4}{y^{12}}$ **6.** 1   **7.** $\dfrac{1}{c^3}$ **8.** 14

9. $49s^{18}t$ **10.** $\dfrac{4}{b^{12}}$ **11.** $\dfrac{16a^{12}}{9b^6}$

12. a. 4.3×10^{10} **b.** 0.000 0056

13. a. $2.4192 \times 10^8 \, \text{m}^3$ **b.** $8.83008 \times 10^{10} \, \text{m}^3$

14. $5x^3 - 7x^2 + 4x + 11$ **a.** 3   **b.** 5

15. $24w^2 - 3w - 4$ **16.** $15x^3 - 7x^2 - 2x + 1$

17. $-10x^5 - 2x^4 + 30x^3$ **18.** $8a^2 - 10a + 3$

19. $4y^3 - 25y^2 + 37y - 15$ **20.** $4 - 9b^2$

21. $25z^2 - 60z + 36$ **22.** $100 - 9w^2$

23. $y^3 - 11y^2 + 32y - 12$

24. Perimeter: $12x - 2$; area: $5x^2 - 13x - 6$

25. $-3x^6 + \dfrac{x^4}{4} - 2x$ **26.** $2y - 7$

27. $w^2 - 4w + 5 + \dfrac{-10}{2w + 3}$

28. $2x^2 + x + 4 + \dfrac{-2x + 9}{x^2 - 2}$

Chapters 1–5 Cumulative Review Exercises, pp. 418–419

1. $-\dfrac{35}{2}$ **2.** 4   **3.** $5^2 - \sqrt{4}$; 23   **4.** $x = \dfrac{28}{3}$

5. No solution   **6.** Quadrant III   **7.** y-axis

8. The measures are $31°$, $54°$, and $95°$.

9. a. 12 in.   **b.** 19.5 in.   **c.** 5.5 hr

d.

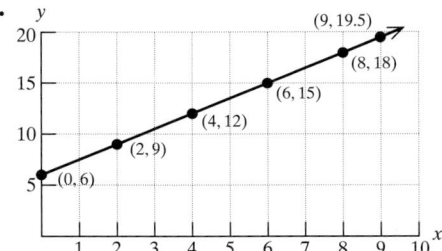

10. $(-3, 4)$

11. $[-5, \infty)$

12. $5x^2 - 9x - 15$ **13.** $-2y^2 - 13yz - 15z^2$

14. $16t^2 - 24t + 9$ **15.** $\dfrac{4}{25}a^2 - \dfrac{1}{9}$

16. $-4a^3b^2 + 2ab - 1$ **17.** $4m^2 + 8m + 11 + \dfrac{24}{m - 2}$

18. $\dfrac{c^2}{16d^4}$ **19.** $\dfrac{2b^3}{a^2}$ **20.** 2.788×10^{-2}

Chapter 6

Chapter Opener Puzzle

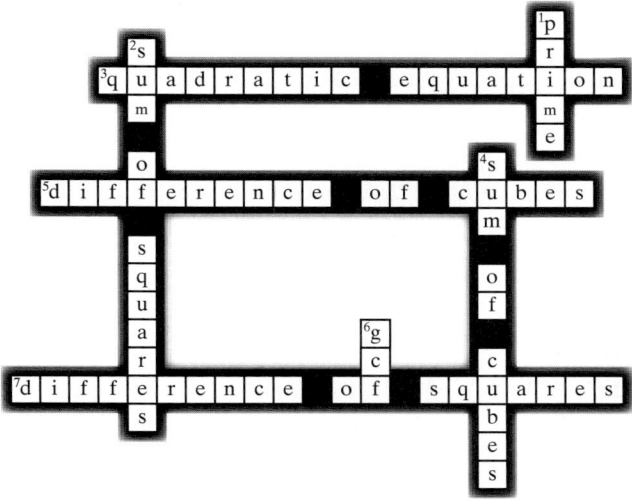

Section 6.1 Practice Exercises, pp. 429–431

3. 7   **5.** 6   **7.** y **9.** 4   **11.** ab **13.** $4w^2z$

15. $2xy^4z^2$ **17.** $(x - y)$ **19. a.** $3x - 6y$

b. $3(x - 2y)$ **21.** $4(p + 3)$ **23.** $5(c^2 - 2c + 3)$

25. $x^3(x^2 + 1)$ **27.** $t(t^3 - 4 + 8t)$ **29.** $2ab(1 + 2a^2)$

31. $19x^2y(2 - y^3)$ **33.** $6xy^5(x^2 - 3y^4z)$

35. The expression is prime because it is not factorable.

37. $7pq^2(6p^2 + 2 - p^3q^2)$ **39.** $t^2(t^3 + 2rt - 3t^2 + 4r^2)$

41. a. $-2x(x^2 + 2x - 4)$ **b.** $2x(-x^2 - 2x + 4)$

43. $-1(8t^2 + 9t + 2)$ **45.** $-15p^2(p + 2)$

47. $-q(q^3 - 2q + 9)$ **49.** $-1(7x + 6y + 2z)$

51. $(a + 6)(13 - 4b)$ **53.** $(w^2 - 2)(8v + 1)$

55. $7x(x + 3)^2$ **57.** $(2a - b)(4a + 3c)$

59. $(q + p)(3 + r)$ **61.** $(2x + 1)(3x + 2)$

63. $(t + 3)(2t - 5)$ **65.** $(3y - 1)(2y - 3)$

67. $(b + 1)(b^3 - 4)$ **69.** $(j^2 + 5)(3k + 1)$

71. $(2x^6 + 1)(7w^6 - 1)$ **73.** $(y + x)(a + b)$

75. $(vw + 1)(w - 3)$ **77.** $5x(x^2 + y^2)(3x + 2y)$

79. $4b(a - b)(x - 1)$ **81.** $6t(t - 3)(s - t^2)$

83. $P = 2(l + w)$ **85.** $S = 2\pi r(r + h)$

87. $\dfrac{1}{7}(x^2 + 3x - 5)$ **89.** $\dfrac{1}{4}(5w^2 + 3w + 9)$

91. For example, $6x^2 + 9x$

93. For example, $16p^4q^2 + 8p^3q - 4p^2q$

Section 6.2 Practice Exercises, pp. 435–437

3. $4xy^5(x^2y^2 - 3x^3 + 2y^3)$ **5.** $(a + 2b)(x - 5)$

7. $(x + 8)(x + 2)$ **9.** $(z - 9)(z - 2)$

11. $(z - 6)(z - 3)$ **13.** $(p + 8)(p - 5)$

15. $(t + 10)(t - 4)$ **17.** Prime

19. $(n + 4)^2$ **21.** a   **23.** c

25. They are both correct because multiplication of polynomials is a commutative operation.

27. The expressions are equal and both are correct.

29. Descending order   **31.** $(y + 20)(y - 8)$

33. $(t + 8)(t + 9)$ **35.** $(q - 30)(q + 40)$

37. $3(x - 12)(x + 2)$ **39.** $8p(p - 1)(p - 4)$
41. $y^2z^2(y - 6)^2$ **43.** $-(x - 4)(x - 6)$
45. $-(m + 2)(m - 3)$ **47.** $-2(c + 2)(c + 1)$
49. $xy^3(x - 4)(x - 15)$ **51.** $12(p - 7)(p - 1)$
53. $-2(m - 10)(m - 1)$ **55.** $(c + 5d)(c + d)$
57. $(a - 2b)(a - 7b)$ **59.** Prime   **61.** $(q - 7)(q + 9)$
63. $(x + 10)^2$ **65.** $(t + 20)(t - 2)$ **67.** The student forgot to factor out the GCF before factoring the trinomial further. The polynomial is not factored completely, because $(2x - 4)$ has a common factor of 2.   **69.** $x^2 + 9x - 52$
71. $7, 5, -7, -5$ **73.** For example: $c = -16$

Section 6.3 Practice Exercises, pp. 447–449

3. $3ab(7ab + 4b - 5a)$ **5.** $(n - 1)(m - 2)$
7. $6(a - 7)(a + 2)$ **9.** a   **11.** b
13. $(2w - 1)(w + 3)$ **15.** $(2a + 3)(a + 2)$
17. $(7y - 5)(y + 2)$ **19.** $(3z - 4)(2z + 3)$
21. $(2p + 5)(5p - 2)$ **23.** Prime   **25.** $(7x - 4)(x - 2)$
27. $(8q - 1)(q + 4)$ **29.** $(12y - z)(y - 6z)$
31. $3(c - 8)(c - 3)$ **33.** $u^6(3u - 1)(u - 4)$
35. $-(x + 2)(x + 5)$ **37.** $-3z^2(3z - 4)(2z + 1)$
39. $10(w + 10z)(6w - 5z)$ **41.** $4y^2 - 28y + 49$
43. a. $x^2 + 12x + 36$ is a perfect square trinomial.
b. $x^2 + 13x + 36 = (x + 9)(x + 4)$;
$x^2 + 12x + 36 = (x + 6)^2$ **45.** $(y - 4)^2$ **47.** $(6p + 5)^2$
49. $(5m - 3n)^2$ **51.** $(w - 2)^2$ **53.** $4(3p - 1)^2$
55. $(2y + 9)(2y + 1)$ **57.** $-5(t - 2)(t - 3)$
59. $(w - 8)(w + 3)$ **61.** $(4p - 1)(p - 2)$
63. $(4r + 3)^2$ **65.** Prime   **67.** $(3p - q)(5p + 2q)$
69. $(4a - 3b)(a + 2b)$ **71.** $(6x + y)(3x - 2y)$
73. Prime   **75.** $(8n + 3)(2n + 1)$
77. $(6x - 5)(2x - 1)$ **79.** $(y + 4)(y - 10)$
81. $(b + 7)(b - 1)$ **83.** $(t + 10)^2$
85. $(p - 9q)(p - 4q)$ **87.** Prime
89. $(x - 18y)(x + y)$ **91.** $(z - 12)(z - 3)$
93. $y(3y^2 - y + 12)$ **95.** $2(w + 7)(w + 3)$
97. $q^2(p - 3)(p - 11)$
99. a. $(x - 15)(x + 2)$ **b.** $(x - 10)(x - 3)$
101. a. $(x - 9)(x - 1)$ **b.** $(x + 9)(x + 1)$
103. $(y^2 - 3)(y^2 + 7)$ **105.** $(p^2 - 8)(p^2 - 5)$

Section 6.4 Practice Exercises, pp. 456–457

3. $(y + 5)(8 + 9y)$ **5.** $12, 1$ **7.** $-8, -1$ **9.** $5, -4$
11. $9, -2$ **13.** $(x + 4)(3x + 1)$ **15.** $(w - 2)(4w - 1)$
17. $(m + 3)(2m - 1)$ **19.** $(4k + 3)(2k - 3)$
21. $(2k - 5)^2$ **23.** Prime   **25.** $(3z - 5)(3z - 2)$
27. $2(7y + 4)(y + 3)$ **29.** $(6y - 5z)(2y + 3z)$
31. $25y^2 - 70y + 49$ **33. a.** $4x^2 - 20x + 25$ is a perfect square trinomial.   **b.** $4x^2 - 25x + 25 = (4x - 5)(x - 5)$;
$4x^2 - 20x + 25 = (2x - 5)^2$ **35.** $(t - 8)^2$ **37.** $(7q - 2)^2$
39. $(2x + 9)^2$ **41.** $2(4x + 5y)^2$ **43.** $4(c + 1)^2$
45. $(3y + 4)(3y + 1)$ **47.** $(5p - 1)(4p - 3)$
49. $(3u - 2v)(2u - 5v)$ **51.** $(4a + 5b)(3a - b)$
53. $(h + 7k)(3h - 2k)$ **55.** Prime
57. $(8w + 1)(2w + 1)$ **59.** $(q - 1)^2$
61. $-2(a - 1)(a - 9)$ **63.** $(m + 3)(m - 2)$
65. $2(10y + 1)(y - 4)$ **67.** $w^3(w - 4)(w - 7)$
69. $r(4r - 5)(r + 2)$ **71.** $4q(q - 5)(q + 4)$
73. $b^2(a + 10)(a + 3)$ **75.** $-1(m - 2)(m + 17)$
77. $-2(h - 9)(h - 5)$ **79.** $(m^2 + 3)(m^2 + 7)$
81. No. $(5x - 10)$ contains a common factor of 5.

Section 6.5 Practice Exercises, pp. 465–466

3. $(3x - 1)(2x - 5)$ **5.** $5xy^5(3x - 2y)$
7. $(x + b)(a - 6)$ **9.** $(5x + 3)^2$ **11.** $x^2 - 25$
13. $4p^2 - 9q^2$ **15.** $(x - 6)(x + 6)$
17. $(w - 10)(w + 10)$ **19.** $(2a - 11b)(2a + 11b)$
21. $(7m - 4n)(7m + 4n)$ **23.** Prime
25. $(y + 2z)(y - 2z)$ **27.** $(a - b^2)(a + b^2)$
29. $(5pq - 1)(5pq + 1)$ **31.** $(c^3 - 5)(c^3 + 5)$
33. $(5 - 4t)(5 + 4t)$ **35.** $x^3, 8, y^6, 27q^3, w^{12}, r^3s^6$
37. $(a + b)(a^2 - ab + b^2)$ **39.** $(y - 2)(y^2 + 2y + 4)$
41. $(1 - p)(1 + p + p^2)$ **43.** $(w + 4)(w^2 - 4w + 16)$
45. $(x - 10)(x^2 + 10x + 100)$
47. $(4t + 1)(16t^2 - 4t + 1)$
49. $(10a + 3)(100a^2 - 30a + 9)$
51. $\left(n - \dfrac{1}{2}\right)\left(n^2 + \dfrac{1}{2}n + \dfrac{1}{4}\right)$ **53.** $(a + b^2)(a^2 - ab^2 + b^4)$
55. $(x^3 + 4y)(x^6 - 4x^3y + 16y^2)$ **57.** Prime
59. $(x^2 - 2)(x^2 + 2)$ **61.** Prime
63. $(t + 4)(t^2 - 4t + 16)$ **65.** Prime
67. $4(b + 3)(b^2 - 3b + 9)$ **69.** $5(p - 5)(p + 5)$
71. $(\frac{1}{4} - 2h)(\frac{1}{16} + \frac{1}{2}h + 4h^2)$ **73.** $(x - 2)(x + 2)(x^2 + 4)$
75. $(q - 2)(q^2 + 2q + 4)(q + 2)(q^2 - 2q + 4)$
77. $\left(\dfrac{2x}{3} - w\right)\left(\dfrac{2x}{3} + w\right)$ **79.** $(2x + 3)(x - 1)(x + 1)$
81. $(2x - y)(2x + y)(4x^2 + y^2)$
83. $(3y - 2)(3y + 2)(9y^2 + 4)$ **85.** $(27k + 1)(3k + 1)$
87. $(k + 4)(k - 3)(k + 3)$ **89.** $2(t - 5)(t - 1)(t + 1)$
91. $y(y - 6)$ **93.** $(2p - 5)(2p + 7)$
95. $(-t + 2)(t + 6)$ or $-1(t - 2)(t + 6)$
97. $(-2b + 15)(2b + 5)$ or $-1(2b - 15)(2b + 5)$

Section 6.6 Practice Exercises, pp. 469–471

3. Look for a perfect square trinomial: $a^2 + 2ab + b^2$ or $a^2 - 2ab + b^2$.
5. Look for a difference of squares: $a^2 - b^2$, a difference of cubes: $a^3 - b^3$, or a sum of cubes: $a^3 + b^3$.
7. a. Difference of squares   **b.** $2(a - 9)(a + 9)$
9. a. None of these   **b.** $6w(w - 1)$
11. a. Nonperfect square trinomial   **b.** $(3t + 1)(t + 4)$
13. a. Four terms-grouping   **b.** $(3c + d)(a - b)$
15. a. Sum of cubes   **b.** $(y + 2)(y^2 - 2y + 4)$
17. a. Nonperfect square trinomial   **b.** $3(q - 4)(q + 1)$
19. a. None of these   **b.** $6a(3a + 2)$
21. a. Difference of squares   **b.** $4(t - 5)(t + 5)$
23. a. Nonperfect square trinomial   **b.** $10(c^2 + c + 1)$
25. a. Sum of cubes   **b.** $(x + 0.1)(x^2 - 0.1x + 0.01)$
27. a. Perfect square trinomial   **b.** $(8 + k)^2$
29. a. Four terms-grouping   **b.** $(x + 1)(2x - y)$
31. a. Difference of cubes   **b.** $(a - c)(a^2 + ac + c^2)$
33. a. Nonperfect square trinomial   **b.** Prime
35. a. Perfect square trinomial   **b.** $(b + 5)^2$
37. a. Nonperfect square trinomial   **b.** $-p(p + 4)(p + 1)$
39. a. Nonperfect square trinomial   **b.** $3(2x + 3)(x - 5)$
41. a. None of these   **b.** $abc^2(5ac - 7)$
43. a. Nonperfect square trinomial   **b.** $(t + 9)(t - 7)$
45. a. Four terms-grouping   **b.** $(b + y)(a - b)$
47. a. Nonperfect square trinomial   **b.** $(7u - 2v)(2u - v)$
49. a. Nonperfect square trinomial   **b.** $2(2q^2 - 4q - 3)$
51. a. Sum of squares   **b.** Prime

53. a. Nonperfect square trinomial   **b.** $(3r + 1)(2r + 3)$
55. a. Difference of squares   **b.** $(2a - 1)(2a + 1)(4a^2 + 1)$
57. a. Perfect square trinomial   **b.** $(9u - 5v)^2$
59. a. Nonperfect square trinomial   **b.** $(x - 6)(x + 1)$
61. a. Four terms-grouping   **b.** $2(x - 3y)(a + 2b)$
63. a. Nonperfect square trinomial
b. $x^2y(3x + 5)(7x + 2)$
65. a. Four terms-grouping   **b.** $(4v - 3)(2u + 3)$
67. a. Perfect square trinomial   **b.** $3(2x - 1)^2$
69. a. Nonperfect square trinomial   **b.** $n(2n - 1)(3n + 4)$
71. a. Difference of squares   **b.** $(8 - y)(8 + y)$
73. a. Nonperfect square trinomial   **b.** Prime
75. $\left(\dfrac{4}{5}p - \dfrac{1}{2}q\right)\left(\dfrac{16}{25}p^2 + \dfrac{2}{5}pq + \dfrac{1}{4}q^2\right)$
77. $(a^4 + b^4)(a^8 - a^4b^4 + b^8)$
79. a. The quotient is $x^2 + 2x + 4$.   **b.** $(x - 2)(x^2 + 2x + 4)$
81. $x^2 + 2x + 4$ **83.** $2x + 1$
85. a. $a^2 - b^2$ **b.** $(a - b)(a + b)$ **87.** 4891

Section 6.7 Practice Exercises, pp. 480–484

3. $4(b - 5)(b - 6)$ **5.** $(3x - 2)(x + 4)$
7. $4(x^2 + 4y^2)$ **9.** Neither   **11.** Quadratic
13. Linear   **15.** $x = -3, x = 1$ **17.** $x = \dfrac{7}{2}, x = -\dfrac{7}{2}$
19. $x = -5$ **21.** $x = 0, x = -\dfrac{1}{3}, x = -1$
23. The equation must have one side equal to zero and the other side factored completely.   **25.** $y = 8, y = -1$
27. $w = 8, w = 2$ **29.** $x = -\dfrac{1}{4}, x = 3$
31. $a = \dfrac{7}{2}, a = -\dfrac{7}{2}$ **33.** $t = -5$
35. $n = 0, n = -\dfrac{1}{3}, n = -1$ **37.** $x = \dfrac{1}{2}, x = 10, x = -7$
39. $t = 0, t = 6$, or $t = -6$ **41.** $y = -3, y = \dfrac{3}{2}$
43. $n = \dfrac{1}{3}, n = -\dfrac{1}{3}$ **45.** $d = 0, d = -2, d = 4$
47. $h = -13$ **49.** $q = -1, q = 4$ **51.** $x = 0, x = 6$
53. $k = \dfrac{3}{4}, k = -3$ **55.** $p = 3, p = -2$ **57.** $z = -\dfrac{4}{7}$
59. $w = 0, w = \dfrac{2}{3}$ **61.** $d = 0, d = -\dfrac{1}{4}$
63. $t = -2, t = 4, t = -4$ **65.** $w = -7, w = 5$
67. $k = 4, k = 2$ **69.** The numbers are $-\dfrac{9}{2}$ and 4.
71. The numbers are 5 and -4.
73. The numbers are 6 and 8, or -8 and -6.
75. The numbers are 0 and 1, or 9 and 10.
77. The painting has length 12 in. and width 10 in.
79. a. The picture is 13 in. by 6 in.   **b.** 38 in.
81. The base is 10 cm and the height is 25 cm.   **83.** 4 sec
85. 0 sec and 4 sec   **87.** Given a right triangle with legs a and b and hypotenuse c, then $a^2 + b^2 = c^2$.   **89.** $c = 5$ m
91. $b = 12$ yd   **93.** The height is 9 km.   **95.** 19 yd
97. 10 m

Chapter 6 Review Exercises, pp. 489–491

1. $3a^2b$ **2.** $x + 5$ **3.** $2c(3c - 5)$ **4.** $-2yz$ or $2yz$
5. $2x(3x + x^3 - 4)$ **6.** $11w^2y^3(w - 4y^2)$
7. $t(-t + 5)$ or $-t(t - 5)$ **8.** $u(-6u - 1)$ or $-u(6u + 1)$
9. $(b + 2)(3b - 7)$ **10.** $2(5x + 9)(1 + 4x)$
11. $(w + 2)(7w + b)$ **12.** $(b - 2)(b + y)$
13. $3(4y - 3)(5y - 1)$ **14.** $a(2 - a)(3 - b)$
15. $(x - 3)(x - 7)$ **16.** $(y - 8)(y - 11)$
17. $(z - 12)(z + 6)$ **18.** $(q - 13)(q + 3)$
19. $3w(p + 10)(p + 2)$ **20.** $2m^2(m + 8)(m + 5)$
21. $-(t - 8)(t - 2)$ **22.** $-(w - 4)(w + 5)$
23. $(a + b)(a + 11b)$ **24.** $(c - 6d)(c + 3d)$
25. Different   **26.** Both negative   **27.** Both positive
28. Different   **29.** $(2y + 3)(y - 4)$
30. $(4w + 3)(w - 2)$ **31.** $(2z + 5)(5z + 2)$
32. $(4z - 3)(2z + 3)$ **33.** Prime   **34.** Prime
35. $10(w - 9)(w + 3)$ **36.** $3(y - 8)(y + 2)$
37. $(3c - 5d)^2$ **38.** $(x + 6)^2$ **39.** $(v^2 + 1)(v^2 - 3)$
40. $(x^2 + 5)(x^2 + 2)$ **41.** The trinomials in Exercises 37 and 38.   **42.** $5, -1$ **43.** $-3, -5$ **44.** $(c - 2)(3c + 1)$
45. $(y + 3)(4y + 1)$ **46.** $(t + 12)(t + 1)$
47. $x(x + 5)(4x - 3)$ **48.** $w(w + 5)(w - 1)$
49. $(p - 3q)(p - 5q)$ **50.** $2(4v + 3)(5v - 1)$
51. $10(4s - 5)(s + 2)$ **52.** $ab(a - 6b)(a - 4b)$
53. $2z^4(z + 7)(z - 3)$ **54.** $(3m - 1)(3m + 2)$
55. $(3p + 2)(2p + 5)$ **56.** $(7x + 10)^2$ **57.** $(3w - z)^2$
58. The trinomials in Exercises 56 and 57.
59. $(a - b)(a + b)$ **60.** Prime
61. $(a + b)(a^2 - ab + b^2)$ **62.** $(a - b)(a^2 + ab + b^2)$
63. $(a - 7)(a + 7)$ **64.** $(d - 8)(d + 8)$
65. $(10 - 9t)(10 + 9t)$ **66.** $(2 - 5k)(2 + 5k)$
67. Prime   **68.** Prime   **69.** $(4 + a)(16 - 4a + a^2)$
70. $(5 - b)(25 + 5b + b^2)$ **71.** $(p^2 + 2)(p^4 - 2p^2 + 4)$
72. $\left(q^2 - \dfrac{1}{3}\right)\left(q^4 + \dfrac{1}{3}q^2 + \dfrac{1}{9}\right)$ **73.** $6(x - 2)(x^2 + 2x + 4)$
74. $7(y + 1)(y^2 - y + 1)$ **75.** $2(c^2 - 3)(c^2 + 3)$
76. $2(6x - y)(6x + y)$ **77.** $(p + 3)(p - 4)(p + 4)$
78. $(k - 2)(2 - k)(2 + k)$ or $-1(k - 2)^2(2 + k)$
79. $(6y + 1)(y - 2)$ **80.** $3(p - 1)^2$
81. $x(x - 6)(x + 6)$ **82.** $(k - 7)(k - 6)$
83. $(c - 2d)(7a - b)$ **84.** $q(q - 4)(q^2 + 4q + 16)$
85. $4(2h^2 + 5)$ **86.** Prime   **87.** $m(m - 8)$
88. $(x + 4)(x + 1)(x - 1)$ **89.** $3st(4s + t)(s - 4t)$
90. $5q(p^2 - 2q)(p^2 + 2q)$ **91.** $3(3a - 1)(2a + 5)$
92. $w^2(w + 8)(w - 7)$ **93.** $n(2 + n)(4 - 2n + n^2)$
94. $14(m - 1)(m^2 + m + 1)$
95. $(x - 3)(2x + 1) = 0$ can be solved directly by the zero product rule because it is a product of factors set equal to zero.
96. $x = \dfrac{1}{4}, x = -\dfrac{2}{3}$ **97.** $a = 9, a = \dfrac{1}{2}$
98. $w = 0, w = -3, w = -\dfrac{2}{5}$ **99.** $u = 0, u = 7, u = \dfrac{9}{4}$
100. $k = -\dfrac{5}{7}, k = 2$ **101.** $h = -\dfrac{1}{4}, h = 6$
102. $q = 12, q = -12$ **103.** $r = 5, r = -5$
104. $v = 0, v = \dfrac{1}{5}$ **105.** $x = 4, x = 2$

106. $t = -\dfrac{5}{6}$ **107.** $s = -\dfrac{2}{3}$ **108.** $y = \dfrac{2}{3}, y = 6$

109. $p = \dfrac{11}{2}, p = -12$ **110.** $y = 0, y = 7, y = 2$

111. $x = 0, x = 2, x = -2$
112. The height is 6 ft, and the base is 13 ft.
113. The ball is at ground level at 0 and 1 sec.
114. The ramp is 13 ft long.
115. The legs are 6 ft and 8 ft; the hypotenuse is 10 ft.
116. The numbers are -8 and 8.
117. The numbers are 29 and 30, or -2 and -1.
118. The height is 4 m, and the base is 9 m.

Chapter 6 Test, p. 491

1. $3x(5x^3 - 1 + 2x^2)$ **2.** $(a - 5)(7 - a)$
3. $(6w - 1)(w - 7)$ **4.** $(13 - p)(13 + p)$
5. $(q - 8)^2$ **6.** $(2 + t)(4 - 2t + t^2)$
7. $(a + 4)(a + 8)$ **8.** $(x + 7)(x - 6)$
9. $(2y - 1)(y - 8)$ **10.** $(2z + 1)(3z + 8)$
11. $(3t - 10)(3t + 10)$ **12.** $(v + 9)(v - 9)$
13. $3(a + 6b)(a + 3b)$ **14.** $(c - 1)(c + 1)(c^2 + 1)$
15. $(y - 7)(x + 3)$ **16.** Prime
17. $-10(u - 2)(u - 1)$ **18.** $3(2t - 5)(2t + 5)$
19. $5(y - 5)^2$ **20.** $7q(3q + 2)$
21. $(2x + 1)(x - 2)(x + 2)$ **22.** $(y - 5)(y^2 + 5y + 25)$
23. $(mn - 9)(mn + 9)$ **24.** $16(a - 2b)(a + 2b)$

25. $(4x - 3y^2)(16x^2 + 12xy^2 + 9y^4)$ **26.** $x = \dfrac{3}{2}, x = -5$

27. $x = 0, x = 7$ **28.** $x = 8, x = -2$

29. $x = \dfrac{1}{5}, x = -1$ **30.** $y = 3, y = -3, y = -10$

31. The tennis court is 12 yd by 26 yd.   **32.** The two integers are 5 and 7 or -5 and -7.   **33.** The base is 12 in., and the height is 7 in.   **34.** The shorter leg is 5 ft.

Chapters 1–6 Cumulative Review Exercises, p. 492

1. $\dfrac{7}{5}$ **2.** $t = -3$ **3.** $y = \dfrac{3}{2}x - 4$

4. There are 10 quarters, 12 nickels, and 7 dimes.
5. $[-4, \infty)$

6. a. Yes   **b.** 1   **c.** $(0, 4)$ **d.** $(-4, 0)$
e.

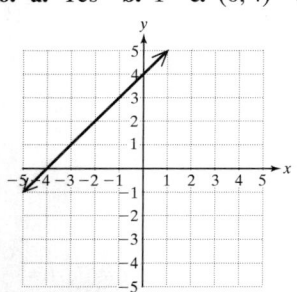

7. a. Vertical line   **b.** Undefined   **c.** $(5, 0)$
d. Does not exist   **8.** $y = 3x + 14$ **9.** $(5, 2)$

10. $-\dfrac{7}{2}y^2 - 5y - 14$ **11.** $8p^3 - 22p^2 + 13p + 3$

12. $4w^2 - 28w + 49$ **13.** $r^3 + 5r^2 + 15r + 40 + \dfrac{121}{r - 3}$

14. c^4 **15.** 1.6×10^3 **16.** $(w - 2)(w + 2)(w^2 + 4)$
17. $(a + 5b)(2x - 3y)$ **18.** $(2x - 5)(2x + 1)$
19. $(y - 3)(y^2 + 3y + 9)$ **20.** $x = 0, x = \dfrac{1}{2}, x = -5$

Chapter 7

Chapter Opener Puzzle

	Column 1		Column 2
	$\dfrac{3}{2}\;\boxed{+}\;\dfrac{1}{2} = 2$		$\dfrac{3}{10}\;\boxed{\cdot}\;\dfrac{5}{6}\;\boxed{\div}\;\dfrac{1}{8} = \dfrac{1}{8}$
	$\dfrac{3}{4}\;\boxed{\div}\;\dfrac{1}{8} = 6$		$\dfrac{1}{2}\;\boxed{+}\;\dfrac{1}{2}\;\boxed{\div}\;\dfrac{3}{4} = \dfrac{7}{6}$
	$\dfrac{5}{7}\;\boxed{\cdot}\;\dfrac{14}{3} = \dfrac{10}{3}$		
	$\dfrac{2}{3}\;\boxed{-}\;\dfrac{1}{2} = \dfrac{1}{6}$		

Section 7.1 Practice Exercises, pp. 501–504

3. a. A number $\dfrac{p}{q}$, where p and q are integers and $q \neq 0$.

b. An expression $\dfrac{p}{q}$, where p and q are polynomials and

$q \neq 0$.   **5.** $-\dfrac{1}{8}$ **7.** $-\dfrac{1}{2}$ **9.** 0   **11.** Undefined

13. a. \$196   **b.** \$168   **c.** \$145.60
15. $\{h \,|\, h \text{ is a real number and } h \neq 4\}$
17. $\left\{y \,|\, y \text{ is a real number and } y \neq -\dfrac{7}{3}, y \neq -3\right\}$
19. $\{c \,|\, c \text{ is a real number and } c \neq 6, c \neq -1\}$ **21.** The set of all real numbers   **23.** The set of all real numbers

25. For example: $\dfrac{1}{x - 5}$

27. For example: $\dfrac{1}{(x + 1)(x - 4)}$

29. a. Undefined   **b.** $\dfrac{2}{3}$ **31. a.** $\dfrac{9}{5}$ **b.** $\dfrac{9}{5}$

33. a. $\{x \,|\, x \text{ is a real number and } x \neq 1\}$ **b.** 2
35. a. $\{r \,|\, r \text{ is a real number and } r \neq 2\}$ **b.** $r + 2$

37. a. $\left\{a \,|\, a \text{ is a real number and } a \neq 0, a \neq \dfrac{3}{4}\right\}$ **b.** $\dfrac{2a}{4a - 3}$

39. a. $\left\{b \,|\, b \text{ is a real number and } b \neq \dfrac{5}{2}, b \neq -\dfrac{5}{2}\right\}$

b. $\dfrac{4}{2b + 5}$

41. a. $\{t \,|\, t \text{ is a real number and } t \neq -5, t \neq 4\}$ **b.** $\dfrac{t - 2}{t - 4}$

43. $\dfrac{5}{c^2}$ **45.** $\dfrac{4a^3}{5}$ **47.** $-\dfrac{5s^2}{r^3 t}$ **49.** 2   **51.** $\dfrac{c - 1}{c + 2}$

53. $\dfrac{1}{9(n + 2)}$ **55.** $\dfrac{(p - 3)^2}{p + 2}$ **57.** $\dfrac{1}{2c - 3}$ **59.** $\dfrac{3}{x - 5}$

61. $\dfrac{3(p+2)}{q-2}$ **63.** $\dfrac{5}{z-7}$ **65.** $b+8$

67. Cannot simplify **69.** $\dfrac{h+3}{h+4}$ **71.** $\dfrac{x-7}{2x-5}$

73. $\dfrac{4}{(t-2)(t+2)}$ **75.** $\dfrac{p-q}{p+q}$ **77.** $\dfrac{m}{m+1}$

79. $\dfrac{3}{2(x-y)}$ **81.** $\dfrac{1}{ax-b}$ **83.** $\dfrac{2c-d}{5c-2d}$

85. They are opposites. **87.** -1 **89.** -1

91. $-\dfrac{1}{3}$ **93.** -1 **95.** Cannot simplify

97. $\dfrac{t-4}{4+t}$ **99.** $-\dfrac{7+b}{b-3}$ **101.** $y+3$

103. $\dfrac{x-5}{x^2-5x+25}$

Section 7.2 Practice Exercises, pp. 508–509

1. To multiply fractions, we multiply the numerators and multiply the denominators. To divide the fractions we would multiply the first fraction by the reciprocal of the second fraction.

3. $\dfrac{5}{14}$ **5.** 9 **7.** $\dfrac{7}{5}$ **9.** 6 **11.** $\dfrac{1}{2s}$ **13.** $\dfrac{10}{a}$

15. $\dfrac{3(a+4)}{2}$ **17.** $\dfrac{p}{6}$ **19.** $-\dfrac{10}{3}$ **21.** $-\dfrac{x-y}{x+y}$

23. $\dfrac{3}{2c^2}$ **25.** $6a^3b$ **27.** $\dfrac{16}{15}$ **29.** $\dfrac{16a^2}{15}$

31. $\dfrac{b(3+b)}{3(b+1)}$ **33.** $\dfrac{5(x+y)}{2x-y}$ **35.** $\dfrac{2t-1}{t}$

37. $\dfrac{5t+1}{5t-1}$ **39.** 4 **41.** $\dfrac{4r}{2r+3}$ **43.** $\dfrac{1}{5}$

45. $\dfrac{z-4}{z-1}$ **47.** $\dfrac{p+1}{p-3}$ **49.** $\dfrac{q-7}{q+4}$ **51.** $\dfrac{y+6}{y+1}$

53. $-\dfrac{(p-2)(p+1)}{4(4+p)}$ **55.** $\dfrac{4h-1}{3h-2}$ **57.** $\dfrac{2}{y}$

59. $\dfrac{x+2}{2}$ **61.** $\dfrac{2m+n}{2}$ **63.** $\dfrac{1}{3}$ **65.** $\dfrac{t}{5(t-1)}$

67. r^2-rs+s^2

Section 7.3 Practice Exercises, pp. 515–518

3. $\{x\,|\,x$ is a real number and $x \neq 1, x \neq -1\}$; $\dfrac{3}{5(x-1)}$

5. $\dfrac{a+5}{a+7}$ **7.** $\dfrac{2}{3y}$ **9.** 36 **11.** 6

13. $15p$ **15.** $12xyz^3$ **17.** $w^2+8w+12$ **19.** -20

21. $5x-10$ **23.** -6 **25.** a, b, c, d **27.** x^5 is the lowest power of x that has x^3, x^5, x^4 as factors.

29. The product of unique factors is $(x+3)(x-2)$.

31. Because $(b-1)$ and $(1-b)$ are opposites; they differ by a factor of -1. **33.** 45 **35.** 16 **37.** 63

39. $9x^2y^3$ **41.** w^2y **43.** $(p+3)(p-1)(p+2)$

45. $9t(t+1)^2$ **47.** $(y-2)(y+2)(y+3)$

49. $3-x$ or $x-3$ **51.** $\dfrac{6}{5x^2}$; $\dfrac{5x}{5x^2}$

53. $\dfrac{24x}{30x^3}$; $\dfrac{5y}{30x^3}$ **55.** $\dfrac{10}{12a^2b}$; $\dfrac{a^3}{12a^2b}$

57. $\dfrac{6m-6}{(m+4)(m-1)}$; $\dfrac{3m+12}{(m+4)(m-1)}$

59. $\dfrac{6x+18}{(2x-5)(x+3)}$; $\dfrac{2x-5}{(2x-5)(x+3)}$

61. $\dfrac{6w+6}{(w+3)(w-8)(w+1)}$; $\dfrac{w^2+3w}{(w+3)(w-8)(w+1)}$

63. $\dfrac{6p^2+12p}{(p-2)(p+2)^2}$; $\dfrac{3p-6}{(p-2)(p+2)^2}$

65. $\dfrac{1}{a-4}$; $\dfrac{-a}{a-4}$ or $\dfrac{-1}{4-a}$; $\dfrac{a}{4-a}$

67. $\dfrac{8}{2(x-7)}$; $\dfrac{-y}{2(x-7)}$ or $\dfrac{-8}{2(7-x)}$; $\dfrac{y}{2(7-x)}$

69. $\dfrac{1}{a+b}$; $\dfrac{-6}{a+b}$ or $\dfrac{-1}{-a-b}$; $\dfrac{6}{-a-b}$

71. $\dfrac{-9}{24(3y+1)}$; $\dfrac{20}{24(3y+1)}$ **73.** $\dfrac{3z+12}{5z(z+4)}$; $\dfrac{5z}{5z(z+4)}$

75. $\dfrac{z^2+3z}{(z+2)(z+7)(z+3)}$; $\dfrac{-3z^2-6z}{(z+2)(z+7)(z+3)}$; $\dfrac{5z+35}{(z+2)(z+7)(z+3)}$

77. $\dfrac{3p+6}{(p-2)(p^2+2p+4)(p+2)}$; $\dfrac{p^3+2p^2+4p}{(p-2)(p^2+2p+4)(p+2)}$; $\dfrac{5p^3-20p}{(p-2)(p^2+2p+4)(p+2)}$

Section 7.4 Practice Exercises, pp. 524–527

1. $\{x\,|\,x$ is a real number and $x \neq 6, x \neq -6\}$

3. a. $\dfrac{1}{6}$, 0, undefined, $-\dfrac{1}{4}$, undefined

b. $(a-6)(a+2)$; $\{a\,|\,a$ is a real number and $a \neq 6, a \neq -2\}$

c. $\dfrac{a-1}{a-6}$ **5.** $\dfrac{6t-1}{(t+1)(t-6)}$ **7.** $\dfrac{8}{3}$ **9.** $\dfrac{2}{3}$ **11.** $\dfrac{b+9}{b-3}$

13. 6 **15.** $\dfrac{6p+5}{2p+1}$ **17.** $2w+1$ **19.** $k-3$

21. 5 **23.** $y+7$ **25.** $\dfrac{1}{x+3}$ **27.** $\dfrac{1}{x+4}$

29. $\dfrac{36}{t+3}$ **31.** $\dfrac{1}{12p}$ **33.** $\dfrac{10b-7a^2}{6a^2b^2}$ **35.** $\dfrac{q^2-2p}{p^2q^3}$

37. $\dfrac{1}{2}$ **39.** $\dfrac{11}{5(c-4)}$ **41.** $\dfrac{5h-32}{(h-5)(h+5)}$ **43.** $\dfrac{5}{24}$

45. $\dfrac{4}{2w-1}$ or $\dfrac{-4}{1-2w}$ **47.** $\dfrac{10x+3y}{(x+y)^2}$

49. $\dfrac{2r-5}{7}$ or $\dfrac{-2r+5}{-7}$ **51.** $\dfrac{4m+1}{m-2}$ or $\dfrac{-4m-1}{2-m}$

53. $\dfrac{3(5y-3)}{y(y-1)}$ **55.** $\dfrac{2y^2+35x}{14xy^2}$ **57.** $\dfrac{12-5q}{3q(q-2)}$

59. $\dfrac{3}{z+4}$ **61.** $\dfrac{2b}{(b+1)(b-1)}$

63. $\dfrac{-x(x+11)}{(x+1)(x+4)(x-3)}$ **65.** $\dfrac{8y^2-7y+5}{(3y+1)(2y-3)(y+1)}$

67. $\dfrac{-1}{3q-2}$ **69.** $\dfrac{x-y}{x+y}$ **71.** $\dfrac{-2}{x-y}$ **73.** $\dfrac{13t+2}{2t^2}$

75. $\dfrac{1}{n+6}$ **77.** $\dfrac{12}{p}$ **79.** $n+\left(5 \cdot \dfrac{1}{n}\right)$; $\dfrac{n^2+5}{n}$

81. $\dfrac{1}{m}-\dfrac{3m}{7}$; $\dfrac{7-3m^2}{7m}$ **83.** $\dfrac{2m^2+1}{(m-1)^2(m^2+m+1)}$

85. $\dfrac{-17t^2 - 6t + 2}{(2t + 1)(4t - 1)(t - 5)}$

87. $\dfrac{-5n + 17}{2(3n + 1)(n - 3)}$ or $\dfrac{5n - 17}{2(3n + 1)(3 - n)}$ **89.** $\dfrac{2p + 3}{3p + 4}$

91. $\dfrac{2}{m}$

Chapter 7, Problem Recognition Exercises— Operations on Rational Expressions p. 528

1. $\dfrac{-2x + 9}{3x + 1}$ **2.** $\dfrac{1}{w - 4}$ **3.** $\dfrac{y - 5}{2y - 3}$

4. $\dfrac{7}{(x + 3)(2x - 1)}$ **5.** $-\dfrac{1}{x}$ **6.** $\dfrac{1}{3}$ **7.** $\dfrac{c + 3}{c}$

8. $\dfrac{x + 3}{5}$ **9.** $\dfrac{a}{12b^4c}$ **10.** $\dfrac{2a - b}{a - b}$ **11.** $\dfrac{p - q}{5}$

12. 4 **13.** $\dfrac{10}{2x + 1}$ **14.** $\dfrac{w + 2z}{w + z}$ **15.** $\dfrac{3}{2x + 5}$

16. $\dfrac{y + 7}{x + a}$ **17.** $\dfrac{1}{2(a + 3)}$ **18.** $\dfrac{2(3y + 10)}{(y - 6)(y + 6)(y + 2)}$

19. $(t + 8)^2$ **20.** $6b + 5$

21. $\{x \mid x \text{ is a real number and } x \neq -1\}$

22. $\{x \mid x \text{ is a real number and } x \neq 3, x \neq -3\}$

Section 7.5 Practice Exercises, pp. 534–536

3. $\left\{ a \mid a \text{ is a real number and } a \neq \dfrac{3}{2}, a \neq -5 \right\}; \dfrac{1}{2a - 3}$

5. $\dfrac{3(2k - 5)}{5(k - 2)}$ **7.** $\dfrac{1}{3xy(x + y)}$ **9.** $\dfrac{2b - 9a^2}{3a(5 - 4ab)}$

11. $\dfrac{4a}{5}$ **13.** $\dfrac{3}{2}$ **15.** $\dfrac{27y}{10x^4}$ **17.** $\dfrac{p^2q^2}{2w^2}$

19. $\dfrac{10}{23}$ **21.** $1 + b$ **23.** $\dfrac{5(k + 5)}{k + 1}$ **25.** 6

27. $\dfrac{3(4 + p^2)}{2(p - 3)(p + 3)}$ **29.** $\dfrac{2(3 + 2m^2)}{m(6 - 5m)}$ **31.** $\dfrac{3w + 5}{4w + 7}$

33. $\dfrac{p + 3}{p + 2}$ **35.** $\dfrac{3}{2m}$ **37.** $\dfrac{10}{\frac{2}{5} - \frac{1}{4}}; \dfrac{200}{3}$ **39.** $\dfrac{\frac{3}{5} - \frac{1}{2}}{4}; \dfrac{1}{40}$

41. a. 51.4 mph average   **b.** 51.4 mph   **c.** Because the rates going to and leaving from the destination are the same, the average rate is unchanged. The average rate is not affected by the distance traveled.

43. $\dfrac{-3w + 20}{4(w + 5)}$ **45.** $-\dfrac{(y + 3)(y - 2)}{(y + 1)(y - 3)}$ **47.** $\dfrac{5}{3}$

Section 7.6 Practice Exercises, pp. 544–547

3. $\dfrac{2}{4x - 1}$ **5.** $5(h + 1)$ **7.** $\dfrac{(x + 4)(x - 3)}{x^2}$

9. $b = 3$ **11.** $k = \dfrac{5}{11}$ **13.** $y = \dfrac{1}{3}$

15. a. $\{z \mid z \text{ is a real number and } z \neq 0\}$ **b.** $5z$ **c.** $z = 5$

17. a. $\{x \mid x \text{ is a real number and } x \neq 2, x \neq -3\}$

b. $(x - 2)(x + 3)$ **c.** $x = 7$

19. $x = -\dfrac{21}{8}$ **21.** $b = 4$ **23.** $x = -\dfrac{1}{6}$

25. $m = 4, m = -2$ **27.** $b = 8$

29. $t = -4; (t = 0 \text{ does not check.})$

31. $n = 3$ **33.** No solution; $(q = -5 \text{ does not check.})$

35. $w = 5$ **37.** $x = -9; x = -1$

39. $x = -7; (x = \dfrac{3}{2} \text{ does not check.})$ **41.** No solution; $(x = -3 \text{ does not check.})$ **43.** $y = 2; (y = -4 \text{ does not check.})$ **45.** $a = -5; (a = 5 \text{ does not check})$

47. $w = 7, w = -2$ **49.** The number is 7.

51. The number is -3.   **53.** $a = \dfrac{FK}{m}$ **55.** $R = \dfrac{KE}{I}$

57. $r = \dfrac{E - IR}{I}$ or $r = \dfrac{E}{I} - R$ **59.** $r = \dfrac{C}{2\pi}$

61. $w = \dfrac{V}{lh}$ **63.** $m = \dfrac{T}{g - f}$ or $m = \dfrac{-T}{f - g}$

65. $w = \dfrac{n}{1 - Pn}$ or $w = \dfrac{-n}{Pn - 1}$ **67.** $P = \dfrac{A}{1 + rt}$

69. $b = \dfrac{fa}{a - f}$ or $b = \dfrac{-fa}{f - a}$

Chapter 7, Problem Recognition Exercises— Comparing Rational Equations and Rational Expressions, p. 547

1. $\dfrac{y - 2}{2y}$ **2.** $x = 6$ **3.** $t = 2$ **4.** $\dfrac{3a - 17}{a - 5}$

5. $\dfrac{4p + 27}{18p^2}$ **6.** $\dfrac{b(b - 5)}{(b - 1)(b + 1)}$ **7.** $h = 5$

8. $\dfrac{2w + 5}{(w + 1)^2}$ **9.** $x = 7$ **10.** $m = 5$

11. $\dfrac{3x + 14}{4(x + 1)}$ **12.** $t = \dfrac{11}{3}$

Section 7.7 Practice Exercises, pp. 556–560

3. Expression; $\dfrac{m^2 + m + 2}{(m - 1)(m + 3)}$ **5.** Expression; $\dfrac{3}{10}$

7. Equation; $p = 2$ **9.** $p = 95$ **11.** $z = 1$

13. $a = \dfrac{40}{3}$ **15.** $x = 40$ **17.** $y = 3$ **19.** $z = -1$

21. $a = 1$ **23. a.** $V_f = \dfrac{V_i T_f}{T_i}$ **b.** $T_f = \dfrac{T_i V_f}{V_i}$

25. Toni can drive 297 mi on 9 gallons of gas.

27. They would produce 1536 lb.

29. The opponent received 4336 votes.

31. This represents 262.5 miles.   **33. a.** 4 cm   **b.** 5 cm

35. $x = 3.75$ cm; $y = 4.5$ cm

37. The height of the pole is 7 m.

39. The light post is 24 ft high.

41. The speed of the current is 2 mph.

43. The wind speed is 35 mph.

45. Shanelle skis 10 km/hr and Devon skis 15 km/hr.

47. The speeds are 45 mph and 60 mph.

49. Kendra flew 180 mph to Cincinnati and 150 mph back.

51. Sergio rode 12 mph and walked 3 mph.

53. $\dfrac{1}{2}$ of the room   **55.** $5\frac{5}{11}$ $(5.\overline{45})$ minutes

57. $22\frac{2}{9}$ $(22.\overline{2})$ min   **59.** $3\frac{1}{3}$ $(3.\overline{3})$ days

61. 48 hr   **63.** 16 min

65. There are 40 smokers and 140 nonsmokers.

67. There are 240 men and 200 women.

Chapter 7, Review Exercises, pp. 566–568

1. a. $-\dfrac{2}{9}, -\dfrac{1}{10}, 0, -\dfrac{5}{6}$, undefined

b. $\{t \mid t$ is a real number and $t \neq -9\}$

2. a. $-\dfrac{1}{5}, -\dfrac{1}{2}$, undefined, $0, \dfrac{1}{7}$

b. $\{k \mid k$ is a real number and $k \neq 5\}$

3. a, c, d

4. $\{x \mid x$ is a real number and $x \neq \frac{5}{2}, x \neq 3\}$; $\frac{1}{2x - 5}$

5. $\{h \mid h$ is a real number and $h \neq -\frac{1}{3}, h \neq -7\}$; $\frac{1}{3h + 1}$

6. $\{a \mid a$ is a real number and $a \neq 2, a \neq -2\}$; $\frac{4a - 1}{a - 2}$

7. $\{w \mid w$ is a real number and $w \neq 4, w \neq -4\}$; $\frac{2w + 3}{w - 4}$

8. $\{z \mid z$ is a real number and $z \neq 4\}$; $-\frac{z}{2}$

9. $\{k \mid k$ is a real number and $k \neq 0, k \neq 5\}$; $-\frac{3}{2k}$

10. $\{b \mid b$ is a real number and $b \neq -3\}$; $\frac{b - 1}{2}$

11. $\{m \mid m$ is a real number and $m \neq -1\}$; $\frac{m - 5}{3}$

12. $\{n \mid n$ is a real number and $n \neq -3\}$; $\frac{1}{n + 3}$

13. $\{p \mid p$ is a real number and $p \neq -7\}$; $\frac{1}{p + 7}$

14. y^2 **15.** $\dfrac{u^2}{2}$ **16.** $v + 2$ **17.** $\dfrac{3}{2(x - 5)}$

18. $\dfrac{c(c + 1)}{2(c + 5)}$ **19.** $\dfrac{q - 2}{4}$ **20.** $-2t(t - 5)$

21. $4s(s - 4)$ **22.** $\dfrac{1}{7}$ **23.** $\dfrac{1}{n - 2}$ **24.** $-\dfrac{1}{6}$

25. $\dfrac{1}{m + 3}$ **26.** $\dfrac{-1}{(x + 3)(x + 2)}$ **27.** $-\dfrac{2y - 1}{y + 1}$

28. $5x + 5$ **29.** $2y + 4$ **30.** $6w - 24$

31. $2r + 6$ **32.** $s^2 - 6s + 8$ **33.** $u^2 - 5u - 6$

34. $a^2 b^3 c^2$ **35.** $xy^2 z^4$ **36.** $(p + 2)(p - 4)$

37. $q(q + 8)$ **38.** $(m - 4)(m + 4)(m + 3)$

39. $(n - 3)(n + 3)(n + 2)$ **40.** $2t - 5$ or $5 - 2t$

41. $3k - 1$ or $1 - 3k$ **42.** $c - 2$ or $2 - c$

43. $3 - x$ or $x - 3$ **44.** 2 **45.** 2 **46.** $a + 5$

47. $x - 7$ **48.** $\dfrac{-y - 18}{(y - 9)(y + 9)}$ or $\dfrac{y + 18}{(9 - y)(y + 9)}$

49. $\dfrac{t^2 + 2t + 3}{(2 - t)(2 + t)}$ **50.** $\dfrac{m + 8}{3m(m + 2)}$ **51.** $\dfrac{3(r - 4)}{2r(r + 6)}$

52. $\dfrac{p}{(p + 4)(p + 5)}$ **53.** $\dfrac{q}{(q + 5)(q + 4)}$ **54.** $\dfrac{1}{2}$

55. $\dfrac{1}{3}$ **56.** $\dfrac{a - 4}{a - 2}$ **57.** $\dfrac{1}{4z}$ **58.** $\dfrac{w}{2}$

59. $\dfrac{8}{y}$ **60.** $y - x$ **61.** $-(b + a)$ **62.** $-\dfrac{2p + 7}{2p}$

63. $-\dfrac{k + 10}{k + 4}$ **64.** $x = -8$ **65.** $y = -2$

66. $h = 0$ **67.** $w = 2$ **68.** $t = -2$ **69.** $p = 3$

70. $y = -11, y = 1$

71. No solution; $(z = 2$ does not check.)

72. The number is 4.   **73.** $h = \dfrac{3V}{\pi r^2}$ **74.** $b = \dfrac{2A}{h}$

75. $m = \dfrac{6}{5}$ **76.** $a = 19.2$ **77.** 10 g

78. Ed travels at 60 mph, and Bud travels at 70 mph.

79. Together the pumps would fill the pool in 16.8 min.

80. $x = 11; y = 26$

Chapter 7, Test, pp. 568–569

1. a. $\{x \mid x$ is a real number and $x \neq 2\}$ **b.** $-\dfrac{x + 1}{6}$

2. a. $\{a \mid a$ is a real number and $a \neq 0, a \neq 6, a \neq -2\}$

b. $\dfrac{7}{a + 2}$ **3.** b, c, d   **4.** $\dfrac{y + 7}{3(y + 3)(y + 1)}$

5. $-\dfrac{1}{5(9b + 1)}$ **6.** $\dfrac{1}{w + 1}$ **7.** $\dfrac{t + 4}{t + 2}$

8. $\dfrac{x(3x + 5)}{(3x + 4)(x - 2)}$ **9.** $\dfrac{1}{m + 4}$ **10.** $a = \dfrac{8}{5}$

11. $p = -2; (p = 1$ does not check.)

12. $c = 1$ **13.** No solution. $(x = 4$ does not check.)

14. $r = \dfrac{2A}{C}$ **15.** The number is $-\dfrac{2}{5}$.   **16.** $y = -8$

17. $1\frac{1}{4}$ (1.25) cups of carrots

18. The speed of the current is 5 mph.   **19.** 2 hr

20. $a = 5.6; b = 12$ **21. a.** $15(x + 3)$ **b.** $3x^2 y^2$

Chapters 1–7, Cumulative Review Exercises, pp. 569–570

1. 32   **2.** 7   **3.** $y = \dfrac{10}{9}$

4.

Set-Builder Notation	Graph	Interval Notation
$\{x \mid x \geq -1\}$	(graph, point at -1)	$[-1, \infty)$
$\{x \mid x < 5\}$	(graph, point at 5)	$(-\infty, 5)$

5. The width is 17 m and the length is 35 m.

6. The base is 10 in. and the height is 8 in.

7. $\dfrac{x^2 y z^{17}}{2}$ **8. a.** $6x + 4$ **b.** $2x^2 + x - 3$

9. $(5x - 3)^2$ **10.** $(2c + 1)(5d - 3)$

11. $\left\{x \mid x \text{ is a real number and } x \neq 5, x \neq -\dfrac{1}{2}\right\}$

12. $\dfrac{x - 3}{x + 5}$ **13.** $\dfrac{1}{5(x + 4)}$ **14.** -3 **15.** $y = 1$

16. $b = -\dfrac{7}{2}$

17. a. x-intercept: $(-4, 0)$; y-intercept: $(0, 2)$ **b.** x-intercept: $(0, 0)$; y-intercept: $(0, 0)$

18. a. $m = -\dfrac{7}{5}$ **b.** $m = -\dfrac{2}{3}$ **c.** $m = 4$ **d.** $m = -\dfrac{1}{4}$

19. $y = 5x - 3$

20. One large popcorn costs $3.50, and one drink costs $1.50.

Chapter 8

Chapter Opener Puzzle

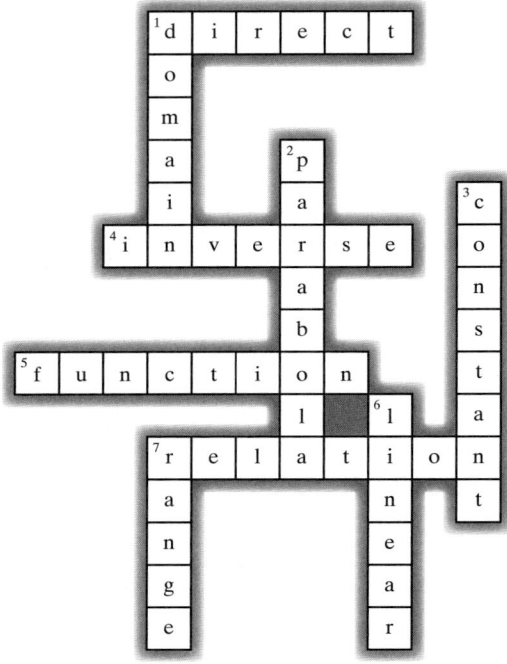

Section 8.1 Practice Exercises, pp. 577–580

3. $\{(A, 1), (A, 2), (B, 2), (C, 3), (D, 5), (E, 4)\}$

5. $\{(64, 37.99), (128, 42.99), (256, 49.99), (512, 74.99)\}$

7. Domain $\{A, B, C, D, E\}$; range $\{1, 2, 3, 4, 5\}$

9. Domain $\{64, 128, 256, 512\}$; range $\{37.99, 42.99, 49.99, 74.99\}$

11. Domain $[-5, 3]$; range $[-2.1, 2.8]$

13. Domain $[0, 4.2]$; range $[-2.1, 2.1]$

15. Domain $(-\infty, 2]$; range $(-\infty, \infty)$

17. Domain $[-4, \infty)$; range $[-4, -2] \cup (2, \infty)$

19. Domain $(-\infty, \infty)$; range $\{-2\}$

21. Domain $\{-3, -1, 1, 3\}$; range $\{0, 1, 2, 3\}$

23. Domain $[-4, 5)$; range $\{-2, 1, 3\}$

25. a. 2.85   **b.** 9.33   **c.** Dec.   **d.** Nov.   **e.** 7.63   **f.** {Jan., Feb., Mar., Apr., May, June, July, Aug., Sept., Oct., Nov., Dec.}

27. a. 31.876 million or 31,876,000   **b.** The year 2012

29. a. For example: {(Julie, New York), (Peggy, Florida), (Stephen, Kansas), (Pat, New York)}   **b.** Domain {Julie, Peggy, Stephen, Pat}; range {New York, Florida, Kansas}

31. $y = 2x - 1$ **33.** $y = x^2$

Section 8.2 Practice Exercises, pp. 588–593

3. a. $\{(-2, -4), (-1, -1), (0, 0), (1, -1), (2, -4)\}$

b. $\{-2, -1, 0, 1, 2\}$ **c.** $\{-4, -1, 0\}$

5. Function   **7.** Not a function   **9.** Function

11. Not a function   **13.** Function   **15.** Not a function

17. -11 **19.** 1   **21.** 2   **23.** $6t - 2$ **25.** 7

27. 4   **29.** 4   **31.** $6x + 4$ **33.** $-4x^2 - 8x + 1$

35. $-\pi^2 + 4\pi + 1$ **37.** 7   **39.** $-6a - 2$

41. $|-c - 2|$ **43.** 1   **45.** 7   **47.** -18.8

49. -7 **51.** 2π **53.** -5 **55.** 4

57. a. 2   **b.** 1   **c.** 1   **d.** $x = -3$ **e.** $x = 1$

f. $[-3, 3]$ **g.** $[-3, 3]$

59. a. 3   **b.** $H(4)$ is not defined because 4 is not in the domain of H.   **c.** 4   **d.** $x = -3$ and $x = 2$ **e.** All x on the interval $[-2, 1]$ **f.** $(-4, 4)$ **g.** $[2, 5]$

61. The domain is the set of all real numbers for which the denominator is not zero. Set the denominator equal to zero, and solve the resulting equation. The solution(s) to the equation must be excluded from the domain. In this case, setting $x - 2 = 0$ indicates that $x = 2$ must be excluded from the domain. The domain is $\{x \mid x$ is a real number and $x \neq 2\}$.

63. $(-\infty, -6) \cup (-6, \infty)$ **65.** $(-\infty, 0) \cup (0, \infty)$

67. $(-\infty, \infty)$ **69.** $[-7, \infty)$ **71.** $[3, \infty)$ **73.** $\left(-\infty, \frac{1}{2}\right]$

75. $(-\infty, \infty)$ **77.** $(-\infty, \infty)$

79. a. $h(1) = 64$ and $h(1.5) = 44$ **b.** $h(1) = 64$ means that after 1 sec, the height of the ball is 64 ft. $h(1.5) = 44$ means that after 1.5 sec, the height of the ball is 44 ft.

81. a. $d(1) = 11.5$ and $d(1.5) = 17.25$ **b.** $d(1) = 11.5$ means that after 1 hr, the distance traveled is 11.5 mi. $d(1.5) = 17.25$ means that after 1.5 hr, the distance traveled is 17.25 mi.

83. a. $P(0) = 0\%$ $P(5) = \dfrac{100}{3} \approx 33.3\%$

$P(10) = \dfrac{200}{3} \approx 66.7\%$ $P(15) = \dfrac{900}{11} \approx 81.8\%$

$P(20) = \dfrac{800}{9} \approx 88.9\%$ $P(25) = \dfrac{2500}{27} \approx 92.6\%$.

If Brian studies 25 hours he will get a score of 92.6%.

b. $A(0, 0)\ B\left(5, \dfrac{100}{3}\right) C\left(10, \dfrac{200}{3}\right) D\left(15, \dfrac{900}{11}\right) E\left(20, \dfrac{800}{9}\right)$

$F\left(25, \dfrac{2500}{27}\right)$

85. $(4, \infty)$ **87.** $\{5, -3, 4\}$ **89.** $\{0, 2, 6, 1\}$ **91.** -7

93. 0 and 2   **95.** 6

97.

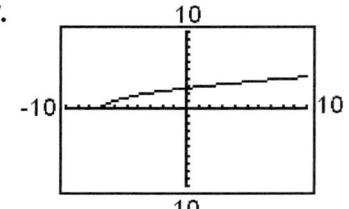

99. a. $h(1) = 64$ **b.**

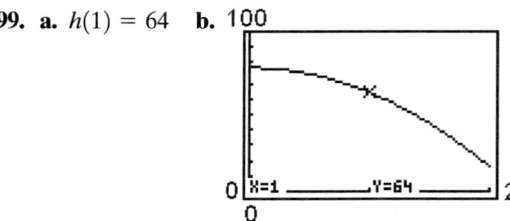

Section 8.3 Practice Exercises, pp. 601–606

3. a. Yes   **b.** $\{7, 2, -5\}$ **c.** $\{3\}$

5. a. $g(2) = -2, g(4) = 2, g(5) = 1$, and $g(3)$ cannot be evaluated because $x = 3$ is not in the domain of g.

b. $(-\infty, 3) \cup (3, \infty)$

7. $V(2) = -64$ means that 2 sec after release, the object is falling at 64 ft/sec. $V(5) = -160$ means that 5 sec after release, the object is falling at 160 ft/sec.

9. m is the slope, and $(0, b)$ is the y-intercept.

11. Domain $(-\infty, \infty)$; range $(-\infty, \infty)$

13.

x	$g(x)$
-2	2
-1	1
0	0
1	1
2	2

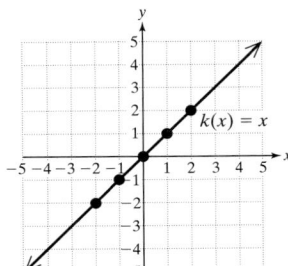

15.

x	$k(x)$
-2	-2
-1	-1
0	0
1	1
2	2

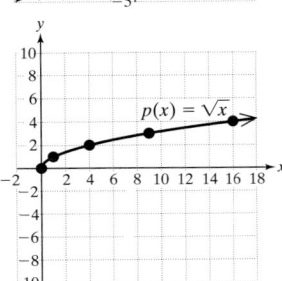

17.

x	$p(x)$
0	0
1	1
4	2
9	3
16	4

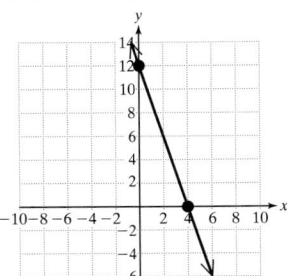

19. Quadratic   **21.** Linear   **23.** Constant
25. None of these   **27.** Linear   **29.** None of these

31.

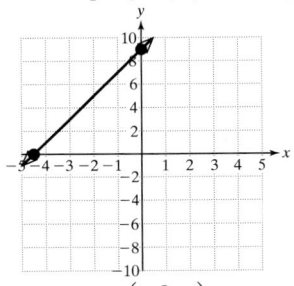

x-intercept: $(4, 0)$; y-intercept: $(0, 12)$

33.

x-intercept: $\left(-\dfrac{9}{2}, 0\right)$; y-intercept: $(0, 9)$

35.

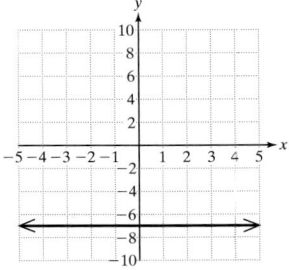

x-intercept: none; y-intercept: $(0, -7)$

37.

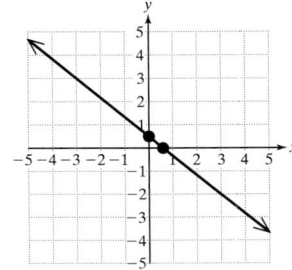

x-intercept: $\left(\dfrac{3}{5}, 0\right)$; y-intercept: $\left(0, \dfrac{1}{2}\right)$

39. a. $x = 1$ **b.** $f(0) = -1$ **41. a.** None   **b.** $f(0) = 1$
43. a. $x = 1$ and $x = -1$ **b.** $f(0) = 1$
45. a. $(-\infty, \infty)$ **b.** $(0, 1)$ **c.** iii
47. a. $(-\infty, \infty)$ **b.** $(0, -2)$ **c.** v
49. a. $[-4, \infty)$ **b.** $(0, 2)$ **c.** i
51. a. $(-\infty, -1) \cup (-1, \infty)$ **b.** $(0, 1)$ **c.** x
53. a. $(-\infty, \infty)$ **b.** $(0, 3)$ **c.** ix
55. a. $(-4, -2) \cup (0, 3)$ **b.** $(-2, 0)$ **c.** $(3, \infty)$
57. a. $(-\infty, -1) \cup (1, \infty)$ **b.** $(-1, 1)$ **c.** Never constant
59. a. $(0, \infty)$ **b.** $(-\infty, 0)$ **c.** Never constant
61. a. $(0, \infty)$ **b.** $(-\infty, 0)$ **c.** Never constant
63. a. Never increasing   **b.** $(-\infty, 0) \cup (0, \infty)$ **c.** Never constant

65. Yes; A function is decreasing on an interval I if for all $a < b$ on I, $f(a) > f(b)$. In this case, if k is decreasing on $(-2, 0)$, then $k(-1.5) > k(-0.5)$ must be true.

67.

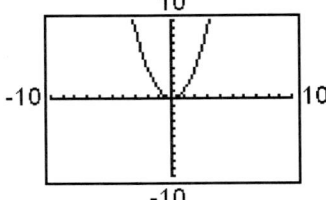

69.

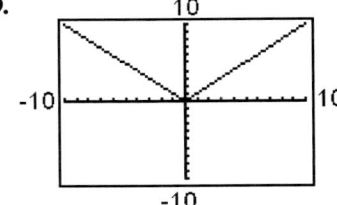

71.

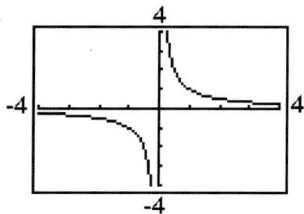

Section 8.4 Practice Exercises, pp. 612–614

3. -1 **5.** $x = -2$ **7.** $[-4, \infty)$

9. a. Increase   **b.** Decrease   **11.** $T = kq$

13. $W = \dfrac{k}{p^2}$ **15.** $Q = \dfrac{kx}{y^3}$ **17.** $L = kw\sqrt{v}$

19. $k = \dfrac{9}{2}$ **21.** $k = 512$ **23.** $k = 1.75$ **25.** $Z = 56$

27. $L = 9$ **29.** $B = \dfrac{15}{2}$ **31.** 355,000 tons   **33.** 42.6 ft

35. 18.5 A   **37.** 1.25 Ω   **39.** 20 lb   **41.** 2224 lb

43. a. $A = kl^2$ **b.** The area is 4 times the original.
c. The area is 9 times the original.

Chapter 8 Review Exercises, pp. 618–621

1. For example: {(Peggy, Kent), (Charlie, Laura), (Tom, Matt), (Tom, Chris)}

2. Domain $\left\{\dfrac{1}{3}, 6, \dfrac{1}{4}, 7\right\}$; range $\left\{10, -\dfrac{1}{2}, 4, \dfrac{2}{5}\right\}$

3. Domain $[-3, 9]$; range $[0, 60]$
4. Domain $[-4, 4]$; range $[1, 3]$

5. Domain $\{-3, -1, 0, 2, 3\}$; range $\left\{-2, 0, 1, \dfrac{5}{2}\right\}$

6.   **7.**

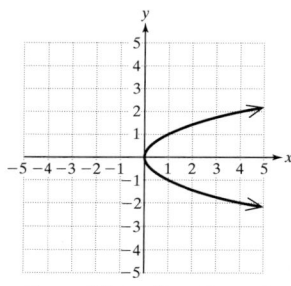

8. a. Not a function   **b.** $[1, 3]$ **c.** $[-4, 4]$
9. a. Function   **b.** $(-\infty, \infty)$ **c.** $(-\infty, 0.35)$
10. a. Function   **b.** $\{1, 2, 3, 4\}$ **c.** $\{3\}$
11. a. Not a function   **b.** $\{0, 4\}$ **c.** $\{2, 3, 4, 5\}$
12. a. Not a function   **b.** $\{4, 9\}$ **c.** $\{2, -2, 3, -3\}$
13. a. Function   **b.** $\{6, 7, 8, 9\}$ **c.** $\{9, 10, 11, 12\}$
14. -4 **15.** 2   **16.** 2   **17.** $6t^2 - 4$ **18.** $6b^2 - 4$
19. $6\pi^2 - 4$ **20.** $6\square^2 - 4$ **21.** 20   **22.** $(-\infty, \infty)$
23. $(-\infty, 11) \cup (11, \infty)$ **24.** $[8, \infty)$ **25.** $[-2, \infty)$
26. a. \$98   **b.** \$123   **c.** \$148

27.   **28.**

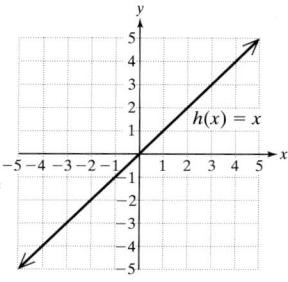 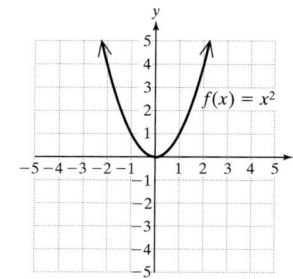

29.   **30.**

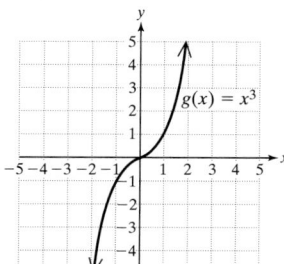 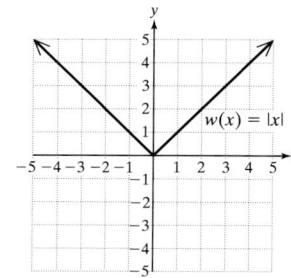

31.   **32.**

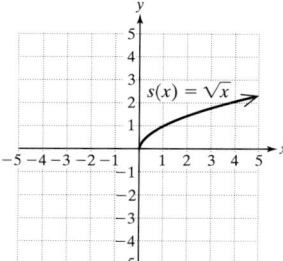 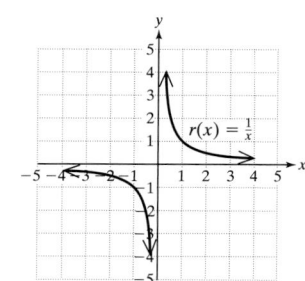

33. The function is constant over its entire domain, $(-\infty, \infty)$. It is never increasing or decreasing.

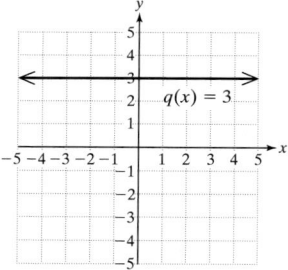

34. The function is increasing over its entire domain, $(-\infty, \infty)$. It is never constant or decreasing.

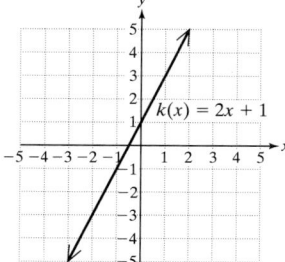

35. x-intercept: $\left(\dfrac{7}{4}, 0\right)$; y-intercept: $(0, -7)$

36. x-intercept: $\left(\dfrac{9}{2}, 0\right)$; y-intercept: $(0, 9)$

37. a. $b(0) = 4.5$. In 1985 consumption was 4.5 gal of bottled water per capita. $b(7) = 9.4$. In 1992 consumption was 9.4 gal of bottled water per capita.
b. $m = 0.7$. Consumption increased by 0.7 gal/year per capita.
38. -1 **39.** 1   **40.** $x = 0$ and $x = 4$
41. There are no values of x for which $g(x) = -4$.
42. $(-4, \infty)$ **43.** $(-\infty, 1]$ **44.** $(-4, 2)$
45. $(2, \infty)$ **46. a.** $r(4) = 0, r(5) = 2, r(8) = 4$ **b.** $[4, \infty)$
47. a. $h(-3) = -\dfrac{1}{2}, h(-1) = -\dfrac{3}{4}, h(0) = -1, h(2) = -3,$
$h(3)$ is undefined, $h(4) = 3$ **b.** $(-\infty, 3) \cup (3, \infty)$
48. a. $F = kd$ **b.** $k = 3$ **c.** $d = 1\frac{2}{3}$ ft
49. $y = 256$ **50.** $y = 12$ **51.** 48 km

Chapter 8 Test, pp. 622–623

1. a. Not a function   **b.** $\{-3, -1, 1, 3\}$ **c.** $\{-2, -1, 1, 3\}$
2. a. Function   **b.** $(-\infty, \infty)$ **c.** $(-\infty, 0]$
3. To find the x-intercept(s), solve for the real solutions of the equation $f(x) = 0$. To find the y-intercept, find $f(0)$.

4. 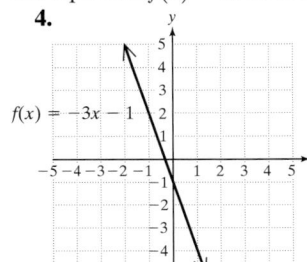   **5.**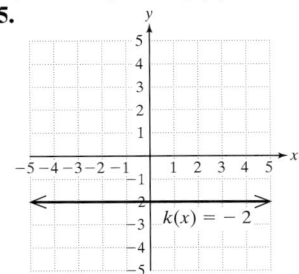

6. 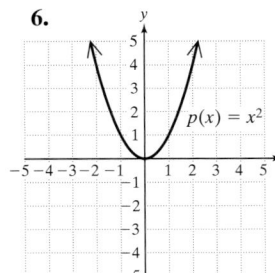   **7.**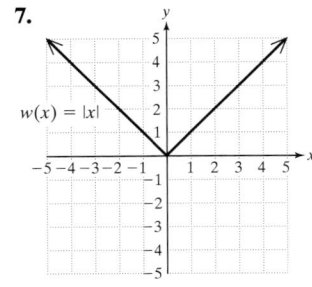

8. $(-\infty, -7) \cup (-7, \infty)$ **9.** $[-7, \infty)$ **10.** $(-\infty, \infty)$
11. a. $r(-2) = 9, r(-1) = 4, r(0) = 1, r(2) = 1, r(3) = 4$
b. $(-\infty, \infty)$
12. a. $s(0) = 36$. In 1985 the per capita consumption was 36 gal. $s(7) = 47.2$. In 1992 the per capita consumption was 47.2 gal.   **b.** $m = 1.6$. Per capita consumption increases by 1.6 gal/year.
13. 1   **14.** 2   **15.** $[-1, 7]$ **16.** $[-1, 4]$ **17.** False
18. $(6, 0)$ **19.** $x = 6$ **20.** All x in the interval $[1, 3]$
and $x = 5$ **21.** $(3, 4)$ **22.** $(-1, 1) \cup (4, 7)$ **23.** $(1, 3)$
24. Quadratic   **25.** Linear   **26.** Constant
27. None of these
28. x-intercept: $(-12, 0)$; y-intercept: $(0, 9)$ **29.** $x = \dfrac{ky}{t^2}$
30. 3.3 sec

Chapters 1–8 Cumulative Review Exercises, pp. 623–624

1. $t = -\dfrac{12}{7}$ **2.** 14
3. $\left(-\infty, \frac{1}{6}\right]$ **4.** 1474 in.^{3} **5.** $m = \pm\dfrac{2}{5}$
6. a. To find the x-intercepts, let $f(x) = 0$ and solve for x. To find the y-intercept, find $f(0)$.
b. $(0, 2)$ **c.** $\left(-\dfrac{2}{3}, 0\right)$
7. $(4, 0)$ **8.** The numbers are 24 and 36.
9. $\dfrac{x^2 y^2}{z^5}$ **10.** $p^3 + 3p^2 - 17p + 6$
11. $p^2 - 2p + 7$ **12.** $(k + 1)(2k - 1)$
13. $k = -1, k = \dfrac{1}{2}$ **14.** $(-\infty, 15) \cup (15, \infty)$
15. $x = -1$; ($x = 15$ does not check.)
16. $\dfrac{2}{2x + 1}$ **17. a.** 8620 students   **b.** The year 2011
18. Domain $\{12, 15, 18, 21\}$; range $\{a, c, e\}$; yes   **19. a.** 1
b. 33   **20.** $3500

Chapter 9

Chapter Opener Puzzle
Math is 2 good 2 B 4 got 10

Section 9.1 Practice Exercises, pp. 631–633
3. No solution; Inconsistent system   **5.** The speeds are 65 mph and 58 mph.   **7.** $(4, 0, 2)$ is a solution.
9. $(1, 1, 1)$ is a solution.   **11.** $(1, -2, 4)$
13. $(-1, 2, 0)$ **15.** $(-2, -1, -3)$ **17.** $(1, -4, -2)$
19. $(1, 2, 3)$ **21.** $(-6, 1, 7)$ **23.** $\left(\dfrac{1}{2}, \dfrac{2}{3}, -\dfrac{5}{6}\right)$
25. $\left(\dfrac{1}{2}, 2, -\dfrac{1}{2}\right)$ **27.** $(6, -3, -2)$ **29.** $(0, 4, -4)$
31. $(1, 0, -1)$ **33.** $(-20, 8, 0)$ **35.** $(-9, 5, 5)$
37. Inconsistent system   **39.** $(1, 3, 1)$ **41.** $\left(-\dfrac{1}{2}, \dfrac{1}{3}, \dfrac{1}{6}\right)$
43. Dependent system   **45.** $(1, 0, 1)$ **47.** $(0, 0, 0)$
49. Dependent system

Section 9.2 Practice Exercises, pp. 637–640
1. $(-7, 0, 8)$ **3.** Inconsistent system   **5.** $67°, 82°, 31°$
7. $25°, 56°, 99°$ **9.** 10 cm, 18 cm, 27 cm   **11.** 11 in., 17 in., 26 in.   **13.** The fiber supplement has 3 g; the oatmeal has 4 g; and the cereal has 1.5 g.   **15.** Kyoki invested $2000 in bonds, $6000 in mutual funds, and $2000 in the money market account.   **17.** $1500 at 5%, $2000 at 5.5%, $1000 at 4%   **19.** 148 adult tickets, 51 children's tickets, 23 senior tickets   **21.** 24 oz peanuts, 8 oz pecans, 16 oz cashews   **23.** Vanderbilt 6100, Baylor 12,200, Pace 8900   **25.** $a = -9, b = 5,$ and $c = 4$. The parabola is given by $y = -9x^2 + 5x + 4$.   **27.** $a = 2, b = -1,$ and $c = -5$. The parabola is given by $y = 2x^2 - x - 5$.   **29.** $a = -3, b = 4, c = 0$. The parabola is given by $y = -3x^2 + 4x$.
31. $a = 1, b = 4, c = -1$. The parabola is given by $y = x^2 + 4x - 1$.

Section 9.3 Practice Exercises, pp. 646–649

3. $\left(12, \dfrac{1}{2}\right)$ **5.** $(1, 3, -3)$ **7.** 3×1, column matrix

9. 2×2, square matrix   **11.** 1×4, row matrix

13. 2×3, none of these   **15.** $\begin{bmatrix} 1 & -2 & | & -1 \\ 2 & 1 & | & -7 \end{bmatrix}$

17. $\begin{bmatrix} -2 & 1 & | & -1 \\ 3 & 1 & | & 7 \end{bmatrix}$ **19.** $\begin{bmatrix} 5 & 0 & 2 & | & 17 \\ 8 & -1 & 6 & | & 26 \\ 8 & 3 & -12 & | & 24 \end{bmatrix}$

21. $\begin{aligned} -2x + 5y &= -15 \\ -7x + 15y &= -45 \end{aligned}$ **23.** $x = 0.5, y = 6.1, z = 3.9$

25. a. -13 **b.** 0 **27.** $\begin{bmatrix} 1 & 1 & | & 7 \\ 0 & 1 & | & -2 \end{bmatrix}$

29. $\begin{bmatrix} -7 & 2 & | & 19 \\ 9 & 6 & | & 13 \end{bmatrix}$ **31.** $\begin{bmatrix} 1 & 3 & | & -5 \\ 0 & 8 & | & 2 \end{bmatrix}$

33. a. $\begin{bmatrix} 1 & 2 & 0 & | & 10 \\ 0 & -9 & -4 & | & -47 \\ -3 & 4 & 5 & | & 2 \end{bmatrix}$

b. $\begin{bmatrix} 1 & 2 & 0 & | & 10 \\ 0 & -9 & -4 & | & -47 \\ 0 & 10 & 5 & | & 32 \end{bmatrix}$ **35.** $(-3, -2)$ **37.** $(5, 4)$

39. Infinitely many solutions; $\{(x, y) \mid 2x + 5y = 1\}$; Dependent system   **41.** $(1, 2)$

43. $(2, 2)$ **45.** No solution; Inconsistent system

47. $\left(7, 2, -\dfrac{3}{2}\right)$ **49.** $(13, 20, 5)$ **51.** True   **53.** True

55. Multiply row 3 by 2. Replace row 3 with the result.

57. Multiply row 2 by 4 and add to row 3. Replace row 3 with the result.   **59.**

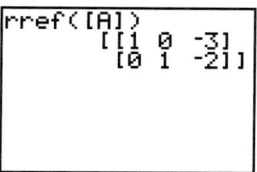

```
rref([A])
       [[1 0 -3]
        [0 1 -2]]
```

61.   **63.**

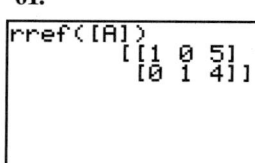

```
rref([A])
       [[1 0 5]
        [0 1 4]]
```

```
rref([A])
       [[1 0 0 7    ]
        [0 1 0 2    ]
        [0 0 1 -1.5]]
```

Section 9.4 Practice Exercises, pp. 658–661

3. 16   **5.** 1   **7.** 4   **9.** 6   **11.** 26   **13.** 6

15. 46   **17. a.** 30   **b.** 30   **19.** Choosing the row or column with the most zero elements simplifies the arithmetic when evaluating a determinant.   **21.** 12

23. -15 **25.** 0   **27.** $8a - 2b$ **29.** $4x - 3y + 6z$

31. 0   **33.** $\mathbf{D} = 16; \mathbf{D}_x = -63; \mathbf{D}_y = 66$ **35.** $(2, -1)$

37. $(4, -1)$ **39.** $\left(\dfrac{23}{13}, \dfrac{9}{13}\right)$ **41.** Cramer's rule does not apply when the determinant $\mathbf{D} = 0$.   **43.** No solution; Inconsistent system   **45.** $(0, 0)$

47. Infinitely many solutions; $\{(x, y) \mid x + 5y = 3\}$; Dependent system   **49.** $x = 1$ **51.** $z = \dfrac{1}{2}$

53. $y = \dfrac{16}{41}$ **55.** $(3, 2, 1)$ **57.** Cramer's rule does not apply.   **59.** $x = -19$ **61.** $w = 1$ **63.** 36

65. a. 2   **b.** -2 **c.** $x = -1$

Chapter 9 Review Exercises, pp. 666–667

1. $(2, -1, 5)$ **2.** Inconsistent system   **3.** Dependent system   **4.** $(-1, -2, 2)$ **5.** $(1, -3, -2)$ **6.** $(-1, 4, 0)$

7. 5, 12, and 13 ft   **8.** The pumps can drain 250, 300, and 400 gal/hr.   **9.** She invested $6000 in Fund A, $4000 in Fund B, and $2000 in Fund C.   **10.** The angles are $13°, 39°,$ and $128°$.   **11.** $y = x^2 - 8x + 15$ **12.** $y = -x^2 - 2x + 3$

13. 3×3 **14.** 3×2 **15.** 1×4 **16.** 3×1

17. $\begin{bmatrix} 1 & 1 & | & 3 \\ 1 & -1 & | & -1 \end{bmatrix}$ **18.** $\begin{bmatrix} 1 & -1 & 1 & | & 4 \\ 2 & -1 & 3 & | & 8 \\ -2 & 2 & -1 & | & -9 \end{bmatrix}$

19. $\begin{aligned} x &= 9 \\ y &= -3 \end{aligned}$ **20.** $\begin{aligned} x &= -5 \\ y &= 2 \\ z &= -8 \end{aligned}$ **21. a.** 4

b. $\begin{bmatrix} 1 & 3 & | & 1 \\ 0 & -13 & | & 2 \end{bmatrix}$ **22. a.** $\begin{bmatrix} 1 & 2 & 0 & | & -3 \\ 0 & -9 & 1 & | & 12 \\ -3 & 2 & 2 & | & 5 \end{bmatrix}$

b. $\begin{bmatrix} 1 & 2 & 0 & | & -3 \\ 0 & -9 & 1 & | & 12 \\ 0 & 8 & 2 & | & -4 \end{bmatrix}$ **23.** $(1, 2)$ **24.** $(-3, 6)$

25. $(1, 3, -2)$ **26.** $(6, 1, -1)$ **27.** -11 **28.** -60

29. 1   **30.** 12   **31.** 18   **32.** 0   **33.** -30

34. 48   **35.** 7   **36.** 0   **37.** -40 **38.** 4

39. $(3, 2)$ **40.** $(1, 4)$ **41.** $\left(\dfrac{29}{11}, \dfrac{6}{11}\right)$ **42.** $\left(-\dfrac{14}{5}, -\dfrac{18}{5}\right)$

43. $(1, -2, 3)$ **44.** $(5, -2, 0)$ **45.** Infinitely many solutions; $\{(x, y) \mid 2x - y = 1\}$, Dependent system

46. No solution, Inconsistent system

Chapter 9 Test, pp. 667–668

1. $(16, -37, 9)$ **2.** $(2, -2, 5)$ **3.** No solution

4. The sides are 5 ft, 15 ft, and 23 ft   **5.** $5500 was invested in Fund A. $4000 was invested in Fund B, and $2000 was invested in Fund C.   **6.** Joanne 142 orders; Kent 162 orders; Geoff 200 orders   **7.** $y = -x^2 - 6x - 8$

8. For example: $\begin{bmatrix} 2 & 1 \\ 0 & -4 \\ 2.6 & 7 \end{bmatrix}$ **9. a.** $\begin{bmatrix} 1 & 2 & 1 & | & -3 \\ 0 & -8 & -3 & | & 10 \\ -5 & -6 & 3 & | & 0 \end{bmatrix}$

b. $\begin{bmatrix} 1 & 2 & 1 & | & -3 \\ 0 & -8 & -3 & | & 10 \\ 0 & 4 & 8 & | & -15 \end{bmatrix}$ **10.** $(6, -1)$ **11.** $(2, -4, 3)$

12. 7   **13.** 20   **14.** $y = -\dfrac{40}{11}$ **15.** $y = 8$

16. Infinitely many solutions; $\{(x, y) \mid y = 3x\}$, Dependent system   **17.** $(1, 0, 4)$ **18.** The system is dependent.

Chapters 1–9 Cumulative Review Exercises, pp. 668–669

1. No solution   **2.** $-\dfrac{19}{12}$ **3.** $\dfrac{a}{36b^2}$ **4.** $\left(-\dfrac{7}{5}, \infty\right)$

5. Slope $\dfrac{5}{2}$; x-intercept $(3, 0)$; y-intercept $\left(0, -\dfrac{15}{2}\right)$

6.

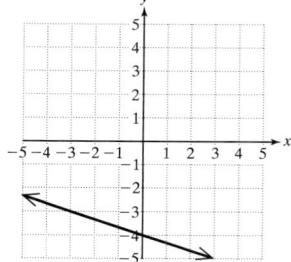

7. $y = -4x + 4$ **8.** Inconsistent system; No solution
9. $(a - 8)(a + 8)$ **10.** $(a - 4)(a^2 + 4a + 16)$
11. $4x^2 - x + 5 + \dfrac{-5}{x - 3}$ **12.** $30x^2y^4$ **13.** $\dfrac{x - 4}{x + 3}$
14. a. $g(0) = -6$ **b.** $g(1)$ is undefined. **c.** $g(-6) = 0$
15. $(2, 0, -1)$ **16.** 2×3
17. a. $f(-1) = 3$ **b.** $f(6) = 4$ **c.** $x = 0$ and $x = 4$
d. $(-\infty, \infty)$ **e.** $[0, \infty)$ **18.** $(1, 1)$
19. 14 **20.** $(1, 1)$

Chapter 10

Chapter Opener Puzzle

d	i	v	i	s	i	o	n
1	6	4	2	7	3	8	5

Section 10.1 Practice Exercises, pp. 678–681

3. $(-\infty, 2]$ **5.** $\left(-\infty, \dfrac{1}{18}\right)$ **7.** $(-0.4, 0.2)$
9. a. $[-1, 5)$ **b.** $(-2, \infty)$
11. a. $(-1, 3)$ **b.** $\left(-\dfrac{5}{2}, \dfrac{9}{2}\right)$
13. a. $(0, 2]$ **b.** $(-4, 5]$
15. $[-2, 3]$
17. $(3, 6)$
19. $(-\infty, 2]$
21. $[8, \infty)$
23. No solution
25. $-4 \le t$ and $t < \frac{3}{4}$
27. The statement $6 < x < 2$ is equivalent to $6 < x$ and $x < 2$. However, no real number is greater than 6 and also less than 2.
29. The statement $-5 > y > -2$ is equivalent to $-5 > y$ and $y > -2$. However, no real number is less than -5 and also greater than -2.
31. $\left[\dfrac{5}{2}, 7\right)$
33. $(-6, 6]$
35. $(2, 8)$
37. $\left[-\dfrac{10}{3}, -\dfrac{7}{3}\right]$
39. $\left[-\dfrac{1}{2}, \dfrac{3}{2}\right)$
41. $(-\infty, -4) \cup (-2, \infty)$
43. $(-\infty, -2) \cup [2, \infty)$
45. $(-\infty, 2]$
47. $[-6, \infty)$
49. $(-\infty, \infty)$

51. $\left(-\infty, -\dfrac{4}{3}\right) \cup (2, \infty)$
53. $(-\infty, -4] \cup (4.5, \infty)$
55. a. $(-10, 8)$ **b.** $(-\infty, \infty)$
57. a. $(-\infty, -8] \cup [1.3, \infty)$ **b.** No solution
59. $\left(-\dfrac{11}{2}, \dfrac{7}{2}\right]$ **61.** $\left(-\dfrac{13}{2}, \infty\right)$ **63.** $(-\infty, -17)$
65. $(-\infty, 1) \cup (5, \infty)$
67. a. $4800 \le x \le 10,800$ **b.** $x < 4800$ or $x > 10,800$
69. a. $2.0 \times 10^5 \le x \le 3.5 \times 10^5$
b. $x < 2.0 \times 10^5$ or $x > 3.5 \times 10^5$
71. All real numbers between $-\dfrac{3}{2}$ and 6
73. All real numbers greater than 2 or less than -1

Section 10.2 Practice Exercises, pp. 690–693

3. $(-\infty, -4) \cup \left(\dfrac{1}{3}, \infty\right)$ **5.** $\left[3, \dfrac{16}{5}\right]$ **7.** $[-1, 3]$
9. a. $(-2, 0) \cup (3, \infty)$ **b.** $(-\infty, -2) \cup (0, 3)$
c. $(-\infty, -2] \cup [0, 3]$ **d.** $[-2, 0] \cup [3, \infty)$
11. a. $(-1, 1]$ **b.** $(-\infty, -1) \cup [1, \infty)$
c. $(-\infty, -1) \cup (1, \infty)$ **d.** $(-1, 1)$
13. a. $x = 4, x = -\frac{1}{2}$ **b.** $(-\infty, -\frac{1}{2}) \cup (4, \infty)$ **c.** $(-\frac{1}{2}, 4)$
15. a. $x = -10, x = 3$ **b.** $(-10, 3)$ **c.** $(-\infty, -10) \cup (3, \infty)$
17. a. $p = -\dfrac{1}{2}, p = 3$ **b.** $\left[-\dfrac{1}{2}, 3\right]$ **c.** $\left(-\infty, -\dfrac{1}{2}\right] \cup [3, \infty)$
19. $(-1, 7)$ **21.** $(-\infty, -2) \cup (5, \infty)$ **23.** $[-5, 0]$
25. $[4, 8]$ **27.** $\left(-\infty, -\dfrac{1}{5}\right) \cup (1, \infty)$ **29.** $[-3, 2]$
31. $(-11, 11)$ **33.** $\left(-\infty, -\dfrac{1}{3}\right] \cup [3, \infty)$
35. $(-\infty, -3] \cup [0, 4]$ **37.** $(-2, -1) \cup (2, \infty)$
39. a. $x = 7$ **b.** $(-\infty, 5) \cup (7, \infty)$ **c.** $(5, 7)$
41. a. $z = 4$ **b.** $[4, 6)$ **c.** $(-\infty, 4] \cup (6, \infty)$
43. $(1, \infty)$ **45.** $(-\infty, -4) \cup (4, \infty)$
47. $\left(2, \dfrac{7}{2}\right)$ **49.** $(5, 7]$ **51.** $(-\infty, 0) \cup \left[\dfrac{1}{2}, \infty\right)$
53. $(0, \infty)$ **55.** $(-\infty, \infty)$ **57.** No solution
59. No solution   **61.** $\{0\}$ **63.** $(-\infty, -12) \cup (-12, \infty)$
65. $\{-12\}$ **67.** No solution   **69.** $(-\infty, \infty)$
71. Quadratic; $[-4, 4]$ **73.** Quadratic; $(-\infty, \infty)$
75. Linear; $(-\frac{5}{2}, \infty)$ **77.** Rational; $(-\infty, -1)$
79. Polynomial (degree > 2); $(0, 5) \cup (5, \infty)$
81. Polynomial (degree > 2); $(2, \infty)$
83. Rational; no solution   **85.** Quadratic; $(-2, 2)$
87. Quadratic; $(-\infty, -2] \cup [1, \infty)$
89. Rational; $(-\infty, -2] \cup (5, \infty)$
91. Linear; $(-\infty, 5]$
93. Quadratic; $(-\infty, \infty)$
95. $(-\infty, 0) \cup (2, \infty)$

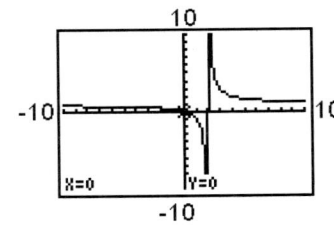

97. $(-1, 1)$

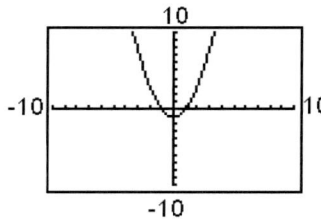

99. $\{-5\}$

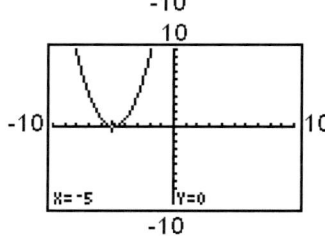

101. No solution

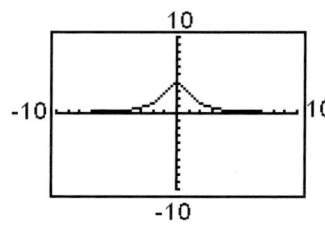

Section 10.3 Practice Exercises, pp. 698–700

3. $(-\infty, -2] \cup [2, \infty)$ **5.** $(-\infty, -1) \cup \left[\dfrac{1}{2}, \infty\right)$

7. $(-1, 0) \cup (8, \infty)$ **9.** $q = 10, q = -10$

11. $x = 23, x = -23$ **13.** $y = \dfrac{5}{8}, y = -\dfrac{5}{8}$

15. No solution   **17.** $p = 0$ **19.** $x = \dfrac{5}{4}, x = -\dfrac{7}{4}$

21. $w = 15, w = -21$ **23.** $t = 8, t = -7$

25. No solution   **27.** $x = 0, x = 6$

29. $x = -\dfrac{6}{5}, x = -\dfrac{8}{5}$ **31.** $x = -\dfrac{1}{5}$

33. No solution   **35.** $k = \dfrac{3}{2}$ **37.** $h = -3, h = \dfrac{3}{2}$

39. $x = -\dfrac{1}{2}, x = \dfrac{3}{2}$ **41.** $x = 0, x = -\dfrac{10}{3}$

43. $y = -8, y = \dfrac{6}{5}$ **45.** $a = -\dfrac{2}{9}$ **47.** $p = -10, p = \dfrac{6}{5}$

49. $n = \dfrac{11}{12}$ **51.** $n = -2.775, n = -0.375$

53. No solution   **55.** $|x| = \dfrac{7}{2}$ **57.** $|x| = 9$

59. $x = \sqrt{3}, x = 0$ **61.** $w = 11 + 3\sqrt{2}, w = 5 - 3\sqrt{2}$

63. $x = 4\sqrt{5}, x = -\dfrac{4\sqrt{5}}{5}$

65. $x = 7, x = 1$

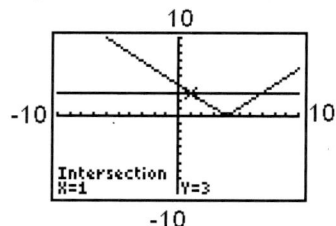

67. No solution

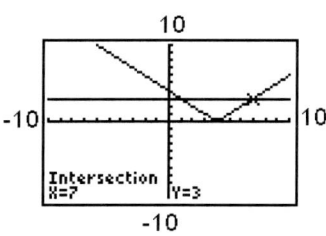

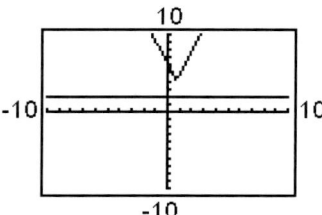

69. $x = -1$

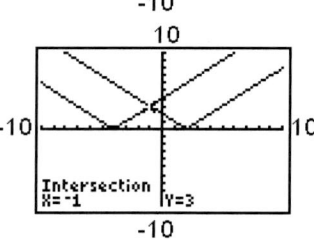

71. $x = -5, x = 1$

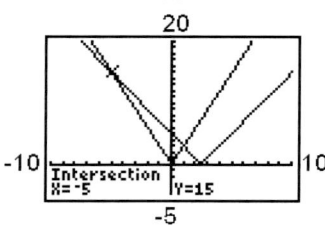

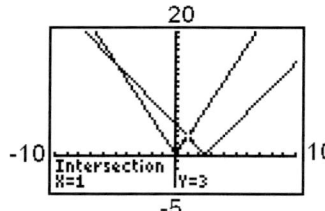

Section 10.4 Practice Exercises, pp. 709–712

3. $x = \dfrac{4}{7}, x = \dfrac{6}{7}$ **5.** $[4, \infty)$

7. $(-\infty, 3) \cup (6, \infty)$

9. a. $a = 4, a = -4$

b. $(-\infty, -4) \cup (4, \infty)$

c. $(-4, 4)$

11. a. $w = 4, w = -8$

b. $(-\infty, -8) \cup (4, \infty)$

c. $(-8, 4)$

13. a. No solution

b. $(-\infty, \infty)$

c. No solution

15. a. No solution

b. $(-\infty, \infty)$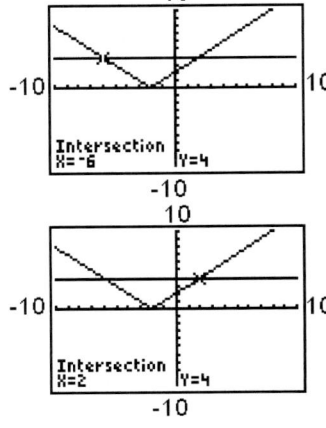

c. No solution

17. a. $p = -3$

b. $(-\infty, -3) \cup (-3, \infty)$

c. No solution

19. $(-\infty, -6) \cup (6, \infty)$

21. $[-3, 3]$

23. $(-\infty, \infty)$

25. $(-\infty, -2] \cup [3, \infty)$

27. No solution

29. $[-10, 14]$

31. $\left(-\infty, -\dfrac{5}{4}\right] \cup \left[\dfrac{23}{4}, \infty\right)$

33. $\left(-\dfrac{21}{2}, \dfrac{19}{2}\right)$

35. $(-\infty, \infty)$

37. No solution

39. $\{-5\}$

41. $(-\infty, 6) \cup (6, \infty)$

43. $(-\infty, 2] \cup [4, \infty)$

45. $(4, 8)$

47. $(-10.5, 4.5)$

49. $|x| > 7$ 　　**51.** $|x - 2| \le 13$ 　　**53.** $|x - 32| \le 0.05$

55. $\left|x - 6\frac{3}{4}\right| \le \dfrac{1}{8}$

57. $1.99 \le w \le 2.01$ or, equivalently in interval notation, $[1.99, 2.01]$. This means that the actual width of the bolt could be between 1.99 cm and 2.01 cm inclusive.

59. b 　　**61.** a

63. $(-\infty, -6) \cup (2, \infty)$

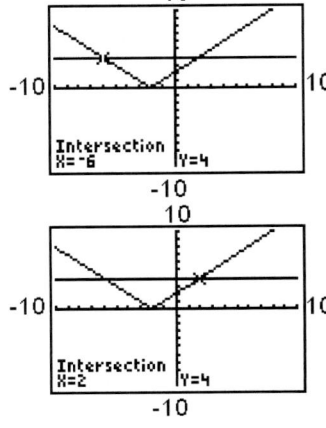

65. $(-7, 5)$

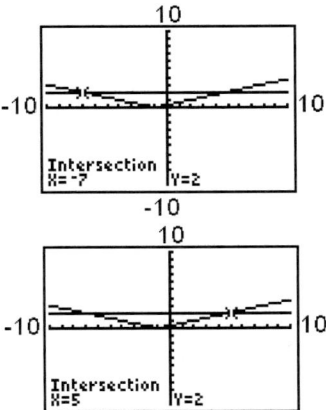

67. No solution

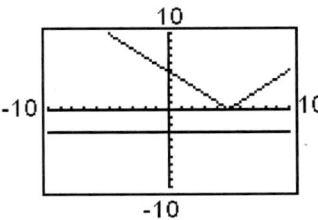

69. $(-\infty, \infty)$

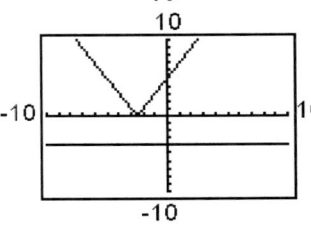

71. $x = -\dfrac{1}{6}$

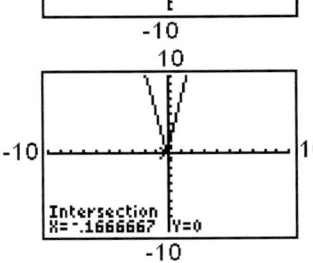

Chapter 10 Problem Recognition Exercises— Equations and Inequalities, pp. 712–713

1. Quadratic; $(-\infty, -9) \cup (-1, \infty)$ 　　**2.** Linear; $a = -26$

3. Rational; $[-8, -2)$ 　　**4.** Quadratic; $x = -\frac{1}{4}, x = 2$

5. Absolute value; $\left(-\frac{1}{3}, 1\right)$ 　　**6.** Quadratic; no solution

7. Linear; $(-\infty, -10)$ 　　**8.** Quadratic; $[-4, 1]$

9. Quadratic; $\left(-\infty, -\dfrac{1}{3}\right) \cup [2, \infty)$

10. Rational; $x = -\frac{11}{3}$ 　　**11.** Linear; $\left[\dfrac{16}{7}, \infty\right)$

12. Absolute value; no solution

13. Polynomial; $[-5, -\frac{1}{2}] \cup [3, \infty)$

14. Quadratic; $(-\infty, \infty)$ 　　**15.** Rational; $(-\infty, -1) \cup (2, \infty)$

16. Quadratic; $x = 4, x = 1$

17. Absolute value; $(-\infty, -8) \cup (-8, \infty)$

18. Quadratic; $(-6, 6)$

19. Polynomial; $y = 2, y = -2, y = -\frac{5}{3}$

20. Absolute value; $x = 12, x = -4$

21. Linear; $b = 0$ 　　**22.** Rational; $(-\infty, 0] \cup (5, \infty)$

23. Absolute value; $x = -\frac{1}{2}$ 　　**24.** Polynomial; $\left(-\frac{3}{2}, 0\right)$

Section 10.5 Practice Exercises, pp. 720–726

3. $[1, 3)$ **5.** $(-\infty, -4] \cup (-2, \infty)$

7. a. No **b.** No **c.** Yes **d.** Yes

9. a. No **b.** Yes **c.** Yes **d.** No

11. $\leq$ **13.** $<$ **15.** $\geq, \geq$

17.

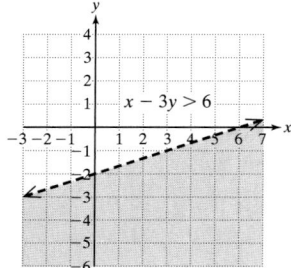

19.

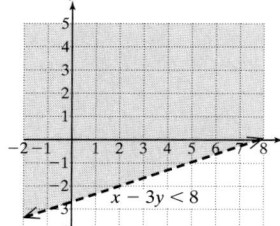

21.

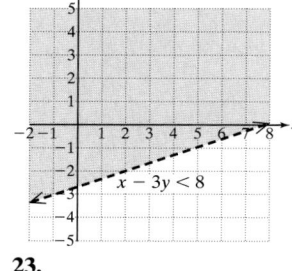

23.

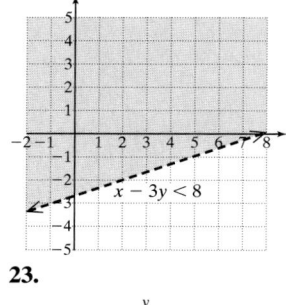

25.

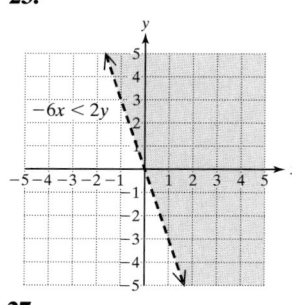

27.

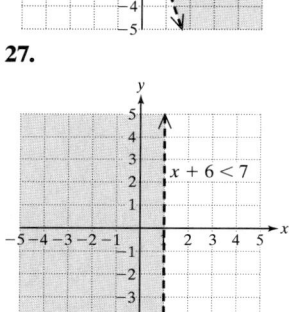

29.

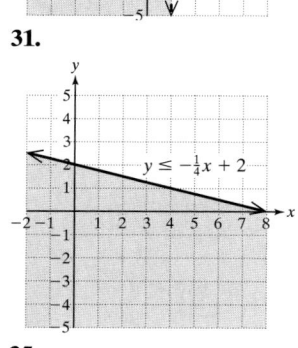

31.

33.

35.

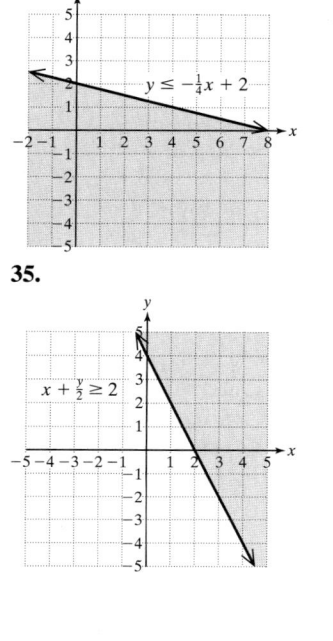

37.

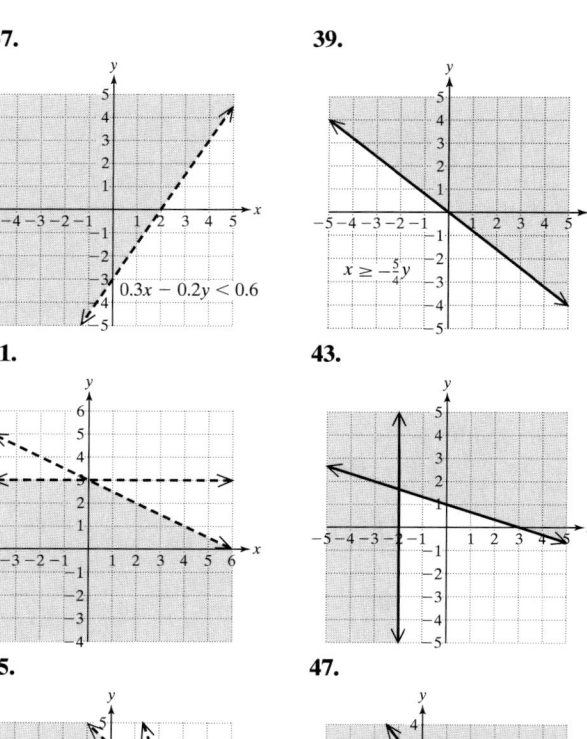

39.

41.

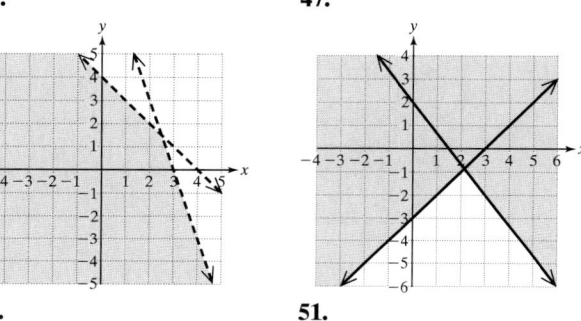

43.

45.

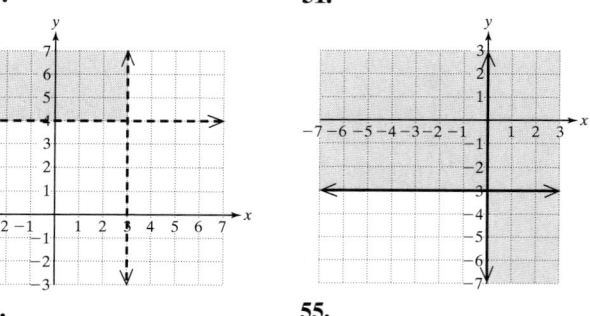

47.

49.

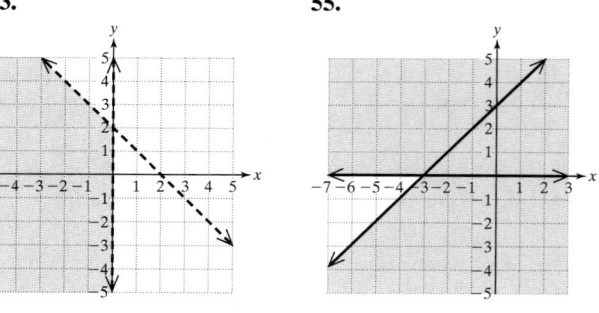

51.

53.

55.

57.

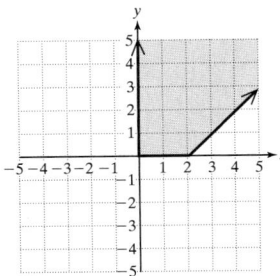

59.

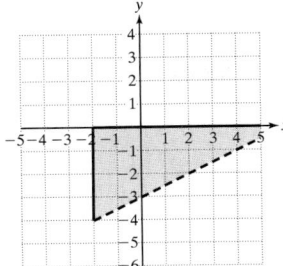

61.

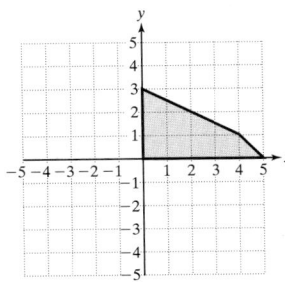

63. a. $x \geq 0, y \geq 0$ **b.** $x \leq 40, y \leq 40$ **c.** $x + y \geq 65$
d.

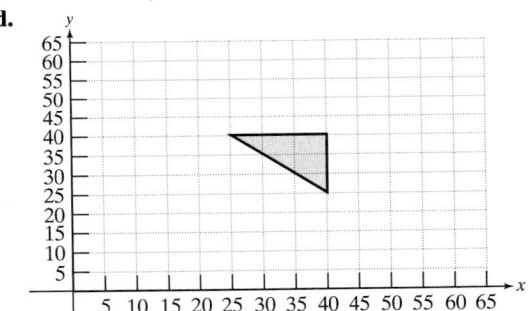

e. Yes; The point $(35, 40)$ means that Karen works 35 hours and Todd works 40 hours.   **f.** No; The point $(20, 40)$ means that Karen works 20 hours and Todd works 40 hours. This does not satisfy the constraint that there must be at least 65 hours total.

Chapter 10, Review Exercises, pp. 731–735

1. $\left(-\dfrac{11}{4}, 4\right]$ **2.** $[-5, 2)$ **3.** No solution

4. No solution   **5.** $(-\infty, 6] \cup (12, \infty)$

6. $(-\infty, -2) \cup [2, \infty)$ **7.** $\left(-\infty, \dfrac{1}{2}\right)$ **8.** $(-5, \infty)$

9. $\left[0, \dfrac{4}{3}\right]$ **10.** $[-7, -2)$

11. All real numbers between -6 and 12
12. a. $140 \leq x \leq 225$ **b.** $x < 140$ or $x > 225$
13. a. $125 \leq x \leq 200$ **b.** $x < 125$ or $x > 200$
14. a. Yes   **b.** No   **c.** Yes
15. a. The solution is the intersection of the two inequalities. Answer: $\{x \,|\, -2 \leq x \leq 5\}$.   **b.** The solution is the union of the two inequalities. Answer: the set of all real numbers.
16. a. $(-2, 0)(2, 0)$ are the x-intercepts.   **b.** On the interval $(-2, 2)$ the graph is below the x-axis.   **c.** On the intervals $(-\infty, -2)$ and $(2, \infty)$ the graph is above the x-axis.
17. a. $x = 0$; $(0, 0)$ is the x-intercept.   **b.** $x = 2$ is the

vertical asymptote.   **c.** On the intervals $(-\infty, 0]$ and $(2, \infty)$ the graph is on or above the x-axis.   **d.** On the interval $[0, 2)$ the graph is on or below the x-axis.
18. $(-2, 6)$ **19.** $(-\infty, \infty)$ **20.** $(-\infty, -2) \cup [0, \infty)$
21. $(-\infty, -1] \cup (1, \infty)$ **22.** $(-2, 0) \cup (5, \infty)$
23. $(-2, 0) \cup \left(\dfrac{5}{2}, \infty\right)$ **24.** $[-3, -1]$
25. $(-\infty, -5) \cup (1, \infty)$ **26.** $(3, \infty)$ **27.** $(-3, \infty)$
28. $\{-5\}$ **29.** $(-\infty, -2) \cup (-2, \infty)$
30. $x = 10, x = -10$ **31.** $x = 17, x = -17$
32. $x = 1.3, x = 7.4$ **33.** $x = -0.44, x = 2.54$
34. $x = 5, x = -9$ **35.** $x = 3, x = 1$
36. No solution   **37.** No solution
38. $x = \dfrac{3}{7}$ **39.** $x = -\dfrac{5}{4}$
40. $x = 6, x = \dfrac{4}{5}$ **41.** $x = -\dfrac{1}{2}$
42. Both expressions give the distance between 3 and -2.
43. $|x| > 5$ **44.** $|x| < 4$
45. $|x| < 6$ **46.** $|x| > \dfrac{2}{3}$
47. $(-\infty, -14] \cup [2, \infty)$ ⟵———□   □———⟶
                                              -14 2
48. $[-11, -5]$ ———□——————□———⟶
                        -11 -5
49. $\left(-\infty, \dfrac{1}{7}\right) \cup \left(\dfrac{1}{7}, \infty\right)$ ⟵——————✕——————
                                                    $\frac{1}{7}$
50. $\left(-\infty, -\dfrac{1}{5}\right) \cup \left(-\dfrac{1}{5}, \infty\right)$ ⟵———————✕———————
                                                      $-\frac{1}{5}$
51. $\left[-2, -\dfrac{2}{3}\right]$ ———□———□———⟶
                             -2 $-\frac{2}{3}$
52. $\left[0, \dfrac{6}{5}\right]$ ———□———□———⟶
                       0 $\frac{6}{5}$
53. $(2, 22)$ ———(———)———⟶
                     2 22
54. $(-12, 0)$ ———(———)———⟶
                      -12 0
55. $(-\infty, \infty)$ ⟵————————⟶
56. $(-\infty, \infty)$ ⟵————————⟶
57. $(-\infty, \infty)$ ⟵————————⟶
58. $(-\infty, \infty)$ ⟵————————⟶
59. No solution
60. No solution
61. If an absolute value is less than a negative number, there will be no solution.
62. If an absolute value is greater than a negative number, then all real numbers are solutions.
63. $0.17 \leq p \leq 0.23$ or, equivalently in interval notation, $[0.17, 0.23]$. This means that the actual percentage of viewers is estimated to be between 17% and 23% inclusive.
64. $3\frac{1}{8} \leq L \leq 3\frac{5}{8}$ or, equivalently in interval notation, $[3\frac{1}{8}, 3\frac{5}{8}]$. This means that the actual length of the screw may be between $3\frac{1}{8}$ in. and $3\frac{5}{8}$ in., inclusive.

65.

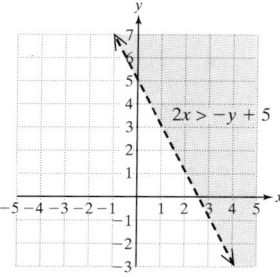

$2x > -y + 5$

66.

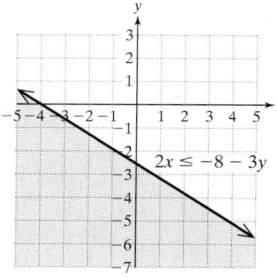

$2x \leq -8 - 3y$

67.

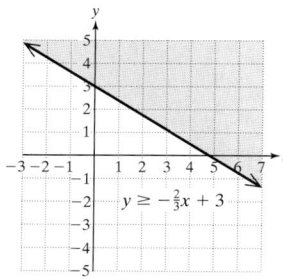

$y \geq -\frac{2}{3}x + 3$

68.

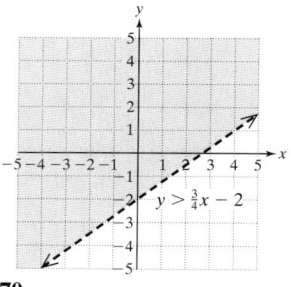

$y > \frac{3}{4}x - 2$

69.

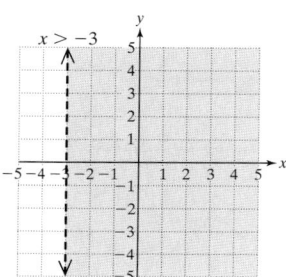

$x > -3$

70.

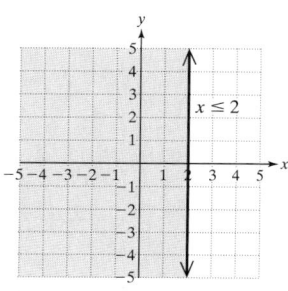

$x \leq 2$

71.

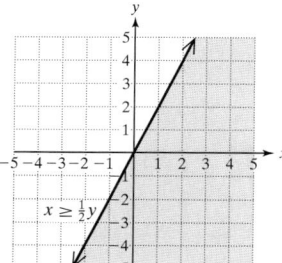

$x \geq \frac{1}{2}y$

72.

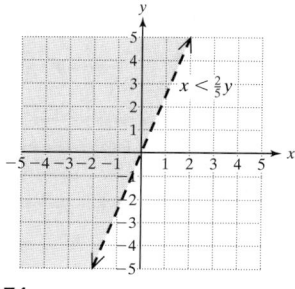

$x < \frac{2}{5}y$

73.

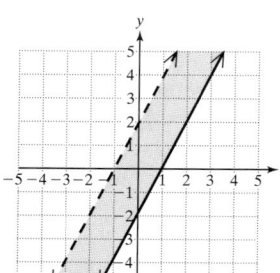

74.

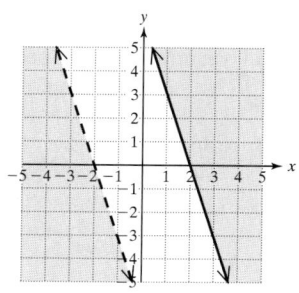

75.

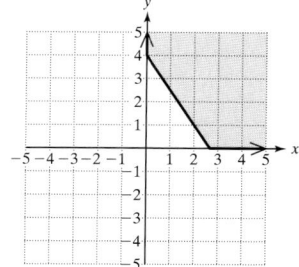

76.

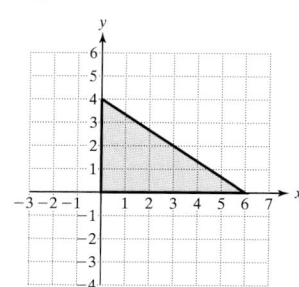

77.

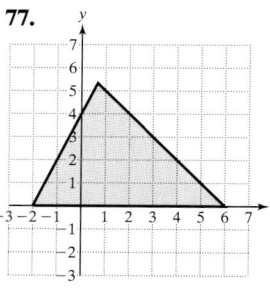

78. a. $x \geq 0, y \geq 0$ **b.** $x + y \leq 100$ **c.** $x \geq 4y$

d.

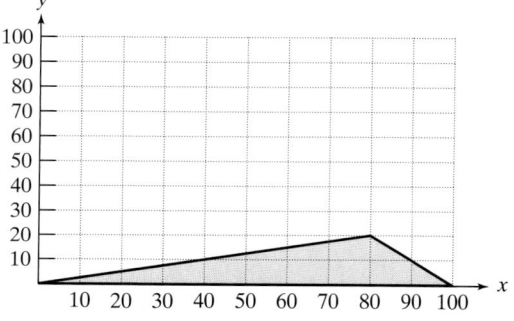

Chapter 10, Test, pp. 735–736

1. $\left[-\dfrac{1}{3}, 2\right]$ **2.** $(-\infty, -24] \cup [-15, \infty)$

3. $[-3, 0)$ **4.** $(-\infty, \infty)$ **5.** No solution

6. a. $9 \leq x \leq 33$ **b.** $x < 9$ or $x > 33$

7. $\left[\dfrac{1}{2}, 6\right)$ **8.** $(-5, 5)$ **9.** $(-\infty, -3) \cup (-2, 2)$

10. $\left(-3, -\dfrac{3}{2}\right)$ **11.** No solution   **12.** $\{-11\}$

13. $x = 10, x = -22$ **14.** $x = -8, x = 2$

15. a. $(7, 0)(-1, 0)$ are the x-intercepts.   **b.** On the interval $(-1, 7)$ the graph is below the x-axis.   **c.** On the intervals $(-\infty, -1)$ and $(7, \infty)$ the graph is above the x-axis.

16. No solution   **17.** $\left(-\infty, -\dfrac{1}{3}\right) \cup \left(\dfrac{17}{3}, \infty\right)$

18. $(-18.75, 17.25)$ **19.** $(-\infty, \infty)$

20. $|x - 15.41| \leq 0.01$

21.

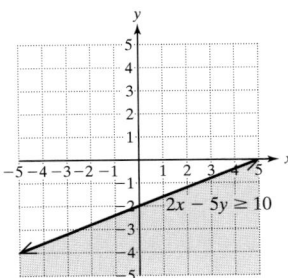

$2x - 5y \geq 10$

22.

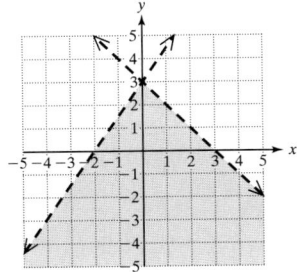

23.

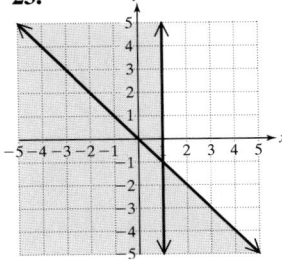

24. a. $x \geq 0, y \geq 0$ **b.** $300x + 400y \geq 1200$

c.

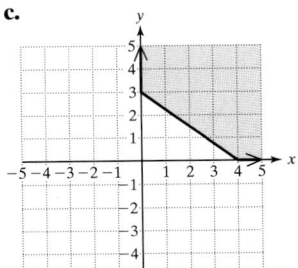

Chapters 1–10, Cumulative Review Exercises, pp. 736–738

1. $x^2 - x - 13$ **2.** $m = -\dfrac{3}{2}, m = 1$

3. a. $p = 0, p = 6$ **b.** $(0, 6)$ **c.** $(-\infty, 0) \cup (6, \infty)$

4. a. $y = 14, y = -10$ **b.** $(-10, 14)$

c. $(-\infty, -10) \cup (14, \infty)$

5.

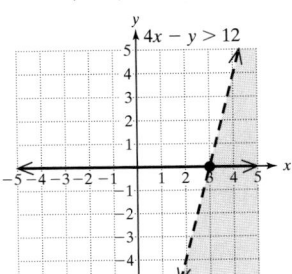

$4x - y > 12$

6. a. $t(1) = 9$. After one trial the rat requires 9 min to complete the maze. $t(50) = 3.24$. After 50 trials the rat requires 3.24 min to complete the maze. $t(500) = 3.02$. After 500 trials the rat requires 3.02 min to complete the maze.
b. Yes. The limiting time is 3 min.   **c.** More than 5 trials

7. a. $\left(-\infty, -\dfrac{5}{2}\right] \cup [2, \infty)$

b. On these intervals, the graph is on or above the x-axis

8.

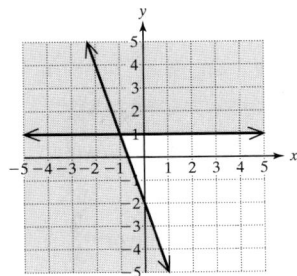

9. $-7x + 13$

10. Approximately $\$1.5 \times 10^6$ per restaurant

11. a. Quotient: $2x^2 - 5x + 12$; remainder: $-24x + 5$

12. a. $b_1 = \dfrac{2A - hb_2}{h}$ or $b_1 = \dfrac{2A}{h} - b_2$ **b.** 10 cm

13. 75 mph   **14.** $57°, 123°$

15. \$8000 in the 6.5% account; \$5000 in the 5% account

16. Neither   **17.** $\left[\dfrac{13}{3}, 5\right)$

18. x-intercept $(1, 0)$; no y-intercept; undefined slope

19. x-intercept $(8, 0)$; y-intercept $(0, 4)$; slope $-\dfrac{1}{2}$

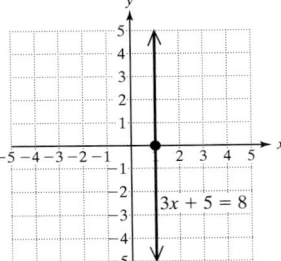

$3x + 5 = 8$

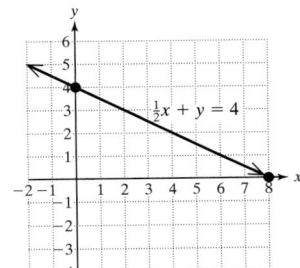

$\frac{1}{2}x + y = 4$

20. $y = -\dfrac{2}{3}x - \dfrac{13}{3}$ **21.** $(-1, 3, -2)$

22. a. 4×3 **b.** 3×3

23. Plane: 500 mph; wind: 20 mph

24. a. Quadratic   **b.** $P(0) = -2600$. The company will lose \$2600 if no desks are produced.   **c.** $P(x) = 0$ for $x = 20$ and $x = 650$. The company will break even (\$0 profit) when 20 desks or 650 desks are produced.

25. $\{x \mid x \leq 50\}$ or $(-\infty, 50]$ **26.** $-\dfrac{xy}{x + y}$

27. $-\dfrac{a + 1}{a}$ **28.** $\dfrac{2x + 5}{(x - 5)(x - 2)(x + 10)}$

29. $(6, 0)$ **30.** $3(x - 2)^2(x + 2)$

Chapter 11

Chapter Opener Puzzle

2	1	^a4	5	6	^b3
5	6	3	^c2	^d1	4
^e3	^f5	1	4	^g2	^h6
4	ⁱ2	6	3	^j5	^k1
1	3	^m5	6	4	2
ⁿ6	4	2	1	3	5

Section 11.1 Practice Exercises, pp. 748–752

3. a. $8, -8$ **b.** 8 **c.** There are two square roots for every positive number. $\sqrt{64}$ identifies the positive square root.
5. a. 9 **b.** -9
7. There is no real number b such that $b^2 = -36$.
9. 7   **11.** -7 **13.** Not a real number
15. $\dfrac{8}{3}$ **17.** 0.9   **19.** -0.4 **21. a.** 8 **b.** 4
c. -8 **d.** -4 **e.** Not a real number **f.** -4
23. -3 **25.** $\dfrac{1}{2}$ **27.** 2   **29.** $-\dfrac{5}{4}$
31. Not a real number   **33.** 10   **35.** -0.2
37. $|a|$ **39.** a **41.** $|a|$ **43.** x^2 **45.** $|x|y^2$
47. $-\dfrac{x}{y}$ **49.** $\dfrac{2}{|x|}$ **51.** -92 **53.** 2   **55.** xy^2
57. $\dfrac{a^3}{b}$ **59.** $-\dfrac{5}{q}$ **61.** $3xy^2z$ **63.** $\dfrac{hk^2}{4}$ **65.** $-\dfrac{t}{3}$
67. $2y^2$ **69.** $2p^2q^3$ **71.** 9 cm   **73.** 13 ft   **75.** 5 mi
77. 25 mi   **79. a.** Not a real number **b.** Not a real
number **c.** 0 **d.** 1 **e.** 2 Domain: $[2, \infty)$
81. a. -2 **b.** -1 **c.** 0 **d.** 1 Domain: $(-\infty, \infty)$
83. $[-5, \infty)$ **85.** $(-\infty, \infty)$ **87.** b   **89.** d
91. $q + p^2$ **93.** $\dfrac{6}{\sqrt[3]{x}}$ **95.** 8 in.   **97.** 8.3066
99. 3.7100   **101.** 15.6525   **103.** -0.1235
105.

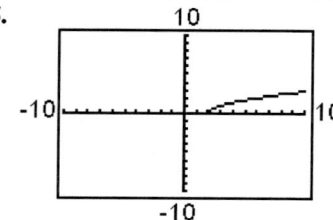

107.

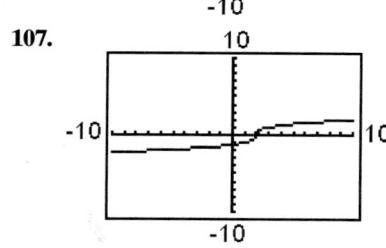

Section 11.2 Practice Exercises, pp. 757–759

3. a. 3 **b.** 27   **5.** 5   **7.** 3   **9.** 12   **11.** -12
13. Not a real number   **15.** -4 **17.** $\dfrac{1}{5}$
19. $a^{m/n} = \sqrt[n]{a^m}$; The numerator of the exponent represents the power of the base. The denominator of the exponent represents the index of the radical.
21. a. 8 **b.** -8 **c.** Not a real number   **d.** $\dfrac{1}{8}$ **e.** $-\dfrac{1}{8}$
f. Not a real number
23. a. 125 **b.** -125 **c.** Not a real number   **d.** $\dfrac{1}{125}$
e. $-\dfrac{1}{125}$ **f.** Not a real number   **25.** $\dfrac{1}{512}$
27. 27   **29.** $-\dfrac{1}{81}$ **31.** $\dfrac{27}{1000}$
33. Not a real number   **35.** -2 **37.** -2 **39.** $\dfrac{1}{9}$
41. 6   **43.** 10   **45.** $\dfrac{3}{4}$ **47.** 1   **49.** $\dfrac{9}{2}$
51. $\sqrt[3]{q^2}$ **53.** $6\sqrt[4]{y^3}$ **55.** $\sqrt[3]{x^2y}$ **57.** $\dfrac{1}{\sqrt[5]{qr}}$
59. $x^{1/3}$ **61.** $10b^{1/2}$ **63.** $y^{2/3}$ **65.** $(a^2b^3)^{1/4}$
67. $\dfrac{1}{x}$ **69.** p **71.** y^2 **73.** $6^{2/5}$ **75.** $\dfrac{4}{t^{5/3}}$
77. a^7 **79.** $\dfrac{25a^4d}{c}$ **81.** $\dfrac{y^9}{x^8}$ **83.** $\dfrac{2z^3}{w}$ **85.** $5xy^2z^3$
87. $x^{13}z^{4/3}$ **89.** $\dfrac{x^3y^2}{z^5}$ **91. a.** 10 in.   **b.** 8.5 in.
93. a. 10.9%   **b.** 8.8%   **c.** The account in part (a).
95. $\sqrt[6]{x}$ **97.** $\sqrt[8]{y}$ **99.** $\sqrt[15]{w}$ **101.** 3
103. 0.3761   **105.** 2.9240   **107.** 31.6228

Section 11.3 Practice Exercises, pp. 764–766

1. a^2b **3.** $\dfrac{1}{x^2y^5}$ **5.** $\sqrt[7]{x^4}$ **7.** $y^{9/2}$ **9.** $x^5\sqrt{x}$
11. $q^2\sqrt[3]{q}$ **13.** $a^2b^2\sqrt{a}$ **15.** $-x^2y^3\sqrt[4]{y}$ **17.** $2\sqrt{7}$
19. $4\sqrt{5}$ **21.** $3\sqrt[3]{2}$ **23.** $5b\sqrt{ab}$ **25.** $3a^3b\sqrt{2b}$
27. $-2x^2z\sqrt[3]{2y}$ **29.** $2wz\sqrt[4]{5z^3}$ **31.** x **33.** p^2
35. 5   **37.** $\dfrac{1}{2}$ **39.** $\dfrac{5\sqrt[3]{2}}{3}$ **41.** $\dfrac{5\sqrt[3]{9}}{6}$ **43.** $15\sqrt{2}$
45. $-30\sqrt{3}$ **47.** $5x^2y\sqrt{y}$ **49.** $3yz\sqrt[3]{x^2z}$ **51.** $\dfrac{2}{b}$
53. $\dfrac{2\sqrt[5]{x}}{y^2}$ **55.** $\dfrac{5x\sqrt{2xy}}{3y^2}$ **57.** $2a^7b^4c^{15}d^{11}\sqrt{2c}$
59. $\dfrac{1}{\sqrt[3]{w^6}}$ simplifies to $\dfrac{1}{w^2}$ **61.** $\sqrt{k^3}$ simplifies to $k\sqrt{k}$
63. $2\sqrt{41}$ ft   **65.** $6\sqrt{5}$ m   **67.** $90\sqrt{2}$ ft ≈ 127.3 ft
69. The path from A to B and B to C is faster.

Section 11.4 Practice Exercises, pp. 770–772

3. $-xy\sqrt[4]{x^3}$ **5.** $\dfrac{b^2}{2}$
7. $(3^5x^{15}y^{10})^{1/5}$ simplifies to $3x^3y^2$ **9.** $y^{11/12}$
11. a. Not *like* radicals **b.** *Like* radicals
c. Not *like* radicals
13. a. Both expressions can be simplified by using the distributive property. **b.** Neither expression can be simplified because the terms do not contain *like* terms or *like* radicals.

15. $9\sqrt{5}$ **17.** $\sqrt[3]{t}$ **19.** $5\sqrt{10}$
21. $8\sqrt[4]{3} - \sqrt[4]{14}$ **23.** $2\sqrt{x} + 2\sqrt{y}$
25. Cannot be simplified further

27. Cannot be simplified further   **29.** $\dfrac{29}{18}z\sqrt[3]{6}$

31. $0.70x\sqrt{y}$ **33.** Simplify each radical: $3\sqrt{2} + 35\sqrt{2}$.
Then add *like* radicals: $38\sqrt{2}$ **35.** 15 **37.** $8\sqrt{3}$
39. $3\sqrt{7}$ **41.** $-\sqrt{2}$ **43.** $\sqrt[3]{3}$ **45.** $-5\sqrt{2a}$
47. $8s^2t^2\sqrt[3]{s^2}$ **49.** $6x\sqrt[3]{x}$ **51.** $14p^2\sqrt{5}$ **53.** $-\sqrt[3]{a^2b}$
55. $33d\sqrt[3]{2c}$ **57.** $2a^2b\sqrt{6a}$ **59.** $5x\sqrt[3]{2} - 6\sqrt[3]{x}$
61. False, $\sqrt{9} + \sqrt{16} \neq \sqrt{9 + 16}, 7 \neq 5$ **63.** True
65. False, $\sqrt{y} + \sqrt{y} = 2\sqrt{y} \neq \sqrt{2y}$

67. $\sqrt{48} + \sqrt{12}$ simplifies to $6\sqrt{3}$

69. $5\sqrt[3]{x^6} - x^2$ simplifies to $4x^2$
71. The difference of the principal square root of 18 and the square of 5
73. The sum of the principal fourth root of x and the cube of y
75. $9\sqrt{6}$ cm ≈ 22.0 cm   **77.** $x = 2\sqrt{2}$ ft
79. a. $10\sqrt{5}$ yd   **b.** 22.36 yd   **c.** $\$105.95$

Section 11.5 Practice Exercises, pp. 778–780

1. $3pq^3\sqrt{6p}$ **3.** $-2ab\sqrt{5bc}$ **5.** $\dfrac{q^2}{p^{1/8}}$ **7.** $2\sqrt[3]{7}$ **9.** $\sqrt[3]{21}$

11. $2\sqrt{5}$ **13.** $4\sqrt[4]{4}$ **15.** $8\sqrt[3]{20}$ **17.** $-24ab\sqrt{a}$
19. $6\sqrt{10}$ **21.** $6x\sqrt{2}$ **23.** $10a^3b^2\sqrt{b}$
25. $-24x^2y^2\sqrt{2y}$ **27.** $6ab\sqrt[3]{2a^2b^2}$ **29.** $12 - 6\sqrt{3}$
31. $2\sqrt{3} - \sqrt{6}$ **33.** $-3x - 21\sqrt{x}$ **35.** $-8 + 7\sqrt{30}$
37. $x - 5\sqrt{x} - 36$ **39.** $\sqrt[3]{y^2} - \sqrt[3]{y} - 6$
41. $9a - 28\sqrt{ab} + 3b$
43. $8\sqrt{p} + 3p + 5\sqrt{pq} + 16\sqrt{q} - 2q$ **45.** 15
47. $3y$ **49.** 6 **51.** 709 **53. a.** $x^2 - y^2$
b. $x^2 - 25$ **55.** $3 - x^2$ **57.** 4 **59.** $64x - 4y$
61. $29 + 8\sqrt{13}$ **63.** $p - 2\sqrt{7p} + 7$
65. $2a - 6\sqrt{2ab} + 9b$ **67.** True
69. False; $(x - \sqrt{5})^2 = x^2 - 2x\sqrt{5} + 5$
71. False; 5 is multiplied by 3 only.   **73.** True   **75.** 39
77. $6x$ **79.** $3x + 1$ **81.** $x + 19 - 8\sqrt{x + 3}$
83. $12\sqrt{5}$ ft^2 **85.** $18\sqrt{15}$ in.^{2} **87.** $\sqrt[4]{x^3}$
89. $\sqrt[15]{(2z)^8}$ **91.** $p^2\sqrt[6]{p}$ **93.** $u\sqrt[6]{u}$ **95.** $\sqrt[6]{(a + b)}$
97. $\sqrt[6]{x^2y}$ **99.** $\sqrt[4]{2^3 \cdot 3^2}$ or $\sqrt[4]{72}$ **101.** $a + b$

Section 11.6 Practice Exercises, pp. 786–788

3. $12y\sqrt{5}$ **5.** $-18y + 3\sqrt{y} + 3$ **7.** $64 - 16\sqrt{t} + t$

9. -5 **11.** $\dfrac{\sqrt{5}}{\sqrt{5}}$ **13.** $\dfrac{\sqrt[3]{x^2}}{\sqrt[3]{x^2}}$ **15.** $\dfrac{\sqrt{3z}}{\sqrt{3z}}$

17. $\dfrac{\sqrt[4]{2^3a^2}}{\sqrt[4]{2^3a^2}}$ **19.** $\dfrac{\sqrt{3}}{3}$ **21.** $\dfrac{\sqrt{x}}{x}$ **23.** $\dfrac{3\sqrt{2y}}{y}$

25. $-2\sqrt{a}$ **27.** $\dfrac{3\sqrt{2}}{2}$ **29.** $\dfrac{3\sqrt[3]{4}}{2}$ **31.** $\dfrac{-6\sqrt[4]{x^3}}{x}$

33. $\dfrac{7\sqrt[3]{2}}{2}$ **35.** $\dfrac{\sqrt[3]{4w}}{w}$ **37.** $\dfrac{2\sqrt[4]{27}}{3}$ **39.** $\dfrac{\sqrt[3]{2x}}{x}$

41. $\dfrac{2x\sqrt[3]{2y^2}}{y}$ **43.** $\dfrac{xy^2\sqrt{10xy}}{10}$ **45.** $\dfrac{2\sqrt{6}}{21}$ **47.** $\dfrac{\sqrt{x}}{x^4}$

49. $\dfrac{\sqrt{2x}}{2x^3}$ **51.** $\sqrt{2} + \sqrt{6}$ **53.** $\sqrt{x} - 23$

55. $\dfrac{4\sqrt{2} - 12}{-7}$ or $\dfrac{-4\sqrt{2} + 12}{7}$ **57.** $4\sqrt{6} + 8$

59. $-\sqrt{21} + 2\sqrt{7}$ **61.** $\dfrac{-\sqrt{p} + \sqrt{q}}{p - q}$ **63.** $\sqrt{x} - \sqrt{5}$

65. $\dfrac{-14\sqrt{a} - 35\sqrt{b}}{4a - 25b}$ **67.** $9 - 2\sqrt{15}$ **69.** $5 - \sqrt{10}$

71. $\dfrac{5 + \sqrt{21}}{4}$ **73.** $-6\sqrt{5} - 13$

75. $\dfrac{16}{\sqrt[3]{4}}$ simplifies to $8\sqrt[3]{2}$

77. $\dfrac{4}{x - \sqrt{2}}$ simplifies to $\dfrac{4x + 4\sqrt{2}}{x^2 - 2}$

79. $\dfrac{\pi\sqrt{2}}{4}$ sec ≈ 1.11 sec   **81.** $\dfrac{2\sqrt{6}}{3}$ **83.** $\dfrac{17\sqrt{15}}{15}$

85. $\dfrac{8\sqrt[3]{25}}{5}$ **87.** $\dfrac{-33}{2\sqrt{3} - 12}$ **89.** $\dfrac{a - b}{a + 2\sqrt{ab} + b}$

Chapter 11, Problem Recognition Exercises—Operations on Radicals, pp. 788–789

1. $3\sqrt{2}$ **2.** $2\sqrt{7}$ **3.** Cannot be simplified further
4. Cannot be simplified further   **5.** $\sqrt{2}$ **6.** $\sqrt{7}$
7. $9 - z$ **8.** $16 - y$ **9.** $8 - 3\sqrt{5}$
10. $-8 + 11\sqrt{3}$ **11.** $-x\sqrt{y}$ **12.** $11ab\sqrt{a}$
13. $-24 - 6\sqrt{6} - 3\sqrt{2}$ **14.** $-80 + 8\sqrt{15} + 16\sqrt{5}$

15. $\dfrac{2\sqrt{x} + 14}{x - 49}$ **16.** $\dfrac{5\sqrt{y} - 20}{y - 16}$ **17.** $3\sqrt{3}$

18. $3\sqrt{5}$ **19.** $\dfrac{\sqrt{7x}}{x}$ **20.** $\dfrac{\sqrt{11y}}{y}$ **21.** $y^2z^5\sqrt{z}$

22. $2q^3\sqrt{2}$ **23.** $3p^2\sqrt[3]{p^2}$ **24.** $5u^3v^4\sqrt[3]{u^2}$ **25.** $x\sqrt{10}$

26. $y\sqrt{3}$ **27.** $20\sqrt{3}$ **28.** $\sqrt{10}$ **29.** $\dfrac{1}{3}$ **30.** $\dfrac{5}{3}$

31. $\sqrt{x} - \sqrt{5}$ **32.** $\sqrt{y} - \sqrt{7}$ **33.** $4x - 11\sqrt{xy} - 3y$
34. $51 + 14\sqrt{2}$ **35.** $8 + 2\sqrt{15}$ **36.** $16 - 2\sqrt{55}$
37. $x - 12\sqrt{x} + 36$ **38.** -88 **39.** $u - 9v$
40. $-3\sqrt{6}$ **41.** $11\sqrt{a}$ **42.** $4x\sqrt{2}$ **43.** 0
44. $5 + \sqrt{35}$ **45.** $a + 2\sqrt{a}$ **46.** $26 - 17\sqrt{2}$

Section 11.7 Practice Exercises, pp. 796–800

3. $\dfrac{3w\sqrt{w}}{4}$ **5.** $3c\sqrt[3]{2c}$ **7.** $4x - 6$

9. $9p + 7$ **11.** $x = 9$ **13.** $y = 3$ **15.** $z = 42$
17. $x = 29$ **19.** $w = 140$ **21.** No solution
23. $a = 7$ ($a = -1$ does not check.)

25. No solution   **27.** $V = \dfrac{4\pi r^3}{3}$

29. $h^2 = \dfrac{r^2 - \pi^2 r^2}{\pi^2}$ or $h^2 = \dfrac{r^2}{\pi^2} - r^2$

31. $a^2 + 10a + 25$ **33.** $5a - 6\sqrt{5a} + 9$
35. $r + 22 + 10\sqrt{r - 3}$ **37.** $a = -3$

39. No solution ($w = \dfrac{1}{2}$ does not check.)

41. $y = 0; y = 3$ **43.** $h = 9$ **45.** $a = -4$

47. $a = \dfrac{9}{5}$ **49.** $h = 2$

51. No solution ($t = 9$ does not check.)

53. $x = -\dfrac{11}{4}$ **55.** No solution ($t = \dfrac{8}{3}$ does not check.)

57. $z = -1$ ($z = 3$ does not check.)

59. $m = \dfrac{1}{3}, m = -1$

61. No solution ($t = 3$ and $t = 23$ do not check.)

63. $y = \dfrac{49}{16}$

65. a. 30.25 ft   **b.** 34.5 m   **67. a.** 305 m   **b.** 460 m
69. a. 12 lb   **b.** $t(18) = 5.1$. An 18-lb turkey will take about 5.1 hr to cook.
71. $b = \sqrt{25 - h^2}$ **73.** $a = \sqrt{k^2 - 196}$
75.

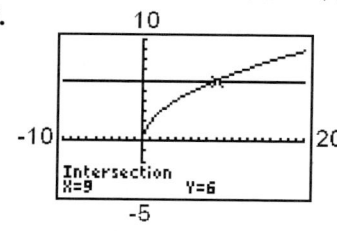

77.

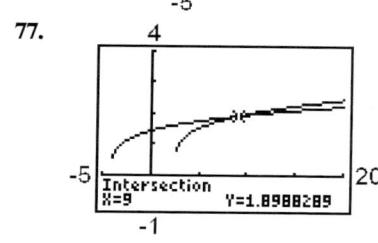

Section 11.8 Practice Exercises, pp. 808–810

3. $3\sqrt{5} - 15\sqrt{2}$ **5.** $9 - x$ **7.** $p = -8$
9. $c = \dfrac{49}{36}$ **11.** $\sqrt{-1} = i$ and $-\sqrt{1} = -1$
13. $12i$ **15.** $i\sqrt{3}$ **17.** $2i\sqrt{5}$ **19.** -60
21. $29i\sqrt{2}$ **23.** $13i\sqrt{7}$ **25.** -7 **27.** -12
29. $-3\sqrt{10}$ **31.** $i\sqrt{2}$ **33.** $3i$ **35.** $-i$
37. 1 **39.** i **41.** 1 **43.** $-i$ **45.** -1
47. $a - bi$ **49.** Real: -5; imaginary: 12
51. Real: 0; imaginary: -6 **53.** Real: 35; imaginary: 0
55. Real: $\dfrac{3}{5}$; imaginary: 1   **57.** $7 + 6i$
59. $\dfrac{3}{10} + \dfrac{3}{2}i$ **61.** $5 + 0i$ **63.** $-1 + 10i$
65. -24 **67.** $18 + 6i$ **69.** $26 - 26i$
71. $-29 + 0i$ **73.** $-9 + 40i$ **75.** $35 + 20i$
77. $-20 + 48i$ **79.** $\dfrac{13}{16}$ **81.** $\dfrac{1}{5} - \dfrac{3}{5}i$
83. $\dfrac{3}{25} - \dfrac{4}{25}i$ **85.** $\dfrac{21}{29} + \dfrac{20}{29}i$ **87.** $-\dfrac{17}{10} - \dfrac{1}{10}i$
89. $-\dfrac{1}{2} - \dfrac{5}{2}i$ **91.** $\dfrac{1}{2} - \dfrac{1}{3}i$ **93.** $1 + 10i$
95. $\dfrac{1 + 2i}{4}$ **97.** $-1 + i\sqrt{2}$ **99.** $-2 - i\sqrt{3}$

Chapter 11, Review Exercises, pp. 816–819

1. a. False; $\sqrt{0} = 0$ is not positive.   **b.** False; $\sqrt[3]{-8} = -2$
2. $\sqrt{(-3)^2} = \sqrt{9} = 3$ **3. a.** False   **b.** True
4. $\dfrac{5}{4}$ **5.** 5   **6.** 6
7. a. 3   **b.** 0   **c.** $\sqrt{7}$ **d.** $[1, \infty)$
8. a. 0   **b.** 1   **c.** 3   **d.** $[-5, \infty)$ **9.** $\dfrac{\sqrt[3]{2x}}{\sqrt[4]{2x}} + 4$
10. a. $|x|$ **b.** x **c.** $|x|$ **d.** $x + 1$
11. a. $2|y|$ **b.** $3y$ **c.** $|y|$ **d.** y **12.** 8 cm
13. Yes, provided the expressions are well defined. For example: $x^5 \cdot x^3 = x^8$ and $x^{1/5} \cdot x^{2/5} = x^{3/5}$
14. n represents the root.
15. Take the reciprocal of the base and change the exponent to positive.

16. -5 **17.** $\dfrac{1}{2}$ **18.** 4   **19.** b^{10} **20.** $\dfrac{16y^{12}}{xz^9}$
21. $x^{3/4}$ **22.** $(2y^2)^{1/3}$ **23.** 2.1544   **24.** 6.8173
25. 54.1819
26. a. \$5278.78   **b.** \$5573.11   **c.** \$43,804.27
d. \$129,655.12   **e.** \$660,236.46
27. See page 760.
28. $6\sqrt{3}$ **29.** $xz\sqrt[4]{xy}$ **30.** $10x$ **31.** $\dfrac{-2x^2y^2\sqrt[3]{2x}}{z^3}$
32. a. The principal square root of the quotient of 2 and x.
b. The cube of the sum of x and 1.   **33.** 31 ft
34. they are *like* radicals.
35. Cannot be combined; the indices are different.
36. Cannot be combined; one term has a radical, but the other does not.   **37.** Can be combined: $3\sqrt[4]{3xy}$
38. Can be added after simplifying: $19\sqrt{2}$
39. $5\sqrt{7}$ **40.** $9\sqrt[3]{2} - 8$ **41.** $10\sqrt{2}$ **42.** $13x\sqrt[3]{2x^2}$
43. False; 5 and $3\sqrt{x}$ are not *like* radicals.
44. False; $\sqrt{y} + \sqrt{y} = 2\sqrt{y}$ (add the coefficients).
45. 6   **46.** $2\sqrt[4]{2}$ **47.** 12   **48.** $-6\sqrt{15} + 15$
49. $4x - 9$ **50.** $y - 16$ **51.** $7y - 2\sqrt{21xy} + 3x$
52. $12w + 20\sqrt{3w} + 25$ **53.** $-2z - 9\sqrt{6z} - 42$
54. $3a + 5\sqrt{5a} - 10$ **55.** $u^2\sqrt[6]{u^5}$ **56.** $\sqrt[4]{4w^3}$
57. $\dfrac{\sqrt{14y}}{2y}$ **58.** $\dfrac{\sqrt{15w}}{3w}$ **59.** $\dfrac{4\sqrt[3]{3p}}{3p}$ **60.** $\dfrac{-\sqrt[3]{4x^2}}{x}$
61. $-\sqrt{15} - \sqrt{10}$ **62.** $-3\sqrt{7} - 3\sqrt{5}$ **63.** $\sqrt{t} + \sqrt{3}$
64. $\sqrt{w} + \sqrt{7}$ **65.** The quotient of the principal square root of 2 and the square of x.
66. $y = \dfrac{49}{2}$ **67.** $a = 31$ **68.** $w = -12$ **69.** $p = 7$
70. $t = 9$ **71.** No solution ($x = -2$ does not check.)
72. $m = \dfrac{1}{2}, m = 4$ **73.** $x = -2$ ($x = 2$ does not check.)
74. $6\sqrt{5}$ m ≈ 13.4 m
75. a. $v(20) \approx 25.3$ ft/sec. When the water depth is 20 ft, a wave travels about 25.3 ft/sec.   **b.** 8 ft
76. $a + bi$, where a and b are real numbers and $i = \sqrt{-1}$
77. $a + bi$, where $b \neq 0$
78. In each case we simplify the expression by multiplying the numerator and denominator by the conjugate of the denominator.
79. $4i$ **80.** $-i\sqrt{5}$ **81.** -15 **82.** $-2i$ **83.** -1
84. i **85.** $-i$ **86.** 0   **87.** $-5 + 5i$ **88.** $9 + 17i$
89. $25 + 0i$ **90.** $24 - 10i$
91. $-\dfrac{17}{4} + i$; Real part: $-\dfrac{17}{4}$; Imaginary part: 1
92. $-2 - i$; Real part: -2; Imaginary part: -1
93. $\dfrac{4}{13} - \dfrac{7}{13}i$ **94.** $3 + 4i$ **95.** $\dfrac{-4 + i\sqrt{10}}{6}$
96. $2 - 4i$

Chapter 11, Test, pp. 819–820

1. a. 6   **b.** -6 **2. a.** Real   **b.** Not real
c. Real   **d.** Real   **3. a.** y **b.** $|y|$ **4.** 3
5. $\dfrac{4}{3}$ **6.** $2\sqrt[3]{4}$ **7.** $a^2bc^2\sqrt{bc}$ **8.** $3x^2\sqrt{2}$
9. $\dfrac{4w^2\sqrt{6w}}{3}$ **10.** $\sqrt[6]{7y^3}$ **11.** $\sqrt[12]{10}$
12. a. $f(-8) = 2\sqrt{3}$; $f(-6) = 2\sqrt{2}$; $f(-4) = 2$; $f(-2) = 0$
b. $(-\infty, -2]$

13. -0.3080 **14.** -3 **15.** $\dfrac{1}{t^{3/4}}$

16. $3\sqrt{5}$ **17. a.** $3\sqrt{2x} - 3\sqrt{5x}$ **b.** $40 - 10\sqrt{5x} - 3x$

18. a. $\dfrac{-2\sqrt[3]{x^2}}{x}$ **b.** $\dfrac{x + 6 + 5\sqrt{x}}{9 - x}$

19. a. $2i\sqrt{2}$ **b.** $8i$ **c.** $\dfrac{1 + i\sqrt{2}}{2}$

20. $1 - 11i$ **21.** $30 + 16i$ **22.** -28 **23.** $-33 - 56i$

24. 104 **25.** $\dfrac{17}{25} + \dfrac{6}{25}i$ **26.** $36 - 11i$

27. $r(10) = 1.34$; the radius of a sphere of volume 10 cubic units is 1.34 units.   **28.** 21 ft   **29.** $x = -16$

30. $x = \dfrac{17}{5}$ **31.** $t = 2$; ($t = 42$ does not check)

Chapters 1–11, Cumulative Review Exercises, pp. 820–821

1. 54   **2.** $15x - 5y - 5$ **3.** $y = -3$ **4.** $(-\infty, -5)$
5. $y = -2x + 5$

6.

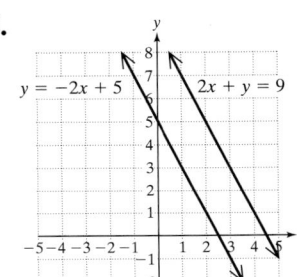

7. $\left(\dfrac{1}{2}, \dfrac{1}{3}\right)$

8. $\left(2, -2, \dfrac{1}{2}\right)$ is not a solution.   **9.** $x = 6, y = 3, z = 8$

10. a. $f(-2) = -10; f(0) = -2; f(4) = 14; f\left(\dfrac{1}{2}\right) = 0$

b. $(-2, -10), (0, -2), (4, 14), \left(\dfrac{1}{2}, 0\right)$

c.

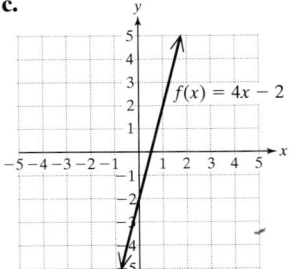

11. Not a function   **12.** $a^4b^8c^2$ **13.** $a^6b^{12}c^4$
14. a. 1.4×10^{-4} **b.** 3.14×10^9
15. $2x^2 - x - 15$; second degree   **16.** $\sqrt{15} + 3\sqrt{2} + 3$
17. $x - 4$ **18.** $-\dfrac{1}{6}$ **19.** $\dfrac{3c\sqrt[3]{2}}{d}$ **20.** $32b\sqrt{5b}$

21. $2 + 3i$ **22.** $y = 10$ **23.** $\dfrac{x - 15}{x(x + 5)(x - 5)}$

24. $\dfrac{-1}{(2a + 1)(a + 2)}$ **25.** $-14x^2 + 26x - 17$

26. $2\sqrt{3} + 2\sqrt{5}$ **27.** $\dfrac{6}{17} + \dfrac{10}{17}i$ **28.** $x = \dfrac{7}{6}, x = -\dfrac{3}{2}$

29. $(x + 3 - y)(x + 3 + y)$ **30.** $(x^2 + 2)(x^4 - 2x^2 + 4)$

Chapter 12

Chapter Opener Puzzle

1. square   **2.** completing   **3.** quadratic
4. discriminant   **5.** parabola   **6.** vertex   **7.** axis

Section 12.1 Practice Exercises, pp. 829–832

3. $y = \pm 2$ **5.** $k = \pm\sqrt{7}$ **7.** $m = \pm 5i$

9. $q = -1, q = -5$ **11.** $y = \dfrac{-3 \pm \sqrt{7}}{2}$

13. $t = -5 \pm 3i\sqrt{2}$ **15.** $w = \pm\dfrac{\sqrt{33}}{3}$

17. $m = -\dfrac{4}{5} \pm \dfrac{\sqrt{3}}{5}$

19. 1. Factoring and applying the zero product rule.
2. Applying the square root property. $x = \pm 6$
21. $n = 9; (x - 3)^2$ **23.** $n = 16; (t + 4)^2$

25. $n = \dfrac{1}{4}; \left(c - \dfrac{1}{2}\right)^2$ **27.** $n = \dfrac{25}{4}; \left(y + \dfrac{5}{2}\right)^2$

29. $n = \dfrac{1}{25}; \left(b + \dfrac{1}{5}\right)^2$ **31.** See page 827.

33. $t = -3, t = -5$ **35.** $x = -3 \pm i\sqrt{7}$
37. $p = -2 \pm i\sqrt{2}$

39. $y = 5, y = -2$ **41.** $a = -1 \pm \dfrac{i\sqrt{6}}{2}$

43. $x = 2 \pm \dfrac{2}{3}i$ **45.** $p = \dfrac{1}{5} \pm \dfrac{\sqrt{3}}{5}$

47. $w = -\dfrac{3}{4} \pm \dfrac{\sqrt{65}}{4}$ **49.** $n = 2 \pm \sqrt{11}$

51. $x = 1, x = -7$ **53. a.** $t = \dfrac{\sqrt{d}}{4}$ **b.** 8 sec

55. $r = \sqrt{\dfrac{A}{\pi}}$ or $r = \dfrac{\sqrt{A\pi}}{\pi}$ **57.** $a = \sqrt{d^2 - b^2 - c^2}$

59. $r = \sqrt{\dfrac{3V}{\pi h}}$ or $r = \dfrac{\sqrt{3V\pi h}}{\pi h}$ **61.** 4.2 ft

63. a. 8%   **b.** 11%   **c.** 14.02%
65. a. 4.5 thousand textbooks or 35.5 thousand textbooks
b. Profit increases to a point as more books are produced. Beyond that point, the market is "flooded," and profit decreases. Hence there are two points at which the profit is $20,000. Producing 4.5 thousand books makes the same profit using fewer resources as producing 35.5 thousand books.

Section 12.2 Practice Exercises, pp. 842–846

3. $x = 2, x = -12$ **5.** $x = 1 \pm 3i$ **7.** $4 - 2\sqrt{5}$
9. $2 - i\sqrt{3}$ **11.** The form $ax^2 + bx + c = 0$

13. $x = \dfrac{-b \pm \sqrt{b^2 - 4ac}}{2a}$ **15.** $b = -\dfrac{1}{5}, b = 3$

17. $t = \dfrac{-3 \pm \sqrt{65}}{4}$ **19.** $n = \dfrac{2 \pm i\sqrt{26}}{5}$

21. $x = \dfrac{2}{3}, x = -\dfrac{1}{4}$ **23.** $k = 3 \pm \sqrt{5}$ **25.** $y = -\dfrac{2}{3}$

27. $m = \dfrac{-3 \pm i\sqrt{7}}{4}$ **29.** $y = -\dfrac{3}{4}, y = -2$

31. $m = -3, m = \dfrac{1}{2}$ **33.** $p = \dfrac{1 \pm i\sqrt{47}}{8}$

35. $w = \dfrac{-7 \pm \sqrt{33}}{2}$ **37.** $y = \dfrac{2}{5}, y = 1$

39. $x = -2 \pm \sqrt{11}$ **41. a.** $(4x + 1)(16x^2 - 4x + 1)$

b. $x = -\dfrac{1}{4}, x = \dfrac{1 \pm i\sqrt{3}}{8}$ **43. a.** $5x(x^2 + x + 2)$

b. $x = 0, x = \dfrac{-1 \pm i\sqrt{7}}{2}$ **45.** $1\dfrac{1}{3}$ ft by 4 ft by 12 ft

47. 2.3 cm and 5.6 cm   **49.** 10.2 and 13.6 ft

51. $t = \dfrac{3 + \sqrt{5}}{2} \approx 2.62$ sec or $t = \dfrac{3 - \sqrt{5}}{2} \approx 0.38$ sec

53. a. 46 mph   **b.** 36 mph
55. a. $4y^2 - 12y + 9 = 0$ **b.** 0   **c.** 1 rational solution
57. a. $5n^2 - 2n + 0 = 0$ **b.** 4   **c.** 2 rational solutions
59. a. $3k^2 + 0k - 7 = 0$ **b.** 84   **c.** 2 irrational solutions
61. a. $2x^2 - 5x + 6 = 0$ **b.** -23 **c.** 2 imaginary solutions
63. x-intercepts: none; y-intercept: $(0, -1)$

65. x-intercepts: $\left(\dfrac{1 + \sqrt{7}}{3}, 0\right)$ and $\left(\dfrac{1 - \sqrt{7}}{3}, 0\right)$;

y-intercept: $(0, 2)$

67. $z = 0, z = -\dfrac{7}{4}$ **69.** $b = \pm i\sqrt{7}$

71. $k = \dfrac{1 \pm i\sqrt{31}}{2}$ **73.** $y = 0, y = -3$

75. $w = \dfrac{5 \pm \sqrt{41}}{2}$ **77.** $b = \dfrac{-1 \pm i\sqrt{11}}{2}$

79. $x = 6, x = -5$ **81.** $k = \dfrac{3}{2}, k = \dfrac{1}{4}$

83. $h = 5, h = -5$
85. a. $x = 5 \pm i\sqrt{2}$ **b.** $x = 5 \pm i\sqrt{2}$ **c.** Answers will vary
87.

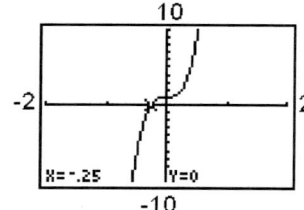

89.

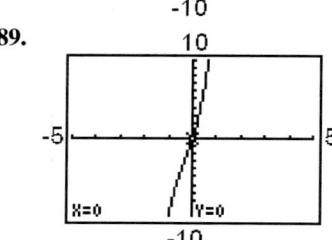

91. a. 12,769 thousand   **b.** 15,507 thousand   **c.** 1989
d.

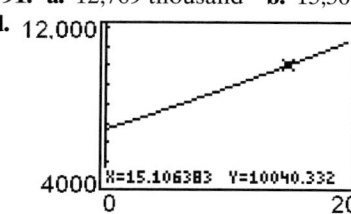

Section 12.3 Practice Exercises, pp. 850–852

3. $x = \dfrac{3}{2} \pm \dfrac{\sqrt{7}}{2}$ **5.** $x = -4$ **7.** $x = 2 \pm \sqrt{26}$

9. a. $u = 7, u = -5$ **b.** $w = 7, w = -1, w = 5, w = 1$

11. a. $u = 3, u = 1$ **b.** $p = -\dfrac{3}{2}, p = 1, p = \dfrac{1}{2}, p = -1$

13. $x = -2, x = 1, x = -3, x = 2$

15. $n = \dfrac{27}{8}, n = -125$ **17.** $p = -32, p = 1$

19. $p = 25, p = 9$ **21.** $t = \dfrac{1}{9}$ ($t = 4$ does not check.)

23. $x = -\dfrac{7}{3}$ **25.** $t = \pm\sqrt{3}, t = \pm 2i$

27. $x = \pm\dfrac{\sqrt{2}}{3}, x = \pm i$ **29.** $y = 4, y = 6$

31. $x = \dfrac{9 \pm \sqrt{73}}{4}$ **33.** $x = 2 \pm \sqrt{3}$

35. $x = \pm 2, x = \pm 2i$ **37.** $x = -\dfrac{7}{4}, x = -\dfrac{3}{2}$

39. $m = \dfrac{1}{2}, m = -\dfrac{1}{2}, m = \sqrt{2}, m = -\sqrt{2}$

41. $x = 2, x = 1, x = -1 \pm i\sqrt{3}, x = \dfrac{-1 \pm i\sqrt{3}}{2}$

43. $x = 5, x = 4$ **45.** $t = 1, t = \dfrac{17}{2}$

47. $x = 64, x = -125$ **49.** $m = \pm\sqrt{2}, m = \pm 2i$
51. $a = \pm 4i, a = 1$ **53.** $x = 4, x = \pm i\sqrt{5}$
55. a. $x = \pm i\sqrt{2}$ **b.** Two imaginary solutions; zero real solutions
c. No x-intercepts   **d.**

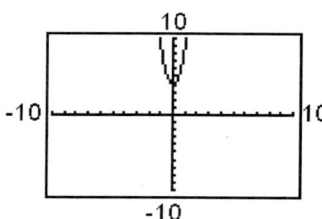

57 a. $x = 0, x = 3, x = -2$ **b.** Three real solutions; zero imaginary solutions   **c.** Three x-intercepts
d.

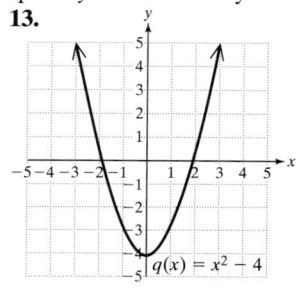

Section 12.4 Practice Exercises, pp. 861–867

3. $y = 3 \pm 2i$ **5.** $t = \dfrac{5 \pm \sqrt{10}}{5}$

7. $x = -27; x = -8$
9. The value of k shifts the graph of $y = x^2$ vertically.
11.

$f(x) = x^2 + 2$

13.

$q(x) = x^2 - 4$

15.

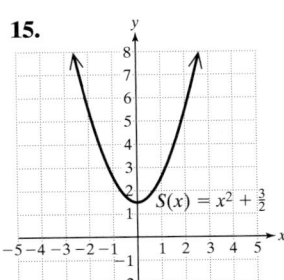

$S(x) = x^2 + \frac{3}{2}$

17.

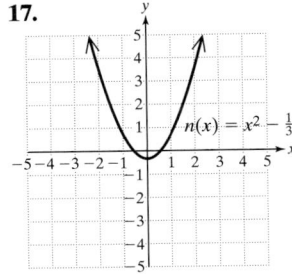

$n(x) = x^2 - \frac{1}{3}$

19.

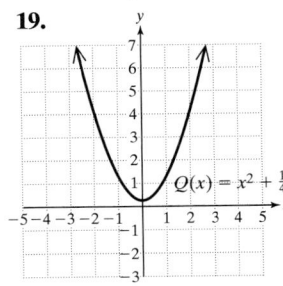

$Q(x) = x^2 + \frac{1}{4}$

21.

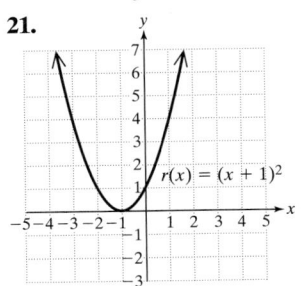

$r(x) = (x + 1)^2$

23.

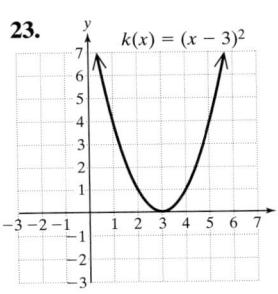

$k(x) = (x - 3)^2$

25.

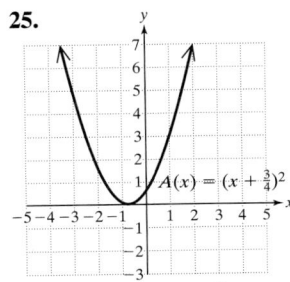

$A(x) = (x + \frac{3}{4})^2$

27.

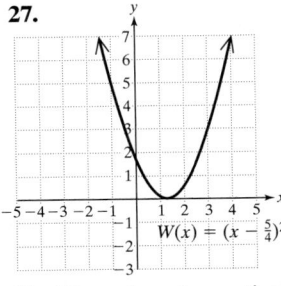

$W(x) = (x - \frac{5}{4})^2$

29.

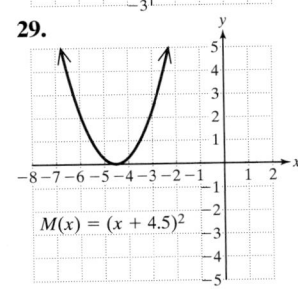

$M(x) = (x + 4.5)^2$

31. The value of *a* vertically stretches or shrinks the graph of $y = x^2$.

33.

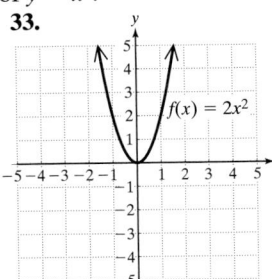

$f(x) = 2x^2$

35.

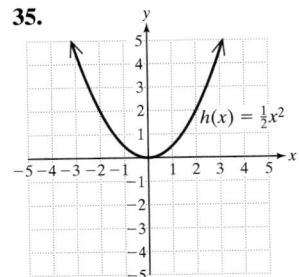

$h(x) = \frac{1}{2}x^2$

37.

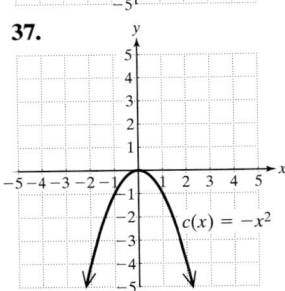

$c(x) = -x^2$

39.

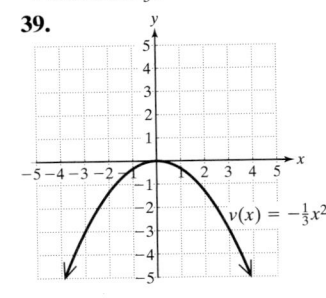

$v(x) = -\frac{1}{3}x^2$

41. d   **43.** g   **45.** a   **47.** e

49.

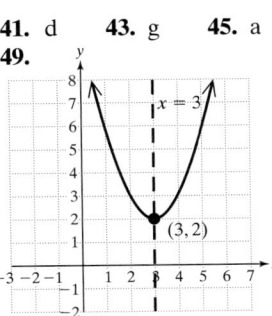

$x = 3$ (3, 2)

51.

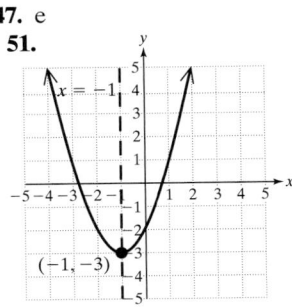

$x = -1$ (−1, −3)

53.

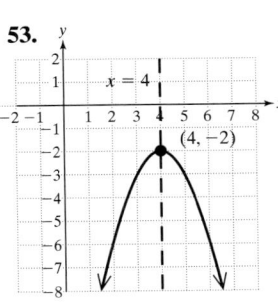

$x = 4$ (4, −2)

55.

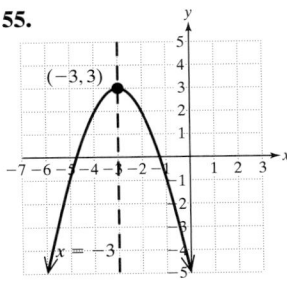

(−3, 3) $x = -3$

57.

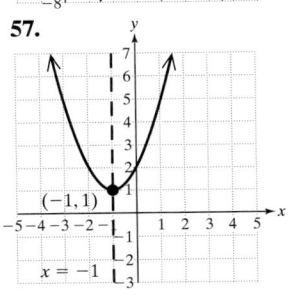

(−1, 1) $x = -1$

59.

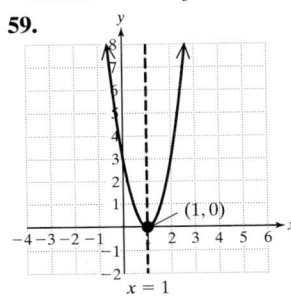

(1, 0) $x = 1$

61.

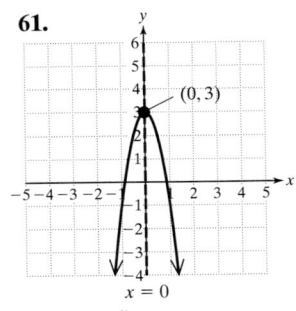

(0, 3) $x = 0$

63.

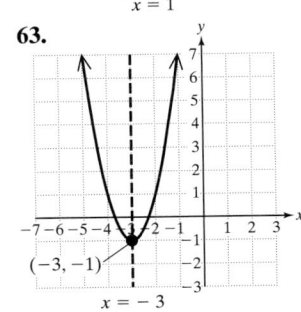

(−3, −1) $x = -3$

65.

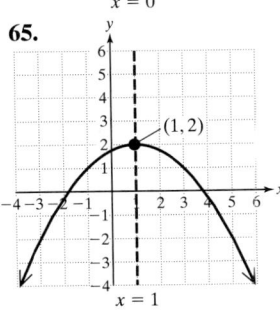

(1, 2) $x = 1$

67. Vertex $(6, -9)$ minimum point; minimum value: -9

69. Vertex $(2, 5)$; maximum point; maximum value: 5

71. Vertex $(-8, 0)$; minimum point; minimum value: 0

73. Vertex $\left(0, \frac{21}{4}\right)$; maximum point; maximum value: $\frac{21}{4}$

75. Vertex $\left(7, -\frac{3}{2}\right)$; minimum point; minimum value: $-\frac{3}{2}$

77. Vertex $(0, 0)$; minimum point; minimum value: 0
79. True   **81.** False
83. **a.** $(60, 30)$ **b.** 30 ft   **c.** 70 ft
85.

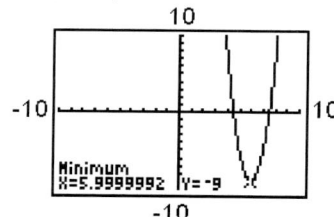

87.

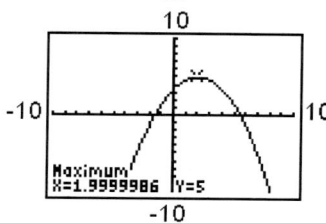

Section 12.5 Practice Exercises, pp. 874–879

3. The graph of p is the graph of $y = x^2$ shrunk vertically by a factor of $\frac{1}{4}$.

5. The graph of r is the graph of $y = x^2$ shifted up 7 units.
7. The graph of t is the graph of $y = x^2$ shifted to the left 10 units.

9. 16   **11.** $\frac{49}{4}$ **13.** $\frac{1}{81}$ **15.** $\frac{1}{36}$

17. $g(x) = (x - 4)^2 - 11; (4, -11)$
19. $n(x) = 2(x + 3)^2 - 5; (-3, -5)$
21. $p(x) = -3(x - 1)^2 - 2; (1, -2)$
23. $k(x) = \left(x + \frac{7}{2}\right)^2 - \frac{89}{4}; \left(-\frac{7}{2}, -\frac{89}{4}\right)$
25. $f(x) = (x + 4)^2 - 15; (-4, -15)$
27. $F(x) = 5(x + 1)^2 - 4; (-1, -4)$
29. $P(x) = -2\left(x - \frac{1}{4}\right)^2 + \frac{1}{8}; \left(\frac{1}{4}, \frac{1}{8}\right)$ **31.** $(2, 3)$
33. $(-1, -2)$ **35.** $(-4, -15)$ **37.** $(-1, 2)$
39. $(1, 3)$ **41.** $(-1, 4)$ **43.** $\left(\frac{3}{2}, \frac{3}{4}\right)$
45. **a.** $\left(-\frac{9}{2}, -\frac{49}{4}\right)$ **b.** $(0, 8)$ **c.** $(-8, 0)(-1, 0)$
d.

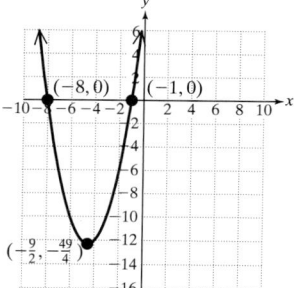

47. **a.** $\left(\frac{1}{2}, \frac{7}{2}\right)$ **b.** $(0, 4)$ **c.** No x-intercepts

d.

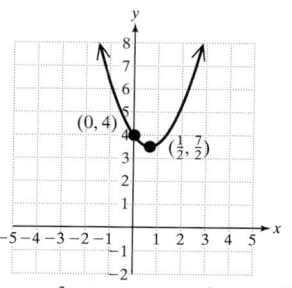

49. **a.** $\left(\frac{3}{2}, 0\right)$ **b.** $\left(0, -\frac{9}{4}\right)$ **c.** $\left(\frac{3}{2}, 0\right)$
d.

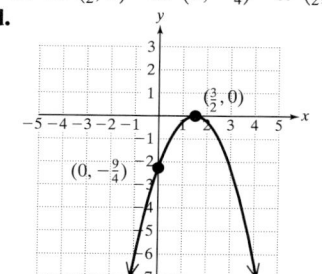

51. Mia must package 10 MP3 players.
53. **a.** 33 psi   **b.** 38 thousand miles
55. **a.** \$450   **b.** $(0, -550)$. If no cookbooks are produced, the organization will lose \$550.   **c.** 50 books or 550 books
d. $(300, 1250)$
e.

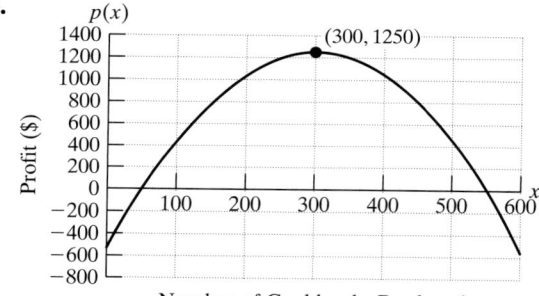

f. 300 books produced will yield a maximum profit of \$1250.
57. 45 mph   **59.** 48 hr
61. **a.** The sum of the sides must equal the total amount of fencing.   **b.** $A = x(200 - 2x)$ **c.** 50 ft by 100 ft
63.

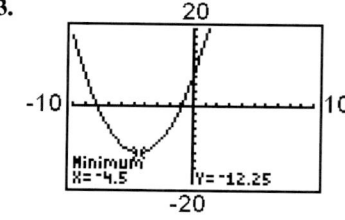

65.

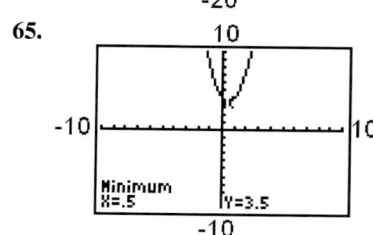

67.

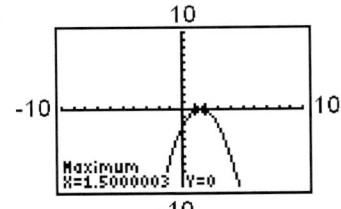

Chapter 12, Review Exercises, pp. 883–886

1. $x = \pm\sqrt{5}$ **2.** $y = \pm 2i$ **3.** $a = \pm 9$

4. $b = \pm\dfrac{i\sqrt{57}}{3}$ **5.** $x = 2 \pm 6\sqrt{2}$ **6.** $x = \dfrac{5 \pm 3i}{2}$

7. $y = \dfrac{1 \pm \sqrt{3}}{3}$ **8.** $m = 4 \pm \sqrt{5}$

9. $5\sqrt{3}$ in. ≈ 8.7 in. **10.** 9 in.

11. $5\sqrt{6}$ in. ≈ 12.2 in. **12.** $n = 64; (x + 8)^2$

13. $n = \dfrac{81}{4}; \left(x - \dfrac{9}{2}\right)^2$ **14.** $n = \dfrac{1}{16}; \left(y + \dfrac{1}{4}\right)^2$

15. $n = \dfrac{1}{25}; \left(z - \dfrac{1}{5}\right)^2$ **16.** $w = -2 \pm 3i$

17. $y = \dfrac{3}{2} \pm i$ **18.** $x = \dfrac{1}{3}, x = -1$

19. $b = \dfrac{1}{2}, b = -4$ **20.** $x = 3 \pm 2\sqrt{3}$

21. $t = 4 \pm 3i$ **22.** See page 838.

23. Two rational solutions **24.** Two rational solutions
25. Two irrational solutions **26.** Two imaginary solutions **27.** One rational solution **28.** Two imaginary solutions

29. $y = 2 \pm \sqrt{3}$ **30.** $m = \dfrac{5 \pm 5i\sqrt{3}}{2}$

31. $a = 2, a = -\dfrac{5}{6}$ **32.** $x = 2, x = \dfrac{4}{3}$

33. $b = \dfrac{4}{5}, b = -\dfrac{1}{5}$ **34.** $k = \dfrac{1}{10}, k = -\dfrac{1}{2}$

35. $x = 2 \pm 2i\sqrt{7}$ **36.** $y = 4 \pm \sqrt{6}$

37. $x = \dfrac{2 \pm \sqrt{22}}{3}$ **38.** $b = 8, b = -2$

39. $y = -7 \pm \sqrt{3}$ **40.** $a = -1 \pm \sqrt{11}$

41. a. 1822 ft **b.** 115 ft/s
42. a. $\approx 53{,}939$ thousand **b.** 2021
43. $x = 49$ ($x = 9$ does not check.)
44. $n = 4, n = 16$ **45.** $y = \pm 3, y = \pm\sqrt{2}$

46. $m = \pm\dfrac{\sqrt{6}}{2}, m = \pm i$ **47.** $t = -243, t = 32$

48. $p = 32, p = 1$ **49.** $t = 3 \pm \sqrt{10}$

50. $m = \dfrac{1 \pm \sqrt{181}}{6}$ **51.** $x = \pm 3i, x = \pm i\sqrt{3}$

52. $x = \pm\sqrt{7}, x = \pm 2$

53.

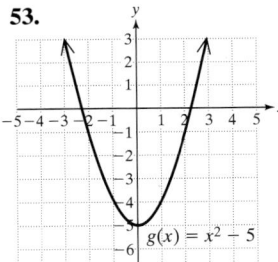

54.

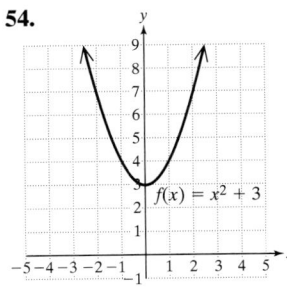

55.

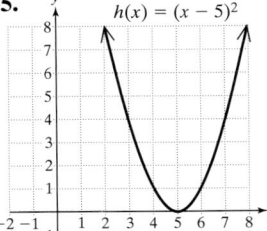

56.

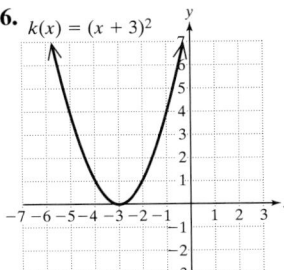

57.

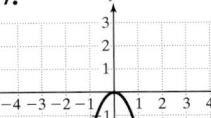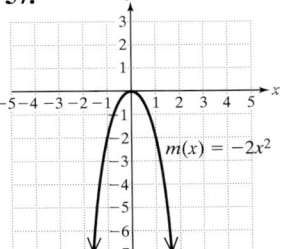

58.

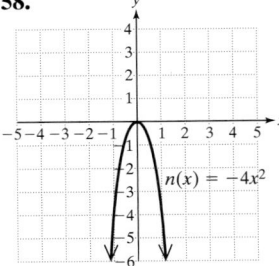

59.

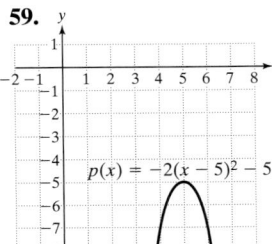

60.

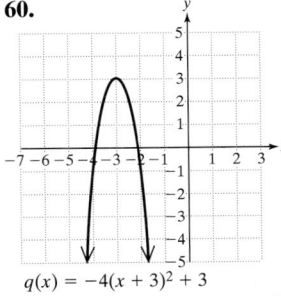

61. $\left(4, \dfrac{5}{3}\right)$ is the minimum point. The minimum value is $\dfrac{5}{3}$.

62. $\left(1, -\dfrac{1}{7}\right)$ is the maximum point. The maximum

value is $-\dfrac{1}{7}$. **63.** $x = -\dfrac{2}{11}$ **64.** $x = \dfrac{3}{16}$

65. $z(x) = (x - 3)^2 - 2; (3, -2)$
66. $b(x) = (x - 2)^2 - 48; (2, -48)$
67. $p(x) = -5(x + 1)^2 - 8; (-1, -8)$
68. $q(x) = -3(x + 4)^2 - 6; (-4, -6)$ **69.** $(1, -15)$

70. $(-1, 7)$ **71.** $\left(\dfrac{1}{2}, \dfrac{41}{4}\right)$ **72.** $\left(-\dfrac{1}{3}, -\dfrac{22}{3}\right)$

73. a. $(-2, 4)$ **b.** $(0, 0) (-4, 0)$

c.

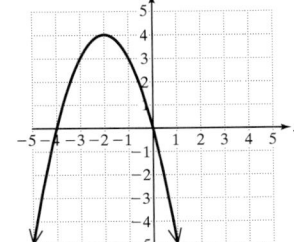

74. 3 sec **75.** 150 meals

Chapter 12, Test, pp. 886–887

1. $x = 2, x = -8$ **2.** $p = 2 \pm 2\sqrt{3}$

3. $m = -1 \pm i$ **4.** $n = \dfrac{49}{4}; \left(d + \dfrac{7}{2}\right)^2$

5. $x = -3 \pm 3\sqrt{3}$ **6.** $x = \dfrac{3 \pm i\sqrt{47}}{4}$

7. a. $x^2 - 3x + 12 = 0$ **b.** $a = 1, b = -3, c = 12$ **c.** -39
d. Two imaginary solutions
   8. a. $y^2 - 2y + 1 = 0$ **b.** $a = 1, b = -2, c = 1$ **c.** 0
d. One rational solution
   9. $x = 1, x = \dfrac{1}{3}$ **10.** $x = \dfrac{-7 \pm \sqrt{5}}{2}$
   11. Height 4.6 ft; base 6.2 ft   **12.** Radius 12.0 ft
   13. $x = 9$ ($x = 4$ does not check.)   **14.** $y = 8, y = -64$
   15. $y = \dfrac{11}{3}, y = 6$ **16.** $p = \pm\sqrt{6}, p = \pm 3$
   17. $y = \dfrac{5 \pm \sqrt{57}}{2}$ **18.** $(4, 0)(2, 0)(0, 8)$; c
   19. $(-4, 0)(3, 0)(-3, 0)(0, -36)$; b
   20. $(-3, 0)(-1, 0)(0, -6)$; d   **21.** $(4, 0)(-3, 0)(0, 0)$; a
   22. 256 ft   **23. a.** 1320 million   **b.** 1999
   24. The graph of $y = x^2 - 2$ is the graph of $y = x^2$ shifted
down 2 units.   **25.** The graph of $y = (x + 3)^2$ is the graph
of $y = x^2$ shifted 3 units to the left.   **26.** The graph of
$y = -4x^2$ is the graph of $y = 4x^2$ opening downward instead
of upward.
   27. a. $(4, 2)$ **b.** Downward   **c.** Maximum point   **d.** The
maximum value is 2.   **e.** $x = 4$
   28. a. $g(x) = 2(x - 5)^2 + 1; (5, 1)$ **b.** $(5, 1)$
   29. 20,000 ft^2

Chapters 1–12, Cumulative Review Exercises, pp. 888–889

   1. a. $\{2, 4, 6, 8, 10, 12, 16\}$ **b.** $\{2, 8\}$ **2.** $-3x^2 - 13x + 1$
   3. -16 **4.** 1.8×10^{10} **5. a.** $(x + 2)(x + 3)(x - 3)$
b. Quotient: $x^2 + 5x + 6$; remainder: 0
   6. $x - 2$ **7.** $\dfrac{2\sqrt{2x}}{x}$ **8.** \$8000 in 12% account; \$2000
in 3% account   **9.** $(8, 7)$ **10. a.** 720 ft   **b.** 720 ft
c. 12 sec   **11.** $x = 3 \pm 4i$ **12.** $x = \dfrac{-5 \pm \sqrt{33}}{4}$
   13. 25   **14.** $2(x + 5)(x^2 - 5x + 25)$
   15.

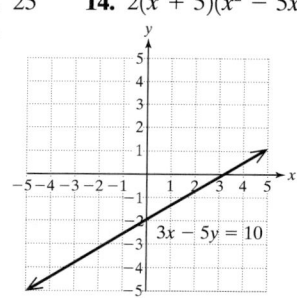

   16. a. $\left(\dfrac{5}{2}, 0\right)(2, 0)$ **b.** $(0, 10)$

   17. Free throws: 565; 2-pt. shots: 851; 3-pt. shots: 30
   18. The domain element 3 has more than one corresponding
range element.
   19.

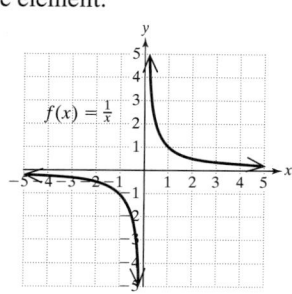

20. $y = 39$ **21. a.** 3   **b.** $\sqrt{2}$ **c.** Not a real number
22. a. $[-4, \infty)$ **b.** $(-\infty, \infty)$ **23. a.** $(-\infty, 2]$
b. $(-\infty, 1) \cup \{2\} \cup [3, 4]$ **c.** 2   **d.** 3   **e.** 2   **f.** $x = -3$
   24. a. $x = 2, x = \dfrac{2}{3}$ **b.** $\left(-\infty, \frac{2}{3}\right) \cup (2, \infty)$
   25. $f = \dfrac{pq}{p + q}$ **26.** $t = 7$ **27.** $\dfrac{y + 1}{y - 3}$
   28.

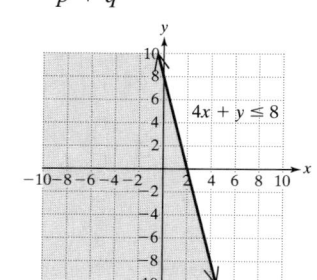

   29. a. $(3, 1)$ **b.** Upward   **c.** $(0, 19)$ **d.** No x-intercept
e.

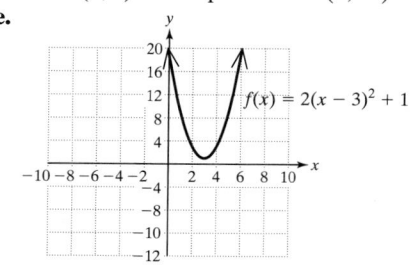

   30. Vertex: $(8, -62)$

Chapter 13

Chapter Opener Puzzle

6	5	3	4^a	2	1^b
4	2	1	6^c	5^d	3
1^e	4^f	5	3	6^g	2^h
3	6^i	2	1	4^j	5^k
5	1	4^m	2	3	6
2^n	3	6	5	1	4

Section 13.1 Practice Exercises, pp. 897–899

   3. $(f - g)(x) = -2x^2 - 3x + 4; (-\infty, \infty)$
   5. $(f + h)(x) = x + 4 + \sqrt{x - 1}; \;\; [1, \infty)$
   7. $(h \cdot k)(x) = \dfrac{\sqrt{x - 1}}{x}; \;\; [1, \infty)$
   9. $(f \cdot k)(x) = \dfrac{x + 4}{x}; \;\; (-\infty, 0) \cup (0, \infty)$
   11. $\left(\dfrac{g}{f}\right)(x) = \dfrac{2x^2 + 4x}{x + 4}; (-\infty, -4) \cup (-4, \infty)$

13. $\left(\dfrac{f}{h}\right)(x) = \dfrac{x+4}{\sqrt{x-1}}$; $(1, \infty)$

15. $(f \circ k)(x) = \dfrac{1}{x} + 4$; $(-\infty, 0) \cup (0, \infty)$

17. $(k \circ f)(x) = \dfrac{1}{x+4}$; $(-\infty, -4) \cup (-4, \infty)$

19. $(h \circ k)(x) = \sqrt{\dfrac{1}{x} - 1}$ or $\sqrt{\dfrac{1-x}{x}}$; $(0, 1]$

21. $(g \circ k)(x) = \dfrac{2}{x^2} + \dfrac{4}{x}$; $(-\infty, 0) \cup (0, \infty)$ **23.** No

25. $(f \circ g)(x) = 25x^2 - 15x + 1$; $(g \circ f)(x) = 5x^2 - 15x + 5$

27. $(f \circ g)(x) = |x^3 - 1|$; $(g \circ f)(x) = |x|^3 - 1$

29. $(h \circ h)(x) = 25x - 24$ **31.** 0   **33.** -64 **35.** 2

37. 1   **39.** $\dfrac{1}{64}$ **41.** 0   **43.** Undefined   **45.** -2

47. 2   **49.** 0   **51.** 1   **53.** 0   **55.** Undefined

57. -1 **59.** 2   **61.** 2   **63.** -1 **65.** -2

67. 3   **69.** -6 **71.** -1 **73.** 4   **75.** 0

77. 2   **79. a.** $P(x) = 3.78x - 1$ **b.** \$188

81. a. $F(t) = 0.2t + 6.4$; F represents the amount of child support (in billion dollars) not paid.   **b.** $F(0) = 6.4$ means that in 2000, \$6.4 billion of child support was not paid. $F(2) = 6.8$ means that in 2002, \$6.8 billion of child support was not paid. $F(4) = 7.2$ means that in 2004, \$7.2 billion of child support was not paid.   **83. a.** $(D \circ r)(t) = 560t$; This function represents the total distance Joe travels as a function of time that he rides.   **b.** 5600 ft

Section 13.2 Practice Exercises, pp. 907–910

3. Yes   **5.** No   **7.** Yes   **9.** $g^{-1} = \{(5, 3), (1, 8),$ $(9, -3), (2, 0)\}$ **11.** $r^{-1} = \{(3, a), (6, b), (9, c)\}$

13. The function is not one-to-one.   **15.** Yes

17. No   **19.** Yes   **21.** $h^{-1}(x) = x - 4$

23. $m^{-1}(x) = 3(x + 2)$ **25.** $p^{-1}(x) = -x + 10$

27. $f^{-1}(x) = \sqrt[3]{x}$ **29.** $g^{-1}(x) = \dfrac{x^3 + 1}{2}$

31. a. 1.2192 m, 15.24 m   **b.** $f^{-1}(x) = \dfrac{x}{0.3048}$

c. 4921.3 ft   **33.** False   **35.** True   **37.** False

39. True   **41.** $(b, 0)$ **43.** For example: $f(x) = x$

45. a. Domain $\{x \mid x \le 0\}$, range $\{y \mid y \ge -4\}$

b. Domain $\{x \mid x \ge -4\}$, range $\{y \mid y \le 0\}$

47. a. $\{x \mid -2 \le x \le 0\}$ **b.** $\{y \mid 0 \le y \le 2\}$

c. $\{x \mid 0 \le x \le 2\}$ **d.** $\{y \mid -2 \le y \le 0\}$

e.

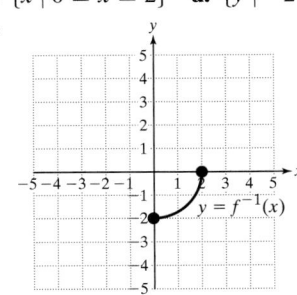

49. a. $\{x \mid 2 \le x \le 5\}$ **b.** $\{y \mid 0 \le y \le 3\}$

c. $\{x \mid 0 \le x \le 3\}$ **d.** $\{y \mid 2 \le y \le 5\}$

e.

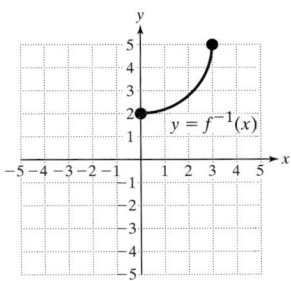

51. a. $(f \circ g)(x) = 5\left(\dfrac{x+2}{5}\right) - 2 = x$

b. $(g \circ f)(x) = \dfrac{(5x - 2) + 2}{5} = x$

53. a. $(f \circ g)(x) = \dfrac{\sqrt[3]{27x^3}}{3} = x$

b. $(g \circ f)(x) = 27\left(\dfrac{\sqrt[3]{x}}{3}\right)^3 = x$

55. a. $(f \circ g)(x) = (\sqrt{x+3})^2 - 3 = x$

b. $(g \circ f)(x) = \sqrt{(x^2 - 3) + 3} = x$

57. $p^{-1}(x) = \dfrac{3 - 3x}{x + 1}$ **59.** $w^{-1}(x) = \dfrac{4 - 2x}{x}$

61. $m^{-1}(x) = \sqrt{x + 1}$ **63.** $g^{-1}(x) = -\sqrt{x + 1}$

65. $v^{-1}(x) = x^2 - 16$, $x \ge 0$

67. $u^{-1}(x) = x^2 - 16$, $x \le 0$

69. $k^{-1}(x) = \sqrt[3]{x + 4}$

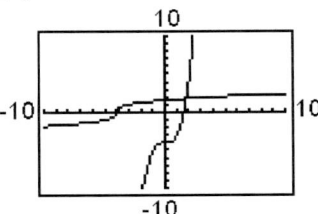

71. $m^{-1}(x) = \dfrac{x + 4}{3}$

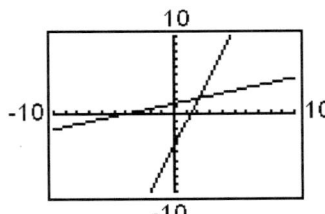

Section 13.3 Practice Exercises, pp. 917–921

3. $-2x^2 + 2x - 3$ **5.** $\dfrac{3x - 1}{2x^2 + x + 2}$ **7.** $6x^2 + 3x + 5$

9. $\left\{(14, -13), \left(-\dfrac{1}{2}, \dfrac{1}{2}\right), (30, 6), \left(0, -\dfrac{5}{3}\right), (1, 0)\right\}$

11. 25   **13.** $\dfrac{1}{1000}$ **15.** 6   **17.** 8   **19.** 5.8731

21. 1385.4557   **23.** 0.0063   **25.** 0.8950

27. a. $x = 2$ **b.** $x = 3$ **c.** Between 2 and 3

29. a. $x = 4$ **b.** $x = 5$ **c.** Between 4 and 5

31. $f(0) = 1, f(1) = \dfrac{1}{5}, f(2) = \dfrac{1}{25}, f(-1) = 5, f(-2) = 25$

33. $h(0) = 1, h(1) = \pi \approx 3.14, h(-1) \approx 0.32,$ $h(\sqrt{2}) \approx 5.05, h(\pi) \approx 36.46$

35. $r(0) = 9, r(1) = 27, r(2) = 81, r(-1) = 3, r(-2) = 1,$ $r(-3) = \dfrac{1}{3}$ **37.** If $b > 1$, the graph is increasing. If $0 < b < 1$, the graph is decreasing.

39.

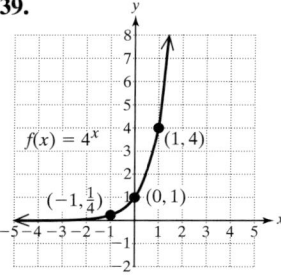

41.

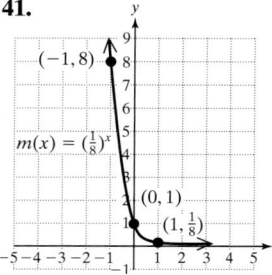

43.

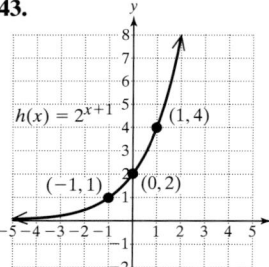

45.

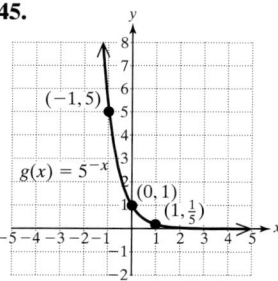

47. a. 0.25 g   **b.** ≈ 0.16 g   **49. a.** 758,000   **b.** 379,000
c. 144,000   **51. a.** $A(t) = 141,340,000(1.015)^t$
b. $A(46) \approx 280,000,000$ **53. a.** \$1640.67   **b.** \$2691.80
c. $A(0) = 1000$. The initial amount of the investment is
\$1000. $A(7) = 2000$. The amount of the investment doubles
in 7 years.

55.

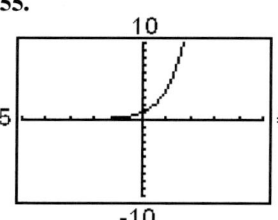

57.

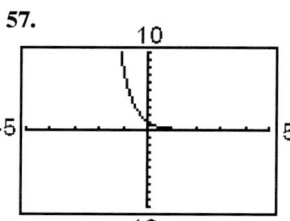

59.

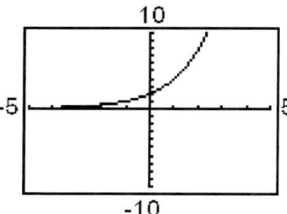

61.

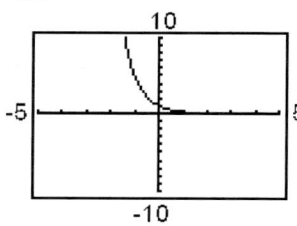

63. a.

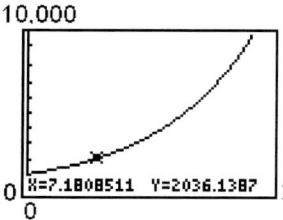

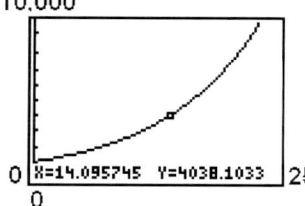

b. 7 yr, 14 yr, 21 yr

Section 13.4 Practice Exercises, pp. 930–935

3. i

5. a. $g(-2) = \dfrac{1}{9}, g(-1) = \dfrac{1}{3}, g(0) = 1, g(1) = 3, g(2) = 9$

b.

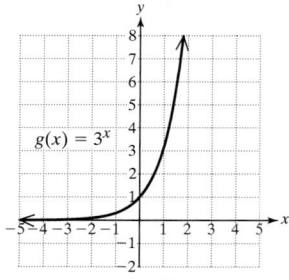

7. a. $s(-2) = \dfrac{25}{4}, s(-1) = \dfrac{5}{2}, s(0) = 1, s(1) = \dfrac{2}{5}, s(2) = \dfrac{4}{25}$

b.   **9.** $b^y = x$

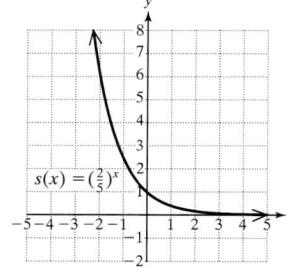

11. $\log_{10} 1000 = 3$ **13.** $\log_8 2 = \dfrac{1}{3}$ **15.** $\log_8\left(\dfrac{1}{64}\right) = -2$

17. $\log_b x = y$ **19.** $\log_e x = y$ **21.** $\log_{5/2}\left(\dfrac{2}{5}\right) = -1$

23. $125^{2/3} = 25$ **25.** $25^{-1/2} = \dfrac{1}{5}$ **27.** $2^7 = 128$

29. $b^y = 82$ **31.** $2^x = 7$ **33.** $\left(\dfrac{1}{2}\right)^6 = x$ **35.** 3

37. -4 **39.** $\dfrac{1}{3}$ **41.** -1 **43.** 3   **45.** 0   **47.** 4

49. $\dfrac{1}{3}$ **51.** 2   **53.** 4   **55.** -1 **57.** -3

59. 0.7782   **61.** 0.4971   **63.** -1.5051 **65.** -2.2676
67. 5.5315   **69.** -7.4202 **71. a.** Slightly less than 2
b. Slightly more than 1   **c.** $\log 93 \approx 1.9685, \log 12 \approx 1.0792$

73. a. $f\left(\dfrac{1}{64}\right) = -3, f\left(\dfrac{1}{16}\right) = -2, f\left(\dfrac{1}{4}\right) = -1, f(1) = 0,$
$f(4) = 1, f(16) = 2, f(64) = 3$
b.

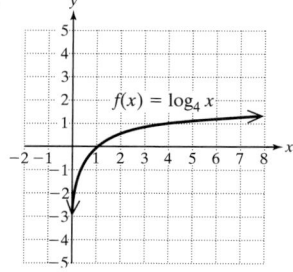

75. $3^y = x$

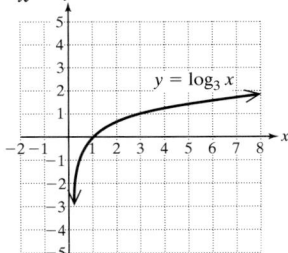

x	y
$\frac{1}{9}$	-2
$\frac{1}{3}$	-1
1	0
3	1
9	2

77. $\left(\dfrac{1}{2}\right)^y = x$

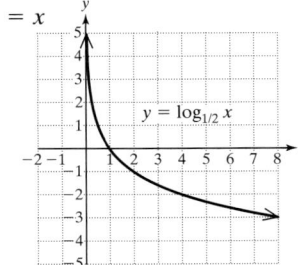

x	y
4	-2
2	-1
1	0
$\frac{1}{2}$	1
$\frac{1}{4}$	2

79. $(5, \infty)$ **81.** $(-1.2, \infty)$ **83.** $(-\infty, 0) \cup (0, \infty)$
85. ≈ 7.35
87. a.

t (months)	0	1	2	6	12	24
$S_1(t)$	91	82.0	76.7	65.6	57.6	49.1
$S_2(t)$	88	83.5	80.8	75.3	71.3	67.0

b. Group 1: 91; Group 2: 88   **c.** Method II
89. Domain: $(-6, \infty)$; asymptote: $x = -6$

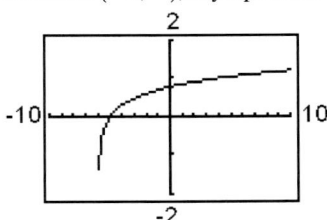

91. Domain: $(2, \infty)$; asymptote: $x = 2$

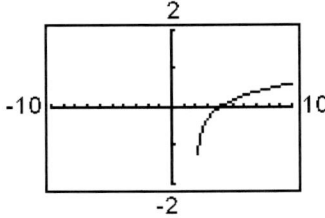

93. Domain: $(-\infty, 2)$; asymptote: $x = 2$

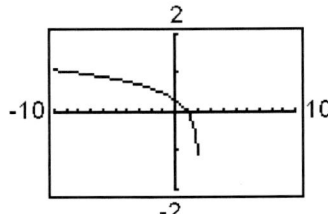

Section 13.5 Practice Exercises, pp. 940–943

3. 4   **5.** $\dfrac{1}{6}$ **7.** 0.9031   **9.** 5.0475   **11.** a

13. c   **15.** For example: $\log_3 3 = 1$

17. For example: $6^{\log_6 4} = 4$ **19.** 1   **21.** 5   **23.** 2
25. 3   **27.** 0   **29.** 5   **31.** 1   **33.** 5   **35.** 3
37. p **39.** 2   **41.** Expressions a and c are equivalent.
43. Expressions a and b are equivalent.
45. $\log_2 y - \log_2 z$ **47.** $\log_6 x + \log_6 y + \log_6 z$
49. $\dfrac{1}{3} \log_7 z$ **51.** $\log_2 x - \log_2 y - \log_2 z$
53. $\log a + \dfrac{1}{3} \log b - \log c - 2 \log d$
55. $\log (a + 1) - \log b - \dfrac{1}{3} \log c$
57. $\dfrac{3}{4} \log_5 w + \dfrac{1}{4} \log_5 z - \dfrac{1}{2} \log_5 x$
59. $-4 \log_3 z$ **61.** $1 - \log_x y - \dfrac{1}{2} \log_x z$
63. $\log (CABIN)$ **65.** $\log_2 \left(\dfrac{\sqrt[3]{x}}{y^5 z^3} \right)$ **67.** $\log_5 \left(\dfrac{a}{\sqrt{bc^3}} \right)$
69. $\log_3 (z^{5/2})$ **71.** $\log_2 (2b^2)$ or $1 + \log_2 b^2$
73. $\log \left(\dfrac{\sqrt[4]{a + 1}}{b^2 c^4} \right)$ **75.** $\log_x (p + 2)$ **77.** 1.386
79. 3.218   **81.** 3.401   **83.** 2.119   **85.** 8.316
87. a. $R = \log I - \log I_0$ **b.** $\log I$
89. a. Domain: $(-\infty, 1) \cup (1, \infty)$ **b.** Domain: $(1, \infty)$

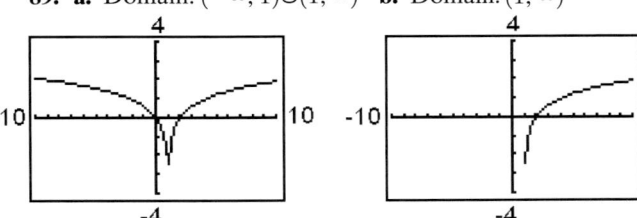

c. They are equivalent for all x in the intersection of their domains, $(1, \infty)$.

Section 13.6 Practice Exercises, pp. 951–957

3.

x	g(x)
-3	125
-2	25
-1	5
0	1
1	$\frac{1}{5}$
2	$\frac{1}{25}$
3	$\frac{1}{125}$

$g(x) = \left(\frac{1}{5}\right)^x$

5.

x	r(x)
0.5	-0.30
1	0
5	0.70
10	1.00

$r(x) = \log x$

7.

x	y
-5	0.05
-4	0.14
-3	0.37
-2	1
-1	2.72
0	7.39

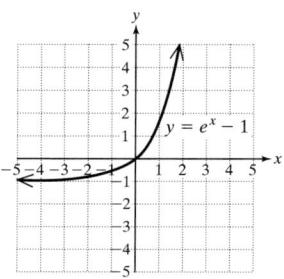

Domain: $(-\infty, \infty)$

9.

x	y
-4	-0.98
-3	-0.95
-2	-0.86
-1	-0.63
0	0
1	1.72

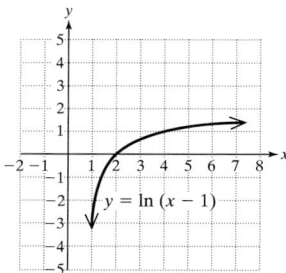

Domain: $(-\infty, \infty)$

11. a. \$7152.26   **b.** \$7740.30   **c.** \$8711.07   **d.** \$10,190.52
An investment grows more rapidly at higher interest rates.
13. a. \$22,161.83   **b.** \$22,321.96   **c.** \$22,358.78
d. \$22,376.76   **e.** \$22,377.37   More money is earned at a
greater number of compound periods per year.
15. a. \$13,498.59   **b.** \$18,221.19   **c.** \$24,596.03
d. \$33,201.17   **e.** \$60,496.47   More money is earned over a
longer period of time.

17.

x	y
1.25	-1.39
1.50	-0.69
1.75	-0.29
2	0
3	0.69
4	1.10
5	1.39

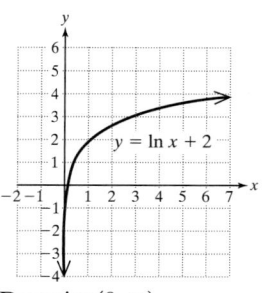

Domain: $(1, \infty)$

19.

x	y
0.25	0.61
0.5	1.31
0.75	1.71
1	2.00
2	2.69
3	3.10
4	3.39

Domain: $(0, \infty)$

21. a.

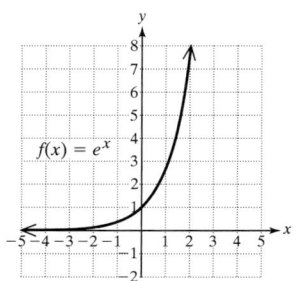

b. Domain: $(-\infty, \infty)$; range: $(0, \infty)$

c.

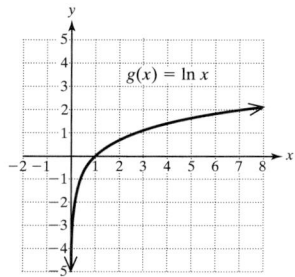

d. Domain: $(0, \infty)$; range: $(-\infty, \infty)$
23. 2   **25.** x **27.** $5-3x$ **29.** 4   **31.** $\ln (w^2 z)$
33. $\ln \sqrt[3]{\dfrac{a^4}{b}}$ **35.** $\ln \left(\dfrac{y^3}{xz}\right)$ **37.** $\ln \left(\dfrac{\sqrt{c} \cdot a}{b^2}\right)$
39. $\dfrac{1}{3} \ln a - \dfrac{1}{3} \ln b$ **41.** $\dfrac{1}{2} \ln c + 1$
43. $\dfrac{1}{2} \ln a + \dfrac{1}{2} \ln b - 3 \ln c$ **45.** $\dfrac{1}{2} \ln 2 + \dfrac{1}{2} \ln a + \dfrac{1}{2} \ln b$
47. a. 2.3023   **b.** 2.3023   **c.** They are the same.
49. 1.4650   **51.** 2.0437   **53.** -0.7104 **55.** -2.3219
57. -3.7227 **59.** -2.0589 **61. a.** 12.6 years
b. 8.7 years   **c.** 17.4 years   **63. a.** 91 deaths   **b.** Sept. 5:
349 deaths; Sept. 10: 459 deaths; Sept. 20: 570 deaths

65. a.–b.

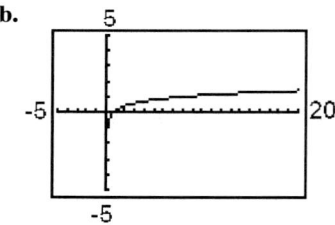

c. They appear to be the same.
67. a.–b.

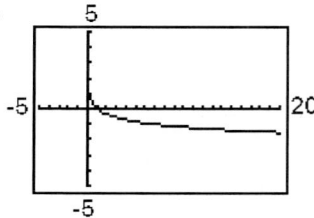

c. They appear to be the same.
69.

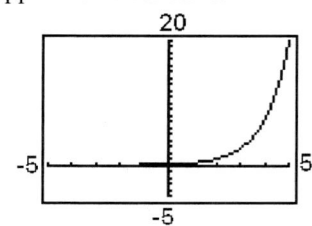

Chapter 13 Problem Recognition Exercises—Logarithmic and Exponential Forms p. 957

	Exponential Form	Logarithmic Form
1.	$2^5 = 32$	$\log_2 32 = 5$
2.	$3^4 = 81$	$\log_3 81 = 4$
3.	$z^y = x$	$\log_z x = y$
4.	$b^c = a$	$\log_b a = c$
5.	$10^3 = 1000$	$\log 1000 = 3$
6.	$10^1 = 10$	$\log 10 = 1$
7.	$e^a = b$	$\ln b = a$
8.	$e^q = p$	$\ln p = q$
9.	$(\frac{1}{2})^2 = \frac{1}{4}$	$\log_{1/2} (\frac{1}{4}) = 2$
10.	$(\frac{1}{3})^{-2} = 9$	$\log_{1/3} 9 = -2$
11.	$10^{-2} = 0.01$	$\log 0.01 = -2$
12.	$10^x = 4$	$\log 4 = x$
13.	$e^0 = 1$	$\ln 1 = 0$
14.	$e^1 = e$	$\ln e = 1$
15.	$25^{1/2} = 5$	$\log_{25} 5 = \frac{1}{2}$
16.	$16^{1/4} = 2$	$\log_{16} 2 = \frac{1}{4}$
17.	$e^t = s$	$\ln s = t$
18.	$e^r = w$	$\ln w = r$
19.	$15^{-2} = \frac{1}{225}$	$\log_{15} (\frac{1}{225}) = -2$
20.	$3^{-1} = p$	$\log_3 p = -1$

Section 13.7 Practice Exercises, pp. 967–972

3. a.

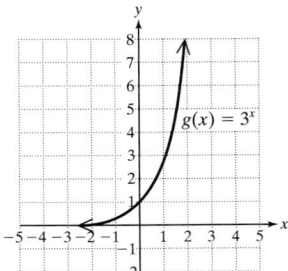

b. Domain: $(-\infty, \infty)$; range: $(0, \infty)$

5.

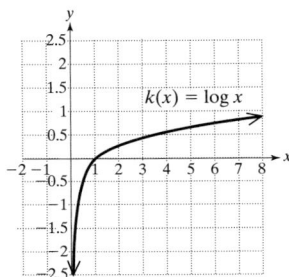

b. $x = 0$ **c.** Domain: $(0, \infty)$; range: $(-\infty, \infty)$

7. $\log_b [x(2x + 3)]$ **9.** $\log_b \left(\dfrac{x + 2}{3x - 5}\right)$ **11.** $x = 4^9$

13. $q = \sqrt{10}$ **15.** $x = e^{19}$ **17.** $x = \dfrac{1}{10}$

19. $x = 10$ **21.** $b = e$ **23.** $b = 64$ **25.** $a = 29$

27. $t = \dfrac{8}{5}$ **29.** $x = -2$ **31.** $x = -1$

33. $h = 3$ ($h = -3$ does not check.) **35.** $x = \dfrac{1}{2}$

37. $x = \dfrac{5}{4}$; ($x = -1$ does not check.)

39. $y = \dfrac{5}{3}\left(y = -\dfrac{1}{2}$ does not check.$\right)$

41. No solution ($h = 3$ does not check.)
43. 10^7 (or 10,000,000) times more intense
45. $10^{-5.5}$ W/m^2 **47.** $x = 4$ **49.** $x = -3$

51. $x = \dfrac{1}{3}$ **53.** $x = \dfrac{4}{3}$ **55.** $x = \dfrac{7}{4}$ **57.** $x = \dfrac{1}{3}$

59. $x = \dfrac{2}{5}$ **61.** $x = -1$ **63.** $y = \dfrac{\ln 39}{\ln 6} \approx 2.045$

65. $x = \ln 0.3151 \approx -1.155$
67. $p = \log 16.8125 \approx 1.226$

69. $k = \dfrac{\ln 4}{0.03} \approx 46.210$ **71.** $n = \dfrac{\ln 20}{0.05} \approx 59.915$

73. $x = \dfrac{\ln 2}{\ln 2 - \ln 7} \approx -0.553$

75. a. 12,921 thousand (or 12,921,000) people
b. The year 2015 ($t \approx 13.9$) **77. a.** 300 bacteria
b. ≈ 357 **c.** ≈ 40 min **79.** 11 years

81. a. 3.1 g **b.** 415.9 days **83.** $t = \dfrac{\ln 2}{r}$

85. a. 6.8 mCi **b.** 13.9 hr **87.** $z = 16, z = \dfrac{1}{2}$
89. $x = 1, x = e^2$
91.

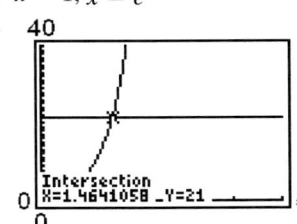

Chapter 13, Review Exercises, pp. 978–981

1. $(f - g)(x) = 2x^3 + 9x - 7; (-\infty, \infty)$
2. $(f + g)(x) = -2x^3 - 7x - 7; (-\infty, \infty)$
3. $(f \cdot n)(x) = \dfrac{x - 7}{x - 2}; (-\infty, 2)\cup(2, \infty)$
4. $(f \cdot m)(x) = (x - 7)\sqrt{x}; [0, \infty)$
5. $\left(\dfrac{f}{g}\right)(x) = \dfrac{x - 7}{-2x^3 - 8x}; (-\infty, 0)\cup(0, \infty)$
6. $\left(\dfrac{g}{f}\right)(x) = \dfrac{-2x^3 - 8x}{x - 7}; (-\infty, 7)\cup(7, \infty)$
7. $(m \circ f)(x) = \sqrt{x - 7}; [7, \infty)$
8. $(n \circ f)(x) = \dfrac{1}{x - 9}; (-\infty, 9)\cup(9, \infty)$

9. $\sqrt{32}$ or $4\sqrt{2}$ **10.** $\dfrac{1}{8}$ **11.** -167 **12.** -10

13. a. $(2x + 1)^2$ or $4x^2 + 4x + 1$ **b.** $2x^2 + 1$

c. No, $f \circ g \neq g \circ f$ **14.** $\dfrac{1}{4}$ **15.** -3 **16.** 0

17. -1 **18.** 4 **19.** 1 **20.** No **21.** Yes

22. $\{(5, 3), (9, 2), (-1, 0), (1, 4)\}$ **23.** $q^{-1}(x) = \dfrac{4}{3}(x + 2)$

24. $g^{-1}(x) = (x-3)^5$ **25.** $f^{-1}(x) = \sqrt[3]{x} + 1$

26. $n^{-1}(x) = \dfrac{4}{x} + 2$

27. $(f \circ g)(x) = 5\left(\dfrac{1}{5}x + \dfrac{2}{5}\right) - 2 = x + 2 - 2 = x$

$(g \circ f)(x) = \dfrac{1}{5}(5x - 2) + \dfrac{2}{5} = x - \dfrac{2}{5} + \dfrac{2}{5} = x$

28. The graphs are symmetric about the line $y = x$.

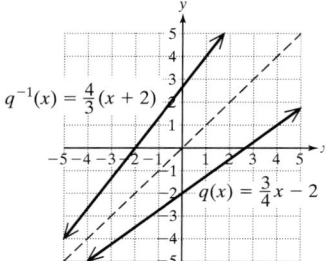

$q^{-1}(x) = \dfrac{4}{3}(x + 2)$

$q(x) = \dfrac{3}{4}x - 2$

29. a. Domain: $\{x \mid x \geq -1\}$; range: $\{y \mid y \geq 0\}$
b. Domain: $\{x \mid x \geq 0\}$; range: $\{y \mid y \geq -1\}$
30. $p^{-1}(x) = (x - 2)^2, x \geq 2$

31. 1024   **32.** $\dfrac{1}{36} \approx 0.028$ **33.** 2   **34.** 10

35. 8.825   **36.** 16.242   **37.** 1.627   **38.** 0.681
39.   **40.**

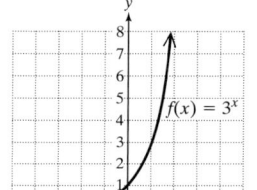

$f(x) = 3^x$

$g(x) = \left(\tfrac{1}{4}\right)^x$

41.   **42.**

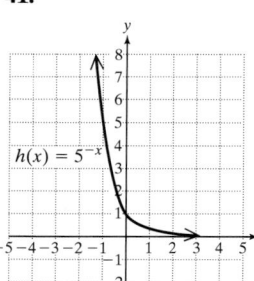

$h(x) = 5^{-x}$

$k(x) = \left(\tfrac{2}{5}\right)^{-x}$

43. a. Horizontal **b.** $y = 0$ **44. a.** 15,000 mrem
b. 3750 mrem **c.** Yes   **45.** -3 **46.** 0   **47.** 1
48. 8   **49.** 4   **50.** 4   **51.** 5   **52.** -1
53.   **54.**

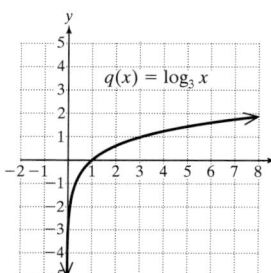

$q(x) = \log_3 x$

$r(x) = \log_{1/2} x$

55. a. Vertical asymptote **b.** $x = 0$ **56. a.** 2.5 **b.** 9.5
57. 1   **58.** 6   **59.** 0   **60.** 7   **61. a.** $\log_b x + \log_b y$
b. $\log_b\left(\dfrac{x}{y}\right)$ **c.** $p \log_b x$ **62.** $\log_b\left(\dfrac{\sqrt[4]{x^3 y}}{z}\right)$

63. $\log_3\left(\dfrac{\sqrt{ab}}{c^2 d^4}\right)$ **64.** $\log 5$ **65.** 0   **66.** b   **67.** a

68. 148.4132   **69.** 14.0940   **70.** 32.2570   **71.** 57.2795
72. 1.7918   **73.** -2.1972 **74.** 1.3424   **75.** 1.3029
76. 3.3219   **77.** 1.9943   **78.** -0.8370 **79.** -3.6668
80. a. \$33,361.92 **b.** \$33,693.90 **c.** \$33,770.48
d. \$33,809.18   **81. a.** $S(0) = 95$; the student's score is 95 at
the end of the course. **b.** $S(6) \approx 23.7$; the student's score is
23.7 after 6 months. **c.** $S(12) \approx 20.2$; the student's score is
20.2 after 1 year. **d.** The limiting value is 20.   **82.** $(-\infty, \infty)$
83. $(-\infty, \infty)$ **84.** $(-\infty, \infty)$ **85.** $(0, \infty)$ **86.** $(-5, \infty)$

87. $(7, \infty)$ **88.** $\left(-\infty, \dfrac{4}{3}\right)$ **89.** $(-\infty, 5)$ **90.** $x = 125$

91. $x = \dfrac{1}{49} \approx 0.02$ **92.** $y = 216$ **93.** $y = 3^{1/12} \approx 1.10$

94. $w = \dfrac{1001}{2} = 500.5$ **95.** $w = 9$ **96.** $p = 5 \,(p = -2$

does not check.)   **97.** $t = \dfrac{190}{321} \approx 0.59$ **98.** $x = -1$

99. $x = \dfrac{4}{7}$ **100.** $a = \dfrac{\ln 21}{\ln 4} \approx 2.1962$

101. $a = \dfrac{\ln 18}{\ln 5} \approx 1.7959$ **102.** $x = -\ln(0.1) \approx 2.3026$

103. $x = -\dfrac{\ln(0.06)}{2} \approx 1.4067$ **104.** $n = \dfrac{\log 1512}{2} \approx 1.5898$

105. $m = \dfrac{\log\left(\dfrac{1}{821}\right)}{-3} \approx 0.9714$

106. $x = \dfrac{3 \ln 2}{\ln 7 - \ln 2} \approx 1.6599$

107. $x = \dfrac{5 \ln 14}{\ln 14 - \ln 6} \approx 15.5734$ **108. a.** 1.09 μg
b. 0.15 μg **c.** 16.08 days   **109. a.** 150 bacteria
b. ≈ 185 **c.** ≈ 99 min   **110. a.** $V(0) = 15,000$; the initial
value of the car is \$15,000. **b.** $V(10) = 3347$; the value of
the car after 10 years is \$3347. **c.** 7.3 years **d.** Yes. The
limiting value is 0.

Chapter 13, Test pp. 982–983

1. $\left(\dfrac{f}{g}\right)(x) = \dfrac{x - 4}{\sqrt{x + 2}}$ **2.** $(h \cdot g)(x) = \dfrac{\sqrt{x + 2}}{x}$

3. $(g \circ f)(x) = \sqrt{x - 2}$ **4.** $(h \circ f)(x) = \dfrac{1}{x - 4}$

5. 0   **6.** $-\dfrac{3}{2}$ **7.** $\dfrac{1}{4}$ **8.** Undefined

9. $\left(\dfrac{g}{f}\right)(x) = \dfrac{\sqrt{x + 2}}{x - 4}$; domain: $[-2, 4) \cup (4, \infty)$
10. A function is one-to-one if it passes the horizontal line
test.   **11.** b   **12.** $f^{-1}(x) = 4x - 12$
13. $g^{-1}(x) = \sqrt{x} + 1$

14.

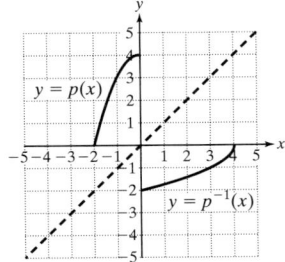

15. a. 4.6416   **b.** 32.2693   **c.** 687.2913

16.

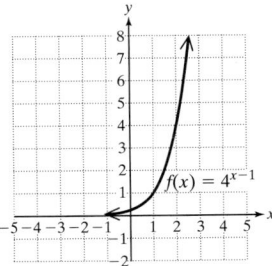

17. a. $\log_{16} 8 = \dfrac{3}{4}$ **b.** $x^5 = 31$

18.

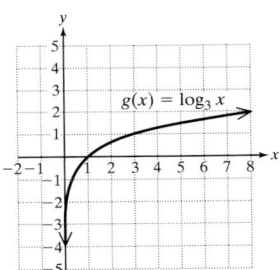

19. $\dfrac{\log_a n}{\log_a b}$ **20. a.** 1.3222   **b.** 1.8502   **c.** -2.5850

21. a. $1 + \log_3 x$ **b.** -5 **22. a.** $\log_b(\sqrt{x}y^3)$

b. $\log\left(\dfrac{1}{a^3}\right)$ or $-\log a^3$ **23. a.** 1.6487   **b.** 0.0498

c. -1.0986 **d.** 1   **24. a.** $y = \ln(x)$ **b.** $y = e^x$

25. a. $p(4) \approx 59.8$; 59.8% of the material is retained after 4 months.   **b.** $p(12) \approx 40.7$; 40.7% of the material is retained after 1 year.   **c.** $p(0) = 92$; 92% of the material is retained at the end of the course.   **26. a.** ≈ 9762 thousand (or 9,762,000) people   **b.** The year 2020 ($t \approx 20.4$)

27. a. $P(0) = 300$; there are 300 bacteria initially.

b. $\approx 35{,}588$ bacteria   **c.** $\approx 1{,}120{,}537$ bacteria

d. $\approx 1{,}495{,}831$ bacteria   **e.** Yes. The limiting value appears to be 1,500,000.

28. $x = 25$ ($x = -4$ does not check.)   **29.** $x = 32$

30. $x = e^{2.4} - 7 \approx 4.023$ **31.** $x = -7$

32. $x = \dfrac{\ln 50}{\ln 4} \approx 2.822$ **33.** $x = \dfrac{\ln 250}{2.4} \approx 2.301$

34. a. $P(2500) = 560.2$; at 2500 m the atmospheric pressure is 560.2 mm Hg.   **b.** 760 mm Hg   **c.** 1498.8 m

35. a. $\$2909.98$ **b.** 9.24 years to double

Chapters 1–13 Cumulative Review Exercises, pp. 984–986

1. $-\dfrac{5}{4}$ **2.** $-1 + \dfrac{p}{2} + \dfrac{3p^3}{4}$

3. Quotient: $t^3 + 2t^2 - 9t - 18$; remainder: 0

4. $|x - 3|$ **5.** $\dfrac{2\sqrt[3]{25}}{5}$ **6.** $\dfrac{4d^{1/10}}{c}$

7. $(\sqrt{15} - \sqrt{6} + \sqrt{30} - 2\sqrt{3})\,\text{m}^2$

8. $-\dfrac{7}{29} - \dfrac{26}{29}i$ **9.** $42°, 48°$ **10.** $m = \dfrac{2}{3}$

11. 4.8 L   **12.** 24 min   **13.** $(7, -1)$

14. $w = -\dfrac{23}{11}$ **15.** $x = \dfrac{c + d}{a - b}$ **16.** $t = \dfrac{\sqrt{2sg}}{g}$

17. $T = \dfrac{1 - \left(\dfrac{V_0}{V}\right)^2}{k}$ or $\dfrac{V^2 - V_0^2}{kV^2}$ **18.** $(7, 0), (3, 0)$

19. a. $-30t$ **b.** $-10t^2$ **c.** $2t^2 + 5t$

20. $q = 0$ **21. a.** $x = 2$ **b.** $y = 6$ **c.** $y = \dfrac{1}{2}x + 5$

22. $40°, 80°, 60°$ **23.** $\{(x, y)\,|\,-2x + y = -4\}$

24. a. vi   **b.** i   **c.** v   **d.** x   **e.** ii   **f.** ix   **g.** iv   **h.** viii   **i.** vii   **j.** iii

25. $\left[\dfrac{1}{2}, \infty\right)$ **26.** $f^{-1}(x) = \dfrac{1}{5}x + \dfrac{2}{15}$ **27.** 40 m^3

28. $-\dfrac{x^2 - 5x + 25}{(2x + 1)(x - 2)}$ **29.** $1 + x$

30. a. Yes; $x \neq 4, x \neq -2$ **b.** $x = 8$ **c.** $(-\infty, -2) \cup (4, 8]$

31. The numbers are $-\dfrac{3}{4}, 4$.

32. $x = -4$ ($x = -9$ does not check.)

33. $(-\infty, \infty)$

34. a. $P(6) = 2{,}000{,}000$, $P(12) = 1{,}000{,}000$, $P(18) = 500{,}000$, $P(24) = 250{,}000$, $P(30) = 125{,}000$

b.

c. 48 hr

35. a. 2   **b.** -3 **c.** 6   **d.** 3

36. a. 217.0723   **b.** 23.1407   **c.** 0.1768   **d.** 3.7293   **e.** -0.4005 **f.** 2.6047

37. $x = \ln 100 \approx 4.6052$

38. $x = 3$ ($x = -9$ does not check.)

39. $\log\left(\dfrac{\sqrt{z}}{x^2 y^3}\right)$ **40.** $\dfrac{2}{3}\ln x - \dfrac{1}{3}\ln y$

Chapter 14

Chapter Opener Puzzle

$$\frac{f}{6} \ \frac{a}{3} \ \frac{c}{5} \ \frac{t}{7} \ \frac{o}{1} \ \frac{r}{4} \ \frac{s}{2}$$

Section 14.1 Practice Exercises, pp. 992–996

3. $3\sqrt{5}$ **5.** $\sqrt{34}$ **7.** $\dfrac{\sqrt{73}}{8}$ **9.** 4   **11.** 8

13. $\sqrt{13}$ **15.** $\sqrt{42}$ **17.** Subtract 5 and -7. This
becomes $5 - (-7) = 12$.   **19.** $y = 13, y = 1$
21. $x = 0, x = 8$ **23.** Yes   **25.** No
27. Center $(4, -2); r = 3$ **29.** Center $(-1, -1); r = 1$

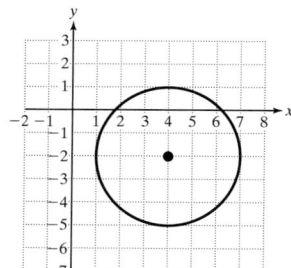

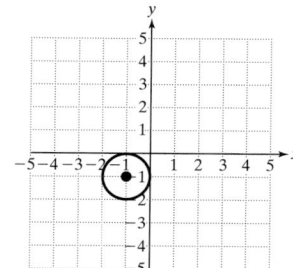

31. Center $(0, 5); r = 5$ **33.** Center $(3, 0); r = 2\sqrt{2}$

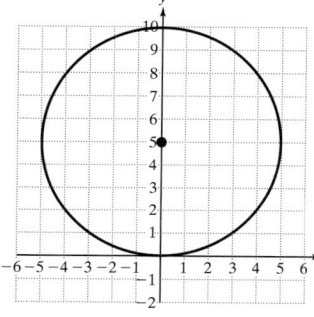

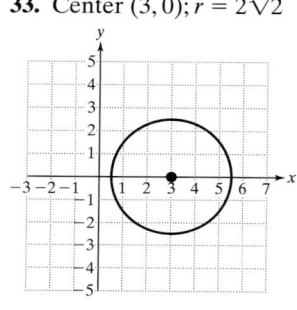

35. Center $(0, 0); r = \sqrt{6}$ **37.** Center $\left(-\dfrac{4}{5}, 0\right); r = \dfrac{8}{5}$

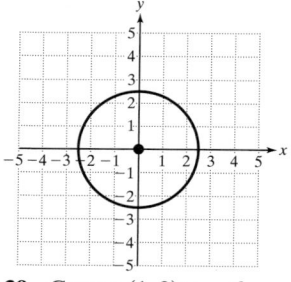

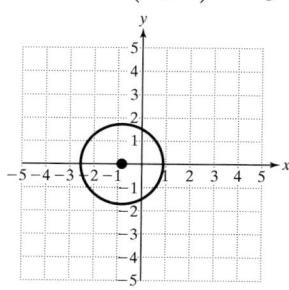

39. Center $(1, 3); r = 6$ **41.** Center: $(0, 3); r = 2$

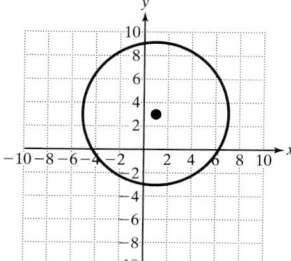

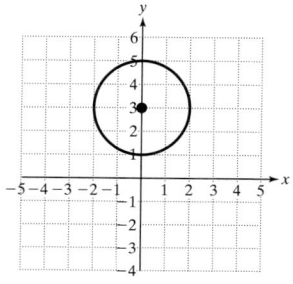

43. Center $(0, -3); r = \dfrac{4}{3}$ **45.** Center: $(6, -6); r = 1$

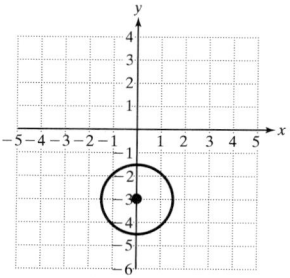

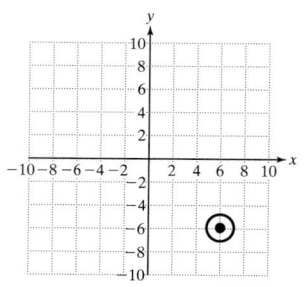

47. Center: $(0, 0); r = 4$ **49.** $x^2 + y^2 = 4$

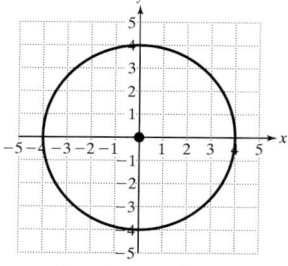

51. $x^2 + (y - 2)^2 = 4$ **53.** $(x + 2)^2 + (y - 2)^2 = 9$
55. $x^2 + y^2 = 49$ **57.** $(x + 3)^2 + (y + 4)^2 = 36$
59. $x^2 + (y - 3)^2 = 4$
61. $(x - 4)^2 + (y - 4)^2 = 16$

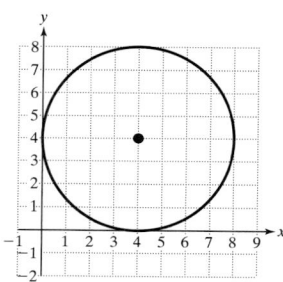

63. $(x - 1)^2 + (y - 1)^2 = 29$
65.

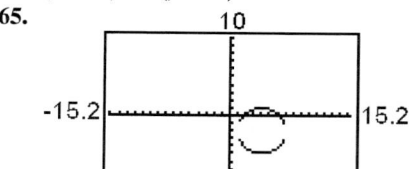

67.

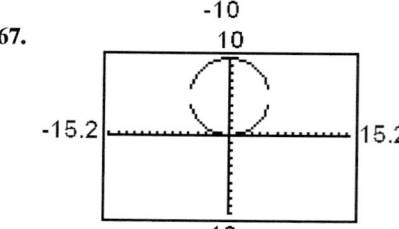

69.

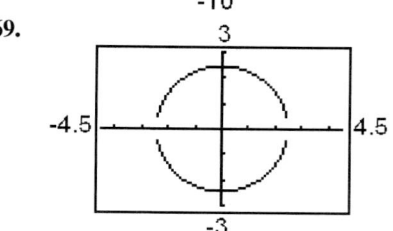

Section 14.2 Practice Exercises, pp. 1003–1006

3. 5

5. Center: $(0, -1)$; radius: 4

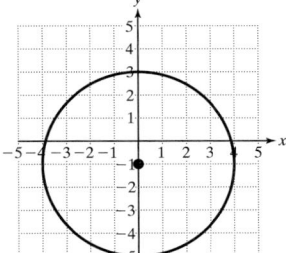

7. Center: $(3, -3)$; radius: 1   **9.**

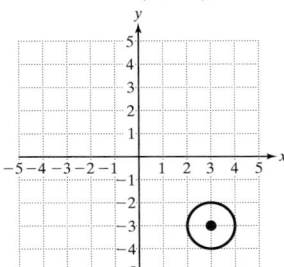
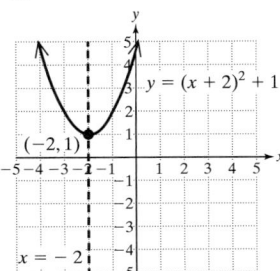

11.   **13.**

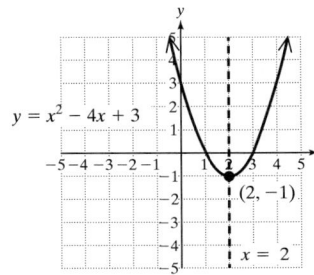
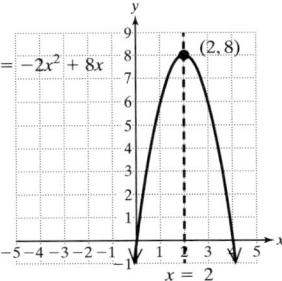

15.   **17.**

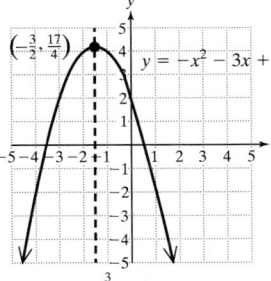
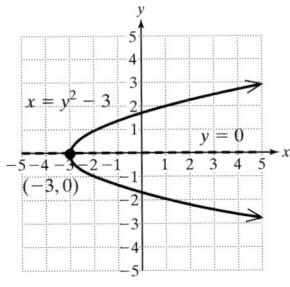

19.   **21.**

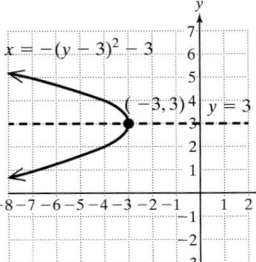
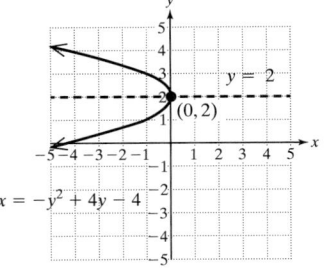

23.

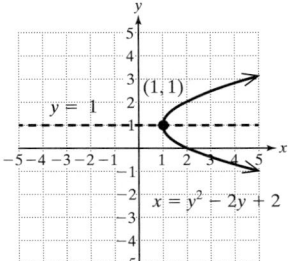

25. $(2, -1)$ **27.** $(5, -1)$ **29.** $(2, 8)$ **31.** $\left(\dfrac{3}{2}, -\dfrac{1}{4}\right)$

33. $(33, 4)$ **35.** For a parabola whose equation is written in the form $y = a(x - h)^2 + k$, if $a > 0$, the parabola opens upward; if $a < 0$, the parabola opens downward. For a parabola written in the form $x = a(y - k)^2 + h$, if $a > 0$, the parabola opens right; if $a < 0$, the parabola opens left.

37. Vertical axis of symmetry; opens upward

39. Vertical axis of symmetry; opens downward

41. Horizontal axis of symmetry; opens right

43. Horizontal axis of symmetry; opens left

45. Vertical axis of symmetry; opens downward

47. Horizontal axis of symmetry; opens right

Section 14.3 Practice Exercises, pp. 1011–1015

3. Center: $(8, -6)$; radius: 10

5. Vertex: $(-3, -1)$; $x = -3$

7. $\left(x - \dfrac{1}{2}\right)^2 + \left(y - \dfrac{5}{2}\right)^2 = \dfrac{1}{4}$

9.   **11.**

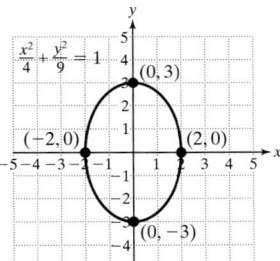
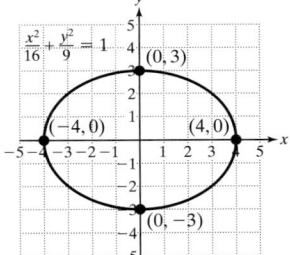

13.

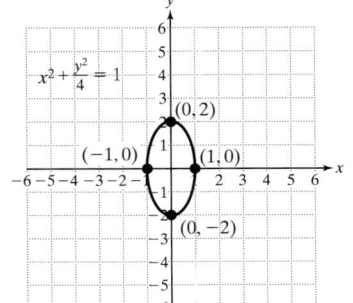

15.

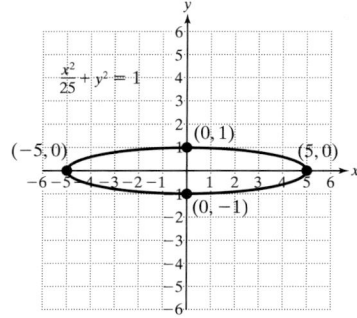

17. Center: $(4, 5)$

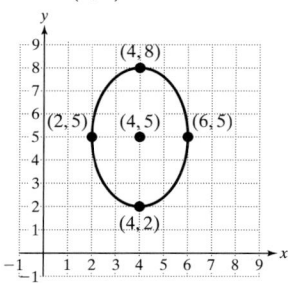

19. Center: $(-1, 2)$

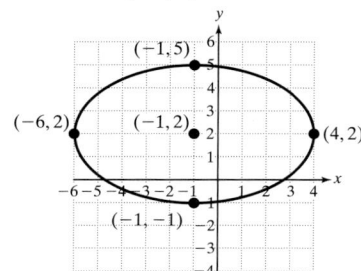

21. Center: $(2, -3)$

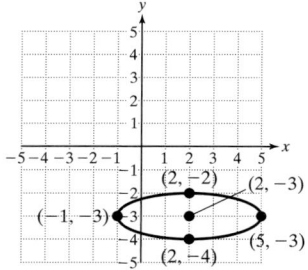

23. Center: $(0, 1)$

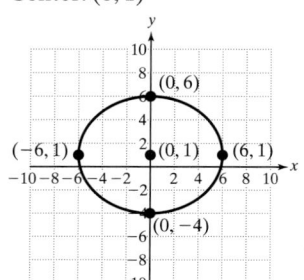

25. Vertical　　**27.** Horizontal
29. Horizontal　　**31.** Vertical

33.

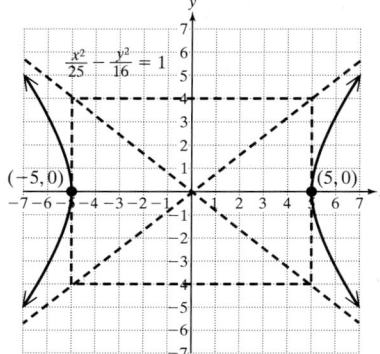

35.

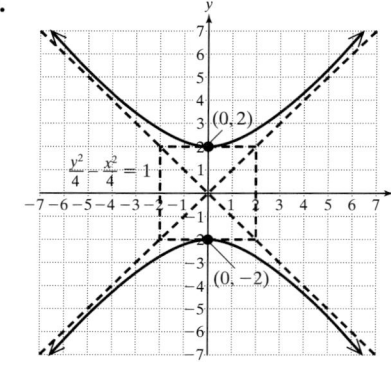

37.

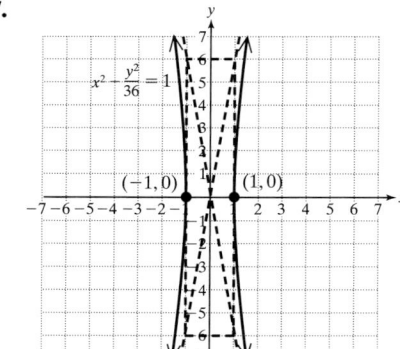

39.

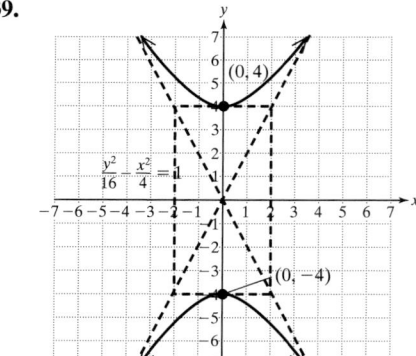

41. Hyperbola　　**43.** Ellipse　　**45.** Ellipse
47. Hyperbola　　**49.** Ellipse　　**51.** Hyperbola
53. 49 ft

55.

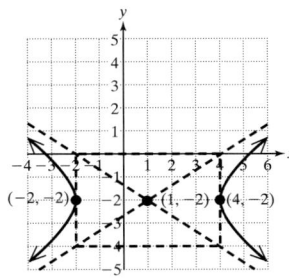

57.

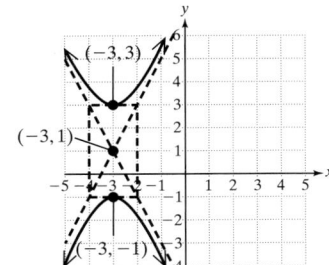

Section 14.4 Practice Exercises, pp. 1021–1023

3. $7\sqrt{2}$ **5.** $(x + 5)^2 + (y - 3)^2 = 64$ **7.** Hyperbola
9. Ellipse   **11.** Circle   **13.** Parabola
15. Zero, one, or two   **17.** Zero, one, or two
19. Zero or one   **21.** Zero, one, two, three, or four
23. $(-3, -5)(2, 0)$ **25.** $(\sqrt{5}, 2\sqrt{5})(-\sqrt{5}, -2\sqrt{5})$

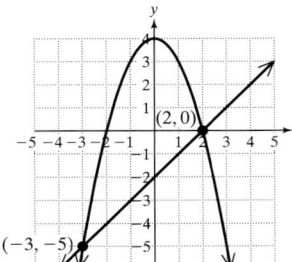
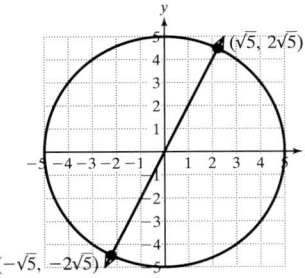

27. $(\sqrt{3}, 3)(-\sqrt{3}, 3)$

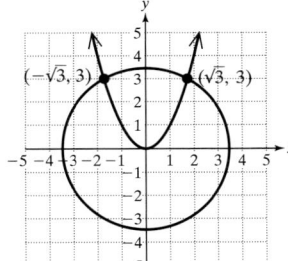

29. $(5, \sqrt{5})$ **31.** $(0, 0)(1, -1)$ **33.** $(-2, 4)$
35. $(0, 0), (3, 9)$ **37.** $(3, 2)(3, -2)(-3, 2)(-3, -2)$
39. $(0, 3)(0, -3)$ **41.** $(2, 1)(-2, 1)(2, -1)(-2, -1)$
43. $(\sqrt{2}, 0)(-\sqrt{2}, 0)$ **45.** $(2, 0)(-2, 0)$
47. $(3, 0)(-3, 0)$ **49.** $(4, 5)(-4, -5)$ **51.** 3 and 4
53. 5 and $\sqrt{7}$, -5 and $\sqrt{7}$, 5 and $-\sqrt{7}$, or -5 and $-\sqrt{7}$
55.

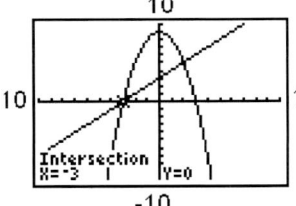
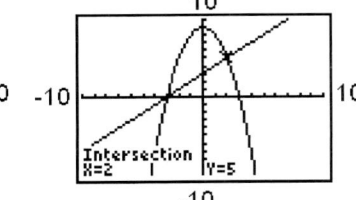

57.

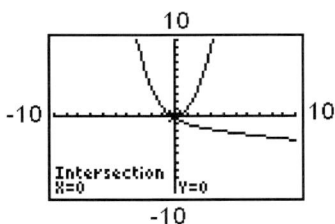

59.

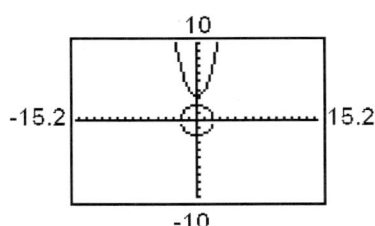

Section 14.5 Practice Exercises, pp. 1027–1032

1. k   **3.** e   **5.** i   **7.** j   **9.** a   **11.** d
13. False   **15.** True
17. a.

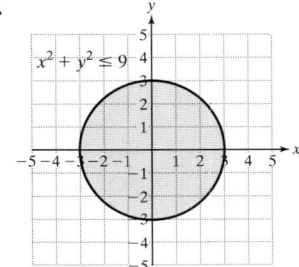

b. The set of points on and "outside" the circle $x^2 + y^2 = 9$
c. The set of points on the circle $x^2 + y^2 = 9$

19. a.

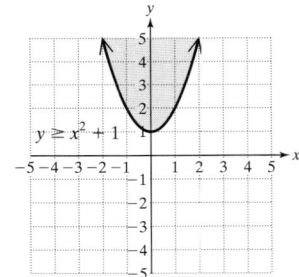

b. The parabola $y = x^2 + 1$ would be drawn as a dashed curve.

21.

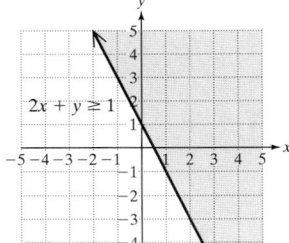

23.

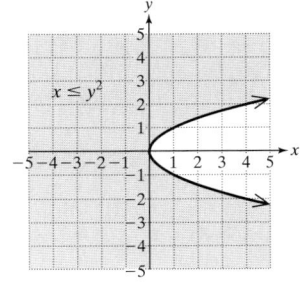

25.

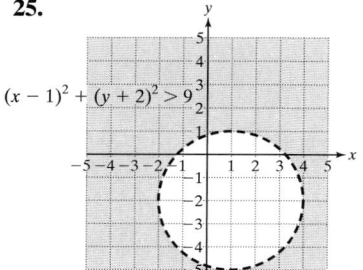

$(x - 1)^2 + (y + 2)^2 > 9$

27.

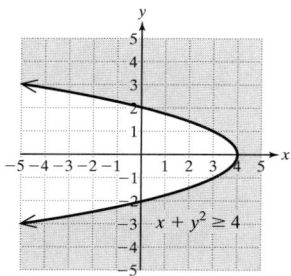

$x + y^2 \geq 4$

29.

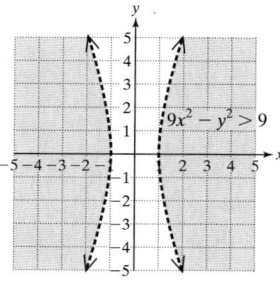

$9x^2 - y^2 > 9$

31.

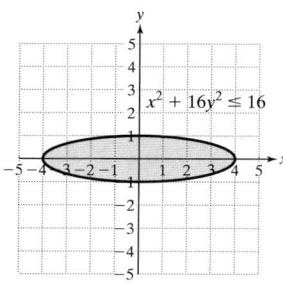

$x^2 + 16y^2 \leq 16$

33.

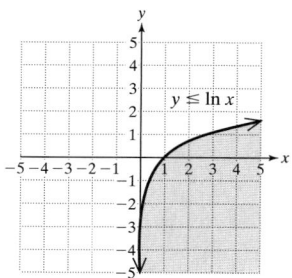

$y \leq \ln x$

35.

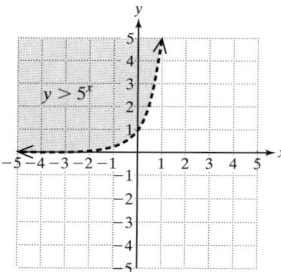

$y > 5^x$

37.

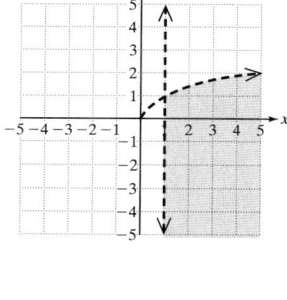

39.

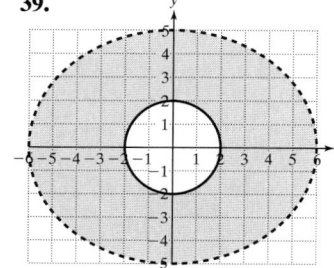

41.

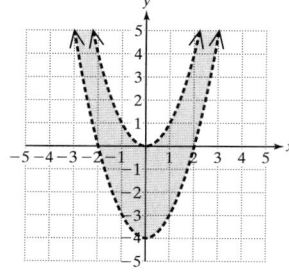

43.

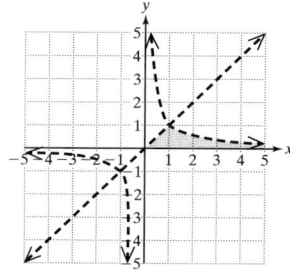

45. No Solution

47.

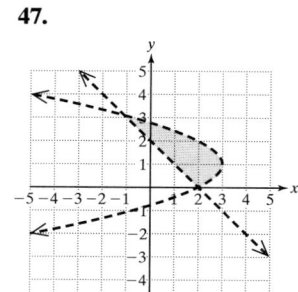

49.

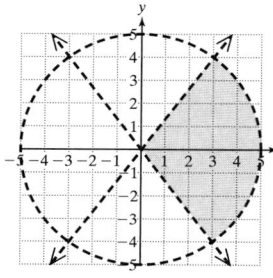

51.

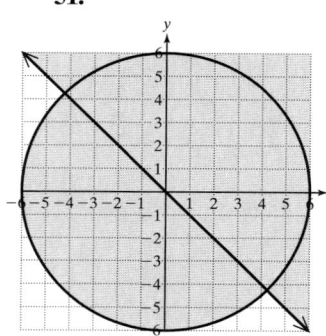

53.

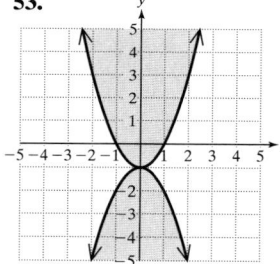

Chapter 14 Review Exercises, pp. 1037–1040

1. $2\sqrt{10}$ 2. $\sqrt{89}$ 3. $x = 5$ or $x = -1$
4. $x = -3$ 5. Collinear 6. Collinear
7. Center $(12, 3)$; $r = 4$ 8. Center $(-7, 5)$; $r = 9$
9. Center $(-3, -8)$; $r = 2\sqrt{5}$
10. Center $(1, -6)$; $r = 4\sqrt{2}$
11. a. $x^2 + y^2 = 64$ b. $(x - 8)^2 + (y - 8)^2 = 64$
12. $(x + 6)^2 + (y - 5)^2 = 10$
13. $(x + 2)^2 + (y + 8)^2 = 8$
14. $\left(x - \dfrac{1}{2}\right)^2 + (y - 2)^2 = 4$
15. $(x - 3)^2 + \left(y - \dfrac{1}{3}\right)^2 = 9$ 16. $x^2 + y^2 = \dfrac{49}{4}$
17. $x^2 + (y - 2)^2 = 9$

18. Vertical axis of symmetry; parabola opens downward
19. Horizontal axis of symmetry; parabola opens right
20. Horizontal axis of symmetry; parabola opens left
21. Vertical axis of symmetry; parabola opens upward

22.

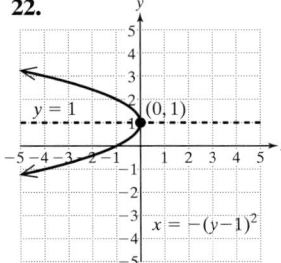

23.

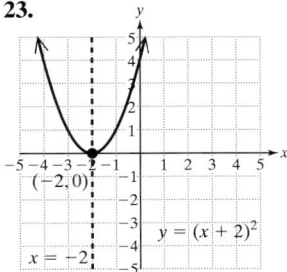

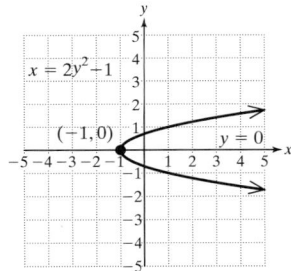

26. $y = (x - 3)^2 - 4$; vertex: $(3, -4)$; axis of symmetry: $x = 3$ **27.** $x = (y + 2)^2 - 2$; vertex: $(-2, -2)$; axis of symmetry: $y = -2$

28. $x = -4\left(y - \dfrac{1}{2}\right)^2 + 1$; vertex: $\left(1, \dfrac{1}{2}\right)$;

29. axis of symmetry: $y = \dfrac{1}{2}$

$y = -2\left(x + \dfrac{1}{2}\right)^2 + \dfrac{1}{2}$; vertex: $\left(-\dfrac{1}{2}, \dfrac{1}{2}\right)$;

axis of symmetry: $x = -\dfrac{1}{2}$

30.

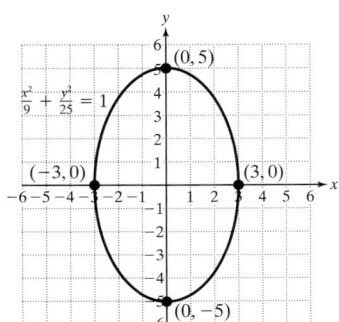

31.

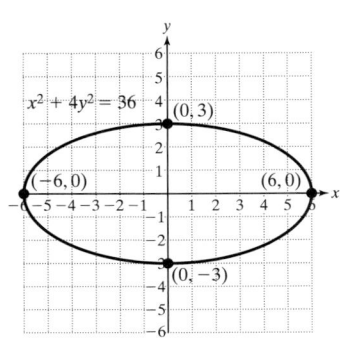

32. Center: $(5, -3)$

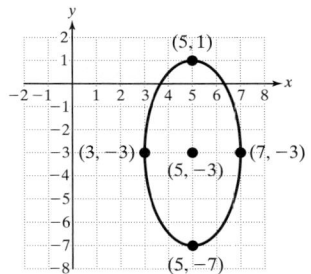

33. Center: $(0, 2)$

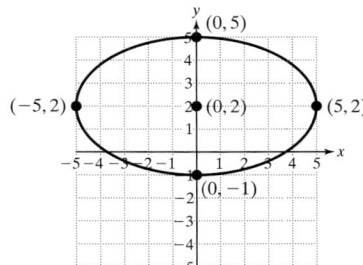

34. Horizontal   **35.** Vertical   **36.** Vertical
37. Horizontal
38.

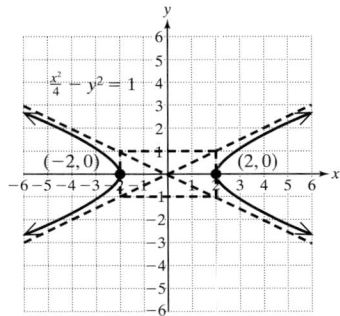

39.

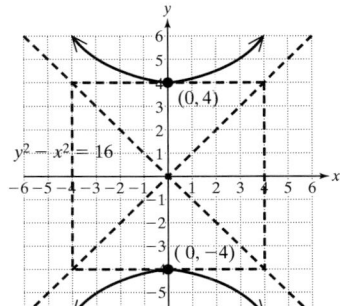

40. Hyperbola   **41.** Ellipse   **42.** Ellipse
43. Hyperbola
44. **a.** Line and parabola

b.

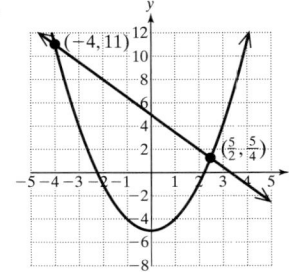

c. $\left(\dfrac{5}{2}, \dfrac{5}{4}\right), (-4, 11)$

45. a. Line and parabola

b.

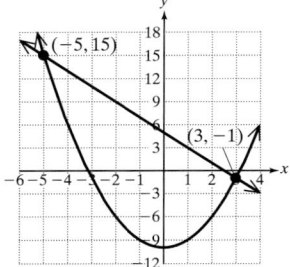

c. $(-5, 15), (3, -1)$

46. a. Circle and line

b.

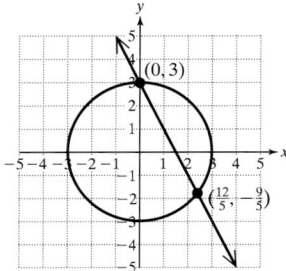

c. $(0, 3), \left(\dfrac{12}{5}, -\dfrac{9}{5}\right)$

47. a. Circle and line

b.

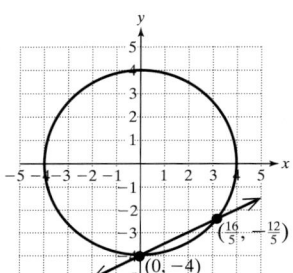

c. $(0, -4), \left(\dfrac{16}{5}, -\dfrac{12}{5}\right)$

48. $(0, -2)\left(\dfrac{16}{9}, \dfrac{14}{9}\right)$ **49.** $\left(-\dfrac{7}{5}, \dfrac{13}{5}\right)(-5, -1)$

50. $(8, 4)(2, -2)$ **51.** $(2, 4)(-2, 4)(\sqrt{2}, 2)(-\sqrt{2}, 2)$

52. $(3, 1)(3, -1)(-3, 1)(-3, -1)$

53. $(6, 5)(-6, 5)(6, -5)(-6, -5)$

54.

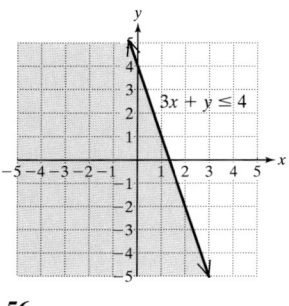

55.

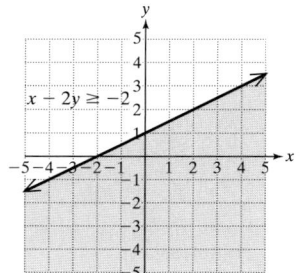

56.

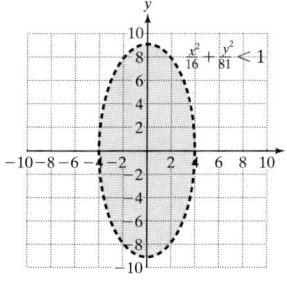

57.

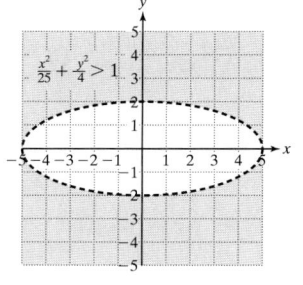

58.

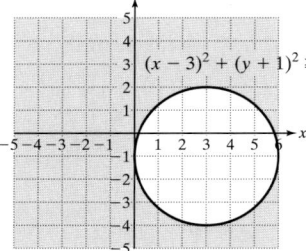

59.

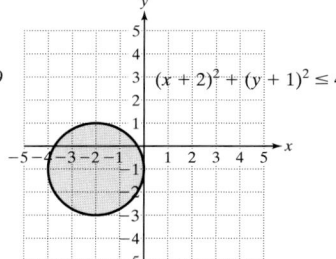

60.

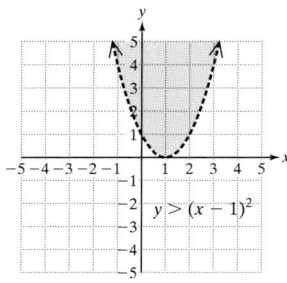

61.

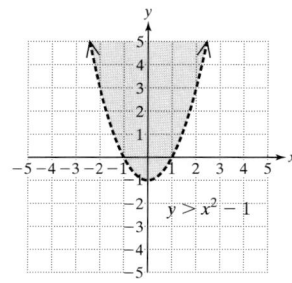

62.

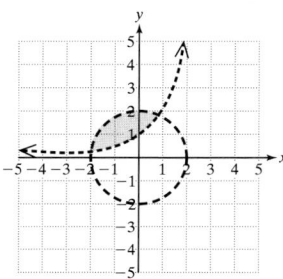

63.

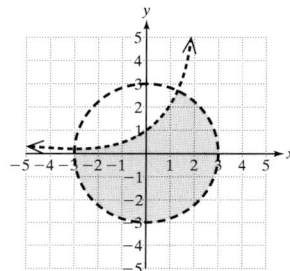

64.

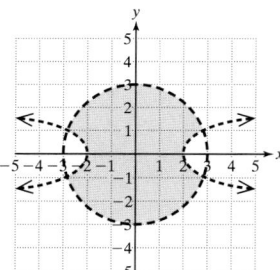

Chapter 14 Test, pp. 1041–1042

1.

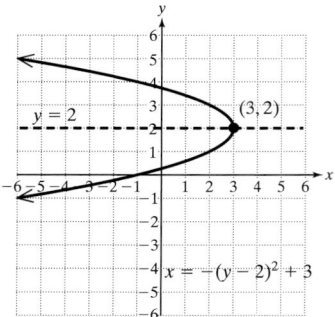

2. Center: $\left(\dfrac{5}{6}, -\dfrac{1}{3}\right)$; $r = \dfrac{5}{7}$

3.

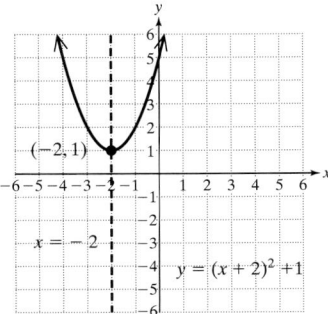

4. $2\sqrt{10}$ **5.** Center: $(0, 2)$; $r = 3$

6. a. $\sqrt{5}$ **b.** $x^2 + (y - 4)^2 = 5$

7.

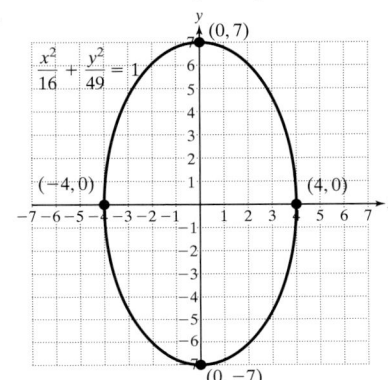

8.

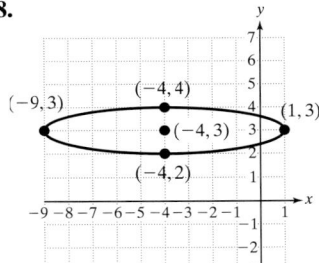

9.

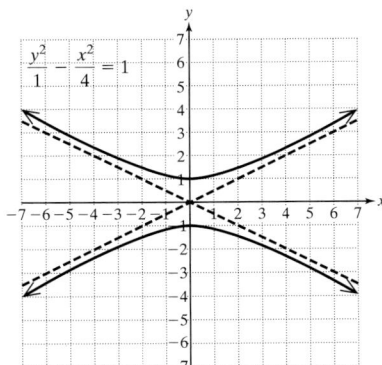

10. a. $(-3, 0)(0, 4)$; ii   **b.** No solution; i

11. The addition method can be used if the equations have corresponding *like* terms.

12. $(2, 0)(-2, 0)$

13.

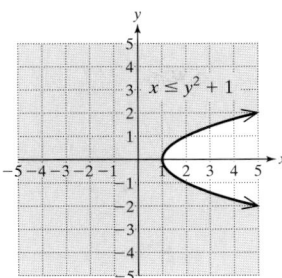

14.

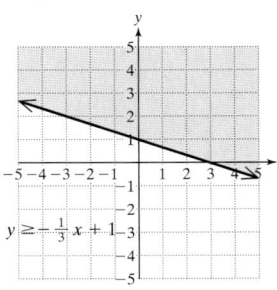

15.

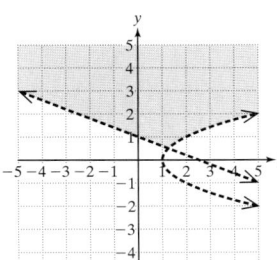

16.

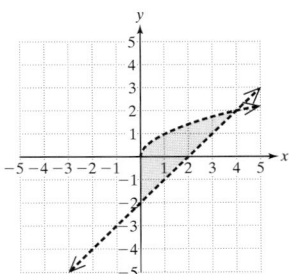

Chapters 1–14 Cumulative Review Exercises, pp. 1042–1044

1. All real numbers

2. $\left(\dfrac{10}{3}, \infty\right)$

3. $10, 15$ **4. a.** $(-5, 0), (0, 3)$ **b.** $m = \dfrac{3}{5}$

c.

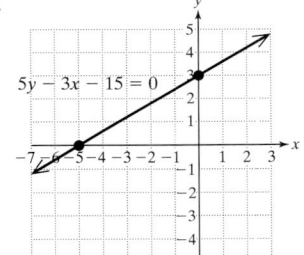

5. a. $m = 20$ **b.** $y = 20x + 390$ **c.** $590 billion

6. slope: $\dfrac{3}{4}$; y-intercept: $\left(0, -\dfrac{3}{2}\right)$

7. 12 dimes, 5 quarters   **8.** $(2, -3, 1)$

9. a. $\begin{bmatrix} 1 & -2 & | & -8 \\ 0 & 1 & | & 2 \end{bmatrix}$ **b.** $\begin{bmatrix} 1 & 0 & | & -4 \\ 0 & 1 & | & 2 \end{bmatrix}$

10. $\left(\frac{5}{2}, \frac{3}{2}\right)$ **11.** $(6, 3)$

12. $f(0) = -12; f(-1) = -16; f(2) = -10; f(4) = -16$

13. $g(2) = 5; g(8) = -1; g(3) = 0; g(-5) = 5$

14. $z = 32$ **15.** $(g \circ f)(x) = x + 7; x \geq -1$

16. a. -5 **b.** $(x + 1)(x^2 + 1); -5$ **c.** They are the same.

17. $(x + y)(x - y - 6)$ **18.** $x^3 - 2x^2 - 7x - 4$

19. $x = 6, x = \dfrac{3}{2}$ **20.** $\dfrac{a - 1}{a + 2}$ **21.** $\dfrac{-x^2 - x - 4}{(x + 3)(x - 2)}$

22. $x = 0, x = -1$ **23. a.** No solution   **b.** $x = -11$

24. a. $-30 + 24i$ **b.** $\dfrac{12}{41} + \dfrac{15}{41}i$ **25.** $2\sqrt{7}$ m

26. a. 17.6 ft; 39.6 ft; 70.4 ft   **b.** 8 sec

27. $w = -\dfrac{1}{5}$; $w = \dfrac{1}{10} \pm \dfrac{i\sqrt{3}}{10}$ **28.** $x = 2 \pm \sqrt{11}$

29. $(-5, -36)$ **30. a.** $(-3, 0), (1, 0)$ **b.** $(0, 3)$ **c.** $(-1, 4)$

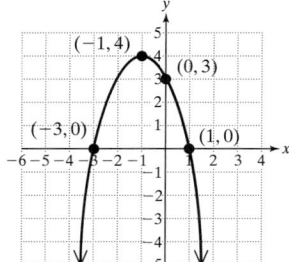

31. $(-1, 19)$ **32.** $\left(-\infty, \dfrac{1}{2}\right] \cup \left[\dfrac{9}{2}, \infty\right)$

33. $\log_8 32 = \frac{5}{3}$ **34.** $x = \dfrac{2}{3}$ **35.** $h^{-1}(x) = \sqrt[3]{x + 1}$

36. $x^2 + (y - 5)^2 = 16$

37. Yes, the circle can be tangent to the parabola.

38. $(0, -4)$

39.

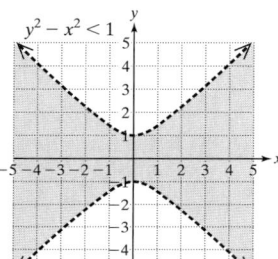

40.

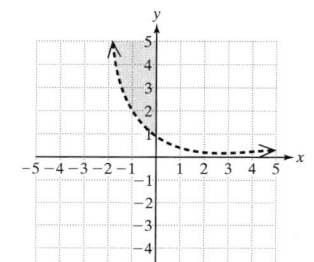

Beginning Algebra Review

Review A, Practice Exercises, pp. R-7–R-9

1. $\left\{-6, -\dfrac{7}{8}, -0.\overline{3}, 0, \dfrac{4}{9}, 7, 10.4\right\}$

3. $\{-6, 0, 7\}$ **5.** $\{1, 8\}$ **7.** $\{-\pi, \sqrt{3}\}$

9. 5 is greater than -2 **11.** -3 is greater than or equal to -4 **13.** 10 is between 6 and 12

15. 2 is not equal to -5 **17.** $6 \cdot 3$ **19.** $20 \div 5$

21. $|-5|$ **23.** $2 + |-8|$ **25.** $4 \cdot \dfrac{1}{3}$ **27.** -2

29. 19   **31.** 16   **33.** -33 **35.** -4.3 **37.** -16

39. -19 **41.** 5.6   **43.** $-\dfrac{9}{8}$ **45.** $-\dfrac{20}{9}$ **47.** -27

49. 35   **51.** 0   **53.** -8.61 **55.** $\dfrac{5}{6}$ **57.** 7

59. -6 **61.** 0   **63.** Undefined   **65.** $\dfrac{1}{2}$ **67.** 9

69. 8   **71.** 2   **73.** -16 **75.** 5   **77.** 15

79. $-\dfrac{8}{3}$ **81.** -16 **83.** -11 **85.** 460 in.^{2}

87. Associative property of multiplication

89. Distributive property of multiplication over addition

91. Commutative property of addition

93. 1; for example $1(3) = 3$

95. a. reciprocal   **b.** $\dfrac{1}{4}$ **97.** $5a + 45$ **99.** $-7y + 7z$

101. $2b + 3c - 5$ **103.** $-2x + 3y - 8$ **105.** $7a + 14$

107. $5y - 25$ **109.** $-w + 12u$ **111.** $14g - 48h$

Review B, Practice Exercises, pp. R-13–R-15

1. $x = -21$ **3.** $t = 60$ **5.** $p = -20$ **7.** $x = 3$

9. $x = -\dfrac{3}{2}$ **11.** $b = 23$ **13.** $w = -\dfrac{7}{2}$

15. $y = -\dfrac{10}{7}$ **17.** $x = -\dfrac{19}{3}$ **19.** $a = 5.5$

21. $t = 2.3$ **23.** $x = 0$ **25.** $y = 2$ **27.** $m = 9$

29. $y = -33$ **31.** $\frac{1}{2}$ **33.** $-20, -19, -18$

35. 4.5%   **37.** $2100 in the 4.2% account, $1400 in the 3.8% account

39. Car: 64 mph; truck: 60 mph

41. $(-\infty, -2)$

43. $[-6, \infty)$

45. $(-2, \infty)$

47. $\left(\dfrac{3}{4}, \infty\right)$

49. $\left(-\dfrac{7}{4}, 1\right]$

51. $(-1, 7)$

53. $\left(-\infty, \dfrac{35}{3}\right)$

55. $(-\infty, -0.2]$

Review C, Practice Exercises, pp. R-21–R-25

1.–7.

9. IV   **11.** III   **13.** I   **15.** II

17.

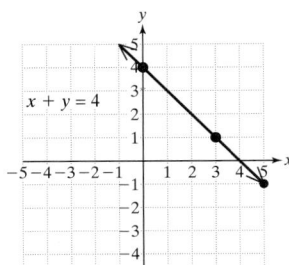

x	y
3	1
5	−1
0	4

19.

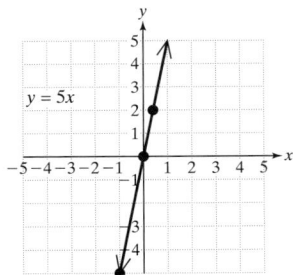

x	y
0	0
$\frac{2}{5}$	2
−1	−5

21.

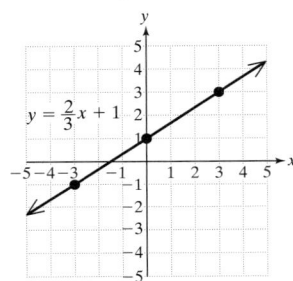

x	y
3	3
−3	−1
0	1

23. x-intercept $(6, 0)$; y-intercept $(0, -9)$

25. x-intercept $(-3, 0)$; y-intercept $(0, -15)$

27. x-intercept $(4, 0)$; y-intercept $(0, 2)$

29. x-intercept $(0, 0)$; y-intercept $(0, 0)$

31. x-intercept $\left(-\frac{9}{2}, 0\right)$; y-intercept $(0, -3)$

33. x-intercept $\left(\frac{5}{6}, 0\right)$; y-intercept $(0, -5)$

35. $x = 3$; vertical   **37.** $y = -\frac{1}{3}$; horizontal

39. $x = 0$; vertical

41. $y = \frac{4}{5}x - \frac{6}{5}$; slope is $\frac{4}{5}$; y-intercept $\left(0, -\frac{6}{5}\right)$

43. $y = 3x + \frac{3}{2}$; slope is 3; y-intercept $\left(0, \frac{3}{2}\right)$

45. $x = 1$; undefined slope; no y-intercept

47. $y = -\frac{1}{4}x$; slope is $-\frac{1}{4}$; y-intercept $(0, 0)$

49. $y = 4$; slope is 0; y-intercept is $(0, 4)$

51.

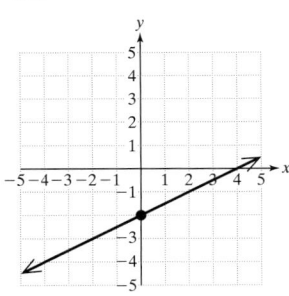

53.

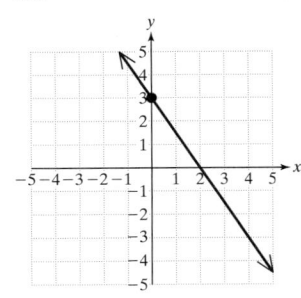

55.

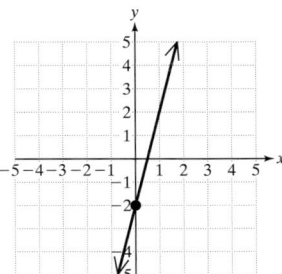

57.

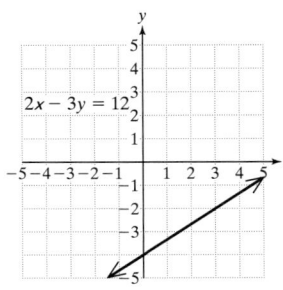

59.

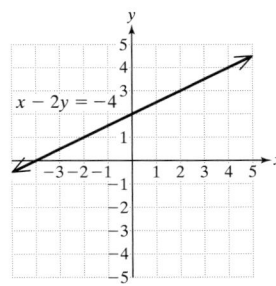

61.

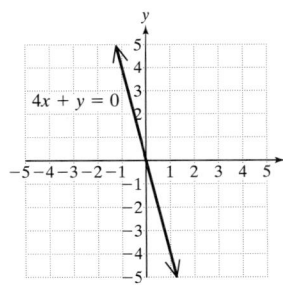

63.

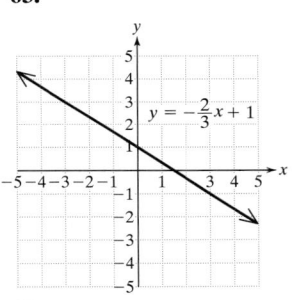

65.

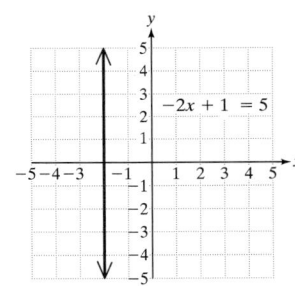

67.

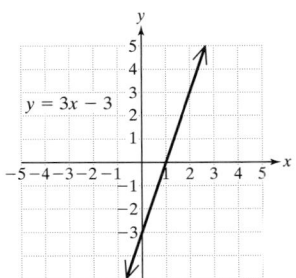

69. Parallel   **71.** Neither   **73.** Perpendicular

75. Parallel   **77.** $y = -3x - 17$; $3x + y = -17$

79. $y = 4x + 8$; $4x - y = -8$ **81.** $y = 2x - 6$;

$2x - y = 6$ **83.** $y = -2x$; $2x + y = 0$ **85.** $x = 4$

Review D, Practice Exercises, pp. R-29–R-31

1.

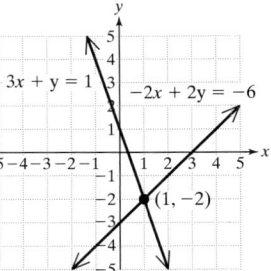

3.

5. $(1, -1)$ **7.** $(1, 0)$ **9.** $(1, -7)$ **11.** $(2, -1)$
13. No solution; inconsistent system   **15.** $(5, -7)$
17. Infinitely many solutions; $\{(x, y)\,|\,y = 2x - 14\}$;
dependent system   **19.** $(2200, 1000)$ **21.** $(2, 0)$
23. $\left(\dfrac{1}{2}, -\dfrac{5}{2}\right)$ **25.** He invested \$8000 at 6% interest and
\$2000 at 4.5% interest.   **27.** The angles are $52°$ and $38°$.
29. Denisha is 26 and her sister is 22.

Review E, Practice Exercises, pp. R-36–R-37

1. 6^7; multiplication of like bases   **3.** y^3, division of like
bases   **5.** $\dfrac{w^4}{256}$; power of a quotient   **7.** q^{30}; power rule

9. 1   **11.** 2   **13.** $\dfrac{1}{25}$ **15.** $\dfrac{7}{6}$ **17.** $\dfrac{1}{r^2}$ **19.** b^2

21. $27x^6$ **23.** $3w^2z^2$ **25.** $\dfrac{b^2 c^6}{a^4 d^8}$ **27.** $\dfrac{1}{wz^3}$

29. 3.05×10^6 **31.** 2.51×10^{-5} **33.** 8.96×10^7
35. 3.9×10^{-4} **37.** 3.6×10^4 **39.** 2.1×10^{16}
41. 2.19×10^{-2} **43.** 8.0×10^{-12}
45. $13p^3 - 4p^2 + 2p + 11$ **47.** $-n^2 - n + 8$
49. $-10u^2v + 16u^2 + 2uv^2$ **51.** $4p^4 - 20p^3 + 8p^2 + 32p$
53. $-6x^2 - 17xy - 5y^2$ **55.** $16b^2 - 16b - 12$
57. $x^3 - 6x^2y + 14xy^2 - 15y^3$ **59.** $9h^2 - 64$

61. $16x^2 - 40x + 25$ **63.** $\dfrac{1}{9}t^4 - 81$

65. $0.04x^4 - 1.2x^2 + 9$ **67.** $2a^2b - 6ab + b - 3$

69. $x^2 - 5x + 11$ **71.** $2p^2 - 7p + 4 + \dfrac{10}{p - 5}$

73. $-5m^4 + 8m^3 - 4m^2 - 4m + 1$

75. $4a^2 + 6a + 9 + \dfrac{18}{2a - 3}$

Review F, Practice Exercises, pp. R-40–R-41

1. $a^2 - b^2 = (a + b)(a - b)$
3. $a^3 + b^3 = (a + b)(a^2 - ab + b^2)$
5. a. Difference of squares   **b.** $(t - 10)(t + 10)$
7. a. Sum of cubes   **b.** $(y + 3)(y^2 - 3y + 9)$
9. a. Trinomial (non-perfect square trinomial)
b. $(d - 4)(d + 7)$ **11. a.** Trinomial (perfect square
trinomial)   **b.** $(x - 6)^2$ **13. a.** Four terms—grouping
b. $(ax + b)(2x - 5)$ **15. a.** Trinomial (nonperfect square
trinomial)   **b.** $(2y - 1)(5y + 4)$ **17. a.** Difference of
squares   **b.** $10(p - 8)(p + 8)$ **19. a.** Difference of cubes

b. $z(z - 4)(z^2 + 4z + 16)$ **21. a.** Trinomial (nonperfect
square trinomial)   **b.** $b(b + 5)(b - 9)$ **23. a.** Trinomial
(perfect square trinomial)   **b.** $(3w + 4x)^2$ **25. a.** Four
terms—grouping   **b.** $10(2x + a)(3x - 1)$ **27. a.** Four
terms—grouping   **b.** $(x - 3)(x + 3)(x + 4)$
29. a. Difference of squares   **b.** $(w^2 + 4)(w - 2)(w + 2)$
31. a. Difference of cubes   **b.** $(t^2 - 2)(t^4 + 2t^2 + 4)$
33. a. Trinomial (nonperfect square trinomial)
b. $(4p - 1)(2p - 5)$ **35. a.** Trinomial (perfect square
trinomial)   **b.** $(6y - 1)^2$ **37. a.** Four terms—grouping
b. $(x + 4)(x - y)$ **39. a.** Sum of squares   **b.** $2(x^2 + 25)$
41. a. Trinomial (nonperfect square trinomial)
b. $s^2(3r - 2)(4r + 5)$
43. a. Trinomial (nonperfect square trinomial)
b. $(x - 3y)(x + 11y)$ **45. a.** Sum of cubes
b. $(m^2 + n)(m^4 - m^2n + n^2)$ **47. a.** Trinomial
(nonperfect square trinomial)   **b.** $(a + b)(x - 4)(x + 3)$
49. a. None of these   **b.** $x(x - 4)$

51. $x = -\dfrac{5}{2}$ or $x = 3$ **53.** $x = 0$ or $x = 5$

55. $w = -3$ **57.** $z = 4$ or $z = -2$

59. $x = \dfrac{5}{3}$ or $x = -\dfrac{1}{2}$ **61.** $x = \dfrac{3}{2}$ or $x = 4$

63. $w = 1$ or $w = -1$

Review G, Practice Exercises, pp. R-47–R-49

1. $\{x\,|\,x$ is a real number and $x \neq -4\}$

3. $\left\{x\,|\,x$ is a real number and $x \neq \dfrac{3}{5}$ and $x \neq -\dfrac{3}{5}\right\}$

5. $\{t\,|\,t$ is any real number$\}$ **7.** $\dfrac{x^3}{4y}$ **9.** $\dfrac{t + 2}{2}$

11. $\dfrac{y + 4}{y + 2}$ **13.** $\dfrac{y^3}{10x}$ **15.** $\dfrac{1}{2}$ **17.** $3ab^2c$

19. $\dfrac{p(p - 6)}{4}$ **21.** $-\dfrac{t^2}{t - 7}$ **23.** $8x^3y^{10}$

25. $(x - 4)(x + 3)(x - 3)$ **27.** $-\dfrac{1}{3x}$ **29.** $\dfrac{5x^2 + 2y}{4x^3y^2}$

31. $x - 7$ **33.** $\dfrac{17x}{2(2x - 1)}$ **35.** $\dfrac{x^2 - 3x + 6}{(x + 4)(x - 3)(x - 1)}$

37. $\dfrac{-3}{y + 3}$ **39.** -1 **41.** $\dfrac{3b + 4a}{7b - a}$ **43.** $2(4x + 1)$

45. $u - v$ **47.** $\dfrac{y + 5}{2y - 9}$ **49.** $x = 10$ **51.** No solution

53. $x = -2, x = -4$ **55.** $x = 7$

Application Index

CONSUMER APPLICATIONS

DISTANCE/SPEED/TIME

ENVIRONMENT

Subject Index

Properties of Real Numbers

Commutative Property of Addition	$a + b = b + a$
Commutative Property of Multiplication	$ab = ba$
Associative Property of Addition	$(a + b) + c = a + (b + c)$
Associative Property of Multiplication	$(ab)c = a(bc)$
Distributive Property of Multiplication over Addition	$a(b + c) = ab + ac$
Identity Property of Addition	0 is the **identity element for addition** because $a + 0 = 0 + a = a$
Identity Property of Multiplication	1 is the **identity element for multiplication** because $a \cdot 1 = 1 \cdot a = a$
Inverse Property of Addition	a and $(-a)$ are **additive inverses** because $a + (-a) = 0$ and $(-a) + a = 0$
Inverse Property of Multiplication	a and $\frac{1}{a}$ are **multiplicative inverses** because $a \cdot \frac{1}{a} = 1$ and $\frac{1}{a} \cdot a = 1$ (provided $a \neq 0$)

Sets of Real Numbers

Natural numbers: $\{1, 2, 3, \dots\}$

Whole numbers: $\{0, 1, 2, 3, \dots\}$

Integers: $\{\dots -3, -2, -1, 0, 1, 2, 3, \dots\}$

Rational numbers: $\{\frac{p}{q} | p$ and q are integers and q does not equal 0$\}$

Irrational numbers: $\{x | x$ is a real number that is not rational$\}$

Application Formulas

Sales tax = (cost of merchandise)(tax rate)

Commission = (dollars in sales)(commission rate)

Simple interest = (principal)(rate)(time): $I = Prt$

Distance = (rate)(time): $d = rt$

Compound interest: $A(t) = P\left(1 + \dfrac{r}{n}\right)^{nt}$

Continuous compound interest: $A(t) = Pe^{rt}$,

where $A(t)$ = balance of account after t years, P = principal, r = annual interest rate, t = time in years, n = number of compound periods per year

Proportions

An equation that equates two ratios is called a proportion:

$$\frac{a}{b} = \frac{c}{d} \quad (b \neq 0, d \neq 0)$$

The cross products are equal: $ad = bc$.

Exponential Functions

A function defined by $y = b^x$ $(b > 0, b \neq 1)$ is an exponential function.

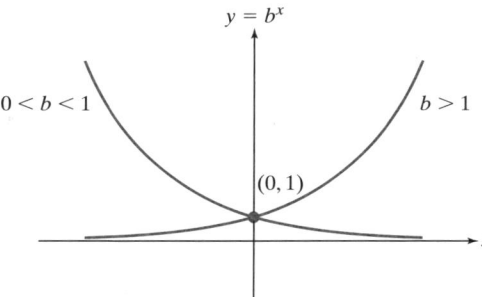

Logarithmic Functions

A function defined by $y = \log_b(x)$ is a logarithmic function.

$y = \log_b(x) \Leftrightarrow b^y = x \quad (x > 0, b > 0, b \neq 1)$

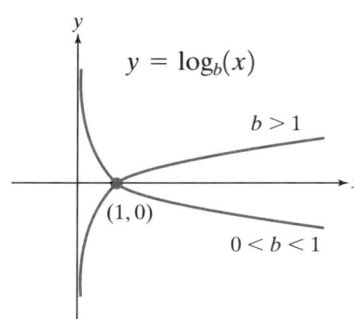